Advances in Cryogenic Engineering

VOLUME 33

A Continuation Order Plan is available for this series. A continuation order will bring delivery of each new volume immediately upon publication. Volumes are billed only upon actual shipment. For further information please contact the publisher.

A Cryogenic Engineering Conference Publication

Advances in Cryogenic Engineering

VOLUME 33

Edited by
R. W. Fast
Fermi National Accelerator Laboratory
Batavia, Illinois

SPRINGER SCIENCE+BUSINESS MEDIA, LLC

The Library of Congress cataloged the first volume of this title as follows:

Advances in cryogenic engineering. v. 1–
New York, Cryogenic Engineering Conference; distributed
by Plenum Press, 1960–
v. illus., diagrs. 26 cm.
Vols. 1– are reprints of the Proceedings of the Cryogenic Engineering
Conference, 1954–
Editor: 1960- K. D. Timmerhaus.

1. Low temperature engineering—Congresses. I. Timmerhaus, K. D.,
ed. II. Cryogenic Engineering Conference.

TP490.A3 660.29368 57-35598

Proceedings of the 1987 Cryogenic Engineering Conference held
June 14–18, 1987, in St. Charles, Illinois

ISBN 978-1-4613-9876-9 ISBN 978-1-4613-9874-5 (eBook)
DOI 10.1007/978-1-4613-9874-5

Originally published by Plenum Press, New York in 1988
Softcover reprint of the hardcover 1st edition 1988

CONTENTS

Foreword .... *xvii*
Dedication .... *xviii*
Russell B. Scott Memorial Awards .... *xix*
1986 Cryogenic Engineering Conference Board .... *xx*
Acknowledgements .... *xxi*

Applications of Superconductivity—Magnets

MAGNETIC RESONANCE IMAGING IN 1987 1
D. E. Andrews

THE PAST, PRESENT AND FUTURE OF MRI SUPERCONDUCTING MAGNET DEVELOPMENT 9
P.A. Rios and R. L. Rhodenizer

DYNAMIC PROCESSES IN STRING OF FULL SCALE SUPERCONDUCTING MAGNETS WITH FIELD 5 T 19
A. I. Ageyev, A. M. Andriishchin, V. I. Gridasov, E. M. Kashtanov, B. V. Kaz'min, S. N. Logachev, K. P. Myznikov, M. B. Nurusheva, A. P. Orlov, M. V. Pryima, A. N. Shamichev, V. V. Sytnik, V. A. Vasiliev, L. M. Vasiliev, O. M. Veselov, A. N. Yerokhin, N. N. Yarygin, and S. I. Zinchenko

DESIGN OF A STRONGLY CURVED SUPERCONDUCTING BENDING MAGNET FOR A COMPACT SYNCHROTRON LIGHT SOURCE 25
B. Krevet, H. O. Moser, and C. Dustmann

TESTING OF A 3 TESLA SUPERCONDUCTING MAGNET FOR FOR THE AMY DETECTOR AT TRISTAN 33
K. Tsuchiya, S. Terada, T. Omori, A. Maki, Y. Doi, T. Haruyama, T. Mito, and K. Asano

SUPERCONDUCTING MAGNETS FOR FUSION APPLICATIONS 41
C. D. Henning

AN INDUSTRIAL SUPERCONDUCTING HIGH GRADIENT MAGNETIC SEPARATOR 53
J. A. Selvaggi, P. Vander Arend, and J. Colwell

13.4 T GENERATION BY SUPERFLUID COOLING 61
M. Wake, K. Maehata, and M. Kobayashi, A. Sato, Y. Yamada, S. Murase, S. Nakamura, and Y. Kamisada

THE POTENTIAL IMPACT OF DEVELOPING HIGH T_C SUPERCONDUCTORS ON SUPERCONDUCTIVE MAGNETIC ENERGY STORAGE (SMES) 69
Y. M. Eyssa, M. K. Abdelsalam, R. W. Boom, and G. E. McIntosh

Applications of Superconductivity—Electronics, Rectifiers

SUPERCONDUCTIVITY: THE ULTIMATE ELECTRONIC TECHNOLOGY......... 77
M. Nisenoff

CRYOGENIC REQUIREMENTS FOR MEDICAL INSTRUMENTATION............. 87
R. E. Sarwinski

INSTRUMENTATION FOR CLINICAL APPLICATIONS OF NEUROMAGNETISM.. 97
D. S. Buchanan, D. Paulson, and S. J. Williamson

HIGH-RESOLUTION MEASUREMENTS OF BIOMAGNETIC FIELDS............. 107
J. P. Wikswo, Jr.

EFFICIENCY ASPECTS IN SUPERCONDUCTING RECTIFIERS.................. 117
G.B.J. Mulder, H.H.J. ten Kate, H.J.G. Krooshoop, and L.J.M. van de Klundert

Applications of Superconductivity—Magnet Stability, Coil Protection

EFFECT OF EPOXY CRACKING ON STABILITY OF IMPREGNATED WINDINGS RELATED TO THERMAL AND MECHANICAL PROPERTIES..... 125
S. Nishijima, T. Yamashita, K. Takahata, T. Okada, T. Fukutsuka, K. Matsumoto, and M. Hamada

LOCAL TEMPERATURE RISE AFTER QUENCH DUE TO EPOXY CRACKING IN IMPREGNATED SUPERCONDUCTING WINDINGS 135
S. Nishijima, K. Takahata, and T. Okada

HELIUM PRESSURE RISE OF SUPERCONDUCTING SUPER COLLIDER DIPOLE MAGNETS FOLLOWING A QUENCH..................... 143
M. A. Hilal, S. D. Peck, E. A. Ibrahim, V. N. Karpenko and W. R. Hassenzahl

MODELING OF PROXIMITY COUPLING BETWEEN SUPERCONDUCTING FILAMENTS IN SUPERCONDUCTING DIPOLE MAGNETS 149
M. A. Green

AC LOSS AND STABILITY OF A FORCE-COOLED SUPERCONDUCTING COIL WITH A BIAS PULSED SUPERCONDUCTING MAGNET................ 159
N. Ohuchi, Y. Makida, J. Yamamoto, and Y. Murakami

QUENCH DETECTION BY FLUID DYNAMIC MEANS IN CABLE-IN-CONDUIT SUPERCONDUCTORS 167
L. Dresner

THERMAL ANALYSIS OF THE FORCED COOLED CONDUCTOR FOR THE TF SUPERCONDUCTING COILS IN THE TIBER II ETR DESIGN.................. 175
J.A. Kerns, D.S. Slack, and J. R. Miller

SIMULATION OF NONSTATIONARY THERMAL REGIMES IN CRYOGENIC LOOPS FOR TOKAMAK-15 ELECTROMAGNETS 183
A. F. Volkov, L. B. Dinaburg, V. V. Kalinin, A. B. Konstantinov, and V. N. Krushev

THE TRANSIENT STABILITY OF LARGE SCALE SUPERCONDUCTORS COOLED IN SUPERFLUID HELIUM 187
J. Waynert, Y. Eyssa, and X. Huang

EVOLUTION OF NORMAL ZONES AND TEMPERATURE IN A SUPERCONDUCTOR COOLED INTERNALLY WITH HELIUM-II 195
E. R. Canavan and S. W. Van Sciver

PROTECTION SYSTEM FOR SUPERCONDUCTING MAGNETIC ENERGY STORAGE (SMES) 203
G.E. McIntosh, Y. M. Eyssa, M. K. Abdelsalam, R. W. Boom, T. A. Gallagher, R. N. Poirier, J. M. Shah, J. R. Bilton, T. F. Garrity, and M. A. Hilal

Applications of Superconductivity—Cryogenic Techniques

THE TWENTE HIGH-CURRENT CONDUCTOR TEST FACILITY, FIRST RESULTS ON CRITICAL CURRENT AND PROPAGATION IN TWO CABLES 211
H.H.J. ten Kate, W. Uyttewaal, B. ten Haken, and L.J.M. van de Klundert

SHORT SAMPLE CRITICAL CURRENT MEASUREMENTS USING A SUPERCONDUCTING TRANSFORMER 219
E.M.W. Leung, H. G. Arrendale, R. E. Bailey, and P. H. Michels

SSC MAGNET CRYOSTAT SUSPENSION SYSTEM DESIGN 227
T. H. Nicol, R. C. Niemann, and J. D. Gonczy

IMPROVED DESIGN FOR A SSC COIL ASSEMBLY SUSPENSION CONNECTION 235
E. T. Larson, J. A. Carson, T. H. Nicol, R. C. Niemann, and R. A. Zink

CRYOGENIC SUPPORT THERMAL PERFORMANCE MEASUREMENTS 243
J. D. Gonczy, W. N. Boroski, T. H., Nicol, R. C. Niemann, J. G. Otavka, and M. K. Ruschman

SSC DIPOLE MAGNET CRYOSTAT THERMAL MODEL MEASUREMENT RESULTS 251
R. C. Niemann, W. N. Boroski, J. D. Gonczy, T. H. Nicol, J. G. Otavka, M. K. Ruschman, and R. J. Powers

OPERATING EXPERIENCE OF THE IFSMTF VAPOR-COOLED LEAD SYSTEM 259
J. W. Lue, D. T. Fehling, W. A. Fietz, M. S. Lubell, J. N. Luton, S. W. Schwenterly, S. S. Shen, R. E. Stamps, D. H. Thompson, C. T. Wilson, and T. Kato

OPERATION OF LIQUID HELIUM LEVEL SENSORS AT ELEVATED PRESSURES 267
A. F. Zeller, H. Laumer, and J. A. Nolen

DESIGN AND TESTING OF ELECTRICAL INSULATION FOR SUPERCONDUCTING COILS....271
S. W. Schwenterly

QUENCH DETECTION OF SUPERCONDUCTING MAGNET BY DUAL-CORE OPTICAL FIBER....283
O. Tsukamoto, K. Kawai, Y. Kokubun, and T. Takao

Insulation

THEORY AND TECHNIQUE FOR REDUCING THE EFFECT OF CRACKS IN MULTILAYER INSULATION FROM ROOM TEMPERATURE TO 77 K....291
Q. S. Shu, R. W. Fast, and H. L. Hart

CRACK COVERING PATCH TECHNIQUE TO REDUCE THE HEAT FLUX FROM 77 K TO 4.2 K THROUGH MULTILAYER INSULATION....299
Q. S. Shu, R. W. Fast, and H. L. Hart

SYSTEMATIC ANALYSIS OF CHARACTERISTICS FOR DIFFERENT TYPES OF MULTILAYER INSULATIONS....305
M. Taneda, T. Ohtani, M. Okuda, and J. Tsukuda

TEST OF MULTILAYER INSULATIONS FOR THE USE IN THE SUPERCONDUCTING PROTON-RING OF HERA....313
H. Burmeister, W. Eschricht, G. Horlitz, D. Sellmann, and J. D. Ye

THERMAL PERFORMANCE OF CANDIDATE SSC MAGNET THERMAL INSULATION SYSTEMS....323
T. Ohmori, W. N. Boroski, J. D. Gonczy, R. C. Niemann, M. K. Ruschman, T. Taira, K. Takahashi, A. Yamamoto, and H. Hirabayashi

PROTOTYPE OF A SEMIFLEXIBLE MULTI-LAYER INSULATED ENCLOSURE FOR CRYOGENIC POWER CABLES AND PIPELINES....333
F. Schauer, M. Hubmann, H. Pirklbauer, and G. Kropatsch

DESIGN, DEVELOPMENT, AND TEST OF SHUTTLE/CENTAUR G-PRIME CRYOGENIC TANKAGE THERMAL PROTECTION SYSTEMS....341
P. N. MacNeil, J. E. England, and R. H. Knoll

Heat Transfer to Liquid Helium and Nitrogen, Flow Regimes

SOUND INDUCED ENHANCEMENT OF HEAT TRANSFER FROM A SOLID INTO LIQUID HELIUM....349
E. Bodegom, J. A. Nissen, L. C. Brodie, and J. S. Semura

FILM BOILING TO A PLATE FACING DOWNWARD....355
R. F. Barron and A. R. Dergham

ANGULAR DEPENDENCE OF BOILING HEAT TRANSFER MECHANISMS IN LIQUID NITROGEN....363
C. Beduz, R. G. Scurlock, and A. J. Sousa

BOILING ON A CRYOGENICALLY COOLED PULSED CONDUCTOR............371
C. J. Crowley, S. S. Kang, and P. H. Rothe

NEW HEAT TRANSFER AND FRICTION FACTOR DESIGN DATA FOR PERFORATED PLATE HEAT EXCHANGERS............383
R. H. Hubbell and C. L. Cain

AN INVESTIGATION INTO FLOW REGIMES FOR TWO-PHASE HELIUM FLOW............391
J. C. Theilacker and C. H. Rode

Heat and Mass Transfer in He II, Superfluid Pumps

METASTABLE HEAT FLOW PHENOMENA AND DROOPING PRESSURE DIFFERENCES HE II USING MULTIPLE CHANNEL CONFIGURATIONS............399
B.P.M. Helvensteijn and S. W. Van Sciver

HEAT TRANSPORT PROPERTIES OF PRESSURIZED AND SATURATED HE II IN THE VICINITY OF T_λ............407
M. Fouaidy and M. X. Francois

APPLICATION OF A SQUARE FUNCTION HEAT PULSE TO FORCED FLOW He II............417
A. Kashani and S. W. Van Sciver

NOISELESS FILM BOILING IN SUPERFLUID HELIUM ON A HEATED THIN (32 μm) WIRE............425
F. Jebali, J. Maza, M. X. Francois, and F. Vidal

TURBULENT TRANSPORT OF He II IN ACTIVE AND PASSIVE PHASE SEPARATORS USING SLIT DEVICES AND POROUS MEDIA............431
S.W.K. Yuan, J. M. Lee, and T.H.K. Frederking

OBSERVATION OF TWO-PHASE HELIUM FLOWS IN A HORIZONTAL PIPE....441
E. Sauvage-Boutar, C. Meuris, J. Poivilliers, and M. X. Francois

PRESSURE DROP AND TEMPERATURE RISE IN He II FLOW IN ROUND TUBES, VENTURI FLOWMETERS AND VALVES............449
P. L. Walstrom and J. R. Maddocks

THE DESIGN OF FOUNTAIN EFFECT PUMPS............457
S.W.K. Yuan and T.C. Nast

OPERATING CHARACTERISTICS OF ISOCALORIC FOUNTAIN-EFFECT PUMPS............465
P. Kittel

OPERATIONAL CHARACTERISTICS OF LOOPS WITH HELIUM II FLOW DRIVEN BY FOUNTAIN EFFECT PUMPS............471
A. Hofmann, A. Khalil, and H. P. Krämer

PERFORMANCE TEST OF A LABORATORY PUMP FOR LIQUID TRANSFER BASED ON THE FOUNTAIN EFFECT............479
W.E.W. Chen, T.H.K. Frederking, and W. A. Hepler

LAB TESTS OF A THERMOMECHANICAL PUMP FOR SHOOT..................487
M. J. DiPirro and R. F. Boyle

EXPERIMENTS ON TRANSFERRING HELIUM II WITH A THERMOMECHANICAL PUMP..................497
G. L. Mills, A. R. Urbach, A. J. Mord, B. H. Brandreth, L. A. Hermanson, and H. A. Snyder

CHARACTERIZATION OF A CENTRIFUGAL PUMP IN He II..................507
J. G. Weisend II and S. W. Van Sciver

PERFORMANCE OF A SMALL CENTRIFUGAL PUMP in He I AND He II..................515
P. R. Ludtke, D. E. Daney, and W. G. Steward

PUMP PERFORMANCE REQUIREMENT FOR THE LIQUID HELIUM ORBITAL RESUPPLY TANKER..................525
J. H. Lee and Y. S. Ng

Refrigeration for Superconducting Systems

CRYOGENIC CHARACTERISTICS OF THE TOPAZ THIN SUPERCONDUCTING SOLENOID..................533
A. Yamamoto, T. Mito, H. Kimura, T. Haruyama, H. Inoue, H. Yamaoka, E. Shiba, H. Kichimi, Y. Doi, T. Ohba, and T. Ohmori

PRESSURE DROP IN FORCED TWO PHASE COOLING OF THE LARGE THIN SUPERCONDUCTING SOLENOID..................543
T. Haruyama, T. Mito, Y. Doi, and A. Yamamoto

CRYOGENIC SYSTEM OF A 3 TESLA SUPERCONDUCTING SOLENOID FOR THE AMY PARTICLE DETECTOR AT TRISTAN..................551
Y. Doi, T. Haruyama, T. Mito, K. Tsuchiya, S. Terada, T. Ohmori, M. Maki, O. Araoka, M. Tadano, S. Suzuki, H. Suzuki, Y. Kondo, M. Kawai, N. Yamashita, and Y. Ibaraki

PERFORMANCE TEST OF THE HERA 3 × 6500 W HELIUM REFRIGERATION PLANT..................559
M. Clausen, C. Gerke, G. Horlitz, H. Lierl, K.-D. Nowakowski, S. Rettig, W. Stahlschmidt, P. Beurer, H. Egolf, H. Herzog, F. Langenecker, and B. Ziegler

CRYOGENIC EXPERIENCE AT THE HERA MAGNET MEASUREMENT FACILITY..................569
H. R. Barton, Jr., M. Clausen, G. Kessler, and S. Rettig

CRYOGENIC TESTING AT THE SSC STRING TEST FACILITY..................577
J. C. Theilacker

THE MIRROR FUSION TEST FACILITY CRYOGENIC SYSTEM —PERFORMANCE, MANAGEMENT APPROACH, AND PRESENT EQUIPMENT STATUS..................585
D. S. Slack and W. C. Chronis

THE TORE SUPRA 300 W—1.75 K REFRIGERATOR REPORT..................591
G. M. Gistau, M. Bonneton, and J. W. Mart

SYSTEM DESIGN AND VERIFICATION TEST RESULTS OF CRYOGENIC SYSTEM FOR DEMONSTRATION POLOIDAL COIL 599
E. Tada, T. Kato, T. Hiyama, K. Kawano, M. Hoshino, and S. Shimamoto

CRYOPUMP OF THE NEUTRAL BEAM INJECTOR FOR JIPPT-IIU 607
Y. Ohtu, S. Kataoka, T. Ohi, S. Kitagawa, Y. Oka, O. Kaneko, T. Kuroda, and K. Sakurai

CRYOGENIC SYSTEM FOR TRISTAN SUPERCONDUCTING RF CAVITY 615
K. Hara, K. Hosoyama, Y. Kimura, Yuuji Kojima, Yuzo Kojima, S. Mitsunobu, M. Morimoto, H. Nakai, S. Noguchi, T. Ogitsu, Y. Sakamoto, T. Nakazato, S. Kawamura, K. Matsumoto, and S. Saito

CEBAF'S CRYOGENIC SYSTEM 623
P. Brindza, W. Chronis, C. Rode, J. P. Kelley, and M. Shea

DESIGN CONCEPTS FOR THE ASTROMAG CRYOGENIC SYSTEM 631
M. A. Green and S. Castles

A CRYOGENIC SYSTEM FOR FUSION NEUTRON IRRADIATION AND IN-SITU MEASUREMENTS ON SUPERCONDUCTORS 639
P. A. Hahn and M. W. Guinan

Cold Compressors

COLD COMPRESSION OF HELIUM FOR REFRIGERATION BELOW 4 K 647
H. Quack

TESTS OF COLD HELIUM COMPRESSORS AT FERMILAB 655
T. J. Peterson and J.D. Fuerst

OPERATING EXPERIENCES AND TEST RESULTS OF SIX COLD HELIUM COMPRESSORS 663
D. P. Brown, R. J. Gibbs, A. P. Schlafke, J. H. Sondericker, and K. C. Wu

THE 300 W—1.75 K TORE SUPRA REFRIGERATOR COLD CENTRIFUGAL COMPRESSORS REPORT 675
G.M. Gistau, Y. Pecoud, and A. E. Ravex

Refrigeration and Liquefaction; Dilution Refrigerators

REFRIGERATION FOR ELECTRONICS: SUMMARY OF A PANEL SESSION 683
R. K. Kirschman, P. J. Kerney, R. C. Longsworth, J. R. Olson, and R. V. Schurter

4 K GIFFORD MC MAHON/JOULE-THOMSON CYCLE REFRIGERATORS 689
R. C. Longsworth

AN INVESTIGATION INTO THE MECHANICS OF JOULE-THOMSON VALVE PLUG FORMATION 699
L. Wade, C. Donnelly, E. Joham, K. Johnson, R. Phillips, E. Ryba, B. Self, and R. Stanton

LOW-COST, COMPACT DILUTION REFRIGERATOR: OPERATION FROM 200 TO 20 mK 707
P. R. Roach and K. E. Gray

CONTROL OF THE INTERFACE BETWEEN ^{3}He-RICH AND ^{4}He-RICH PHASES USING ELECTRIC FIELDS 713
U. E. Israelsson, H. W. Jackson, and D. Petrac

Magnetic Refrigeration; Materials for Magnetic Refrigeration

MAGNETIC REFRIGERATION: A REVIEW OF A DEVELOPING TECHNOLOGY 719
J. A. Barclay

RECENT PROGRESS IN MAGNETIC REFRIGERATION STUDIES 733
T. Hashimoto, T. Yazawa, R. Li, T. Kuzuhara, K. Matsumoto, H. Nakogome, M. Takahashi, M. Sahashi, K. Inomata, A. Tomokiyo, and H. Yayama

AN ERICSSON MAGNETIC REFRIGERATOR FOR LOW TEMPERATURE 743
K. Matsumoto, T. Ito, and T. Hashimoto

ANALYSIS OF MAGNETIC REFRIGERATION WITH EXTERNAL REGENERATION 751
S. R. Jaeger, J. A. Barclay, and W. C. Overton, Jr.

ROTARY RECUPERATIVE MAGNETIC HEAT PUMP 757
L. D. Kirol and M. W. Dacus

OPTIMAL TEMPERATURE–ENTROPY CURVES FOR MAGNETIC REFRIGERATION 767
C. R. Cross, J. A. Barclay, A. J. DeGregoria, S. R. Jaeger, and J. W. Johnson

THE MAGNETOCALORIC EFFECT OF SOME RARE EARTH METALS 777
G. Green, W. Patton, and J. Stevens

PREPARATION AND FABRICATION OF RARE EARTH MAGNETIC MATERIALS 785
B. J. Beaudry

MEASURED PROPERTIES OF GdNi FOR MAGNETIC REFRIGERATION APPLICATIONS 791
C. B. Zimm, W. F. Stewart, J. A. Barclay, C. K. Campenni, W. Overton, C. Olsen, D. Harding, R. Chesebrough, and W. Johanson

Cryocoolers; Refrigeration for Space Applications

CURRENT DEVELOPMENTS IN NASA CRYOGENIC COOLER TECHNOLOGY 799
S. H. Castles

HIGH PRESSURE RATIO CRYOCOOLER WITH INTEGRAL EXPANDER AND HEAT EXCHANGER 809
J. A. Crunkleton, J. L. Smith, Jr., and Y. Iwasa

DEVELOPMENT OF A SPACE QUALIFIED SURFACE TENSION CONFINED LIQUID CRYOGEN COOLER (STCLCC) 819
S. H. Castles and M. E. Schein

SMALL TURBO-BRAYTON CRYOCOOLERS 827
H. Sixsmith, J. Valenzuela, and W. L. Swift

SMALL VUILLEUMIER COOLER 837
H. Yoshimura and M. Kawada

SMALL SCALE FREE DISPLACER STIRLING CRYOCOOLER 845
C. Heiden and G. Reich

DEVELOPMENT AND EXPERIMENTAL TEST OF AN ANALYTICAL MODEL OF THE ORIFICE PULSE TUBE REFRIGERATOR 851
P. J. Storch and R. Radebaugh

PULSE TUBE WITH AXIAL CURVATURE 861
Y. Zhou, W. X. Zhu, and Y. Sun

SORPTION CRYOGENIC REFRIGERATION—STATUS AND FUTURE 869
J. A. Jones

NEW DESIGN OF AN ADIABATIC DEMAGNETIZATION CRYOSTAT FOR SPACE APPLICATION 879
J. Yamamoto, A. Sato, and M. Sahashi

Cryogenic Applications—Space Science and Technology

BAYONET FOR SUPERFLUID HELIUM TRANSFER IN SPACE 885
G. E. McIntosh, D. S. Lombard, D. L. Martindale, and M.J. DiPirro

THE SUPERFLUID HELIUM ON-ORBIT TRANSFER (SHOOT) FLIGHT EXPERIMENT 893
M. J. DiPirro and P. Kittel

CRYOGENIC AND THERMAL DESIGN FOR THE SUPERFLUID HELIUM ON-ORBIT TRANSFER (SHOOT) EXPERIMENT 901
J. H. Lee, S. Maa, W. F. Brooks, and Y. S. Ng

ACQUISITION SYSTEM TESTING WITH SUPERFLUID HELIUM 909
J. E. Anderson, D. A. Fester, and M. J. DiPirro

DIRECT LIQUID CONTENT MEASUREMENT APPLICABLE FOR He II SPACE CRYOSTATS 917
M. Wanner

THERMAL PERFORMANCE OF THE COSMIC BACKGROUND EXPLORER SUPERFLUID HELIUM DEWAR, AS BUILT AND WITH AN IMPROVED SUPPORT SYSTEM 925
R. A. Hopkins and D. A. Payne

DESIGN AND TEST OF A MODIFIED PASSIVE ORBITAL DISCONNECT STRUT (PODS-IV) 935
I. E. Spradley and R. T. Parmley

UNIQUE CRYOGENIC FEATURES OF THE GRAVITY PROBE B (GP-B) EXPERIMENT 943
R. T. Parmley

AN ADIABATIC DEMAGNETIZATION COOLED BOLOMETER SYSTEM 955
L. Lesyna, T. Roellig, M. Werner, and P. Kittel

Commercial Cryogenic Plants

EXPANSION TURBINES AND REFRIGERATION FOR GAS SEPARATION AND LIQUEFACTION 963
L. C. Kun

NITROGEN PRODUCTION FOR EOR 975
R. F. Pahade and J. H. Ziemer

RECOVERY OF VALUABLE HYDROCARBONS USING DEPHLEGMATOR TECHNOLOGY 983
D. P. Bernhard and H. C. Rowles

OPTIMIZATION OF THE OPERATION OF A GAS TERMINAL 991
C. Fuge, P. Eisele, and B. D. Whitehead

Properties of Cryogenic Fluids

NEW MEASUREMENTS OF THE TENSILE STRENGTH OF LIQUID ^{4}He 999
J. A. Nissen, E. Bodegom, L.C. Brodie, and J. S. Semura

SIMULTANEOUS PRESSURE AND TEMPERATURE MEASUREMENTS ON HELIUM IN THE HIGH SPEED ROTATING FRAME 1005
R. M. Igra, M. G. Rao, and R. G. Scurlock

HELIUM ADSORPTION ON ACTIVATED CARBONS AT TEMPERATURES BETWEEN 4 AND 76 K+ 1013
I. Vázquez, M. P. Russell, D.R. Smith, and R. Radebaugh

HIGH PRESSURE ADSORPTION ISOTHERMS OF HELIUM ON ACTIVATED CHARCOAL 1023
L. Duband, J. Chaussy, and A. Ravex

TEMPERATURE DEPENDENCE OF THE COHESION PARAMETER FOR CALCULATING BINARY VLE VALUES FOR SYSTEMS CONTAINING HELIUM AND NEON 1031
Y. Adachi, H. Sugie, and B.C.-Y. Lu

EVIDENCE OF UNRELIABILITY OF FACTORY ANALYSES OF ARGON IMPURITY IN OXYGEN 1039
F. Pavese, D. Ferri, and D. Giraudi

A NEW SQUARE-ROOT-TYPE PSEUDO-CUBIC EQUATION OF STATE 1045
M. Kato and T. Kiuchi

Cryogenic Instrumentation and Data Acquisition; Control Systems

REVIEW OF INSTRUMENTATION FOR SUPERCONDUCTING MAGNETS 1053
J. A. Zichy

CRYOGENIC INSTRUMENTATION OF AN SSC MAGNET TEST STAND 1063
K. McGuire, J. Strait, M. Kuchnir, and A. McInturff

BEHAVIOR OF TURBINE AND VENTURI FLOWMETERS IN SUPERFLUID HELIUM 1071
D. E. Daney

SONIC STANDING WAVE GAS DENSITY MONITOR 1081
R. J. Walker

AN ELECTRONIC BALANCE FOR WEIGHING FOAMS AT CRYOGENIC TEMPERATURES 1089
R. O. Voth and J. D. Siegwarth

ELECTRO-OPTIC SAMPLER FOR CHARACTERIZATION OF DEVICES IN A CRYOGENIC ENVIRONMENT 1097
D. R. Dykaar, R. Sobolewski, J. M. Chwalek, T. Y. Hsiang, and G. A. Mourou

CONTROL SYSTEM FOR HELIUM REFRIGERATORS OF TRISTAN DETECTOR MAGNETS 1105
T. Mito, T. Haruyama, M. Tadano, Y. Kondo, A. Yamamoto, N. Kimura, and Y. Doi

EXPERIENCE WITH A PROCESS CONTROL SYSTEM FOR LARGE SCALE CRYOGENIC SYSTEMS 1113
M. Clausen, K. H. Mess, Chr. Gerke, and S. Rettig

Miscellaneous Cryogenic Applications and Techniques

CRYOGENIC DESIGN OF THE D-ZERO LIQUID ARGON COLLIDER CALORIMETER 1121
G. T. Mulholland, K. J. Krempetz, R. D. Luther, R. H. Wands, and K. J. Weber

CRYOGENIC PULSED POWER TRANSFORMERS 1129
J. D. Rogers, P. W. Eckels, D. T. Hackworth, E. J. Shestak, and S. K. Singh

AEROSPACE GENERATORS USING HIGH-PURITY CRYOGENIC ALUMINUM ROTOR CONDUCTOR 1137
T. A. Keim

LOW LOSS LIQUID HELIUM TRANSFER SYSTEM, USING A HIGH PERFORMANCE CENTRIFUGAL PUMP AND COLD GAS EXCHANGE 1147
H. Berndt, R. Doll, U. Jahn, and W. Wiedemann

A MINIATURE CRYOGENIC HIGH VACUUM VALVE 1153
J. D. Siegwarth and R. O. Voth

TEST AND STUDY FOR A SMALL-SIZED ENVIRONMENTAL SIMULATOR OF SPACE RADIANT COOLERS 1161
J. Han, G.-D. Cui, and G.-T. Hong

A CRYOGENIC SOURCE OF POLARIZED DEUTERONS 1167
A. A. Belushkina, V. P. Ershov, V. V. Fimushkin, G. I. Gaj, L. S. Kotova, Yu. K. Pilipenko, V. B. Shutov, V. V. Smelyansky, A. I. Valevich, and I. V. Zhigulin

Indexes

AUTHOR INDEX 1171

SUBJECT INDEX 1176

FOREWORD

The 1987 joint Cryogenic Engineering Conference/International Cryogenic Materials Conference was held at the Pheasant Run Resort, St. Charles, Illinois from June 14 to 18. Fermi National Accelerator Laboratory, located a few kilometers from Pheasant Run, was the host for this conference. There is a great deal of cryogenic research and development underway at Fermilab and many applications of cryogenic materials and systems are in routine, daily use at the Tevatron. The technical program for the joint conference had over 300 invited and contributed papers from many different countries.

The CEC board and I have tried to dramatically shorten the publication time of this volume of **Advances in Cryogenic Engineering**. In order to help meet the goal of the February publication, I asked the reviewers to complete their reviews before leaving Pheasant Run, after the conference. I would like to thank all of the reviewers for their prompt and throughtful reviews. I very much appreciate the authors following the prescribed format and responding quickly to my requests for revisions.

R.W. Fast
Editor

DEDICATION

The chairman of the Cryogenic Engineering Conference is highly visible during the three or four days every two years when the conference is held. His activities on behalf of the CEC are much less apparent during the eighteen months prior to the conference, but nevertheless are very important, since the success of the conference depends on his leadership and enthusiasm during this planning phase. The conference chairman, who is also the president of Cryogenic Engineering Conference, Inc., presides over the approximately semi-annual CEC board meetings, develops the conference budget, solicits contributions from government and industry, and coordinates conference activities with the chairman of the International Cryogenic Materials Conference, which meets jointly with the CEC. The conference chairman serves the board ex officio for a period of two years after his conference and usually chairs the awards committee during that time.

The list of CEC chairmen reads like "Who's Who in Cryogenics." Nine of the conference chairmen have been awarded Russell B. Scott Memorial Awards for the best papers presented at the CEC, and one has been awarded the Samuel C. Collins Award for outstanding achievements in cryo-science and technology.

Two of the more recent CEC chairmen retired from full-time participation in cryogenics in 1986: Jack E. Jensen and Michael J. Hiza, Jr. Jack earned his B.S. and M.S. in mechanical engineering from the University of Colorado, Boulder and worked for several years at the Boulder Laboratories of the National Bureau of Standards. He joined Brookhaven National Laboratory in 1957, where he remained for most of his career. He has recently been associated with CVI, Inc. Jack co-authored, with J.W. Dean, the paper "Supercritical Helium Refrigeration for Superconducting Power Transmission Cable Studies," which was presented at the 1975 conference and awarded the Scott Award in 1977. Elected to the CEC board in 1971, Jack was the technical program chairman for the 1973 conference and chaired the 1975 conference in Kingston, Ontario. He served the board for many years as its representative to the U.S. National Committee of the International Institute of Refrigeration.

Mike was associated with NBS-Boulder for his entire 31-year career. He holds the B.S. and M.S. degrees in chemical engineering from the University of Colorado, Boulder. In 1966, he co-authored with A.J. Kidnay and presented at the CEC a paper entitled "High Pressure Adsorption Isotherms of Neon, Hydrogen and Helium at 76°K," which was recognized by the Scott Award at the 1967 conference. He published over fifty other papers in the field of phase equilibrium and fluid properties. Mike was elected to the CEC board in 1972, was the technical program chairman in 1975 and chairman of the 1977 conference in Boulder, Colorado.

After retiring from the board, both Jack and Mike continued to serve the CEC by presenting papers, chairing sessions, and reviewing papers for **Advances in Cryogenic Engineering.**

It is in recognition of their dedication to the Cryogenic Engineering Conference and their contributions to cryogenic science and engineering that Volume 33 of **Advances in Cryogenic Engineering** is gratefully dedicated to Michael J. Hiza, Jr. and Jack E. Jensen.

RUSSELL B. SCOTT MEMORIAL AWARDS

The Russell B. Scott Memorial Awards honor the first head of the Cryogenic Engineering Laboratory of the National Bureau of Standards in Boulder. Mr. Scott was the founder of the Cryogenic Engineering Conference, the first of which was held in 1954. He is the author of the best-selling classic, **Cryogenic Engineering**. He retired from the National Bureau of Standards in 1965 after a 37 year career with the Bureau. Mr. Scott died in 1967.

The objectives of the Scott Memorial Awards are to provide an incentive for high quality papers at the Cryogenic Engineering Conference and to provide recognition to those authors who, in the judgment of the CEC board, presented the best papers at the conference. The award consists of a certificate and a $500 honorarium.

The 1987 Scott Awards—for the best papers delivered at the 1985 CEC—were presented at the 1987 Awards Luncheon to Ray Radebaugh, James Zimmerman, David R. Smith, and Beverly Louie (NBS-Boulder) for their paper "A Comparison of Three Types of Pulse Tube Refrigerators: New Methods for Reaching 60 K" and to Herman H.J. ten Kate, Willem Nederpelt, Paul Juffermans, Frank van Overbeeke, and Louis J.M. van de Klundert (Twente University of Technology, The Netherlands) for their paper "A New Type of Superconducting Direct Current Meter for 25 kA."

1986 CRYOGENIC ENGINEERING CONFERENCE BOARD

ACKNOWLEDGEMENTS

The Cryogenic Engineering Conference Board wishes to thank these generous sponsors who contributed to the success of the 1987 Conference:

Air Products and Chemicals, Inc.
Alabama Cryogenic Engineering, Inc.
Astronautics Corporation of America
Ball Aerospace Systems Division
Birmingham Associates
Brookhaven National Laboratory
Creare, Inc.
Cryogenic Consultants, Inc.
CTI—Cryogenics Division, Helix Technology Corp.
CVI, Inc.
Fermi National Accelerator Laboratory
Gary Wheaton Bank of Batavia, Illinois
Harza Engineering Co.
Hitachi, Ltd.
Hughes Aircraft Co.
Ishikawajima—Harima Heavy Industries Co. (IHI)
Janis Research Co., Inc.
Koch Process Systems, Inc.
Lake Shore Cryotronics, Inc.
Massachusetts Institute of Technology
Meyer Tool and Manufacturing Co.
Mycom Corp.
NASA—Ames Research Center
National Bureau of Standards
Rotoflow Corp.
State of Illinois—Department of Commerce and Community Affairs
Sulzer Bros., Inc.
SURA/CEBAF
Teledyne—Wah Chang Albany
Union Carbide Corp.—Linde Div.
University of California—Los Angeles
University of Wisconsin—Madison
US DOE Division of High Energy Physics
US DOE Office of Fusion Energy
Westinghouse R & D Center

MAGNETIC RESONANCE IMAGING IN 1987

David E. Andrews

Oxford Superconducting Technology

Carteret, New Jersey

ABSTRACT

The superior performance of superconducting magnets for magnetic resonance imaging has led to the production of over 1000 magnets for this application. Further growth in this industry will depend on reducing system costs, extending medical applications, and easing the present siting problem. New magnet designs from Oxford address these issues. Compact magnets are economical to build and operate. A 4 Tesla whole body magnet for research in Magnetic Resonance Spectroscopy (MRS) has been successfully tested. Active-Shield magnets, by drastically reducing the fringing field, will allow MRI systems with superconducting magnets to be located in previously inaccessible sites.

MRI IN 1987

As of the spring of 1987, approximately 1100 one-meter bore superconducting magnets have been manufactured for whole body magnetic resonance imaging (MRI). These magnets comprise about 90% of the MRI magnets built, with the balance utilizing resistive and permanent magnet technology. Superconducting magnets have achieved pre-eminence in this application for several reasons:

1. They provide strong magnetic fields economically. Superconducting magnets in routine clinical use today range in field strength from 0.35 Tesla to 2 Tesla. Resistive and permanent magnets have only been practical up to the lower end of this range. Although the optimum field for imaging is the subject of much debate, fields below 0.35 Tesla clearly do not provide as high signal-to-noise data as higher fields. The MRI system design can

convert high signal-to-noise into improved spatial revolution, faster image acquisition, and/or reduced image "graininess".

2. Superconducting magnets in general provide a larger homogeneous volume, allowing imaging over a wider field of view. Weight and power consumption constraints lead to compromises in the useful imaging volume in permanent and resistive magnet designs.

3. Superconducting magnets provide magnetic fields of unsurpassed temporal stability. Operating in the persistent mode and isolated from thermal or electrical transients, stability of better than 0.1 ppm hour is routinely achieved. Good temporal stability is as important as good spatial uniformity in achieving the image quality the industry has come to expect.

4. Superconducting magnets for MRI are highly reliable and simple to operate. Because they operate in the persistent mode and use sealed off vacuum vessels, they are unaffected by power interruptions. The efficient cryogenic performance (specifications on helium consumption are now typically < 0.4 ppm/hour) allows operation in most cases without on site refrigeration systems. The only maintenance requirement is periodic topping up of the cryogens.

REQUIREMENTS FOR FURTHER GROWTH OF MR

Impressive as the last five years' progress has been, the installed base of MRI remains nearly an order of magnitude below that of the competing technology of x-ray Computed Tomography (CT). Several advances are necessary in order for the installed base to approach that of CT.

The price of an MR system is currently about three times that of CT. It must be brought down if the industry is to grow to its full potential. Although the magnet price is only about 20% of the selling price of an MR system, the increasingly cost conscious industry demands magnets that are both economical to build and economical to operate.

Specific medical applications of MRI are to date less widespread than are applications of CT. This is partly due to the newness, and great flexibility of the technology. Different pulse sequences are required to highlight different pathologies, and protocols to establish the optimal approach to different clinical symptons are still being worked out. In addition, there are technical issues in imaging the chest and abdomen, due to image artifacts induced by respiratory and cardiac motion. For this reason, images of the head and spine, which are easily immobilized and where MRI's soft tissue contrast makes it clearly superior to CT, account for about 70% of the scans performed today. New, faster imaging techniques which minimize motion artifacts promise to extend applications of MRI in the body.

A particular class of applications of magnetic resonance to medicine is Magnetic Resonance Spectroscopy (MRS). This technique utilizes MR's capability to identify particular chemical components via their chemical shift. It allows tracing of biochemical reactions in vivo. In principle it could be used to trace the metabolism of pharmaceuticals, assess the effectiveness of cancer therapy, assess cardiac tissue viability following an infarct, and diagnose specific metabolic disorders. In practice MRS is not yet an important clinical tool, despite a great deal of effort. The principle reason for this is the weak signal-to-noise available from the generally low concentration of compounds of interest. Improved signal-to-noise would result from higher field whole body magnets than the 2 Tesla units generally available today. The ultimate success of MRS in becoming a routinely used diagnostic test, rather than a difficult to use research device, will have an important effect on the total market requirements for Superconducting MR magnets.

A final impediment to the larger utilization of MRI equipment is the difficulty end users have in finding suitable sites for the magnets. An early concern that environmental steel would cause unshimmable field gradients was largely unfounded. A very real problem is the impact of the magnet's fringe field on the clinical environment. Public access to fields above five gauss must be controlled. For a 1.5 Tesla magnet, this means an area approximately 12 m x 10 m is required, and as many as five stories in a multistory hospital are affected. Reductions in this volume are possible by use of steel shielding, but the fact remains that MR installations often require special purpose buildings or extensive site rennovations to accommodate the equipment. The cost and red tape associated with such projects continue to slow the diffusion of the technology.

NEW MAGNETS FROM OXFORD

Oxford Superconducting Technology and Oxford Magnet Technology are bringing new magnet designs to the marketplace in 1987 which address the issues raised above.

Compact Magnets

In the interest of economies of scale, Oxford has had a uniform cryostat design for all field strengths from 0.35 Tesla to 2 Tesla for the past three and one half years. Over six hundred magnets have been shipped in the Unistat configuration during this time. To further reduce manufacturing costs, and to take advantage of weight and size reducing design improvements, Compact magnets were introduced in 1986 and are now in full production. Compact magnets have been built with field strength from 0.35 to 1.5 Tesla. Their lighter weight (under 4140 kg for 0.5 Tesla), makes them easier to use on mobile vans as well as in static installations.

Figure 1. Compact magnet on shaker table test

Their suitability for the mobile environment has been verified with tests on shaker tables (Figure 1). They can be ramped to 0.5 Tesla in under 10 minutes. Their helium consumption without refrigeration is the best in the industry at 0.35 liters per hour. Addition of a Gifford-McMahon two stage refrigerator can reduce this further, and eliminate liquid nitrogen consumption. Although the outside dimensions of Compacts are significantly smaller than Unistats, the bore size of the Compact is a full one meter, for compatibility with existing gradient coil and RF coil designs, and compatibility with possible future gradient coil enhancements. In summary, Compact magnets provide a number of economies. They serve the mobile MRI market, which allows costs to be spread among a number of hospitals. They are interchangeable between mobile and static installations, simplifying manufacturing planning. Their smaller size allows savings on material costs. Their efficient use of cryogens reduces operating costs. These economies are achieved without other performance compromises.

4 Tesla Magnet

A new tool for Magnetic Resonance Spectroscopy (MRS) research, to supplement the 2 Tesla High Homogeneity magnets presently in service, is now available. Tests have been successfully completed on a 4 Tesla whole body magnet, which will be shipped this quarter. Tests of the 4T magnet is the culmination of a three year development effort. Design challenges included:

a) Maintaining the required dimensional accuracy of the windings in the presence of the very large intercoil forces.
b) Providing for the safe dissipation of the 35 megajoules of stored energy in event of a quench.
c) Dealing with the total system weight of 43 metric tons.

The magnet reached 4.2 Tesla without quenching. An intentionally induced quench at 4 Tesla verified the protection scheme. A key specification for a magnet whose primary use will be MRS is the homogeneity. Measured values for this magnet were 6 ppm on a 40 cm diameter spherical volume (dsv), and 0.2 ppm on a 15 cm dsv.

Early clinical applications of MRS will most likely be done at 1.5 and 2.0 Tesla, where a significant installed base is available. Research at 4 Tesla should accelerate the development of this potentially revolutionary medical technique.

Active-Shield Magnet

The magnet development likely to have the largest impact on the MRI industry is the Active-Shield magnet. These magnets utilize superconducting shielding coils, at a larger radius than the primary coils, and in series with them, to drastically reduce the fringe field.[1]

Active-Shield magnets have been under development since 1983. A US patent was filed in 1984 and issued in 1986. In 1985 a 1.5 Tesla prototype was constructed. In order to reduce the cancellation of the central field by the shielding coils, they were placed at a much larger radius than the primary coils. This led to a rather unwieldy cryostat. Design effort since has focussed on a magnet whose physical size is comparable to existing magnets. In 1986, 0.7 and 1.0 Tesla prototypes were tested, with coil designs compatible with smaller cryostats. Shipments of 0.5 Tesla production units began in the spring of 1987 (see Figure 2) and tests of the first 1 Tesla and 1.5 Tesla production units are imminent.

SUMMARY

Magnet technology has kept pace with other technical innovations in MRI, the first large scale commercial application of superconductivity. Further growth of the

Fig. 2. Active-Shield magnet

The shielding performance of Active-Shield magnets is generally much superior to steel shielded magnets. A reduction of the volume within the 5 gauss envelope of 20 to 30 is achieved. Typical steel shielded designs achieve factors of 3 to 5, and in addition can be extremely heavy. Designs in excess of 50 tons are not uncommon for 1.5 tesla magnets, which leads to site restrictions.

industry will not be hindered by requirements on the magnets.

ACKNOWLEDGEMENTS

The work described here was performed by the author's colleagues at Oxford Superconducting Technology, Carteret, New Jersey, and Oxford Magnet Technology, Eynsham, England.

REFERENCE

1. D.G. Hawksworth, I.L. McDougall, J.M. Bird, D. Black, Considerations in the design of MRI magnets with reduced stray fields, presented at 1986 Applied Superconductivity Conference, Baltimore, MD.

THE PAST, PRESENT AND FUTURE OF MRI SUPERCONDUCTING MAGNET DEVELOPMENT

P.A. Rios

General Electric Company
Corporate Research and Development
Schenectady, NY 12345

and

R.L. Rhodenizer

General Electric Company
Medical Systems Group
Florence, SC 29501

ABSTRACT

MR magnets have quickly moved from applications in laboratory spectroscopy instruments to application in magnetic resonance imaging (MRI) in hospitals. The level of effort in universities and medical instrument companies directed at MRI and spectroscopy for diagnostics is substantial, and as a result, the magnet performance requirements are changing with time. Generally, these changing requirements are in the field strength, the field quality, and the interaction of the magnet with other system components. The application in hospitals, where cryogenics and magnets are not otherwise in use, brings its own set of demands to magnet design and performance. It is desirable to minimize the support necessary for cryogens by means of active coolers or liquefiers. Mobile MRI systems are also being used that impose their own set of demands on the magnet and cryogenic subsystem. Magnetic shielding that facilitates installation in the vicinity of other medical equipment is necessary in many installations.

Future developments will depend on the success of current research. Proposals have been made for MR systems that will require as much as 10 T in a 1 m bore for research. The applications that will prove to be successful in a clinical environment will ultimately depend on the cost and effectiveness of the procedure. The magnet, which is a major element in terms of both cost and performance, will play an important role.

INTRODUCTION

For over 30 years nuclear magnetic resonance (NMR) has been used in laboratories to perform chemical analyses. Instruments to perform magnetic resonance spectroscopy (MRS) analyses have been commercially available from several companies. The instruments have evolved to the point where ancillary equipment is available that makes it possible to introduce samples automatically in the instrument and record the resulting data in computer files.

MRS has used increasingly higher fields (systems operating at over 10 T are available commercially), but magnets for chemical analysis have only to accommodate maximum samples of approximately 0.025 m diameter. Superconducting windings were introduced into MRS magnets in order to achieve the required high field levels.

More recently, magnetic resonance imaging (MRI) has become a valuable clinical tool. Although MRI is performed at lower fields than MRS, the medical applications generally require that the patient's body be accommodated in the magnet. The magnet technology that has to be used to satisfy this much larger size is substantially different from that of small magnets, since structural problems, thermal problems, and interactions with the environment become much more important in large magnets. Issues related to magnet protection in the event of a transition to normal of a superconducting winding and safety are also crucial, since the stored energy in a whole-body magnet can be substantial.

In the last three years, the MRI United States market has increased fourfold to $510 million in 1987[1]. The capabilities of the instruments are rapidly improving, and an increasingly diverse group is being exposed to the technique, so that new applications are created.

In this paper, the authors will trace the development of magnetic resonance magnets from the chemical spectroscopy applications to the medical applications and will discuss the relevant technological issues that have been addressed in the development of the magnets. A perspective on what the authors believe to be directions for the future is discussed at the end of the paper.

MAGNETS FOR MAGNETIC RESONANCE

The physical principles of magnetic resonance have been covered in a variety of papers and books[2,3] and will not be covered here. The magnet provides the static field that determines the resonance frequency of the chemical species being analyzed. The requirements that are placed on the magnet are first of all one of providing a uniform and accurate magnetic field that is constant with time.

Since it is generally not possible to attain the specified levels of field uniformity in a magnet by carefully designing and fabricating the magnet windings, additional sets of coils are provided that are separately excited to correct for field nonuniformities inherent in the magnet itself or caused by magnetic materials in the environment.

MRS MAGNETS

The earliest application of NMR was for the chemical analyses of solid, liquid, or gaseous materials. Such analyses are based on the absorption of radio frequency radiation in samples that are placed in a uniform background magnetic field. Spectrometers for this purpose were introduced in 1954 and are currently available commercially from several sources.

High resolution NMR spectrometers utilize superconductive magnets to generate the background magnetic fields. Typically, the magnet axis is aligned in the vertical orientation and samples are introduced into the test region with sophisticated sample holders, which often allow the sample to be spun about the vertical axis to reduce the negative effects of spatial field inhomogeneities. Magnets are currently available with bore sizes up to 0.025 m and with field strengths exceeding 10 T. To resolve fine structure in the measured spectrum, the uniformity of the background field over the sample is in the range of 0.01 to 0.1 ppm. Temporal stability (usually better than 0.1 ppm/hr) is achieved through persistent operation with the magnet current trapped in a closed loop through a superconductive switch.

Although high resolution magnets are not physically large by present day standards for superconductive devices, some larger systems are used in horizontal arrangements for studies on bigger test specimens, e.g., large mammals. Typically these devices allow chemical shift imaging in addition to spectroscopy with bore-mounted gradient coils provided for spatial resolution of the rf signals. Field strengths in these systems are usually somewhat lower than for the high resolution spectrometers, up to the range of 5 to 7 T, but magnet cryostat, free bore sizes are larger, typically several tens of centimeters. Such a system is shown in Figure 1.

DEVELOPMENT OF MAGNETS FOR MEDICAL IMAGING

Recent advances in computer and imaging technologies have made MRI possible. In the early 1980's MRI units were installed in selected hospitals where research was conducted to investigate the potential for clinical efficacy of the technique. Initially, imaging was performed in electromagnets using resistive windings, but these magnets have limited field capability in sizes compatible with whole bodies without the expenditure of excessive power. Imaging at higher fields led to the development of superconducting magnets. Permanent magnets provide a third alternative at moderate fields.

The selection of an operating field level is a complex issue that requires tradeoffs among many parameters and components. As happens with such complex issues, different organizations have taken different approaches, and today there are commercially available MRI systems that operate anywhere from 0.1 T to 1.5 T. Consequently, magnets that use resistive windings, permanent magnets and superconducting magnets are all being used in hospitals today.

It is desirable to operate the magnet at a high field level in order to increase the signal-to-noise-ratio, which can translate into faster scanning or improved resolution. Field is proportional to frequency, and operation at high frequency means that shielding currents that are induced in the body result in greater attenuation and phase shift. RF power requirements also rise with frequency. Greater field uniformity has a beneficial effect both in terms of RF power deposition as well as in terms of image quality, but in general the required allowable field nonuniformity as a fraction of the central field decreases at least proportionately with field strength. Consequently, the magnet performance becomes increasingly difficult to attain, not only because of the challenges inherent in operating at high field, but also because the requirements in terms of field quality also become increasingly difficult.

Operation at higher fields opens up a new possibility: imaging and spectroscopy can be performed at a single field and would allow the physician to perform chemical spectrosocpy on a patient after having utilized imaging to identify an abnormality.

Figure 1. Horizontal MRS magnet.

RESISTIVE MAGNETS

Resistive magnets for MRI have been designed using circular coils, with or without an iron yoke. Figure 2 shows a resistive magnet designed for whole-body imaging. Designs have been developed using iron, which utilize windowframe geometry or shaped poles. Since the supporting structure for the coils are generally accessible in resistive magnets, provision for adjustment of the coil position (shimming) can be done, at least to correct major nonuniformities. Correction coils can be added to more accurately adjust the field uniformity.

The effective use of iron in these magnets has the advantages that the power requirement can be reduced since the iron can enhance the field by 10% to 25%, and at the same time reduce the fringe field.

Practical, resistive MRI magnets for whole-body imaging will require tens of kilowatts of power, and in general must be water cooled. Since dimensional stability is critical to field uniformity and temporal stability, it is necessary to control the winding temperature and the coolant temperature. It has been suggested that resistive magnets can be turned on only when the magnetic field is required for imaging in order to conserve electric power; however, in practice it is difficult to control the thermal transient so that imaging can be effectivley done within a short time of energizing the magnet.

The power requirement for resistive magnets is strongly dependent on field level and magnet bore size. Gordon and Timms[4] give estimates of typical power requirements for resistive magnets that utilize four- and six-coil designs. Their curves are reproduced in Figure 3. Since the magnet is connected to an active power supply while imaging is being performed, an accurate and stable power supply is required to satisfy the temporal stability requirement.

It is clear, that resistive magnets can be applied to MRI only at moderate field levels. Although the economic tradeoffs will depend on local conditions, i.e., cost of electricity, superconducting magnets are generally more attractive at fields much above 0.2 T in typical United States installations. Rios and Laskaris[5] show comparative costs for resistive electromagnets, permanent magnets, and superconductive magnets. Figure 4 shows this comparison, which includes capitalized cost of consumables for shielded magnets.

PERMANENT MAGNETS

MRI magnets that use permanent magnets, in spite of being generally heavier than electromagnets, offer a major advantage in relation to resistive electromagnet in that they do not require excitation power or a coolant supply. In principle, once they are fabricated and the magnet is magnetized, they will operate with very little attention.

Figure 2. Resistive magnet for whole-body MRI. (Courtesy of Oxford Superconducting Technology, Carteret, NJ)

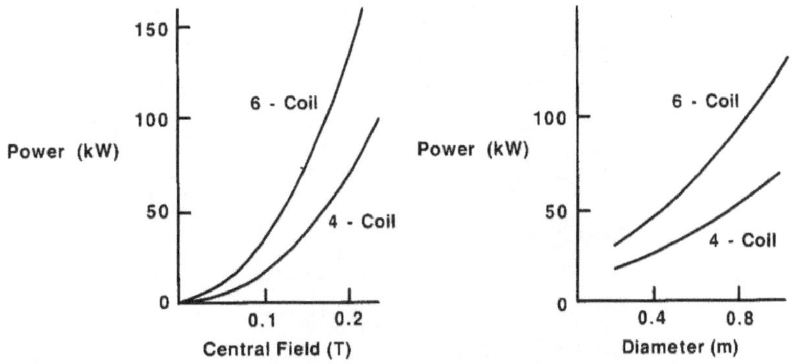

Figure 3. Typical power requirements for resistive magnets. (a) Power requirement vs. central field for a magnet with a 0.8-meter-diameter bore (b) Power requirements vs. bore diameter for magnets operating at 0.15 T.

Magnets that utilize ferrite and higher performance materials such as neodymium-iron have been considered, and MRI systems that operate at 0.3 T are commercially available today, for example, from Intermagnetics General and Fonar. (See Figure 5).

In designing an MRI magnet using permanent magnets, the cost of the magnet material must be carefully traded off against magnetized volume and field. . As field and/or magnetized volume result in rapidly increasing weight and cost of permanent magnet material, it is essential that iron be used to minimize leakage field and that the magnetized volume be minimized. Therefore, practical MRI magnets that use permanent magnets exhibit very little leakage field, and the space available for the imaging coils and the patient is usually smaller than that in superconducting magnets.

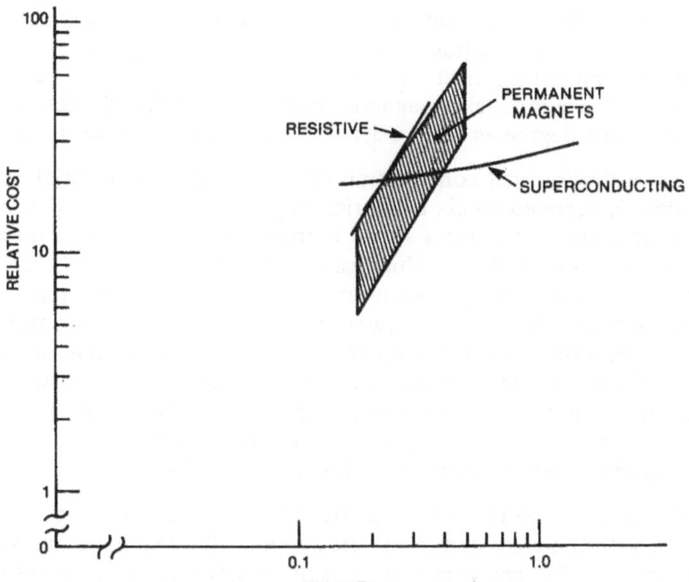

Figure 4. Relative costs for MRI magnet technologies.

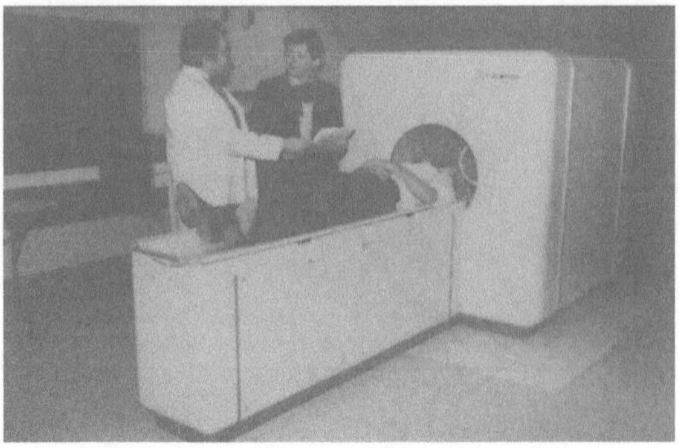

Figure 5. Permanent MR magnet. (Courtesy of Intermagnetics General Corp., Guilderland, NY)

Zijlstra[6] discusses magnets that allow for 1 meter access hole diameter, a common bore size for superconducting magnets, and concludes that systems of reasonable weight and cost can be used under 0.2 T. The region for permanent magnets shown in Figure 4 is bounded at the top by MRI magnets using neodymium-iron and at the bottom by MRI magnets using ferrites, which are considerably heavier. These curves in general agrees with Zijlstra's conclusions.

SUPERCONDUCTING MAGNETS

Superconducting magnets offer unique capabilities in whole-body MRI. Superconducting windings can have nearly unlimited ampere-turn capability and can therefore produce relatively high fields in the required large volume at moderate cost. Since the magnet is electrically lossless, power supplies are only needed for initial excitation, and the magnet may continue to operate in the persistent mode.

Superconducting windings can be used in designs similar to those of magnets using resistive windings or permanent magnets—with iron yokes to reduce the fringe field. There is no disadvantage inherent to superconducting magnets in terms of fringe field. On the other hand, the equipment to accommodate the use of liquid helium, refilling with liquid helium or the application of active refrigerators does add complexity and cost that makes these magnets attractive only if the central field exceeds 0.2 T in typical US installations. Figure 4 reflects this assertion. Since these magnets are typically used only at the higher field levels, magnetic shielding of the fringe field is usually not carried out to the same degree as with resistive electromagnets or permanent magnets.

Figure 6 shows the internal construction of a superconducting MRI magnet. The main magnet field is supplied by six cylindrical superconducting coils that are arranged about the plane of symmetry inside a support structure. Other main coil arrangements may be used, either fewer coils (in which case the nonuniformity of the field will increase), or solenoidal coils with thickened ends. The coils may be wound on the outside of the support structure and overwound with with tape or filaments to retain them in place. In general, the main magnet is housed in a liquid helium tank that provides cooling at 4.2 K. Correction coils may be superconducting, in which case they are also placed in the liquid helium container, preferably inboard of the main magnet. Resistive correction coils may be used in the warm bore of the magnet, or a combination of resistive and superconducting correction coils may be used.

When superconducting correction coils are used, they can also be operated in the persistent mode, with the leads removed. However, if the magnet is operated at 1.5 T or above and is used for spectroscopy as well as imaging it is necessary to make corrections for the field nonuniformity introduced by the patient's body. In that case, provision must be made to easily make minor corrections in field uniformity.

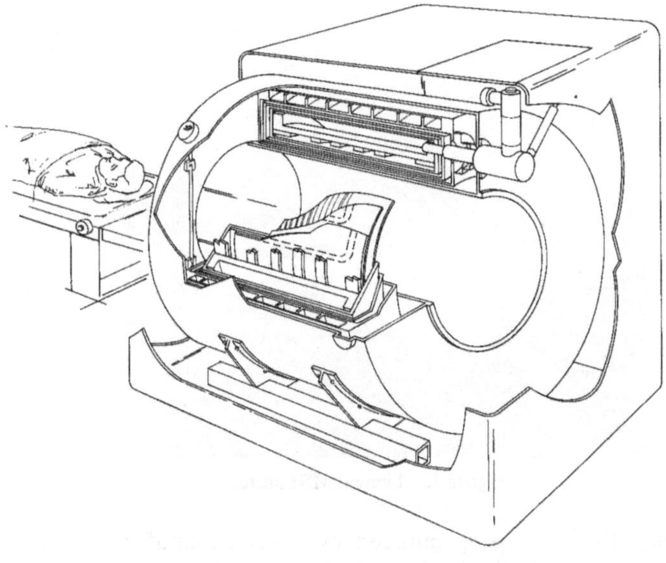

Figure 6. Superconducting MR magnet.

In Figure 6 the leads are shown entering the cryostat at the end of the cryostat. Many commercial magnets have leads that enter at the top to take advantage of thermal stratification. The helium vessel is supported on struts made from glass-reinforced epoxy, and shielded by one or more thermal shields. Figure 6 shows a vapor-cooled shield and a liquid nitrogen-cooled shield. Active refrigerators may be used to cool one or more shields, or to reliquefy the helium in order to minimize service calls for refilling. This feature may be very important if the unit is to be located where liquid helium is not readily available or expensive.

CLINICAL MRI

Commercially available MRI systems have now been available for several years and are no longer considered as research tools but as viable diagnostic devices. The majority of the equipment installed today utilize superconductive magnets with fields in the range of 0.5 T to 2 T. The imaging devices are operated by hospital technicians who are relatively unskilled in cryogenics or superconductive technology and are maintained by service personnel specially trained by the manufacturer. A typical MRI suite is shown in Figure 7.

From the user's point of view, the equipment must be available nearly full time at peak performance with little interuption for service or repair. This implies that the magnet deliver a rated uniform field which is stable over time and which can be maintained with little or no intervention. Typical performance values for high quality imaging would require that field non-uniformity over the imaging field of view be less than 10 ppm and that the non-uniformity be stable over time to better than 0.1 ppm per hour. It is desirable that service intervals be as long as possible with times of 3 months being typical. For cryogen service this implies relatively large storage volumes in the cryostat for the cryogenic fluids (both liquid nitrogen and liquid helium) coupled with high cryostat thermal performance. Liquid loss rates are generally better than 0.5 liters per hour for helium and less than 1 liter per hour for nitrogen.

To ease cryogen service needs, various refrigeration schemes may be considered to reduce or eliminate cryogen losses. Commercially available, Gifford-McMahon shield coolers are used in several cases to reduce losses. Gifford-McMahon recondensers or

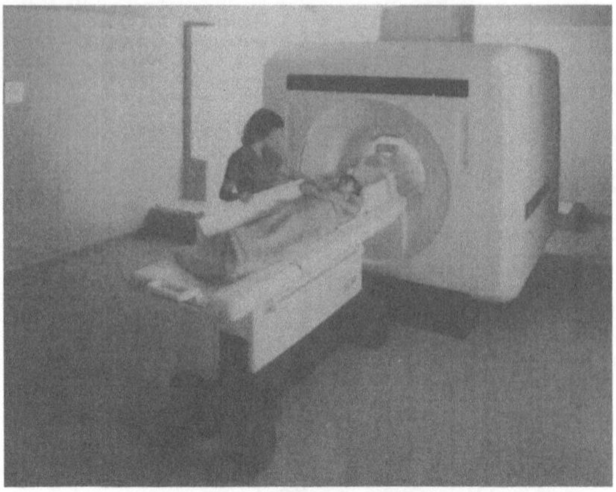

Figure 7. Typical MRI suite.

Claude cycle reliquefiers are being pursued by several manufacturers to significantly reduce the loss of cryogenic fluids, but as of today, the reliability of such systems makes their. commercial application questionable.

Siting is normally a major concern for the installation of an MRI system. One concern is shielding for the stray magnetic fields from superconductive magnets which may exceed acceptable levels in areas adjacent to the imaging suite. Solutions for this problem generally involve the use of iron shields or possibly active shilds, either close to the magnet or in the walls of the room. In either case such shielding is a significant cost factor in the installation and requires close integration of the facility with the magnet design.

A unique siting situation is the use of superconductive magnets in MRI systems installed in mobile vans. This application further emphasizes the need for compact, light weight, thermally efficient designs for practical use. These systems must also tolerate the mechanical loads associated with transport over roads for distances exceeding 100,000 miles and thus presents the magnet designer with the challange of balancing robust mechanical design with high cryostat thermal performance. A mobile installation is shown in Figure 8.

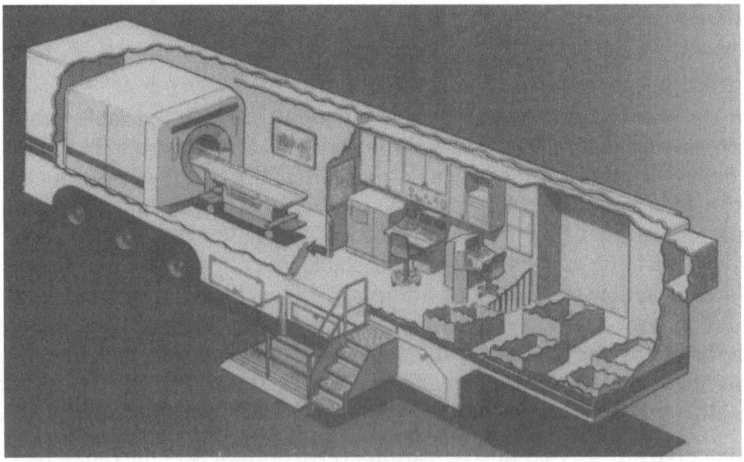

Figure 8. Mobile MRI installation.

CLINICAL SPECTROSCOPY

Investigators continue to search for biological and clinical applications of spectroscopy. Such applications derive clinical information from in vivo MR spectral data. Typically nuclei other than hydrogen are used including ^{31}P,^{13}C,^{19}F,^{23}Na.

It is often the goal to complete spectroscopy measurements in the same magnet used for imaging. This places some additional constraints on the system design. For example, as spectroscopy generally requires relatively high field strengths, the magnet must perform to the upper limits of field normally associated with imaging, e.g., 1.5 T. In addition, field uniformity needs for high resolution spectroscopy exceed those of imaging by, typically, a factor of ten. Generally, however, this is for substantially smaller volumes so that imaging magnets are usually compatible with this requirement.

Research on clinical applications of MR is just beginning. Several plans for research systems have been proposed that will require large, high field, high performance magnets. Typical among these is the proposed "decatesla" which is based on a 10 T, one-meter bore system. It is expected that spectroscopy will be a viable clinical tool in the near future.

OTHER APPLICATIONS

As mentioned earlier in this paper, the exposure of scientists and engineers in diverse fields to magnetic resonance and the capabilities of MRI and MRS are rapidly increasing. One can expect that the cost of compact computers and power electronics will continue to drop in the future, as it has for the past decade. Compact, economical magnets, the other remaining major component in the system, can open up applications of magnetic resonance that are now not practical.

In medical applications MRI had been a relatively expensive instrument limited to hospitals or groups of hospitals with substantial resources. Installations are being made today at self-standing imaging centers and community hospitals[7]. Lower cost systems might address different applications where whole-body magnets and imaging coils are not necessary, as in the imaging of extremities, and can place the instrument within the reach of a broader group of practitioners.

Since the magnet and its associated environmental shielding today represent a major element of cost, it is necessary to move toward very compact systems that retain the patient's space; system tradeoffs are crucial. These tradeoffs will be between performance and cost and between magnet component costs and electronic component costs. Because space is at a premium, there will be increasing electromagnetic coupling between the magnet and other components of the system.

For applications where small magnets are possible, i.e., extremities, resistive windings, and permanent magnets will compete with superconducting magnets. For superconducting magnets to be viable, new, simpler concepts that reduce first-cost and reliable, economical cryorefrigerators that can substanitally reduce the service requirements of the magnet are essential.

In the past five years, superconducting magnet manufacturers have undergone a transformation as a result of the need to produce magnets in quantities and the need to install magnets in an environment where the user may not be familiar with cryogenics. They have had to improve the interface of the magnet with the end-user and to improve the reliability of the magnet to minimize its impact on the system. Compact helium liquefiers must still undergo that transformation if they are to be successfully applied to these systems.

By performing imaging and spectroscopy in animals, medical and pharmaceutical information can be obtained without sacrificing the laboratory animals. Because of similarities between the systems of humans and mammals, it is desirable to provide the capability for using the instrument with large mammals. Magnets that would be desirable for this might have a working bore of 0.5 m and operate at 5 T to 7 T.

It is technically feasible to apply proton imaging, as it is done today, in nonmedical applications. Water infiltration in solids is one area that is amenable. Some examples of applications are measurements of moisture content in soils and concrete. Other potential applications in which moisture or water content is important include drying processes, oil-well core measurements, and moisture measurements in composites and adhesives.

Magnets that provide fields of view outside their bore such as used in the above application are needed when it is not practical to surround the sample with the magnet. One example of such an application, oil well logging, has been reported on Jackson, Brown, and Koelle, who have proposed opposing magnets to provide a toroidal uniform field region away from the centerline of the magnets[8].

Ultimately, magnetic resonance offers the potential for process control—homogeneity of compounds, monitoring of chemical reactions, monitoring of mixtures, monitoring of polymerization—which will require rugged, reliable magnets and refrigerators at moderate cost. Development of solids imaging techniques could open up a variety of NDE applications in composites and ceramics for application in industry. Possibly the newly discovered ceramic superconductors will make it possible to reduce magnet costs by elimination of the vacuum insulation and the use of simpler refrigerators.

CONCLUSION

Magnets for MRS and MRI are now well established. Superconducting magnets represent the major fraction of these magnets, for spectroscopy, chemical shift imaging, and medical proton imaging, but resistive windings, permanent magnets, and hybrid magnets are being used. The transition has been made from a laboratory device to serially produced magnets that provide high-quality magnetic field characteristics and a suitable interface for the user. Current publications show that emphasis is now being placed on reducing cost, size, and environmental shielding of medical magnets.

As the use of magnetic resonance has exploded, new applications are constantly being considered. Imaging and spectroscopy techniques are advancing rapidly. Compact, rugged, and low-cost magnets are needed to make possible broader application in medicine and pharmaceuticals, as well as industrial applications in process monitoring and NDE.

REFERENCES

1. *Diagnostic Imaging*, Vol. 8, No. 11, November 1986, p. 113.
2. L. Kaufman, L.E. Crooks, and A.R. Margulis, *Nuclear Magnetic Resonance in Medicine*, Igaku-Shoin, New York (1981).
3. P. Mansfield and P.G. Morris, *NMR Imaging in Biomedicine*, Academic Press, New York, (1982).
4. R.E. Gordon and W.E. Timms, "Magnet Systems Used in Medical NMR," *Computerized Radiology*, Vol.8, No.5, pp. 245-261, (1984).
5. P.A. Rios and E.T. Laskaris, "Magnets for Magnetic Resonance in Medical Diagnostics: From the Laboratory to a Product," *Proceedings of the 9th International Conference on Magnet Technology*, SIN, CH-5234 Villigen, Switzerland, pp 231-235 (1986).
6. H. Zijlstra, "Permanent Magnet Systems for NMR Tomography," *Philips Journal of Research*, Vol 40, No. 5, pp 259-288 (1985).
7. B. Teal, "Marketing MR in a Community Hospital," *Administrative Radiology*, November (1986).
8. "Western Gas Sands Project Los Alamos NMR Well Logging Tool Development," Progress Report LA-9151-PR, Los Alamos National Laboratory, March 1982.

DYNAMIC PROCESSES IN STRING OF FULL SCALE SUPERCONDUCTING MAGNETS WITH FIELD 5 T

A.I.Ageyev, A.M.Andriishchin, V.I.Gridasov, E.M.Kashtanov, B.V.Kaz'min, S.N.Logachev, K.P.Myznikov, M.B.Nurusheva, A.P.Orlov, M.V.Pryima, A.N.Shamichev, V.V.Sytnik, V.A.Vasiliev, L.M.Vasiliev, O.M.Veselov, A.N.Yerokhin, N.N.Yarygin, S.I.Zinchenko

Institute for High Energy Physics
Serpukhov, USSR

ABSTRACT

The system design for a string of full-scale superconducting magnets for UNK is given. The magnets have iron at room temperature. The system uses single phase flow through the magnet and two-phase return-flow cooling through the outer wall. Measured transient temperatures are reported for pulsed operation at design conditions. Measured cryogenic and electrical transients of the system are reported for the quench of a single magnet.

INTRODUCTION

IHEP simultaneously with model magnets for UNK studies the features of the common work of connected-in-series full-scale superconducting dipoles[1,2]. For this purpose the force-circulating test facility based on the 500 W helium refrigerator operating at 4.5 K is used. Presently, the dynamic characteristics of the cryogenic and electric schemes when simulating the accelerator cycles and emergency operational modes of a two-magnet string are studied. Next stage of the work will be conducted with a four-magnet string.

THE CRYOGENIC SCHEME

The cryogenic scheme of the facility provides cooldown, cooling and warm up of a magnet string as well as helium removal during a superconducting-to-normal transition. A possibility is envisaged to perform study on heat exchange and hydraulics, determine heat influxes in string elements and dynamic heat releases into the magnets, simulate various schemes of cooldown and warmup techniques.

The cryogenic system of the test facility consists of the KGU-400 liquefier, satellite refrigerator and force-circulating block[3]. The system determines the initial parameters of the helium flow: consumption rate, vapour contents, temperature and pressure. Helium flow having the required parameters leaves the circulation block to go into the main block of the string, fig. 1, and cool the superconducting wires connec-

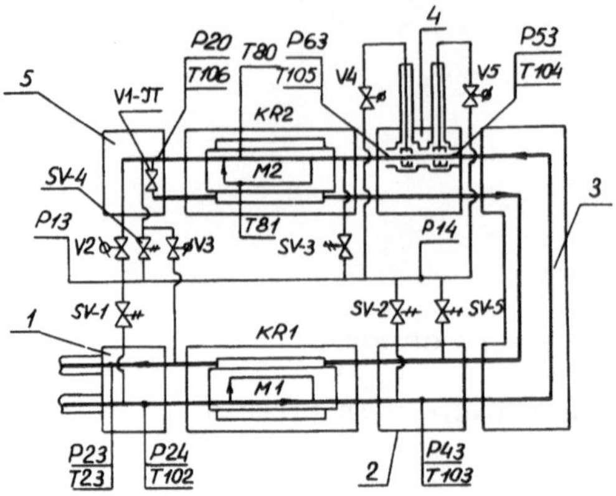

Fig. 1. The layout of the string and helium flows: 1 - main block, 2 - intermediate block, 3 - connecting block, 4 - safety lead block, 5 - return block, KR1, KR2 - cryostats with dipole magnets M1, M2, SV(1-5) safety valves, V1-JT - throttle valve, V2,... V5 - regulation valves, T102, ... T106 - temperature sensors, P20, ... P63 - pressure gauges.

ting the coils with the power source. Then this flow passes through the 1st SC magnet (M1), intermediate block 2, connecting block 3, safety lead block 4, the 2nd SC magnet (M2) and goes into the return block 5. Single-phase subcooled helium is supplied into the string in the mode similar to the UNK operating conditions. In M1 and M2 magnets the single-phase flow splits into the inner flow and by pass one having about equal consumption rates. The inner flow cools a superconducting coil. Heat removal from the coil warms the flow thus increasing its outlet temperature. The bypass flow passes through the channel on the outside of the coil bandage and gets cooled by the colder two-phase counterflow. Heat exchange between the flows takes place through the walls of the helium vessel. After the cryostat, the inner and by-pass flows blend resulting in the decrease of the temperature of the mixed flow. In the return block 5 single-phase helium is turned into two-phase helium with the help of the throttle valve V1 and counterflowing through the superconducting magnets removes heat from the single-phase flow and goes into the force-circulating block. The system makes it possible to cool the string with single-phase or two-phase helium.Valve V1 and V2 are used to choose the operating mode while valve V3 is used during cooldown. Valves V4 and V5 are used to cooldown safety leads. Safety valves SV (1-6) are designed for helium removal when quenches occur.

ELECTRIC SCHEME POWER SUPPLY AND PROTECTION OF THE STRING

The circuit is presented in fig. 2. The pulsed power supply provides ramp at the 125 A/sec rate current in the following cycling mode: 40 s rise, 40 s flattop, 40 s drop. In the operating mode current flows through the coils of magnets M1, M2. In this case the thyristor key TK4 is opened and the automatic switch AK is closed, the switches TK1, TK2 are locked. The status of the magnets is controlled with quench detectors (QD) by the signals from the potential outputs PLM1, PLM2, from magnets M1 and M2. If a quench occurs the QD records the active voltage in the magnet[4], the registration threshold being 1 B.

In the course of study magnet M2 was forced to go normal by pulsed heat input with the help of a tape heater TH2 placed on the 2nd shell of the coil. A pulse from the heater causes a large part of the coil to go to the normal state thus preventing the superconductor from overheating. Simulta-

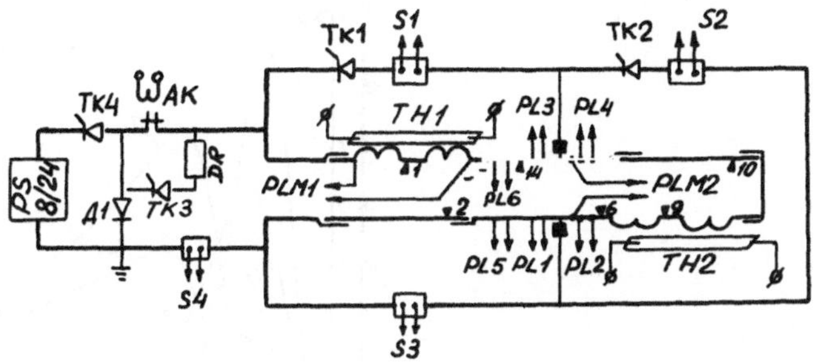

Fig. 2. Electric scheme of the string. PS 8/24 (8 kA, 24 V) - pulsed power supply; TK1,... TK4 -thyristor keys; AK - automatic switch; DR-damp resistor of the stored energy; S(1-4) - shunts to control the current; TH1, TH2 - tape heaters; PH (∇1, ... ∇10, ∇14) point heaters; PL1, ... PL6 - potential terminals; PLM1, PLM2 - potentials leads from the magnets.

neously, magnet M1 remains to be superconducting. The thyristor key TK4 is shut off and TK3 is opened. Energy is removed from magnet M1 in the planned operating mode and onto damp resistance DR. The switch TK2 is opened to provide shunting magnet M2 through safety leads connected between the magnets. A copper cylinder-shaped quench stopper having radiators for cooling purposes provides electric contact between the leads and superconducting cable. Point heaters ∇(1,..., 14) make it possible to cause a quench of any section of the string. The quench propagation process is monitored using the voltage readings from potential pairs PL1-PL6.

TEST RESULTS OF THE STRING

The test programme was aimed primarily at the study of nonstationary cryogenic and electric processes in the string with one magnet quenched. The cooling mode was close to the operating conditions of the superconducting ring of UNK: the temperature of the inlet helium flow was 4.3 K, the pressure was 2.5 bar, a consumption rate of 90 g/sec. The current ramp rate was 125 A/sec. Magnet M2 was quenched intentionally with the help of heater TH2 at currents ranging in the string up to 6 kA when the pressure in the contour was reduced to 1.1 bar. In this range of currents no quench propagation over the cable through the quench stopper was observed. Magnet M1 retained superconducting properties.

Figure 3 shows the temperature and pressure changing in the helium flow when magnet M2 goes normal at 5 kA.Curve E_i describes variation in the instantaneous power released into the coil of M2.As seen,the current damps in magnet M2 within about 1 sec, while that of the string falls with a time constant of 10 sec. Curves T105, T104 and T103 show the temperature growth of the flow at the input port of the QS, its output port and close to dipole M1, respectively. As seen, the helium flow temperature raises to 7 K before the QS to fall rapidly after it with the temperature disturbance in the flow not increasing 0.4 K before dipole M1. Curves P63, P53 and P43 present the pressure front propagation of the flow from M2 to M1. The pressure is seen to reach its maximum value for each contour cross section basically simultaneously with the peak of instant power releasing into the coil of dipole M2. The maximum pressure measured at the input port of the QS is 0.72 MPa. According to the calculated estimates, the pressure developed inside magnet M2 can be as large as 1.0MPa being the maximum tolerable value for some elements of the cryostat. This imposes certain contraints on the value of the operating current in the string cooled by single-phase helium.

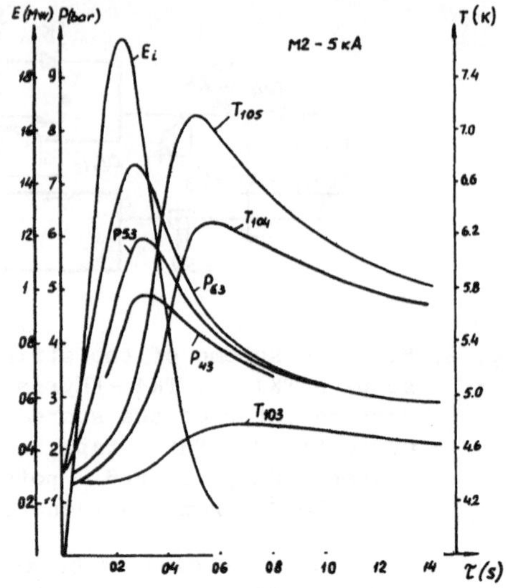

Fig. 3.

The dynamic characteristics of the string cryogenic system (E_i is instant power of the energy by magnet M2; T105, T104, T103 - the temperatures and P63, P53, P43 are the flow pressures in the specific points of the string, Fig. 1).

Figure 4 shows the development of helium flow temperatures in the cryostat of M2 when the string is powered with the cycles of 5 kA, 40-40-40 s trapezoidal currents. During temperature measurements the helium consumption decreases to 50 g/sec. The inner single-phase helium flow cooling the coil is warmed cycle-wise due to ac loss in the magnet (curve A). In this case the temperature change is not more than 0.05 K. The bypass flow temperature (curve B) does not vary because it exchanges the heat with the two-phase counterflow (curve C). To compare time-dependent temperature and current change the dashed lines in the figure show two cycles of current.

CONCLUSIONS

The results on the studies performed confirm that the chosen cooling scheme and that of electric protection of the string of 5 T magnets are correct. The observed apprecialbe helium pressure growth during emergency quenches of the magnets requires that the strength of the cryostat elements should be increased for them to operate in nominal cycles of 6 kA currents. These reasonings suggest that the operating current of the magnet should be cut to 5 kA. The dynamic characteristics will be studied further with the help of a four-magnet string of the same design.

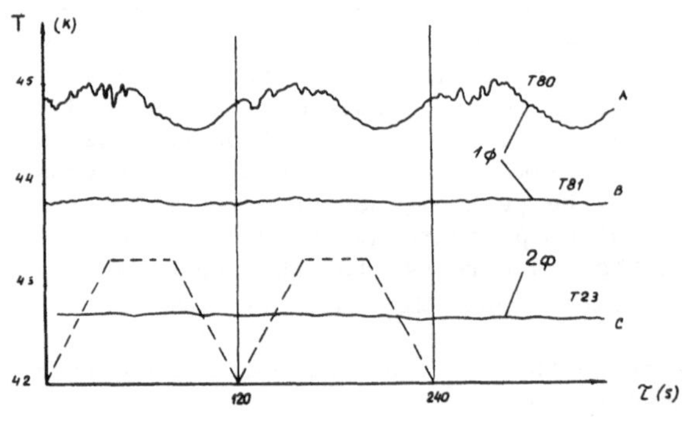

Fig. 4.

Variation in the helium flow temperature during the cyclewise operation of the string: T81 - the bypass flow temperature, T80 the temperature of the flow passing through the coil, T23B - the temperature of two-phase flow leaving the string.

REFERENCES

1. A.I.Ageyev et al. Development of Superconducting Dipoles for UNK. Proc. of Workshop on Superconducting Magnets and Cryogenics. BNL, 1986, p. 13.
2. A.I.Ageyev et al. "Advances in Cryogenic Engineering", v. 31, Plenum Press, New York (1986), p. 167
3. A.I.Ageyev et al. Test Facility for Full-Scale Superconducting Magnet for UNK. Proc. X Int. Cryog. Eng. Conf. Butterworth, Guildford, UK, 1984.
4. G.M.Antonichev et al. Study of Quench Processes in UNK Magnets. Proc. of ICFA Workshop on Superconducting Magnets and Cryogenics, BNL, N.Y., 1986, p. 239.

DESIGN OF A STRONGLY CURVED SUPERCONDUCTING BENDING MAGNET FOR A COMPACT SYNCHROTRON LIGHT SOURCE

B. Krevet, H. O. Moser

Kernforschungszentrum Karlsruhe

C. Dustmann

Brown Boveri & Cie.
Federal Republic of Germany

ABSTRACT

A design of a 90° bending magnet for a synchrotron radiation source with a characteristic wavelength of 0.2 nm is presented which involves a multilayer coil with circular cross-section optionally surrounded by a C-shaped cold iron shielding, and a cold bore. The usefulness of this design is checked by dynamic aperture calculations including nonlinear effects through 3rd order for the lattice of the proposed synchrotron radiation source at Karlsruhe.

INTRODUCTION

Compact electron storage rings are receiving growing attention mainly for their potential use as synchrotron radiation sources in x-ray lithography[1]. The compactness of these rings depends, at least partly, on the use of superconducting bending magnets which are characterized by a comparably small bending radius, or a strong curvature, and a relatively large bore inside

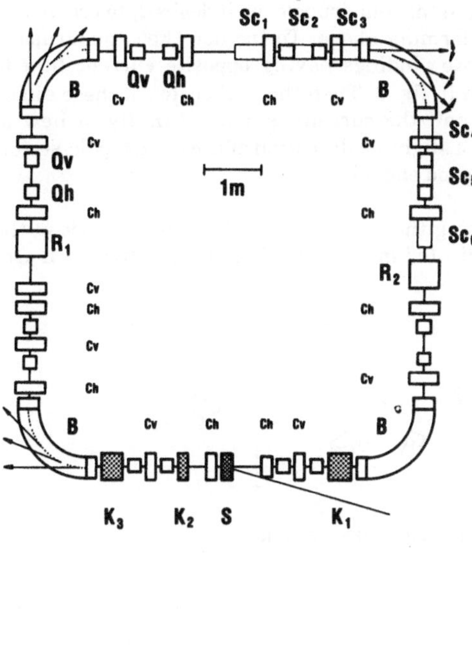

Fig. 1. Schematic of the planned Karlsruhe synchrotron light source for in-depth x-ray lithography. B superconducting bending magnets; Q_h, Q_v horizontally and vertically focusing quadrupoles; C_h, C_v horizontal and vertical corrector magnets; K 1-3 injection kickers; S septum, R_1, R_2 radiofrequency cavities; Sc 1-6 vertical scanning dipoles.

the coils. Fig. 1 shows the lattice of the compact electron storage ring planned at Karlsruhe[2] to be used as a synchrotron light source for in-depth x-ray lithography according to the LIGA process[3]. It is based on 4 superconducting dipole magnets with 90° bending angle. Their field strength and bending radius are 4 T and 1.2 m, respectively. With an electron energy of 1.4 GeV the design wavelength of $\lambda_c = 0.2$ nm is obtained.

Superconducting bending magnets are well established in large accelerators. They must produce extremely homogeneous fields and must have sufficient mechanical restraints to ensure that no movement of conductors is possible which would result in field errors or in quenching of the magnets. Both requirements are met by magnets with a circular cross-section for which a proven technology has been developed at Fermilab[4] and DESY[5]. For the position of conductors the cos(θ) current distribution is used. Preloading the coils is easily provided by radial pressure exerted by clamps.

Applying this technology to compact synchrotron light sources, two additional problems arise. First because of the large bending angle the magnet must be bent, too. Second to let the synchrotron radiation exit there should be a slot extending all over the circumference. Both problems require a modification of the cos(θ) current distribution, the latter in addition a modification of the mechanical structure. This paper describes the layout of a curved dipole magnet built from coils having circular cross-section with the specifications needed for compact storage rings.

MAGNET DESIGN AND 3-DIMENSIONAL FIELD CALCULATIONS

The effect of the slot for the synchrotron radiation is significant. To have sufficient space for the mechanical support this slot should be at least 2 cm wide, allowing the windings to start at an angle of 9°. This requires a complete modification of the cos(θ) distribution. On the other hand the influence of the curvature turns out to be small, allowing to treat it as a perturbation of a two dimensional design. Hence the design is started with a two dimensional optimization program which uses the formula for the vector potential A of a p-pole magnet as given by [6,7]

$$A(r,\theta) = \frac{2\mu_0}{\pi} J \cdot \sum_n \frac{p \cdot r^n \cos(n\theta)}{n^2} \sin(n\phi) \int_{a_1}^{a_2} \frac{da}{a^{n-1}}$$

where $n = 2k + 1, k = 1, 2, \ldots$ and r, θ are the cylindrical coordinates, a_1 and a_2 are the inner and outer radii of the sector with an angular width ϕ. J is the current density. This enables a direct calculation of the multipoles arising from the arc sectors. Starting with a crude approximation for the design a 3-dimensional calculation is carried out as described below. The results of this start design for a straight and a curved magnet ($\rho = 1.2$ m) are displayed in fig. 3. They show that the main influence of the curvature is a quadrupole term resulting in a field deviation less than 1 % in the good field region which is 4 x 2 cm^2. To compensate this quadrupole or, if desired, to create any other, a quadrupole shell is constructed using the formula above. To include these quadrupole-sectors in the dipole coil it is desirable to get rid of those windings having opposite current flow to the dipole by adding suitable dipole sectors as shown in fig. 2. Then the multipoles of these dipole sectors must be taken into account but only with half the current density. Finally, a new 3-dimensional calculation is performed which requires a slight modification of the quadrupole-sector. This procedure is repeated until the field accuracy is good enough.

All field calculations for the coils are done using the law of Biot and Savart. In case of the curved sectors two different methods are used as indicated in fig. 3. Each sector is fitted with arcs

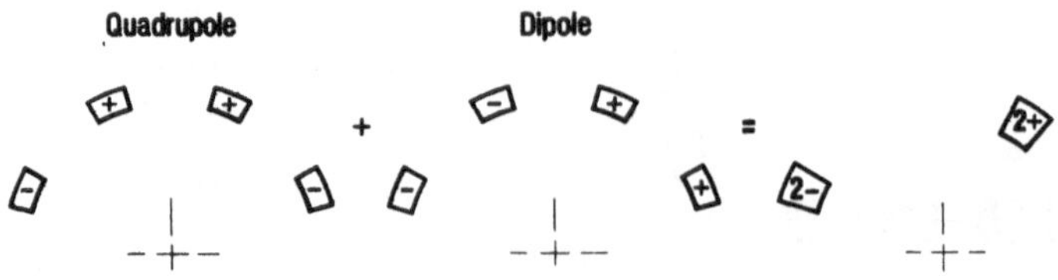

Fig. 2. Combining a quadrupole with a dipole.

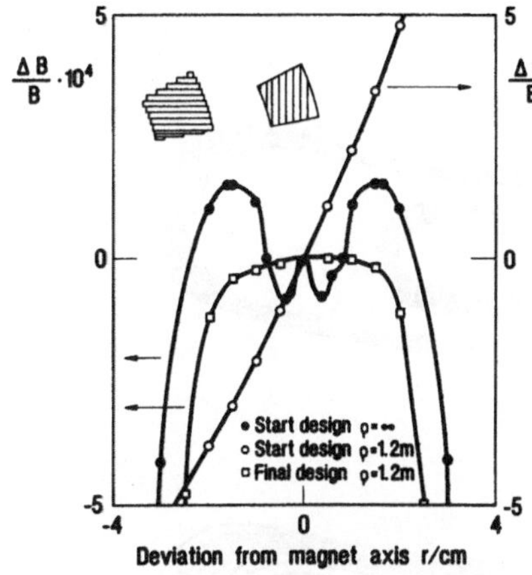

Fig. 3. Field deviation for the curved and straight start design and for the final design E 54.

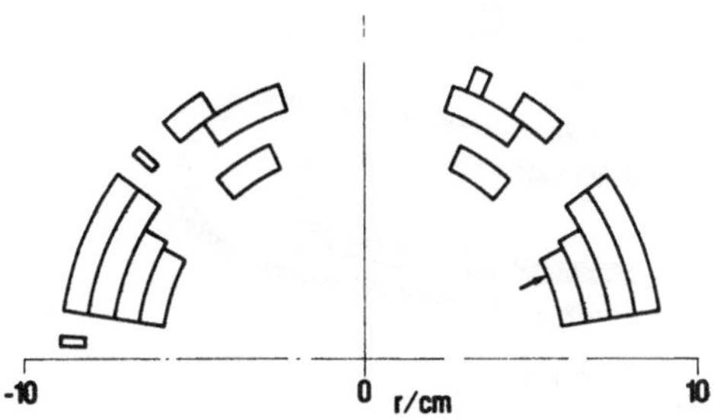

Fig. 4. Cross-section of the upper half of dipole E 54. The unsymmetric sectors in the 4th shell compensate the quadrupole field caused by curving the straight quadrupole-free dipole magnet.

having rectangular cross-section. For the horizontal arcs a routine based on published formulae[8] is used showing no divergences in the vicinity of the conductor. The integration over the arc-angle was performed by gaussian quadrature in steps less than 10°. With a similar fit the field calculation of the straight magnet was performed using the TOSCA-routine for straight bars. A TOSCA-routine for arcs is used in which the arcs are split off in infinitely thin cylinders perpendicular to the plane of the design orbit. For each of these cylinders the fields are calculated by a power-expansion of the incomplete elliptical integrals. This routine was slightly modified to fit the surface of the sectors. On the scale used in fig. 3 both methods give identical results. The fields originating from the winding heads at the ends are calculated with a filament-method.

RESULTS

Dipole Coil

Fig. 4 shows the final layout for the design E 54. The magnet consists of 4 layers each 7.5 mm thick. The inner radius is 6 cm. To produce a field of 4 T an overall current density of 360 A/mm² is

Table 1: Conductor characteristics

Material	Nb-Ti
Cross section of conductor length in mm	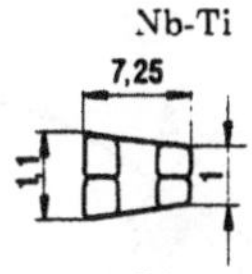
Number of superconducting strands	24
Mean diameter of strand /mm	1.05
Cabling length /mm	65 - 95
Superconductor ratio	1.8
Critical current density at 4.3 K, 6 T /A	4392
Strand parameters:	
Critical current at 4.3 K, 6 T after twisting /A	183
Number of filaments	636
Diameter of filaments /mm	0.014

Table 2: Magnet characteristics

Bending angle		90°
Bending radius /m		1.2
Nominal field /T		4
Good field region /mm²		40 x 20
Number of layers		4
Inner radius /mm		60
Thickness of layer /mm		7.5
Overall current density A/mm²		360
Operating current /A		3232
Maximum field on winding /T		5.2
Number of turns:	Layer 1	32
	Layer 2	22
	Layer 3	50
	Layer 4	49

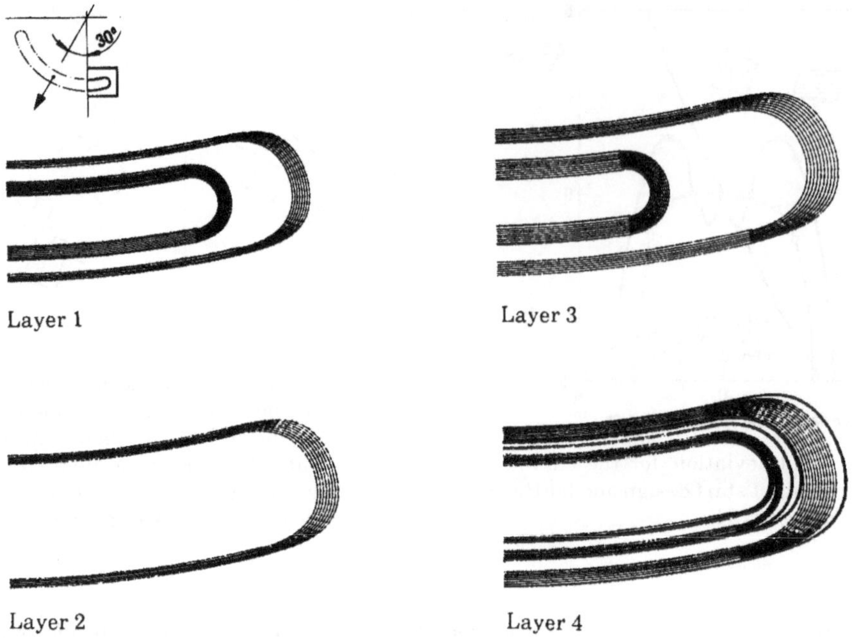

Fig. 5. Top view of the winding heads of magnet E 54. The four layers are represented separately. Only a region extending from an angle of 30° referred to the 45° symmetry plane up to the end is shown.

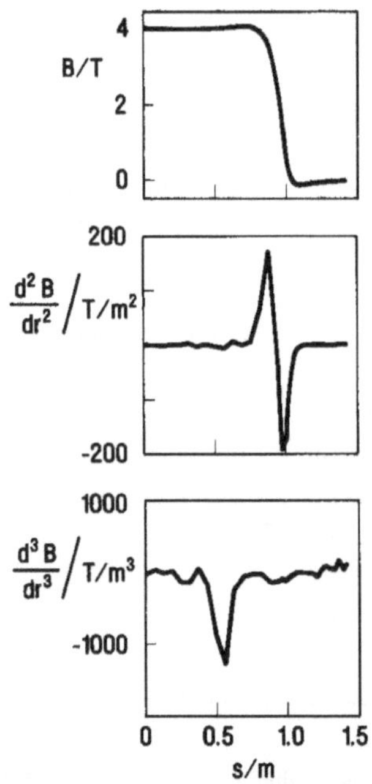

Fig. 6. Total field, sextupole and octupole components of dipole E 54 versus path length s along reference orbit.

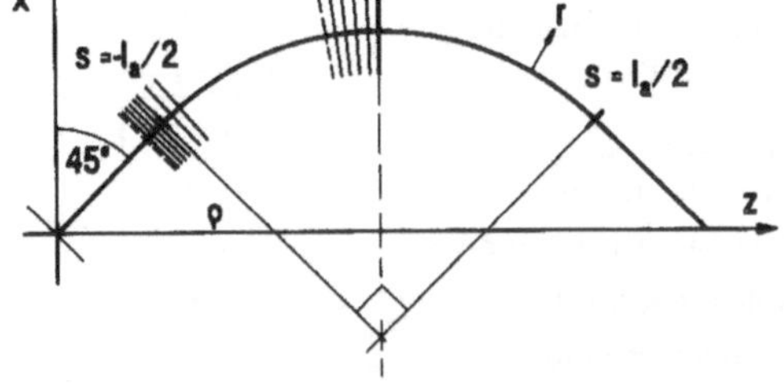

Fig 7 Geometry used for computing magnetic field in and transfer map of a 90° bending magnet with mid-plane symmetry. Actually, the bending radius ρ is 1.2 m, the length of the leading and trailing straight sections 0.61 m.

needed. The maximum field of 5.2 T arises at the position indicated in fig. 4 in the end-region of the magnet at an azimuthal angle of 41° measured from the midpoint. Assuming this is the value to be at 80 % on the load line of the magnet a total current of 4392 A at 6 T and 4.3 K is specified for the conductor. Table 1 and table 2 summarize the characteristics of the magnet and the conductor on which this layout is based The cable has the cross-section of a keystone with a mean width of 1.05 mm, and a height of 7.25 mm. The insulation is 0.075 mm thick. A conductor with these specifications has been fabricated by Vacuumschmelze/Hanau.

The integral multipoles are reduced by suitable choice of the spacers in the winding heads. The geometry of the winding heads is derived from the Fermilab and HERA magnets. If the magnet is thought to be straightened and the resulting cylinder is developed onto a plane, then the winding heads form semicircles. For the configuration shown in fig. 5 the dipole field and the sextupole and octupole components are plotted in fig. 6 versus the position s on the reference orbit. The details of the geometry used are shown in fig. 7. The relative contribution of the integral sextupole and octupole are 7×10^{-6} and 6×10^{-5} at r = 1 cm, respectively.

Force calculation and mechanical design

The forces are computed by a code which simulates the conductor by small current elements and calculates the Lorentz force on each element using the field map (fig. 8) in the conductor region produced by the codes described above. The resulting force distribution is shown in fig. 9 where the numbers indicate the azimuthal line force per sector in units of kN/m .

These forces are supported by applying the principle of prestressing the coil. The prestressing force has to overcompensate the magnetic force under all operating conditions. Fig. 10 shows the cross section of the coil support structure which forms also the helium containment. The coil is supported at 42° by 36 bolts and at 9° by a continuous clamp.

In the upper part of the coil the magnetic stress is 19 N/mm^2 and in the lower part it is 13 N/mm^2. Therefore a prestress of 30 N/mm^2 and 20 N/mm^2 respectively has been chosen to give a safety margin for fabrication tolerances and cool-down. This prestress results in a pressure of 120 N/mm^2 and a shear stress of 150 N/mm^2 in the bolts, a bending of the wedges between the bolts of less than 0.01 mm, an equivalent stress of 250 N/mm^2 in the clamp and a clamp deformation of less than 0.04 mm. As structural material Inconel 706 has been chosen because of its high yield strength (τ_{02} = 1050 N/mm^2) and its fracture toughness at low temperature.

The upper and the lower part of the structure are assembled together with half of the coils in the following sequence by means of a hydraulic assembly fixture: the inner parts of the structure (1) are bolted together and are mounted on the assembly fixture, then the coils (2) and the outer parts of the structure (3) are put into position. Now the prestress is applied by hydraulic pressure and the bolts are fixed so that the prestress is frozen. The structure is removed from the assembly fixture and sealed by welding in order to act as a leak-tight helium vessel, too. The conductor bus work is done at the coil ends in special boxes which are also needed for the He inlet and outlet.

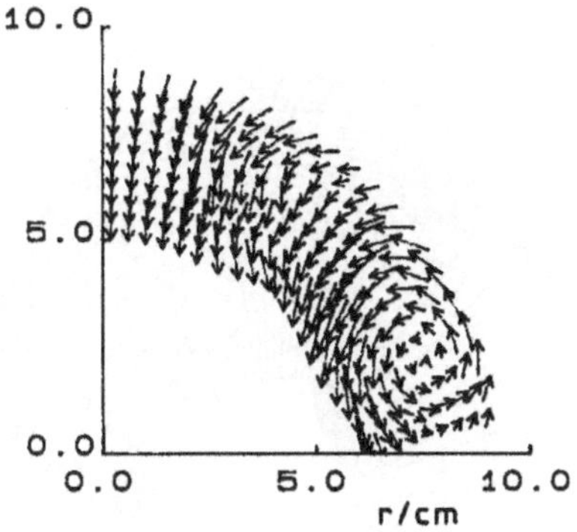

Fig. 8. Magnetic field map in the coil.

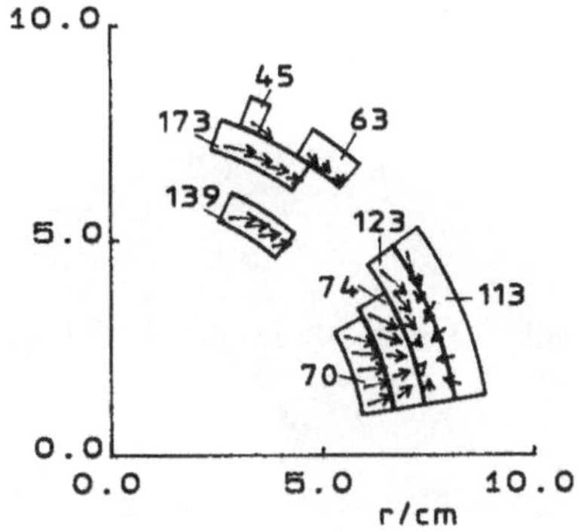

Fig. 9. Force distribution (azimuthal line force in kN/m per sector).

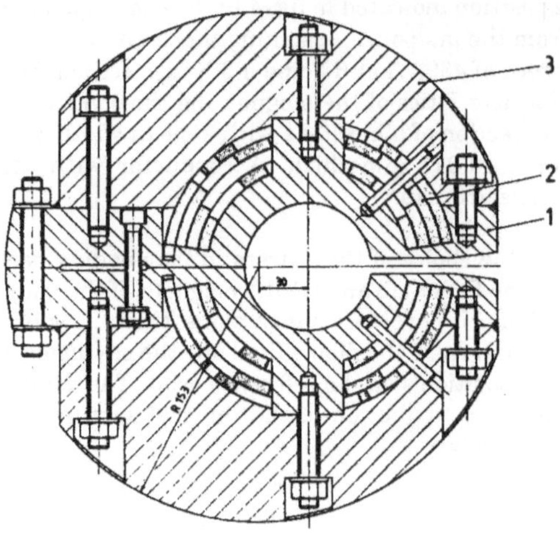

Fig. 10. Cross section of the coil support structure.

Iron shielding

TOSCA was used to compute the iron contribution to the field. Fig. 11 shows the mesh for the iron shield which is at a temperature of 4.2 K and has a C-type cross section with an inner radius of 20 cm. The iron contribution is 12 % at 4 T. Fig. 12 shows the fields obtained from magnetisation integration in the good field region for two different currents. Since this iron shield creates a quadrupole field which is, however, small enough to be compensated by the lattice quadrupoles its use is not yet decided. In any case shielding is desirable to reduce stray fields, which, at a distance of 1 m from the magnet, are still as high as 0.023 T. The reduction factor is 0.4 at the outer and 0.1 at the inner radius, respectively.

Beam optical performance

As a criterion for the usefulness of a specific magnet design the dynamic aperture of a storage ring lattice equipped with four of these magnets is chosen. The dynamic aperture, i. e. that cross-section around the closed orbit whithin which the particle trajectories are confined for a sufficiently large number of turns is found by tracking particles 10^4 times around the ring. Use is made of program MARYLIE 3.0[9] and SCB[10] which permit to carry out a symplectic tracking including real fields and non-linear effects through 3rd order. Fig. 13 shows the dynamic aperture in the middle of a long straight section for the standard lattice with bending magnets E 54 (fig. 1). The diagram contains also bars indicating the magnitude of 10 standard deviations of the horizontal and vertical beam size usually required for a satisfactory beam life time. Since the dynamic aperture exceeds the 10σ-bars by far the E 54 design is acceptable.

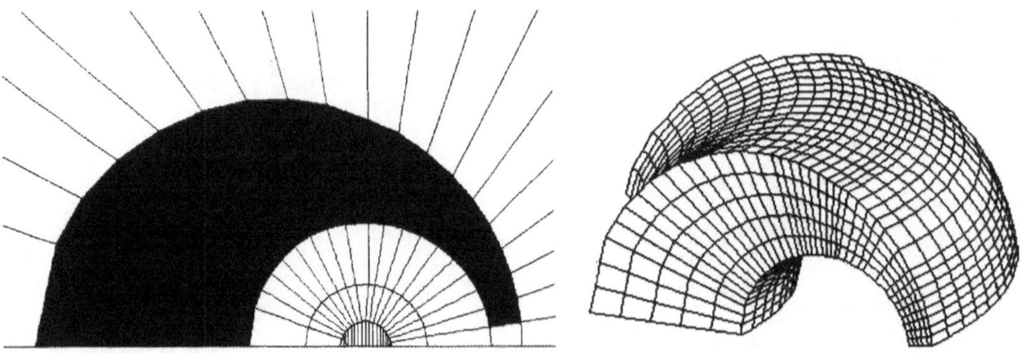

Fig. 11. Part of the TOSCA-mesh and perspective view of the iron.

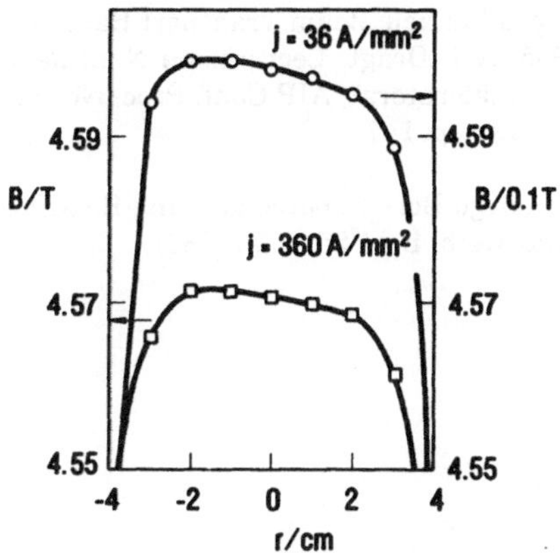

Fig. 12. Magnetic field with iron shield for two different currents.

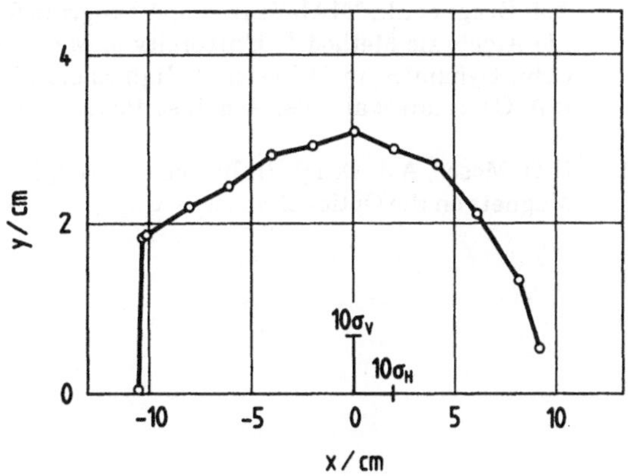

Fig. 13. Dynamic aperture with standard lattice and magnet E 54.

CONCLUSION

The technology developed for straight accelerator magnets can be applied to strongly curved bending magnets having a slot along the circumference. The design presented has an acceptable dynamic aperture when used within the standard lattice of the Karlsruhe compact synchrotron light source.

REFERENCES

1. for references see H. O. Moser, B. Krevet, A. J. Dragt, Nonlinear beam optics with real fields in compact storage rings, in: "Proc. Part. Acc. Conf.", Washington, D. C., IEEE Conference Records 87 CH 2387-9, IEEE Service Center, Piscataway, NJ (1987), submitted to Nucl. Instr. and Meth. B.

2. D. Einfeld et al., "Entwurf einer Synchrotronstrahlungsquelle mit supraleitenden Ablenkmagneten für die Mikrofertigung nach dem LIGA-Verfahren", KfK 3976, Kernforschungszentrum, Karlsruhe, FRG (1986).

3. E. W. Becker et al., Fabrication of microstructures with high aspect ratios and great structural heights by synchrotron radiation lithography, galvanoforming, and plastic moulding (LIGA process), Microelectronic Engineering 4: 35 (1986).

4. R. Palmer, A. V. Tollestrup, "Superconducting magnet technology for accelerators", FNAL-TM-1251, Fermilab, Batavia, Illinois, (1984).

5. S. Wolff, The Superconducting Magnet System for HERA, in: "Proc. 9th Intl. Conf. Magnet Technology", C. Marinucci, P. Wegmuth, eds., Egloff Offsetdruck, Wettingen, Switzerland, Zürich (1985), p. 62.

6. "Superconducting Magnet Systems", H. Brechna, ed., Springer-Verlag, Berlin, Heidelberg, New York (1973).

7. R. J. Lari, "Calculation of the harmonic content of the magnetic field produced by a given distribution of parallel conductors", R.JL-7, Argonne National Laboratory, Argonne, Illinois, (1966).

8. A. Schleich, A. Segessemann, Calculation of the Electromagnetic Forces Acting on the Coils of an Electromagnet, in: "Proc. 3rd Intl. Conf. Magnet Technology", Editorial Committee MT3, DESY, ed., Adam Curtze KG, Hamburg, FRG (1971), p. 198.

9. A.J. Dragt et al., "MARYLIE 3.0, A Program for Charged Particle Beam Transport Based on Lie Algebraic Methods", University of Maryland, 1985; A.J. Dragt, Lectures on Nonlinear Orbit Dynamics, in: "Physics of High Energy Particle Accelerators", AIP Conf. Proc. No. 87, R.A. Carrigan et al., eds., Am. Inst. Phys., New York, (1982), p. 147.

10. H.O. Moser, A.J. Dragt, Influence of Strongly Curved Large-Bore Superconducting Bending Magnets on the Optics of Storage Rings, Nucl. Instr. and Meth. B 24/25: 877 (1987).

TESTING OF A 3 TESLA SUPERCONDUCTING MAGNET FOR THE AMY DETECTOR AT TRISTAN

K. Tsuchiya, S. Terada, T. Omori, A. Maki, Y. Doi,
T. Haruyama and T. Mito

KEK, National Laboratory for High Energy Physics
Tsukuba, Ibaraki, Japan

K. Asano

Hitachi Ltd.
Hitachi, Ibaraki, Japan

ABSTRACT

A 3 tesla superconducting magnet was constructed and installed in an experimental hall at TRISTAN. Cooldown and excitation tests of the magnet were carried out with a dedicated cryogenic system. The coil has a 2.39 m inner diameter and is 1.54 m in overall length with a radial thickness of 0.1 m. The rated current is 5 kA and the stored energy is 40 MJ. The refrigerator and 17-tonne magnet cold mass were cooled to 4.4 K in 7 days. After that, the magnet was energized to the design current of 5 kA and the mechanical stress of the coil supports was measured. The stress on the supports was well below the allowed maximum. The discharging characteristics of the magnet were also measured and it was confirmed that the magnet was very stable and reliable.

INTRODUCTION

The AMY particle detector[1] is one of three colliding beam detector systems for studying electron-positron collisions in the TRISTAN main ring. Compared with other particle detectors used for colliding beam machines, it has very different features. It requires a high magnetic field. The high field provides the ability to do precise momentum measurement of produced particles in a small volume, allowing us to keep the size of the detector small enough to accommodate low-beta quadrupole magnets. Furthermore, the small size makes the detector well suited for muon detection.

The schematic side view of the AMY magnet is shown in Fig. 1. A 2.39 m inner diameter × 1.54 m long superconducting solenoid is surrounded by a thick hexagonal iron return yoke with two end plates which guides the magnetic flux and serves as a hadron absorber. The magnet dewar, which is connected to the cryostat containing the solenoid, is placed at one side of the iron yoke and the vacuum pump unit for the cryostat at the other.

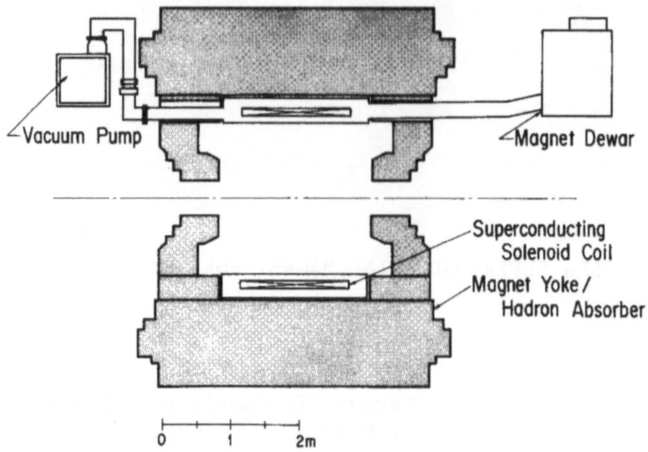

Fig. 1. Schematic side view of the AMY magnet.

The design of the magnet[2] was started in 1983. After about a three year construction period, the magnet was installed in the AMY experimental hall and tests of the performance of the magnet system, including the dedicated refrigerator, were made. This paper gives a brief description of the magnet and test results.

MAGNET

In order to fulfil the basic requirement, that the magnet be safe and reliable when operating at a high magnetic field, the design was based on a cryostable coil cooled by the pool boiling method. The main parameters are shown in Table 1.

The conductor of the coil is composed of 7 strands of NbTi Rutherford type cable, a work-hardened copper (1/2 hard) housing and a high purity aluminum (RRR > 3500) strip. The cable and the aluminum strip are inside of the copper housing and all three components are soldered together. The copper housing works as a mechanical support member and the aluminum as a stabilizer. Owing to this configuration, the conductor can withstand up to a 12.6 kg/mm^2 compressive force with a resistivity of 0.69×10^{-10} Ωm at 4.2 K and zero field.

The design field, 3 tesla at the magnet center, is produced by an 8 layer helically wound coil. The coil fabrication method was as follows:

Table 1. AMY Superconducting Magnet

Current	(A)	5000
Central field	(T)	3.0
Peak field in the windings	(T)	4.0
Coil inner diameter	(m)	2.386
Coil outer diameter	(m)	2.584
Coil length	(m)	1.540
Total ampere turns	(AT)	5.4×10^6
Inductance	(H)	3.2
Stored energy	(MJ)	40

the conductor was wound with a tension of 400 kg on a mandrel of 25 mm thick 304L stainless steel, which became the inner wall of the liquid helium vessel. In the winding process, numbers of 1.6 mm thick molded glass-fiber-reinforced plastic spacers were placed between the turns and an axial pressure of 0.8 kg/mm^2 was applied every 20 turns to keep the winding section tight. An important consideration was to keep the conductor surface clean, in order not to reduce the heat transfer coefficient. Between the layers, 2.5 mm thick GFRP strips were placed. The space factor for the side surface of the conductor and for the up and down surfaces are 50% and 80%, respectively.

The electrical connection between layers was made by standard soldering technics at the ends of the coil. The length of this electrical joint was about 420 mm and the resistance of the joint estimated from sample test results was about 2×10^{-9} Ω at 4.2 K which correspond to 50 mW of Joule heating at the rated current of 5000 A.

After completion of the winding, a layer of fiber-reinforced plastic was formed on the surface of the coil and machined in order to make a smooth surface for shrink fitting. After that, a 36 mm thick 304L stainless steel cylinder, the outer cylinder of the liquid helium vessel, was heated to about 150° C and the coil was inserted into the cylinder. The amount of expansion of the radius of the cylinder was about 0.85 mm which corresponds to a contact pressure of 0.37 kg/mm^2. At the top of the inner surface of the cylinder a special groove was made so that helium gas would not accumulate and cause improper cooling.

The cryostat of the magnet consists of two parts, a horizontal cylindrical vessel containing the superconducting coil and a magnet dewar which has a pair of current leads and three control valves. The liquid helium storage capacity of the cryostat is about 500 L. The vacuum vessel is made from 304 stainless steel. The thermal radiation shield is fabricated from aluminum alloy A1100 sheet and cooled with liquid nitrogen.

The support system must support both the weight of the helium vessel containing the superconducting coil and the axial and radial forces due to the electromagnetic force coming from any non-symmetrical assembly of the solenoid within the iron return yoke. The decentering force was estimated to be 3.0 tonnes/mm and 1.1 tonnes/mm in the axial and radial directions, respectively. Assuming that the maximum offset of the coil relative to the yoke is 15 mm, the support system was designed to withstand 45 tonnes and 16 tonnes in the axial and the radial directions, respectively. The support system consists of six axial support rods located at the end opposite from the magnet dewar side and twelve radial support rods at each end of the helium vessel. The arrangement of these rods are shown in Fig. 2. The support rods were made from Inconel 718 which has a high ultimate tensile strength, 142 kg/mm^2 at room temperature. The lengths of the axial and the radial support rods were 290 mm and 345.5 mm, respectively. The estimated conduction heat leak of the LHe vessel through these supports was 0.82 W for each axial support and 0.36 W for each radial support.

ELECTRICAL SYSTEM

A schematic of the electrical circuit is shown in Fig. 3. The system is composed of a rectifier using thyristors, active and passive filters, a DC breaker and a protection resistor. The capacity of this power supply is 5 kA and 10 V. The protection and interlock system can initiate a slow discharge, τ = 1600 sec, by firing an SCR, or a fast discharge, τ = 21 sec, by opening the DC circuit breaker. Quench detection is done by the usual bridge-balance method. The typical quench threshold used in the

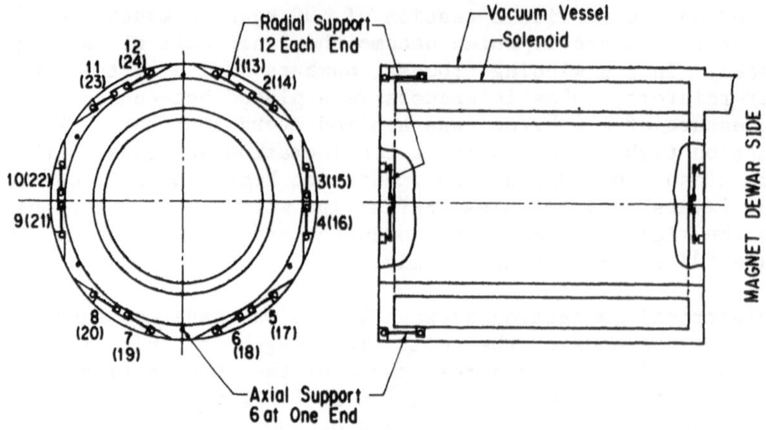

Fig. 2. Schematic diagram of the support rod arrangement. The number in the side view corresponds to the rod number.

excitation test was 300 mV and 500 msec. Furthermore, the pressure of the cryostat was always monitored and if it rose above 0.5 kg/cm^2, the protection system would be triggered.

TEST OF THE MAGNET

Cooldown

The helium flow circuit for cooling the magnet is shown in Fig. 4. During cooldown, helium gas from the precooling line contained in the refrigerator passed through the transfer line and the control valve TCV-701, entered into the bottom of the cryostat and then flowed back to the refrigerator through the return line from the magnet dewar. In order to keep the temperature difference within the coil less than 70 K, the helium gas temperature and the flow rate in the precooling line were controlled by mixing warm and cold gas in the refrigerator. This was done by a process control computer and electrically operated valves. After the magnet was cooled down, liquid helium in the storage dewar was transferred to the cryostat, first through PCV-702 and then through LCV-703.

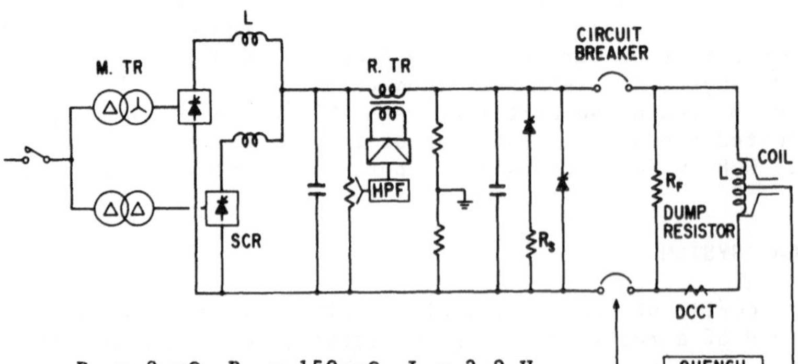

$R_S = 2\ m\Omega$, $R_F = 150\ m\Omega$, $L = 3.2$ H

Fig. 3. Electrical circuit of the AMY solenoid.

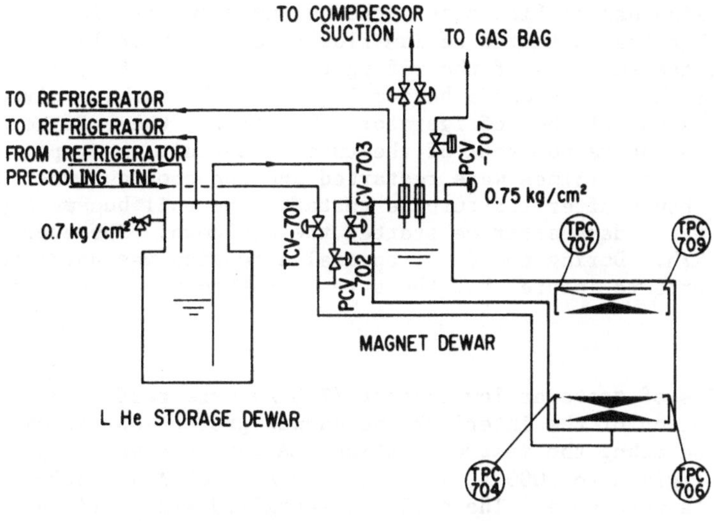

Fig. 4. Helium flow schematic.

The temperature of the coil was monitored with platinum-cobalt resistors at several locations. Typical positions of the temperature sensors are shown in Fig. 4. The temperature of the thermal radiation shield was measured with copper-constantan thermocouples. Another consideration during cooling of the magnet was the thermal stress of the support rods. To monitor the stress we attached strain gauges on the rods and the output signals were read by a computer.

The cooldown curve of the magnet is shown in Fig. 5. On December 6 we started the cooldown of the magnet with cold helium gas of temperature about 70 K lower than the magnet. The temperature sensor TPC-704 showed the lowest temperature and TPC-709 showed the highest. The pressure of the cryostat was maintained between 0.25 and 0.40 kg/cm^2. Since the

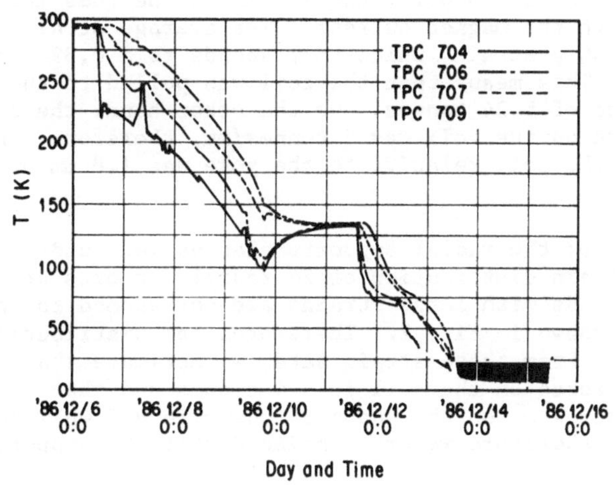

Fig. 5. Temperature trend of the coil during the cooldown period. Due to refrigerator trouble the cooldown was paused from December 9 through 11.

precooling line has no flow meter, the helium flow rate in the line was not known, but the estimated helium flow rate was about 15 g/sec. On December 9, the turbines of the refrigerator were started, but soon it became clear that there was a helium leak from the turbine circuit to the insulation vacuum of the refrigerator. Therefore, the refrigerator was warmed up and the vacuum seal of the turbine circuit was repaired. On December 11, the turbines were restarted and the cooldown was continued. Twenty-four hours after the turbine restart, the coil became superconducting. This was 7 days after we started the cooldown. The average cooldown rate was 2 K/h. During cooldown, special attention was payed to the radial supports to insure that the stress on them was evenly distributed.

Excitation

A series of tests at low current (200 A) were performed to verify proper operation of the interlock and quench protection systems and to calibrate and debug the instrumentation. After this was completed, the coil was energized to 1000 A, 2000 A, 3000 A, 4000 A and 5000 A and various tests were done. The coil was energized with a linear ramp rate of $\sim$167 A/min over most of the ramping time. After the coil reached the preset current, the strain of the support rods was measured and slow discharge and fast discharge tests were performed. In order to measure the charging and discharging characteristics, various signals (coil voltages, coil current, bridge-balance signal of the quench detector etc.) and data from the cryostat (pressure, temperatures, helium level in the cryostat and strains of the support rods) were recorded using a tape recorder and a computer.

The magnet inductance was accurately determined from the magnet charging curve, $L = \int Vdt/I$, where V is the magnet charging voltage and I is the magnet current, and the field was measured with a calibrated hall probe. In Fig. 6 the inductance and the central field of the magnet vs current are shown. The results are in good agreement with a calculation using the program POISSON.

Strain of the support rods

Although a few strain gauges attached to the support rods did not work properly, we could still estimate the electromagnetic forces on the coil. In case of axial support, the strain on the rods increased almost quadratically with the magnet current. The average strain at 5000 A excitation was 130 μ strain, which corresponds to a 0.89 tonne tension force on a rod. This means that the coil was pulled to the magnet dewar side with a force of 5.34 tonnes. On the other hand, the calculated axial decentering force on the coil was 3 tonnes/mm, therefore, the estimated axial offset of the coil relative to the yoke was 1.8 mm on the dewar side.

The strains of the radial supports also varied quadratically with the current. The strain distribution of 24 radial supports at 5000 A is shown in Fig. 7. The rods with large strains are correspond to the vertical ones, which are shown in Fig. 2. Therefore, the distribution of strain is very reasonable. From these strain data, we estimate that the electromagnetic force on the coil is 6.4 tonnes and that the direction is downward. The stresses on the support rods were all well below the allowed maximum, therefore we are convinced that the support system is very safe.

Fast discharge of the magnet

Even though the coil was designed to be cryostable, the magnet should be protected from an unexpected quench by extracting the stored energy

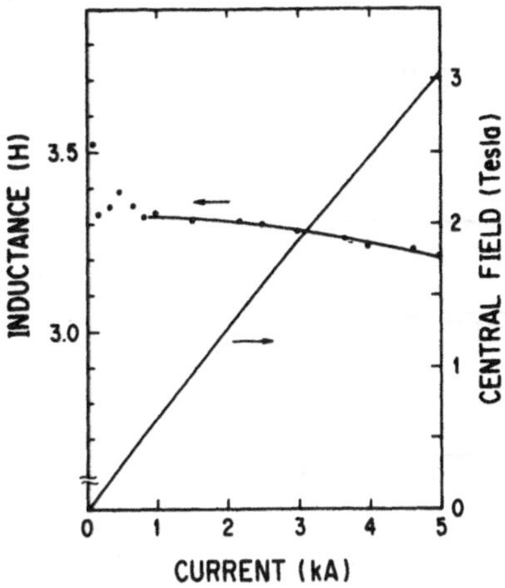

Fig. 6. Measured inductance and central field of the magnet vs current.

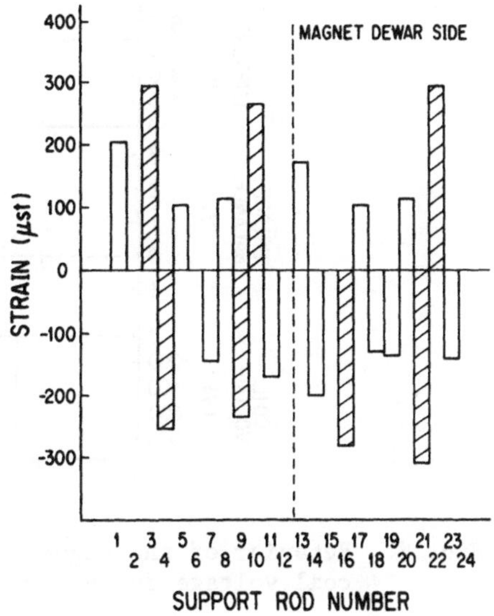

Fig. 7. Strain distribution of the radial support rods at 5000 A excitation.

quickly. Therefore we tried a fast discharge test at various currents. The main objectives of the test were to see (1) whether the coil or a part of the coil becomes normal when the current decays quickly (2) whether an electrical breakdown occurs in the coil (3) whether the cryostat is strong enough against the pressure rise during the fast discharge and the electromagnetic force due to induced eddy currents.

The first two objectives were studied by analyzing the current decay, voltage in the coil and the bridge signal of the quench detector. At low current, 2000 A, the current decayed exponentially and the time constant was in good agreement with that expected, $\tau = L/R = 21$ sec. However, the time constant became shorter, $\tau = 16.5$ sec, when the starting current was 5000 A. This phenomena was simulated using a program which was a modified version of QUENCH, and we found that it was due to the temperature change of the protection resistor. The calculated maximum temperature of the resistor was about 300° C if we assumed all of the stored energy of the coil was dumped into the resistor. Figure 8 shows the magnet current, the coil voltage and the bridge-balance signal of the quench detector resulting from the fast discharge at 5000 A. Although there is an apparent signal in the bridge-balance, we couldn't confirm that it was real.

Also of interest in the fast discharge test was the pressure rise of the cryostat. This is due to the eddy current induced in the stainless steel wall of the helium vessel and the conductor. The cryogenic interlock system, when fast discharge occurred, operated as follows: the helium line of the magnet was isolated from the storage dewar and the refrigerator and the valve PCV-707 was opened to reduce the pressure rise of the cryostat. However, the pressure increased somewhat because of the finite capacity of the piping system. In the test, the pressure increase of the magnet dewar was 0.17 kg/cm^2 and 0.38 kg/cm^2 at 4000 A and 5000 A fast discharge, respectively. In Fig. 9 the behavior of the pressure and the helium level in the magnet dewar during a fast discharge at 5000 A is shown. The amount of liquid helium loss due to a 5000 A fast discharge was 90 $\sim$ 100 L.

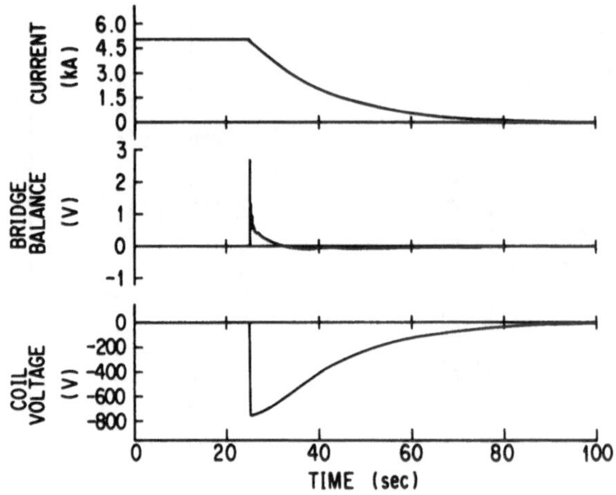

Fig. 8. Behavior of the magnet current, bridge balance signal, and coil voltage during fast discharge at 5000 A.

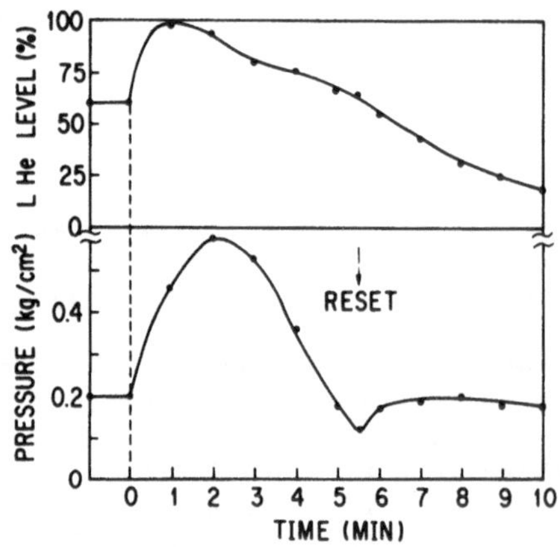

Fig. 9. Trend of magnet dewar pressure and liquid helium level after fast discharge at 5000 A.

CONCLUSION

The AMY 3 tesla superconducting magnet was installed in the TRISTAN experimental hall and tested. As expected it was stable and reliable.

- the cooldown time in normal operation is about 5 days and the magnet is ready to excite within one additional day.
- the stress on the support rods is well below the allowed maximum.
- even if a fast discharge occurs at 5000 A, the magnet is stable and the liquid helium loss is only 100 liters.
- the recovery time after a fast discharge is very short, about 40 min.
- the total heat leak of the liquid helium vessel is 35 W.

REFERENCES

1. C. Back, A. Bodek, H. Budd et al., TRISTAN EXP-003 KEK (1984).
2. K. Tsuchiya, S. Terada, A. Maki et al., A 3 Tesla Superconducting Solenoid for the AMY Particle Detector at TRISTAN, IEEE Trans. Magn. MAG-22 (1987) to be published.

SUPERCONDUCTING MAGNETS FOR FUSION APPLICATIONS*

C. D. Henning

Magnetic Fusion Energy
Lawrence Livermore Laboratory
Livermore, California

ABSTRACT

Fusion magnet technology has made spectacular advances in the past decade: to wit, the Mirror Fusion Test Facility and the Large Coil Project. However, further advances are still required for advanced economical fusion reactors. Higher fields to 14 T and radiation-hardened superconductors and insulators will be necessary. Coupled with high rates of nuclear heating and pulsed losses, the next-generation magnets will need still higher current density, better stability and quench protection. Cable-in-conduit conductors coupled with polyimide insulations and better steels seem to be the appropriate path. Neutron fluences up to 10^{19} neutrons/cm^2 in niobium tin are achievable. In the future, other amorphous superconductors could raise these limits further to extend reactor life or decrease the neutron shielding and corresponding reactor size.

INTRODUCTION

The application of superconducting magnets to fusion experiments has been limited, but increasing rapidly. Since 1970, when the Baseball II[1] and IMP[2] magnets at LLNL and ORNL respectively, were first used, the T-7 system[3] at the Kurchatov (USSR) has been in operation. More recently, the MFTF magnet system[4,5] at LLNL and the TriAM-1M in Japan[6] have been completed. Soon the Tore Supra[7] at Cadarach, France and the T-15[8] at Kurchatov will be nearing completion. These fusion magnet systems represent significant advancements in magnet technology starting from small confinement experiments to near-reactor sizes. The Large Coil Project[9] has further advanced the technology and enhanced an industrial base for magnet construction.

Now we are on the brink of nuclear application of fusion magnets. Currently, the U.S., U.S.S.R., Japan and the European Community are discussing the cooperative design of an International Thermonuclear Experimental Reactor (ITER). The impetus for this design effort came from a joint statement issued by President Reagan and Chairman Gorbachev at the November 1986 Geneva Summit Conference. These leaders called for the

*Work performed under the auspices of the U.S. Departmet of Energy by the Lawrence Livermore National Laboratory under contract number W-7405-ENG-48.

"...widest practical development of international cooperation (in fusion)." This statement resulted in a team from the Department of Energy meeting with representatives from the U.S.S.R. in April 1986. In October 1986, the U.S. formally proposed that the European communities, Japan, the U.S.S.R., and the U.S. collaborate in producing a detailed design for the next major device in fusion energy development. The final design will provide enough detail for the participating nations to make an informed decision on whether to construct and test the device. If the ITER is constructed, it could confirm the feasibility of fusion power by demonstrating physics and plasma performance as well as test components and materials under integrated reactor conditions.

Lawrence Livermore National Laboratory is leading the U.S. effort on the ITER. While compromises in design with other nations are expected, we can project some of the necessary conditions for superconducting magnet applications from the Tokamak Ignition/Burn Experimental Reactor Study (TIBER)[10]. This design (Figure 1) incorporates physics and engineering characteristics that could lead to a compact, steady-state, current-driven reactor to be used for testing components and materials in a nuclear environment similar to that expected in a power reactor. In fusion reactors, plasma cycling causes time-dependent changes in environmental testing conditions (for example, tritium-concentration profiles, eddy-current effects, thermal conditions, and failure or fracture modes). Because we are primarily interested in the effects of long-pulse reactor conditions on components, the often dominant short-pulse effects must be minimized so that testing results can be accurately extended to reactor conditions. For this reason, TIBER is designed with enough auxiliary current-drive power to maintain steady-state operations. A summary of the major parameters is given in Table 1.

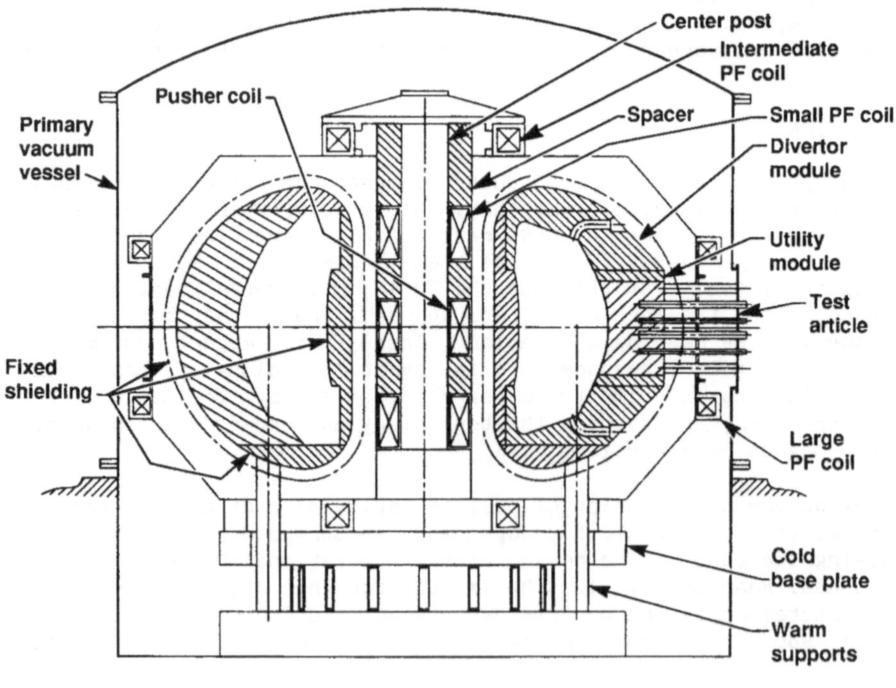

Fig. 1. TIBER - a Tokamak Ignition/Burn Experimental Reactor

Table 1. TIBER Design Parameters

Fusion Power	314 MW
Major Radius	3.0 m
Minor Radius	.83 m
Elongation	2.4
Burn Time	Steady State
Average Neutron Wall Loading	1.3 MW/m^2
Peak Neutron Wall Loading	2.0 MW/m^2
Fluence Goal	>3 MW/m^2
Availability Goal (Testing Phase)	30%
Tritium Consumption (Testing Phase)	4.5 kg/yr
Surface Heat Flux	
First Wall (Uniform)	0.25 MW/m^2
Divertor (Peak)	2.2 MW/m^2
Test Area	
Frontal Area of Test Spaces	1.2 m x 2 m
Number of Test Spaces	8

ADVANCES IN MAGNET TECHNOLOGY

The size and cost of the next generation tokamak fusion experiments with ignited plasmas will be largely determined by the size characteristics of the center post region. Starting from the radial dimensions and current density of the central ohmic heating coil, the toroidal field coil current density and inner neutron shield control the tokamak inner plasma radius. Since plasma minor radius and aspect ratio are usually minimized (consistent with fusion power and physics scaling) the entire tokamak size is greatly influenced by the magnet current density, field, and radiation tolerance.

Current Density. The winding pack current density in the TIBER II design for both the poloidal and toroidal field coils was 4 KA/cm^2 at 14 T and 12 T respectively[11]. This performance is above traditional standards, since cryogenically stable conductors have current densities about half of that projected for Cable In Conduit Conductors (CICC) as used in TIBER. Here the conductor stability is determined largely by the enthalpy of the trapped helium rather than the boiling heat transfer of liquid helium from the conductor surface. Not only does it result in high current density, but the conductor is much more radiation tolerant because it is not affected by copper resistance increases due to neutron damage.

CICC Coil operating experience has been bolstered tremendously by small scale test and development activities at a number of laboratories in the U.S., Japan, and Europe. Figure 2 gives some perspective of what has been achieved along with the goals of present development activities. Operating current density over the cable space, (i.e., inside the conduit of the CICC) is plotted versus the maximum field at the windings. The lower point shown for the ORNL Nb-T coil corresponds to a stability margin well in excess of 100 mJ/cm^3. The solenoid also operated quite stably up to the critical current, where the stability margin was measured to be 56 mJ/cm^3.

Other operating points for the Westinghouse LCT coil[12] are shown plus the demonstrated performance of the MIT 12 T coil, which was tested in the LLNL, High Field Test Facility (HFTF)[13]. The lower of the two MIT 12 T points represents the only point where stability of the coil was tested. The coil stably absorbed a pulse of >200 mJ/cm^3 without consequence.

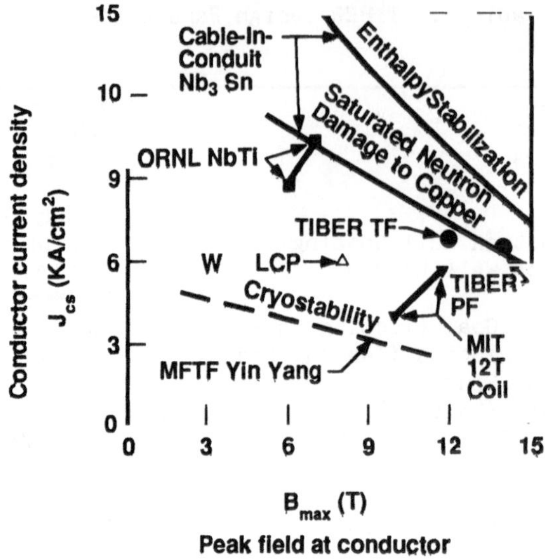

Fig. 2. Superconductor Stability - Higher conductor current densities through enthalpy stabilization in cable in conduit conductor are possible even with saturated neutron damage to the copper.

The upper enthalpy stabilization curve in Fig. 2 is a projection of the performance that is obtainable with state-of-art MF-Nb_3Sn in a CICC design. The projected improvement is possible because the performance of a CICC is tied directly to the increased current margin (or alternatively, increased temperature margin) provided by recent improvements in the critical currents of MF-Nb_3Sn wires. The middle curve takes account of the effects of radiation damage to the copper stabilizer in transforming the estimated performance boundary to one appropriate for the TIBER II TF coil environment, where shielding is much reduced.

Strain Effects. In a CICC conductor, the strain in the Nb_3Sn filaments, and thus the critical current at operating conditions, is determined by a complex interaction between the sheath and the cable inside from formation at around 1000 K, through cooldown to the operating temperature around 4 K, and charging to full field and current. An experimental program is in progress at LLNL to examine the conductor/sheath interaction through a series of electromechanical tests[14]. In the first series of tests, the critical currents of model CICCs were measured, with the axial load applied to the sheath increased by small increments until the measured critical current was observed to pass through a maximum. The ratio I_{ci}/I_{cm} of initial or "no-load" critical current to maximum critical current was observed to vary with fraction of the internal space available for helium in otherwise identical specimens. These data are plotted in Fig. 3, along with other values for similar conductors inferred from data in the literature[15,16]. A factor f_{He} has been plotted to account for the observation that, the greater the He fraction, the less rigidly linked are the conductor strands to the compressive influence of the sheath during cooldown.

An equally important factor in tokamak design is cyclic failure. In Fig. 4 the stress in the poloidal field coil sheath has been calculated to vary from 100 MPa to 420 MPa. The net 370 MPa cyclic stress amplitude then predicts a 50,000 cycle life limit (with a safety factor of two) for an initial 10 percent thickness flaw to propagate completely through the

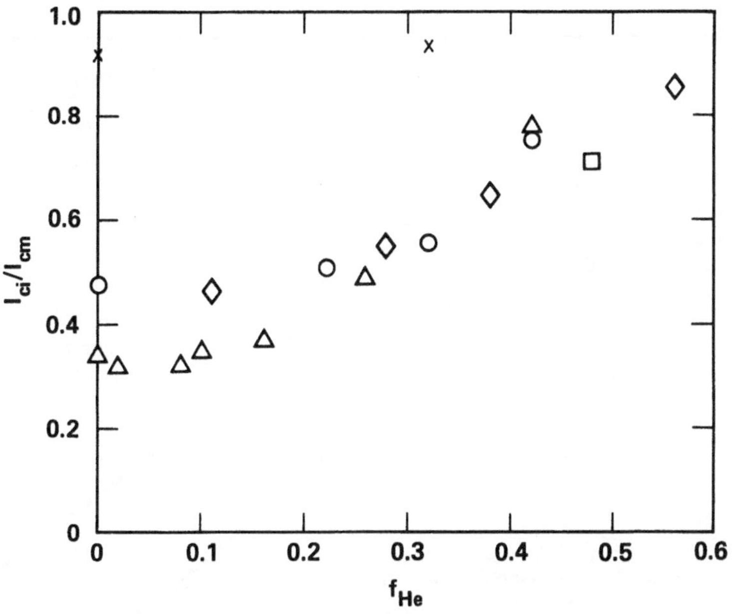

Fig. 3. Ratio of initial (unloaded) critical current of CICC samples to maximum critical current as a function of void or helium fraction inside the conduit.

sheath. This does not constitute catastrophic failure, but only leaks helium into the magnet case as a warning of limited remaining magnet life.

Radiation Damage. Figure 5 is representative of the effects of high-energy neutron fluence on Nb_3Sn conductors, both pure and modified[17]. The samples tested were irradiated with 14.1-MeV neutrons at room temperature and then tested at 4.2 K. The lower horizontal axis in Fig. 5 accounts for the "softer" shielded neutron-energy spectrum calculated for TIBER II. Initial increases in current density are explained by the damage giving rise to increased normal-state resistivity of the superconductors, resulting in an increase in the upper critical field. In the modified materials, the upper critical field has already been optimized by additions of a third element to increase the normal-state resistivity. Therefore, any further increase due to damage might be expected to be small. Eventually, at high fluence a decrease in performance is caused by a drop in critical temperature with increased damage. For unmodified Nb_3Sn, it is clear that the conductor will retain its performance beyond the 10^{19} n/cm^2 fluence in TIBER II. Modified Nb_3Sn is useful for attaining the high field performance in the center poloidal field coils, where the high-field windings are essentially completely shielded by the TF system and the outer, lower field turns.

For radiation-hardened magnets the insulation must retain adequate strength and electrical properties after irradiation to end-of-life fluences above 10^{19} n/cm^2 and gamma dosages above 10^{10} rads, especially in those portions of the magnet adjacent to shield-penetrations for diagnostics on plasma-heating systems. Irradiation-damage studies of organic materials are usually done using gamma (γ) irradiation alone, if possible, or in combination with neutron-fluences on the order of 10^{16} n/cm^2, which are thought to cause negligible additional damage[18]. High γ dosages cause breakage and/or rearrangement of bonds, so that scission, cross-linking of adjacent long-chain polymers (with decreased ability to

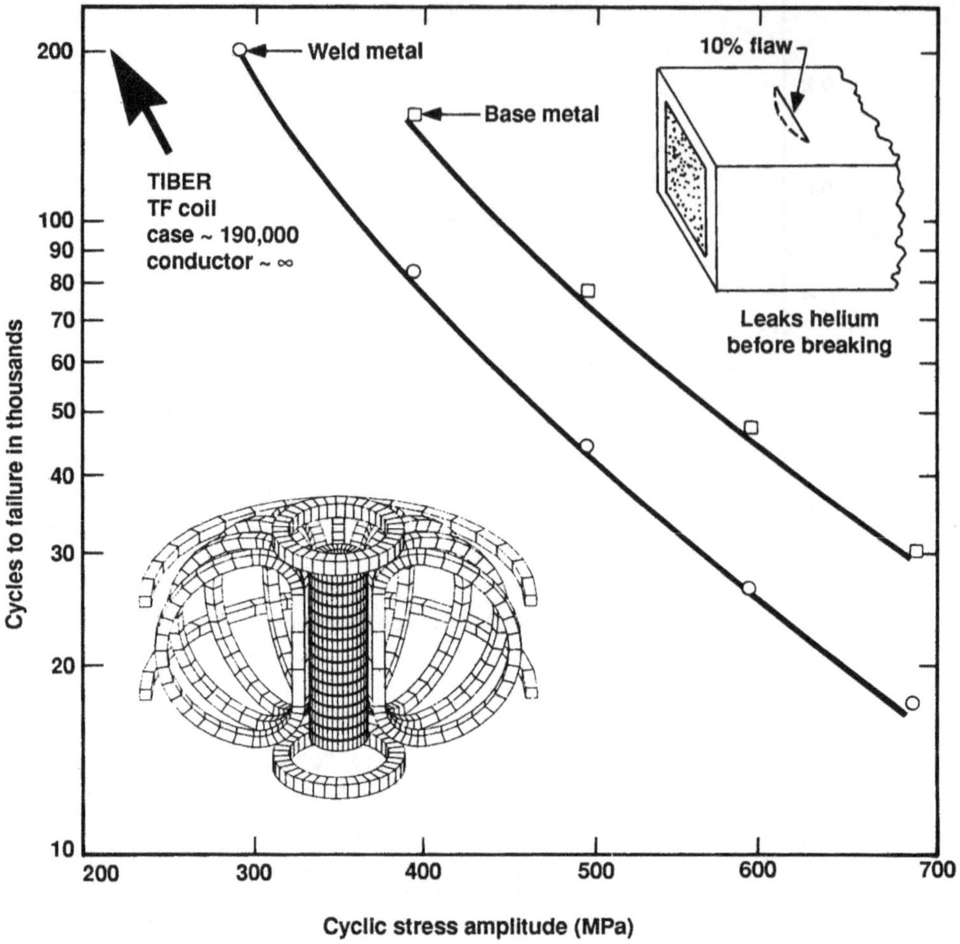

Fig. 4. Cyclic lifetime of an initial crack in the conductor sheath is calculated from crack growth data for JBR-75 steel.

slide over each other under load), and creation of free radials (fission fragments) and gases occurs.

For the load-bearing insulation in superconducting magnets, the two main classes of organic materials that have excelled are based on the use of a glass-fiber filament woven into a two-dimensional cloth, with layers of cloth being bonded together into a three-dimensional mass with either an "epoxy-type" or "polyimide type" binder. The details of the manufacture and basic materials science of such composites are described in Refs. 19 and 20, for example.

Much information on the response of these and other fiberglass-reinforced composites to γ irradiation to doses below 10^8 rads is available and is summarized in Refs. 21, 22, and 23. More recent work has concentrated on determination of the effects of higher γ dosages, in the range 10^8 - 10^{10} rads, on both the electrical properties (such as resistance and breakdown voltage and mechanical properties such as failure strengths under various static modes of loading such as bending, compression, tension[24-30] and fatigue modes of loading[31-33]. Usually,

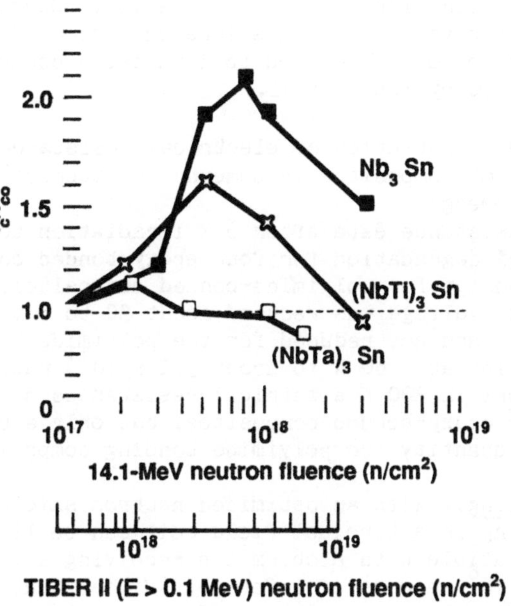

Fig. 5. Critical current vs. neutron fluence. Data were taken using a 14.1-MeV neutron source. The lower horizontal axis accounts for the "softer" neutron energy spectrum expected in TIBER II.

Table II. Neutron irradiation data for organic insulators (irradiated to 10^8 Gy or 10^{10} rad)

Material	Unirradiated strength Compression (MPa)	Unirradiated strength Flexure (MPa)	Irradiated strength Compression (MPa)	Irradiated strength Flexure (MPa)
G-10	886	1100	65	95
G-10 BF		990		108
G-11	826	1113	62	110
Epikote 828 (epoxy)	513	165	125	25
Stycast 2850 (epoxy)	570	262	50	57
Spaulrad (polyimide)	680	990	400	640
Norplex (polyimide)	900	690	900	445
Vespal (polyimide)	250	320	255	320

irradiation is done at 4-9 K; the specimens are warmed to room temperature during removal from the irradiation source and then are cooled to 77 K and tested. This procedure is not representative of real operation, where the radiation damage and mechanical loading occur simultaneously at 4-5 K but is useful for determination of generic behavior. Summarized in Table II are the results of cryogenic irradiations on various fiberglass-reinforced composites, as ratios of irradiated to irradiated compressive and flexure strength as reported by Mauer et al.

The database on retention of electrical resistance and electrical breakdown voltage of composites is somewhat scattered[22,24,34], but the following trends emerge:

- 300 K resistance data after 5 K irradiation to 10^{10} rads showed 30 to 80% degradation for four epoxy-bonded composites, but little change for polyimide-bonded composites. Dielectrical breakdown voltage was reduced about 30 to 50% for the epoxy materials and not reduced for the polyimide bonded materials.
- Irradiation at ~ 60 K to about 3.2×10^{11} rads resulted in reductions in 300 K electrical-resistances of about 10^4 to 10^9 for five epoxy-bonded composites, but only a six-fold reduction in this quantity for polyimide bonding composite.

<u>Nuclear Heating</u>. With an optimized neutron shield of about 50 cm, the neutron heating in a toroidal field coil can be limited to 5 mW/cm^3. This rate is compatible with niobium tin receiving a fluence of 10^{19} n/cm^2 corresponding to 10^8 seconds of lifetime. Higher heating rates up to 100 mW/cm^3 are possible, but the reactor lifetime would be correspondingly reduced. For TIBER II the 5 mW/cm^3 heating rate at the inside turns is low enough that a full pancake can be cooled without intermediate helium connections.

Perhaps contrary to intuition, the cold helium in this scheme is injected into the windings at the outermost turns (one inlet at the outer end of each half-pancake). The helium from both halves is exhausted at the same opening on the innermost and most highly heated turn of the pancake. The exhaust port can be located along the outer leg of the "Dee" where space is available. With the helium injected at relatively low pressures, there is significant Joule-Thompson expansion that gives some added cooling and stabilization of temperature in the more highly heated portions of the windings. This can be seen in the path traced out on the TS diagram for the 72 kW total heat load case shown in Fig. 6. The numbers on the curve represent conditions in particular turns of the winding, counting in from the outside.

With temperature distribution in the flow path determined, the conductor can be designed to give the desired performance, i.e., stability margin at the desired current density, in the region of the windings where conditions are most severe, i.e., the highest fluid temperature and field. For the heat loads expected in TIBER II, there are options that hold the fluid temperature in the 12 T region of the TF coils to 5.3 K or less.

CONCLUSION

The advancement toward fusion energy has brought new requirements on superconducting magnet technology. Not only does performance (field and current density) need to be improved, but nuclear effects must be accommodated. Neutron damage and nuclear heating will be prime considerations in the next generation of fusion experiments.

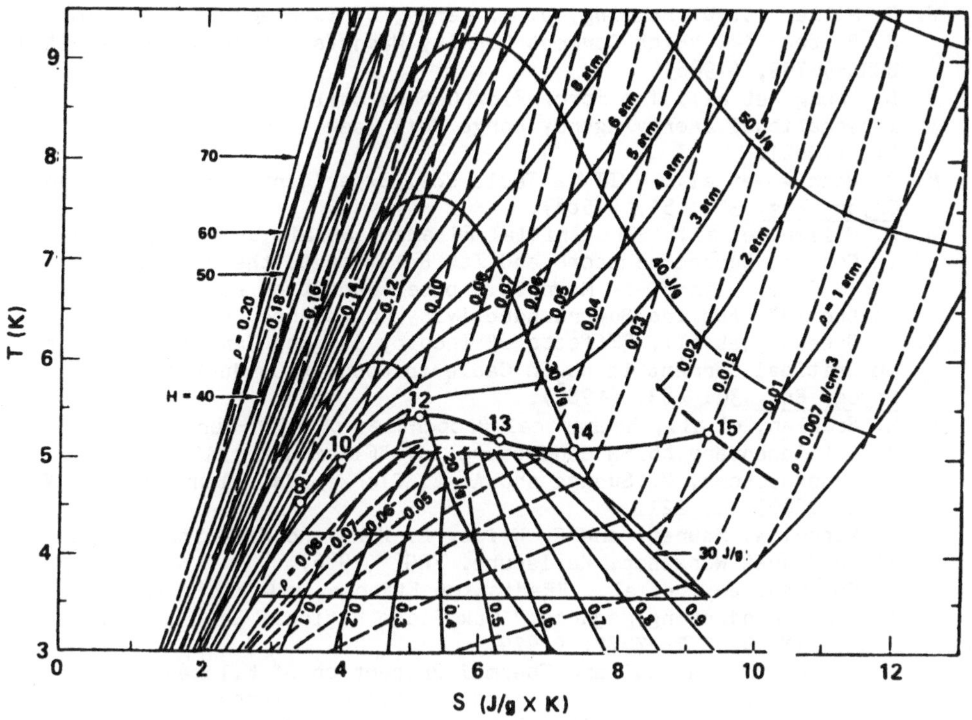

Fig. 6. Thermodynamic path taken by the helium in a TIBER II TF coil corresponding to a total radiation heat load of 72 kW to all coils. The numbers on the curve are turn numbers, helium is injected at the outside (Turn 1) at 4.5 K, 5 atm, and 3000 $g \cdot s^{-1}$.

REFERENCES

1. C. D. Henning, et al., "Large Superconducting Baseball Magnet, Part II," 8th Int. Cong. of Refrigeration, International Institute of Refrigeration, Washington, D.C., Aug. 27 - Sept. 3, 1971.
2. K. R. Efferson, et al., "The IMP Superconducting Coil System," 4th Symposium on Engineering Problems of Fusion Research, Naval Research Laboratory, Washington, D.C., April 20-23, 1971.
3. D. Ivanov, Private Communication, Kurchatov Institute, Moscow, (1977).
4. C. D. Henning, et al., "Mirror Fusion Test Facility Magnet System - Final Design Report," UCRL-52955, (1980).
5. K. H. Krause, T. A. Kozman, J. L. Smith, R. J. Horan, "MFTF-B PACE Tests and Final Cost Report," UCID-20819, (1986).
6. Hitachi Corporation - TriAM-1M Magnet System for Kyushu University, (1986).
7. R. Aymar, et al., "TORE SUPRA-Status Report," Magnetics, MAG-17. No. 5: 1911, (1981).
8. B. Kadomtsev, Private Communication, (1986).
9. P. N. Haubenreich, et al., "Procurement and Testing of Superconducting Toroidal Field Coils in the Large Coil Program," Seventh Symposium on Engineering Problems of Fusion Research, Knoxville, (1977).
10. C. D. Henning, B. G. Logan, et al., "TIBER II Tokamak Ignition/Burn Experimental Reactor," 1986 Status Report, UCID-20863, (1986).

11. J. R. Miller, C. D. Henning, J. A. Kerns, D. S. Slack, L. T. Summers, J. P. Zbasnik, "High Current Density Magnets for INTOR and TIBER," UCRL-95759, (1986).
12. J. L. Young, et al., The Force Flow Cooled Coils for the International Energy Agency Large Coil Task," Adv. Cryog. Eng. 27: 11, (1982).
13. M. O. Hoenig, et al., "MIT 12 Tesla Coil Experimental Results," Adv. Cryog. Eng. 31: 151, (1986).
14. J. R. Miller, et al., "The Initial Filament Strain State of Cable-in-Conduit Superconductors and Its Relation to the Design of Large-Bore, High-Field Magnets," paper submitted for presentation at the 1986 Applied Superconductivity Conference, Baltimore, MD.
15. M. M. Steeves, et al., "Effects of Incoloy 903 and Tantalum Conduits on Critical Current in Nb_3Sn Cable-in-Conduit Conductors," Adv. Cryog. Eng. 30: 883, (1982).
16. R. M. Scanlan, et al., Mechanical Properties of High-Current Multifilamentary Nb_3Sn Conductors in: "Filamentary A15 Superconductors, M. Suenga and A. F. Clark, Eds. Plenum Press, New York (1980), p. 221.
17. R. Flukiger, W. Maurer, and F. Weiss, P. Hahn and M. Guinan, High-Field Magnet Workshop, Karlsruhe, FRG, (1986).
18. R. P. Coltman, Jr., et al., "Radiation Effects on Organic Insulators for Superconducting Magnets," Oak Ridge National Laboratory, Oak Ridge, TN, ORNL TM-7077, (1979).
19. M. B. Kasen, "Mechanical and Thermal Properties of Filamentary-Reinforced Structural Composites at Cryogenic Temperatures-1: Glass-Reinforced Composites," Cryogenics, 15:327-49, (1975).
20. J. R. Benzinger, "The Manufacture and Properties of Radiation-Resistant Laminates," presented at the ICEC Conference, San Diego, CA, August (1981).
21. R. Evans and J. T. Morgan, "A Review of the Effects of Ionizing Radiation on Plastic Materials at Low Temperatures," Rutherford Laboratory, Didcot Berkshire England presented at the ICEC Conference, San Diego, CA, August 1981.
22. M. Van de Voorde, Ed., "Low Temperature Irradiation Effects on Materials and Components for Superconducting Magnets for High Energy Physics Applications," CERN, Geneva, Switzerland, Report CERN 77-03, (1977).
23. J. Guess, et al., "A Survey of Radiation Damage Effects in Superconducting Magnet Components and Systems," Oak Ridge National Laboratory, Oak Ridge, TN, ORNL TM-5787, (1975).
24. H. Weber, E. Kubasta, W. Steiner, H. Benz, and K. Nylund, "Low Temperature Neutron and Gamma Irradiation of Glass Fiber Reinforced Composites," Journal of Nuclear Materials, 115:11-15, (1983).
25. R. E. Schmunk, G. R. Imel, and Y. O. Harker, "Irradiation Studies of Magnet Insulator Materials," Journal of Nuclear Materials, 115:11-15, (1983).
26. G. R. Imel, D. V. Kelsey, and E. H. Ottewitte, "The Effects of Irradiation on TFTR Coil Materials," Journal of Nuclear Materials, 85-86:367-71, (1979).
27. S. Takamura and T. Kato, "Mechanical Properties of Organic Insulators for Superconducting Magnets After Low Temperature Irradiations," Paper CP-9 presented at the ICEC Conference, Colorado Springs, CO, August 1983.
28. S. Takamura and T. Kato, "Low Temperature Irradiation Effects on Mechanical Properties of Epoxy Used in Superconducting Magnets," Cryogenics, pp. 215-219, (1978).

29. R. R. Coltman, Jr., C. E. Klabunde, R. H. Kernohan, and C. J. Long, "Effects of Radiation at 5° K on Organic Insulators for Superconducting Magnets," IEEE Procedures in Magnetics, MAG.-15:1694-1698, (1979).
30. R. R. Coltman, Jr., and C. E. Klabunde, "Mechanical Strength of Low Temperature Irradiated Polyimides: A Five-to-Ten End Improvement in Dose-Resistance with Epoxies," Journal of Nuclear Materials, 103-104: 717-722, (1981).
31. R. Schmunk and H. Berkel, "Tests in Irradiated Magnet Insulator Materials," Paper presented at the 3rd Fusion Materials Conference, Albuquerque, NM, (1983).
32. R. Schmunk, "Irradiation and Testing of Spaulrad-S for Fusion Magnet Applications," Idaho National Engineering Laboratory, Idaho Falls, ID, unpublished manuscript, (1983).
33. S. Nishijima, S. Ueta, and T. Okada, "The Effects of Low Temperature Irradiation Effect on the Cryogenic Fatigue Resistance of Epoxy Resin Used in Superconducting Magnets," Cryogenics, pp. 312-313, (1981).
34. C. J. Long, R. Kernohan, and R. R. Coltman, Jr., "Radiation Effects on Insulators for Superconducting Magnets" and "Non Metallic Materials and Composites at Low Temperatures," A. F. Clark, R. P. Reed, and G. Hartwig (Editors), Plenum, New York (1979), pp. 141-153.

AN INDUSTRIAL SUPERCONDUCTING HIGH GRADIENT MAGNETIC SEPARATOR

J.A. Selvaggi

Eriez Magnetics, Erie, Pennsylvania

P. Vander Arend

Cryogenics Consultants, Inc., Allentown, Pennsylvania

J. Colwell

Huber Corporation, Wrens, Georgia

ABSTRACT

An industrial superconducting high gradient magnetic separator was designed, fabricated, assembled, tested and placed in operation. It is used to process clay. It has been operating continuously since the early part of May 1986. The warm bore is 87 inches in diameter. The maximum field of 2 tesla can be generated in less than 60 seconds with equal time for decay. Tests were conducted on heat leaks, ramp losses, steady state losses, liquefier capability, protection schemes, power supply operation, and liquid nitrogen requirements. The iron enclosure was designed to insure a uniform and linear field to 2 tesla. The cryogenic system was designed for present and future expansion requirements. The conductor is niobium-titanium and cryostable. Comparison with conventional water-cooled units will be made. Discrepancies between design and operating characteristics will be presented.

INTRODUCTION

Present high gradient magnetic separators are almost exclusively used by the clay processing companies. The popular units operate with a magnetic field up to 2 tesla. The system is composed of an iron-clad solenoid with a bore diameter to accept an 84 inch by 20 inch high canister (see Fig. 1). The canister is packed with magnet stainless steel wool and a density varying between 6 and 7%. Clay slurry is passed through the canister and paramagnetic particles are captured by the high gradient resulting from the wool situated in a uniform magnetic field. Proper design of the iron structure insures uniformity to 2 tesla. The process is cyclical. After a period of time the wool accumulates enough material resulting in objectionable contaminants passing through. At a defined level of contamination the solenoid is de-energized, the wool flushed with clear water, and the process resumed. The system is completely automated.

The conventional units utilized water-cooled conductors to generate the magnetic field. Power consumption ranges from 600 KW to 270 KW. It was decided to replace the copper conductors with a superconducting system based initially on the savings in electric power consumption. The Huber Corporation decided on the superconducting system for their next HGMS.

Fig. 1. Canister

Since a unit of this size with a closed loop liquid helium system has never been manufactured, there was no model to gain pertinent data as to possible subtle design requirements. The various engineering disciplines were assembled. The technology to design and manufacture each part of the system was available. The fully integrated system was new.

ANALYSIS

The system was designed to generate 2 tesla when fully iron-clad and effectively uniform throughout the bore (see Fig. 2). With a coil 20" high and sufficient iron, the behavior is equivalent to a coil of infinite length (see Fig. 3). Such a coil has an internal uniform field but tapers linearly from maximum field to zero field from the inside diameter to the outside diameter. The actual coil height was 20" but the requirements of the cryostat resulted in pole pieces, top and bottom, of 3-5/16". These short pole pieces result in a slight field distortion, on center, in the radial direction.

The superconductor is niobium-titanium, 1008 turns, cabled, cryostable, with a copper to superconductor ratio of 39/1. Since this system operates 24 hrs./day with extreme minimum shutdown per year, operational reliability was crucial. Avoidance of a quench was one of the major design criteria. It was designed for 10^6 amp-turns with two full cycles per hour for a total of 10^6 cycles. The ramp time to 2 tesla is 60 seconds with equal time for decay. During testing the times were reduced to 45 seconds. The only observable instrument recording change was a slightly higher helium boil-off rate.

The current flow from the power supply to the magnet is carried from the warm (room temperature) environment to the cold (4.2 K) environment, where the magnet is located, by the two conventional power leads. To function properly, a flow of cold helium gas is maintained in the standpipe power leads to disperse the heat generated internally. If at any current level the voltage drop exceeds 80 mV (Reference voltage), one or both of two sensing voltage drop protective relays, one for each lead, will trigger an alarm, open one or two solenoid values to increase cold helium gas flow, and operate the power supply in the full discharge mode.

The iron enclosure was designed to operate at 1.5 tesla, excepting that the two pole pieces operated between 2 to 2.1 tesla. The iron saturates at 2.14 tesla. Therefore the only deviation from linearity was due to the pole pieces. 40,425 amp-turns per inch or 808,500 amp-turns for a gap of 20" is required. Due to some iron saturation the amp-turns required for 2 tesla was 820,000. This insured linearity from 0 to 2 tesla. The iron enclosure is symmetrical both in the axial and radial direction from its center.

Fig. 2. Iron Enclosure

Fig. 3. Superconducting Coil

The calculated axial force on the coil was 30,000 lbs./in.; the radial 10,000 lbs./in; the maximum hoop stress is 17,000 psi. Due to the symmetry of the iron enclosure and the coil in its cryostat, careful attempts were made to center the coil in the iron. No tests have been run to check these positions, but the radial force constant was questioned.

TABLE 1

Stored Energy	3.53 MJ
Operating Current at 2 T	840 A
Maximum Current	999
Inductance	10 Henry
Ramp Rate	1 min up, 1 min down
Charging Voltage	140 V
Ampere-turns (designed maximum)	1.0×10^6
Coil Type	Superconducting Solenoid
Orientation	Vertical Axis
Cooling	Liquid Helium Bath, 4.5 K (nominal)
Heat Load (calculated)	
Steady State Operation	
Vapor-cooled Leads	4.0 litres/hr.
For a 1-min ramp	
Coil	900 J
Cryostat	4200 J
Liquid Helium Capacity	200 L
Conductor	
Material	NbTi/Copper
Configuration	Cable
Coil Configuration	Double pancake wound
Number of Layers	28
Number of Turns per Layer	36
Winding I.D.	93 in.
Winding O.D.	102.8 in.
Coil Height	20 in.
Coil Weight	5800 lbs.
Axial Force Constant	30,000 lbs./in. (calculated)
Radial Force Constant	10,000 lbs./in. (calculated)
Maximum Hoop Stress	17,000 psi

The behavior of energized coils is to enclose maximum flux or to maximize inductance. It appeared that radial movement would reduce both. A scaled model was built and suspended in the iron enclosure both in the radial and axial direction. Offsetting the coil in the radial direction while axially centered always resulted in the coil returning to the center. Exaggerated offsets behaved the same. Offsetting in the axial direction caused the coil to move in the same direction when energized due to increase in flux-linkages. Since these tests were run after all designs were completed and fabrication in progress, no changes were made. The compensation for the calculated forces increased the cost of the unit.

The calculated value of inductance, assuming no iron saturation, was 10 henries. Since R is zero in the superconducting state, $L = VT/I$, where V is the applied constant voltage, T is the time in seconds, and I is current. The measure data was about 10.2 henries. The stored energy when operating at 2 tesla is 3.51×10^6 joules.

POWER SUPPLY

The power supply is not one but two power supplies connected in parallel, controlled by a common master current regulator (see Fig. 4). There is a high voltage and a low voltage power supply. The power supplies work together to form an integrated system capable of fast field force on, steady state operation at optimum power factor and minimal input power consumption, plus the ability to fast field discharge the magnet's stored energy back into the power source.

Each power supply consists of an isolation power transformer and a 6 SCR full wave AC to DC converter. The low voltage power supply is used to sustain the magnet field intensity after it has been forced on by the high voltage power supply. It is current regulated providing an adjustable output current from 0-1100 ADC. It is rated to provide up to 12 VDC -- however, it generally operates at a much reduced point compensating only for system losses. System losses are only due to the conventional conductors from the power supply to the coil. With the low voltage power supply operating at a maximum voltage of 12 VDC vs the high voltage power supply at 200 VDC, the maximum peak ripple voltage is 13 during 95% of the operating cycle. This low peak ripple voltage results in less helium boil-off of the liquefier (less power loss).

The high voltage power supply is used briefly at two different times during an operating cycle. At the start of the cycle the magnet accepts current very slowly due to its inductance. Being current regulated and unable to achieve setpoint due to the inductance, it immediately goes to full voltage (200 VDC). This forces the current to build in the magnet many times faster than could be achieved by a power supply designed only to operate at its steady state voltage of 5 VDC. Once the current achieves setpoint the voltage from the high voltage power supply backs off rapidly to 0 VDC and the low voltage power supply picks up the load, operating at better power factor, efficiency, and lower losses than the high voltage could ever hope to.

The high voltage power supply is not used again until the end of the operating cycle. At turn off the high voltage SCR's are momentarily turned on to allow current to flow in a forward direction, then the master current regulator is locked off removing the reference control signal from both the high and low voltage control amplifiers. The low voltage SCR's turn off and are no longer used. The high voltage SCR's gate signals are left on, however, and are phase shifted to a time where current cannot flow from the source to the magnet during the negative portion of the sine wave. To the magnet it appears as if the coils were open circuited. The coil voltage rises to the transformer's secondary voltage value and with the SCR's gates phase shifted, energy from the coils is discharged back through the transformer and into the source. The higher the transformer secondary voltage, the faster discharge will occur. Discharge voltage and thus discharge time can be controlled by the position of the SCR gates in the discharge region.

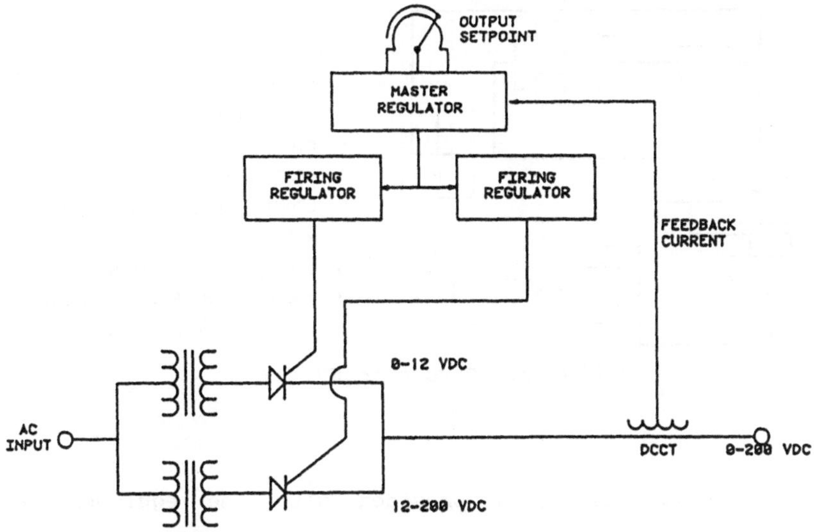

Fig. 4. Simplified One-Line Diagram

To back up the discharge circuit two bootstrap SCR's are used with individual sensing circuits. Should the discharge circuit fail the energy stored in the coils would force the DC voltage to reverse and climb toward infinity. Before damage can occur the bootstrap circuit senses the situation and shorts across the coils. If the bootstrap fires, discharge of energy will occur slowly because it occurs at only approximately 2 VDC.

CRYOGENIC SYSTEM FOR THE MAGNET

The cryogenic system (see Fig. 5) provides the following functions:

a.) Complete system cool-down from ambient temperature to 4 K.
b.) Steady state operation, including periodic removal of the field for regeneration of the bed.
c.) Provide full redundancy for failure of various components to assure 100% availability.
d.) Provide the capability to warm up the system.

The following general comments can be made:

Utilities

Utilities for the system are:

a.) Cooling Water
b.) Power
c.) Liquid Nitrogen
d.) Helium Gas

Consumption of these utilities is as follows:

a.) Cooling water is circulated through a cooling tower. Makeup of water amounts to 25 gpm.
b.) Power consumption for a one-year operating period is 645,000 kwh.
c.) Liquid nitrogen consumption for a one-year operating period is 20.7 litres, including start-up.
d.) Helium consumption for a one-year operating period is .934 litres/hr. including start-up and gas storage leaks.

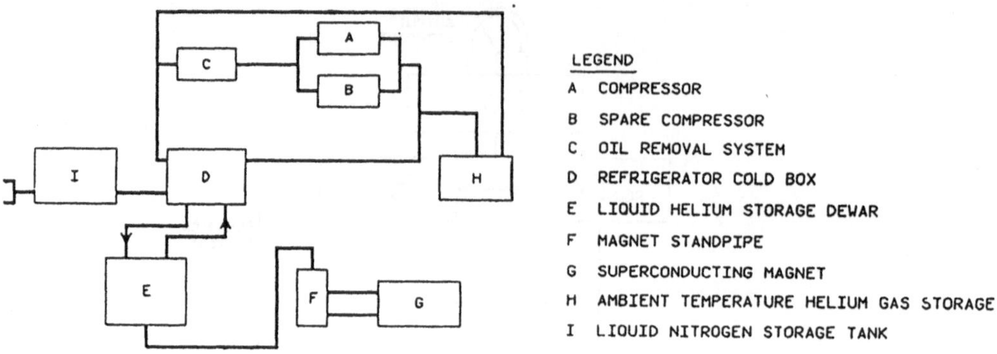

Fig. 5. Cryogenic System

The process guarantees provided with the equipment were as follows:

a.) Cooling Water -- 20 gpm
b.) Power -- 438,000 kwh
c.) Liquid N_2 Consumption -- 10.265 litres/hr.
d.) Helium Gas Loss -- minimal

Description of Process

The helium refrigerator (D) (see Fig. 6) employs liquid nitrogen precooling and a reciprocating expander for production of refrigeration below 80 K. The mixture of liquid and gas produced at the J-T valve flows to the liquid helium storage tank (E), where gas and liquid separate. The vapor returns to the refrigerator.

Liquid helium flows from the storage tank (E) to the magnet standpipe (F). Flow rate is controlled by a superconducting level sensor which controls a throttle valve in the liquid line.

The magnet is housed in a cryostat (G) which is maintained at full level by means of a gravity driven circulation system. As a result of this arrangement, magnet liquid inventory is relatively small.

Power leads for the magnet enter the standpipe (F) and are cooled by a flow of helium gas from the standpipe, through the leads, out to compressor suction.

The magnet and cryostat (G) is equipped with a helium gas cooled shield, with gas from this shield flowing to the 80 K temperature level of the refrigerator.

Liquid nitrogen is supplied from a so-called "customer station" (I) (tank supplied and maintained by the supplier of the liquid nitrogen). Flow is controlled by a throttle valve activated by a differential level control.

The system has two oil-flooded screw compressors (A-B), each capable of providing the necessary flow for system operation. Only one of the compressors operates.

Oil is removed in an oil removal system (C), consisting of two stages of coalescing and one stage of vapor adsorption on charcoal.

The gas storage vessel (H) is a propane tank with a volume of 15,000 gallons in which gas can be stored at pressures between atmospheric and 250 psig.

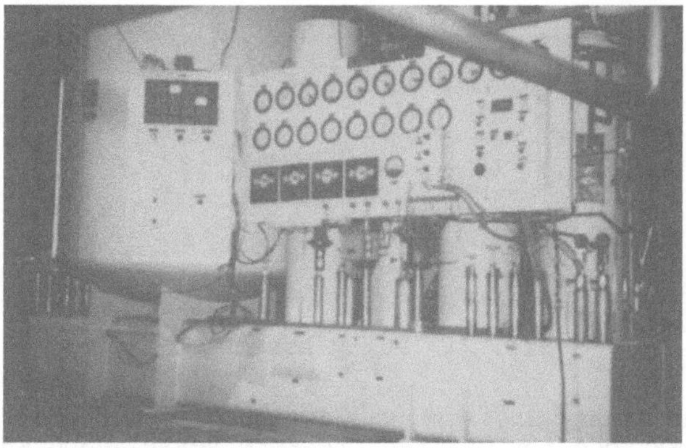

Fig. 6. Helium Liquefier

REDUNDANCY

The system is designed to provide continuous refrigeration to the magnet. Only power failure is expected to be a reason for downtime. In order to assure 100% on-stream time, various components back up other components. For instance:

1.) Loss of the expander of the refrigerator will be offset by consumption of liquid helium. The vaporized helium is warmed in the refrigerator cold box exchanger and compressed and stored in gas storage vessel (H) for later liquefaction.

2.) A loss of the refrigerator will be offset with consumption of liquid helium, which can then be compressed and stored in vessel (H) for later reliquefaction. The liquid helium storage tank (E) contains as much as 4,000 litres of liquid which is sufficient to keep the magnet in operation for one week without use of the refrigerator cold box.

3.) The spare compressor provides backup for the operating compressor.

4.) Cold box contamination can be taken care of in eight (8) hours, during which time liquid helium will be used for magnet operation.

SYSTEM DESIGN PHILOSOPHY

Since the system is part of a commercial production process, reliability and redundancy were considered to be the most important parameters.

This has led to some overdesign and overcapacity; for instance:

1.) The refrigerator can produce as a liquefier 25-30 litres/hr. of liquid helium when using a single compressor. This provides the capability of reliquefying warm gas at a relatively high rate. On the other hand, it results in somewhat high power consumption during steady state since the screw compressor does not turn down very efficiently.

2.) The liquid helium storage tank can hold as much as 4,000 litres of liquid. This resulted in a slower start-up from warm conditions and a somewhat higher heat leak.

3.) The magnet support system was designed for high forces in case the magnet was not absolutely centered. These forces were given by the coil manufacturer and resulted in an extra heat leak of some 4-5 watts.

Fig. 7. Fully-Assembled System

4.) The magnet cryostat was designed in accordance with the ASME Code for unfired pressure vessels and was also designed to withstand the pressure generated from a quench. As a result eddy currents in the shell during ramping are relatively high.

As a result of the conservative design, it appears that the refrigeration system will be capable of handling a second magnet system, operating in parallel with the existing system. It also means that a magnet at a different location could be provided with a less conservative cryogenic system. This would lead to savings in power and liquid nitrogen.

OPERATING HISTORY

At the request of the Huber company the equipment was assembled in Bethlehem, Pennsylvania for a full-scale cryogenic system start-up from scratch with magnet operation to full field and ramps at the desired frequency.

Cool-down, magnet filling and magnet operation took place during the month of December, 1985. The acceptance consisted of 48 hours of operation and was carried out during the first week of January, 1986. The equipment was then disassembled and transported to Wrens, Georgia.

Assembly started in February, 1986 and cool-down was started in May, 1986. Plant operation was started on Friday, May 16, 1986 (see Fig. 7). The plant has been in operation since this time. Following are some significant facts of this one year of operation.

1.) Plant crew training was carried out at Wrens, Georgia and not during the initial test at Bethlehem, Pennsylvania.

2.) Power failures (due to thunderstorms primarily) totaled 5. Plant downtime contributed to power failures was less than one hour.

3.) Downtime for other reasons was 100 hours.

4.) Onstream time for the first year of operation was 8500 hours. This is 99% of available time.

5.) Man-hours for plant operation were 8500 hours.

CONCLUSIONS

The superconducting high gradient magnet separator has been successfully operating for at least one year. The increased liquid nitrogen consumption can be attributed to start-up requirements and nitrogen storage tank losses. The helium consumption can be attributed to start-up consumption and warm end gas leaks. Gas leaks have been reduced. 1987-1988 data is being accumulated and will be compared with data presented in this paper. The calculated heat leak was 9.17 watts but measured at 14.37 watts. The increase is due to dewar losses (3% per day instead of 1.2% as calculated) and radial support losses to counteract offset radial forces.

13.4 T GENERATION BY SUPERFLUID COOLING

M.Wake, K.Maehata, M.Kobayashi

KEK National Laboratory for High Energy Physics, Oho,305 Japan

A.Sato, Y.Yamada, S.Murase

Toshiba Research and Development Center,Kawasaki,210 Japan

S.Nakamura, Y.Kamisada

Showa Wire and Cable Co.,Kawasaki,210 Japan

ABSTRACT

A magnet with NbTi/Ta and NbTi conductors was constructed to be operated in a superfluid cryostat. The operation experiment was successful and it achieved 13.4 T. This is the highest magnetic field ever achieved by alloy conductors. The advantage of the superfluid cooled alloy conductor magnet for the purpose of high field application for particle accelerators are encouraged from the result of the test operation.

INTRODUCTION

It is known that alloy conductors can have higher current carrying capacities than compound conductors at the field range of less than 12 T [1] if the operation temperature is as low as 1.8K. Since the mechanical strength of compound material has little expectation of drastic improvement, it may be necessary to use alloy conductors for the high field application of dipole magnets for high energy particle accelerators, because mechanical stress in such magnets have to be very large to avoid trainings. The disadvantage of the superfluid cooling is in its inefficiency of cryogenic system and this must be overcome by some means. Although, the magnet construction feasibility itself has been proved to some extent by the success of the 10 Tesla dipole magnet [2], the superiority of the superfluid cooled alloy conductor magnet could be extended up to 12 T if the current density is the only limiting factor of the magnetic field. Small magnets with ternary alloy conductors were prepared for a series of test operations. The test result of NbTi/Hf magnet which produced 12.2 T was already reported [3] with the comparison data of various ternary alloy conductors. This report describes the test result of NbTi/Ta magnet which achieved the highest magnetic field of 13.4 T with the improved superfluid cooling equipment at 1.55 K.

MAGNET STRUCTURE

The magnet is composed of two coils. The outer coil is wound with 0.5 mm × 1 mm NbTi rectangular cable while the inner coil is made of 0.4 mmϕ NbTi/Ta cable. The outer coil has inner diameter of 60 mm, outer diameter of 134.3 mm and length 260 mm. The inner coil has inner diameter of 10 mm, outer diameter of 54 mm and length 120 mm. The main parameters of these two coils are listed in Table 1. The material composition of the conductors are Nb-64.7Ti-7.2Ta for

Table1. Magnet Parameters

Parameters	Outer Coil	Inner Coil
Bore Space (mm)	60	10
Winding Inner Diameter(mm)	65.2	20.0
Winding Outer Diameter (mm)	134.3	54.0
Winding Length (mm)	260.	120
Number of Turns	14863	11703
Inductance (H)	5.48	0.68
Field to Current Ratio (T/A)	0.0670	0.118
Material (atomic%)	Nb-63Ti	Nb-64.7Ti-7.2Ta
Filament Diameter (μm)	17.6	15
Number of filaments	925	295
Copper to Superconductor Ratio	0.99	1.4
Conductor Size(mm)	$0.5\times1.0\times0.5^R$	0.4ϕ

Fig.1. Photo of the magnet. The inner coil (right) was inserted into the outer coil (left).

inner coil and Nb-63Ti for outer coil. Both conductors were drawn with large reduction ratio after the heat treatment to increase the current density. The structures of the coils are simple multi-layer solenoids. No extra cooling channel was considered but the impregnation of epoxy in the coil was avoided. Due to the very low viscosity of superfluid, helium can easily penetrate into the coil through narrow channels. Large heat capacity of liquid helium largely contribute for the improvement of the stability of the magnet. Both magnets were fixed by a G10 holder and mounted in a superfluid (HeII) bath. The excitation current of the coils were supplied separately using two independent power supplies. The current was fed through the normal helium bath to the HeII bath using 1.2 mm × 2.4 mm 1:1 NbTi/Cu conductor. Since the region between normal and HeII bath is in the insulation vacuum, the cooling in the current path has to be carefully monitored. The current paths were tested up to 200A without magnet prior to the experiment. The magnetic field in the center of the magnet was monitored by a Hall probe. Figure 1 is the photo of the magnet.

SUPERFLUID CRYOSTAT

The magnet was immersed in a superfluid helium bath at a pressure of one atmosphere. The superfluid vessel is separated by a vacuum insulation jacket from the normal liquid helium environment in the cryostat. The communication channel between the HeII bath and the normal helium bath which is normally closed is opened as the safety valve when the magnet quenches. As is shown

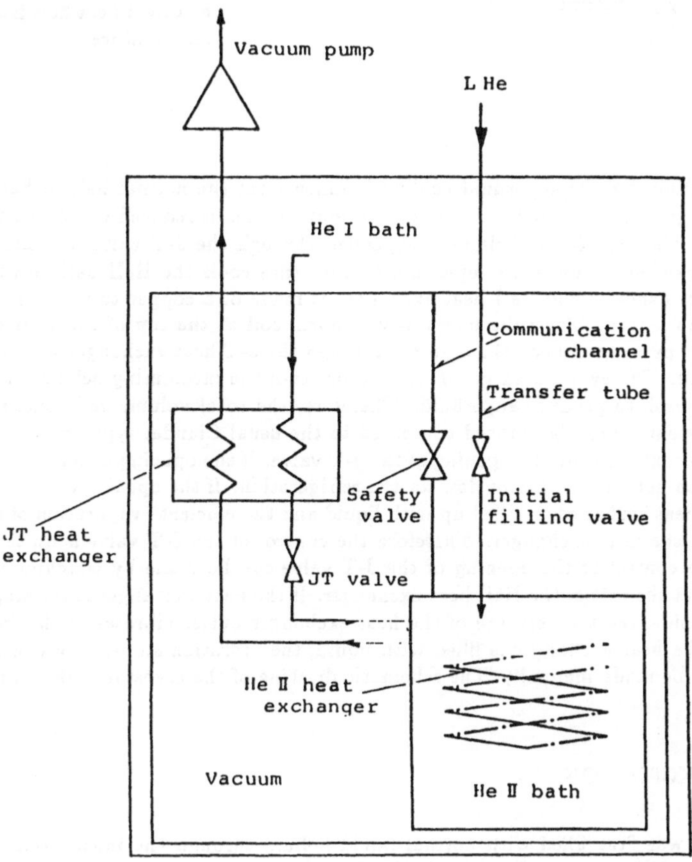

Fig.2. Flow diagram of the system. There is no evaporation pot in the system. The refrigeration is controlled only by the flow.

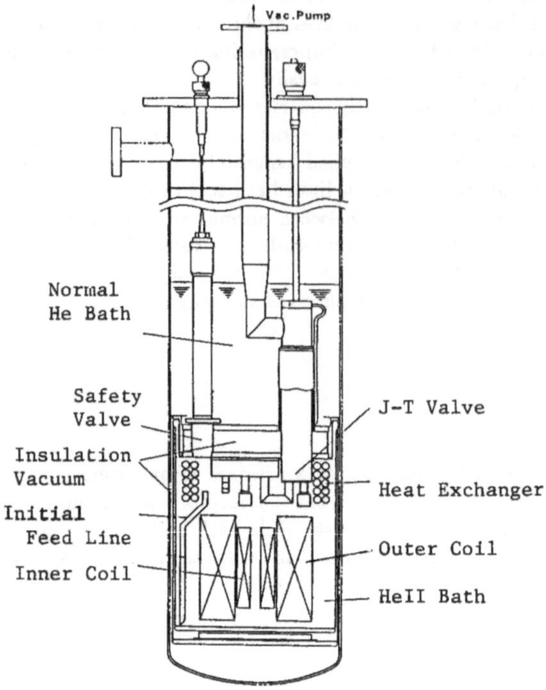

Fig.3. Arrangement in the cryostat. The jacket of the HeII bath is vacuum insulated and sorrounded by the 4.2 K normal He. Therefore, there is no direct heat flow from room temperature.

in Fig. 2, liquid helium for the superfluid cooling is taken from the normal helium bath through the J-T heat exchanger where the temperature of the liquid helium is reduced down to 2.2 K by the returning gas flow. Then the liquid helium is expanded through the J-T valve into the HeII heat exchanger. The expanded helium with superfluid temperature cools the HeII bath down to 1.8 K through this heat exchanger. The HeII heat exchanger is made of a copper tube of 14 mm O.D. , 12 mm I.D. and 8 m length. The tube forms 5-turn spiral coil at the top of the HeII bath. The evaporated helium is pumped outside the cryostat through the J-T heat exchanger by a $7000L$/min rotary vacuum pump. The evaporated gas is used to pre-cool the incomming helium. This system has no evaporation bath to produce superfluid. Therefore, the total volume and consequently the cold mass of the system is greatly reduced compared to the usual Claudet type cryostat. The flow rate of the system is controlled by the opening of the J-T valve. If the opening of the J-T valve is too small, the system can not have sufficient flow for the refrigeration. If the opening of the J-T valve is too large, the HeII heat exchanger is filled up with liquid and the efficient evaporation of the helium can not be made in the heat exchanger. Therefore the control of the J-T valve is quite important in this system. The control of the opening of the J-T valve can be made by watching the carbon resistance attached at the exit of the HeII heat exchanger. If the heat exchanger is working normally with mist helium inside, the temperature of the heat exchanger outlet vibrates at the frequency of vacuum pump. If the heat exchanger is filled with liquid, the vibration stops. The control itself is very stable and can be made manually. The schematic drawing of the cryostat is shown in Fig. 3.

COOLING AND EXCITATION

The HeII bath was first filled with normal liquid helium through the initial feed line. After isolating the HeII bath from normal helium bath, the vacuum pump was started to make up the flow in the cooling line. The above adjusting method of J-T valve was able to bring down the temperature of the HeII bath to T_λ in 23 minutes. Another 28 minutes was necessary to bring down the temperature to 1.7K. Figure 4 is the cool down curve of the HeII bath. The refrigeration power

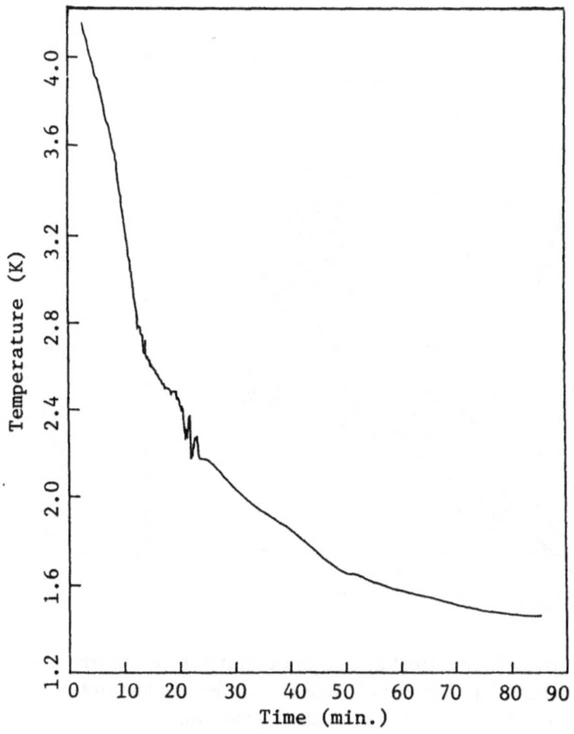

Fig.4. Cool down curve. The oscillations around 2.2 K are due to the difficulty in the operation of the J-T valve. The rapid change of the flow because of the superfluid transition affects the operation of the J-T valve. Once it reaches to the steady state, the operation of J-T valve become easy.

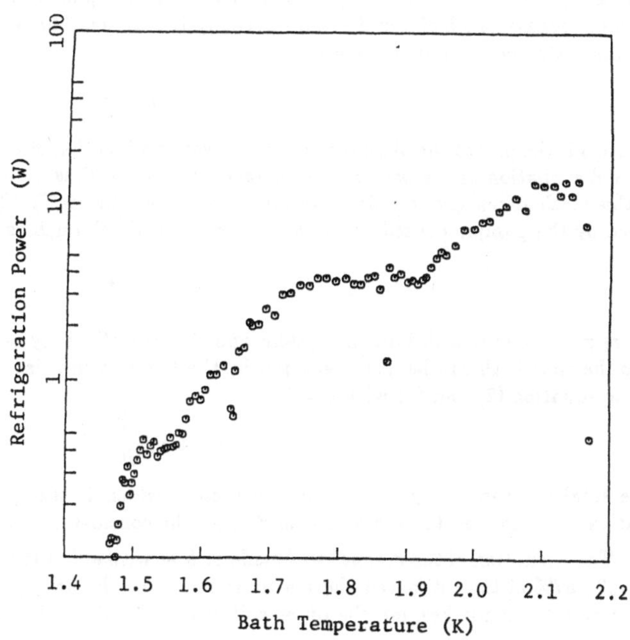

Fig.5. Apparent cooling power.

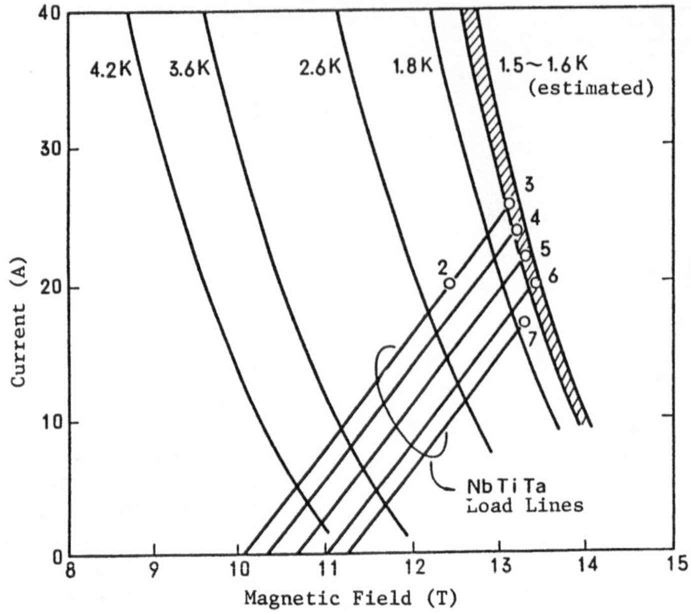

Fig.6. Quenches on the load line. The vertical axis is the current in the inner coil. Circles indicate the quenches. The numbers beside the circles are the run numbers. The maximum field of 13.4 T was obtained in the 6th run.

of the cryostat can be calculated from the cooling curve. If the heat capacity of liquid helium in the HeII bath and the magnet are C_{He} and C_{mag}, respectively, the cooling power, Q, is expressed by:

$$Q = -\{C_{He} + C_{mag}\}\frac{dT}{dt} \tag{1}$$

The heat capacity of the magnet, 2 J/K at 4.2 K for example, is negligible compared to that of liquid helium, because 23.1 L of liquid helium in the HeII bath has 14000 J/K heat capacity at 4.2 K. Then, the cooling power is simply given by:

$$Q = -C_{He}\frac{dT}{dt} \tag{2}$$

Figure 5 is the plot of the observed cooling power as a function of the temperature. Since the temperature distribution in the bath at the temperature below T_λ is quite homogeneous, the measurement in this method can give a reliable result in the region below T_λ. The cooling power should also be noted by the pumping speed, q and the latent heat of helium, L as:

$$Q = fqL - r \tag{3}$$

where r is the heat leak into the system and f is the efficiency of the refrigeration. The heat leak into the HeII bath can be also measured by the temperature rise after stopping the vacuum pump. Using equation (2) and (3) with $q = 0$,

$$r = C_{He}\frac{dT}{dt}. \tag{4}$$

The total refrigeration generated in the system which is the sum of the cooling power and the heatload is estimated to be 5 W at 1.8 K after the consideration of the above heat leaks.

The excitation of the magnet was made at first with only the outer coil. The outer coil produced 10.18T, 87% of the estimated short sample limit, for the first quench. Then the excitation of the inner coil was made keeping the outer coil field at 10.05T. The second run tripped when the field reached at 12.4 T. This was found to be a trip due to the noise. The third run reached at 13.18T and had quench in the inner coil. The runs after fourth were made by increasing the outer coil current with pre-excited inner coil. The excitation was made 7 times in total. The maximum excitation field was obtained in the 6th run, in which the current of 20.1 A was kept in the inner coil and the current in the outer coil was slowly increased. The inner coil made quench when the outer coil current reached 164.4 A. The central field just before quench was 13.4 T which is higher than the reported highest field [4]. These currents correspond to 96% of the estimated short sample limit. The short sample limit was extrapolated from 2.2K data. Figure 6 shows the load line and quench points during the experiment. The temperature of the magnet in these runs are between 1.55 K and 1.60 K.

CONCLUSION

The excitation of the ternary alloy magnet was very successful. The maximum field obtained by the alloy conductor magnet with superfluid cooling reached 13.4 T. The refrigeration power of the pressurized superfluid system was close to the designed value. It is possible to build high field dipole magnet beyond 12 T for future high energy accelerators.

REFERENCES

1. M.Wake et al., Use of superconducting alloys in high field magnet, in "High Field Magnetism", M.Date,ed. ,North-Holland Pub.Co:, Amsterdam (1983), p.339.
2. T.Shintomi et al., The construction and test results of a 10T dipole magnet, IEEE trans. Nucl. Sci. NS-32: 3719 (1985).
3. M.Wake et al., 1.8 K test of binary and ternary superconductors, IEEE trans. on Magnetics MAG-19: 552 (1983).
4. K.Noto et al., NbTiHf-high field superconducting magnet in 1.8K- pressurized liquid HeII, in "Proc. Tenth Intl. Cryo. Engr. Conf." Butterworth, Guildford, UK (1984), p.181.

THE POTENTIAL IMPACT OF DEVELOPING HIGH T_C SUPERCONDUCTORS ON SUPERCONDUCTIVE MAGNETIC ENERGY STORAGE (SMES)

Y. M. Eyssa, M. K. Abdelsalam,
R. W. Boom and G. E. McIntosh

Applied Superconductivity Center
University of Wisconsin-Madison
Madison, Wisconsin

ABSTRACT

The discovery of superconductivity with $T_c > 77$ K (liquid nitrogen boiling temperature) is potentially of great importance for large scale electric utility applications such as the transmission and storage of electrical energy. Superconducting magnetic energy storage (SMES) is already a promising technology for electric utility load leveling. Therefore, it is useful to assess SMES with oxide superconductors cooled by inexpensive and plentiful liquid nitrogen (LN_2) instead of NbTi cooled by the more expensive liquid helium. Liquid nitrogen cooling will significantly reduce the refrigeration energy requirements, and, especially, would help make small size SMES units more economic. The paper presents the impact of LN_2 on efficiency, design and economics of SMES.

INTRODUCTION

The object of this article is to evaluate the impact of recent discoveries[1,2] of high T_c superconductors on SMES technology for electric utility load leveling.[3,4] SMES energy is stored in the magnetic field of a superconductive coil cooled to liquid helium temperature. Electromagnetic axial and radial forces are generated. To avoid costly cold structure, the radial forces are transferred through low heat leak struts to warm bedrock structure. The axial forces are contained internally by aluminum structure which also is vailable to absorb the stored energy in case of an emergency dump situation.[3,5,6]

Mechanical thermal insulating struts transfer radial forces from the cold coil (1.8 K) to the warm rock structure (300 K). The struts have two heat intercepts at 77 K and 28 K to minimize refrigeration power.[7-10] Other major heat loads to the system are radiation, current leads and ac losses. Radiation and electric lead heat loads are reduced by heat intercepts. The intercepts at 77 K and 28 K are metallic shields of 1100 aluminum for struts and radiation shields. Helium gas is optimally used to cool the electrical leads. Heat loads in the helium dewar include ac losses, conductor resistive joints, and friction between structure components.

The major part of the SMES operating cost (about 40%) corresponds to refrigeration energy required to maintain the SMES coil at operating temperature. The daily energy at room temperature to provide refrigeration is listed as a percentage of stored energy in Table 1 for 1.8 K He II SMES units of different sizes. For large units refrigeration represents only a small fraction of daily energy delivered in contrast to the case for small units with higher percentages. For units storing less than 100 MWh, it is important to operate at low current because the lead loss is proportional to current. For this reason smaller units (E_s < 20 MWh) are difficult to justify for utility applications even if their capital cost is acceptable. This evaluation is not applicable to SMES units which are cycled several times per day.

HIGH T_C SUPERCONDUCTORS

Current densities of 10^5 A/cm^2 at 77 K and zero tesla have been reported for the copper oxide conductors.[1] Even though these new oxide materials do not yet carry high current densities in high fields, it is still interesting to assess the impact of developing LN_2 temperature superconductors on SMES. Three areas are addressed: refrigeration power requirements, SMES design and economics.

REFRIGERATION POWER

There are four major heat loads in an SMES unit: (1) electrical lead cooling, (2) mechanical strut cooling, (3) thermal radiation heat load and (4) ac, frictional and other losses in the dewar. The first three are minimized by intercepting heat at selected higher temperatures and by optimum design.[7-10] The fourth can be minimized through better conductor and structure design.

The two extremes discussed here are for heat intercepts at an infinite number of locations between the warm and the cold end temperatures (T_H and T_c) and for no heat intercepts. Actual efficiencies in percent of Carnot are 20% at 1.8 K, 30% at 28 K and 40% at 78 K. Efficiencies are interpolated for other temperatures. The formulas in Table 2 derived

Table 1. Daily Room Temperature Refrigeration Energy as Percentage of Energy Stored, E_s for a 1.8 K SMES unit. Both 50 kA and 230 kA Leads are Considered. Units have a β = 0.02 Aspect Ratio and a Midplane Field = 3.73 tesla.

E_s(MWh)	LEAD (230 kA)	LEAD (50 kA)	STRUTS	RADIATION	AC LOSS	TOTAL (230 kA)	TOTAL (50 kA)
10.0	71.91	15.69	8.58	5.39	2.60	88.48	32.26
20.0	35.95	7.84	6.81	4.28	2.06	49.11	21.00
50.0	14.38	3.14	5.02	3.15	1.52	24.08	12.83
100.0	7.19	1.57	3.98	2.50	1.21	14.88	9.26
200.0	3.60	0.78	3.16	1.99	0.96	9.70	6.89
500.0	1.44	0.31	2.33	1.46	0.71	5.94	4.81
1000.0	0.72	0.16	1.85	1.16	0.56	4.29	3.73
2000.0	0.36	0.08	1.47	0.92	0.44	3.19	2.91
5000.0	0.14	0.03	1.08	0.68	0.33	2.23	2.12
10000.0	0.07	0.02	0.86	0.54	0.26	1.73	1.67

Table 2. Mathematical Formulas For Optimum Upper and Lower Limits of Refrigeration Power

	Lower Limit (Infinite number of heat intercepts)	Upper Limit (no heat intercepts)
(1) Electrical lead (one lead)		
P/I (WA^{-1})	$2\int_{T_c}^{T_H} \frac{\eta}{T}\{\rho k T_H f(1 - \frac{\eta' f T^2}{\eta T_H})\}^{1/2} dT$	$\eta f_c\{\int_{T_c}^{T_H} \rho k dT\}^{1/2}$
IL/A (Am^{-1})	$\int_{T_c}^{T_H} \{\frac{k}{\rho f}(\frac{T_H}{T^2} - \frac{\eta'}{\eta T_H})\}^{1/2} dT$	$\int_{T_c}^{T_H} \frac{k}{\{\int_{T}^{T_H} \rho k dT\}^{1/2}} dT$
(2) Struts (support) or superinsulation		
PL/A (Wm^{-1})	$\int_{T_c}^{T_H} \{k(\frac{\eta T_H}{T^2} - \eta' f)\}^{1/2} dT$	$\eta f_c \int_{T_c}^{T_H} k dT$
(3) Cold end losses (Actual work)		
P/Q_c	ηf_c	ηf_c

f = $(T_H/T - 1)$ and f_c = $(T_H/T_c - 1)$ are Carnot efficiencies
η = a multiplier to account for deviation from carnot
η' = is derivative of η
ρ = is electrical resistivity
k = thermal conductivity

for the two cases: optimally cooled (infinite number of heat intercepts) and no cooling (no heat intercepts) refer to: 1-the optimum values of room temperature refrigeration power per ampere, P/I and the optimum current x length ÷ cross-sectional area, IL/A of electrical leads, 2-the optimum room temperature refrigeration power x length ÷ cross-sectional area, PL/A of mechanical supports or multilayer superinsulation, 3-the cold end losses resulting from ac or friction losses. Figures 1 and 2 show the optimum values of P/I and IL/A for 300 RRR copper leads. Figure 3 shows the values of PL/A for G-10 mechanical struts for the two limits. PL/A for superinsulation have similar curves as those of Fig. 3. Eddy current losses are significantly lower at higher operating temperatures due to the high resistivity of the matrix in addition to lower Carnot efficiencies at high T_c's.

As shown in Figs. 1 and 3 the refrigeration power difference between the two cooling schemes is very small at values of $T_c > 77$ K compared to $T_c = 4.2$ K. We may conclude that for SMES or other superconducting magnets operating at high temperatures there is no need for continuous or discrete cooling for either electrical leads, low conductivity mechanical

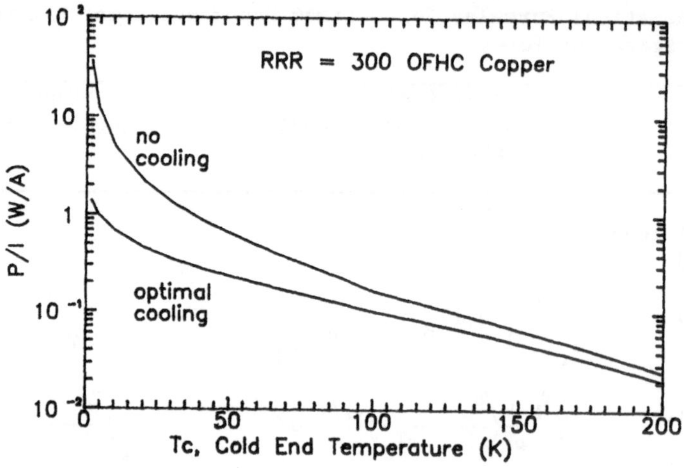

Fig. 1. Lead losses P/I (room temperature refrigeration power/current).

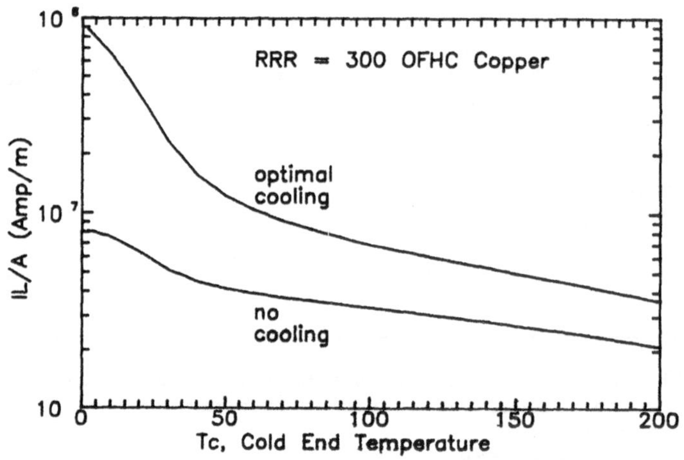

Fig. 2. Optimum IL/A for the two cases represented in Fig. 1. L is the current lead length. A is the cross-section area.

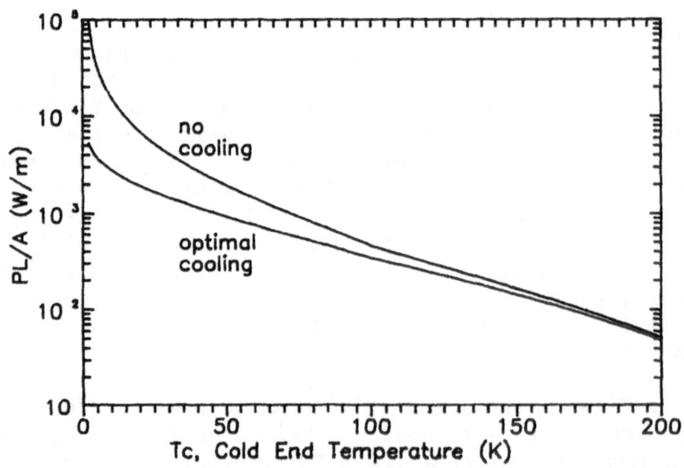

Fig. 3. PL/A vs. T_c for G10 struts. P is room temperature refrigeration power, L = strut length, A = strut cross-section area.

supports or superinsulation. Cold end ac losses which are inversely proportional to normal conductor resistivity, such as eddy current and coupling losses, are extremely small and can be neglected compared to lead, strut or radiation load.

Table 3 lists the room temperature daily refrigeration energy for a 77 K SMES unit. As shown, units as small as 10 MWh have small daily refrigeration energy compared to the energy stored.

STABILITY

SMES conductors in liquid helium are stabilized by mounting NbTi in high purity aluminum to give a low resistance alternate current path for recovery of superconductivity following energy disturbance. At liquid nitrogen temperature an alternate current path may also be required (for protection) because the ceramic oxide has resistivities several orders of magnitude larger than copper or aluminum. Enthalpy absorption by current carrying structure would also be needed to limit temperatures below melting points of affected materials in the conductor and system. A major factor that would enhance the stability of high T_c superconductors is the very high specific heat of metals at 77 K which is 3 orders of magnitudes higher than at 4K. Large energy disturbances should not raise the conductor temperature beyond one or two degrees. The use of copper or aluminum with high T_c superconductors should be aimed toward protection rather than stability.

PROTECTION

Protection of large SMES system is achieved by absorbing the energy in the cold aluminum axial structure during an emergency dump.[3,5,11] First the large liquid helium inventory is removed in a short time by pressurized 28 K helium gas. This expensive system is required to save the costly helium.[11] For SMES units cooled with LN_2 at 77 K, the liquid dumping system should be very simple and much cheaper. There will not be need for a large amount of liquid nitrogen for cooling as it is the case in SMES at 1.8 K. This eliminates the need for dump reservoir, piping

Table 3. Daily room temperature refrigeration energy as a percentage of energy stored, E_s for a 77 K SMES unit. Both 50 kA and 230 kA leads are considered. Units have a $\beta = 0.02$ aspect ratio and a midplane field = 3.73 tesla.

E(MWh)	2 LEADS (230 kA)	2 LEADS (50 kA)	STRUTS	RADIATION	AC LOSS	TOTAL (230 kA)	TOTAL (50 kA)
10.0	10.79	2.35	1.14	0.44	0.00	12.37	3.93
20 0	5.39	1.18	0.91	0.35	0.00	6.65	2.43
50.0	2.16	0.47	0.67	0.26	0.00	3.08	1.40
100.0	1.08	0.24	0.53	0.20	0.00	1.81	0.97
200.0	0.54	0.12	0.42	0.16	0.00	1.12	0.70
500.0	0.22	0.05	0.31	0.12	0.00	0.65	0.48
1000.0	0.11	0.02	0.25	0.09	0.00	0.45	0.36
2000.0	0.05	0.01	0.20	0.07	0.00	0.32	0.28
5000.0	0.02	0.00	0.14	0.06	0.00	0.22	0.20
10000.0	0.01	0.00	0.11	0.04	0.00	0.17	0.16

and cryogenic valves discussed in reference 11 for a liquid helium dump system.

DEWAR AND STRUCTURE

The dewar or cryostat for a liquid nitrogen-cooled SMES is considerably less complex than for 1.8 K helium. Refrigeration is supplied in the form of make-up liquid from a nitrogen liquefier with the heat load reflected in venting through back pressure control valves. It may be feasible to build a non-magnetic cold box which could be located adjacent to the coil with only warm gas lines leading to and from compressors located in a lower field environment. A single lay-up of insulation will be sufficient and support struts will not require heat intercepts. There will be strong motivation to build a simple, reliable insulated structure with minimum cryogenic technology requirements. For instance, it may be cost effective to use non-vacuum purged perlite insulation instead of a conventional evacuated dewar. use of a non-vacuum insulation system reduces the liquid container and vacuum jacket cost as well because it eliminates 1 atm of pressure differential. Another possibility is to combine struts and insulation into powder-filled fiberglass/epoxy honeycomb. This configuration would greatly simplify the struts and uniformly distribute radial loads but it would probably require a vacuum because of cracks and fit-up problems.

Structure in SMES coils is designed to accommodate cooldown, operational and warmup forces in case of energy dump situation. Structure has to be metallic preferably aluminum to act as an internal energy dump during energy scamp. Several designs have been proposed for liquid helium operation include rippled and segmented structure.[12,13] The idea behind ripples or segmented structure is to allow for zero or small cooldown stresses, otherwise the 0.43% contraction in aluminum results in 350 MPa (50 ksi) cooldown stress. Because cooldown forces at 77 K is not much different than at liquid helium temperature, we do not expect that the basic structure of a SMES unit should be different for liquid nitrogen superconductors. Developing superconductors that work at temperatures higher than 77 K may lead to change in design philosophy. A continuous structure that is stressed in tension due to cooldown forces may be preferred as long as the cooldown stresses are much smaller than yield stress of aluminum. Other operational stresses such as axial and radial magnetic stresses have to be added to the cooldown stress in this case.

ECONOMICS

Previous work on SMES at UW shows that low aspect ratio coils ($\beta = 0.02$) has an advantage in terms of cost, lower heat leak through struts and lower pressure on rock support.[3,12] Table 4 breaks down the cost of all component of a 1.8 K He II-cooled 5500 MWh unit. The cost analysis are based on data from reference 13 applied to the UW non-rippled design concept.[12] The table also shows how the cost items scales with the stored energy. The cost analysis considered here does not include any fixed charges which may result in underestimating the cost of small units. The purpose of the table is only to show approximately the impact of using high T_c materials on SMES as function of size. Table 5 lists cost items vs. stored energy. The helium related and the refrigerator cost shown in Table 4 are added in Table 5 to indicate the potential savings in case of high T_c SMES. As shown small unit sizes have higher percentage of helium related cost compared to large units. Developing high T_c SMES can significantly improve the economics of small

Table 4. Capital Cost Factors. Reference Unit: 1.8K Helium II-Cooled Two Layer, B = 4T, E_o = 5500 MWh, β = 0.02, P = 10^6 kW, I_o = 228 kA, Magnet System Cost* C_o = \$910 x 10^6, Power Cost** = \$82.4 x 10^6. Costs are scaled from reference 13.

Cost Component	Scaling Factor	Non-rippled Design
I. $E^{2/3}$ Related		
Helium related (dewar, dump system, shields, etc.)	$C_1\ [E/E_o]^{2/3}$	$C_1 = 0.0900C_o$
Others (conductor, trench, etc.)	$C_2\ [E/E_o]^{2/3}$	$C_2 = 0.5050C_o$
II. E Related		
(axial structure)	$C_3[1.0834 - 0.0834\ (E_o/E)^{1/3}]E/E_o$	$C_3 = 0.3580C_o$
III. Refrigerator***	$C_4\ [0.95\ (E/E_o)^{2/3} + .05\ I/I_o]^{0.7}$	$C_4 = 0.0465C_o$
IV. Power Related		
(power system)	C_pP	C_p = 82.4 \$kW**

* Includes 41% for contingency and allowance for funds during construction.[13]

** Equivalent to (\$50/kW) x VI, where V is the purchased voltage rating (6 kV) and I is the purchased current rating (228 KA) with an additional 20.5% for contingency.

*** Refrigeration room temperature power input is 6 MW for the reference unit.

Table 5. Helium II Related Costs, Conductor Related Costs, and Total Cost vs. Size for 1.8 K SMES Units.

E(MWh)	Helium Related+ Cost (% of Total)	Conductor Related Cost (% of Total)	Total Cost (% of Reference Unit)
10.0	49.735	16.501	1.637
20.0	41.662	18.735	2.288
50.0	33.115	20.705	3.814
100.0	28.234	21.455	5.843
200.0	24.435	21.663	9.186
500.0	20.567	21.282	17.224
1000.0	18.211	20.515	28.225
2000.0	16.181	19.711	46.861
5000.0	13.874	18.293	93.007
5500.0	(Reference Unit)	------	100.0
10000.0	12.398	17.187	157.144

energy stored units in addition to the improvement in efficiency indicated earlier. It is early to estimate the cost of the new high T_c materials to realize the savings or the extra cost over the conductor cost listed in Table 5.

CONCLUSION

High T_c SMES are to be expected to be more economically attractive to utility applications specially for small units. Thermal analytical analysis shows no need for heat intercepts to reduce the refrigeration power. High T_c conductors would impact the dewar more than the structure design of SMES. SMES should be more reliable at 77 K than 1.8 K.

REFERENCES

1. "The New York Times," National News, May 11, 1987.
2. A. L. Robinson, An oxygen key to the new superconductors, Science, 236:1063 (1987).
3. R. W. Boom, Y. M. Eyssa, G. E. McIntosh and S. W. Van Sciver, Cryogenic aspects of inductor-converter superconductive magnetic energy storage, in "Proceedings of the Ninth International Cryogenic Engineering Conference," Butterworth & Co., Surrey, U.K. (1982), p.731.
4. M. K. Abdelsalam, "Operational aspects of superconductive mag netic energy storage," paper presented at the 1987 Intermag Conference, Tokyo, Japan (1987). To be published in IEEE Trans. Mag. (1988).
5. Y. M Eyssa, G. E. McIntosh, M. A. Hilal, A. Khalil and B. Nilsson, An energy dump concept for large energy storage coils, in: "Proceedings of the 9th Symposium on Engineering Problems of Fusion Research," IEEE Pub. No. 81CH1715-2 NPS, (1981) p. 1412.
6. J. Waynert, Temperature and voltage analysis of a protection scheme for large energy storage coils, in "Advances in Cryogenic Engineering,"Vol. 31, Plenum Press, New York (1986), p. 399.
7. M. A. Hilal, Optimization of current leads for superconducting systems, IEEE Transactions on Magnetics MAG-13(1):690 (1977).
8. M. A. Hilal and R. W. Boom, Optimization of mechanical supports for large superconductive magnets, in "Advances in Cryogenic Engineering," Vol. 22, Plenum Press, New York (1977), p. 224.
9. M. A. Hilal and Y. M Eyssa, Minimization of refrigeration power for large cryogenic systems, in "Advances in Cryogenic Engineering," Vol. 25, Plenum Press, New York (1980), p. 350.
10. A. Khalil and G. E. McIntosh, Liquid neon heat intercept for superconducting energy storage magnets, in "Advances in Cryogenic Engineering," Vol. 27, Plenum Press, New York (1982), p. 587.
11. G. E. McIntosh, et. al., Protection system for superconducting magnetic energy storage unit, in "Advances in Cryogenic Engineering," Vol. 33, Plenum Press, New York (1988).
12. R. W. Boom, Y. M. Eyssa, Y. Huang and G. E. McIntosh, "Two layer non-rippled superconductive magnetic energy storage systems," in "Proceedings of the Eleventh International Cryogenic Engineering Conference," Butterworth, Surrey, U.K. (1987), p. 464.
13. R. J. Loyd, "Design Improvement and Cost Reductions for a 5000 MWh Superconducting Magnetic Energy Storage Plant, Part 2," LASL Report LA-10668-MS, October 1985.

SUPERCONDUCTIVITY: THE ULTIMATE ELECTRONIC TECHNOLOGY

M. Nisenoff

Naval Research Laboratory
Washington DC 20375-5000, USA

ABSTRACT

The unique properties of the superconducting state can provide an electronic technology with properties not obtainable from any other known technology. Superconducting electronic devices and circuits can exhibit very low loss, zero frequency-dispersion signal transmission, very high Q-value resonators and filters and quantum limited electromagnetic sensors for radiation from dc through the millimeter wave region and into the visible and ultra-violet region. Very high performance signal and data processing systems of interest to both the military and civilian communities can be built using this technology. The impact of the newly discovered high temperature superconducting materials can revolutionize electronic technology as it would bring these very interesting properties and device behavior to an operating temperature where (a) refrigeration requirements are greatly reduced, and (2) where hybrid semiconductor-superconductor circuits can be built which make use of the best features of each technology.

INTRODUCTION

Superconductivity, due to its quantum mechanical nature, can provide the civilian and military communities with the ultimate in electronic device and circuit technology. Superconductivity can be employed to provide very low loss, zero frequency-dispersion signal transmission lines and passive circuit components while electromagnetic sensors and digital gates can have characteristics limited by the Heisenberg Uncertainty Principle. These is no other know electronic technology that can provide any of these features, let alone all of them.

Prior to December 1986, all known superconductors exhibited transition temperatures below 23 K. These very low temperatures required the use of liquid helium, which is moderately expensive, or the use of energy intensive closed cycle refrigeration system. Thus superconductivity remained in the laboratory or were used only in very special situations, such as radio astronomy, where the ultimate in performance was required and where cost was not an issue.

However, the recent breakthrough in superconducting materials has resulted in materials with transition temperatures as high as about 100 K with, as yet, unsubstantiated reports of transition temperatures as high as 150 K. These materials, when suitably fabricated into useful conductors or electronic devices and circuits, will be capable of operation in the temperature range near 77 K. Providing cryogenic environments in this temperature region can be achieved using liquid nitrogen, which is relatively inexpensive—a liter of nitrogen costs less than a liter of milk—or with the use of very small, lightweight and inexpensive closed cycle coolers. Thus, superconductivity can now come out of the laboratory and be economically used in the civilian and military communities.

BACKGROUND

Superconductivity is a quantum mechanical phenomena that is found in many element, compounds and alloys when they have been cooled to cryogenic temperatures. In 1911, superconductivity was discovered in mercury at a temperature of 4.2 K. Over the next 75 years, additional superconducting materials were discovered with ever increasing transition temperatures. In 1973, the compound of niobium and germanium, Nb_3Ge, was discovered to be superconducting with a transition temperature of 23 K. See Fig. 1. Over this span of time, the highest know transition temperature increased with a rate of about 0.3 K/year. The compound Nb_3Ge held the record for the highest known transition temperature, T_c until the end of 1986. At that time, the compound La-Ba-Cu-O was prepared in single phase form and exhibited a T_c of about 30 K. Within the next three *months* other ceramic copper oxide compounds were studied and T_c values as high as 120 K were reported. During this period of time, the rate of increase of T_c was of the order of 0.3 K/day, a rate about 1,000 greater than prior to December 1986. This spectacular rate of increase in T_c and the ability to obtain superconductivity at liquid nitrogen temperatures has initiated a *revolution* in how the engineer and scientist and the layperson, as well, think about superconductivity. More about these spectacular results and its impact on electronic technology will be discussed below.

The superconducting state, by virtue of its quantum mechanical origin, has very interesting properties that can yield electrical and electronic devices and components with unique characteristics. The best know property, which resulted in the use of the term "superconductivity" to describe this phenomena, is the property of zero electric resistance. When materials have been cooled to temperatures below their *transition temperature*, the specimen has zero electrical resistance, that is, electrical current can flow with no heat dissipation. If an electrical current flows in an all superconducting current, it will continue to flow indefinitely without diminishing in magnitude. (If the current does decrease, it should take more than 10,000,000 years before the magnitude of the current has dropped to 23 percent of the original value. This assumes, of course, that the sample is continuously maintained below its T_c.) The electrical resistance of a superconductor is more than 16 orders of magnitude smaller than high purity copper at room temperature.

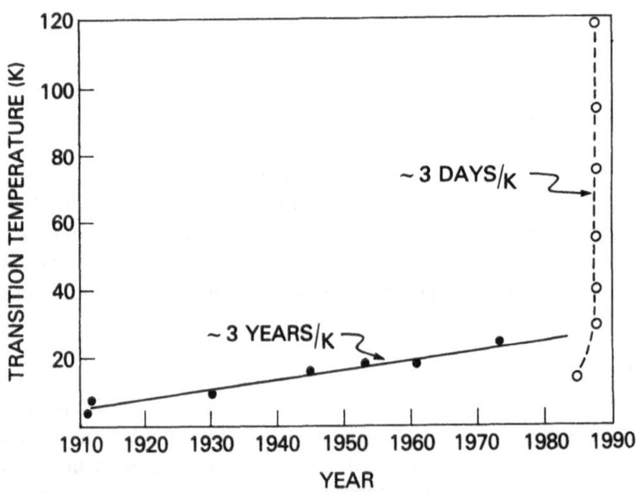

Fig. 1. Highest known superconducting transition temperature as function of time. The filled circle and solid line represent data for elemental, alloy and compound super conductors while the open circles and the broken line are for oxide superconductors.

A second property of s superconductor is magnetic flux exclusion. If a superconductive specimen at a temperature above T_c is exposed to a magnetic field, when it is cooled through the transition temperature, the magnetic field is *expelled* from the interior of the specimen. See Fig. 2. (Strictly speaking, the magnetic field is expelled from the interior of the specimen, but does penetrate slightly into the specimen to a depth of the order of a fraction of a micrometer.) If the specimen is in the form of a hollow cylinder or washer-shaped sample, the field is expelled from both the inner and outer surface of the specimen with the magnetic field trapped in the bore of the sample having discrete values with magnitude equal to the magnetic flux quantum divided by the area of the bore. The magnetic flux quantum is equal to the ratio of two fundamental physical quantities, Planck's constant, h, divided by twice the charge on the electron, 2e. This quantum of magnetic flux has the value of 2.07×10^{-15} webers (2.07×10^{-7} gauss-cm^2). For example, if the bore has a area of 1 cm^2, the trapped magnetic field will be multiples of 2.07×10^{-11} tesla (2.07×10^{-7} gauss) which is a very small value of magnetic field. The quantization of magnetic flux is used in creating magnetic shields to provide stable magnetic field environments and for EMI protection.

A third property of the superconducting state is associated with the current flowing across a structure consisting of two superconducting electrodes separated by a thin insulating barrier layer, whose thickness is of the order of 1 nanometers. As the temperature of the superconductor is lowered through its transition temperature, the electron energy states at the Fermi Level are split into two bands, the states below the Fermi level are associated with the bound Cooper pairs, which can move through the specimen with no electrical resistance while the state above the Fermi level are associated with normal electrons. As the temperature is lowered, this energy gap, the spacing between the top of the energy states for the Cooer pairs and the bottom of the band for normal electrons, gradually opens. At absolute zero of temperature, the energy gap is approximately proportional to the superconducting transition temperature. In Table I, the correspondence between transition temperature and energy gap expressed in volts and in frequency are shown. The semiconductor-style energy band diagram for a superconductor-insulator-superconductor structure for zero applied voltage across the device is shown in the upper left portion of Fig. 3. As a bias voltage is applied to the SIS structure, the bands begin to be offset. As long as the filled states near the top of the energy band associated with the Cooper pairs are opposite to the energy gap in the other electrode, no current can flow. However, once the top of the filled Cooper band is opposite to the bottom of the normal electron band (see lower left hand portion of Fig. 3), Cooper pairs will tunnel across the barrier region and will break up into two normal electrons resulting in a normal (dissipative) current flow across the device. As this bias condition is exceeded, there will be an abrupt increase in current flow. The current-vs.-voltage characteristics for such a device is shown in the right

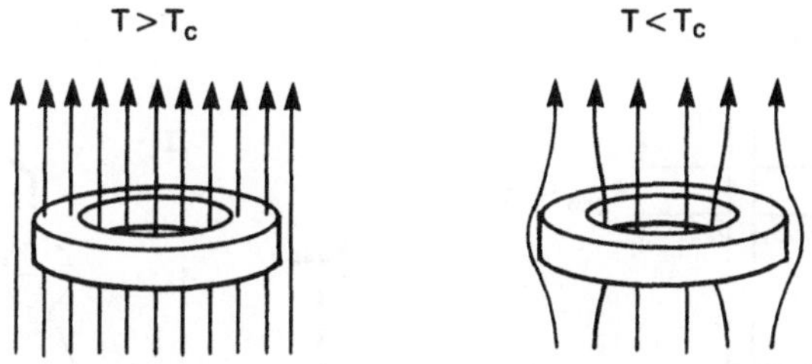

Fig. 2. Magnetic flux expulsion from a superconducting ring cooled in a constant magnetic field. The left hand figure illustrates the magnetic flux threading the ring when the temperature is above the superconducting transition temperature while the right hand figure illustrate the magnetic flux lines when the temperature is below the transition temperature of the ring.

Table I. Comparison of Transition Temperature of a Superconductor with the Energy Gap Expressed in Millivolts and Frequency

Transition Temperature (K)	1	10	15	30	90	120
Energy Gap (mV)	0.3	4.3	8.6	8.6	26	35
(THz)	0.068 (68 GHz)	1.03	2.1	4.1	12.3	16.4

hand side of Fig. 3. Note the very abrupt increase in current at the gap voltage, V_{gap}. This very sharp non-linearity in the I-V curve of an SIS structure is used to detect microwave and millimeter wave radiation with very high efficiency. (See below).

For completeness, two other concepts about the superconducting state will be mentioned. The *critical magnetic field* is the largest magnetic field to which a superconductor can be exposed and still remain in the superconducting state. The critical magnetic field, like the critical temperature, is an intrinsic property of the particular superconductor. In Fig. 4, the critical magnetic field is plotted as a function of temperature for a number of superconductors. Note that the shape of these curves are similar and that as T_c increases, the value of the magnetic filed also tends to increase. Assuming this general scaling between T_c and critical field the new ceramic oxide superconductors which have T_c near 100 K, should have a value for its critical magnetic field in excess of about 200 Tesla (2 Megaguass)!

The other crucial parameter is the *critical current density* which is the largest current density that a superconducting specimen can support while remaining in the superconducting (zero dissipation) state. The critical current density is an extrinsic property and depends very strongly on the processing and past history of the specimen. Typical values of critical current density for a suitably prepared sample at temperatures well below T_c and 10^6 to 10^7 amperes per square centimeter. (As the temperature is raised toward T_c, the maximum current density decreases, going to zero at T_c.)

REFRIGERATION

These properties of the superconducting state were known before December 1987. A variety of sensors, devices and circuits had been fabricated using the "low" temperature superconductor then known, that is, those with T_c less than 23 K. However, one of the primary obstacles to the wide spread use of superconducting electronic technology based on the use of materials with T_c's of less than 23 K was the very expensive cryogenic refrigeration systems that are required to cool these materials below their transition temperatures. Table II lists some of the properties of liquid cryogens and cryogenic refrigeration systems required to provide

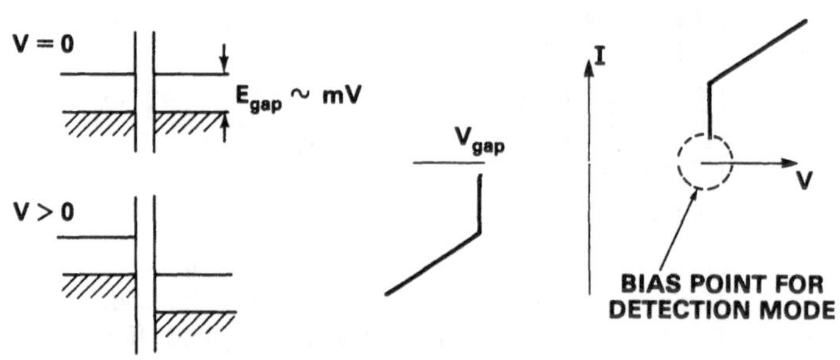

Fig. 3. Energy band diagram for superconducting tunnel junction for zero applied bias (upper left hand trace) and for finite bias (lower left hand trace) and current-vs.-voltage characteristic for superconducting tunnel junction device.

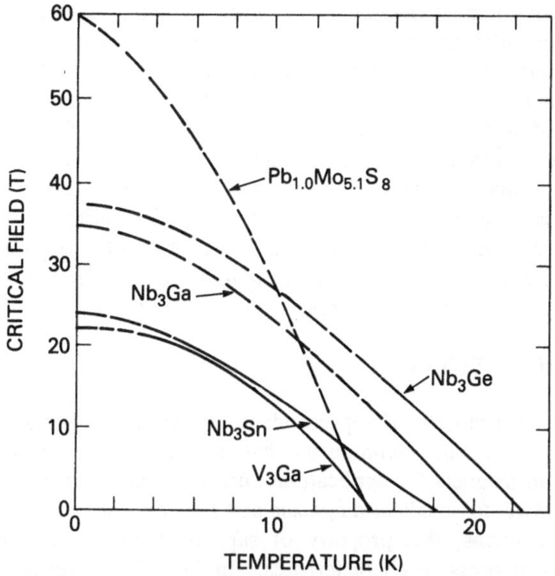

Fig. 4. Plot of critical magnetic field as function of temperature for several superconductors.

temperatures in the 4 K to 77 K range. The properties of a superconductor are temperature independent at temperatures well below T_c but begin to change as the temperature is raised. At about 1/2 of T_c, the parameters of the superconductor are close to their $T = 0$ K values. At about 2/3 of T_c, the parameters have decreased slightly. At still higher temperature, approaching the transition temperature, the parameters change rapidly with temperature. Thus, for practical reasons, one would prefer to operate superconductors at temperatures near or below 1/2 T_c but devices could also operate up to about 2/3 T_c before the strong temperature dependence would be a serious problem. The third column of Table II shows the values of T_c for which the indicated bath temperature would be about 2/3 T_c. Thus liquid helium is a suitable bath for superconductors with T_c of 6 K or greater, liquid hydrogen for materials with T_c of 30 K or higher while liquid nitrogen would provide a suitable cryogenic environment for superconductors with T_c greater than about 120 K. The fourth column shows the relative heat of vaporization for the various cryogens. Note that the heat of vaporization of nitrogen is about 60 times greater than that of liquid helium. That is to say, for a given open cycle dewar system, if one liter of liquid helium lasts for one day, then one liter of liquid nitrogen would last for about 60 days. These extended hold times for a nitrogen dewar relative to a helium dewar greatly relaxes the logistic requirements for replenishing liquid cryogens. In addition, there is a very large saving in cost: a liter of liquid helium costs between $5 and $10 per liter while liquid nitrogen costs only about $0.25 per liter. Thus, the combined greater heat of vaporization and the lower price of nitrogen results in a reduction in a factor of about 1200 in the *daily* cost of providing the required environment for a superconducting device or system operating near 77 K compared to one operating at 4 K.

Table II. Cryogenic Refrigeration Considerations

Gas	Liquid Cryogen			Cryocooler	
	Boiling Point	$T_{op} = 2/3\,T_c$	Relative Heat of Vaporization	Relative C.O.P Carnot Eff.	(W/W)
Helium	4.2 K	6 K	1	1	1000
Hydrogen	20.4 K	30 K	11.6	5.1	195
Neon	27.1 K	40 K	38.5	7.0	143
Nitrogen	77 K	114 K	59.4	24.3	42

The cryogenic environment required for a superconducting device can also be provided by closed cycle cryogenic refrigeration systems (''cryocooler''). According to Carnot's Law, a cryocooler requires more energy the lower the cold temperature. In column 5 of Table II, the relative Coefficient of Performance, the inverse of the ideal Carnot Efficiency, is shown. An ideal cryocooler providing a given amount of cooling at 77 K would require 24 times less input power than would be required to provide 1 watt of cooling at 4 K. The last column of the table, presents typical values of input power at room temperature required to produce one watt of cooling at the indicated temperature. A 4 K cryocoolers requires about 1,000 watts of input power for each watt of cooling at 4 K while a 77 K cryocooler requires between 30 and 40 watts per each watt of cooling at 77 K. *One of the advanages of high temperature superconductor with T_c near 100 K is the greatly reduced cryogenic refrigeration requirements.*

ELECTRONIC APPLICATIONS

The zero resistance property of superconductivity has many applications in electrical and electronic technologies. Its most common use has been in various forms of conductors which have been used to build magnets for applications such as nuclear particle accelerators, Magnetic Resonance Imaging (MRI) for medical diagnosis and for very intense research magnets. In low power electronic applications, this property of superconductivity can provide passive circuit components with very impressive characteristics. In Fig. 5, the surface resistance of several normal metal conductors at 300 K and 4 K are shown along with the intrinsic electrical losses of a superconductor at several values of reduced temperature, that is the ratio of the operating temperature to the transition temperature. Normal conductor losses vary as the square root of the frequency while the surface resistance of a superconductor varies as the square of the frequency, and exponentially with the reduced temperature. It can be seen that a superconductor operating at a reduced temperature of about 0.55 (that is lead with a T_c = 7 K operating at 4 K or Y-Ba-Cu-O which has a T_c = 90 K operating at 50 K) the surface loss at 1 GHz is smaller by five orders of magnitude than copper at room temperature, while at 100 GHz, the advantage is only about a factor of 30. At a reduced temperature of 0.12 (that is, operating a 90 K superconductor at 10 K), the losses for a superconductor are more than six orders of magnitude smaller at 1 GHz and about four orders of magnitude smaller at 100 GHz. Using superconductor for microwave filters and resonant circuits, Q-values of greater than 10^6 to 10^9 can be achieved. The very low loss property of a superconductor can be used to build staggered tuned

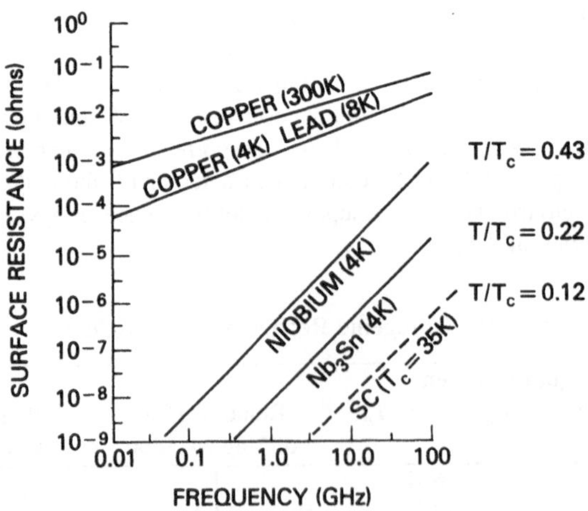

Fig. 5. High frequency surface resistance for several metals at room temperature and 4 K and for several superconductors materials at 4 K.

multiple filters with very steep frequency roll-off characteristics and very high out-of-band isolation.

This low loss property can also be used to build microwave frequency striplines with very interesting properties as shown in Fig. 6. This figure illustrates the attenuation and phase velocity for striplines fabricated from cooper and from a superconductor operating at a reduced temperature of 0.55. The upper curve shows that the attenuation for superconductor striplines is orders of magnitude smaller than for the copper lines for all frequencies below the frequency corresponding to the energy gap, E_g. The phase velocity is constant independent of frequency from very low frequencies up to about 1/3 or 1/2 of the gap frequency. Thus superconducting striplines can transmit pulse with very low attenuation and no diepersion for pulses with frequency components below about 1/3 to 1/2 of the gap frequency. Thus, not only does the signal amplitude not diminish after propagating along a superconducting stripline, but all of the frequency components will travel with the same velocity and thus there will be very little "spreading" of the signal. These properties of superconducting passive components have very important implications in both high frequency analog signal and high speed digital interconnects.

The use of superconducting signal interconnects may made a very major impact on interconnecting conventional semiconductor devices and chips. Most silicon-based devices experience carrier freeze-out in the temperature range below about 30 K, while many GaAs devices can function down to about 4 K. Thus superconducting interconnects fabricated with low temperature superconductors such as niobium and niobium nitride could be used with GaAs devices but not with silicon chips. However, interconnects fabricated with the newly discovered superconductors with T_c above 90 K can be operated at 30 to 70 K and thus can be employed to interconnect both silicon devices and chips. *The potential for interconnecting semiconductor devices and chips with superconducting interconnects, taking advantage of the best features of each technology, can have an immense impact on electronic technology.*

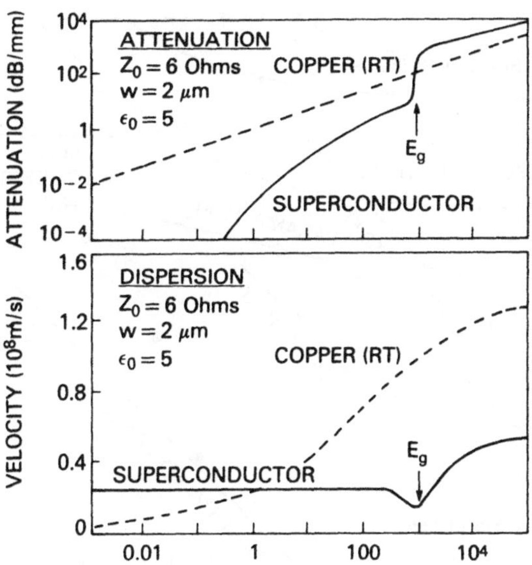

Fig. 6. Attenuation and dispersion versus. frequency for microstrip lines fabricated with superconducting lead and with room temperature lead.

HIGH FREQUENCY DETECTION

The very sharp non-linearity of the current-vs.-voltage curve of a superconductor-insulator-superconductor tunnel device (such as illustrated in Fig. 3) can be used for quantum limited detection of microwave and millimeter wave radiation. This is illustrated in Fig. 7 conversion gain, the influence of the IF amplifier on the overall receiver noise temperature is diminished. Receivers at 36 GHz and 94 GHz have been built with superconducting mixers operated such that they show converison gain and have demonstrated receiver noise temperatures of 24 K or less. These are the lowest noise temperatures reported for any millimeter wave receiver obtained with any technology. It is anticipated that this level of performance—close to quantum limited mode—should be achievable well above 100 GHz up to about 1/2 the gap frequency from which the mixers are fabricated. (See Table I.)

DIGITAL LOGIC AND MEMORY DEVICE TECHNOLOGY

Superconducting tunnel device can be operated as very fast gates, by switching from the superconducting to the normal state. The speed-power figure of merit for several technologies are shown in Fig. 8. Superconducting gates and High Electron Mobility Transistors (HEMT) can switch in times near 10 picoseconds. However, the power dissipation for the superconducting device is two to three orders of magnitude smaller than for the HEMT. Thus higher packing densities should be achievable for Josephson gates. The ultimate switching time for the where the noise temperature (left hand scale) and the noise figure (right hand scale) for a number of detector technologies are plotted as a function of signal frequency. In the 1 to 10 GHz region, field effect devices with gate lengths of the order of 0.5 micrometers to 0.25 micrometers can provide very low noise amplifiers. With increasing signal frequency, the noise increases with the frequency and at frequencies greater than 30 to 50 GHz, all conventioanl devices, both room temperature and cryogenically cooled, have degraded performance. At frequencies up to about 100 GHz, superconducting SIS mixers have demonstrated, to within a factor of two, quantum limited detection, that is, for each incident microwave or millimeter wave photon, one electron is generated in the external circuit in which the mixer is embedded. No conven-

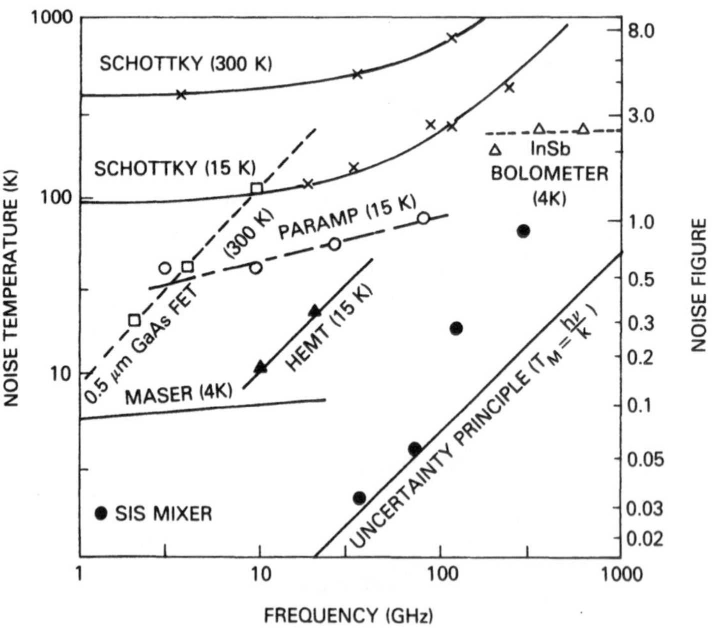

Fig. 7. Comparison of noise characteristics of various mixes technologies at room temperature and cryogenic temperatures.

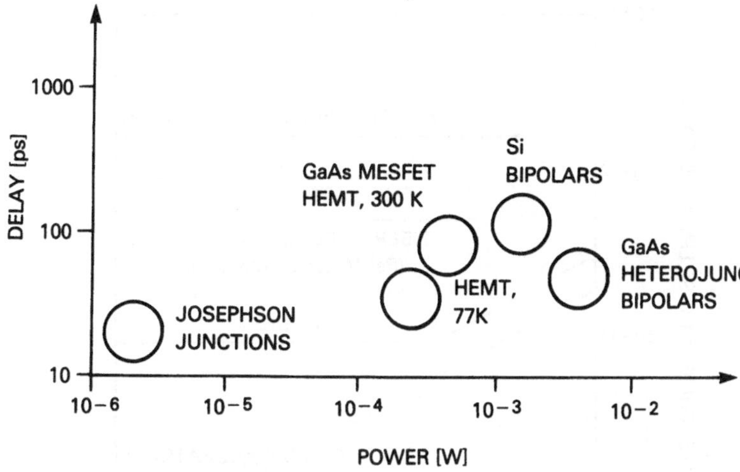

Fig. 8. Comparison of gate delay-power dissipation characteristics for several digital technologies.

tional technology can approach this level of quantum-limited operation. Furthermore, since the mixer process is quantum-mechanical in nature, under suitable conditions, SIS devices can exhibit conversion *gain*, a property not found in any other mixer technology. If the mixer has HEMT device depends on the transit time for electrons between the source and the drain. Thus very small dimensions, of the order of 0.25 micrometers was used to achieve these very fast switching times. However, the minimum feature size for the Josephson device was 2.5 micrometers. Thus Josephson technology still can be extended to shorter switching times before the limits of fabrication will be an obstacle. In principle, since the operation of the Josephson device is based on quantum mechanical tunneling, the ultimate limitation on switching time should be given by the Heisenberg Principle which, for a material with a transition temperature of 10 K, would be about 0.1 picosecond while, for a 100 K material, the limiting switching time should approach 0.01 picosecond. To fabricate an FET with comparable switching times would require gate dimensions of 25 Angstroms or about 5 lattice spacing—a very demanding requirement for the lithography. Thus superconducting digital gates should approach quantum limited operation within the domain of present status of device processing.

MAGNETIC FIELD SENSORS

The electrical impedance of a circuit containing SIS devices in an otherwise totally superconducting circuit, can be very sensitive to low frequency magnetic fields. These circuits are commonly referred to as Superconducting Quantum Interference Devices or SQUID. The magnetic field sensitivity of various magnetometer technologies are shown in Fig. 9. The performance achieved for the best SQUID magnetometer in the high frequency "white noise" region is within a factor of 2 of the value corresponding to an energy change of the order of Planck's constants—the quantum of energy. All competing technologies currently have intrinsic magnetic field noise levels of three to four orders of magnitude above this ultimate quantum limits. (The source of the low frequency magnetic field noise, which varies inversely with the square root of frequency, is not understood at the present time.) Superconducitng magnetometers have found widespread use in ultra-sensitive laboratory measurement instrumentation, for geophysical exploration and in biomagnetic studies of the human body.

SUMMARY

Superconductivity can provide electronic sensors with ultimate sensitivity limited by the Uncertainty Principle. Furthermore, the low electrical losses in superconductors can be used for very low loss, zero frequency-dispersion signal transmission lines or interconnects and for

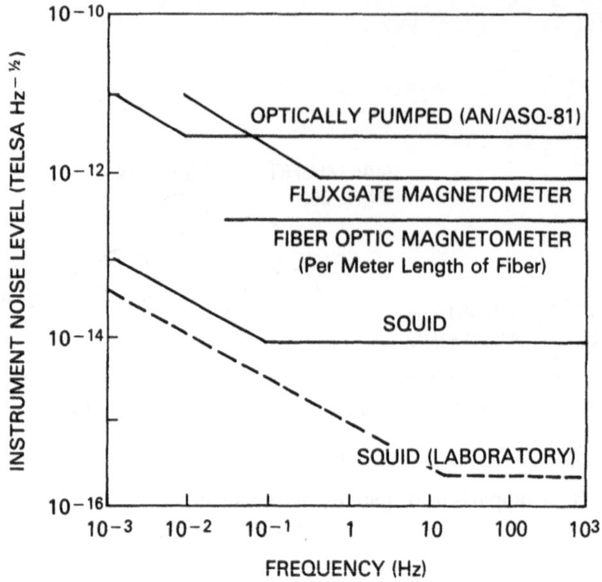

Fig. 9. Comparison of magnetic field sensitivities versus signal frequency for several magnetometer technologies.

very high Q-value filters and resonators. No other electronic technology can provide comparable characteristics.

The advent of the recently discovered high temperature superconducting materials with transition temperatures near 100 K will expedite the introduction of superconductivity into wide spread use. The elevated operating temperature for the new materials will require much more energy efficient cryogenic refrigeration systems. In addition, it will permit hybrid semiconductor-superconductor circuits and systems which can employ the best features, both technical and economical, of each technology. Thus, the advent of high temperature superconductivity can have a revolutionary impact on all areas of electronic technology.

CRYOGENIC REQUIREMENTS FOR MEDICAL INSTRUMENTATION

Raymond E. Sarwinski

Advanced Cryo Magnetics, Inc.

San Diego, California

ABSTRACT

Cryogenic instrumentation is becoming more common in the hospital environment. Led by MRI magnets, cryocoolers as well as large amounts of liquid helium and liquid nitrogen are being used in both mobile and fixed installations. Liquid helium is also necessary in the use of SQUID instrumentation. Different methods of attaining low temperatures are compared as to their technical advantages and disadvantages, cost and reliability. Possible new high temperature superconductors would have a considerable impact on dewar systems as well as on changing the emphasis of cryogenic refrigerators and liquid cryogens to temperatures associated with liquid neon, nitrogen or even freon.

INTRODUCTION

Over the past fifteen years, a group of important diagnostic instruments using cryogenic liquids have made or are making their way from the research laboratory to the hospital environment. In most cases, these instruments involve superconductivity and must operate at liquid helium temperatures. At the same time, liquid nitrogen may be used for thermal shielding.

These instruments and their dependence on cryogenic liquids will be described. Included will be Magnetic Resonance Imaging, Positron Emission Tomography and Computed Tomography scanners as well as a number of SQUID based instruments. The advantages and disadvantages of the common sources of refrigeration will also be reviewed. The effect of high temperature superconductivity on magnet and SQUID systems will be explored.

SOURCES OF REFRIGERATION

There are five practical ways to obtain cryogenic temperatures. These include transfer from storage dewars, transfer from an on-site liquifier, mechanical cryocoolers, Joule-Thomson and thermo-electric devices. Of these, the Joule-Thomson and the thermo-electric devices are more likely to be used only at high temperatures.

Storage Dewars

Over the years the usual way of obtaining low temperatures in the laboratory was to transfer liquid helium and liquid nitrogen from storage containers. It is not surprising that this method was used in the first commercial hospital instruments. The advantages of using an open bath cryostat include the fact that the instrument's cryostat design is straight-forward and the work and equipment necessary to produce the cryogenic liquids is maintained off-site. The end user simply orders what is needed and the product is delivered. Of course, a price must be paid for this convenience. That price is paid in terms of cost per litre, the need of a skilled technician to transfer the liquid efficiently and the dependence on the supplier for timely delivery of the product.

In many parts of the world, the cost per litre is high and scheduled delivery is not always assured. On transfering, the cost may be increased even more since even the most skilled technician can make a mistake. The method used in transferring liquid helium hasn't changed significantly over the years. In fact, the design of the transfer line hasn't changed in the last fifty years. That design didn't really have to since it's present form is inexpensive to manufacture and it's design adequate for the average graduate student to use in the laboratory. But for use in industry, the transfer line should be of sturdier construction, making use of bayonet couplings and valves. At least one storage dewar manufacturer has started in this direction. Efficient, standard bayonet couplers and valves built into storage dewars and transfer lines would lead to better transfers and less wear and tear on equipment and personnel.

On-Site Liquifiers

Small liquifiers are available for a cost starting at $120,000. They can easily produce the necessary refrigeration for even the largest hospital instrument. Five watts at 4 K and 60 watts at 80 K is typical. To produce this refrigeration, the compressor bank needs about 20 kW of electric power. The noisy compressors are located in their own room and the liquifier itself is located at a remote location. Liquified gases are passed to the instrument through long transfer lines. Liquifiers are relatively complex machines and need to be monitored in operation and maintained at regular intervals. Since reliability and quick repair turn-around is a prime concern in the operation of every hospital instrument, major parts should be kept on site or quickly available. The on-site liquifier allows operation in any part of the world.

Gifford-McMahon Cryocoolers

Large numbers of this type of mechanical cryocooler have been used in vacuum pumps for years and are quite reliable. The bare cryocoolers come in a variety of configurations. A single stage machine can produce 50 watts of cooling at 80 K. A two stage machine can produce refrigeration of 50 watts at 80 K and 10 watts at 15 K. Input power requirements for these machines range between 1.7 and 3.0 kW while the cost of a bare machine can range between $11,000 and $18,000. These machines include a cold head consisting of a valve motor and displacer weighing about 20 pounds. The cold head is connected by flexible gas lines to a compressor unit weighing between 150 and 300 pounds. The system operates at a low speed of about one Hertz and only requires maintenance every 10,000 hours of operation. A separate Joule-Thomson (J-T) loop can be added to a two stage cryocooler to produce refrigeration at 4 K. These machines require an input power of over 5 kW and cost upwards of $50,000. The cold head weighs about 30 pounds and the compressor about 500 pounds. Adding the

J-T loop increases the complexity and need for maintenance, while reducing the reliability.

Joule-Thomson Refrigerators

It is possible to operate a number of J-T stages in a cascade configuration not relying on mechanical cryocooler precooling. The precooling is done by J-T stages at higher temperature. To operate at 4 K, a helium J-T must be precooled to about 25 K. This can be done by using a hydrogen J-T. To operate at 25 K, a hydrogen J-T must be precooled to about 80 K. This can be done by using a nitrogen J-T. A nitrogen J-T can be operated from room temperature but is sometimes precooled to 140 K by using a freon J-T. The greatest heat load can be on the freon J-T. The problem is that four systems are now running at the same time. This means four separate gas loops and four separate compressor systems are required. J-T machines have no cold moving parts and so cause little vibration. They do require compressors which may be operated some distance away. They need counterflow heat exchangers, which are easy to make, and a high impedance pressure dropping valve (the J-T valve) located where cooling is to be produced. This valve is usually the weak link since the J-T valve is prone to blockage. A large amount of gas continually flows through this valve. If the gas is not extremely clean, gas impurities, oil from the compressor and general debris can clog the valve.

Thermo-Electric Coolers

This type of device seems to be the ideal cooler. With the flip of a switch, cooling is available. Thermo-electric coolers only need to reject some heat at room temperature. Unfortunately, not much cooling is available and present day devices will not produce very low temperatures. A multistaged device can produce a few mW of cooling at 130 K. For some instrument applications, this may be sufficient. Not much cooling is needed from the final stage of thermoelectric devices since these are multistaged devices with small temperature differences between stages.

DESCRIPTION OF INSTRUMENTS

To best understand how to choose a refrigeration system, it is helpful to look at the personnel involved in the operation of a clinical instrument. The DIAGNOSTICIAN is the person who interprets the data taken on the instrument. This individual is supported by the TECHNICIAN who actually takes the data, positions the instrument, pushes the right buttons, etc. If a malfunction occurs, the technician is the first to know. The technician then alerts the SITE ENGINEER. The site engineer maintains the equipment, checks software for bugs, interfaces with the instrument manufacturer and transfers the cryogens. The site engineer may maintain instruments at a number of sites and is always on call. The site engineer is trained in all aspects of instrument repair as well as the handling of cryogens. While the site engineer is usually an employee of the manufacturer, some installations are now hiring a MAINTENANCE ENGINEER who would be an employee of the hospital. The maintenance engineer would take over most of the duties of the site engineer, especially in the transfer of cryogens. For small installations, not as many people are involved and the task of cryogenic maintenance is more visible.

Magnetic Resonance Imaging

MRI machines are the largest and most common cryogenic instruments found in hospitals. At this time, there are about 600 in operation worldwide in both fixed and mobile installations.

A MRI device can give a two dimensional picture of the hydrogen distribution in a biological sample[1,2]. While hydrogen density may not differentiate between blood and muscle, or between gray and white matter, use of the longitudinal relaxation time (T1), gives a much greater contrast sensitivity than that achieved by any other method of imaging.

An installation consists of a superconducting magnet providing uniform fields between 0.35 and 2 Tesla. The magnet vessel is toroidal in shape having an inner bore diameter of 1 meter, a length of 2 meters and a weight between 5 and 10 tons. It is enclosed by an electrically screened room. The magnet power supply, rf generator and gradient coil supplies are located outside the room. Signals are routed to a computer in another room outside of the fringing field of the magnet. In a mobile installation, all of the equipment is mounted in an aluminum, air-ride van (complete with air conditioning).

A MRI installation is large in terms of the space it takes up, cost, and the number of people involved in the operation. The full complement of personnel described above are usually used at a MRI site. With a site engineer in attendance, any cryogenic problem is not as noticeable.

Almost all MRI magnets in use today, are operated in simple open cryostat systems. The magnet is immersed in a liquid helium vessel holding between 300 and 1500 litres. This vessel is surrounded by a radiation shield cooled to about 30 K by the evolving helium gas. The 30 K shield is in turn surrounded by a shield thermally connected to a reservoir of liquid nitrogen. This reservoir may contain between 30 and 300 litres of liquid nitrogen. Super-insulation is used to reduce the radiation heat leak between shields and room temperature. The liquid helium boil-off can vary between 0.3 litres per hour for a stationary system with retractable current leads to 1 litre per hour for an active mobile system with permanently installed current leads. Since the helium volume is so large, liquid helium need only be transferred every month or so.

Liquid nitrogen boil-off may vary between 0.5 and 1 litre per hour. If the liquid nitrogen reservoir is small, liquid nitrogen is transferred daily by an automatic system from an external storage dewar. At best, liquid nitrogen is transferred at the same time as the liquid helium. The advantages and disadvantages of the stored cryogen system described previously are multiplied because of the size of these cryostats. Disadvantages that stand out include the need for a skilled technician, cost and availability of liquid helium and the fact that a mistake in transfer can cost over $2500 (by loosing 500 litres at $5 per litre).

In many parts of the world, some of these problems are intolerable. Liquid helium and nitrogen may not be readily available or at least quite expensive. In this case, the open dewar liquid cryogen system must be modified.

An on-site liquifier may be used[3], however, it represents a substantial investment, requires siting and maintenance but does free the installation from purchasing both liquid helium and nitrogen. In one form, the magnet system remains the same open dewar system but with the remote liquifier either replenishing the liquid helium or using the incoming liquid helium from the liquifier to reduce the helium boiloff to zero. An additional circuit from the liquifier, circulates 70 K gas through the liquid nitrogen reservoir, reducing its net evaporation to zero. The advantage of this form is that during maintenance or disruption of service, such as breakdown or power failure, the MRI system can remain in operation until repairs are effected. Another advantage of keeping the liquid cryogens in quantity is that the temperature stability of the magnet is maintained. A change in temperature of the 80 K shield will cause a change in it's electrical resistivity, changing it's eddy current shielding effect and necessitating a re-tuning of the electronics. In those cases where cryogens are not stored in bulk, some form of temperature regulation may be necessary.

Use of the on-site liquifier changes problems of obtaining and transferring liquid cryogens to those of maintenance and reliability of the liquifier. Within a year, at least two companies will offer small on-site liquifiers for use with MRI magnets.

Another alternative to the open dewar method is the use of Gifford-McMahon cryocoolers[4]. The simplest application is to use a single stage machine to maintain the liquid nitrogen shield. The cryocooler may have an attachment to reliquify the gas in the liquid nitrogen reservoir, reducing liquid loss to zero. The liquid nitrogen system remains closed and operates at a slight positive pressure. For maintenance or repair, the attachment may be removed without affecting the operation of the magnet.

The mechanical cooler may be mounted directly to the 80 K shield with no liquid nitrogen involved. The shield temperature will not change rapidly due to the stabilizing effect of the heat capacity of the aluminum shield. The cryostat designer must be innovative so that dewar vacuum need not be broken when the cooler must be shut down and removed for maintenance. Both recondensing and direct shield cooling have been used with a high degree of success.

A slightly more expensive and mechanically more complicated approach is to use a two stage cryocooler. The first stage operates the 80 K shield while the second stage cools the 30 K shield to an even lower temperature, perhaps as low as 15 K. It is not unreasonable to expect such a system, equipped with removable leads, to have a helium boil-off of 0.2 to 0.3 litres of liquid helium per hour. A 1000 litre reservoir operating down to 20% would last for 4 months between helium refills. Since these coolers are quite reliable, the only problem may be in making good thermal contact between the cooler and the shields in such a way that the cooler could be easily removed when necessary. A number of solutions have been proposed and most work to some degree.

The two stage cryocooler with J-T loop may also be used to reduce liquid cryogen boil-off[5]. This machine can provide either reliquification at or cooling of the 80 K stage, cooling at the 15 K stage and sufficient cooling at 4 K to reliquify helium, reducing all boil-offs to zero. In principle, this sounds quite good, but these machines are expensive and introduce even more complication into the thermal coupling scheme. Due to previously mentioned problems with J-T loops, the reliability of this alternative is not as high as that of the simpler systems. Besides thermal coupling and removal problems, clever schemes are needed to cool the magnet current and instrumentation leads previously cooled by the helium boil-off gas. If more cooling is needed, two cryocoolers may be installed.

Cryocoolers have little effect on the imaging system. They have been operated in high fringing fields. They do not affect magnetic field homogeneity and vibration, though felt and heard, does not seem to bother the operation of the MRI instrument. The one and two stage machines have been on the market for a long time and have proven reliability.

When discussing cryocoolers, it always seems necessary to mention cascade Joule-Thomson coolers. The invitation of a compact, vibration free, long life refrigerator is very strong even though it is not the most efficient of machines. Multi-staged or even large single stage coolers have not been developed for MRI applications as of yet.

Even further away in development for MRI instruments, are the thermo-electric coolers. They have neither the cooling power nor the ability to operate at low enough temperatures.

Positron Emission Tomography

PET is another advanced technology technique used to gain information about body chemistry and physiology. It is not only useful as a diagnostic tool but has great potential in the study of chemical defects as related to both mental and metabolic abnormalities[6]. For decades, it was recognized that the position of certain positron emitting radionuclides could be determined with precision and sensitivity by positioning detectors around a source. The direction of the two photons emitted on the annihilation of a positron and an electron are correlated. Rings of detectors can determine the position of the event precisely by measuring the angle and time of flight after the event. Computers resolve the events into images.

The radiopharmaceuticals needed as source materials have relatively short half-lives ranging from minutes to hours. It is therefore advantageous to produce these materials as near as possible to where they will be used. This is done using small mobile cyclotrons. The cyclotrons can use superconducting magnets to advantage. PET and MRI magnets can have some of the same cryogenic problems. All of the comments made about MRI magnets can be applied to PET magnets. An additional problem with PET magnets, is that any external refrigeration supplement must be mounted in such a way as to be effective while penetrating heavy iron and lead shielding.

X-Ray Scanners

Certain crystals, when cooled to low temperatures, make sensitive, low noise X-ray detectors. They operate well at temperatures between 80 K and 130 K. A line of detectors is mounted in a stationary full, or a rotating partial ring surrounding the source. The source of cooling must be free of vibration and should provide cooling for at least five days. For the more difficult rotating ring dewar, this can be done with a simple, horizontal, liquid nitrogen dewar having a capacity of 10 liquid litres of nitrogen. The dewar can be filled by a relatively inexperienced technician at the beginning and end of each week.

Mechanical cryocoolers are ruled out for detector ring cooling because of their vibration. J-T coolers having about 5 watts of cooling capacity could be used but have not yet been developed. The J-T exchanger and valve would be mounted on the rotating sensor mount and be coupled through two small diameter flexible lines to the compressor positioned in another room.

If the operating temperature could be raised to about 160 K, a bank of thermo-electric coolers could be used. The connection between the moving sensors and the room would then consist of a pair of heavy, flexible, low voltage current leads. A whisper fan would be mounted outboard to cool the heat rejection area. Thermo-electric coolers are a feasible solution to intermediate temperature needs.

SQUID Instruments

SQUID is an acronym for Superconducting QUantum Interference Device. These devices are accepted to be the most sensitive magnetic field detectors available. For over 15 years, SQUID based instruments have been used in research institutions to measure physiological magnetic fields[7]. In recent years, these non-invasive machines have become available to the clinical diagnostician.

One of the benefits of using this type of instrument is that no contact with the patient is necessary. Information has been gathered on many of the major organs of the body, including the heart, brain, liver and lungs[8,9,10,11]. Clinical instruments located at NYU, MIT, NIH, UCLA, Simon Fraser, Helsinki and Rome measure human iron stores in the liver, the location of the foci of epileptic seizures and magnetic particles in the lung.

The basic SQUID instrument consists of one or more signal channels, each associated with a superconducting pickup coil connected to a SQUID and operated in a liquid helium bath. The pickup coil is usually some order of gradiometer, the dewar is non-metallic and the whole system is run in a shielded room. The liquid helium dewar contains 10 to 15 litres of liquid helium and must be refilled once a week. Vaopr cooled radiation shields are used, eliminating the need for liquid nitrogen.

All SQUID instruments are now operated in open dewar systems. A technician is needed to transfer helium. As mentioned above, a careful redesign and standardization of the transfer process would result in more efficient and calmer transfers.

Helium is delivered to the instrument site in 50 to 100 litre storage dewars. The transfer line is inserted into the storage container and the instrument dewar, the storage dewar is pressurized to slightly above atmospheric pressure with gaseous helium and helium is transferred until a meter indicates the transfer is complete. The transfer line is removed and the storage dewar parked in some corner until it is used again. Boil-off in the storage dewar amounts to less than a litre per day for a 100 litre dewar. Instrument dewar boil-off is about 1.5 litres per day. Therefore, a 100 litre storage dewar can service a 15 litre instrument dewar for about 3 to 4 weeks, taking into account the efficiency of the transfers.

As on-site liquifier would not relieve the necessity to transfer every week. It could only provide an on-site source of liquid helium. One way of freeing the instrument from using liquid helium would be the use of a modified Gifford-McMahon cryocooler with a 4 K J-T stage. Direct coupling of a mechanical cryocooler to reliquify the helium or even to reduce the boil-off presents a very difficult problem due to the acoustic and magnetic noise produced by the valve motor and the reciprocating motion of the cryocooler displacer. The cold head would have to be mounted off the instrument and at some distance from it. The 4 K J-T stage would have to be mounted on the instrument and connected to the rest of the cryocooler through an insulated transmission line. Such a system could provide direct cooling of the instrument or serve as a recondenser for the instrument dewar helium bath.

Another solution is to use a cascade J-T system as previously described. The J-T produces almost no vibration and has no moving parts to interact magnetically with the SQUID pickup coils. Unfortunately, there are no such commercial or even experimental machines available at this time. Another point of consideration is that even if a commercial machine were available, it would represent a substantial investment compared to the cost of the instrument. A site engineer would still be necessary to monitor and perform maintenance on the system.

HIGH TEMPERATURE SUPERCONDUCTORS

Because of recent progress in the development of high temperature superconductors, it is appropriate to speculate on the future of cryogenic instruments. Nearly all the instruments discussed here depend on superconductivity. The superconducting magnets and SQUIDS need a 4 K environment to function but if that temperature were raised to 80 K, their cryogenic requirements would be quite different.

High Temperature Superconducting Magnets

To be practical, magnet wire at 80 K would need a high critical field, a high critical current and the proper normal matrix for stabilization. Joints having very low resistance would have to be made to realize near persistent operation. The wire would also have to be flexible enough to wind. MRI magnet systems using such wire would be easier to construct. Their intermediate 30 K shield could be eliminated. For an open bath dewar containing 1500 litres, the heat leak into the bath would still be 30 to 50 watts and the boil-off would still be between .6 and 1 litre per hour. If the wire behaved properly by allowing magnet operation in the vapor, the dewar would have to be refilled every 2 to 3 months. Since it is much easier to transfer liquid nitrogen than liquid helium, the burden of using cryogenic liquids would be greatly relieved.

We would see a boom in the development of small nitrogen liquifiers to support those locations that need on-site liquification. These relatively small liquifiers should be less expensive than helium liquifiers and mechanically more reliable. The electric power requirements for a nitrogen liquifier are much less than those for a helium liquifier.

Gifford-McMahon cryocoolers already exist for use as reliquifiers. If the magnet would not need the thermal ballast of a liquid nitrogen bath, existing mechanical cryocoolers could operate a dry system down to 50 K without modification. If great care is taken in the design of the insulation system, two stage cryocoolers could even take care of the cryogenic needs of magnets operating in a closed liquid neon bath at 30 K.

High Temperature SQUIDS

The future does not look as bright for SQUIDS at high temperatures as it does for magnets. It is believed that 80 K SQUIDS will always be electrically more noisy than 4 K SQUIDS. Many of the SQUID instruments mentioned here are limited in sensitivity by SQUID noise even when operating at 4 K. The 80 K SQUID, though not as quiet as the 4 K SQUID, could still be part of a very sensitive instrument, especially one where it could be operated in high magnetic fields or in the very low frequency range down to dc. 80 K wire could also be used in the construction of pickup coils coupled to 4 K SQUIDS. In some instruments, it is necessary to locate the pickup loops as near to the warm outer surface of the dewar as possible. If those loops were made of 80 K wire, they could be located much nearer to the dewar outside. In another variation of today's magnetometer or gradiometer, the 80 K pickup coils could be located at some distance from the 4 K SQUID and connected to it through liquid nitrogen transmission lines.

CONCLUSION

Hundreds of diagnostic instruments using liquid cryogens are in use in hospitals and clinics. Most use open bath dewars into which liquid nitrogen and liquid helium must be transferred. In almost all cases involving large systems, site engineers maintain the apparatus and accept the task of transferring the liquid cryogens. The cost and complication of liquid transfers may be reduced by using mechanical cryocoolers as shield coolers and/or reliquifiers. Single and double stage coolers are the most reliable and operate well without affecting instrument operation. Mechanical cryocoolers with piggy-back J-T loops are more expensive, less reliable, but necessary for some locations.

At this time, the difficult problem of coupling cryocoolers to SQUID systems has not been solved. Mechanical coolers are too noisy and no commercial cascaded J-T refrigerators have been developed. An immediate solution to part of the problem is the standardization and redesign of the current techniques of transferring cryogenic liquids.

The development of usable high temperature superconducting wire will greatly affect the design and operation of instruments based on superconducting magnets. The SQUID instruments will not benefit as much as magnets and will still have to wait for a better, vibration free, low noise, efficient and reliable refrigerator to be developed.

REFERENCES

1. P. A. Bottomley, NMR imaging techniques and applications, Rev. Sci. Instrum. 53(9):1319 (1982).
2. T. F. Budinger and P. C. Lauterbur, Nuclear magnetic resonance technology for medical studies, Science, 226:288 (1984).
3. F. F. Murray, K. F. Hwang and W. D. Markiewicz, Refrigerator operating experience on whole body MRI magnet systems, in: "Advances in Cryogenic Engineering," Vol. 31, Plenum Press, New York (1986).
4. R. C. Longsworth, Interfacing small closed cycle refriderators to liquid helium cryostats, Cryogenics, 24:175 (1984).
5. R. C. Longsworth, 4 K refrigerator and interface for MRI cryostats, in: "Advances in Cryogenic Engineering," Vol. 31, Plenum Press, New York (1986).
6. G. L. Brownell et al, Positron tomography and nuclear magnetic resonance imaging, Science, 215:619 (1982).
7. R. Hari et al, Biomagnetism in the study of brain functions, in: "Biomagnetism:Applications and Theory," H. Weinberg, G. Stroink and T. Katila, eds., Permagon Press, New York, (1985).
8. G. M. Brittenham et al, Diagnostic assessment of human iron stores by measurement of hepatic magnetic susceptibility, Il Nuovo Cimento, Vol. 2D, N.2:567 (1983).
9. S. J. Williamson and L. Kaufman, Analysis of neuromagnetic signals, in: "Handbook of electroencephalography and clinical neurophysiology," A. Remond and A. Gevins, eds., Elsevier, Amssterdam, (1986).
10. S. Barbanera et al, Use of a superconducting instrumentation for biomagnetic measurements performed in a hospital, IEEE Trans on Mag, Vol. 17, N.1:849 (1981).
11. D. Cohen et al, Ferrimagnetic particles in the lung-the relaxation process, IEEE Trans on Biomed Eng, Vol. BME-31, N.3:274 (1984).

INSTRUMENTATION FOR CLINICAL APPLICATIONS OF NEUROMAGNETISM

D.S. Buchanan, D. Paulson

Biomagnetic Technologies, inc.
San Diego, California

and

S.J. Williamson

Neuromagnetism Laboratory
Department of Physics
New York University, New York, New York

ABSTRACT

Measurements of the magnetic field of the human brain in a clinical setting require a higher level of performance from the instrumentation than is generally acceptable for laboratory research. We describe several significant advances that are intended for such neuromagnetic applications as well as for a broader range of biomagnetic studies. We report the development of a closed-cycle refrigerator capable of sustaining a SQUID-based sensor without introducing significant deterioration of the noise level. This eliminates the need for liquid cryogens and permits the sensor to be operated in various orientations, including inverted. The performance of a new type of magnetically shielded room is then evaluated for neuromagnetic studies. It has the advantages of being pre-fabricated and of providing a large interior for convenient clinical studies. Its ceiling supports a versatile gantry that holds one or two sets of magnetic sensors. This arrangement, when used with a magnetic system for precisely determining the sensor positions with respect to the patient's head, is feasible for precise localization of neural sources within the brain. We end with an example of the kinds of clinical studies that are now being carried out with the aid of neuromagnetic measurements.

INTRODUCTION

The technology for magnetic studies of the brain has greatly advanced since the first observation of neuromagnetic fields by Cohen.[1,2] It is now recognized that the dominant source of magnetic fields measured outside the scalp is the pattern of intracellular currents in active neurons. Since magnetic fields in the frequency domain of interest are little affected by the electrical properties of the cranium, they emerge from the head without distortion. One important advantage of magnetic studies is the possibility of determining the three-dimensional location of active neural regions in the brain through measurements and analysis of the field pattern measured over the scalp. These methods are described in recent reviews[3-5] and in a textbook.[6] In many cases it is possible to localize a confined source with a precision of a few millimeters. As the magnetic fields of interest range in strength from about 10 fT to 1000 fT, the only field sensor having the required sensitivity and small sampling volume is the SQUID (superconducting quantum interference device). The need for a dewar to contain the liquid helium that maintains the SQUID below its superconducting transition temperature imposes significant contraints on how and where such a sensor may be used.

The past three years have seen several multi-sensor systems developed for neuromagnetic applications.[7-9] One motivation for introducing an array of sensors within a single dewar is to speed neuromagnetic measurements by reducing the number of times the dewar must be moved from one location to another over the scalp to record a field pattern. While virtually all SQUIDs presently sense the fields of interest by means of a flux transformer connected to a detection coil wound of superconducting wire, considerable effort is being invested in the development of techniques for fabricating thin-film coils with various planar geometries. One interest in using planar first-order and higher-order gradiometers is to improve the spatial resolution whereby the net magnetic flux from a more distant source is reduced in favor of the closest-lying source. Another motivation is to improve the precision of the coil's field balance, i.e. matching the area-turns ratios of component coils, so that for a given geometry the sensor is insensitive to relatively uniform fields from distant noise sources. Still another motivation is the desire to simplify the process of fabricating such coils, in anticipation of the time a few years hence when arrays of 100 or more detection coils will be used to measure simultaneously the field over the entire scalp.

CRYOSQUID

During the past decade there has been continuing interest[10-13] in developing a closed-cycle refrigerator for cooling SQUID-based magnetometers for a variety of applications, including neuromagnetometry. Such a device is desirable for several reasons, including: reduction of operating costs; possible use in a remote, untended environment; the unavailabilty of liquid helium in certain circumstances; safety; and the convenience of not having to transfer helium every two or three days.

The system we describe is intended to be used for neuromagnetometry both by itself and in conjunction with other neuromagnetometers. Toward that end we established several design goals. The system should be able to operate at or near current dewar-based noise levels, which means a noise level of approximately 20 $fT/Hz^{1/2}$. No cryogenic liquids, either He or N_2, should be required for operation. It should operate for at least one month between shut-downs. It should be able to operate over a range of orientations, including nearly up-side down. And it should also be capable of operating either inside or outside a magnetically shielded room. We have succeeded in developing such a system that meets these goals. To acknowledge the method of cooling the sensor we call this device "CryoSQUID".

We have constructed a prototype hybrid system incorporating a Gifford-McMahon (GM) cycle and a Joule-Thomson (JT) refrigerator designed to provide cooling of a standard BTi DC SQUID and a single second-order gradiometer detection coil of conventional design (Fig. 1). This

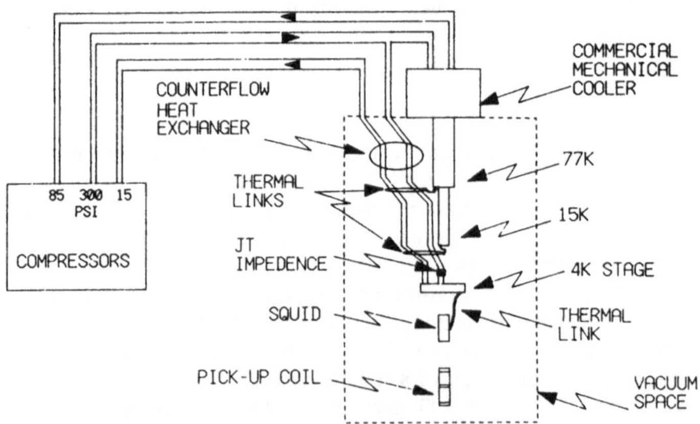

Fig. 1. Schematic for gas flow lines and thermal links in the CryoSQUID system.

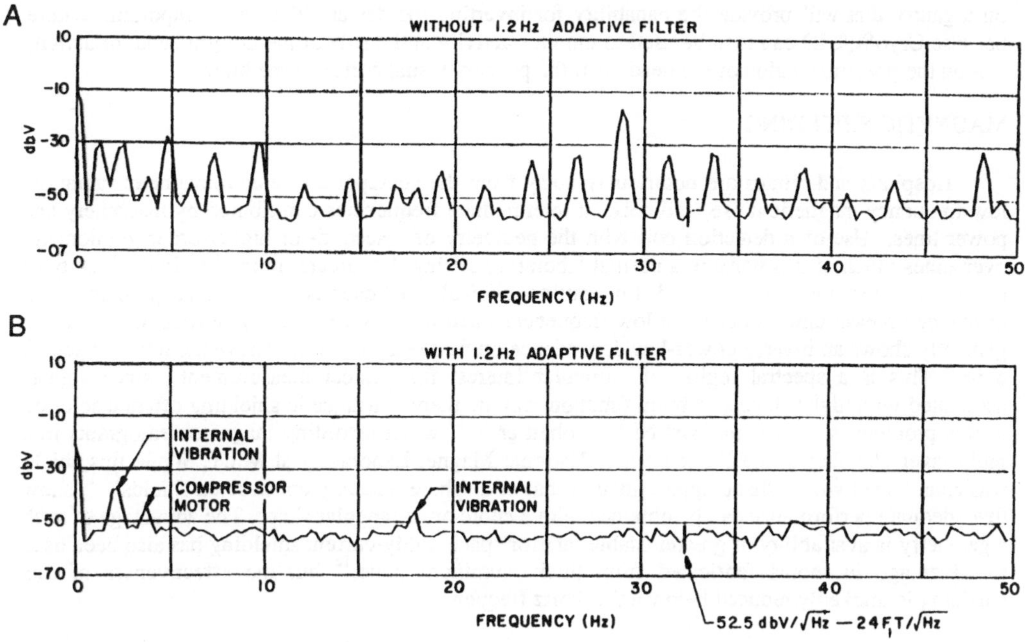

Fig. 2. Noise spectral density of CryoSQUID when the magnetic sensor is placed in a magnetic shield. A) System operating without adaptive filter. B) System operating with the adaptive filter that subtracts a sample of the previously measured noise from the record.

system has operated successfully at equivalent noise levels of 25 $fT/Hz^{1/2}$ over a bandwidth of 1-500 Hz, with minor exceptions where noise peaks are found at a few undamped internal vibrational frequencies of the system (Fig 2). The system is designed around an off-the-shelf, two-stage GM cooler having approximately one watt of cooling at 15 K. To this is added a helium gas JT loop and a suspension system that is designed to minimize the vibration transmitted to the SQUID and detection coil from the GM stage. The JT is run with an input pressure of 2 MPa and a return pressure of 100 kPa. The GM uses the same input pressure and a return pressure of 570 kPa. Two standard helium compressors provide the gas supplies. The current prototype weighs approximately 23 kg and has a cool-down time from room temperature of approximately 34 hours.

The use of a JT cycle for the low temperature stage eliminates the noise problems associated with the temperature fluctuations of an all mechanical design.[12] It also provides the lower temperature (less than 5.5 K) required for low-noise operation of the BTi DC SQUID. Most of the tests performed so far have been done while running the JT loop in an open cycle from a helium storage bottle.

While the suspension system reduces the vibration experienced by the SQUID and pickup coil it does not eliminate it. This is evident in Fig. 2A where at 4.8 Hz the equivalent noise is 550 $fT/Hz^{1/2}$. These data were taken while running the system in a mu-metal can to shield the system from ambient field noise. Results outside of the mu-metal give similar results for the vibrational noise. Because of the stable characteristics of the vibrational noise most of it can be effectively eliminated by the use of a computer-based adaptive filter as shown in Fig. 2B. Here the noise characteristics were monitored and averaged by computer with a reference synchronized to the GM basic period. When the resulting averaged noise is subtracted from the time series data, a relatively flat spectrum is obtained.

We are in the process of building an improved version of this device. This will reduce the weight of the dewar and contents to less than 15 kg, and it should eliminate unwanted internal vibrations. Furthermore, we estimate it should reduce the cool-down time to less than 18 hours. The white noise level of the SQUID system should be reduced to 20 $fT/Hz^{1/2}$ through the elimination of the present excessive rf shielding within the dewar. This version is also being designed to operate on a gantry that will provide the capability for inverting the dewar. This is an important feature, because CryoSQUID can then be used to monitor activity at the side of the upright head, or activity low on the posterior region of the head, near the primary visual cortex of the brain.

MAGNETIC SHIELDING

Hospitals and clinics are notoriously noisy from the standpoint of electromagnetic radiation, low-frequency magnetic noise, and noise at intermediate frequencies contributed by machinery and power lines. Use of a detection coil with the geometry of a second- or higher-order gradiometer overcomes much of this noise in a normal laboratory setting; but greater reliability in the quality of noise reduction may be required for routine clinical applications. A special problem with unshielded measurements occurs at low frequencies, below a few hertz, where environmental noise generally shows an inverse power-law dependence on frequency and rises above the intrinsic sensor noise. This is a spectral region of particular interest for clinical measurements, since signals associated with higher levels of brain function may lie there.[14] Magnetic shielding offers a solution to this problem, as first demonstrated by Cohen *et al.*[15] when recording magnetocardiograms in a multi-layer chamber at the Francis Bitter National Magnet Laboratory at M.I.T. While this shield was constructed with a shape approximating that of a sphere, recently constructed shields[15-17] show that adequate performance can be obtained with a cubic or rectangular shape. The advantage of such a geometry is availability of greater usable interior space. Eddy-current shielding has also been used to advantage in rooms fabricated from thick aluminum plate,[18] but the effectiveness of this shielding is markedly reduced below a few hertz frequency.

We report here the performance of a particularly large magnetically shielded room (MSR) recently installed in the Neuromagnetism Laboratory at New York University.[19] Similar MSRs have been erected during the past half-year at three other institutions. The interior has floor dimensions of 3 m × 4 m and a height of 2.4 m. Such a generous space is important for patient comfort and safety, especially when the patient may need constant attention by a physician. This room consists of an inner shield of mu-metal™ mounted on 8-mm thick aluminum plate that serves as an eddy-current magnetic and radio-frequency shield. The montage is supported by a stiff aluminum framework of 15-cm thickness, and the outer surface of the framework is covered by a second mu-metal shield. The ceiling of the MSR supports aluminum railings which may be used to suspend one or two gantries for holding the dewar containing the magnetic sensors. A single door when closed provides magnetic continuity for the two layers of magnetic shielding. Figure 3 shows the front of this room before cosmetic panels were attached to its surfaces. Access ports are provided for air circulation, optical fibers, and filtered electrical leads.

The attenuation provided by this MSR was evaluated by R.T. Johnson and J.R. Marsden by placing an electromagnet about 5 m or more from the wall of the room with its axis vertical. A SQUID magnetometer positioned at the center of the room detected the interior field strength when fields of various frequencies were applied. Figure 4 illustrates the deduced attenuation over the frequency range of primary interest. After a 60-Hz shaking field was applied, the steady earth's field is attenuated by a factor of 10^3. However, shielding is less effective for very low-frequency fields, being only about 30 db. The effect of eddy-current shielding becomes apparent above about 10^{-1} Hz, and the attenuation rises to a value of about 10^4 at 10^2 Hz with a tendancy toward saturation at higher frequencies. The shielding is not quite as effective as for MSRs having three[17] or more[15] separated layers of magnetic shielding, but the present room is considerably less expensive and has greater interior space.

The improvement in noise level that the MSR provides can be appreciated by examining the spectrum in Fig. 5. This shows the field spectral density measured by J. Shang and B. Schwartz of a SQUID sensor placed near the center of the room and oriented vertically. This sensor, which is one of five mounted on a probe suspended in a single dewar,[8] has a second-order gradiometer as its detection coil, with a baseline between adjacent coils of 4 cm. In the absence of the MSR the low-frequency ambient noise first became apparent over the intrinsic sensor noise at a frequency of 25 Hz. With the MSR it becomes apparent at the considerably lower frequency of about 1 Hz.[20] This substantial improvement is considered very satisfactory although not state-of-the-art. The noise level can be further reduced by improving the area-turns balance of the detection coils, which are known to be out of balance by a few parts in 10^4. This dewar contains three other SQUID sensors in

Fig. 3.

Magnetically shielded room at the Neuromagnetism Laboratory of the Departments of Physics and Psychology at New York University.

addition to the five used to detect neuromagnetic fields. These three have magnetometer detection coils that are oriented to monitor three orthogonal components of the ambient field, so that their outputs can be properly scaled and subtracted from the signal to further improve the noise level. We are now in the process of evaluating how effective this procedure is to further enhance sensitivity within the MSR.

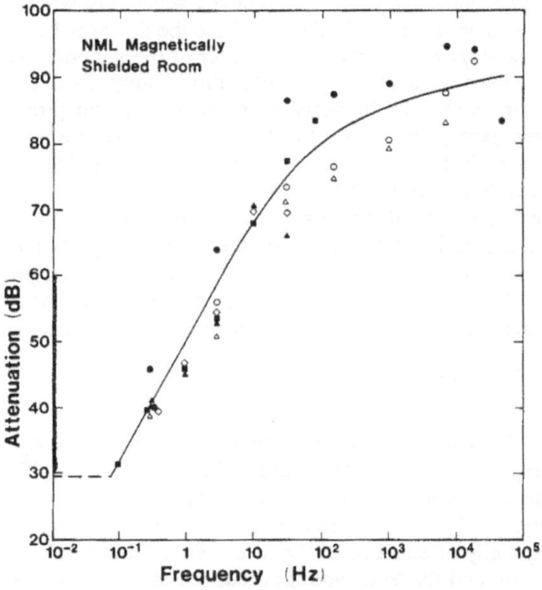

Fig. 4. Attenuation at various frequencies of the MSR shown in Figure 3. Different symbols indicate measurements with the field coil at positions in front or to the side of the room, and at different distances from it.

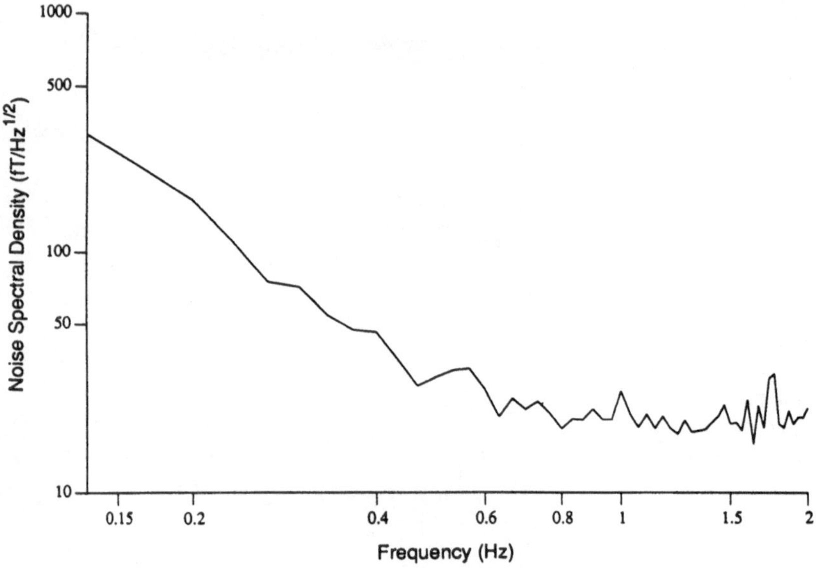

Fig. 5. Field noise indicated by a SQUID sensor placed near the center of the MSR. The high-frequency noise is sensor system noise, and corresponds to a field spectral density of about 20 $fT/Hz^{1/2}$. Environmental noise becomes apparent below about 1 Hz.

DEWAR GANTRY

To determine where neural sources lie within the head it is necessary to measure accurately the position and orientation of each field sensor with respect to landmarks on the patient's scalp. Traditionally the dewar was placed in the desired location, and the position of its tail with respect to convenient landmarks was measured across the scalp. But this procedure has inherent inaccuracies, due principally to the irregular shape of the head. One advance was to align the patient's head within a reference framework, and to move the dewar accurately with respect to this framework.[3,8] An example of such a setup is shown in Fig. 3, where the dewar is held in a carriage that moves so that the dewar's axis always points toward the center of the patient's head. This has an advantage when the head is modeled as a sphere for computing source locations, for the field component provided by the sensors is exactly the radial component. Another procedure is to use a computer-controlled mechanized gantry that moves the dewar to a pre-determined position and orientation in space.[20] In Fig. 6 a different approach is shown, where two independent gantries permit the operator to move each dewar by hand to the desired location. Independent movement is provided along two orthogonal horizontal directions and the vertical direction, with rotation allowed about the vertical axis and the horizontal axis where the gantry supports the dewar. Friction holds the dewar in place when the operator releases it, and a secure lock is provided by compressed gas brakes that secure all these degrees of freedom. The dewar can also be rotated about its own axis if it is desired that the individual sensors within the dewar be placed in a particular orientation.

PROBE POSITION INDICATOR

The gantry just described is augmented by a magnetic system that determines the position of each sensor with respect to the patient's head. This is called the "Probe Position Indicator". It consists of a transmitter mounted on the bottom of the main section of each dewar, four receivers, an electronic controller, and computer software to interpret the output. Each transmitter and receiver contains three orthogonally oriented coils. An ac current at about 10 kHz is sent through each of the transmitter coils in turn, and the three voltages induced in the coils of each receiver are measured. The position and orientation of a given receiver is computed from the nine amplitudes of the voltages it provides.

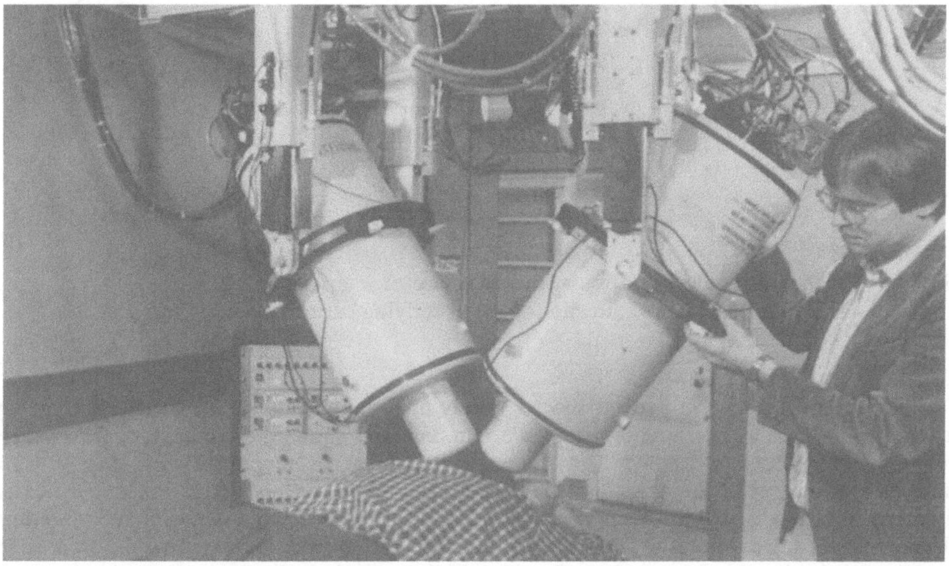

Fig. 6. A pair of dewars, each containing a probe with 7 SQUID sensors, supported by gantries over a subject in the MSR at the Center for Neuromagnetism of the NYU Medical Center.

When recording neuromagnetic data, a patient wears a headband similar to an athlete's sweatband, and from one to three receivers are attached to the band. The fourth receiver, which has a short pointer attached, is used by the operator to point to landmarks on the scalp, so that the positions of these receivers are established with respect to the landmarks. It is necessary to have only one receiver on the headband if there is assurance that it will not move or tip from its original orientation during measurements. However, it is more reliable to use two or three receivers, since inter-comparison of their independent measurements provides an indication of how stable the arrangement is. Between neuromagnetic recordings the operator can determine the position of the the sensors with respect to the subject's head whenever desired.

Measurements carried out by J. Shang have shown that the accuracy of this Probe Position Indicator is generally better than ±2 mm when operating in the shielded room depicted in Fig. 6, provided the tails of the dewars are close to the scalp. Accuracy decreases if the dewars are moved further than about 20 cm away. In practice, with this system, the accuracy in determining sensor positions is limited by how reproducibly the landmarks can be identified by the operator and not by the accuracy of the Probe Position Indicator itself.

CLINICAL APPLICATIONS

Ongoing clinical research at the University of California at Los Angeles[21,22] and at the Istituto di Neurochirurgia of the Università degli Studi di Roma[3,23] have revealed a number of significant applications of magnetic localization in studies of tumors and epilepsy. The presence of a tumor, even those too small to be resolved by x-ray CT or magnetic resonance images, may produce abnormal magnetic activity that can be localized.[3] However, most of the investigations reported to date have concentrated on epilepsy, which is a disorder characterized by the abnormal electrical discharge of neurons within the brain that result ultimately in behavioral seizures. Localized discharges, or focal epilepsy, is the most common, and it is estimated that some 800,000 individuals in the United State alone have such a disorder. Most can be treated effectively with medication; however, this is insufficient for as many as 360,000 of them. For perhaps 15% of the latter group, surgical removal of the epileptic brain tissue may be considered. Obtaining a precise localization of the focal region is of paramount importance in deciding whether to intervene surgically.

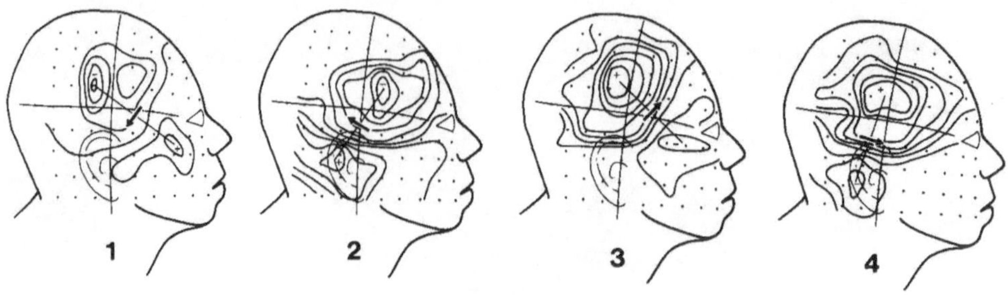

Fig. 7. Isofield contours for a four-wave sequence of interictal epileptic activity. The arrows show the position and direction of the underlying current sources. From Ref. 21.

The discharges seen in patients have a variety of forms. Some can be localized to a single position, while others have a more complex origin. Figure 7 shows the sequence of field patterns from a complex series of discharges in a young male patient, as reported by Barth *et al.*[21] Two different sources generate biphasic discharges in an interleaved sequence. The first map indicates a discharge in the right anterior temporal lobe with the current directed downward and posteriorly. The second, recorded 16 milliseconds after the first, indicates a discharge from a different source lying 1 cm behind and below the first. The third discharge is from the first source with the current reversed, and the last discharge is identified as originating from the second source, with reversed current. Viewed magnetically, it is often possible to unravel such complex spatial and temporal discharge patterns that cannot be interpreted from the electroencephalogram alone.

In this particular example, the intracranial locations of the first and second sources are reasonable from the brain's anatomy and observed pathology. Both sources lie at the edges of a large region of scar tissue within the right temporal lobe. Figure 8 shows x-ray CT scans on which the computed source positions are shown by crosses. At the borders of scarring, functioning neurons are often disturbed, thus producing intermittent epileptic discharges. Accumulating evidence such as this, together with a large body of data obtained in studies of normal brain activity, provides verification of existing procedures for neuromagnetic localization.

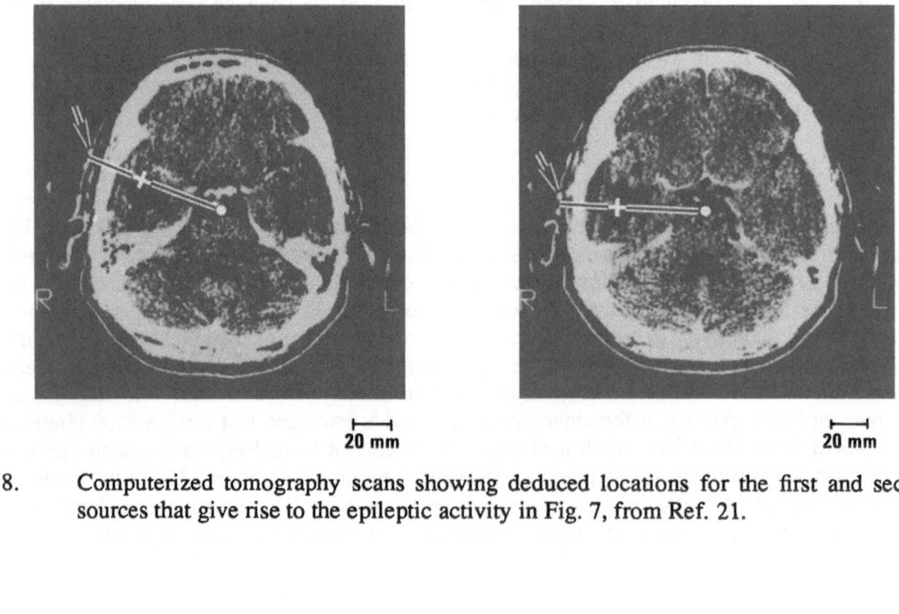

Fig. 8. Computerized tomography scans showing deduced locations for the first and second sources that give rise to the epileptic activity in Fig. 7, from Ref. 21.

ACKNOWLEDGEMENTS

We express our sincere appreciation to L. Kaufman, D.B. Crum, R.J. Ilmoniemi, R.T. Johnson, J.R. Marsden, B. Schwartz, and J. Shang for their help and informative discussions. Research at New York University is supported in part by Air Force Office of Scientific Research Contracts AFOSR-84-0313 and F49620-86-C-0131.

REFERENCES

1. D. Cohen, Magnetoencephalography: evidence of magnetic fields produced by alpha rhythm currents, Science 161:784 (1968).
2. D. Cohen, Magnetoencephalography: detection of the brain's electrical activity with a superconducting magnetometer, Science 175:664 (1972).
3. G. L. Romani and L. Narici, Principles and clinical validity of the biomagnetic method, Med. Prog. thru. Tech. 11:123 (1986).
4. R. Hari and R. J. Ilmoniemi, Cerebral magnetic fields, CRC Critical Rev. in Biomed. Eng. 14:93 (1986).
5. S. J. Williamson and L. Kaufman, Analysis of neuromagnetic signals, in: "Handbook of Electroencephalography and Clinical Neurophysiology," A. Gevins and A. Rémond, eds., Elsevier, Amsterdam (1987), Chapter 14.
6. S. J. Williamson, G. L. Romani, L. Kaufman, and I. Modena, "Biomagnetism: An Interdisciplinary Approach," Plenum Press, New York (1983).
7. R. Ilmoniemi, R. Hari, and K. Reinikainen, A four-channel SQUID magnetometer for brain research, Electroenceph. clin. Neurophysiol. 58:467 (1984).
8. S. J. Williamson, M. Pelizzone, Y. Okada, L. Kaufman, D. B. Crum, and J. R. Marsden, Magnetoencephalography with an array of SQUID sensors, in: "Proceedings of the 10th International Cryogenic Conference - ICEC10", Butterworth, London (1984), p. 339.
9. G. L. Romani, R. Leoni, and C. Salustri, Multichannel instrumentation for biomagnetism, in: "SQUID '85: Superconducting Quantum Interference Devices and their Applications," H. D. Hahlbohm and H. Lübbig, eds., Walter de Gruyter, Berlin (1985), p. 919.
10. J. E. Zimmerman and R. Radebaugh, Operation of a SQUID in a very low-power cryocooler, in: "NBS Special Publication 508," National Bureau of Standards (1978).
11. J. E. Cox and S. A. Wolf, Magnetic and vibrational characteristics of a close cycle refrigerator, in: "NBS Special Publication 508," National Bureau of Standards (1978).
12. D. B. Sullivan, J. E. Zimmerman and J. T. Ives, Operation of a Practical SQUID Gradiometer in a Low-Power Stirling Cryocooler, in: "NBS Special Publication 607," National Bureau of Standards (1981).
13. E. Tward and R. Sarwinski, A closed cycle cascade Joule Thomson refrigerator for cooling Josephson junction devices, in: "NBS Special Publication 698," National Bureau of Standards (1985).
14. Y. Okada, L. Kaufman, and S. J. Williamson, Hippocampal formation as a source of endogeneous slow potentials, Electroenceph. clin. Neurophysiol. 55:417 (1982).
15. A. Mager, The Berlin magnetically shielded room (BMSR), Section A: Design and construction, in: "Biomagnetism," S.N. Erné, H. D. Hahlbohm, and H. Lübbig, eds., Walter de Gruyter, Berlin (1981), p. 51.
16. M. Ibuka, H. Hosomatsu, and S. Naito, A SQUID magnetometer using a niobium thin-film microbridge, IEEE Trans. Instrum. and Meas. IM-30:251 (1981).
17. V.O. Kelhä, J. M. Pukki, R. S. Peltonen, A. J. Penttinen, R. J. Ilmoniemi, and J. J. Heino, Design, construction, and performance of a large-volume magnetic shield, IEEE Trans. Magn. MAG-18:260 (1982).
18. J. E. Zimmerman, SQUID instruments and shielding for low-level magnetic measurements, J. Appl. Phys. 48:702 (1977).
19. Fabricated by Vacuumschmelze GmbH, Hanau, Federal Republic of Germany.
20. J. Vrba, M. Burbank, H. Ensing, A. Fife, E. Heijster, C. Marshall, J. McCubbin, D. McKenzie, M. Tillotson, K. Watkinson, H. Weinberg, and P. Brickett, Integrated biomagnetic robotic system, in: "Biomagnetism: Applications and Theory," H. Weinberg, G. Stroink, and T. Katila, eds., Pergamon Press, New York (1984), p. 52.

21. D. Barth, W. Sutherling, J. Engle, Jr., and J. Beatty, Neuromagnetic evidence of spatially distributed sources underlying epileptiform spikes in the human brain, Science 223:293 (1984).
22. J. Beatty, D.S. Barth, and W. Sutherling, Magnetically localizing the sources of epileptic discharges within the human brain, Naval Res. Rev. 2:20 (1984).
23. G.B. Ricci, Clinical magnetoencephalography, Nuovo Cimento 2D:517 (1983).

HIGH-RESOLUTION MEASUREMENTS OF BIOMAGNETIC FIELDS

John P. Wikswo, Jr.

Living State Physics Group
Department of Physics and Astronomy
Vanderbilt University
Nashville, Tennessee

ABSTRACT

Attempts to understand the fundamental determinants of magnetic fields produced by bioelectric activity have been hindered by the low spatial resolution and 1 cm or greater Dewar thickness of conventional SQUID biomagnetometers. To overcome this and provide data for our model studies of the relationship of bioelectric and biomagnetic fields, we have built both superconducting and semiconductor magnetometers with small, room-temperature, toroidal pickup coils. We can readily measure cellular action currents in isolated nerve and muscle preparations, and are evaluating this instrument for use during neurosurgical repair of damaged peripheral nerves in humans. To analyze the 2 and 3-dimensional current distributions in the exposed, in vivo heart and brain, we are considering high-resolution SQUID systems, hopefully with coil diameters of only several millimeters and a similar coil-to-animal spacing.

INTRODUCTION

Over the past 25 years, a number of different types and configurations of magnetometers have been used to record the magnetic fields associated with biological electrical activity. In their pioneering measurements in 1963, Baule and McFee[1] used a pair of coils each with 2 million turns of copper wire wound on a ferrite core to record the magnetic field of the human heart. Cohen[2] used a similar coil to measure the magnetoencephalogram (MEG) from currents flowing in the brain. In 1970, Cohen, Edelsack and Zimmerman[3] achieved substantial increases in signal-to-noise ratio by using a SQUID magnetometer. Most of the biomagnetic studies since then have used SQUIDs to assess the utility of biomagnetic measurements for the clinical diagnosis of electrophysiological disorders, such as are associated with heart attacks and epilepsy, or to explore noninvasively the response of the brain to a variety of well controlled stimuli such as visual patterns or auditory tone bursts.

While studies on human volunteers have certain advantages, they provide only limited information regarding the relation between biomagnetic fields and cellular electrophysiology. There is little opportunity to use conventional, highly-invasive electrophysiological techniques to confirm the predictions of the models used for data analysis. Furthermore, the size of the magnetometer pickup coils and the several centimeter distance between the magnetometer and the biological sources results in a low spatial resolution not conducive to understanding the cellular basis of the magnetic signals. The inability to conduct controlled biochemical interventions limits the assessment of how various factors govern the production of these magnetic fields. To circumvent these shortcomings and gain some much-needed basic information regarding biomagnetic fields, we have concentrated on measurements using isolated, one-dimensional tissue preparations from frogs, crayfish, and rats. We can place the magnetometer immediately adjacent to the tissue being studied, control many experimental parameters, and obtain an independent and detailed description of the properties and location of the biological electric current sources. These experiments require magnetometers with as high a spatial resolution as possible, which presents a new set of instrumentation problems not addressed by the large SQUID arrays being developed for human studies.

MEASUREMENTS IN ONE DIMENSION

Our first recording of the magnetic field associated with an action potential, i.e. a nerve impulse, propagating along a frog sciatic nerve provides a clear example of how conventional SQUID biomagnetometers cannot provide the necessary spatial resolution. Figure 1 shows the magnetic field pattern associated with a propagating nerve impulse. The magnetic field falls off as $1/r$ close to the nerve, consistent with the finite spatial extent of the depolarization currents, $1/r^2$ further from the nerve, consistent with the dipole nature of the depolarization currents, and eventually as $1/r^3$, as expected for a quadrupolar current source. To measure this field, we placed an isolated frog sciatic nerve bundle next to the face of a SQUID magnetometer with a 1.5 cm × 2.7 cm elliptical pickup coil 15 mm from the nerve[4,5]. The signal, shown in Fig. 2a, was barely detectable after averaging 1024 times.

The key to successful measurement of the magnetic field from such a weak and highly-localized current source is to place the magnetometer as close as possible to the tissue. Possibly the easiest way is to thread the nerve through a room-temperature, ferrite-cored, pickup coil so that the nerve serves as the primary to this transformer, and the copper wire wound on the core serves as the secondary. A number of different techniques can be used to sense the current or voltage induced in this coil by the magnetic field as it propagates through the toroid. Initially, we wrapped several turns of wire around the outer tail of our SQUID magnetometer Dewar and connected this coil to the toroidal pickup coil[4,5]. The SQUID was then used simply as an inductively-coupled ammeter, and we obtained the data in Fig. 2b. We subsequently improved the sensitivity and noise rejection by more closely matching the toroidal pickup coil to the SQUID, which was housed in a cryogenic dip probe that could be operated in a liquid helium storage Dewar[6].

There are three sources of noise in a SQUID magnetometer with a room-temperature pick-up coil: The Johnson noise in the copper pickup coil, the

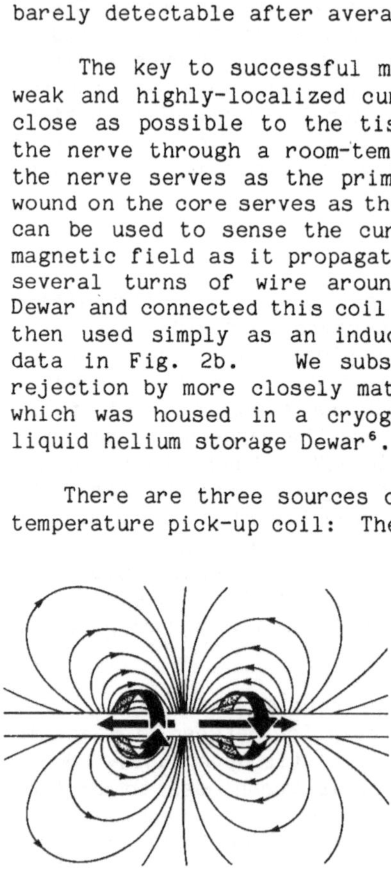

Fig. 1. The magnetic field (dark bands) and extracellular electric field (thin lines) associated with a nerve action potential propagating from right to left. The effective current sources are represented by a pair of axial current dipoles, with the leading dipole depolarizing the nerve membrane and the oppositely-directed one repolarizing it.

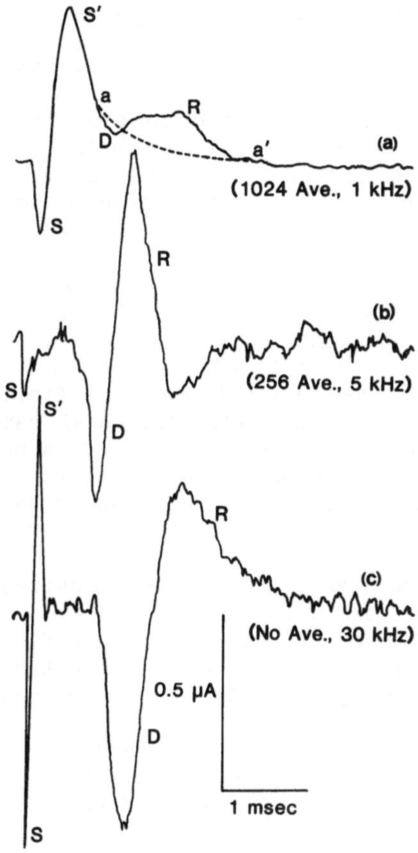

Fig. 2. a) The SQUID output with an isolated frog sciatic nerve in a conducting medium placed as close as possible to the pickup-coil. S and S' are stimulus artifact, the deviation from the line aa' is due to the magnetic field of the nerve, and D and R are the depolarization and repolarization phases[5]. b) The SQUID output when the frog nerve was coupled to the SQUID by a small, room-temperature toroidal coil[4]. c) The corresponding output of the toroidal pickup-coil/semiconductor current-to-voltage converter[7,8].

Johnson noise from the rf-shielding transformer within the probe, and the intrinsic SQUID noise. Since the Johnson noise in the toroidal pickup coil was dominant in our system, we recognized that it was unnecessary to use an amplifier as quiet as a SQUID. However, since the toroidal pickup coils must be matched to the limited spatial extent of the propagating action potential, it is difficult to fabricate small toroids whose impedance is greater than 50 to 100 Ohms. This in turn places a severe limitation on the use of conventional semiconductor amplifiers, since their voltage noise is usually excessive when the amplifier is connected to such a low source impedance. Our solution to this problem was to develop a low noise current-to-voltage converter with 10 parallel input stages designed specifically to reduce the input voltage noise of the amplifier[7]. Figure 2c shows the excellent signal-to-noise ratio that can be obtained with this instrument, which has an input current noise of only 120 $pA/Hz^{1/2}$ and is now being used to study propagation in single axons, bundled nerves, cardiac muscle, and skeletal muscle[8,9] and for evaluation of peripheral nerve performance during neurosurgery[10].

MEASUREMENTS IN TWO AND THREE DIMENSIONS

While our studies on isolated one dimensional preparations have been successful, the experimental techniques have we developed are not readily extended to measurements of the more intriguing 2 and 3 dimensional current sources in the heart, brain and other organs. If we are to extend our basic research in biomagnetism to these more challenging problems, we will need the sensitivity, coil configuration, and noise rejection of a SQUID gradiometer, but the SQUID biomagnetic instruments presently in use have pick-up coils that are 1.5 to 3 cm in diameter and typically 1 to 2 cm from the outside of the liquid helium Dewar that contains the magnetometer. Smaller coil diameter and spacing are not necessary for non-invasive studies of the human heart or brain due to the substantial thickness of tissue and bone that separates the Dewar from the organ being studied, but in animal studies, all intermediate bone and tissue can be removed to study the exposed heart, brain, and spinal cord with higher spatial resolution and with invasive electrical validation of the magnetic results. In such studies, it will be most worthwhile to have smaller coils located closer to the tissue than possible with conventional SQUID systems.

As an example of the benefits of placing small coils close to the source, we will consider a simulated measurement of the magnetic field from nerve impulses propagating down the spinal cord of a rat. The field pattern would have the same general shape as shown in Fig. 1. Figure 3 shows the calculated isofield contours for the vertical magnetic field component recorded at several distances from an in vivo, horizontal rat spinal cord. The field measured at 16 mm from the cord is three orders of magnitude smaller than that recorded at 1 mm; the pattern at 16 mm is so spread out that it is difficult to imagine using it for localization of spinal cord damage or chronic pain foci.

The requirement that the SQUID be placed near the nerve is emphasized in the plot of the peak-normal-field versus distance above the nerve in Fig. 4. The solid curve demonstrates the $1/r$ to $1/r^2$ to $1/r^3$ fall-off of the neuromagnetic field with distance. The graph also shows the operating region of two SQUID systems: a high-resolution system with coils 2 mm from the spinal cord and 50 $fT/Hz^{1/2}$ noise indicated by (a), and the other a state-of-the art MEG system with pick-up coils 10 mm from the cord and 20 $fT/Hz^{1/2}$ noise by (b). The closest possible coil-to-animal spacing locates the vertical line, while the system noise in a 10 kHz bandwidth determines

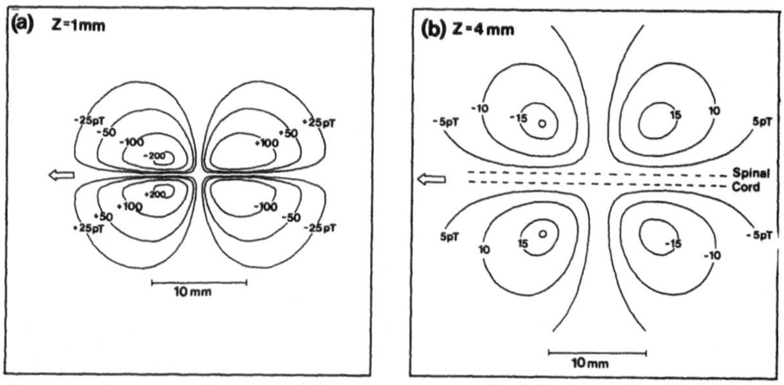

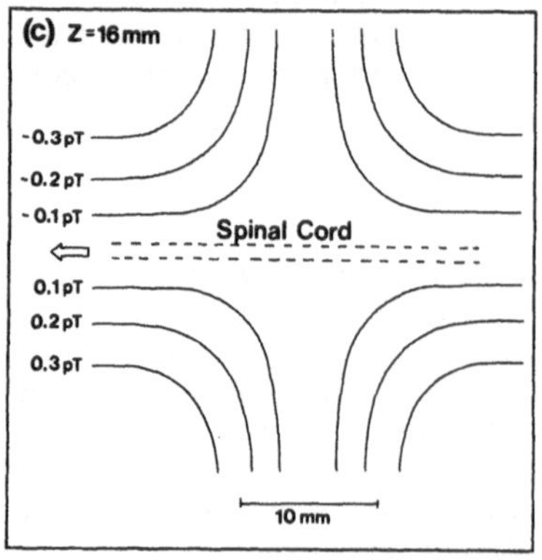

Fig. 3. The calculated isofield contours for three distances above a rat spinal cord. The arrow shows the direction of propagation of the pattern. The field values are in picoTesla.

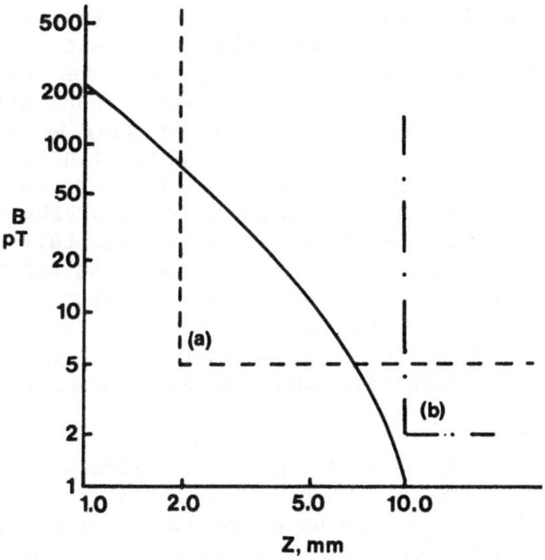

Fig. 4. The calculated peak value of the normal component of the magnetic field from an intact rat spinal cord as a function of distance from the center of the spinal cord. The noise and spatial resolution limits of the high-resolution magnetometer are shown by corner (a) while those for conventional biomagnetic SQUIDs are shown by (b).

the horizontal one. The high resolution system will detect the spinal cord signal with a 20-to-1 signal to noise ratio while the conventional one will have a 1 to 1 signal-to-noise ratio. This is an excellent example of how the loss in absolute field sensitivity accompanied by the use of small pickup coils is amply compensated for by the large field strengths close to the source.

HIGH RESOLUTION SQUID SYSTEMS

With present SQUID magnetometers, the pickup coils reside in the liquid helium space inside the Dewar, so that the spacing between the pickup coils and the room-temperature experimental animal is determined by the thickness of the inner and outer Dewar walls, the superinsulation and the radiation shield, and by the thermal contraction of the inside of the Dewar when it is cooled to cryogenic temperatures. Freake, Swithenby and Thomas[11] recently demonstrated that a 5 mm spacing could be achieved, as shown in Fig. 5.

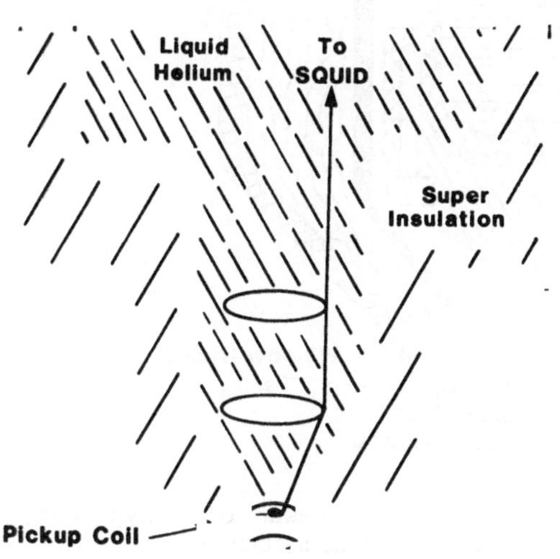

Fig. 5. A schematic representation[11] of a special fiberglass Dewar with conical internal and external tails, a second-order gradiometer with a 2 mm diameter pickup coil next to the end of the Dewar, and a 5 mm coil-to-animal spacing. The 90 $fT/Hz^{1/2}$ noise originates primarily from Johnson noise in the rf-shield (not shown) and can possibly be reduced by up to a factor of 3.

We are examining a more radical approach to achieve high resolution and a coil-to-animal spacing of only 2 mm. If the SQUID and pickup coils are located in the vacuum space between the Dewar walls, the inner wall need not be thin and thereby structurally weak. While a Dewar is generally designed to minimize the thermal input over the entire surface, we can accept somewhat higher local thermal input as long as the area with less-than-ideal insulation is minimized. The only requirement is that the coils be mounted on a substrate with sufficiently high thermal conductivity to keep them superconducting. Figure 6 shows a cross-section of the tail of a Dewar that would meet these requirements. The thermal shield and super-insulation will minimize the radiative heat input to the coil system. The use of the new high-temperature superconductors for the pickup coils could allow them to operate at a higher temperature and would reduce both the required thermal conductivity of the substrate and the associated heat load into the helium space.

Rather than a use thick outer vacuum wall, a thin vacuum window would separate the magnetometer face coil from the experimental preparation. A beryllium copper or other metal foil could be used for the window, but Johnson noise may prove to be a limitation. Alternatively, it could be formed from plastic. In any case, the thickness and mounting of the window must be chosen to minimize any bowing inward from atmospheric pressure. Crucial in this design is the use of a mechanism to adjust the separation between the pickup coils and the window to account for thermally-induced changes in the length of the inner Dewar wall, for upon warming the expansion of the inner Dewar could perforate the vacuum window. Whether a coil-to-animal spacing of less than 2 mm can be achieved will be determined by a variety of trade-offs between helium boil-off, noise, and construction cost.

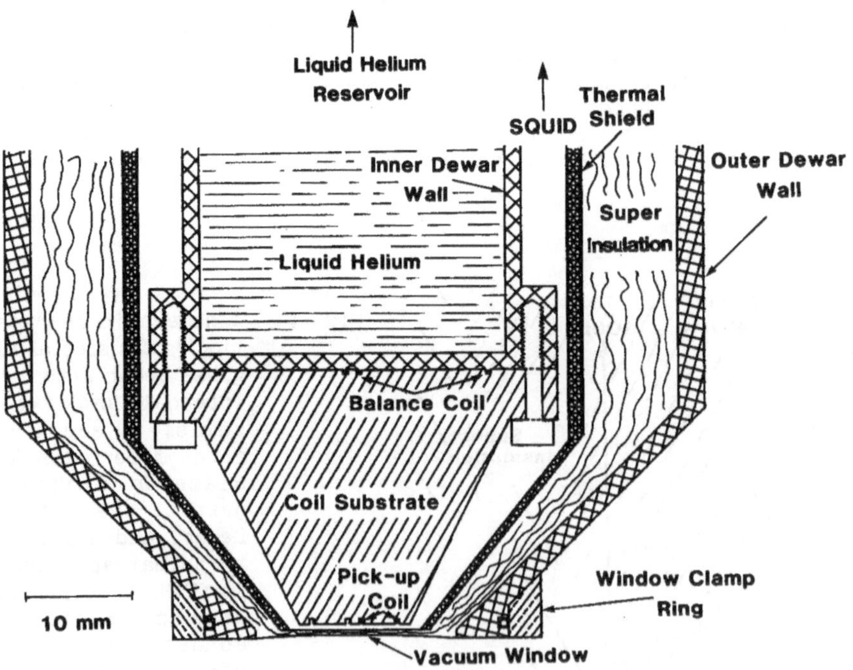

Fig. 6. A cross-section through the tail of the Dewar of a hypothetical high-resolution, four-channel, SQUID gradiometer.

Our studies so far suggest that it is possible to construct a high-resolution SQUID system with four independent first-order asymmetric gradiometers with 3 mm diameter face coils 1 to 2 mm from the animal, and an expected noise level of 50 $fT/Hz^{1/2}$. In optimizing this system, the first question that must be addressed is what size pick-up coils are needed, and how close the coils should be to the animal. Based on experience with SQUID systems developed for recording the MCG and MEG, as a general guide the coil diameter should be approximately the same as the coil-to-animal spacing. For now we will assume that cryogenic coils can be placed as close as 1 mm to the animal. Thus we can consider coils as small as 1 mm.

There is, of course, no single coil diameter that is best for general use. It would, however, not be advantageous to use a large, closely-placed coil to measure fields from a compact source, since the average field over a large coil would be very low. The situation is even worse when a large coil is placed some distance from a compact source. Clearly 3 mm diameter coils used to record the isofield contours in Fig. 3a or 3b will provide more accurate localization and a more detailed and accurate representation of the action currents than 15 mm coils used to record the field in Fig. 3c.

The choice between building a magnetometer or gradiometer is governed to a large extent by the need for noise rejection. If second-order gradiometers were used, it might be possible to avoid the need for a shielded room, but there are severe problems in balancing such a multichannel system with millimeter coils. If the shielded room were to have an excellent shielding factor, it might be possible to use simple pick-up loops so that the system would operate as a magnetometer, but since the cost of building such a room is excessive, we have ruled out simple magnetometers. The first-order gradiometer provides the middle ground. The decision then has to be made as to the actual coil geometry. Millimeter diameter symmetric coils would be difficult, if not impossible to balance adequately, whereas with the asymmetric design the upper coil can be 1 cm or larger in diameter so that permanent balance tabs can be installed and adjusted. The choice of turns and asymmetry ratios is governed by the need to match the coil to the SQUID sensor. Figure 7a shows the calculated coil inductance of asymmetric gradiometers wound from 2 mil wire as a function of face coil diameter, where the face coil has 4, 9, 16, 25, or 36 turns and the cancelling coil has one turn[12]. The horizontal dashed line shows the 2 μH SQUID inductance. A 16 turn face coil 3 mm in diameter with a single turn cancelling coil has an inductance of 2 μH. Figure 7b shows the effective input noise of the various gradiometer geometries, assuming a dc-SQUID sensor noise of 1.5×10^{-12} $A/Hz^{1/2}$. The 3-mm / 16-turn gradiometer has a noise of 50 $fT/Hz^{1/2}$, and the 16-turn system is quieter than the other configurations in the 2 to 5 mm range of diameters. Obviously the system noise could be halved by going to coils with a 5 mm diameter, but all simulations conducted so far indicate that the compromise in spatial resolution would be too great. While this sensitivity will be about one-third of that of the best conventional dc-SQUID systems, Figures 3 and 4 show that the amplitude of the field 1 mm from the source can be as much as 50 times that at 1 cm.

The four-channel gradiometer pick-up coil is shown in Fig. 8. A 3 cm baseline was chosen because such a short baseline will help reject external noise, yet will not contribute to significant signal loss or distortion because the biological current sources will be highly localized and close to the face coil. The axes of the gradiometers are not parallel, since the face-coils need to be located close to each other. This may complicate the balancing procedure somewhat, but this geometry is deemed necessary for the anticipated studies. The spacing between the four gradiometers would be determined after their mutual inductance and cross-talk is calculated.

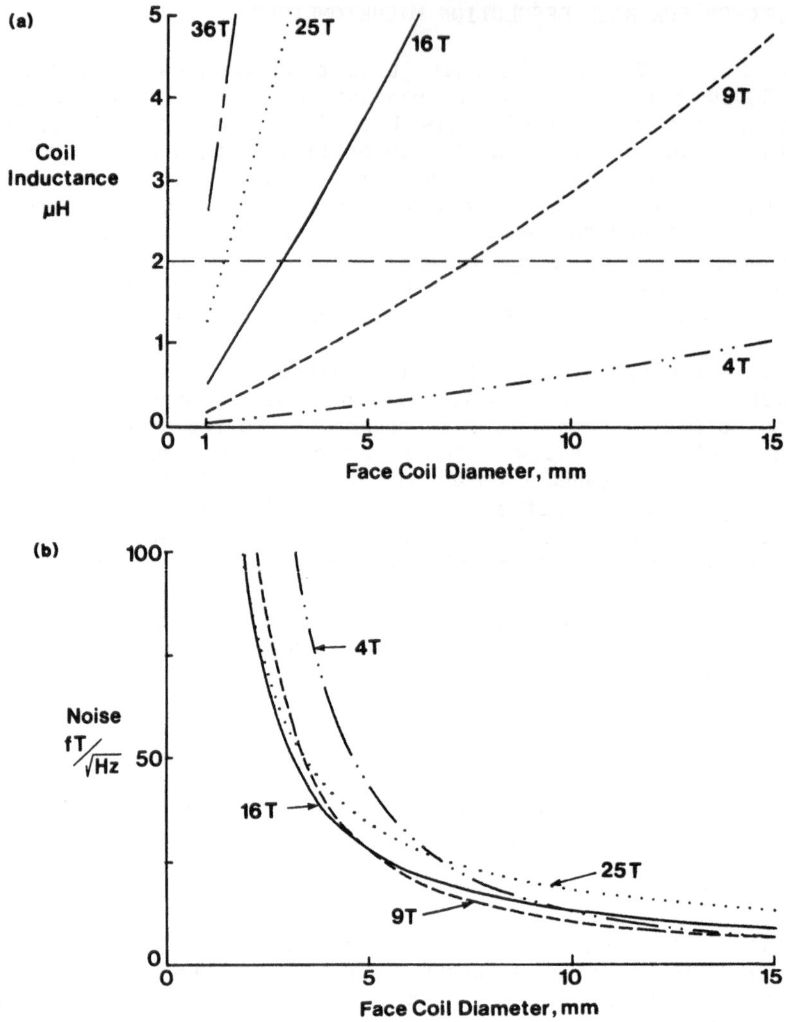

Fig. 7. Plots of the coil inductance (a) and effective noise (b) as a function of face coil diameter for five asymmetric gradiometer configurations with differing numbers of turns on the face coil.

APPLICATIONS

Our studies using one-dimensional animal preparations have shown that the external electric and magnetic fields associated with the nerve action potential contain similar information about the transmembrane action potential of the nerve, but for several reasons the magnetic measurements are more accurate[9]. When electric measurements of the action potential and magnetic measurements of the action current are combined, it is possible to obtain accurate values for the electrical conductivity of the axoplasm within the nerve. Experimental and mathematical studies of septated nerves and cardiac muscle suggest that this approach might also provide important information on electrical behavior of cell-to-cell coupling. The mathematical models that we have developed to analyze two-dimensional current

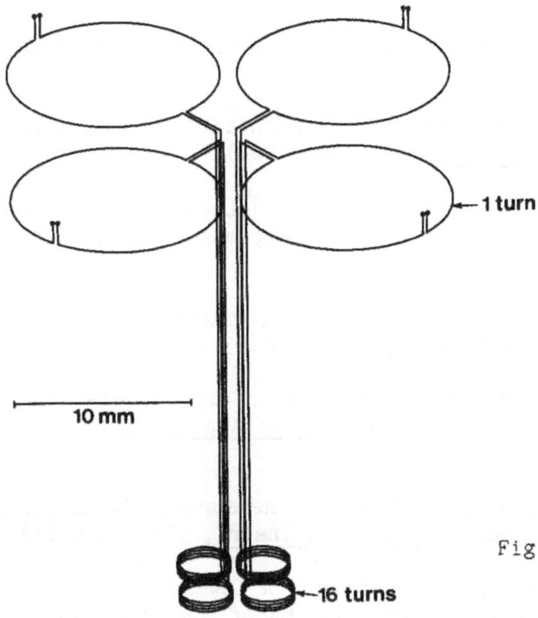

Fig. 8. An isometric view of the pick-up coils for the high-resolution SQUID magnetometer.

patterns in cardiac muscle[13] have also provided us with new insights into the role of the electrical anisotropy of the tissue in determining both the electric and magnetic fields, and these models need to be tested experimentally.

The analysis presented in this paper suggests that it is indeed feasible to construct a high-resolution biomagnetometer with four independent pick-up coils located behind a thin vacuum window so that the coils will be an order of magnitude closer to the sample than in present biomagnetic SQUIDs. The area of each pick-up coil will be two orders of magnitude smaller than in present SQUIDs, but will have a field sensitivity comparable to existing rf SQUIDs. Any loss in sensitivity of our proposed system, as compared to existing ones with dc SQUIDs, is more than compensated for by the increase in field amplitude obtained with the decreased coil-to-animal spacing.

This instrument would allow us to make a variety of biomagnetic measurements that were not previously possible, including studying how the anisotropic and bisyncytial properties of cardiac muscle affect cardiac depolarization currents and propagation of cardiac activation, and how the MEG, modeled equivalent sources, and actual bioelectric current sources in animal cortex are related in both the normal and epileptic brain. The dc sensitivity of the magnetometer should greatly facilitate the study of cardiac ischemia and epilepsy spread in the cortex. We could examine fields from the spinal cord, peripheral nerves, and skeletal muscle. Other applications would include the mapping of corrosion currents in small samples[14], and non-destructive evaluation of metallic structures[15,16].

ACKNOWLEDGEMENTS

The suggestions of Duane Crum regarding SQUID design are appreciated. This research is supported by NIH Grants NS 19794 and NS 24751, ONR Contract N00014-82-K-0107, and Vanderbilt University.

REFERENCES

1. G. Baule and R. McFee, Detection of the magnetic field of the heart, Am. Heart J. 65:95 (1963).

2. D. Cohen, Magnetoencephalography: Evidence of magnetic fields produced by alpha-rhythm currents, Science, 161:784 (1968).

3. D. Cohen, E.A. Edelsack, and J.E. Zimmerman, Magnetocardiograms taken inside a shielded room with a superconducting magnetometer, Appl. Phys. Lett., 16:278 (1970).

4. J.P. Wikswo, Jr., J.P. Barach, and J.A. Freeman, Magnetic field of a nerve impulse:First measurements, Science, 208:53 (1980).

5. J.P. Barach, J.A. Freeman, and J.P. Wikswo, Jr., Experiments on the magnetic field of nerve action potentials, J. of Appl. Phys., 51:4532 (1980).

6. J.P. Wikswo, Jr., Improved instrumentation for measuring the magnetic field of cellular action currents, Rev. Sci. Instr., 53:1846 (1982).

7. J.P. Wikswo, Jr., P.C. Samson and R.P. Giffard, A low-noise, low input impedance amplifier for magnetic measurements of nerve action currents, IEEE Trans. Biomed. Eng., BME-30:215 (1983).

8. F.L.H. Gielen, B.J. Roth, and J.P. Wikswo, Jr., Capabilities of a toroid-amplifier system for magnetic measurements of current in biological tissue, IEEE Trans. Biomed. Eng., BME-33:910 (1986).

9. B.J. Roth and J.P. Wikswo, Jr., The magnetic field of a single nerve axon: A comparison of theory and experiment, Biophys. J., 48:93 (1985).

10. J.P. Wikswo, Jr., G.S. Abraham, and V.R. Hentz, Magnetic assessment of regeneration across a nerve graft, in Biomagnetism Theory and Applications, H. Weinberg, G. Stroink and K. Katila, Eds., Pergamon Press, New York, pp. 88 (1985).

11. S.M. Freake, S.J. Swithenby, and I.M. Thomas, A miniature SQUID gradiometer for the detection of quasi-dc ionic current flow in developing organisms, to appear in Proc. 6th Int. Conference on Biomagnetism, Tokyo (1987).

12. J.P. Wikswo, Jr., Optimization of SQUID differential magnetometers, AIP Conf. Proc., 44:145 (1978).

13. N.G. Sepulveda and J.P. Wikswo, Jr., Electric and magnetic fields from two-dimensional anisotropic bisyncytia, Biophys. J., 51: 557 (1987).

14. J.G. Bellingham and M.L.A. MacVicar, SQUID technology applied to the study of electrochemical corrosion, in "Proc. Applied Superconductivity Conf.," Baltimore (1986).

15. H. Weinstock and M. Nisenoff, Nondestructive evaluation of metallic structures using a SQUID gradiometer, in "3rd Int. Conf. on Supercond. Quantum Devices," H.-D. Hahlbohm and H. Lübbig, eds., Berlin (1985).

16. H. Weinstock, T. Erber, and M. Nisenoff, Threshold of Barkenhausen emmision and onset of hysteresis in iron, Phys. Rev. B, 31: 1535 (1985).

EFFICIENCY ASPECTS IN SUPERCONDUCTING RECTIFIERS *

G.B.J. Mulder, H.H.J. ten Kate, H.J.G Krooshoop and L.J.M. van de Klundert

Department of Applied Physics
University of Twente
Enschede, The Netherlands

ABSTRACT

The subject of this paper is the overall efficiency of thermally and magnetically controlled superconducting rectifiers. Such devices can be used to energize high-current superconducting loads, or alternatively, to make up the loss in a load which has been energized previously by means of (demountable) current leads. Both applications can result in a considerable saving of helium consumption. An analysis is made of the losses occurring in the rectifier components. The presented formulae are useful when designing a high-efficiency rectifier. Also, for an existing rectifier, the influence of parameters such as primary waveform, the primary amplitude, the frequency and the fraction of pumping time can be studied. To illustrate the theory, it is used to determine the overall efficiency of two experimental rectifiers constructed at the University of Twente.

INTRODUCTION

The use of superconducting rectifiers can be avoided by applying the conventional technique of high-current input leads in combination with a high-current power supply. However, above a certain current level, we expect a superconducting rectifier to be more economical than its conventional counterpart. The sum of heat leak and dissipation of current input leads increases approximately linearly with the design current of such leads.[1] A theoretical lower limit for a single counterflow lead is about 0.6 W/kA, but in practice[1] this value ranges from 1 to 3 W/kA. The efficiency of a rectifier on the other hand, is almost independent of the maximum secondary current.[2,3] Beside the efficiency, there may be other advantages of a superconducting rectifier system such as low cost and weight, which will not be discussed here.

We will restrict ourselves to transformer-type full-wave superconducting rectifiers with either magnetically or thermally controlled switches (see Fig. 1). A detailed treatment of the operating principles of

* Supported by F.O.M., The Netherlands Foundation for Fundamental Research on Matter, Utrecht, The Netherlands.

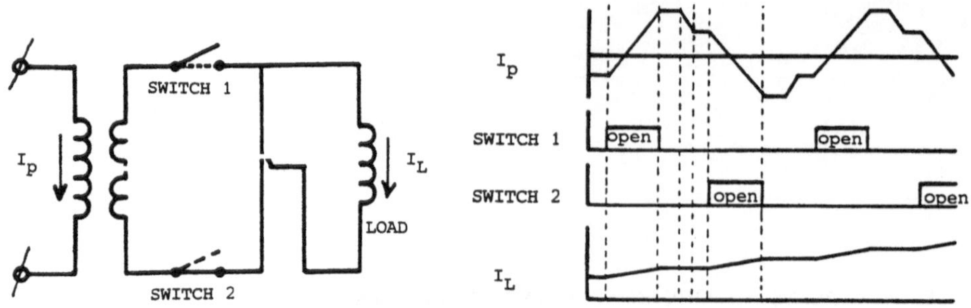

Fig. 1. Operation diagram of a full-wave superconducting rectifier

such rectifiers is given elsewhere.[3,4] The theoretical behaviour of a sc. rectifier depends on the waveform of the control signals and in particular on the shape of the primary current. From a variety of primary waveforms which result in a correct operation of the rectifier, we have selected three different waveforms which seem to be the most suitable for a high-efficiency rectifier. As shown in Fig. 2, they have a linear variation of the primary current during the pumping or commutation stages and a constant current during the time intervals for activation or recovery of the switches. The three waveforms are characterized by :

A: The time intervals for pumping, commutating, opening and closing the switches are constant. Consequently, the operating frequency is also constant. The current rates and corresponding voltages involved are variable, that is the pump voltage decreases and the commutation voltage increases during the process of energizing a magnet.

B: The current rates for pumping and commutation are constant, but in general not equal. The frequency becomes variable. This waveform implies a simplification of the primary power supply because a) the primary voltage is constant and b) the ratio of peak primary power to average primary power can be relatively low, for example less than 10.

C: This is the case of a loss make-up rectifier. The rectifier is used to compensate the losses in a persistent mode circuit, which means that the secondary current remains perfectly constant. The equations for the average power and loss of this type of rectifier are in fact equal to the instantaneous power and loss of the rectifier with waveform A.

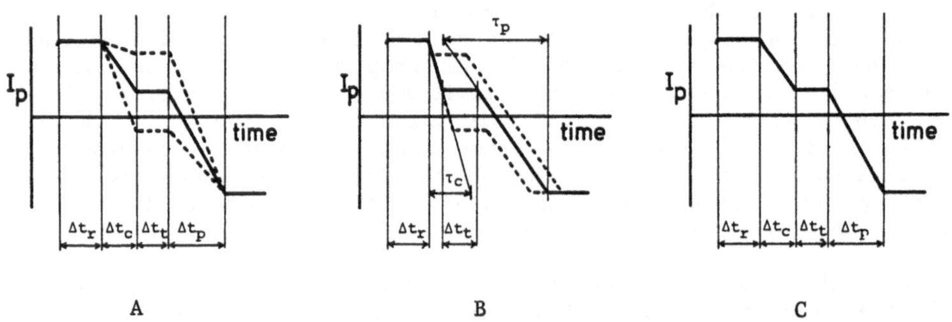

Fig. 2. Three possible waveforms for the primary current:
case A : constant time intervals --> constant frequency.
case B : constant current rates --> variable frequency.
case C : loss make-up rectifier --> constant load current.

THEORY

If a rectifier is used to energize a magnet (cases A and B), the efficiency η_w can be defined as

$$\eta_w = \frac{W_l}{W_l + \sum_i W_i} = \frac{1}{1 + \sum_i W_i/W_l}$$

where W_l is the energy stored in the magnet and $\sum_i W_i$ is the summation of the energy dissipated in all cryogenic parts of the rectifier. This definition refers only to the cryogenic environment and does not take into account the losses in the magnet itself. If the rectifier is used as a loss make-up device (case C), one should use the power efficiency η_p,

$$\eta_p = \frac{P_l}{P_l + \sum_i P_i} = \frac{1}{1 + \sum_i P_i/P_l}$$

where P_l and ΣP_i are the average output power and the average dissipation. Subsequently, the relevant heat sources in the cryogenic part of the rectifier system will be treated. One can distinguish:

* Reverse current loss in the switch elementsduring the pumping step,
* Loss due to resistive commutation or imperfect inductive commutation,
* The dissipated energy involved with opening and closing a switch,
* Eddy current and hysteresis loss in transformer windings and structure,
* Dissipation in the electrical connections between the components.

Reverse current loss

During the pump step, the full secondary voltage, that is twice the load voltage, appears across one of the two switches. The resulting reverse current loss depends on the type of switch.

* For magnetically controlled switches we can assume the resistance during the open-state to be constant and equal to the full normal resistance.
* In a thermally controlled switch, after triggering with a short heater pulse, the switch keeps itself in the open-state by means of the reverse current loss. At a sufficiently high voltage V_{sw} across the switch, the resistance is constant, namely the full normal resistance. This situation should be avoided if a fast thermal recovery is required. At a reduced voltage V_{sw}, the resistance becomes proportional to V_{sw}. Now the reverse current loss is not sufficient to open the switch completely and a stationary situation of coexisting normal and superconducting regions occurs. In this case, the theory of normal zone propagation predicts a constant leakage current I_{leak} through the switch equalling the minimum propagation current of the conductor for this specific cooling situation. Indeed, this effect of a voltage-independent leakage current has been measured in about ten experimental switches.

To summarize, the reverse current loss depends on whether the switch develops its full normal state resistance or not. In the first case the dissipation is proportional to V_{sw}^2 whereas in the latter case it is proportional to V_{sw}.

Commutation loss

The objective of the commutation step is to transfer the load current from one half of the secondary circuit to the other half. This can be achieved either inductively by generating the appropriate step in the primary current, or resistively by just opening a switch carrying the full load current. In the latter case it is advisable to shorten the primary of the transformer during commutation in order to reduce the loss. If the primary is short-circuited, only $\frac{1}{2} L_{sec} I_l^2 (1-k^2)$ is dissipated in the switch element instead of the complete secondary energy $\frac{1}{2} L_{sec} I_l^2$. With inductive transfer, the commutation loss is avoided almost completely.

Depending on the accuracy of the control electronics, only a fraction α of the load current flows through the switch when it has to be opened. The commutation loss occurring every half-cycle can therefore be described by $\alpha^2(1-k^2)\ \frac{1}{2}\ L_{sec}I_1^2$ where α equals 1 for the resistive commutation mode and α is small (say, less than 0.05) for the inductive commutation mode. It is clear that a transformer coupling constant k close to 1 is important, in particular for sc. rectifiers using the resistive commutation mode.

Switch triggering loss in thermal or magnetic switches

During the rectifier operation, a switching loss occurs which is the same every half-cycle:

* The energy necessary to open a thermally controlled switch. After triggering, the reverse current through the switch keeps it in the open-state and a stationary heat-flow occurs. The enthalpy associated with this stationary temperature distribution comes partly from the heater trigger pulse and partly from the initial ohmic dissipation in the switch element. Every half-cycle, the enthalpy difference is lost in the helium bath, when the switch recovers.
* A.c. loss resulting from the alternating control field of a magnetically controlled switch. The hysteresis loss in the control coil and the switch is constant every half-cycle. The same applies for the eddy current loss provided the control field rates are kept constant.

Dissipation in the transformer

An accurate evaluation of the dissipation in the transformer is quite difficult. A basic problem is that the primary and both secondary parts of the transformer generate a magnetic field acting on the filaments which is a combination of a d.c., an a.c.and a rotating field. In literature, only the separate cases of alternating or rotating fields have been dealt with.[5] Moreover, the loss per cycle is a function of the actual load current and therefore it changes during the operation of the rectifier.

Here we have chosen a simple approach which yields approximate results. The following assumptions were made:

a) The primary and both secondary coils generate magnetic fields which have locally the same orientation so there are no rotating field components. The local field change is then proportional to the change in primary current, that is $\Delta B = i\hat{I}_pK_c(x,y,z)$ for the commutation interval and $\Delta B = (1-i)\hat{I}_pK_p(x,y,z)$ for the pumping interval. This condition is in practice well satisfied for toroidal and long solenoidal transformers.

b) The local hysteresis loss is proportional to the field change ΔB. This corresponds to a rectangular magnetization loop of the filaments, which is valid if ΔB is much larger than the penetration field of the filaments and if the critical current density is constant.

c) The eddy current loss for a field change ΔB in Δt seconds is proportional to $(\Delta B)^2/\Delta t$. This is correct provided that 1) the matrix resistance is constant, 2) the return currents through the outer filaments do not exceed the local critical current and 3) Δt is much larger than the time constant of the eddy currents.

After integration of the local dissipation over the coil body one finds a total transformer loss per half-cycle described by

$$Q = i\hat{I}_pC_{hc} + (1-i)\hat{I}_pC_{hp} + (i\hat{I}_p)^2C_{ec}/\Delta t_c + ((1-i)\hat{I}_p)^2C_{ep}/\Delta t_p \ .$$

The loss constants, C_{hc}, C_{hp}, C_{ec} and C_{ep} depend on the geometry of the transformer and the quality of the primary and secondary conductors. In a given transformer they need to be determined only once, either by measurement or by integration of existing loss formulae[5] over the volume of the windings. C_{hc} and C_{ec} describe the hysteresis loss and eddy current loss during the commutation step, whereas C_{hp} and C_{ep} are related to the pumping step. For example, to obtain C_{hc}, the hysteresis loss involved with a primary current sweep from $\hat{I}_p$ to $-\hat{I}_p$ while the secondary coils are short-circuited (commutation stage), has to be divided by $\hat{I}_p$. The other constants are obtained in a similar way.

Dissipation in the joints

In the sc. rectifier, electrical connections have to be made between the sc. wires or cables of the rectifier components. Dissipation in these joints contributes to the inefficiency of the rectifier. According to the experiments of ten Kate,[3] joint resistances vary from 1 to 10 nΩ for currents between 1 and 25 kA. We assume the resistance of the joints to be constant, though ten Kate has shown that this is only approximately true. The dissipation in the joints then equals $R_j I_1^2$, where R_j is the effective resistance of the rectifier circuit.

RESULTS AND DISCUSSION

The equations describing the scaled loss contributions W_i/W_1 or P_i/P_1 mentioned above are summarized in table 1. They can be obtained as follows. For rectifiers employing primary waveform A or B, the losses (which depend on the actual secondary current) have to be integrated over the time required to energize the magnet from zero to I_1 and scaled afterwards with the corresponding stored energy. For the loss make-up rectifier with waveform C, the dissipated energy per cycle (which is now constant) has to be divided by the energy delivered to the magnet during the same cycle.

The obtained results were used to calculate the loss and overall efficiency of two experimental rectifiers employing primary waveform A:

1) A 5 Hz magnetically controlled rectifier described elsewhere.[6]
2) The same rectifier, but equipped with thermally controlled switches which limit the rectifier operating frequency to about 2 Hz.

Both 1 and 2 are suitable for a secondary current of at least 1 kA. With regard to the average output power, which is proportional to $f\cdot\hat{I}_p^2$, rectifier 1 is superior. The power amounts to approximately 100 W at f = 1 Hz and $\hat{I}_p$ = 30 A. Relevant data for these rectifiers are collected in table 2.

Fig 3a and 3b show how the losses are distributed over the rectifier components in the case f = 1 Hz and $\hat{I}_p$ = 30 A. For these rectifiers the dissipation in the joints and the commutation loss appear to be negligible. Due to the application of wires with CuNi-matrix, the eddy current loss in the transformer is also very small. On the other hand, the hysteresis loss in the transformer, which is identical for both rectifiers, is quite substantial. Comparing the switch triggering loss of the rectifiers, a difference by a factor of 5 is observed. In rectifier 1, this switching loss dominates the other contributions. The poor value of C = 1.4 J per switching action is the result of filament hysteresis loss in the control coil and switch element. An improvement can be achieved by using technologically more advanced wires with smaller filament diameters (presently 8.6 μm). The reverse current loss is observed to be higher in the thermal switches than in the magnetic switches. They both have approximately the same full normal state resistance of 0.8 Ω, however, the voltage across the switches in the thermally controlled rectifier is insufficient to develop the full normal resistance.

The loss contributions (given in table 1) all depend in a different way on the frequency, the primary amplitude and the normalized secondary current. Consequently, the distribution of the losses over the rectifier

Table 1.
Scaled loss contributions
W_i/W_1 or P_i/P_1 in a rectifier
employing waveform A, B or C.

		WAVEFORM A CONSTANT FREQUENCY	WAVEFORM B VARIABLE FREQUENCY	WAVEFORM C [redacted] LOSS MAKE-UP RECTIFIER
1)	Reverse current loss when R_{sw} is constant.	$\frac{f\,L_{sec}}{\delta_p\,R_{sw}}\left[\frac{4(2-i)}{i}\right]$	$\frac{f_o\,L_{sec}}{\delta_{po}\,R_{sw}}\cdot\frac{8}{i}$	$\frac{f\,L_{sec}}{\delta_p\,R_{sw}}\left[\frac{4(1-i)}{i}\right]$
2)	Reverse current loss when I_{leak} is constant.	$\frac{I_{leak}}{k\,\hat{I}_p\sqrt{L_p/L_{sec}}}\cdot\frac{2}{i}$	$\frac{I_{leak}}{k\,\hat{I}_p\sqrt{L_p/L_{sec}}}\cdot\frac{2}{i}$	$\frac{I_{leak}}{k\,\hat{I}_p\sqrt{L_p/L_{sec}}}\cdot\frac{1}{i}$
3)	Loss due to resistive or imperfect inductive commutation.	$\alpha^2(1-k^2)\left[\frac{2}{i^2}\ln\left(\frac{1}{1-i}\right)-\frac{2+i}{i}\right]$	$\alpha^2(1-k^2)\left[\frac{2}{i^2}\ln\left(\frac{1}{1-i}\right)-\frac{2+i}{i}\right]$	$\alpha^2(1-k^2)\left[\frac{i}{1-i}\right]$
4)	Switch triggering loss.	$\frac{C}{k^2L_p\hat{I}_p^2}\left[\frac{1}{i^2}\ln\left(\frac{1}{1-i}\right)\right]$	$\frac{C}{k^2L_p\hat{I}_p^2}\left[\frac{1}{i^2}\ln\left(\frac{1}{1-i}\right)\right]$	$\frac{C}{k^2L_p\hat{I}_p^2}\left[\frac{1}{2i(1-i)}\right]$
5)	Hysteresis loss in the transformer.	$\frac{C_{hp}}{k^2L_p\hat{I}_p}\left[\frac{1}{i}+\frac{C_{hc}}{C_{hp}}\left(\frac{1}{i^2}\ln\left(\frac{1}{1-i}\right)-\frac{1}{i}\right)\right]$	$\frac{C_{hp}}{k^2L_p\hat{I}_p}\left[\frac{1}{i}+\frac{C_{hc}}{C_{hp}}\left(\frac{1}{i^2}\ln\left(\frac{1}{1-i}\right)-\frac{1}{i}\right)\right]$	$\frac{C_{hp}}{k^2L_p\hat{I}_p}\left[\frac{1}{2i}+\frac{C_{hc}}{C_{hp}}\left(\frac{1}{2(1-i)}\right)\right]$
6)	Eddy current loss in the transformer.	$\frac{C_{ep}\,f}{k^2L_p\delta_p}\left[\frac{2-i}{i}+\frac{\delta_pC_{ec}}{\delta_cC_{ep}}\left(\frac{2}{i^2}\ln\left(\frac{1}{1-i}\right)-\frac{2+i}{i}\right)\right]$	$\frac{C_{ep}\,f_o}{k^2L_p\delta_{po}}\left[\frac{2}{i}+\frac{\delta_{po}C_{ec}}{\delta_{co}C_{ep}}\left(\frac{2}{i^2}\ln\left(\frac{1}{1-i}\right)-\frac{2}{i}\right)\right]$	$\frac{C_{ep}\,f}{k^2L_p\delta_p}\left[\frac{1-i}{i}+\frac{\delta_pC_{ec}}{\delta_cC_{ep}}\left(\frac{1}{1-i}\right)\right]$
7)	Dissipation in the joints.	$\frac{\bar{R}_j}{f\,L_{sec}}\left[\frac{2}{i^2}\ln\left(\frac{1}{1-i}\right)-\frac{2+i}{i}\right]$	$\frac{2\bar{R}_j}{f_oL_{sec}}\left[\left(\frac{1+\delta_{co}-\delta_{po}}{i^2}\right)\left(\ln\left(\frac{1}{1-i}\right)-\frac{i^3}{3}-\frac{i^2}{2}-i\right)+\frac{i}{3}\right]$	$\frac{\bar{R}_j}{f\,L_{sec}}\left[\frac{i}{1-i}\right]$

Table 2. Data of rectifiers 1 and 2.

L_p	0.184	H	C_{hc}	$7.08\cdot10^{-3}$	J/A	Rectifier 1:		
L_{sec}	$0.232\cdot10^{-3}$	H	C_{hp}	$1.56\cdot10^{-2}$	J/A	R_{sw}	0.8	Ω
L_l	0.0135	H	C_{ec}	$6.4\cdot10^{-7}$	Js/A^2	C	1.4	J
k	0.975		C_{ep}	$1.9\cdot10^{-6}$	Js/A^2	Rectifier 2:		
δ_p	0.3		$\bar{R}_j$	$\approx 10^{-8}$	Ω	I_{leak}	4.0	A
δ_c	0.2		α	0.03		C	0.3	J

components is also a function of f, $\hat{I}_p$ and i. Fig. 4 shows the influence of the frequency on the efficiency of the magnetically controlled rectifier. With increasing frequency, the reverse current loss becomes more pronounced than the other contributions, the overall efficiency drops and the optimum in the efficiency shifts to a higher value of i. For the thermally switched rectifier (below 1.3 Hz) the efficiency is independent of the frequency because of the constant leakage current in the switches. The influence of the primary amplitude on the efficiency of rectifier 2 is shown in Fig. 5. With increasing amplitude, the switch triggering loss becomes less important and the efficiency increases. For rectifier 1 there is a similar dependence on $\hat{I}_p$.

The two rectifiers treated above serve as an illustration of the obtained results. It is not the intention of this paper to give an extensive comparison between thermally and magnetically controlled rectifiers, but merely to give a general method to determine the rectifier efficiency. The comparison between fast thermal or magnetic switches is still studied at the moment at the University of Twente.

LIST OF SYMBOLS

C Switch triggering loss per half-cycle.
C_{hc}, C_{hp} Constants describing the hysteresis loss in the transformer.
C_{ec}, C_{ep} Constants describing the eddy current loss in the transformer.

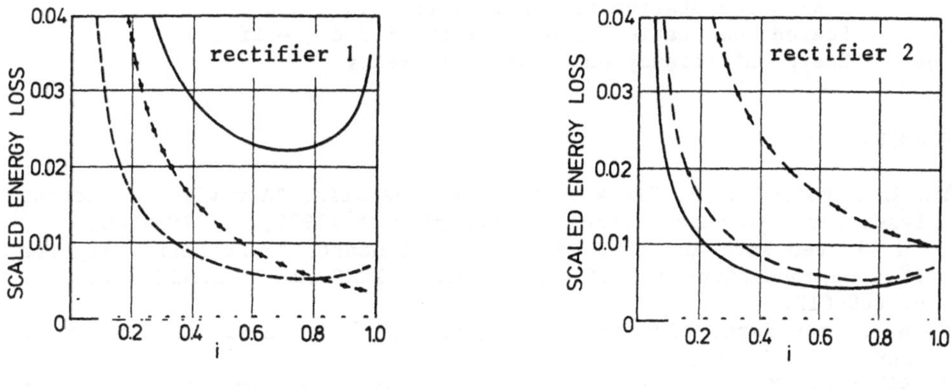

(a) (b)

Fig.3. Loss contributions in the magnetically (a) and thermally (b) controlled rectifier at f=1 Hz and $\hat{I}_p$ = 30 A.
—— switch triggering loss
-+-+- transformer eddy current loss
- + - reverse current loss
- - - transformer hysteresis loss

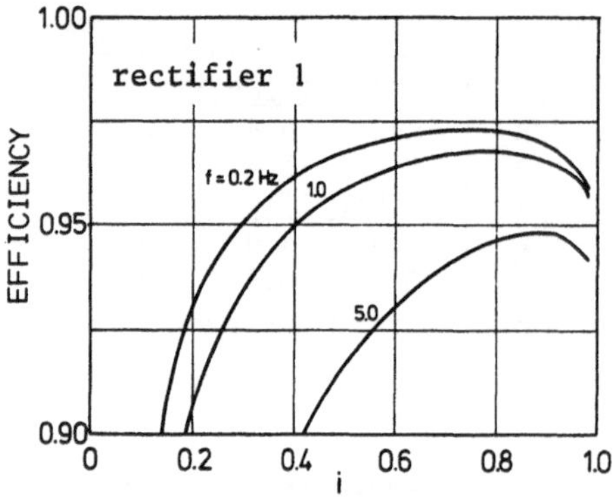

Fig.4. Influence of the frequency on the efficiency of rectifier 1 (at $\hat{I}_p$=30 A).

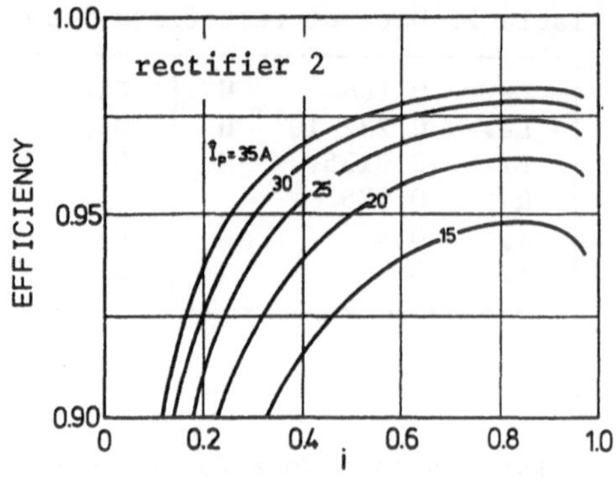

Fig.5. Influence of the primary amplitude on the efficiency of rectifier 2 (at f=1 Hz).

f	Frequency of rectifiers A and C (= $0.5/(\Delta t_r+\Delta t_c+\Delta t_t+\Delta t_p)$).
f_o	Initial frequency of rectifier B (= $0.5/(\Delta t_r+\Delta t_t+\tau_p)$).
i	Normalized secondary current (= I_l/I_{max}).
I_{max}	Maximum achievable secondary current (= $2k\,\hat{I}_p\sqrt{L_p/L_{sec}}$).
$\hat{I}_p$	Amplitude of the primary current.
I_l	Secondary current through the load.
I_{leak}	Reverse leakage current through the switch.
k	Coupling constant of the transformer.
L_p, L_{sec}	Primary and secondary inductance of the transformer.
R_{sw}, R_j	Switch resistance and effective joint resistance.
Δt_t, Δt_r	Triggering and recovery times of the switches (see Fig. 2).
Δt_p, Δt_c	Times required for pumping and commutating. (see Fig. 2).
τ_p, τ_c	Times for the full pump step and commutation step (see Fig. 2).
V_{sw}	Voltage across the switch
α	Constant describing the accuracy of the commutation step.
δ_p	Scaled pumping time for rectifiers A and C (= $2f\Delta t_p$).
δ_{po}	Scaled pumping time for rectifier B (= $2f_o\tau_p$).
δ_c	Scaled commutation time for rectifiers A and C (= $2f\Delta t_c$).
δ_{co}	Scaled commutation time for rectifier B (= $2f_o\tau_c$).
η_w, η_p	Energy efficiency and power efficiency.

REFERENCES

1. Yu. L. Buyanov, A. B. Fradkov and I.Yu. Shebalin, "A review of current leads for cryogenic devices", Cryogenics 15 (1975), p. 193-200.
2. H. H. J. ten Kate and L. J. M. van de Klundert, "Feasibility aspects of superconducting rectifiers", Proc. ICEC-10 Helsinki (1984), p. 809-812.
3. H. H. J. ten Kate, "Superconducting rectifiers", Thesis University of Twente, Enschede (1984).
4. J. Sikkenga, M. Groenenboom and H. H. J. ten Kate, "Operation modes of superconducting rectifiers", Proc. MT-9 Zürich (1985), p. 852-855.
5. C. Y. Pang, "Losses in type-II superconducting wire due to alternating and rotating fields", Thesis University of Cambridge (1980).
6. G. B. J. Mulder, H. J. G. Krooshoop, H. H. J. ten Kate and L. J. M. van de Klundert, "A high-power magnetically switched superconducting rectifier operating at 5 Hz", Proc. ASC-86 Baltimore (1986).

EFFECT OF EPOXY CRACKING ON STABILITY OF IMPREGNATED WINDINGS

RELATED TO THERMAL AND MECHANICAL PROPERTIES

S. Nishijima, T. Yamashita, K. Takahata, T. Okada
ISIR Osaka University, Ibaraki, Osaka, Japan

T.Fukutsuka, K.Matsumoto, M.Hamada
Kobe Steel LTD, Kobe, Hyogo, Japan

ABSTRACT

The anisotropy of thermal conductivity induced by epoxy cracking in impregnated windings have been studied in order to estimate the stability change. Epoxy cracking or debonding have been found to release the stored elastic energy as the form of heat which could cause the premature quench of the impregnated windings. Furthermore it would degrade not only the macroscopic mechanical properties but also thermal properties. The change of macroscopic mechanical properties was measured in terms of the load-displacement curves of the windings and the stress analysis was made by finite element method. The change of heat dissipation rate was measured using the heater and thermo-couples which were installed in the windings. The temperature distribution was also measured. The stability of the windings are discussed based on mechanical and thermal properties and the effect of induced epoxy cracks will be studied.

INTRODUCTION

The impregnation of magnet winding with epoxy resin has been used to make a high performance superconducting coil, such as a large forced flow cooled coil and a coil with a complicated shape. The purpose of the epoxy impregnation is to prevent the wire motion, to enhance the rigidity of the winding, and to assemble the windings in one body.[1] Nevertheless, there are some cases in which the impregnation does not work efficiently due to the epoxy cracking, the debonding or the slipping at wire-epoxy interface.[2] When such mechanical fracture occurs, the stored elastic energy is released as heat in the coil.[3] The mechanical fracture causes the decrease of rigidity and hence the increase of deformation. The thermal macroscopic conductivity is also degraded. Training behavior and/or degradation may be brought by these processes.[4-6]

To solve this problem, it is required to understand the mechanical behavior accurately considering the compositeness of impregnated coil and to estimate the change of mechanical behavior induced by the fracture. It is also necessary to clarify the mechanism of degradation of coil performance due to the mechanical fractures.

Actual impregnated windings have complicated structural composition composed with conductor, insulator, impregnant, structural material and others. It is, therefore, difficult to estimate the mechanical behaviors and superconducting properties of the windings precisely in the magnet. In this work, the basic behaviors are examined using the a model simulating the impregnated coil. Changing the amount of mechanical failure by loading, the mechanical and thermal properties of the model were investigated. The effects of the cracks on stability of the impregnated coil are also discussed.

EXPERIMENTALS

The NbTi wire-epoxy composite model beam examined in this work is illustrated in Fig.1. It was vacuum impregnated and cured at 373K for two hours. Epoxy resin was Epikote 828 with Jephermine D-230 hardener with a ratio 10:3. The NbTi wire used here is 0.5mm in diameter and copper-super ratio of 0.13. It contains 61 filaments of 0.042mm in diameter twisted at 10mm. The critical currents are 150A and 100A in external field of 5T and 7T, respectively.

The composite model beam had a point heater(Manganin wire: 3.74Ω at liquid helium temperature) in the center and six thermocouples (AuFe-chromel) are located around the heater. The leads were lined along the wires to prevent the heat diffusion through them.

The force was applied to the model beam by the three points bending test in liquid helium to introduce mechanical fractures such as cracks or debondings. The heat generation induced by crack initiation, the change of thermal and mechanical properties of the model beam were examined. Figure 2 presents the schematic diagram of the experimental apparatus.

EXPERIMENTAL RESULT AND DISCUSSION

Load-displacement curves and thermocouple outputs corresponded to the displacement were presented in Fig.3. In this experiment, the load was applied to the model beam up to 932kg (maximum flexural stress: σ_{max}=50.84kg/mm^2, maximum shear stress: τ_{max}=6.35kg/mm^2) and removed. This process is called "EX1" in this work. The load was applied again up to the same level as EX1 of 932kg and removed (named "EX2"). The load drops were observed on load-displacement curve of EX1. They must be caused by the introduction of mechanical fractures. The displacement did not return to zero even when the load was removed. This might be the another evidence of crack introduction. On the second loading i.e. in EX2, the load-displacement curve traced on that of unloading process in EX1 at first and came to deviate from a certain load level. In the case of EX2, no introduction of large fracture might occur because no sudden load drop was observed. The spikes of temperature rise were observed with the load drops on the load-displacement curve in EX1. It was found that these (temperature rises) are due to a heat generation induced by mechanical fracture in the model beam. No spike was detected in EX2.

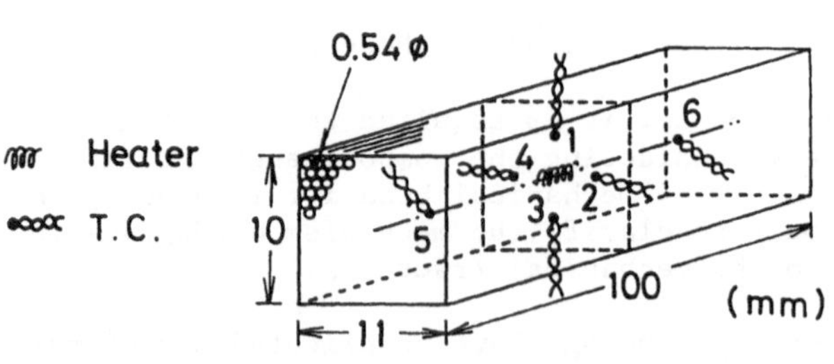

Figure 1. NbTi wire-epoxy composite model beam

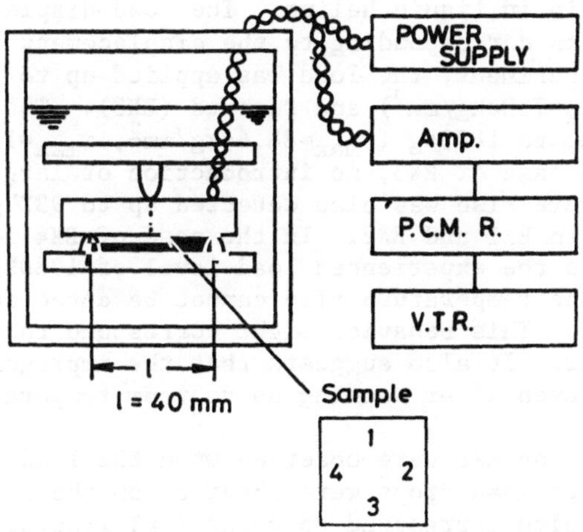

Figure 2. Schematic diagram of the experimental apparatus

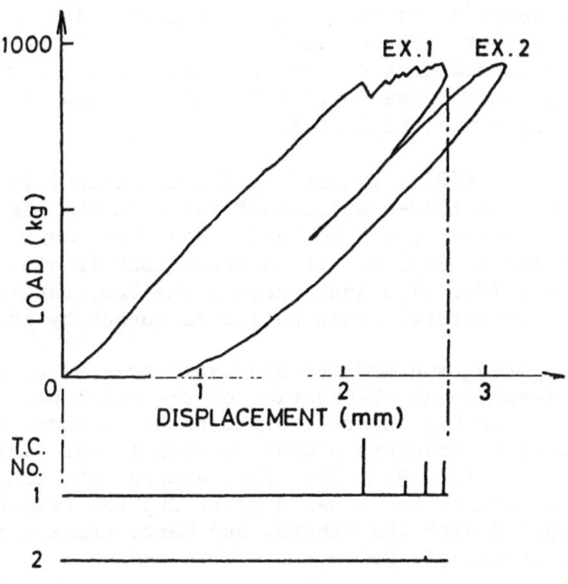

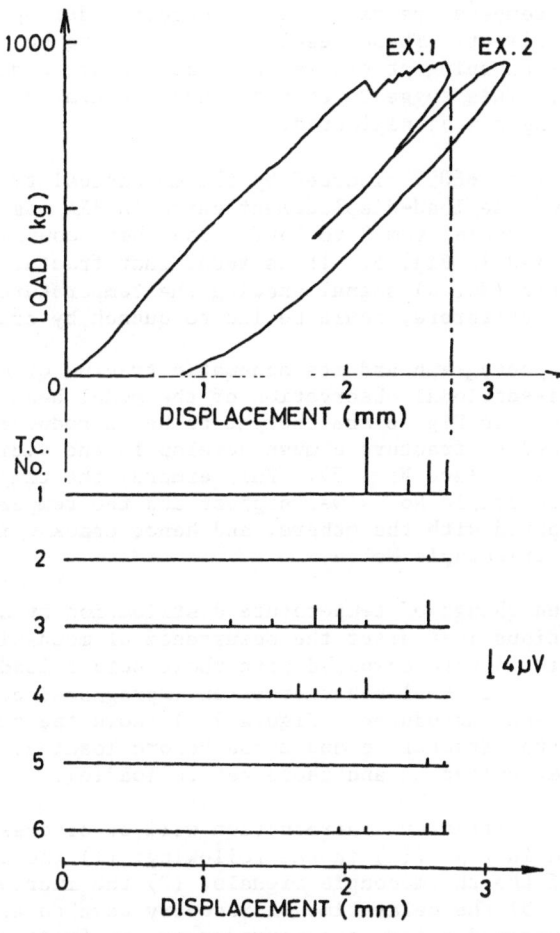

Figure 3. Load-displacement curves and thermocouple outputs

This model beam once was warmed up to room temperature after EX1 and EX2 and was loaded again in liquid helium. The load-displacement curves and thermocouple outputs corresponding to the displacement are presented in Fig. 4. In this experiment, the load was applied up to 1080kg (σ_{max}=58.91kg/mm^2, τ_{max}=7.36kg/mm^2) and removed (EX3). The load was subsequently applied up to 1625kg (σ_{max}=88.64kg/mm^2, τ_{max}=11.08kg/mm^2) and removed (EX4). In the case of EX3, no introduction of large fracture appeared. No temperature rise was also detected up to 932kg to which load had been applied in EX1 and EX2. In the case of EX4 no temperature rise was detected up to the experienced load level of 1080kg in EX3. The results suggest that the temperature rise cannot be detected within the experienced load level. This behavior might correspond to the training of the impregnated coil. It also suggests that the impregnated coil would remember the training even after warming up to room temperature.

In EX4, a number of spikes were observed when the load exceeded 1080kg, and three sudden load drops were observed on the load-displacement curve. These drops, which correspond to mechanical fracture in the model, were named a, b, and c, respectively. It is shown that these fractures a, b, c correspond to the spikes of the temperature rise. The results indicate that a number of cracks and debondings occur when the applied load exceeds the maximum experienced level.

The spike of the temperature rise was not detected during the unloading process in every experiment. These results show that the heat generation induced by frictional slipping at cracks and debonded areas may be too small to be detected. This suggests that the heat generation induced by the frictional slipping can be neglected.

The temperature rise (660μV) induced by the mechanical fracture c detected at the end of the load-displacement curve in EX4 was much larger than the others, which range from 2 to 15μV. The thermocouple outputs at fracture c are magnified in Fig. 5. It is found that fracture c caused the thermocouple (No. 4) signal showing the temperature rise of 46K. The impregnated coil, therefore, could be led to quench by crackings.

Figure 6 shows a photograph and its schematic tracing of cracks obtained in the cross-sectional observation of the model beam after the experiment. The cracks in Fig. 6 are thought to be introduced by fracture c because cracks caused by fracture c must develop in the vicinity of thermocouples No. 1, 2, 4 (see Fig. 5). Furthermore, the temperature rise detected by thermocouple No. 4 was highest and the temperature decreased slowly compared with the others, and hence crack c is initiated in the vicinity of thermocouple No. 4.

Figure 7 shows the change of temperature distribution by a heater. Temperature distributions just after the occurrence of mechanical fractures such as a, b, c in Fig. 4 were compared with those before loading. This experiment simulates the thermal behavior of an impregnated coil when mechanical fractures are introduced. Figure 7(a) shows the temperature distributions just after fracture c and those before loading. Figure 7(b) also shows those after unloading and those before loading.

First, the effect of fracture introduction will be discussed. Differences are found in Fig. 7(a) in the following: (1) the delay time of the initial rising of the thermocouple signals, (2) the increase of the maximum temperature, (3) the delay time of recovery down to 4.2K. These phenomena were not observed before the introduction of fracture c, that is, the temperature rise started just after the drive of the heater pulses, and was confirmed to be caused by fracture c. They may originate

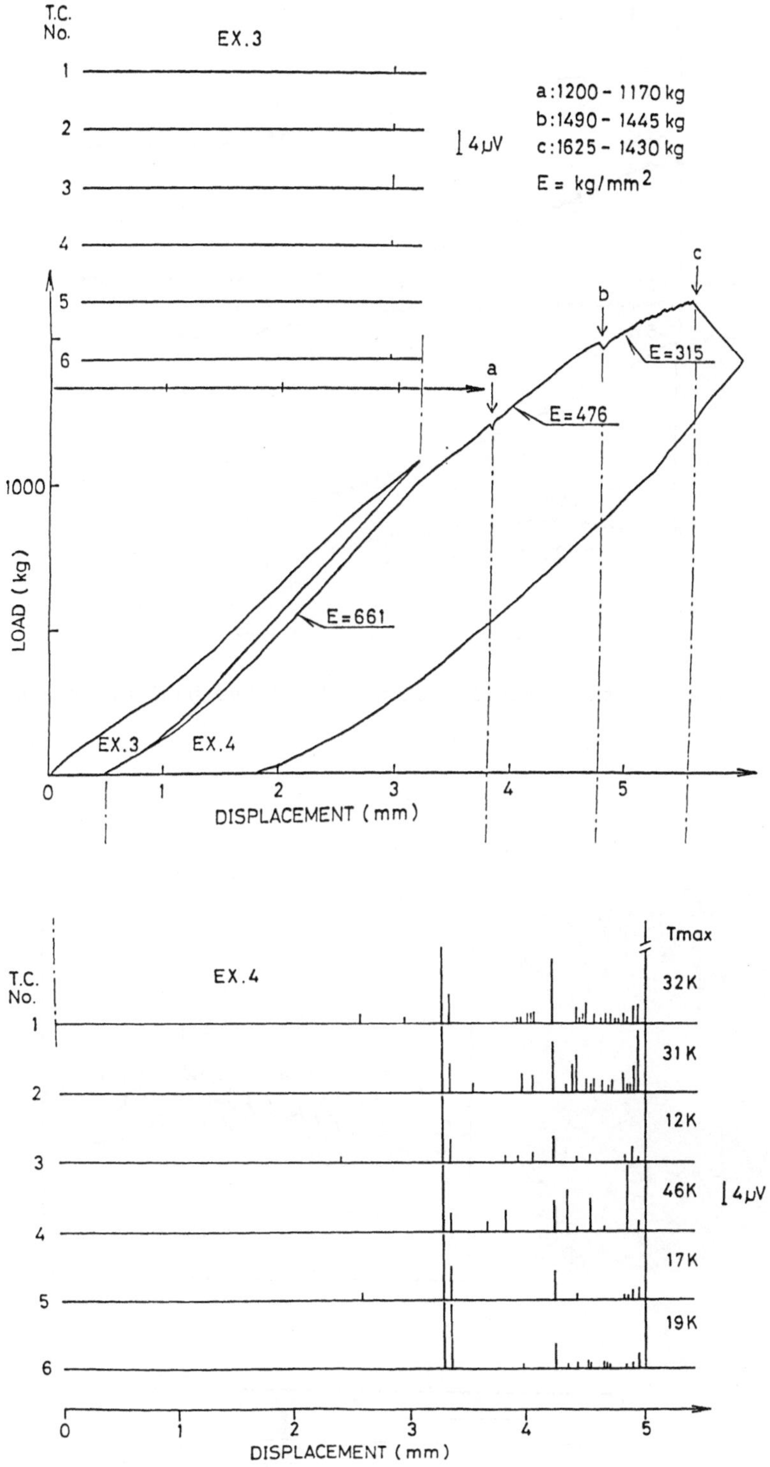

Figure 4. Load-displacement curves and thermocouple outputs corresponding to the displacement in EX3 and EX4.

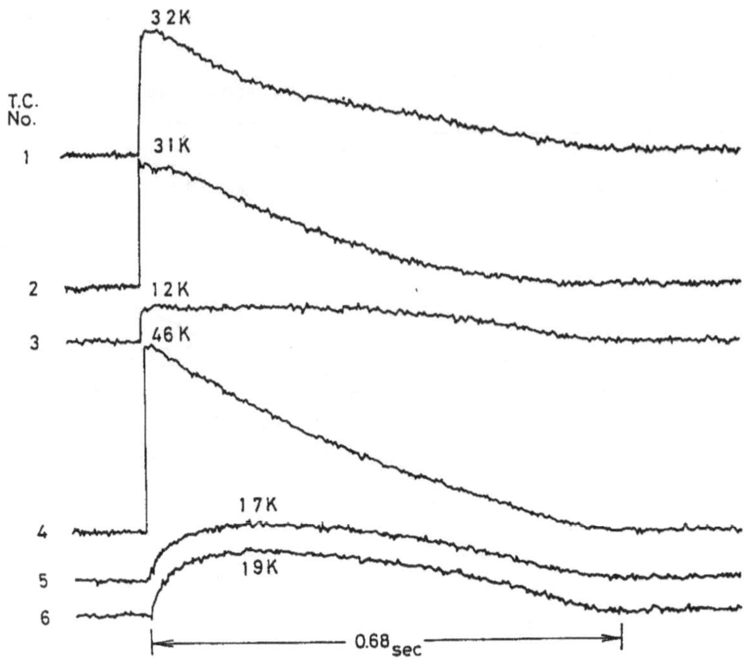

Figure 5. Magnified thermocouple outputs due to mechanical fracture c

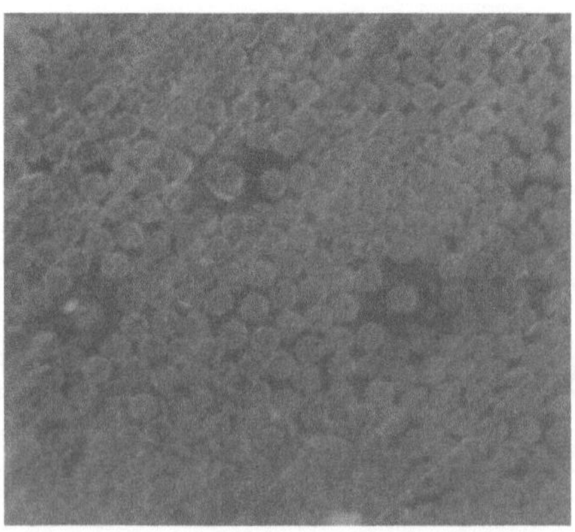

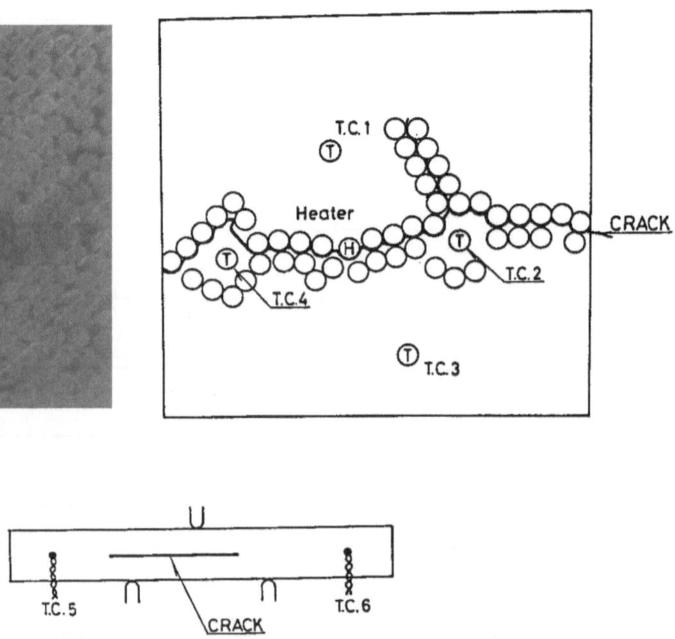

Figure 6. Schematic figure of crack obtained by the sectional observation of model beam after the experiment

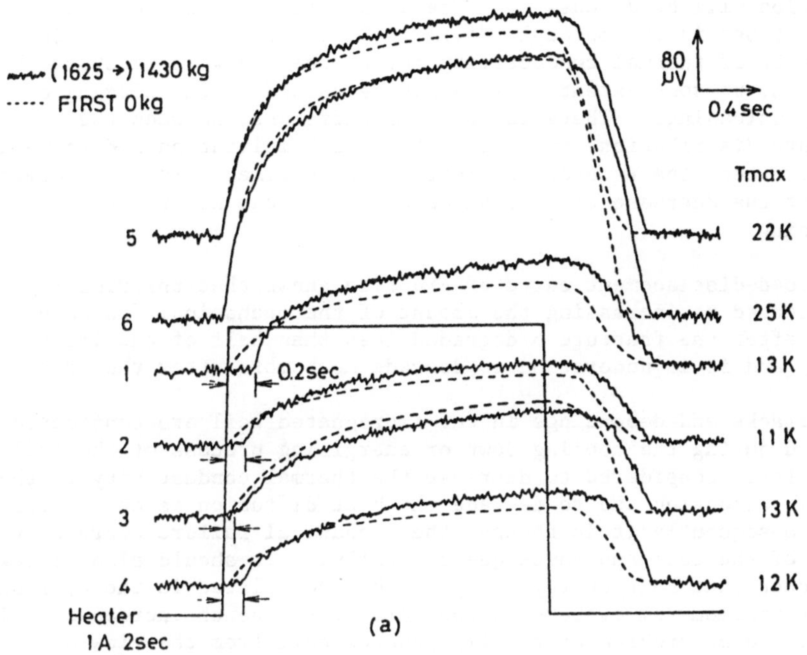

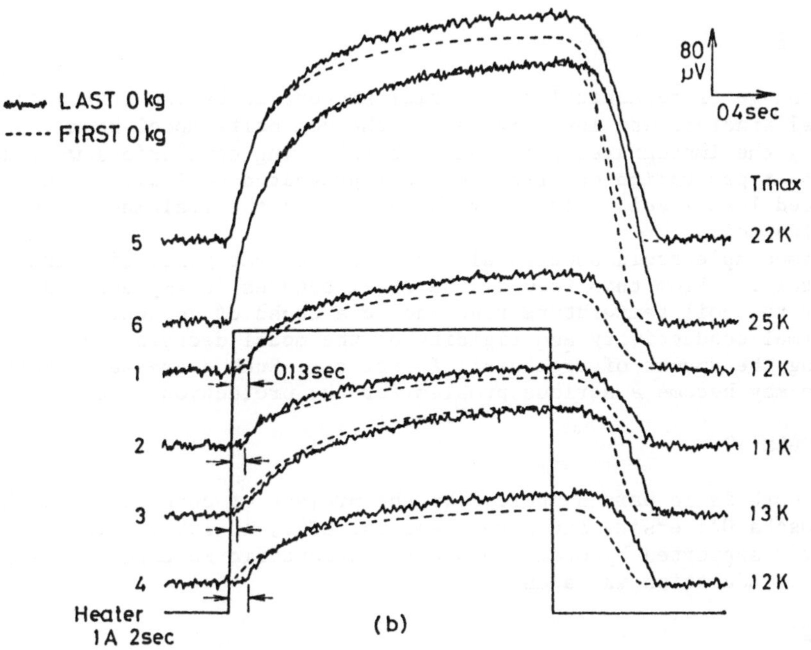

Figure 7. Induced temperature rise caused by heater. Temperature distributions obtained under loaded (a) and unloaded condition (b) are compared with those before loading.

from the decrease of the thermal conductivity due to the cracks in the vertical direction to the wire axis.

Secondly, the effect of unloading after mechanical fracture introduction will be discussed. The delay time was recovered to some extent (0.07sec in thermocouple No.1). It is, therefore, considered that the decrease of thermal conductivity due to the crack is recovered by the load removal to some extent. This may be caused by the closing of the cracks by unloading. There is not much difference between the temperature distributions just after the crack initiation and those after unloading except the recovery of initial delay time. It is, therefore, found that the decrease of thermal conductivity can not recover completely by unloading.

The load-displacement curve of EX4 also shows that the rigidity of the beam decreased as increasing the amount of the mechanical fracture. The rigidity after the fracture b degraded less than half of the initial rigidity, and it is understood to degrade much more after the fracture c.

The cracks and debondings in the impregnated coil are considered to be introduced during the cooling down or energizing process of the coil. It is, therefore, considered to decrease the thermal conductivity in the transverse direction and to degrade the heat diffusion as energizing the coil. Consequently it is thought the mechanical failure decrease the regidity of the coil and cause quench easily. It should also be taken into account that heat energy is apt to be accumulated in the coil and hence the maximum temperature in the coil after quench increases. It might come to be problem of the impregnated coil from the view point of coil protection.

CONCLUSIONS

The change of mechanical and thermal properties by the introduction of mechanical fracture was investigated on the composite model beam simulating the impregnated coil, and the following conclusions were drawn.

(1) No heat generation occurred in the impregnated coil within the experienced load level. This may correspond to the training behavior of impregnated coil.

(2) Thermocouple could successfully detect the heat generation induced by the fracture. From this result, the crack occures in impregnated coil may cause the coil temperature rise and be a cause of quench.

(3) Thermal conductivity and rigidity of the model decreased as increasing the amount of mechanical fracture. Such decrease of heat diffusion may become a serious problem for the protection of coil.

ACKNOWLEDGEMENT

This work is in part supported by the project of cooperative work between Osaka University and Kobe Steel Co. LTD., in 1987. This work is also partly supported by Grant in Aid for Scientific Research No.61050042, Ministry of Education in Japan.

REFERENCES

1. M. N. Wilson, Some Basic Problems in Superconducting Magnet Design, IEEE Trans. on Magnetics, MAG-17:1815(1981)
2. O. Tsukamoto and Y. Iwasa, Epoxy Crackings in the Epoxy-Impregnated Superconducting Winding, IEEE Trans. on Magnetics, MAG-21:377(1985)
3. P. F. Smith, B. Colyer, A Solution to the 'Training' Problem in Superconducting Magnets, Cryogenics 15:201(1975)

4. Y. Yasuda and Y. Iwasa, Stress-Induced Epoxy Cracking and Energy Release at 4.2K in Epoxy-Coated Superconducting Wires, Cryogenics 24:423 (1984)
5. S. Nishijima, T. Okada, N. Yabuta, K. Matsumoto, M. Hamada and T. Horiuchi, New Proposal to Reduce Training for Potted Superconductive Coil in Relation to Composite Stress/Strain Effect, IEEE Trans. on Magnetics, MAG-17:2055(1981)
6. S. Nishijima, K. Shibata, T. Okada, K. Matsumoto, M. Hamada and T. Horiuchi, An Attempt to Reduce Training Using Filled Epoxy as an Impregnating Material, IEEE Trans. on Magnetics, MAG-19:216 (1983)

LOCAL TEMPERATURE RISE AFTER QUENCH DUE TO EPOXY CRACKING IN IMPREGNATED SUPERCONDUCTING WINDINGS

Shigehiro Nishijima, Kazuya Takahata, and Toichi Okada

ISIR Osaka University, Ibaraki, Osaka, Japan

ABSTRACT

Relationships between epoxy cracking and the temperature rise in a coil after a quench have been investigated in epoxy-impregnated single-layer superconducting coils. The acoustic emission (AE) method was used to locate the cracking areas. The temperature rise of the wire was observed with voltage taps in the wire. The AE sources were localized in a certain area and a temperature rise was observed in that area. The temperature rise became greater as the quench number increased because the temperature rise itself caused further debonding and additional cracking. Areas where the epoxy was deliberately debonded during impregnation also exhibited temperature rises.

INTRODUCTION

The epoxy impregnation technique has been widely used in superconducting magnets with high current density. The premature quenching caused by the wire motion can be prevented by this technique. The training behavior, however, persists in the impregnated magnets.[1,2] Premature quenches are brought about by releasing the previously stored energy in the form of epoxy cracking. The stored energy arises from the difference of thermal contraction between the wire and the epoxy. Consequently, even impregnated magnets must be designed taking into account the quench in order not to cause burn-out of the wire due to an excessive temperature rise.

Until now, the protection of magnets has been investigated using analysis of the normal zone propagation velocity and the current decay time.[3] It is, however, also necessary to consider that inhomogeneous heat diffusion will be enhanced when epoxy cracking occurs in the magnet. If the magnet has been trained through several premature quenches, one must pay attention to the epoxy cracking.

In this paper, the behavior of epoxy cracking is examined using AE measurements, and the effects of epoxy cracking on the temperature rise of the wire after quench are investigated using an epoxy-potted short superconducting wire and single-layer coil.

EXPERIMENTAL

In this work, experiments were done using two types of samples. In the first experiment, an epoxy-potted short sample (named "Sample A") was fabricated. The experimental setup is illustrated schematically in Fig. 1 (a). The superconducting wire used in this work is multifilamentary Nb-Ti-Zr-Ta composite with a copper matrix of 0.35 mm diameter. The normal zone was induced locally by a manganine heater wound around the wire under a transport current of 100 A. Five potential taps were attached to the wire to monitor the normal zone behavior and the temperature rise of the wire. Two PZT type piezoelectric sensors which have resonant frequency of 140 kHz were attached to the epoxy block with vacuum grease. The coincidence method was used in order to reduce the noise. The AE may be generated by the wire and flux motions in addition to epoxy cracking.[4,5] In this experiment, no external magnetic field was applied to the coil in order to detect only the AE induced by the epoxy cracking.

In other experiments, two epoxy-potted single-layered coils (called "Sample B" and "Sample C") were fabricated. The experimental setup is shown in Fig. 1 (b). The wire was wound on a bakelite bobbin of 30 mm diameter. The number of turns was 10. The coil was impregnated with epoxy resin. The thickness of the outside epoxy layer was approximately 3 mm. 70 quenches were induced in the same manner as in the first experiment. The transport current was also set at 100 A and cut off when the voltage difference between the two ends of the coil reached 10 V at each quench. Three potential taps were attached to the wire to examine the temperature of the wire as shown in Fig. 1 (b), and the voltage differences between taps 1 and 2 or 2 and 3 were measured. The heater was located between taps 1 and 2. The measured voltage differences were converted into the average temperature between the taps considering the temperature dependence of the electrical conductivity of copper. Four AE sensors were attached to the bobbin to locate the AE sources.

RESULTS AND DISCUSSION

Short wire

Figure 2 shows the change of voltage differences between the taps and the AE signals at the quench of Sample A. The voltage differences increased with increasing temperature of the wire. The AE signals were not detected just after the normal zone generation, but appeared after the temperature of the wire increased enough. The temperature was estimated to be more than 100 K at the time when the AE appeared. This AE feature indicates that AE signals are generated by epoxy cracking. The thermal stress is produced by the difference of thermal contraction between the wire and epoxy during the initial cooling-down process. The temperature rise of the wire may enhance the difference of thermal stress and trigger the epoxy cracking.[6,7] The data obtained, information, and calculation method established in the preliminary experiment were applied to the coil simulation experiments described below.

Potted coil (B)

Figure 3 shows the change of voltage difference between taps 1 and 2 and the AE signals at the first quench of Sample B. The voltage difference increased with increasing temperature of the wire. It fell to zero when the transport current was cut, presented as "breaker on" in the figure. The interval between the normal zone generation and the

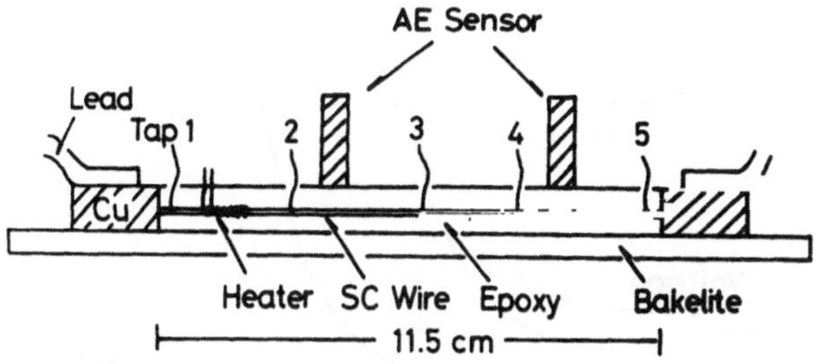

(a)

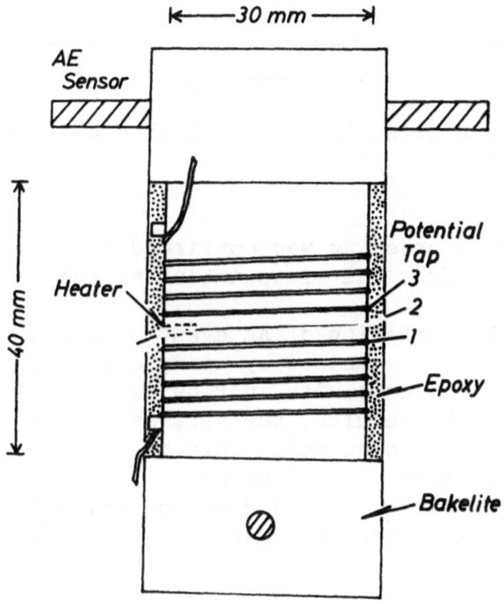

(b)

Fig. 1. Experimental setup. (a) Epoxy-potted superconducting wire (Sample A) and (b) coils (Samples B and C).

current cut-off was approximately 115 msec. The maximum voltage difference, $V_{m,n}$ in the nth quench, was measured and the maximum temperature of the wire, $T_{m,n}$, was calculated at 110 msec after the normal generation. The temperature of the wire $T_{m,n}$s was 184 K and 144 K between taps 1 and 2 and 2 and 3, respectively in Fig. 4. The AE signals appeared just after the normal zone generation. This feature differs from that in the quench of Sample A. The reason for this is apparently that an inhomogeneous thermal stress was produced during the cooling down and epoxy cracking was induced. The AE events continued to occur even after current cut-off. This AE may be caused by epoxy cracking due to variations in the thermal stress distribution and/or frictional motion of epoxy and wire in the epoxy cracking.[6]

Figure 4 shows the change of $T_{m,n}/T_{m,1}$ with quench number. The degree of temperature rise of the wire increased as the quench number increased. This can be explained in terms of the hindrance of heat diffusion caused by the epoxy cracking. The degree of temperature rise is enhanced as quench proceeds and hence has the possibility to burn out the

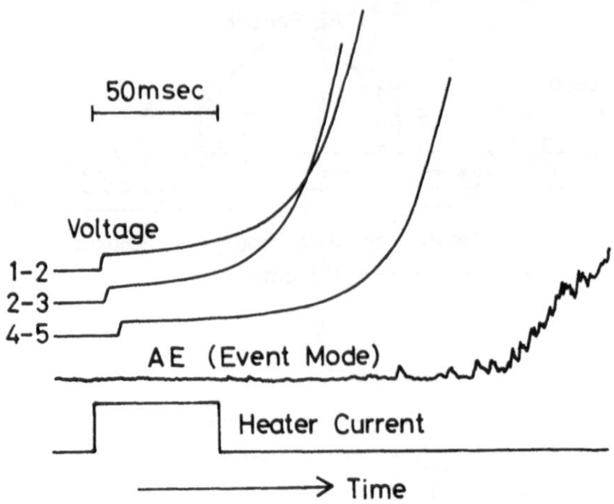

Fig. 2. Voltages between the taps and AE signals at the quench of Sample A.

wire in the coil. Epoxy cracking was confirmed to be an important factor in the impregnated coil from the viewpoint of magnet protection.

Figure 5 shows the variation of AE event counts after quench with quench number. If AE is caused by epoxy cracking, the AE event counts are expected to decrease with increasing quench number because of the Kaiser effect. The results, however, show that an unexpected increase of AE event counts occurred. This may be attributed to an increase of temperature rise. This increase of thermal stress may induce further epoxy cracking. The sudden increase of AE count may be induced by further epoxy cracking due to the increase of thermal

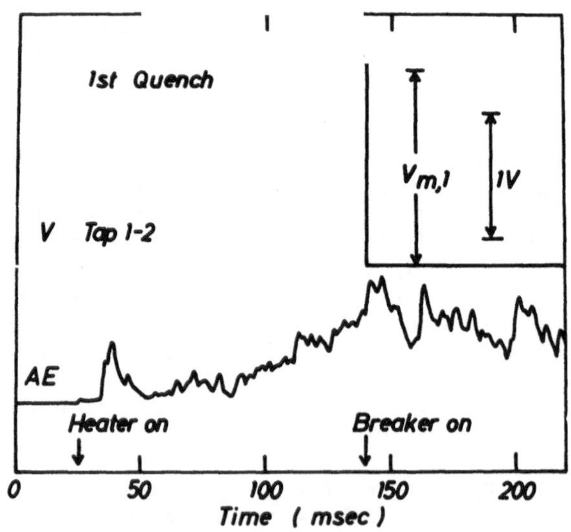

Fig. 3. Voltage between taps 1 and 2 and AE signals at the first quench of Sample B.

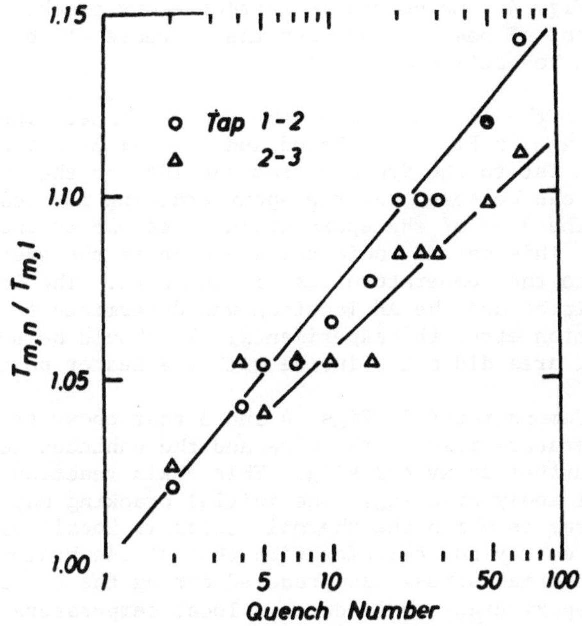

Fig. 4. Variation of $T_{m,n}/T_{m,1}$ with quench number (Sample B).

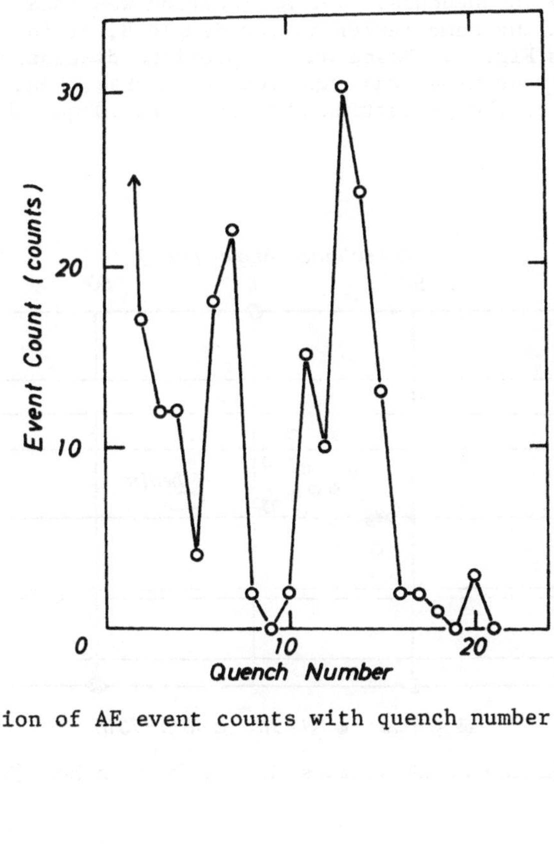

Fig. 5. Variation of AE event counts with quench number (Sample B).

stress. The sudden increase of AE count corresponds to the temperature rise as shown in Fig. 4. Consequently further epoxy cracking can be detected in terms of AE measurement, and the increase of temperature rise after quench could be estimated.

Figure 6 shows the AE location on the coil surface. The surface is expanded into a plane in Fig. 6. Closed and open circles indicate the AE locations from the 1st to the 9th and from the 10th to the 70th quenches, respectively. It can be seen that the epoxy cracking is localized in a certain region. The area of the epoxy cracking spread as the quench number increased. This spread indicates a change in the thermal stress distribution due to the temperature rise in the wire. The local epoxy cracking was confirmed and the AE location was determined to be accurate by visual observation after the experiments. It should be noted that the epoxy cracking area did not coincide with the heater position.

It has been demonstrated in Figs. 4 and 5 that epoxy cracking enhances the temperature rise of the wire and the enhanced temperature rise results in further epoxy cracking. This chain reaction is responsible for local epoxy cracking. The initial cracking may be produced in the area in which the thermal stress is localized. In the test coil, this area may not coincide with that of the heater because an inhomogeneous thermal stress was produced during the cooling process. The local epoxy cracking leads to a local temperature rise. One has to take these phenomena into account in designing the protection for impregnated superconducting magnets.

Potted coil (C)

This coil (Sample C) was used to examine the correlation between AE occurrence and burn-out in the coil. Figure 7 shows the location of AE sources during 70 quenches. The AE location was localized in a certain region in the same manner as for Sample B, at the coordinates (0 deg, 27 mm) in Fig. 7. Based on the previous discussion, the most dangerous area for burn-out was expected to be this area. The coil was excited without the protection circuit after 70 quenches. The wire

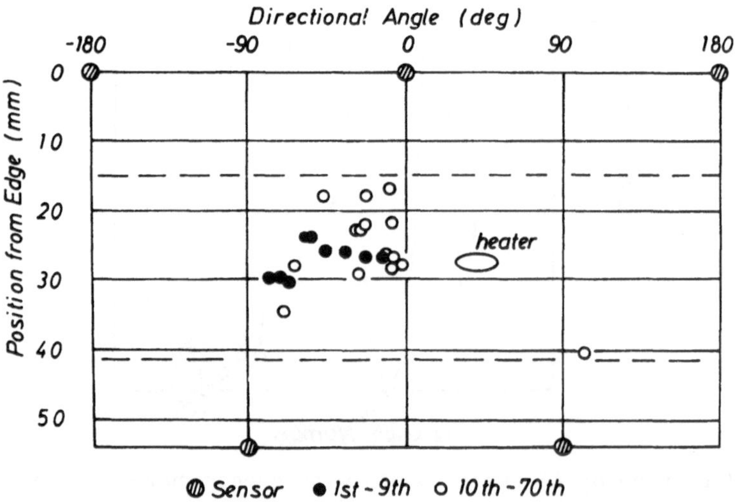

Fig. 6. Location of AE sources during 70 quenches (Sample B).

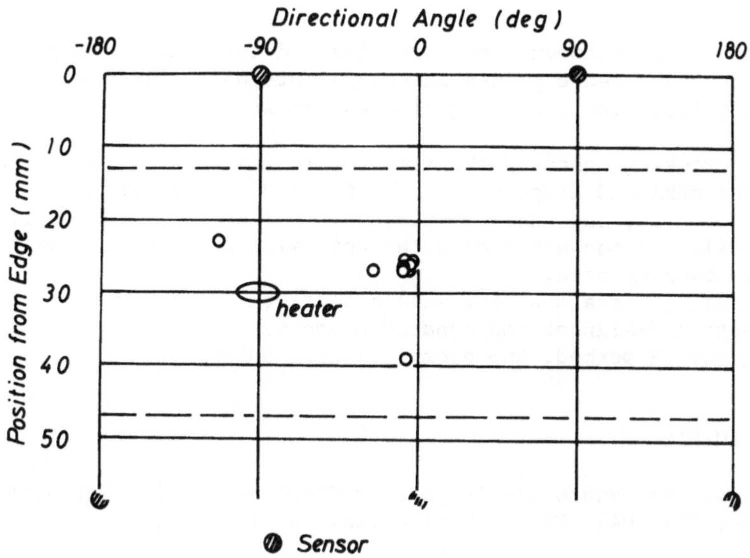

Fig. 7. Location of AE sources during 70 quenches (Sample C)

was then burned out. Figure 8 shows the location of AE sources and the burn-out area. The burn-out area coincided with just the region in which the AE sources were localized during 70 quenches as shown in Fig. 7. This result also indicates that local epoxy cracking corresponds to the local temperature rise and hence has the possibility of making the coil burn out. By means of AE measurements, the possible place for burn-out can be located in an impregnated coil.

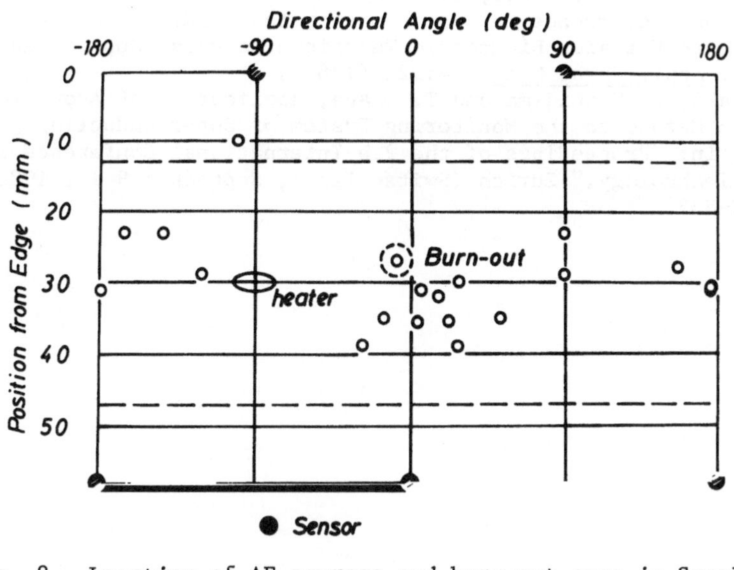

Fig. 8. Location of AE sources and burn-out area in Sample C at the last quench without a protection circuit.

CONCLUSIONS

Relationships between epoxy cracking and temperature rise of wire after quench have been examined on epoxy-impregnated single-layer superconducting coils. The following conclusions are drawn:

(1) Epoxy cracking enhanced the temperature rise of wire after quench, and the enhanced temperature rise induced further epoxy cracking.
(2) Epoxy cracking and enhanced temperature rise occurred locally in the coil. In the worst case the coil will burn out in the local epoxy cracking area.
(3) The local temperature rise should be taken into account in the protection design of impregnated magnets.
(4) Using the AE method, the epoxy cracking behavior can be examined.

ACKNOWLEDGEMENTS

This work was supported in part by Grant in Aid for Scientific Research No. 61050042, Ministry of Education of Japan.

REFERENCES

1. Y. Iwasa, Experimental and Theoretical Investigation of Mechanical Disturbances in Epoxy-Impregnated Superconducting Coils, Cryogenics 25:304 (1985).
2. S. Nishijima, K. Shibata, T. Okada, K. Matsumoto, M. Hamada and T. Horiuchi, An Attempt to Reduce Training Using Filled Epoxy as an Impregnating Material, IEEE Trans. on Magnetics, MAG-19:216 (1983).
3. M.N. Wilson, Some Basic Problems in Superconducting Magnet Design, IEEE Trans. on Magnetics, MAG-17:815 (1981).
4. O. Tsukamoto, J.F. Maguire, E.S. Bobrov and Y. Iwasa, Identification of Quench Origins in a Superconductor with Acoustic Emission and Voltage Measurements, Appl. Phys. Lett. 39:172 (1981).
5. H. Nomura, M.N.L. Sinclair and Y. Iwasa, Acoustic Emission in a Composite Copper NbTi Conductor, Cryogenics 20:283 (1980).
6. O. Tsukamoto and Y. Iwasa, Correlation of Acoustic Emission with Normal Zone Occurrence in Epoxy-Impregnated Windings: An Application of Acoustic Emission Diagnostic Technique to Pulse Superconducting Magnets, Appl. Phys. Lett. 44:922 (1984).
7. H. Iwasaki, S. Nishijima and T. Okada, Application of Acoustic Emission Method to the Monitoring System of Superconducting Magnet, in: "Proceedings of the 9th International Conference on Magnet Technology," Zurich (Switzerland), September 9-13, 1985, pp. 830-833.

HELIUM PRESSURE RISE OF SUPERCONDUCTING SUPER COLLIDER DIPOLE MAGNETS FOLLOWING A QUENCH

M.A. Hilal, S.D. Peck, and E.A. Ibrahim
General Dynamics Space Systems Division
San Diego, California

and

V.N. Karpenko and W.R. Hassenzahl
Lawrence Berkeley Laboratory
Berkeley, California

ABSTRACT

Heat transfer to the helium fluid within the magnet windings during and following a quench results in high-pressure gas propagating along the windings. A quasi-steady state approach[1] can be used for cryostable magnets with a relatively short vent line. For long adiabatic dipole magnets, the maximum pressure rise depends strongly on the magnet length. To determine the helium pressure distribution in the magnet, it is necessary to simultaneously solve the time dependent continuity, momentum, and energy equations for helium in the flow channels, the heat conduction equation in the windings, and the magnet current decay equation. General Dynamics has developed the code QMAG to do this. An early version of the code was applied to a full-length Superconducting Super Collider (SSC) dipole magnet for the extreme case of having helium flow only around the bore tube. Some of the results of that study are reported in this paper. We also report on the code modifications made to model a large cooling tube connected to the bore tube through the regularly spaced gaps in the collar and yoke laminations surrounding the windings. Furthermore, results of a study of a long dipole magnet model are reported in this paper to show the effect of the cooling tube internal flow on the bore tube pressure. The quench pressure results for SSC dipole magnets using the modified code will be published later.

INTRODUCTION

Following a magnet quench, helium is expelled from the windings and is forced to flow in the cooling channel along the magnet. It is important to determine the maximum pressure induced during magnet quench to ensure magnet structural integrity. The problem is complicated since it is difficult to accurately model the helium flow as it is expelled from the windings and to account for the mixing and fluid expansion processes. A model with some simplifying assumptions about flow from the windings was made so that the helium flow could be analyzed. The computer code QMAG was developed at General Dynamics to implement this model.[2] Helium flow in a single cooling channel adjacent to the windings is assumed in the QMAG code. This led to pessimistic predictions of quench pressures for SSC dipole magnets, since helium flow in the outer cooling tubes was not considered. We present in this paper the results of a modified model where helium flow to a large cooling tube is considered.

MATHEMATICAL MODEL

Figure 1 is a schematic of a dipole magnet quench pressure model. The model consists of the windings and two flow channels. To simplify the mathematical description of the model, the following assumptions are made:

1) Helium flow is one-dimensional.
2) The helium within the windings has the same temperature as the conductor.
3) No helium mixing is assumed within the windings.
4) Helium within the windings has the same longitudinal pressure distribution as the helium pressure in the bore tube channel.
5) No helium flows axially through the windings.
6) No helium is directly expelled to the outer cooling channel, since a stainless steel membrane covers the outside surface of the windings.
7) Velocity of the helium fluid expelled from the windings or flowing to the outer channel is normal to the longitudinal velocity.
8) Helium flows from the bore tube channel to the outside cooling channel through uniformly spaced channels.
9) The outside cooling channel has a large cross-section and helium flow to this channel does not cause any significant pressure rise.
10) Heat transfer to the outside cooling channel is negligible during the quench time duration.

The first seven assumptions were previously implemented[1] in developing the QMAG code and will not be discussed here. Assumption 8 is valid for long dipole magnets excluding the dipole ends. We plan to test the effect of Assumption 9 and the results will be published later. The outside tube in the SSC magnet is separated from the windings by the collar and yoke assemblies, which will effectively insulate the windings during the time periods of interest here; hence Assumption 10 is expected to be true.

The continuity equation for helium flow in the bore tube channel is

$$\frac{\partial \rho}{\partial t} = -\frac{\partial(\rho V)}{\partial x} + m_s + m_i$$

where m_s represents the rate of helium mass expelled from the windings and m_i is the internal helium flow rate from the bore tube to the outside cooling channel. Evaluation of both terms follows.

The momentum equation is

$$\frac{\partial(\rho V)}{\partial t} = -\frac{\partial(\rho V^2)}{\partial x} - \frac{\partial P}{\partial x} - \rho F$$

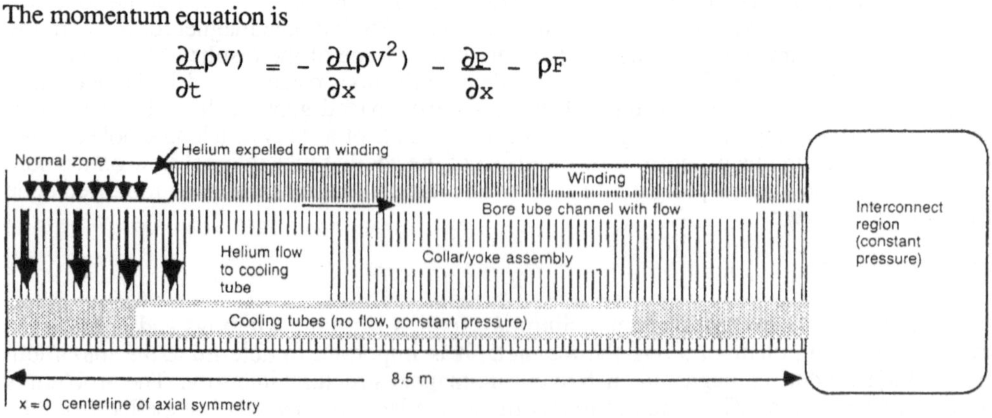

Figure 1. A conceptual sketch of the dipole magnet quench pressure model.

and the energy equation is

$$\frac{\partial}{\partial t}[\rho(u + V^2/2)] = - \frac{\partial}{\partial x}[\rho V(u + V^2/2)] - \frac{\partial(PV)}{\partial x} + q_t$$

with $$q_t = q_s + q_c + q_i,$$

where ρ is the density of the helium in the bore tube,
V is the velocity of the helium in the bore tube,
F is a function of the friction factor,
u is the internal energy of the helium in the bore tube,
q_t is the total power added to the helium in the channel,
q_s is the energy transferred from the winding due to helium expulsion,
q_c is the convection heat transfer,
and q_i is any imposed surface heating.

No modification to the heat conduction equation or the current decay equation as they were used in an earlier study[2] is necessary.

The model used to determine m_s and q_s is also discussed elsewhere[2], but we think it useful to include it in this paper also since it represents one of the basic assumptions of our approach. Figure 2 represents the winding volume per unit channel volume. Volume V consists of the windings and the helium fraction trapped within the windings. It is assumed that helium flows from V to the cooling channel without mixing with helium flowing from the other radial sections of the windings. Mixing takes place only in the cooling channel. It is difficult to predict the effect of this model on the results, but it is an essential assumption to simplify the mathematics. Based on this assumption m_s and q_s are given by

$$m_s = - \int_{\Delta V} r \frac{\partial \rho}{\partial t} dV$$

$$q_s = - \int_{\Delta V} r \frac{\partial(\rho u)}{\partial t} dV$$

where r is the fraction of the winding node occupied by helium.

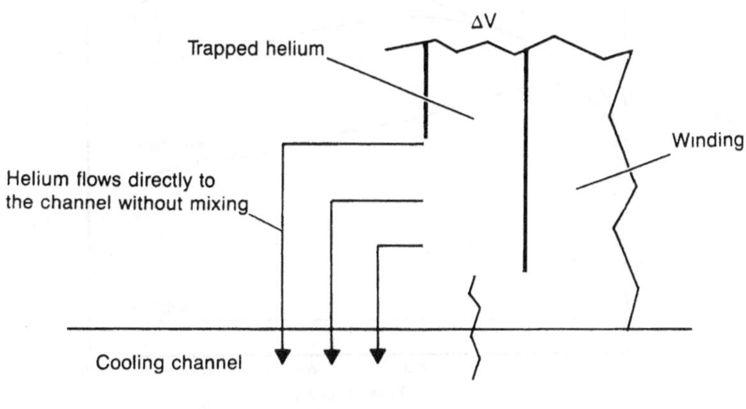

Figure 2. Model of the energy transfer from the windings. No mixing prior to flow in the channel is assumed.

The helium mass flow rate per unit length between the flow pipes, m_i, is calculated from

$$m_i = \beta\,(P - P_{out}),$$

where P is the helium pressure distribution along the inner flow pipe (bore tube), P_{out} is the helium pressure distribution along the outer flow pipe, and ß is given by

$$\beta = \frac{1}{2}\left[\frac{\rho A^2 d}{fL(P - P_{out})}\right]^{0.5}$$

where d is the hydraulic diameter of the connecting flow passage, f is the Fanning friction factor, L is the separation length between channels, and A is the flow area per unit length.

In general ß depends on the flow rate between cooling tubes. For laminar flow the above expression for ß reduces to $A\rho d^2/32\mu L$ (where μ is the viscosity), and the determination of m_r becomes a linear problem. The results reported hereafter were obtained assuming laminar flow between cooling tubes.

RESULTS

Figure 3 shows the pressure as a function of time at several stations along the bore tube of an SSC dipole following a quench. This figure is reproduced from the initial study of the SSC quench pressure rise[3]. QMAG at that time did not consider helium flow to an outer cooling tube, so this figure represents the case of zero ß. These results also are for a void fraction in the windings of only 1%, lower than the SSC design value of approximately 3%. The predicted maximum pressure for these conditions is 46 atmospheres. Given a void fraction of 3%, the maximum pressure would be higher than the structural capability of the bore tube. This result prompted the present modifications to QMAG so advantage could be taken of the venting benefit of the outer cooling tube.

The results in this paper are presented primarily to show the effect of ß on the maximum pressure. The values assumed for ß do not necessarily reflect a specific design, but rather were chosen to determine the significant range over which the exact magnitude of ß is important. The effect of ß on maximum pressure is shown in Figure 4 for void fractions of 2%

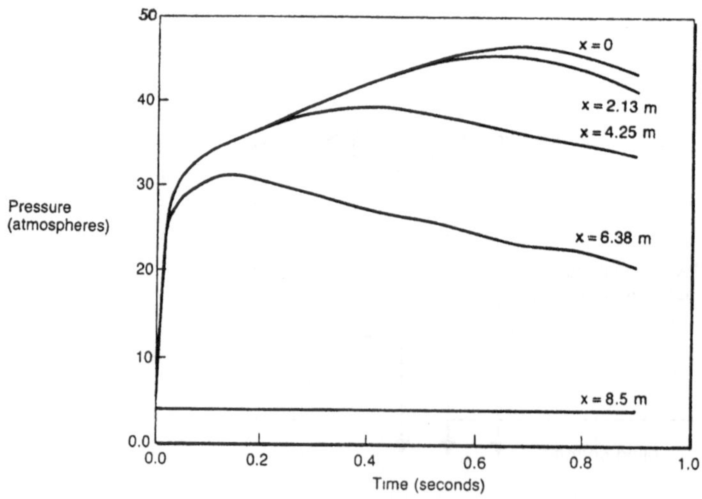

Figure 3. Pressure vs. time for a global quench of an SSC dipole with 1% void fraction.

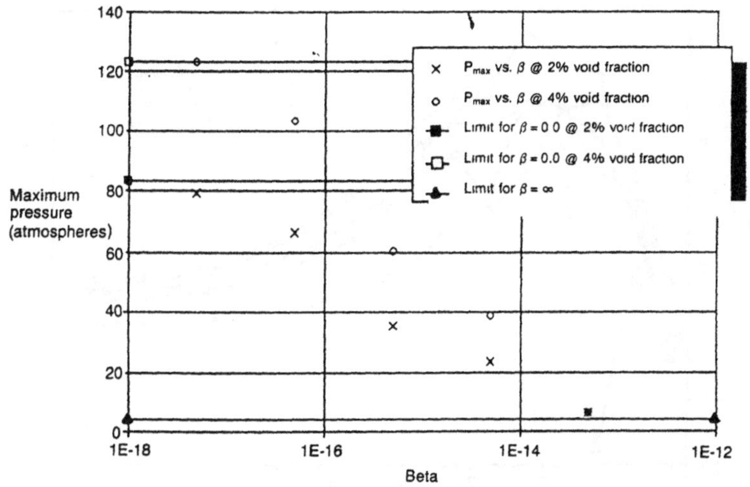

Figure 4. Maximum pressure is sensitive to β in the range 10^{-17} to 10^{-13} seconds.

and 4%. Upper limits on pressure calculated for zero ß occur at 123 atm for 4% void fraction and 82 atm for 2% void fraction. These values of pressure are reached for values of ß smaller than 10^{-17} sec. On the other hand, for infinite ß one would expect a negligible pressure drop between the bore tube and cooling tube. This condition occurs for ß larger than about 10^{-13}. In other words, for ß larger than 10^{-13}, the axial pressure distribution due to a quench is governed by the cooling tube geometry and not the bore tube geometry. The appropriate value of ß for the SSC dipoles has not yet been determined.

Figures 5 and 6 show the pressure and velocity distribution along the channel for a value of ß of 5×10^{-13} sec. Figure 7 shows the pressure rise versus time for the dipole magnet central node. The results shown in these figures are for a case where the quench starts as a two-meter-long normal zone located at the axial midpoint of the coil.

CONCLUSIONS

It is important to include the flow between flow pipes in determining dipole magnet quench pressure. Assuming all the helium ejected from the winding is trapped in the bore tube leads to overly pessimistic predictions of peak pressure. The plenum formed by the large cooling tube can be effective in absorbing the hot helium from the winding and minimizing the pressure rise during quench.

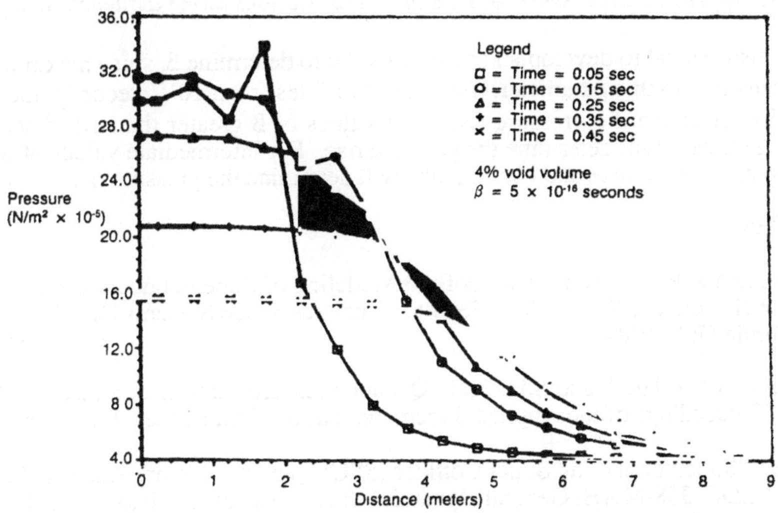

Figure 5. Approximate pressure profile during quench for a typical dipole.

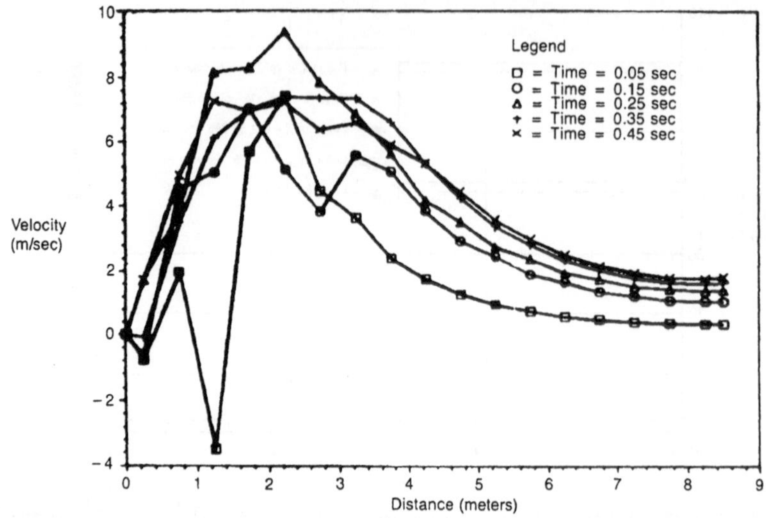

Figure 6. Approximate velocity profile during quench for a typical dipole.

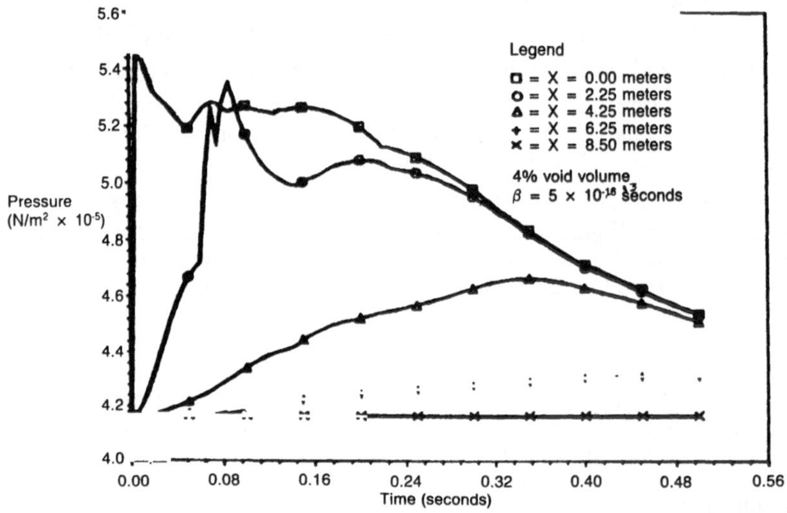

Figure 7. Approximate pressure vs. time at several stations along a quenching dipole.

It is also critical to develop an accurate model to determine ß, since maximum quench pressure is sensitive to this parameter. For values of ß less than 10^{-17} seconds, the bore tube geometry will determine the pressure rise. For values of ß greater than 10^{-13} seconds, the cooling tube geometry will determine the pressure rise. For intermediate values of ß, the flow passage between the bore tube and cooling tube will determine the pressure rise.

REFERENCES

1. M.A. Hilal and K.R. Johnson, "Thermofluid Modeling of Superconducting Magnets During a Quench," GDC-ERR-84-901, General Dynamics Space Systems Division, San Diego, California (July 1984).

2. E.A. Ibrahim, M.A. Hilal, and S.D. Peck, Quench pressure analysis of adiabatically stable magnets, in: "Proceedings of the Applied Superconductivity Conference," Baltimore, 1986.

3. M.A. Hilal, "Superconducting Super Collider (SSC) Quench Pressure Analysis Final Report," SSC-CDG-338-MAH, General Dynamics Space Systems Division, San Diego, California (2 June 1986).

MODELING OF PROXIMITY COUPLING BETWEEN SUPERCONDUCTING FILAMENTS IN SUPERCONDUCTING DIPOLE MAGNETS

M.A. Green

Lawrence Berkeley Laboratory
Berkeley, California

There is strong experimental evidence for superconducting current coupling between filaments in fine filament multifilamentary Nb-Ti which has been proposed for the SSC. The coupling between the fine filaments appears to be a Type II superconducting coupling. As a result, a.c. loss and magnetization can increase as one reduces the filament size in a fine filamentary conductor. This paper presents a computer model which fits experimental data for proximity superconducting coupling between filaments in a superconductor matrix. Using this model, one can predict the effect of proximity coupling on the field uniformity of a dipole magnet for the SSC.

BACKGROUND

The field generated by circulating currents flowing within the superconductor of a superconducting dipole was observed as early as 1970.[1] This magnetic field was observed to be rich in higher multipoles even though the coil configuration in which the circulating currents flowed did not generate higher multipoles when it was powered by a transport current. It was recognized from the very beginning that these higher multipoles would adversely affect the operation of a superconducting storage ring at injection if the field at injection was low enough.

The SSC requires very uniform magnetic fields in its dipole magnets at injection, better than 1 part in 10,000 (1 unit), over its useful aperture of 10 mm in radius.[2] Cost considerations dictate that the injection induction for the SSC dipoles be low (about 0.33 T when the full field of the SSC magnet is 6.6 T). At an injection induction of 0.33 T, the circulating currents in the superconducting filaments generate a field error of 17 to 20 units when the average superconductor filament diameter is 20 microns.[3] The magnitude of the field generated by circulating currents goes down as the filament diameter goes down. It was thought that reducing the filament diameter would go a long way toward reducing the field generated by the circulating currents.

Two effects have made the reduction of filament diameter as a means of reducing the circulating current field less attractive. The first is the "so-called" surface current which can be attributed to an extension of the lower critical field H_{c1}. The H_{c1} effect places a lower limit on the field generated by magnetization or circulating currents. The second effect which has been observed is coupling due to a form of Type II tunneling between filaments within the strands. As spacing between filaments is reduced, the magnetization observed <u>increases</u> as the filaments carry the circulating currents which are coupled across the filaments by superconducting currents. Unlike a.c. loss coupling between filaments, the proximity effect coupled currents do not decay with time.

This paper explains a computer modeling method which has been used to predict how proximity coupling will affect the magnetic field in an SSC dipole at injection. Computer results generated using the LBL-SCMAGØ4 computer code are presented.

BASIC THEORY, MAGNETIZATION AND DOUBLET STRENGTH

The field generated by circulating currents in a single filament of superconductor can be represented by the classical hydrodynamic doublet equation[4] which takes the following form in the complex plane:[5]

$$H''^{*}(Z) = \frac{\Gamma e^{i\alpha}}{2\pi i(Z - Zc)^2} \qquad (1)$$

where $H''^{*}(Z)$ is the complex conjugate of the field $H''(Z)$ at a point Z. This field is generated by a current doublet with a strength Γ and a doublet angle α at a location Zc. Γ is the product of the circulating current I and the average distance between the current d. The doublet angle α is the angle of the magnetic flux line through the conductor (mostly due to the magnetic field generated by the transport current) minus $\pi/2$. Both Γ and α are functions of the previous flux history of the superconducting filament.

Equation 1 can be used as it is or it can be expanded in a Taylor series about the origin Z = 0. This series will take the following general form (with no iron):

$$H''^{*}(Z) = \sum_{N=1}^{\infty} a''_N \, Z^{N-1} \qquad (2)$$

the general form of the multipole coefficient a''_N is as follows:

$$a''_N = \frac{-\Gamma e^{i\alpha}}{2\pi i} N \, Z_c^{-(N+1)} \qquad (3)$$

which applies for all multipoles N (N = 1 is dipole, N = 2 is quadrupole and so on). The radius of convergence is $|Z_c|$.

One can extend the theory to the case where there is an infinitely permeable circular iron shell which has its center at Z = 0. One can use the method of images to do this calculation. The method is described further in Reference 5. The power series expansion of the image currents doublets is included in the SCMAGØ4 code.

The doublet strength factor is proportional to the superconductor magnetization. On the filamentary level, the relationship between the doublet strength factor Γ and magnetization takes the following form:

$$\Gamma_f = \frac{\pi D_f^2 M_f}{4} \tag{4a}$$

where D_f is the filament diameter (m); M_f is the superconductor filament magnetization (Am^{-1}); and Γ_f is the product of the doublet current I and the doublet distance d. On the strand level, the relationship between magnetization and Γ is as follows:

$$\Gamma_s = \frac{\pi D_s^2 M_s}{4} \tag{4b}$$

where D_s is the strand diameter (m); M_s is the superconducting strand magnetization (Am^{-1}); and Γ_s is the strand doublet strength factor (Am).

The magnetization of the superconductor in an SSC dipole has several sources: 1) There is magnetization due to bulk current circulation within the filaments. These currents are responsible for a.c. losses at the filamentary level when the field in the filament changes. These a.c. losses per cycle are frequency independent. 2) There is magnetization due to H_{c1}.[6] This magnetization is maximum at H_{c1}, and it is zero when $H = 0$ or as H approaches H_{c2}. This magnetization generates no a.c. loss. 3) There is magnetization due to coupled eddy currents across the filaments. These currents are proportional to dB/dt, and they are inversely proportional to the square of the twist pitch of the strand.[7] These currents generate an a.c. loss per cycle which is frequency dependent. 4) There is magnetization due to superconducting tunneling between filaments. This magnetization is dB/dt independent, and it does generate a.c. losses per cycle which is frequency independent. 5) There is magnetization due to eddy current coupling due to coupling between strands of a cable. This coupled eddy current magnetization behaves like coupled eddy current magnetization between filaments. The SCMAGØ4 code calculates the magnetization and field which are generated by the first four items.

The first two item on the magnetization list occur at the filamentary level so that M_f in Equation 4a is the sum of these two items. For a fully penetrated superconductor with a constant J_c, the bulk current magnetization M_{f1} will take the following approximate form:

$$M_{f1} \simeq \frac{2}{3\pi} D_f J_c [1 - \delta] \tag{5}$$

where D_f is the filament diameter; J_c is the critical current density of the superconductor in filament (this should come from magnetization measurements rather than short sample measurements); and δ is the fraction of the filament current carrying capacity which carries transport current or coupled currents. The second filamentary term is the H_{c1} effect magnetization M_{f2} which has the following form in the SCMAGØ4 code:

$$M_{f2} \simeq H_{c1} - \frac{\ell N[(H - H_{c1}/2)\phi]}{\ell N[(H_{c2} - H_{c1}/2)\phi]} H_{c1} \qquad \text{for } H > H_{c1} \tag{6}$$

for $H \leq H_{c1}$ the magnetization $M_{f2} = H$ where H_{c2} is the upper critical field; H_{c1} is the lower critical field; and $\emptyset = \lambda/(2.07 \times 10^{-15})$ where λ is the penetration depth of the conductor.

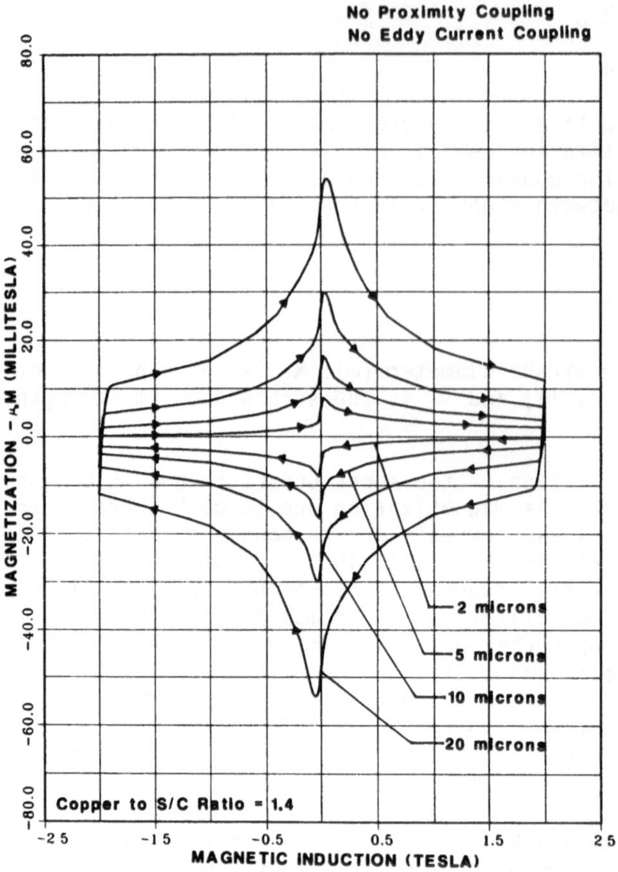

Fig. 1. Magnetization versus magnetic induction and filament diameter (no proximity coupling).

Figure 1 shows magnetization curves generated by the SCMAGØ4 code for a conductor with a copper to superconductor ratio of 1.4 and various filament sizes from 2 microns to 20 microns. There is no coupling at the strand level shown in Figure 1. One will note that the curves are skewed. This skewing is the same for all filament diameters. The skew is observed in magnetization loops which are generated when the field change goes through zero.[8,9] The skewed behavior can be attributed to H_{c1} currents and vortex currents.[6] This skew does not die out until the field on the filament reaches H_{c2}.

Items 3 and 4 on the list of contributing magnetizations occur at the strand level. To first order, these two magnetizations can be summed to get M_s in Equation 4b. The magnetization due to coupled eddy current takes the following form provided the conductor has some twist and the rate of flux change is moderate.[7]

$$M_{s1} \approx \frac{\mathring{B}}{\rho_e}\left(\frac{L}{2\pi}\right)^2 \tag{7}$$

where L is the twist pitch of the superconductor; $\mathring{B}$ is the rate of flux change; and ρ_e is the effective transverse resistivity of the matrix which is defined as follows:

$$\frac{1}{\rho_e} = \frac{1}{\rho_t} + \frac{2W}{D_s \rho_M} + \frac{D_s W}{2\rho_M}\left(\frac{2\pi}{L}\right)^2 \qquad (7a)$$

where ρ_t is the interfilament resistivity which is defined as:

$$\rho_t = \frac{6.56 \times 10^{-16}}{W} + \rho_M \qquad (7b)$$

where W is the interfilamentary spacing; ρ_M is the matrix material bulk resistivity including magnetoresistivity; and D_s is the strand diameter. The proximity coupling magnetization takes the following approximate form:[10]

$$M_{s2} = \frac{2}{3\pi}\,\frac{D_s J_c}{r + 1}\, e^{-\varepsilon} \qquad (8)$$

where D_s and J_c are previously defined; r is the copper to superconductor ratio; and ε is a function of W, H and ρ_t. The value of ε derived for use in the SCMAGØ4 code was fit to the Brookhaven

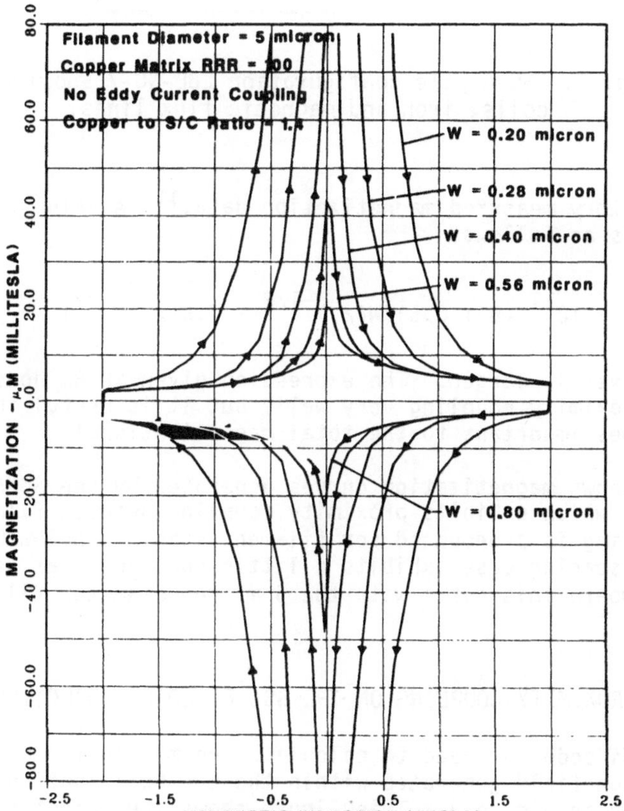

Fig. 2. Magnetization versus magnetic induction and filament spacing W with proximity coupling (copper matrix RRR = 100).

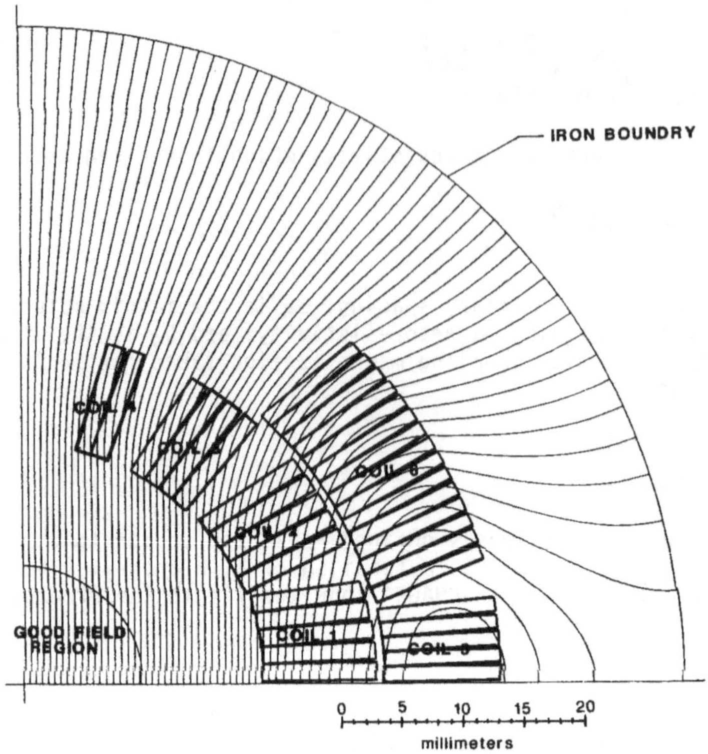

Fig. 3. SSC dipole configuration LBL-NC-7 showing coils, iron and magnetic flux lines.

National Laboratory measured magnetization data.[11] A value ε which fits the data is as follows:

$$\varepsilon = 7.18 \times 10^{11} (B + 0.5)(W(\rho_t)^{-1/2} - 7.8 \times 10^{-12}) \qquad (8a)$$

or zero, whichever is larger. The expression given in 8a does not predict the onset of proximity coupling very well, but it is useful in the region where M_{S2} becomes important to the total magnetization.[12]

Figure 2 shows magnetization curves generated by the SCMAGØ4 code for a conductor which exhibits proximity coupling between filaments. The proximity coupling is pronounced for filament spacings below 0.8 microns. The 0.8 micron spacing case exhibits a little coupling near zero induction. Compare this curve with the 5 micron diameter filament in Figure 1.

THE EFFECT OF PROXIMITY COUPLING ON THE SSC FIELD AT INJECTION

The SCMAGØ4 code was used to calculate the multipole components of the magnetization field generated within the LBL-NC 7 configuration of the SSC dipole shown in Figure 3. This figure shows the direction of the magnetic flux lines within the coil which determines the doublet angle α in various parts of the coil. Figure 3 shows that the flux lines change direction in coil 5 of the magnet and that the field is lowest in coils 5 and 6 of the magnet. Since proximity coupling is strongest at low fields, it is these outside coil blocks which have the largest effect on the proximity coupling magnetization seen in the good field region of the magnet.

Figure 4 shows the ratio of sextupole to dipole at a radius of 10 mm as a function of central induction and filament spacing. The multipole ratio is presented for the central field going down from 6.6 T to zero and back up again. In a superconductor without proximity coupling, most of the 5 micron filaments are fully penetrated by the flux change at a central field of 0.08 T going up. One can see in Figure 4 that when the filament spacing is 0.2 microns, flux penetration of the second flux change does not occur until the central induction change reaches about 0.5 T going up. The shape of the 0.2 micron spacing curve is very different when the field goes up as compared to when the field goes down. This difference reflects what is happening in the low field region of the magnet.

Figure 5 shows the sextupole component at SSC injection as a function of filament diameter D and the ratio of filament spacing W to diameter for a Residual Resistance Ratio (RRR) 100 copper matrix conductor. The dashed line indicates what the sextupole ratio would be with no proximity coupling. This curve does not approach zero even when the filament diameter approaches zero. This is due to the H_{c1} surface current effect.

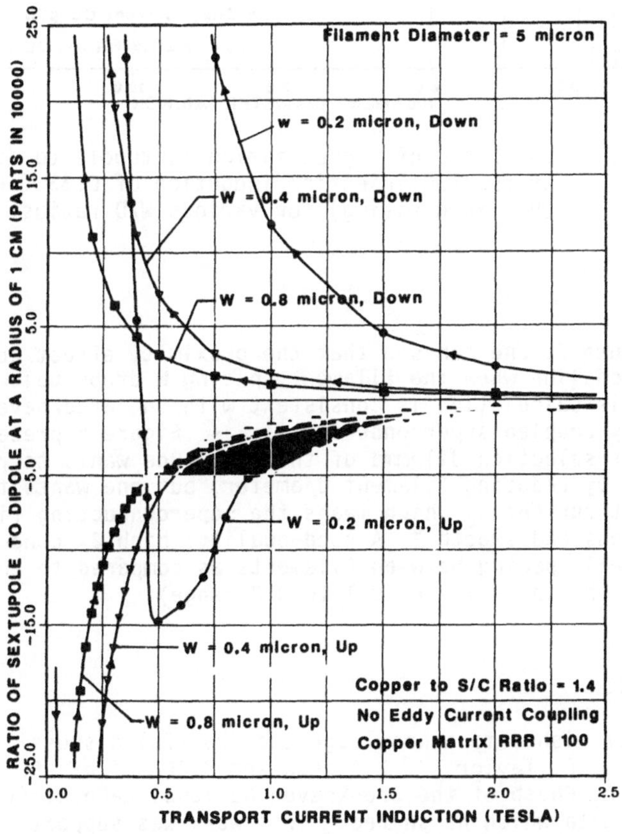

Fig. 4. The ratio of magnetization sextupole to dipole as a function of transport current induction, induction change direction and filament spacing W.

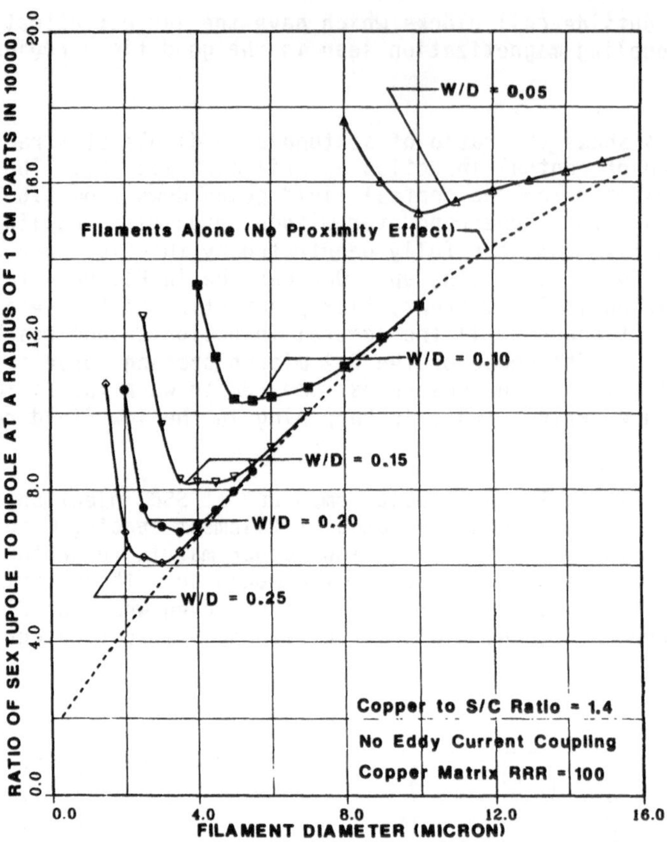

Fig. 5. The ratio of magnetization sextupole to dipole at the SSC injection induction of 0.33 T (with the field rising) for various W/D ratios.

From Figure 5, one can see that the proximity effect curves depart from the dashed line when the filament spacing W drops below 0.8 to 1 micron. This calculation is consistent with the Brookhaven measurements of a proximity coupled superconductor.[11, 12] Figure 5 presents the superconductor selection dilemma of the SSC. One wants to reduce magnetization by reducing filament diameter, but one wants to maintain a high-transport current J_c which means the superconducting filaments have to be continuous and smooth.[9] A good-quality, high J_c conductor has a relatively small spacing between filaments as compared to the filament diameter (W/D should be in the 0.1 to 0.2 range).

ACKNOWLEDGMENTS

The author wishes to acknowledge the fruitful discussions he has had with W. Gilbert, C. Taylor, J. Peterson and S. Caspi of LBL; and the author thanks A. Ghosh of the Brookhaven National Laboratory for the magnetization data which he shared. This work was supported by the Office of High Energy Physics of the United States Department of Energy under Contract Number DE-AC03-76SF00098.

REFERENCES

1. M. A. Green, IEEE Transactions on Nuclear Science NS-18 (1971), p.664.

2. Superconducting Super Collider Conceptual Design Report SSC-SR-2020 (1986).

3. M. A. Green, "Control of the SSC Fields Due to Superconductor Magnetization in the SSC Magnets", to be published in IEEE Transactions on Magnetics MAG-23 (2) (1987).

4. L. M. Milne-Thomson, Theoretical Hydrodynamics, McMillan Company, N.Y. (1960), p. 204.

5. M. A. Green, "Residual Fields in Superconducting Magnets", published in the Proceedings of the MT-4 Conference at Brookhaven National Laboratory (1972), p. 339.

6. W. J. Carr, IEEE Transactions on Magnetics MAG-21 (2), p. 335.

7. M. N. Wilson, Superconducting Magnets, Clarendon Oxford Press, Oxford, UK (1983), pp. 174-181.

8. W. J. Carr and C. R. Wagner, Advances in Cryogenic Engineering Vol. 30, Plenum Press (1984), p. 923.

9. A. K. Ghosh and W. B. Sampson, Advances in Cryogenic Engineering Vol. 32, Plenum Press (1986), p. 809.

10. E. W. Collings, "Stabilizer Design Considerations in Ultrafine Filamentary Cu/Ti-Nb Composites", Proceedings of the 6th Workshop on Niobium-Titanium Superconductors, University of Wisconsin, Madison, WI (1986).

11. A. K. Ghosh et al, "Anomalous Low Field Magnetization in Fine Filament Nb-Ti Superconductor", to be published in IEEE Transactions on Magnetics MAG-23 (2) (1987).

AC LOSS AND STABILITY OF A FORCE-COOLED SUPERCONDUCTING COIL WITH A BIAS PULSED SUPERCONDUCTING MAGNET

Norihito Ohuchi, Yasuhiro Makida, Junya Yamamoto and Yoshishige Murakami

Laboratory for Applied Superconductivity
Osaka University
Suita, Osaka, Japan

ABSTRACT

Stability of a force-cooled hollow superconductor coil for the AC loss was studied from a point of view of a cooling characteristic of supercritical helium (SHE). The coil was operated in the fluctuating magnetic field (maximum $\dot{B}$=3.75 T/s) generated by a bias pulsed superconducting magnet. In this experimental system, the maximum AC loss in the hollow conductor coil during operation reached 1.82 W/m and the normal transition was observed due to the accumulation of the AC loss induced by the repeated operation of the bias pulsed magnet. The results from our analysis are as follows. (1) The required energy for the normal transition was almost decided by the heat capacity of SHE from the initial temperature to the critical temperature of the coil. (2) The operating mode of a force-cooled pulsed magnet is predominantly decided by the transit time and the heat capacity of SHE.

INTRODUCTION

The force-cooling method has been used in several superconducting magnets[1,2], and there exists the magnet which has operated for long duration[1]. It has been considered that the force-cooled magnet has the superiority to the bath-cooled magnet for both allowable magnetic force and electrical insulation. Then, the force-cooled magnet is expected to be applied to a pulsed magnet. In this application, the AC loss generates in the whole coil and the one changes the hydrodynamic condition of supercritical helium (SHE). The change of the flow condition induced by the thermal load makes the stability theory of the force-cooled magnet complicated. Several works about stability of the force-cooled superconductor have already been reported[3-7]. We have studied on a hollow superconductor coil (SHETEM2a) and reported its safety operating region when the constant thermal load by a heater wire was uniformly applied on an entire length[6]. Especially, the heat capacity of SHE from the initial condition to the critical temperature of the superconductor played the important role, with deciding the safety region against the thermal disturbance. As an advanced step, we installed a hollow superconductor coil (SHETEM2b) inside a pulsed superconducting magnet and analyzed the influence of the AC loss induced in the conductor on the stability of force-cooled magnet with a practical model.

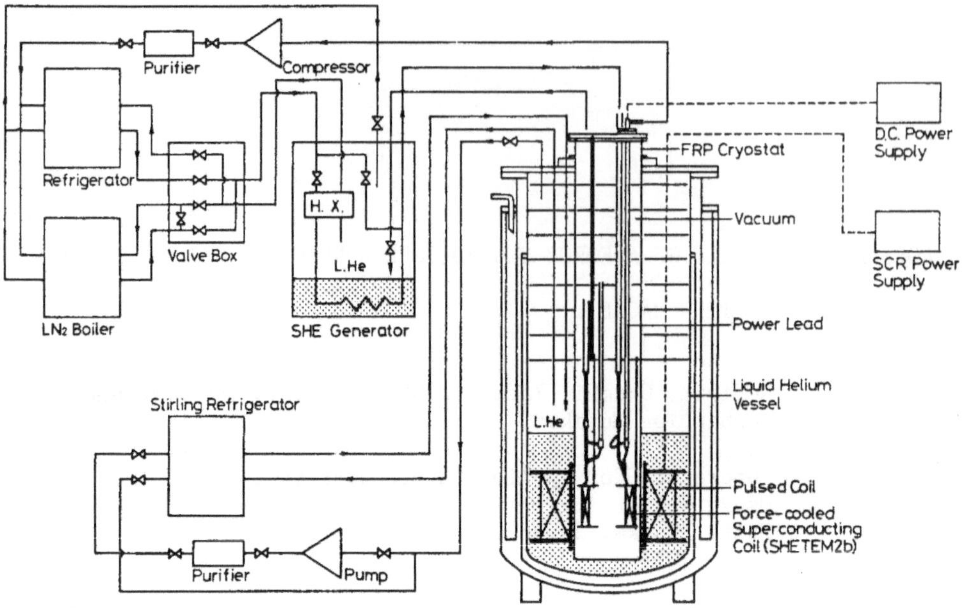

Figure 1. Schematic view of the experimental system.

EXPERIMENTAL APPARATUS

Figure 1 shows the schematic view of the experimental system. The force-cooled hollow conductor coil (SHETEM2b) was put into a vacuum vessel within the bore of the bath-cooled pulsed magnet (0.5 MJ pulsed magnet[9]), and the vessel was made of FRP to keep off the eddy current loss during operation of the bias pulsed magnet. The force-cooling system is able to supply SHE of 4.5 K, 1.0 MPa and 1.0 g/s of the mass flow rate, constantly. The cross sectional view of the hollow conductor is shown in Fig. 2. The conductor consists of 2 superconductors (NbTi), 2 heater wires, 5 electrical taps, copper wires and a copper tubing as a cooling channel. The copper ratio for the whole cross section area of the conductor is Cu/NbTi = 44/1. Then, the stability of the hollow superconductor coil was analyzed with the eddy current loss at the copper fraction. Electrical taps are mounted non-inductively on the conductor at intervals of 11 meters in order to observe electrical resistance due to the normal transition. This solenoid coil has 79.5 turns (26.5 turns times 3 layers) and the self-inductance is 1.1 mH. The constant of the magnetic field to a transport current is 0.48×10^{-3} T/A. Because FRP sheets exist between layers and the conductor is covered with Kapton tapes, heat conduction between turns is very poor. In case of SHE of 5.0 K, 0.8 MPa and 1.0 g/s, the pressure drop in the conductor was 0.01 MPa. The mass flow rate was measured at the inlet side with an orifice flow meter. As shown in Fig. 3, the temperature and the pressure of SHE were measured at the inlet and the outlet of the coil with the carbon resistance thermometers and the strain gage pressure sensors. The outer pulsed magnet can be operated at the rate of 5 T/s and the maximum center magnetic field of 5 T. The inner diameter and the height of this magnet are 310 mm and 255 mm. Then, the pulsed magnet can apply the almost uniform magnetic field to the whole of SHETEM2b because the outer diameter and the height of SHETEM2b are 187 mm and 165 mm, respectively. Details of these superconducting coils are shown in Tab. 1.

Table 1. Specification of SHETEM2b and 0.5 MJ pulsed magnet.

Conductor (SHETEM2b)		Copper Tubing	
length	44 m	outer diameter	4 mm
diameter	5.62 mm	inner diameter	3 mm
cross section area	22.8 mm^2	Coil	
void fraction	31 %	outer diameter	187 mm
Superconductor (NbTi)		inner diameter	145 mm
diameter	0.81 mm	height	165 mm
copper ratio		number of turns	79.5
Cu/CuNi/NbTi = 2/0.4/1		self-inductance	1.1 mH
critical current (4.2 K, 2 T)		Bias Pulsed Magnet	
	480 A	(0.5 MJ Pulsed Magnet)	
Heater Wire (CuNi)		outer diameter	494 mm
diameter	0.7 mm	inner diameter	310 mm
Copper Wire (RRR:188)		height	255 mm
diameter	0.81 mm	central field	5 T

PRELIMINARY EXPERIMENT WITH HEATER

Thermal characteristics and the required energy for the normal transition of the hollow superconductor coil (SHETEM2a[8,10]) from a standpoint of the flow effect of SHE, which was heated along the entire length, have been reported. The conductor of SHETEM2b is half as long as SHETEM2a, and these conductors have the same composition, excepting electrical taps of SHETEM2b. As the preliminary experiment, the required energy for the normal transition in a bias magnetic field was estimated with the heater wire in order to know thermal characteristics of SHETEM2b. Moreover, the influence of the thermal load on the mass flow rate which was the important parameter for the stability of the force-cooled magnet was studied.

Required energy for the normal transition

The required energy for the normal transition (Q_j, J/m) was measured when SHETEM2b was heated in a bias magnetic field (0 ∼ 3.5 T). The thermal load was uniformly given to the entire conductor by the heater wire. Q_j was estimated by the following equation (1).

$$P_\eta = Q_j/V\int_{T_i}^{T_{cr}} Cp\cdot\rho dT, \qquad (1)$$

where T_i: initial temperature of SHE (K), T_{cr}: critical temperature of the

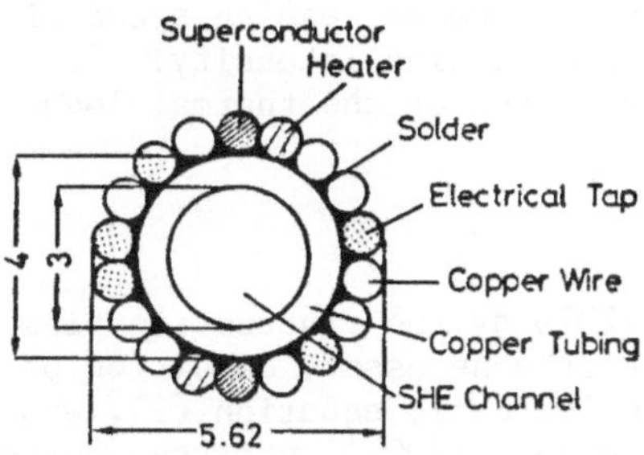

Figure 2. Cross sectional view of the hollow conductor.

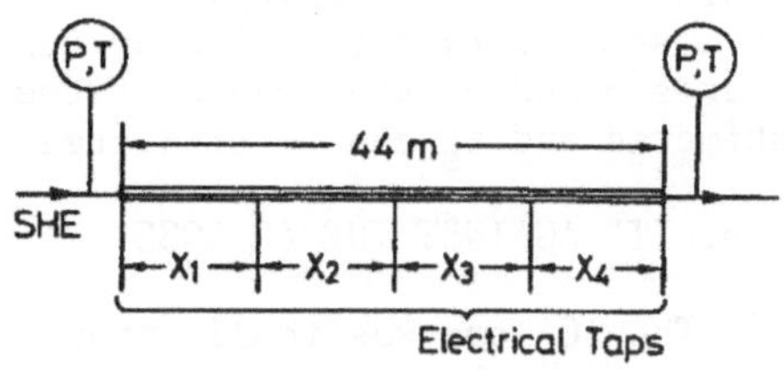

Figure 3. Positions of sensors and electrical taps. SHE passed through the coil from the innermost layer.

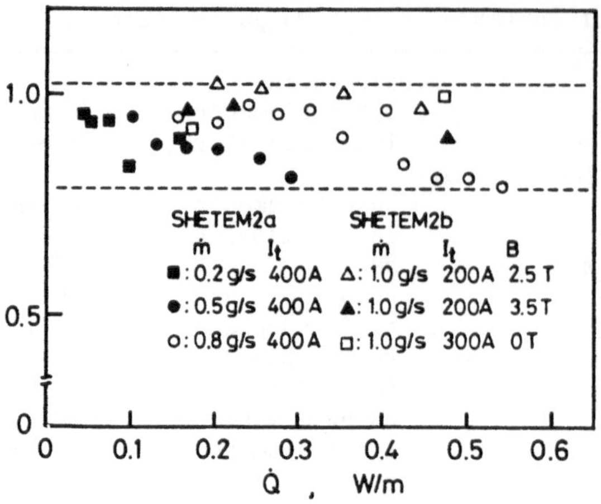

Figure 4. The relation between P_η(required energy for the normal transition/heat capacity of SHE) and $\dot{Q}$ (heating intensity) when the whole conductor was heated by a heater wire. $\dot{m}$, I_t and B are the mass flow rate, a transport current of SHETEM2b and the bias magnetic field.

coil (K), Cp: specific heat capacity of SHE (J/kg•K), ρ: density of SHE (kg/m^3) and V: volume of SHE in the conductor of a unit length (m^3/m) and P_η: the ratio of Q_j to the energy needed to increase the temperature of SHE in the conductor of a unit length from the initial temperature to the critical temperature of the coil. The value of Cp and ρ were used at the initial pressure. The condition of SHE before heating was 5.3 K of temperature, 0.83 MPa of pressure and 1.0 g/s of the mass flow rate. The experimental results are shown in Fig. 4, with the results obtained by SHETEM2a. In the figure, $\dot{m}$, I_t and B are the mass flow rate, transport current and the bias magnetic field, respectively. $\dot{Q}$ (W/m) of the abscissa means the heating intensity. According to Fig. 4, P_η was from 0.8 to 1.02 and it was confirmed that Q_j could be calculated from the heat capacity of SHE. Then, as the necessary condition of the safety operation of the force-cooled magnet, the equation (2) should be satisfied.

$$Q_t < V\int_{T_i}^{T_{cr}} Cp\cdot\rho dT, \qquad (2)$$

where, Q_t: the thermal load given to the conductor during the transit time of SHE.

Influence of the disturbance on the mass flow rate

As shown in the equation (2), the transit time of SHE (t_s) is the important parameter for the stability of the force-cooled superconducting magnet. Figure 5 shows the change of the mass flow rate by the stepwise heat input. The numerals in the figure are the run number. During heating, the mass flow rate decreased by the expansion of SHE and the increase of the flow impedance, however, that recovered the initial value within 10 s after heating. As shown in Fig. 6, the decreasing speed of the mass flow rate is a linear function of the heating intensity. From figures 5 and 6, the change of the mass flow rate for the thermal load is estimated and t_s can be predicted.

STABILITY AGAINST THE AC LOSS

The AC loss was really induced in SHETEM2b by the fluctuating bias field. From the results of this experiment, the necessary condition of the safety operating region of the force-cooled coil, equation (2), was verified. As shown in Fig. 7, a constant current (200 A) was transported to SHETEM2b and the bias magnetic field ($\dot{B}$ = 1.25-3.75 T/s, B_{max} = 2.5 T, t_i = 1-12 s) was changed repeatedly with the 0.5 MJ pulsed magnet. The AC

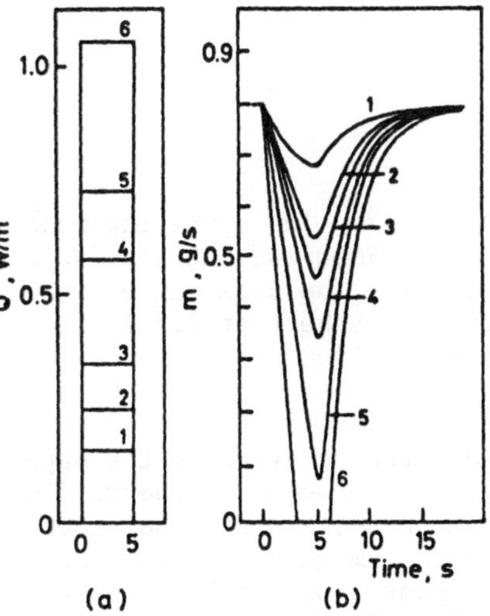

Figure 5. Changes of the mass flow rate (shown in (a)) when SHETEM2b was heated by a heater wire(b). The numerals in the figure are the run number.

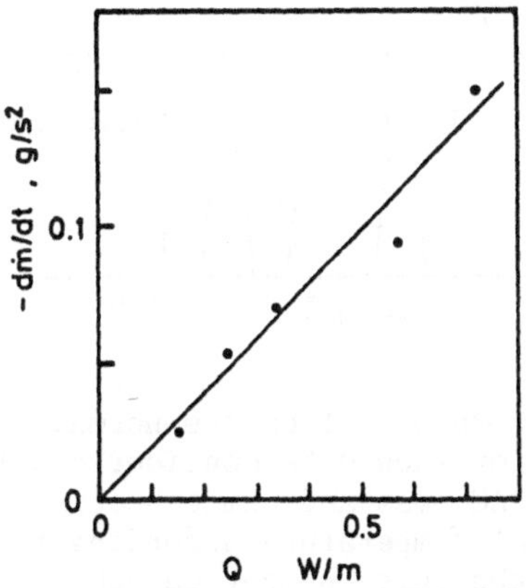

Figure 6. Decreasing rate of the mass flow rate against the thermal load.

loss of SHETEM2b for the different magnetic field sweep rate is shown in Tab. 2. The AC loss was acquired by the enthalpy change of SHE passing through the coil when the pulsed magnet was solo driven. These values are well agreed with the calculation. The critical temperature of SHETEM2b is 7.2 K at 200 A of I_t and 2.5 T of B. It was neglected for the distribution of the self-generated field in the coil (0-0.095 T) to affect that of its critical temperature. When the conditions of SHE were 5.3 K and 0.83 MPa, the right side of the equation (2) is 8.83 J/m. From the equation (2), in case of $\dot{B}$ = 2.5 T/s and t_i = 1 s, it is expected that the normal zone appears after the fifth pulse. Figure 8 (a,c,e) and (b,d,f) show the bias magnetic field and the appearance of electric voltage due to the normal transition at X_3, respectively. The numerals in the figure are the number of pulses. The mass flow rate of SHE was 1.0 g/s and t_s at the initial condition was about 41 s. The results for $\dot{B}$ = 2.5 T/s are shown in Fig. 8 (c,d). The normal transition was not observed before the fourth pulse. At the fourth pulse, the normal region slightly appeared when the magnetic field was increasing, however, this region turned to the superconducting state during the constant magnetic field (B = 2.5 T). By the AC loss of the fifth pulse, the normal transition was observed while driving the pulsed magnet. The thermal load of ohmic heating was about 8 J/m and it was considered that the temperature rise of SHE was equal to about 2 K. In this case, the normal zone returned to the superconducting state owing to the increase of the critical temperature when the bias magnetic field became 0 T. At the sixth pulse, the superconducting state

Table 2. AC loss of SHETEM2b.

$\dot{B}$ T/s	AC loss W/m	Thermal load for one pulse J/m	t_p s
1.25	0.23	0.91	4.00
2.50	0.91	1.82	2.00
3.75	1.82	2.43	1.33

where, t_p: heating duration for one pulse.

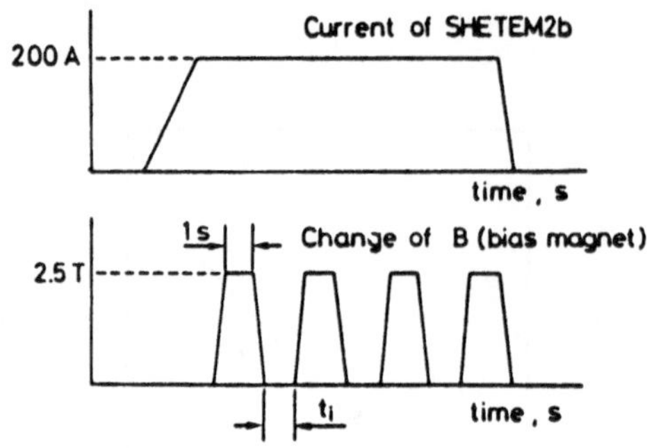

Figure 7. The operating mode of SHETEM2b and the bias pulsed magnet. The bias magnetic field was kept at 2.5 T for 1 s.

did not recover from the normal. In this way, the behavior of the superconductor should be considered with the relation between the thermal load by the AC loss and ohmic heating at the normal zone, and the change of the critical temperature according to the fluctuating magnetic field. However, the pulse number for the first normal transition can be estimated by the equation (2). The pulse number for $\dot{B} = 3.75$ T/s by the equation (2) was 4, but the normal zone appeared at the third pulse. This is due to the heat transfer between the conductor and SHE. The heat transfer of forced-flow is described by the following equations with the heat transfer coefficient of Dittus-Boelter.

$$q = h \times \Delta T, \tag{3}$$

$$h = 0.023(\lambda^{0.6}C_p^{0.4}/\eta^{0.4})(\dot{m}^{0.8}/D^{0.2}), \tag{4}$$

where, q: heat flux (W/cm^2), h: heat transfer coefficient (W/cm^2•K), λ: thermal conductivity (W/cm•K), η: viscosity (g/cm•s), D: equivalent diameter (cm) and ΔT: temperature difference (K). The AC loss of 1.82 W/m is equal to 19.3 mW/cm^2 as the heat flux to SHE. After two pulses, the temperature of SHE increased to 6.35 K. The mass flow rate decreased to 0.76 g/s by the AC loss. In this case, h was 21.0 mW/cm^2•K and ΔT was

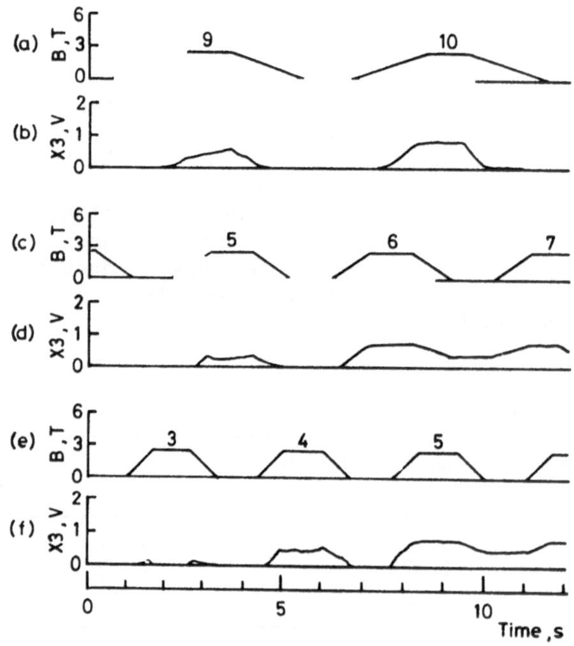

Figure 8. The appearance of the normal zone at X_3 for the different changing rate of the bias magnetic field. The operating rates of the bias magnetic field are 1.25 T/s (a), 2.5 T/s (c) and 3.75 T/s (e). The numerals in the figure are the pulse number.

Table 3. Experimental results for the different interval between pulses.

	t_i, s	X_1	X_2	X_3	X_4	Np
	2	N	N	N	N	4
$\dot{B}$ = 3.75 T/s	4	S	S	N	N	4
$\dot{m}_i$ = 1.0 g/s	8	S	S	S	N	4
(t_{si} = 41 s)	10	S	S	S	N	4
	1	N	N	N	N	5
	2	N	N	N	N	5
$\dot{B}$ = 2.50 T/s	4	N	N	N	N	5
$\dot{m}_i$ = 0.8 g/s	6	S	N	N	N	5
(t_{si} = 51 s)	10	S	S	N	N	5
	12	S	S	S	N	5

where, S: superconducting condition, N: normal condition, Np: the number of pulses required for the normal transition, $\dot{m}_i$: the mass flow rate of SHE at the initial condition and t_{si}: transit time of SHE for $\dot{m}_i$.

0.92 K from equations (3) and (4). Then, the conductor temperature exceeded the critical temperature and the normal zone appeared.

In Tab. 3, the experimental results of the different time interval are shown. The pulse number of the normal transition (N_p) agreed with that estimated by the equation (2). With extending the time interval between pulses, the normal transition occurred at the down stream. This is due to the temperature distribution of SHE along the flow direction formed by the intermittent thermal load. In case of $\dot{B}$ = 3.75 T/s and t_i = 10 s, or $\dot{B}$ = 2.5 T/s and t_i = 12 s, the section of the normal transition was limited at X_4. For the time interval longer than that shown in Tab. 3, the normal zone did not appear. In other words, the interval between operations of the force-cooled superconducting pulsed magnet should be decided by the transit time of SHE through the coil and the heating density of the AC loss. As mentioned above, the equation (2) is the necessary condition of the safety operation of the pulsed force-cooled magnet and the large mass flow rate of SHE makes the superconducting coil stable for the AC loss. Reversely, from the operating mode of the pulsed magnet and the heating density of the AC loss, the transit time of SHE is decided, and concretely the flow velocity of SHE and the length of one cooling channel are restricted.

CONCLUSION

The influence of the AC loss on the stability of the force-cooled superconducting magnet was studied in a practical process with the bias pulsed magnet. As a result, in case the AC loss was generated in a force-cooled superconducting magnet, the important parameter for the stability was the heat capacity of SHE from the initial temperature to the critical temperature of the coil.

The necessary condition for the safety operation of the force-cooled pulsed superconducting magnet is that the thermal load which is applied to the conductor during the transit time of SHE through the coil should be smaller than the energy needed to increase the SHE temperature to the critical temperature of the coil. Under the condition that the operating mode of the force-cooled pulsed magnet is already decided, the flow velocity of SHE and the length of one cooling channel are restricted.

ACKNOWLEDGMENT

The authors would like to thank Sumitomo Electrical Industries, Ltd. for the production of the hollow coil. This work has been partially supported by Grant-in-Aid for Scientific Research from the Ministry of Education (No. 61050047).

REFERENCE

1. I. Horvath and G. Vecsey, Ten years experience in operation of the superconducting muon channel at the swiss institute for nuclear research, in:"Proc. MT-9," SIN, Zurich (1985), p. 174.
2. J. L. Young, C. J. Heyne, P. Komarek, H. Krauth, G. Vecsey and C. Marinucci, The forced flow cooled coil for the international energy agency large coil task, in:"Advances in Cryogenic Engineering," Vol. 27, Plenum Press, New York (1982), p. 11.
3. J. R. Miller, Empirical investigation of factors affecting the stability of cable-in-conduit superconductors, Cryogenics 25:552 (1985).
4. M. C. M. Cornelissen and C. J. Hoogendoorn, Thermal stability of superconducting magnets: dynamic criteria, Cryogenics 25:3 (1985).
5. L. Dresner, Parametric study of the stability margin of cable in conduit superconductors: theory, IEEE Trans. Mag., MAG-17:753 (1981).
6. A. Y. Lee, Cryogenic recovery analysis of forced-flow supercritical helium-cooled superconductors, in:"Advances in Cryogenic Engineering," Vol. 23, Plenum Press, New York (1978), p. 235.
7. D. Junghans, Stability of force-cooled superconductors Part 2: experiment, Cryogenics, 23:227 (1983).
8. N. Ohuchi et al, Flow effect of supercritical helium gas on the stability of a hollow superconducting coil (SHETEM2a) with thermal load, presented at Applied Superconductivity Conference in Baltimore, LH-14 (1986).
9. Y. Murakami et al, Experiments and analyses of thermal characteristics and stress/strain distribution of a 0.5 MJ pulsed coil, in:"Advances in Cryogenic Engineering," Vol. 29, Plenum Press, New York (1984), p. 167.
10. N. Ohuchi et al, Cooling characteristics and flow instability of a supercritical helium cooled hollow superconducting coil, in:"Proc. Ninth Intl. Cryo. Engr. Conf.," Butterworth, Guildford, UK (1986), p. 419.

QUENCH DETECTION BY FLUID DYNAMIC MEANS IN CABLE-IN-CONDUIT SUPERCONDUCTORS*

Lawrence Dresner

Oak Ridge National Laboratory

Oak Ridge, Tennessee

ABSTRACT

The tight confinement of the helium in cable-in-conduit superconductors creates protection problems because of the substantial pressure rise that can occur during a quench. But the same pressure rise offers the useful possibility of a non-electrical means of detecting incipient quenches by monitoring the outflow from the various hydraulic paths of the magnet. If the method is to work, (1) the signal must be large enough to be detected unambiguously at an early enough time, and (2) the signal must not depend too strongly on the length, Joule power density, or rate of growth of the initial normal zone (because these things are not entirely within our control). This paper explores by calculation the degree to which these conditions can be met. The Westinghouse coil for the Large Coil Task (LCT) is used as the basis for illustrative examples.

INTRODUCTION

The Westinghouse (WH) coil of the International Fusion Superconducting Magnet Test Facility (IFSMTF) at Oak Ridge National Laboratory is one of six large magnets of different design that operate in a torus. If one of its neighbors is dumped, the rapidly changing magnetic flux induces large voltages in the WH coil. These voltages can only be compensated for imperfectly, and the WH protection system responds to them as to an incipient quench. Such false positives may require a dump of the WH coil if we do not wish to risk a period of rapidly changing flux during which the protection system is temporarily blinded.

For this reason, a non-electrical means of detecting incipient quenches is desirable. Monitoring the outflow from the various hydraulic paths may provide such a means, but if it is to work, the following conditions must be met:

(1) The signal must be large enough to be detected unambiguously.
(2) The signal must not depend too strongly on the length, Joule power density, or rate of growth of the initial normal zone (since these are beyond our control).

The purpose of this paper is to explore by calculation the degree to which these conditions are met.

*Research sponsored by the Office of Fusion Energy, U.S. Department of Energy, under contract DE-AC05-84OR21400 with Martin Marietta Energy Systems, Inc.

BASIC EQUATIONS*

The key to making the calculations is the very large length-to-diameter (L/D) ratio of the helium volume inside the conductor. The length of the hydraulic paths in the Westinghouse IFSMTF magnet is 120 m, whereas the hydraulic diameter of the helium-filled interstices is only 0.4 mm; the L/D ratio is thus 3×10^5. During a quench, the Joule heating raises the pressure of the helium. The helium tries to relieve this pressure by expanding, but its expansion is opposed by friction with the wires and the walls and by the inertia of the fluid. Because of the very large L/D ratio, the pressure gradient in the helium is almost entirely expended in overcoming friction, and accordingly *we neglect the inertia of the fluid.* This simplification enables us to obtain formulas that show explicitly the dependence of the expulsion velocity on the various parameters of the conductor.

The flow equations (continuity, momentum, energy) for a heated pipe are

$$\frac{d\rho}{dt} + \rho\frac{\partial v}{\partial x} = 0 \tag{1a}$$

$$\rho\frac{dv}{dt} = -\frac{\partial p}{\partial x} - \rho F \tag{1b}$$

$$\rho\frac{d}{dt}\left(e + \frac{v^2}{2}\right) = -\frac{\partial}{\partial x}(pv) + Q \tag{1c}$$

where F, the frictional force per unit mass, is given by $F = 2fv^2/D$. (Symbols are defined at the end of the paper.) The frictional force appears in the momentum equation (1b) just as any external force would, but not in the energy equation (1c) because the work done by the fluid against the frictional force is not removed from the fluid (as it would be if the work were against an external force) but is returned to it as heat.

If we multiply (1b) by v and subtract it from (1c), we find, after using (1a) and the second law of thermodynamics, $T\,ds = de + p\,d\tau$, that

$$T\frac{ds}{dt} = \frac{Q}{\rho} + Fv \tag{2}$$

The term Fv on the right-hand side of (2) represents entropy production due to irreversible conversion by friction of kinetic energy to internal energy. Had the term $-\rho Fv$ been present on the right-hand side of (1c), as it would have been had F been an external force, then the term Fv would not appear in (2).

The basic assumption of this work is that the frictional forces greatly dominate inertial forces in a long, narrow tube. This means that the left-hand side of (1b) is very much less than either term on the right. In other words, the pressure gradient expends itself in overcoming friction, not in accelerating the fluid. Hence, we set $dv/dt = 0$ in (1b). We can eliminate the derivative of ρ from (1a) using the thermodynamic identify $d\rho = dp/c^2 - (B\rho/c_p)T\,ds$ so that

$$\frac{1}{\rho}\frac{d\rho}{dt} = \frac{1}{\rho c^2}\frac{dp}{dt} - \frac{B}{c_p}\left(\frac{Q}{\rho} + Fv\right) \tag{3}$$

*This section is quoted nearly verbatim from Ref. 1.

Using (1a,b) and (3), we find

$$\frac{\partial v}{\partial x} + \frac{1}{\rho c^2}\frac{\partial p}{\partial t} = \frac{B}{c_p}\left[\frac{Q}{\rho} + Fv\left(1 + \frac{c_p}{Bc^2}\right)\right] \tag{4}$$

Finally, consulting NBS-621, we find that Bc^2/c_p is always close to 1, so the bracketed quantity on the right-hand side of (4) is always close to 2.

Further Reduction of the Equations

We use (1b), with its left-hand side set equal to zero, and (4) to calculate the expulsion velocity at the ends of a long, slender pipe caused by a thermal perturbation near its center. We simulate the effect of the thermal perturbation and the subsequent growth of a nonrecovering normal zone by a pressure p in the central plane $x = 0$ that rises proportionally to a power n of the time t: $p(0,t) = p_0 t^n$. So throughout the pipe, i.e., for $x > 0$, the Joule power density, Q, is zero. When $Q = 0$, the right-hand side of (4) is proportional to v^3. Early enough, when the velocity v is still small, we expect v^3 to be small compared with $\partial v/\partial x$. Accordingly, we drop the entire right-hand side of (4). Later, when we have solved the resulting equations, we shall use the solution to evaluate the right-hand side of (4) and the term $\partial v/\partial x$. Comparing them will give a limit on the elapsed time t, below which the right-hand side of (4) is negligible compared with $\partial v/\partial x$.

With these approximations, (1b) and (4) become

$$\frac{\partial p}{\partial x} = -\frac{2f}{D}\rho v^2 , \quad \frac{\partial v}{\partial x} + \frac{1}{\rho c^2}\frac{\partial p}{\partial t} = 0 \tag{5}$$

Early, when not much helium has been expelled from the pipe, the helium remains on the high-density side of the pseudo-critical curve and behaves like a liquid. We therefore take the physical properties ρ and c to be constants. This done, we henceforth interpret p as the *pressure rise* above ambient pressure. Finally, to simplify the appearance of the equations, we work in a special system of units in which $\rho = c = D/4f = 1$. Then (5) becomes

$$\frac{\partial p}{\partial x} + \frac{v^2}{2} = 0 , \quad \frac{\partial v}{\partial x} + \frac{\partial p}{\partial t} = 0 \tag{6}$$

To these we add the boundary and initial conditions

$$p(x,0) = v(x,0) = 0 ; \quad p(\infty,t) = v(\infty,t) = 0 ; \quad p(0,t) = p_0 t^n \tag{7}$$

Strictly speaking, the second boundary condition refers to a semi-infinite pipe. But we can use the solution corresponding to it to estimate the expulsion velocity from a pipe of finite length L for small enough times: if the pressure rise $p(L,t)$ is small compared with the driving pressure rise $p_0 t^n$, the expulsion velocity can be approximated by $v(L,t)$. This puts another condition on the elapsed time.

SIMILARITY SOLUTION*

Equation (6) is invariant to the one-parameter family of one-parameter groups of affine transformations

$$p' = \lambda^\alpha p , \quad v' = \lambda^\gamma v , \quad t' = \lambda^\beta t , \quad x' = \lambda x , \quad 0 < \lambda < \infty \tag{8}$$

*A general reference to the material in this section can be found in Ref. 2.

where the constants α, β, and γ are related by the two linear equations $\alpha - 2\beta = -3$, $\gamma - \beta = -2$. The number λ is the group parameter that labels the transformations of a particular group; any one of the numbers α, β, and γ can be used to label the various groups in the family.

Solutions of (6) that are themselves invariant to all the transformations of a particular group must have the form

$$p = t^{\alpha/\beta} P(x/t^{1/\beta}) , \quad v = t^{\gamma/\beta} V(x/t^{1/\beta}) \tag{9}$$

where P and V are as yet undetermined functions of the single argument $y = x/t^{1/\beta}$. The numbers α, β, and γ are determined by the boundary and initial conditions: substituting (9) into (7) gives

$$\alpha = 3n/(2-n) , \quad \beta = 3/(2-n) , \quad \gamma = (2n-1)/(2-n) \tag{10a}$$

$$P(0) = p_0 , \quad P(\infty) = 0 , \quad V(\infty) = 0 \tag{10b}$$

If we substitute (9) into the partial differential equations (6), we obtain the following two *ordinary* differential equations (ODEs) for P and V:

$$\dot{P} + V^2/2 = 0 , \quad \beta\dot{V} + \alpha P - y\dot{P} = 0 \tag{11}$$

Here $\dot{P}$ is an abbreviation for dP/dy. Since we are mainly interested in V, we eliminate P from (11) and obtain the following second-order ODE for V:

$$\beta\ddot{V} - \tfrac{1}{2}(\alpha - 1)V^2 + yV\dot{V} = 0 \tag{12}$$

The ODEs (11) and (12) are themselves invariant to the associated group of affine transformations

$$P' = \mu^{-3}P , \quad V' = \mu^{-2}V , \quad y' = \mu y , \quad 0 < \mu < \infty \tag{13}$$

The quantities $u = y^2V$ and $w = y^3\dot{V}$ are both invariant to (13) [$\dot{V}' = \mu^{-3}\dot{V}$ follows immediately from (13)]; they are called an invariant and a first differential invariant, respectively. According to a theorem of Lie's,[3] using an invariant and a first differential invariant as new independent and dependent variables reduces (12) to a first-order differential equation:

$$\frac{dw}{du} = \frac{(3\beta - u)w + \frac{1}{2}(\alpha - 1)u^2}{\beta(2u + w)} \tag{14}$$

Figure (1a) shows schematically the direction field of (14) for $0 < n < 1/2$ and Fig. (1b) shows the direction field of (14) for $1/2 < n < 2$. (For $n > 2$, the problem posed by (5) and (7) is not solved by a similarity solution.) Some typical integral curves are sketched in each figure. Only the fourth quadrant is shown because, y and V being positive and $\dot{V}$ being negative, $u > 0$ and $w < 0$. There are two singular points, the origin (0,0) and the point $R : (6, -12)$. Note carefully that the coordinates of the point R are independent of α, β, and γ.

The integral curve of (14) that we seek passes through the origin, for when $y = 0$, $u = y^2V$ and $w = y^3\dot{V}$ are both zero. The integral curves emanating from the origin are of three types: those that intersect the locus of zero slope, those that intersect the locus of infinite slope, and the separatrix that divides them. The separatrix passes through the singular point R. As we approach R along the separatrix, $y \to \infty$.* Since $u_R = 6$, $V = u/y^2 \approx 6/y^2$ near R. So the separatrix is the integral curve we want, for the V it determines obeys the boundary condition $V(\infty) = 0$.

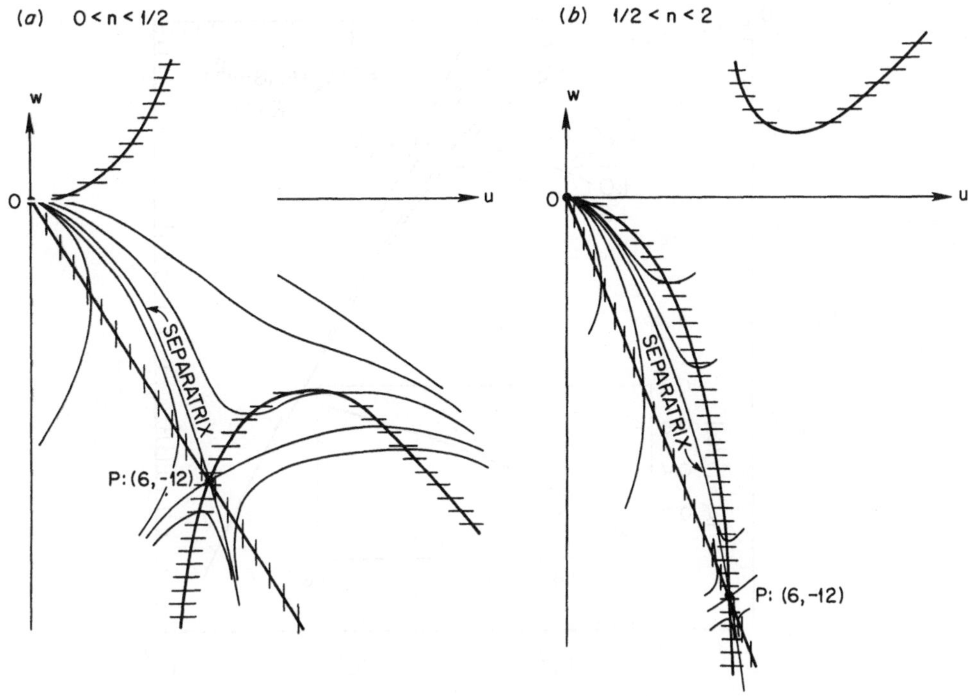

Fig. 1. The direction field of Eq. (14).

The separatrix is a relation of the form $w = F(u)$ or equivalently $y^3\dot{V} = F(y^2V)$, i.e., a first-order ordinary differential for V in terms of y. The general solution of a first-order ODE is a one-parameter family of curves. Since the differential equation is invariant to the group (13), so must be the entire family of solution curves. This means that each solution curve is carried into another by the transformations (13). The *common* asymptote of all the solution curves, $V = 6/y^2$, is carried into itself since y^2V is an invariant of the group (13).

Any value of y can correspond to any point on the separatrix except 0 and R. Suppose we have a set of three values y, V, $\dot{V}$ that map onto a point (u, w) on the separatrix, i.e., for which $u = y^2V$ and $w = y^3\dot{V}$. Any image μy, $\mu^{-2}V$, $\mu^{-3}\dot{V}$ of y, V, $\dot{V}$ will also map onto the same point (u, w), but with a different value of y. We can find points (u, w) on the separatrix by using $u_s = 6 - \epsilon$, $w_s = -12 - \epsilon\dot{w}_R$ as starting values for a numerical integration in the (stable) direction $P \to O$. To the values u, w we append a *guess* for y and thereby obtain starting values y, V, $\dot{V}$ for a (stable) backward integration of (12). If the value obtained for $P(0) = -\beta\dot{V}(0)/\alpha$ [follows from (11)] does not equal p_0 (which will generally be the case), we scale the calculated curve $V(y)$ according to (13) to obtain the curve we want. Shown in Fig. 2 are curves of $V(y)$ normalized to $V(0) = 1$ obtained in this way for $n = 0$, 1/4, 3/8, 1, and 3/2. Shown in the insert is the quantity $A = -\dot{V}(0) = \alpha P(0)/\beta$ as a function of n when $V(0) = 1$. A useful approximation to the curves in Fig. 2 is

$$V = 6/(6 + 6Ay + y^2)\ , \quad A = 0.529n^{0.703} \tag{15}$$

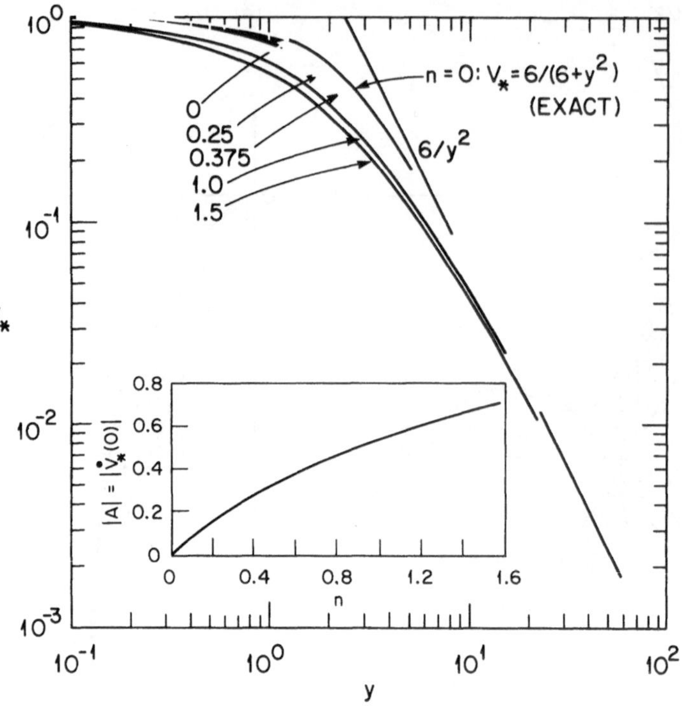

Fig. 2. Curves of $V(y)$ normalized to $V(0) = 1$ for various n. The insert shows the quantity $A = -\dot{V}(0) = \alpha P(0)/\beta$ when $V(0) = 1$ as a function of n.

Henceforth, we shall denote the curve $V(y)$ normalized to 1 at $y = 0$ as $V_*(y)$. Then $V_*(0) = 1$ and $\dot{V}_*(0) = -A$.

RESTRICTIONS ON THE ELAPSED TIME

There are two restrictions on the elapsed time. The first arises from the condition $v^3 \ll |\partial v/\partial x|$, which can be written $t^n \ll -\dot{V}/V^3$. We evaluate the right-hand side for $y = 0$. Since $\dot{V}(0) = -(\alpha/\beta)P(0) = -np_0$, the value of μ that transforms $V_*(y)$ into $V(y)$ is $\mu = (A/np_0)^{1/3}$. Then $V(0) = \mu^{-2}V_*(0) = (np_0/A)^{2/3}$ so that $t^n \ll -\dot{V}(0)/V^3(0) = A^2/np_0$. It appears from the numerical results that the quantity $-\dot{V}/V^3$ is monotone increasing, so t^n is always $\ll -\dot{V}/V^3$.

*To see how this comes about, differentiate the equation $u=y^2V$ to get $y(du/dy)=2u+w$. Near R, this becomes $dy/y=du/(2+\dot{w}_R)(u-6)$, where $\dot{w}_R$ is the slope of the separatrix at R. Thus near R, y varies as $(6-u)^{2+\dot{w}_R}$. We can find $\dot{w}_R$ from (14) by using l'Hospital's rule: $\dot{w}_R=[\beta-6-\sqrt{(\beta-6)^2+24\beta(\alpha+1)}]/2\beta$. From this formula we see that $\dot{w}_R<-2$, which shows that $y\to\infty$ as $u\to 6$ from below.

This last result is correct in special units. To write it in any units, e.g., mks units, we must make the various terms dimensionless by introducing powers of ρ, c, and $D/4f$, all of which are numerically equal to 1 in special units. Since $p_0 t^n$ is a pressure, we can write in ordinary units $t^n \ll A^2 \rho c^2/np_0$.

The second restriction on the time comes from the requirement that the pressure rise at the end of a hydraulic path be small compared with the driving pressure rise. This is the same as the requirement that we be able to use the asymptotic forms $6/y^2$ and $6/y^3$ for V and P, respectively ($P = 1/2 \int_y^\infty V^2 dy = 6/y^3$ if $V = 6/y^2$). Figure 2 shows the curves $V_*(y)$, which are normalized to $V_*(0) = 1$, on a log-log plot. If we transform $V_*(y)$ by (13) we can get $V(y)$ having any value at the origin, depending on the value of μ we use. Such a transformation on a log-log plot is simply translation of the curves $V_*(y)$ parallel to lines of slope -2, i.e., parallel to the line $6/y^2$, which maps into itself. The requirement on t can then be written

$$\left(\frac{np_0}{A}\right)^{2/3} = V(0) \gg \frac{6}{y^2} = \frac{6t^{2/\beta}}{L^2} \quad \text{or} \quad \frac{np_0}{A} \gg \frac{6^{3/2} t^{2-n}}{L^3} \tag{16}$$

Equation (16) is correct in special units. In ordinary units, it becomes

$$t \ll \left(\frac{4}{6\sqrt{6}} \frac{n}{A} \frac{fL^3}{Dc^2} \frac{p_0}{\rho c^2}\right)^{1/(2-n)} \tag{17}$$

NUMERICAL EXAMPLE

The following numerical data are appropriate to the WH coil: $\rho = 150\ \mathrm{kg \cdot m^{-3}}$, $c = 290\ \mathrm{m\ s^{-1}}$ (10 atm, 4.0 K), $D = 4 \times 10^{-4}$ m, $L = 60$ m (half-length of hydraulic path), $f = 0.01$. According to Ref. 1, the pressure in an initially small, nonrecovering normal zone grows proportionally to the time so that $n = 1$. In Ref. 1, the proportionality constant p_0 is given as 35 atm/s for the WH coil. With these data, the limits on the time turn out to be 1.0 s and 3.0 s, respectively. The time it takes for a sonic signal to travel from the center to the end of a hydraulic path is 60/290 = 0.21 s; before this much time elapses no disturbance at the center of a path can be detected at the ends. The expulsion velocity is given by $v = t^{\gamma/\beta}$ $6t^{2/\beta}/L^2 = 6t/L^2$ in special units, or

$$v = \frac{3}{2} \frac{c^2 Dt}{fL^2} \tag{18}$$

in ordinary units. At $t = 0.5$ s, $v = 70$ cm/s, corresponding to an outward mass flow of 9.3 g/s. This is nearly five times the ambient flow of 2 g/s in a hydraulic path and so should certainly be large enough to detect unambiguously, if we monitor each flow path individually. Even if we monitor only the total (48 parallel paths), it will increase by nearly 10% after 0.5 s and 20% after 1 s, which should be enough to be clearly noticeable.

CONCLUDING REMARKS

Whether the increased outflow is large enough to be detected unambiguously (condition 1 of the introduction) can be determined for a particular design using (18) and the restrictions on the time mentioned earlier. The independence of the asymptotic formula (18) of the parameters p_0 and n means that the outflow velocity does not depend strongly on the details of the nonrecovering normal zone. In this way, condition 2 of the introduction is fulfilled. It should not be forgotten that p_0 and n enter the formulas restricting the time interval during which (18) is valid.

List of Symbols

A	$-\dot{V}(0) = \alpha P(0)/\beta$
B	volume coefficient of thermal expansion, $\rho(\partial\tau/\partial T)_p$
c	sonic speed
c_p	specific heat at constant pressure
D	hydraulic diameter
e	specific internal energy
f	Fanning friction factor
F	frictional force per unit mass, $2fv^2/D$
L	half-length of pipe
n, p_0	constants in power law for driving pressure; see Eq. (7)
$P(y), V(y)$	functions defined in Eq. (9)
Q	Joule power density
s	specific entropy
t	time
T	temperature
u	y^2V; $w = y^3\dot{V}$
$V_*(y)$	the curve $V(y)$ normalized to $V(0) = 1$
x	distance down the tube from the center
y	$x/t^{1/\beta}$
$\alpha, \beta, \gamma, \lambda, \mu$	constants in the transformation formulas (8) and (13)
ρ	density
τ	specific volume, ρ^{-1}

REFERENCES

1. L. Dresner, Thermal expulsion of helium from a quenching cable-in-conduit conductor, in: "Ninth Symposium on Engineering Problems of Fusion Research," IEEE, New York (1981), pp. 618–621 (1981).
2. L. Dresner, "Similarity Solutions of Nonlinear Partial Differential Equations," Pitman Publishing, Inc., Marshfield, Mass. (1983).
3. A. Cohen, "An Introduction to the Lie Theory of One-Parameter Groups," G. E. Stechert and Co., New York, 1931.

THERMAL ANALYSIS OF THE FORCED COOLED CONDUCTOR FOR THE TF SUPERCONDUCTING COILS IN THE TIBER II ETR DESIGN

J. A. Kerns, D. S. Slack, and J. R. Miller

Lawrence Livermore National Laboratory
University of California
Livermore, California

ABSTRACT

The Tokamak Ignition/Burn Experimental Reactor (TIBER) is being designed to provide nuclear testing capabilities for first wall and blanket design concepts. The baseline design for TIBER II is to provide steady-state nuclear burn capabilities. These objectives must be met using reactor relevant components, such as state-of-the-art current drive schemes coupled with superconducting toroidal field (TF) and poloidal field (PF) coils. The design is also constrained to be cost effective, which forces us to make the machine as small as possible. This last constraint limits the nuclear shielding in TIBER. Therefore, the TF coils will have a high nuclear heat load of up to 4.5 kW per coil. The cooling scheme and the thermal analysis for this design are presented.

INTRODUCTION

The Tokamak Ignition/Burn Experimental Reactor (TIBER), a fusion reactor design study, is a steady-state machine in which the plasma current is driven by neutral beams or other means. TIBER uses toroidal field (TF) and poloidal field (PF) coils wound with cable-in-conduit conductors (CICC) made of Nb_3Sn superconductors to produce the high fields needed for confining and controlling the plasma. The TF coils will be subjected to high nuclear heating. Design margins in TIBER allow a possible total nuclear heat load of 72 kW to the 16 TF coils, although estimates of the actual nuclear heat load are approximately 27 kW. The helium in the CICC must remove the nuclear heat and maintain the superconductor at low enough temperatures to provide adequate operating stability margins.

We evaluated the TF coil design operating with the high heat loads. Our analysis is based on solving coupled differential equations, which describe the evolution of helium pressure p and temperature T_b vs distance s along the flow paths. Appropriate models are used to describe the fluid friction and the nuclear heat generation. Once the helium temperature is known along s, the current-sharing temperature T_{cs} and stability margin ΔH are determined. The following sections will describe the physical model, the analytical models, and the results of the calculations.

COIL DESCRIPTION

The geometry of the D-shaped coil must be known so that s along the coil can be related to the heat flux and magnetic field generated within the torus. For example, the peak nuclear heat load on the coils is specified as a function of the poloidal angle θ, which is measured from the plasma center located at a radius of 3.1 m. We define θ to be zero at the midplane on the low-field side of the TF coil. The geometrical relations expressing s as a function of θ and turn number i are straightforward but tedious; and therefore, they are not be presented here although they are included in the analysis.

The TF coil conductor for TIBER was designed using a separate calculation, which is described by Miller et al.[1] This calculation optimizes the fractions of conduit, helium, conductor, and copper in the conductor for specific operating conditions. Pertinent parameters of the TF conductor design are given in Table 1. This design is based on a single point along the helium flow path through the CICC. As such, the appropriateness of this design must be checked along the entire flow path of the helium.

The conductor is pancake wound to form the TF coil. Helium inlets are connected to the conductor at the crossover turn in the low-field region inside the TF coil. From there, the helium flows through both halves of the pancake to outlet connections outside the torus. Therefore, the flow direction for the helium is from the "hot" or heated inner turns to the "cold" outer turns. We distinguish between the inner and outer turns because the neutron flux decays exponentially through the windings. Most heating occurs in the turns nearest the plasma inside the torus. The nuclear heat per unit volume generated in each turn can be determined if we multiply the heat going into the coil as a function of θ and turn number N_t by the exponential loss per turn, that is:

$$q'''(\theta,i) = q'''(\theta)\exp\left(\frac{-\Delta r}{\varepsilon}(i-1)\right). \tag{1}$$

HELIUM-FLOW ANALYSIS

The following equations govern the state of the supercritical helium flowing in a heated channel[2]:

$$\frac{dp}{ds} = \frac{\dfrac{-2G^2 f}{\rho D} + \dfrac{4q''G\,\beta}{\rho D(C_p - v^2\beta)}}{1 - \dfrac{G^2}{\rho}(K + \beta\phi)} \tag{2}$$

$$\frac{dT}{ds} = \frac{\dfrac{4q''}{GD(C_p - v^2\beta)} - \dfrac{2G^2 f\phi}{\rho D}}{1 - \dfrac{G^2}{\rho P}(K + \beta\phi)} \tag{3}$$

where

$$\phi = \frac{\psi C_p + v^2 K}{C_p - v^2\beta}$$

Table 1. TF Conductor Parameters

Operating current	(I)	36.16 kA
Conductor area	(A_{cond})	904×10^{-6} m^2
Cable space area	(A_{cs})	510×10^{-6} m^2
Fraction of conductor	(f_{cond})	0.6
Fraction of copper	(f_{Cu})	0.6
Wire diameter	(d_w)	1.0×10^{-4} m
Cooled perimeter	(P_{He})	1.18 m
Mean length of a turn	(L)	18.55 m

The helium properties are local variables that must be evaluated along the flow path. An analytical solution to Equations 2 and 3 is impossible because of rapid and significant variation of the helium properties. We solve these coupled differential equations using a Runge-Kutta fourth-order numerical integration scheme. The helium properties are calculated at each increment using HEPROP subroutines by Hands,[3] with modifications by Arp.[4] The only variables that we still lack are the friction factor f and the heat flux q into the helium.

The friction factor for cable conductors has been experimentally investigated by Lue et al.[5] Their results showed that the CICC has a friction factor approximately two to three times greater than that for smooth tubes. The increased friction results from the tortuous flow produced by the twisted, cabled conductor inside the conduit. Hooper[6] used the results of this experiment to write an empirical relation for the friction factor given by

$$\ell n\ f = 13.15\ Re^{-0.36} - 4.338 \quad (40 < Re < 10^4), \tag{4}$$

where Re is the Reynolds number.

The peak nuclear heat load vs θ was estimated by El-Guebaly.[7] This calculation shows a broad maximum in the peak heating that occurs along the straight leg of the torus for θ between 120° and 240°. The heating on the high-field section is approximately three orders of magnitude higher than heating on the low-field section. This difference is primarily due to relatively thin nuclear shielding on the high-field sections, which is thicker on the low-field side. Another effect is that the surface area of the torus increases with radius, so the neutron flux must decrease with r. The shape of the heating function is periodic with minima at θ equal to 0 and 2π and a maximum at π. This shape can be approximated by an eleventh-order Fourier series. The comparison between data points taken from Ref. 7 compared with the analytical curve is shown in Fig. 1. The heat flux into the helium q'' is found by multiplying q''' by the area of the conductor and dividing by the cooled cable perimeter. For this analysis, we assumed only the variation of q'' and q''' with Θ and turn number. The normalization is determined separately, to set the total heat into the TF coils at either 27 or 72 kW.

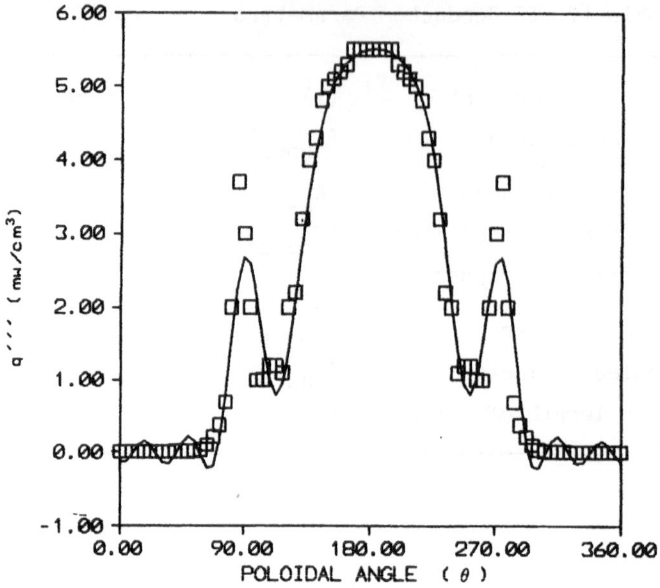

Fig. 1. Comparison between discrete data points taken from Ref. 7 and an eleventh order Fourier series.

The analytical heat flux equation, friction factor, and helium properties are combined with Eqs.(2) and (3) to predict p and T_b as a function of s. The inlet temperature is set at 4.5 K, and the inlet pressure is adjusted so the entire flow remains single phase. We also maintained the outlet pressure greater than 0.2 MPa to simplify the refrigeration system. These requirements resulted in an inlet pressure of 2.5 MPa for a heat input of 72 kW. The inlet pressure was reduced to 2.0 MPa for the 27-kW heat input. As helium enters the conduit, the nuclear heating and the Joule-Thompson (JT) effect cause the temperature to rise initially. As the flow continues through the conduit, the nuclear heating diminishes and the JT expansion cools the helium. This cooling can produce a maximum in T_b; for example, when the total heat load is 27 kW, the maximum temperature is 7.3 K.

Since T_b nearly doubles from inlet conditions, we must carefully determine T_b as a function of s (Fig. 2). Large temperature increases occur in the first several turns of the coil and have a step-like appearance. This shape is caused by the coil heating that is localized to the inboard straight-leg region. The step-like pattern is seen only in the first few turns along the helium flow path because the heating decays exponentially with distance into the coil.

Temperature, Current, and Stability Margin

The TF conductor must be designed with adequate margins to stabilize it against energy releases of about 100 mJ/cm^3 within the coil. If adequate margins are not provided, these perturbations can quench the coil. We define the temperature margin as:

$$\Delta T_m = T_{cs} - T_b \tag{5}$$

where T_{cs} is the current-sharing temperature. We estimate the current-sharing temperature for Nb_3Sn by:

$$T_{cs} = T_c(B)\left[1 - \frac{I}{A_{cs} f_{cond} (1 - f_{Cu}) J_{c0}(B)}\right] \tag{6}$$

where:

$$T_c(B) = T_{c0}\left(1 - \frac{B}{B_{c20}}\right); \tag{7}$$

$$J_{c0}(B) = \frac{\left[111\left(1 - \frac{B}{B_{c20}}\right)\right]^2}{B^{1/2}} \times 10^6 ; \tag{8}$$

$T_{c0} = 18$ K; $B_{c20} = 27.8$ T .

The θ variation of B around the torus can be approximated with sufficient accuracy using the first five terms of a Fourier cosine series in which the even terms are zero. The field is assumed to drop linearly to zero with turns into the winding (a good approximation in the inboard leg). The full dependence is then:

$$B = \frac{N_t - (i - 1)}{N_t}\left[(B_{max} - B_{min})\, F(\theta) + B_{min}\right] \tag{9}$$

where F(θ) is the Fourier series; B_{max} is 11.91 T; and B_{min} is 6.49 T.

The flow direction through the coil is radially outward because we feel that the large swings in ΔT_m and the absolute minimum, which occur in

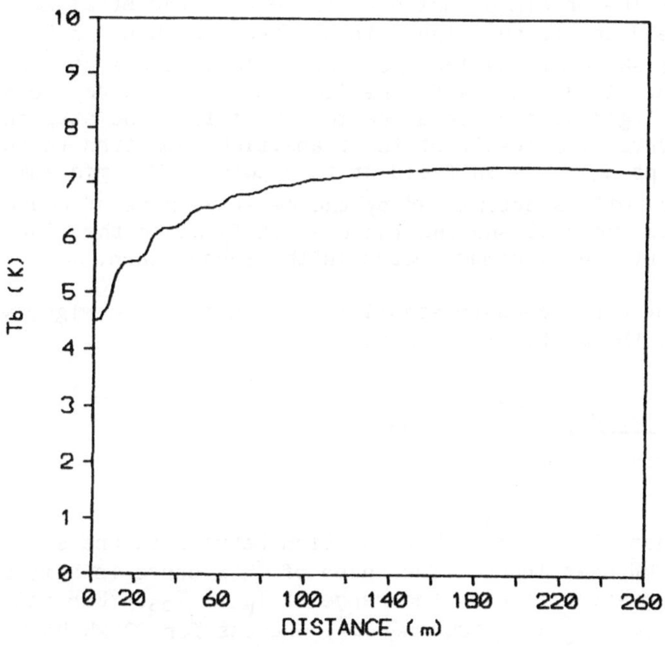

Fig. 2. Helium temperature along the flow path.

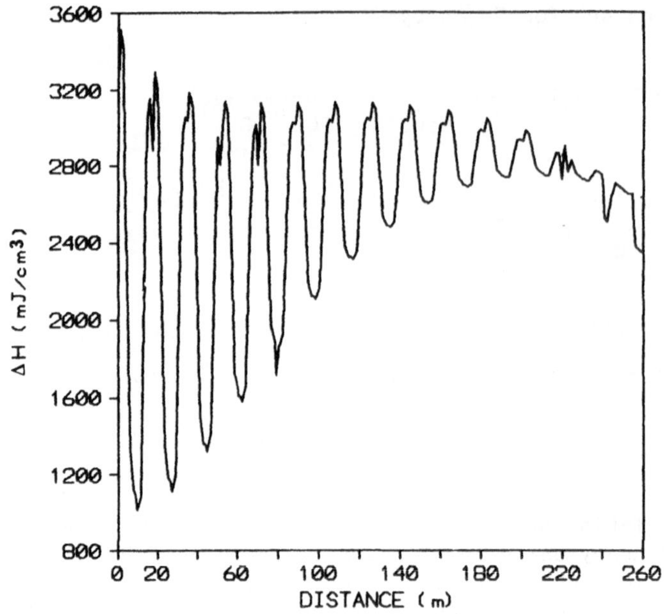

Fig. 3. Temperature margin along the flow path for heat input of 27 kW.

the inner turns, can be better controlled when p and ρ are high and T_b is low. The characteristic shape of ΔT_m for radially outward flow is shown in Fig. 3 for 27-kW heat input. The periodic nature of the curve results from the helium flow path around the torus in which B changes by almost a factor of 2. The change of B causes T_{cs} to be small in the high-field regions of the coil. As T_{cs} decreases, T_b increases so that the ΔT_m must decrease. The smallest value of ΔT_m occurs near the end of the straight leg where both the B-field and T_b are high. This effect is shown in Fig. 3 by the nonsymmetrical shape of the lobes of the periodic curve at the minimum in the first turn. At distances further into the coil where the nuclear heating is negligible, the lobes become symmetric. The step on the upper part of the curve is a result of the transition from turn to turn that causes an abrupt decrease in field at this point. The minimum value of ΔT_m throughout the coil is determined by the relative rate of decrease in B radially across the coil and the increase in T_b along the flow path. For 72-kW heat input, the minimum occurs in the second turn.

To guarantee an adequate stability margin in the design, we have specified that the stability margin, defined by

$$\Delta H = \frac{1 - f_{cond}}{f_{cond}} \int_{T_B}^{T_{cs}} \rho \, C_p \, dT \,, \qquad (10)$$

be not less than 300 mJ/cm^3. The relation between ΔH and s is shown in Fig. 4 for 27-kW heat input. The shape of this curve is similar to Fig. 3, since the integration is over the range of T_b to T_{cs}. For this heat input, the minimum value of ΔH is 1000 mJ/cm^3, whereas for 72-kW heat input, the minimum is 300 mJ/cm^3.

The ratio of operating current to the critical current (I/I_c) is also calculated. This ratio is small when T_{cs} is large, and it goes to one as T_{cs} approaches T_b. Therefore, the relation between the current ratio and s is 180° out of phase with our temperature and stability margins. The largest value of I/I_c is 0.52 for 27-kW heat input and 0.79 for 72-kW heat input.

CONCLUSIONS

A TF coil conductor has been designed for TIBER to operate at a nominal heat input of 27-kW. The design can tolerate a total heat input of 72 kW. The design was verified in calculations using details of the helium temperature and pressure as a function of flow distance through the winding pack. Details of the friction and nuclear heat input into the coil are included in the calculations, as are the compressibility effects in the helium. The temperature and stability margins and the current ratio were calculated over the length of the flow path, and the results verified that the stability margins we established are adequate for operating conditions in TIBER.

NOTATION

A_{cond}	Conductor area	f_{cond}	Conductor fraction in arcs
A_{cs}	Cable space area	f_{Cu}	Copper fraction in dw
B	Magnetic field	G	Mass flow
C_p	Specific heat	ΔH	Stability margin
D	Hydraulic diameter	I	Operating current
d_w	Wire diameter	I_c	Critical current
f	Friction factor	i	Turn number

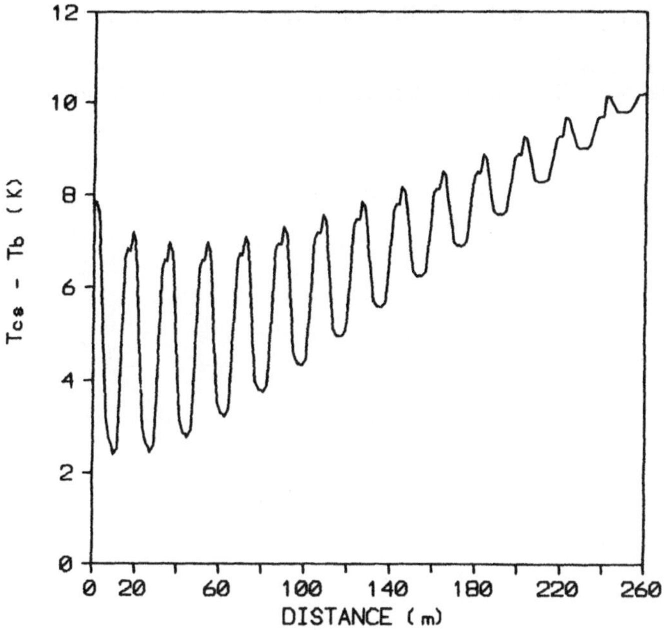

Fig. 4. Stability margin along the flow path for heat input of 27 kW.

J_c	Critical current density	Re	Reynolds number
K	Compressibility coefficient	s	Distance
N_t	Total turns in flow path	T	Temperature
p	Pressure	T_b	Bulk temperature
P_{He}	Cooled perimeter	T_{cs}	Current-sharing temperature
q''	Heat flux	ΔT_m	Temperature margin
q'''	Heat per unit volume		

GREEK SYMBOLS

β	Expansion coefficient	ρ	Helium density
ψ	Joule-Thompson coefficient		

REFERENCES

1. J.R. Miller et al., "High Current Density Magnets for Intor and TIBER," Lawrence Livermore National Laboratory Report UCRL-95759, Livermore, CA (1986).

2. V. Arp, Adv. in Cryo. Eng. 17:342-351 (1972).

3. B.A. Hands, "HEPROP--computer code for the thermodynamic and thermophysical properties of helium," third ed., Oxford University Engineering Laboratory Department Report 1289/79 (1979).

4. V. Arp., National Bureau of Standards, private communication (1987).

5. J.W. Lue, J.R. Miller, and J.C. Lottin, IEEE Trans. Mag. 15:53-55 (1979).

6. R.J. Hooper, Fusion Engineering Design Center memo FEDC-M-84-E/M-016, (March 5, 1984).

7. L. El-Guebaly, University of Wisconsin, private communication (1987).

Work was performed under the auspices of the U.S. Department of Energy by Lawrence Livermore National Laboratory under contract number W-7405-48.

SIMULATION OF NONSTATIONARY THERMAL REGIMES IN CRYOGENIC LOOPS FOR TOKAMAK-15 ELECTROMAGNETS

A.F.Volkov, L.B.Dinaburg, V.V.Kalinin,
A.B.Konstantinov, V.N.Khrushev

D.V.Efremov Scientific Research Institute of
Electrophysical Apparatus, USSR

ABSTRACT

Electromagnetic system (EMS) of Tokamak-15 contains superconducting toroidal field coil (TFC) and cryoresistive nitrogen cooled coils for plasma equilibrium and shap control[1]. EMS coils of Tokamak-15 are manufactured from hollow forced cooling conductors. The results of numerical simulation of some nonstationary thermal and hydrodynamic problems associated with forced cooling conductor application in Tokamak-15 EMS coils are outlined in the paper.

INTRODUCTION

The first problem of numerical simulation consisted in providing conditions for simultaneous and uniform cool down of all coils for Tokamak-15 under moderate thermo-mechanical stress in EMS structure. The second problem determined conditions of coil effective cooling under dynamic thermal loads of working cycle of the Tokamak-15. The third problem was concerned with protection of cryogenic loop of TFC against pressure rise under extreme thermal disturbances due to quenches and EMS cryostat failure.

ANALYSIS

For coils with forced cooling cool down is determined greatly by heat-exchange intensity between cryogen flows in adjacent turns and layers of the coil with thermal contact. This heat-exchange is essential for two--layered type coils.It results in considerable decrease in cool down rate. Two-layered coil elements may be designed as pancakes and cylindrical turns. Winding of pancakes is used in TFC and circular cryoresistive coils of Tokamak-15. The cylindrical two-layered winding is characteristic of the central solenoid (CS) (Fig.1). It is in this coil that interlayered heat-exchange is the most active[2]. Fig.2 shows results of temperature field calculation at CS cool down. A hollow aluminium CS conductor is 54*54 mm in cross-section. The cooling channal is 26 mm in diameter. The conductor in one layer is 135 m in length. It is clear from Fig.2a that flow outlet temperature becomes considerably lower that the maximal flow temperature in the coil. As a result flow cooling capacity, Qa, proportional to temperature flow difference at the inlet and outlet of the coil, is less than the maximal one, Q, proportional to difference between the maximal flow temperature in the coil and the at inlet. The experimen-

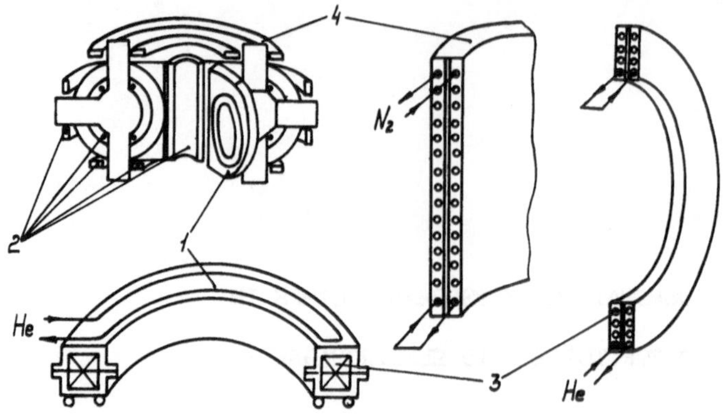

Fig. 1. Layout and principle of Tokamak-15 coils cooling.
1 - superconductive coil (TFC); 2 - cryoresistive coils;
3 - TFC pancakes; 4 - central colenoid (CS)

tal simulation of cool down has been carried out, taking into account that estimated thermal resistance of electrical insulation of coils may differ from the real. A copper conductor of 11.5*11.5 cross-section with a cooling channal 8 mm in diameter has been investigated. The conductor is wound as a two-layered cylindrical section. The composition and method of heat treatment for the electrical insulation are similar to those in CS. Fig.3a shows temperature fields when flow passes both layers connected in series. The regime of temperature jump has been investigated. Flow temperature was 50K lower than the initial temperature of an experimenal section. Fig.3b shows experimental data on changes in the ratio of real cooling capacity of cryogen flow Qa to the maximal one of flow Q for connections in series (line 2) and in parallel (line 1). Numerical simulation results and experiment were in a good agreement. It confirmed the possibility of applying the developed calculation technique to simulate cooling processes of full scale coils. The heat exchange between layers has been found to increase the CS cool down time by a factor of 1.5.

In the operating regim the cryoresistive coils are cooled by nitrogen flow pressurized to 0.8 MPa and cooled to 78 K. Coils are heated within 7 second due to ohmic losses. An interval between current pulses last 600 s. The numerical simulation was to determine the nitrogen flow rate to recover the temperature of coils by the start of the following current pulse. Fig.2b shows temperature fields in CS cooling channals. As early as after the 3rd pulse (1228s) the cyclic repeating dynamic fields are observed. In other coils the heat exhange between layers influences negligibly

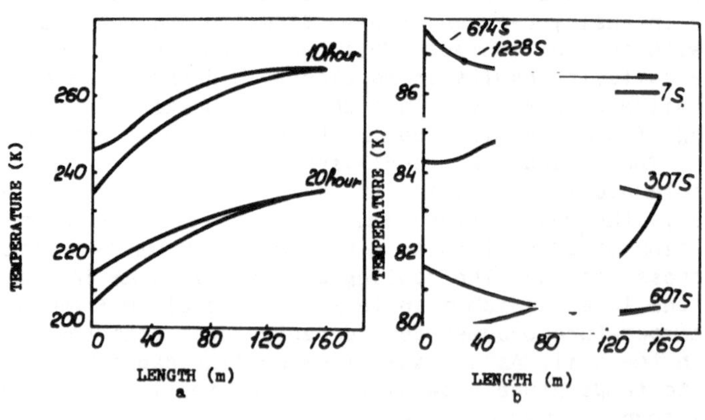

Fig. 2. Temperature fields in CS (calculation)
a - cooling down; b - operating regim

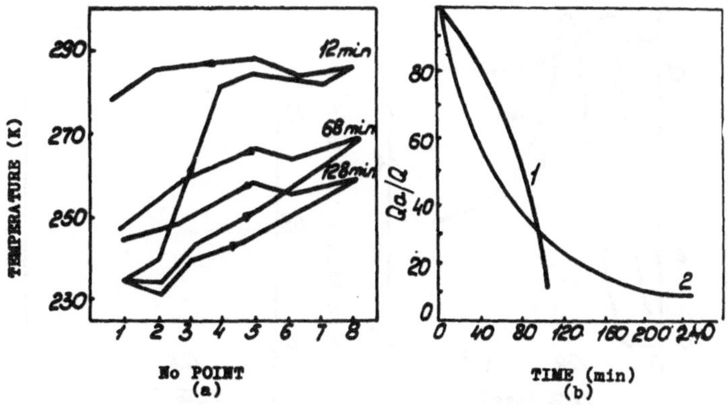

Fig. 3. Temperature field of CS experimental section (experiment)
a - temperature; b - relative cooling capacity

the cool down time. The principal difference lies only in changing temperature fields in coils. For example, in TFC two-layered pancakes (Fig.4a) the temperature increase and decrease are observed along every turn. Fig.4b shows cool down computation result for TFC, the weight of which is 260 tons. The regime of linear cryogen flow temperature decrease at the inlet of TFC is considered. Into the calculation procedure a condition, providing moderate value of thermomechanical stresses in TFC, is introduced. According to this condition the flow heating should not exceed 40K. At the initial helium temperature decrease rate of 5 K/hour its value limiting the heating flow regulates cool down in about 20 hours. From this moment the temperayure decrease rate at the inlet amounts to 4.3 K/h on the average. Fig.4c shows the rate of helium temperature decrease at the inlet and outlet in cooling down the test block of TFC[3].

TFC in operating regime is cooled by supercritical (0.5 MPa) or two-phase (0.16 MPa) helium. Helium flow is provided by cryoejectors[4]. TFC is effected by steady-state heat load (900W), as well as by dynamic cyclic

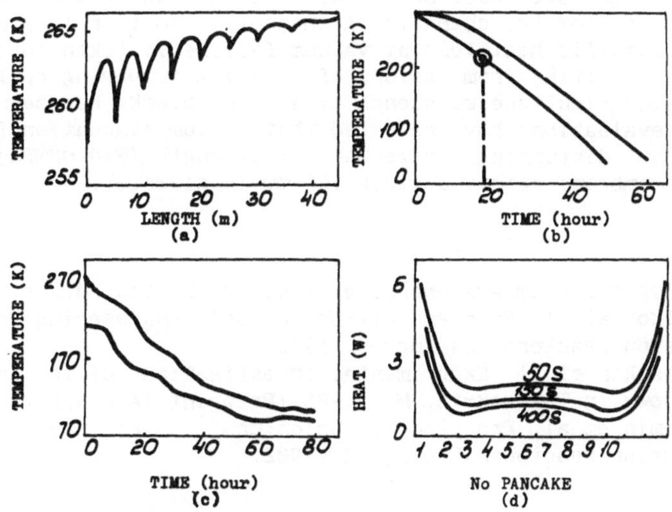

Fig. 4. Superconductive coil cooling (TFC)
a - temperature field during cool down (calculation); b - cool down rate (calculated); c - cool down rate (experimental);
d - heat load in TFC block pancakes

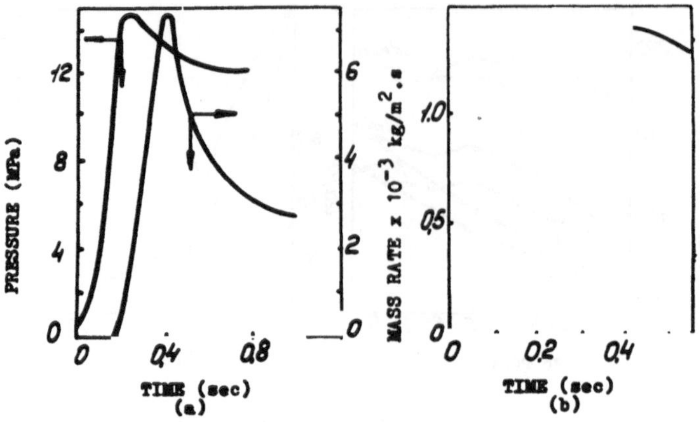

Fig. 5. Helium evacuation TFC cooling channels
a - pancake; b - block case

heat load due to losses under variable magnetic field (300 W). To reduce direct thermal influence on the conductor the mechanical cases of each TFC block (24 in number) are provided with cooling channels. This channels are arranged along the perimeter of side plates of cases. Side plates of cases being in close thermal contact with bounday pancakes of block are considerably affected by heat. Evaluation results of heat load distribution between 12 pancakes of TFC block are shown in Fig.4d. The heat load into the boundary pancakes is seen to be 2.5-3 times as high as into the central ones. During EMS operation extreme situations, resulting in the large transient heat disturbances in TFC, are possible. Fig.5a shows calculation results of pressure increase dynamics and helium output rate from the conductor channels at quench. This quench was simulated by momentary heating of the whole conductor (340 m) up to 30 K, transport current being 5.3 KA. The total helium flow evacuated from the TFC is proportional to output mass flow rate, chanel cross-section area in the conductor and the number of blocks in quenches. Extreme heat loads due to vacuum failure in cryostat affect primarily helium in cooling channels of the block cases. Heat flow penetrating the pancakes of the block is not great during relatively long period, as massive cases are slowly heated. Fig.5b shows helium evacuation dynamics from the channels. A channel is 18 m in length and 14 mm in diameter. Specific heat load at vacuum failure is taken to be 10^4 W/m^2. The situation resulting from failure of a helium supplying collector of the TFC coil and simultaneous quench of all TFC blocks has been investigated. The evaluations have revealed that helium evacuation from the cryostat without disturbing its mechanical strength (P=0.02MPa) is possible with 8 membrane safety devices 150 mm in diameter.

REFERENCES

1. V. P. Belyakov. Atomnaya energiya, 1982, V.55, 2nd Ed., p.101-108.
2. A. F. Volkov et al. Proc.2nd All-Union Conf. Engineering Problems of Fusion reactors, Leningrad, 1982.
3. I. O. Anashkin et al. Experimental investigation of TFC test block cool down in Tokamak-15. M., 1986 (Preprint IAE: 4320/10).
4. N. N. Barmin et al. Proc.2nd All-Union Conf. on Engineering Problems of Fusion Reactors, Leningrad, 1982.

THE TRANSIENT STABILITY OF LARGE SCALE SUPERCONDUCTORS COOLED IN SUPERFLUID HELIUM

J. Waynert

Astronautics Technology Center
Madison, Wisconsin

Y. Eyssa and X. Huang

Applied Superconductivity Center
University of Wisconsin-Madison
Madison, Wisconsin

ABSTRACT

This paper discusses the transient stability of large scale, cylindrical and rectangular, monolithic superconductors cooled by He II. The analysis assumes a long normal zone exists such that end effects may be neglected. The time-dependent joule heating of the conductor is calculated based on the magnetic diffusion of the transport current into the stabilizing matrix. The effects of cool-down and cyclic strain, and magnetoresistivity, on the high purity stabilizing material are considered. Results for transient heat transfer in the superfluid for both geometries are presented. The effects of using temperature dependent thermodynamic properties for the superfluid are discussed. Both the heat generation and cooling as functions of time are used to discuss conductor design and the associated stability margin.

INTRODUCTION

It has been proposed that large scale superconducting magnets, for example Torus Supra[1], or large scale superconducting magnetic energy storage (SMES) devices[2], use subcooled superfluid helium to make use of its enhanced heat transfer properties. Another more recent application is for energy storage and power conditioning in space applications[3] in which transient stability of the conductor rather than cryostability is being considered in an effort to reduce the conductor mass. To ensure the high reliability desired for these devices and to obtain the maximum use of the increased heat transfer properties of the superfluid, requires a thorough knowledge of the stability of the conductors. Previous analyses have lacked either the detailed knowledge of the time-dependent heat generation in the conductor and heat transfer to the superfluid or have not had the benefit of recently developed stability models.

Transient stability refers to the ability of a superconductor in which a section has been driven normal (non-superconducting) to recover its superconducting properties before steady-state conditions have been established. The argument for recovery is that the energy of the thermal disturbance plus the normal-state joule heating energy must, at some time, be balanced by the time-dependent available enthalpy of the helium. The argument would be applied to a situation in which the normal, steady-state joule heating exceeds the cooling capability of the helium as shown in fig. 1. This paper considers a variation in which the conductors are designed to be cryostable[4] but the conductors in the inner layers of the inductor only have access to the large helium volume via channels.

The transient stability model begins with an energy disturbance, say the small movement of a conductor against the supporting structure during a load change, which deposits thermal energy in the conductor raising its temperature above the current sharing temperature. The current must diffuse throughout the conductor until a uniform current density is achieved. The result is heat generated in excess of the steady state joule power loss because of the initially high values of current density.

The joule heat flux generated is compared to the maximum allowable heat flux in the superfluid. The area under the appropriate heat flux curve yields the energy deposited (joule heating) or removed (helium cooling) per unit surface area. If at some point in time the two energies are equal, the conductor may recover at that time. See fig. 1. This is refered to as the equal-area criterion by Seyfert <u>et. al.</u>[5].

Two conductor and cooling bath geometries are considered. In one case, the conductor is rectangular with one surface exposed to a channel of He-II. In the other case, the conductor is cylindrical with half its circumference exposed to an axially symmetric cylindrical bath. See fig. 2. In both cases, the superconducting filaments are assumed to be on the outer surface of the conductor.

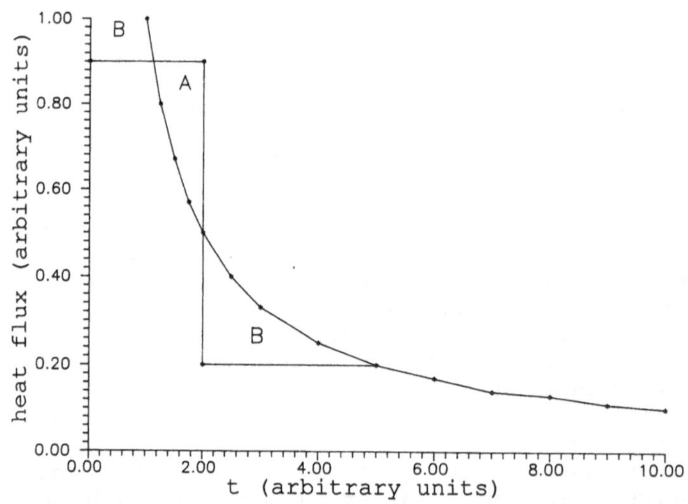

Fig. 1. The cooling capability of superfluid helium is represented by the smooth curve. The upper horizontal line represents the energy deposited by a thermal disturbance, while the lower one represents the steady-state joule heating. The conductor may recover when the excess heating (area A) equals the excess cooling (area B).

It was thought that the cylindrical geometry offered several advantages over the rectangular geometry. First, the current should diffuse faster over the same distance for the cylindrical geometry since the area is decreasing as the current diffuses inward. Second, the cylindrical geometry offers increasing helium enthalpy as the heat pulse travels radially outward. The actual results are discussed in the heat transfer section and the excess heat generation section.

In the conclusions section, the results are summarized, the stability margins are discussed, and the implications for conductor design are presented.

HEAT TRANSFER IN SUPERFLUID HELIUM

Heat transport in superfluid helium does not obey Fourier's linear law of conduction. Instead, a nonlinear, empirically determined form has been proposed by Gorter and Mellink[6] for heat transport in linear channels:

$$q'' = -k(dT/dx)^{1/3} \tag{1}$$

Here k is a temperature dependent (frequently taken as constant) conductance parameter and q" is the heat flux in W/cm^2. Fig. 1 shows the conductor as a heat source at temperature T_c and the He II bath at initial temperature T_b. The variables x and r for linear and cylindrical geometries respectively, are also shown.

The continuity equation, with the appropriate form of the divergence, is used with equation 1 to obtain the equation for the time-dependent temperature distribution.

$$\text{div } q'' = -C(\delta T/\delta t) \tag{2}$$

The result is used in equation one, evaluated at x = 0, to derive the heat flux at the conductor - helium interface.

Several choices of boundry conditions are possible. The isothermal boundry condition assumes the right or outer boundry is at a fixed temperature as if in contact with a large helium reservoir or a refrigerated surface. Or there might be zero heat flux if the boundry is insulated, an adiabatic boundry. The helium boundry conditions at the conductor interface can also be for constant heat flux or, as is more representative in the case of a normal zone heat source, at a clamped temperature of T_L, the lambda temperature. Here it is assumed that for large scale superconducting applications, the superfluid would be at atmospheric pressure. The appearance of a normal zone in the conductor would probably be accompanied by a thin film of He-I at the surface. Thus, the superfluid surface essentially coincides with the conductor surface and is at the lambda temperature.

The transient heat balance or continuity equation for superfluid helium has been solved by several people for various boundry conditions[7,8]. The results of a finite difference approach show that for times less than the thermal diffusion time constant T_D, the transient solution closely follows the solution for an infinite channel with a clamped temperature T_L, i.e.

$$q'' = 2.62t^{-1/4} \qquad W/cm^2 \tag{3}$$

with t in seconds. This result is valid for $t \ll T_D$ since the outer boundry is unaware of the existence of the thermal pulse. It is also approximately independent of whether temperature dependent or constant values are used for the conductance parameter and specific heat. Fig. 2, curve a, illustrates the cooling capability q_H, of the superfluid helium for the very short times associated with transient stability for the linear geometry.

The cooling curves shown in fig. 3 for the cylindrical geometry were obtained by replacing x by r in equation 1 and using the divergence in cylindrical coordinates in the continuity equation. The resulting differential equation for the temperature distribution was solved numerically using the finite difference technique. Results are shown for an effectively infinite bath, curve a and an isothermal outer boundry at a radius of 9 cm. In both cases the conductor radius is 3 cm. Differences in the calculated heat fluxes using the temperature dependent versus the temperature independent values for the thermal properties is typically 0.25 W/cm^2 (temperature dependent values being less). This difference is relatively small whereas the computation time is several orders of magnitude longer for the temperature dependent case.

As might be expected, the heat flux handling capability of the superfluid is greater for the cylindrical geometry for the dimensions chosen. For the short times considered for transient stability, the heat flux is limited by the available helium enthalpy. Since more helium is available in the cylindrical geometry, the maximum heat flux at all times is greater than for the linear channel.

EXCESS HEAT GENERATION IN THE CONDUCTOR

The conductor being analysed is a large monolithic, aluminum-stabilized, superconducting conductor, capable of carrying 228.8 kA. Such a conductor has been proposed for use in a magnetic energy storage

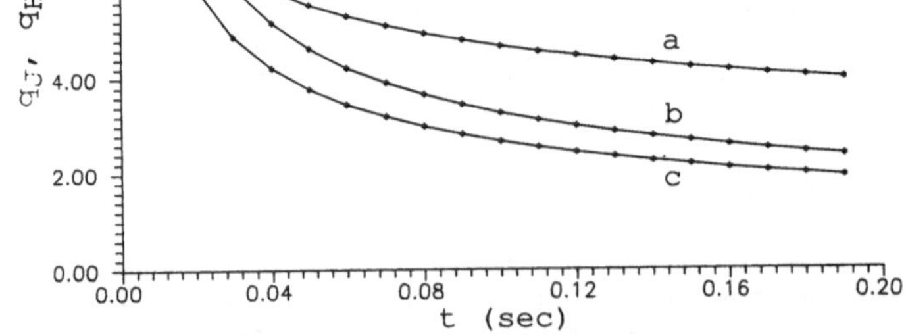

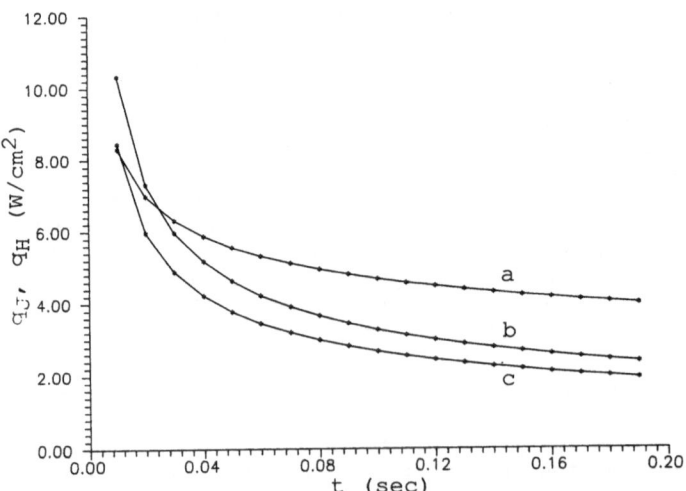

Fig. 2. Curve a is the maximum allowable heat flux in an infinite channel. Curves b and c are the joule heat per unit surface area in an aluminum conductor 3.0 X 9.42 cm carrying 228.8 kA with RRR's of 400 and 600 respectively. Flux is through the 9.42 cm surface.

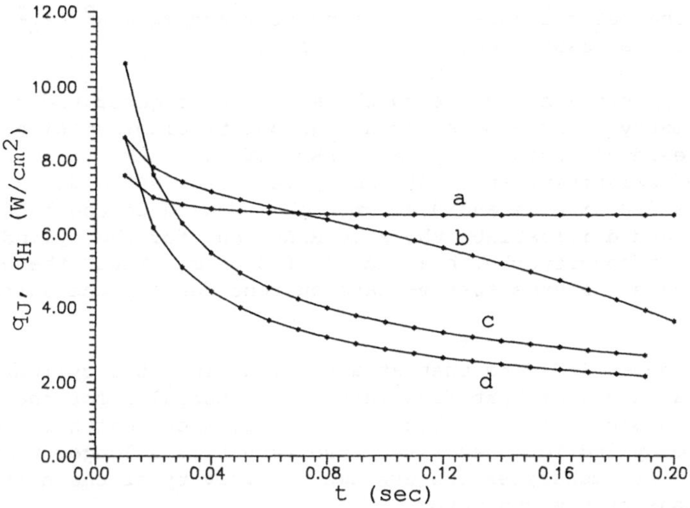

Fig. 3. Curve a is the maximum allowable heat flux in an infinite outer-radius cylinder with the inner radius of the helium at 3.0 cm. Curve b is for an adiabatic boundry at 9.0 cm. Curves c and d (RRR = 400, and 600 respectively) are the joule heat per unit surface area in an aluminum conductor of 3.0 cm radius carrying 228.8 kA.

device in conjuction with electric utility diurnal load variations. The analysis considers both rectangular and cylindrical conductors with the superconducting filaments on the outer surface. If the conductor experiences a thermal disturbance which deposits enough energy to raise the temperature of the conductor above the current sharing temperature, the current in the superconducting filaments will diffuse throughout the stabilizer. The current diffusion takes time. During this time, the value of the current density near the filaments is greater than it will be in the steady state. The result is an excess of joule heat q_J being generated until the steady state is reached. Also, a long normal zone is assumed so that end effects may be neglected.

To find the amount of excess heat generated, the diffusion equation for the current density J, is solved for the appropriate geometry. The conductor is approximately at a uniform temperature because the thermal time constant is much less than the magnetic diffusion time constant. Thus, J is a function of position and time only. It is assumed that at t=0 the current density is described by a delta function distribution which is non-zero only at the conductor surface. Then the integral of $J^2\rho$ (ρ is the resistivity) over the cross-sectional area of the conductor yields energy per unit length being generated as a function of time. The results for the surface heat flux for the rectangular conductor of width 2a, a height h and magnetic diffusivity D carrying a current I, is

$$q''(t) = (I^2\rho/ah^2)[1 + 2\Sigma\exp(-2n^2\pi^2Dt/a^2)] \quad (4)$$

and for the cylindrical geometry

$$q''(t) = (I^2\rho/2\pi^2R^3)[1 + \Sigma\exp(-2t/T_n)] \quad (5)$$

where R is the outer radius of the conductor and $T_n = R^2/Da_n{}^2$ and a_n is n'th zero of the Bessel function of order 1.

Fig. 2, curves b and c show the excess heat generated for the channel geometry. Curve b is for a high purity aluminum with an effective residual resistivity ratio RRR, of 400 (RRR = room temperature resistivity/resistivity at 4.2K) and curve c, for RRR = 600. The conductor is 3.0 cm wide and 9.42 cm high (exposed to the bath). Fig. 3, curves c and d illustrate the same RRR's but for the cylindrical geometry for a conductor with a radius of 3.0 cm. Thus, these two conductors have the same steady-state current density and surface heat flux.

The results indicate that at any given time, the cylindrical conductor has a higher heat flux than the rectangular for the conductor dimensions chosen. This is because the diffusion length is twice as long for the cylindrical geometry. The result is a longer time constant. This emphasizes the extreme sensitivity of the heat flux to the dimensions of the conductor.

EQUAL AREA CRITERION

The joule heat energy deposited per unit surface area from t=0 to some arbitrary time t, is found by integrating equations 4 and 5. The results for the rectangular and cylindrical geometries are respectively

$$E_J(t) = (I^2\rho/ah^2)[t + a^2/6D - (a^2/\pi^2 D)\Sigma(1/n^2)\exp(-2n^2\pi^2 Dt/a^2)] \quad (6)$$

$$E_J(t) = (I^2\rho/2\pi^2 R^3)\{t + \Sigma(T_n/2)[1-\exp(-2t/T_n)]\}. \quad (7)$$

The results can be compared to the allowable enthalpy capacity of the superfluid helium E_H, found by integrating its maximum heat flux. In the case of the channel geometry, the heat flux for an infinite channel is used for simplicity. Numerical integration of the finite difference results for q" is used for the cylindrical geometry.

Figs. 4 and 5 illustrate the difference E_J-E_H for the two geometries and the the two RRR's. According to the equal area criterion, the conductor can first recover at a time when the curve crosses the x-axis. (All the conductors presented appear to eventually

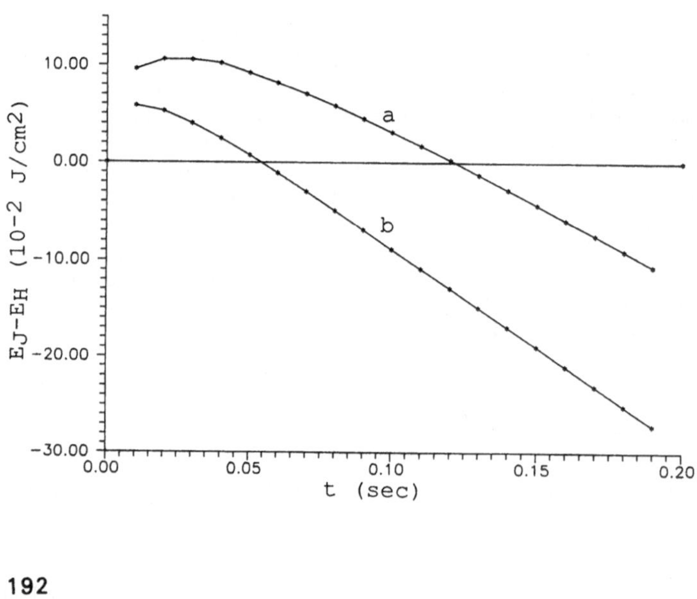

Fig. 4.

The difference between the heat energy deposited E_J and the energy removed by the superfluid E_H for channel geometry. RRR = 400, and 600 for curves a and b respectively.

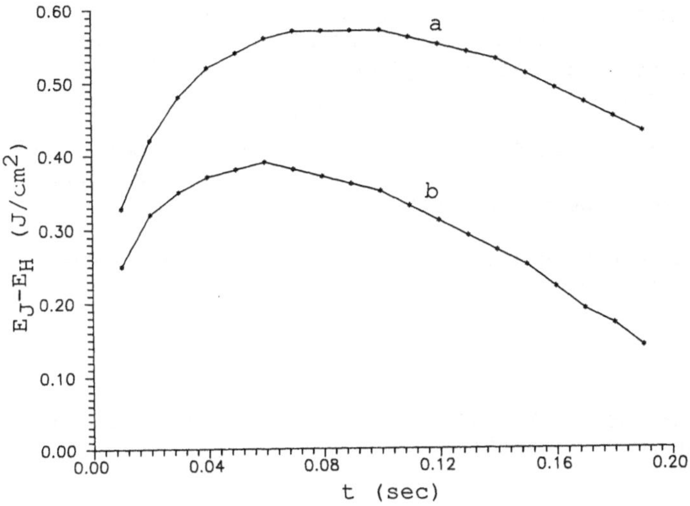

Fig. 5.

The difference between the heat energy deposited E_J and the energy removed by the superfluid E_H for cylindrical geometry. RRR = 400, and 600 for curves a and b respectively.

recover <u>if enough helium is available</u> since they have been designed to be cryostable.) For the conductor dimensions chosen, this happens only in the case of the rectangular conductor. Of course the disturbance used represents the minimum effect of a normal zone and does not account for the energy deposited to raise the conductor into the current sharing regeme. For example, the energy needed to raise the conductor temperature from 2 to 8 K is about 0.035 J/cm^2. A stability margin of 0.035J/cm^2 corresponds to a conductor movement of 80 microns if the coefficient of friction is 0.5. From the graph of fig. 4, recovery could occur after about 145 msec for RRR = 400.

As seen in equations 6 and 7, a controlling factor in the behavior of the heating curves is the time constant. Greater transient stability is achieved when smaller time constants are used. This means using the highest possible RRR aluminum. Aluminum is chosen over copper for the stabilizer since it is easier to obtain lower resistivity aluminum and the magnetoresistivity of aluminum is much less than that of copper.

The RRR values of 400 and 600 chosen for the example reflect the degradation of the electrical resistivity from monotonic and cyclic strain, and the magnetoresistivity. An initial RRR of 10,000, decreases to about 7370 due to cool-down strain (0.4%). A cyclic strain range of 0 to 0.1% further degrades the resistivity to an effective RRR of 930. Finally, a magnetic field of 3.5 T further reduces the RRR to an effective RRR of about 450[9].

CONCLUSIONS

The results of this study indicate that large, monolithic superconductors cooled by superfluid helium may achieve transient stability. Greater stability margins are obtained by reducing the excess joule heat energy being produced by the diffusing transport current. This is accomplished by minimizing the distance filament currents must diffuse to achieve uniform current density and/or reducing the steady-state current density.

By increasing the aspect ratio of rectangular conductors (height exposed to He bath/width), it is possible to reduce the current diffusion length dramatically, thereby decreasing the time required to recover from a thermal disturbance. Similar results can probably be obtained in cylindrical conductors by increasing the outer radius and creating a hollow region in the interior. In either case, extreme attention must be paid to the conductor design to achieve reasonable stability margins. Expectations need to be verified experimentally, especially since no consideration has been given to heat transfer effects at the helium-conductor interface.

Temperature dependent values for thermal conductance parameter and the specific heat yield superfluid heat fluxes about 4% less than those obtained from using averaged thermal properties.

Transiently stablized conductors offer the prospect of a reduced conductor mass which makes superconducting magnetic energy storage a more attractive option for space-based applications.

ACKNOWLEDGEMENTS

This work was supported by the United States Department of Energy under contract number DE-AC03-85SF15934.

REFERENCES

1. R. Aymar et al, Conceptual Design of a Superconducting Tokamak: Torus II Supra, IEEE Trans. on Magnetics, 15:542 (1979).

2. R. W. Boom et al, Cryogenic Aspects of Inductor-Converter Superconductive Magnetic Energy Storage, in "Proceedings of ICEC-9", Butterworth, Buildford, UK (1982), p. 731.

3. M. K. Abdelsalam, Y. M. Eyssa, Pulsed Magnetic Energy Storage for Space Applications, presented at Applied Superconductivity Conference, Baltimore, MD, 1986, to appear in Trans. on Magnetics.

4. Y. M. Eyssa, X. Huang, J. Waynert, Heat Transfer in Helium II for Two-Layer Energy Storage Magnets, presented at Applied Superconductivity Conference, Baltimore, MD, 1986, to appear in Trans. on Magnetics.

5. P. Seyfert, J. Lafferranderie, L. Claudet, Time-dependent Heat Transport in Subcooled Superfluid Helium, Cryogenics, 22:401 (1982).

6. C. J. Gorter, J. H. Mellink, On the Irreversible Processes in Liquid Helium II, Physica XV:285 (1949).

7. P. Seyfert, Practical Results of Heat Transfer to Superfluid Helium, Stability of Superconductors, Intern. Inst. of Ref. Comm. A 1/2 (1981) p. 53.

8. L. Dresner, Transient Heat Transfer in Superfluid Helium, in "Advances in Cryogenic Engineering," Vol. 27, Plenum Press, New YOrk (1982), p. 411.

9. K. T. Hartwig, G. S. Yuan, P. Lehman, Strain Resistivity at 4.2 K in Pure Aluminum, in: "Advances in Cryogenic Engineering - Materials,", Vol. 32, PLenum Press, New YOrk (1986), p. 405.

EVOLUTION OF NORMAL ZONES AND TEMPERATURE IN A SUPERCONDUCTOR COOLED INTERNALLY WITH HELIUM-II

E. R. Canavan and S. W. Van Sciver

Applied Superconductivity Center
University of Wisconsin-Madison
Madison, Wisconsin

ABSTRACT

In an effort to understand the stability mechanisms of internally cooled superconductors in the He II regime, measurements are made on a hollow composite superconductor open at both ends to a bath of subcooled He II. The vacuum insulated conductor, 3 meters long and 1.1 millimeters internal diameter, carries a transport current in a transverse background field of up to 11 Tesla. Heaters mounted on the conductor at its center are used to initiate local disturbances. The propagation or collapse of these disturbances is then observed using a series of temperature sensors and voltage taps attached symmetrically about the heater. The stability margin of the conductor is mapped out, and, using a simple theory, the pressure rise and thermal expulsion velocity are derived from the voltage and temperature profiles.

INTRODUCTION

A magnet cooled internally with Helium-II would retain all the advantages of the internally cooled design, but, for fields up to 13 Tesla, brittle A-15 compounds could be replaced by NbTiTa, thus avoiding many fabrication difficulties. For space applications, superconducting magnets will probably use an internally cooled design with He-II because pool boiling is not effective without gravity and because helium will most likely be stored below the lambda point to reduce the weight of the dewar structure. In addition to the improved superconductor properties available at reduced temperatures, the extremely high effective thermal conductivity of He-II means that the entire enthalpy reserve between the bath temperature and the lambda temperature in the region of a thermal disturbance is available for absorbing heat. Further, after a temperature excursion from which the conductor recovers, the energy in the warmed section of liquid diffuses away in a time much shorter than the residence time of the liquid in the flow path. Unfortunately, while a great deal of work has been done on internally cooled conductors, particularly internally cooled cabled superconductors (ICCS), very little of this work has investigated the region below 4.2 K[1]. Also, most of the experiments have dealt with disturbances which deposit heat uniformly over the entire length of the conductor. While the consequences of a disturbance which drives an entire flow path normal all at once are the most severe, it is much more likely that a sudden heat input from, for example, conductor motion will occur only over a small region. There is reason to believe that there should be a significant difference between these two situations. When an entire flow path goes normal, the strong transient flow generated by the expanding helium greatly increases the heat transfer in the outer regions of the conductor. However, when the disturbance is local, the initial transient flow is not as strong, and in fact the expansion of the hot gas is what causes the normal zone to spread. The experiment described in this paper measures the stability against local disturbances of a superconductor cooled internally with Helium-II.

DESCRIPTION OF THE EXPERIMENT

The conductor used in this experiment is wound into a single solenoid and mounted inside a vacuum can, as shown in Fig. 1. The ends of the conductor are attached to large copper tubes (I.D. = 1.3 cm) which pass through feed-throughs at the top of the can and are open to the bath. The vacuum can fits into the 10 cm bore of a NbTi magnet. The magnet, designed for operation at 9 Tesla in normal helium, will achieve over 11 Tesla in subcooled He-II. The magnet and the experiment fit inside a vacuum insulated can which is cooled below the temperature of the main bath with a heat exchanger containing He-II at saturated vapor pressure.

The conductor is shown in the inset of Table 1. It consists of a copper tube with six composite superconductor wires bonded to its circumference with lead-tin solder. Some of its parameters are described in Table 1. Although the fraction of copper is very high compared to a typical ICCS, most of it is the low conductivity copper of the tube. Because of this, the resistance per unit length is equal, at 7 Tesla, to that of a conductor with half the volume of 100 RRR OFHC copper. In this way, the problem of avoiding buckling in a small thin walled tube of soft copper was circumvented. In

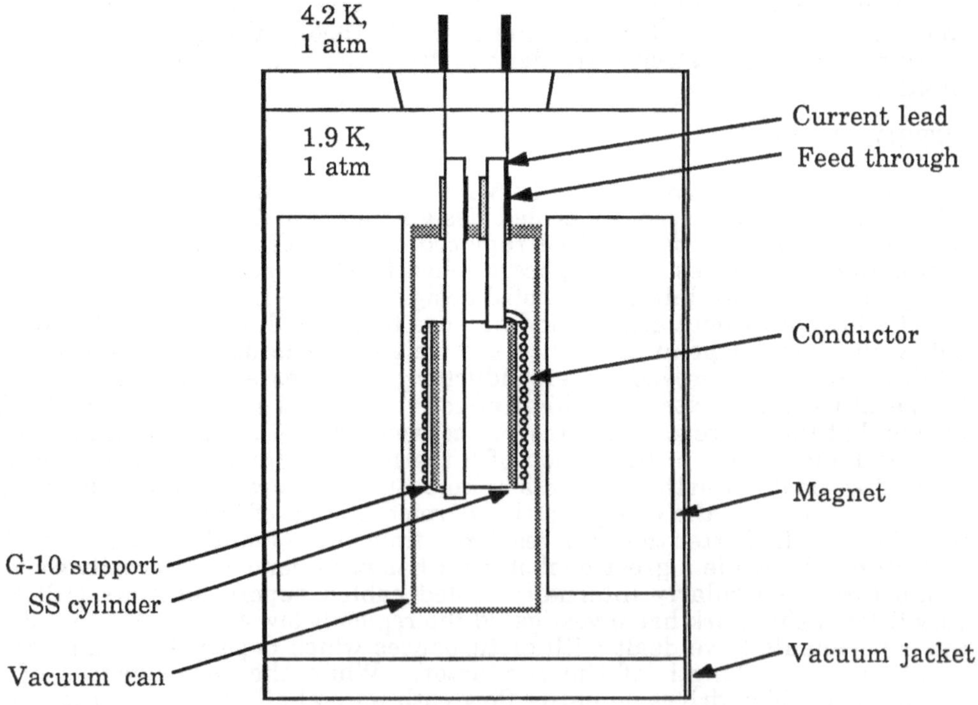

Fig. 1. Schematic diagram of the experiment.

Table 1

Conductor Parameters

Outer diameter	1.9 mm	
Inner diameter	1.1mm	
Wire diameter	0.51 mm	
Number of composite wires	6	
Fraction of helium = A_{He}/A_{Total}	0.23	
Overall Cu : S/C ratio	11.9	
Conductor length	3.188 m	

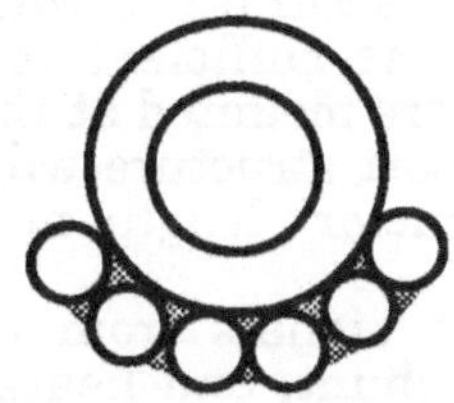

Sensor positions

Voltage Tap	Position [m]	Temperature Sensor	Position [m]
1...... {	-1.500	1	-0.600
	+1.500	2	-0.244
2...... {	-0.600	3	-0.130
3...... {	-0.250	4	-0.075
4...... {	-0.125	5	0.000
5...... {	-0.075	6	+0.075
6...... {	+0.075	7	+0.125
7...... {	+0.125	8	+0.250
8...... {	+0.250	9	+0.600
	+0.600		

spite of its relatively high resistivity, the thermal diffusion time across half the tube's circumference is estimated to be only 50 μs, much smaller than other important time constants of the experiment. While the conductor does not seem to bear much resemblance to a force cooled ICCS, it should provide an effective model and yet still allow certain simplifications. Although the liquid in the conductor is stagnant, many experiments have shown that the steady state flow velocity has a relatively small effect on stability because the strong transient flow generated when a section of the conductor turns normal is much larger than typical background velocities. Furthermore, though the helium flow path is a simple cylindrical tube, because the inner diameter is so small, the ratio of the wetted conductor perimeter to the cross sectional area of the helium space is approximately the same as in experiments using the cable in conduit configuration, most notably the triplex cable used by Lue, Miller and Dresner.[2] Thus, for the same current and resistance per unit length, the heat flux into the helium as well as the time to vaporize the adjacent helium should be about the same. The Fanning friction factor often used for cable-in-conduit conductors, 0.013, is substantially higher than that for a tube of this roughness (0.0075 at Re = 10^5), so for the same hydraulic diameter and expulsion velocity, the pressure drop in the tube will be about 40% lower. However, flow in tubes is better characterized and the friction factor is known to a higher degree of accuracy. This geometry has the further advantage that it allows the conductor temperature to be easily monitored.

Table 1 also shows the location of voltage taps and temperature sensors. The temperature sensors used in this experiment are 75 Ohm, 1/8 Watt Allen

Bradley resistors which are ground flat until they reach a resistance of 100 Ohms. Before mounting, they are thermally cycled ten times, then calibrated in field against the vapor pressures of helium, neon, and nitrogen. They are then bonded directly to the conductor with a thin layer of epoxy. The calibration curves of Allen Bradley resistors do continue to shift with repeated thermal cycling. This can be partially accounted for by subtracting the offset from the known bath temperature, but the calibration curves are still only accurate to within approximately two percent. Fortunately, this accuracy is sufficient for the present measurements. The temperature sensors are mounted at the position of the voltage taps, or as close to them as the support structure will allow. A sensor is also mounted at the center of the conductor's length, next to the 10 cm and 20 cm resistive heaters.

The signals from the temperature sensors, voltage taps, transport current shunt, and heater voltage taps are sent through a set of isolating amplifiers and read by a multiplexed A/D directly into a microcomputer. A D/A is used to control the power supply which drives the heater. In a typical shot of the experiment, the computer will start the A/D, wait a short period to obtain a baseline from which to calculate offsets, then fire a 10 millisecond heater pulse and continue to collect data for a set period. It then converts the resistor signals to temperature and displays the data in graphical form, as well as storing it for later analysis. If the coil quenches, a voltage comparator reading the signal from the full coil voltage tap shuts off the current before the normal zone propagates to the ends of the conductor. In this way, very severe heating is avoided and the conductor cools down to bath temperature in less than ten minutes. For each coil current, the reference voltage of the quench detector is turned to a small value, and a series of pulses fired to find the maximum pulse energy for which the conductor recovers. The energy limit is determined to within a half a percent. The reference voltage is then returned to its higher setting and data sets recorded for pulse energies just above and below the stability boundary. Some data is also taken with heat pulses five and ten percent above this limit.

RESULTS

Stability Boundary

Because an internally cooled conductor is a metastable system, its stability is characterized by the largest excursion from which it will return to its initial state. The usual measure of this is the stability margin, the largest amount of heat per unit volume of metal deposited suddenly and uniformly over the conductor from which it can recover. Unfortunately, for two reasons this quantity is not appropriate for describing the conductor used in the present experiment. Firstly, the heat pulse is not applied uniformly, but rather only over a small region. Secondly, the volume fraction of copper is much higher than that of a practical conductor, even though the resistance per unit length, and thus the Joule heating power per unit length, is in the proper range. Since the intent of this experiment is more the understanding of the mechanisms of stability rather than the design of an optimum conductor, the measure used here is the heat deposited in the conductor per unit volume of helium under the heater. This stability limit at 7, 9, and 11 Tesla is shown in Fig. 2.

The most striking feature of the stability diagram is the shape of the stability curve at 9 Tesla. The stability limit is multi-valued for currents between 119 and 133 Amperes. This phenomenon has been observed before, but only in experiments using uniform heating and bath temperatures well above the lambda point. Basically, the flat part of the curve at high current is determined by transient heat transfer in the helium, either transient Gorter-Mellink diffusion below the lambda line, or normal thermal diffusion

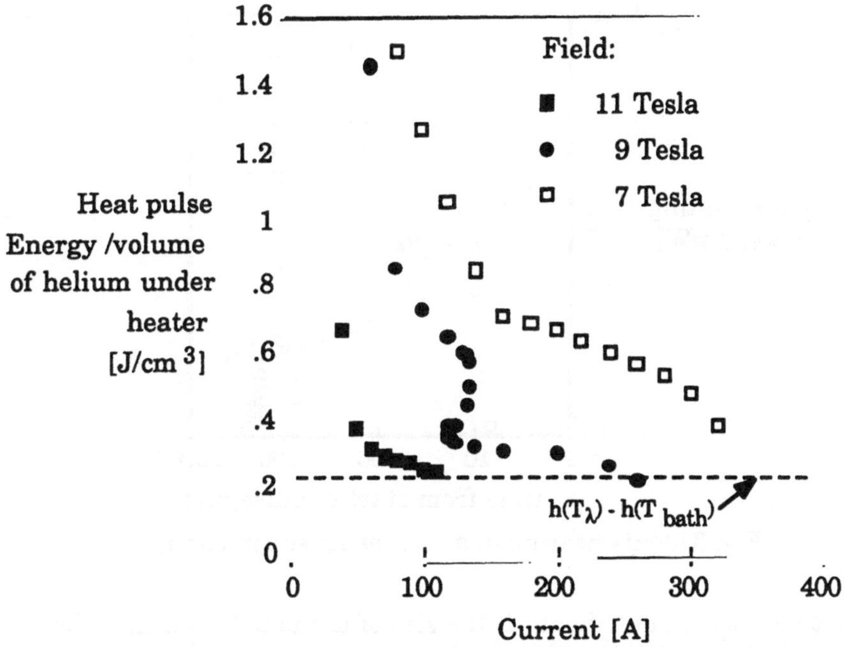

Fig 2. Stabiltiy limit as a funtion of current.

above it.[2] At somewhat lower currents, stability is determined by convective heat transfer enhanced by the strong transient flows induced when bubbles form over the heated region and expand rapidly. In the transition between these two regimes, it is possible to have a situation where, at a given current (⇒ current sharing temperature, Joule heating power), energies in a certain range are too large to be absorbed by the conduction mechanism, but not large enough to induce flow sufficient to cause stabilization, while at a somewhat higher level the necessary flow is induced. Unfortunately, although Dresner has developed a scaling law for the current at which the transition between the two regimes occurs,[3] it is not known why in some configurations the transition region is multi-valued, and in others it is smooth.

Because the effective thermal conductivity of He-II is very high, it is expected that, even for very short heat pulses, the temperature profile of the helium in the heated section of the tube will be approximately constant in the radial direction until the temperature of the liquid at the wall reaches the lambda line. Beyond this point, a sharp temperature gradient developes because the thermal conductivity of He-I is quite poor. Thus, one would expect that in the high current region, where stability is controlled by conduction in the radial direction, the limiting stability would be given by just the difference in enthalpy per unit volume of helium between the bath and the lambda temperatures. This value is shown as a dashed line in Fig. 2. The 9 and 11 Tesla data do approach it at high currents. Part of the additional stability observed in the high current region may be due to end effects.

A third stability regime has been observed in this experiment. At very low currents, the conductor can recover even if the transient flow generated by the heat pulse is not strong enough to cause immediate recovery. If this happens, a bubble is left at the center of the tube, with the uncooled conductor

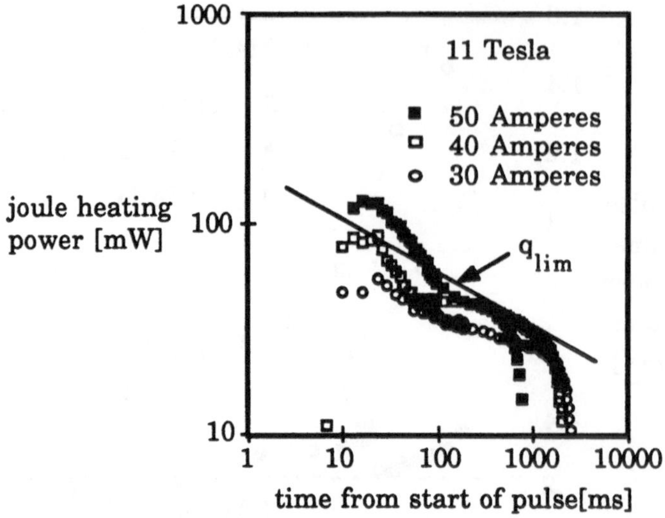

Fig. 3. Joule heating in a recovering superconductor.

next to it producing Joule heat. If the size of the bubble, which is determined by the size of the initial pulse, is small enough, and if the Joule heating power per unit length is low enough, the power the normal section emits will be less than the power which can be conducted away from its ends by the He-II. Although the limiting power for steady state Gorter-Mellink conduction is only 21 milliwatts, the limits to transient conduction at short times are much higher. The limiting heat flux, q, as a function of time, t, is given by Dresner's solution of the Gorter-Mellink diffusion equation for an infinitely long channel with a clamped temperature boundary condition and constant properties[4]:

$$q_{lim} = \left(\frac{3\sqrt{3}}{8} \right)^{1/6} K^{3/4} S^{1/4} (T_\lambda - T_{bath})^{1/2} t^{-1/4} \tag{1}$$

where K and S are constants. This curve is shown in Fig. 3, along with the Joule heating power observed in the conductor after the addition of heat pulses with energies slightly below the stability limit. It can be seen that, except for a small initial spike, the Joule power remains just below the limiting power, so heat is removed and the normal zone contracts. A slightly larger initial normal zone will at some point cross the limiting curve and grow unstably. This description in some ways resembles the "equal area" model developed for cooling channels in pool boiling magnets in He-II[5]. This "axial conduction" regime, so called to distinguish it from the radial conduction important at high currents, shows up most clearly in the 11 Tesla data. At 7 Tesla, transient convection, as evidenced by strong Joule heating peaks which rapidly drop off, is still the dominant mechanism except at the lowest currents.

Thermal Expulsion Velocity and Pressure Rise

Once a normal zone starts to propagate, a great deal of Joule heat is dumped into the adjacent helium vapor, causing it to expand rapidly. As this warm, low density gas expands, it drives more of the conductor normal, increasing the total Joule heating power. This rapid expansion, besides quenching the conductor, has two other serious effects in a large conductor.

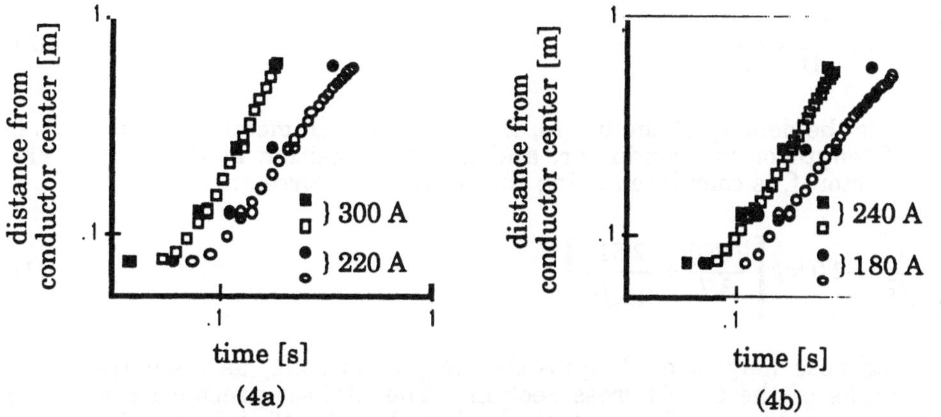

Fig. 4. Propagation of the normal zone front and the thermal front. The solid symbols show the time at which the conductor temperature at that position crosses the lambda line; the hollow symbols show the position of the normal zone front.

First, the expanding gas rapidly expels cold helium out the end of the tube, possibly causing damage to the refrigeration equipment. Second, the pressure inside the conductor may be large enough to cause structural damage.

Dresner proposed a simple "piston in a pipe" model for the expansion of a local normal zone[6]. The data from the unstable region confirm some of the assumptions he used. The first assumption, that the boundary of the warm, low density helium travels with the same velocity as the normal zone front, is supported by Fig. 4. This figure shows the position of the normal zone front, as derived from voltage tap measurements, as a function of time. It also shows the position of several temperature sensors as a function of the time at which the temperature crosses 4.2 K at that location. It can also be noted from the data that the position of the normal zone front is proportional to time to some power n. Applying a fitting routine to the data sets shown on the figure gives values for n of 1.30, 1.31, 1.60 and 1.83 at currents of 180, 220, 240, and 300, respectively. Because the equations derived from this simple model are still intractable with analytical techniques, Dresner postulates that the position of the boundary is proportional to t^n, then uses the equations from the model to compute the exponent. Obviously, this was the proper form to assume. Unfortunately, it is necessary to make the additional assumption that the final pressure is much greater than the initial pressure. As shown below, the estimated pressure rise in this experiment never exceeded one atmosphere, so it is not surprising that the calculated value of n, 4/3, does not match the values computed from the data, although they are reasonably close considering the simplicity of the model.

Using the assumptions of the above model, the velocity of the vapor front can be found by numerically differentiating the position of the normal zone front. Since at low temperature and small pressure rises, the helium in the tube is essentially incompressible, this is also the expulsion velocity. The highest velocity, observed at the highest current, 322 Amperes, was 7.1 meters per second. The pressure rise, ΔP, across the liquid region can be calculated from this velocity, V, using:

$$\Delta P = f \frac{(L-x)}{D} \frac{\rho V^2}{2} \qquad (2)$$

where ρ is the density of the liquid, D is the inner diameter of the tube, L is the half length of the conductor, and x is the position of the front. The friction factor, f, is calculated using the rough tube correlation:

$$\frac{1}{\sqrt{f}} = -2.0 \log \left[\frac{(e/D)}{3.7} + \frac{2.51}{Re\sqrt{f}} \right] \qquad (3)$$

The roughness height, e, is approximately 5 microns, as measured from micrographs of the tube's cross section. The highest pressure rise found was 44.6 kilopascals. Because of the low density of the gas phase, the pressure rise in the central region is very small relative to the pressure rise through the liquid.

SUMMARY

This paper describes an experiment which measures the stability of a superconductor cooled internally with He-II against local disturbances. The stability curves show three regimes. The first regime, observed at high currents and 9 and 11 Tesla, is controlled by radial conduction into the helium and, as predicted by theory, gives a stability limit approximately equal to the enthalpy available in the helium under the heater between the bath and lambda temperatures. The second region, observed at 7 and 9 Tesla, is controlled by thermally induced flows which enhance convection. At 9 Tesla, stability is multivalued in the transition between the two regimes. The third region, observed at low currents, is controlled by transient conduction in the axial direction. Voltage and temperature measurements from conductor quenches confirm the assumptions of Dresner's "piston in a pipe" model. Using the model, the maximum thermal expulsion velocity and pressure rise in the tube are computed from the data.

ACKNOWLEDGMENTS

Work funded jointly by the National Science Foundation under grant MEA-8310770 and the Department of Energy under grant DEAC02-82ER52077

REFERENCES

1. J. C. Lottin and J. R. Miller, Stability of internally cooled superconductors in the temperature range 1.8 to 4.2 K, IEEE Tran. Mag. MAG-19(3):439 (1983).
2. J. W. Lue, J. R. Miller, and L. Dresner, Stability of cable-in-conduit superconductors, J. Appl. Phys. 51(1):772 (1980).
3. L. Dresner, Parametric study of the stability margin of cable-in-conduit superconductors: theory, IEEE Tran. Mag. MAG-17:753 (1981).
4. L. Dresner, Transient heat transfer in superfluid helium -- part II, in: "Advances in Cryogenic Engineering," Vol. 29, Plenum Press, New York (1984), p. 323.
5. P. Seyfert, J. Lafferranderie, and G. Claudet, Time-dependent heat transport in subcooled superfluid helium, Cryogenics 22:401 (1982).
6. L. Dresner, Protection considerations for force-cooled superconductors, in: "Proc. Eleventh Sym. on Fusion Engr.," Austin, Texas (1985) p. 1218.

PROTECTION SYSTEM FOR SUPERCONDUCTING MAGNETIC ENERGY STORAGE (SMES)

G. E. McIntosh, Y. M. Eyssa, M. K. Abdelsalam, R. W. Boom

Applied Superconductivity Center
University of Wisconsin-Madison
Madison, Wisconsin

T. A. Gallagher, R. N. Poirier, J. M. Shah

CBI Industries, Inc.
Plainfield, Illinois

J. R. Bilton, T. F. Garrity

EBASCO Services, Inc.
New York, New York

M. A. Hilal

General Dynamics Corp.
San Diego, California

ABSTRACT

A unique cryogenic system is needed to dump 10^6 to 10^7 litres of 1.8 K helium from a superconducting magnetic energy storage (SMES) unit in 5 to 20 seconds after the detection of an internal fault. The dump system includes quick-acting closures suitable for superfluid helium, refrigerated storage to receive and hold liquid helium without flashing losses, and a gas supply to maintain a positive pressure in the cryostat. The protection system transfers stored electromagnetic energy at low voltage into thermal energy absorbed by internal metallic axial structure. A rapid helium dump helps obtain more uniform final temperatures in the axial structure and conductor turns. Recommendations for specific hardware development are discussed.

INTRODUCTION

Large 1000-5000 MWh (3.6-18 GJ) SMES units have energy densities of about 100-200 J/g. An emergency energy discharge requires safe dissipation of this energy. The Applied Superconductivity Center (ASC) of the University of Wisconsin has addressed this problem in several publications.[1-4] The stored energy should be absorbed inside the magnet, not dissipated in an external dump resistor which requires megavolts. The mass of the structure, conductor and heat sink materials should be enough to achieve the desired final temperature.[2,3] The coil turns consist of Al-NbTi composite conductors, 2 to 6 cm in diamter, wound in electrical contact with axial structure of aluminum alloy. The structure, which is larger in cross-sectional area by a factor of six, carries current and absorbs energy during a discharge. The energy must be dissipated uniformly over the SMES mass for optimum use of structure and heat sink materials. An excessive temperature gradient during the energy dump can lead to undesired thermal stresses and large imbalance between the inductive and ohmic voltage distribution in the winding. The superfluid helium must be dumped quickly to avoid excessive temperature and voltage gradients. Shorting sections of the coil in parallel during dump contributes to uniform temperature distribution.[1,2,4]

During a dump several thousand cubic meters of superfluid helium must be transferred from the SMES to refrigerated storage in 5 to 20 seconds, with little or no increase in cryostat pressure and total helium recovery. The helium dump system requires high reliability and low standby heat leak, particularly in respect to the He II-normal helium interface at each valve. The helium dump/recovery system involves several uncommon design techniques as demonstrated herein.

ENERGY DUMP CONSIDERATIONS

Transferring energy to an external discharge resistor sized at several times the normal resistance of the superconductive coil would subject the coil to several megavolts, which is unacceptable. Internal energy absorption can limit temperatures to about 400 K. During energy dump, the current will divide between the conductor and structure inversely proportional to resistances because the magnetic diffusion time is very small compared to the current decay time of the coil. The thermal conductance of the helium gas between the conductor and the structure is h = 1000 (T/150) W/m^2 where T is the average temperature of the conductor and the structure. Although real contact between the conductor and the structure results in higher values of h, most of the thermal resistance is in the thick structural aluminum alloy. The electric time constant of the dump is long enough to allow the larger energy dissipated per unit volume of the conductor (due to its lower resistance) to diffuse to the large thermal capacity of the structure. The present UW two-layer design has the conductor in good thermal and electrical contact with the structure.

As now visualized, the dump procedure is started by shorting the coil leads externally before forcing the 1.8 K He II out of the helium vessel in a short time (τ_d). Warm 28 K helium gas pressurizes liquid helium out the bottom. The uncovered turns below the liquid level become normal in a fraction of a second since the enthalpy of aluminum is extremely small below the NbTi critical temperature. During dump all covered turns remain superconducting. At the end of the dump, the top of the coil is at a higher temperature than the bottom. The difference in temperature is proportional to the dump time. As a result, the final coil temperature difference between the top and the bottom can be high, resulting in undesired thermal stresses and high voltages. Table 1 lists the final temperature of the top and the bottom of the coil for different dump times. Figure 1 is a plot of the temperatures of conductor and structure at the top and bottom of the coil for a dump time of 10 s. As shown, the temperature of the conductor is slightly higher than that of the structure. The conductor temperature is close to that of the structure even though the energy is dissipated at a higher rate in the conductor due to its lower resistivity. The temperatures of the top and the bottom of the coil are almost equal for dump time < 20 s. For larger dump times, the temperature difference can be substantial as shown in

Table 1. Final Coil Temperature of the 5500MWh SMES Unit.

τ_{dump} (s)	T_{top} (K)	T_{bottom} (K)
10	370	350
20	378	348
60	410	315
100	445	285

Table 1. The voltage distribution along the shorted coil during a 100 second dump time is shown in Fig. 2. Turn to ground voltages as high as 8 kV arise from the imbalance between inductive and ohmic voltages in the solenoid. For short dump times the temperature of the coil is more uniform resulting in smaller turn to ground voltages.

Thermal stresses and voltage problems can be eliminated for longer dump times. The solenoid can be divided into a number of sections with a small current lead coming from each as shown in Fig. 3. These sections are joined to small cold mechanical switches which are connected as shown and closed only in a dump situation. Figure 3 shows a 25 kA switch[5] to connect the shorted coil sections in parallel. A small current transfer equalizes voltages for a more uniform temperature distribution.

Table 2 lists the maximum temperatures of conductor or structure at different times during a 100 s helium dump for eight sections connected in parallel.

HE II DUMP SYSTEM

The proposed SMES superfluid helium dump system is shown schematically in Fig. 4. Principal elements include:

1. Independently insulated and evacuated reservoirs held at 0.11721 MPa and 4.424 K.

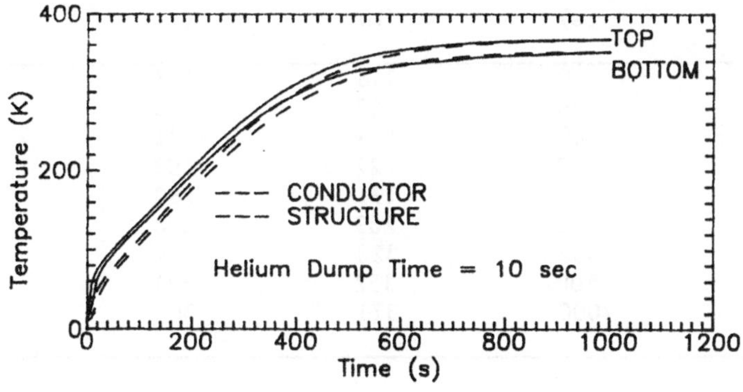

Fig. 1. Conductor and structure temperature rise in a 5500MWh SMES coil.

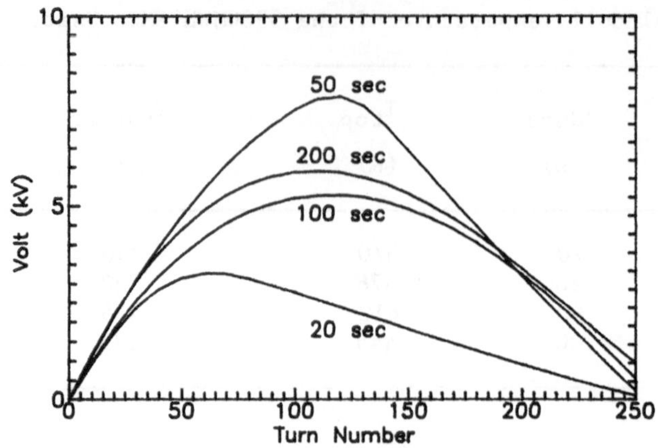

Fig. 2. Voltage vs. turn (height) vs. time after dump for a 5500MWh SMES coil. The dump time is 100 s.

2. Dump pipes 0.324 m OD connecting cryostat and reservoir.
3. Superfluid isolation valves in each pipe.
4. Gas supplies at 28 K, 1.723 MPa gas supply to maintain pressure and displace He II from the cryostat.
5. Pressure regulators and superfluid isolation valves for the gas supply.

The reservoir size is based on the total volume of He II in the cryostat and the volume of liquid condensed from vapor originally in the reservoir. Sizing is complicated by the fact that helium density decreases from 0.147 kg/L as superfluid at 1.8 K and 0.1013 MPa to about 0.143 kg/L at final conditions of 0.01 MPa and 2.5 K. The minimum volume of the reservoir is 17 to 18% larger than the volume of He II.

Standby conditions in the reservoir contribute about 16% to its size. In turn, the standby conditions are predicated on the desire to

Table 2. Maximum temperatures of conductor and structure (T_c, T_s) at different times during a dump.

Time (s)	T_c (K)	T_s (K)
0	1.8	1.8
40	87	63
80	114	91
100	127	103
200	191	160
300	269	228
400	323	284
500	352	321
1000	371	368

balance the pressure on each side of the dump isolation valves. With atmospheric pressure at the top of the cryostat and an allowance for hydrostatic pressure, the liquid-side total pressure at the valves is about 0.1172 MPa (17 psig) for which the normal helium saturation temperature is 4.381 K. Maintaining the reservoir slightly warmer at 4.43 K prevents condensation until superfluid helium enters.

Balanced pressure across the isolation valves is an important part of superfluid technology. The objective is to have tight seals at the top and bottom of an insulated plug with minimal differential pressure. The lower seal prevents superfluid creep and resultant heat leak and the upper seal restricts flow of normal helium into the isolation area where it would condense at the lower seal if not restricted. A concept design of the isolator valve is shown in Fig. 5. It utilizes upper and lower elastomer lip seals successfully[6] used in He II service with a close-fitting G-11CR isolator plug. Estimated heat leak of the nominal 0.3 m valve is 5×10^{-3} W from 4.43 to 1.8 K.

Dump pipes are sized based on an acceptable pressure drop of 2550 Pa (0.37 psi) for a maximum length of 6.1 m. At a flow rate of 310.61 kg/s per pipe, the specified pressure drop is met with pipes having an ID of 0.315 m (12 IPS, sch. 5). Flow work per pipe is 5.4×10^3 W or about 25×10^3 J per pipe for a complete dump which increases internal energy of the full reservoir by 0.35%.

Helium gas must be supplied to replace the volume of liquid drained from the cryostat. The quantity required is uncertain because there will be condensation at the liquid interface and some expansion due to heating from exposed normal turns. (This problem is being treated analytically by one of the co-authors and an experimental investigation is underway at UW.) Since there is no gas in the cryostat initially, it is clear that a quantity must be supplied at the outset to get the dump started with the total amount determined by the balance between condensation and vapor expansion. It is currently assumed that it is necessary to displace 125% of the liquid volume with helium vapor at 0.10477 MPa and the supply temperature of 28 K.

The physical gas supply containment consists of a 0.315 m ID aluminum pipe (12 IPS, sch. 10) around the periphery of the SMES. Stored gas is maintained at 28 K by contact with the cryostat cold thermal radiation shield and at approximately 1.723 MPa which corresponds to the helium refrigerator compressor outlet pressure. The system is completed by a number of parallel lines leading from storage to the cryostat. Each line has a pressure control valve (PCV) to deliver gas at slightly above atmospheric pressure and a superfluid shutoff valve similar to the larger liquid dump valves.

The final element of the dump arrangement is the liquid supply/relief system consisting of 4.4 K, 0.119 MPa normal helium supply reservoir, relief devices and thermally isolating plug valves separating the normal and superfluid helium. The plug valves do not seal tightly so that pressure differentials between normal and superfluid reservoirs are equalized by leaks. However, for this application the top works of each plug valve is spring loaded to relieve into the normal helium supply manifold at 0.122 to 0.1358 MPa (3 to 5 psig). In turn, the normal helium manifold relief valves are set to vent at 0.14 MPa (5.61 psig). This is the main cryostat over-pressure protection circuit as well as an element of the dump system. However, installation of a number of parallel cold burst discs is being considered as a backup.

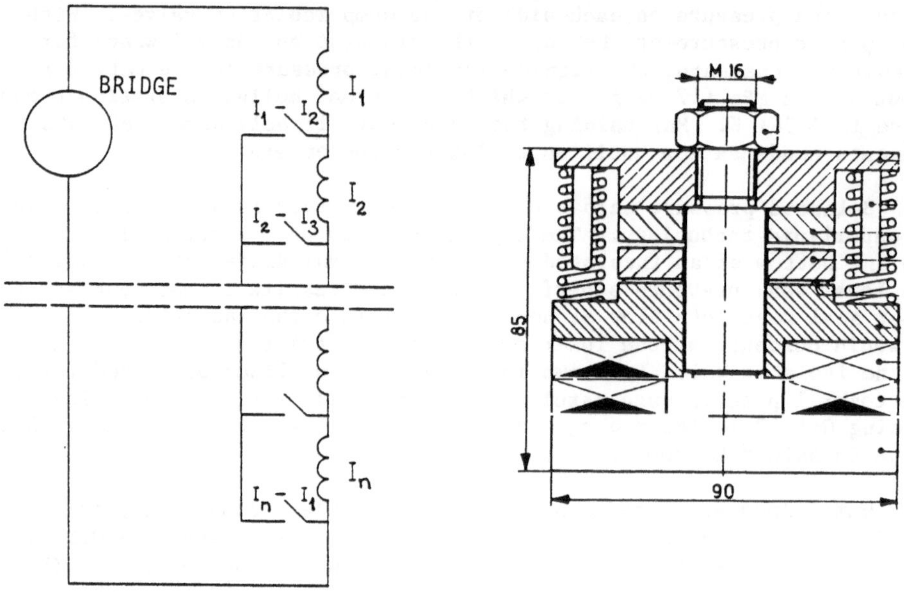

Fig. 3. Shorting mechanism required to equate axial temperature and voltage gradient in case of a long helium dump time. Switch dimensions are in mm.

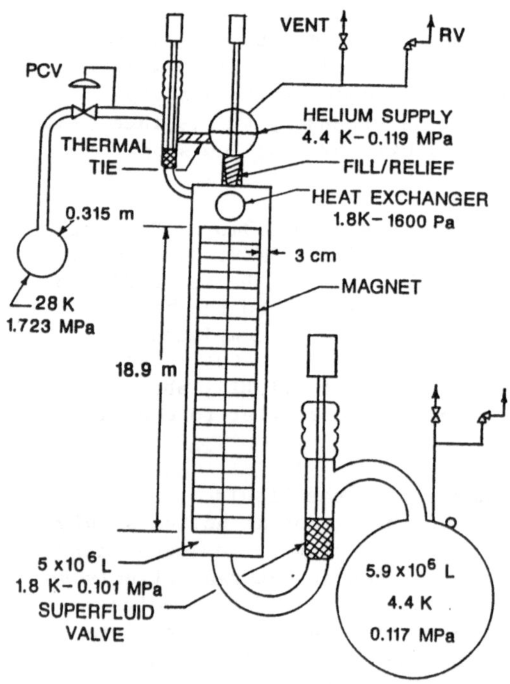

Fig. 4. Helium dump system.

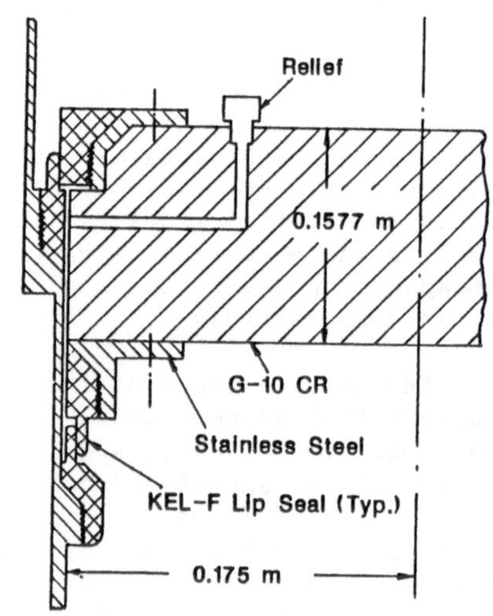

Fig. 5. Superfluid helium isolation valve.

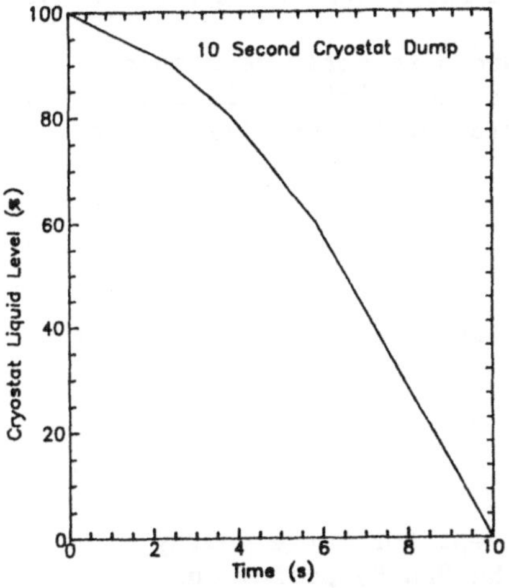

Fig. 6. Cryostat liquid level vs. time.

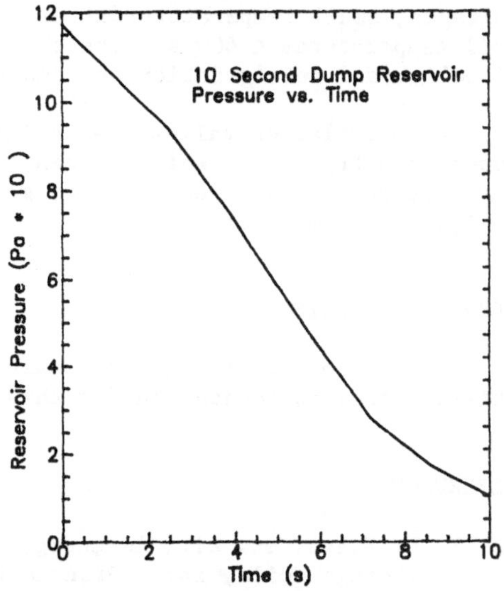

Fig. 7. Reservoir pressure vs. time.

PROTECTION SYSTEM OPERATION

An unacceptable helium usage rate signals the need for an emergency dump. The first action is to disconnect the SMES from the utility system. Then the operation is: shorting switches are closed; helium is expelled; current flows in the structure warming it up; and all energy is absorbed in 10 minutes. At the conclusion, all of the helium is in the refrigerated reservoir at approximately 10^4 Pa (0.1 atm) and 2.5 K. Windings and structure are surrounded by helium gas at vent pressure of 0.1358 MPa and 370 K. Helium in the reservoir will gradually rise to equilibrium conditions at 4.4 K and the cryostat will be ready for cool-down by forced circulation of helium gas at atmospheric temperature and pressure.

During a 10 second dump, the projected cryostat liquid level and reservoir pressure are shown as a function of time in Figs. 6 and 7. The reservoir pressure begins to fall as soon as He II enters and the rate continues almost linearly to a final pressure of 0.01 MPa (0.1 atm). The actual ending pressure may be lower than 0.01 MPa if the reservoir volume is greater than the minimum required or it may rise somewhat if the dump valves are closed slowly. The cryostat liquid level drops gradually at first but accelerates later as the internal pressure drop decreases and the differential pressure increases due to falling reservoir pressure. Plots of conductor and structure temperature vs. time for 10 second helium dumps are shown in Fig. 1. With a 10 second dump, top and bottom temperatures are within 20 K and the maximum is about 370 K. A 100 second dump (Table 1) results in a temperature difference of 150 K and a maximum temperature of 445 K.

CONCLUSIONS

An SMES protection system which utilizes the internal mass of conductor and cold structure to absorb stored electric energy is described. Dump times of 10 to 20 second helium yield low coil to ground voltages, small temperature difference from top to bottom and maximum coil temperatures < 400 K. The energy dump is accomplished without loss of helium and refrigeration is conserved with reservoir storage at 4.4 K.

Dump isolation valves are designed around a recently developed superfluid lip seal. While proven in another superfluid application, manufacture and extended-term cryogenic testing of the specific valve design is recommended.

ACKNOWLEDGEMENTS

The authors gratefully acknowledge support of the Wisconsin Electric Utility Research Foundation for this work.

REFERENCES

1. Y. M. Eyssa, et. al., An energy dump concept for large energy storage coils, in: "Proceedings of the 9th Symposium on Engineering Problems of Fusion Research," IEEE Pub. No. 81CH1715-2 NPS, (1981), p. 456.
2. R. W. Boom, et. al., Cryogenic aspects of inductor-converter superconductive magnetic energy storage, in: "Proceedings of the Ninth International Cryogenic Engineering Conference," Butterworth & Co., Surrey, U.K., (1982), p. 731.
3. J. Waynert, Temperature and voltage analysis of a protection scheme for large energy storage coils, in: "Advances in Cryogenic Engineering," Vol. 31, Plenum Press, New York, (1986), p. 399.
4. R. W. Boom, Y. M. Eyssa, Y. Huang and G. E. McIntosh, Two-Layer Non-rippled Superconductive Magnetic Energy Storage Systems, in: "Proceedings of the Eleventh International Cryogenic Engineering Conference," Butterworth, Surrey, U.K. (1987), p. 464.
5. H. H. ten Kate, B. ten Kate and L. J. Van de Klundert, An experimental mechanical switch for 3 kA driven by superconducting coils, in: "Advances in Cryogenic Engineering," Vol. 31, Plenum Press, New York (1986), p. 243.
6. G. E. McIntosh, D. S. Lombard, D. L. Martindale and M. J. DiPirro, "Bayonet for Superfluid Helium Transfer in Space," paper FC-4, 1987 Cryogenic Engineering Conference, June 14-18, 1987.

THE TWENTE HIGH-CURRENT CONDUCTOR TEST FACILITY, FIRST RESULTS ON CRITICAL CURRENT AND PROPAGATION IN TWO CABLES

H. H. J. ten Kate, W. Uyttewaal,
B. ten Haken and L. J. M. van de Klundert

University of Twente
Dept. of Applied Physics
Applied Superconductivity Centre
Enschede, the Netherlands

ABSTRACT

A conventional measurement of the critical current and other relevant properties of superconducting cables within the field range of 0 to 8 tesla is complicated because of the large current supply required, the strong magnetic forces on the conductor, the minimum bending diameter of the cable and the high running costs of a conventional large scale facility. In our laboratory we have put into use a new small-size conductor test facility in which cables can be investigated concerning their critical current, a.c. losses, stability and propagation phenomena in the field range of 0 to 7.5 tesla. The test current is generated by a single-cycle superconducting rectifier which is designed to generate the cable current in the range to 25 kA nominal with an absolute maximum of 40 kA. In this low-cost facility we started with the investigation of several cables. In this paper we report on the first successful experiments during sample tests of two high-current conductors which are a monolithic conductor for the French TORE SUPRA fusion project and an 11-strand cable which is applied in the generator development program of Siemens. Critical current and propagation velocities were measured and analyzed over the mentioned field range.

INTRODUCTION

The application of high-current superconducting cables in the various large-scale magnet projects for fusion and SMES, and in generators and certain accelerator magnets makes it necessary to test small-scale models, and prototypes as well as the full-size cables. Because of the large critical current in these types of superconducting cables (which ranges from 1 to 50 kA in the field of 1 to 8 tesla), the large cross-section and the large minimum bending diameters, investigation in the ordinary solenoidal small-bore magnets is not possible. For such types of measurements there exist only a few appropriate facilities which can include the presence of a large magnet, a heavy duty power supply for the current range indicated above and a relatively large liquefier power to make up the current lead loss of approximately 2.5 W/kA. The high running costs and the long cooling times of such facilities were the main motivation for us to develop a

magnet system with the following properties:
* compact magnet, low stored energy, field range 0-7.5 tesla,
* cooling time of magnet and sample holder 15 and 1 hours respectively,
* replaceable sample holder when magnet is at 4 K,
* test current up to 40 kA supplied by a superconducting rectifier,
* possible tests are critical current, propagation velocity and a.c. losses,
* constant requirement of liquid helium between 1.5 and 2 L/hr.

SYSTEM COMPONENTS

Magnet system

The magnet system consists of two solenoids which are connected in anti-series thus generating a quadrupole field with a maximum between both coils at a radius of 10 cm, see figure 1. The torus shaped test volume has a section of $30*10\ cm^2$. The field homogeneity is better than 1.5% in a sphere of 2 cm. Note that in this geometry the circuit of the test conductor and the magnet are magnetically decoupled since the field of the magnet is in radial direction. The magnet system can generate a field of 7.8 tesla at 105 A and has a maximum stored energy of 184 kJ. More details about the magnet itself are given in reference 1.

Cryogenic power supply

The cryogenic power supply provides the current in the sample and is an essential part of the facility, see figure 2. It is actually a single cycle superconducting rectifier[2] which consists of a superconducting transformer and a thermally activated superconducting switch, see figure 3. The sample conductor is connected to this circuit by means of two soldered joints which can easily be opened and closed by built-in heater elements. Both the sample and the cryogenic power supply are connected to the insert structure which can be replaced while the magnet is kept cold.

Operating cycle. The operating cycle to generate a sweep of the current in the sample is as follows, see figure 4: the switch is opened and the primary current is decreased to -Ip; the switch is closed after which

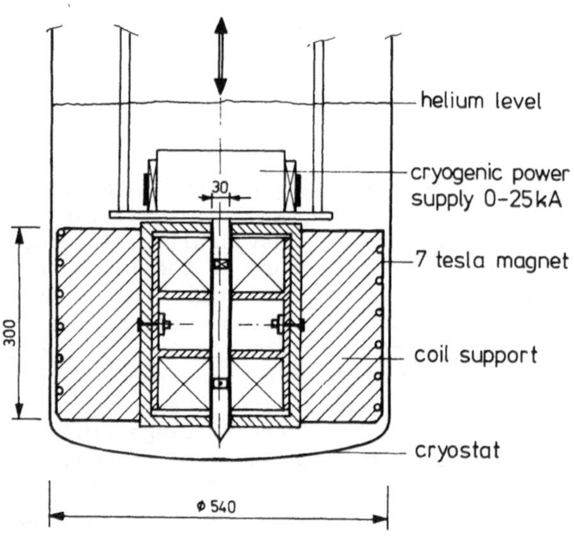

Fig. 1.

Longitudinal section of the magnet part of the facility showing the magnet system, the test volume and the sample holder connected to the cryogenic power supply and insert structure.

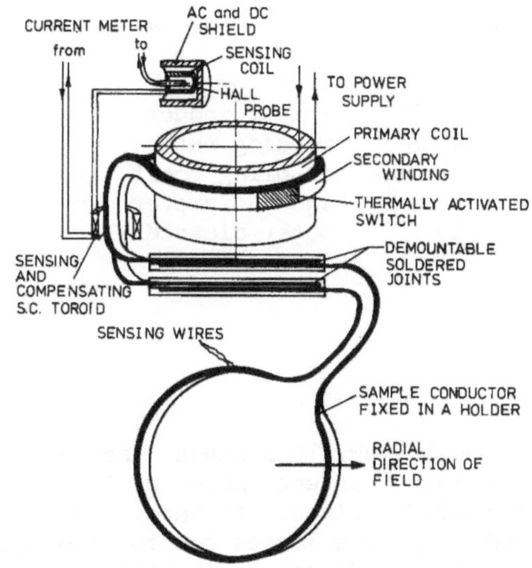

Fig. 2.

Impression of the cryogenic part of the facility showing the sample holder, the superconducting rectifier, the connections and the current meter.

given in table 1. The coils of the transformer are wet-wound using Stycast 2850FT. The secondary winding and the switch are made of a round braided cable with a high-resistivity matrix[3]. The thermally activated superconducting switch in the secondary circuit is part of the single secondary transformer winding.

Current amplification. The current amplification, Ips of the transformer depends on the inductance of the test sample according to Ips = M/(Lsec+Lsample), where M is the mutual inductance of the transformer and Lsec the inductance of the secondary winding. The factor Ips ranges from measuring can start; the primary current can now be swept up to +Ip in order to generate the secondary current with a prescribed shape. The desired shape of the test current is obtained by operating the transformer in the feed-back mode so that the circuit becomes a current supply rather than a voltage source. The specifications of the rectifier circuit are 1250 to 2500 depending on the inductance of the test conductor in the

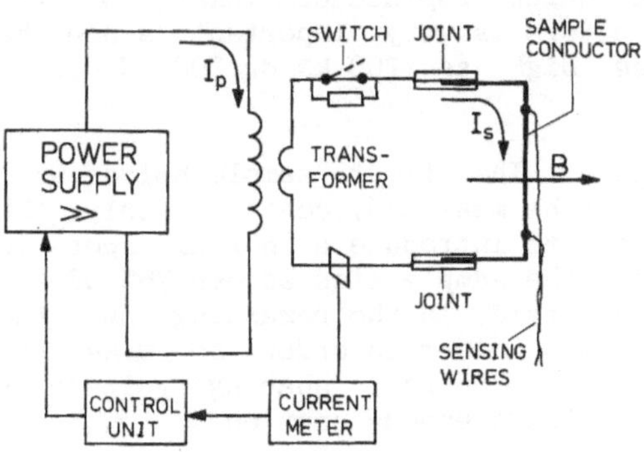

Fig. 3. Circuit of the single cycle superconducting rectifier to generate the test current.

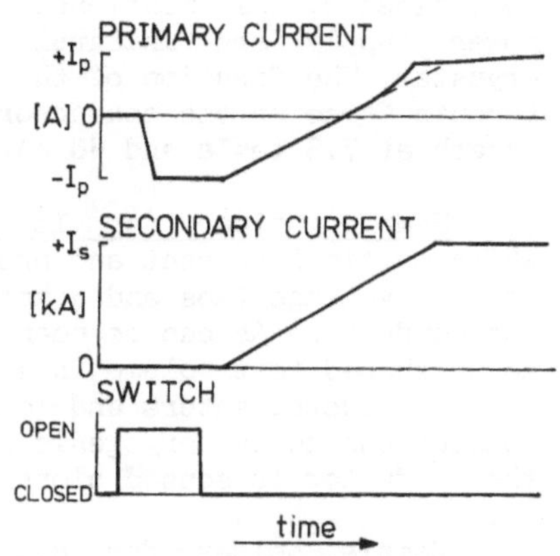

Fig. 4. Operating cycle of the cryogenic power supply.

Table 1. Specifications of the rectifier circuit.

primary coil		secondary coil		transformer	
inductance	= 10.1 H	inductance	= .3 μH	mutual induct.	= 1.4 mH
turns	= 7555	max. current	= 40 kA	winding ratio	= 1
max. current	= 75 A	cable size	= $40*6\ mm^2$	coupling coeff.	= .80
wire diam.	= .33 mm	length	= .65 m	inner diameter	= 160 mm
Cu/NbTi	= 1.35	strands	= 720	outer diameter	= 192 mm
filaments	= 367	Cu30Ni/NbTi	= 1.1	length	= 72 mm
wire length	= 4.2 km	strand diam.	= .3 mm	max.field (50 A)	= 4.2 T

sample holder, which lies between 0.3 and 0.9 μH. This means that a sample current of 25 kA can be attained by a primary current sweep of 10 to 20 A, provided the resistance of the secondary circuit is zero. Additional primary current is required to make up the losses due to the joint resistances and current sharing in the sample. Moreover, in the case of an investigation of the normal zone propagation velocity, the sample current has to be kept constant long enough in order to measure the voltage rise across the normal spot. This also requires an extra contribution from the primary current sweep.

Electrical connections. The electrical connections between the cryogenic power supply and the sample are soldered joints each of which has a resistance of about 1 nΩ. Consequently the decay time of the secondary circuit is approximately 500 seconds providing the primary current is constant. The connections were designed to facilitate a fast and easily repeatable replacement of test samples. They consist of two copper connecting parts (sizes $220*50*10\ mm^3$) having slots in which the cable terminals are soldered. Both parts can easily be connected by a soldering procedure using an electrical heating element which is permanently attached to the permanent connecting part of the current supply. Obviously two solders with a different melting temperature should be used.

Sample holders

Two types of sample holders were developed, one to provide critical current and propagation measurements and a second one for a.c. loss measurements. As mentioned before, the sample holder and the cryogenic power supply are integral with the quick replaceable insert of the cryostat. The fixation of the test conductors is very important since the Lorentz force on the conductor can be as high as 200 kN or 3000 N per cm length at 7.5 tesla and 40 kA.

Sample holder for Ic measurements. The first sample holder with which critical current and propagation can be measured, contains only the sample, voltage taps and a heating element to introduce a normal spot in the conductor. As can be seen in figure 2 the sample lies across 75% of the torus shaped test volume in a homogeneous field. In the remaining part the test conductor enters and leaves the sample holder in order to make the connections to the cryogenic power supply. The current sharing voltage of the conductor is sensed along 45 cm in the homogeneous section.

Sample holder for a.c. loss measurements. In order to facilitate loss measurements, the sample holder has a more complicated construction, see figure 5. In this case it also includes a.c. coils that generate the alternating field on the conductor, pick-up and compensating coils to sense the voltage due to magnetisation loss, voltage taps on the test conductor

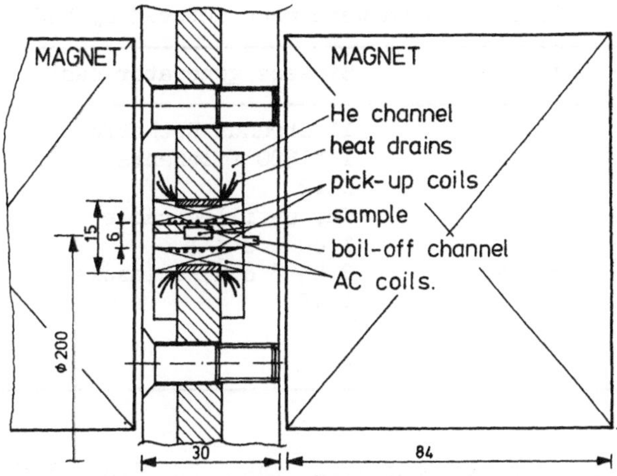

Fig. 5.

Detail of the longitudinal section of the sample holder that provides a.c. loss measurements.

to sense the voltage due to transport current loss and finally a housing to detect the helium evaporation. The holder is made in such a way that it can take a large variety of cable sizes without changing the a.c. field coils and the voltage detection system. The a.c. field coil system consists of an outer and inner coil which are connected in anti-series to minimize the inductance and so maximize the field amplitude or field rate. The coil system can generate a field of 30 mT/A at the conductor in a background field of 7.5 T. As an example, when using a 100V/10A power amplifier the achievable field amplitude is 0.3 T at 25 Hz with a field rate of approximately 75 T/s. Note that the direction of the a.c. field produced by these coils is perpendicular to the main field. With this facility it is possible to investigate the effect of this field geometry on the losses of the conductor. However, it can be expected that this effect is small because the main role of the large stationary field is to reduce the critical current density and to decrease the matrix losses due to magneto- resistivity.

Pick up coils. The set of pick-up and compensating coils are located under the first layer and on the last layer of the outer and inner field coil respectively. Since the length of these coils is not infinite compared to the width of the test conductor, the calculated loss using the measured voltages of these coils is somewhat smaller than the true loss. Therefore, a correction factor has to be taken into account which has to be calculated for various sizes of test conductors. For practical conductors in this facility the correction amounts to a few percent.

The direct current meter

A serious handicap of using a transformer to induce the sample current is the current measurement which is much more complicated. Conventional methods to measure a stationary current in a closed superconducting circuit fail or are not accurate enough. Therefore we developed especially for this kind of circuit a new type of current meter which enables us to measure a direct current in presence of large stray-fields with an accuracy of 0.1%, see reference 4. This type of direct current meter is successfully applied in this facility.

CRITICAL CURRENT OF TWO CABLES

As an initial test of the facility we investigated two different types of high-current superconducting cables. Their specifications are listed in table 2. The critical current was measured by applying the resistivity

Table 2. Specifications tested conductors, both were manufactured by VAC.

property		TORE SUPRA conductor	Siemens generator cable
type	=	monolith	11 strands Rutherford
superconductor	=	10164 filaments NbTi	11*1500 filaments NbTi
filament dia. [μm]	=	23.5	23.2
matrix	=	Cu with CuNi barriers	Cu, unfilled cable
Cu:CuNi:NbTi	=	2.20 : 0.33 : 1	1.60 : 0 : 1
cable size [mm^2]	=	2.8 * 5.6, uninsulated	2.45 * 8.5, uninsulated
strand dia. [mm]	=	-	1.45
twist length [mm]	=	40	16
cable pitch [mm]	=	-	80

criterion of 10^{-14} Ωm. The current rate was less than 10 A/s. Both conductors showed a clear current sharing state at magnetic fields above 3 tesla. The measured critical currents of both conductors versus the applied magnetic field are shown in figure 6. The generator cable can carry a much larger current than the TORE SUPRA conductor. In order to investigate the pinning force as well as the current density one has to consider the self-field contribution to the total field on the filaments. The maximum self-field of the TORE SUPRA and the Siemens generator cable are in this test geometry 0.09 and 0.06 T/kA respectively. The maximum pinning force versus the total magnetic field that includes a self-field contribution, is shown in figure 6b. It is obvious that the 11-strand Rutherford cable is much better optimized than the monolithic conductor.

A good procedure to evaluate a critical current measurement is to check the validity of the Kim relation. This states that in the field range below about 5 tesla the critical current density of the superconductor can be described by $J_c(B) = J_0 * B_0 / (B_0 + |B|)$, where J_0 and B_0 are constants. Figure 7 shows the critical current density and its inverse value versus

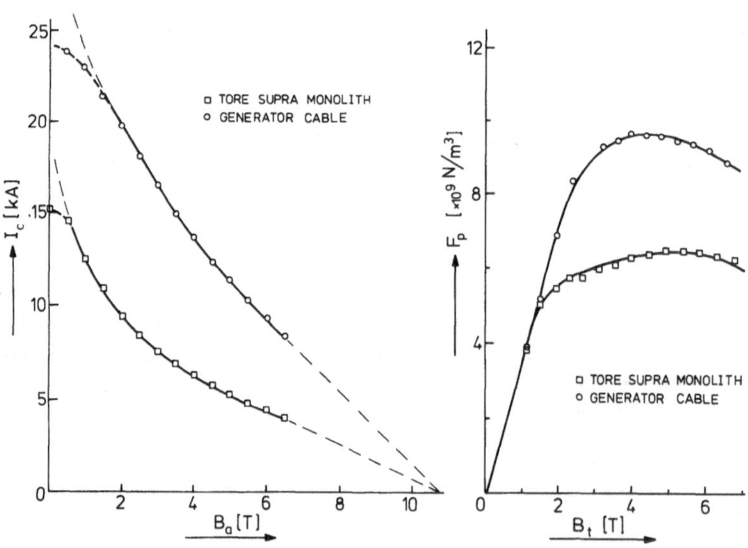

Fig. 6. (a) Critical current $I_c(Ba)$ and (b) the pinning force $F_p(Bt)$ versus the total magnetic field of the TORE SUPRA and the Siemens generator conductor.

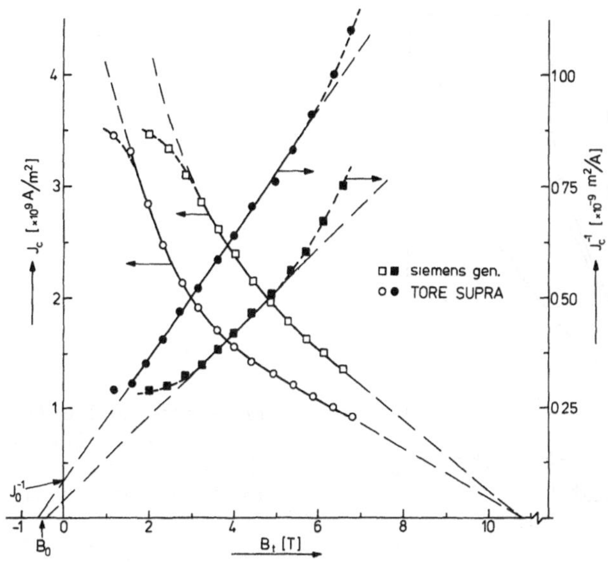

Fig. 7. Critical current density and its inverse value versus the total magnetic field on the filaments. Same conductors as in figure 6.

the total magnetic field on the filaments. The critical current density of the conductors can be described by the following B_0 and J_0 provided B <5 T:
* TORE SUPRA monolith : B_0 = 0.6 T and J_0 = 12000 A/mm^2;
* Siemens (11-strand Rutherford) cable: B_0 = 0.4 T and J_0 = 25000 A/mm^2.
Above 5 tesla the critical current density falls linearly with the field up to the second critical field of approximately 10.8 tesla.

It can be clearly observed that the maximum current at low fields is less than the critical current due to self-field instability of the conductors. Its criterion sets a maximum to the term J_c*D, where D is the characteristic size or diameter of the conductor. Note that in this case the unaffected current density is limited to about 3000 A/mm^2. Further, it can be concluded that the critical current density in the monolithic conductor is very low, as can be illustrated by the critical density found at 5 T:
* TORE SUPRA monolithic conductor: J_c = 1280 ± 20 A/mm^2 at 5 T;
* Siemens generator cable : J_c = 1900 ± 20 A/mm^2 at 5 T.
The TORE SUPRA conductor can carry the working current of 1400 A at 9 T.

NORMAL ZONE PROPAGATION VELOCITY

The normal zone propagation velocity in the TORE SUPRA conductor as a function of the current and magnetic fields of 1, 3 and 5 tesla was determined by measuring the resistance rate after a normal zone was introduced by a heating pulse, see figure 8. Moreover, the results obtained are compared to calculated velocities using the models of Turck and Dressner as described in references 5 and 6. The conductors are glued to the sample holder and then covered with a thin insulating layer of STYCAST resin which results in a minimum propagation current of about 1 kA. In the case of the TORE SUPRA conductor with its mixed matrix, the measured and calculated velocities do not fit very accurately but the models can be used to estimate the velocity within about 30%. The limited accuracy of the models mentioned above in the case of superconductors having (partly) a high-resistivity matrix was already concluded in an earlier study of non-copper matrix superconductors, see reference 6.

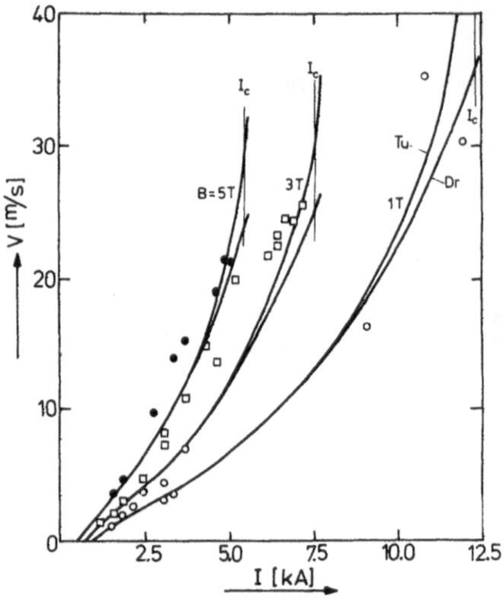

Fig. 8.

Measured and calculated propagation velocity in the TORE SUPRA conductor versus current at 1, 3 and 5 tesla.

CONCLUSION

The first experiments in the test facility as presented here showed that it is an effective tool to investigate high-current superconducting cables on a small scale and with acceptable costs concerning critical current, propagation phenomena as well as a.c. losses. The facility provides test currents up to 40 kA in a field range of 0 to 7.5 tesla which enables the user to test cables under realistic circumstances rather than using small-scale models or extrapolation of questionable results obtained at elevated temperatures in order to reduce the test current.

ACKNOWLEDGEMENT

The authors would like to thank FOM, the Netherlands Foundation for Fundamental Research on Matter, who partly supports the investigations, and Vaccumschmelze and Siemens who kindly supplied us with the cable samples.

REFERENCES

1. H. H. J. ten Kate et al, A test facility for high-current superconducting cables up to 25 kA at 7 tesla, in: "Proc. 11th. Int. Cryog. Eng. Conf., Berlin, April 1986, "Butterworth, Guildford, UK (1986), p. 500.
2. H. H. J. ten Kate, "Superconducting Rectifiers", thesis University of Twente, April 14, 1984, the Netherlands (1984).
3. H. H. J. ten Kate et al, A 25 kA, 0.5 kW thermally switched superconducting rectifier, in: "Proc. MT-8, Grenoble, 1983, Journal de Phys. C1, Vol. 45, (1984) p. 471.
4. H. H. J. ten kate et al, A new type of superconducting current meter for 25 kA, in: "Advances in Cryogenic Engineering", Vol. 31, Plenum Press, N.Y. (1986), p. 1309.
5. B. Turck, About the propagation velocity in superconducting composites, Cryogenics, Vol. 24, (1980) p. 145.
6. H. H. J. ten Kate et al, Longitudinal propagation velocity of the normal zone in superconducting wires, in: "Proc. ASC 1986, Baltimore, October 3, 1986.

SHORT SAMPLE CRITICAL CURRENT MEASUREMENTS USING A SUPERCONDUCTING TRANSFORMER

E. M. W. Leung, H.G. Arrendale, R.E. Bailey and P. H. Michels

General Dynamics Corporation, Space Systems Division

P.O. Box 85990, San Diego, CA 92138

ABSTRACT

We have developed a superconducting transformer (SCT) for measurement of the critical current of large composite superconductors for high field and high current applications such as those used in fusion magnets. It consists of a 700-turn superconducting primary and a one-turn secondary, of which the sample conductor forms a part. When an electrical current of up to 100 amperes is introduced into the primary, a current calculated to be greater than 25,000 amperes is induced in an opposite direction in the secondary. A 20-centimeter diameter, 7.5-Tesla magnet provides the background magnetic field. The secondary current is measured using both a Rogowski coil and a Hall probe. Calibration of the device is achieved by measurement against a superconductor critical current Standard Reference Material (SRM #1457) obtained from the National Bureau of Standards. Advantages associated with this approach include elimination of an expensive high current, low ripple power supply, and the reduction in liquid helium boiloff as a result of being able to use a pair of current leads at least two orders of magnitude below that of the test current. The mathematical theory, design, and test results associated with our device will be presented.

INTRODUCTION

Critical current (short sample current) measurements for composite superconductors, having a high current carrying capacity at high magnetic fields, have always been difficult. The short sample data obtained for the same conductor by various laboratories, due to different sample holder configuration and effects like differential thermal contraction between the holder and the sample, strain imposed on the conductor, joint length allowed for current transfer, mounting techniques, self-field effect, and electric field normal criterion, could vary up to 30%[1] from the average measured value. Things have improved since the work of Clark, Goodrich, and Fickett[2], leading to the introduction of a critical current measurement standard[3] and a Standard Reference Material (SRM #1457[4] by the National Bureau of Standards. Unfortunately, this standard only covers composite superconductors with a current carrying capacity below 600A. While the above error-causing factors become more prominent for a larger, higher-current composite superconductor, the conventional experimental setup for critical current measurement, using a low ripple DC power supply and vapor cooled leads, becomes ecomomically unacceptable and technically difficult. A pair of 20KA commercially available current leads consumes 56L of the \$5/L liquid helium per hour of operation. A high current power supply with low ripple output can also be prohibitively expensive. We came across the concept of using a superconducting transformer[5], which had been applied to low current short sample measurement. A mathematical model (the constant voltage charge mode) for the operation of such a transformer was developed and presented in a previous paper[6] by the author. In this paper, the constant current charge rate case of the theory, comparison of the model to actual measurements, and detailed design of our apparatus is presented. Experience associated with the calibration (against NBS SRM #1457) and actual usage of the superconducting transformer for measurement of a medium current composite superconductor is discussed and the viability of this alternative critical current measurement method assessed.

THEORY

Qualitative

A superconducting transformer (SCT) consists of a superconducting primary with multiple turns and a single-turn superconducting secondary, wholly or partially made of the sample conductor to be tested. While charging the primary with a small current of the magnitude of a few hundred amperes, a much larger current, depending on the turns ratio between the primary and the secondary, of the magnitude of several tens of thousand amperes, can be induced in the secondary in a direction opposite to the direction of the primary current flow, in accordance with the Lenz's Law of magnetodynamics. The magnitude of the current generated in the secondary can be determined by a precalibrated Rogowski coil or a Hall sensor. In theory, when the value of the secondary current exceeds that of the critical current for the sample, the secondary will quench and the classical voltage-current (V-I) characteristic curves can be generated at different background fields with a X-Y chart recorder or a fast oscilloscope. By calibrating the SCT against a critical current measurement SRM, the accuracy and sensitivity of this device can be obtained.

Quantitative

Fig. 1 represents the schematic based on which the mathematical model of the superconducting transformer was developed. Two general equations, in accordance with the Kirchhoff's law, can be written for the circuitry:

$$L_1 \frac{dI_1}{dt} + M \frac{dI_2}{dt} + R_1 I_1 = E \quad [1]$$

$$L_2 \frac{dI_2}{dt} + M \frac{dI_1}{dt} + R_2 I_2 = 0 \quad [2]$$

where L_1 = self inductance of the primary
L_2 = self inductance of the secondary
R_1 = resistance of the primary
R_2 = resistance of the secondary
M = mutual inductance between the primary and the secondary
I_1 = current flowing in the primary

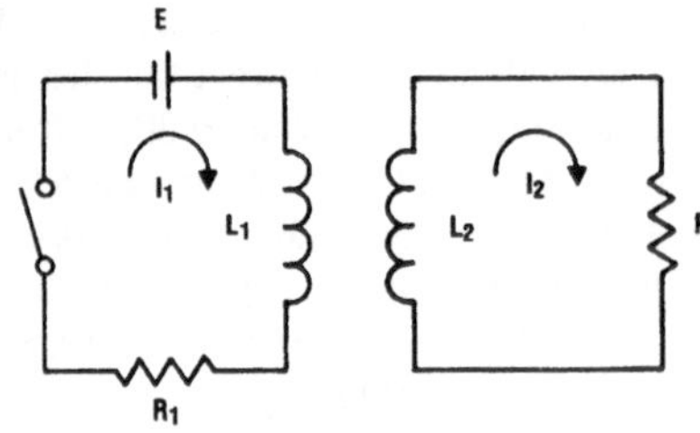

Fig. 1. Superconducting transformer mathematical model.

I_2 = current flowing in the secondary
E = charging voltage applied to the primary
k = $(M^2/L_1L_2)^{1/2}$
t = time

Constant current charge rate charging mode. In this case, $dI_1/dt = C$, a constant, by definition; $E = E(t)$, is a function of time.

Equations [1] and [2] become

$$L_1 C + M \frac{dI_2}{dt} + R_1 I_1 = E(t) \qquad [3]$$

$$L_2 \frac{dI_2}{dt} + MC + R_2 I_2 = O \qquad [4]$$

Applying the initial condition of $I_2 = 0$ when $t = 0$ to equation [4] gives an expression for I_2 as

$$I_2 = \left(\frac{MC}{R_2}\right) [\exp(-R_2 t/L_2) - 1] \qquad [5]$$

Putting the result of [5] into [3] gives

$$E(t) = C \{ L_1 [1 - k^2 e^{-(R_2 t/L_2)}] + R_1 t \} \qquad [6]$$

Constant voltage charging mode. The derivation of the expressions governing this case was given in a previous paper by the author[6] and will not be repeated here. For completeness, the final formats of the expressions are given below:

$$I_1 = \left(\frac{E}{R_1}\right) \left\{ \frac{1 - e^{-at}\cosh(bt) + [(a^2-b^2)L_2 - aR_2] e^{-at}\sinh(bt)}{bR_2} \right\} \qquad [7]$$

$$I_2 = \frac{-[(a^2-b^2)ME] e^{-at}\sinh(bt)}{bR_1R_2} \qquad [8]$$

and

$$I_2 \text{ reaches a maximum at } t = \left(\frac{1}{b}\right) \tanh^{-1}\left(\frac{b}{a}\right) \qquad [9]$$

where

$$a = \frac{R_1L_2 + R_2L_1}{2(L_1L_2 - M^2)}$$

and

$$b = \frac{[(R_1L_2 - R_2L_1)^2 + 4M^2R_1R_2]^{1/2}}{2(L_1L_2 - M^2)}$$

The above analytic expressions were checked using a numerical circuit computer analysis code by the name of SPICE[7]. They give almost identical results. Two simple BASIC codes were then written on an IBM PC for experimental simulation.

EXPERIMENTAL SETUP

Our critical current measurement setup, using a superconducting transformer, is illustrated in Fig. 2. The transformer is coupled mechanically to a 7.5T, 20-cm-bore background solenoid magnet in such a way that their main field direction is orthogonal to that of the magnet. This assembly is immersed in the liquid helium housed in a 0.66-m ID, 2.72-m deep test dewar, installed in a pit built with concrete reinforced with a nonmagnetic magnesium-based alloy. A differential copper-constantan thermocouple, a calibrated Allen Bradley carbon resistor, and a pressure gauge monitor the pressure and temperature of the liquid helium test environment. Fig. 3 shows the location of the F. W. Bell Model #BHT921 cryogenic Hall sensor and a lab-fabricated Rogowski coil for measuring the secondary current, with respect to the primary and secondary. The error in the Hall sensor secondary

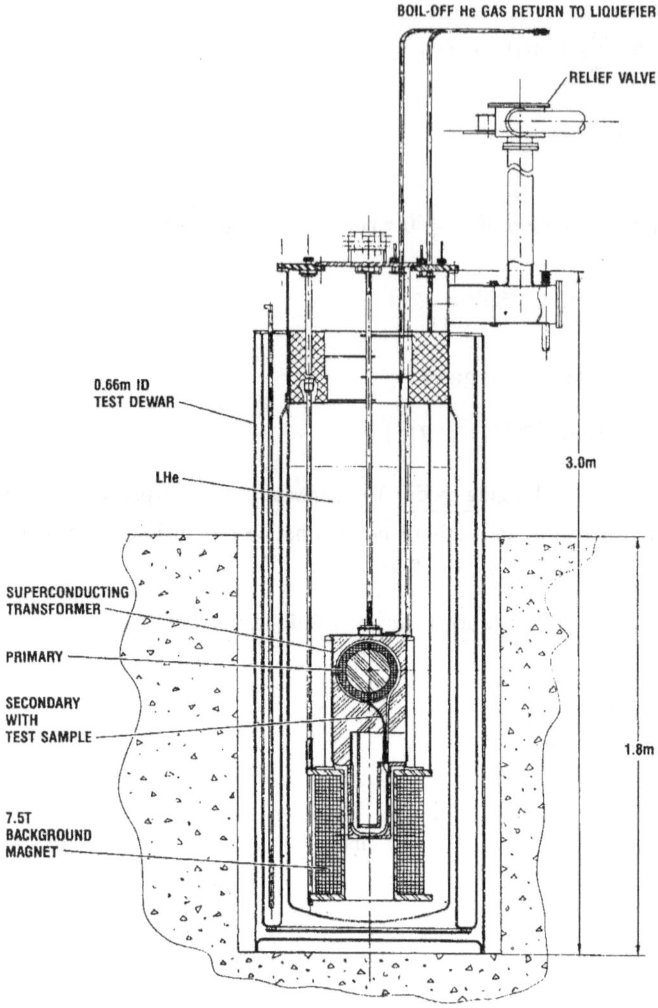

Fig. 2. Assembly for SCT test.

current reading coming from the effect of the primary current is calculated and correctable while the Rogowski coil only measures the current in the secondary. The inverse broken tennis racket configuration (Fig. 3) of the secondary enables two samples to be measured at the same time. The required overlapping lengths for the joints in the secondary are calculated using expressions obtained from reference 8. Maximum bending strains are 2.8% and 0.5%, respectively, for the Elmo Bumpy Torus (EBT[9]) and NBS SRM #1457 samples. The instrumentation setup for the test assembly is presented in Fig. 4. The advantages of this arrangement are that only a 10V, 100A DC power supply is needed instead of a high-current 25000A one and that a 160A rated vapor-cooled lead can be used instead of a 25000A rated one.

TEST PROCEDURE

We applied the SCT measurement method to two conductor samples, first the Elmo Bumpy Torus (EBT) conductor and then the NBS SRM #1457 conductor. In our first test, the constant current charge rate mode was used. Only a limited amount of data was taken. Basically it was a checkout of the system electronics. We did not attempt to obtain any V-I characteristics curves for the conductor. The short sample current values obtained were hand integration of the output of the Rogowski coil. By the time we ran the second test, we had programmed our Hewlett Packard Model

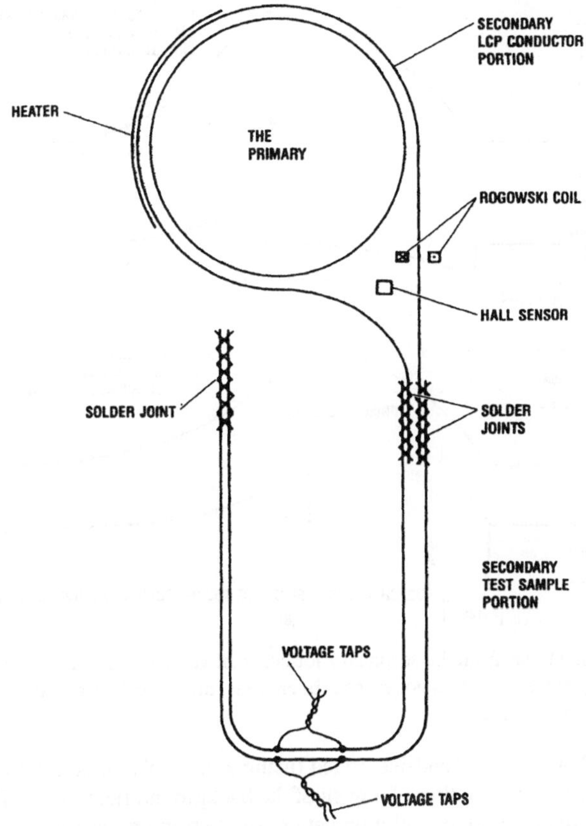

Fig. 3. A Rogowski coil and a cryogenic hall sensor were used to measure the induced secondary current.

1000E computer to give us a constant voltage mode charge. The results obtained were more repeatable than in the first test. A typical output of the Rogowski coil, responding to a constant voltage charge of the primary of the transformer, is shown in Fig. 5. The Rogowski coil was first calibrated at room temperature. Because of the calibration's insensitivity to temperature change, the Hall sensor

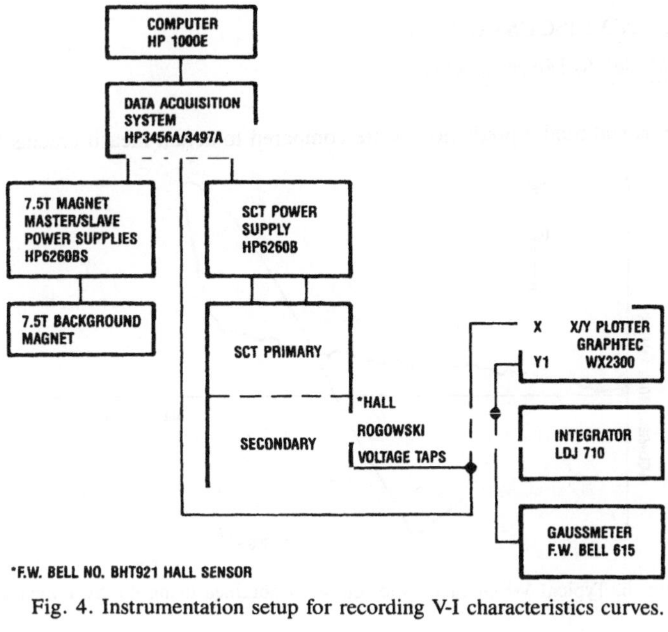

Fig. 4. Instrumentation setup for recording V-I characteristics curves.

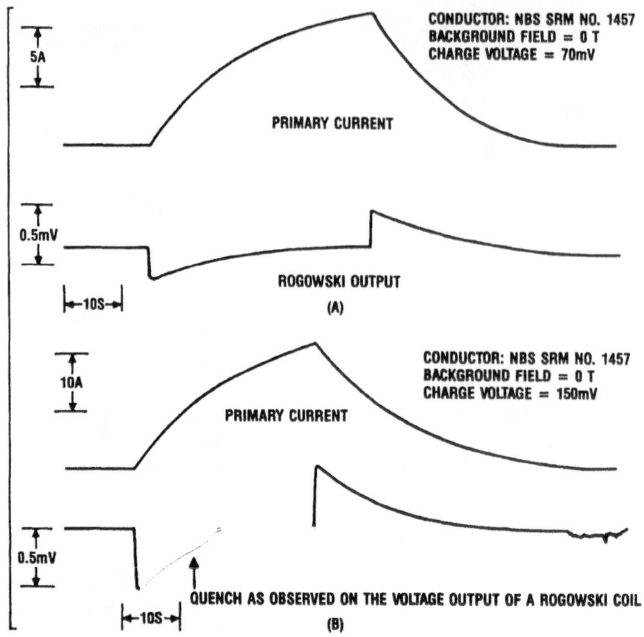

Fig. 5. Typical Rogowski output from the secondary for constant voltage charging of the primary, for the case of the induced secondary current (a) does not exceed (b) exceeds, the critical current of the sample conductor at a given background magnetic field.

output was calibrated against the hand-integrated (using a polar plannimeter) output of the Rogowski coil at its below-quench point condition for each of the background field levels. The peak value of the secondary current attained and time evolution pattern of the primary current are compared to predictions based on theory. The voltage measured across the sample is then plotted in real time against the output from the Hall sensor on a X-Y recorder to obtain the classical V-I characteristics curves for different charge voltages and background field values. Fig. 6 represents a typical V-I curve obtained for the NBS SRM #1457 conductor. A more direct method is to plot the voltage across the sample against the integrated output of the Rogowski coil at a given time on the X-Y recorder. We had to use the first method because our integrator did not respond well enough to the fast superconducting-normal transition.

TEST RESULTS AND DISCUSSION

Comparison Of Model To Measurements

The mathematical model predictions were compared to actual measurements for the constant

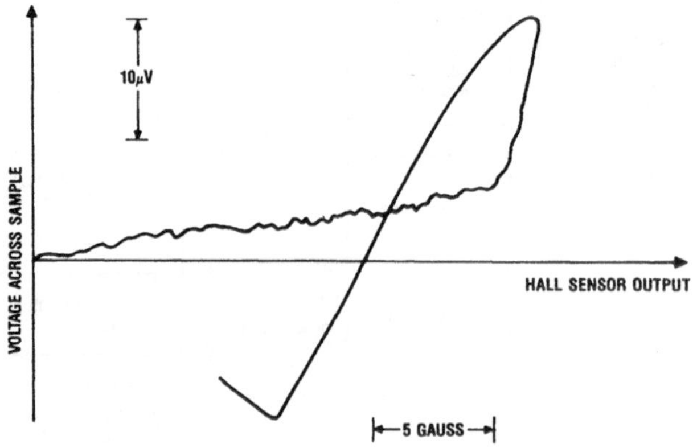

Fig. 6. Typical V-I characteristics curve as obtained using the SCT method.

Table 1. Comparison of time evalution patterns of primary current at constant charge voltage of 70 mV.

Time (in S)	Primary Current (in A)	
	Measured	Calculated
0	0	0
5	3.5	3.2
10	5.8	5.4
15	7.5	7.0
20	8.8	8.3
25	9.8	9.3
30	10.5	10.2
35	11.1	10.9

voltage charge mode. Table 1 summarizes the time evolution pattern of the primary current of the transformer with the constant charge voltage set at 70 mV and at zero background magnetic field. The values for the various parameters chosen are R = 7 mohm, R = 1.65E-07 ohm, L = 0.1696 H, L = 1.27E-06 H, and k = 0.5. We tried different methods to measure k, without much success because of the complicated shape of the secondary. The peak secondary current attained and the time at which it occurs for a 100mV constant voltage charge was calculated to be 560A, 23.7s while actual test data gave 595A and 24s. The test data, as evidenced from Table 1, matched fairly well with theory.

Results From The EBT Conductor Test

Table 2 summarizes the test results obtained using the SCT measurement method as compared to data extrapolated from known and reliable measurements[10] of the critical current of the EBT conductor at a field of 7T. This initial data was encouraging enough for us to apply the SCT method to the standard reference material from NBS.

Results From The NBS SRM #1457 Conductor Test

At each background magnetic field level, we took critical current measurements using the SCT at a minimum of three different constant charge voltages. The spread of the test results were then plotted in Fig. 7 as I's against the critical current values ⊙ given for the standard reference material[4]. Values for the critical current at magnetic fields B = 0, 7, and 7.5T were deduced from data that comes with the standard reference material at B = 2, 4, 6, and 8T. The electric field criterion used for determining the short sample currents was 0.5 microvolts/cm in all cases.

CONCLUSIONS

Calibration of our Superconducting Transformer(SCT) was performed using the NBS critical current measurement standard reference material number 1457. Test data suggested that the SCT method can be a viable alternative to the conventional approach in the measurement of critical current of composite superconductors. This is especially useful for high current and high background magnetic field conditions. Further development of the concept using a conductor with a bigger

Table 2. Comparison of results for the EBT conductor with previously obtained data using the conventional test method.

Magnetic Field (T)	Short Sample Data (in A) Obtained Using		Deviation
	Conventional Method*	SCT Method	
6.5	3,891	4,641	+19.2%
5	5,606	6,574	+17.3%
3	8,473	7,757	− 8.4%
1	15,691	15,305	− 2.6%

*Extrapolated number from actual measurement of 3,000 ± 150 A at 7.5T (reference 10).

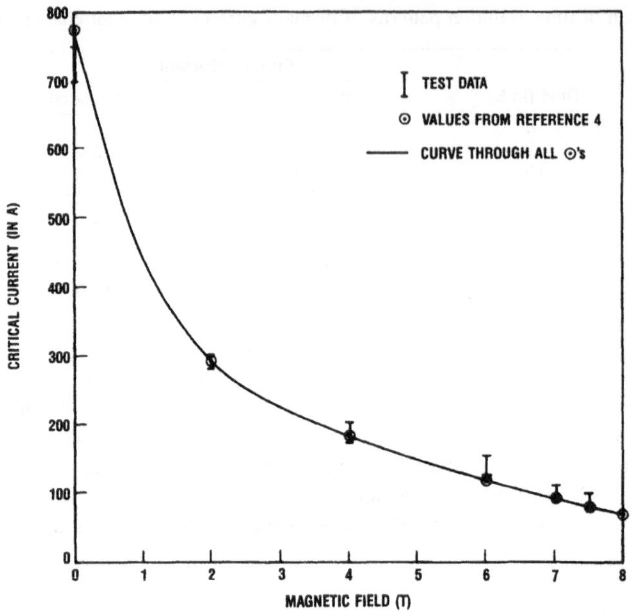

Fig. 7. It is evident that the SCT method can be a viable alternative to the conventional critical measurement method for composite superconductors.

current carrying capacity is recommended. The theory developed for this type of single shot superconducting transformer compares well with the measurements.

ACKNOWLEDGEMENTS

The authors wish to express their thanks to Mr. Edward McCaslin of General Dynamics Space Systems Division for the careful work that went into the assembly of the test apparatus; Mr. Robert A. Johnson, Head of our Energy Programs Department, for his continuing support in the project; and Dr. Lorentz Goodrich of the National Bureau of Standards, for helpful discussions and encouragement.

REFERENCES

1. Lorentz Goodrich, NBS, private communication.
2. A. F. Clark, L. F. Goodrich, F. R. Fickett, "Development of Standards for Superconductors," NBSIR 80-1642, 82-1678.
3. "Standard Test Method for D. C. Critical Current of Composite Superconductors," ASTM Designation B714-82.
4. L. F. Goodrich et al, Critical current measurements on an NbTi superconducting wire standard reference material, NBS Special Publication #260-91.
5. J. R. Purcell, H. DesPortes, Short sample testing of very high current superconductors, *Review of Scientific Instruments,* vol. 44, no. 3 (1973).
6. E. M. W. Leung, The birth of an applied superconductivity test laboratory, "Proc. of Nineth International Magnet Technology Conference," 1985, p. 862.
7. A. R. Newton, SPICE version 2.G user's guide, U. C. Berkeley, 1981.
8. J. W. Ekin, Current transfer in multifilamentary superconductors, I. Theory, *J. Appl. Phys. 49,* p. 3406(1978).
9. S. L. Ackerman et al, Design and construction details of the 7.4T superconducting metastable magnet for the Elmo Bumpy Torus proof-of-principle program, "Proc. of the Eighth International Magnet Technology Conference," 1983.
10. General Dynamics Space Systems Division, EBT program data file.

SSC MAGNET CRYOSTAT SUSPENSION SYSTEM DESIGN

T.H. Nicol, R.C. Niemann and J.D. Gonczy

Fermi National Accelerator Laboratory
Batavia, Illinois

ABSTRACT

The design of the cryostat for the Superconducting Super Collider (SSC) dipole magnets has largely been driven by the design of the cold mass suspension and anchor systems. Rigorous structural requirements in combination with low allowable heat loads have resulted in a suspension system that represents a significant departure from current superconducting magnet design practice both in performance concept and materials selection. This paper presents a summary of the suspension and anchor system designs being employed in the SSC.

INTRODUCTION

The suspension system in a superconducting magnet performs two essential functions. First, it resists internally and externally generated structural loads imposed on the cold mass assembly ensuring that the position of that assembly is stable over the operating life of the magnet. Second, it serves to insulate the cold mass from heat conducted from the outside world.

To satisfy the first function the normal operating stresses in the suspension system must be low enough to avoid creep in the component materials yet sufficient reserve strength must exist to handle loads imposed during shipping and handling of the magnet assembly, seismic excitations, and internally generated quench loads. To satisfy the second function in some optimal way we must size the suspension components to just meet the structual requirements and utilize materials that offer a good compromise between mechanical strength and thermal impedance. Table 1[1] summarizes the structural and thermal loads considered during the design of the suspension system for the SSC.

Table 1. SSC Dipole Structural and Thermal Load Summary

Shipping and handling loads:	vertical	2.0 G
	lateral	1.0 G
	axial	1.5 G
Seismic load guidelines:	Nuclear Regulatory Guide 1.61 vertical and horizontal spectra scaled by 0.3	
Maximum axial quench load:		11360 kg
Budgeted conduction heat loads per magnet:	80 K	7.20 W
	20 K	0.82 W
	4.5 K	0.12 W

DESIGN DEVELOPMENT

Support Post

Several candidate support systems were considered during the early development stages of the SSC.[2] A reentrant post assembly was ultimately selected for its superior mechanical and thermal performance. Fig. 1 is a cross section through a typical SSC support post. Its design has been previously described in some detail.[2,3] Briefly, it consists of two concentric composite tubes; one operating between 300 K and 80 K, the other between 80 K and 4.5 K. The connection between these inner and outer tubes is via a metallic tube located in the annular space between them. It connects the top of the outer tube to the bottom of the inner tube and allows the inner tube to reenter the outer, increasing the total conductive heat path and accounting for its 'folded' construction. All of the metal to composite joints are effected by sandwiching the composite tubes between metallic discs and rings. Clamping forces are generated by shrink fits at all joints.[2,3] No other mechanical or chemical bonds are made.

Of interest here is some detail regarding the size and material selection for the inner and outer composite tubes. Given the notation in Fig. 2, consider three components of stress acting at points 1 and 2; the points of maximum bending stress in tubes 1 and 2, respectively. The three stresses are induced by: 'g' forces acting at the cold mass centerline, denoted by F_g, quench forces acting at the top of the post, denoted by F_q, and the weight of the cold mass, denoted by W. Given this notation, the total stress acting at point 1 is given by

$$\sigma_1 = \frac{M_{g1}d_1}{2I_1} + \frac{M_{q1}d_1}{2I_1} - \frac{W}{A_1} \qquad (1)$$

where M_{g1} = bending moment at point 1 due to F_g
M_{q1} = bending moment at point 1 due to F_q
W = weight of the cold mass assembly (per support)
d_1 = diameter of tube 1
I_1 = section modulus of tube 1
A_1 = cross sectional area of tube 1

Similarly for point 2,

$$\sigma_2 = \frac{M_{g2}d_2}{2I_2} + \frac{M_{q2}d_2}{2I_2} - \frac{W}{A_2} \qquad (2)$$

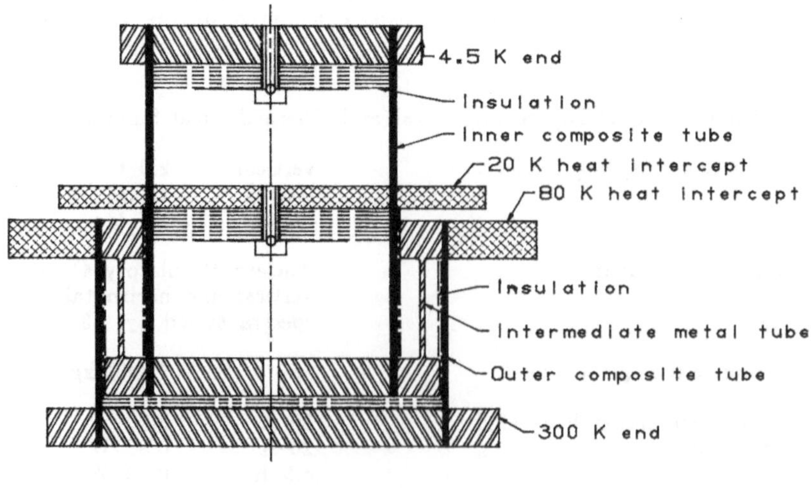

Fig. 1. Cross section through an SSC support post.

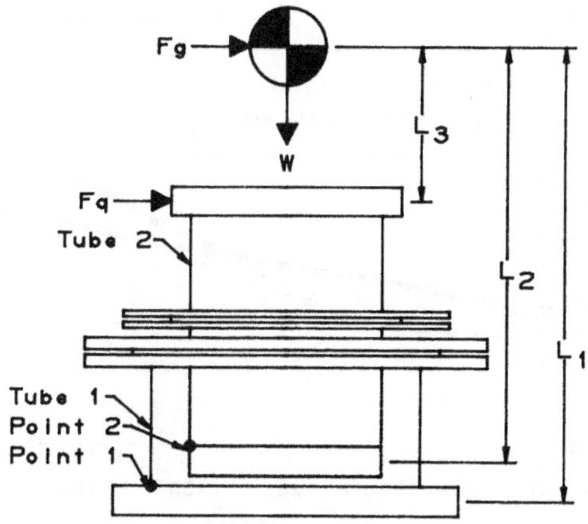

Fig. 2. SSC support post design analysis notation.

Bending moments M_{g1}, M_{q1}, M_{g2}, and M_{q2} are related to F_g, F_q, L_1, L_2, and L_3 by

$$\begin{aligned} M_{g1} &= F_g(L_1) \\ M_{q1} &= F_q(L_1\text{-}L_3) \\ M_{g2} &= F_g(L_2) \\ M_{q2} &= F_q(L_2\text{-}L_3) \end{aligned}$$

So (1) and (2) may be rewritten as

$$\sigma_1 = \frac{F_g L_1 d_1}{2I_1} + \frac{F_q(L_1\text{-}L_3)d_1}{2I_1} - \frac{W}{A_1} \tag{3}$$

and

$$\sigma_2 = \frac{F_g L_2 d_2}{2I_2} + \frac{F_q(L_2\text{-}L_3)d_2}{2I_2} - \frac{W}{A_2} \tag{4}$$

To produce an optimum design using (3) and (4), σ_1 and σ_2 must be equated to the ultimate strength for the materials used of tubes 1 and 2. However, a thin walled tube may fail due to elastic instability (local buckling) at a stress below the ultimate. Elastic instability in thin wall tubes is determined by[4]

$$\sigma_{ei} = \frac{2Et}{\sqrt{3}\,\sqrt{1-\nu^2}\,d}$$

where σ_{ei} = stress at which elastic instability occurs
E = elastic modulus
t = wall thickness
ν = Poisson's ratio
d = tube diameter

Our own tests show that the above expression must be derated by approximately 1.5 to agree with results on actual tube material. For tubes 1 and 2 this yields

$$\sigma_{ei1} = \frac{2E_1 t_1}{1.5\,\sqrt{3}\,\sqrt{1-\nu_1^{2}}\,d_1} \tag{5}$$

and

$$\sigma_{ei2} = \frac{2E_2 t_2}{1.5\,\sqrt{3}\,\sqrt{1-\nu_2^{2}}\,d_2} \tag{6}$$

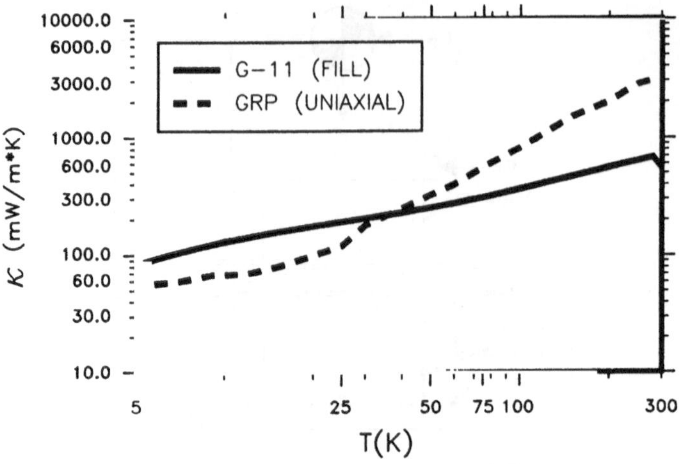

Fig. 3. Thermal conductivity of G-11CR and GRP tube material.

The complete optimization process entails use of (3), (4), (5), and (6). In principle we could optimize with respect to any aspect of the post geometry. In our case the tube lengths are determined more by other cryostat components than by optimization criteria and in the current configuration the structural and thermal performance parameters are not very sensitive to small changes in the tube diameters. Our final optimization is with respect to the thicknesses of tubes 1 and 2, t_1 and t_2, respectively. Note that although they do not appear explicitly in (3) and (4), t_1 and t_2 are implicit in the expressions for I and A.

Given the properties for candidate materials, a general post geometry, and specified load constraints, we solve equations (3) and (4) for t_1 and t_2 respectively. Next, we replace the left sides of (3) and (4) with (5) and (6), respectively; again solving for t_1 and t_2.

The materials selected by this process were NEMA G-11CR for tube 1 and graphite reinforced plastic (GRP) for tube 2. GRP exhibits structural performance superior to that of G-11CR, but has significantly higher thermal conductivity between 300 K and 40 K (see Fig. 3[5,6]). This makes it undesirable for use in tube 1 which operates between 300 K and 80 K, but potentially attractive for tube 2 operating between 80 K and 4.5 K.

The SSC cryostat contains five support posts; five being the number that limits sag in the cold mass assembly to less than 0.25 mm. Using this, the conduction heat loads to 80 K, 20 K, and 4.5 K in a dipole magnet, due to the support posts only, are 10.52 W, 1.60 W, and 0.08 W, respectively. Table 2 lists the input parameters used in the optimization process used to determine the configuration of the current SSC support post. Table 3 lists the resulting heat loads to 80 K, 20 K, and 4.5 K for two post configurations; one utilizing G-11CR and GRP, the other utilizing only G-11CR. The current SSC support post is that represented by the first column. Note that the use of GRP reduces the 4.5 K heat load dramatically without significant impact on the 80 K and 20 K loads.

Table 2. Current SSC Support Post Input Parameters

F_g = 1450 kg	W = 1450 kg
F_q = 3860 kg	SF = 2.0 (safety factor)
L_1 = 376 mm	L_2 = 351 mm
L_3 = 152 mm	
E_1 = 27.6 GPa	E_2 = 68.9 GPa
ν_1 = 0.2	ν_2 = 0.2
d_1 = 179 mm	d_2 = 127 mm
σ_{u1} = 276 MPa	σ_{u2} = 413 MPa

Table 3. Comparison Of Two Optimized Post Configurations

	Outer G-11CR Inner GRP	Outer G-11CR Inner G-11CR
t_1	2.77 mm	2.77 mm
t_2	3.28 mm	5.11 mm
Q_{80}	2.103 W	2.038 W
Q_{20}	0.320 W	0.370 W
$Q_{4.5}$	0.015 W	0.030 W

Anchor System

The five support posts used in each SSC cryostat will share vertical and lateral loads induced by shipping and handling and seismic excitations. Thermal contraction of the cold mass assembly during warmup and cooldown necessitates axial sliding between the cold mass and each of the four outboard posts so they cannot contribute to axial load restraint. The center post is attached rigidly to the cold mass assembly to ensure correct axial position within the cryostat vacuum vessel. Given no other axial restraint, this means that the center post would see the complete axial load induced during shipping and handling, seismic disturbances, and differential pressures which can occur during magnet quenching. A single post is incapable of handling these loads alone. Utilizing a 'strong' post at the center would impose intolerable heat loads on the cryogenic systems. We require a separate means of effecting axial restraint.

One of the early anchor schemes employed in prototype SSC cryostats is shown in Fig. 4. It consists of two tubular struts connected via pinned joints at 300 K to the base of the center post and at 4.5 K to the cold mass assembly. The struts share any axial load imposed on the cold mass; one reacting in tension, the other in compression. As an anchor, this system works well. Its impact on the cryostat as a whole, however, is not ideal. First, it adds two more suspension components conducting heat from 300 K to 4.5 K, increasing the heat loads to all intercept stations. Second, it penetrates the thermal shields at 80 K and 20 K, complicating assembly and necessitating elliptical holes which are difficult to insulate. Even a small crack in the insulating layers produces an increase in radiative heat load.

Ideally one would like an anchor system with negligible thermal impact on the cryogenic systems and which introduced no perturbations into other cryostat components (like the shield holes in the case of the strut system).

Recall from a previous discussion that the sliding post/cold mass attachments implies that the fixed center post is the only one capable of resisting axial loads. Recognizing that the bending strengths of all five posts could be combined to effectively act as a single axial restraint, we have chosen to connect the 4.5 K ring of each post to that of each adjacent post with axial tie bars.

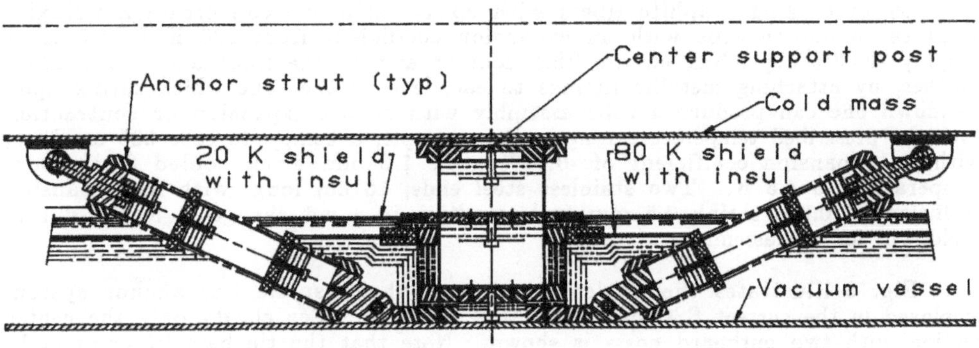

Fig. 4. Strut type axial anchor system used on SSC prototypes.

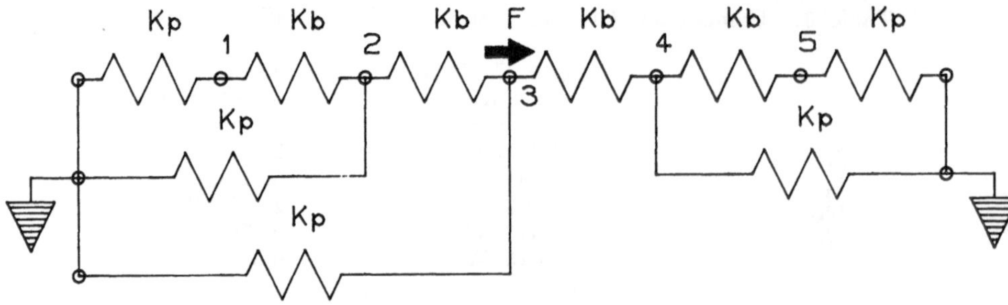

Fig. 5. Equivalent spring diagram of the post/tie bar anchor system.

To understand the effectiveness of such a scheme, we must understand the degree to which an axial load applied at the center post is shared by the remaining four posts. Specifically, we need to know the reaction forces at the top of each post, given a force applied at the center. Fig. 5 is an equivalent spring diagram of the post/tie bar system used to determine the stiffness of the total system. Points 1 through 5 represent the post/tie bar attachment points, point 3 being the top of the center post. The k_p's represent the bending stiffness of each support posts. The k_b's represent the axial stiffness of each tie bar. Grounded points represent the 300 K attachment of each post to the cryostat vacuum vessel.

If the tie bars were infinitely stiff one could expect the posts to share an axial load equally, i.e., for five posts, each would see one-fifth of the total. Obviously this does not represent a feasible solution. As a more realistic example, let k_p and k_b equal 1000 and 2000 kg/mm, respectively. Reaction force analysis on the system in Fig. 5 shows that 35.4% of the load goes into the center post, 19.4% goes into each of the next two outboard posts, and 12.9% goes into the two outermost posts. The total system stiffness, k_t, for this example, is 2820 kg/mm. The system efficiency may be defined as the ratio between the actual and theoretical maximum stiffnesses, $k_t/5k_p$). Using this definition, the efficiency of the system in this example is 56.4%. Increasing k_b by a factor of four yields a system whose efficiency is 69.7% so clearly one would like to make k_b as high as possible.

The remaining hurdle in the design of the post/tie bar anchor system relates to the material selection for the tie bars themselves. Materials commonly selected for use in superconducting magnet cryostats; glass composites, stainless steel, and aluminum, for example, all exhibit decreases in length when cooled from 300 K to 4.5 K. Selecting such a material for the anchor tie bars would result either in very high tensile loads in the tie bars themselves or very high bending moments in the post assemblies because of the shrinkage which occurs during cooldown. Unlike most materials, however, graphite fibers exhibit a negative coefficient of thermal expansion meaning that they grow when cooled. In most graphite composites this effect is masked by the resin system which shrinks upon cooldown, particularly if the bulk of the fibers are oriented off-axis from the measurement direction.

By pultruding graphite fibers with epoxy resin one can create a uniaxial graphite composite tube with an expansion coefficient from 300 K to 4.5 K of roughly -0.03% depending on the fiber content and on the fiber and epoxy used. Further, by attaching metallic fittings to each end of the tube which shrink upon cooldown one can produce a tube assembly with no net expansion or contraction over the prescribed temperature range. For example, a composite tube 325 cm long, with an expansion coefficient of -0.03% grow 1.0 mm when cooled from room temperature to 4.5 K. Two stainless steel ends, 16 cm long, with an expansion coefficient of 0.3%, shrink 0.5 mm each resulting in a net change in length during cooldown, for the assembly, of zero.

Fig. 6 illustrates the implementation of the post/tie bar anchor system employed in the current SSC dipole cryostat assembly. For clarity only the center section with two outboard posts is shown. Note that the tie bars fit completely

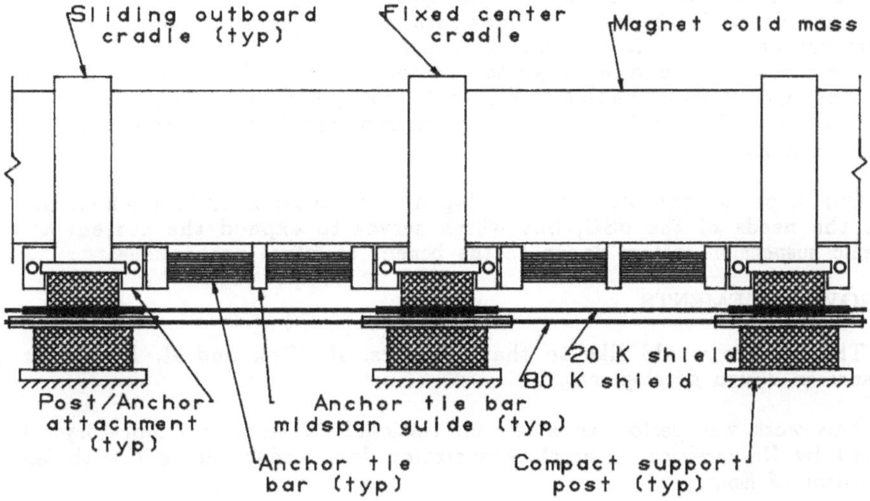

Fig. 6. Partial view of the SSC post/tie bar anchor system.

inside the 20 K shield, eliminating the need for shield penetrations. In addition, because both ends of each tie bar assembly operate at 4.5 K, there is no added conduction heat load. Metallic fittings on each end of the tie bar assemblies are attached by the same shrink fit method used in support post fabrication.[1,2] Also shown in Fig. 6 is a guide located midway along each tie bar. The guide is a loose fitting collar that does not restrict axial movement of the tie bar assembly, but that prevents column buckling when the tie bar is loaded in compression.

The tie bars used in the present configuration are 305 cm long tubes with an outside diameter of 64.5 mm and a wall thickness of 6.35 mm. The material is a uniaxial GRP pultrusion with and elastic modulus of 124 GPa. Using earlier notation, the calculated anchor component and system parameters are

$$k_p = 1618 \text{ kg/mm}$$
$$k_b = 3690 \text{ kg/mm}$$
$$k_t = 4764 \text{ kg/mm}$$
$$eff = 58.9 \text{ \%}$$

The fixed center post sees 34.0% of the total axial load, the two outboard posts nearest the center see 19.5% each, and the outermost two posts see 13.5% each. Note that the value for F_q listed in Table 2 is 34% of the maximum quench load listed in Table 1 and represents the maximum load seen by the center support post during a magnet quench.

CONCLUDING REMARKS

The suspension system for SSC magnets has evolved toward its current configuration as the result of many design iterations, some based on established practice, others developed as complete new concepts.[2] The post/tie bar system is a combination of old and new. Support posts in some form have been used for many years, but none, to our knowledge, have been developed into devices which afford such low heat loads and high structural stiffnesses and strengths. Further, none have been developed using shrink fit bonds at all composite to metal joints nor have they played such an integral role in the anchor system performance. We are also unaware of the use of uniaxial graphite pultrusions in combination with metallic ends to effect axial links which are structurally sound and which exhibit little or no change in length during the temperature variations seen in superconducting magnets.

The current post/tie bar suspension system design meets the static structural requirements set forth for SSC magnets and exceeds the required thermal performance at 4.5 K. It constitutes an assembly which requires a minimum of added perturbations to other cryostat components and lends itself well to easy fabrication and mass-production. Yet to be completed is the analysis and testing related to its performance in a dynamic environment like that experienced during seismic excitations.

Our hope is that we have developed a suspension system which not only serves the needs of the SSC, but which serves to expand the current state of cryogenic suspension system design to the benefit of future applications.

ACKNOWLEDGEMENTS

The authors would like to thank Messrs. R. Zink and R. Dixon for their assistance in design development.

This work was performed at Fermi National Accelerator Laboratory which is operated by Universities Research Association Inc. under contract with the U.S. Department of Energy.

REFERENCES

1. SSC Central Design Group, "Superconducting Super Collider Magnet System Requirements," SSC-100, October 1986.
2. T.H. Nicol et al, A suspension system for superconducting super collider magnets, in: "Proceedings of the Eleventh International Cryogenic Engineering Conference," Butterworths, Surrey, UK (1986).
3. J.D. Gonczy et al, Cryogenic support thermal performance measurements, in: "Advances in Cryogenic Engineering" Vol. 33, Plenum Press, New York (to be published).
4. R.J. Roark and W.C. Young, "Formulas for Stress and Stain," Fifth Edition, McGraw Hill, New York (1975), p. 428.
5. J.R. Bensinger, Properties of cryogenic grade laminates, IEEE Conference, Boston (1979).
6. M. Takeno et al, Thermal and mechanical properties of advanced composite materials at low temperatures, in: "Advances in Cryogenic Engineering" Vol. 32, Plenum Press, New York (1986).

IMPROVED DESIGN FOR A SSC COIL ASSEMBLY SUSPENSION CONNECTION

E.T. Larson, J.A. Carson, T.H. Nicol, R.C. Niemann, R.A. Zink

Fermi National Accelerator Laboratory
Batavia, IL

ABSTRACT

Close control of the alignment of magnets for the proposed Superconducting Super Collider (SSC) high energy physics research facility is essential for the success of this small bore accelerator. The connection of the magnet coil assembly to the cryostat suspension system presents many challenges to the cryostat designer. The resulting design must withstand shipping and seismic loads, allow axial contraction of the coil assembly, position the coil assembly center line within a 0.25 mm radius, provide rotational adjustment of the coil, resist axial quench loads, provide a bearing assembly which is tolerant to high vacuum, high radiation and cryogenic conditions, and fit within the stringent geometric constraints of the cryostat assembly. A coil assembly suspension connection which meets these criteria is described. Measurements of the effective friction coefficient, static load deflection, and component stresses are compared to the predicted performance. Experiences with a full length model of the coil assembly connection are presented.

INTRODUCTION

Dipole magnets of the proposed Superconducting Super Collider (SSC) will require an integrated suspension system which meets the demanding criteria of low heat leak, high structural strength, precise alignment, high reliability and low cost required for this system. This paper outlines the design considerations, and test results for an improved connection between the reentrant post[1] and the dipole cold mass assembly.[2]

To date all dipole model magnets have been assembled with a cold mass connection which uses insert pins to penetrate the outer shell and contact the iron yoke within the cold mass assembly as shown in Figure 1. The insertion pins act both as fiducial and structural support members, and require a seal weld on the outer shell for each pin penetration. Parallel rods with sleeve bearings support and guide a tie bar which attaches to both insertion pins locking the cold mass in rotation and lateral position.

Difficulty in positioning the insertion pins accurately in their welded sockets, along with a lack of rotational freedom initiated the search for an improved design. Areas of the original design requiring improvement are:

- A Providing angular position adjustment of cold mass during assembly
- B Increasing structural support to withstand dynamic loads
- C Improving bearing performance, lower wear rate, reduced stick slip binding
- D Avoiding multiple weld penetrations in the cold mass outer shell.

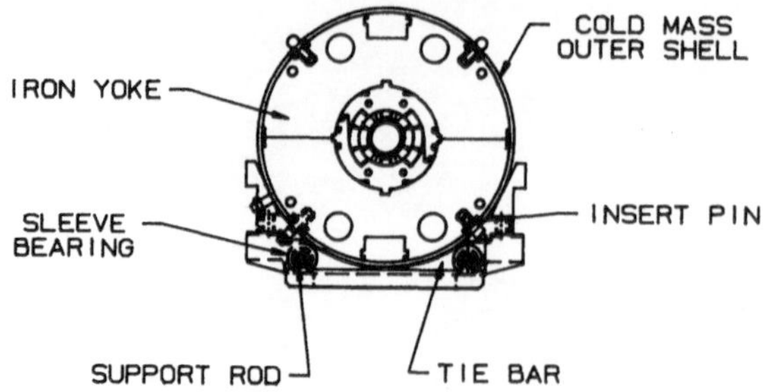

Fig. 1. Present cold mass slide connection.

An improved cold mass slide connection for SSC dipole magnets has been designed and preliminary testing completed. To date it has successfully passed loading and slide tests at or above design criteria.

DESIGN CONSIDERATIONS

A cold mass connection must perform the following:

A Position the magnet bore center repeatably and accurately
B Support static and dynamic loads
C Allow for axial contraction during cyclic cold mass cooldowns
D Fit within the close geometry of the cryostat
E Withstand the cryogenic and radiation environments associated with superconducting accelerator magnets

The typical layout of the improved coldmass suspension system is shown in Figure 2. Four of the five connections to the cold mass are of a sliding design, allowing both rotational and axial degrees of freedom. The center connection is fixed during assembly and locks the angular and axial position of the cold mass. Four anchor tie bars will be installed between the five reentrant posts to share the axial loads transfered from the fixed center connection to the center post.[3]

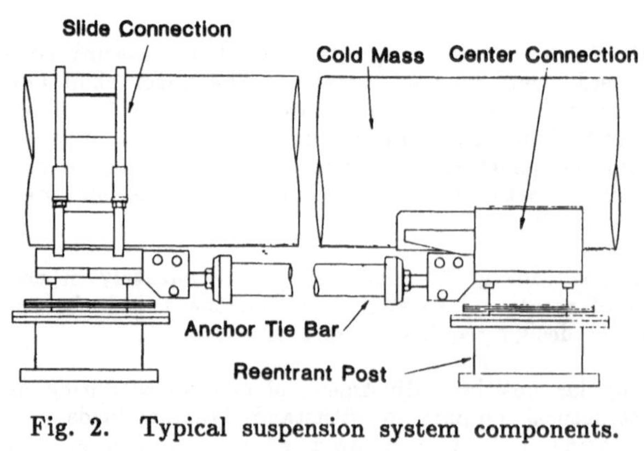

Fig. 2. Typical suspension system components.

Figure 3 shows the main components of the improved cold mass slide connection. The base sits on top of the 4 K ring of the reentrant post assembly with a split flange ring locking the base to the top of the ring. The cradle arms of the lower half contain bearing blocks which retain removable bearing pads. The outer shell of the cold mass contacts these pads to establish the position of the magnet assembly. The upper half of the connection has an identical bearing support which retains the cold mass during shipping, quench or seismic loads. The four points which contact the outer shell are located to avoid the longitudal weld seams which close both halves of the outer shell and the cut out in the iron yoke, see Figure 4. The outside diameter of the outer shell at these points is not concentric due to fabrication distortions, and therefore is not a reliable surface for alignment.

The outer surface of the coldmass at suspension points will be used as a fiducial for locating the magnet bore center. Using the surface of the cold mass as a fiducial requires referencing the outside diameter to the magnet bore center. Errors in the collared coil assembly and gaps between the collars and iron yoke are the same for any fiducial system referenced to the iron yoke. Efforts are centered on reducing the error between the outside iron yoke surface and the stainless steel outer shell surface. Two proposals for achieving this are discussed later.

Magnet alignment is critical in three directions; vertical height (y), lateral position (x), and angular position (θ). Height and lateral position are to be established by alignment targets placed on complete cold mass connection/post subassemblies. Shims establish height while lateral position will be achieved by bolt clearances. When the five cold mass connection/post subassemblies are aligned in x and y the cold mass will be set into the lower connection half, measured to determine the angular offset of its average vertical magnetic plane, rotated to compensate for this offset and fixed at the center. The critical dimension in the lower half of the cold mass connection is the radius created by the bearing insert. Tolerances in the counter bore or post assembly can be removed from alignment

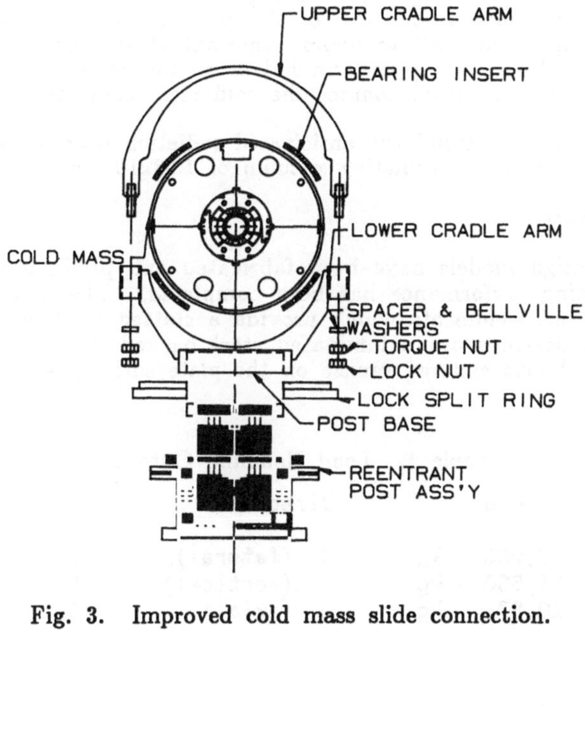

Fig. 3. Improved cold mass slide connection.

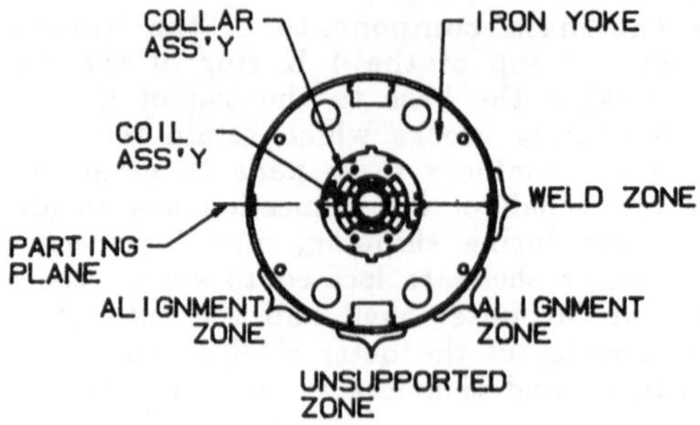

Fig. 4. Cold mass alignment zones.

error by shims and lateral positioning. The upper half of the slide connection gives structural strength for dynamic loads, but does not establish alignment, therefore it does not have high fabrication tolerances.

The magnet assembly is subjected to shipping, quench and seismic loads which are transfered to suspension components through the cold mass connections. Load handling criteria are shown in Table 1.[4] The five connections share the lateral and vertical loads, while the center connection alone transfers the axial load. The only axial load the slide connections see is the result of axial forces transfered through the bearing pads during cooldown.

During the 300 to 4 K cooldown the 17 meter coldmass shrinks 51 mm overall. The center connection being fixed causes the supports farther from center to see greater relative axial motion. The connections 3.454 meters from center see 10 mm axial motion while the connections 6.908 meters from center experience 20 mm axial contraction. This axial motion must be permitted while minimizing the axial reaction load in the bearings; and subsequent bending moment imposed on the reentrant post assembly. Reducing axial reaction loads also prevents large post deflections and potential binding or misalignment. The design is based on 70 thermal cycles occuring during the life of a magnet.[5]

The SSC cryostat cross section, Figure 5, has small clearance in the vertical direction due to insulation requirements, post geometry, etc. For this reason the cold mass is positioned as close to the top of the post as practical with a minimum thickness in the upper arm half to avoid thermal shield interference. Insulating standoffs are installed on the upper arm halves to minimize thermal conduction if the 20 K insulation shield should contact the cold mass connection.

The design must withstand the anticipated radiation over a twenty year life of 6×10^6 rad, and operate in an insulating vacuum of 1.33×10^{-4} Pa.

TEST EXPERIENCE

Prototype design models have been fabricated and preliminary testing of the structural and sliding performance has been completed. The model assemblies are wrought stainless steel weldments which provide a slotted land for the bearing pad to set into. The bearings are a laminated steel backed plate with a lead/teflon mixture impregnated into sintered bronze on the plate surface, which are rolled into

Table 1. Load Handling Criteria

Load	Direction	G's
7,250 kg	X (lateral)	1
14,500 kg	Y (vertical)	2
10,875 kg	Z (axial)	1.5

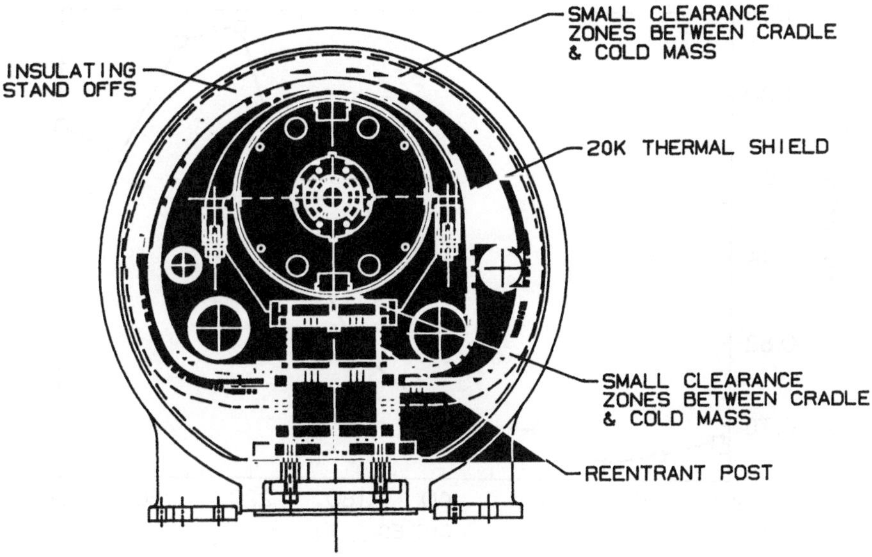

Fig. 5. SSC cryostat cross section.

final form. This self lubricating bearing material, like other bearings, requires a surface finish of 0.50μm or better for optimum performance. The bearing is rated for service at 4 K, and has low outgassing in high vacuum. Higher radiation resistant bearing materials such as polyimides and molybdebum disulfide coatings are being investigated, but because of availability and ease of fabrication laminated steel bearings have been chosen for prototype models.

The slide connection deflection under static loading has been measured and the results are shown in Table 2. Values represent the cold mass center deflection relative to the slide connection base, with post deflections subtracted out. Induced stresses are less than 25% of material yield point. The outside of the cold mass is 9 mm from the post top ring, while the upper arm has a 10 mm gap between its top point and the thermal shield avoiding interference. The two halves of the connection interlock in a guided slot to increase strength during lateral loading. The halves are bolted together with stainless steel bellville washers acting as tensioning springs and preloaded to a total of 550 kg for the four springs. This allows for any differential thermal contraction between the connection and cold mass, and provides a uniform preload in the bearings.

A full length cold mass has been tested in static ambient conditions to monitor bearing response to cyclic axial displacement at full load. A double acting hydraulic cylinder was used to measure axial load while moving the cold mass in the axial direction. Substituting a slide connection at the normally fixed center support allows axial displacement of 25 mm in each direction. The surface finish of the cold mass was improved to 0.50μm at the five support locations. Figure 6 shows the axial actuation force vs equivelant thermal cycles for the full length cold mass installed with 3 mm center to end sagitta. The 70 cycle design life, and

Table 2. Slide Connection vs Load.

Load	Direction	Deflection
4,500 kg	-Y Vert. Down	0.8 mm
3,000 kg	+Y Vert. Up	1.8 mm
2,250 kg	X Lateral	1.3 mm

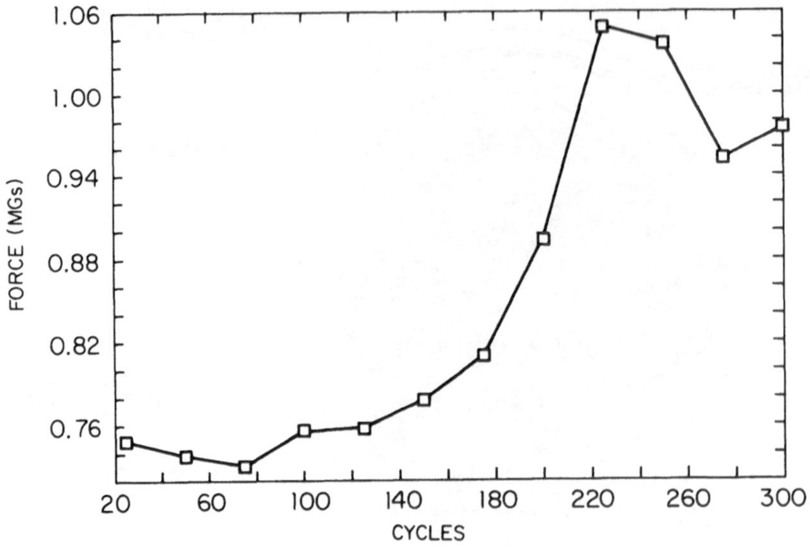

Fig. 6. Full length test data.

endurance factor of 4 result in test runs with a minimum of 280 cycles. The peak measured force of 1050 kg is distributed over five supports resulting in 210 kg per post. Initial cycles are 25% lower than this. The post has been designed to withstand 1450 kg loads without the assistance of anchor tie bars. By monitoring the deflection of the post assemblies and comparing them it was possible to determine axial load distribution. As expected the distribution was essentially equal throughout the testing, averaging 1 mm peak deflection at the top post ring. This deflection will be reduced considerably when the anchor tie bar system is installed.[3]

TEST PLANS

Several areas of the design are undergoing further investigation, they include:

A Measurement of connection response at cryogenic and vacuum conditions
B Measurement of bearing materials at cryogenic and vacuum conditions
C Alternate connection fabrication methods (High production)
D Cold mass outer shell fabrication techniques

A short magnet model is being fabricated to allow cryogenic testing of the connection and alternate bearing materials. Cooldown and transient thermal response will be measured, and provisions will be made to allow axial actuation through an external hydraulic cylinder. Bearing life, change in coefficient of friction and post deflection will be key areas of investigation.

Alternate bearing materials must withstand high radiation, have structural strength at cryogenic temperatures, low outgassing at high vacuum, and low friction coefficient. Polyimide bearings with 15% molybdenum disulfide added will be tested in the short magnet model to compare performance with laminated steel backed bearings. Low friction coatings and other surface treatments will be applied to stainless steel bearing pads and tested as well.

The complete slide connection support structure is being fabricated as a sand casting in test batches to determine feasability for high production. Tests for structural strength and integrity, vacuum outgassing, and fracture toughness at cryogenic temperatures will be performed on test parts to qualify the method. Economic and performance comparisons will be made between castings and welded assemblies.

Two techniques for fabricating the cold mass outer shell are being investigated. Using the external surface of the cold mass as a fiducial requires referencing the outer shell surface to the magnet bore center. Investigations are centered on reducing the tolerance between the outside iron yoke surface and the cold mass outer shell surface. One proposal for achieving this relies on minimizing the thickness variation in the steel used for forming the outer shell, and creating sufficient tension in the outer shell during longitudal weld seam shrinkage to eliminate gaps between the iron yoke and shell. The other method relies on machining the outside diameter of the shell after forming and welding, referencing the cutting head to a contact point on the iron yoke surface. Both methods must be capable of producing high quality surface finishes to enhance bearing performance. Tests of each method are being developed and implemented to allow quantitative comparison.

CONCLUSIONS

A slide connection for SSC magnets which addresses all the design criteria has been built and successfully passed structural and sliding tests. Completion of the remaining test program will verify the integrity of the connection for use in future model magnets.

ACKNOWLEDGEMENT

The authors wish to acknowledge the contributions of J. Tweed and M. Ruschman in the laboratory evaluations.

The work as presented was performed at Fermi National Accelerator Laboratory which is operated by Universities Research Association Inc., under contract with the U.S. Department of Energy.

REFERENCES

1. R.C. Niemann, et al., Design, construction and performance of a post type cryogenic support, in: "Advances in Cryogenic Engineering," Vol. 31, Plenum Press, New York (1986), p. 73.

2. C.E. Taylor, et al., A 6.4 tesla dipole magnet for the SSC, in: "Advances in Cryogenic Engineering," Vol. 31, Plenum Press, New York (1986), p. 25.

3. T.H. Nicol, R.C. Niemann and J.D. Gonczy, SSC magnet cryostat suspension system design and performance, in: "Advances in Cryogenic Engineering," Vol. 33, Plenum Press, New York (to be published).

4. SSC Central Design Group, Superconducting super collider magnet system requirements, "SSC-1000," October 1986.

5. SSC Central Design Group, Superconducting super collider magnet system requirements, (to be published).

CRYOGENIC SUPPORT THERMAL PERFORMANCE MEASUREMENTS

J.D. Gonczy, W.N. Boroski, T.H. Nicol, R.C. Niemann,
J.G. Otavka, and M.K. Ruschman

Fermi National Accelerator Laboratory
Batavia, Illinois

ABSTRACT

The stringent refrigeration requirements of the Superconducting Super Collider (SSC) and the premium nature of radial space in the SSC cryostat have led to the development of a reentrant tube cryogenic support. Thermal shrink fitting techniques are used to assemble the support. The thermal performance of two cryogenic support models is presented. The geometry of each model, its instrumentation, and experimental test arrangement in a Heat Leak Test Facility are described. Heat leak and temperature profile measurements made with a primary heat intercept temperature controlled between 10 K and 40 K are presented. Heat leak values to 4.5 K were measured by means of a heatmeter. Heat leak values to the primary and secondary heat intercepts were derived using the measured temperature profiles and component material properties. Presented are thermal performance measurements of copper cable connections used to heat sink the primary and secondary heat intercepts to their respective thermal radiation shields. Temperature measurements also were made on identical model supports installed in a full length (17.5 meters long) SSC dipole magnet cryostat thermal model. The thermal performance of the cryogenic supports for the two measurements is compared.

INTRODUCTION

The SSC cryostat development program has resulted in an innovative design for the main structural support member for the SSC magnet cold mass, and the primary and secondary thermal radiation shields.

The Compact Cryogenic Support (CCS) shown in Fig. 1 employs metallic end and heat intercept connections which are securely joined to composite tubes by the clamping pressure generated from designed dimensional interferences between components at the joint locations. Thermal shrink fitting techniques are used to assemble the cryogenic support. The shrink fitted joint provides a tightly clamped connection between the composite tube and the metallic components, and results in a support that maintains its structural integrity for load conditions in tension, compression, bending, and torsion for thermal cycles between 300 K and 4.5 K.

By proper selection of metallic materials, e.g., stainless steel for the inner disc and aluminum for the outer ring, a shrink fitted junction becomes stronger as it becomes colder due to the added clamping afforded by the differential thermal contraction of the junction components. Selection of the composite material for the tubular elements of the CCS is determined by concurrent consideration of stress, deflection, heat leak, creep, and installation geometry. The tubular material must allow for the development of flexure and torsional stiffness while maintaining a small cross-sectional area that is required to limit conduction heat leak.

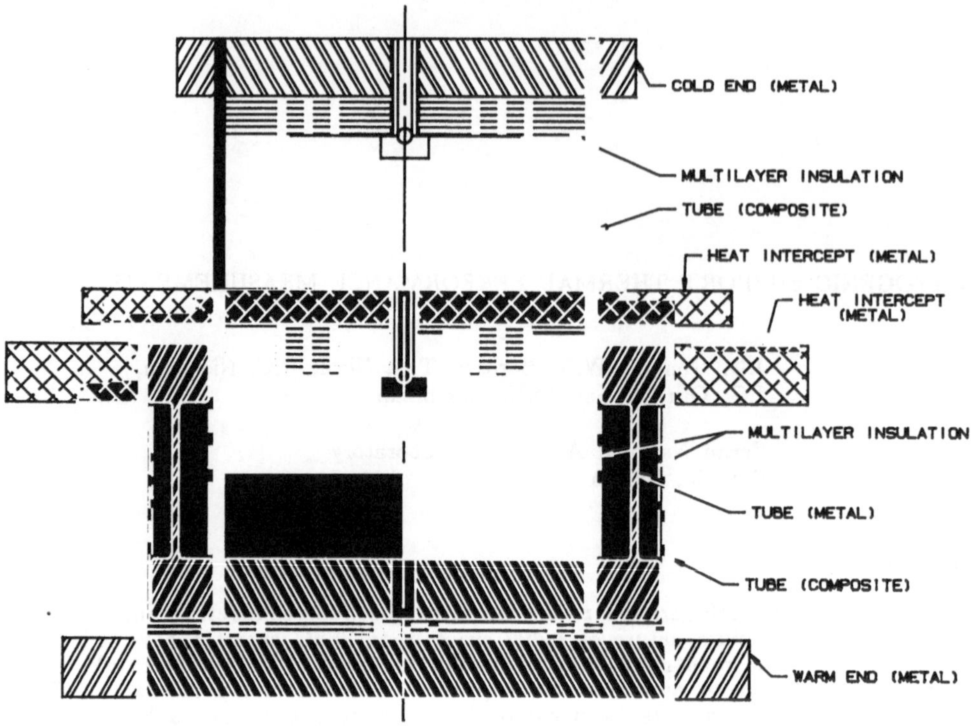

Fig. 1. Compact cryogenic support cross-section.

Also by its nature, a CCS is subject to internal thermal radiation. Internal radiation can significantly affect the thermal performance of the support and must be controlled. Effective control can be achieved by use of multilayer insulation internal to the CCS structure.

As the SSC support requirements became better defined, and the understanding of the structural and thermal performances of the support known, a more substantial CCS was developed, and the earlier model was replaced.[1,2] Structural and thermal evaluation of the Compact Cryogenic Support continues as working models of the design are optimized.[3]

COMPACT CRYOGENIC SUPPORT MODEL GEOMETRY

The 4x5 CCS and the 5x7 CCS are designated by the diameters (measured in inches) of the innermost and outermost composite tube elements. The material comprising the composite tubes in the 4x5 CCS model is G-10CR. In the 5x7 CCS, G-11CR is used. The wall thickness of the composite tubes in both models is 1.58 mm.

In each model, a stainless steel middle tubular element bridges the outer and inner composite tubes and accomplishes the reentrance of the conductive heat path within the outer tube, thereby adding length to the path without compromising axial height. The assembled height of each model is 213 mm.

The two models are identical in concept, differing in construction primarily in the size of their corresponding components, and with the 5x7 CCS model having a change in the shape of the stainless steel tubular element.

By design, the CCS has four distinct temperature stations which are defined by the CCS thermal connections. They include a 300 K connection to the room temperature vacuum vessel, an 80 K outer shield support and heat intercept connection, a 20 K inner shield support and heat intercept connection, and a connection to the 4.5 K magnet cold mass.

Multilayer insulation (MLI) internal to the CCS is used to limit radiant heat transfer between components. Radial heat transfer, i.e., radiation between the outer composite tube and the stainless steel middle tube, and between the stainless steel middle tube and the inner composite tube is reduced by MLI spiral wrapped on the corresponding cold areas of temperature boundary surfaces. Thermal radiation in the axial direction is controlled by wafers of MLI secured to the metal discs by tubular G-10 fasteners which allow evacuation of the annular space of the inner composite tube.

THERMAL PERFORMANCE EVALUATIONS IN THE HEAT LEAK TEST FACILITY

CCS temperature profile and heat leak measurements were performed in the Heat Leak Test Facility (HLTF).[4] Figure 2 illustrates the HLTF and shows the test geometry for the thermal evaluation of the 5x7 CCS with a cold mass slide connection.[5]

The HLTF provides corresponding thermal connections to the CCS temperature stations, and also provides the capability for controlling the inner shield connection at a temperature between 10 K and 40 K. Copper heat sink straps on the CCS heat intercepts are used to thermally anchor the intercepts to their appropriate HLTF thermal connection.

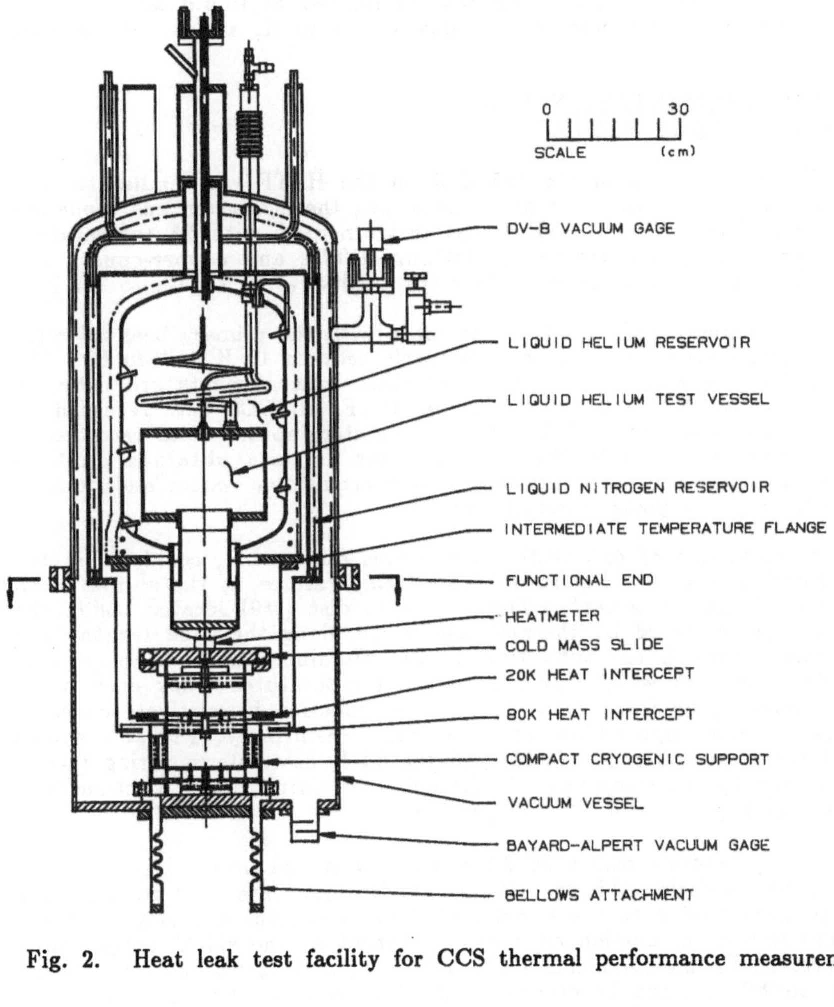

Fig. 2. Heat leak test facility for CCS thermal performance measurements.

Carbon resistors are used to monitor HLTF temperatures near 4.5 K; platinum resistance temperature detectors (RTDs) and cryogenic linear temperature sensors (CLTS) are used at temperatures near 80 K.[6] A four lead-wire system is employed for the sensors. The sensors are connected in series with a constant current source whose polarity is manually controlled. A current of 10 μA of is used for measurements near 4.5 K. For increased sensitivity, 100 μA is used for temperatures near 80 K. Sensor potential leads are connected to a high impedance data logger.

A bellows attachment to the HLTF vacuum vessel provides an effective load, due to atmospheric pressure, of approximately 273 kg on the CCS test geometry, and allows for thermal motion of the CCS during thermal cycles.

Insulating vacuum in the Heat Leak Test Facility is initiated by a turbomolecular pumping station. System vacuum is monitored by a Bayard-Alpert ionization gage and controller and is recorded by the data acquisition data logger.

The primary method of determining heat leak in the HLTF is by measurement using a heatmeter.[7] The heatmeter employs a thermally resistive reference section which is sandwiched between thermally conductive ends. For measurements to helium temperatures, a pair of carbon resistors sense the temperature across the reference section. For measurements to 80 K, platinum resistors are used. A calibration heater is located at the warm end of the heatmeter. Calibration of the heatmeter is accomplished by equating the temperature difference across the reference section to the value of heat flow through the heatmeter as generated by the calibration heater. Heatmeter calibration was performed at liquid nitrogen and liquid helium temperatures for ranges of 0 - 3.0 watts to 80 K, and 0 - 0.250 watt to 4.5 K.

4x5 CCS TEST INSTRUMENTATION AND THERMAL PERFORMANCE RESULTS

The installation geometry of the 4x5 CCS in the HLTF was similar to that shown in Fig. 2, but without the slide attachment, i.e., the heatmeter was connected directly to the 4x5 CCS. The 4x5 CCS was instrumented with 23 temperature sensors which included carbon resistors, platinum RTDs and copper-constantan thermocouples. The sensors were located along the conductive heat path to 4.5 K.

Thermal measurements on the CCS were made with the primary heat intercept controlled at different steady state temperature levels between 10 K and 40 K. The temperature profiles generated by the results of these measurements are listed in Table 1. Subsequent to the measurements in the HLTF, all CCS data recorded by the HLTF data logger was converted for use as computer input. Data recorded in the HLTF was then processed in the same manner as data obtained on CCS performance measurements using a computerized data acquisition system and done in a full length SSC magnet cryostat thermal model.[8]

Graphic representation of 4x5 CCS temperatures during LN_2 cooldown in the HLTF is illustrated in Fig. 3. Figure 3a shows temperatures at the shrink fitted junction of the 20 K heat intercept. The aluminum disc (#9) located inside the inner composite tube is cooled by thermal conduction along the tube length, and also by conduction through the tube wall to the aluminum heat intercept ring (#10). The closeness with which the component temperatures track each other illustrates that the clamping force at a shrink fitted junction is relatively constant for thermal excursions imposed on the junction. Similarly, Fig. 3b shows temperatures of the 80 K lower disc (#15) and the 80 K heat intercept ring (#16). Temperature tracking by the components is through two shrink fitted junctions and includes the thermal impedance of the stainless tube.

The CCS temperature stations at 20 K and 80 K are used to structurally support the thermal radiation shields. Copper cable straps are used to heat sink each CCS temperature station to the shield that it supports, thereby employing the CCS temperature station to function as a heat intercept at the shield temperature. Figure 4a features HLTF measurements on a 5x7 CCS, and shows the temperature profile along an 80 K heat intercept heat sink strap. Figure 4b shows

Table 1. 4x5 CCS Temperature Profile Measurements

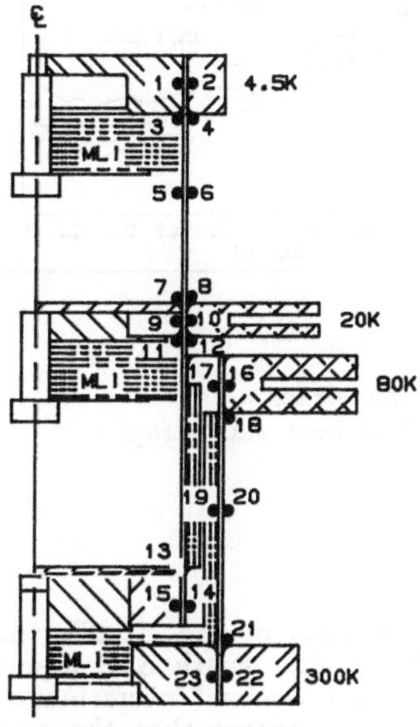

SENSOR LOCATION ALONG CONDUCTION HEAT PATH

4x5 CCS TEMPERATURE PROFILE MEASUREMENTS (K)						
SENSOR NUMBER	SENSOR TYPE	DESIGN TEMP. (K)	INTERCEPT TEMPERATURE			
			10K	20K	30K	40K
1	CARBON RES.	4.5	5.3	6.8	7.9	9.4
2	CARBON RES.	4.5	5.4	6.9	8.1	9.7
3	CARBON RES.	4.9	5.4	6.9	8.1	9.7
4	CARBON RES.	4.9	5.6	7.7	9.3	11.6
5	CARBON RES.	13.6	8.2	14.7	19.9	27.5
6	CARBON RES.	13.6	8.4	15.4	21.1	29.6
7	CARBON RES.	20.0	8.7	15.4	19.4	24.0
8	CARBON RES.	20.0	11.0	24.3	35.1	49.8
9	CARBON RES.	20.0	10.1	21.2	29.2	40.8
10	CARBON RES.	20.0	10.0	20.9	28.8	40.1
11	CARBON RES.	20.0	13.6	25.1	34.2	46.7
12	CARBON RES.	20.0	13.9	27.0	37.4	52.3
13	Pt RTD	79.0	80.6	80.9	81.9	82.2
14	Pt RTD	79.1	82.8	82.9	83.6	83.6
15	Pt RTD	79.1	82.7	82.9	83.5	83.5
16	Pt RTD	80.0	80.1	80.0	80.3	80.2
17	Pt RTD	80.0	OPEN	OPEN	OPEN	OPEN
18	Pt RTD	80.8	91.5	91.5	92.5	92.5
19	Cu/CON Tc	200.0	182.5	182.8	188.6	188.9
20	Cu/CON Tc	200.0	183.1	183.6	189.2	189.5
21	Cu/CON Tc	299.7	286.2	287.7	296.9	297.5
22	Cu/CON Tc	300.0	290.8	292.5	302.0	302.6
23	Cu/CON Tc	300.0	291.2	292.8	302.3	302.9

measurements made in the SSC thermal model, and profiles the 5x7 CCS temperatures along a 20 K heat intercept heat sink strap.

5x7 CCS TEST INSTRUMENTATION AND THERMAL PERFORMANCE RESULTS

Thermal performance measurements of the 5x7 CCS in the HLTF used the test geometry as shown in Fig. 2. Carbon resistors and platinum RTDs provided the temperature sensor instrumentation along the conductive heat path.

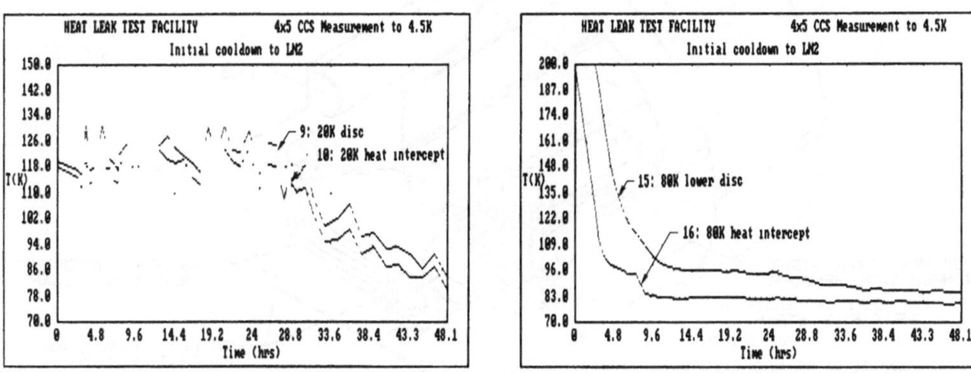

(a) (b)

Fig. 3. 4x5 CCS component temperatures during thermal cycle. (a) at 20 K intercept, (b) at 80 K intercept.

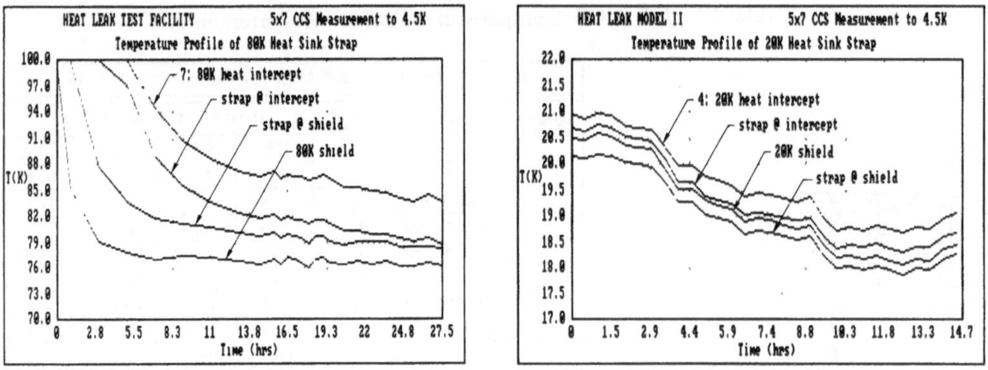

(a) (b)

Fig. 4. 5x7 CCS temperature profile of heat sink straps.
(a) at 80 K heat sink strap (b) at 20 K heat sink strap.

In an experimental test arrangement separate from that of the HLTF, the thermal performance of the 5x7 CCS was also evaluated. A full length SSC dipole magnet thermal model (SSC-TM) was constructed and its thermal performance evaluated.[9] The thermal model is identical to magnet models except that the cold mass contains a simulated coil assembly. In the thermal model, (see Fig. 5) the cold mass and shields are supported relative to the vacuum vessel at five points along their lengths with each 5x7 CCS supporting a structural load of approximately 1455 kg. Temperature measurements along the 5x7 CCS were limited to the metal rings at the CCS temperature stations.

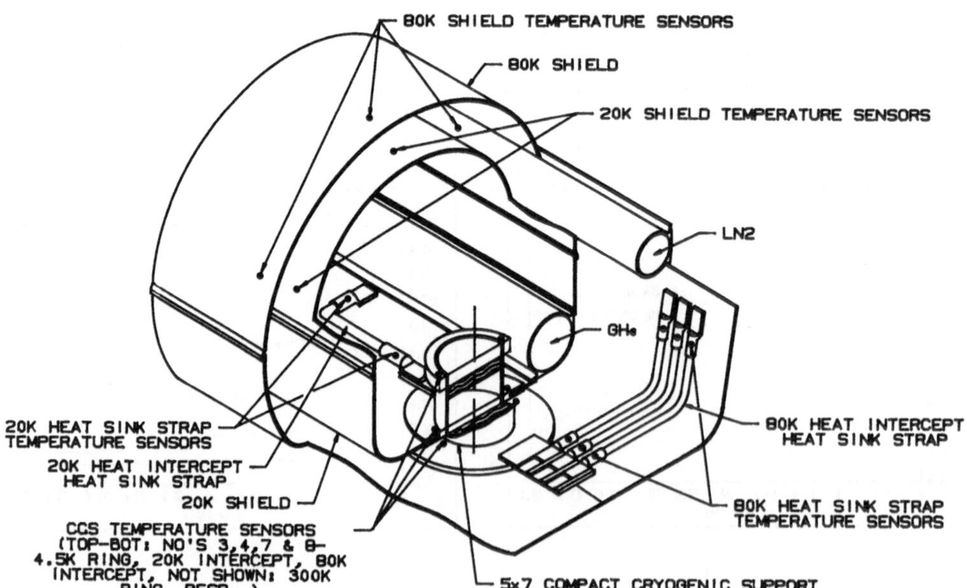

Fig. 5. 5x7 CCS installation in a full length SSC dipole magnet thermal model.

Table 2. 5x7 CCS Thermal Performance Results From The HLTF And SSC-TM

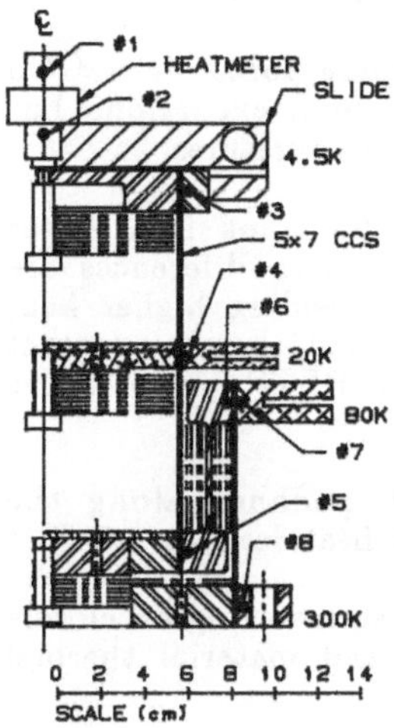

5x7 CCS TEMPERATURE PROFILE MEASUREMENTS (K) IN THE HEAT LEAK FACILITY AND SSC THERMAL MODEL									
SENSOR NUMBER	DESIGN TEMP. (K)	10K HEAT INTERCEPT		20K HEAT INTERCEPT		30K HEAT INTERCEPT		35K HEAT INTERCEPT	
		HLTF	SSC-TM	HLTF	SSC-TM	HLTF	SSC-TM	HLTF	SSC-TM
1	4.5	4.8	-	5.1	-	5.3	-	5.5	-
2	-	5.3	-	6.5	-	7.2	-	7.7	-
3	4.5	7.9	6.6	12.3	9.2	14.3	10.7	15.8	11.9
4	10.0	11.4	12.2	-	-	-	-	-	-
	20.0	-	-	22.7	22.9	-	-	-	-
	30.0	-	-	-	-	28.6	29.4	-	-
	35.0	-	-	-	-	-	-	33.3	34.6
5	79.1	84.2	-	84.9	-	84.7	-	85.8	-
6	80.0	80.0	-	80.7	-	80.4	-	81.4	-
7	80.0	83.1	83.6	83.7	84.7	83.3	84.0	84.3	84.2
8	300.0	286.9	293.6	286.2	289.1	286.1	290.5	285.1	287.6
HEAT LEAK MEASUREMENTS (mW) BY HEATMETER & MATERIAL PROPERTIES									
QHM		5.5	-	18.6	-	27.9	-	34.3	-
Q3		4.7	7.2	17.9	22.7	26.7	33.9	34.5	43.5
Q4		139.1	137.2	112.3	109.9	93.2	86.5	79.9	68.2
Q7		1077.1	1121.6	1048.0	1098.3	1094.8	1123.2	1089.0	1110.5

The thermal performance results of the 5x7 CCS obtained in the two facilities are compared in Table 2. Shown in Table 2 are temperature profile and heat leak measurements with the CCS at similar thermal conditions in the two facilities. In the HLTF, heat leak values to 4.5 K were obtained by direct measurement using the heatmeter as well as by calculation using the measured temperature profiles and the material properties for G-11CR thermal conductivity. Heat leak values to the remaining CCS temperature stations were derived using only the temperature profiles and material properties. In the thermal model, the heat leak values to the CCS temperature stations were derived by temperature and material properties alone.

DISCUSSION

The fabric reinforcement in G-11CR is a plain weave E-glass cloth having interlaced threads in the warp (length) and fill (width) directions of 43 threads per inch (16.9 per centimeter) and 32 threads per inch (12.6 per centimeter), respectively. Tubular G-11CR has the warp threads wrapped circumferentially about the tube and the fill threads along the tube length. Normal to the fabric weave is along tube radii. The CCS conductive heat path is along the tube length and is therefore in the fill direction.

G-11CR thermal conductivity (κ) values are reported by Kasen et al.[10] for the warp, and normal to the fabric weave directions; however, little published data could be found for thermal conductivity for the fill direction. Equation 1 was used with the referenced measurements of warp and normal thermal conductivity to derive values for thermal conductivity for the fill direction.

$$\kappa_{fill} = \kappa_{warp} - \left((1 - 32/43)\ (\kappa_{warp} - \kappa_{normal}) \right) \qquad (1)$$

Thermal impedance is offered by the CCS cold mass slide connection, and is realized as a reduction to the CCS conductive heat leak. This fact is made apparent by the steady-state temperature level above 4.5 K at which the CCS connection to the cold mass slide equilibrates. Calculations show a heat leak reduction of 33% under the 273 kg load in the HLTF, and an 18% reduction under the 1455 kg load in the SSC-TM.

CONCLUSIONS

- **Supports can be designed, built, and operated as required by the SSC.**
- **The CCS shrink fit design assumption that the clamping force in a CCS shrink fitted junction is nearly constant for junction thermal excursions has been verified by the tracking of the junction components temperatures.**
- **Temperature profile and heat leak measurements in the Heat Leak Test Facility and in the SSC thermal model compare very favorably; differences are largest in the measurements to 4.5 K (lower temperatures yielding higher heat leak values) and can be attributed to better mechanical and thermal contact due to higher loading in the thermal model than in the HLTF (1455 kg vs 273 kg, respectively).**
- **The cold mass slide connection presents a thermal impedance along the conduction heat path of the CCS. The result is a lower heat leak to 4.5 K.**
- **Heat leak measurements to 4.5 K in the HLTF by the heatmeter agree closely with calculated values using the measured temperatures and material thermal properties, and agree well with design values.**

ACKNOWLEDGEMENTS

The authors express our sincere appreciation to Messrs. Ray Hanft, Moyses Kuchnir, Peter Limon, Paul Mantsch, Peter Mazur and Mike McAshan for their dedicated interest, time, criticism and constructive evaluation of these measurements and methods.

The work as presented was performed at Fermi National Accelerator Laboratory which is operated by Universities Research Association Inc., under contract with the U.S. Department of Energy.

REFERENCES

1. R.C. Niemann et al., The cryostat for the 6 T magnet option, in: "Advances in Cryogenic Engineering" Vol. 31, Plenum Press, New York (1986), p. 63.
2. T.H. Nicol, R.C. Niemann and J.D. Gonczy, A suspension system for superconducting super collider magnets, in: "Proc. Eleventh Intl. Cryo. Conf.," Butterworth, Guilford, UK (1986), p.533.
3. T.H. Nicol, R.C. Niemann and J.D. Gonczy, SSC magnet cryostat suspension system design and performance, in: "Advances in Cryogenic Engineering," Vol. 33, Plenum Press, New York (to be published).
4. J.D. Gonczy et al., Heat leak measurement facility, in: "Advances in Cryogenic Engineering" Vol. 31, Plenum Press, New York (1986), p. 1291.
5. M. Kuchnir, Pulsed current resistance thermometry, in: "Advances in Cryogenic Engineering" Vol. 29, Plenum Press, New York (1984), p. 879.
6. E.T. Larson et al., Improved design for a SSC coil assembly suspension connection, in: "Advances in Cryogenic Engineering," Vol. 33, Plenum Press, New York (to be published).
7. M. Kuchnir, J.D. Gonczy and J.L. Tague, Measuring heat leak with a heatmeter, in: "Advances in Cryogenic Engineering," Vol. 31, Plenum Press, New York (1986), p. 1285.
8. R.C. Niemann et al., The cryostat for the SSC 6 T magnet option, in: "Advances in Cryogenic Engineering," Vol. 31, Plenum Press, New York (1986), p. 63.
9. R.C. Niemann et al., SSC dipole magnet cryostat thermal model measurement results, in: "Advances in Cryogenic Engineering," Vol. 33, Plenum Press, New York (to be published).
10. M.B. Kasen et al., Mechanical, electrical, and thermal characterization of G-10CR and G-11CR glass-cloth/epoxy laminates between room temperature and 4 K, in: "Advances in Cryogenic Engineering Materials," Vol. 26, Plenum Press, New York (1979), p. 235.

SSC DIPOLE MAGNET CRYOSTAT THERMAL MODEL MEASUREMENT RESULTS

R.C. Niemann, W.N. Boroski, J.D. Gonczy, T.H. Nicol, J.G. Otavka and M.K. Ruschman

Fermi National Accelerator Laboratory
Batavia, Illinois

R.J. Powers

Powers Associates, Inc.
Swampscott, Massachusetts

ABSTRACT

Thermal performance of the conceptual design SSC dipole magnet cryostat has been experimentally evaluated. A full scale thermal model was constructed and open cycle thermal performance measurements were made. Details of the measurement program, measurement results and a comparison of predicted and measured performance are presented. The measurement methods and improvements of them for possible follow-on evaluations are discussed.

INTRODUCTION

The SSC dipole magnet development program includes the design, construction and testing of superconducting magnet models.[1] The model program is structured to provide information and to gain experience with design features, fabrication, handling and operational performance, both magnetic and cryogenic.

As a precursor to the construction of actual magnet models, a full length (17.5m) thermal model was built. The objectives of the thermal model program were to utilize and improve the magnet production facility and manufacturing procedures, to evaluate the cryostat design from a production standpoint, to gain experience in magnet handling and transportation, to monitor the transient thermal and structural responses of the cryostat and to measure the heat leaks to the cold mass and the thermal shields.

CRYOSTAT

The cryostat design has been previously described in detail[2] and thus only its major features are presented herein. The cryostat general arrangement is as shown by Fig. 1 and 2.

Cold Mass Assembly

The cold mass assembly consists of the beam tube, magnetic correction elements, collared coils, laminated iron yoke and outer helium containment shell.

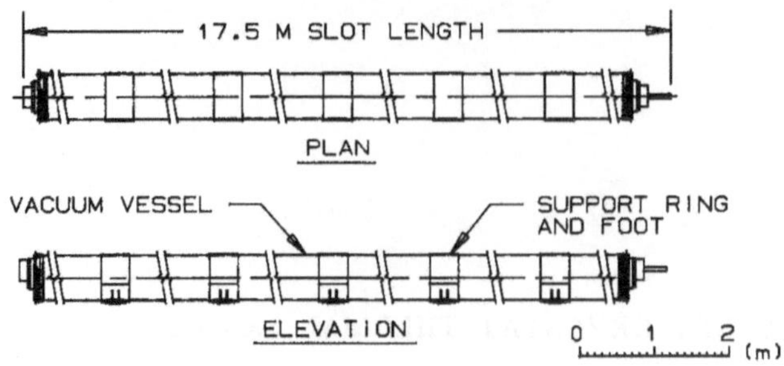

Fig. 1. Cryostat plan and elevation views.

The cold mass components are joined together forming a leak-tight and structurally rigid assembly.

Cryogenic Piping

The cryostat assembly contains all piping that interconnects the magnet refrigeration system throughout the circumference of the accelerator rings. A five pipe system is employed for cryogenic and safety reasons.

Thermal Shields

Thermal shields maintained independently at 20 and 80 K surround the cold mass assembly. The shields absorb the radiant heat flux and provide heat sink stations for the suspension system. The shields are supported by and thermally connected to the cold mass assembly supports.

Insulation

Insulation is installed on the external surfaces of the inner and outer shields. The insulation consists of blankets of flat, reflective radiation shields with mat spacers.[3] One blanket is installed on the inner shield and four are installed on the outer shield.

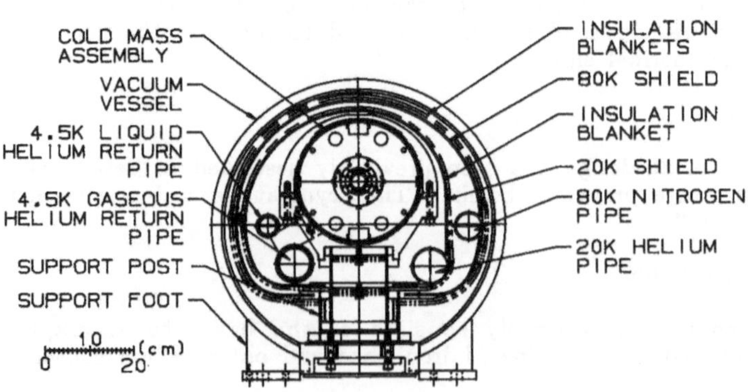

Fig. 2. Cryostat cross section at a suspension point.

Suspension System

The cold mass and shields are supported relative to the vacuum vessel at five points by the suspension system.[4] The system incorporates a reentrant type support post. The insulating sections of the support are fiber reinforced plastic tubing with attached metallic end connections and heat intercepts.

Vacuum Vessel

The vacuum vessel provides containment for insulating vacuum and the cold mass connection to ground. The magnet assembly procedure incorporates insertion of the complete internal sub-assembly into the vacuum vessel. The support post-vacuum vessel connection determines the alignment of the cold mass assembly relative to the vacuum vessel.

Interconnections

Mechanical and electrical interconnections between adjacent magnets are required at the magnet ends. It is essential that the connections be straightforward to assemble and disassemble, compact and reliable. The design facilitates assembly and disassembly operations in the SSC tunnel.

THERMAL MODEL

The thermal model[5] is identical to magnetic models except that the cold mass assembly contains a simulated collared coil assembly and the model ends have been reconfigured for an open cycle heat leak measurement. The model has been instrumented to evaluate thermal and structural performance. Included are temperature sensors to monitor cooldown, warmup and steady state conditions and strain gauges to monitor the performance of the suspension system during onsite transit, cooldown and operation. Construction of the thermal model employs the fabrication facility, procedures and the components of the magnetic models.[6]

MEASUREMENT METHODS

The measurement system is shown by Fig. 3.

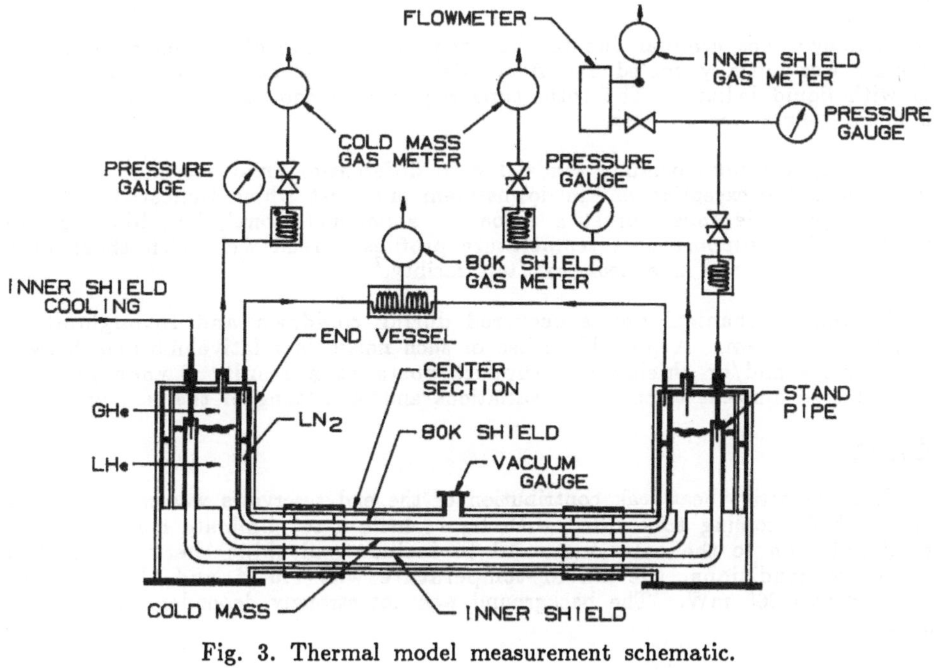

Fig. 3. Thermal model measurement schematic.

The measurements were performed in an open cycle mode. The cold mass was supplied liquid helium by reservoirs at each end. The inner shield was supplied helium from an external dewar. Since the inner shield temperature was an operational variable, its value was controlled by varying the flowrate through the shield. The outer shield was supplied liquid nitrogen by reservoirs at each end. The external piping was equipped to measure the flowrates of the cold mass, inner shield and outer shield gas streams.

After cooldown, initial filling and general stabilization of the three systems, the temperature of the inner shield was regulated to a selected test value. After regulation, temperatures and heat leaks at each system were monitored to establish equilibrium. The equilibration process for each operating point required several days. Once at equilibrium, steady state heat leak data was taken.

Cold Mass

The cold mass heat leak was determined from boiloff measurements. Since the measurement includes the heat leaks of both the cold mass and the end reservoirs, a separate measurement of the end reservoir heat leak was made and then subtracted from the total measured heat leak.

Inner Shield

The inner shield heat leak was determined from the shield gas stream heating. The flowrate through the shield and the helium ΔT along the shield were monitored.

Outer Shield

The outer shield heat leak was determined from boiloff measurements. Like the cold mass measurements, an end reservoir heat leak subtraction was employed.

RESULTS

The experimental program extended for ~3 months and involved 23 data taking runs as identified by the inner shield temperature. The following results have been screened to exclude transient periods, upset conditions and operational problems.

Cooldown

Cooldown was gradual due to the open cycle nature of the operation. The cold mass was initially cooled and filled with liquid nitrogen, evacuated and then filled with liquid helium. The total time required to cool and fill the cold mass with liquid helium was 294 hrs.

The support post bending loads due to differential axial thermal contraction were low with the exception of the downstream end post which indicated a load of 4800N. A probable cause for such a load is a nonoperational; i.e., binding, cold mass slide. The support post temperature profiles agreed well with those of an identical post measured in a heat leak test facility.[7]

Random mechanical noises occurred during cooldown and throughout the measurement program. A probable cause of such noises is relative motions between the cold mass and/or shields and support posts as a result of vacuum vessel motions due to ambient temperature variations and/or settling of the test pad.

Cold Mass

The subtractive heat leak contribution of the end reservoirs was measured with the inner shield cooling tube filled with liquid helium to eliminate conduction and thermal radiation to the cold mass and to include end shine thermal radiation. Under these conditions, the shield temperature was 7.6 K and the measured background was 865 mW. The background was not strongly dependent on reservoir liquid level.

Calibration heaters in the cold mass were employed to evaluate the accuracy of the measurement system. The heaters were energized with the inner shield at ~20 K. The measured heat leak increases corresponding to heater power levels of 200 mW and 398 mW were 197 mW and 376 mW, respectively.

The cold mass heat leak at the cryostat design point; i.e., inner shield operating at 20 K, was 140 ±40 mW as compared to the predicted 128 mW and the budgeted 300 mW. The cold mass heat leak vs inner shield temperature is given by Fig. 4.

Considerable differences exist between measured and predicted heat leaks at several experimental points. Factors that can contribute to these differences are as follow:

- The center section heat leak (140 mW) is small relative to the balance of the apparatus (865 mW). This unbalance amplifies end vessel effects.
- Thermal communication exists between the inner shield cooling circuit and the end vessel helium reservoirs as a result of the inner shield supply and return piping passing through the cold mass helium reservoirs. Even though the piping was insulated, transients in inner shield supply conditions could be seen to affect the reservoirs.
- Level instability (sloshing) occurred occasionally in the end vessels. The sloshing was most often associated with changes in the inner shield circuit operation.
- The predicted sensitivity to insulating vacuum is high. At the design inner shield operating temperature of 20 K, the residual gas (helium) conduction at 10^{-6} torr is 43 mW and at 10^{-5} torr is 430 mW. The insulating vacuum at the vacuum vessel midspan port ranged from 1.3 X 10^{-6} to 2.9 x 10^{-6} torr during the span of the measurements. The location and installation of the vacuum gauge was demonstrated by diagnostic measurements to inaccurately relate changes in the insulating vacuum that could correspond to changes in the outgassing rate of the mild steel vacuum vessel due to changes in ambient temperature. Consequently, vacuum was not monitored frequently during most of the data taking.
- Variations in atmospheric pressure result in temperature changes of the liquid which influence the apparent heat leak. Atmospheric pressure was not monitored frequently during most of the data taking. Where possible, the heat leak was corrected for changes of liquid temperature and pressure with time. The cold mass end to end temperature variation during operation was characteristically ±30 mK.

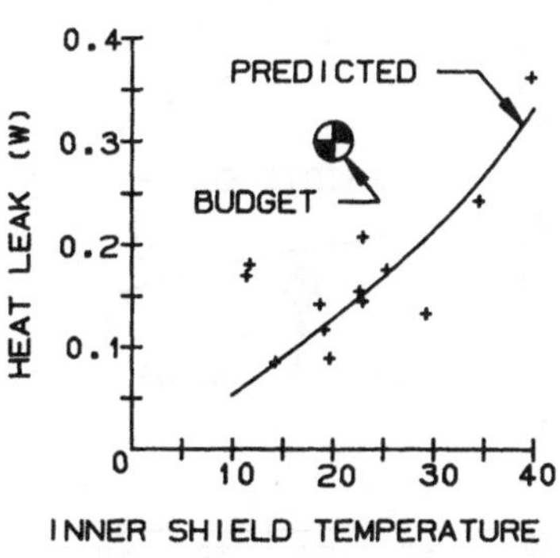

Fig. 4. Measured cold mass heat leak.

The inner shield heat leak at the cryostat design point was 5.0 ±0.4 W as compared to the predicted 2.77 W and the budgeted 2.5 W. The inner shield heat leak vs inner shield temperature is given by Fig. 5.

A corroboration of the inner shield heat leak measurement was provided by performing inner shield boiloff measurements during the cold mass background measurement. With the shield cooling tube filled with liquid helium, the heat leak as measured by boiloff was 5.64 W.

The factor of two difference between measured and predicted heat leaks is felt to be due to thermal shorts between the 20 and 80 K systems, by locally compressed insulation between the shields and by an insufficient number of layers of insulation. Shield and support temperature monitors indicate the possibility of shorts. An autopsy of the thermal model to investigate the existence of such thermal shorts has not been conducted for programmatic reasons.

Outer Shield

The subtractive heat leak contribution of the end reservoirs was made with the center section removed and the reservoirs connected together. The nominal background heat leak was ~24 W. The background was found to vary with level; i.e., ~3 W for a level change from "full" to "half" and accordingly a variable subtraction was employed. The outer shield heat leak at the cryostat design point was 19 ±2 W compared to the predicted 23.3 W and the budgeted 25 W. The outer shield heat leak vs inner shield temperature data is given by Fig. 6. The less than predicted measured heat leaks are felt to be associated with the thermal shorts that are suspected exist between the inner and outer shields and their connections. The scatter in the measured heat leak is felt to be associated with the sensitivity of the insulation systems to insulating vacuum[8] in the higher pressure ranges; i.e., $>10^{-4}$ Torr. As noted earlier, the insulating vacuum, while not accurately measured, showed changes with ambient temperature.

IMPROVEMENTS FOR FOLLOWON MEASUREMENTS

Methods

The measurements of the cold mass and outer shield heat leaks were indirect since they included the measurement of an associated background which in both cases was large. Direct measurements that involve devices such as heatmeters[9] would improve the measurements and should be evaluated for their accuracy and suitability to the model measured.

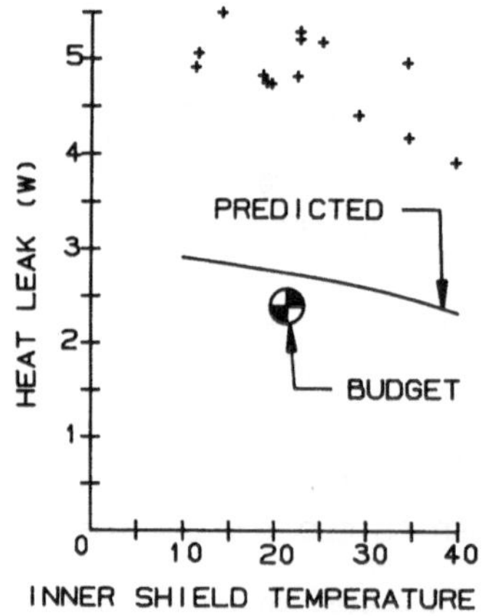

Fig. 5. Measured inner shield heat leak.

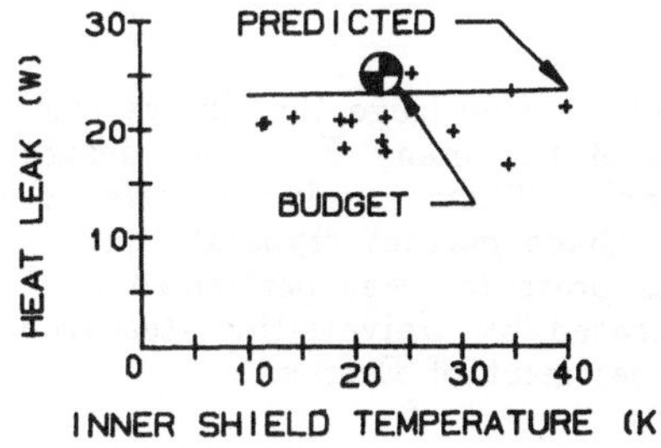

Fig. 6. Measured outer shield heat leak.

Thermal Model

For the same experimental method, the model requires several changes. The backgrounds should be reduced significantly by incorporating low heat end vessels. The inner shield cooling circuit should be totally separate from the cold mass circuit. Placement of pressure and vacuum measurement transducers should be improved.

Instrumentation

All transducer signals should be directly inputed to the computer driven data acquisition system to permit total on-line data processing and evaluation of transients, upset conditions, etc. Pressure and vacuum monitoring should be expanded to include end vessels, interior regions of the model, etc.

Protocol

Background measurements should be expanded to establish the effects of operating pressure, insulating vacuum and liquid level. Transient performance criteria should be evaluated and updated.

Component Measurements

Measurements of cryostat components that contribute to the heat leak; i.e., insulation, support posts, etc. should continue to be made. Operation at off-design conditions should be .included.

Analytical Model

The analytical model for the cryostat should continue to be improved in the areas of transient and off-design conditions. Model inputs should include factors generated by component evaluations. Correlations of the physical and analytical model should be ongoing.

CONCLUSIONS

- The measurement methods are appropriate.
- The model, instrumentation and protocol can be improved in a straightforward manner to improve measurement results.
- The cryostat, with the possible exception of a cold mass slide, performed well during cooldown and steady state operation.
- With the exception of the inner shield, the measured heat leaks were within budget. The temperature monitors on the model indicated possible shorts between the inner and outer shields.
- The thermal measurements indicate that the superconducting magnets for the SSC can be built within the heat leak budgets as required for accelerator operation.

ACKNOWLEDGEMENTS

The authors gratefully acknowledge the sincere interest, extensive contributions and diligent performance of the many Fermilab design, magnet production and engineering test personnel. Their combined efforts resulted in the design, construction and test of a dipole magnet cryostat that answers the cryogenic needs of the SSC. The work as presented was performed at Fermi National Accelerator Laboratory which is operated by Universities Research Association Inc., under contract with the U.S. Department of Energy.

REFERENCES

1. SSC Central Design Group, Conceptual design of the superconducting super collider, "SSC-SR-2020," March 1986.
2. R.C. Niemann, et al, Superconducting super collider magnet cryostat, cryogenic properties, "Processes and Applications," No. 251, Vol. 82, p. 136, (1986).
3. T. Ohmori, et al, Thermal performance of candidate SSC magnet cryostat thermal insulation systems, "Adv. Cryo. Engr.," 33, Plenum Press, New York (to be published).
4. T.H. Nicol, et al, SSC magnet cryostat suspension system design and performance, "Adv. Cryo. Engr.," 33, Plenum Press, New York (to be published).
5. R.C. Niemann, et al, Design, construction and test of a full scale SSC dipole magnet cryostat thermal model, "Applied Superconductivity Conference," September 30 - October 3, 1986 in Baltimore, Maryland (to be published).
6. N.H. Engler, et al, SSC long dipole magnet model construction experience, "Adv. Cryo. Engr.," 33, Plenum Press, New York (to be published).
7. J.D. Gonczy, et al, Cryogenic support thermal performance measurements, "Adv. Cryo. Engr.," 33, Plenum Press, New York (to be published).
8. T. Ohmori, et al, Thermal performance of candidate SSC magnet cryostat thermal insulation systems, "Adv. Cryo. Engr.," 33 Plenum Press, New York (to be published).
9. M. Kuchnir, et al, Measuring heat leak with a heatmeter, "Adv. Cryo. Engr.," 31, (1986) Plenum Press, New York.

OPERATING EXPERIENCE OF THE IFSMTF VAPOR-COOLED LEAD SYSTEM*

J. W. Lue, D. T. Fehling, W. A. Fietz, M. S. Lubell,
J. N. Luton, S. W. Schwenterly, S. S. Shen, R. E. Stamps,
D. H. Thompson, C. T. Wilson, and T. Kato†

Oak Ridge National Laboratory
Oak Ridge, Tennessee

ABSTRACT

The International Fusion Superconducting Magnet Test Facility (IFSMTF) uses six pairs of vapor-cooled leads (VCLs) to introduce electric power to six test coils. Each VCL is housed in a dewar outside the 11-m vacuum vessel and is connected to the coil via a superconducting bus duct; the various VCLs are rated at 12 to 20 kA. Heat loss through the leads constitutes the single largest source of heat load to the cryogenic system. Concerns about voltage breakdown if a coil quenches have led to precautionary measures such as installation of a N_2-purged box near the top of the lead and shingles to collect water that condenses on the power buses. A few joints between power buses and VCLs were found to be inadequate during preliminary single-coil tests. This series of tests also pointed to the need for automatic control of helium flow through the leads. This was achieved by using the resistance measurements of the leads to control flow valves automatically. By the time full-array tests were started, a working scheme had developed that required little attention to the leads and that had little impact on the refrigerator between zero and full current to the coils. The operating loss of the VCLs at full current is averaging at about 7.4 g/s of warm flow and 360 W of cold-gas return load. These results are compared with predictions that were based on earlier tests.

INTRODUCTION

The Large Coil Task (LCT) is now testing six large, D-shaped coils arranged in a toroidal array at the IFSMTF.[1] Each coil is powered by a separate dc power supply. Six pairs of VCLs rated at 12 to 20 kA are used as interfaces between the room-temperature power bus and the liquid-helium-cooled superconducting bus to minimize heat load. Nevertheless, because of the large conductor cross section needed to carry the current and the large number of VCLs used, this system still constitutes the single largest source of heat load to the cryogenic system.

---

*Research sponsored by the Office of Fusion Energy, U.S. Department of Energy, under contract no. DE-AC05-84OR21400 with Martin Marietta Energy Systems, Inc.

†Japan Atomic Energy Research Institute, Tokai, Japan

Ground insulation for the VCL lies very close to the top of the lead, where it is susceptible to moisture accumulation that weakens the voltage breakdown withstand capability. During the course of the test program, several steps were taken to arrive at a satisfactory solution.

The heat load of the VCL is taken by both the flow through the lead and by a cold vapor return that goes through the heat exchangers of the refrigerator cold box. The heat load has been monitored during various test modes. The results are analyzed to establish optimum operating parameters for later runs.

INSULATION CONSIDERATIONS

Each VCL is housed in a dewar outside the 11-m vacuum vessel, as shown in Fig. 1. Test coil dump voltage as high as 2.5 kV is expected during some of the tests. Ground insulation between the VCL and the lead dewar is provided by inserting a G-10 insulating flange beneath the stainless steel flange at the top of the dewar. Plastic sleeves are used to cover the flange bolts. Some of these sleeves were found to have slight mechanical damage and arc damage that resulted in low breakdown voltage during some of the high-voltage testing of the coil and VCL system routinely done before operation. No in-service faults were encountered.

Moisture accumulation (both ice and water) on the cold VCL results in water dripping on and around the insulating flange, providing a current path from the VCL to the grounded dewar. This problem was remedied by the addition of a nitrogen-purged PVC enclosure, which prevented moisture accumulation on the insulating flange. The exposed VCL top and the flexible jumper were still cold enough at times to cause

Fig. 1. Typical VCL dewar and power bus assembly.

moisture accumulation, which dripped onto the electrical bus supports. Fiberglass "shingles" were added to catch and divert the dripping water to a "gutter-like" assembly where the water was collected and drained into a safe area.

INSTRUMENTATION AND FLOW CONTROL

Liquid helium for the VCL dewar is supplied by either of two storage dewars. Figure 2 is the simplified VCL helium-flow diagram. Liquid level is continuously monitored by a superconducting level probe with low- and high-level limit outputs; the probe controls the opening and closing of the inlet valve. This automatic level control has been working very well, and the liquid consumption rate can be calculated by the level drop rate data.

Part of the helium boiloff flows through the VCL to intercept the heat conducted from the room-temperature bus and the resistive heating generated in the VCL itself. This warm gas return is directed to the compressor suction. During the single-coil tests, the increase in the demand of this warm return from zero current to full current for a coil was found to be large enough to cause a rise in compressor suction pressure and upset refrigerator performance. An auxiliary compressor, added to handle the VCL warm gas return, worked well during the full-array tests. It also eased the problem of large boiloff during a coil quench.

The remaining helium boiloff is taken cold and joined with boiloffs from the pool-boiling (PB) coils for return to the cold box. There it is used to cool incoming gas through the counterflow heat exchangers. Thus, the cold vapor return constitutes a much lower refrigeration load to the refrigerator than the warm gas return. By manually closing the supply valve, this flow arrangement also allows cold vapor from the test coils to backflow through the cold return valves to cool a VCL when no current is expected in that coil.

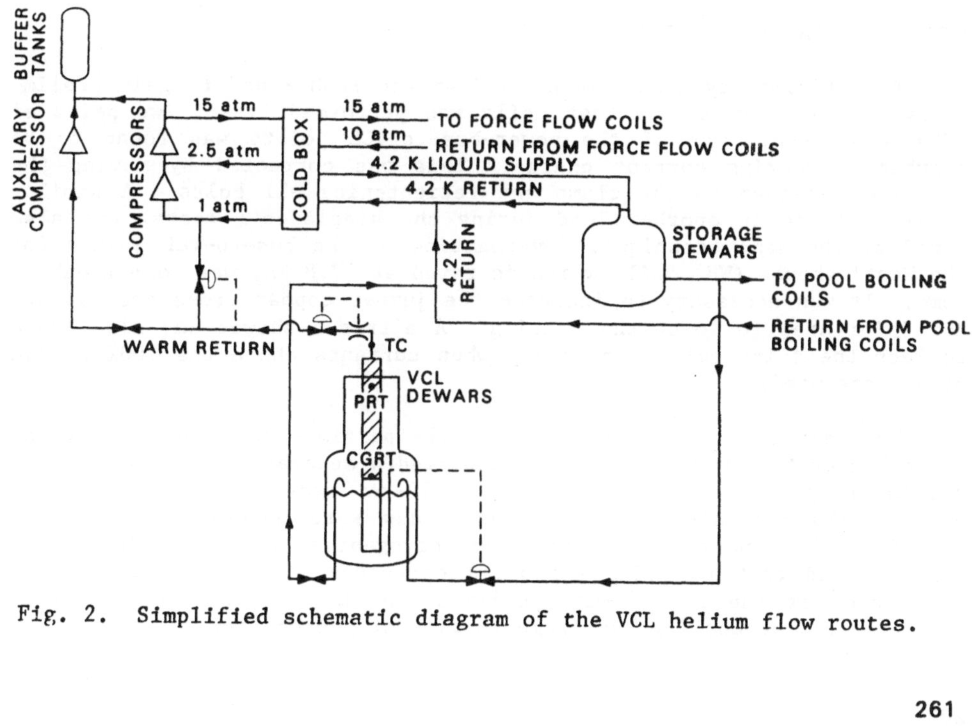

Fig. 2. Simplified schematic diagram of the VCL helium flow routes.

In addition to the liquid level probe, the VCL is also instrumented with a carbon-glass resistance thermometer (CGRT) near the vapor entrance, a platinum resistance thermometer (PRT) just inside the top flange, and a thermocouple (TC) in the warm gas return line just beyond the lead. Voltage taps across the lead are used to obtain an analog reading of resistance. This is done by performing an analog division of VCL voltage by coil current. Both the voltage and current signals are fed to a specially designed electronic module via isolation amplifiers. The lead resistance meter contains a limiter circuit that causes the meter to indicate zero resistance at very low currents to prevent the erroneous readings that occur when the denominator input to the divider circuit is near zero. All temperature measurements, helium level, lead resistance, and VCL warm gas flow are monitored by the facility data acquisition system. Alarm levels are selected for each of these signals. A voting scheme is used to trigger a coil dump if two or more of these alarm levels are exceeded.

Because the flow demand of the VCLs depends on coil current, it is necessary to vary the warm gas flow through the lead to achieve optimum operation of the refrigerator and to prevent too much moisture condensation on top of the lead. Adjusting 12 valves manually while the full array is being charged is too demanding for operators, to say the least. The lead resistance measurement is used as the process variable to control flow automatically. Each of the 12 warm return flow control valves is connected to a manual station, which can set the valve position or allow the facility programmable logic controller (PLC) to control the valve. When the station is in "manual" mode, the PLC tracks the manual setting. When control is transferred to "auto," the valve will be controlled by the PLC, starting at the initial position set by the operator. A minimum lead resistance value was set in the register. The PLC takes over valve control when the VCL resistance reads above this minimum value. Opening of the valve is controlled by deviation of the lead resistance from a preset value. The PLC releases control when VCL resistance reads below the minimum set value and reverts the valve setting to the initial position. This automatic flow control scheme, which was implemented shortly before the full-array tests began, worked well, and the VCL flow control has required little attention since then.

SUMMARY OF EARLIER TEST DATA ON VCLs

A partial-array test[2] to shake down the IFSMTF and to gain preliminary data on two of the test coils was performed first; two pairs of VCLs were used. One of the power-bus-to-VCL joints was found to be overheating during current charge. This was corrected by moving the bolt holes closer to the clamp and retightening all bolts. A similar problem, found in another lead during the single-coil tests, was also cured by the same technique. Overheating of the bus-to-VCL joints for the Westinghouse (WH) coil, which is rated at 17.8 kA, was more troublesome. It was necessary to increase the jumper copper cross section and add joint area to lower the heating. A nitrogen shower was also added to keep the joint cold, especially when currents above the 100% design point were used.

The heat load of the VCLs used in the partial-array tests was found to be higher than expected.[3] This prompted a separate lead test outside the facility. The test used a pair of 12-kA leads similar to the ones used in the partial-array test; the leads were housed in a pair of facility VCL dewars joined together cryogenically. This "twin-lead test" showed that the VCLs performed correctly (i.e., the self-demand loss rate met the 1.8-L/h-kA specification and exceeded the 2 min of full current, no cooling requirements without overheating).

The measured helium boiloff rate as a function of VCL flow can be fed to a refrigeration performance curve to find the cryogenic margins of various operating modes for the coil tests. Refrigerative performance of the IFSMTF refrigerator[4] can be summarized by the measured refrigeration power vs liquefaction rate curve shown in Fig. 3. The refrigerator can deliver about 1.5 kW of refrigerative power or liquefy about 380 L/h of liquid helium or produce any combination of refrigerative power and liquefaction rate in between. The warm VCL flow of the leads constitutes liquefaction load to the refrigerator, and the cold vapor return constitutes refrigeration load to the refrigerator. Scaling to differently rated leads and summing all heat loads of the 12 VCLs resulted in the three load lines shown in Fig. 3; these represent predicted loads in standby and at full charges of a full-array test, based on either partial-array test data or twin-lead test data. In these plots, it was assumed that the facility and test coils generate an additional heat load of 600 W in standby and 800 W in charged-up conditions.[4] When the load line lies inside the refrigeration curve, excess refrigeration or liquefaction power is left from the refrigerator, and the liquid inventory will be replenished. When the load line lies beyond the refrigeration curve, not enough refrigeration is provided by the refrigerator, and the liquid inventory of the storage dewar will be depleted. The cryogenic margin of the system depends not only on the mode of operation but also on how the heat load is being handled. That is, the efficiency of the refrigeration system also depends on how much of the heat load is taken by the warm VCL flow. If the 12 VCL heat loads are as low as the twin-lead test results indicate, the VCLs can be operated with very little warm flow and can rely on the heat exchangers in the cold box to handle the refrigeration load. If heat loads are as

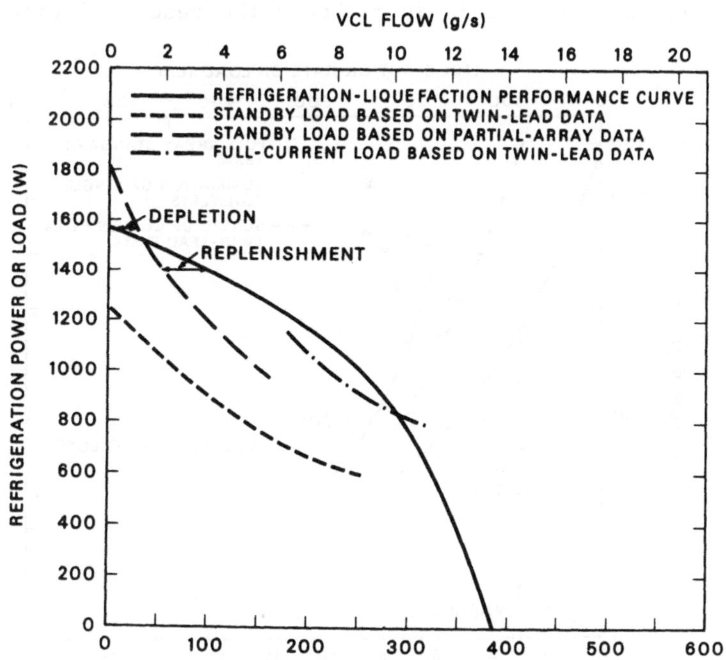

Fig. 3. Performance curve of the IFSMTF refrigerator. The refrigeration and liquefaction loads of the VCLs based on the partial-array and twin-lead test data are incorporated to determine possible cryogenic margins of the system.

high as the partial-array results indicate, higher warm flow is necessary and advantageous. Figure 3 also indicates that the margin for the refrigerator is very small. Any significant heat loss in the VCL system at full charge higher than scaled from the twin-lead test will cause a deficiency in refrigerative power.

FULL-ARRAY RESULTS

In the full-array tests (standard-1), one of the coils was designated as the test coil and the other five coils were background coils. The test coil was charged to 100% of its design current. The background coils were charged to 60–95% of their design currents to help produce a midplane field of 8 T at the test coil. The variation in background coil currents was necessary because of different ampere-turns in each coil and because of the desire to produce minimum out-of-plane loads.

Implementation of automatic VCL flow control has made the operation much easier. To minimize the impact on compressor suction pressure, VCL flow was made somewhat higher from the start. The resulting heat loads at standby and at full currents are shown in Fig. 4. Each data point represents one of the six standard-1 tests. The boiloff rate vs VCL flow of the leads at zero and full current as scaled from the twin-lead test were drawn in for comparison. Also plotted are the single-coil test results, which were summations of losses at each of the six single-coil tests. All zero current standby data seem to lie on a curve somewhat higher than the twin-lead results, similar to the partial-array data. The difference gets smaller at higher VCL flows. The loss rates at full currents were also somewhat higher than those shown in the scaled curve for the twin-lead test. They cluster together rather than lying on a trend curve. This is probably the result of varying back-

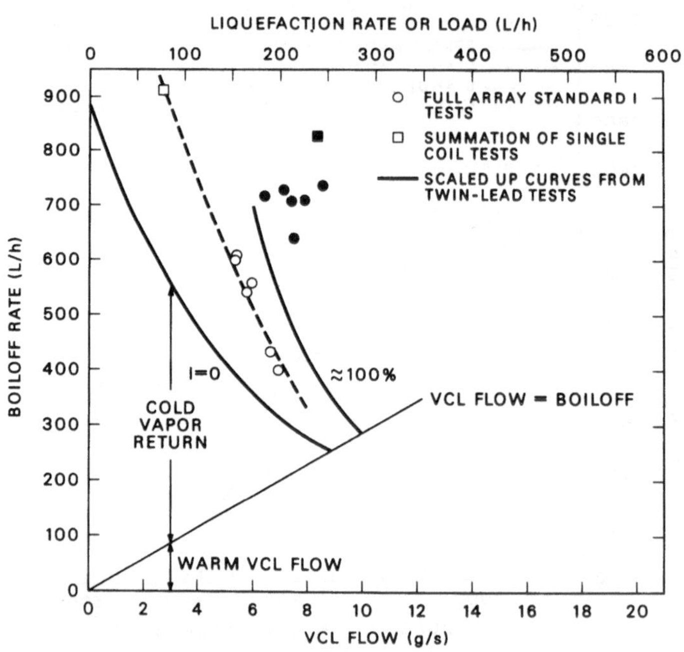

Fig. 4. Total helium boiloff rates at standby (open circle) and at full charge (closed circle) during full-array tests. Scaled-up curves from the twin-lead test are drawn for comparisons.

ground coil currents and different lengths of time at full currents when the data were taken. On the average, the 12 pairs of VCLs were operated with a warm flow of 7.4 g/s (which constitutes a liquefaction load of 210 L/h) and a cold return of 490 L/h (which constitutes a refrigeration load of 360 W) at full charges.

The VCL flow requirements at full current are more or less dictated by the self-heating demand of the leads. The flow is controlled automatically by regulating to a selected lead resistance. VCL flow at standby can, however, be changed more freely. By inserting the standby data of the standard-1 test into a refrigeration performance curve similar to Fig. 3, one can expect to find optimum operating points. Two sets of data points are shown in Fig. 5. One represents a background load of 600 W at zero current (and 800 W at full current), and the other represents a background load of 1000 W at zero current (and 1200 W at full current).

These curves can be used to measure liquid replenishment rate as a function of VCL flow during standby conditions. The results are shown in Fig. 6. For a 600-W background load, Fig. 6 shows a broad optimum operating range of 3.7–5.5 g/s for total VCL flow. The maximum possible liquid replenishment rate is about 120 L/h. For a 1000-W background load, liquid inventory is depleted until VCL flow is more than 6.0 g/s. Even then, the replenishment rate can barely reach 20 L/h. This is understandable because, when the background loss is high, a large amount of cold vapor is already returning to the heat exchangers of the cold box. The exchangers cannot take any more gas from the VCLs. Thus most lead losses should be directed through the warm return. The measured liquid inventory losses of the six standard-1 tests are averaging about 35 L/h in standby and 280 L/h at full current. Thus, the background heat loads are closer to 1000 W than the 600 W estimated earlier, or the refrigerator performance has been degraded.

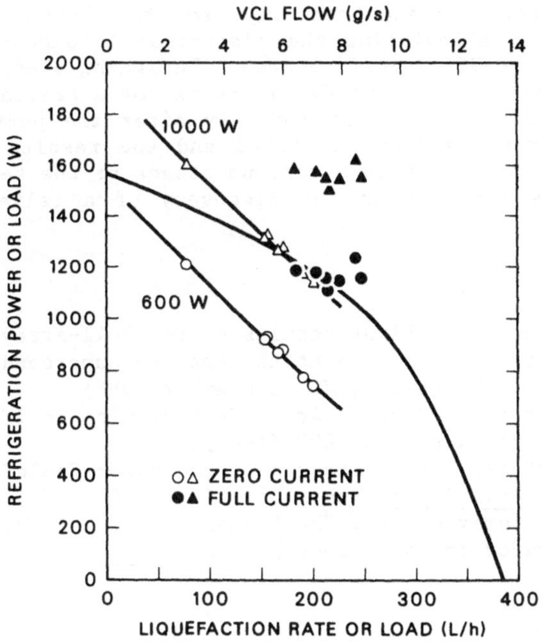

Fig. 5. Total refrigeration and liquefaction loads at two different background heat loads.

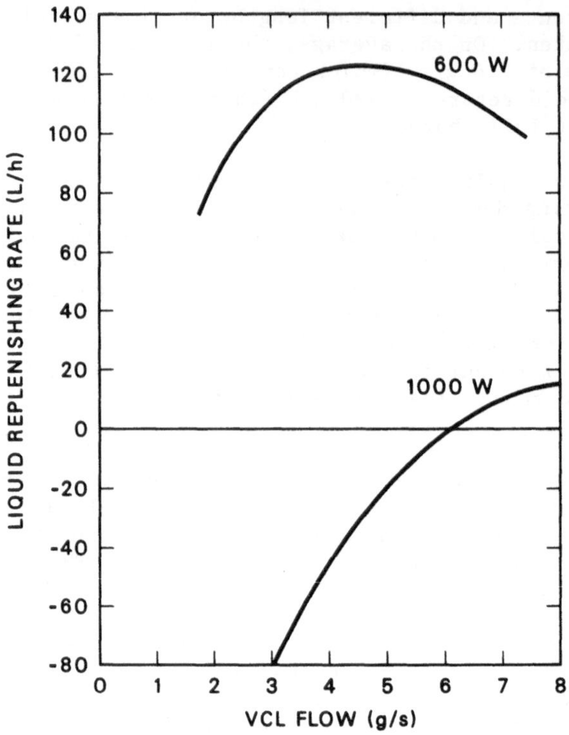

Fig. 6. Liquid helium inventory replenishing and depleting rate as a function of VCL flow at two different background heat loads.

CONCLUSION

The six pairs of VCLs constitute one of the most challenging problems, both electrically and thermally, for the IFSMTF. Much attention has been given to alleviating the electrical breakdown problem at the room-temperature-to-VCL interface area. Designing electrical joints for high currents and high current densities is not a trivial task. Thermal loads of the VCLs can be minimized only after the performance of both VCL and refrigerator systems is found and the results are integrated. All problems are solved through various stages of the test program. The test experiences resulted in the discovery of a reliable and optimal operating mode.

REFERENCES

1. S. S. Shen et al., "First Results of the Full-Array LCT Coil Tests," paper presented at the Applied Superconductivity Conference, Baltimore, MD (September 1986).
2. J. W. Lue et al., "Partial-Array Test Results in IFSMTF," Fusion Technol. 8(1), pt. 2A: 807 (1985).
3. J. W. Lue et al., "Test Results of the Vapor-Cooled Leads for the IFSMTF," Adv. Cryog. Eng., 31: 217 (1986).
4. S. W. Schwenterly et al., "Performance of the IFSMTF Helium Refrigerator in Partial-Array Tests," Adv. Cryog. Eng., 31: 617 (1986).

OPERATION OF LIQUID HELIUM LEVEL SENSORS AT ELEVATED PRESSURES

A.F. Zeller, H. Laumer and J.A. Nolen

National Superconducting Cyclotron Lab
Michigan State University
East Lansing, Michigan

ABSTRACT

Operation of standard liquid helium level sensors depends on the difference in heat transfer between the gas and liquid phases. At increased pressure inadequate cooling by the liquid causes the resistive zone to penetrate below the liquid surface when the liquid level rises. Thus the level detector records the minimum level and does not follow any subsequent rise in the level. Measurements of indicated levels as a function of heater current at pressures up to 0.2 MPa (29 psia) were made using an American Magnetics sensor to determine the proper heater currents. Reduced heater currents were found to permit operation at pressures up to nearly the critical point.

INTRODUCTION

The normal operational pressure of the 2500 L liquid helium (LHe) supply dewar for cyclotron operation at NSCL is between 0.16 and 0.18 MPa, corresponding to liquid temperature of 4.7 to 4.9 K. In this pressure range, the standard American Magnetics Inc. (AMI) level detector, which is based on the Efferson design[1], "sticks". The detector faithfully follows the surface when the level is falling, but does not indicate a rising level. These Efferson-type detectors are a thin superconducting wire being heated at the top by a short bifilar wound heater. With the proper heater current the difference in heat transfer between the liquid and gas phases causes a resistive zone above the liquid while the wire remains superconducting below the surface. The level is indicated by the voltage drop across the wire. At higher pressures, and hence, temperatures, the difference in heat transfer decreases until at 0.228 MPa (18.3 psig), 5.20 K, the critical point is reached (the density of the liquid and gas are equal). Thus at some pressures below 0.228 MPa the resistive heating of the normal zone will be more than can be removed by the reduced heat transfer, and the resistive zone will be driven below the meniscus. At dewar pressures of 0.16-0.18 MPa, the resistive zone is about in equilibrium with the surface, so it reads properly when the level is decreasing. However, a rising level does not produce enough additional cooling to cause the resistive zone to shrink, so the meter "sticks" at the lowest level. The true level can be read, of course, by turning the heater off and back on.

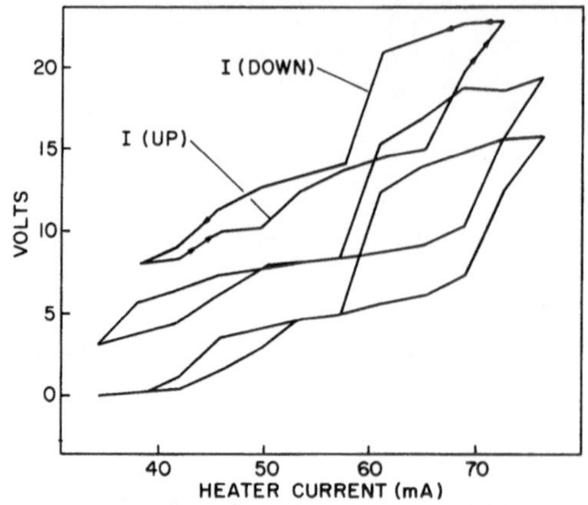

Fig. 1. Level sensor voltage while ramping heater current up and down at dewar pressures of 0.199 (bottom curve), 0.201 and 0.205 (top) MPa.

RESULTS

Tests were done in a 500 L dewar using a standard AMI 46 cm sensor inserted to about one half its length into the liquid. The voltage drop was measured as a function of heater current. The current was increased until the dramatic rise in voltage signaled that the wire was completely normal, then the current was decreased slowly to zero. During the decreasing current phase, the normal zone would suddenly retreat to the interface, but at a current below where it had run away on the increasing ramp. The operating range is then defined by the overlapping section of the two curves. This behavior is discussed by Efferson[1] and by Eckels et al.[2] who studied the effects at sub-atmospheric pressures. Measured curves for three pressures are shown in fig. 1 for pressures from 0.199 to 0.201 MPa. The curve for the highest pressure has no overlap region, within the 3 mA step size limit of the power supply, and hence no operating range. Curves tend to have longer operating range as the pressure is decreased, as shown in fig. 2. The line represents a reasonable operating point as a function of pressure. It would appear that a constant current of approximately 58 mA would suffice for all pressures, but operation at such a low current at low pressure would greatly reduce the response time, so

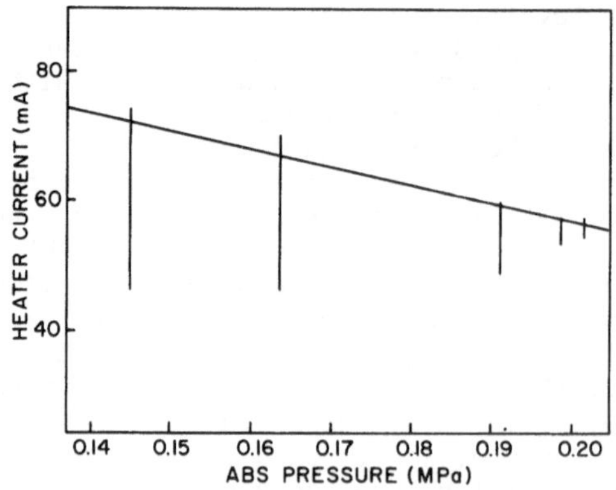

Fig. 2. Operating ranges as a function of pressure. The line represents an appropriate operating range for each pressure which insures a fast response, although the most efficient point is not very sensitive to small pressure changes.

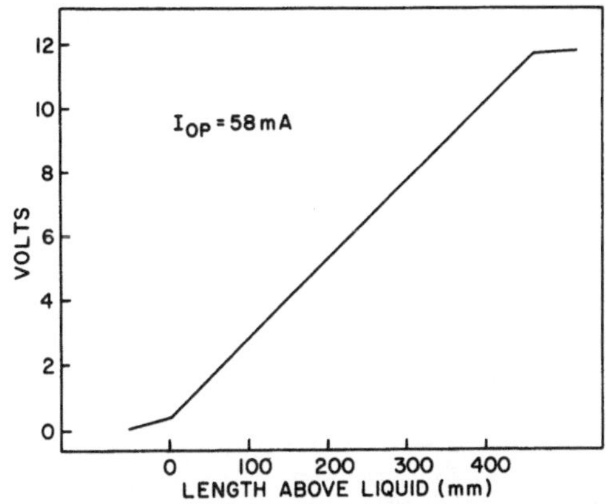

Fig. 3. Voltage versus length at an operating current of 58 mA at 0.200 MPa for a standard AMI sensor, 46 cm long. The curve was measured by pulling the probe up and then lowering it down over the length of the sensor, so it measured relative to an arbitrary zero when sensor is fully under the liquid.

currents should be kept at the upper end of the operating range. Operating at 58 mA at a pressure of 0.200 MPa produced a linear response for the full 46 cm active length of the sensor, both for rising and falling levels; this is shown in fig. 3. This curve was measured by raising and lowering the sensor in the dewar, so the abscissa is relative to an arbitrary zero. The plateau regions on either end (fully immersed and fully exposed) result from resistive changes due to partial exposure of the short heater and increased resistance due to warmer helium gas, respectively.

In initial tests in the 500 L dewar, the pressure was changed by regulating the return gas flow. It was found that a relatively sharp temperature gradient is maintained in the dewar even at pressures above the critical point. The level sensor may as a result still indicate a level. This effect is, of course, not observed if the fluid in the dewar is disturbed, such as during filling.

DISCUSSION

In October 1986, the standard AMI heater power supply output of 72 mA was reduced to 65mA. Since then the sticking of the level has not occurred, and reliable readings are obtained. It is possible that operation, at or above 0.205 MPa (15 psig) is achievable with a different NbTi wire or a change in the insulation, since the heat transfer characteristics depend strongly upon surface properties. Certainly a lower current is needed, but a pulsed mode may be required to monitor falling helium levels if a satisfactory overlap cannot be found.

ACKNOWLEDGEMENTS

This work supported by the National Science Foundation under grant number PHY 8611210. Helpful discussions with Dr. K. Efferson are acknowledged.

REFERENCES

1. K.R. Efferson, A superconducting (Nb-Ti) liquid helium level detector, in: "Advances in Cryogenic Engineering", Vol. 15, Plenum Press, New York (1969), p. 124.
2. P.W. Eckels et al, Superconducting generator cooling system simulation, Cryogenics 25:471 (1985).

DESIGN AND TESTING OF ELECTRICAL INSULATION FOR SUPERCONDUCTING COILS*

S. W. Schwenterly

Oak Ridge National Laboratory
Oak Ridge, Tennessee

ABSTRACT

This paper reviews and summarizes the dielectric properties of liquid and gaseous helium and nitrogen, vacuum, and solid insulation materials for practical insulation design for superconducting coils. The influence of electrode geometry on breakdown voltage is discussed, and features that require special care are pointed out. Appropriate room-temperature high-voltage testing is considered. Finally, the paper examines the implications for coil insulation design of a new class of superconductors that can operate at temperatures approaching 77 K.

INTRODUCTION

Large superconducting magnets contain considerable amounts of stored energy, which is resistively converted to heat during a quench or emergency dump. In some cases, most of the energy is dissipated in an external dump resistor. Because this resistor must be large enough to dissipate the energy before normal zones in the coil overheat, potentials of several thousand volts can appear across the magnet terminals. If the magnets are part of a fusion confinement system, they must also withstand voltage transients induced by plasma disruptions.

Large energy storage coils may contain so much energy that external dissipation would require excessive voltage levels. In this case, the energy is absorbed in the winding; however, even with zero terminal voltage, very large internal voltages may appear. These coils are also subject to switching surges and lightning transients from the power grid.

Some applications, such as accelerators, require long strings of magnets connected in series. Quenching of one magnet produces a steep-fronted voltage surge, which propagates into adjacent magnets and can produce very nonuniform voltage distributions. Most of the quench voltage might appear across the first 10 or 20 turns of the adjacent magnet.

The voltage withstand requirements of a particular coil system are thus strongly influenced by the application. Once the designer has determined the maximum voltages generated in normal and emergency operation, safety factors must be chosen

*Research sponsored by the Office of Fusion Energy, U.S. Department of Energy, under contract DE-AC05-84OR21400 with Martin Marietta Energy Systems, Inc.

to determine test voltages during manufacture and final acceptance of the coil. An additional factor must be added in choosing the ultimate design voltage to give reasonable assurance of passing the tests. Hence, the design voltage may easily reach an order of magnitude above the operating voltage. The safety factors chosen will depend on the particular application, consequences of coil failure, ease of coil repair, and the amount of previous experience with the selected insulation system.

Careful consideration must therefore be given to electrical insulation from the outset of the magnet design. This paper gives a broad overview of insulation breakdown properties. Basic requirements of coil insulation are considered, questions of design and testing are discussed, and areas of the coil that require particular attention are pointed out.

INSULATION REQUIREMENTS

Room-temperature insulation properties play a much larger role in the design of superconducting coils than might first be expected. Of course, materials must be selected that are appropriate for use at low temperatures; however, during production and before acceptance the magnet must be tested at room temperature at several times the operating voltage to verify adequate safety factors. As far as electrical stresses are concerned, these room-temperature tests will probably be the most stringent conditions the magnet will ever experience and may have the strongest influence on insulation design.

Several other properties in addition to dielectric strength are important. The selected material must have the mechanical strength to withstand coil winding loads, Lorentz forces during operation, and differential thermal contraction loads during cooldown. It should not become too brittle at low temperatures. High thermal conductivity may be important to enhance conductor stability or to improve thermal contact with temperature sensors, which must be isolated from the conductor. Certain applications in fusion power systems and particle accelerators require resistance to radiation damage.

PROPERTIES OF DIELECTRICS

The practical breakdown strength of a material is influenced by contaminants and flaws and, in the case of liquids, by the presence of vapor bubbles. The practical strength can easily be an order of magnitude below the so-called intrinsic strength of the material, which is determined under ideal laboratory conditions. External factors such as electrode size and shape, surface condition, gap length, and time of voltage application can work to further reduce the practical breakdown strength. The following section summarizes some representative practical breakdown strengths for various materials and indicates the effects of external factors.

Breakdown Strengths

Gases. Breakdown in gases occurs by an electron multiplication process, in which electrons drifting across the gap collide with neutral gas atoms or molecules and ionize them to produce new electrons, which add to the discharge. The collision rate is density-dependent, and for gases that are not too dense Paschen's Law applies. Thus, the breakdown voltage is a function of Pd or ρd, the product of pressure or density and electrode gap. Figure 1 shows Paschen curves for several gases.[1–4] Note that these voltages are measured in uniform electric fields with clean, smooth electrodes. Each gas has a characteristic curve with a minimum voltage of a few hundred volts. As Pd is raised above its minimum-voltage value, the electrons cannot pick up enough energy between collisions to ionize another atom unless the applied voltage is increased. If Pd goes below the minimum-voltage value, the breakdown voltage again rises because the electrons must now gain more energy from the field between increasingly less frequent collisions to have a high enough probability of

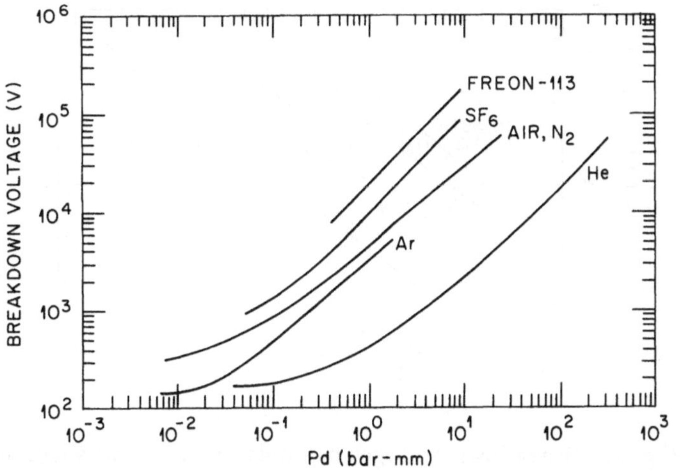

Fig. 1. Paschen curves for various gases.

producing enough extra electrons to sustain the discharge. Eventually, breakdown voltages characteristic of vacuum will be reached.

In Fig. 1, the gas temperature is assumed to be 20°C (293 K). Over the range where the gas is ideal, breakdown voltages for other temperatures can be obtained by displacing the curves to the left or right by the ratio T(K)/293. Helium gas is a relatively poor dielectric, with breakdown voltages at least a factor of 10 below nitrogen or air to the right of the minimum. Sulfur hexafluoride (SF_6) is roughly a factor of 3 better than nitrogen. The highest breakdown voltages are obtained with the halocarbons, for example, Freon-113 ($C_2Cl_3F_3$). These materials are best used in mixtures with nitrogen to reduce the amount of carbon deposit created in a breakdown.

As the gas temperature approaches the critical point and the density at a given pressure rises above ideal gas values, breakdown voltages tend to fall below the Paschen curve. Gerhold's comprehensive review of breakdown characteristics of cryogenic gases and liquids gives details.[5]

Vacuum. For the purposes of electrical insulation, a "vacuum" is a gas at a low enough pressure that the mean free path between molecules is larger than the electrode separation. Generally a pressure below 10^{-5} mbar is necessary. In this case, electron avalanches are impossible and breakdown is triggered by field emission of electrons from imperfections on the electrodes. Breakdown voltages are strongly dependent on the surface condition of the electrodes, and a given electrode configuration can often be conditioned by a series of low-current discharges to several times the initial breakdown voltage. Reported values of vacuum breakdown voltage vs gap are scattered over a wide band,[6] the lower edge of which is shown in Fig. 2. Because a gap of 1 mm will have a minimum breakdown voltage of over 20 kV, vacuum breakdown characteristics should not seriously limit the design.

Liquids. The principal liquids for cryogenic coil insulation are liquid helium and nitrogen. Reported liquid helium breakdown voltages in particular are distributed very widely, and Fig. 2 again shows only the lower edge of the band.[7,8] Breakdown in cryogenic liquids seems to be triggered by discharges in vapor bubbles, and voltages may be up to an order of magnitude higher if boiling is suppressed by minimizing heat leaks or pressurizing the liquid. However, even the minimum values shown will in most cases provide good insulation safety factors. Much higher breakdown voltages have been measured in liquid nitrogen.[8,9] As will be discussed later, it is usually the

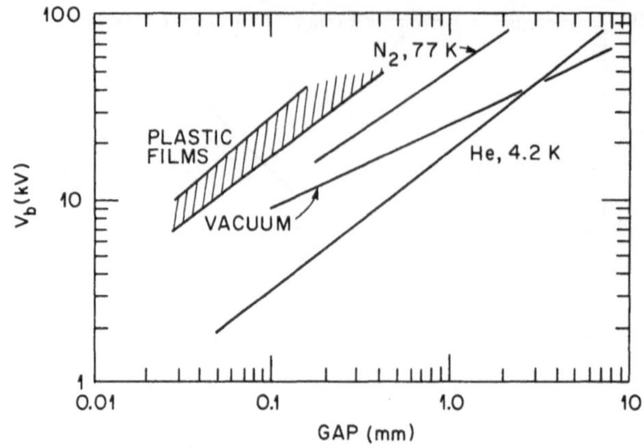

Fig. 2. Breakdown voltages of solids, liquids, and vacuum.

breakdown strength of the gas at some higher temperature rather than the liquid breakdown strength that will determine the size of the gaps.

Solids. Breakdown strengths of solids are usually well above those of liquids and gases. Most of the common plastic films, such as polyethylene, polycarbonate, Teflon, nylon, Kapton, and Mylar, have been investigated. In general, there is only a slight increase in breakdown strength in cooling to cryogenic temperatures. For nonporous materials, the room-temperature values can generally be used with confidence if no low-temperature value is available. Several sets of measurements for various plastic films at cryogenic temperatures lie in the band shown in Fig. 2.[10–12] These data were obtained by ramping up the voltage over a few minutes. For film thicknesses below 0.1 mm, single-layer breakdown strengths are over 200 kV/mm. Thicknesses over 0.1 mm generally require multiple layers, and the breakdown strength falls rapidly to reach typical values of 100 kV/mm at 1 mm thickness. For porous insulation materials, breakdown strengths depend much more on the impregnant and are a factor of 2 or more below these values.[13] Discharges at gaps can lead to long-term degradation, and operating stresses should be kept well below the expected breakdown stress. This is particularly important because solid insulation is not self-healing after a breakdown. In ac applications, a material with a low dissipation factor should be used to reduce the heat load on the refrigerator. Polyethylene and Teflon have the lowest dissipation factors at 4.2 K ($\sim 10^{-6}$), whereas the nylon, Kapton, and Mylar factors are a hundred times as great.

In some cases, a solid dielectric and a liquid or gas are used in parallel, such as in a standoff insulator or in an electrical break in a coolant supply line. For this arrangement, breakdown usually occurs first across the surface of the solid. This so-called flashover can appear at less than 50% of the breakdown voltage for the liquid or gas alone.[14–16] Flashover through the interfaces between solid layers occurs at even lower voltages, and electric stresses parallel to the surface in layered dielectrics are typically kept to only about 10% of the perpendicular stresses.

External Factors

Electrode Geometry. If the electrodes have regions where the radius of curvature is smaller than the gap, the surface electric field will be increased in these areas. The "enhancement factor" f is defined by the relation

$$f = \frac{E_{\max}}{V/d} ,$$

where V is the applied voltage and d is the gap. Figure 3 plots values of f vs the ratio of gap to electrode radius for several common geometries.[17] References 18 and 19 also give useful data for many configurations. For gases, when f is less than about 5, the breakdown voltage can be predicted fairly accurately by dividing the uniform-field Paschen curve breakdown voltages from Fig. 1 by f. If f is greater than 5, breakdown is in general preceded by a corona discharge near the sharply curved electrode, and space charge in the gap makes the breakdown voltage difficult to estimate. For liquid helium and nitrogen, the values in Fig. 2 can be similarly derated by the factor f. In a vacuum, the situation is different. Experiments in moderately divergent fields at gaps below 1 mm suggest that the divergent field actually increases the vacuum breakdown voltage.[20] However, the effect decreases for larger gaps, and in any case large enhancement factors should be avoided to prevent damage if the vacuum fails. For solid dielectrics, the breakdown strength of the material is so high that sharp-edged electrodes may not affect the short-term breakdown voltage. However, discharges at the edges can eventually erode the insulation, particularly with ac and repeated impulse voltages.

Gap Length and Electrode Area. The relation between breakdown voltage and gap is not generally linear. For many materials,

$$V_{br} = \alpha d^n ,$$

where α is a constant. For liquids and solids, n is about 0.8, as seen in Fig. 2. For vacuum, n is less than 0.5, although values up to 0.8 have been observed with small, clean electrodes.[20] For gases, n is about 0.9 for values of Pd well to the right of the minimum (Fig. 1).

When the volume of the insulation under stress is scaled up at constant thickness, the greater statistical probability of defects also leads to lower breakdown voltages. Consideration of the statistical distribution of defects leads to the relation[21]

$$E_{br} = \frac{\beta}{(Ad)^m} \quad \text{or} \quad V_{br} = \frac{\beta d^{1-m}}{A^m} ,$$

where Ad is the volume of insulation and β is a constant. Generally the value of m ranges from 0.05 to 0.15.[21] Thus the gap exponent $1 - m$ could be between 0.85 and 0.95, and much of the gap dependence previously discussed could also result from

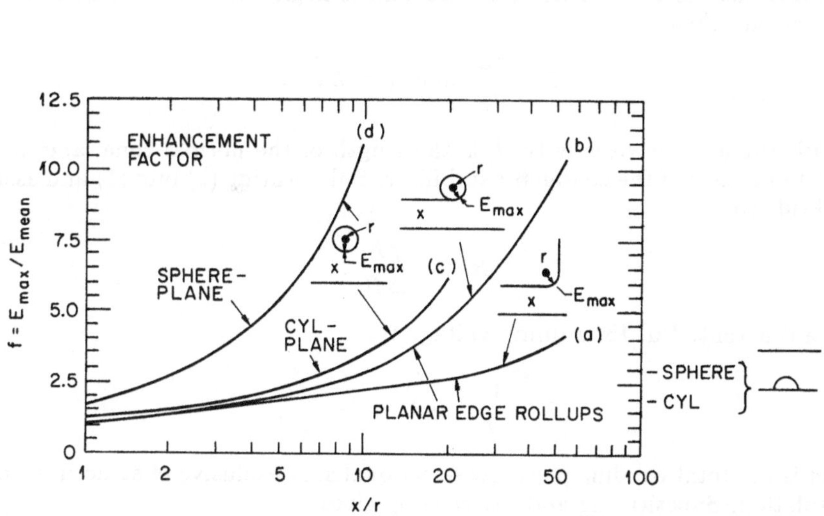

Fig. 3. Geometric enhancement factors vs gap-radius ratio; f for half-sphere and half-cylinder on plane is constant at 3 and 2, respectively.

statistical factors. If m is taken to be 0.1, scaling up the insulation area by a factor of 10^4 causes a reduction to only 40% of the test voltage with small electrodes.

Time. Even if no degradation processes affect the insulation, statistical factors lead to a reduction in breakdown strength with increasing time of voltage application according to

$$E_{br} = \frac{\gamma}{t^p} ,$$

where γ is a constant. At room temperature p is about 0.1,[21] but at cryogenic temperatures p can fall below 0.01.[22] This can be important for applications with continuously stressed insulation.

INSULATION DESIGN

To prevent costly modifications late in the design cycle, electrical insulation requirements should be fully considered at the start of conceptual design of the coil. Usually the required magnetic field distribution and the physical envelope available will be known at the outset. The structure required to support the windings can then be designed to fit the physical envelope, leading to a definition of the space available for the winding pack. At this point, the superconductor, stabilizer, coolant, and insulation begin competing for space. In many cases, a conductor will already have been chosen because it was left over from another program. If not, electrical insulation considerations would suggest as large a conductor with as high a current rating as possible, since this would decrease the number of turns and lead to a lower dump voltage. This is easily demonstrated. If the required field, shape, and size of the magnet are known, the stored energy E and the amp-turns $N\ I$ are both predetermined. In a dump, the proportion of stored energy dissipated in the winding is given by

$$E\left(\frac{\bar{r}}{\bar{r}+R}\right) = v\Delta H , \tag{1}$$

where $\bar{r}$ is the average resistance of the winding during the dump, R is the dump resistor value, v is the conductor volume, and ΔH is the allowed enthalpy rise per unit volume in the conductor, which is determined by the allowable thermal expansion and temperature rise. Heat transfer to the coolant is neglected. The winding resistance and volume are then

$$\bar{r} = \frac{\bar{\rho}L}{a_c} \quad \text{and} \quad v = La_c , \tag{2}$$

where $\bar{\rho}$ is the average resistivity, L is the length of the normal zone, and a_c is the cross-sectional area of the conductor stabilizer. Substituting (2) into (1) and assuming $\bar{r} \ll R$ leads to

$$Ra_c^2 = \frac{\bar{\rho}E}{\Delta H} . \tag{3}$$

This is a constant. But for a dump voltage V,

$$R = \frac{V}{I} \quad \text{and} \quad a_c = \frac{A}{N} , \tag{4}$$

where A is the total winding pack cross-sectional area exclusive of structure, coolant, and insulation. Substituting and rearranging gives

$$V = \left[\frac{\bar{\rho}\ N\ I\ E}{\Delta H}\right]\frac{N}{A^2} . \tag{5}$$

The part of Eq. (5) inside the brackets is a constant because NI is fixed by the required field. Thus, the dump voltage can be decreased by reducing the number of turns and particularly by increasing the area of the winding pack. The resulting large, high-current conductor has many disadvantages, however. A higher-current power supply will be more expensive, and larger current leads will produce a larger heat load, requiring a more-expensive refrigerator. The reduced surface area per unit volume of conductor lowers heat transfer to the coolant and compromises stability. The area A is usually limited by other equipment near the coils. Large conductors are more difficult to bend and wind properly, and multiple smaller conductors would probably have to be wound in parallel. The final coil design should be an optimized compromise among these factors. Most likely, a process of successive approximations will be needed. For each configuration, the designer must determine the maximum electrical and mechanical stress on the insulation based on the total dump scenario and all internal and external transients. Provisions for cooling channels and perhaps sensor leads may be necessary. The final design voltage will be the product of the operating voltage, the design safety factor of perhaps 4 or 5 (to be verified by the insulation tests), and a further safety factor of perhaps 2 (based on the test voltages). Consequently, the design voltage V_d will be about an order of magnitude above the operating voltage.

Prospective insulation materials can be selected on the basis of desired mechanical, thermal, and radiation resistance properties. Breakdown voltage data then lead to the required thickness of each material appropriate for V_d. The effects of insulation area, gap length, contamination, and conductor surface condition must be carefully considered in this process. Certain areas of the winding deserve special care, as shown in Fig. 4. Electric field enhancements can occur at the winding pack corners (a), at the ends of layers or pies (b), and at sensor leads coming through the insulation. Conductor edges in these areas should be well rounded or shielded, and extra insulation should be added if possible. Gaps in layered insulation (c) should be avoided because these will see increased electric field and could trigger a breakdown. Similarly a filler material should be used at junctions between solid insulation and conductor surfaces in areas of high electrical stress to fill small voids caused by imperfections in the mating surfaces. Curable RTV silicone rubber works well for this purpose. Solid insulation should not meet a conductor at an acute angle (d) because discharges at the junction could trigger a flashover. Instrument and power feedthroughs and electrical breaks in coolant supply lines are particularly susceptible to flashover. References 23 and 24 give useful information about the design of these components. In general, filled or fiber-reinforced epoxies or plastics should be used for large pieces of solid insulation to prevent cracking during cooldown. Instrument leads should be divided into cables at similar voltage, and leads at widely differing voltages should not be mixed on the same feedthrough. The same care must be taken in routing the leads

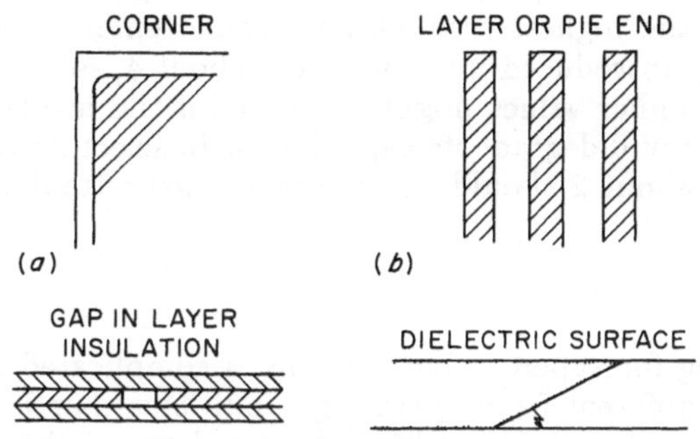

Fig. 4. Areas of insulation needing special attention.

from the coil to the instrumentation readouts. The coil should be fabricated in a clean room or enclosure to exclude dust and contaminants.

The configuration of the winding pack can be critically important in equipment that must withstand rapid voltage surges, such as transformers in a commercial power network, or magnets connected in long strings on an accelerator. Forsyth and Meth[25] recommend that each turn face its neighbor with as large an area as possible and face the nearest ground plane with as small an area as possible to spread out the initial surge over many turns. Thus, the ground capacitance should be minimized by small area and large insulation thickness. Electrostatic shields or divider networks can also be added to grade the voltage distribution more favorably.

Once the required insulation thickness has been determined, the overall size of the winding pack can be calculated and compared with the available space inside the support structure. Several iterations of the design may be necessary before a properly optimized version is obtained.

INSULATION TESTING

Because the insulation design voltage is determined by the test voltage, development of the insulation design and the testing procedures must go hand-in-hand. The insulation test procedure must:

1. give confidence that the coil will perform under operating conditions,
2. put the coil at minimum risk of failure during testing,
3. provide a method of locating defects, and
4. properly document the results.

The test procedure should also specify whether any sensor leads are to be grounded, floating, or tied to the conductor, as well as the polarity or type (dc, ac, impulse, etc.) of voltage to be applied. A resistance of several MΩ is usually placed between the voltage source and coil to limit current in a breakdown. The current should be monitored as the voltage is ramped up. Any increase above a linear dependence on voltage could indicate a defect.

Selection of Safety Factors

Various methods can be used to choose appropriate safety factors. One method would be to use existing test standards. For example, IEEE Standard 95-1977 suggests a test voltage that is the higher of

$$V_{\text{test}} = [2V_{\text{oper}} + 1000\ V] \quad \text{or} \quad 2000\ V\ .$$

This test was developed for commercial electrical equipment operating at power frequencies and may not be appropriate for a magnet, which must withstand transient high voltages only during a dump. Common practice may be a better guideline. A scan of recent literature on superconducting coil testing yields safety factors ranging from 2 to 5. Since the highest credible field enhancement factor from conducting particles accidentally introduced into the coil is about 4, safety factors in this range are recommended. Higher values might be chosen if the insulation must withstand special stresses that could degrade its capabilities. In selecting the design voltage, an additional factor of about 2 should be applied to ensure that the coil will pass the tests.

Test Conditions

Insulation testing for superconducting coils is complicated by the fact that test conditions are very different from operating conditions. Furthermore, during a coil dump, both the coil voltage and the breakdown voltage of the coolant are changing rapidly as the coolant is warmed and vented. The point of minimum safety factor for *solid* insulation will be at the start of the dump at maximum voltage, whereas the minimum safety factor in areas insulated by *coolant* may occur some time after the start. This is because the coil voltage falls roughly exponentially, but as the coolant warms and perhaps vaporizes, its density and breakdown voltage at first may fall faster. Very little is known about heat transfer between the conductor and coolant during a dump, particularly for pool-boiling designs. The most conservative assumption would put the coolant temperature equal to the conductor temperature to minimize the coolant density and breakdown voltage. Depending on experience with a given design, less-conservative assumptions might be made. Whatever the assumptions are, they will eventually lead to a choice of voltage, coolant temperature, pressure, and density where the minimum safety factor occurs, and the electrode gaps or surface flashover distances can be chosen to ensure that this safety factor equals or exceeds the desired value. The test program must then be laid out to verify the voltage withstand capabilities of all solid insulation, surface flashover paths, and gaps between conductors in the coil.

Choosing the test conditions will often lead to a dilemma for helium-cooled coils. It is desirable to perform tests at several stages of coil construction. When the coil case or conductor is not closed, atmospheric air is the most convenient medium for testing. However, air is equivalent in dielectric strength to 1-bar helium at roughly 14 K. Because the coolant temperature could go much higher than this in a dump, air has an unrealistically high dielectric strength, and a gap that could survive the test in air might break down in cold helium gas during a dump. On the other hand, if higher test voltages are applied, the solid insulation might be overstressed. If the coil is temporarily enclosed by bagging or some other means, it could be purged with helium. However, if the gaps in the coil were designed to operate in helium at 100 K, then room-temperature helium could withstand only 1/3 of this voltage. Proper gap lengths could be safely verified by testing at a lower voltage, but the solid insulation would then be understressed. Higher density and breakdown voltage could be obtained by pressurizing to perhaps 3 bar, but the case would then have to be closed more elaborately. The best compromise for thorough testing of both the solid insulation and the gaps and flashover paths would be to test first in air at full voltage and then in 1-bar helium at some scaled-down voltage. If testing only in air is possible, the solid insulation will have to be overdesigned to allow for stressing of all gaps and flashover paths to their full safety factors. Because of the high dielectric strength of solid insulation, this may not cause difficulties.

Another consideration stems from the time effect mentioned in the previous section. Insulation tests are usually conducted at various stages of completion of the coil. In many designs, it is impossible to test the most recently completed section without also stressing already-tested sections. The sections tested early will spend much longer at test voltage than those tested last. If the coil design requires minimal safety factors, it is possible that a breakdown could occur in a section completed earlier, an accident which could entail a major unwinding of the coil for repairs. In this case, it is best to set up a test program where the voltage for successive tests decreases by a certain interval or percentage. The size of the stepdown will be determined by the number of intermediate tests and the allowable difference between the initial and final test voltages.

HIGH-TEMPERATURE SUPERCONDUCTORS

The recent discovery of a new class of superconducting materials offers the prospect of coils that can operate at liquid nitrogen temperature or higher. The use of liquid or pressurized nitrogen as the coil coolant will probably lead to an insulation design that differs considerably from that of a helium-cooled coil. For example, the ratio $\bar{\rho}/\Delta H$ in Eq. (5) may change. For a given allowable thermal expansion in a dump, a copper-stabilized coil rising 100 K above helium temperature could rise only 50 K above liquid nitrogen temperature. This is caused by the lack of thermal expansion for about the first 50 K of temperature rise above 4 K. The corresponding allowable ΔH for a high-temperature coil would be about the same as for helium temperatures—roughly 100 J/cm^3. However, the average resistivity $\bar{\rho}$ will double from 50 to 100 mΩ-cm, so $\bar{\rho}/\Delta H$ will also double. This will affect the choice of N, A, and I.

The use of nitrogen as the coolant eases the insulation design problem. Although the properties of any solid insulation in the coil will not be appreciably different, any gaps and flashover paths can be decreased because of the greater breakdown strength of liquid and gaseous nitrogen over helium. However, if this is done the room-temperature testing is complicated because with 1-atm air or nitrogen the solid insulation cannot be fully stressed without overstressing the gaps (the opposite situation to the room-temperature test problem with a helium-cooled coil). Either the gaps and flashover paths must be overdesigned, or a gas with higher breakdown strength than nitrogen must be used. SF_6, commonly used for high-voltage gas-insulated apparatus, has roughly 3 times the breakdown strength of nitrogen. However, the addition of only 20% SF_6 to pure nitrogen will raise the breakdown strength of the mixture to 80% of the strength of pure SF_6.[26] Freon is another possible additive. The addition of 10% Freon-113 to nitrogen will more than double the pure nitrogen breakdown strength.[4] Reference 27 gives considerable background on dielectric gas mixtures.

It could be argued that normal zones in a high-temperature superconductor would be nearly impossible because heat capacities would be so high that no credible disturbance could cause a large enough temperature rise. However, fast discharges would still be necessary to protect the coil from failures of the refrigerator or other equipment. Also, high-voltage insulation would still be required to handle the external voltage transients faced by the coil.

CONCLUSIONS

A successful cryogenic coil insulation design must consider the breakdown performance of solids, liquids, gases, and vacuum from room temperature down to operating temperature. Insulation breakdown strengths must be derated for factors such as damage during installation, electrode geometry, total size, and lifetime, and once test voltages are chosen, an additional safety margin must be provided to give confidence that the insulation will pass the tests. The operating voltage should therefore be kept an order of magnitude below the ultimate design voltage. The new high-temperature superconductors will ease but not eliminate the problem of high-voltage insulation for superconducting coils.

REFERENCES

1. H. Winkelnkemper et al., Breakdown of gases in uniform electric fields, *Electra* 52:67 (1977).
2. J. D. Cobine, "Gaseous Conductors," McGraw-Hill, New York (1941), pp. 143–165.
3. E. Husain and R. S. Nema, Electric stress at breakdown in uniform field for air, nitrogen, and sulfur hexafluoride, in: "Proc. Fourth Intl. Symp. on Gaseous Dielectrics," Pergamon, New York (1984), p. 168.
4. E. Gockenbach, The dielectric strength of some freons and their mixtures with nitrogen and sulphur hexafluoride, in: "Proc. First Intl. Symp. on Gaseous Dielectrics," CONF-780301, Oak Ridge National Laboratory, Oak Ridge, Tennessee (1978), p. 355.
5. J. Gerhold, Dielectric breakdown of cryogenic gases and liquids, *Cryogenics* 19:571 (1979).
6. M. J. Mulcahy et al., Designed experiments on high voltage vacuum breakdown, in: "Proc. Second Intl. Symp. on Insulation of High Voltages in Vacuum," Massachusetts Institute of Technology, Cambridge (1966), p. 177.

7. B. Fallou et al., High voltage dielectric behavior of liquid and hypercritical helium, in "Low Temperatures and Electric Power," Pergamon, New York (1970), p. 377 (in French).
8. K. N. Mathes, Dielectric properties of cryogenic liquids, *IEEE Trans. on Elec. Ins.* EI-2:24 (1967).
9. P. H. Burnier, J. L. Moreau, and J. P. Lehmann, The dielectric strength of cryogenic fluids and solid insulators, in: "Advances in Cryogenic Engineering," Vol. 15, Plenum Press, New York (1970), p. 76.
10. P. Chowdhuri, Some characteristics of dielectric materials at cryogenic temperatures for HVDC systems, *IEEE Trans. on Elec. Ins.* EI-16:40 (1981).
11. M. M. Menon et al., Dielectric strength of liquid helium impregnated plastic tapes, in: "Annual Report—1975 Conf. on Electrical Insulation and Dielectric Phenomena," National Academy of Sciences, Washington, DC (1978), p. 277.
12. G. Bogner, Cryopower transmission studies in Europe, *Cryogenics* 15:79 (1975).
13. S. J. Rigby and B. M. Weedy, Liquid nitrogen-impregnated tape insulation for cryoresistive cable, *IEEE Trans. on Elec. Ins.* EI-10:1 (1975).
14. I. Ischii and T. Noguchi, Surface flashover strength in super-critical helium, in: "Annual Report—1980 Conf. on Electrical Insulation and Dielectric Phenomena," National Academy Press, Washington, DC (1980), p. 397.
15. R. Hawley, Solid insulators in vacuum: A review, *Vacuum* 18:383 (1968).
16. J. Wańkowicz, Flashover voltage of spacer insulators in a vacuum at 290-6K, *Cryogenics* 23:482 (1983).
17. D. B. Hopkins, Design considerations and data for gas-insulated high voltage structures, in: "Proc. Sixth Symp. on Eng. Probs. of Fusion Research," IEEE, New York (1976), p. 435.
18. A. Bouwers and P. G. Cath, The maximum electrical field strength for several simple electrode configurations, *Philips Tech. Rev.* 7:270 (1941).
19. H. Ryan and C. A. Walley, Field auxiliary factors for simple electrode geometries, *Proc. IEE* 114:1529 (1967).
20. M. Rabinowitz, Electrical breakdown in vacuum: new experimental and theoretical observations, *Vacuum* 15:59 (1965).
21. J. Artbauer and J. Griač, Some factors preventing the attainment of intrinsic electric strength in polymeric insulations, *IEEE Trans. on Elec. Ins.* EI-5:104 (1970).
22. A. Bulinski, J. Densley, and T. S. Sudarshan, The ageing of electrical insulation at cryogenic temperatures, *IEEE Trans. on Elec. Ins.* EI-15:83 (1980).
23. J. Gerhold, Design criteria for high voltage leads for superconducting power systems, *Cryogenics* 24:73 (1984).
24. F. Schauer, A capacitance-graded cryogenic high voltage bushing for vertical or horizontal mounting, *Cryogenics* 24:90 (1984).
25. E. B. Forsyth and M. Meth, "Electrical insulation requirements for superconducting magnets used in a large system," BNL 29968, Brookhaven National Laboratory, Upton, New York (1981).
26. M. J. Mulcahy et al., A review of insulation breakdown and switching in gas insulation, *Insulation/Circuits* 16:55 (1970).
27. D. W. Bouldin et al., A current assessment of the potential of dielectric gas mixtures for industrial applications, in: "Proc. Fourth Intl. Symp. on Gaseous Dielectrics," Pergamon, New York (1984), p. 204.

QUENCH DETECTION OF SUPERCONDUCTING MAGNET BY DUAL-CORE OPTICAL FIBER

O. Tsukamoto, K. Kawai, Y. Kokubun and T. Takao

Faculty of Engineering
Yokohama National University
Tokiwadai, Yokohama, Japan

ABSTRACT

We are developing a quench detecting technique using a dual-core optical fiber. The fiber has two single-mode optical cores in one fiber. Using this technique, we have demonstrated that a temperature rise of 1.0 K at 4.2 K, so that, a quench in a superconducting magnet was detectable. We verified that an electromagnetic force-stress to the optical fiber did not deteriorate the sensitivity of the quench detection. A quench detector using this optical method is immune from electromagnetic noises and free from troubles caused by high voltage tension. Problems arising when applying this technique to a large scale magnet and a possible improving technique in the instrumentation are discussed.

INTRODUCTION

Usually, to detect quenches in superconducting magnets, signals from voltage taps attached to conductors are monitored. This method, however, causes many troubles. Small voltage signals must be detected to reliably protect the magnets from damages caused by a quench. An abrupt conductor motion produces a voltage spike large enough to frequently trigger the quench detectors. A superconducting magnet itself is a large pick-up coil and the quench detectors are susceptible to electromagnetic noises, such as field fluctuations from other magnets and line noises. For example, toroidal magnets in a tokamak fusion reactor are subject to fluctuations in magnetic field caused by plasma disruptions, and by field perturbations of poloidal and other toroidal magnets. Therefore, it is generally considered to be very difficult to detect quenches in toroidal magnets. Pulsed magnets are subject to high voltages of several kV. It is very difficult to detect the small voltage due to a normal zone in such high-voltage magnets. Lead-wires from the voltage taps often cause arcs and insulator break-down in the windings. Therefore, it is important to develop new quench detection methods which use sensors that are electrically insulated from the magnet conductors.

We have developed a technique to detect a temperature rise at 4.2 K by using a dual-core optical fiber[1]. When a part of the fiber is warmed, the length of the optical path of each core in the fiber changes differently, because the refractive index of each core has a slightly different temperature dependence. Applying laser light to one end of the dual-core fiber, a temperature rise can be detected by observing shifts in the interference fringes formed at the other end of the fiber. Using this technique, a quench in a superconducting magnet can be detected by detecting temperature rises of the winding due to Joule heating in the normal zone. This optical method uses no probes electrically connected to the magnet conductors and has many advantages over the conventional voltage-measuring quench detection method.

We performed a basic experiment and demonstrated that a temperature rise of 1.0 K could be detected at 4.2 K. Sensitivity of this technique can be improved by using a longer fiber and a more sensitive detector of the interference fringe shifts. Using this technique, it was demonstrated that a quench was detectable[2]. The optical fiber placed in a magnet winding is subject to stresses by electromagnetic force. There is a possibility that the interference fringes shift due to the electromagnetic stress. In the experiment, it was also demonstrated that the electromagnetic force-stress did not shift the interference fringes.

In the optical arrangement of the quench detector an image-fiber is used to optically transfer the interference fringe pattern to the room temperature region from the cryogenic region. An image-fiber is not so flexible to thread though a narrow and complicated pass, and it is hard to make the optical system disconnectable. We are developing a new optical device which can transfer the fringe-shift signal with two single-core fibers. A system using this new device is well-applicable to a large scale magnet.

PRINCIPLE

As is shown in the figure 1, optical phase ϕ of a single-core and single-mode fiber changes with changes in temperature and mechanical stress. When the temperature of the fiber changes, the refractive index n of the optical core changes, so that, the optical length of the core changes. Generally, at the cryogenic temperature below 20 K, the thermal expansion coefficient of glasses is zero[3], and the length of the fiber L is supposed to be unchanged by the temperature changes in that range. Therefore, the optical phase change $\Delta\phi$ caused by the temperature change ΔT is only due to the change in the refractive index, as is indicated by following equation

$$\Delta\phi = \alpha L \Delta T \qquad \alpha = \frac{2\pi}{\lambda_0}\left(\frac{\partial n}{\partial T}\right) \tag{1}$$

Mechanical stress affects the optical phase in two ways, affecting L and n. $\Delta\phi$ due to change in mechanical stress $\Delta\varepsilon$ is expressed as follows

$$\Delta\phi = \beta L \Delta\varepsilon \qquad \beta = \frac{2\pi}{\lambda_0}\left(\frac{\partial L}{\partial \varepsilon} + \frac{\partial n}{\partial \varepsilon}\right) \tag{2}$$

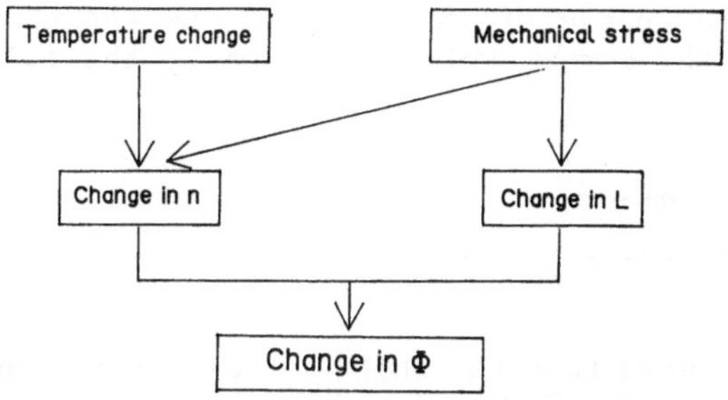

Fig. 1 Effect of temperature and mechanical stress on optical phase ϕ of single mode and single core fiber at cryogenic temperature. L and n are length and refractive index of optical fiber respectively.

Figure 2 shows the schematic arrangement for a temperature measurement at cryogenic temperature using a dual-core optical fiber. Laser light is introduced into the cryogenic region by a single-core and single-mode optical fiber. At the junction of the single-core and the dual-core fibers, the laser light is equally split to two cores of the dual-core fiber. Interference fringes are formed on an end surface of the image-fiber at the interference section. The fringe pattern is transferred from the cryogenic region to the room temperature region through the image-fiber. The shifts of the fringes due to the temperature changes of the dual-core fiber are detected at room temperature.

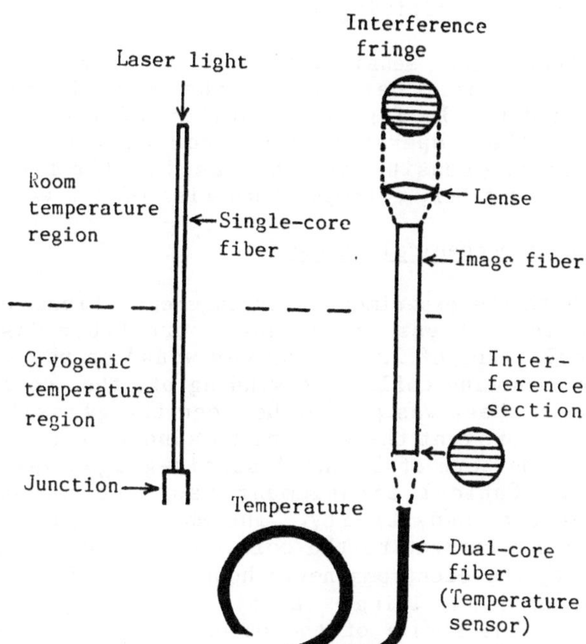

Fig. 2 Schematic arrengement for an optical temperature measurement using a dual-core optical fiber.

The temperature change of the fiber causes the change in the phase difference between the cores, which is given by the following equation, considering Eq. 1

$$\Delta\phi_1 - \Delta\phi_2 = (\alpha_1 - \alpha_2)\Delta T\, L$$
$$\alpha_j = \frac{2\pi}{\lambda_0}\left(\frac{\partial n_j}{\partial T}\right) \quad j = 1, 2 \tag{3}$$

This equation shows that the shift in the interference fringes, that is, $(\Delta\phi_1 - \Delta\phi_2)$ is proportional to ΔT and L.

The dual-core fiber is placed in windings of a magnet and subject to mechanical stress caused by electromagnetic force. When the fiber is subject to mechanical stress, the change in the phase difference between the two cores is given by

$$\Delta\phi_1 - \Delta\phi_2 = (\beta_1 - \beta_2)\Delta\varepsilon\, L$$
$$\beta_j = \frac{2\pi}{\lambda_0}\left(\frac{\partial L}{\partial\varepsilon} + \frac{\partial n_j}{\partial\varepsilon}\right) \quad j = 1, 2 \tag{4}$$

As is shown in the next section, experiments performed at the liquid helium temperature shows that the interference-fringe shift due to the temperature change is more dominant than that due to the change of the mechanical stress. This characteristics of the dual-core fiber is favorable to its use as a quench detector.

EXPERIMENT

Temperature sensitivity

Temperature sensitivity of a dual-core optical fiber was measured at 4.2 K region. Details of the experiment are described in the reference 1. In figure 3, number of the fringe shift was plotted against the temperature rise from 4.2 K for various value of L. The temperature sensitivity increased as the fiber was longer. With the 3.0 m long fiber, a temperature rise of 1 K was detectable.

Effect of mechanical stress

With the experimental arrangement illustrated in Fig. 4, effect of mechanical stress on the dual-core fiber was examined. A dual-core optical fiber of 3.0 m long was wound on the outermost layer of a test superconducting coil. The winding of the coil was epoxy-impregnated. A kapton sheet was placed between the glass-fiber bobin and innermost layer to prevent the winding to adhere to the bobin. The coil produced 1 T in the bore at 50 A. A strain-gauge was placed on the outermost layer. Optical arrangement was same as was used to measure the temperature sensitivity. The whole sample was put in liquid helium. Applying a current to the coil, a longtudinal stress was applied to the fiber by the electromagnetic hoop stress of the winding, and the shift in the interference fringe was measured. Figure 5 shows an experimental result. The strain of the outer surface increased with the increase of the magnet current. However, no fringe shift was observed during the increase and decrease of the current. It was demonstrated that, to the stress level of the experiment, the value of Eq. 4 is negligibly small compared to the value of Eq. 3.

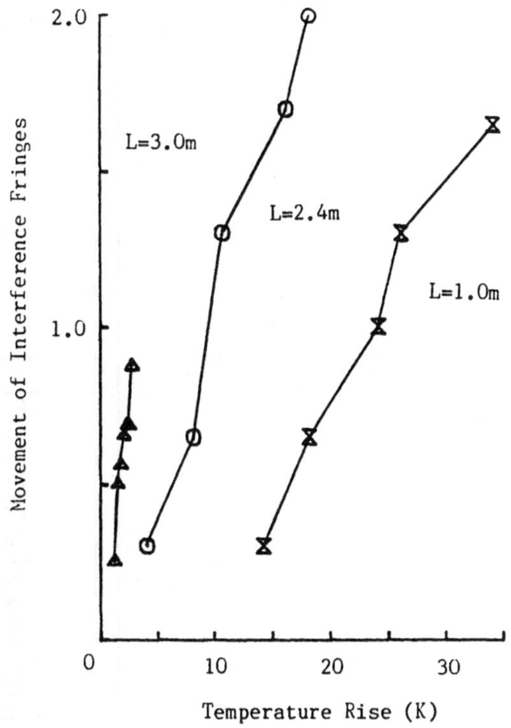

Fig. 3 Temperature rise vs. number of fringe shift at 4.2 k for various length (L) of the dual-core optical fiber.

Quench detection

An experiment was performed with a small superconducting magnet, 30 mm outer diameter, 80 mm long, to demonstrated the quench was detectable by this optical method[2]. A ditch was made on an outer surface of a glass-fiber-epoxy bobin and a dual-core optical fiber 5.0 m long was wound there. A superconducting wire was wound tightly on the bobin and the optical fiber. The magnet was designed to produce 2 T in the center of the bore at 30 A. The optical arrangement was same as is shown in Fig. 4. In the experiment, the magnet current and terminal voltage were recorded by a transient recorder, and the fringe pattern was recorded by a video recorder.

During a charge-up of the magnet, the fringe pattern did not move until a quench occurred. A typical quench event is shown in Fig. 6. The quench event occurred during a charge-up. Rising of the terminal voltage due to the quench and shift in the fringe pattern occurred simultaneously. Voltage fluctuations before the quench was due to the current changes during charge-up.

CONCLUSIONS AND DISCUSSIONS

It was demonstrated that a quench of a magnet was detectable by the optical method using a dual-core optical fiber. The optical method was less sensitive than the voltage method in this experiment where electromagnetic environment of the magnet was quiet and the charging rate was low. The optical method, however, is still preferable in the case that a magnet is subject to high voltage and large electromagnetic noises.

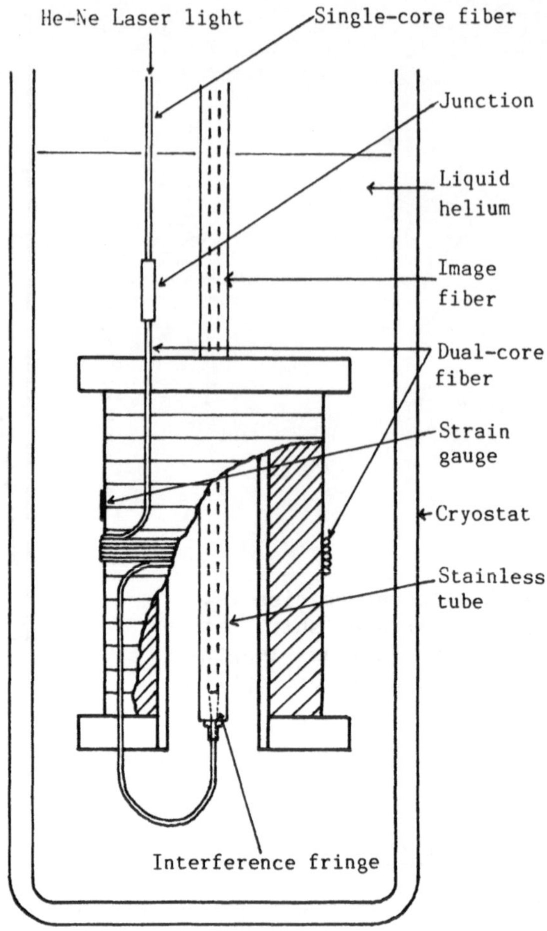

Fig. 4 Experimental arrengement for investigating mechanical stress effect on dual-core fiber.

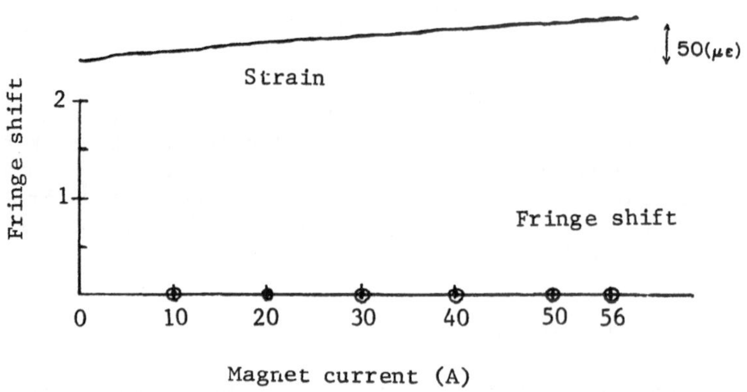

Fig. 5 Experimental result.

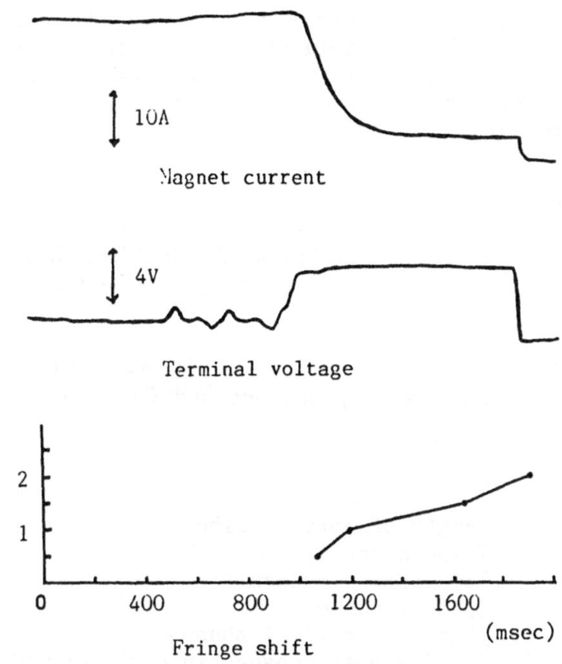

Fig. 6
Typical quench event.

In the optical quench detection system, the arrangement to transfer the fringe pattern to room temperature region is rather complicated. There is an arrangement using only a dual-core fiber, where both ends of the fiber are at room temperature and the middle part of the fiber is the temperature sensing section at cryogenic temperature. Application of a laser light to this arrangement and forming of interference fringes are at room temperature. This arrangement although simple is not practical because the temperature measurement is greatly affected by fluctuations in the temperature distribution in the boundary area between room and cryogenic temperatures.

In the system, an image-fiber is used to transfer the fringe pattern. The arrangement for forming the fringe pattern at the end of the image-fiber is delicate and hard to be taken apart. We are developing a new optical system shown in Fig. 7 to transfer the fringe pattern shift. A single-mode slab optical wave-guide (Aint-Resonant Reflection Optical Wave-guide ; ARROW) is inserted between the dual-core fiber and two single mode fibers. Light from the dual-core connected one end of the wave-guide forms one dimensional interference fringe at the other end. Two single-core fibers are connected to this end with a distance between the fibers corresponding to the phase angle of $\pi/2$ in

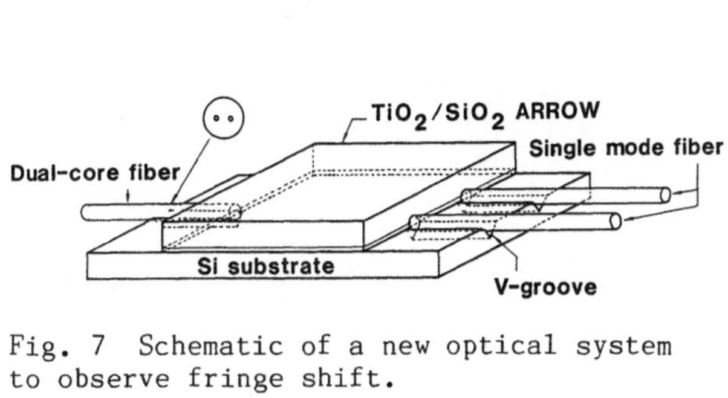

Fig. 7 Schematic of a new optical system to observe fringe shift.

the interference fringe. With this system, amount and direction of the shift in the fringe pattern are detectable and only a pair of flexible single-core fiber can transfer the signal instead of a stiff image-fiber. The performance of this system has been basically demonstrated at liquid helium temperature.

ACKNOWLEDGEMENT

We wish to thank Dr. K. Inada, Mr. O. Fukuda and Mr. K. Sanada of Fujikura Ltd. for useful discussions and supplying us dual-core fibers and other optical fibers.

Work is supported Grant-in-Aid for Engineering Research, 'The ministry of Education, Science and Culture.

NOMENCLATURE

L	Length of optical fiber
T	Temperature
ΔT	Temperature change
ϕ	Optical phase
$\Delta\phi$	Change in optical phase
$\Delta\phi_1, \Delta\phi_2$	Optical phase change in each core of dual-core fiber
n	Refractive index of optical core
n_1, n_2	Refractive index of each core of dual-core fiber
ε	Mechanical stress
$\Delta\varepsilon$	Change in mechanical stress
λ_0	Wave length of light

REFERENCES

1. O. Tsukamoto, Y. Kokubun and T. Toyama, Detection of temperature rise at 4.2 K by using a dual-core optical fiber - a optical method to detect a quench of a superconducting magnet, in: "Advances in Cryogenic Engineering", Vol. 31, Plenum Press, New York (1986), P1269

2. O. Tsukamoto et al, Quench detection of superconducting magnet by dual-core optical fiber, IEEE Trans. on Mag., MAG-23, (1987), (to be published)

3. G. K. White, in: "Experimental Techniques in Low-Temperature Physics", Oxford at The Clarendon Press (1968), P377

THEORY AND TECHNIQUE FOR REDUCING THE EFFECT OF CRACKS IN MULTILAYER INSULATION FROM ROOM TEMPERATURE TO 77 K*

Q.S. Shu,[†] R.W. Fast, and H.L. Hart

Fermi National Accelerator Laboratory
Batavia, Illinois

ABSTRACT

Cracks or gaps in multilayer insulation blankets can significantly increase the heat load to a cryogenic system. Our experiments gave the mean equivalent thermal conductivity of a narrow crack to be 3 to 5 W/m-K between room temperature and 77 K. The heat flux through a crack was found to be ~150 W/m^2. The dependence of the heat load on crack width, geometry, properties of the cold surface under the crack, the depth of the crack, and overall vacuum pressure were systematically studied. Aluminized Mylar patches covering the cracks were found to be very effective in reducing the heat load. Using the optimum number and distribution of patches determined in our experiments, it is possible to eliminate the effect of cracks on the overall heat load. In order to understand the mechanism of heat transfer through cracks, the temperature distributions in the multilayer insulation adjacent to cracks were measured. A theoretical model has been developed to explain this "black crack" phenomena and provide more quantitative heat leak estimates.

INTRODUCTION

Multilayer insulation (MLI) has been studied for a long time and many aspects of its performance are now very predictable. However, there are still serious problems associated with cracks or penetrations in the MLI system which require a better fundamental explanation. Previous experimental studies have indicated that the degradation of the thermal performance of an MLI system due to cracks is much worse than previously thought.[1] Moreover, in large cryogenic devices, e.g., Superconducting Super Collider (SSC)[2] which has 8600 magnets with a total surface area of $2.3 \times 10^6\ m^2$, there are almost always some assembly joints, gaps, overlaps and penetrations between prewrapped MLI blankets. The economic significance of reducing the effects of cracks in MLI systems is obvious, since most of the electric power consumed in large superconducting devices is in the refrigeration system.

Our goal in the investigation was not to study MLI materials themselves, but to focus on the basic heat transfer mechanism of cracks: why the effects are so serious, and how the effects can be reduced. The so-called Enhanced Black Cavity Theoretical Model was developed to explain the unexpectedly large heat load caused by the cracks.[3] Experimentally, several patch methods were used to reduce the effects of cracks. The optimum number and optimum distribution of these patches along cracks in a MLI blanket were determined. A

*Work sponsored by Universities Research Association under contract with the U.S. Department of Energy.
†Visiting Scientist from Cryogenics Laboratory, Zhejiang University, Hangzhou, China.

good patch system can improve the thermal performance of a MLI blanket with cracks to about that of a blanket without.[4] There are many different combinations of crack dimensions, patch material and distribution, which could have been tested. Several typical combinations were chosen for this study, but the theoretical model and experimental results are of general applicability, regardless of the MLI materials used.

EXPERIMENTAL ARRANGEMENT

The apparatus has been described in detail in a previous paper.[5] The inner copper plate, with an area of 2.26 m^2, was refrigerated to 77 K by a thermosiphon tube connected to the liquid nitrogen supply and boil-off vessel. The temperature of the outer (warmer) box was automatically maintained at 277 K. The insulation vacuum was measured by a cold cathode gauge mounted directly on the warm box inside the vacuum space. Individual MLI layers were hung vertically on the inner plate. Accurately sized cracks of various widths and lengths were cut in the fluffy MLI blanket using a razor blade and an adjustable Micarta frame. The cracks were of zero width, called slits; of finite width, called slots; and with equal length and width, called square holes. An optical transit was used to measure the actual dimensions of the cracks. To avoid errors associated with layer density, one 30-layer MLI blanket was used for the seven runs to investigate the effects of slot width on thermal performance, by starting with the narrowest slot and progressively enlarging it. Another blanket was used for testing heat load reduction methods. Flat, double-aluminized Mylar (DAM, 500 angstroms each side) and crinkled single-aluminized Mylar (SAMC, 300 angstroms each side) were used as patch materials. The patches were centered over the slots and square holes with a 10-mm overlap and secured with tape. In order to understand the heat transfer mechanism in a MLI blanket with cracks, copper-constantan thermocouples were mounted with aluminum tape to layers 5, 10, 15, 20, 25, 30 and to the cold plate and warm box. The junction was 3 mm from the edge of the slot and almost in the longitudinal center. All of the thermocouple wires were led out of the cryostat to a scanner without joints or splices to avoid the errors due to thermovoltage. The heat sinking of the wires was done carefully.

RESULTS OF CRACK EXPERIMENTS

Increase in Heat Load

The heat load, Q_0, from a 277 K copper box to a black painted fin, with an area $A =$ 2.24 m^2, through 30 layers of MLI without cracks is 1.4 W and the corresponding heat flux, Q'_0, (Q_0/A) is 0.63 W/m^2. The total heat load, Q_C, from the same warm box to the same fin through 30 layers of MLI with fourteen slots (each nominally 4 × 254 mm^2, and total measured area A_S = 0.0114 m^2) is 2.74 W. The increment of heat load, which is the heat load through the slots, $(Q_S = Q_C - Q_0)$ is 1.34 W. The heat load through the slot per unit slot area labeled as the heat flux through the slot, Q_S/A_S = 117.5 W/m^2. This flux, Q'_S, is about 200 times larger than the heat flux through a similar MLI blanket without cracks, *i.e.*, $Q'_S/Q'_0 \cong 200$.

Effect of Crack Width on Heat Flux

In order to investigate the effect of slot width on heat flux, the heat load through a 30-layer MLI blanket was measured for crack widths of 0, 2, 4, 6, 9, 15 and 30 mm.

One-dimensional slit. The measured data with one-dimensional slits shows clearly that such slits do not increase the heat flux through a MLI blanket. They do, however, simplify evacuation of the MLI blanket.

Two-dimensional slot. For slots, Q_C/A is a monotonically increasing function of slot width. The slope of the curve up to a slot width of about 2 mm was 33.5 W/m^3, for widths between 2 and about 9 mm the slope was 243 W/m^3; above 9 mm it was 179 W/m^3. Figure 1 shows the Q_S/A_S and the equivalent thermal conductivity as a function of the slot width. The heat flux increases rapidly with width to a broad maximum of about 150 W/m^2 at 9 mm, then decreases slowly and approaches the bare plate value of 24.7 W/m^2.[5] The maximum value is more than 200 times the heat flux through a 30-layer MLI blanket without slots.

Effect of Crack Geometry on Heat Flux

To investigate the effect of crack geometry on heat flux, square holes were cut in the 30-layer MLI blanket in the same geometric distribution and the same area as the 4 mm slots. The overall heat load increment caused by the square holes is 1.57 W; and the heat flux through the slot is 149 W/m^2, which is slightly larger than that caused by the 4-mm slots.

Effect of the Properties of the Cold Surface Under Slots

The effect on the heat flux of the properties of the cold surface under the slots was tested with two runs using the same MLI blanket with fourteen 4×256 mm^2 slots. For one run the cold copper surface was painted black; in the other it was covered by 3M #425 aluminum tape. The experimental data shows that the increment of heat flux due to the slots is almost independent of the cold surface emissivity: the values were 1.34 W for the black painted surface and 1.40 W for the aluminum taped surface.

Effect of Slot Depth on Heat Flux

In order to study the effect of slot depth on heat flux, fourteen 6-mm wide, 245-mm long slots were cut on both a 30-layer and a 90-layer MLI blanket. The measured data shows that the heat flux through the slots is 130 W/m^2 for the 30-layer blanket and 139 W/m^2 for the 90-layer blanket. Evidently the deeper the slots, the more heat goes through the slots.

The Effect of Overall Vacuum on Heat Flux

Figure 2 shows the overall heat flux through MLI blankets with various slot widths as a function of overall vacuum pressure. It can be seen that slots in an MLI blanket provide less thermal protection under poor vacuum conditions.

THEORETICAL MODEL

In order to explain the unexpectedly large heat transfer through a crack in a MLI blanket

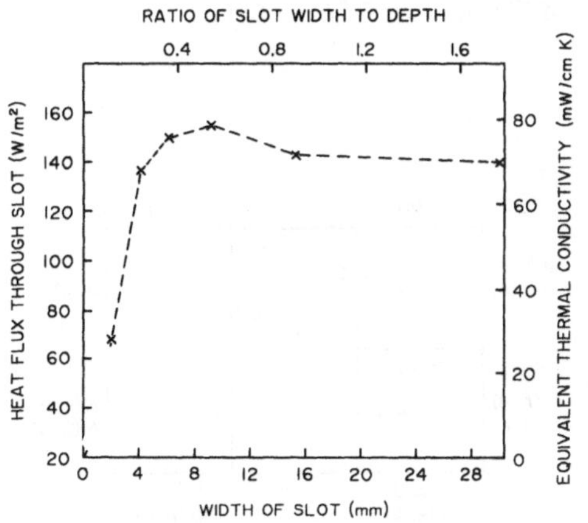

Fig. 1. Heat flux through slot and equivalent thermal conductivity of slots as a function of slot width for exposed slots.

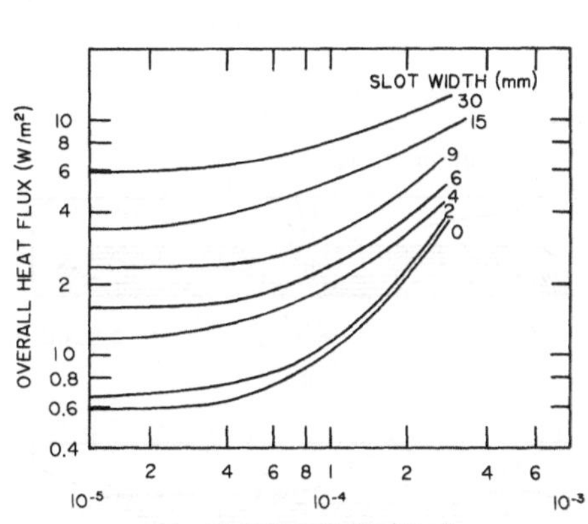

Fig. 2. Heat flux as a function of overall vacuum level for exposed slots of various widths.

and to optimize procedures for reducing this significant effect, a theoretical model, the so-called Enhanced Black Cavity Model was developed.[3] The steps in the model analysis can be briefly described as follows:

1. A crack in a MLI blanket was first considered as a cavity in a low emissivity enclosure, with the edges of the cut MLI acting as a wall which absorbed all the incident radiant energy in the sandwich structure of the MLI edges.
2. The multi-reflection of radiation flux between the shiny MLI Mylar and the polished inner surface of the vacuum jacket greatly enhances the radiant flux into the crack.
3. The temperature distribution in the region near the crack affects the heat transfer through the MLI blanket, which in turn increases the heat flux to the crack.
4. The relative dimensions of a crack also influence its absorptivity; this is considered here as a cavity effect.

Figure 3 shows that radiant fluxes from the warm external environment enter directly and indirectly into a crack and are totally absorbed by the structure of the MLI and the cold surface after successive reflections inside. Figure 4 is a simplified example, showing one black surface DC (represented as a slot in the MLI) surrounded by three shiny surfaces. The radiation energy from point 1 of plane C reflected by A to the slot, appears to come from an image point 1′ in an image plane $C(A)$. In this case there are thirteen additional image surfaces, all of which emit additional radiation to the slot greatly and enhancing the transmitted energy flux. So if Q_E is the radiation energy directly emitted from warm box, then the total increment of heat flux due to the slot is

$$Q_T = \eta Q_E \tag{1}$$

where

$$\eta = \eta_R \, \eta_T \, \eta_C \tag{2}$$

and η is the total enhancement factor, which unifies the enhancement factor for multi-reflection η_R, the enhancement factor for temperature distribution change η_T, and the cavity factor $\eta_C \cdot \eta$ can be determined by a combination of theoretical calculations and experimental results. Once η is determined, the total enhancement of the heat flux due to cracks can be calculated using Eq. (3) from Reference 3:

$$Q_T = \eta \sum_{i=1}^{N} \varepsilon_2 \alpha T_2^4 \left[S_i \left(F_{S_i - A} + F_{S_i - B} + F_{S_i - C} \right) \right] \tag{3}$$

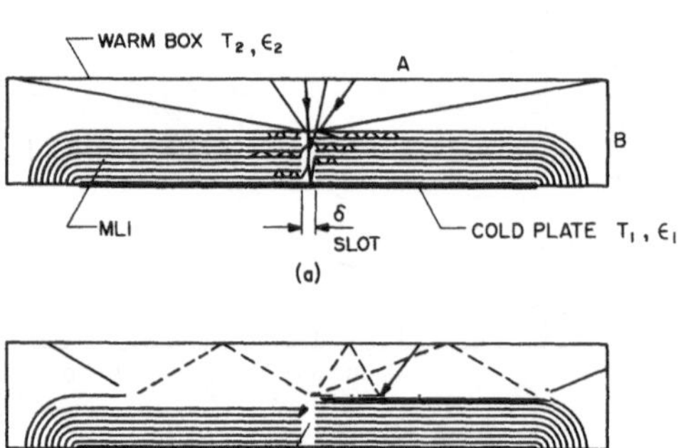

Fig. 3. Radiation flux entering crack in MLI; (a) directly, (b) by reflection.

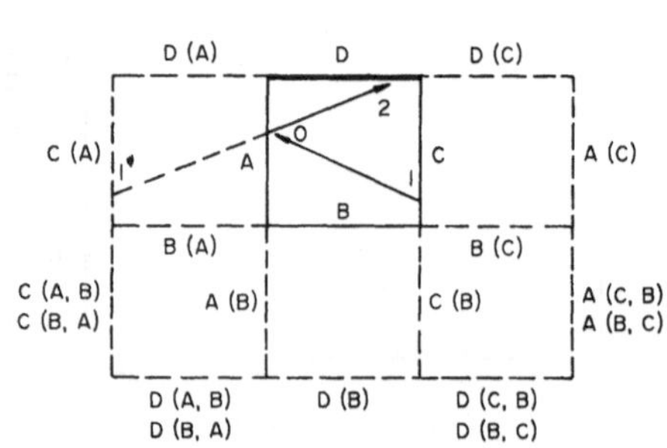

Fig. 4. Theoretical approximation of a slot in a MLI blanket.

where N is the number of slots, α is the Stefan-Boltzman constant; $\mathcal{E}_2$ is the emissivity of warm box; T_2 is the temperature of warm box; S_i is the area of slot i, and F is the corresponding view factor. Figure 5 is an example of the calculated of heat load through the crack for our experimental apparatus.

CRACK COVERING TECHNIQUE USING PATCHES

Most of the results of the work to reduce the effects of cracks by using patches to cover them are shown in Fig. 6. We optimized the number and distribution of the patches, and studied two different patch materials.

Optimum Patch Distribution

Four data runs were made to find the optimum patch distribution. DAM patches were used, and as shown in Fig. 6d-g, could reduce the heat flux to approximately that of a MLI blanket without cracks. Locating a few patches in the outer (warm) half of the blanket, is as effective as a uniform distribution of patches and better than patches in the inner (cold) half. Figure 6m shows that locating all patches on the top of the crack is not effective.

Optimum Number of Patches

Having quantitatively determined that patches in the warm part of the blanket were very effective in reducing the effect of cracks, the number of patches was investigated. The results are shown in Figure 6h-m. It appears that between four and six patches is the minimum necessary to achieve a significant reduction in the heat flux, and the effect becomes marginal above about six. The optimum, for this geometry at least, is six patches.

Patch Material

Figure 6 shows that 2 × 500 angstrom DAM may be more effective as a patch material than 300-angstrom SAMC. For an identical, uniform distribution of patches, i.e. Fig. 6g and i, it is quite clear that the DAM is preferable.

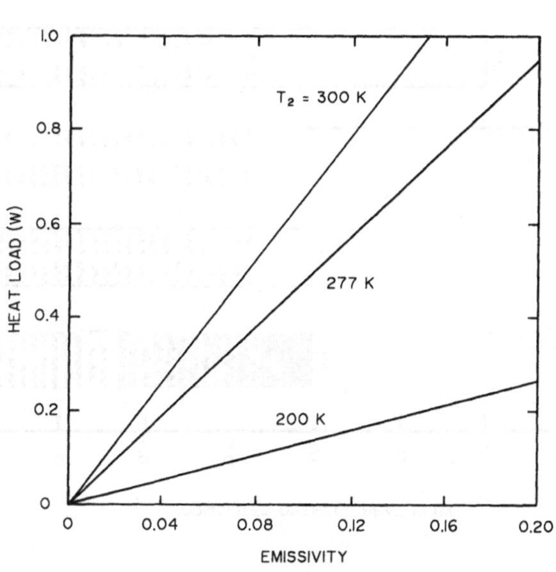

Fig. 5. Heat load as a function of the temperature and emissivity of warm surface.

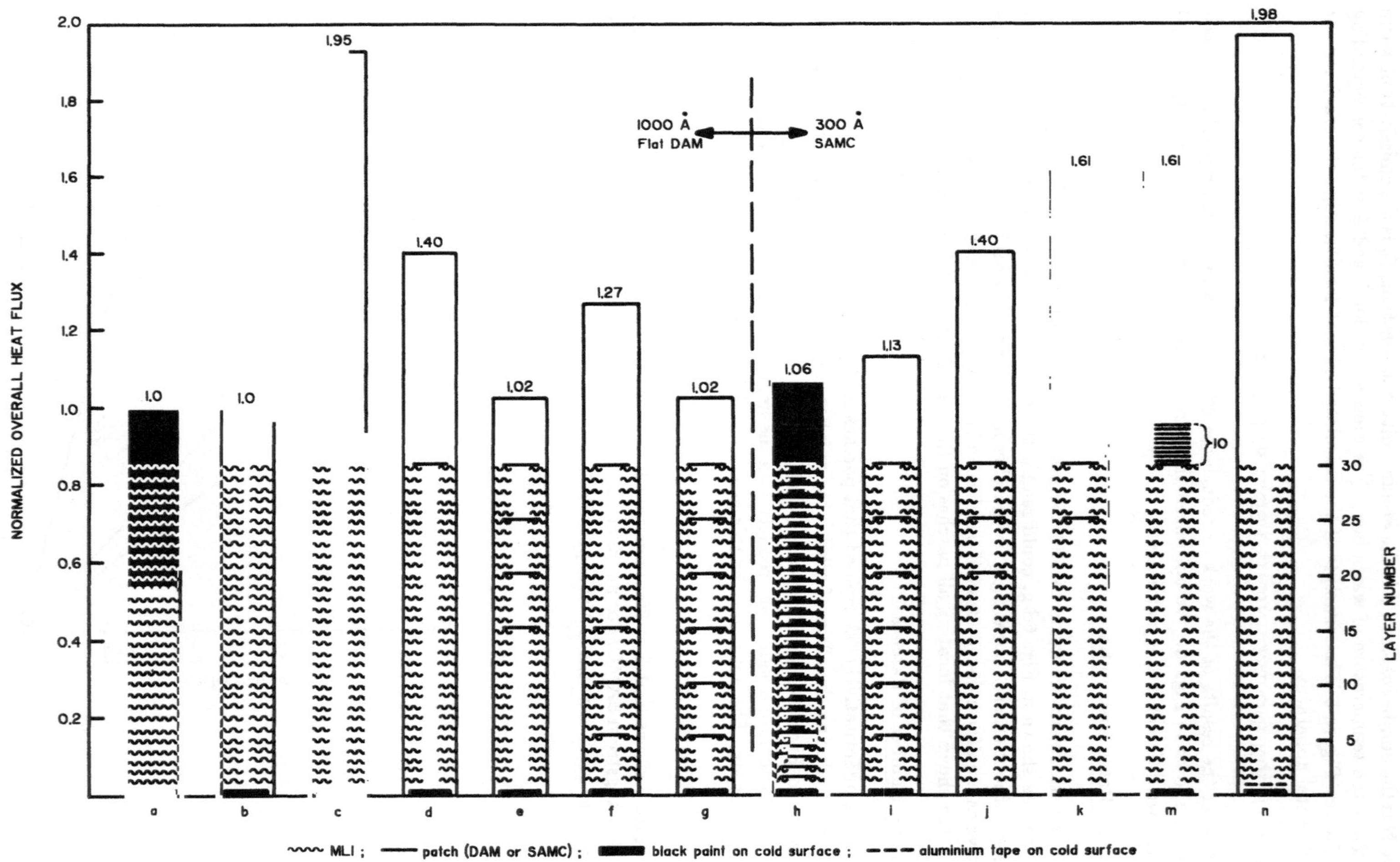

Fig. 6. Graphic summary of experimental patch study. a, no cracks; b, one-dimensional slits; c to m, various patch geometries; n, no patches.

Patches on Square Hole

The square holes in the 30-layer MLI blanket were covered by four patches, located at the 15th, 20th, 25th and 30th layers. This reduced the overall heat flux from 1.57 W/m^2 to 0.83 W/m^2, which is the same as that for DAM patches on slots of the same area.

TEMPERATURE DISTRIBUTION AND EQUIVALENT THERMAL CONDUCTIVITY

The temperature distribution in MLI near cracks is a sensitive function of the crack width and the patch locations among the layers. This provides considerable information on the heat transfer mechanism through the insulation blanket. A typical temperature distribution, Fig. 7, shows that the presence of cracks changes the entire temperature distribution. The local equivalent thermal conductivity in the vicinity of the cracks was calculated from the temperature distribution, and is shown in Fig. 8. From these data, the temperature in the last patch (on the warmest layer) is an indication of the improvement in thermal performance resulting from the patches. Qualitatively, the higher the temperature of the last patch, the smaller the heat flux transmitted through the crack.

CONCLUSIONS

The quantitative effects of cracks in MLI blankets on the heat load to a cryogenic device are dependent on the MLI material, the geometry of the device and the care with which the MLI is applied. Nevertheless, the experimental results and theoretical model presented here lead to conclusions which should be of general interest. These can be summarized as follows:

1. A significant increase in the heat flux will be caused by cracks. The mean equivalent thermal conductivity of a narrow crack between room temperature and 77 K is 3-8 W/m-K. The heat flux through a crack is a function of the aspect ratio of the crack, with a maximum of about 150 W/m^2.

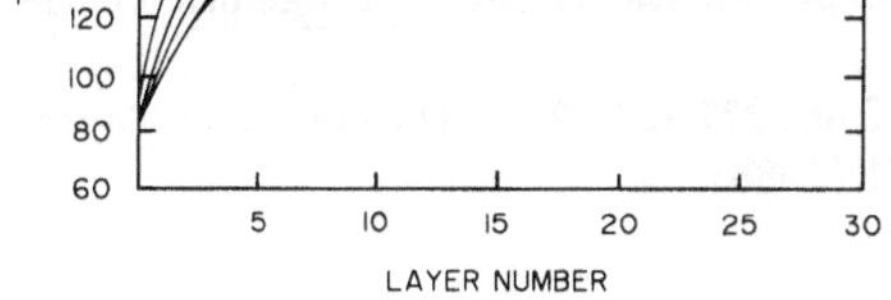

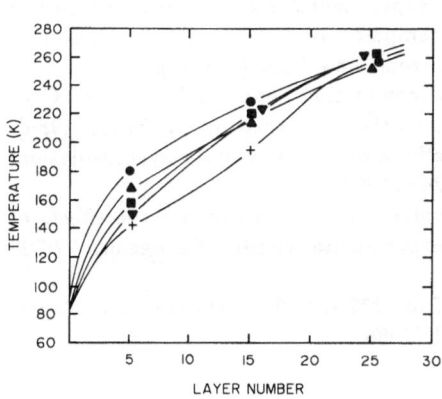

Fig. 7. Temperature distributions for slots with DAM patches ● , slots without patches; ▲ , single patches on layer 30; +, patches on layers 15, 20, 25, 30; ■ , patches on layers 5, 10, 15, 30; ▼, patches on layers 5, 10, 15, 20, 25, 30.

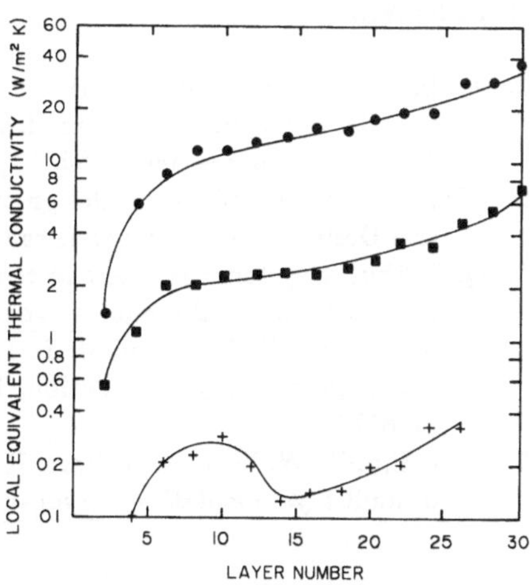

Fig. 8. Local equivalent thermal conductivity for slots with DAM patches as a function of depth in blanket. ● , slots without patches; +, patches after layers 15, 20, 25, 30; ■ , patches after layers 5, 10, 15, 30.

2. According to the Enhanced Black Cavity Model, the unexpectedly large heat transfer can be attributed to: (a) the crack acting like a black cavity and, (b) the multi-reflection of the radiation flux outside the crack. The temperature distribution in a blanket with cracks corresponds to greater heat transfer. There is a geometric factor, determined by the dimensions of the crack, which parameterizes the black cavity effect. Once this enhancement factor is determined, the heat increment due to the cracks can be calculated.

3. The use of aluminized Mylar patches to cover the crack at each layer is good, but may not be easy to install. On the other hand, the use of flat, DAM, 1000 angstrom patches on a few layers will give almost the same improvement as patches between each layer. Placing the patches in the outer (warmer) half of the blanket is much better than in the inner (colder) half. Putting patches on the outside of the crack is not very effective.

4. Mylar with more aluminizing (1000 angstroms) is better as a patch material. Crinkled Mylar with 300-500 angstrom single aluminizing is easier to install since spacers are not required.

5. Reducing the emissivity of the cold surface under a narrow crack does not significantly effect the heat flux.

6. The presence of cracks and patches changes the temperature distribution in the MLI blanket near the cracks. The temperature of the layer closest to the cold surface increases and that of the layer closest to the warm surfaces decreases as the width of the slots increases. Patches have the effect of lowering the temperature of the MLI blanket near a crack below that for an exposed slot. The higher temperature of the last (warmest) patch indicates the ability of a patch system to protect the cold surface against incoming radiation.

7. The local equivalent thermal conductivity of a crack is a sensitive function of the distribution and number of patches along the crack. It is a minimum around the first patch from the cold surface, and a maximum a few layers after the first patch.

REFERENCES

1. Q. S. Shu, R. W. Fast, and H. L. Hart, An experimental study of heat transfer in multilayer insulation systems from room temperature to 77 K, in: "Advances in Cryogenic Engineering," Vol. 31, Plenum Press, New York (1986), p. 455.
2. "Preliminary Report on the Design of the Superconducting Super Collider," SSC Central Design Group, Universities Research Association, Berkeley, California (1986).
3. Q. S. Shu, A systematic study to reduce the effects of cracks in multilayer insulation, Part 1: theoretical model, Cryogenics 27:249 (1987).
4. Q. S. Shu, R. W. Fast, and H. L. Hart, A systematic study to reduce the effects of cracks in multilayer insulation, Part 2: experimental results, Cryogenics 27:298 (1987).
5. Q. S. Shu, R. W. Fast, H. L. Hart, Heat Flux from 277 K to 77 K through a few layers of multilayer insulation, Cryogenics 26:671 (1986).

CRACK COVERING PATCH TECHNIQUE TO REDUCE THE HEAT FLUX FROM 77 K TO 4.2 K THROUGH MULTILAYER INSULATION*

Q.S. Shu,† R.W. Fast, and H.L. Hart

Fermi National Accelerator Laboratory
Batavia, Illinois

ABSTRACT

The effects of cracks in a multilayer insulation blanket on the heat load from 77 K to 4.2 K is a serious problem, but less so than from room temperature to 77 K. The technique of patching the cracks and the enhanced black cavity model developed to reduce heat flux through cracks in a MLI blanket to a 77 K surface were applied to the 77-4.2 K situation. The optimized patch covering technique is also very effective in this temperature region. The heat load through cracks in a MLI system between 77 K and 4.2 K, calculated by means of the enhanced black cavity model, is in agreement with the experimental data.

The temperature distributions were measured, and the corresponding equivalent thermal conductivity of the cracks were deduced. Both of these quantities were quite different from those measured at higher temperatures.

INTRODUCTION

The theory and techniques to reduce the effects of cracks in MLI insulation on the heat load from room temperature to 77 K have been investigated experimentally.[1,2] It was a conclusion of that study that cracks in a MLI blanket seriously degrade the thermal performance and cause an unexpectedly large heat load through the insulation system. It was also noted that an optimized crack-covering technique could improve the thermal performance of a MLI system with cracks to that without cracks in that temperature region. It is well known that there are three heat transfer mechanisms in MLI systems: residual gas molecular conduction, radiation, and solid conduction between the adjacent layers. The relative importance of each is a function of temperature. Because of the low heat of vaporization of liquid helium, any degradation of the MLI on superconducting devices can have serious operational consequences. It was obvious, therefore, that the effect of cracks on the heat load from 77 K to 4.2 K should be examined and patches applied to the cracks to achieve reduced heat leak.

Since we believed that the basic results of the earlier study might be generally applicable to the 77-4.2 K temperature region, only a single crack width and a single patch geometry were tested between 77 K and 4.2 K.

---

*Work sponsored by the Universities Research Association, under contract with the U.S. Department of Energy.

†Visiting Scientist on leave from Cryogenic Laboratory, Zhejiang University, Hungzhou, People's Republic of China.

EXPERIMENTAL ARRANGEMENT

The experimental arrangement was the same as that described in an earlier paper.[3] The central fin, wrapped with 30 layers of MLI, was refrigerated to 4.2 K by liquid helium through a thermosiphon tube from the boiloff vessel, which was guarded by another liquid helium vessel. The outer box was maintained at 77 K by two thermosiphon tubes from a liquid nitrogen vessel. The accuracy of the wet test meter used for the helium gas boiloff is ±0.2%. The correction factor to convert the boiloff data to evaporation rates is 1.154. The thermocouples to measure the temperature distribution in the MLI blanket were Chromel/gold 0.07% iron and were installed in the same way as before.[2]

Fourteen slots, 254 mm long by 4 mm wide, were cut in the MLI blanket as shown in Figure 1. The patch material was 1000 angstrom double aluminized flat 6.35-μm Mylar.

EXPERIMENTAL RESULTS

Effect of Slots on Heat Load

The heat load from the 77 K copper box to the 4.2 K black painted fin (area $A = 2.24\ m^2$) through a 30-layer MLI blanket without cracks, Q_0, was 29.9 mW and the corresponding heat flux, Q'_0, (Q_0/A) was 13.2 mW/m^2. The total heat load measured with slots in the MLI, Q_C, was 32.42 mW. The heat load through slots defined as $Q_S = Q_C - Q_0$ was 2.52 mW. Since total area of the slots A_S was 0.014 mm^2, the heat flux through the slots Q'_S defined as Q_S/A_S was 180 mW/m^2. The heat flux through slots was therefore about 14 times that through a MLI blanket without slots ($Q'_S/Q'_0 = 14$).

Patch Covering Techniques

To quantitatively verify the patch covering technique at the lower temperatures, the slots on layers 15, 20, 25 and 30 were covered with patches. The measured heat load was 30.3 mW, which is approximately equal to the original heat load through the MLI blanket without slots.

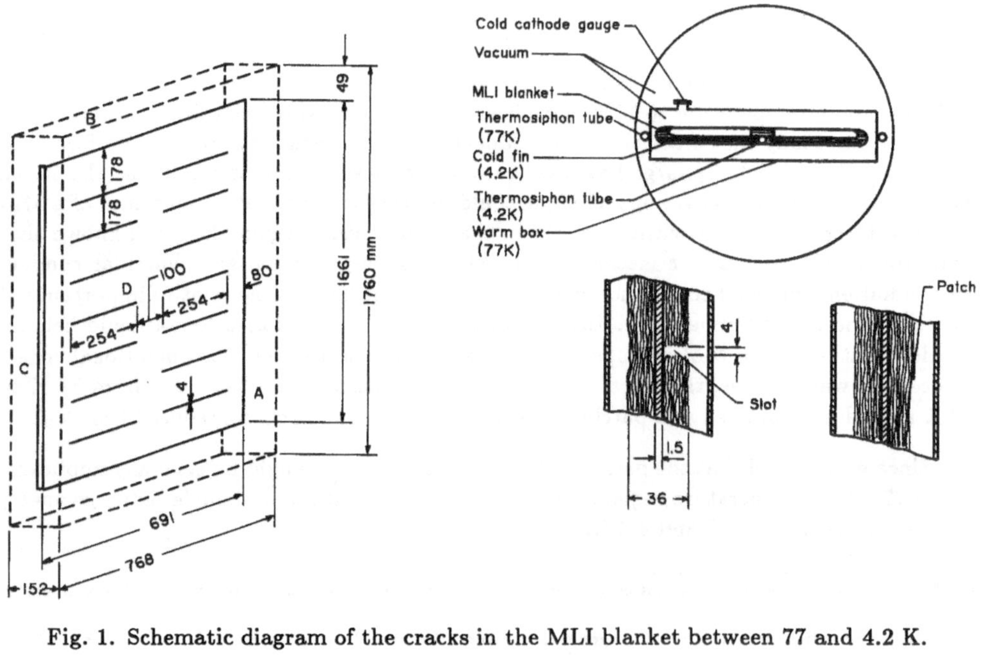

Fig. 1. Schematic diagram of the cracks in the MLI blanket between 77 and 4.2 K.

Temperature Distribution and Thermal Conductivity

Figure 2 shows the temperature distribution in the 30-layer MLI blanket at a vacuum pressure of 5×10^{-8} torr for three different arrangements: (1) without slots, (2) with slots, and (3) with patches covering the slots on layers 15, 20, 25 and 30. Figure 3 gives the equivalent thermal conductivity of the slots, defined as $Q_S \cdot \Delta N/A_S\Delta T_N$, where $\Delta T_N/\Delta N$ is the local temperature gradient. It can be seen that the thermal conductivity in either the inner or outer portions of the MLI blanket is less than that observed in layers 12 to 22. The maximum conductivity of the slots is more than one order of magnitude greater than the conductivity of the blanket without slots. The values with patches covering the slots are about the same as without slots over the entire temperature range. Figure 4 shows the equivalent thermal conductivity as a function of temperature in a MLI blanket.

Temperature Difference Between Outer Layer and Warm Box

The earlier experiments,[2] from room temperature to 77 K, showed that the temperature difference between the outermost layer and the warm box was less than 2 K. However, the 77 to 4.2 K data showed a temperature difference of about 7 K.

CALCULATION OF HEAT LOAD THROUGH CRACKS

The heat load from the 77 K box to the 4.2 K surface through cracks can be calculated by means of the enhanced black cavity model.[6] The following calculation shows how the model

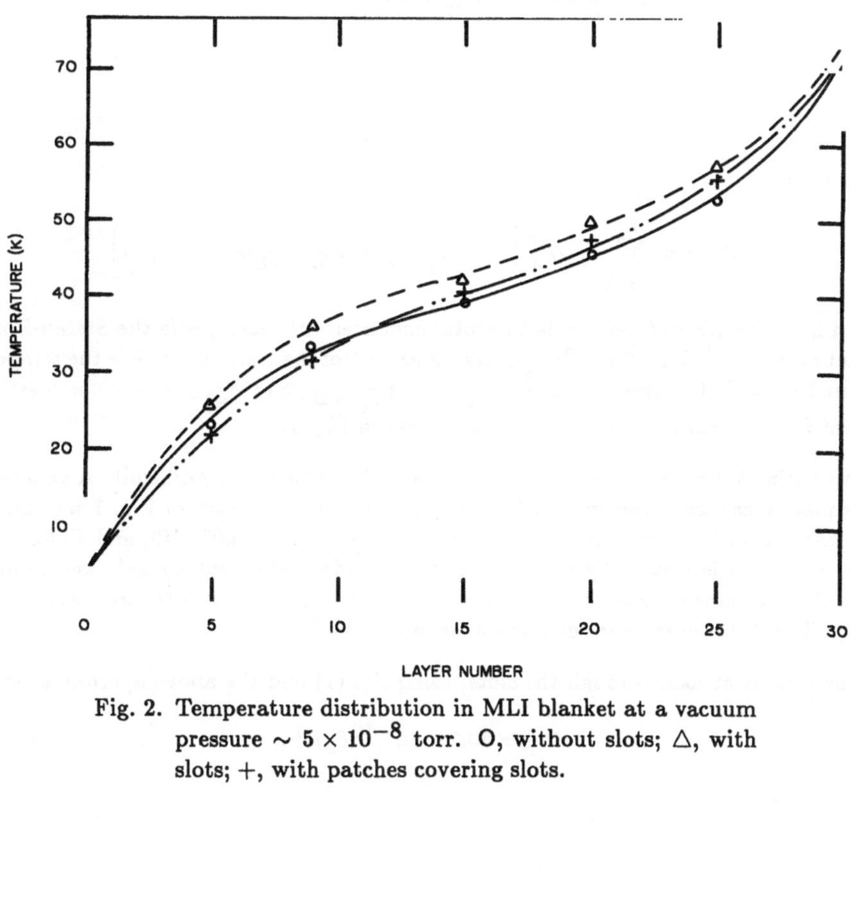

Fig. 2. Temperature distribution in MLI blanket at a vacuum pressure ~ 5×10^{-8} torr. O, without slots; Δ, with slots; +, with patches covering slots.

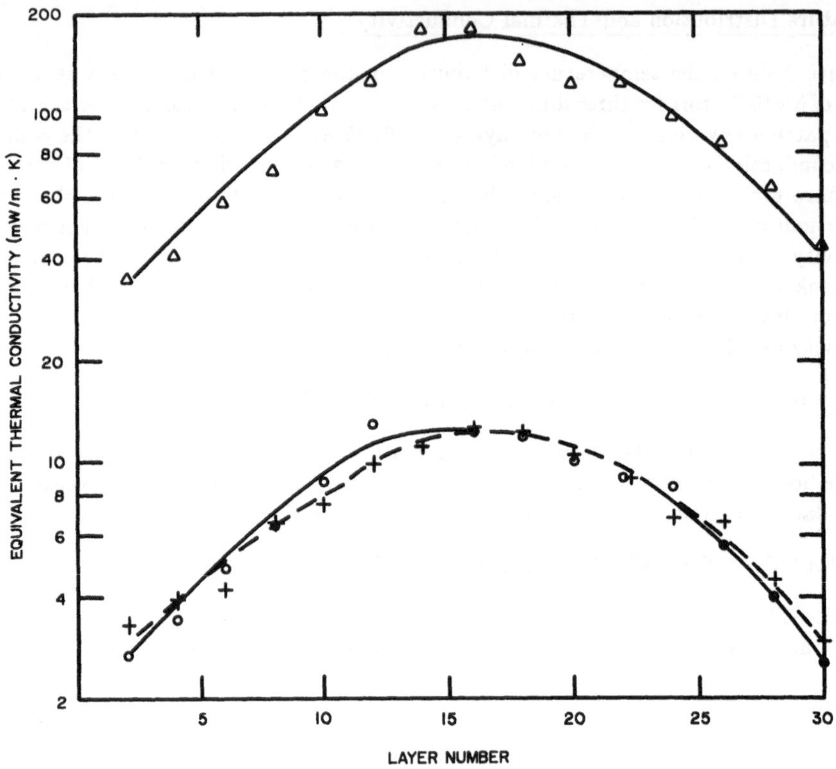

Fig. 3. Equivalent thermal conductivity as a function of depth in a MLI blanket. O, without slots; Δ, with slots; +, with patches covering slots.

may be used.

$$Q_S = \eta \sum_{i=1}^{n} \mathcal{E}_2 \alpha T_2^4 \left[S_i \left(F_{S_i-A} + F_{S_i-B} + F_{S_i-C} \right) \right] \tag{1}$$

where n is the number of slots, η is the total enhancement factor; α is the Stefan-Boltzman constant 5.67×10^{-8} Wm^{-2} K^{-4}; $\mathcal{E}_2$ is the emissivity of the warm box; T_2 is the temperature of warm box; S_i is the area of slot i. F_{S_i-A}, F_{S_i-B}, and F_{S_i-C} are the view factors from slot i to the warm walls A, B, and C, shown in Fig. 1.

Evaluating the quantity inside the brackets of Eq. (1) is usually difficult. A general series approximation can be found in Ref. 5. A solution for the geometry of Fig. 1 was calculated earlier and are as follows: $S(F_{S-A}) = 0.013$, $S(F_{S-2B}) = 0.000042$, and $S(F_{S-2C}) = 0.0014$. Each slot is viewed by both end and both side plates, but by only one front plate. The total enhancement factor can be determined by a combination of theoretical calculations and experimental results. For our apparatus, η was 3.8.[6]

The total heat load through the slots, using Eq. (1) and the above approximation

$$Q_S = 30.7 \times 10^{-10} \mathcal{E}_2 T_2^4 \, . \tag{2}$$

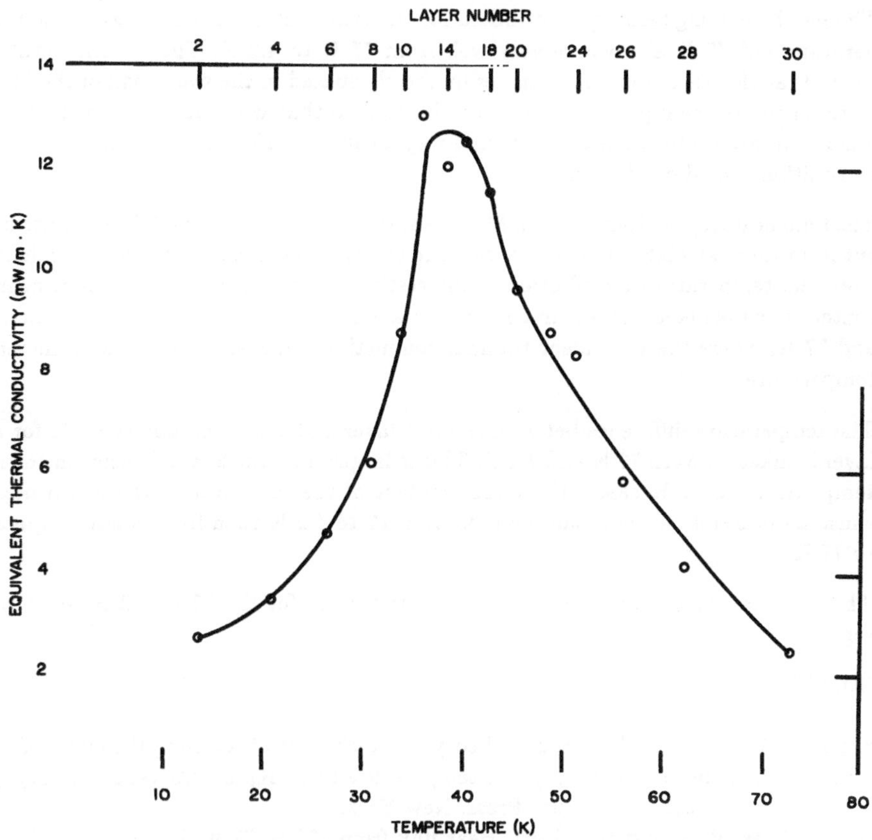

Fig. 4. Equivalent thermal conductivity as a function of temperature in a MLI blanket.

Assuming $\mathcal{E}_2 = 0.03$, the total heat load through the slots from the 77 K warm box to the 4.2 K surface is 3.04 mW. If $\mathcal{E}_2 = 0.02$, then $Q_S = 2.1$ mW; the dependence of Q_S on the emissivity is $100\mathcal{E}_2$.

DISCUSSION AND CONCLUSION

From the above experiments, it is obvious that the theory and the technique developed to reduce the effects of cracks in a MLI system between room temperature and 77 K is useful for the 77 to 4.2 K temperature region. The basic conclusions are:

1. The heat flux through cracks in a 30-layer MLI blanket between 77 K and 4.2 K is 180 mW/m^2, which is fourteen times the flux through the same MLI blanket without cracks.

2. Compared to the case of a MLI blanket between room temperature and 77 K, where the heat flux through cracks is nearly two hundred times that through the blanket, the effect from 77 K to 4.2 K is less. One of the reasons is that the heat transfer in a MLI system between room temperature and 77 K at very low vacuum pressure is primarily by radiation, while solid conduction plays a more important role between 77 K and 4.2 K.

3. The patch covering technique optimised for the temperature region between room temperature and 77 K also works very well from 77 K to 4.2 K. The experimental data shows that the use of just four patches uniformly spaced in the warm half of the blanket restored the thermal performance of the blanket to that without cracks. On the other hand, the careless installation of patches might cause a serious increase in heat load due to additional solid conduction.

4. The temperature distributions in a MLI blanket between 77 K and 4.2 K are quite different from those at higher temperatures. The equivalent thermal conductivity calculated from the temperature distribution is interesting: It has a broad maximum near the center of the blanket. This is in contrast to a MLI blanket between room temperature and 77 K, where the equivalent thermal conductivity always increases with increasing temperature.

5. The temperature difference between the last layer and the warm box is ~7 K for a 30-layer blanket between 77 K and 4.2 K. This is larger than the 2-K difference in the room temperature to 77 K case. Therefore, contact of the last layer to the warm box will cause a more serious heat load increase from 77 to 4.2 K than from room temperature to 77 K.

6. The enhanced black cavity model is also satisfactory for the 77 to 4.2 K temperature region.

REFERENCES

1. Q.S. Shu, R.W. Fast, and H.L. Hart, Theory and technique of reducing the effect of cracks in multilayer insulation from room temperature to 77 K, in: "Advances in Cryogenic Engineering," Vol. 33, Plenum Press, New York.
2. Q.S. Shu, R.W. Fast, and H.L. Hart, Heat flux from 277 to 77 K through a few layers of multilayer insulation, Cryogenics 26:671 (1986).
3. E.M.W. Leung et al., Techniques for reducing radiation heat transfer between 77 and 4.2 K, in: "Advances in Cryogenic Engineering," Vol. 25, Plenum Press, New York, p. 489.
4. Q.S. Shu, R.W. Fast, and H.L. Hart, Systematic study to reduce the effects of cracks in multilayer insulation. Part 2: experimental results, Cryogenics 27:298 (1987).
5. R. Siegel, "Thermal Radiation Heat Transfer," NASA Report SP-164, U.S. Government Printing Office, Washington (1971).
6. Q.S. Shu, Systematic study to reduce the effects of cracks in multilayer insulation. Part 1: theoretical model, Cryogenics 27:249 (1987).

SYSTEMATIC ANALYSIS OF CHARACTERISTICS FOR DIFFERENT TYPES OF MULTILAYER INSULATIONS

M. Taneda, T. Ohtani, M. Okuda, J. Tsukuda

Mechanical Engineering Research Laboratory
Kobe Steel, Ltd.
Kobe, Japan

ABSTRACT

Thermal characteristics for different types of multilayer insulation(MLI) were experimentally studied to correlate these characteristics with various constructions, shapes, and sizes. In this study the heat flux and the temperature distribution of MLI between 300K and 77K were measured, using a cylindrical calorimeter. In addition, the effects of layer density and packing pressure were observed. Then, a simple MLI model was used to analyse the contributions of radiation, solid conduction and gas conduction to the thermal performance of MLI as observed in this experiment. Then an effort was made to obtain a correlation that would indicate the performance of all types of MLI used in this study. Four MLI types were tested: type A (flat double aluminized polyester film + nylon net spacer with a larger mesh); type B (flat double aluminized polyester film + polyester net spacer with a smaller mesh); type C (crinkled double aluminized polyester film [larger crinkle] + no spacer); and type D (dimpled double aluminized polyester film [smaller crinkle] + no spacer). Experimental results showed that, at the same layer density, the heat flux of MLI with spacer was greater than that of MLI without spacer. Furthermore, at the same packing pressure, the heat flux of MLI of smaller size was greater than that of larger size.

INTRODUCTION

Multilayer insulation (MLI) is often used for cryogenic equipments, and the performance of MLI has been studied for a long time. Few efforts seem to have been made, however, to analyze the relation between diferent types of MLI. Little is known, for example, about the difference between MLI with large mesh spacer and MLI with small mesh spacer, or between MLI using deep crinkling and MLI with shallow crinkling. Therefore comparison tests were made on the different types of MLI, in particular in regard to spacer mesh size and reflector projection, in order to correlate the characteristics of MLI with various constructions, shapes, and sizes.

EXPERIMENTAL APPARATUS AND METHOD

A cylindrical calorimeter was used to mesure the heat flux of the MLI. Fig.1 shows the design of the experimental apparatus, which is similar to that described by Kropschot[1] and Scurlock[2] The inner vessel was constructed with a center test chamber and with upper and lower guard chambers, all made of stainless steel. Copper skirts were inserted between the test chamber and the guard chambers, and soldered to the guard chamber. A distance of 2 mm was left between the test chamber and the skirt. MLI was wound on the test chamber and upper and lower guard chamber in thirty layers, as shown in Fig.1. Copper constantan thermocouples were inserted between every fifth and sixth layer. The temperature ranged from 300K (room temperature) to 77K (the temperature of liquid nitrogen). The upper and lower guard chambers were filled with liquid nitrogen. The vaporization rate of liquid nitrogen in the test chamber was measured by a wet flow meter. The pressure in the guard chamber was maintained at about 50 mmH_2O, higher than that in the test chamber, in order to prevent the vaporized gas from recondensation through the guard chamber. The pressure in the vacuum chamber was measured with an ionization gauge at the center of the cryostat.

Because of the large amount of MLI wound around the test chamber, the distance between the surface area of the test chamber and the surface area of the outermost MLI film was significant. Therefore the logarithmic mean of the two surface was used to evaluate the performance of the MLI regarding the heat flux per unit area.

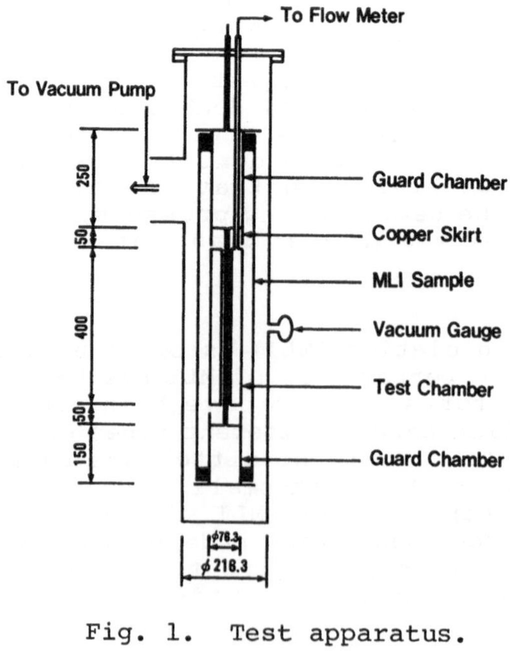

Fig. 1. Test apparatus.

TYPES OF MLI

Four types of MLI were used in this experiment, represented here as types A, B, C and D. In order to study the size effects of MLI, Table 1 shows the various details of type and size. Types A and B had both reflectors and spacers. Types C and D had only reflectors (crinkled or dimpled). The size of net spacer of type A was larger than that of type B, i.e., the mesh density of the net spacer of type A was smaller and the diameter of the thread was larger. The size of crinkle of type C was larger than that of type D, i.e., the height of the crinkle for type C was greater than that of type D, and the density of crinkle was lower.

EXPERIMENTAL RESULT

Layer density and heat flux

The heat flux was measured for each type of MLI, using thirty layers of MLI in each case. Fig.2 shows the experimental results. The temperature range was from 300K to 77K. Test pressure was 1x10^{-} Torr. The results showed that at low layer density, heat flux of types A,C and D were about 0.3 W/m2. The heat flux of type B would also be expected to be the same value. In this layer density region, heat flux is mainly governed by radiation heat transfer.

It is furthermore important to notice that the heat flux of types A and B(which had spacers) was larger than that of types C and D (which had no spacers) at the same layer

Table 1. Detail of MLI Tested in This Study

Type	Reflector	Spacer
A	Double Aluminized Polyester Film Shape : Flat Thickness : 25 μm	Nylon Net Mesh Density : 0.44 1/cm^2 Thread Diameter : 0.3 mm
B	Double Aluminized Polyester Film Shape : Flat Thickness : 25 μm	Polyester Net Mesh Density : 80 1/cm^2 Thread Diameter : 0.05 ~ 0.1 mm
C	Double Aluminized Polyester Film Shape : Crinkled Thickness : 25 μm Height of Crinkle : ~ 600 μm Density of Crinkle : 12~13 1/cm^2	No Spacer
D	Double Aluminized Polyester Film Shape : Dimpled Thickness : 25 μm Height of Dimple : ~ 300 μm Density of Dimple : 20 1/cm^2	No Spacer

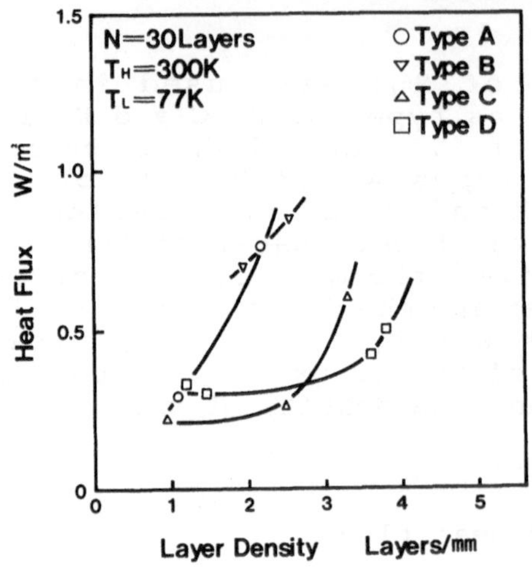

Fig. 2. Heat flux vs. layer density.

density. This may be the result of the fact that when spacers are used, a larger force is needed to maintain a constant density, and this force increases the area of contact which in turn leads to increase solid conduction heat transfer.

Packing pressure and heat flux

In experiments to obtain the relation between packing pressure and heat flux, it was not possible to determine the packing pressure directly. Therefore a packing test was carried out as shown in Fig.3. Thirty layers of MLI were stacked on a flat table, covered with a flat plate. A load varying in weight up to a few kilogram was imposed on the top of this pile. By measuring the total height of the MLI and the weight of the load, it was possible to determine the correlation between the packing pressure and the layer density. Fig.4 shows the relation between packing pressure and MLI height (30 layers). On the basis of these results, Fig.2 was redrawn as Fig.5, where packing pressure was used as the horizontal axis. As shown in this figure, the heat flux of types B and D (i.e. those of smaller size) were larger than that of types A and C (the larger size) at the

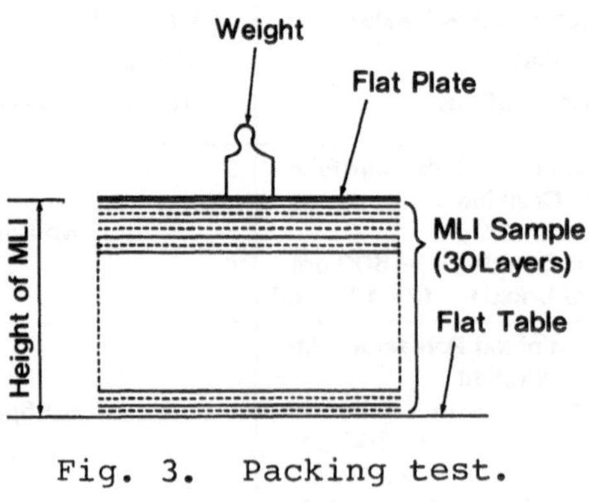

Fig. 3. Packing test.

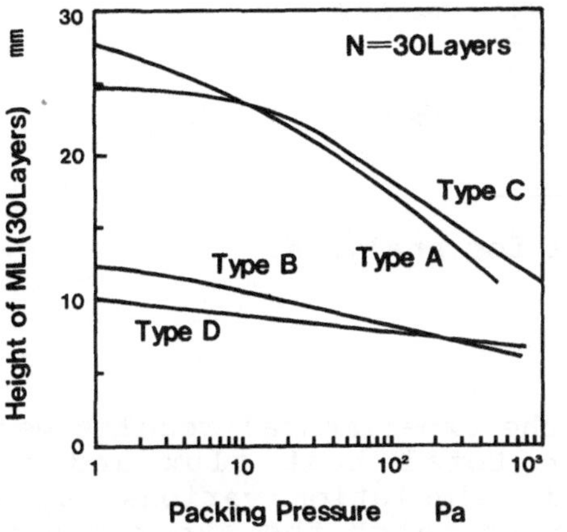

Fig. 4. Packing pressure vs. height of MLI (30 layers).

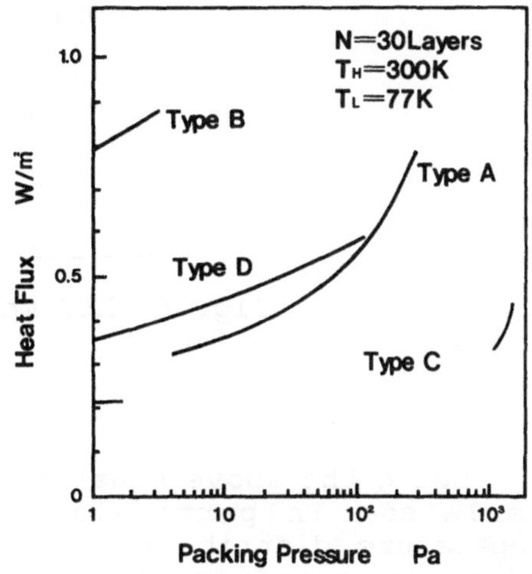

Fig. 5. Packing pressure vs. heat flux.

same packing pressure. Presumably this is because types B and D had more contact points and therefore greater contact area, leading to an increase in solid conduction heat transfer.

Temperature distribution

The temperature distribution of MLI is a very important factor in making it possible to analyze the heat flux through MLI in terms of radiation and solid conduction separately. First, an MLI model was desined. Then the radiation heat flux and solid conduction heat flux were calculated for each separate layer. Here, gas conduction heat flux was ignored because the test pressure was 1×10^{-6} Torr, so that the rate of gas conduction heat flux was very low. Fig.6 shows the MLI model which was used for the analysis. This model shows the two reflectors and the pillars between them. In this model, the total heat flux, Q is described by this equation,

$$Q=\pi D_i H\{E_i\sigma(T_{i+1}^4+T_i^4)+K_i(T_{i+1}-T_i)\}$$

$$E_i=\frac{1}{\frac{1}{\varepsilon_i}+\frac{D_i}{D_{i+1}}\left(\frac{1}{\varepsilon_{i+1}}-1\right)}$$

$$K_i=\frac{A}{R_{1,i}+R_{2,i+1}}$$

where Di is the i-th reflector's diameter, σ is the Stefan-Boltzmann constant, ε_i is the i-th reflector's emissivity, H is vertical length, A is contact area, R_i represents contact thermal resistance between spacers and reflectors, and T_i is the i-th reflector's temperature. The first term on the right hand represents radiation heat flux, and second term presents solid conduction heat flux. The effect of the thermal conductivity of the spacers was ignored, since it was very small compared to R_i.

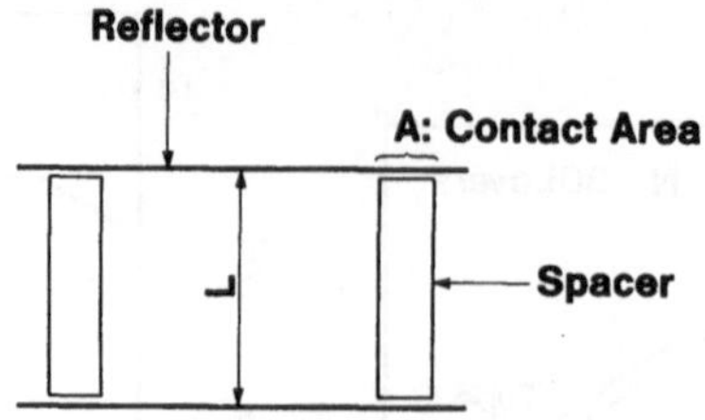

Fig. 6. MLI model for analysis.

Using the above equation, the experimental results were simulated, in particular, the total heat flux and the temperature distribution. In this simulation various number values were applied to ε and K to match the experimental results. Fig.7 shows both the experimental and the simulated results for temperature distribution for type A. This figure shows that, if the packing pressure increases, solid conduction heat flux also increases, with the result that the temperature distribution goes down. This emissivity were reported by Cunington et al[3]. For types B and D, this emissivity also showed a close correlation with the experimental results. For type C, however, the emissivity was about 40% less than that for other types. It is surmised that the reason for this is the fact that the shape of the type C reflector is really uneven, i.e., it should not be assumed to be flat, as shown in Fig.6. In this simulation K was assumed to be constant across the whole temperature range. But from Fig.7 it can be observed that the simulated results for temperature distribution are lower than the

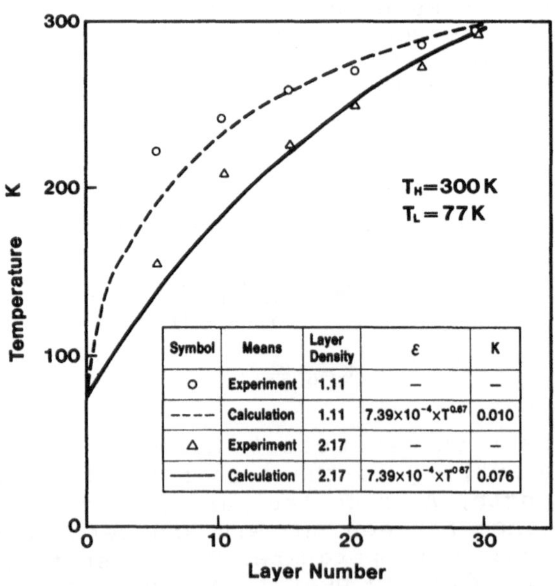

Fig. 7. Temperature distribution of MLI for Type A.

experimental results in the lower temperature range. It may be necessary to reconsider the temperature dependence of K.

SUMMARY OF RESULTS

The results of the experiments, then, can be summarized as follows.

1. At the same layer density, the heat flux of MLI with spacers (types A and B) was greater than that without spacer (types C and D).

2. At the same packing pressure, the heat flux of MLI of smaller sizes (types B and D) was greater than that for larger sizes (types A and C).

3. The emissivity for the flat reflector was the same as that reported, but the apparent--though not true--emissivity for the reflector with deep crinkling was lower.

REFERENCES

1. R.H. Kropschot, Cryogenics 1:171 (1961).
2. R.G. Scurlock and B. Saull, Cryogenics 5:303 (1976).
3. G.R. Cunnington and C.L. Tien, AIAA Paper No. 69-607 (1969-6).

experimental results in the lower temperature range. It may be necessary to reconsider the temperature dependence of K.

SUMMARY OF RESULTS

The results of the experiments, then, can be summarized as follows:

1. At the same layer density, the heat flux of MLI with spacers (types A and B) was greater than that without spacers (types C and D).

2. At the same packing pressure, the heat flux of MLI of smaller sizes (types B and D) was greater than that for larger sizes (types A and C).

3. The emissivity for the flat reflector was the same as that reported, but the apparent thermal emissivity for the reflector with deep crinkling was lower.

REFERENCES

1. R.H. Kropschot, Cryogenics [illegible] (1961).
2. R.G. Scurlock and B. Saull, Cryogenics 8:303 (1976).
3. G.R. Cunnington and C.L. Tien, AIAA Paper No. 69-607 (1969).

TEST OF MULTILAYER INSULATIONS FOR THE USE IN THE SUPERCONDUCTING PROTON-RING OF HERA

H. Burmeister, W. Eschricht, G. Horlitz,
D. Sellmann

Deutsches Elektronen-Synchrotron DESY
Hamburg, Germany

Ye Jia-ding

Cryogenic Laboratory, Chinese Academy of
Sciences, Beijing, China

ABSTRACT

For large cryogenic distribution systems the choice of the optimal thermal insulation is of great importance. In order to select a design of multilayer insulation, which meets all thermodynamical and mechanical requirements of both the 6.4 km helium transferline and the magnet ring of the HERA-superconducting proton accelerator, several kinds of organic and inorganic shield materials in connection with different kinds of spacer materials have been tested. A short description of the testing facility is given. Results of the heat loss rate measurements from ≈ 300 K to ≈ 77 K and gamma ray irradiation tests are reported.

INTRODUCTION

For the thermal insulation of the superconducting magnets, and the more than 6 km transferline connecting the central he-refrigerator with the magnets in the ringtunnel of the new accelerator HERA at DESY, a high-grade and radiation resistant multilayer insulation is necessary. A radiation dose of 5×10^5 Gy/a for elements in the ringtunnel is caused mainly by the electron accelerator synchrotron radiation. This source is about two orders of magnitude more intense than the proton accelerator, so the usability of organic materials appears to be uncertain. To attain the specified total loss rates of about 1.3 W/m² in the shield circuit between 300 K and 60 K, the insulation of the shield should not have losses of more than 1 W/m². This is not easy to achieve with a complete inorganic insulation. Gamma irradiation experiments with aluminized mylar foils prove the usability of this material up to a radiation dose of 5×10^6 Gy. So we investigated in the main aluminized mylar foils in our loss rate tests.

IRRADIATION EXPERIMENT

For realistic irradiation conditions, the foil samples were exposed in a small vacuum vessel to gamma radiation (6×10^3 Gy/h) from the positronconverter of the LINAC II at DESY. The total absorbed dose was $5.7 \pm 1.2 \times 10^6$ Gy, determined by RPL-glass dosemeters. The gas developed by the irradiated mylar has been analyzed with a mass spectrometer. The irradiated foil was subject to a strain and tensile test.

The gas, originated from the irradiated mylar foil, is in the main hydrogen (≈ 70%), carbon dioxide (≈ 20%) and carbon-monoxide (≈ 10%) (fig. 1). In small trace-concentrations methane and some heavier hydrocarbons are detected.

This points at a radiation-induced cross-linkage of the irradiated polyethylenterephtalate and coincides with the results of the strain and tensile test. The elastic modulus and the ultimate stress are scarcely effected by the applied radiation dose, but the failure strain is reduced by a factor of two. The foil becomes brittle, but is still usable as an insulation material. This result corresponds with other measurements[1].

MULTILAYER INSULATION TESTS

In the test-cryostat (fig. 2) liquid nitrogen in the gauging tube is evaporated by the heat absorption of the cylindrical gauging surface. The gas mass flow is determined in a volumetric displacement flowmeter (fig. 3). Liquid nitrogen from the reservoir keeps the gauging tube completely filled. The temperature of the gauging surface and of several layers

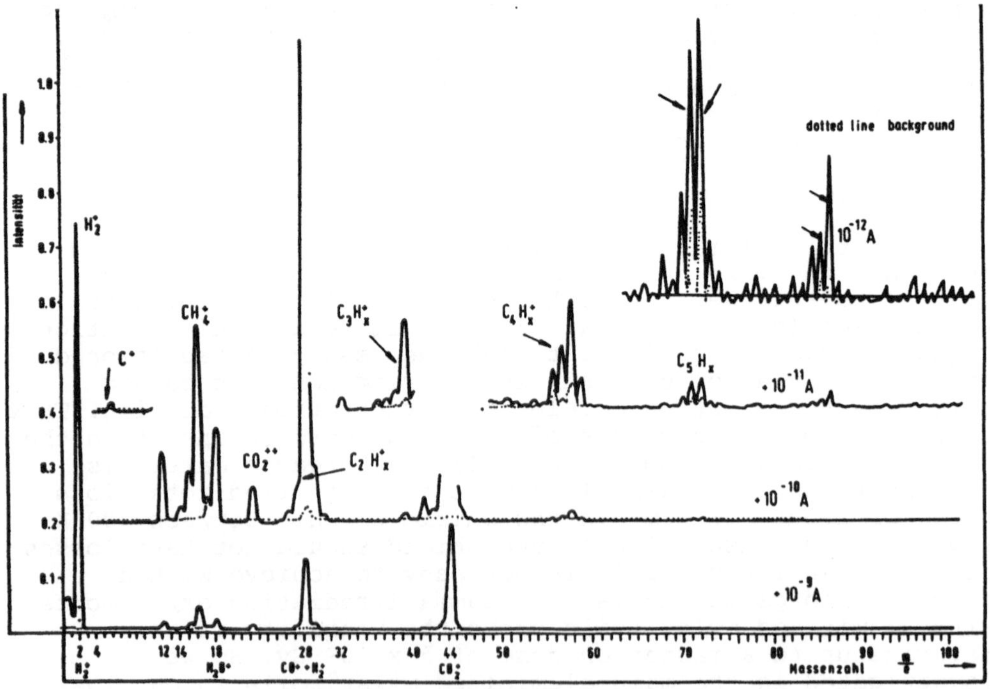

Fig. 1. Mass-spectrogram of gas evolved during irradiation of Mylar.

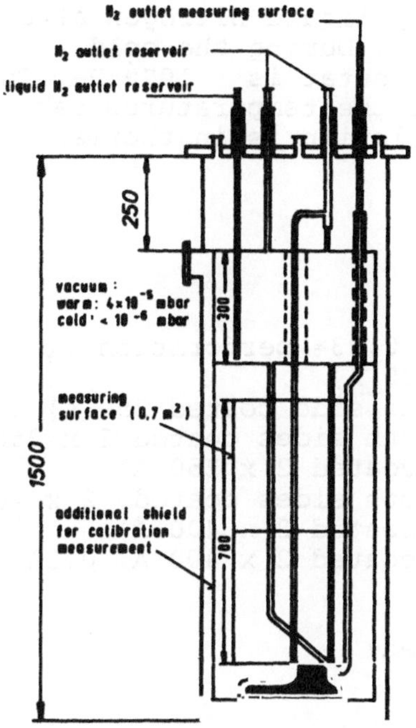

Fig. 2. Test cryostat.

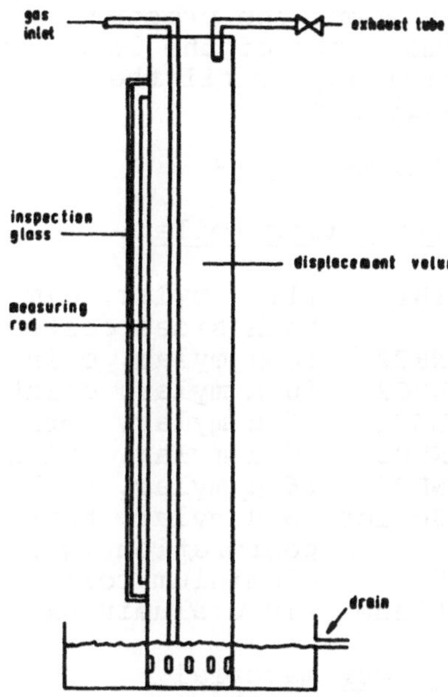

Fig. 3. Volumetric displacement flowmeter.

of the multilayer insulation under test is measured by platinum-resistance thermometers.

The volumetric displacement flowmeter has a volume of about 44 litres to compensate short-time fluctuations in the evaporation rate by a sufficient measuring time.

A calibration measurement for which the whole gauging surface was surrounded by a shield cooled to 77 K, determined the background of our test assembly to 0.11 W/m². A calibration with an electric heater on the gauging surface yields an excellent correlation between the electric heater power and the measured values.

Tests made later on with an additional shielding of the annular clearance between the insulation under test and the normal shield by an aluminized mylar foil, showed a reduction of the measured loss rates by 0.44 W/m². This is an effect of heat radiation entering the edge front of the insulation after reflection on the inner surface of the shields. The preceeding measurements are corrected by these values.

The insulations are spiral wound on the gauging cylinder by hand and fixed to the surface by three plastic punches on the upper end of the cylinder. This technique produces an increasing pressure on the inner layers, proportional to the total number of layers, which seems to be realistic as a test for an insulation for a long and complex transferline and the 9 m dipole magnets.

The platinum resistance thermometers are fixed on the reflecting foils of the insulation by aluminized tape.

The cryostat is cooled down by liquid nitrogen after pumping the vacuum to ≈ 4 x 10^{-3} Pa. During the cold measurement the pressure in the cryostat is 10^{-4} Pa. The measurements of the loss rates and the temperatures take several days until the tested insulation is in thermal equilibrium.

TESTED MATERIALS

A) reflecting foils

1. IHI (12 µ mylar, dimples, ≈ 0.03% perforation (ϕ 2 mm) both sides coated 2 x 400 Å)
2. NRC2 (6 µ mylar, crinkled, one side coated 250 Å)
3. NRC2 (6 µ mylar, crinkled, both sides coated 2 x 250 Å)
4. NRC2 (6 µ mylar, both sides coated 2 x 250 Å)
5. NRC2 (6 µ mylar, crinkled, both sides coated, 2 x 400 Å)
6. NRC2 (6 µ mylar, both sides coated 2 x 400 Å)
7. Jehier (6 µ mylar, both sides coated 2 x 400 Å. 0.1% perforation (ϕ 2 mm))
8. 9 µ aluminium foil
9. Linde (10 µ aluminium, embossed)

B) spacer material

1. glass net (Vitrulan SD 4220 C/51, mesh size 4 x 5 mm, δ ≈ 80 g/m², finish)
2. glass net (Interglas 01868, mesh size 5 x 5 mm, δ ≈ 32 g/m², finish)
3. glass silk (Interglas 02034, no finish
4. carbon paper (China Scientific Instruments & Materials Corp., δ ≈ 32 g/m²)
5. polyamide tulle (Jehier superinsulation)
6. glass silk (Linde superinsulation)

RESULTS

A comparison between the temperature behaviour of the different layers of some multilayer insulations and ideal radiation shields (fig. 4) shows clear advantages for a spacer material between the reflecting foils, even for reflecting foils like the IHI, which is provided for use without a spacer.

Likewise this figure shows that for glass silk as spacer the aluminized mylar foils have advantages compared to the aluminium foil. On the other hand, the both sides thicker coated mylar foils are better than the thinner coated ones and obtain the lowest loss rates.

Different kinds of spacer-materials, especially a carbon covered glass paper, are tested with aluminized mylar foil (fig. 5). The absorbent carbon was expected to improve the vacuum between the reflecting foils and so to decrease the loss rates. Because the multilayer insulation in the HERA-transferline and magnets will not be heatable, the carbon paper in our test is not baked out, too. So the use of the carbon paper in the warm layers of the insulation is senseless. Our tests show that carbon paper as a spacer for both sides coated (2 x 400 Å) mylar has no advantages compared to a glassnet spacer[2]. Because of the high degassing rate of

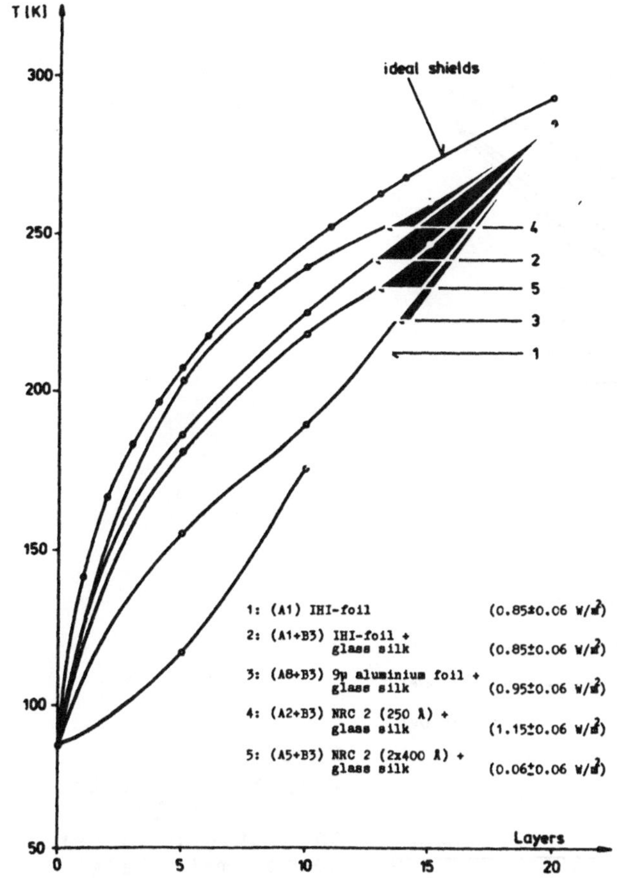

Fig. 4.
Temperature distribution in 20-layer insulations (several reflecting foils).

the carbon paper, it prolonges the pumping time by a factor of two to eight, according to the applied number of carbon paper layers (10-40).

Our tests with different numbers of layers for the same kind of multilayer insulation show, that from 20 to 30 layers only a small improvement of the loss rate is possible and the increase from 30 to 40 layers has no perceptible effect on the loss rate [3, 4] (fig. 6). To explain the effect which prevents the theoretically possible improvement by a factor of two for an increase in the numbers of layers from 20 to 40, we evaluate the fraction of the total heat-flux through the insulation due to radiation heat transport.

The emissivity of the foil was calculated by the temperatures of the outer insulation layer and the cryostat vacuum vessel, and the loss rate of the insulation. Assuming a constant emissivity for all layers, it is possible to roughly evaluate the fraction of radiation heat transport of the total heat-flux in the region between two temperature measured layers. This evaluation shows for 20 layer insulations the highest fraction of radiation heat transport in the outer layers and a continuous decrease of this fraction in the inner layers (fig. 7). For the 40 layer insulation we observe a similar characteristic, but for the innermost layers (1-5) we watch a slight increase of the fraction of radiation heat transport (fig. 8).

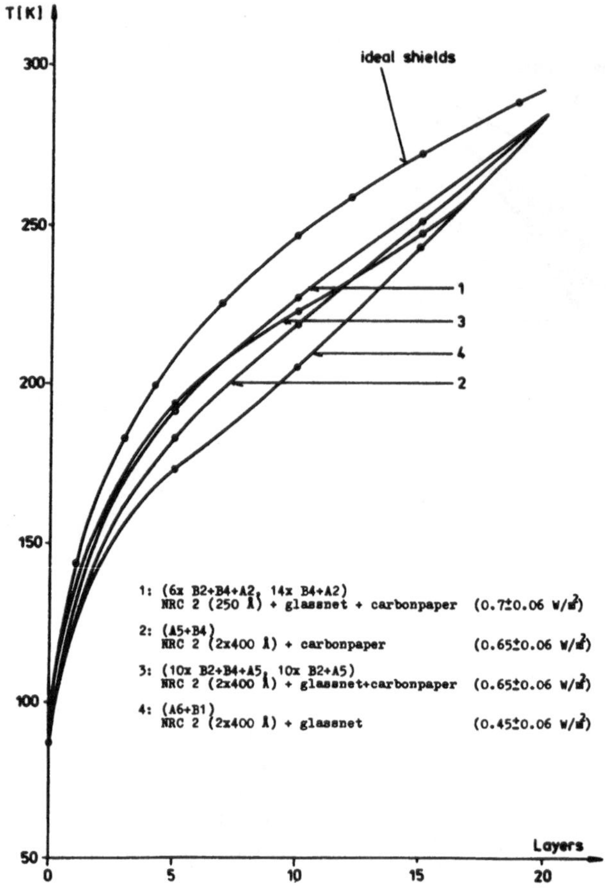

Fig. 5. Temperature distribution in 20-layer insulations (several spacer materials).

It seems that the pressure of the insulating vacuum between the layers of this insulation has risen to a level where the cryopump effect of the cold gauging surface is sufficient for effectively reducing the heat conduction by gas, leading to an increase in the fraction of radiation heat transport. This means that an increase in the number of layers for an insulation of some 20 layers causes a deterioration of the insulating vacuum in an increasing region of the insulation[5]. So there is hardly an increase in the numbers of "effective" layers. The solid heat conduction cannot explain the observed increase of the fraction of radiation heat conduction in the innermost insulation layers. An influence of a perfo-ration of the reflection foil to improve the vacuum on the temperature distribution and the loss rates of the multilayer insulation is not measurable (fig. 9). This may depend on the geometry of our testing facility. Here the evacuation of the space between the layers is possible through the edge front of the insulation as well as through the perforated layers.

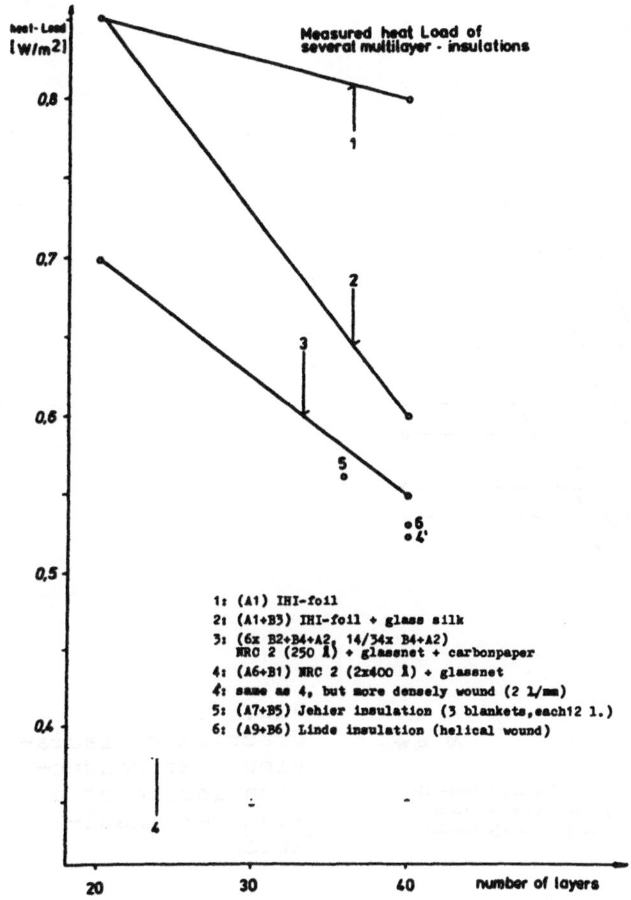

Fig. 6.
Heat load of several multilayer insulations.

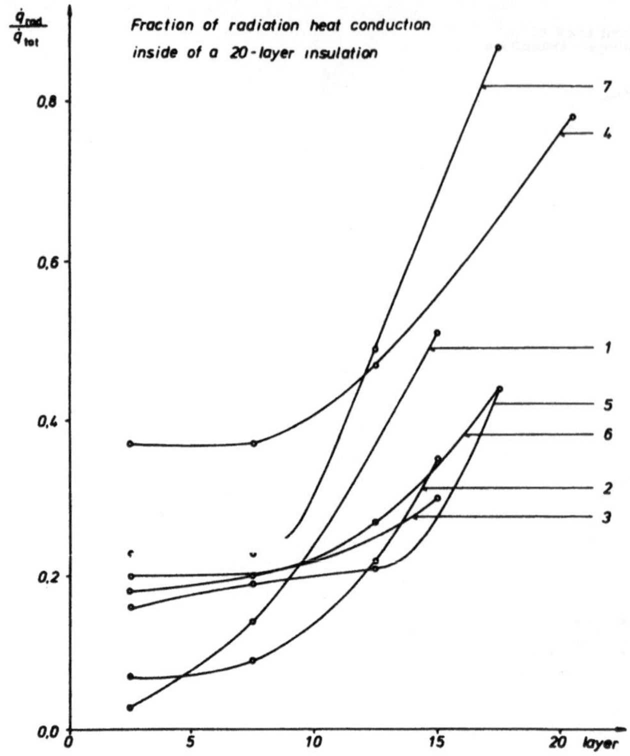

1: (A1) IHI foil
2: (A8+B3) 9µ alu-foil+glass-silk
3: (A1+B3) IHI foil + glass-silk
4: (A2+B4+B2) NRC2 (250 Å)+carbonp.+glassnet
5: (A5+B3) NRC2 (2x400Å)+glass-silk
6: (A5+B4) NRC2 (2x400Å)+carbonp.
7: (A5+B2) NRC2 (2x400Å)+glassnet

Fig. 7.
Fraction of radiation heat conduction inside of a 20-layer insulation.

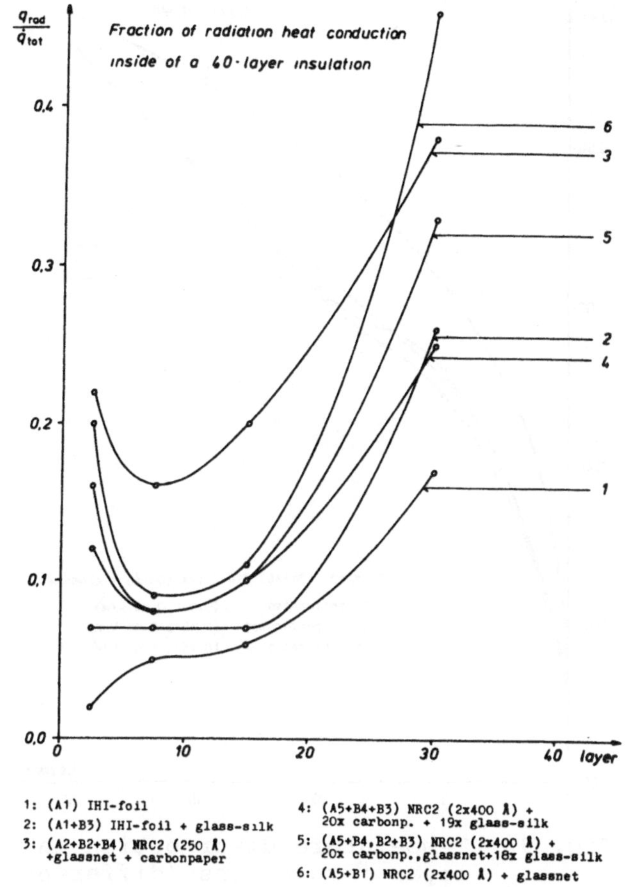

Fig. 8. Fraction of radiation heat conduction inside of a 40-layer insulation.

CONCLUSION

The gamma ray irradiation test shows the usability of aluminized mylar foils up to radiation dose of 5 x 10^6 Gy, which is enough for the use in the HERA transferline and magnets.

Our loss rate measurements show the best results for both side thickly aluminium coated (2 x 400 Å) mylar foils with a glass net spacer. This combination has even with a quite dense winding (≈ 1.8 layer/mm) very low loss rates (≈ 0.55 W/m² for 300 K - 77 K) (fig. 6).

The use of carbon paper as an adsorbing spacer has, without baking out, no advantages, and prolonges the time for pumping down the vacuum.

A perforation of up to 1% of the reflecting foils shows no effect on the loss rates in our testing facility, but this result may not be transferable to bigger coherent areas of multilayer insulations.

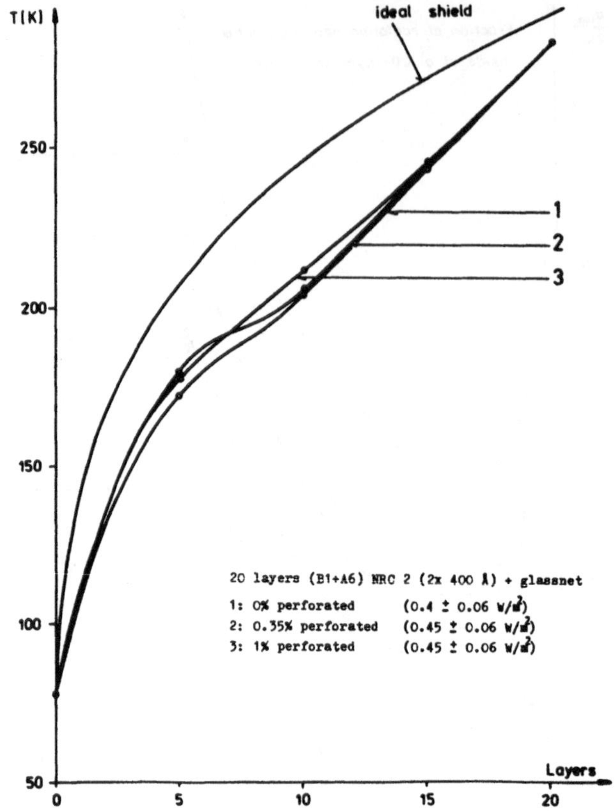

Fig. 9. Temperature distribution in 20-layer insulations (different perforations).

The dependence of the measured loss rate on the number of layers differs increasingly from the theoretical behaviour with an increasing number of layers. Insultions with 30 and 40 layers have the same loss rates. It can be concluded from the temperature distribution inside the insulation, that the vacuum between the layers deteriorates with the increase of the number of layers, so the number of "effective" layers does not increase any more.

REFERENCES

1. W. Schnabel, "Kunststoffe, ihr Wesen und Verhalten gegenüber ionisierender Strahlung" Hahn Meitner Institut für Kernforschung Berlin (1962)
2. T.L. Halaczek, J. Rafalowicz, Heat trasport in self-pumping multilayer insulation, Cryogenics 26: 373 (1986)
3. T. R. Gathright, P. A. Reeve, Effect of multilayer insulation on radiation heat transfer at cryogenic temperatures, TRIUMF University of Victoria (Canada 1986)
4. Q. S. Shu, R. W. Fast, H. L. Hart, Heat flux from 277 to 77 K through a few layers of multilayer insulation, Cryogenics 26: 671 (1986)
5. R. S. Mikhalchenko, N. P. Pershin, I. G. Sidorenko, Heat and mass transfer in multilayer insulations, Sixth ICEC

THERMAL PERFORMANCE OF CANDIDATE SSC MAGNET THERMAL INSULATION SYSTEMS

T. Ohmori*1, W.N. Boroski, J.D. Gonczy,
R.C. Niemann and M.K. Ruschman

Fermi National Accelerator Laboratory*2
Batavia, Illinois

T. Taira and K. Takahashi

Ishikawajima-Harima Heavy Industries Co., Ltd. (IHI)
Yokohama, Japan

A. Yamamoto and H. Hirabayashi

National Laboratory for High Energy Physics (KEK)
Tsukuba, Japan

ABSTRACT

The thermal performance of three candidate thermal insulation systems for the SSC magnet cryostat has been evaluated experimentally in a heat leak test facility. The systems evaluated were multilayer insulation (MLI) blankets consisting of aluminized polyester film with fiberglass mat spacer, aluminized polyester film with spunbonded nylon spacer and aluminized dimpled perforated polyester film without spacer. Performance between 300 and 80 K with good and degraded insulating vacuum was studied. The inter-layer heat transfer coefficient h_{itl} and the effective thermal conductivity K_{eff} of MLI blanket were used to compare the different insulation systems and to predict the performance of the SSC magnet insulation.

INTRODUCTION

The proposed Superconducting Super Collider high energy physics research accelerator facility utilizes about 10,000 superconducting devices for particle beam acceleration and control. The magnet cryostats for these devices must have very low refrigeration loads in order to control the capital and operating costs of the refrigeration system. The insulation required for such a large number of magnet cryostats is extensive and thus the thermal performance of the insulation must be optimized to reduce the cost of materials and their installation.[1,2]

*1 Visiting scientist from IHI Co., Ltd. Isogoku, Yokohama 235, Japan
*2 Operated by Universities Research Assn. Inc., under contract with the U.S. Department of Energy

In operation, the magnet cryostat can experience good vacuum, e.g., 1.3 x 10^{-4} Pa (~10^{-6}Torr) when it is cooled down after installation, and also degraded vacuum, e.g., when it is heated during a magnet quench. These are important transient phenomena having different time constants and require study not addressed in this paper. Of equal importance is the steady-state thermal performance of the candidate insulation systems operating with good to degraded vacuum. In this series of measurements, the steady-state thermal performance of the MLI systems were studied in a heat leak test facility[3] between temperatures of 300 K and 80 K and pressures from 10^{-4}Pa to 1 Pa.

The reported studies address the MLI material properties and the fabrication and installation of the MLI blankets. Since heat transfer through an insulation blanket is inherently a function of the MLI material properties and the blanket layer density, accurate measurements of material properties and layer density are essential for obtaining reliable MLI thermal performance data and an optimum cryostat design. Careful measurements were therefore conducted on the physical dimensions of the installed MLI test samples as well as on the surface resistance of the aluminum layer on the polyester film substrate.

HEAT LEAK TEST FACILITY AND TEST METHODS

The heat leak test facility employs a heat meter[4] to measure the heat transfer through the insulation (Fig. 1). The heatmeter measurement is not affected by atmospheric pressure changes. This method is an improvement over boil-off calorimetry which requires special pressure control valving to measure the heat leak for a short period of time without the additional complication associated with a change in atmospheric pressure.

The sample MLI blanket is fabricated around the side and the bottom of a cylindrical cold plate constructed of OFHC copper. The cold plate is 394 mm in diameter and 305 mm in height. Thermal communication between the cold plate and the surfaces of the liquid nitrogen reservoirs, either by thermal radiation or by gas conduction, is limited to a negligible amount with the use of MLI in the space separating the surfaces. The MLI in this space is not shown in Fig. 1 so as not to be confused with the MLI pictured under test. A calibration heater embedded in the cold plate was employed and confirmed that no significant heatflux was bypassed around the heatmeater for both good and degraded vacuum conditions.

The insulating vacuum pressure (pressure of the space between insulation blanket and hot plate) was controlled by means of a turbo-molecular pump and an isolation valve. Vacuum levels higher than 8 x 10^{-3} Pa were obtained with a fully opened isolation valve and was measured with a an exposed ion gage. Degraded vacuum was controlled by closing the valve and was measured with a thermocouple gage. Degraded vacuum lower than 7 x 10^{-1} Pa was obtained by introducing nitrogen gas into the vacuum space. Hot plate temperature is controlled with an embedded heater and promotes the stability of the insulating vacuum. Control of the hot plate temperature is especially important to the study of MLI thermal performance in a degraded vacuum where the performance can be strongly dependent on insulating vacuum pressure.

CANDIDATE INSULATION SYSTEMS

The three candidate MLI systems evaluated were insulation blankets consisting of (1) double aluminized Mylar (DAM) films with fiberglass mat (FGM) spacers (FGM/DAM), (2) double aluminized Mylar films with Cerex spunbonded nylon (CSN) spacers (CSN/DAM) and (3) dimpled and perforated (DP) double aluminized Mylar films without spacers (DP-DAM). The blanket structures are as follows and their geometry is illustrated in Fig. 2:

1) FGM/DAM -- The MLI test sample consists of one blanket of FGM/DAM. The blanket construction is a stacked assembly of alternating DAM and FGM layers stacked to a height of 11 layers of DAM films (Scharr; DAM 25.4 μmt) and 10 layers of polyester bonded fiberglass mat spacers (Nicofibers Inc.; SURMAT 254 μmt, fiber diameter is 18 to 21 μm). The fiber glass mat binder adheres to DAM film so that the blanket structure requires no materials to fasten the layers to each other.

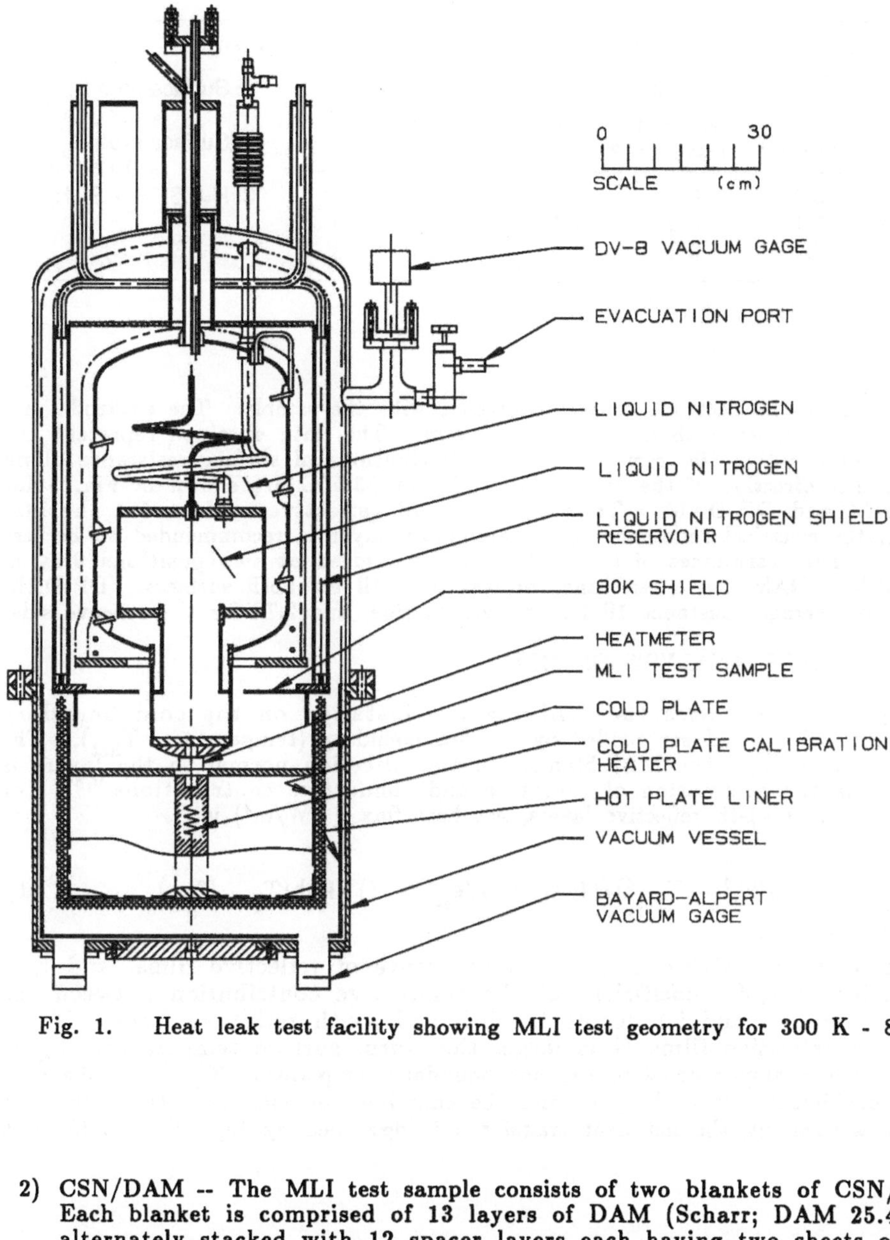

Fig. 1. Heat leak test facility showing MLI test geometry for 300 K - 80 K

2) CSN/DAM -- The MLI test sample consists of two blankets of CSN/DAM. Each blanket is comprised of 13 layers of DAM (Scharr; DAM 25.4 μmt) alternately stacked with 12 spacer layers each having two sheets of CSN (James River Corp.; CEREX 81.3 μmt, 6/6 nylon fibers 21 μm in diameter). The MLI layers were secured together by 16 nylon tag-pins through the blanket.
3) DP-DAM -- The MLI test sample consists of two blankets of DP-DAM. Each blanket contains of 10 layers of DP-DAM (IHI; Dimple 12 μmt; 2 μm diameter perforations for every 100 x 100 μm area) and three DAM layers. The DAM layers are located at the outer most blanket surfaces and in the middle of the blanket. The blanket employs 6 nylon tag-pins to secure the MLI layers.

The layer density of blanket as averaged from the side and bottom parts of the cold plate is listed in Table 1.

Surface resistance of the aluminum coating on the polyester film was measured on sample films by the four electrode method in order to avoid measurement errors

Table 1. Total Thickness H of MLI Blanket Sample

Test Sample	Blanket thickness H (mm): H_s, mm side of cold plate *1	H_b, mm bottom of cold plate *2	H, mm mean value *3	N layers	N/H layers /mm	Note
FGM/DAM	5.4	10.0	6.5	11	1.7	*1: Surface area $S_s = 0.377\ m^2$
CSN/DAM	9.1	8.0	8.9	26	2.9	*2: Surface area $S_b = 0.123\ m^2$
DP-DAM	6.4	3.6	5.7	26	4.6	*3: $H=(S_s H_s + S_b H_b)/S$
						* $S = S_s + S_b$

due to variations in electrode contact resistance with the sample. The electrodes are constructed of copper with a width of 100 mm. The data obtained represent the resistance of a 100 x 100 mm area. The distributions of surface resistance along the transverse direction of the original roll film at 23.5°C are shown in Fig. 3 for both DAM and DP-DAM. To obtain a good reflective surface for thermal radiation, the electrical resistance of the aluminum layer is recommended to be less than 1Ω.[5] The resistances of both sides were measured at two positions 9.14 m apart. Scharr DAM has a resistance of less than 1Ω for both surfaces. DP-DAM film has an average resistance 1Ω for the convex side, and 2.7Ω for the concave side.

HEAT TRANSFER ANALYSIS OF MLI

Figure 4 illustrates the MLI blanket installed on the cold boundary (temperature: T_{cold}) and surrounded by the hot boundary (temperature: T_{hot}). The heat transfer through the MLI blanket in the direction normal to the layers is described as the summation of radiative and conductive contributions.[6,8] For adjacent i-th and i-1-th reflective layers, net heat flux q (W/m^2) is

$$q = \sigma(T_i^4 - T_{i-1}^4)/(1/\epsilon_i + 1/\epsilon_{i-1} - 1) + h(T_i - T_{i-1}) \qquad (1)$$

where ϵ is the emissivity, T is the temperature of reflective films, and h is effective heat transfer coefficient of the conductive contribution between the reflective films. i and i-1 denote the i-th and i-1-th reflective films. If the number of reflective films N is large, the warm surface temperature T_N of the insulation is almost equal to the hot boundary temperature T_{hot}.[9] If the heat transfer coefficients for each space and the emissivity of each reflective films are regarded as constant, the net heat transfer q is described by Eq. (2) by adding N

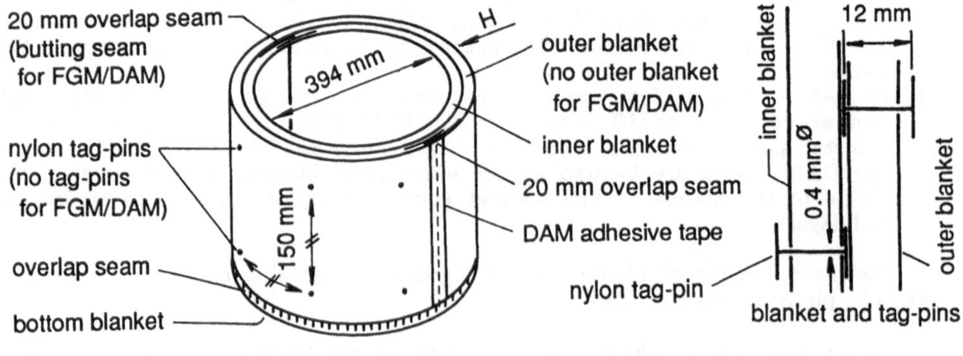

Fig. 2. Geometry of MLI blanket test sample for the heat leak test facility.

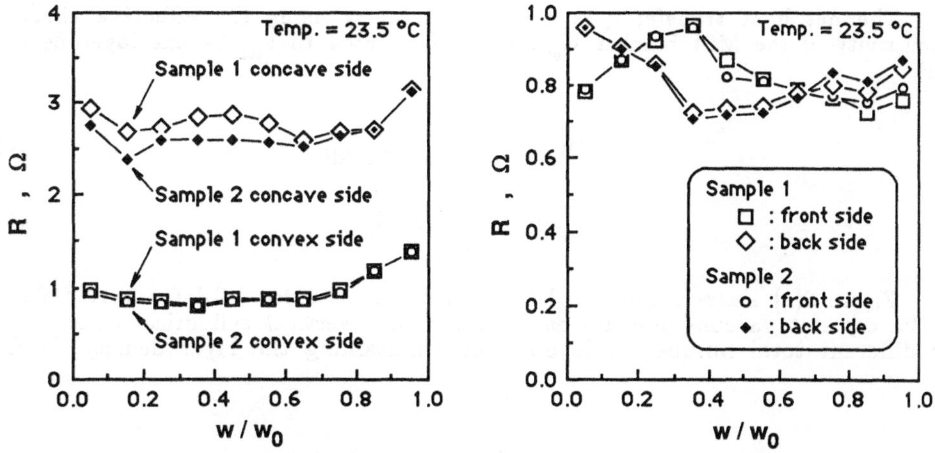

(a) DIMPLE (w_0 = 1000 mm)   (b) DAM (w_0 = 1067 mm)

Fig. 3. Surface resistance of aluminized polyester film.

sets of Eq. (1) for each insulation effective. Thus,

$$Nq = \sigma(T_{hot}^4 - T_{cold}^4)/(2/\epsilon - 1) + h(T_{hot} - T_{cold}) \tag{2}$$

By using the inter-layer heat transfer coefficient h_{itl} (W/m²K) between adjacent reflective films, Eq. (2) is written as

$$Nq = h_{itl}(T_{hot} - T_{cold}) \tag{3}$$

$$h_{itl} = 4\sigma\overline{T}^2\overline{T} \;/\; (2/\epsilon - 1) + h \tag{4}$$

where $\overline{T}^2$ is the mean square temperature defined as $(T_{hot}^2+T_{cold}^2)/2$ and $\overline{T}$ is the mean temperature defined as $(T_{hot}+T_{cold})/2$. The emissivity ϵ is the radiative property of the reflective film. The heat transfer coefficient h is a parameter which describes the fabrication condition of the MLI blanket, for example, layer density N/H or compression of the blanket. If MLI blankets are fabricated with the same materials and have the same fabrication conditions, Nq and h_{itl} become identical for the same boundary temperatures even if the total number of reflective films N is different. Then the net heat transfer q is inversely proportional to N.

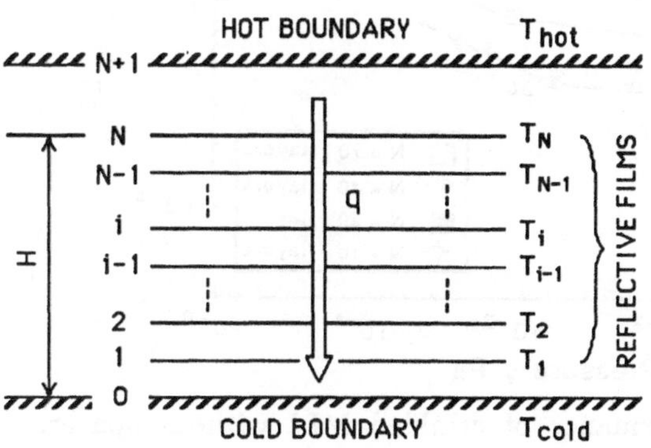

Fig. 4. Heat transfer through the MLI blanket.

The net heat transfer q is also evaluated by using the effective thermal conductivity of the MLI blanket K_{eff}, which is related to h_{itl} by the layer density N/H.

$$q = (K_{eff}/H)(T_{hot} - T_{cold}) \quad (5)$$

$$K_{eff} = h_{itl} / (N/H) \quad (6)$$

Figure 5(a) shows the thermal performance of crinkled MLI for 3×10^{-5} Pa to 0.2 Pa cryostat vacuum pressure as measured in a vertical cylindrical calorimeter[7] for different total number of layers, but maintaining the layer density, N/H,

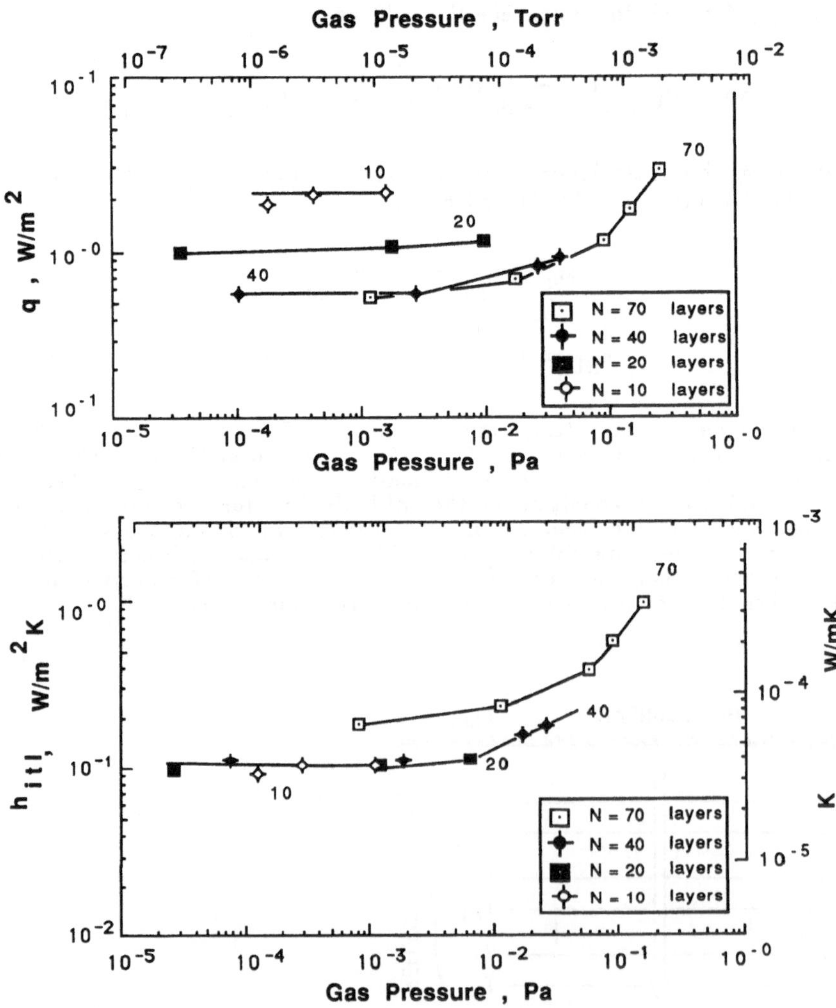

Fig. 5. Thermal performance of crinkled MLI without spacer.

identical at 2.8 layers/mm.[10] If the thermal performance of the MLIs are expressed in terms of h_{itl} or K_{eff}, the results for the different number of layers, i.e., 10, 20, and 40 lie on the same curve as shown in Figure 5(b). The data for 70 layers has a higher heat transfer coefficient than the others. The guard reservoir height L of 400 mm is designed to avoid lateral heat transfer from the cryogen fill/vent ports to the main reservoir. The thickness of this insulation H is 25 mm so that the ratio L/H is 16 for this case.[7] The higher heat flux for N=70 is felt to be due to the compression of the insulation. However, from the data between N=10 and 40, the proportionality of heat flux to N^{-1} is very good. It can be said that h_{itl} and K_{eff} become constant and are characteristic properties of thermal insulation performance for MLI.

If the total number of layers N is small, the space between the hot boundary and the outermost surface of MLI contributes to the insulation performance. Then the left side of Eq. (2) must be (N+1)q. In this situation, the characteristic properties of thermal insulation performance are h_{itl} (or (N+1)q) and N/H.

$$(T_{hot} = 285\ K,\ T_{cold} = 77\ K,\ N/H = 2.8\ mm^{-1})$$

TEST RESULTS AND DISCUSSION

Thermal performance data between 300 K and 80 K for candidate MLI blankets are listed in Table 2 and are shown in Fig. 6. The values for h_{itl} and K_{eff} remain essentially constant under good vacuum condition and increase significantly under degraded vacuum condition; i.e., $>10^{-2}$ Pa. The performance difference between each blanket decreases in the degraded vacuum region. For the entire range of vacuums studied, DP-DAM has the lowest h_{itl} and K_{eff}.

The best vacuum recorded was 1.8 x 10^{-4} Pa for DP-DAM, 1.1 x 10^{-3} Pa for FGM/DAM and 1.2 x 10^{-3} Pa for CSN/DAM. Cryostat vacuum better than 1.33 x 10^{-4} Pa (10^{-6} Torr) could not be attained, because FGM/DAM blankets are employed for the liquid helium and liquid nitrogen reservoirs of the facility. After filling the reservoirs with liquid nitrogen, the vacuum improved to better than 1.33 x 10^{-3} Pa (10^{-5} Torr) in 50 hours for DP-DAM, 280 hours for FGM/DAM and 400 hours for CSN/DAM. The FGM/DAM insulation recorded better vacuum with faster evacuation speeds than CSN/DAM because the former has fewer layers to evacuate than does the CSN/DAM.

The inter-layer heat transfer coefficient h_{itl} for DP-DAM was also measured between 293 K and 78 K by the vertical cylindrical calorimeter and was evaluated as 5.95 x 10^{-2} W/m^2K for a sample without tag-pins.[5] The layer density of the sample is 4.02 layers/mm and rather small. The results are better by factor of 1.17 than the results obtained by the heat leak test facility because DP-DAM has 6 tag-pins and 6 flat films of DAM. From the analysis of conduction heat transfer

Table 2. Candidate Thermal Insulation Systems for The SSC Magnet 80 K Shield*

Insulation	Data obtained in the heat leak test facility (T_{hot} = 300 K , T_{cold} = 80 K)					Design 80 K insulation system		
System	H	N/H	Nq	h_{itl}	K_{eff}	N_t	H_t	w_t
	mm	mm^{-1}	W/m^2	W/m^2K	W/mK	layers	mm	kg/m^2
FGM/DAM	6.5	1.69	26.7	0.122	7.19 x 10^{-5}	44	26.0	3.37
CSN/DAM	8.9	2.93	35.8	0.163	5.55 x 10^{-5}	59	20.1	3.27
DP-DAM	5.7	4.55	15.9	0.0722	1.59 x 10^{-5}	26	5.7	0.452

* Insulation heat leak budget for 80 K shield = 0.61 W/m^2

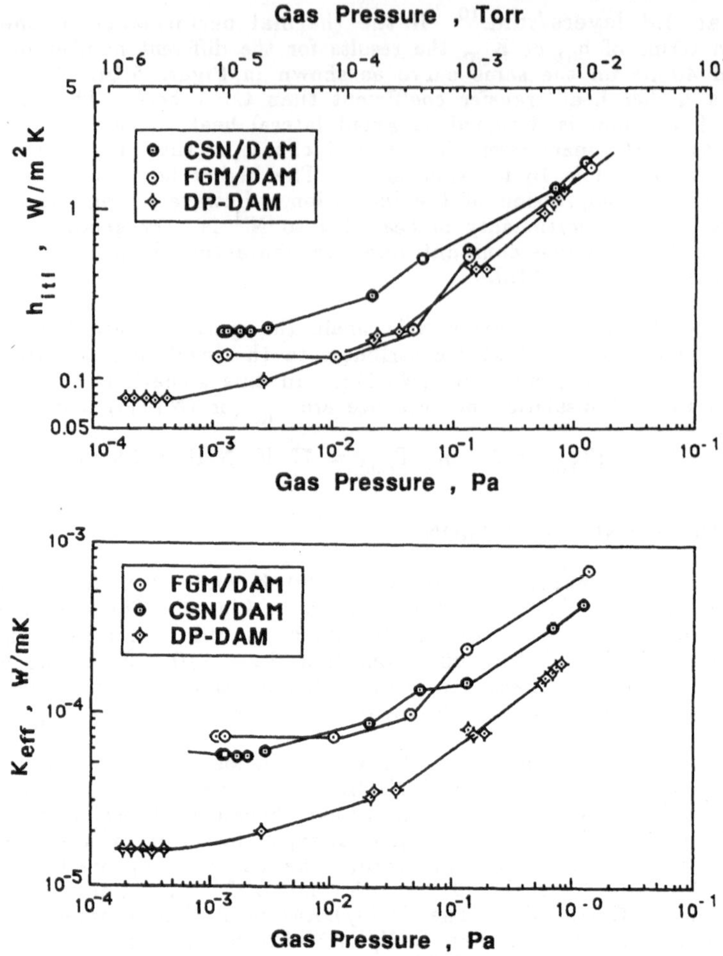

Fig. 6. Thermal performances of candidate insulation systems.

through the nylon tag-pin, the total heat flux increase by 6 tag-pins is estimated to be 5.3% for DP-DAM. The total heat flux increase by the 16 tag-pins for CSN/DAM is estimated to be 4.5% which is smaller than in the DP-DAM insulation because the blanket is thicker than the DP-DAM insulation.

The three candidate thermal insulation systems as designed for the SSC magnet cryostat 80 K shield are listed in Table 2. From the data obtained in the heat leak test facility, the total number of reflective layers N_t of the insulation system which has the same layer density as the MLI blanket test sample required to meet the 80 K heat leak budget, and the corresponding thickness H_t and specific weight w_t are as estimated in Table 2.

The total number of layers N_t is a measure of material and assembly cost of MLI blanket. DP-DAM has the lowest N_t with FGM/DAM second and CSN/DAM third. This ordering is similar to that for the parameter h_{itl}. The material costs for CSN/DAM and FGM/DAM are at present less than DP-DAM, but the fabrication cost of a blanket for the former is higher than DP-DAM.

The total thickness H_t must be smaller than the available installation space in order to avoid mechanical compression of the insulation. DP-DAM has the smallest H_t while CSN/DAM is second and FGM/DAM is third. This ordering is similar to that for the effective thermal conductivity K_{eff}. The installation space for the 80 K and 20 K thermal insulation is especially restricted in the interconnection region between adjacent magnets because the cold mass and other cryogenic pipings are connected by bellows. If a FGM/DAM system is employed, it can encounter high blanket compression which increases the heat leak. The half cell of the magnet string, which consists of 5 dipole magnets, 1 quadrupole magnet and 1 spool piece,[11] has mobile pumping systems at its ends. The vacuum pumpout space around each insulated assembly can only be assured continuity by using thin insulation.

The specific weight w_t of the insulation blanket must be small in order to avoid self-compression of the insulation due to its weight; particularly on a horizontal cylindrical shield. DP-DAM is the lightest blanket because the N_t is the smallest and the insulation requires no spacer material.

The inter-layer heat transfer coefficient h_{itl} of DP-DAM without tag-pins between 78 and 4.2 K measured by the cylindrical calorimeter is 7.93 x 10^{-3} W/m^2K.[5] The thermal insulation system of DP-DAM for the 20 K shield is designed to meet the heat leak budget of 0.075 W/m^2. This system has 22 dimpled layers with a total thickness of 5.5 mm($N/H = 4.02 mm^{-1}$). The CSN/DAM system can be considered as potential material for the 20 K shield, only if its thermal performance h_{itl} and K_{eff} between 80 and 20 K are much better than that for the DP-DAM system.

CONCLUSIONS

The thermal performances from 300 to 80 K of three candidate insulation systems were evaluated and compared experimentally. DP-DAM has the lowest inter-layer heat transfer coefficient h_{itl} between adjacent reflective films and the lowest effective thermal conductivity K_{eff} between 300 and 80 K both in good and degraded insulating vacuum. The insulation system for the SSC 80 K shield utilizing DP-DAM has the smallest total number of layers and the smallest total thickness.

The thermal insulation system for the 20 K shield can be designed with DP-DAM using the data obtained with the cylindrical calorimeter. As the heat transfer mechanism through MLI between 80 and 20 K is different from that between 300 and 80 K, performance studies of CSN/DAM and DP-DAM will be continued in this temperature region.

REFERENCE

1. R.C. Niemann, et al, The cryostat for the SSC 6T magnet option, in: "Adv. Cryo. Engr.," Vol. 31, Plenum Press, New York (1985), p. 63.
2. R.C. Niemann, et al, Superconducting super collider magnet cryostat, in: "6th Intersociety Cryogenic Symposium," Miami, Florida, Nov. 2-7, 1986.
3. J.D. Gonczy, et al, Heat leak measurements facility, in: "Adv. Cryo. Engr.," Vol. 31, Plenum Press, New York (1986), p. 1291.
4. M. Kuchnir, et al, Measuring heat leak with a heat meter, in: "Adv. Cryo. Engr.," Vol. 31, Plenum Press, New York (1986), p. 1285.
5. T. Ohmori, et al, Multilayer insulation with aluminized dimpled polyester film, in: "Proc. Eleventh Intl. Cryo. Engr. Conf.," Butterworth, Guilford, UK (1986), p. 567.
6. C.L. Tien and G.R. Cunnington, Cryogenic insulation heat transfer, in: "Adv. in Heat Transfer," Vol. 9, Academic Press, New York (1976), p. 39.
7. ibid, p. 403.
8. N. Inai, Trans, Japan Soc. Mech. Eng., Ser.B, 43:217 (1977), P.217,
9. Q.S. Shu, et al, Heat flux from 277 to 77 K through a few layers of multilayer insulation, Cryogenics, 26:671 (1986).
10. T. Ohmori, et al, Thermal performance of evacuated multilayer insulation, in: Proc. 27th Meeting of Cryogenic Association of Japan, Tokyo (1981), p. 60, Cryogenic Association of Japan.
11. SSC Central Design Group, Superconducting Super Collider Conceptual Design, SSC-SR-2020, March 1986, p. 331.

PROTOTYPE OF A SEMIFLEXIBLE MULTI-LAYER INSULATED ENCLOSURE FOR CRYOGENIC POWER CABLES AND PIPELINES

F. Schauer, M. Hubmann, H. Pirklbauer, and G. Kropatsch

Anstalt für Tieftemperaturforschung

Graz, Austria

ABSTRACT

A 21 m long prototype of a semiflexible multi-layer insulated cryogenic envelope was built and tested to prove the suitability and economy of such a thermal insulation system for superconducting power transmission cables and liquid gas pipelines. The enclosure consists basically of an outer rigid tube at ambient temperature, a corrugated inner tube containing the cryogenic fluid, and superinsulation within the evacuated annulus. The flexible inner pipe is pulled into loosely connected rigid "spacer tubes" which are held coaxially to the outer tube by synthetic ropes. These spacer tubes serve also as winding cores for the insulation.

For the prototype which includes two 30°-bends, the cheapest possible insulation and structure materials were employed and the simple assembly procedures put to test. The inner cold cryogen duct and outer tube diameters are 100 and 273 mm respectively. Tests with liquid nitrogen and liquid helium as cryogens yielded satisfactorily low thermal losses of ≤0.6 W/m.

INTRODUCTION

The cryogenic enclosure of a superconducting power transmission cable is a major cost factor which influences significantly the economy of the whole system. From cost estimates[1] and experience with the most advanced prototype cable systems[2,3] follows that the only chance to build a superconducting power transmission line competitively would be the omission of He-gas or LN_2 cooled intermediate temperature shields.

The envelopes for cryogenic power cables or, by the same token, cryogenic gas pipelines which have been developed so far are either rigid with thermal contraction compensation elements[3-5] or fully flexible, consisting of corrugated tubes[2,6]. Both systems have their advantages: The thermal losses of the rigid system can be kept very low due to the possibility of using tension-loaded spacers such as thin plastic rope assemblies which are installed far apart from each other in axial direction. The large spacer distances in turn facilitate automatic winding of the superinsulation. Additionally, large tube diameters can be produced economically. On the other hand, fully flexible systems are much easier to lay in the field, special installations are needed neither for compensation of thermal contraction nor for bends. Further, the necessary joints which may bring about additional heat leaks and require low temperature welding seams are greatly reduced with the latter system.

The new "semiflexible" enclosure[7] which can be used for cryogenic power cables as well as for pipelines combines the main advantages of both the completely rigid and the fully flexible systems at minimal costs. The rigid "spacer tubes" -4- (Fig. 1) are held by low-loss spacer ropes (or rods) -5-. The large distances between the spacers -5- allow simple automatic winding of the superinsulation onto the spacer tubes. The inner corrugated pipe -2- at operating temperature needs to be joined in distances of up to several 100 meters only. The rigid parts of the envelope can be produced in transportable units within a factory. In the field, only the outer tubes -1- have to be welded together. Bends of up to several degrees are easily provided for simply by oblique connecting the outer tube sections, sharper bends are made by standardized elbows.

LAYOUT OF THE PROTOTYPE

The actual prototype consists of two 6 m long straight sections, -A- and -B-, two 0.8 m long, identical 30°-knees $-K_1-$ and $-K_2-$ with a bend radius of 1.6 m, a 3.4 m long inclined section -C-, and finally another 4,2 m long horizontal section -D-. The difference of level between the terminations $-T_1-$ and the elevated $-T_2-$ is 1.5 m. The system is designed for a test pressure of 20 bar. Fig. 2 shows the finished line.

The bore diameter of the corrugated inner cryogen duct -2- is 100 mm and lies thus in a relevant order of magnitude both for superconducting power transmission cables (~180 mm for a 1 GVA, 110 kV 3-phase cable) and cryogen transfer lines[4,5]. This corrugated stainless steel tube (110 mm o. D., 0.5 mm wall) was kindly provided by the Kabelmetal Electro GmbH, Hannover, FRG. The rigid spacer tubes -4- of the straight sections as well as of the knees (e.g. 4A, 4K), and the outer tubes -1- are commercially available carbon steel (St 37) pipes with standard dimensions (ϕ273/5 mm and ϕ127/4 mm; not optimized). The sleeves -6- for the loose telescopic connections of the spacer tubes consist of carbon steel too. The brittleness of the cold carbon steel is of no concern since the mechanical loads are static only and stress concentrations are avoided.

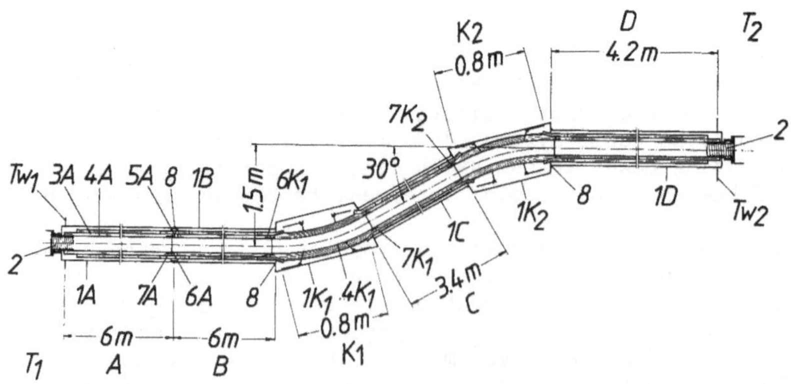

Fig. 1. Enclosure prototype: 1 - outer tubes; 2 - inner corrugated tube; 3 - superinsulation; 4 - spacer tubes; 5 - spacer ropes; 6 - steel sleeves; 7 - multi-layer sleeves; 8 - welding seams; A thru D - straight sections; K_1, K_2 - 30°-bends; T_1, T_2 - terminations; Tw_1, Tw_2 - termination walls.

Fig. 2. The finished enclosure.

The spacer -5A- consists simply of a Kevlar® rope attached at both ends to the outer tube -1A- such as to form an U-shaped loop in which the spacer tube -4A- is lying. The spacer plane is inclined with respect to the axis by an angle of 60° (Fig. 1) in order to reduce heat conduction and to minimize thermal radiation through the lead-in holes of the superinsulation layers. In a real pipeline, every other straight tube section would contain such spacers at each end. The inclination of the spacer rope plane would then also serve to fix the spacer tubes in axial direction. Apart from small axial forces during pulling in the inner corrugated tube, these spacers are only loaded by weight. During transport, the spacer tubes would have to be fixed by some provisional device. The spacer ropes and superinsulation layers are protected by metallic sheets from the welding arc at the welding seams -8-.

The outer tubes -1K- of the elbows are larger (ϕ400/4 mm) to accommodate the bent spacer tubes -4K- (ϕ127/3 mm, St37) and for ease of assembly. Forces in all directions are to be expected at the elbows, especially due to the cryogen pressure. Therefore, three ropes are arranged radially. Such spacers would be used also at straight, inclined pipeline sections, or where larger pull forces have to be supported.

The spacer tube -4B- of section -B- is supported telescopically by the sleeves -6A- of section -A- and -6K_1- of the first knee. The design of the straight sections -C- and -D- corresponds to section -B-, no separate spacer assemblies are required for the spacer tubes.

The superinsulation of the straight sections consists of 100 layers (2/mm) of 9 μm thick pure aluminum foil as it is commonly used in packing industry. A 32 μm thick polyester fibre fleece is inserted between the aluminum foils. This latter material is widely used as a dielectric component within electrical machinery. More expensive aluminized polyester foils are wrapped within the knees, for the overlapping superinsulation sleeves -7- (Fig. 1), and within the terminations. A carbon filled glass fibre paper is wound as a getter directly onto the cold spacer tubes.

The conductive and much of the radiative heat leakage to the line bore -RM_1- (Fig. 3) via the terminations is intercepted by separate guard

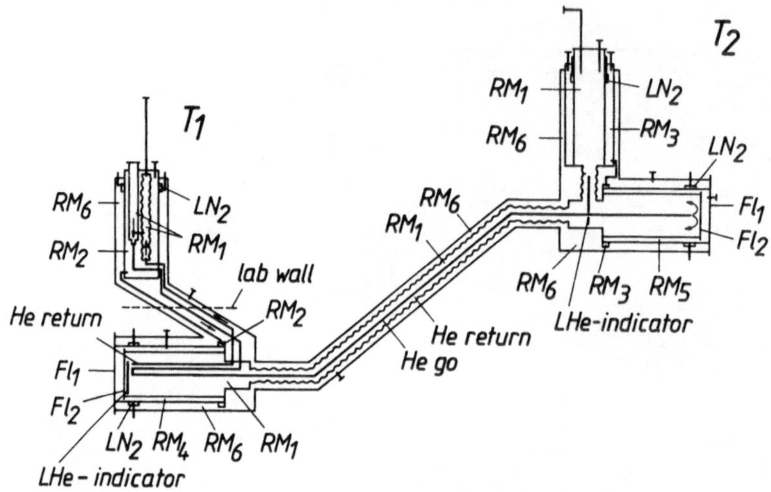

Fig. 3. Schema of the whole setup: T_1, T_2 - terminations; RM_1 - inner cryogen room; RM_2, RM_3 - guard rooms; RM_4, RM_5, RM_6 - vacuum rooms; Fl_1 - warm flange; Fl_2 - cold flange.

rooms (-RM_2-, -RM_3-) containing the same cryogen as the inner tube. The thermal insulation space -RM_6- of the envelope is extended to insulate also most of the terminations. The chambers -RM_4- and -RM_5- between the removable flanges -Fl_1- and -Fl_2- are evacuated separately. Thus the main vacuum does not have to be broken for access to the inner room -RM_1-.

The He go line within the inner corrugated tube -2- supplies cold He during cooling down from the refrigerator at T_1 to the termination -T_2-. This go line consists essentially of two concentric corrugated Cu-tubes (10 mm i. D., 53 mm o. D., respectively) and a 10 mm thick cable paper insulation within the annulus. The paper serves as a thermal insulator between the He go and return streams.

The vertical part of the termination -T_1- is erected within the lab as indicated in Fig. 3, the rest of the line lies outside (Fig. 2).

ASSEMBLY OF THE PROTOTYPE

The construction procedure was simple and relevant also for long lines. The straight, sandblasted spacer tubes were wrapped with superinsulation, inserted into the sandblasted outer tubes, and held there temporarily by provisional spacers. These assemblies were covered with plastic sheets and stored in a laboratory hall for 10 months while the indoor works on the terminations were carried on as far as possible. The vacuum-exposed surfaces were not cleaned any more before final assembly.

Both knees were completely assembled within the laboratory. This procedure would generally be preferable also with all the straight sections. However, in this special case the proceedings were somewhat different because one end of section -A- had to be joined to a termination. Therefore, the spacer rope -5A- was assembled only after connecting section -A- to the termination -T_1- at the final place of installation.

The laying work proceeded without major problems. Firstly, the outer tube -1A- was welded to the termination wall -Tw_1-. Then the spacer -5A- was installed simultaneously with the superinsulation sleeve -7A-. Section -B- was then telescoped to section -A- and arc-welded along the seam -8-. The other bent and straight sections were assembled correspondingly.

One mishap occurred during the assemblage of the inclined section -C- when the loose superinsulation pack slid somewhat towards the lower level knee $-K_1-$. The corresponding wider gap at the upper end was subsequently covered by the sleeve $-7K_2-$. However, the insulation either continued to slide, or the sleeve was dislocated or somewhat damaged during subsequent works as indicated later on by a slight additional heat input at this spot. In future lines, such problems may easily be avoided.

After assembly of this rigid thermal insulation system proper, the corrugated inner tube was pulled into the system. The maximal pulling force, mainly caused by the knees, was on the order of 800 N.

The outdoor works took altogether seven weeks until everything was sealed. Most of this time was needed for the terminations and instrumentation. The partly installed tubes and terminations were covered with plastic sheets overnight and during the weekends. Several rainy days and thunderstorms occurred within this period. The infiltration of humidity into the vacuum space was minimized by slowly blowing dry nitrogen gas therein during these times. The final leak tests, plumbing works, and evacuation took another month until cooling down could be started.

EXPERIMENTS

LN_2 - cooling

The prototype was filled with LN_2 within six hours while a vacuum pump was running at each end. The vacuum improved considerably during cooling from 8×10^{-5} mbar to 3×10^{-6} mbar at the terminations. After a few days, the pumps were switched off for more than a month until the end of the LN_2 - experiments. During this period, the vacuum readings oscillated between 3×10^{-6} mbar and 10^{-5} mbar, depending on the liquid level within the system and the outside temperatures (in summer 1986). It took six days until the spacer tubes -4- reached their equilibrium values of 110 K (Fig. 4). Time constants of several days for the superinsulation layers (depending mainly on the layer number) to reach their temperature equilibrium are to be expected with any such insulation system.

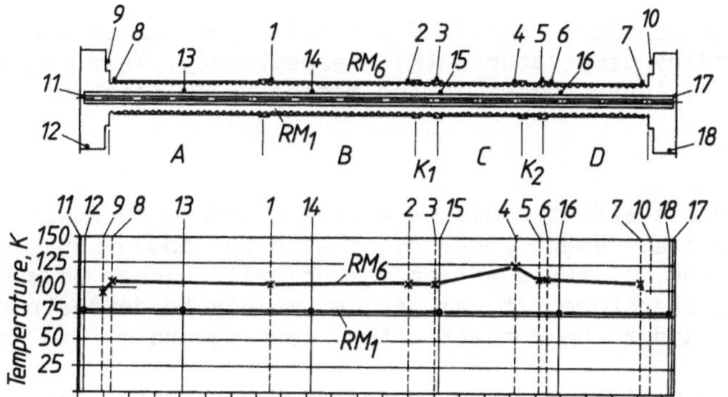

Fig. 4. Temperature distribution within the LN_2-filled enclosure: 1 thru 10 - temperature sensors within the vacuum room RM_6; 11 thru 18- temperature sensors within the cryogen room RM_1. Each sensor consists of a Cu-constantan thermocouple and a carbon resistor.

The spacer tube thermocouple No.4 indicated a temperature of 122 K. This is interpreted as being due to the mentioned injury of the insulation. This assumption was supported by the subsequent LHe experiments.

The loss measurements were started three weeks after cooling down. Three methods were employed: The rate of descend of the LN_2 - level at the termination $-T_2-$ was measured with a float gauge, the evaporation rate with a flow meter, and the integral evaporation rate with a gas counter. Condensation effects within the measuring chamber $-RM_1-$ were avoided by increasing the pressures of the screening rooms $-RM_2-$ and $-RM_3-$ to values slightly above the corresponding pressures within $-RM_1-$. Such effects could have been caused mainly by the LN_2 of the guard room $-RM_2-$ at the lower level termination $-T_1-$ (Fig. 3).

Three measurement cycles, each lasting more than 30 hours, were performed. The results obtained with the three methods per cycle agreed well and yielded a loss rate of 14.5±0,5 W. Taking into account the radiation losses at the terminations, the losses relative to the corrugated inner tube were ~0.6 W/m.

The heat leak indicated by sensor No. 4 (Fig. 4) was estimated by putting up a heat balance differential equation for this spot. From this follows an additional leak of 2 W. Without the insulation injury, the losses of the cryogenic envelope would thus be ~0.5 W/m.

With reference to a mean insulation diameter of 180 mm, this value corresponds to a loss rate of 0.88 W/m^2 (1.25 W/m^2 with respect to the spacer tube diameter). With an apparatus according to references 8 or 9, almost the same loss rate of 0.86 W/m^2 with respect to the mean diameter (1 W/m^2 relative to the cold surface) was achieved with 20 layers only. The layer density was in both cases 2/mm. This result agrees well with other experiments known from literature indicating an almost constant heat flux for a total layer number of more than 20 to 30 layers[10-12]. In future lines only 30 layers may be employed without deteriorating the insulation quality significantly. The measured 20-layer-value agrees reasonably with the calculated loss rate of 0.7 W/m^2 resulting from the semiempirical equation[13]

$$q = \frac{44.8\cdot 10^{-9}\, n^{2.56}}{N}(T_h^2 - T_c^2) + \frac{16.2\cdot 10^{-12}}{N}(T_h^{4.67} - T_c^{4.67}) \quad W/m^2, \qquad (1)$$

with n = layer density (cm^{-1}), N = total layer number and T_h, T_c = hot and cold boundary temperatures (K), respectively. This equation is valid for a low total layer number in the region below constant heat flux.

LHe - cooling

Considerable effort would have been required to achieve with LHe the same loss measurement accuracy as with the above LN_2 experiments. According to Equ. 1, there are only marginally higher losses to be expected at 4.2 K. Therefore, it was decided mainly to demonstrate the operation with LHe and to be content with a less accurate loss measurement.

After having finished the described experiments, the LN_2 was evaporated with a heater at the lowest point of room $-RM_1-$, and the remaining gas sucked off. Thereafter, the interior of the system was flushed with pure He for several hours. By the time the cooling down with the He refrigerator began, the temperatures of the inner tube and the spacer tubes were on the order of 140 K each.

When the inner pipe temperature was about 50 K, a small cold leak of the main vacuum room -RM6- appeared probably within a termination at one of the numerous brazing or welding seams. However, the vacuum could easily be held at $3x10^{-6}$ mbar with a small pump. The problem was irrelevant regarding these experiments and was thus not investigated further.

A temperature of 7 K was reached with the available refrigerator power. The cooling was continued by filling LHe from a storage tank into the termination $-T_1-$. The liquid accumulation was observed with a LHe indicator at T_1 until the liquid reached the inclined section -C-. Because of the high vapor portion (>80% gas/volume due to the overall losses) towards the gas outlet, only a wavering amount of LHe could be detected by the indicator reaching from the vertical part of T_2 down to the go line level (Fig.3).

The LHe-temperature of 4.25 ± 0.05 K was achieved along the whole length of the inner tube and at the bottom of T_2 (sensor No.18, Fig. 4). The spacer tubes reached stable temperature values of 72 K. Again, the sensor No. 4 showed a higher temperature of 93 K.

Two loss measurement cycles using a gas flow meter were performed within two hours. At the beginning of each cycle, the LHe supply to T_1 was stopped. The evaporation rate was on the order of 6 cm^3/s LHe. This rate allowed enough time to perform the measurements without having to worry about LHe drying out somewhere within the inner tube bore volume of 160 L. A vapor flow velocity <1 cm/s at the end of the corrugated inner tube was estimated. According to the Baker diagram, this means stratified two-phase flow. Therefore, the losses of the straight section -D- might not be represented completely by the measured evaporation rate.

The losses, corrected for the calculated additional 2 Watts at the injury, came out to be 14 W. Assuming that only half of the losses of section -D- were available for evaporating the LHe, the overall losses are estimated to be 16 W. This corresponds to ~ 0.6 W/m of the line proper without the terminations and agrees reasonably with the LN_2-experiments.

SUMMARY AND CONCLUSION

A new type of an enclosure for cryogenic power transmission cables and gas pipelines was demonstrated to be feasible. The advantages of this system are low material, manufacturing, and laying costs, as well as good thermal insulation quality. The loss rate of the prototype with an inner cryogen duct diameter of 100 mm is ≤ 0.6 W/m for cryogen temperatures down to 4,2 K.

Special attention was paid to choose the cheapest possible materials for the superinsulation and the rigid cold and warm tubes. The potential for automatically wrapping the superinsulation was demonstrated. The economy of a future optimized version of such a line may still be improved considerably by reducing the number of insulation layers by a factor of three and choosing smaller rigid tube dimensions.

Two spacer assemblies were developed and tested: A very simple one for horizontal line sections to support the weight of the inner parts and moderate axial forces only, the other type for larger axial forces and curved line sections. Standardized elbows can be used without any significant complication of the laying works. No special care was taken with regard to cleanliness of the tube surfaces exposed to the vacuum. For example, the sandblasted carbon steel pipes were stored within the laboratory for 10 months without further cleaning them before assembly.

The outdoor works were performed under realistic conditions, including several rainy days and thunderstorms during this time. In spite of this, the vacuum performance at the LN_2 - experiments was excellent.

In order to study the long term behavior of the prototype, the permanent vacuum will be continually observed in future. One vacuum measurement was done already six months after finishing the experiments, the reading was still 3.5×10^{-2} mbar. A LN_2-cooling experiment without using vacuum pumps is intended to be performed from time to time.

The measured prototype heat load would correspond to ~ 1 W/m thermal losses of a 1 GVA, 110 kV three phase superconducting power transmission cable. If the new high T_c-superconductors should turn out to be technically feasible for superconducting electrical energy transport at LN_2-temperatures, the presented enclosure would be well suited for such lines too.

The heat losses of long LHe transfer lines are usually required to be an order of magnitude lower than achievable without intermediate thermal shields. The inner corrugated tube of the described envelope would in such an application advantageously contain one or more separately vacuum-insulated LHe ducts surrounded by LN_2 or cold He gas serving as the thermal shield.

This enclosure may also be an alternative to other thermally insulated pipelines for cryogenic fluids such as He gas, LH_2, LN_2, or LNG. Obviously, a trade-off considering capital and operating costs can only be established for a particular case.

ACKNOWLEDGMENTS

The authors revere the memory of the late Prof. Klaudy, one of the pioneers of superconductor-technology. He was the main initiator of this project.

The work was supported by the Österreichische Nationalbank, the Verband der Elektrizitätswerke Österreichs, and the Österreichische Elektrizitätswirtschafts-AG.

REFERENCES

1. E. B. Forsyth, Cryogenics 17:3 (1977).
2. P. A. Klaudy and J. Gerhold, IEEE Trans. Magn., MAG-19, 3:656 (1983).
3. E. B. Forsyth and R. A. Thomas, Cryogenics 26:599 (1986).
4. I. Horvath et al, in "Proc. Eleventh Intl. Cryo. Engr. Conf.", Butterworth, Guildford, UK (1986), p. 522.
5. C. Rode et al, in: "Advances in Cryogenic Engineering", Vol. 27, Plenum Press, New York (1982), p. 769.
6. H. Blessing et al, in: "Advances in Cryogenic Engineering", Vol. 27, Plenum Press, New York (1982), p. 761.
7. F. Schauer, in "Proc. Tenth Intl. Cryo. Engr. Conf.", Butterworth, Guildford, UK (1984), p.731.
8. R. H. Kropschot, Cryogenics 1:171 (1961).
9. R. G. Scurlock and B. Saull, Cryogenics 16:303 (1976).
10. Q. S. Shu, R. W. Fast, and H. L. Hart, Cryogenics 26:671 (1986).
11. E. M. W. Leung et al, in: "Advances in Cryogenic Engineering", Vol. 25 Plenum Press, New York (1980), p. 489.
12. T. R. Catright and P. A. Reeve, in: "Proc. Ninth Intl. Conf. on Magn. Technology", SIN, Villigen, Switzerland (1985), p. 696.
13. T. C. Nast, D. J. Frank, and I. E. Spradley, in "Proc.Eleventh Intl. Cryo. Engr. Conf.", Butterworth, Guildford, UK (1986), p. 561.

DESIGN, DEVELOPMENT, AND TEST OF SHUTTLE/CENTAUR G-PRIME CRYOGENIC TANKAGE THERMAL PROTECTION SYSTEMS

Peter N. MacNeil and James E. England

General Dynamics/Space Systems Division
San Diego, California

Richard H. Knoll

NASA-Lewis Research Center
Cleveland, Ohio

ABSTRACT

The thermal protection systems (TPS) for the Shuttle/Centaur were designed to provide fail-safe thermal protection during prelaunch, launch ascent, and on-orbit operations as well as during potential abort, where the Shuttle and Centaur would return to Earth. The TPS selected used a helium-purged polyimide foam beneath three radiation shields for the liquid-hydrogen (LH2) tank and radiation shields only for the liquid-oxygen (LO2) tank. A double-walled vacuum bulkhead separated the two tanks. The LH2 tank had one 1.9 cm-thick layer of foam on the forward bulkhead and two layers on the larger-area sidewall. Full scale tests of the flight vehicle in a simulated Shuttle cargo bay gave total prelaunch heating rates of 25.9 and 12.9 kW for the LH2 and LO2 tanks, respectively. Calorimeter tests on a representative sample of the LH2 tank sidewall TPS indicated that the measured unit heating rate would rapidly decrease from the prelaunch rate of $\approx 300\ W/m^2$ to a desired rate of $< 4\ W/m^2$ once on-orbit.

INTRODUCTION

The Shuttle/Centaur (S/C) G-prime (G') vehicle, using liquid hydrogen (LH2) and liquid oxygen (LO2) as propellants, was designed for use in the Space Transportation System (STS) and afforded a means of significantly increasing the payload capability of the STS[1]. Although the S/C project has been terminated, the design, development, and test history of the cryogenic tankage TPS used for the vehicle are reviewed to possibly benefit future programs requiring the transfer of cryogenics to Earth orbit.

The TPS for the Centaur G' LH2 and LO2 tanks were unique in that they were designed to provide thermal protection during prelaunch, launch-ascent, and on-orbit operations as well as during abort operations, where the Shuttle and its cargo would return to designated landing sites. Other cryogenic stage TPS such as the Shuttle External Tank, the Saturn, and the Atlas/Centaur[2] provide protection primarily during prelaunch and aerodynamic boost. The Titan/Centaur[3] also had to provide thermal protection for an extended time on-orbit, but none of the expendable vehicles had a requirement to return to Earth. In addition to affording thermal protection it was imperative that the Centaur G' TPS not be hazardous to the Shuttle or its crew. This report emphasizes the LH2 tank TPS as it had to meet the design criteria imposed on all S/C TPS[4] plus had the additional requirement, because of the lower temperature of the LH2 tanks, of precluding formation of liquid air or liquid nitrogen on any of its surfaces.

SYSTEM DESIGN REQUIREMENTS

Thermal Requirements for Mission

The S/C G' vehicle[1] (Figure 1) was originally designed to propel the Galileo and Ulysses spacecraft from low-Earth orbit to their respective destinations of Jupiter and a polar orbit about the Sun (via Jupiter). Nominal mission time was 6 hr from STS launch to separation from the orbiter. An additional 3 hr was allowed for deployment delay, giving a total of 9 hr. During this time period, the thermal environment imposed on the vehicle would vary from a room-temperature, gaseous-nitrogen-purged environment at launch to the near vacuum of space with a varying radiant heat flux once on-orbit. Additionally in case of an abort the orbiter cargo bay would be refilled with atmospheric air during its descent to Earth. In order to accomplish the intended missions, in this rather severe and varying environment, the following thermal criteria were imposed:

(1) The TPS shall be designed to prevent liquefaction of gases (air or nitrogen) on its external surfaces.
(2) The prelaunch heating rates of the LH2 and LO2 tanks shall be less than 30.2 and 11.7 kW, respectively.
(3) The maximum on-orbit heating rates through the LH2 tank forward bulkhead and sidewall TPS shall not exceed 0.5 and 0.3 kW, respectively.
(4) The maximum on-orbit heating rates through the LO2 tank sidewall and aft bulkhead shall not exceed 0.051 and 0.053 kW, respectively.

Requirements Imposed by Shuttle Cargo Bay

As a payload in the STS the Centaur G' had to meet the Shuttle safety requirements[5,6] and had to withstand the STS induced environments[7]. Some of the more important safety criteria influencing the TPS design were as follows:

(1) Materials must be noncombustible or self-extinguishing in the upward flame propagation test[8].
(2) The TPS shall have an ultimate safety factor of 1.4 or more and shall be capable of withstanding limit loads without loss of function and ultimate loads without failure.
(3) All systems shall be two-failure tolerant against catastrophic hazards.
(4) Any material exposed to gaseous or liquid oxygen must pass the impact sensitivity tests[8].
(5) All hardware with metalized surfaces shall be electrically bonded per MIL-B-5087. Cargo bay hardware with volume resistivities greater than 10^9 ohm-cm shall not accumulate an electrical charge.

STS-induced structural loads considered in the design of the TPS were launch and emergency landing loads, mechanical loads due to structure-borne or airborne (i.e., acoustical) excitations, loads due to rapid pressure changes in the cargo bay during ascent and abort descent, and localized loads on the TPS due to flow impingement near the cargo bay vent ports.

Finally all materials used in the cargo bay had to meet strict outgassing and cleanliness requirements.

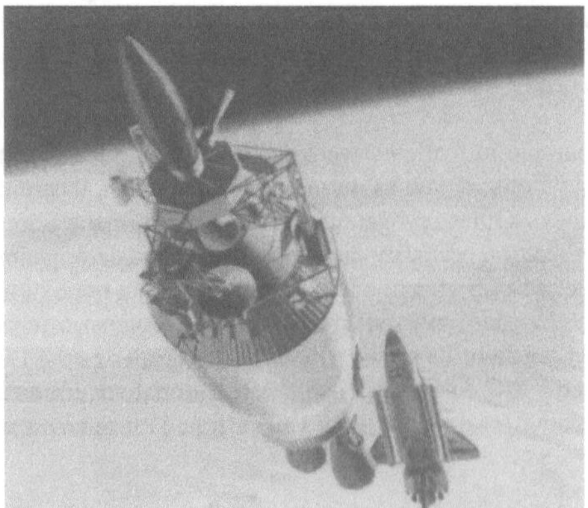

Fig. 1. The Shuttle/Centaur and Galileo spacecraft.

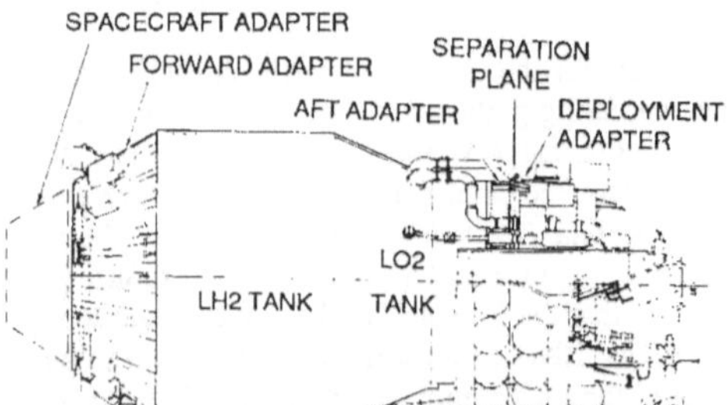

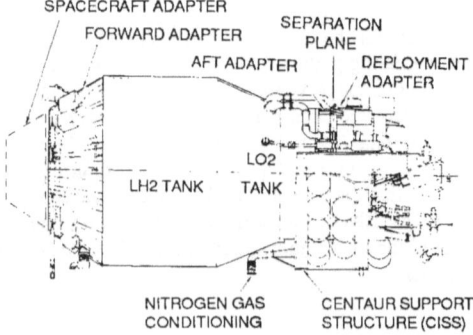

Fig. 2. Centaur in the CISS.

SYSTEM DESIGN

The Centaur vehicle is supported by a cylindrical deployment adapter on its aft end and by a three-point attachment system on its forward end (Figure 2). The Centaur Integrated Support System (CISS) was designed to provide all the necessary structural, fluid, and electrical interfaces for the vehicle. The basic vehicle (Figure 3) had the following key thermal features: (1) the forward adapter's highly conductive aluminum conical section (where most of the avionics were mounted) was isolated from the LH2 tank by the forward adapter's composite cylindrical section; (2) the LH2 and LO2 tanks were separated by a double-walled bulkhead that contained a cryopumped (vacuum) TPS to limit gaseous conduction; and (3) a low-conducting composite adapter was also used between the aft end of the LO2 tank and the warmer aluminum deployment adapter of the CISS.

The helium-purged TPS on the LH2 tank was required to prevent liquefaction of the cargo bay gaseous-nitrogen purge during prelaunch and air during abort operations. The helium purge was contained by the forward adapter purge diaphragm, the forward adapter itself, the innermost sidewall radiation shield, which was external to the foam and acted as a sealed membrane, and the purge plenum, which sealed the system on the aft end of the LH2 tank (Figure 4). The foam insulation was used to achieve sufficient thermal resistance to meet the ground-hold heating criteria and to ensure that all the radiation shield temperatures exceeded the liquefaction temperatures of the surrounding gases. The radiation shields themselves afforded very little thermal protection during ground-hold operations but would provide nearly all the thermal protection once the shields were evacuated on-orbit.

The helium purge would begin roughly 1 hr before the cryogenic propellants were loaded. Prior to this time gaseous nitrogen would be maintained in the purge volume to preclude any damage to the shield surfaces from moisture condensation. At helium purge initiation the insulation blanket ΔP limits were set high enough to allow the ΔP within the purge volume to exceed the nominal 2.75 kPa necessary to open the relief valves. Once the valves were opened, the heavier gaseous nitrogen would be forced out by the lighter helium (Figure 5). After 1 hr the blanket ΔP control limits would be reset (via software) to maintain the nominal ΔP between 0.7 and 2.1 kPa, thus closing the relief valves. At this point the helium purge supply would be governed by the leakage rate of the insulation blanket. These ΔP limits would be generally maintained until just before launch, when the limits would be lowered preparatory to venting the cargo bay at lift-off. This was necessary since the cargo bay itself is slightly pressurized (3.5 to 4.8 kPa) by its gaseous-nitrogen purge. At launch a vent door (not shown) in the forward adapter would be opened, and the pressure within the contained purge volume would then closely follow that of the cargo bay during ascent. In an abort a second vent door would reseal the forward adapter, and the helium purge would be reinitiated.

The LO2 tank used radiation shields only and required no purge since there was no danger of liquefying the gaseous-nitrogen purge or air on the warmer LO2 tank surface. Because of the heat capacity and relatively small surface area of the LO2 tank, little or no thermal protection was needed for the ground-hold phase. The shields were used primarily to afford thermal protection on-orbit. Three shields were used on the tank sidewall and four on the aft bulkhead.

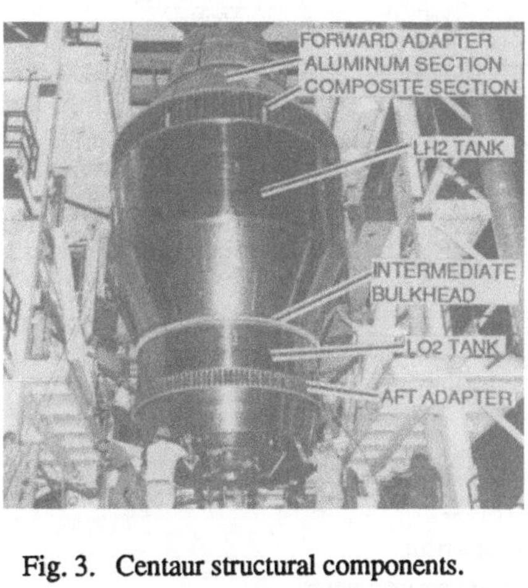

Fig. 3. Centaur structural components.

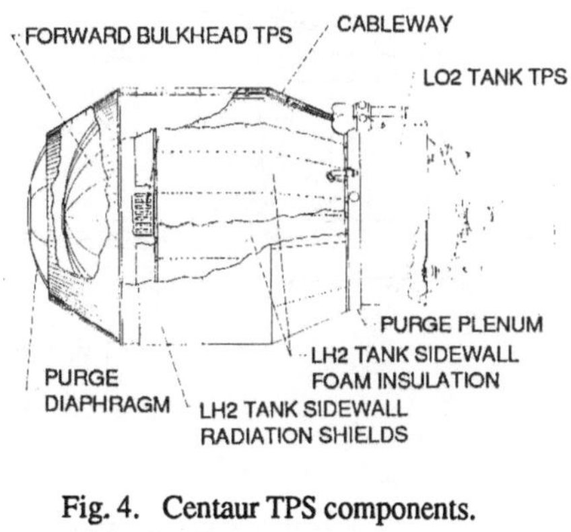

Fig. 4. Centaur TPS components.

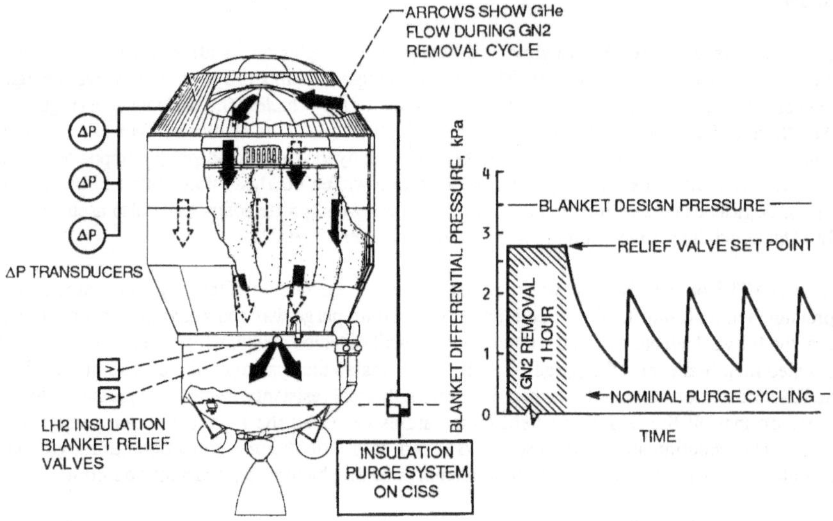

Fig. 5. TPS purge system operation and flow paths.

DESIGN OF LH2 TANK TPS

The purge containment diaphragm on the forward end of the vehicle consisted of two Kevlar-cloth-reinforced shields separated by an embossed Kapton shield (Figure 6). The high-strength Kevlar-reinforced shields were required to withstand a purge system design ΔP of 3.45 kPa. All diaphragm shield surfaces had a vapor-deposited layer of aluminum (VDA) applied to achieve emittances of 0.05 or less. Both of the outer two shields were broadside vented so that they could be rapidly evacuated during ascent to eliminate most of the gaseous conduction heat transfer. The nonvented inner shield provided the required seal for the helium purge. The diaphragm was constructed from gore sectors sewn at their adjoining seams. Each gore sector contained three radiation shields. The sewn joint seam had aluminized (VDA) tape applied both outboard and inboard. The inboard tape had a thermal plastic adhesive to help ensure an adequate seal. A Kevlar strip was sewn to the shields to provide a rigid edge member for attachment to the forward adapter. A silicone sealant was applied to the edge member and the forward adapter to ensure a proper seal.

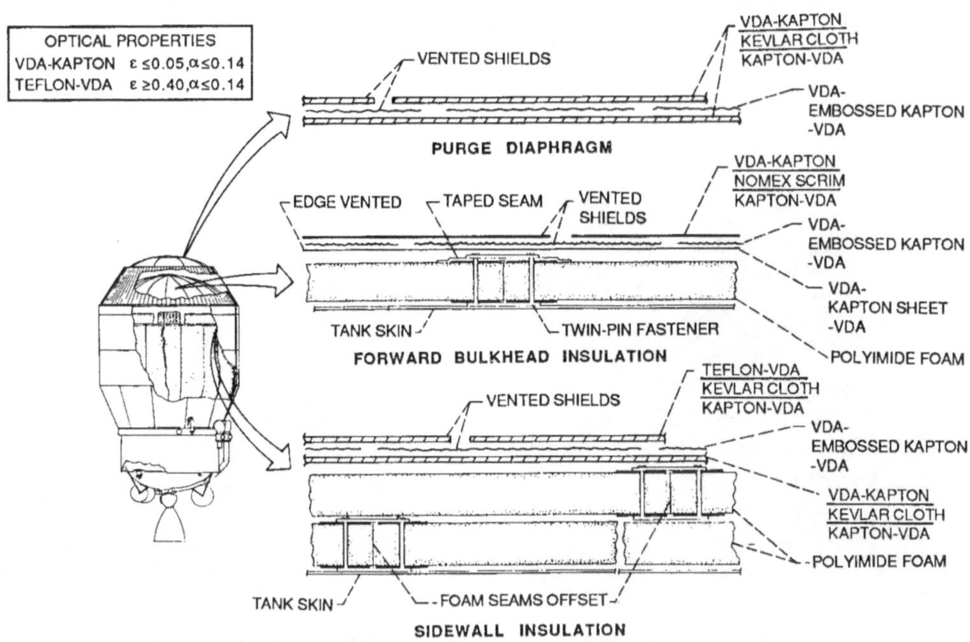

Fig. 6. Primary TPS materials and arrangements.

The forward bulkhead TPS consisted of a 1.9 cm-thick layer of open-cell polyimide foam covered by three radiation shields. The outermost shield is Nomex scrim reinforced. This material was selected for its low weight and rip-resistant features. The higher strength material used on the purge diaphragm was not required here since the forward bulkhead insulation was completely contained within the helium-purged volume and therefore was not exposed to any pressure loading. The center shield was embossed Kapton and the innermost shield was a flat Kapton sheet, and all three shields had a VDA coating on both surfaces. The foam insulation used a typical twin-pin fastener to attach adjacent foam panels. A layer of Tedlar tape was applied over the seam to minimize direct convection paths between the shields and the tank surface. The insulation assembly was held in place by the radiation shields, which were attached with Velcro to the inner surface of the forward adapter. The entire TPS on the forward bulkhead, including the foam, the shield, the fasteners, etc., had a mass of 49.6 kg or about 2.0 kg/m^2.

The LH2 tank sidewall TPS consisted of two layers of polyimide foam (each 1.9 cm thick) covered by three radiation shields. The shields were identical to those used on the purge diaphragm except the outermost laminate material had a 0.013 mm-thick layer of Teflon with an internal VDA coating to achieve a relatively low ratio of solar absorptance to thermal emittance. This low ratio was required to help minimize heating from the Sun and the Earth's albedo while on-orbit. Material specifications for all outboard radiation shield surfaces required that the solar absorptance be less than or equal to 0.14 and the thermal emittance be greater than or equal to 0.4 to give a solar absorptance to thermal emittance ratio of less than 0.35. Each layer of the LH2 tank sidewall foam insulation was composed of several longitudinal panels, the seams of which were offset between inner and outer layers to minimize direct convection currents. The foam panels were held at the fore and aft ends by support channels attached to the forward tank ring and to tank brackets at the aft end of the tank. The foam panels had Kevlar reinforcements bonded to the edges. The LH2 tank sidewall radiation shield was assembled from three circumferential panels and thirteen conical sectors. Sewn joints were used to join adjacent panels then a VDA-coated tape with a thermal plastic adhesive was applied to the inner surface, and a VDA-backed Teflon tape was applied to the outer surface. These tapes approximately matched the optical properties of the shield materials. A Kevlar strip was sewn to the shield to provide a rigid edge for attachment. A sealant was used on the inboard surface of the reinforcing strip to complete the seal (Figure 7). The plenum, constructed of Kevlar/epoxy, seals to the LO2 tank, provided an attachment for the LH2 sidewall shield, and provided mounting locations for the two purge relief valves. The entire inner surface of the plenum was lined with a 1.9 cm-thick layer of polyimide foam. The completed sidewall TPS (including the shields, the foam layers, the fasteners, the retainer rings, etc.) had a mass of 262.8 kg or about 5.93 kg/m^2.

The particular materials selected met the flammability and outgassing requirements of the Shuttle cargo bay and in many cases were the same general materials used extensively in the cargo bay (e.g., aluminized Kapton).

The predicted heat transfer rates (Figure 8) indicate that the forward bulkhead and sidewall are the predominant heat transfer contributors during ground hold, whereas the common vacuum bulkhead would be the major contributor once on-orbit. Shortly after launch the expected heating rates drop by roughly a factor of 100 once the pressures within the insulation approach near-vacuum conditions. The radiation shield temperatures and heating

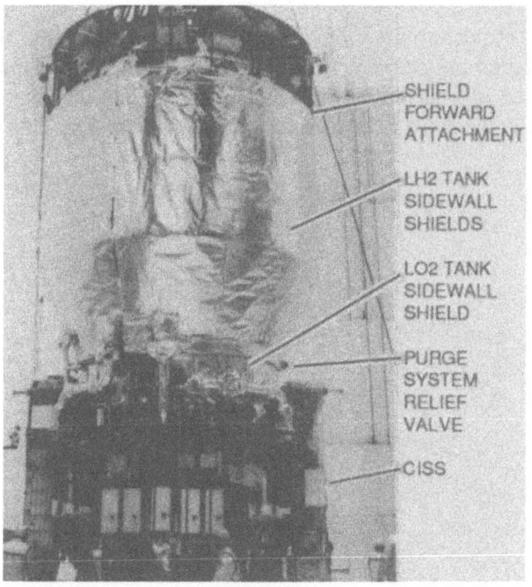

Fig. 7. Complete Centaur in CISS.

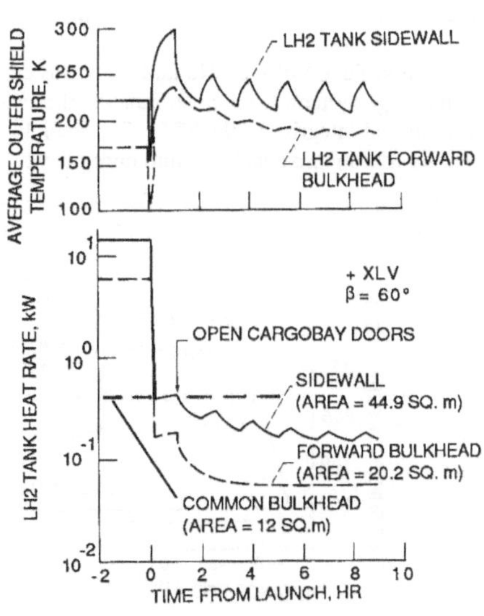

Fig. 8. Tank heating predictions.

rates then rise until the cargo bay doors are opened at 1 hr after launch. At this point the temperatures generally decrease in a cyclic fashion (due to day-night cycles) as steady-state conditions are approached.

DEVELOPMENTAL TESTING

The TPS materials were validated by several small scale tests including tensile strength, coefficient of thermal expansion, purgability, rapid depressurization, heat formability, and flexibility at cryogenic temperatures[4]. These tests showed that the materials chosen were more than adequate and very durable even at cryogenic temperatures.

The performance of the LH2 tank TPS during prelaunch, launch-ascent, and on-orbit was simulated with a 24.1 cm-diameter double-guarded calorimeter. The temperature and heat transfer data taken was quite consistent and showed the expected effects of decreasing pressure. That is, the foam temperatures decreased rapidly, and the radiation shields began to approach a fourth power temperature profile. The heat transfer data indicated that, for the lower pressures expected on-orbit (i.e., 10^{-4} torr or less), heat transfer rates on the order of 3 to 4 W/m^2 will be achievable with external shield temperatures of 258 K.

One of the requirements for the Shuttle cargo bay hardware was that any surface exceeding a volume resistivity of 10^9 ohm-cm be designed to prevent the accumulation of an electrostatic charge. Volume resistivity of the Teflon film on all outboard radiation shields exceeded 10^{15} ohm-cm. Electrostatic charging tests conducted showed the arc energy of a possible discharge was less than 0.0017 mJ which is 10 times less than the minimum to sustain a reaction[9] for the most explosive mixture of hydrogen and air.

There were concerns that high velocity flow near open cargo bay vents could cause flutter or damage to the radiation shields during Shuttle ascent and descent. Since it was not possible to predict the flutter- or flow-induced oscillatory stresses, a test was conducted which subjected the shields to the STS prime contractor's estimated conditions. The TPS was not damaged or degraded in any way by these tests.

During an abort of the Shuttle after lift-off the Centaur propellant tanks would be emptied (some residual propellants would remain) and the LH2 tank TPS helium purge would be reinitiated for a descent and landing. If for some reason the helium purge was unexpectedly terminated or was not reinitiated, air could be ingested into the sidewall TPS and eventually condense on the cold liquid hydrogen tank surface (≈ 18 kg of liquid air). Concerns for this failure scenario were that liquid air (1) could possibly compromise the structural integrity of the radiation shields or (2) could form on the external surfaces of the radiation shields and present a potential hazard to the Shuttle and its crew. All the liquid was assumed to collect on the inner shield (vehicle in horizontal position). As a result of these concerns a small-scale experiment was set up to demonstrate the liquid-air containment capabilities of the sidewall radiation shields (Figure 9). There was no evidence of structural degradation of the shield and no evidence of liquid runoff during the entire test.

As there was a potential of forming liquid air within the LH2 tank TPS after single-point failures testing was planned to prove that the TPS was not impact sensitive in a liquid-air or a gaseous-oxygen environment. These tests were underway when the S/C program was cancelled. If the results of the uncompleted impact testing were unfavorable steps would have been taken to add another level of redundancy to events causing loss of helium purge thereby eliminating the possibility of failure.

A full-scale test of the LH2 tank TPS was conducted: (1) to determine the performance characteristics of the helium purge system, (2) to determine the thermal characteristics and performance of the TPS, and (3) to evaluate the structural integrity of the TPS. The vehicle was enclosed in a shroud during testing to simulate the Shuttle cargo bay during prelaunch. Liquid nitrogen was used in the uninsulated LO2 tank for safety reasons. The TPS on

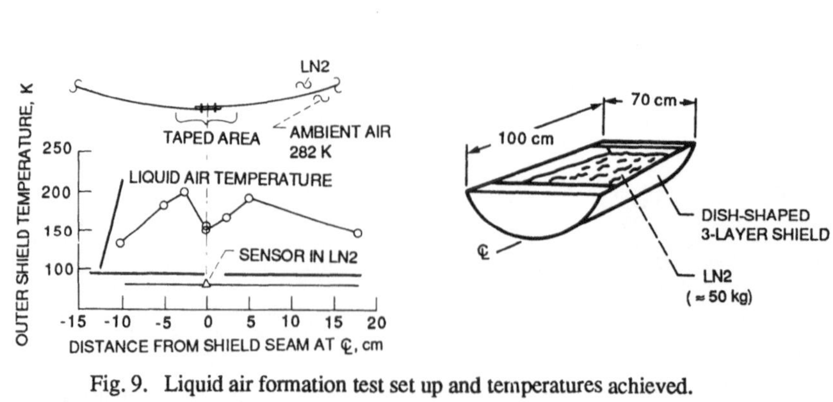

Fig. 9. Liquid air formation test set up and temperatures achieved.

the LH2 tank consisted of the flight design with some minor variations to accommodate the various full-scale vehicle tests planned.

The measured helium leakage rates during purge system tests were well within the leakage goals. The results of two gaseous nitrogen displacement tests indicate that one hour allowed for this cycle is more than sufficient for removing most of the gaseous nitrogen from the system. The two relief valves, which were designed to open between 2.45 and 3.15 kPa, opened at 2.82 and 2.84 kPa. The control system for supplying the helium to the blanket performed flawlessly. Typical pressure histories are given subsequently in the following section.

The measured heat transfer rate of 24.2 kW into the LH2 tank during the thermal performance test was well within the maximum and minimum rates of 22.5 and 25.5 kW predicted for the test article. These rates differed slightly from the flight vehicle predictions because of the various peculiarities associated with the test (e.g., using liquid nitrogen in the LO2 tank and different tank penetrations on the forward end). Typical measured temperatures are compared with flight vehicle data in the following section. Overall, the thermal performance of the system was as expected.

At the conclusion of the purge and thermal performance testing several structural tests were performed on the helium containment shield (innermost shield). The first test consisted of a rapid pressurization to simulate the maximum rate of pressure change expected on the TPS during Shuttle ascent - about 2.1 kPa/sec. The second test imposed a proof pressure of 3.45 kPa on the system, and the final test pressurized the shield until it failed (burst test). The shield system withstood the rapid pressurization tests and the proof pressure test without failure. In the burst test the blanket failed at 4.55 kPa, well above the design proof pressure of 3.45 kPa.

The first flight vehicle and its CISS were mounted in a simulated cargo bay at Cape Kennedy for a terminal countdown demonstration (TCD). In this test the vehicle was tanked and controlled by its onboard computer systems up to a simulated abort just prior to lift-off. Two TCD's were performed. The steady-state temperatures of the TPS are shown in Figure 10 for the various flight temperature transducers used on the vehicle. For reference the temperature data acquired on the full-scale developmental tests (previous section) are included where the sensors are in the same general location. In the forward bulkhead area the lower temperatures measured for the flight vehicle were expected since the shields on the developmental test vehicle were more loosely fitting. The steady-state heat transfer rates for the LH2 and LO2 tanks as determined from boil-off tests were 25.9 and 12.9 kW, respectively.

The LH2 tank insulation blanket purge system performed flawlessly throughout the simulated prelaunch countdown. Figure 11 shows a typical pressure history of the blanket demonstrating the cycling of the blanket at 5 min before the planned launch. Prior to T-5 min the blanket was in its nominal cycle mode. At T-5 min the control band was changed via software to prepare for the events at lift-off. The control band was tightened so that the peak differential pressure was low enough to ensure that the blanket ΔP would not exceed 2.41 kPa when the cargo bay vents were opened at T-41 sec. The average ΔP of the blanket was also raised in preparation for terminating the purge at T-20 sec to ensure that the ΔP would still be positive at lift-off and prevent any backfilling with air. Post-test inspection of the TPS after the two TCD's indicated that no structural damage occurred.

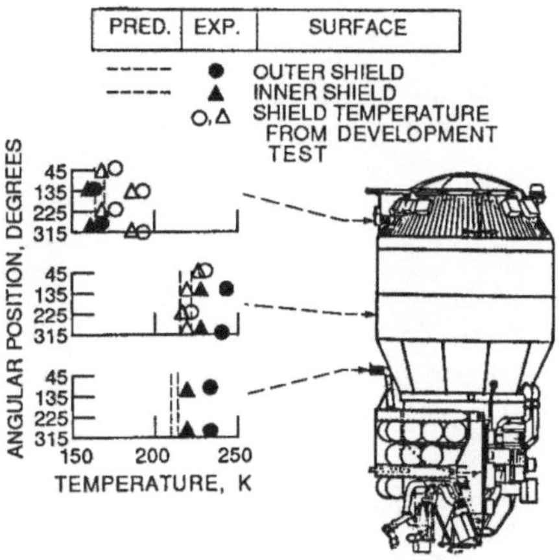

Fig. 10. Flight vehicle TPS prelaunch performance.

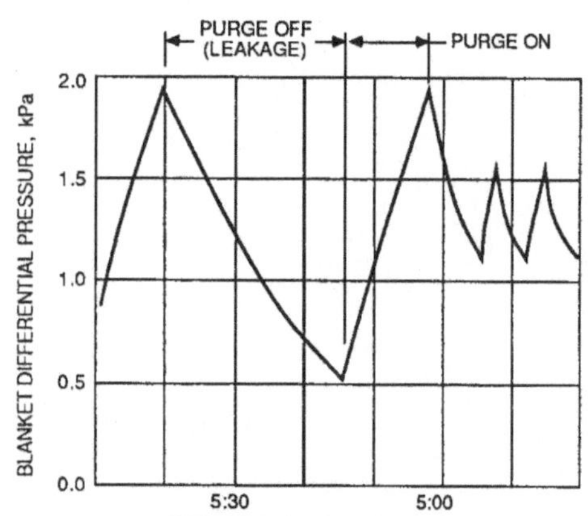

Fig. 11. Helium purge pressure cycling.

CONCLUDING REMARKS

The helium-purged foam and radiation shield blanket concept for the LH2 tank proved to be lightweight, rugged, and reliable. The high-strength Kevlar-cloth-reinforced shields were more than adequate for withstanding all the pressure loads imposed on the system. The measured thermal performance of the LH2 tank TPS during simulated prelaunch conditions was essentially as predicted, and limited calorimeter data indicated that thermal performance on-orbit would have been adequate for the intended missions. The helium purge system for maintaining a helium environment around the LH2 tank during all operations within the atmosphere performed flawlessly.

A considerable amount of development effort was necessarily devoted to safety issues. The basic thermal protection system materials had to, and did, meet strict cargo bay flammability, cleanliness, and outgassing requirements. The basic system design exceeded the structural design requirements for cargo bay payloads and showed no degradation after extensive testing during the terminal countdown demonstrations.

REFERENCES

1. O.F. Spurlock, "Shuttle/Centaur - More Capability for the 1980's," IAF Paper 83-18, Oct. 1983.
2. "Atlas-Centaur AC-19 and AC-20 Performance for the 1969 Mariner Mars Missions," NASA TM X-2278, 1971.
3. R.F. Lacovic, "Thermodynamic Data Report for the Titan/Centaur TC-5 Extended Mission," NASA TM X-73605, 1977.
4. R.H. Knoll, P.N. Mac Neil, and J.E. England, "Design, Development, And Test Of The Shuttle/Centaur G-Prime Cryogenic Tankage Thermal Protection System," NASA TM-89825, 1987.
5. "Safety Policy and Requirements for Payloads Using the Space Transportation System (STS)," NASA TM-85402, 1982.
6. "Payload Ground Safety Handbook," KHB-1700.7, Rev. A.
7. "Shuttle System Requirements For Shuttle/Centaur Stage And Airborne Support Equipment," NASA Johnson Space Center, JSC-07700, Vol. X, Appendix 10.16, Dec. 1985.
8. "Flammability, Odor, And Outgassing Requirements And Test Procedures For Materials In Environments That Support Combustion," NASA TM-84066, 1981.
9. E.C. Magison, "Electrical Instruments In Hazardous Locations," Third Revised Edition, Instrument Society Of America, Pittsburgh, 1978.

SOUND INDUCED ENHANCEMENT OF HEAT TRANSFER FROM A SOLID INTO LIQUID HELIUM

Erik Bodegom, Joel A. Nissen, Laird C. Brodie, and Jack S. Semura

Department of Physics
Portland State University
Portland, Oregon

ABSTRACT

We are reporting on a preliminary study of the enhancement of heat transfer from a germanium crystal heater-thermometer into liquid helium as a result of the application of an ultrasonic pulse. We found upon rapid heating of a germanium crystal in the helium that the total crystal superheat can be substantially lowered by the application of a sound pulse. Application of a short sound pulse, prior to reaching a certain threshold superheat, had little effect on the maximum crystal temperature. Application of the sound pulse after this threshold value was reached resulted in a rapid decrease in the crystal superheat. These results are quite similar to those experiments showing the enhancement of heat transfer due to a light pulse which have been published previously.

INTRODUCTION

The study of the heat transfer from a solid into liquid helium is important from several perspectives. On the one hand there is the engineering aspect of cooling superconductors and on the other hand there is the study of helium and its basic properties. One aspect of heat transfer touching upon both issues is the measurement of the homogeneous nucleation temperature, which we reported elsewhere[1,2]. Typically, in these experiments a single crystal of bismuth immersed in liquid helium was subjected to an electric current pulse. Measuring the magnetoresistance of the bismuth allowed us to deduce the crystal superheat. It was found that if the current pulse was large enough, the temperature reached a maximum (unless, of course, filmboiling occurred) independent of the applied power.

The explanation is that when the liquid helium in contact with the heater-thermometer reaches the homogeneous nucleation temperature, the formation of helium vapor at the surface produces a very efficient mechanism for cooling the crystal. If the heating rate is less, heterogeneous nucleation will play a dominant role (besides convection and conduction) in limiting the crystal superheat. In this case, many hard to control experimental conditions will be important, such as surface roughness and naturally occurring radioactivity. The homogeneous nucleation temperature, however, is a property of the helium itself and is described by the Becker-Doring theory[3].

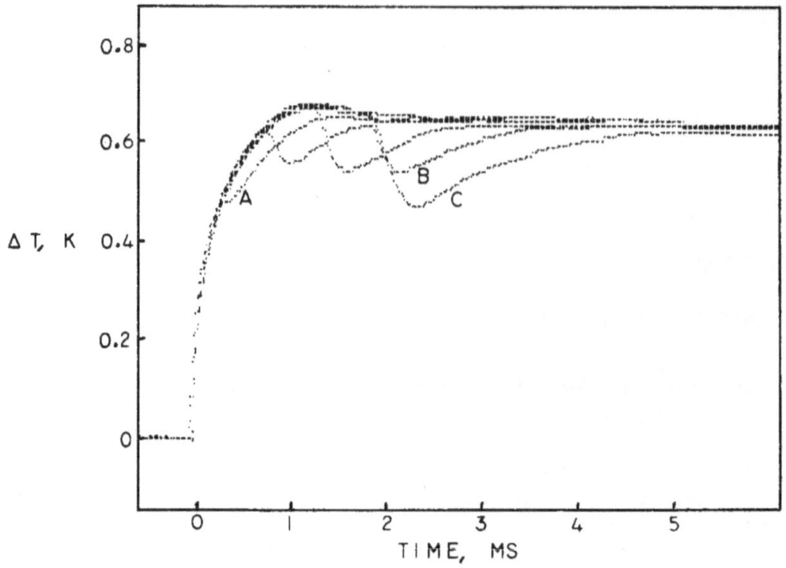

Fig. 1. Effect of a sound pulse (500 kHz) on the crystal superheat. The figure shows a composite record of six separate experiments (T_{bath} = 2.7 K) with a short sound pulse occurring at different times. Curve A shows a situation where the sound just barely gives a noticeable effect. Curve C shows the superheat when the sound pulse is 2½ times as long as in the other cases. The pulse was initiated at the same time as in curve B.

This theory deals with the kinetics of the formation of one phase from another. The formation of droplets of the other phase is associated with the existence of a thermodynamic barrier which has to be overcome in order for the droplet to grow. The rate of formation of stable droplets of the other phase is strongly temperature dependent and leads to an effective kinetic limit of the metastability of the parent phase, which is the nucleation temperature.

In the course of our investigations it was observed that light (either a flash or a steady light) lowered the superheat of the crystal sometimes as much as 75 percent. This light effect has so far not been explained, but one of the unexplored hypotheses is that phonons generated by the light hitting the solid surface induces the enhanced heat transfer.

Another way of reaching the limit of metastability of helium is to stress the liquid. This has been attempted[5,6,7]. For various reasons, however, these measurements fail to reach this limit with the possible exception of our recent experiments[7]. In most of these experiments, ultrasound with large amplitude is used to subject the liquid to large stresses. Because of the difficulty in maintaining an interference-free liquid environment, most of these experiments fail to reach the limit. Thus, although the Becker-Doring theory has been verified in the positive pressure range (by the transient superheating experiments), in the negative pressure range the results until recently have been in disagreement with this theory.

Both the lack of agreement between theory and stretching experiments and the lack of an explanation for the light effect indicates the need to study nucleation of helium I in the presence of a sound field. In this paper we present the results of superheating experiments in the presence of small amplitude sound fields.

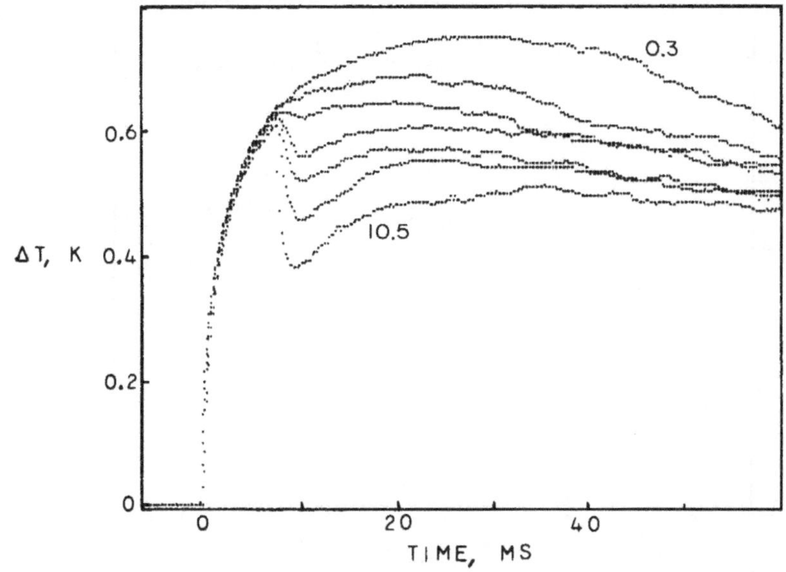

Fig. 2. Effect of the amplitude of the sound pulse (500 kHz) on the crystal superheat. The figure shows a composite record of seven experiments (T_{bath} = 2.8 K) with a sound pulse 2 ms long, initiated 7 ms after the heating pulse started with a varying pressure amplitude (resp. 0.3, 1.1, 1.9, 2.6, 3.7, 6.5, and 10.5 arbitrary units).

EXPERIMENTAL

For the heater-thermometer a commercial germanium crystal is used (Lake Shore Cryotronics GR200A). This crystal is used with the protective can removed so the crystal is in direct contact with liquid helium. This crystal has a surface area of about 4 mm^2. The electrical circuit which is used to apply the heating current and to measure the resistance of the germanium has been described elsewhere[8].

The ultrasound is applied by means of either one of two kinds of PZT piezoelectric crystals, which are submerged in the liquid helium. A thickness mode hemispherical focusing transducer is used in some of the experiments. The transducer is operated at its resonance frequency of 500 kHz. In these experiments the germanium crystal is placed at the focal point of the transducer (inside diameter of the transducer is 1.2 cm). The other one is also a thickness mode transducer. It, however, is a flat transducer (2.5 cm in diameter) with a resonant frequency of 820 kHz. This transducer is either driven at its resonant frequency or used in a standing wave resonant cavity. With the flat transducer, the heater-thermometer is placed on the axis of the piezoelectric crystal at a distance of about 4 mm from the surface.

RESULTS

Figure 1 represents a sequence of experiments showing some of the characteristics of the enhanced heat transfer due to the application of a sound pulse. Here a short burst of ultrasound (0.2 ms duration) is incident on the germanium. On a finer timescale, a time delay of 0.025 ms is observed between the application of power to the piezoelectric crystal and the arrival of the ultrasound as indicated by a lowering in temperature at the crystal, thereby excluding the possibility that the observed effect is due to some electromagnetic effect.

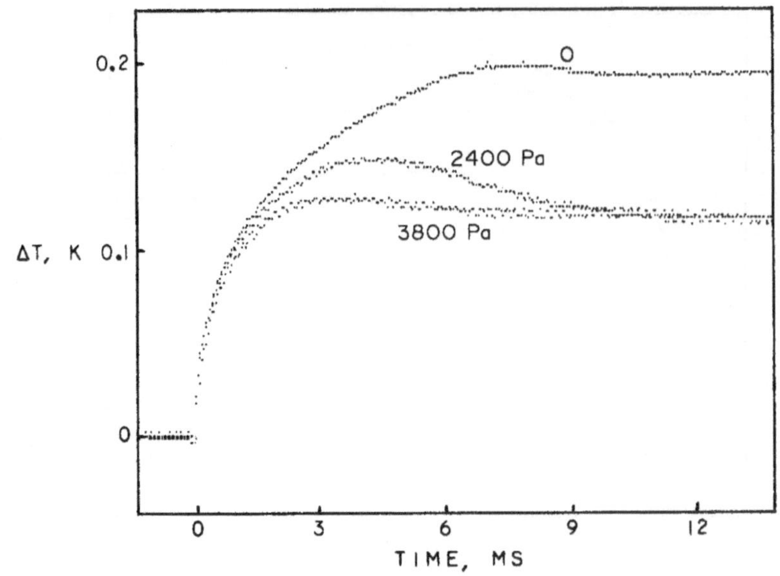

Fig. 3. Effect of a continuous sound field on lowering the crystal superheat. The figure shows three different experiments (T_{bath} = 4.2 K) with a pressure amplitude of 0, 2400, and 3800 Pa, respectively.

In these experiments the ultrasound was applied at different times after the heating pulse was started. The same characteristic effects are observed as in the case of the light effect. If the second pulse (or light pulse) impinges on the crystal slightly before a certain threshold in the crystal superheat is reached, no cooling is noticeable. Upon increasing the delay times and thereby increasing the superheat, the characteristic curves as shown in Figure 1 are obtained. In curve A, the sound is applied just slightly beyond the threshold. The effect of the duration of the sound pulse is indicated by curve B and C. In the latter case, a sound pulse of 0.5 ms duration but of the same amplitude and same delay time as the former was applied to the crystal. It is observed that the rate of cooling, as indicated by dT/dt, is the same in both cases but that the total amount of cooling is more in the case of the longer pulse.

To see the effect of the amplitude of the ultrasound, a series of experiments was done as indicated in Figure 2. In these experiments, short bursts, 2 ms long, were applied to the crystal 7 ms after the heating pulse started. It is clear that increased cooling is associated with larger pressure amplitudes.

Because of the interesting technical aspects, several experiments were done to establish the relationship between the continuous presence of sound and its influence on the crystal superheat. Some of the results are given in Figure 3 which shows three different superheating experiments, one with no sound present and the other two in the presence of a sound field with a pressure amplitude of 0.024 and 0.038 atm, respectively. Note that the larger sound field lowers the superheat temperature by about 0.08 K. This decrease is more than twice as large as the value 0.03 K which can be estimated from homogeneous nucleation theory by assuming an isothermal pressure reduction of 0.038 atm. Therefore, to explain such a significant lowering of the temperature another mechanism besides just the pressure variation must operate. The mechanism that could explain it is that

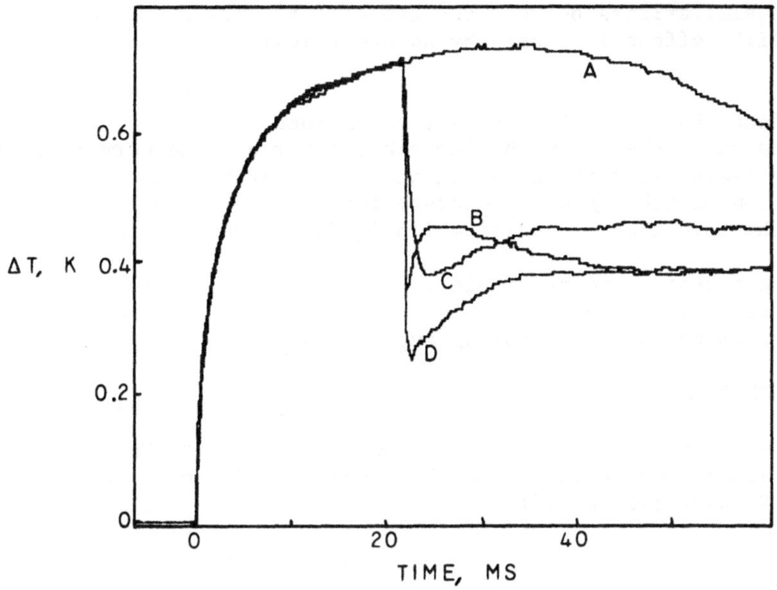

Fig. 4. Comparison of the effect of a light pulse versus a sound pulse (T_{bath} = 2.6 K). Curves A, B, C and D represent four experiments with neither sound nor light, light only, sound only, and both light and sound applied, respectively.

the sound field, besides lowering the energy barrier, activates additional nucleation sites which enhances the formation of vapor and thus provides additional superheat reduction. It is to be noted that the sound effect is qualitatively independent of the acoustical frequency.

In Figure 4, four experiments are shown which stress the analogy between the light effect and the sound effect. Curve A shows the superheat without any extra cooling mechanism applied. Curves B and C give the cooling produced by a light pulse and by a sound pulse 2 ms long, respectively. A small difference can be seen: in the case of the light pulse, the instantaneous cooling is much larger than in the case of the sound pulse. This difference can be explained by the fact that the light flash has a duration of about 40 μs, which is much shorter than the burst of sound. We believe that a shorter pulse, but of larger amplitude, would give a much larger rate of cooling. The remaining curve shows the combined effect of light and sound. It is worth noting that the effects are additive and thereby increase the heat transfer from the solid into the liquid.

DISCUSSION

These experiments have shown that sound can enhance the heat transfer significantly, much like the previously reported light effect. In fact, these experiments show there is a qualitative analogy that is rather striking. First, both the light and the sound effect have the feature of a threshold superheat, before which neither produces much extra heat transfer. Second, both show large cooling rates upon application of a pulse, as given by dT/dt. The only minor difference here is that the cooling rate of a sound pulse is smaller than that of the light pulse with a similar overall change in the temperature. But we think this is at least partly due to the fact that, in order to produce the similar temperature change, our sound pulses had to be of a longer duration. Third, both the light and the sound effect produced intense cooling.

These similarities between the two effects increases the likelihood that the light effect is caused by an acoustical vibration of the heater surface.

The fact that sound fields seem to enhance nucleation might help to explain the sharp breaks in the temperature versus time curves occurring during the measurement of the homogeneous nucleation temperature. Here the formation of bubbles will generate local sound fields that can increase the formation of nuclei which leads to the increased cooling.

From a technical point of view, we have shown that light and ultrasound can be used in combination to reduce the maximum superheat, which might be of importance for cooling superconductors.

ACKNOWLEDGMENTS

This work has been supported by the Portland State University Research and Publications Committee and the Environmental Sciences and Resources program (ESR publication 216).

REFERENCES

1. D. N. Sinha, J. S. Semura, and L. C. Brodie, Homogeneous nucleation in ^{4}He: A corresponding-states analysis, Phys. Rev. A 26:1048 (1982).
2. D. Lezak, L. C. Brodie, J. S. Semura, and E. Bodegom, Homogeneous nucleation temperature of liquid helium three, accepted for publication in Phys. Rev. B.
3. M. Blander and J. L. Katz, Bubble nucleation in liquids, AIChE J. 21:833 (1975).
4. D. Lezak, L. C. Brodie, and J. S. Semura, Light induced cooling of a heated solid immersed in liquid helium I, Advances in Cryogenic Engineering 29:289 (1984).
5. P. L. Marston, Tensile strength and visible ultrasonic cavitation of superfluid ^{4}He, J. Low Temp. Phys. 25:383 (1976).
6. R. D. Finch, T. G. Wang, R. Kagiwada, M. Barmatz, and I. Rudnick, Studies of the threshold-of-cavitation noise in liquid helium, J. Acoust. Soc. Am. 40:211 (1966).
7. J. A. Nissen, E. Bodegom, L. C. Brodie, and J. S. Semura, New measurements of the tensile strength of liquid ^{4}He, submitted to Advances in Cryogenic Engineering 1987.
8. L. C. Brodie, D. N. Sinha, J. S. Semura, and C. E. Sanford, Transient heat transfer into liquid helium I, J. Appl. Phys. 48:2882 (1977).

FILM BOILING TO A PLATE FACING DOWNWARD

Randall F. Barron and Ali R. Dergham

Mechanical Engineering Department
Louisiana Tech University
Ruston, Louisiana

ABSTRACT

An experimental study was conducted to verify an analytical model developed to predict the film boiling heat transfer coefficient from a circular plate facing downward in liquid nitrogen. The film boiling coefficients were measured using a transient technique, in which the entire boiling curve was generated in a single experimental run. The analytical solution was of the form, $Nu = f(Ra)\ Ra^{1/5}$, where Ra is the film boiling Rayleigh number, and the function f(Ra) involved the effect of the finite thickness of the vapor layer at the edge of the plate. The agreement between the analytical model and the experimental results was good for the range of Rayleigh number between 10^{10} and 5×10^{12} for plate diameters ranging from 50 mm to 150 mm and temperature differences between 50 K and 210 K.

INTRODUCTION

Film boiling heat transfer occurs in a variety of cryogenic systems, including heat exchangers, cooldown of components, and pressurization systems. Film boiling on horizontal plates facing upward has been the object of several studies.[1,2,3,4,5] Little analytical or experimental data are available for the case of film boiling to a plate facing downward, however.

The purpose of this study was to develop an analytical correlation for the film boiling heat transfer from a horizontal circular plate facing downward. The analytical model was verified through experiments conducted with liquid nitrogren as the boiling fluid.

ANALYTICAL DEVELOPMENT

The physical system and the coordinate system considered for the analytical development are shown in Fig. 1. A film of vapor is formed on the lower surface of a horizontal circular plate of diameter D. The thickness of the vapor film at any radial distance r is denoted by δ. The liquid in contact with the vapor is assumed to exist as a saturated liquid at T_{sat}.

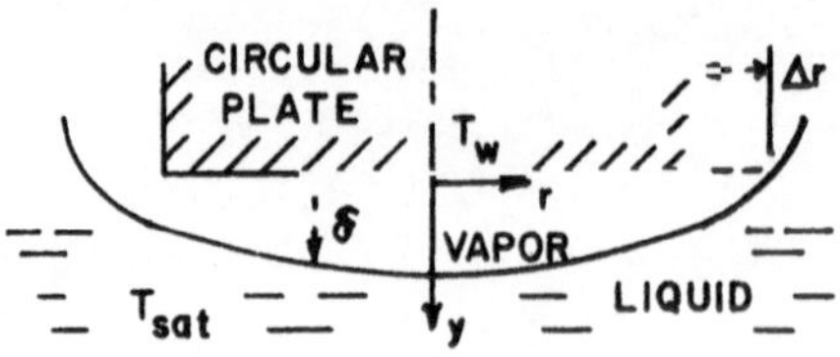

Fig. 1. Coordinate system for film boiling on a plate facing downward.

It has been shown that changes in momentum within the vapor layer are small in comparison with the frictional and bouyancy forces,[6] so the Navier-Stokes equation for the radial direction reduces to

$$-\frac{\partial p}{\partial r} + \mu \frac{\partial^2 u}{\partial y^2} = 0 \qquad (1)$$

The hydrostatic pressure at any point in the vapor layer is given by

$$p = p_0 + g(\rho_L - \rho)\,\delta \qquad (2)$$

Making this substitution into eqn. (1), the governing expression for the velocity distribution within the vapor layer is obtained.

$$\frac{\partial^2 u}{\partial y^2} = \frac{g(\rho_L - \rho)}{\mu}\frac{d\delta}{dr} \qquad (3)$$

The following velocity distribution is obtained by integrating eqn. (3), subject to the conditions that, at the surface of the plate (y = 0), the velocity u = 0, and at the edge of the vapor layer, the shear stress is negligible.

$$u = C_3(\tfrac{1}{2}y^2 - y\,\delta) \qquad (4)$$

where

$$C_3 = \frac{g(\rho_L - \rho)}{\mu}\frac{d\delta}{dr} \qquad (5)$$

The boundary layer integral technique may be used to determine the variation of the vapor film thickness, using eqn. (4) for the velocity distribution and a linear temperature distribution within the vapor layer,

$$T - T_{sat} = \Delta T\,(1 - y/\delta) \qquad (6)$$

The boundary layer integral energy equation for this system is

$$\frac{r\,k\,\Delta T}{d} = \frac{d}{dr}\left[\int_0^\delta r\,\rho\,u\,[i_{fg} + c(T - T_{sat})]\,dy\right] \qquad (7)$$

Carrying out the integration, the following expression is obtained for the vapor layer thickness:

$$\frac{\delta}{r}\frac{d}{dr}\left[r\,\delta^3\frac{d\delta}{dr}\right] = -\frac{3\,D^3}{Ra} \qquad (8)$$

where

$$Ra = \frac{g\ (\rho_L - \rho)\ \rho\lambda\ D^3}{\mu\ k\ \Delta T} = \text{film-boiling Rayleigh number} \tag{9}$$

$$\lambda = i_{fg} + 0.375\ c\Delta T$$

Let us introduce the following dimensionless parameters:

$$\eta = 2\ r/D \quad \text{and} \quad \phi = (\delta/D)^4 (Ra/3)^{4/5}$$

In terms of these dimensionless variables, eqn. (8) becomes

$$\frac{1}{\eta}\frac{d}{d\eta}\left(\eta\frac{d\phi}{d\eta}\right) + \phi^{-\frac{1}{4}} = 0 \tag{10}$$

Eqn. (10) may be linearized by introducing

$$\phi^{-\frac{1}{4}} = (\phi_o)^{-5/4}\ \phi \tag{11}$$

where ϕ_o is the value of the parameter ϕ at the center of the plate. The solution of eqn. (10) may be obtained, subject to the conditions

(a) At $\eta = 0$, $\phi = \phi_o$ and (b) At $\eta = 0$, $d\phi/d\eta = 0$

The solution is

$$\phi = \phi_o\ J_o(2\ B\ r/D) \tag{12}$$

where $$B = (\phi_o)^{-5/8} \tag{13}$$

The local or point Nusselt number may be determined from

$$Nu_x = \frac{h_x D}{k} = \frac{D}{\delta} \tag{14}$$

The overall or average Nusselt number is found by integration,

$$\overline{Nu} = 2\int_o^1 Nu_x\ \eta\ d\eta = \frac{h_c D}{k} \tag{15}$$

Making the substitutions from eqs. (12) and (14), we find

$$\overline{Nu} = 2\ (\phi_o)^{-\frac{1}{4}} \quad (Ra/3)^{1/5} \int_o^1 \frac{\eta\ d\eta}{[J_o(B\ \eta)]^{\frac{1}{4}}} \tag{16}$$

Suppose we let Δr be the distance beyond the edge of the plate where the vapor layer thickness is zero, relative to the plate surface. Then,

$$B(1 + 2\Delta r/D) = 2.40482 \tag{17}$$

The dimensionless ratio $2\Delta r/D$ is a function of the Rayleigh number[7], given by

$$2\Delta r/D = C\ Ra^n \tag{18}$$

The final expression for the average Nusselt number for film boiling to a circular plate facing downward is

$$Nu = f\,(Ra)\; Ra^{1/5} \tag{19}$$

where

$$f(Ra) = \left(\frac{7.854}{1 + C\, Ra^{n}} \right)^{2/5} \int_{0}^{1} \frac{\eta\, d\eta}{[J_0(B\,\eta)]^{\frac{1}{4}}} \tag{20}$$

EXPERIMENTAL APPARATUS

A schematic of the experimental apparatus used in this study is shown in Fig. 2. A more detailed description of the apparatus is given by Dergham.[8] The test plates were flat circular aluminum plates. The following plate diameters and plate thicknesses were used: 50.8 mm (25.4 mm thick), 76.2 mm (25.4 mm thick), 101.6 mm (12.7 mm thick), 127.0 mm (12.7 mm thick), and 152.4 mm (12.7 mm thick). The upper surface and the edge of each plate was insulated with a 12.7 mm thick layer of closed-cell foam insulation. The plates were supported by three thin stainless steel brackets attached at the plate edge by small dowels. The support system was adjustable such that the lower surface of each plate was maintained at a constant depth of 75 mm below the surface of the liquid nitrogen during the experimental runs.

The plate temperature was measured by calibrated copper-constantan thermocouples. The thermocouples were placed in 3 mm diameter holes from the upper (insulated) side of the plate, and a high-conductivity epoxy was used to fill the holes to obtain good thermal contact between the thermocouples and the plate. Because the Biot number for the plates was on the order of 0.02, it was estimated that the difference between the surface temperature and the back-side temperature of the plate was less than 1 K; therefore, only the surface temperature was measured. The thermocouple emf was recorded on a potentiometric strip chart recorder at a chart speed of 1.27 mm/s.

The heat transfer coefficient data were obtained by a transient technique, similar to that used by Merte and Clark.[9] Because the Biot number for the plates was less than about 0.10, the heat transfer problem could be treated as a "lumped-parameter" problem, which yields the following expression for the film-boiling heat transfer coefficient:

$$h_c = - \frac{m\, c_s}{A\,(T_w - T_{sat})} \frac{dT_w}{dt} \tag{21}$$

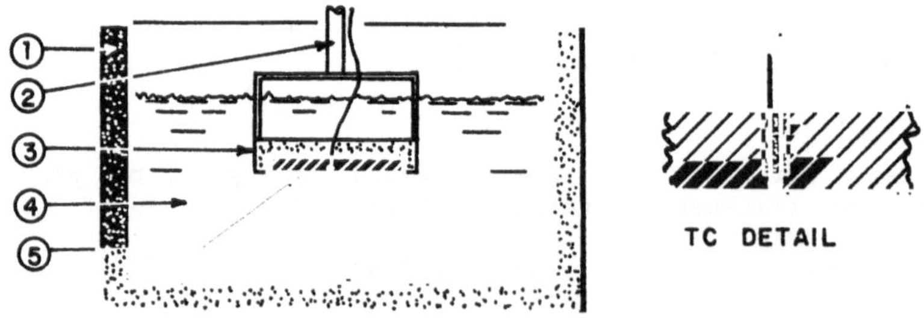

Fig. 2. Experimental apparatus. Legend: (1) insulated container; (2) support structure; (3) plate insulation; (4) liquid nitrogen; (5) test plate.

The temperature data from the strip-chart recorder for each run were fitted to an equation of the form

$$t = C_o + C_1 \ln \theta + C_2[\ln \theta]^2 + C_3[\ln \theta]^3 \quad (22)$$

where: $\theta = T_w - T_{sat}$ The temperature derivative was calculated from the experimental data using the following relationship:

$$\frac{dT_w}{dt} = \frac{\theta}{C_1 + 2\,C_2 \ln \theta + 3\,C_3[\ln \theta]^2} \quad (23)$$

The variation of the specific heat of aluminum with temperature was considered by calculating the specific heat from

$$c_s/R = b_o + b_1\,x + b_2\,x^2 + b_3\,x^3 \quad (24)$$

where R is the gas constant; $x = T_w/\theta_D$; θ_D is the Debye temperature for aluminum (385 K); and the constants b_o, b_1, b_2, and b_3 have the values of 3.5475, -0.4255, -0.05703, and 0.00873, respectively. Eqn. (24) is valid for the temperature range between 75 K and 300 K.

The data were reduced by a computer program, using the measured plate temperatures at appoximately 15-sec intervals in eqs. (20) and (22). All recorded data points were in the stable film-boiling regime.

RESULTS

The experimental values of the film-boiling heat transfer coefficient as a function of the temperature excess are shown in Fig. 3. Because approximately 120 experimental points were obtained, only curves through the experimental points are shown. A complete computer printout of the data is given elsewhere.[8]

From the experimental data, the pool-boiling heat transfer correlation could be written as

$$\overline{Nu} = \left(\frac{0.325}{1 + C\,Ra^n}\right)^{2/5} Ra^{1/5} \quad (25)$$

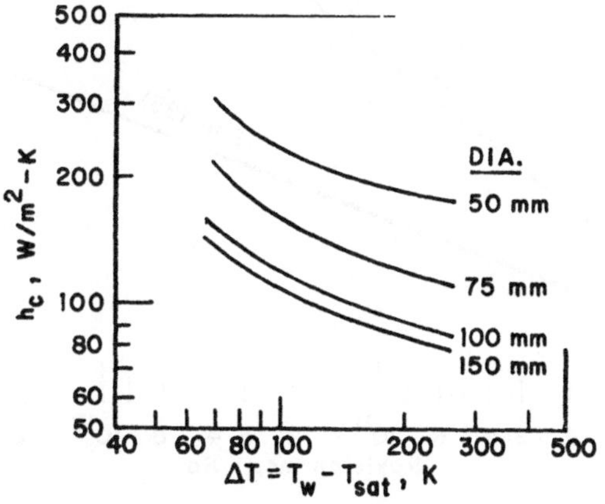

Fig. 3. Film boiling heat transfer coefficients for various plate sizes.

where $C = 1.486 \times 10^6$ and $n = -0.50$, and the vapor-phase properties are evaluated at the mean film temperature. The latent heat of vaporization and the liquid density are evaluated at the fluid saturation temperature. A plot of the experimental correlation is shown in Fig. 4.

Seki et al.[5] gave the following correlation for film boiling on a downward-facing plate in refrigerant R-11:

$$Nu = 0.35\ (i_{fg}/c\Delta T)\ Ra^{0.25} \tag{26}$$

The exponent on the Rayleigh number in eqn. (26) is slightly higher than that obtained in the present correlation; however, the form for the two correlations is not identical. As shown in Fig. 4, the correlation of Seki et al., which was developed from data using refrigerant R-11, predicted heat transfer coefficients which were between 1.5 to 10 times higher than those obtained in the present study.

An analytical study of stable film boiling from a horizontal plate facing downward was made by Farahat and Madbouly[10] for water as the boiling liquid. They did not present a generalized correlation, so no direct comparison could be made with the present analysis; however, the predicted heat flux from their data for water was generally of the same order of magnitude as that from this study.

CONCLUSIONS

An analytical correlation was developed for the film-boiling heat transfer coefficient for a horizontal circular plate facing downward. The analytical correlation was in good agreement with experimental data taken on plates having diameters ranging from 50 mm to 150 mm, for boiling Rayleigh numbers between 10^{10} and 5×10^{12}.

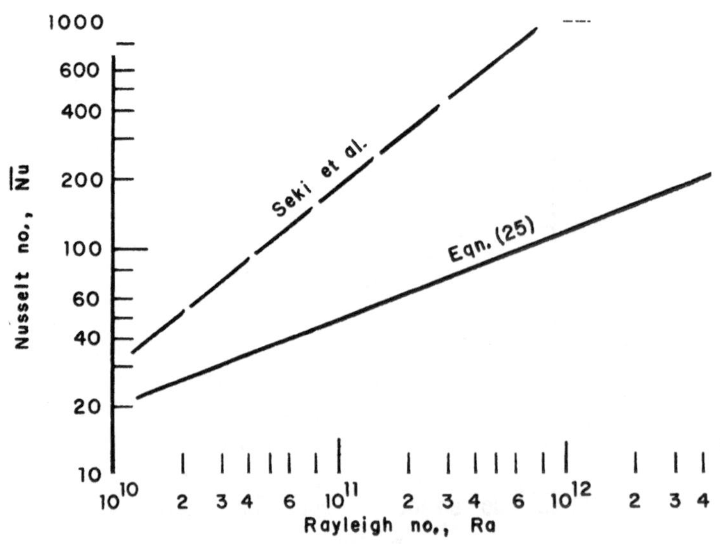

Fig. 4. Heat transfer correlation for stable film boiling to liquid nitrogen from a circular plate facing downward, eqn. (25). The experimental correlation of Seki et al.[5] is also shown, eqn. (26).

NOTATION

A = area of lower surface of plate
b_0, b_1, b_2, b_3 = constants
B = parameter (eqn. 12)
c = vapor specific heat
c_s = specific heat of solid
C_0, C_1, C_2, C_3 = constants
D = plate diameter
g = local acceleration due to gravity
h_c = average film boiling heat transfer coefficient
h_x = local or point heat transfer coefficient
i_{fg} = latent heat of vaporization
$J_o(x)$ = Bessel function of the first kind, order zero
k = vapor thermal conductivity
m = mass of the plate
Nu = Nusselt number, h_cD/k
p = hydrostatic pressure
p_o = hydrostatic pressure at the plate surface
r = radial coordinate
Δr = distance beyond edge of plate where $\delta = 0$
R = gas constant
Ra = film-boiling Rayleigh number, eqn. (9)
t = time
T = vapor temperature
T_{sat} = liquid saturation temperature
T_w = surface temperature of the plate
ΔT = temperature excess, $T_w - T_{sat}$
u = vapor velocity
x = T_w/θ_D
y = coordinate normal to the plate surface

Greek Symbols

δ = vapor film thickness
θ = temperature difference, $T_w - T_{sat}$
θ_D = Debye temperature
λ = effective heat of vaporization, $i_{fg} + 0.375\ c\ \Delta T$
η = parameter, $2\ r/D$
μ = vapor viscosity
ρ = vapor density
ρ_L = liquid density
θ = parameter, eqn. (20)

REFERENCES

1. P. J. Berenson, Film boiling heat transfer from a horizontal surface, ASME J. Heat Transfer 83:351 (1961).

2. E. R. Hosler and J. W. Westwater, Film boiling on a horizontal plate, ARS Journal (1962), p. 553.

3. T. H. K. Frederking, Y. C. Wu, and B. W. Clement, Effects of interfacial instability on film boiling of saturated liquid helium I above a horizontal surface, AIChE J. 12:238 (1966).

4. H. J. Sauer and K. M. Ragsdale, Film pool boiling of nitrogen from flat surfaces, in: "Advances in Cryogenic Engineering," Vol. 16, Plenum Press, New York (1971), p. 412.

5. N. Seki, S. Fukusako, and K. Torikoshi, Experimental study on the effect of orientation of heating circular plates on film boiling heat transfer for fluorocarbon refrigerant R-11, ASME J. Heat Transfer 100:624 (1978).

6. G. Leppert and B. Nimmo, Laminar film condensation on surfaces normal to body or inertial forces, ASME J. Heat Transfer 90:178 (1968).

7. D. K. Edwards and J. C. Haiad, Average Nusselt number on the downward-facing heated plate, Int. J. Heat & Mass Transfer 29:1607 (1986).

8. A. R. Dergham, "Film Boiling of Nitrogen on a Plate Facing Downward," M.S. Thesis, Louisiana Tech University, Ruston, LA (1987).

9. H. Merte and J. A. Clark, Boiling heat transfer data for liquid nitrogen at standard and near-zero gravity, in: "Advances in Cryogenic Engineering," Vol. 7, Plenum Press, New York (1961), p. 546.

10. M. M. Farahat and E. E. Madbouly, Stable film boiling heat transfer from flat horizontal plates facing downwards, Int. J. Heat & Mass Transfer 20:269 (1977).

ANGULAR DEPENDENCE OF BOILING HEAT TRANSFER

MECHANISMS IN LIQUID NITROGEN

C.Beduz, R.G.Scurlock and A.J.Sousa

Institute of Cryogenics
University of Southampton
Southampton, England

ABSTRACT

The overall heat transfer coefficient of aluminium and copper samples with different surface finish have ben measured at atmospheric pressure in liquid nitrogen. The measurements show that the influence of surface orientation is very large at low heat fluxes and for polished surfaces, the maximum value of the heat transfer coefficient is near the horizontal downward facing position. The angular dependence of the overall heat transfer coefficient decreases for surfaces with a large number density of active nucleation sites. Following the use of photographic visualization techniques, a model of high flux boiling heat transfer for downward facing surfaces is developed in which periodic heat transfer is strongly dependent on angle, but weakly dependent on heat flux.

INTRODUCTION

The effect of surface orientation on the nucleate pool boiling heat transfer and on the critical heat flux (CHF) has received little attention to date. The object of this work was to obtain additional experimental information on the angular dependence of the heat transfer characteristics of surfaces with different finish, boiling in liquid nitrogen.

Previous investigations on the effect of orientation on pool boiling using smcoth surfaces[1-5], have shown that the heat transfer is enhanced when an upward facing horizontal surface ($\theta = 0°$) is rotated. The maximum heat transfer coefficient is obtained in the region of $\theta = 170° - 178°$, and is followed by a sharp decrease at $\theta = 180°$. The effect of surface orientation diminishes for increasing heat flux and above a critical value q_i the heat transfer coefficient becomes independent of θ. In all cases the CHF is a maximum at $\theta = 0°$ and decreases with increasing angle.

For a horizontal smooth ($\theta = 0°$) surface, the low number of active nucleation sites at low heat flux produces isolated bubbles which grow and depart without interaction between one another. In this situation the frequency of departure and bubble size are controlled by gravity.

Using the hydrodynamic theory of Zuber[6], Moissis and Berenson[7] showed that when the heat flux is increased beyond a certain value the isolated bubble regime changes into a "slugs-and-columns" regime (interference regime). Stable vapour jets, formed by coalescence of bubbles, provide the escape flow mechanisms for vapour generated from the evaporation of a liquid macrolayer at the base of the columns. The distance between vapour jets is given by the "most dangerous wavelength" of Taylor instability.

$$\lambda = \sqrt{3}\,\lambda_T = 2\pi\,[3\,\sigma/g\,(\rho_\ell - \rho_v)]^{\frac{1}{2}} \qquad (1)$$

The diameter of the vapour jets is taken as $d = \lambda/2$.

Lienhard[8] has pointed out that in the interference regime the vapour removal mechanism is independent of gravity. Therefore for heat fluxes higher than q_i a change in orientation of the surface will not influence the heat transfer. The value of q_i derived by Zuber is

$$q_i = 0.11\;\rho_v\,H_{fg}\,[\,\sigma g/(\rho_\ell - \rho_v)]^{\frac{1}{4}}\;\beta^{\frac{1}{2}} \qquad (2)$$

In this theory the CHF is identified with the heat flux at which the vapour jets break down due to Rayleigh instability. The transition to film boiling is interpreted as a deficiency in vapour removal rather than a deficiency in liquid supply. The value of CHF obtained in this way is

$$q_c = 0.131\,[\sigma g\,(\rho_\ell - \rho_v)/\,\rho_v^{\,2}]^{\frac{1}{4}}\;\rho_v H_{fg} \qquad (3)$$

Katto[9,10] has extended the hydrodynamic instability model and related the CHF with the Helmotz instablility of small feeder jets in the macrolayer existing under the unstable vapour blanket. According to this theory the maximum thickness of the boiling liquid macrolayer is given by

$$\delta_c = \frac{\pi}{2}\,\sigma[(\rho_\ell + \rho_v)/\rho_\ell\,\rho_v]\,(A_v/A_w)^2(\rho_v\,H_{fg}/q)^2 \qquad (4)$$

where

$$A_v/A_w = 0.0584\,(\rho_v/\,\rho_\ell)^{0.2}.$$

However, there is experimental evidence that at high heat flux a large number of unstable dry areas develop and collapse on the boiling surface. The local temperature in these dry patches depends on the lifetime of the dry patches, and on the thermal conductivity and thickness of the surface (these variables are not contained in the hydrodynamic instability theory). Ouwerkerk[11] analysed the stability criteria of existing dry patches taking into account these parameters. He derived an equation relating the critical radius of a circular dry patch $S_f{}^*$ with the heat flux, thermal conductivity and thickness of the plate and the coefficients of heat transfer to liquid and vapour. The local surface temperature of a dry patch of radius larger than $S_f{}^*$ is too high for rewetting to occur. It continues to grow with time and finally a transition to film boiling takes place.

EXPERIMENTAL PROCEDURE

A rotating frame allowed the angular orientation of the surface ($\pm$ 0.5°) to be changed without removal of the sample from the transparent glass dewar used in the experiments. The dimensions of the copper and aluminium samples was 5cm x 5cm x 0.6cm. Four differential copper-constantan thermocouples (Ø = 80μm) were used to measure the mean temperature difference between the heated surface and the bulk of the liquid. A thermofoil heater was glued to the back of the plate, and the sample was potted in a composite resin of low thermal contraction. the heat leaking through the insulation was calculated to be less than 2% of the power dissipated in the heater. The six samples used in the experiments were mirror polished, grooved and rolled, and plasma sprayed on aluminium (1,2,3) and copper (4,5,6).

In sample 2, 0.3mm deep paralle V-shaped grooves 1mm apart were machined on the surface; re-entrant cavities were then formed by rebating the groove lips by a rolling procedure. Sample No.5 was treated in a similar way, but in this case the pitch between grooves was reduced to 0.6mm. The boiling surface of sample 3 was plasma sprayed with aluminium. The 0.12mm thick coating had a porosity of approximately 50%. Sample 6 was bronze sprayed, and the 0.35mmm coating had a porosity of 10%. In samples 2 and 5 the axis of rotation was set parallel to the grooves.

To minimize the effect of aging and hysteresis, the samples were conditioned for one hour at $q = 10^5\ W/m^2$. Then the heat flux was set to a constant value, and the wall superheat was measured at various angles. Repeating this procedure for different heat fluxes, the parametric $q - \Delta T$ curves were obtained without removing the sample from the liquid nitrogen bath.

EXPERIMENTAL RESULTS

The results for the three aluminium samples are plotted in Fig.1. The strong dependence of the heat transfer coefficient of the polished sample with θ at low heat flux decreases for increasing heat flux and the curves appear to merge for q in the region of $8 \times 10^4\ W/m^2$.

As a consequence of having a large number density of nucleation sites, sample 2 exhibits an enhanced heat transfer in comparison with the polished sample. Curves 8 and 9 represent the data obtained with this sample for $\theta = 0°$ and $\theta = 90°$. No change in heat transfer performance was observed for θ varying from 90° to 176°. This small enhancement (25%) seen for up to 90° is gradually reduced for increasing q. The heat transfer characteristics for sample 3 are shown by curve 10 and 11, for $\theta = 0°$ and $\theta = 176°$. This sample revealed no measurable variation of heat transfer with θ.

Figure 2 shows the results for the copper samples. It can be seen that the polished sample presents similar features to those of the polished aluminium sample discussed above. The grooved and coated samples do not show the large improvement obtained with the treated aluminium surfaces. As with the grooved aluminium surface, the wall superheat decreased slightly when θ varied from 0° to 90° and remained constant at larger angles.

The CHF of the six surfaces at $\theta = 0°$ are 19.1, 25.0, 26.5, 17.0, 31.3, $25.0 \times 10^4\ W/m^2$ (samples 1-6) respectively. The angular dependence (Fig.3) shows a steady decrease of the CHF with angle. At $\theta = 176$, the CHF value for all samples is only 30-40% of the value for $\theta = 0°$. Although the q_c/q_{co} values for the copper samples initially fall less rapidly than the values for the aluminium samples, at large angles the curves cross over and no clear trend can be associated with the metal base or surface finish.

DISCUSSION

The boiling curves of the two polished samples show a similar dependence on surface orientation. At low heat flux the heat transfer mechanisms for surfaces facing up (isolated bubble regime) and surfaces

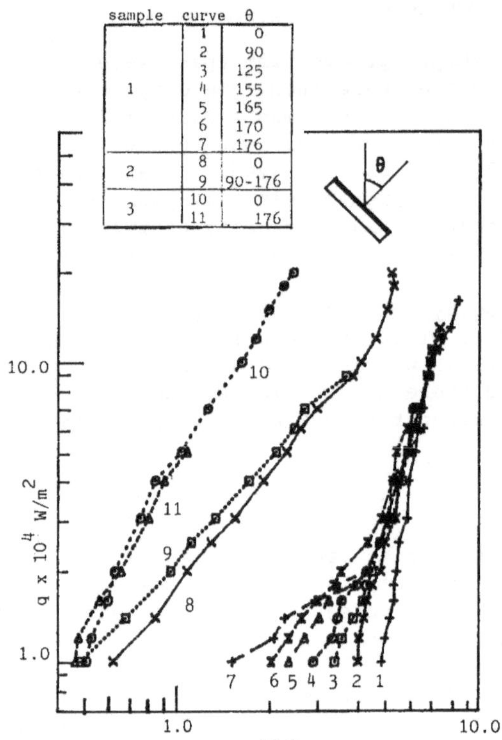

Fig.1. Effect of orientation for the three aluminium samples

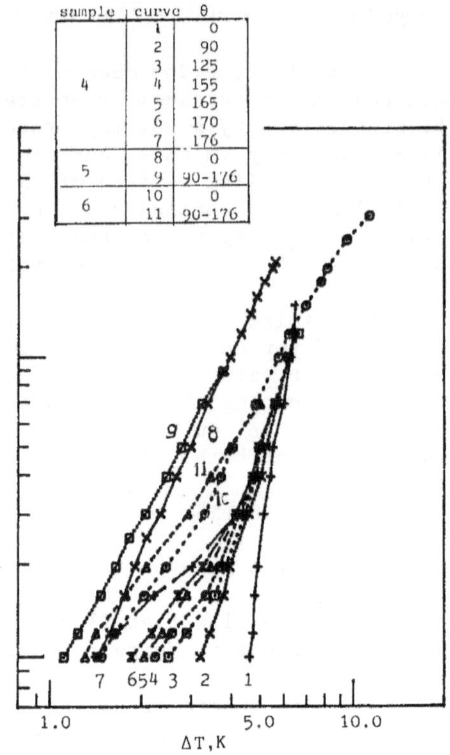

Fig.2. Effect of orientation for the three copper samples

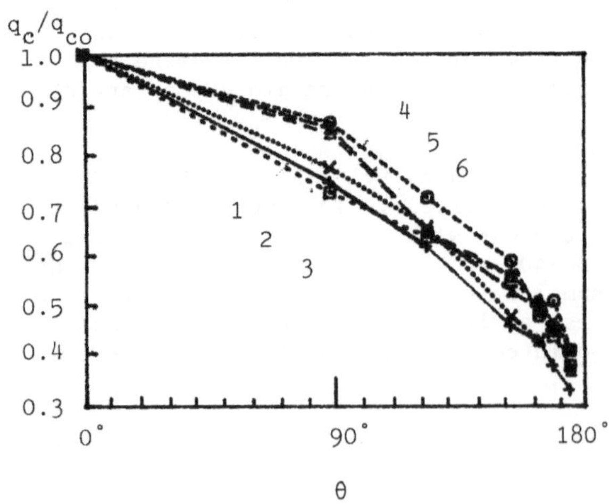

Fig. 3. Dimensionless critical heat flux q_c/q_{co}; curves designated by sample numbers.

facing down are very different. In the latter case, the bubbles originating at an active nucleation site remain in contact with the surface for a relatively long time and grow to large dimensions. Evaporation from the initial liquid macrolayer reduces its thickness and eventually dry patches appear near the trailing edge of the moving bubble. Figure 6 and Fig.7 show the macrolayer and dry patches developed under the large bubbles at $q = 0.2 - 0.8 \times 10^4\ W/m^2$. The residence time of the bubble and the percentage of the surface covered by the macrolayer is reduced at smaller angles and the enhancement of heat transfer produced by the macrolayer evaporation mechanism decreases gradually until θ reaches 90° (vertical position). For θ between 90° and 0° the heat transfer is governed by the growth and departure of individual bubbles. Although the number density of active nucleation points appears to be a maximum at θ = 0°, the heat transfer is a minimum.

At higher heat flux the influence of orientation decreases and in the interference regime the heat transfer becomes independent of angle. Using the Zuber equation to estimate q at the transition we obtain

$$q_i = 8.7 \times 10^4\ W/m^2 \quad \text{for } \beta \simeq 70^\circ$$
$$q_i = 5.7 \times 10^4\ W/m^2 \quad \text{for } \beta = 30^\circ$$

The independence of heat transfer on orientation shown by sample 3 suggests that gravity does not enter in the vapour removal mechanism. A possible explanation is that in this highly porous surface, stable vapour channels develop inside the coating even at low heat flux (see Fig.4). The dimensions and number density of these channels would be determined by the morphology of the coating rather than by hydrodynamic conditions.

Sample 2 and 6 (grooved), and sample 5 (low porosity coating) show a weak angular dependence. In the grooved samples, the location of the active nucleation sites is fixed and the transition to the interference regime is likely to occur at a much lower heat flux than for a polished surface.

At high heat flux our visualization shows the following intermittent mechanism of boiling for θ > 90°. A vapour cloud forms over the entire area, while on the surface patches of boiling liquid of decreasing size can be observed. The vapour blanket grows thicker with time, until buoyancy forces cause the removal of vapour from the upper edge of the plate. Then a liquid wave wets the surface at the trailing edge of the moving vapour cloud and boiling is initiated. After a short time the merging

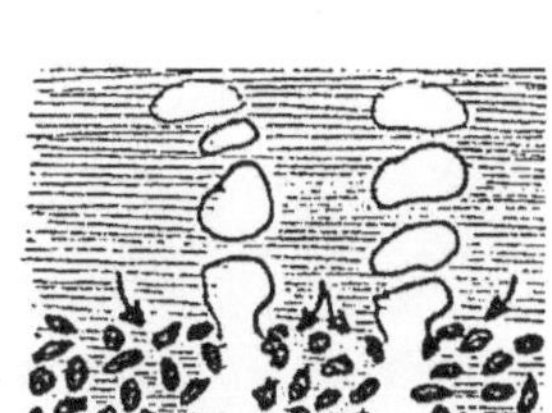

Fig.4. Vapour channels in a porous surface

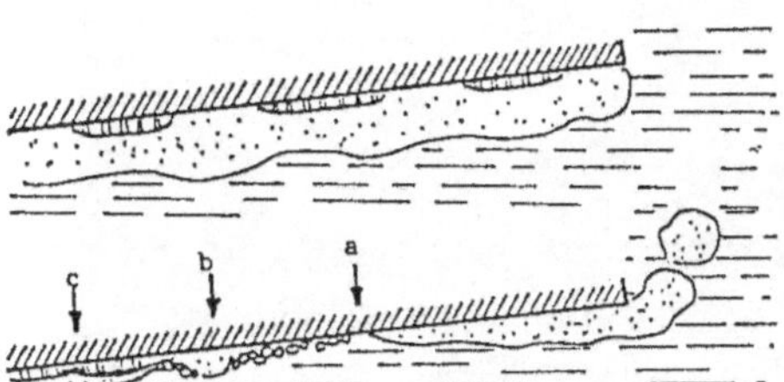

Fig.5. Re-wetting cycle for θ >90°
a: advancing wet contact point
b: dry patch formation
c: vapour cloud initiation

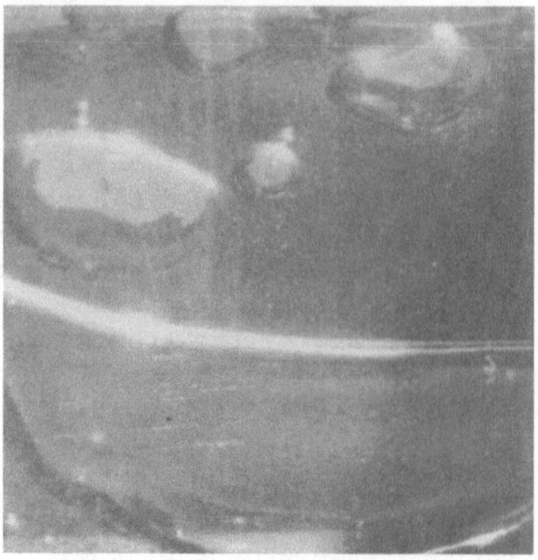

Fig.6. Boiling at $\theta = 178°$, $q = 0.2 \times 10^4$ W/m^2, white areas are dry.

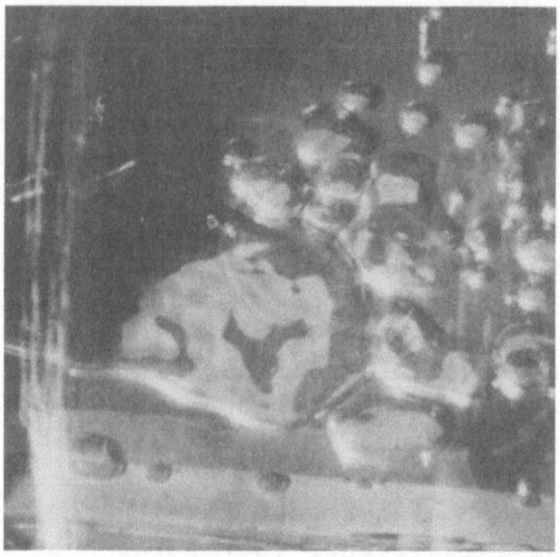

Fig.7. Boiling at $\theta = 178°$ $q = 0.8 \times 10^4$ W/m^2

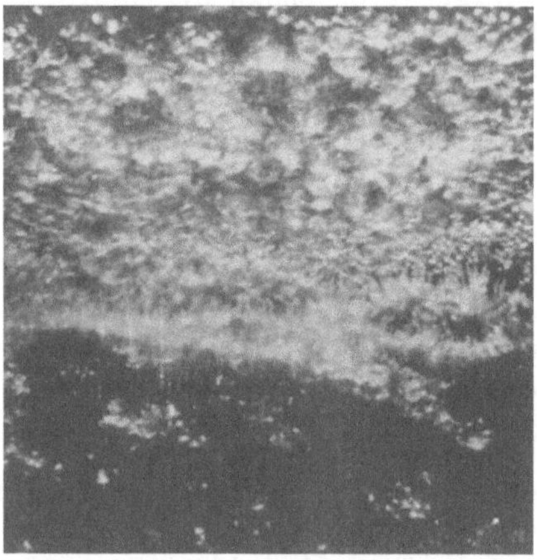

Fig.8. Wet wave-front advancing from top to bottom, $\theta = 170°$, $q = 7.2 \times 10^4$ W/m^2.

Fig.9. Dry patch formation. $\theta = 170°$, $q = 7.2 \times 10^4$W/m^2

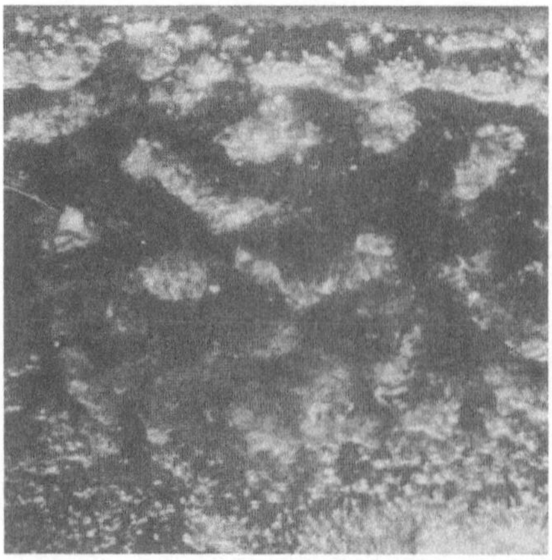

Fig.10. Formation of isolated wet patches, $\theta = 170°$, $q = 4.4 \times 10^4$ W/m^2

bubbles produce a new vapour blanket, advancing across the liquid macro-layer attached to the surface. Irregular evaporation of the macro-layer is seen to take place so that dry patches appear with some periodic structure (as shown in Fig.5 and Fig.8).

The dry patches expand with time, maintaining a periodic form as seen in Figures 8 and 9. The sequence is repeated at a frequency f_r which depends strongly on θ, but weakly on q. The mean distance between patches is comparable with the "most dangerous wavelength" of Taylor instability; the calculated value is λ = 11.4mm while the observed value is in the range 8-15mm.

The calculated CHF value for $\theta = 0°$ in liquid nitrogen using the Zuber formula is $q_c = 16.2 \times 10^4$ W/m^2, which compares well with the polished samples. However, the hydrodynamic theory does not take into account surface characteristics and cannot explain the increased values of CHF measured with the treated samples. From these observations, it appears that the evaporation process in the wet patches results in a decreasing wetted area with time while the thickness of the macrolayer remains constant, rather than a thinning of the liquid macrolayer. Although at present no theory is available on the angular dependence of the CHF, we can use Katto's model to interpret the results. If we assume that the surface is initially covered by a macrolayer of thickness δ_c, with dry patches of diameter $\lambda/2$ spaced at a distance λ, the life-time of the liquid film is determined by

$$\tau_a = {}^1/_{fa} = 0.803 \; \delta_c \left(1 - \frac{A_v}{A_w}\right) \rho_\ell Hfg \frac{1}{q_a} \qquad (5)$$

This equation represents a critical frequency which when compared with f_r determines whether the surface dries out. For a surface with very small heat capacity, the transition to film boiling will occur if f_e is smaller than f_a.

For example, with the polished aluminium surface at $\theta = 176°$ the measured q_c and f_r values were $q_c = 6.1 \times 10^4$ W/m^2 and $f_r = 4.04$ Hz. Using this figure for f_a the above formula yields $q_a = 8.5 \times 10^4$ W/m^2. At this point it should be noted that we have not considered the thermal characteristics of the surface. Applying the theory of Ouwerkerk we see that if the condition $S^*_J < \lambda/\sqrt{2}$ is satisfied, the appearance of unwettable dry patches will promote the transition to film boiling at lower heat fluxes than predicted by the application of the hydrodynamic instability theory. This conclusion is supported by our visualization during the transition to film boiling; a stable dry patch which is not wettted appears at some point, while liquid patches are still present on the rest of the surface. If the thickness of the macrolayer is assumed to be constant, the change in size of the dry patch with time can be caluclated; then Ouwerkerk's model can be applied to obtain the CHF if the dependence of f_r with θ and q is known.

For $\theta < 90°$ the angular dependence of the CHF is likely to be related to the change of frequency of bubble departure (in Katto's model) with θ.

CONCLUSION

The main conclusions from this work are as follows:

1. In the isolated bubble regime, the heat transfer of smooth surfaces shows a large angular dependence. The magnitude of this effect is probably related to the sample size through the residence time

of the vapour blanket (macro-bubble) in the downward facing position. More experimental data are needed to clarify this phenomena.

2. The heat transfer of coated surfaces with large porosity is independent of θ. Vapour removal and liquid supply mechanisms are determined by the geometrical characteristics of the coating.

3. In agreement with previous authors, the CHF of treated surfaces can be much larger than the value predicted by the hydrodynamic instability theory.

4. Our experimental results and flow visualization support the view that a theory of CHF should include the thermal properties of the surface.

ACKNOWLEDGMENTS

This work was carried out with the financial support of BOC Cryoplants Ltd. One of us (AJS) acknowledges the University of Brasilia, and CAPES/MEC, Brazil for a research scholarship.

REFERENCES

1. K.Nishikawa et al, Effects of surface configuration on nucleate boiling heat transfer, ASME J.Heat Transfer, 27:1559 (1984).
2. J.E.S.Vernat, A.C.M.Sousa, and D.S.Jung, Nucleate film boiling heat transfer in R11: the effects of enhanced surfaces, and inclination in "Proc.8th.Int.Heat Transfer Conf", Vol.4, Hemisphere Publ.Corp.,Washington, US (1986), p.2019.
3. I.P.Vishnev et al, Study of heat transfer in boiling of helium on surfaces with various orientations, Heat Transfer, Soviet Research, 8:104 (1976).
4. B.D.Marcus and D.Dropkin, The effect of surface configuration on nucleate boiling heat transfer, Int.J.Heat and Mass Transfer, 6:863 (1963).
5. L.Bewilogua, R.Knoner and H.Vinzelberg, Heat transfer in cryogenic liquids under pressure, Cryogenics, 15:121 (1975).
6. N.Zuber, Nucleate Boiling in the Region of Isolated Bubbles and the Similarity with natural convection, Int.J.Heat and Mass Transfer3, 6:53 (1963).
7. R.Moissis and P.J.Berenson, On the hydrodynamic transition in nucleate boiling, ASME J.Heat Transfer, 85:221 (1963).
8. J.H.Lienhard, On the two regimes of nucleate boiling, ASME J.Heat Transfer, 107:262 (1985).
9. Y.Katto, Critical heat flux in forced convection, in "Proc. ASME Thermal Engr. Joint Conf." Vol3, ASME, New York, p.l. (1983).
10. Y.Katto, Critical heat flux, in "Advances in Heat Transfer", vol.107, p.262 (1985).
11. H.J.Ouwerkerk, Burnout in pool boiling, the stability of boiling mechanisms, Int.J.Heat and Mass Transfer, 55:25 (1972).

BOILING ON A CRYOGENICALLY COOLED PULSED CONDUCTOR

Christopher J. Crowley
Sukhvinder S. Kang
Paul H. Rothe

Creare Incorporated
Hanover, New Hampshire

ABSTRACT

Active cryogenic cooling of pulsed conductors in high power systems offers greatly reduced resistive energy losses. This paper presents unique boiling data for pulsed power inputs at repetition rates of 5 Hz and heating pulses of 2 to 20 ms duration. We have developed an initial analytical model for the transient boiling behavior during pulsed heating and cooldown cycles based in part on a quasi-steady extension of existing correlations and data for steady boiling in the nucleate, transition, and film boiling regimes. Both pool and forced flow boiling are treated. This model includes the concept of a mixed microlayer for treatment of enhanced heat flux to the coolant during rapid heating transients in boiling. Investigated test conditions include those where the power input to the conductor exceeds the value at steady-state critical heat flux by several times.

INTRODUCTION

The work described here is motivated by the need to design thermal management systems for high energy pulsed power systems. The general technology being studied involves the combined elements of:

- CRYOGENIC OPERATION to minimize resistive power losses by exploiting the low resistivity of high purity materials (several orders of magnitude smaller than at room temperature),
- ACTIVE COOLING to remove resistively dissipated energy from the conductors during repeated, rapid, high-energy pulses, and
- COOLABLE GEOMETRY in which a large conductor surface area to volume ratio enhances the ability to cool the conductors.

Potential applications of this technology include the inductors, bus bars, and other components of pulsed, high-power systems.

Relatively little work has been performed on the rapid cyclical boiling behavior of present interest. Table 1 surveys the closely related experimental literature. For our applications we are concerned with boiling from an electrical conductor with:

- Short high-power pulses (2 to 20 ms) followed by relatively longer cooling periods (0.1 to 1 s),
- Heat pulses at high repetition rates, and
- Forced flow of cryogenic coolant (0.1 to 10 m/s).

Table 1. Summary of Pulsed Boiling Experiments in the Literature

REFERENCE	TRANSIENT		TEST SURFACE GEOMETRY		FORCED FLOW	CRYOGEN	PRESSURE
	REPEATED PULSE	OTHER	WIRE	OTHER			
1. Present Interest	Yes	---	Stainless steel wire	---	Wire in crossflow	LN_2	1 to 3 bar
2. Steward[1]	Yes	Single pulse, step heat input	---	Carbon film on quartz substrate	No	LN_2 LHe	0.9 to 3 bar
3. Giarratano[2]	No	Single pulse, step heat input	Platinum wire	Platinum film on quartz	No	LN_2	0.83 bar
4. Giarratano & Frederick[3]	No	Linearly increased surface temperature	---	Carbon film on quartz	No	LHe	0.83 bar
5. Giarratano & Steward[4]	No	Single pulse, step heat input	---	Carbon film on quartz	Forced flow in channel	LHe	1 to 10 bar
6. Yanagi & Akiyama[5]	Yes	Single pulse, step heat input	---	Cu plate	Buoyancy induced flow in parallel plate channel	LN_2 LHe	1 bar
7. Iwasa & Apgar[6]	No	Single pulse, step heat input (Film boiling Regime)	---	Cu plate	Buoyancy induced flow in parallel plate channel	LHe	1 bar
8. Schmidt[7]	No	Single pulse, step heat input	NbTi/Cu wire	---	No	LHe	1 bar
9. Oker & Merte[8]	No	Single pulse, step heat input	No	Gold film on quartz substrate	No	LN_2	1 bar
10. Derewnicki[9]	No	Single pulse, step heat input	Platinum wire	---	No	No (Water)	1 to 10 bar

The pulsed power levels are well beyond the steady state critical heat flux (CHF) in the cryogens of interest. Although the geometry of applied interest is quite complex, heat transfer from single wires is described in the present work so as to relate the findings to prior basic work and develop understanding of the physical processes involved.

PREVIOUS WORK

Exponential heatup of tubes for application to power excursions in nuclear reactors has been studied for example by Sakurai and Shiotsu[10]. Table 1 summarizes more recent experimental investigations which consider the boiling behavior of conductors with step changes in current or single pulses of current. The application for most of these data are the partial quench and recovery processes in superconducting magnets. In the present application, the heat transfer behavior is of interest during transient heatup as well as transient cooldown.

Heatup Period

The heat transfer at the earliest stages of a single pulse was shown to occur by transient conduction into the surface adjacent liquid layer[1,2,11]. Following the conduction controlled period, the heat transfer behavior was modeled[3] as an extension of steady-state nucleate boiling plus sensible heating and vaporization of a thin liquid microlayer on the heated surface. The model is expressed as

$$q_{tr} - q_{ss} = \delta \rho_f c_{pf} (dT/dt) - \rho_f h_{fg} (d\delta/dt) \qquad (1)$$

where δ, the thickness of the microlayer, decreases with time. Following complete evaporation of the layer, transition to film boiling occurs.

Iwasa and Apgar[6] modeled the transient heating process starting from film boiling conditions in terms of the surface heatup rate as

$$q_{tr} - q_{ss} = a(\Delta T)(dT/dt) \quad (2)$$

where $a(\Delta T)$ is the heat capacity of the vapor layer adjacent to the heated surface. The heat capacity was backed out from their experimental data.

Derewnicki[9] measured rapid transient heat transfer from platinum wires to water and reported heat flux levels up to two orders of magnitude greater than steady-state CHF. He speculated that the high heat flux levels were due to convection driven by mechanisms such as high frequency oscillations of the vapor bubbles and thermocapillary forces. He did not formulate mathematical models for these mechanisms.

Cooldown Period

Steward[1] measured the cooldown period during rapid transient boiling in liquid helium and nitrogen for single and repeated heat pulse to a thin-film heater. Schmidt[7] measured the cooldown time of a superconducting wire following a quenching event (loss of superconductivity). Neither of these investigators modeled the cooldown behavior.

Iwasa and Apgar[6] modeled the transient cooldown in film boiling by the same equation as the heatup model (Equation 2). This model yields a step change in the surface heat flux at the start and termination of the heat pulse. Such a behavior has been seen experimentally[5,6]. Bui and Dhir[12] measured transient cooldown in the transition boiling regime. Their results show that heat flux values during transient cooldown are lower than the heatup or steady-state values for this regime.

Repeated Pulses

Yanagi and Akiyama[5] investigated the boiling behavior from a repeatedly pulsed copper plate. Their results show a rapid increase in heat flux at initiation of the heat pulse and a step decrease in surface heat flux when the pulse is terminated and the cooling transient begins. No models were formulated.

The experimental results presented in the present work are unique because they include data both during the heatup and the cooldown periods for conditions where the power input to the conductors exceeds the steady-state values for CHF by several times. The analytical modelling developed in the present study treats the coupled and repetitive transient heat transfer during the heat pulse and also during the cooldown of the conductor. The behaviors modeled are unique to repetitive pulses and boiling from small-diameter wires and are not treated analytically in prior work.

TEST FACILITY AND PROCEDURES

The present transient boiling were data obtained using annealed stainless steel (AISI 302) wires 0.25 and 0.50 mm in diameter. Fig. 1 shows a schematic of the test facility. The test section was a 5 cm ID glass tube flow passage that was located inside an evacuated Plexiglas tube with 15 cm OD. The heater wire was mounted across the diameter of the tube, normal to the flow direction, and held in place by copper alloy prongs. Electrical current was supplied to the wire through the prongs. The test section was transparent to permit the boiling phenomena at the surface of the wire to be observed.

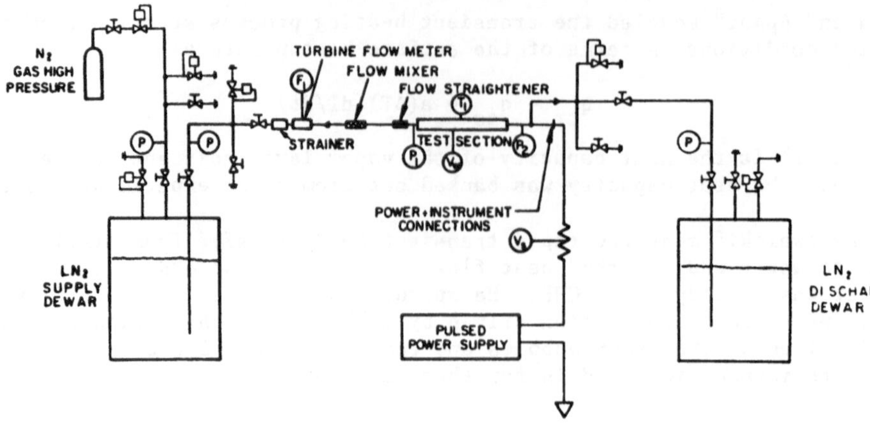

Figure 1. Schematic of Test Facility

The stainless steel wire was its own temperature sensor. Voltage taps were attached to the wire to measure the voltage drop across a known length. This voltage drop was used in conjunction with the measured current (using a standard series resistance) to obtain the wire resistance. The transient wire temperature was determined to within ± 1K from a calibration of the resistance versus the temperature. The measured bulk wire temperature was estimated to be up to 2K higher than the surface temperature of the highest heat flux levels employed. No corrections were applied for this difference. Bulk temperature of the liquid nitrogen coolant, the pressure, and the coolant flow rate were also measured.

The wire was heated with electric current from a pulsed DC power source with input power ranging more than an order of magnitude greater than steady state CHF. A small continuous current was supplied to the wire to measure its resistance and hence its temperature during the cooling period between pulses. Input pulses were at constant current I for fixed duration t_p and were repeated at a frequency of 5 Hz. Sufficient repetitions were performed in each test so that the conductor reached a quasi-steady condition. That is, the conductor cycled between minimum and maximum temperatures which were the same for successive pulses.

Computerized data acquisition with digitization at 10 kHz per channel was used, providing 0.1 ms resolution in the data. A MASSCOMP model MC5600 super-microcomputer along with the commercial software product product IDARS (Integrated Data Acquisition and Reduction System) was used to acquire and reduce the data.

STEADY-STATE BOILING HEAT TRANSFER

Parameters such as conductor material, material surface, and conductor geometry have been found to significantly affect steady-state boiling behavior (Frost[14]). Giarratano[2] has also shown that for very small wire diameters (0.04 mm), the steady-state nucleate and film boiling behavior differs significantly from the models in the literature. Therefore, steady-state boiling experiments were performed with the conductors of interest here to serve as a baseline for the evaluation of transient boiling test results. Our steady-state data showed good agreement with boiling data in the literature. Comparisons against established models for the various boiling regimes show that for the present wire sizes:

- Nucleate boiling can be modelled by the Kutateladze[15] correlation if the coefficient is increased by a factor of approximately 1.6. This correlation has been used successfully in the past for nucleate boiling in liquid nitrogen (Frost[14]).

- Critical Heat Flux (CHF) can be modelled by the superposition of Kutateladze's correlation[15] with corrections for subcooling (Gambill[16]) and forced convection effects (Fand and Keswani[17]). This superposition approach was suggested by Gambill.

- Pool Film Boiling can be modelled by the Breen and Westwater[18] correlation. This correlation has agreed well previously with film boiling data for cryogens (Brentari and Smith[19]).

- Film Boiling with forced flow of liquid can be modeled by the superposition of pool boiling[18] and high velocity flow boiling[20] correlations. The superposition approach suggested by Kutateladze with an exponent of 4 fit the data well.

- Forced convection without boiling can be modeled by the Fand and Keswani[17] correlation.

Comparisons of steady-state free convection and Minimum Film Boiling (MFB) correlations were not made due to insufficient data.

TRANSIENT BOILING TEST RESULTS AND MODELS

Pulsed transient tests were conducted for saturated and subcooled pool boiling conditions over a range of pulse current levels. One set of tests was also conducted with forced crossflow of subcooled liquid. Table 2 summarizes the test matrix for these transient experiments. Fig. 2 illustrates the typical temperature transient (a) and heat transfer behavior on a boiling map (b) for the baseline transient experiments conducted in a saturated pool. The dashed lines on the boiling map indicate the steady-state pool boiling results for reference. It is seen from Fig. 2b that the heat flux is almost constant during heatup. A jump in heat flux to a high value at the beginning of repeated pulses is also seen in previous data[5]. At the end of the pulse, the heat flux rapidly adjusts to a value in the transition boiling regime. During cooldown, the heat flux increases at first, peaking at a value lower than steady-state CHF. It then stays almost constant for a while and then decreases at typical nucleate boiling levels. The test illustrated in Fig. 2 shows conductor cooldown to the coolant temperature between pulses and is referred to as a Fully-Cooled test. When the conductor cycles with a minimum temperature which is above the coolant temperature, it is refered to as an Overheated test. Here we deal primarily with Fully-Cooled tests.

Table 2. Summary of Test Conditions for Transient Boiling (5 Hz)

TEST SERIES	WIRE DIAMETER, d_w (mm)	SUBCOOLING, ΔT_{sub} (K)	FLOW VELOCITY, V_f (cm/s)	PRESSURE, P (bar)	PULSE TIME, t_p (ms)	HEAT PULSE AMPLITUDES TESTED, I/I_{CHF}
SWTR-33	0.25	0	0	0.99	2	2.75, 3.29, 3.68, 4.07
SWTR-26 to 28	0.25	0	0	0.99	6	1.75, 2.16, 2.43
SWTR-22 to 25	0.25	0	0	0.99	10	1.22, 1.49, 1.56, 1.69
SWTR-34	0.25	0	0	0.99	20	0.98, 1.15, 1.31, 1.49
SWTR-30	0.25	4.3 - 6.5	0	2.3 - 3.2	10	1.27, 1.43, 1.66, 2.04
SWTR-31	0.25	5	13	2.3	10	1.29, 1.53, 1.80, 2.09
SWTR-36 (35)	0.50	0	0	0.99	10	1.24, 1.50, 1.61, 1.73
SWTR-37	0.50	0	0	0.99	6	1.54, 1.78, 1.93, 2.17

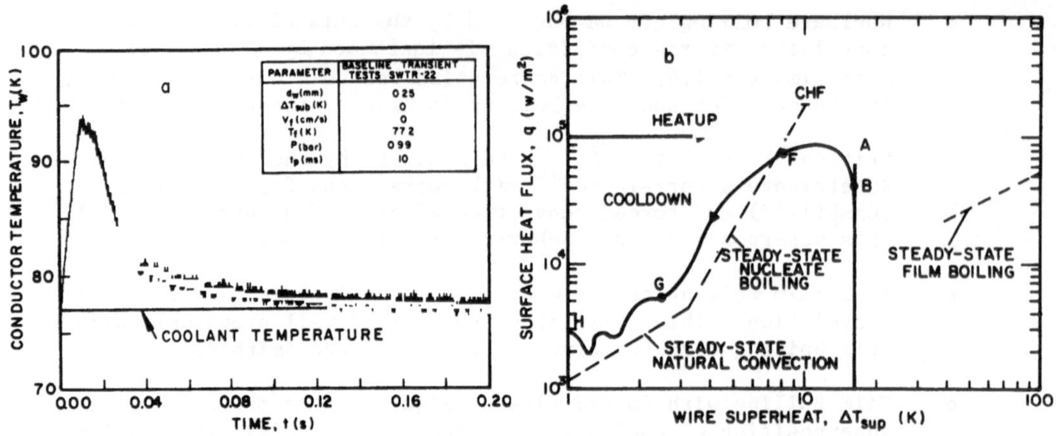

Figure 2. Typical Transient Pulsed Boiling Data

For the present tests, the temperature during the pulse is observed to vary linearly with time. Also, the heat flux during the heating period is almost constant (within +/- 20 percent) and both rate of temperature rise and heat flux increase with increasing pulse current levels. Given a linear temperature increase with time, heat fluxes governed by thermal conduction and convection depend upon time to approximately the one-half power and linearly on time, respectively. Neither result is consistent with the measured heat fluxes, which are approximately constant with time. Nucleate boiling also varies with temperature and therefore with time so that models proposed by Giarratano and Frederick[3] do not describe the observed behavior with repeated heat pulses either.

During transient heatup of a wire in helium[3], a liquid microlayer was hypothesized to account for high measured heat transfer rates. In repeated pulses, we expect that a similar microlayer would rapidly form on the wire surface at the beginning of a pulse. Vapor bubbles remaining on the wire surface from previous pulses would aid the rapid growth of bubbles and creation of the microlayer on the wire surface at much lower superheat values than if the wire were heated for the first time. Possibly, the most significant heat transfer mechanism during repeated pulsed heating is the sensible heatup and vaporization of the microlayer. The heat transfer from the wire surface during the transient heatup can then be written as

$$q = \delta \rho_f c_{pf} (dT/dt) + q_\ell \qquad (3)$$

The first term on the right hand side of this equation models sensible heating of the liquid layer. Its form implies a rapidly mixed microlayer which heats up as rapidly as the wire. Mixing mechanisms like thermocapillary convection[9], bubble agitation or high frequency oscillations are possible. The second term in the equation accounts for the energy removed by evaporation.

Fig. 3 presents the peak measured surface heat flux from the wire as a function of heatup rate. The experimental data are well represented by equation (3) with the coefficients $\delta = 0.07d_w$ and $q_\ell = 0.6 \times 10^5$ W/m^2. Thus, initial microlayer thickness is about 16 μm for the 0.25 mm wire and 32 μm for the 0.5 mm diameter wire. Previously[3], a microlayer thickness of about 4 μm was reported for a 0.025 mm wire in tests with helium. During the pulse, the microlayer thickness decreases from its peak value at a rate corresponding to the vaporization heat flux q_ℓ. Complete microlayer dry-out is predicted in 45 ms and 90 ms for the 0.25 mm and 0.5 mm wires

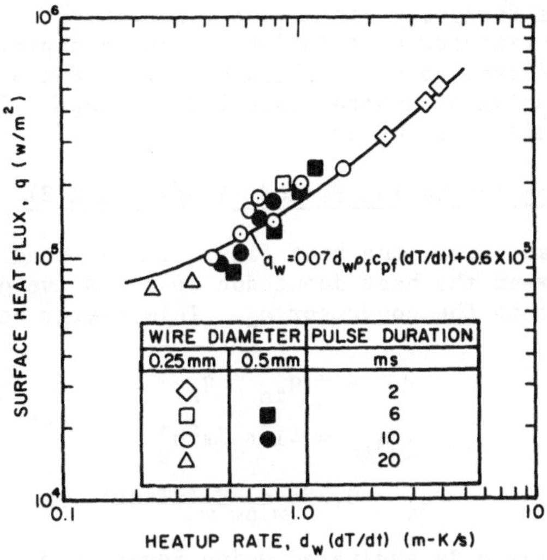

Figure 3. Heat Flux During Pulsed Heatup

respectively. Therefore, complete dryout did not occur in any of our tests since the maximum pulse time period employed was only 20 ms.

A further consequence of an unevaporated microlayer on the wire surface is low temperature gradients in the fluid adjacent to the wire when heatup is ended. A rapid decrease in heat flux could be expected at the beginning of the cooldown period. This behavior, is seen in our results and the results of Yanagi and Akiyama[5].

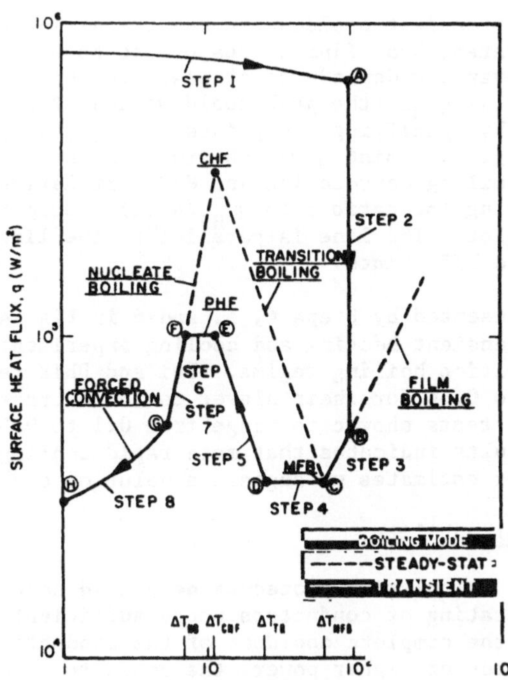

Figure 4. Idealized Map for Repetitive Pulsed Boiling Model

Based upon the foregoing discussion a piecewise model was developed for transient heat transfer calculations. Fig. 4 depicts the idealized transient process (corresponding labels on Fig. 2 are provided for comparison), consisting of general calculation steps. Each of the steps in the process is discussed below.

Heat Transfer During The Heatup Period (Steps 1 and 2)

During the heat pulse, the heat flux q at the conductor surface is the difference between the heat input due to resistive dissipation q_{in} and the heat stored within the conductor q_s. This energy balance can be written as

$$q = q_{in} - q_s \tag{4}$$

$$q_{in} = 4I^2\eta_w/\pi^2 d_w^3 \tag{5}$$

$$q_s = 0.25\rho_w c_{pw} d_w (dT/dt) \tag{6}$$

The surface heat flux q is evaluated using Equation 3. The conductor electrical resistivity η_w and specific heat c_{pw} are functions of temperature. Using finite time steps, equations 3 through 6 can be evaluated iteratively from the beginning of the pulse to yield the conductor temperature at each time step. At the end of the pulse, the heat flux decreases to the film boiling value at the same superheat (Step 2 in Fig. 4).

Heat Transfer During The Cooldown Period

As shown in Fig. 4, the model for transient boiling uses directly the aforementioned steady-state boiling models in the film, nucleate, and convective boiling regimes. The transient process differs significantly from the steady-state only in transition boiling (Steps 4, 5, and 6 in Fig. 4).

Our model for transient cooldown in the transition boiling regime is in three parts: constant heat flux corresponding to minimum film boiling (MFB) in step 4, power law dependence of heat flux on superheat in step 5, and constant heat flux q_{PHF} (the peak cooldown heat flux) in step 6. The model is completed by specifying the points C (q_{MFB}, ΔT_{MFB}) and E (q_{PHF}, ΔT_{CHF}) on Fig. 4. Point C was obtained as the intersection of the steady-state film boiling correlation and MFB correlation[21]. Point E was obtained by specifying the ratio c (= q_{PHF}/q_{CHF}). Step 5 is a straight line on a log-log plot. The line is parallel to the line joining the steady-state CHF and MFB points.

The model represented by Steps 4, 5, and 6 in Fig. 4 is supported by Bui and Dhir's[12] transient heating and cooling experiments for vertical plates in the transition boiling regime. Bui and Dhir report values of c ranging from 0.43 to 0.64 for their slower cooldown transients. For more rapid cooldown, our tests show c to range from 0.1 to 0.5. The difference between the two results indicates that more rapid cooling results in lower values of q_{PHF}. For estimates we suggest a value of c = 0.3.

Limits of Model Applicability

The models and calculation procedure described here apply only to repetitive pulsed heating of conductors where sufficient time is allowed between pulses for the complete cooldown of the conductor. For higher pulse repetition rates or higher power, the conductors do not cool fully between pulses so that after a few pulses quasi-steady overheated operation occurs. At still higher power levels and pulse repetition rates, the conductor temperature increases with every pulse until it fails

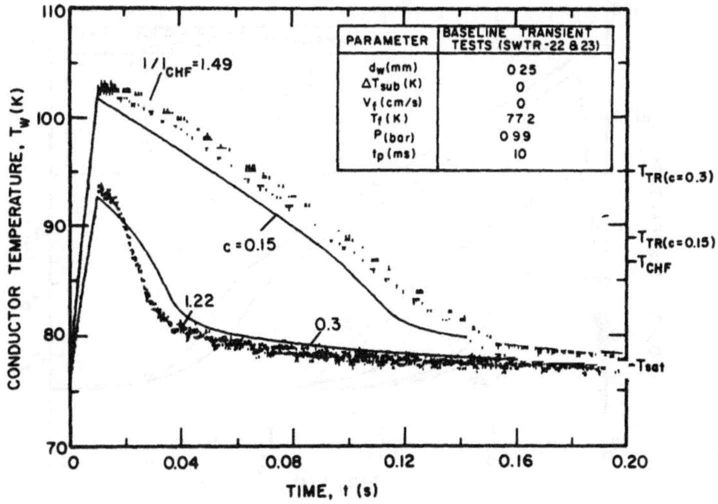

Figure 5. Comparison Of Analysis Predictions With Data For Saturated Pool Boiling

mechanically (burnout). Although these cases can occur, they are not of applied interest for pulsed cryogenic conductors. The present models may be useful there for scoping evaluations.

ANALYSIS COMPARISONS WITH TRANSIENT BOILING DATA

The ultimate objective of the analysis is to be able to predict the transient temperature response of conductors under various conditions. The models described in the previous section were programmed for computerized calculations. Fig. 5 compares the calculated and measured temperature histories for saturated pool boiling conditions and Fig. 6 shows corresponding comparisons for forced convection conditions. The analytical methodology is shown to predict the temperature history of the conductors quite well.

The analysis also allows the transition between Stable, Fully-Cooled and Stable, Overheated behavior to be predicted. For a fixed cooldown period, a maximum permissible conductor heatup ΔT_{max} can be obtained. Equations 3 through 6 combined and integrated using constant properties, yield

$$I^2 = \pi^2 d_w^3/4\eta_w \; [(\delta\rho_f c_{pf} + 0.25\rho_w c_{pw} d_w)\Delta T_{max}/t_p + 0.6x10^5] \qquad (7)$$

For simplicity, we further scale the current by the steady state current at CHF because for the limiting fully-cooled case most of the cooldown occurs at MFB and in the transition boiling region, both of which scale approximately with CHF. Average values of η_w, c_{pw}, and δ were used. This approximate stability model is compared against experimental data in Fig. 7. It correctly predicts the transition from Stable, Fully-Cooled to Overheated behavior.

SUMMARY AND CONCLUSIONS

The boiling behavior during rapid, pulsed heating and cooling transients for a cryocooled conductor was shown to be related in part to the steady-state boiling behavior. No significant delay times were observed in the transitions from one boiling regime to another.

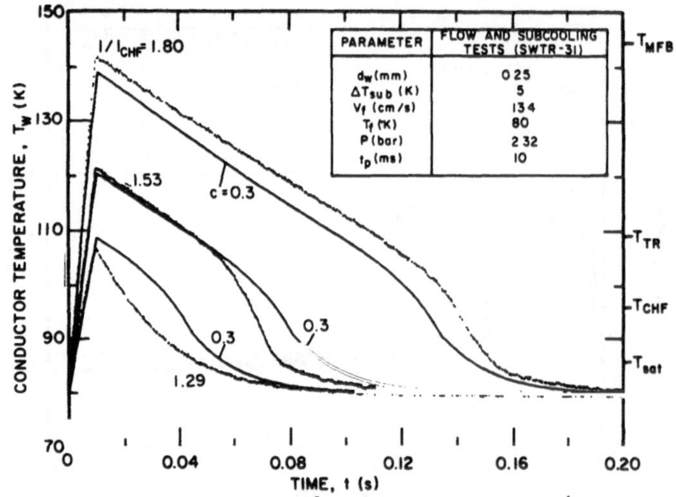

Figure 6. Comparison of Analysis Predictions With Data For Boiling With Forced Flow And Subcooling

During the input current pulse, i.e. the heatup period of the conductor, the heat flux to the coolant remains approximately constant. The magnitude of the heat flux during this period is related to the rate of heatup of the conductor. While the heatup model contains empirical coefficients based upon the experimental data, the form of the model was consistent with heat transfer to a microlayer on the wire surface. The microlayer thickness was directly related to the diameter of the wire.

Heat transfer behavior during the cooldown period of the transients was for the most part close to the steady-state behavior for the film, nucleate, forced convection, and natural convection regimes. In transition boiling, the peak heat flux q_{PHF} reached during transient cooldown was less than the steady-state CHF value but greater than the MFB value. Specifically, the peak heat flux was about 10% to 50% of the CHF value. Transient temperature response predictions using these models were in good agreement with experimental measurements.

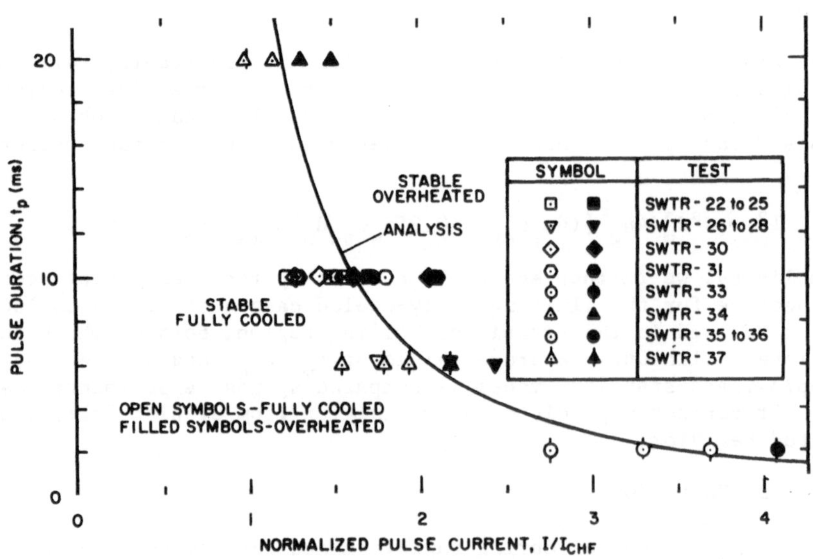

Figure 7. Stability Map For The Pulsed Conductor (5Hz)

The conductors exhibit two important types of temperature cycles:

- Stable, Fully-Cooled--where the conductor cools down to the bulk liquid nitrogen temperature during the cooling period, and
- Stable, Overheated--where the conductor cycles repeatedly between two temperatures above the bulk temperature of the liquid nitrogen coolant.

The transition point between the two types of behavior defines a stability limit important to the use of pulsed conductors to minimize resistive losses. This stability limit was predicted analytically and shown to be in good agreement with the experimental data. Presumably still higher power levels or pulse duration frequencies would lead to an unstable "runaway" situation which is not of applied interest.

ACKNOWLEDGMENT

The work described in this paper has been sponsored by the Aero Propulsion Laboratory, Air Force Wright Aeronautical Laboratories, Aeronautical Systems Division (AFSC), United States Air Force, Wright-Patterson AFB, Ohio 45433-6563.

NOMENCLATURE

c	constant in transient cooldown model
c_p	specific heat at constant pressure (J/kg-K)
d_w	wire or conductor diameter (m)
h_{fg}	latent heat of vaporization (J/kg)
I	current (A)
P	pressure (bar = 10^5 N/m^2)
q	surface heat flux (W/m^2)
q_ℓ	evaporative heat flux in microlayer model (W/m^2)
q_{in}	input heat flux due to resistive dissipation (W/m^2)
q_s	stored heat flux due to temperature rise of conductor (W/m^2)
T_{sat}	fluid saturation temperature (K)
T_w	wire or conductor temperature (K)
ΔT_{sub}	subcooling (K)
ΔT_{sup}	wire superheat ($T_w - T_{sat}$) (K)
ΔT_{max}	maximum wire heatup (K)
t_p	pulse time period (s)
δ	microlayer thickness (m)
η	resistivity (ohm-m)

Subscripts

CHF	critical heat flux
f	liquid phase
MFB	minimum film boiling
PHF	peak heat flux during cooldown in transition boiling
ss	steady-state
TR	transition boiling
tr	transient
w	wire or conductor

REFERENCES

1. W. G. Steward, Int. J. Heat Mass Transfer 21:863 (1978).
2. P. J. Giarratano, Int. J. Heat Mass Transfer 27(8):1311 (1984).
3. P. J. Giarratano and N. V. Frederick, in: "Advances in Cryogenic Engineering," Vol. 25, Plenum Press, New York (1980), p. 455.
4. P. J. Giarratano and W. G. Steward, J. Heat Transfer, Trans. ASME 105(2):350 (1983).
5. H. Yanagi and M. Akiyama, J. Fac. Eng. Univ. Tokyo, Series B 36(1):233 (1981).
6. Y. Iwasa and B. A. Apgar, Cryogenics 267 (1978).
7. C. Schmidt, Appl. Phys. Lett. 32(12):827 (1978).
8. E. Oker and H. Merte, Jr., in: "6th Int. Heat Transfer Conf.," Vol. 1, Hemisphere Publ. Corp., Washington DC (1978) p. 139.
9. K. P. Derewnicki, Int. J. Heat Mass Transfer 28(11):2085 (1985).
10. A. Sakurai and M. Shiotsu, J. Heat Transfer, Trans. ASME 99(4):547 (1977).
11. V. D. Arp et. al., in: "Cryogenic Processes and Equipment 1982," T. H. K. Frederking et al., eds., AIChE Symposium Series 79(224):126 (1983).
12. T. D. Bui and V. K. Dhir, J. Heat Transfer, Trans. ASME 107:756 (1985).
13. J. R. Lewis, in: "Handbook of Stainless Steels," D. Peckner and I. M. Bernstein, eds., McGraw-Hill Book Company, New York (1977).
14. W. Frost, ed., "Heat Transfer at Low Temperatures," Int. Cryogenics Monograph Series, Plenum Press, New York (1975).
15. S. S. Kutateladze, "Fundamentals of Heat Transfer," Edward Arnold Ltd., London, England (1963).
16. W. R. Gambill, Chem. Eng. Prog. Symp. Ser. 59(41):71 (1963).
17. R. M. Fand and K. K. Keswani, Int. J. Heat Mass Transfer 15:1515 (1972).
18. B. P. Breen and J. W. Westwater, Chem. Eng. Prog. 58(7):67 (1962).
19. E. G. Brentari, R. V. Smith, in: "Int. Advnaces in Cryogenic Engrg.," Vol. 10, K. D. Timmerhaus, eds., Plenum Press, New York (1964), p. 325.
20. S. Yilmaz and J. W. Westwater, J. Heat Transfer, Trans. ASME 102:26 (1980).
21. J. H. Lienhard and V. E. Schrock, J. Heat Transfer, Trans. ASME 85:261 (1963).

NEW HEAT TRANSFER AND FRICTION FACTOR DESIGN DATA FOR PERFORATED PLATE HEAT EXCHANGERS

Richard H. Hubbell

Arthur D Little, Inc.
Cambridge, Massachusetts

Christina L. Cain

Flight Dynamics Laboratory
Wright-Patterson AFB, Ohio

ABSTRACT

Perforated plate heat exchangers have been found to have inherently low axial conduction and are therefore excellent candidates for cryogenic applications where an all-metal design is required. A total of three plate cores were tested; two were chemically etched and the other mechanically punched. Hole size, percent open area and plate thickness parameters were varied among the plates. Experimental results were compared to analytical projections and found to differ significantly. The single-blow, transient test technique was used to determine the heat transfer coefficients and the isothermal pressure drop test was used to determine friction factors, as a function of Reynolds number.

INTRODUCTION

In order to obtain design data, it was necessary to test the heat transfer plates of all-metal heat exchangers being developed. Consequently, a test section using the single blow transient testing technique[1], was built and used to test perforated plates being considered for the heat exchangers.

All the perforated plate test sections had 40 copper plates separated by 39 stainless steel e-seal spacers and were designed for a helium working fluid. Two test sections had plates with etched holes (9% open area, .0724 cm hole diameter, and .0457 cm plate thickness; and 24.5% open area, .0749 cm hole diameter, and .0457 cm plate thickness), and one had plates with punched holes (23.6% open area, .0457 cm hole diameter, and .0533 cm plate thickness). Figure 1 shows an assembled test section, an e-seal spacer, and the three plates.

ANALYSIS

The objective of the tests was to compare test sections composed of different plates. Since the N_{tu}/plate dictates the size of the heat exchanger for any required effectiveness, the number of heat transfer

Figure 1. Test core, e-seal spacer, and 3 plate types.

units (N_{tu}) was chosen to compare the plate heat transfer characteristics; the friction factor (f_e) was selected for the pressure drop comparison. Data was then gathered for hole Reynolds numbers ranging from 4 to 666. The Reynolds number used for all calculations (hole Reynolds number) was calculated using a hole hydraulic diameter as the representative length. The mass flow through the core is calculated using the minimum free flow area through the plates. The following properties were measured and used to calculate N_{tu} and f_e: the mass flow rate, the pressure drop across the test section, the temperature upstream and downstream from the test section, and the transient temperature of the test fluid downstream from the test section.

Theoretical Predictions

Heat transfer predictions. The N_{tu}/plate can be determined as a function of plate geometry and hole Reynolds number. The equations used to calculate the N_{tu}/plate assume that $1 \le N_{Re} \le 1000$, $N_{Pr} = 0.7$, and the plate holes have a uniform wall temperature. Two laminar flow cases were considered. One case assumed a developing velocity profile in each hole and the other assumed a fully developed parabolic velocity profile.

Developing Velocity Profile:

$$N_{tu}/\text{plate} = [(1-\sigma)\sigma^{-.232}\ (2.554 + 2.011\ N_{Re}^{.52}\ \sigma^{.121}) + 20.9(t/d)]/N_{Re} + 0.417/[1+.012(N_{Re}\ d/t)^{0.8}]$$

Parabolic Velocity Profile:

$$N_{tu}/\text{plate} = [(1-\sigma)\sigma^{-.232}\ (2.554 + 2.011\ N_{Re}^{.52}\ \sigma^{.121}) + 20.9(t/d)]/N_{Re} + 0.269/[1+.032(N_{Re}\ d/t)^{0.667}]$$

Friction factor predictions. The observed friction factor for a plate can be considered the sum of the effect due to shear forces in the holes, and the entrance and exit effects.

$$f_e = (d/4t)(K_c + K_e) + f$$

Kays and London[2] present K_c, K_e, and f graphically as a function of hole Reynolds number, hole diameter, plate thickness, and plate open area ratio.

Data Reduction

Heat transfer results. The core N_{tu} is a function of the longitudinal heat conduction (thermal conduction in the solid, parallel to the flow direction) and the maximum slope of the temperature vs time plot. The relationship between the longitudinal heat conduction, the maximum rate of change of temperature with time and the N_{tu} is presented in tabular and graphic form by Pucci[3]. Note that the Colburn modulus can be calculated given the heat transfer number (N_{tu}), the Prandtl number (N_{Pr}), and the plate geometry (A_c/A) using the following relationship:

$$j = N_{tu} \, (A_c/A) \, N_{Pr}^{2/3}$$

Friction factor results. The friction factor per plate can be calculated given the mass flow rate and pressure drop across the test section during steady state operation.

$$f_e = (D_h \, \Delta P \, \rho \, A_c^2 \, g_c)/(2 \, n \, t \, \dot{m}^2)$$

$(D_h \, A_c^2)/(2nt)$ is a physical property of the test section, and the temperature, pressure change, and mass flow rate were measured.

TEST METHOD

Description of Apparatus

The test apparatus (Fig 2) consists of compressed air (or helium), a resistance heater followed by a flow straightening section, and the test section. The test section is in a switching box which allows it to be moved quickly into the stream of heated fluid. When the test section is not in the system, a dummy core, which causes the same pressure drop as the test section, is in place. Temperature and pressure readings are taken downstream from the straightening section, and the volumetric flow rate is measured upstream from the staightening section using a rotometer. The outlet temperature is measured after the switching box, and the outlet pressure tap is far enough downstream from the test section (∿7.6 cm) to allow for pressure recovery.

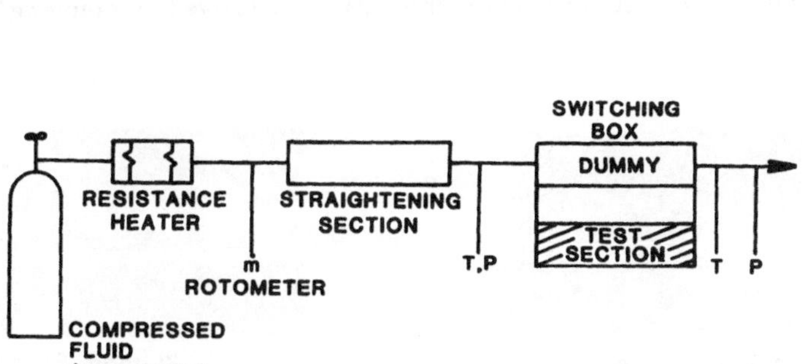

Figure 2. Schematic of test apparatus.

Test Procedure

The test procedure used to collect the data follows:

Heat transfer data. (1) Establish a uniform temperature in the test core by blowing room-temperature compressed air, or helium, through the test apparatus with the test section in place. A uniform temperature (± 0.056 K) is necessary through the entire test core volume to ensure repeatable results. (2) Move the dummy section into the flow to allow the fluid to bypass the test section. (3) Use the resistance heater to increase the fluid temperature about 11 K. (4) Maintain the heated airflow until all the apparatus components reach a stable temperature. The time-temperature plot should vary less than 0.56 K/hr once this steady-state condition is achieved. (5) Next, move the room-temperature test section into the hot air stream, and record the air temperature leaving the test section as a function of time. The pressure drop across the test section and the dummy section must be within 0.254 cm H_2O of each other to maintain a constant flow rate through the apparatus before and after the test section is moved into the stream.

Friction Factor Data. Record the pressure drop across the test section for steady-state, room-temperature fluid as a function of the flow rate.

TEST RESULTS

The two heat exchanger core types (punched, and etched) were tested four ways. The punched plates have a breakout caused by manufacturing (Fig 3). Tests were run with the breakout facing both upstream and downstream. The orientation of the core affected the N_{tu}/plate but did not affect the friction factor. Since the etched plates do not have breakouts, orientation was not considered. The etched plates were tested with both helium and air to insure there was not a significant effect due to the difference in Prandtl numbers (N_{Prair} = .72, N_{PrHe} = .68).

Heat Transfer Data

Figures 4a and 4b present the theoretical and measured heat transfer data. The theoretical results indicate that N_{tu}/plate increases with decreasing d/t ratio. The experimental results indicate that for hole Reynolds numbers above 30, the N_{tu}/plate increases as both d/t and σ decrease. Also, the slope of the N_{tu}/plate vs N_{Re} decreases as d/t increases.

Figure 5 demonstrates the dependence of the heat transfer data on plate orientation (or hole shape). The punched plates were oriented in the core with all the hole breakouts facing the same direction. When the core was oriented so the fluid in the holes followed a converging path,

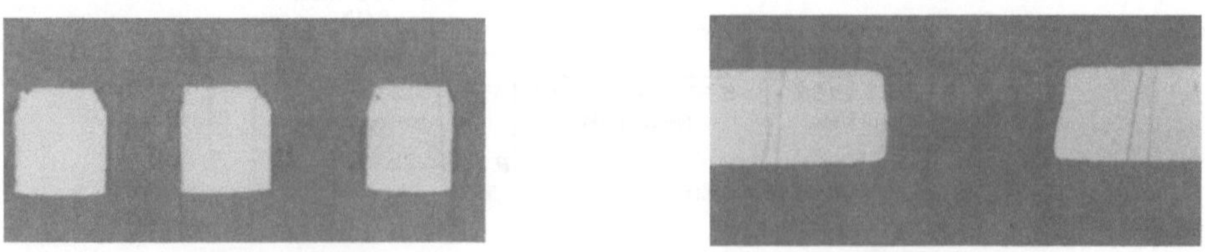

Punched Plate (t=.053 cm, d=.046 cm) — Etched Plate (t=.046 cm, d=.0724 cm)

Figure 3. Photograph of hole cross sections.

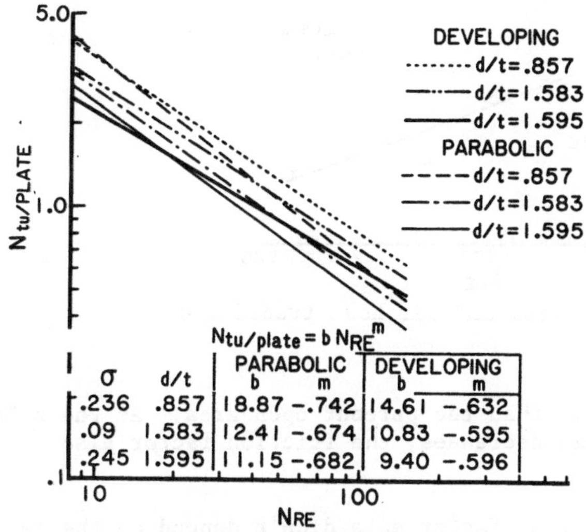

Figure 4a. Theoretical predictions for heat transfer data.

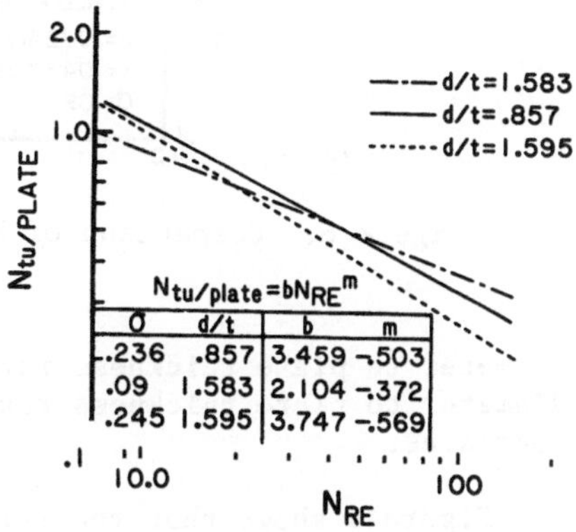

Figure 4b. Experimental heat transfer data.

the N_{tu}/plate for a given N_{Re} was higher than for a core oriented so the fluid followed a diverging path. This small, but measurable, increase in N_{tu}/plate is probably due to less flow separation in the holes.

The heat transfer data didn't depend on the test gas (Figure 6).

Note that the theoretical heat transfer predictions are not accurate. Both the predicted slope and intercepts are off, resulting in predictions that are too high. Also, the model failed to predict that larger d/t ratios result in larger drops in N_{tu}/plate as N_{Re} increases. The discrepancies are probably due to uncertainty over actual fluid flow in the holes. Clearly hole shape affects the heat transfer data and the model assumes cylindrical holes with no flow separation. Actually the holes are not cylindrical and there may be flow separation which accounts for the lower measured N_{tu}/plate values.

Friction Factor Data

Figure 7 presents both the theoretical and measured friction factors as a function of hole Reynolds number. The theoretical results show that the friction factor decreases as the percent open area increases and as the hole diameter to plate thickness ratio decreases. The experimental results indicate that the friction factor is more sensitive to the hole

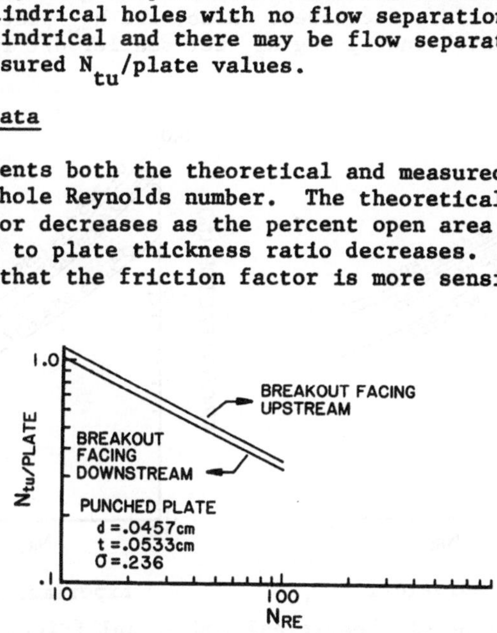

Figure 5. Effect of plate orientation on N_{tu}/plate.

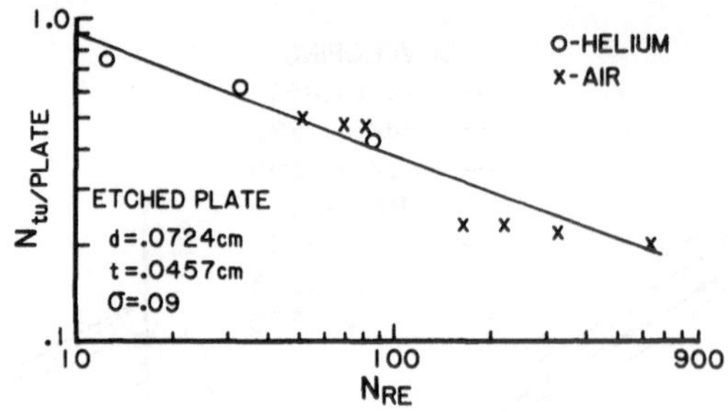

Figure 6. Comparison of helium and air heat transfer data.

diameter to plate thickness ratio than the percent open area. As the hole diameter to plate thickness ratio decreases, the friction factor also decreases.

Figure 8 shows that the friction factor data didn't depend on the test gas (air or helium) or the core orientation.

CONCLUSIONS

1. The apparatus using the single blow transient test method to determine the heat-transfer characteristics of perforated-plate heat exchangers produces results accurate enough for design evaluations.

2. Air can be used to test heat exchangers designed for helium systems.

3. Hole shape affects the heat transfer characteristics. The increase in N_{tu}/plate when the punched plates were tested with the breakout facing upstream indicates that a converging-diverging hole may provide better heat transfer characteristics.

4. Heat transfer characteristics depend strongly on the hole diameter to plate thickness ratio. For hole Reynolds numbers greater than 30, the N_{tu}/plate increases as d/t decreases. Also, the N_{tu}/plate doesn't decrease as quickly with increasing N_{Re} for smaller d/t ratios.

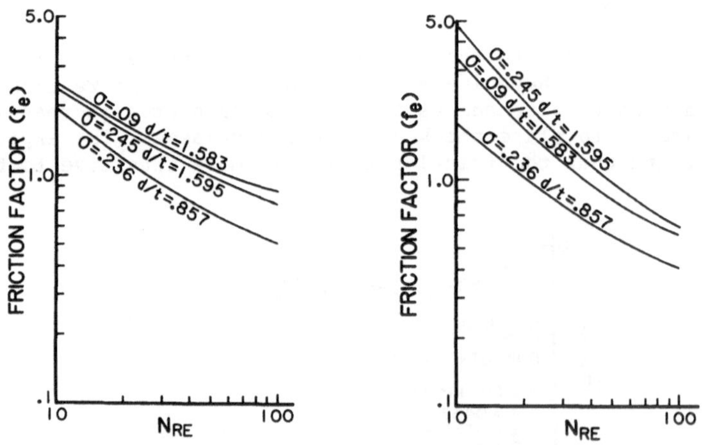

Figure 7. Comparison of theoretical and actual friction factor data.

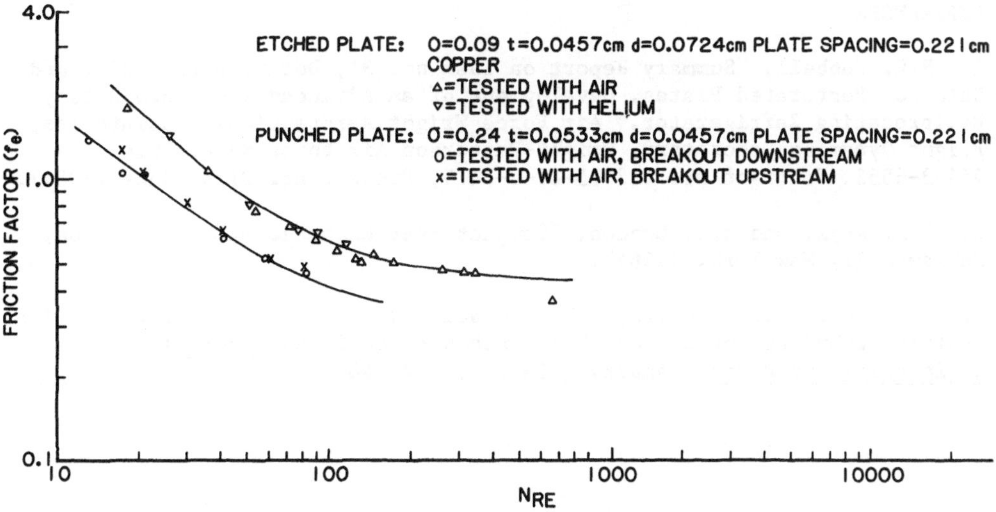

Figure 8. Effect of orientation and test gas on friction factor data.

5. The model used to determine the N_{tu}/plate doesn't completely describe the system. The correlation between the theoretical and experimental results is poor but improves as the hole Reynolds number increases. The correlation is also better assuming a parabolic velocity profile instead of a developing velocity profile in the holes. The difference between the predicted and actual values is probably due to uncertainty over actual flow conditions in the holes.

NOMENCLATURE

- A Total heat transfer surface area.
- A_c Minimum free flow area (plate open area).
- d Hole diameter.
- D_h Hole hydraulic diameter.
- f Friction factor due to wall shear in holes only.
- f_e Friction factor including hole entrance and exit effects.
- g_c Proportionality factor in Newton's second law.
- j Colburn modulus.
- K_c Entrance coefficient.
- K_e Exit coefficient.
- $\dot{m}$ Mass flow rate (based on minimum free flow area).
- n Number of plates in the core.
- N_{Pr} Prandtl number.
- N_{Re} Reynolds number (based on hole diameter).
- N_{tu} Number of heat transfer units.
- ΔP Pressure change across the test section.
- t Plate thickness.
- ρ Fluid density.
- σ Ratio of plate open area to frontal area.

ACKNOWLEDGEMENTS

This research was sponsored by the Air Force Wright Aeronautical Laboratories, Flight Dynamics Laboratory, under contract No. F33615-81-C-3419.

REFERENCES

1. R.H. Hubbell, "Summary Report on Task no. 31, Determination of f and j Data for Perforated Plates, Development of an Advanced Two-Stage Rotary Reciprocating Refrigerator," Air Force Wright Aeronautical Laboratories, Flight Dynamics Laboratory, Wright-Patterson Air Force Base, Ohio 45433-6553, Contract no. F33615-81-C-3419, Project no. 2126, June (1984).

2. W.M. Kays, and A.L. London, "Compact Heat Exchangers," 2nd Edition, McGraw Hill, New York, (1964).

3. P.F. Pucci, C.P. Howard, C.H. Piersall, Jr., The Single-Blow Transient Testing Technique for Compact Heat Exchanger Surfaces, Journal of Engineering for Power, January (1967), pp. 29-40.

AN INVESTIGATION INTO FLOW REGIMES FOR TWO-PHASE HELIUM FLOW

J.C. Theilacker

Fermi National Accelerator Laboratory*
Batavia, Illinois

C.H. Rode

Continuous Electron Beam Accelerator Facility+
Newport News, Virginia

ABSTRACT

The Tevatron accelerator at Fermilab incorporates long two-phase helium passages. During magnet design, the generalized flow map of Baker was used to predict homogeneous flow. Longer than expected magnet time constants led to this investigation. The importance of predicting the flow regime has been amplified with the advent of non-horizontal accelerator designs.

A test setup was constructed at Fermilab to investigate two-phase helium flow regimes for conditions practical in accelerator designs. The setup consisted of a standard Tevatron satellite refrigerator, subcooling dewar, heater, 35 m long transfer line, and a specialized end box. A knife blade on the midplane of the end of the transfer line diverted the flow from the upper and lower halves of the pipe to separate vessels in the end box. The amount of liquid above and below the plane was measured at various total mass flow rates and liquid percentages.

The results show that stratified flow occurs at much higher liquid percentages than predicted by the Baker diagram (several orders of magnitude). We were not able to produce high enough steady state flows to find a boundary to a homogenous flow regime. Stratified flow occurred over all practical conditions for long accelerator magnet systems.

INTRODUCTION

The Tevatron accelerator at Fermilab was designed with continuous two-phase helium heat exchange with the collared coil assembly. This ensures a uniform temperature distribution throughout the magnet strings as long as the two-phase pressure drop is minimized. To achieve good

---

* Operated by Universities Research Association, Inc., under contract with the U.S. Department of Energy.

+ Operated by Southeastern Universities Research Association, Inc. under contact with the U.S. Department of Energy.

radial heat transfer and to minimize system time constants, it was desirable to design for a homogeneous two-phase flow. Future superconducting accelerators (HERA, LEP, SSC, UNK) are being proposed with longer magnet strings (up to 4000 m long) on inclines (up to 1 1/2% grade). Under these conditions it becomes more important to verify the two-phase flow regime.

Flow regime maps are used to design for a specific flow pattern. Two-phase flow can exist in a variety of homogeneous and nonhomogeneous regimes as shown in Fig. 1. Savery[1] reviewed flow maps developed for horizontal as well as vertical channels. One of the more popular horizontal channel charts is that of Baker[2] (Fig. 2). Using this chart, the Tevatron was designed to operate in the froth flow regime.

During the commissioning of the Tevatron, longer than expected time constants were measured for the magnet two-phase circuit. This suggested a nonhomogeneous two-phase flow. As a result, this investigation was made to experimentally locate the boundaries between homogeneous (froth) and nonhomogeneous (stratified, wavy, slug/plug, annular) flows for two-phase helium. Results show that stratified flow occurs at much higher liquid percentages than predicted by the Baker diagram. We were unable to locate the homogeneous flow boundary boundaries over the operating range of the test setup.

SYSTEM DESCRIPTION

The experimental setup is shown schematically in Fig. 3. It consists of the following components:

- 625 watt refrigerator (standard Tevatron satellite refrigerator operating in stand-alone mode)
- 450 liter subcooling dewar
- Transition box
- Adiabatic test line
- Collection box

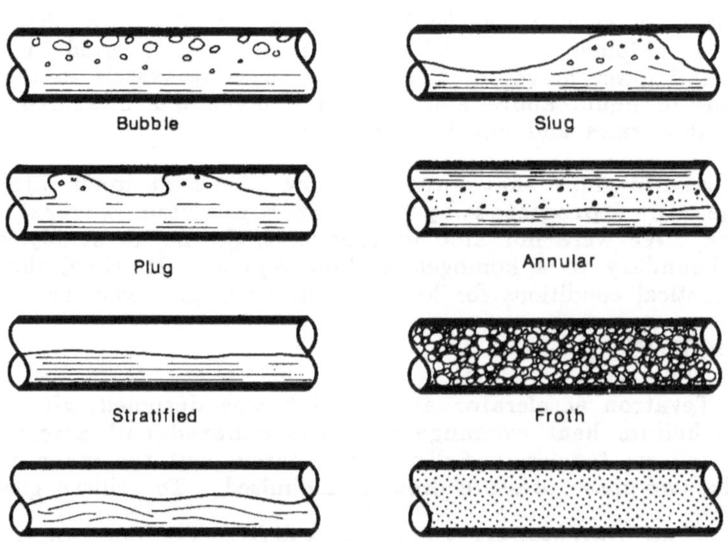

Fig. 1. Two phase flow regimes.

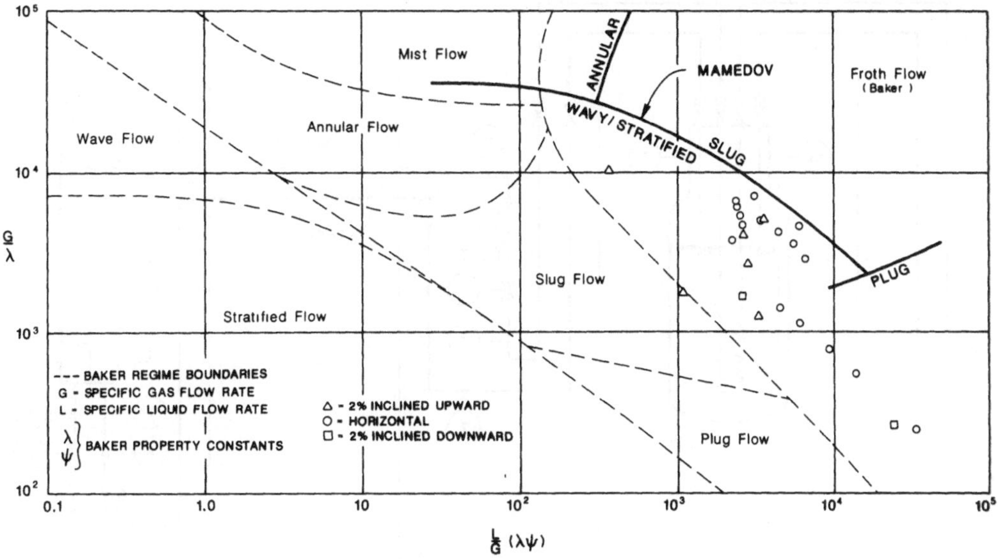

Fig. 2. Baker diagram for horizontal channels.

Output from the refrigerator (typically 1.6 atm) is first subcooled by 1.2 atm boiling liquid in the subcooling dewar. This assured a constant temperature output. Between each test run, the dewar is refilled using the refrigerator. After the dewar is full, the reciprocating expansion engines are turned off to reduce the noise on the pressure and flow measurements.

Subcooled liquid helium then enters the transition box where the following took place:

- Amount of subcooling is measured with a vapor pressure thermometer and pressure measurement.
- Flow rate is measured with a venturi flowmeter
- An electric heater is used to burn off the subcooling and to adjust liquid percentage

The resulting two-phase helium first passes through an 8m long, 23mm ID pipe before entering the 24 m long, 45 mm ID test line. Both lines are made adiabatic by shielding them with the 4.5 K return flow.

At the end of the test line, a knife blade separates the top and bottom halves of the flow and drains each into phase separators. The collected liquid can be measured as a rate of change in liquid levels or by drain venturi flowmeters. The gas flows are combined and pass through a pressure regulating valve, returning to the refrigerator through the test line shield.

This setup was designed to distinguish between stratified flow and other two-phase flow regimes. The specific type of flow regime could not be determined in most cases. For stratified flow with a liquid percentage <80% (the point at which perfectly stratified vapor and liquid phases traveling at the same velocity each occupy half the cross section at 1.6 atm), one would expect to see liquid only in the phase separator of the lower half of the pipe. If however, there where nearly equal amounts of liquid reaching each phase separators, then the regime would not be clear.

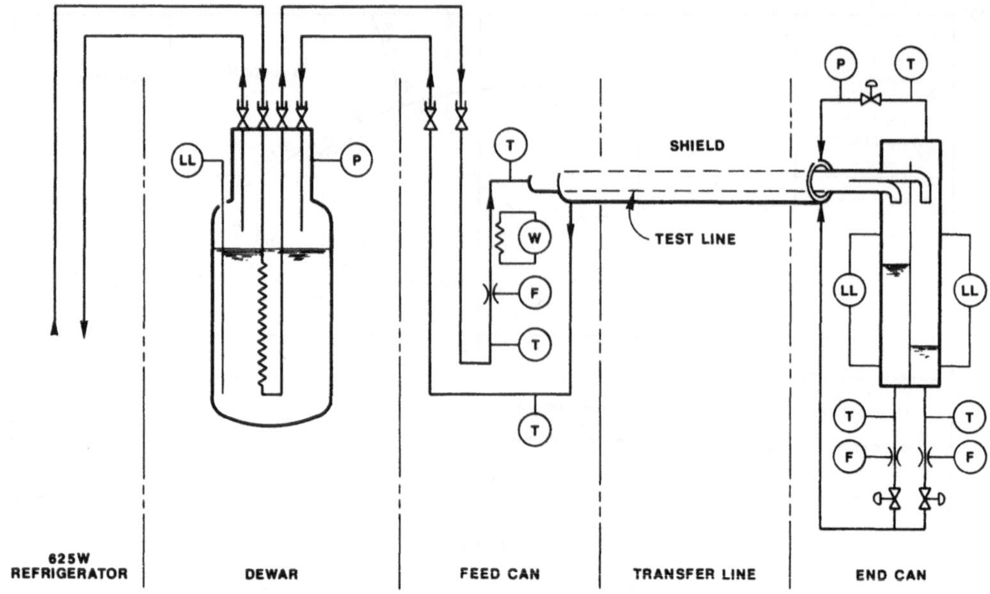

Fig. 3. Two phase test schematic.

In this case, the flow could be homogeneous (froth) or nonhomogeneous (annular).

For circumstances where the liquid flow is predominantly in the lower half of the pipe, again assuming a liquid percentage <80%, then any of the following conditions could be true.

- wavy flow
- slug flow
- gravitational effected annular flow
- stratified flow

To further investigate the characteristics of two-phase helium flow, the test setup was built to allow inclines of ±2%. This angle includes all proposals for large accelerators.

RESULTS

A total of 26 test runs were made; six inclined 2% upward, 18 horizontal, and two inclined 2% downward. For operational convenience, all runs were made at 1.6 atm with the exception of five of the six runs inclined 2% upward which were 1.8 atm. All of the test points indicated a stratified or wavy two-phase flow. Results of the flow geometries are discussed below.

2% Inclined Upward

For the test, the line was first filled with saturated vapor. The line was then filled with a two-phase mixture and the time delay for liquid to reach the phase separators was measured. Test results are shown in Table I.

For a homogeneous froth flow, one would expect a fast time response. On the other hand, the slowest possible time response would be to consider just the liquid flow filling the inclined volume. The calculated delays for these extremes are given in Table 1 columns 4 and 5 respectively. Measured time delays are given in column 6. Column 7 is the ratio of measured (column 6) to liquid only (column 5) time delays. This represents the fraction of the inclined volume which is actually filled with liquid. To cross check the data, we followed the test with a second wave with a few percent gas to "top off" the line (column 8). The top off time should be the difference between columns 5 and 6. Data point #6 was the only point whose top off time was inconsistant with this rule.

Time delays in Table 1 clearly shows a stratified flow for the first few runs. The decreasing ratio in column 7 corresponds to an increase in gas phase flow (i.e., the gas phase requires more of the volume in order to escape). This gas volume is shown in Figure 4a. As an attempt to verify the mechanics of Fig. 4, the gas cross sectional area necessary to result in a pressure drop equal to the liquid head was calculated. Since the liquid head per unit length is constant, one would expect the gas pressure drop to also be constant. As a result, the gas cross sectional area would be nearly constant along the length. The results are shown in Fig. 5. The gas flow point near 40 g/s (#6) deviates from the curve, suggesting that the gas flow may be breaking up the liquid column.

The two 44 g/s data points both appear to be near phase boundaries. The high liquid data point (#5) showed liquid hitting the upper pot 0.4 min after hitting the lower pot (slug or plug flow?). This was the only one of the six points that liquid reached the upper pot. The high gas point (#6) showed an effect that may have been a decreasing frequency wave action hitting the knife.

Table 1. Two-Phase Test Results for 2% Incline Upward.

				Time Delay				
Run	1 Press	2 Flow	3 Quality	4 Froth	5 Liquid	6 Measured	7 Ratio	8 Top Off Time
#	(atm	(g/sec)	% Gas	(Min)	(Min)	(Min)	(Col. 6/5	(Min)
1	1.6	10.5	60	2.8	18.5	16.0	.865	2.2
2	1.8	10.6	45	3.6	12.6	10.5	.832	~2.5
3	1.8	20.8	49	1.8	6.9	5.0	.721	1.5
4	1.8	30.2	51	1.2	5.0	3.0	.600	~1.3
5	1.8	44.	43	0.9	3.0	1.8	.600	1.2
6	1.8	44.	88	0.6	13.9	1.8	.129	1.9

Horizontal

For the eighteen horizontal test runs, subcooled liquid (at 1.6 atm) was first circulated through the test line. The heater was then gradually increased to eliminate the subcooling. From the saturated liquid point, the heater was increased to achieve the desired percent liquid. After the line reached an equilibrium, liquid flow rates for the top and bottom halves of the pipe were measured. These flows were found by measuring the rate of rise in the phase separator liquid levels for low flows, or by venturi flowmeters for high flows. Test results are shown in Table 2.

Included in Table 2 are calculated values of half pipe liquid flow for froth and separated flow regimes. In both cases, values were calculated assuming the liquid and vapor velocities were equal. Column 4 shows that for many runs, no liquid would be expected in the top half of the pipe if a smooth stratified flow exists. Measurements in column 6 shows that wave

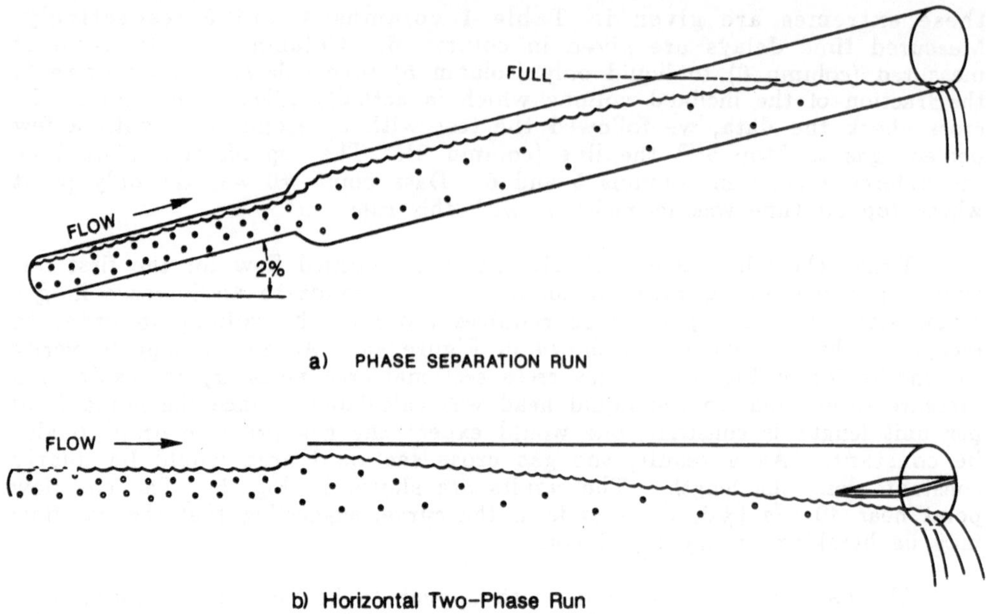

Fig. 4. Liquid separation in test line.

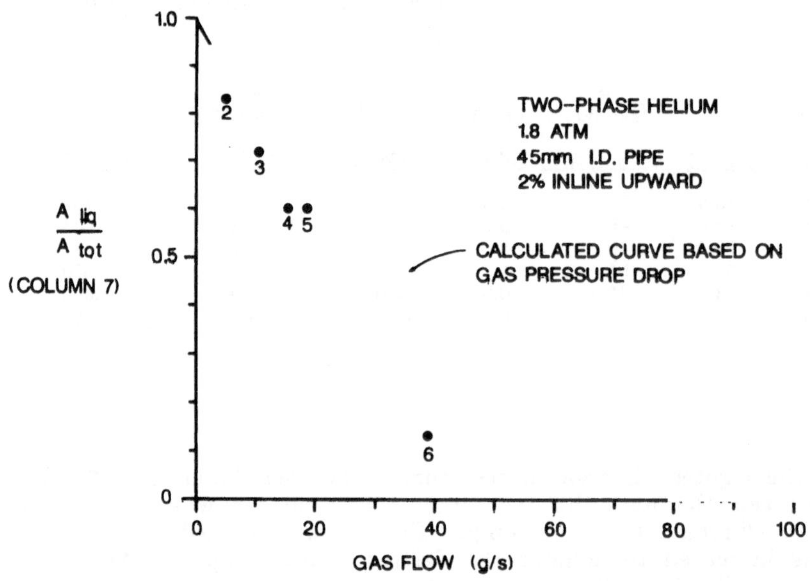

Fig. 5. Liquid flow area results for inclined upward flow.

flow must exist, allowing some liquid above the midplane. Comparing column 6 to 3 shows that a homogeneous froth flow does not exist, as would be predicted by the Baker diagram (Fig. 2). The boundary of the stratified flow regime appears to be shifted at least three orders of magnitude to the right.

Table 2. Horizontal Two-Phase Results

	1	2	3	4	5	6	7	8
			Half Pipe Liquid Flows					
			Calculated			Measured		Gas Vol Ratio
	Mass	Liquid	Froth	Separated Flow				
Run #	Flow (g/s)	Mass (%)	Top (g/s)	Top (g/s)	Bottom (g/s)	Top (g/s)	Bottom (g/s)	(Actual/calc)
7	57.5	94.6%	27.2	21.3	33.1	>10	<44	<0.46
7	57.5	90.1%	25.9	15.1	36.7	>10	<42	<0.77
7	57.5	86.3%	24.8	9.8	39.8	>10	<40	<1.01
7	57.5	80.8%	23.2	2.2	44.3	>10	<36	<1.31
8	31.	57.3%	8.9	0.0	17.8	1.5	16.	0.61
9	42.5	60.7%	12.9	0.0	25.8	2.	24.	0.65
10	48.5	60.4%	14.6	0.0	29.3	≲0.5	28.5	≳0.69
11	53.	59.0%	15.6	0.0	31.3	≲0.5	31.	≳0.68
12	56.5	59.0%	16.7	0.0	33.3	≲0.5	33.	≳0.68
13	56.5	62.2%	19.3	0.0	38.6	≲0.5	36.	≳0.77
14	56.5	72.8%	20.6	0.0	41.1	3.5	37.5	0.77
15	56.5	77.0%	21.8	0.0	43.5	7.5	36.	0.75
16	56.5	81.9%	23.1	3.7	42.6	>10.	<33.5	<0.77
17	56.5	77.0%	21.8	0.0	43.5	8.5	35.5	0.71
18	19.	95.4%	9.1	7.4	10.7	4.	14.	2.27
19	18.5	89.4%	9.3	4.5	12.0	3.	13.5	1.25
20	18.5	18.5%	7.9	2.6	13.2	2.5	13.5	1.01
21	18.5	18.5%	7.3	0.0	14.5	2.	12.5	0.82
22	18.5	18.5%	6.8	0.0	13.3	1.	12.5	0.79
23	78.	78.6%	30.7	0.0	61.3	7.5	52.	0.84
24	73.	65.9%	24.1	0.0	48.1	0.	48.	>0.75
25*	15.2	93.8	7.1	5.3	8.9	1.	12.	2.27
26*	15.2	60.6%	4.6	0.0	9.2	0.	7.5	>0.70

* Inclined 2% downward
** Unable to measure liquid reaching phase separators in points #7 due to a choked venturi

For fixed conditions, the velocity of the liquid is proportional to the pipe cross sectional area occupied by vapor. Theoretical vapor areas are easily calculated for conditions of stratified flow with equal liquid and vapor velocities. Measured vapor areas can be estimated by examining the fraction of the liquid flow which is below the midplane and assuming a smooth stratified flow. The ratio of "measured" vapor area to calculated is shown in column 8 of Table 2. Values less than one indicate either a wavy flow or vapor velocities greater than liquid. The four data points in the lower right of Fig. 2 were the only points where this ratio was greater than one. This implies that the liquid velocity is greater than the vapor, possibly due to gravitational effects as shown in Fig. 4b.

Also shown on Fig. 2 is the two-phase helium data (1.2 atm) of Mamedov et al[4] converted to Baker diagram coordinates. They found the boundary between stratified/wavy flow and intermittent (slug) flow. Their data confirms that we were operating in a stratified or wavy flow regime.

CONCLUSIONS

From the results of this experiment we draw the following conclusion about two-phase helium flow:

- It is not practical to design long continuous two-phase heat exchange accelerator systems in a flow regime other than stratified or wavy.
- Two-phase flow is not suitable for an inclined SSC (due to time constants and control).
- Nonhomogeneous regimes exist for total flow<5g/s-cm^2 (D=45 mm)

Inclined 2% Upward

- For G<2 g/s cm^2: Perfectly separated flow, Gas area predicted by pressure drop.

- For G>2 g/s cm^2: Gas flow area is larger than predicted

Inclined 2% Downward

- Liquid velocity ~0.5 m/s

REFERENCES

1. C.W. Savery, Polyphase flow and transp technol, Century 2-Emerging Technol Conf, San Francisco, CA, ASME, New York, NY, (1980) p. 75-88
2. O. Baker, Simultaneous flow of oil and gas, Oil Gas J., 53:185 (1954)
3. C.H. Rode and J. Theilacker, Preliminary results of fermilab two phase helium tests, FNAL SSC CRYO 85-9, Fermi National Accelerator Laboratory, Batavia, IL., (1985).
4. I.S. Mamedov et al, Two-phase helium flow regimes in horizontal channels, (in Russian) Joint Institute of Nuclear Research (JINR) Report P8-84-156, Dubna (1984).

METASTABLE HEAT FLOW PHENOMENA AND DROOPING PRESSURE DIFFERENCES IN HE II USING MULTIPLE CHANNEL CONFIGURATIONS

B. P. M. Helvensteijn and S. W. Van Sciver

Applied Superconductivity Center
University of Wisconsin-Madison
Madison, Wisconsin

ABSTRACT

Superheating of liquid helium from the He II phase up to 3.0 K is demonstrated in one dimensional multiple channel configurations at pressures below the lamda point pressure. The individual channels have diameters of 10, 50 or 100 μm. Counterflow is produced by heating the lower reservoir which is connected to the bath through the multiple channel. The fountain pressure reaches a plateau when the temperature increases steeply due to turbulence. The observed Gorter-Mellink parameter appears diameter-dependent below 100 μm. Relaxation times related to the decay of the vortex line density may adversely effect He II thermal conductivity during cooldowns.

INTRODUCTION

Heat transfer in He II takes place either in steady state or in a transient mode. In the analysis of transient behavior it is generally assumed that the thermal conductivity of He II is identical to the steady state case.

This paper studies transient heat flow patterns which deviate from those observed in steady state. The metastable phenomena under consideration are:

- superheating above the He II phase where there is only a moderate pressure head between a pumped bath and the volume studied;
- metastable laminar flow;
- relaxation in turbulent flow.

In addition the diameter dependence of the Gorter-Mellink parameter is investigated.

THEORY

Heat transfer in He II in a one dimensional dead end system takes place by counterflow of the normal and superfluid components. This is caused by the fact that the respective densities, ρ_n and ρ_s, are strong functions of the temperature, $\rho_n \sim (T/T_\lambda)^{5.6}$. When the heat flux q is low the flow is laminar. The laminar flow equations relating the temperature and pressure gradients in a tube of circular cross section are respectively[1]:

$$\nabla T_L = \frac{32\eta_n q}{\rho^2 S^2 T d^2} \quad (1)$$

$$\nabla P_L = \frac{32\eta_n q}{\rho S T d^2} \quad (2)$$

In these equations η_n is the normal fluid viscosity; ρ is the density; S is the entropy; d is the diameter of the tube. The relative velocity of the two components which determines whether the flow becomes turbulent, relates to the heat flux as follows:

$$u = \frac{q}{\rho_s S T} \quad (3)$$

Once the relative velocity exceeds the critical velocity (on the order of one centimeter per second), the fluid becomes turbulent and is characterized by a complex array of vortices. The interaction between the normal fluid and the vortices, known as mutual friction, has been adequately described in a phenomenological treatment by Vinen.[2] Mutual friction gives an extra contribution to the temperature gradient:

$$\nabla T_{MF} = \frac{A\rho_n}{S} \left(\frac{q}{\rho_s S T}\right)^3 \quad (4)$$

In Eq. (4), A is known as the Gorter-Mellink mutual friction parameter, a quantity which has roughly a cubic temperature dependence. Beyond the onset of turbulence the temperature gradient is the sum of Eqs. (1) and (4).

Equation (4) does not clarify the relation to the vortices in the liquid. For steady-state flow Schwarz[3] has derived the relation between the vortex line density L per unit volume and the relative velocity, simplified as:

$$L^{1/2} = Zu \quad (5)$$

The parameter Z is a mild function of temperature and is on the order of 10^6 sm^{-2}. Combining the Eqs. (3) through (5) one finds:

$$\nabla T_{MF} = Z' \, L \, u \quad (6)$$

where Z' contains parameters defined above. Equation (6) implies that when the relative velocity exceeds the critical velocity but no vortices have formed yet (L = 0), the system is in what is called metastable laminar heat flow. In addition, when disturbances are suddenly removed the line density may not drop according to Eq. (5), while Eq. (6) still holds. This is a relaxation effect which has the result that in such a case the Gorter-Mellink Eq. (4) does not hold.

To date the most satisfactory analysis of the pressure gradient in turbulent He II is by Childers and Tough.[1,4] The turbulence is suggested to effectively raise the viscosity by an eddy viscosity η_e which varies according to the empirical formula:

$$\eta_e = \kappa\rho \, (\lambda^2 L)^{2/3} \quad (7)$$

where κ is the circulation quantum of a vortex line ($\kappa = h/m \approx 10^{-7}$ m^2/s) and $\lambda = 3 \times 10^{-7}$ m. The turbulent pressure gradient is given by Eq. (2) when the eddy viscosity is added to the normal fluid viscosity.

EXPERIMENT

A can holding up to 500 cm^3 of He II is temperature controlled by means of a heater in the liquid and a vacuum control valve in the pumpline. An external pressure transducer is used to regulate the control valve. A germanium thermometer is inserted in the He II for electronic temperature control, calibration and experimental measurement of the temperature and pressure of the bath. The can is filled by opening a needle valve in the filling line to the outer 4.2 K helium reservoir (see Fig. 1). Because of an imperfect seal of the needle valve experiments have been limited to the temperature range above 1.3 K.

Inside the vacuum can a small chamber is suspended by capillary tubing from the regulated He II reservoir. Thorough purging at room temperature with He gas ensures that the tubing and chamber are free of frozen air. Three sets of glass tubing have been tested. All seals to the glass are made with stycast 2850 FT. Dimensional details are listed in Table 1 where: N is the number of tubes in the set; D is the nominal diameter; d is the actual diameter; OD is the outer diameter; l is the tube length; V is the volume which includes the transducer volume.

Set 1 has all capillaries in one single rod which is sealed at each end to invar flanges, and needs no structural support to hold the chamber and the pressure transducer. Sets 2 and 3 are both sealed at the top and the bottom inside six stainless steel tubes, 0.1 mm wall thickness and 1.6 mm OD. The stainless steel tubes have been punctured, so the annular space is open to vacuum.

The lower chamber is equipped with a manganin wire heater, wound around the outer surface. An Allen Bradley nominal 75 Ω resistance-thermometer is inside the chamber. A differential capacitance pressure

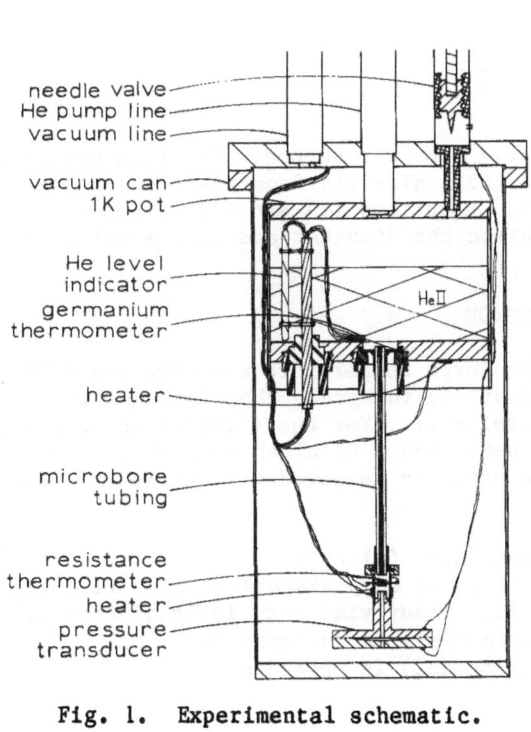

Fig. 1. Experimental schematic.

Table 1. Dimensions of the Capillary Samples.

set	N	D(μm)	d(μm)	OD(mm)	l(cm)	V(cm³)	Support
1	1000	10	9.4	4.0	2.0	0.74	none
2	24	50	47.3	0.08	9.2	0.56	SS
3	6	100	100.6	0.18	9.2	0.42	SS

transducer is attached to the chamber using an indium seal tightened with a lock nut. The heater and resistance thermometers are measured using the standard I-V method. All wiring is wound around a set of thin wall teflon supports to increase the length to over 1 m in order to minimize the heat leak between the experiment chamber and the pumped bath.

The low temperature capacitance pressure transducer has been built to allow accurate pressure measurements in the chamber. The transducer used six miniature bellows. A floating electrode is positioned between three bellows on each side and moves under influence of the difference in pressure inside each set of bellows. In the present experiments, one set of bellows is always open to vacuum. On either side of the moving electrode is a stationary electrode. The outer electrodes are charged out of phase with each other, using a 1 V, 1 kHz input signal. A small preamplifier on top of the cryostat takes the voltage of the center electrode and feeds it to a Stanford Research Systems SR510 lock in amplifier.

Because of long term drifts in the pressure transducer the data were taken in the transient mode. The heat applied to the chamber is in the form of a square pulse of variable duration, typically ~ 200 sec. During this time and following the pulse, the temperature and pressure traces of both the chamber and the bath are monitored. The total heat flow through the tubes Q_{out} is extracted from the data using the simple equation:

$$Q_{in} = Q_{out} + \rho V \frac{dH}{dt} \tag{8}$$

where Q is the energy flow per second. The time derivative of the enthalpy (dH/dt) follows directly from the temperature of the helium within the chamber. Equations (1) and (8) in the laminar region have been used to calculate the diameter and helium volume listed in Table 1.

RESULTS AND DISCUSSION

Because of the large temperature and pressure differences across the tubes, it is necessary to integrate the flow equations for comparison with the experimental data. For the integrations a fifth-order Runge-Kutta routine has been used. Critical velocities have not been taken into account. The data are compared with different models labeled as follows:

L: laminar flow (Eqs. (1) & (2));
t: turbulent flow, no eddy viscosity (Eqs. (1), (2) & (4));
e: turbulent flow, eddy viscosity included (Eqs (1), (2),(4) & (7));
d: data (experimental) in downward mode;
u: data (experimental) in upward mode.

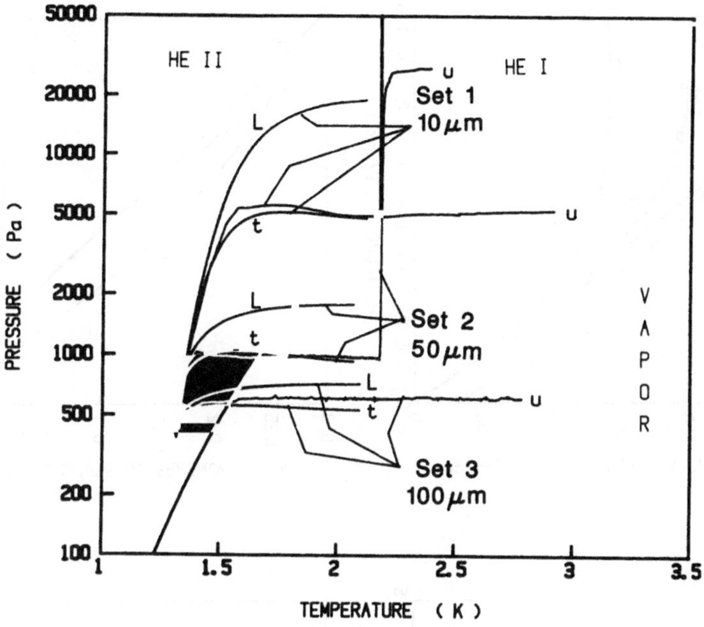

Fig. 2. Superheating from the He II phase.

Superheating

All data on superheating are taken switching the heater on in one single step and monitoring the temperature and pressure changes. In Fig. 2 the helium phase diagram is shown including the measurements on superheating and the theoretical predictions. The data clearly demonstrate superheating beyond the extended He II-He I transition in sets 1 and 3. The duration of the superheating above the lambda temperature was on the order of 5 sec. Close to the saturation line the superheating could last more than 15 minutes. For set 2 the superheating seems to be limited to the lambda transition, which is identical to the findings of Rybarcyk and Tough.[5] However, extrapolating the data suggests that immediately after the onset of boiling the vapor was in equilibrium with liquid at a temperature in excess of 3.0 K. Since in set 2 the heater was located above the thermometer, it is possible that an insulating layer of He I was formed, leaving the volume below it including the thermometer at the He II-He I transition. Further data (not shown) has demonstrated that for sets 1 and 3, the pressure and temperature immediately after leaving the superheated state are in close agreement with saturation conditions.

The data in Fig. 2 show that the pressure increases until it reaches a certain plateau, or even decreases (set 1), while the temperature increases until boiling. This is in accordance with the results of Hammel and Keller[6], and with the general trend in the model calculations. Calculations not including any turbulent contributions show the pressure continuously increasing. The plateau is apparently due to the high temperature rise and the resulting changes in helium properties once the flow becomes turbulent. Temperature and pressure differences measured in experiments in which the heat was either switched

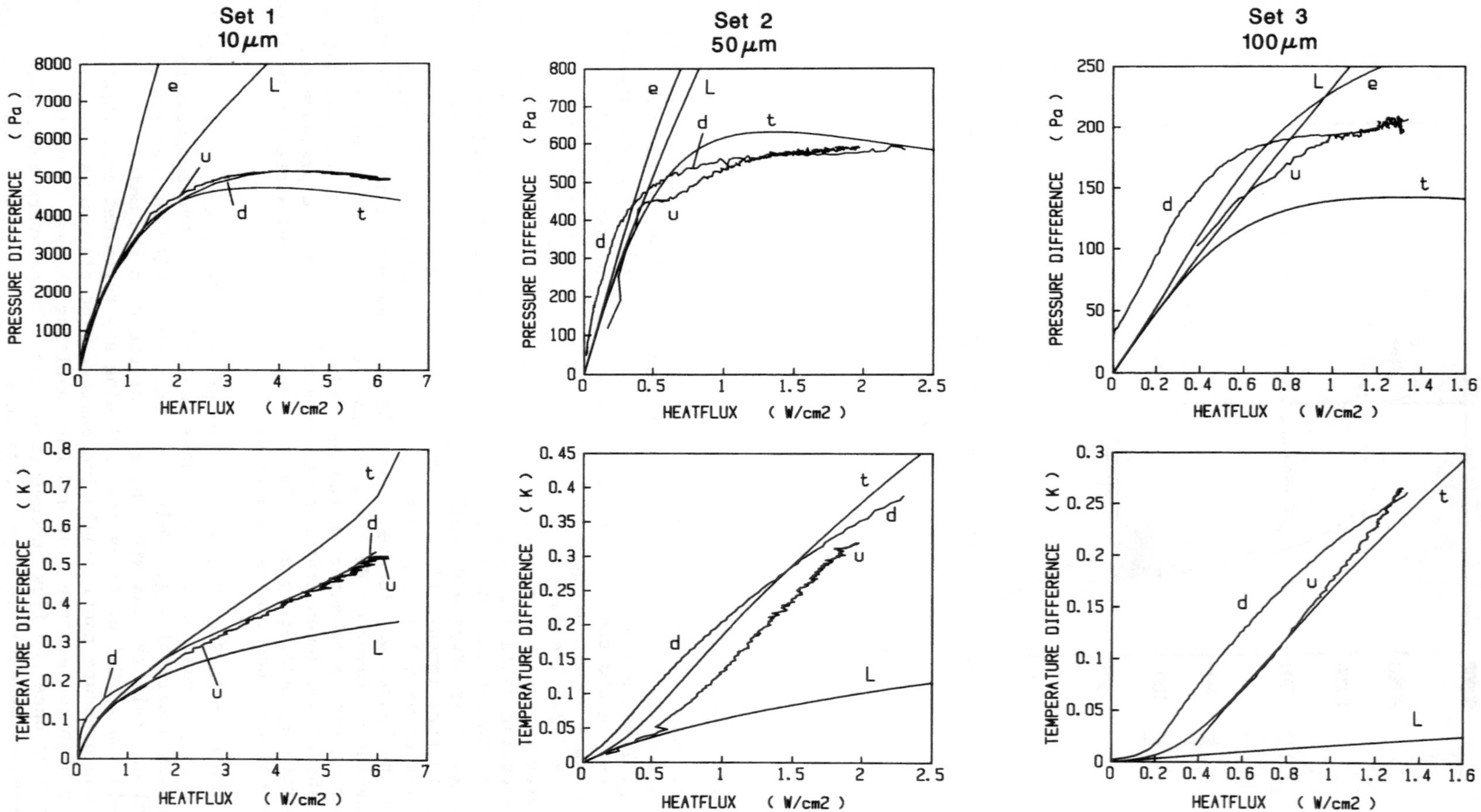

Fig. 3. Temperature and pressure differences versus heat flux.

on or off are plotted versus heat flux in Fig. 3. The heat flux is derived from Eq. (8).

The Gorter-Mellink Parameter

Except for the largest tubing the predicted temperature differences exceed the experimental values (u) with the largest deviation occurring for the smallest diameter tubing. According to Eq. (4) this behavior suggests that the Gorter-Mellink parameter decreases for decreasing diameter. Possibly more significant is that according to Eq. (6) a lower temperature gradient implies a lower vortex density. The observed pattern may be explained as follows. In Schwarz's model[3] it is assumed that annihilation of vortices occurs only at the walls. Thus when the surface to volume ratio is increased, i.e. when the diameter is decreased, the annihilation process is expected to become more important and the vorticity declines.

Eddy Viscosity

Figure 3 shows disagreement between the pressure calculations (t) and the data (u). If we include the eddy viscosity in the calculations the predictions are shifted in the right direction, however, the effect is too large. This may be due to an overestimate of the vortex density based on Eq. (5) or inappropriate values for the λ parameter in Eq. (7). The reasons for the anomalously small pressure differences for set 2 are yet unknown.

Relaxation of Vorticity

The experiments (d) where the heat was abruptly turned off while monitoring the temperature and pressure profile show the following features. At the high heat flux end the downward data start from where one would expect from the data taken turning on the heat. However, for decreasing heat flux both the temperature and pressure differences (d) are considerably higher than in the upward curves (u). This effect is observed more clearly for the larger tubes.

Equation (6) shows that the higher gradients correspond to higher vortex line densities than occur in steady state. This implies that relaxation times related to the annihilation of vortex lines may have a significant effect on the thermal conductivity in transient experiments. The relaxation times are thought to be smaller for smaller tubes because annihilation of vortices occurs at the walls.

Metastable Laminar Heat Flow

A few experiments at low heat flow have been carried out in order to observe metastable laminar heat flow. A successful attempt is demonstrated in Fig. 4.

Contrary to findings by Childers and Tough[1] the observed temperatures and pressures are continuous in time during the transition to turbulent flow. The time derivatives appear not to be continuous indicating a sharp decrease in heat flux at the onset of turbulence (Eq. (8)). However, the resolution of the data is insufficient to state that the heat flux is discontinuous.

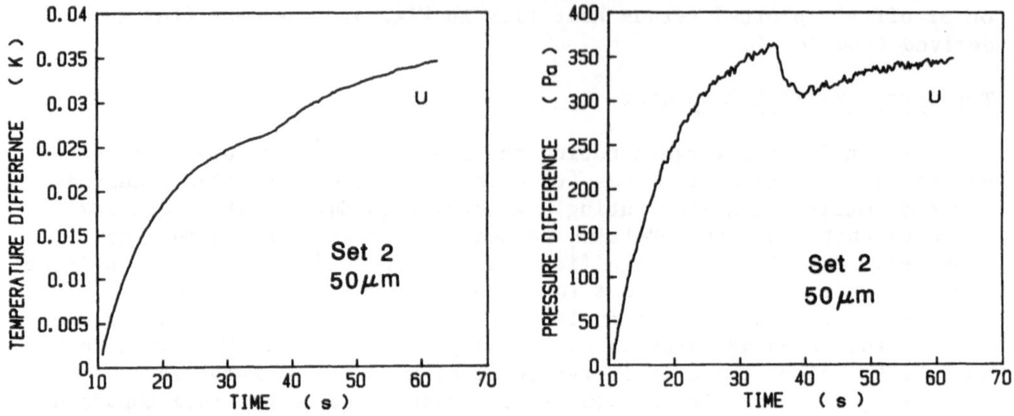

Fig. 4. Temperature and pressure difference as a function of time for metastable laminar heat flow.

CONCLUSIONS

Superheating of liquid helium at low pressures in a non vibration isolated experiment is demonstrated up to 3.0 K for times on the order of 5 s beyond the extended He II-He I transition. When superheating into the metastable He I phase thermal equilibrium is easily lost resulting in erroneous readings.

The fountain pressure reaches a plateau when the temperature rises steeply due to turbulence. The Gorter-Mellink parameter has been shown to decrease for tube diameters below 100 μm. The vortex line density is confirmed to be dependent on history. This may result in metastable laminar heat flow when heating up and reduced thermal conductivity when cooling down. It has been found that in the transition from metastable laminar flow to turbulent flow the temperature and pressure of He II are continuous.

ACKNOWLEDGEMENT

This work is in part supported by the National Science Foundation under grant MEA 8310770. Glass samples provided by Vitro Dynamics Inc.

REFERENCES

1. R. K Childers and J. T. Tough, Helium II Thermal Counterflow: Temperature- and Pressure- Difference Data and Analysis in Terms of the Vinen Theory, Phys. Rev. B, 13:1040 (1976).
2. W. F. Vinen, Mutual Friction in a Heat Current in Liquid Helium II; III. Theory of Mutual Friction, Proc. Roy. Soc., A242:493 (1957).
3. K. W. Schwarz, Generation of Superfluid Turbulence Deduced from Simple Dynamical Rules, Phys. Rev. Lett., 49:283 (1982).
4. R. K. Childers and J. T. Tough, Eddy Viscosity of Turbulent Superfluid ^{4}He, Phys. Rev. Lett., 35:527 (1975).
5. L. J. Rybarcyk and J. T. Tough, Superheating in He II and the Extension of the Lambda Line, J. Low Temp. Phys., 43:197 (1981).
6. E. F. Hammel and W. E. Keller, On the Existence of a Maximum in the Fountain Pressure vs. Temperature Relationship in Liquid Helium II, Physica, 31:89 (1965).

HEAT TRANSPORT PROPERTIES OF PRESSURIZED AND SATURATED HE II IN THE VICINITY OF T_λ

M. Fouaidy and M.X. François

Laboratoire de Thermodynamique des Fluides
Université P. et M. Curie
Orsay, France

ABSTRACT

A precise and quantitative experimental analysis of the heat transport properties of He II confined to a channel has been performed for Heat flux density and bulk temperature for which the transitions He II - He I, He II - vapor and He II - He I - vapor could occur in the neighborhood of the heating source or in the channel itself. The temperature measurements of the heater and of the He I and He II flow in the channel allow a thermohydrodynamic description of the heat flow. Particular attention is given to the λ transition and thus to the case where a He I layer separates the heater from the He II channel. Moreover, the dynamics of the vaporized He I or He II bubbles is analyzed and their role played in the heat transfer mechanism.

INTRODUCTION

Helium II is used as a coolant of many technical devices which require a large temperature uniformity and stability against uncontrolled perturbations. And due to its heat transport properties by means of the so called internal convection it can be used in restricted geometries as a channel to connect the heat source to the refrigerator. Thus, the thermal stability of the cooled devices is determined by the heat transport characteristic of the helium II filled channel specially when the heat flow is large enough to induce a local or broad phase transition into He I or helium vapor. Although several studies involving phase transition have been already made[1-4], the difference in the test cells used by each group or even in the experimental procedure and the lack of knowledge in the hydrodynamic of such multiphase flow are too significant to allow for a precise description of the stationary and transient behavior. The present study considers heat transfer in a channel as regards pressure around P_λ temperature under T_λ and stationary heat fluxes involving phase transitions in the channel.

DESCRIPTION OF THE EXPERIMENT

The experimental cell is shown in Fig. 1a : a stainless cylindrical channel of 10 mm in diameter and 150 mm in length is closed at the bottom by a copper block and connected to the helium thermostated bath at its upper end. Heat is supplied to the copper block by two heaters, the cons-

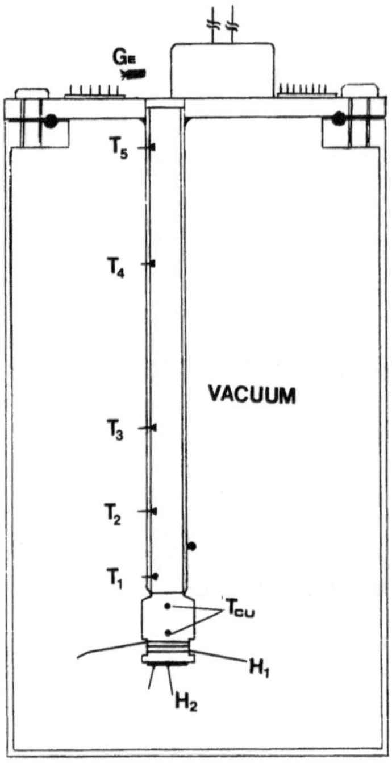

Fig. 1(a) Experimental apparatus - Vertical test channel ; T_i : carbon resistance thermometers ; H_1, H_2 : Constantan wire or strip line heater.

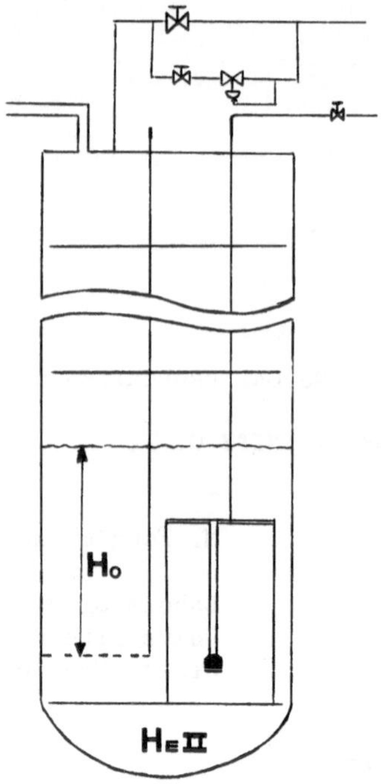

Fig. 1(b) Saturated He II experimental facility ; H_o : liquid helium height over the copper.

tantan wire (H_1) wrapped around and the constantan strip (H_2) stuck to the bottom. We used these two different locations of the heater to be able to detect the influence of the heat flux uniformity on the phase transition at the copper-helium interface. The power supply can be driven with various positive or negative constant slopes (W/s) ranging from 10 μW/s to 0.1 W/s. A stand-by position is available for each value of the heat flux.

Five small carbon thermometers (Allen-Bradley) are located in the channel at 5, 25, 50, 100, 135 mm respectively from the copper surface whose temperature is measured by two carbon thermometers of the same type. The bath temperature is measured by a calibrated germanium thermometer N° 19937 of the lake shore Cryotronics Inc. and by a precise measurement of the vapor pressure (Texas Instrument Fused-quartz-gauge, model 145-01). The bath temperature can be controlled to within $\pm 10^{-4}$ K and the helium level is known to within ± 0.5 mm, giving the value of the local pressure (± 0.7 Pa) and thus the theoretical equilibrium transition temperature to within ± 50 μK. The test channel and the copper block are surrounded by vacuum, and thus all the heat supplied to the copper is transferred to the Helium II channel ; the minimal heat leak by conduction between the copper and the stainless tube can be detected by another thermometer attached to the tube. A diagram of the cryostat used in the present set of experiments is shown in Fig. 1(b).

EXPERIMENTS WITH $P > P_\lambda$, 2 mK $< T_\lambda - T_o <$ 35 mK

In the first part of this study, the local pressure near the heater is greater than P_λ : and one can expect a channel filled with a mixture of He I, vapor and He II after the emergence of the first transition. In order to analyze the subsequent results, we recall (Fig. 2) the well-

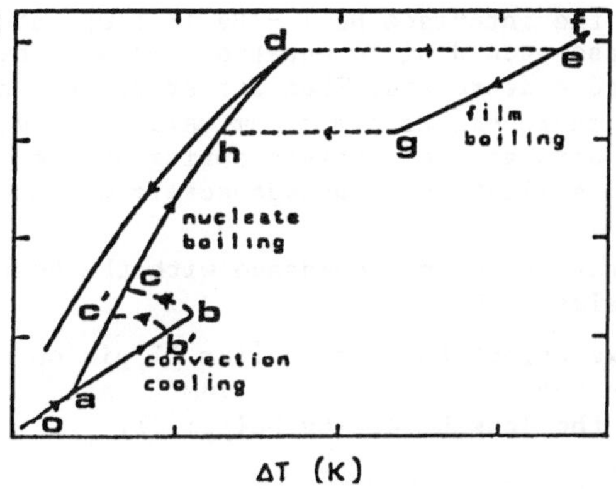

Fig. 2. Typical steady state heat transfer characteristic showing the different possible regimes, the hysteresis, and for B'C', B"C" the metastable states of natural turbulent convection.

known heat transfer characteristics of the classical fluid He I. For an increasing heat flux, the heat transfer mechanism may be the natural laminar or turbulent convection (OAB) type*, followed by the efficient nucleate boiling (CD)**, which becomes a transition DE to the film boiling regime EF***.

For decreasing heat flux, we see that one finds again (H) the nucleate boiling regime after a large hysterisis EG. We also notice that CD is not reversible and that the point B, which characterizes a metastable state of a natural turbulent convection, can be replaced by B' or B", depending on the previous history of the heat transfer. For low heat flux (T_λ-T_o< 10 mK), an unstable nucleate boiling (B'C' or B"C") may occur due to the fact that the mode of heat transfer switches back and forth intermittent between the nucleate boiling and the natural convection. Now if we come back to the experiment, with the heated copper block wetted by He II and connected to the thermostatic bath (T_o) by the channel of length L, we expect that for increasing heat flux the "isobaric" heat flow by internal convection of normal and superfluid helium will increase the temperature T_{F1} of the fluid layer in front of the copper surface according to the Gorter-Mellink law :

$$\int_{T_o}^{T_{F1}} \frac{dT}{f(T)} = LQ^{(m)} \qquad (1)$$

In this equation f(T) and m have to be determined experimentally as m is supposed to diverge close to T_λ [5]. When T_{F1} equals T_λ, that is to say for Q =$Q_\lambda(T_o$,L), the first layer of He II transforms into He I and the solid is now wetted by the classical fluid whose heat transport properties have been recalled. It is thus clear that the next value of the copper temperature will depend on the position of $Q_\lambda(T_o$,L) on the curve (Fig. 2), and that we can adjust this point by varying T_o, or L. Moreover, as soon as the first layer (thickness δ) of He I is nucleated, the He II-filled channel of length L-δ may drive $Q_\lambda(T_o$, L-δ) > $Q_\lambda(T_o$,L). This shows that δ is always controlled by two mechanisms :

* (referred here after as N.C.) - ** (referred here after as N.B.) - *** (referred here after as F.B.).

a) If δ increases, the interface He I - He II temperature T_I(which was T_λ) tends to decrease according to equation (1) and thus the He I layer is cooling again and δ decreases. Then for stationary heat flux greater than Q_λ $(T_o,L-\delta)$, there exists a minimum value of δ imposed by the helium II "piston", which acts as a compressed spring on the interface. This could also lead to oscillatory or pseudooscillatory modes for stationary heat transfer.

b) the thickness δ has to be in accordance with the heat transfer mode in fig. (2) for the value of Q.

In conclusion we expect that the helium II piston will adapt its length to the given heat flux and the helium I part to get its own heat transfer regime in the length left by helium II.

RESULTS FOR $P > P_\lambda$

A typical result of heat transfer characteristics is shown in Fig. (3a, 3b) for the heater, and Fig. (3c, 3d) for four thermometers in the channel T_1(0.5 cm), T_2(2.5 cm), T_3(5 cm), T_4(10 cm). We use the following notations :

regime 1 : copper wetted by Helium II : internal convection in the channel
regime 2 : copper wetted by Helium I in N.C. : channel with He I and He II
regime 3 : copper wetted by Helium I in N.B. : channel filled with He I, Helium vapor and Helium II
regime 4 : copper wetted by Helium vapor F.B. : channel filled with vapor, He I and Helium II.

The transition temperature from one regime to another is noted as T(i,j) where i is the starting point (temperature T_i) and j the final state (temperature T_j). On Fig. 3a, the temperature growth of the copper T_{cu} is a direct reflection of increase in the thermal resistance of the channel ; thus the monotomic parts of the curve $T_{cu}=f(Q)$ have to be connected to a corresponding heat transfer regime and a sharp increase or decrease to a transition between two regimes. On fig. 3b,c the value of the mean temperature of the four thermometers allows us to verify the hydrodynamical model, and the thermal fluctuations give data on the stability of each regime, the presence of bubbles or vapor pockets flow.

DESCRIPTION OF THE SIX PARTS OF THE HEAT TRANSFER CURVE

a) For $T_{cu} \leqslant T(1,2)$, $Q < Q_\lambda(T_o,L)$, the heat is removed by helium II. The values of $Q(T_o,L)$ have been measured down to $T_\lambda - T_o > 2$m K can be represented by a law of the type :

$$Q_\lambda(T_o,L) = b(T_\lambda - T)^x + a$$

with a = - 147 b = 11.613 and x = 1. For $T_\lambda - T_o > 10$ mK ; these results are in agreement with Alher's work[6].

b) At $T_{F1} > T_\lambda$, $T_{cu} > T_1$, a microlayer of He I is formed on the heater surface with no detectable temperature jump. Note that an increase of $10^{-4}.Q_\lambda$ above Q_λ ($\sim$ 10 µW/cm^2) induces a space of 50 microns in equilibrium condition, to which would correspond a ΔT of 2.5 K if the heat transfer was of the diffusion type in He I. The experimental value of ΔT is much lower ; moreover the values of $Q_\lambda(L,T_o)$ correspond to NC regime (Fig. 2). If we plot the numerical results of $Q/Q_\lambda(Q > Q_\lambda)$ versus $\Delta T = T_{cu}(Q > Q_\lambda) - T_\lambda$, the linear curves obtained and reported in Fig. 4, have a slope $k = Q_\lambda^{-1}\{\partial Q/\partial(\Delta T)$, and $kQ_\lambda(T_o,L)$ would give the slope of the curve Fig. 2 for the corresponding value of the heat flux Q, that is the heat transfer coefficient h.

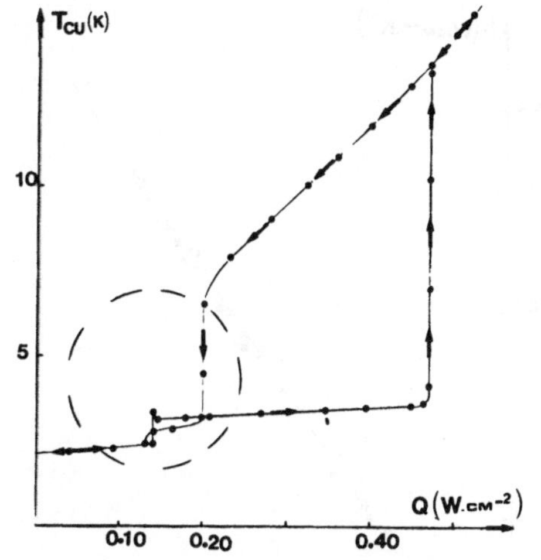

Fig. 3(a) - Steady state characteristic T_{cu} versus the heat flux Q. For T_o=2.147 K, h=26.5 cm and P = 38.4 torr.

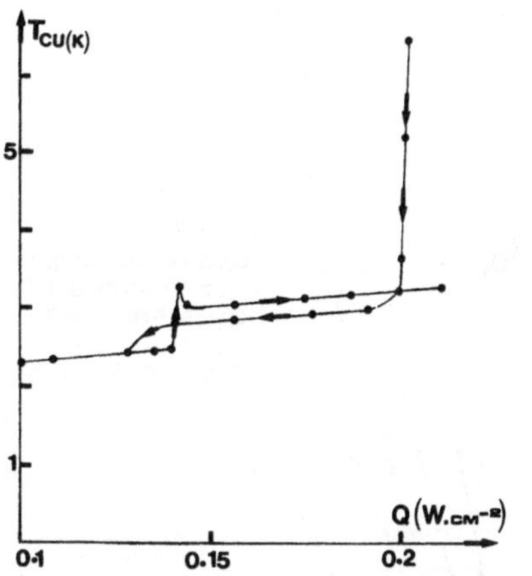

Fig. 3(b) - Enlargment of fig. 3(a) around the transitions T(1,2), T(2,3) from internal convection to natural convection and nucleate boiling ; return : T(3,2), T(2,1).

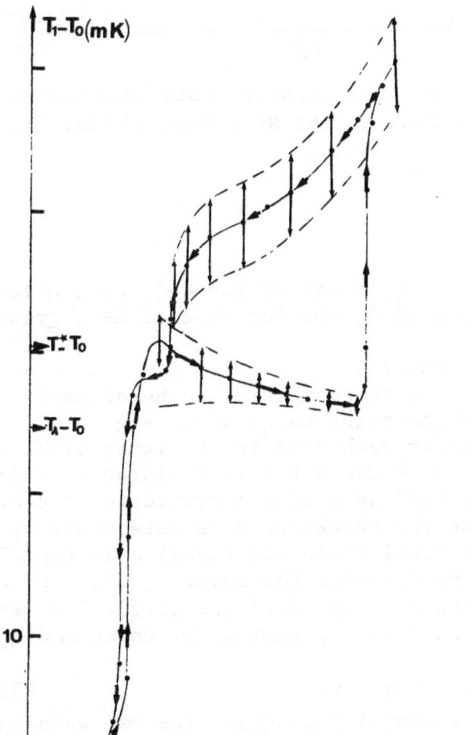

Fig. 3(c) - Mean temperature variations of T_1 versus the heat flow, for the same run as Fig. 3(a). The dashed area gives the fluctuation amplituds.

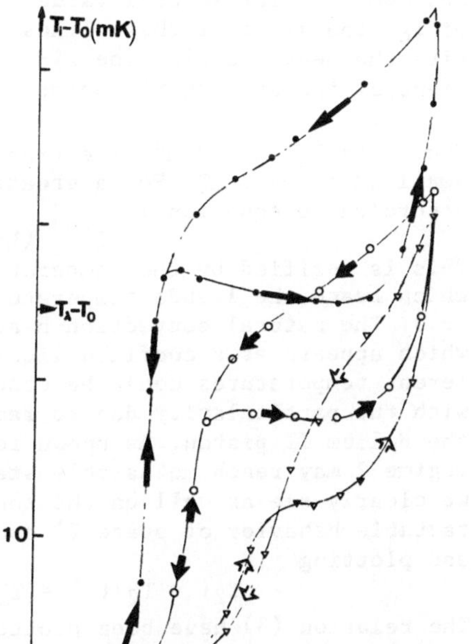

Fig. 3(d) - Mean temperature variations of T_2,T_4,T_5 for the same run as Fif. 3(a) and Fig. 3(c). The fluctuations are not reported but exist as for T_1;

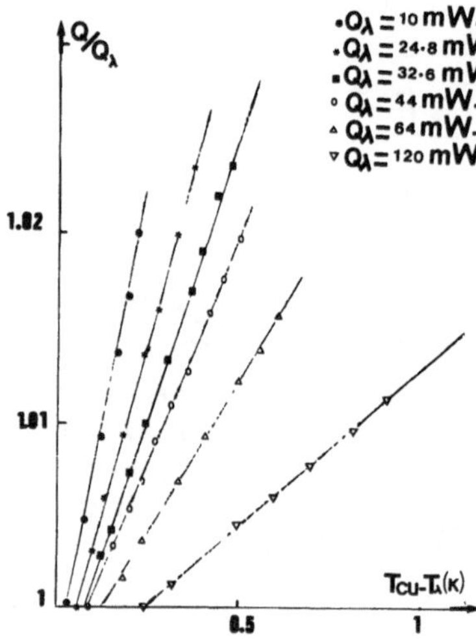

Fig. 4. Details of the heat transfer characteristic close and above Q_λ for several value of T_o. The slope of these lines give the heat transfer coefficient of the arising NC regime

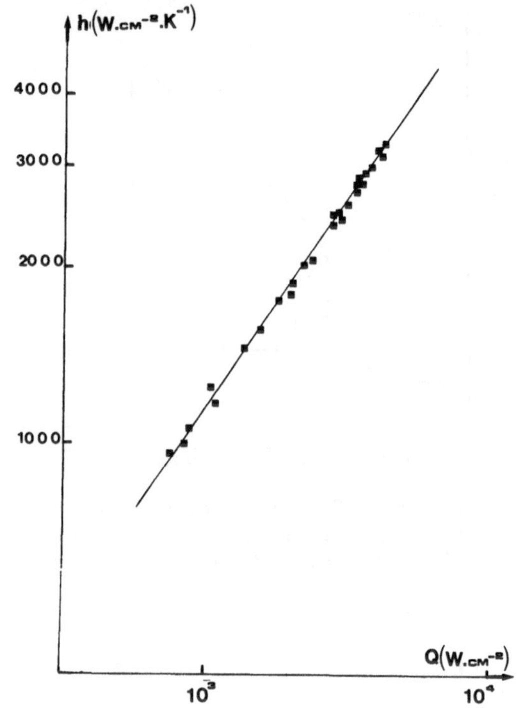

Fig. 5. Heat transfer coefficient versus heat flux in the NB regime of Fig. 3a.

Thus, for T_λ-T_o> 10 mK. the transition $T_{r(1,2)}$ occurs for Q=Q_λ with a very small additional ΔT. For a greater value of Q, the length δ of He I grows according to equation :

$$Q^{(m)}\ (L-\delta) = \text{constant}$$

This is verified by the temperature of the thermometers in the channel which reach the lambda temperature for the right value of Q (Fig. 3(b), (c)). The natural convection heat transfer mechanism in the layer of He I, which appears as a confined liquid medium between two flat plates at different temperatures could be conceptualized as a Bénard convection pattern with the particularity due to fact that the thickness δ is determined by the Helium II piston. As known for classical fluid and recalled in Fig. 1, regime 2 may reach metastable states and describe the curve 2, 2', 3 which we clearly see as well on the thermometers : Fig. 3(b) and (3(c). The metastable behavior of state 2' could identified by fixing the value of Q=$Q_{2'}$ and plotting

$$T_{2'} = T_{2'}(t) = T_{2'}(t=0)\ .\exp(-\ t/\tau) \qquad (3)$$

The relation (3) have been plotted on a semi-log diagram, for two values of $T_{2'}(t=0)$ at the same bath temperature T_o, and we see that the greater the

Table 1. Heat transfer coefficient in the NC regime

Q_λ (W.m^{-2})	100	227	290	440	630	1200
T_λ-T_o(mK)	13.5	14.8	15.47	16.4	18.2	23
$\frac{\partial Q}{\partial(\Delta T)}$ (W.m^{-2}K^{-1})	10	18.2±1	19.1±1	21±1	21.9±0.9	20.7±0.7

initial deviation from the equilibrium, the smaller the time constant. At T_o=2.141 K, q = 0.12 $W.cm^{-2}$, τ ranges from 1 to 7 minutes.

$Q > Q_{NB}$ - The appearance of temperature fluctuations on the copper and the first thermometer are correlated with the beginning of nucleated boiling on the heater surface and the instability of the NC regime. By increasing the heat flux until Q_{FB} we see each thermometer reaching successively the local saturation temperature and increasing frequency in the thermal fluctuation for the typical example of Fig. 7, the heat transfer coefficient h = Q/ΔT, at T = 2,143 K, is fitted by the law : (Fig. 5)

$$h = 8.375\ Q^{0.727}\ W.m^{-2}\ K^{-1}$$

the high value of the exponent of Q[8] could be linked to a very good uniformity of the cavity sizes on the heater surface.

$Q = Q_{FB}$ - The NB regime is suddenly replaced at $T_{(3,4)}$ by the film-boiling regime, and the layer of vapor gives an additional thermal resistance T_4-T_3 to the heat transfer which can be of several Kelvins : (10 Kelvins (Fig. 3(a))). The fluctuations of the copper temperature cease as the copper is decoupled from the unstable vapor-liquid interface by the vapor layer, but not the fluctuations of thermometers in the channel as seen on Fig. 3(b) and (c). The increase of the mean channel temperature profile could be seen as a rise of the local pressure or as the measurement of a mean temperature between non equilibrium superheat vapor and liquid ; but, except for transient heat transfer or stationary heat transfer in restricted channel, no pressure rise of about 2 torr was measured. The heat transfer coefficient is about 0.038 W/m^2. The nucleate boiling peak heat flux Q_{FB} is known to depend on pressure[9], and several correlations have been proposed by various authors. We find ourselves in relative agreement with Bewilogua[10] :

$$Q_{max} = 0.421 + 3.58\ P/P_c - 6.19(\frac{P}{P_c})^2 + 2.21(\frac{P}{P_c})^3$$

in the range of pressure studied; in the case mentioned, Fig. 3(a), where p = 38,4 torr, the correlation gives (495 ± 50) mW/cm^2 and we find 470 mW/cm^2.

$Q < Q_{FB}$ - The film boiling regime can be maintained for much lower values of the heat flux, showing a large hysteresis. The minimum value is temperature and pressure dependent and the experimental results are in agreement with the correlation of Zuber[11] :

$$Q_{min} = \frac{L\rho_v}{2}\left(\frac{\sigma\ \Delta\ \rho}{\rho_v+\rho_L}\right)^{1/4}$$

at T*≃2.18 K, σ = $0.308.10^{-3}$ $N.m^{-1}$, ρ_L # 146 $kg.m^{-3}$

ρ_v=1.235 $kg.m^{-3}$, L = 22.23 $J.g^{-1}$, where T* is the saturation temperature of the helium in front of the heater.

Q_{min} = 1812 $W.m^{-2}$ and Q_{min}(exp) = 1980 $W.m^{-2}$

As Q_{min} is a characteristic of He I, the transition $T_{(4,3)}$ could be $T_{(4,2)}$ or $T_{(4,1)}$ depending of the relative value of Q_{min} with respect to $Q_{NC}(T_{1,2})$ or $Q_{NB}(T_{2,3})$. But one expects that Q_{min} must not be lower than Q_λ as for $Q < Q_\lambda$, the whole channel is in the superfluid state. This is true if we are able to define the lambda transition in such a flow pattern of normal fluid in non equilibrium state. This point is not really clear at the present time.

EXPERIMENTS WITH $P < P_\lambda$, 35 mK < $T_\lambda - T_o$ < 50 mK

The same experiments with saturated and just subcooled He II have been performed with the same test cell (Fig. 1). As the results and the theory are well known for the regime 1 we focus our attention to the tem-

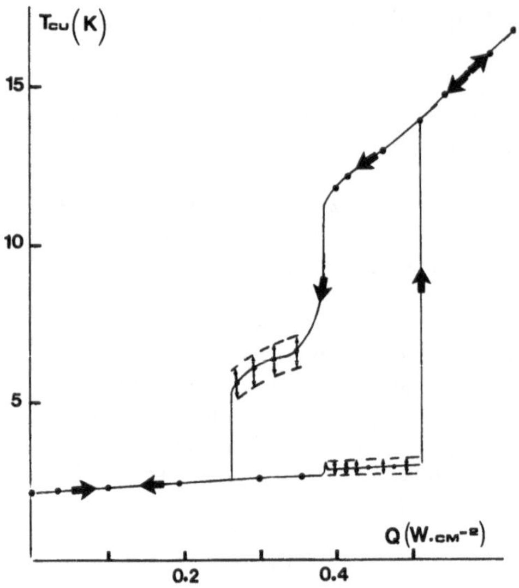

Fig. 6(a) $P < P_\lambda$ - Heat transfer characteristics T_o=2.127K-h=22.6cm. The dashed area gives the amplitude of temperature fluctuations (P = 36.2 Torr).

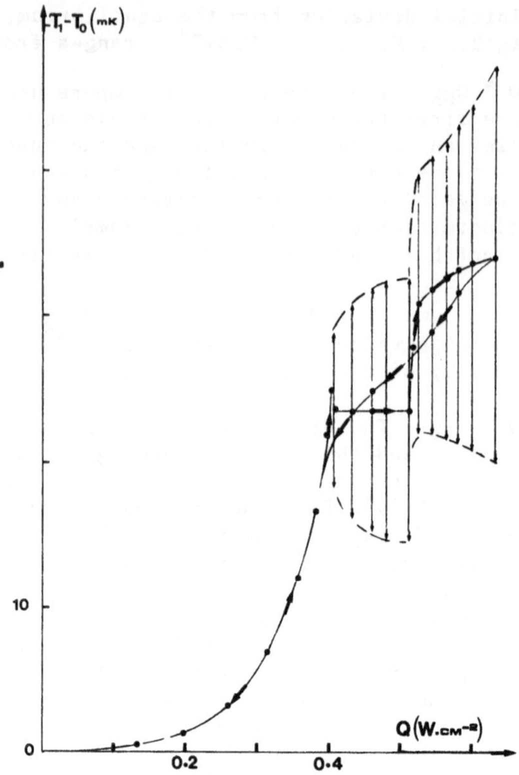

Fig. 6(b) - Temperature variations of T_1 in the channel correlated to the Fig. 6(a).

perature range close to saturation and try to identify the intermediate regime between internal convection and film boiling. We know that if the channel has a constricted part, one can reach supersaturated states for the liquid which relax with a time constant different of that of the heating and thus lead to a pseudo periodic film boiling which is actually a kind of nucleate boiling[12]. The heat transfer curve for T_{bath} = 2.127 K and Δp (hydrostatic) = 23 cm are reported in Fig. 6(a) and in Fig. 6(b) for the temperature profile in the channel. Let us first briefly describe the results.

Kapitza and Gorter Mellink regime - using the expression :

$$Q = h_o \left\{1 + \frac{3}{2}\frac{\Delta T}{T_1} + \frac{(\Delta T)^2}{T_1^2} + \frac{1}{4}\left(\frac{\Delta T}{T_1}\right)^3\right\} \Delta T = h\Delta T$$

with $\Delta T = T_{cu} - T_1$, we find $0.49 < h < 0.67$ W/cm^2.K

The temperature of the copper starts to oscillate with a large amplitude of several 0.1 K, indicating that T_{F1} has reached the local saturation temperature T_{sat}. The fluid, in the channel is not affected by these fluctuations as they are localized in the immediate vicinity of the copper surface.

Increasing the heat input will increase the period and the amplitude of the oscillations which begin to affect the different thermometers in the channel ; a high speed motion picture, made with a pyrex test cell, shows the nucleation of a bubble at a supposed hot point of the surface, which expands rapidly over a small or broad part of the surface as a very thin vapor layer (∿ 1/10 mm) ; the vapor condenses smoothly next, the

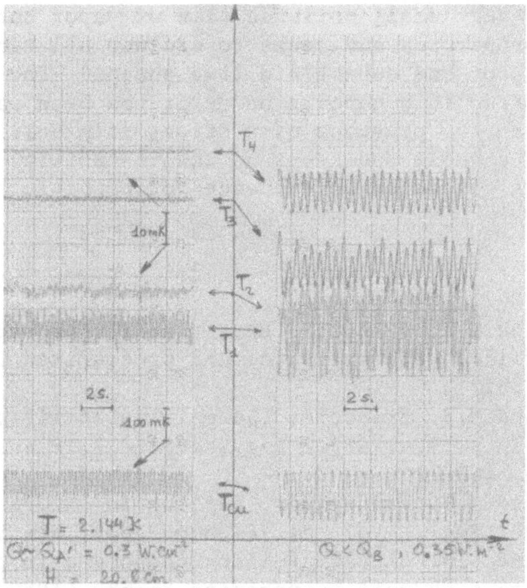

(a) (b)

Fig. 7 - Temperature fluctuations of T_{cu}, T_1, T_2, T_3, T_4, for heat flux in the range Q(A'B) Fig. 6(a) for T = 2.144 K and h = 20.8 cm.

7(a) : Just for $Q > Q_{A'} = 0.3\ W.cm^{-2}$

7(b) : $Q = 0.36\ W.cm^{-2} < Q_B$.

liquid appears to stream on the surface. The copper is wetted by helium II during a long enough time to be cooled as efficiently as in the OA regime, if one excepts the temperature fluctuations of about one Kelvin (for the copper). They are shown in Fig. 7; a part of the cooling process for $Q > Q_{sat} \simeq Q_A$, could be now attributed to the backward and forward motion of the liquid piston[13].

The same regime persists until Q_B. As Q approaches Q_B the pseudo period goes to zero, the amplitude of copper temperature saturates because the wetted time is too short ; the boiling is eventually noisy. Notice that the mean temperature T_1 is equal to $T_{1(sat)}$ and $T_{1(A')}$ is greater than $T_{1(sat)}$.

The transition to film boiling (regime 4) gives a similar result to the preceding case.

T_1 is now surrounded by a mixture of superheated vapor and He II and sees a net increment of the mean measured temperature.

Going back, we see that the film boiling degenerates to a more efficient pseudo nucleate boiling which is different than the AB regime and not well understood ; the mean value of h is $0.08\ W.cm^{-2}.K^{-1}$.

CONCLUSION

As a conclusion, the purpose of the present study is to analyze in details the different heat transfer mechanisms which can occur in He II filled channel connecting the heat source to the cooling source. The experiments have been carried out around P_λ to see the respective influence of the heat transport properties of each phase. One of the main results which could be used not only in stationary conditions, is that the He II piston has a very clear behavior with a very well defined heat transport capacity which controls and stabilizes (in a sense) a large part of the process. A more complete analysis of the transient behavior will be published in an other paper. The very small collapse time of vapor bubbles[14,15] in He II, emphasizes the properties and seems to exclude all kinds of homogeneous flow of helium vapor and superfluid (two phases). The experiments in saturated He II enforced this conclusion as it has been seen that the Helium II piston (even in the presence of a strong film boiling at one of its end) can move as a whole and keeps up the instability but also the cooling. These experiments give also an evidence for possible superheat states of saturated He II.

REFERENCES

1. D. Gentile and M.X. FRANCOIS, Heat transfer properties in a vertical channel filled with saturated and pressurized He II, Cryogenics 21:234 (1981) .
2. D. Gentile and M.X. Francois, Thermal instabilities in an Helium II channel, in : "Adv. Cryo. Eng.," Vol. 27, Plenum Press, New York (1982), p. 468.
3. S.R. Breon and S.W. Van Sciver, Boiling in saturated He II, in : "Adv. Cryo. Eng.," Vol. 31, Plenum Press, New York (1986), p. 465.
4. S.R. Breon and S.W. Sciver, Boiling phenomena in pressurized He II confined to a channel, Cryogenics 26:682 (1986).
5. C.E. Swanson and J. Donnellt, Vortex dynamics and scaling in turbulent counter-flowing Helium II, J. of Low Temperature Physics 61:363 (1985).
6. G. Ahlers, Mutual friction in He II near the superfluid transition, Phys. Rev. Let. 22:54 (1969).
7. O.E. Dwyer, Nucleate-boiling heat transfer, in : "Boiling Liquid Metal Heat Transfer," eds., American Society, Ill, (1976), p. 225.
8. C. Schmidt, Review of steady state and transient heat transfer in pool boiling Helium I, in : "Proc. of Workshop held at Saclay," (1981), p. 17.
9. J.M. Pfotenhauer and R.J. Donnelly, Heat transfer in liquid Helium, in : "Adv. in Heat Transfer," Vol. 17, Ed. Academic Press, Inc, London, (1985), p. 66.
10. L. Bewilogua et al, Heat transfer in cryogenic liquids under pressure, Cryogenics 15:121 (1975).
11. N. Zuber et al, "AECU"-4439, (1959).
12. M.X. François et al, Observation of nucleate boiling in He II and its effect on heat transfer, in : "Proc. of Workshop held at Saclay," (1981), p. 85.
13. M. Fouaidy, Contribution à l'étude des conduits de réfrigération d'Hélium Superfluide pour les flux supercritiques, "Thèse, Université Paris VI," (1987).
14. E. Sauvage, Contribution à l'étude des milieux diphasiques d'Hélium. "Thèse Université Paris VI," (1984).
15. F. Jebali et al., Noiseless film boiling in superfluid helium on a heated 30 μm diameter wire, presented at the same conference.

APPLICATION OF A SQUARE FUNCTION HEAT PULSE

TO FORCED FLOW He II

A. Kashani* and S. W. Van Sciver

Applied Superconductivity Center
University of Wisconsin-Madison
Madison, Wisconsin

ABSTRACT

The present is the third report on a series of experiments involving forced convection heat transfer in He II. A square function heat pulse is applied to the middle of a copper flow tube, 3 mm in diameter and 2 m long, containing He II. The width and amplitude of the pulse are varied in the different experimental runs. The bath temperature is set at three values of 1.65 K, 1.80 K and 1.95 K. The bath pressure is kept under the saturation vapor pressure. The He II flow velocity is varied between zero and 80 cm/s. The fluid temperature is monitored at eight locations along the flow tube. Experimental temperature profiles are then compared with a numerical solution to the He II energy equation.

INTRODUCTION

He II forced convection heat transfer has been the subject of several recent studies.[1-3] The growing interest in this area can be attributed to two large scale applications of He II, i.e., cooling of superconducting magnets and cooling of space-based devices such as infrared telescopes.

In the past two reports have been published on steady state and transient forced convection heat transfer in He II.[2,3] This is an extension of the previous studies which involves application of a heat pulse to flowing He II.

EXPERIMENT

A detailed description of the experimental apparatus has been given in an earlier report.[2] In brief the experiment consists of a motor-driven hydraulic pump and a flow tube. The pump employs two stainless steel welded bellows. Each bellows is connected to one end of the flow tube. A constant He II flow rate in the tube is obtained by compressing one bellows at a constant rate while expanding the other bellows at the same rate.

The copper flow tube is 2 m long and 3 mm in ID. The tube is heated at the midpoint along its length with a strain gauge heater. The He II temperature is monitored at eight locations along the tube using Allen-Bradley carbon resistors. There are four resistors on each side of the heater positioned at 3 cm, 33 cm, 66 cm and 100 cm away from the heater.

The pressure is monitored at both ends of the tube using Siemens pressure transducers, Model KPY-12. Also the He II flow rate is obtained by measuring the vertical displacement of the bellows using a long stroke linear variable displacement transducer.

---

* Sterling Fed. Sys. Inc., Palo Alto, California

Heat transfer measurements are made by first establishing a constant flow in the tube and then applying a square pulse of known amplitude and duration to the heater. A real time data acquisition system is used to collect and store the experimental temperature and pressure traces.

RESULTS AND DISCUSSION

The experimental results can be divided into two parts. The first part involves pulses with durations equal to or greater than 1 s. In the second part the pulse duration is set at values considerably less than 1 s.

Results of the experiments involving long heat pulses are described first. Figure 1 shows the development of the temperature profile in the flow tube after a heat pulse. The bath temperature is set at $T_b = 1.8$ K and the flow velocity is $v = 21$ cm/s. The heat flux, i.e., the heat per unit cross-sectional area of the tube, is $q = 5.71 \mathrm{W/cm}^2$ and the pulse lasts for 3.5 s. The profile during the pulse is similar to the case of a step function heat input. After the heater is turned off, the temperature profile decays while its peak temperature moves with time in the direction of flow.

An increase in the flow velocity results in a lower peak temperature. Also, increasing the flow velocity causes the pulse to decay faster in the tube. These two effects are observed in Fig. 2, where the pulse is the same as in Fig. 1 but the flow velocity is increased to $v = 39$ cm/s.

In Fig. 3 the development of a 1 s pulse at $T_b = 1.8$ K is shown. The pulse heat flux is $q = 5.71 \mathrm{W/cm}^2$ and the flow velocity is $v = 21$ cm/s. For this pulse the profile decays faster than the pulse described above. By comparing Figs. 1 and 3 it is observed that the decrease in the peak temperature, 1 s after the pulse, is smaller for the longer pulse than for the shorter pulse. This is because after the longer pulse more energy has to be carried out of the tube into the bath. Also, after the pulse the profile on the downstream side of the heater is steeper for the shorter pulse. Therefore, internal convection is more effective in reducing the peak temperature for the shorter pulse.

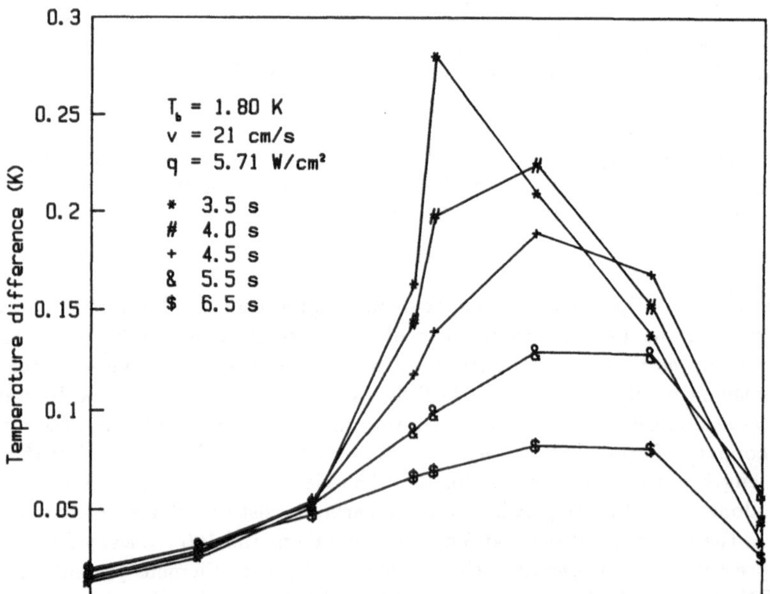

Fig. 1 Development of temperature profile after a 3.5 s square heat pulse.

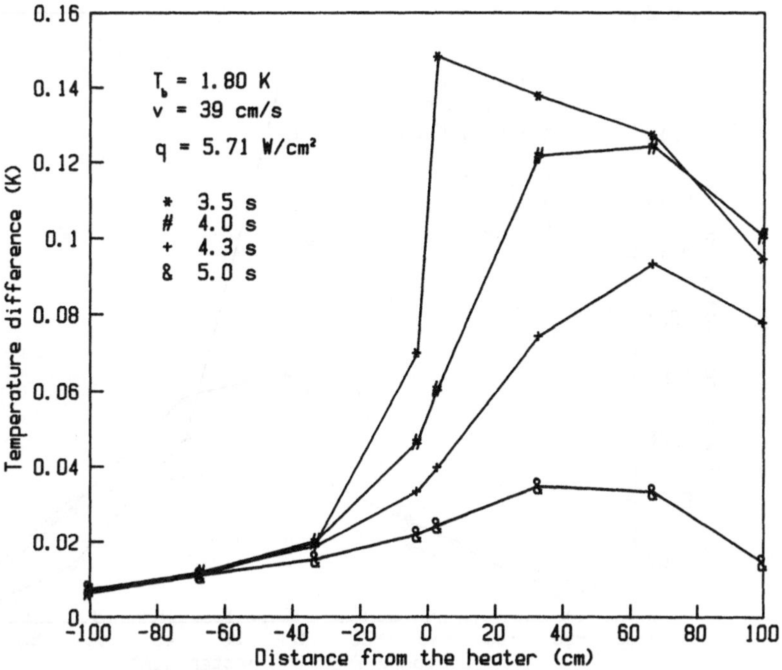

Fig. 2 Effect of increasing flow velocity on development of temperature profile, for the pulse of Fig. 1.

The second series of experiments involve short pulses with widths significantly less than 1 s. Figure 4 shows the development of the temperature profile at $T_b = 1.65$ K, when a heat flux of $q = 8.93\text{W/cm}^2$ is applied at the heater for a duration of 0.1 s. For narrow pulses, it is expected that the peak temperature would travel at a speed close to the flow velocity. Thus, in Fig. 4 where $v = 59$ cm/s the time for the peak temperature to travel between the first and the second thermometers downstream of the heater should be approximately 0.5 s. The first profile in the figure corresponds to the time when the peak occurs at the first thermometer. For the peak to travel at the flow velocity the third profile which is taken 0.5 s after the first one should have a peak at the second thermometer. Since the third profile is fairly flat around the second thermometer it is not clear whether the peak occurs near that thermometer.

Narrow pulses with more energy are required to investigate whether the peak travels at the flow velocity. To achieve this the pulse amplitude is increased by a factor of sixteen, i.e., $q = 125.1\text{W/cm}^2$, but the width of the pulse is reduced to 10 ms. Figure 5 shows the profiles with the new pulse and a flow velocity of $v = 80$ cm/s. For the peak temperature to move with the flow velocity, it should take approximately 0.37 s to travel between the two thermometers on the downstream of the heater. For comparison the time elapsed between the first and the third profiles, in Fig. 5, is about 0.37 s. In the first profile the peak occurs at the first thermometer. On the third temperature profile the peak seems to be very close to the second thermometer. That is, for very narrow pulses the peak temperature appears to travel close to the flow velocity.

An interesting phenomenon occurs in the large amplitude pulse experiments. Immediately after the pulse a jump in the output of the pressure transducers is observed. Figure 6 shows the outputs of the pressure transducers before and after the pulse. The increase in pressure is due to vapor formation at the heater. The expansion of the vapor causes a pressure surge, in the flow tube, which is felt by the pressure transducers after the pulse is

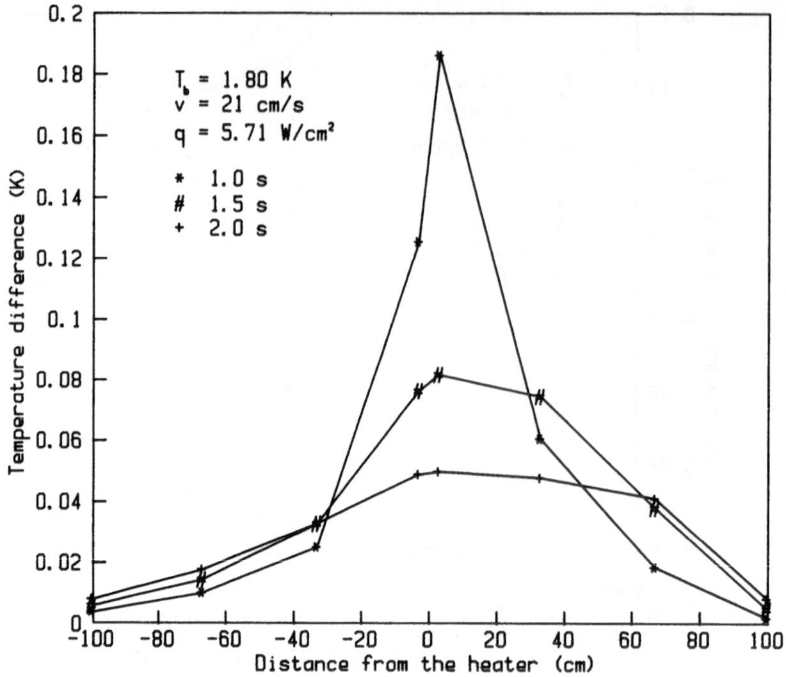

Fig. 3 Development of temperature profile after a 1.0 s square heat pulse.

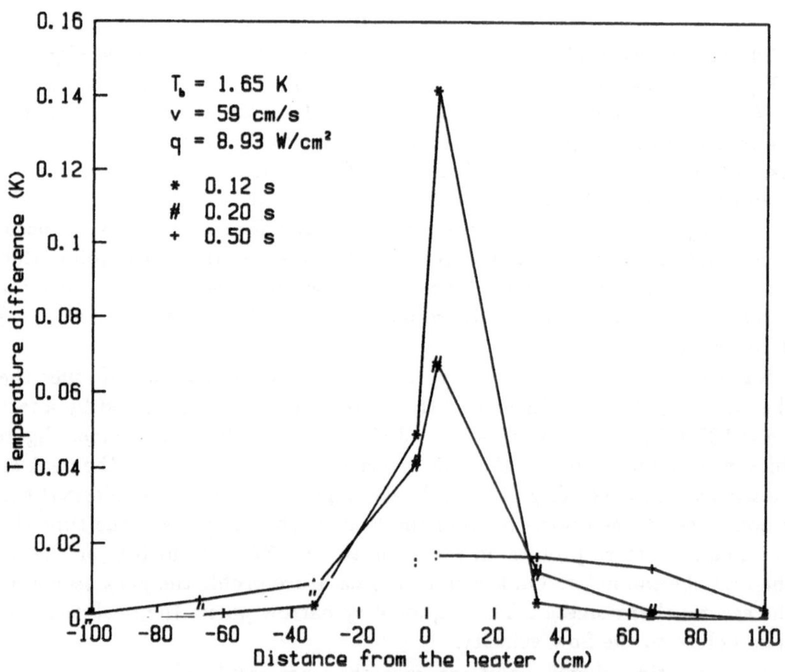

Fig. 4 Development of temperature profile after a 0.1 s square heat pulse.

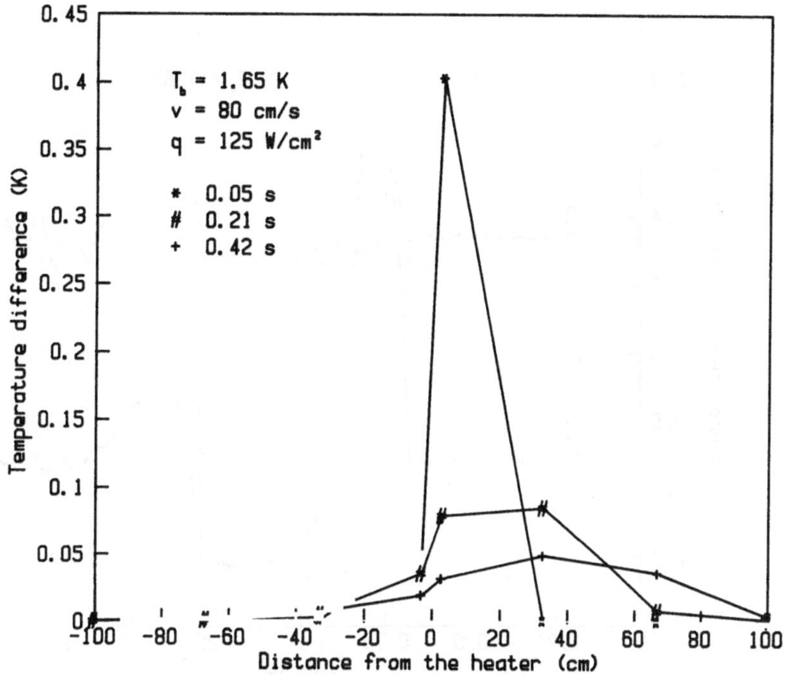

Fig. 5 Development of temperature profile after a 10 ms square heat pulse.

applied. The change in pressure is larger for the transducer upstream of the heater, since the bellows on the downstream side is open to the He II bath. The size of the bubble is smaller than the volume of the tube between the two thermometers closest to the heater. This is because no oscillations are observed in the temperature traces of these thermometers. The energy deposited into the liquid in 10 ms is approximately 0.11 J. The energy required to raise the temperature of the liquid under the heater from 1.65 K up to the λ-temperature is about 0.03 J. Thus, the energy of the pulse deposited in a very short time is sufficient to vaporize some of the liquid under the heater.

The temperature traces of the first two thermometers downstream of the heater are shown in Fig. 7. Initially a dip is observed in the temperature trace of the second thermometer on the downstream side of the heater. The same effect is seen on the outputs of all thermometers except the two closest to the heater. This peculiar observation may be attributed to the sudden increase in the pressure. The rise in pressure occurs so rapidly that it can be considered an isentropic process. For He II an isentropic increase in pressure results in a decrease in temperature. Therefore it is plausible that right after the pulse the pressure rise causes the temperature in the tube to cool below the bath temperature, except for the liquid close to the heater. The dip in the temperature is between 4 mK and 5mK and is present for all bath temperatures.

After the bubble has completely collapsed the temperature in the tube rises slightly because of the heat given off by the bubble. This is observed as a bump on the temperature traces in Fig. 7. The time that the bump appears corresponds closely to the time at which pressure oscillations end.

THEORETICAL MODEL

To compare the experimental temperature profiles with a theoretical model, the He II energy equation is employed. In one dimension (X) and as a function of time (t) this equation is

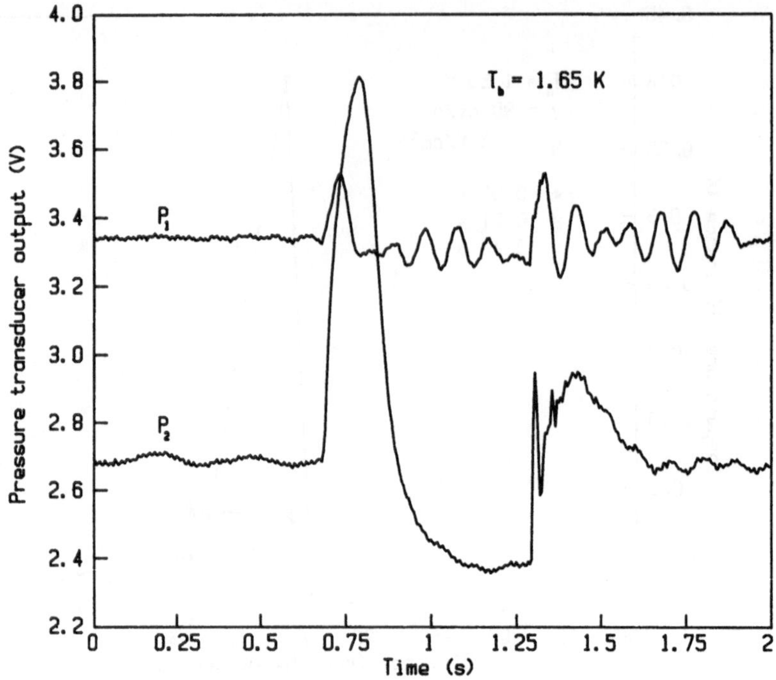

Fig. 6 Pressure transducers outputs after a 10 ms square heat pulse.

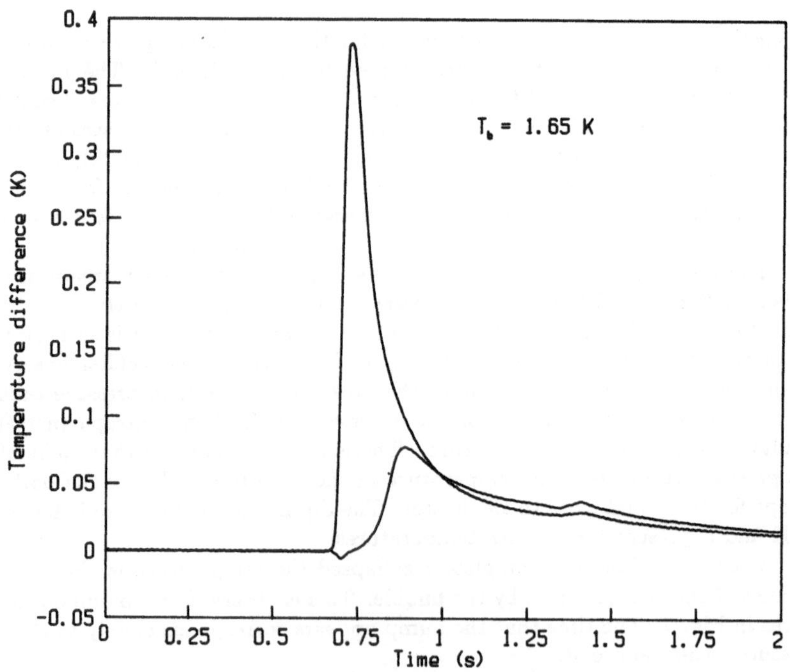

Fig. 7 Temperature traces of the first two thermometers downstream of the heater after a 10 ms heat pulse.

$$\rho C_p \frac{\partial T}{\partial t} + \rho C_p v \frac{\partial T}{\partial X} - \frac{\partial}{\partial X}\left(\frac{1}{f(T)}\frac{\partial T}{\partial X}\right)^{1/3} = q_{gen}(X,t) \tag{1}$$

where for simplicity the change in enthalpy with pressure is neglected. In Eq. (1), T is the temperature, v is the flow velocity, q_{gen} is the heat generation and $f(T)$ is the He II conductivity function. The functional dependence of $f(T)$ and the specific heat (C_P) on temperature can be given as

$$C_p = C_\lambda (T/T_\lambda)^{5.7} \tag{2a}$$

$$f(T) = \frac{A_\lambda}{\rho^2 s_\lambda^4 T_\lambda^3} \frac{1}{\left((T/T_\lambda)^{5.7}(1-(T/T_\lambda)^{5.7})\right)^3} \tag{2b}$$

where $s_\lambda = 1.559$ J/g-K,[4] $A_\lambda = 115$ cm-s/g, $C_\lambda = 8.88$ J/g-K and $T_\lambda = 2.172$ K. A_λ and C_λ are the Gorter-Mellink parameter and the specific heat at the λ-point but their values are chosen to best fit the heat conductivity and the specific heat data from $T_b = 1.6$ K to 2.15 K.[5] The He II density is taken to be a constant at $\rho = 0.1455$ g/cm^3 under the saturation vapor pressure.

In a dimensionless form Eq. (1) can be written as

$$\frac{\partial \theta}{\partial \tau} + \frac{\partial \theta}{\partial x} - \frac{F}{\theta^{5.7}}\frac{\partial}{\partial x}\left(\theta^{5.7}(1-\theta^{5.7})\left(\frac{\partial \theta}{\partial x}\right)^{1/3}\right) = \frac{Q}{\theta^{5.7}} \tag{3}$$

where $\theta = T/T_\lambda$, $\tau = vt/L$, $x = X/L$, and the dimensional quantities F and Q are given by

$$F = \left(\frac{\rho^2 S_\lambda^4 T_\lambda^3}{A_\lambda}\frac{T_\lambda}{L}\right)^{1/3} \Big/ \rho v C_\lambda T_\lambda$$

$$Q = L q_{gen} \Big/ \rho v C_\lambda T_\lambda$$

Equation (3) can be solved numerically using a unique finite difference scheme.[6] To apply Eq. (3) to the problem under consideration, the heat generation term must be set to zero everywhere in the tube except at the midpoint where the heater is located. The initial condition is the bath temperature and the boundary conditions are taken to be the experimental temperatures at the tube ends. Figure 8 shows the numerical temperature profiles, for a pulse of width 3.5 s and a heat flux of $q = 5.71$W/cm^2, to be in close agreement with the experimental temperature profiles. In the solution the parameters F and Q are 0.228 and 4.85, respectively.

For the large amplitude narrow heat pulses bubble formation is observed at the heater. Although the numerical model predicts boiling for this case, it cannot be employed to describe the behavior of the system any further.

CONCLUSIONS

In application of a heat pulse to flowing He II, the decay rate of the pulse is dependent on the input energy of the pulse and the He II flow velocity. For narrow heat pulses the peak temperature appears to travel at a speed close to the the velocity of the fluid. Therefore it seems possible to use narrow pulses for measurement of He II flow velocity.

As in the case of a step function heat input, the He II energy equation appears to predict the experimental temperature profiles, for the case of a square function heat pulse, with good success.

ACKNOWLEDGMENTS

This work is supported by the U.S. Department of Energy under Grant DE-AC02 82ER52077. Special thanks go to E. Dreier for his support during the experimental phase of this study.

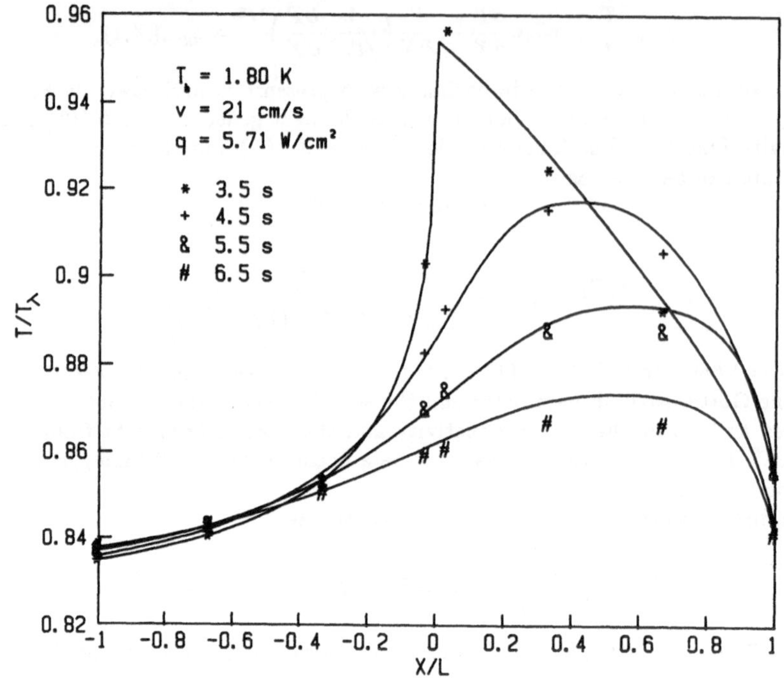

Fig. 8 Comparison of experiment with theory.

REFERENCES

1. R. Srinivasan and A. Hofmann, Investigations on cooling with forced flow of He II, Cryogenics 25:641 (1985).
2. A. Kashani and S. W. Van Sciver, Steady state forced convection heat transfer in He II, in:"Advances in Cryogenic Engineering," Vol 31, Plenum Press, New York (1986), P. 489.
3. A. Kashani and S. W. Van Sciver, Transient forced convection heat transfer in He II, in:"Proc. Eleventh Intl. Cryo. Engr. Conf.," Butterworth, Guildford, UK (1986), P. 654.
4. A. Singsaas and G. Ahlers, Entropy of He II from 1.6 K to the λ-line, Phys. Rev. B 29:4951 (1984).
5. P. E. Dimotakis and J. E. Broadwell, Local temperature measurements in supercritical counterflow in liquid helium, Phys.of Fluids 16:1787 (1974).
6. A. Kashani, "Forced Convection Heat Transfer in He II," Ph.D. Thesis, University of Wisconsin, Madison, Wisconsin (1986).

NOISELESS FILM BOILING IN SUPERFLUID HELIUM ON A HEATED THIN (32 μm) WIRE

F. Jebali, J. Maza*, M.X. François, and F. Vidal*

Laboratoire de Thermodynamique des Fluides
Université P. et M. Curie
Orsay, France

ABSTRACT

We will present here some of our results on the heat transfer in superfluid helium for a heat flux density range high enough to locally evaporate the liquid. In the present paper, we will focus on the case of a thin wire heater (32 μm) placed horizontally in the helium bath. Also, the scenario presented here will be the noiseless (stable) film boiling. For a given vapour film (constant thickness), we will present the experimental dependance of the heat flux density at the vapour-liquid interface as a function of the hydrostatic pressure and of the heater temperature as a function of the heat flux density at the surface heater. These experimental results will be analysed on the ground of the irreversible thermodynamics theories and on the kinetic approaches.

INTRODUCTION

Heat transfer processes in superfluid He in the range of heat fluxes large enough to induce the local evaporation of the fluid is currently of great importance from both the applied and fundamental point of view[1-2].

Although a considerable number of works on this subject have been done in the last decades, it is only recently that it was possible to perform some quantitative analysis of these processes[1-3]. This is a consequence of the increase of the experimental accuracy and the development of new theoretical models for the vaporisation processes in superfluid helium. In particular, the approach of Labuntzov et al[4-8], mainly based on non equilibrium molecular-kinetic considerations gives well defined predictions for noiseless film boiling in He II.

In the last four years, we have mainly studied the flat heater geometries[9-12]. The first aim of this paper is to extend these works to the thin wire heaters. The principal advantage in this case is a better definition of the limit conditions. In spite of the fact that such a geometry has been

* Permanent address : Laboratorio de Fisica de Materiales, Departamento de Fisica de la Materia Condensada, Universidad de Santiago de Compostela, Spain.

already used in a considerable number of works[1-7], to our knowledge the present paper is the first in which the relationship between q_i, h and T_w are presented simultaneously for a given vapour film. Here, q_i is the heat flux density at the vapour film interface, h is the liquid helium depth of the heater and T_w is the heater temperature. To obtain this last parameter, we use as heater a metallic alloy wire (Cu-Pb-Sn) having an important temperature dependence of its electrical resistivity. Another aim of this paper is to analyse these experimental results on the ground of the existing theoretical models for the vaporisation processes in superfluid helium. More precisely, we will use the approach of Labuntzov et al (4-8), mainly based on non-equilibrium molecular-kinetic considerations. This model gives well defined predictions for the relationships between q_i, T_w and h for noiseless film boiling in He II. In addition, we have shown recently[12] that this model is compatible with other general kinetic treatments of vaporisation as well as with the linear irreversible thermodynamic approaches[13-15].

EXPERIMENTAL RESULTS

The present measurements were made by using as heater a thin metallic wire of 32 μm diameter and of nominal composition : 97.97 % Cu, 1.9 % Sn and 0.13 % Pb (all in weights).

As noted in the introduction, the basic property of this type of wire is that its electrical resistance R is temperature dependent. For instance, for T_s = 4.2 K, $(1/R)(dR/dT)_{T_s} = 45 \times 10^{-3}\ K^{-1}$. So, by measuring the wire resistivity it is possible to obtain its temperature variations associated with the vapour film nucleation.

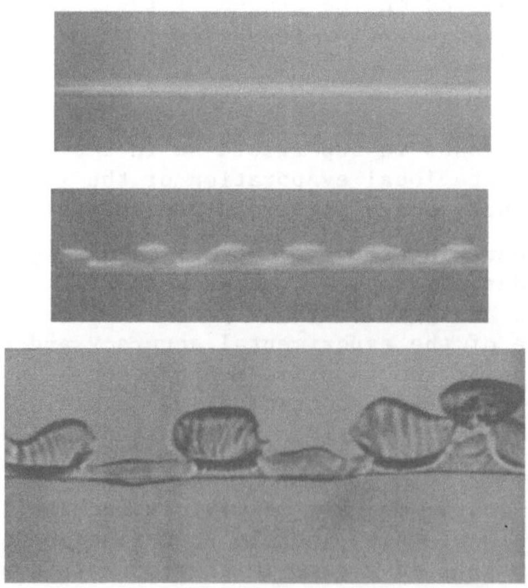

(a)

(b)

(c)

Fig. 1 a) Stable film of vapor around a 32 μm heated wire for heat flux q_w = 16.4 W/cm^2 at T_s = 2.02 K, h = 4.15 cm, e = 0.16 mm and T_w = 126 K ; b) Typical stationary mode of vapor-liquid interface with q_w = 15.3 W/cm^2, T_s = 2.02 K, T_w = 120 K, h = 2.3 cm and wave length 1.5 mm ; c) An example of film boiling in He II. This last photograph has been made with a time exposure of 4×10^{-5}sec.

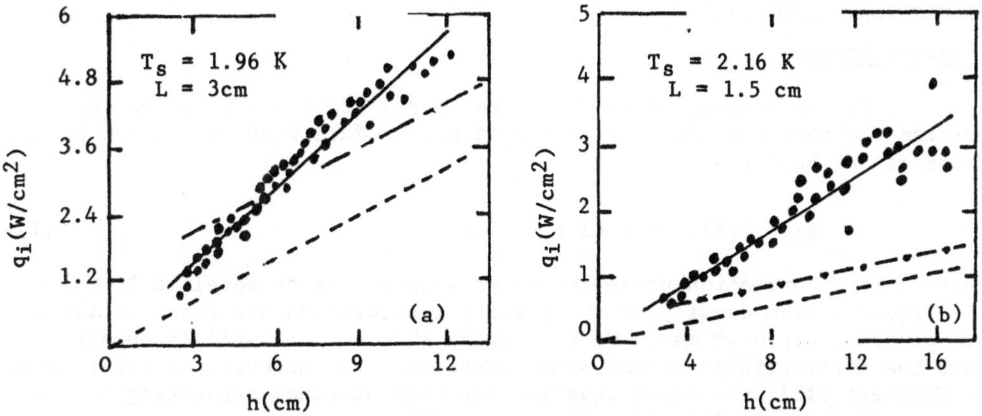

Fig. 2 · Heat flux density (q_i) as a function of the wire depth (h) in liquid helium for a constant vapour thickness (e = 0.1 mm). (a), T_s = 1.96 K ; and (b), T_s = 2.16 K

Three typical photographs of the vapour configuration during heat transfer from the heater wire (32 μm) to He II are shown in Fig. 1. The photographs (a) and (b) have been realized at the same bath temperature T_s, but for different heigh (h) of liquid between the free liquid helium surface and the heating wire.

Figure 2 (a) and (b) show the relationship between the heat flux density through the vapour film interface (q_i) and h for a constant vapour thickness (e) and for two different liquid helium temperatures. In both figures, the solid line is the standard average of the experimental data points, whereas the dashed and the dot-dashed lines are the theoretical predictions obtained as explained in the next section.

Figure 3 (a) and (b) show the measured heater temperature (T_w) as a function of the heat flux density (q_w), at the heater surface (data points). Also, we have represented in that figure the theoretical dependence obtained as indicated in the next Section.

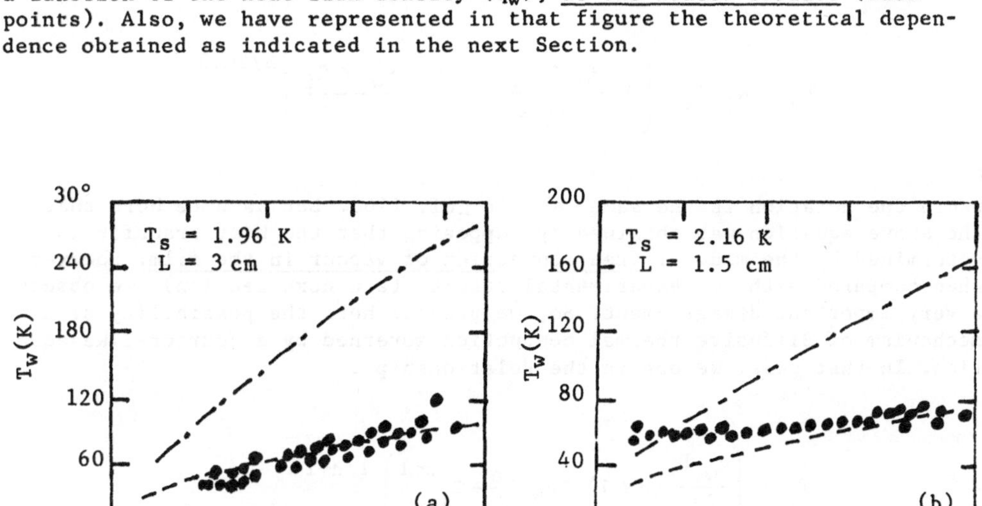

Fig. 3 Heater temperature as a function of the heat flux density at the heater surface. The curves are the theoretical expectations obtained as indicated in the text (section III).

THEORETICAL BACKGROUND

Relation between q_i and h

On the grounds of the Labuntzov model (2, 4 - 8), the relation between the heat flux density at the film interface with the liquid helium depth of the heater, h is

$$q_i = 2.27\,\varepsilon\,\rho_l\,g\,(2\,r\,T_s)^{\frac{1}{2}}\,h\ , \tag{1}$$

where ε is a correction parameter taken values less or equal to 1, ρ_l is the liquid helium density, g the gravity acceleration and r is the ideal gas constant per unit mass. We have shown that equation (1) is compatible with the irreversible thermodynamic approach (12). However, as first noted by Wiechert (15), the above equation has been obtained neglecting the surface tension of the liquid at the vapour film interface.

If the surface tension is take into account, then the pressure of the vapour inside the film will be

$$P_g = P_l + \frac{\sigma}{R_c}\ , \tag{2}$$

where P_l is the liquid pressure at the interface, σ is the surface tension and R_c is the radius of curvature. So, the new relationship between q_i and h take the form :

$$q_i = 2.27\,\varepsilon\,(\rho_l\,g\,h + \frac{2\sigma}{e})\,(2r\,T_s)^{\frac{1}{2}}\ . \tag{2}$$

Relationship between T_w and q_w

The relationship between the temperature T_w and the heat flux density q_w of wire heater has been obtained on grounds of kinetic theory by Labuntzov and Ametistov (16). Such a relation may be written for the average wire temperature :

$$T_w = K_1\,T_s\left[\left(D\,\frac{q_w - q_i}{P\,\rho_l\,g}\,\mu_s\right)^{1/3}\,\frac{q_w + q_i}{\lambda_s\,T_s}\right]^{3/2n+3}\ , \tag{3}$$

where the notation is the same as in Ref. (16). Let us note here that the above equation was obtained by supposing that the heat transfer is determined by the mode of free convection of vapour in the film. However when compared with our experimental results (see next section), we observe a very important disagreement. So, we propose here the possibility of a mechanism of diffusive thermal conduction governed by a Fourier-like equation. In that case, we obtain the relationship :

$$T_w = \left[\frac{q_w\,D}{2a}\,(n+1)\,\mathrm{Log}\,\frac{2e}{D} + T_i^{\,n+1}\right]^{1/n+1}\ , \tag{4}$$

where the exponent n and the constant a are related by $\lambda = aT^n$; λ and T are respectively the vapour thermal conductivity and temperature. T_i is the vapour-liquid interface temperature.

COMPARISON BETWEEN THEORY AND EXPERIMENTS - CONCLUSIONS

We have plotted in Fig. 2 (a and b) the theoretical behavior of q_i as a function of h obtained from Eq. (1) (dashed line). The dot-dashed line has been obtained from Eq. (2), i.e. it takes into account the surface tension. The main conclusions concerning the comparison of these theoretical curves with the experimental data points are :

i) The linear behavior of q_i versus h predicted by the theory (4-8) is well confirmed by the present experiments, for all temperatures studied.

ii) However, we observe a systematic slope difference due, very probably, to the indetermination of R_i, the thermal resistance of heat transfer at the interphase boundary on the liquid side. In fact, detailed measurements of R_i as a function of T_s will be very useful.

iii) The introduction of the correction associated with the superficial tension does not modify the differences of slopes between theory and experiments. However, this correction seems to improve somewhat the agreement between the absolute values.

In Fig. (3), we show the theoretical relationship between the heater temperature and heat flux density as obtained from Eq. (3) (dot-dashed line). We think that the important disagreement of this curve with the experimental results is not due, in this case, to possible systematic errors in the choice of the values of the different parameters arising in Eq. (3), but to the failure of the free convection model. This seems to be confirmed by much better agreement observed when Eq. (4) is used (dashed line). New measurements are under way to confirm this point of view.

ACKNOWLEDGEMENT

J. Maza and F. Vidal acknowledge the financial support from CAICYT PR 620-84 and from an Accion Integrada Hispano-Francesa 1984-1987.

REFERENCES

1. L. DRESNER and S.W. VAN SCIVER, Studies of heat transport to forced flow He II. Cryogenics 26:11 (1986) and references herein.
2. Ye. V. AMETISTOV, Heat transfer with He II noiseless film boiling, Cryogenics 23:179 (1983) and references herein.
3. For early experiments on film boiling in superfluid helium see for instance, H.E. RORSHACH and F.A. ROMBERG, Proceeding of the fifth international conference on low temperature Physics and Chemistry (University of Wiscosin Press, Madison, Wiscosin, 1958), p. 35 ; F. HAENSSLER and L. RINDERER, Helv. Phys. Acta 32:322 (1959).
4. T.M. MURATOVA, D.A. LABUNTZOV, Kinetic analysis of processes of vaporisation and condensation, Teplofizika Vysikikh Temperatur 75:959 (1969).
5. D.A. LABUNTZOV, A.V. LUIKOV, ANBSSR 6:33 (1977).
6. D.A. LABUNTZOV, A.P. KRYNKOV, Analysis of the intensive evaporation and condensation, Int. J. Heat Mass Transfer 22:989 (1979).
7. D.A. LABUNTZOV et al, Investigation of film boiling of superfluid (He II), Teploenergetika 4:18 (1981).
8. D.A. LABUNTZOV and Y.V. AMETISTOV, The theory of He II film boiling on horizontal cylinders, Cryogenics 21:51 (1981).

9. E. SAUVAGE, M.X. FRANCOIS, G. DEFRESNE and F. VIDAL, Heat transfer in superfluid helium with vapour-liquid interface, Phys. Lett. A 111, 430 (1985).
10. J.G.M. ARMITAGE, M.X. FRANCOIS, E. SAUVAGE and F. VIDAL, Quelques résultats expérimentaux sur la création et la dynamique de l'interface d'un film de gaz (He) dans l'Helium Superfluide, "Compte-rendu du Congrès National de Physique," Nice 1985, Les Editions de Physique, Paris (1985), p. 133.
11. G. DEFRESNE, M. FOUAIDY and M.X. FRANCOIS, Thermométrie cryogénique (1,3 k - 40 k) : étalonnage et loi de comportement, "IIème Journées de Cryogénie," Aussois, France, Les Editions de Physique, France (1986), p. 151.
12. J. MAZA, F. MIGUELEZ, C. TORRON, J.A. VEIRA, F. VIDAL, F. JEBALI, M. FOUAIDY, E. SAUVAGE and M.X. FRANCOIS, Some results on noiseless film boiling in superfluid helium with flat heater geometries, "Proceeding of the XVII International Congress of Refrigeration," Vienne (1987), to be published.
13. J.W. CIPOLLA et al, J. Chem. Phys. 61, 69 (1974).
14. V.P. PAO, Phys. Fluids 14:1340 (1971)
15. H. WIECHERT, Boundary conditions for the liquid-vapour interface of helium II, J. Phys. C 9:553 (1976).
16. D.A. LABUNTZOV, E.V. AMETISTOV, The theory of laminar film boiling of helium II, Thermal engineering 29:122 (1982) (Teploenergetika, 29:10 (1982)).
17. F. JEBALI et al, to be published in Phys. Lett. A.

TURBULENT TRANSPORT OF He II IN ACTIVE AND PASSIVE PHASE SEPARATORS USING SLIT DEVICES AND POROUS MEDIA

S.W.K. Yuan

Lockheed Research & Development Division
Palo Alto, California

J.M. Lee

NASA-Ames Research Center
Moffett Field, California

T.H.K. Frederking

University of California
Los Angeles, California

ABSTRACT

The turbulent transport mode of vapor liquid phase separators (VLPS) for He II has been investigated comparing passive porous plug separators with active phase separators (APS) using slits of variable flow paths within a common frame of reference. It is concluded that the basic transport regimes in both devices are identical. An integrated Gorter-Mellink equation, found previously to predict VLPS results of porous plugs, is employed to analyze APS data published in the literature. It is found that the Gorter Mellink flow rate parameter for 9 μm and 14 μm APS slit widths are relatively independent of the slit width having a rate constant of about 9 ± 10%. This agrees with the early heat flow results for He II entropy transport at zero net mass flow in wide capillaries and slits.

INTRODUCTION

Dynamic control of liquid He II vapor-liquid phase separation has been demonstrated using an active phase separator[1-6]. The mass flow through the APS is controlled by actively varying the flow area and path length of an annular slit. Published results of the performance of APS are frequently characterized by graphical plots of the mass flow rate vs. the vapor pressure difference for various slit lengths, e.g. Fig.1.

Most of the APS results in the literature are characterized by a flat section known as the counterflow regime, and a steep section called the "Gorter-Mellink (GM) flow"[1-6]. Figure 1 depicts typical APS results as mass flow

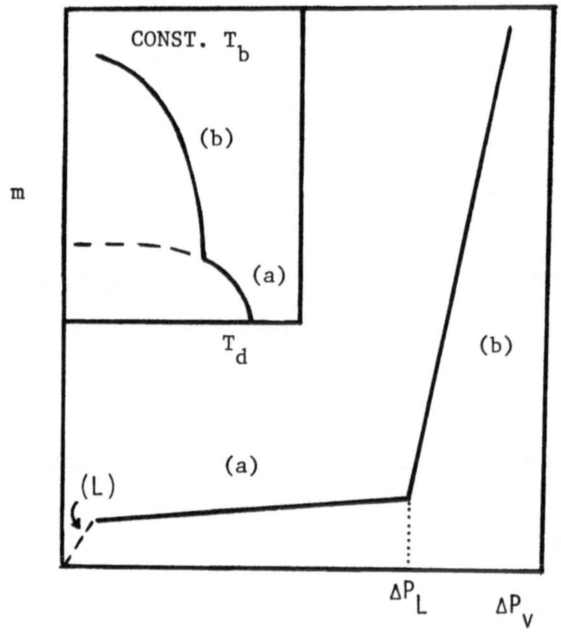

Figure 1. Typical results of active phase separation in mass flow rate vs. vapor pressure difference. Inset: same data in m vs. downstream temperature for a constant bath temperature (schematically).

rate versus vapor pressure difference, ΔP_V. The inset of Fig. 1 presents the data in different coordinates with mass flow rate as a function of downstream temperatures for a constant bath temperature. The curve "a" (Fig. 1) is the flat region observed starting from a finite m value toward the limit, ΔP_L. It is in this region "a" that true VLPS action exists. A frame of reference is used (to correlate the APS data) which has been established for ZNMF transport[7] and passive VLPS systems for porous media[8]. The section "b" of Figure 1 permits a large increase in mass flow rate for a small increment in ΔP_V. In this steeply sloped region "b", liquid breakthrough occurs[2] to the downstream side of the APS. Subsequently a heat exchanger assists in the evaporation of the liquid. Similar breakthrough data in a ceramic plug was reported by Elsner[9]. The dashed section marked "L" is the usual linear range of VLPS documented with porous media. This line starts at the origin of ΔP_V ($\dot{m}$) and is well known in the theoretical treatment provided by the two-fluid model of He II. The two APS regimes "a" and "b" however, are less well understood in view of a lack of a generally accepted nomenclature. It is the purpose of this paper to resolve transport details of section "a" by a quantitative comparison with other VLPS results outside the linear regime "L".

EQUATIONS FOR THE PHASE SEPARATION OF He II

Vapor-Liquid phase separators make use of the fountain effect to prevent excessive liquid loss by the retention of He II within the restricted geometries of the phase separators. The superfluid and normal fluid flow in opposite directions with the normal component carrying the heat which is rejected at the downstream side of the VLPS through evaporation.

Frequent VLPS operation based on porous media was found to be within the non-linear regime[8], corresponding to section "a" (in Fig. 1). Transport in such a system is very close to the zero net mass flow mode, where the heat flux density can be expressed by the conventional Gorter-Mellink equation[7]

$$q_{ZNMF} = \rho_s ST[S\nabla T/\rho_n A_{GM}]^{1/3} \qquad (1)$$

Due to the finite vaporization of He II at the downstream of the VLPS, the heat flux density through a phase separator (based on the two fluid model) can be written as[8]

$$q = q_{ZNMF} / (1 + ST/\lambda) \qquad (2)$$

Recent VLPS work in porous media has indicated that the heat flux density accompanying VLPS-mass flow is pore-size dependent[8]. Looking at Equation (1), it appears to be essential to eliminate the temperature dependency of the A_{GM} coefficient, so that the geometry influence can be observed clearly. This can be done through the substitution of a constant K_i[10], where

$$A_{GM} = [K_i{}^3(\eta_n \rho_s/\rho)]^{-1} \qquad (3)$$

Combining Equations (1) through (3) together with the superficial heat flux density for ZNMF ($q_{ZNMF}=j_o ST$), one gets the mass throughput for the VLPS

$$\dot{m}=j_o A_{tot}=A_{tot} w_{eff} \rho_s (1+ST/\lambda)^{-1} ST/\lambda \qquad (4)$$

$$\text{with} \quad w_{eff}=K_i[(\rho_s/\rho)(\eta_n/\rho_n)S|\nabla T|]^{1/3}$$

RESULTS AND DISCUSSION

For ZNMF in wide ducts, $K_i=K_{GM}\sim 10$ (to first order) was reported[10]. As for VLPS by porous media, it was found that $K_i=K_{VLPS}$ is a strong function of the throughput geometry[8] (e.g. Darcy's permeability, κ)

$$K_{VLPS}=[K_{GM}{}^{-2} + (10^{-7}/\kappa)]^{-1/2} \qquad (5)$$

Figure 2 shows the comparison between Equation (4) and the phase separation results of porous media. These data were fitted by integrating Equation (4) from a constant downstream temperature T_d to the appropriate upstream temperature T_u. Constants K_{VLPS} of 0.4 and 0.26 were used for the 10 μm and 2 μm Mott stainless steel plug respectively.

Figures 3 to 5 represent predictions based on Equation (4) for the flat section "a" (vortex shedding regime) of the active phase separation. These APS data are plotted as T_d versus mass flow rate for constant bath temperatures. The results for a 9 μm slit are presented in Figures 3 and 4 for

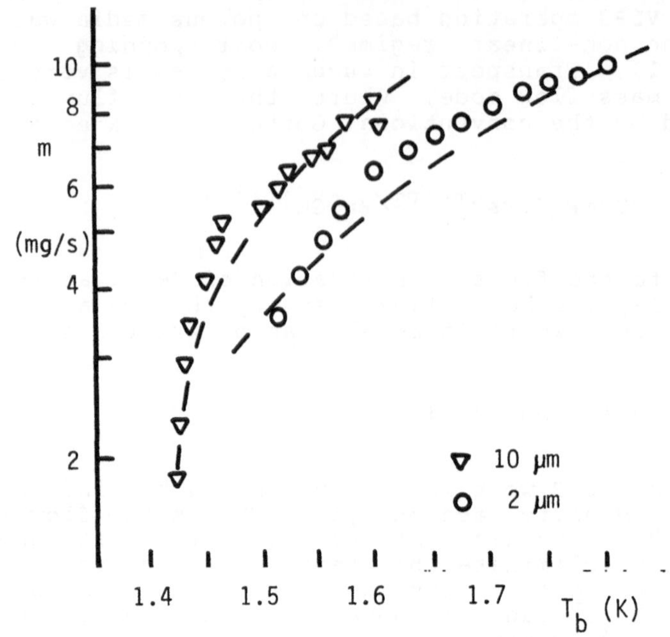

Figure 2. Vapor-Liquid phase separation data of porous plugs; dashed lines represent Equation (4).

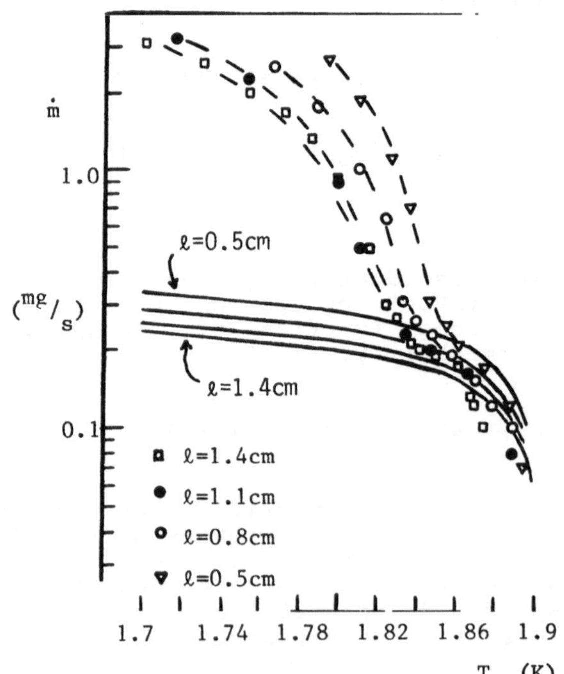

Figure 3. Correlation of the active phase separation data by a 9 μm slit at T_b=1.9 K; solid lines represent Equation (4), dashed lines are drawn to guide the eye.

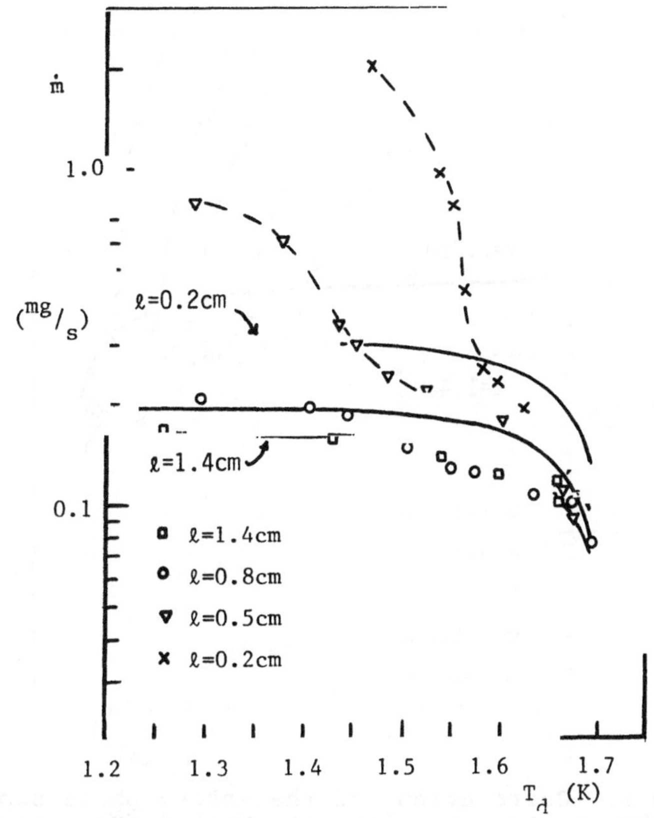

Figure 4. Correlation of the active phase separation data by a 9 μm slit at T_b=1.7 K; solid lines represent Equation (4), dashed lines are drawn to guide the eye.

T_b of 1.9 and 1.7 K respectively. The data tend to agree with Equation (4) up to the critical pressure difference ΔP_L where the onset to liquid breakthrough is initiated. This critical value is higher at low temperatures (also experienced by porous media VLPS experiments[9]) as indicated by Figure 4. The APS results for a 14 μm slit are plotted in Figure 5. K_{VLPS} values of 8 and 10 were obtained for the 9 μm and 14 μm slit respectively.

The K_{GM} constants for the ZNMF of He II in wide ducts and slits are compared with K_{VLPS} values for the APS and the porous media VLPS results in Figure 6. The horizontal axis represents the equivalent Darcy permeability (κ=$D^2/32$ for wide ducts and κ=$d^2/12$ for slits). As we can see from Figure 6, the APS results agree with the ZNMF data of He II in slits and wide capillaries with K_{GM} ~ 10. The porous plug VLPS data however, show a strong geometry dependency according to Equation (5). (Porous media data have been omitted in Fig. 6 for clarity, interested readers are referred to Ref. 8).

Concerning region "b" in Fig.1, it is noted that liquid was reported to enter into the vent line[2]. Thus, the identification of the regime by reference to a Gorter-Mellink function[7] is at variance with the established regime classification[2] as transition regime from onset of liquid breakthrough to breakthrough completion. In addition, the

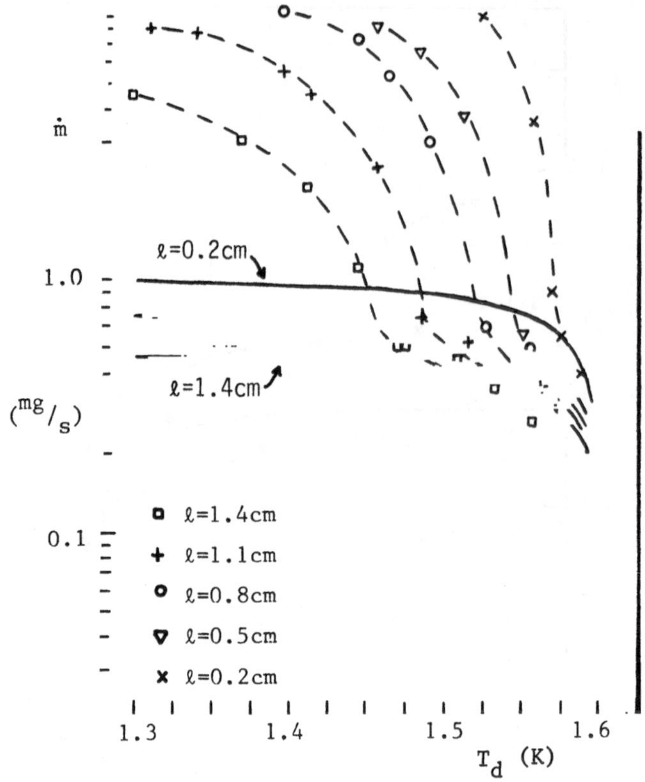

Figure 5. Correlation of the active phase separation data by a 14 μm slit at T_b=1.6 K; solid lines represent Equation (4), dashed lines are drawn the guide the eye.

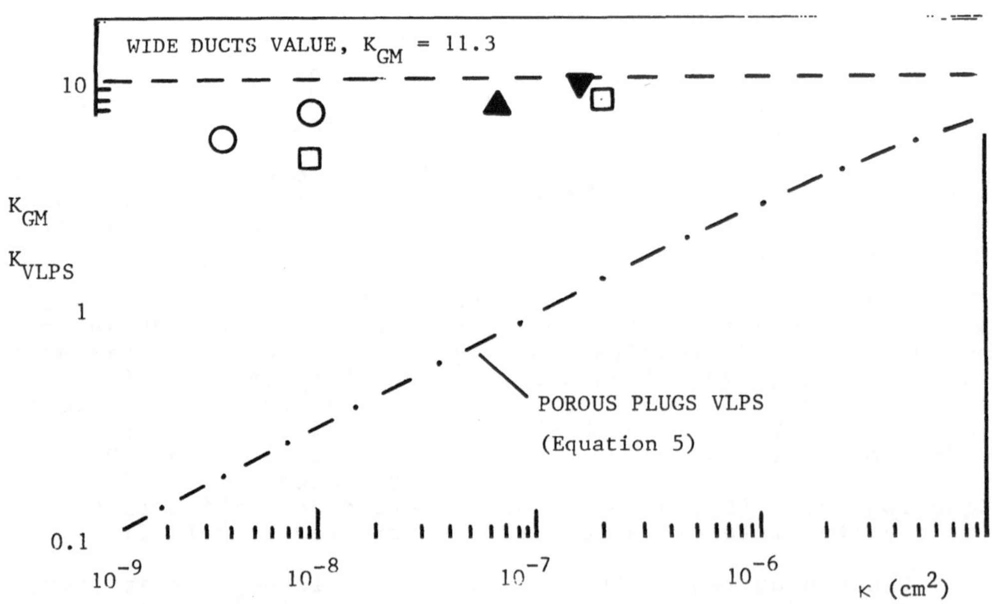

Figure 6. Comparison of rate constants for ZNMF (K_{GM}) with phase separation constants (K_{VLPS}); ZNMF slit data: ○ Keesom et al.[11,12], □ Keller et al.[13]; APS data: ▲ 9 μm and ▼ 14 μm.

Gorter-Mellink coefficient used for "b" had to be made twice as large as the highest value published originally for ZNMF addressed in the treatment of Gorter and Mellink[4].

CONCLUSIONS

It can be concluded that the basic flow regimes (Fig.1) in both the APS and porous media are identical. The normal operating condition of phase separators lies in the vortex shedding regime "a" (as shown in Fig. 7). Laminar transport of He II is possible, however this limit is not easily accessible, especially at low-g. At high flow rates, liquid breakthrough is observed, as the critical pressure difference ΔP_L across the separator is exceeded. The APS results agree with the ZNMF mode of He II in slits and wide ducts (within data scatter), where the constant K_{VLPS} is almost independent of the flow geometry. The porous media VLPS results on the other hand show a strong pore-size or permeability dependency described by Equation (5).

NOMENCLATURE

A_{GM}	Gorter-Mellink coefficient
A_{tot}	Total cross-sectional area of plug
D	Diameter of capillaries
d	Slit width of APS
j	Mass flux density
K_{GM}	Gorter-Mellink transport parameter
K_{VLPS}	Transport parameter in porous plug phase separators
K_i	Size dependant rate constant
$\dot{m}$	Mass flow rate
ΔP_V	Vapor pressure difference

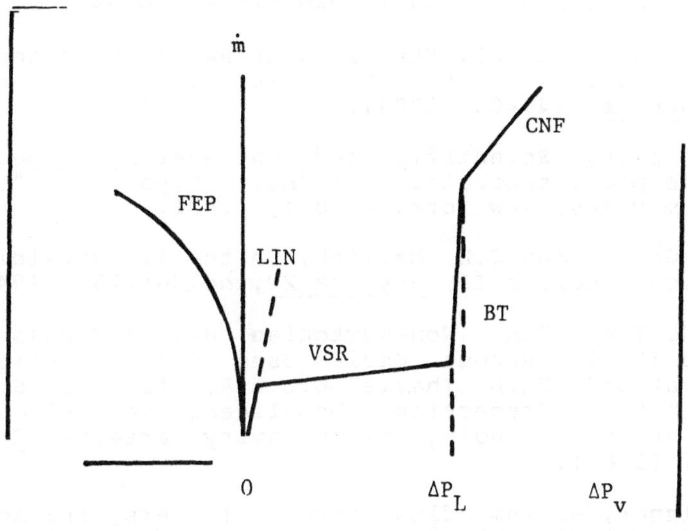

Figure 7. Various transport regimes in phase separation of He II; LIN,linear regime of constant chemical potential; VSR,vortex shedding regime; BT,liquid breakthrough; CNF,classical Newtonian flow above the completion of BT; FEP,fountain effect pumping (schematically).

ΔP_L	Limiting value of ΔP_V at the onset of liquid breakthrough
q	Heat flux density
S	Entropy per unit mass
T	Temperature
w_{eff}	Effective counterflow velocity
η	Shear viscosity
κ	Darcy permeability
ℓ	Slit length of APS
λ	Latent heat of vaporization
ρ	Density

Subscripts

n	normal fluid
o	superficial value
s	superfluid
VLPS	vapor-liquid phase separation
ZNMF	zero net mass flow

REFERENCES

1. K. Luders and G. Klipping, Phase separation under zero g conditions, in:"Proceedings of Eighth ICEC", IPC Science & Technology Press, Guildford (1980), p.14.

2. U. Schotte and H.D. Denner,Heat conduction through narrow channels and phase separation of He II at zero gravity, Z. Phys. B, 41:139 (1981).

3. H.D. Denner et. al, Improved active phase separator for Helium II space cooling systems, in:"Adv. Cryog. Eng.", Vol. 27, Plenum Press, New York (1981), p.1079.

4. I. Arend et. al, Fountain effect inversion: experiments with an active phase separator, in:"Proceedings of Tenth ICEC", Butterworth, Guildford (1982), p.62.

5. H.D. Denner et. al, Stability of He II phase separation - investigations with the active phase separator, Cryogenics, 24:403 (1984).

6. G. Klipping, Scientific and engineering aspects of the active phase separator, in:"Adv. Cryog. Eng.", Vol. 31, Plenum Press, New York, (1985), p.851.

7. C.J. Gorter and J.H. Mellink, On the irreversible process in liquid Helium II, Physica XV, no.3-4:285 (1949).

8. S.W.K. Yuan, "The Non-Newtonian Heat and Mass Transport of He II in Porous Media Used for Vapor-Liquid Phase Separation", Ph.D. Thesis, U.C.L.A., (1985); S.W.K. Yuan and T.H.K. Frederking, Non-linear vapor-liquid phase separation including microgravity effects, Cryogenics, 27:27 (1987).

9. A. Elsner, Helium flow through filters, in:"Adv. Cryog. Eng.", Vol. 18, Plenum Press, New York, (1973), p.141.

10. S.C. Soloski and T.H.K. Frederking, Dimensional analysis and equation for axial heat flow of Gorter-Mellink convection (He II), Int. J. Heat & Mass Transfer, 23:437 (1980).

11. W.H. Keesom and G. Duyckaerts, Mesures sur la conductibilite thermique et l' effet thermomecanique de l' helium liquide II, Physica, 13:153 (1947).

12. J.M. Lee, "Permeabilities of Sintered Porous Metal Plugs and Transport Rates of Vapor-Liquid Phase Separators for He II Vessels", M.Sc. Thesis, U.C.L.A., (1983).

13. W.E. Keller and E.F. Hammel Jr., Heat conduction and fountain pressure in liquid He II, Ann. Phys., 10:202 (1960).

OBSERVATION OF TWO-PHASE HELIUM FLOWS IN A HORIZONTAL PIPE

E. Sauvage-Boutar, C. Meuris, J. Poivilliers, M.X. Francois

CEN/Saclay, I.R.F., DPhPE/STIPE

91191 Gif-sur-Yvette Cédex (France)

ABSTRACT

An experimental apparatus has been built to investigate steady state two-phase helium flow in a horizontal circular tube. This paper reports preliminary results of the hydrodynamic structure of a helium I flow through a tube of 10 mm inner diameter. A glass section with a special optical system allows a study of the flow pattern by observation and high speed photography. The experiment covers mass fluxes from 2 to 7 g/s, vapor quality from 0 to 1, and pressures between 1 and 1.5 atm. The variation of the vapor quality is obtained by evaporation of the liquid by a coax heater soldered to the pipe. The experimental flow regime charts are compared with the theoretical chart of Y. Taitel and A.E. Dukler.

INTRODUCTION

Superconducting systems can be cooled by a forced helium flow at saturation. Very good heat transfer between solid and wetting fluid is obtained in the nucleate boiling regime and the coolant temperature is more or less constant along the pipe.

Single-componant two-phase flows (water - steam for instance) have been studied very extensively for thermal energy transport, and several teams have proposed theoretical models to describe the behaviour of the liquid-gas mixture under steady-state conditions, without singularities. The subject is dealt with in the works of J.M. Delhaye[1]. Unfortunately the similarity factors are not known for two-phase flows, and these results cannot be transposed to the case of light and rather non-viscous coolants such as helium. The important factors are of course the pressure drop terms Δp in horizontal and vertical flows, the role of singularities, and the heat exchange coefficients h. In practice the quantities Δp and h depend not only on the mass flow rate of each phase but also on the spatial distribution of the two which distinguishes the type of flow. This factor plays a major part in horizontal flows, where the effect of gravitation may lead to incomplete wetting of the tube walls by the liquid phase (phase stratification).

The first step therefore is to know the types of flow. These are determined beforehand from charts in the form of two-coordinate diagrams, generally chosen according to the test parameters of the experiment, on which are represented the boundaries between the different flow types. The latter may then be determined for given fluid properties, tube geometry and flow speed of each phase[2,3,4]. The systematic study necessary requires optical observation of the flow, a critical point no doubt responsible for the limited amount of work on helium so far[5]

The present study concerns the development of an experimental device and the determination of flow types for saturated helium flowing under steady-state conditions in a 10 mm-diameter horizontal adiabatic duct, under 1.1 atm pressure (i.e. reduced pressure p/pc = 0.48) and with total mass flow rate between 2 and 7 g/s. The first results obtained are compared with the theoretical findings of Y. Taitel and A.E. Dukler[6] who account for all working parameters: pressure, tube diameter, mass flow rates, ...

EXPERIMENTS

Description of the apparatus (see Fig.1)

Natural flow in a 10 mm-diameter copper tube placed under vacuum is obtained by pressurization of the upstream vessel (1), the downstream vessel (2) being at atmospheric pressure. The helium passes through a venturi tube (3) enabling the mass flow rate to be measured and adjusted by opening or closing of the cold valve (4), which also allows a pressure drop to be created at the input and the flow thus stabilised. The total mass flow is only adjustable within the range 2 - 7 g/s, i.e. a mass flow density between 25.5 and 89 $kgm^{-2}s^{-1}$, i.e. velocities from 0.21 to 0.72 m/s for a liquid flow at saturation and 1.4 to 4.8 m/s for a gas flow at saturation.

An exchanger (5) of boiling helium at a sub-atmospheric pressure obtained by pumping recondenses the gas produced in the transfer line and as necessary subcools the fluid as it leaves the exchanger, giving a pure liquid at the valve outlet.

By means of three resistance heaters welded to the copper tube, (H1): 0.2 m, (H2) : 10 m and (H3) : 0.2 m, distant from the observation point by 0.2 m, 0.5 m and 10.6 m respectively, vapor quality ranging from 0 to 1 may be obtained by variation of the heat flux injected.

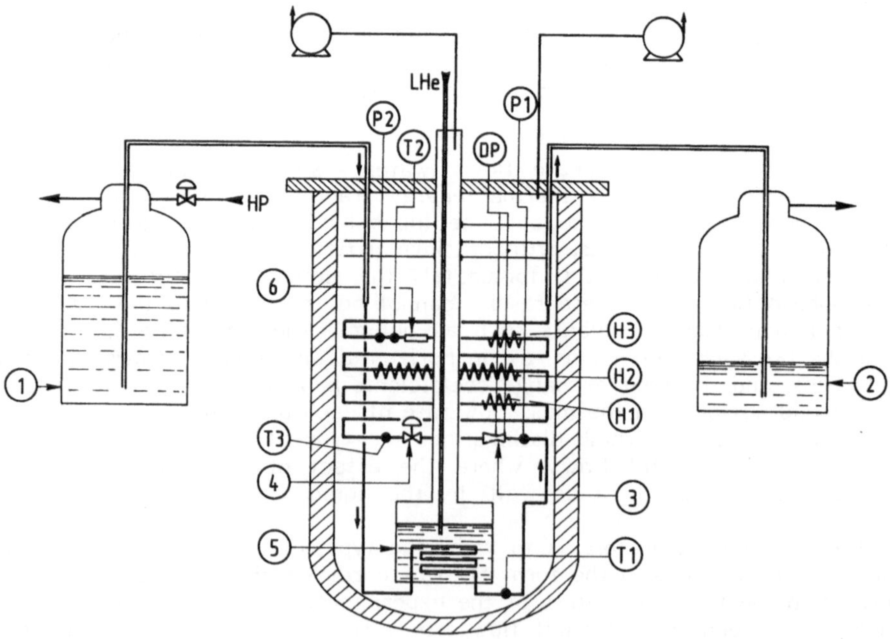

Fig. 1. Diagram of the apparatus (see description in the text).

An observation section consisting of a 5 cm long Pyrex tube (6) is welded on a straight part of the duct through silver-sealed Pyrex-Covar joints. An original optical device has been designed whereby the liquid-gas interfaces are visible on a light background; the flow is lit by an optical fibre. It should be noted that observations in helium are very difficult because of the great similarity between the refractive index of the gas and liquid, the angle of deflection of light by a bubble being 4 degrees.

The different parameters are monitored by means of temperature transducers, carbon resistances (T1), (T2) and (T3), as well as absolute (P1), (P2) and differential (DP) pressure transducers as indicated in Fig.1.

Experimental procedure

For a set mass flow value m, measured by pressure drop at the venturi tube with a precision of about 3%, the temperature of the exchanger is adjusted to give a pure liquid at saturation at the observation point. Heat losses upstream from this point, $\dot{Q}p$, are thus offset by subcooling of the liquid at the exchanger outlet :

$$\dot{m}\,(h_2 - h_1) = \dot{Q}_p$$

where h_2 is the enthalpy of the saturated liquid at the observation point : $h_2 = h(T_{sat}(p_2))$ and h_1 that of the liquid at the exchanger outlet : $h_1 = h(T_1, p_1)$. These losses, due mainly to radiation from the cryostat wall at 110 K, are estimated at about 1.5 W.

The different types of flow are then observed and photographed for a mass fraction of gas $x = \dot{m}_G/\dot{m}$ between 0 and 1 with a precision of 8% :

$$x = \dot{Q}/(L\dot{m})$$

with $\dot{Q}$ the power injected by one of the resistance heaters and L the latent heat of vaporisation of helium at pressure p_2 (pressure drop is negligible over 10 m : $\Delta p < 20$ mbars).

The effects of the different heat inputs, i.e. production of x by different fluxes per unit heated surface $\dot{q}$, and of the adiabatic length before observation are investigated. Heat transfer in the heated zone takes place by convection for heater (H2) ($\dot{q} < 0.05\ 10^4\ W/m^2$) and by nucleate or film boiling for heaters (H1) and (H3) as soon as $x > 0.1$.

PRELIMINARY RESULTS

The main flow regimes observed with other fluids by means of unheated observation sections in horizontal flow are shown on Fig.2.

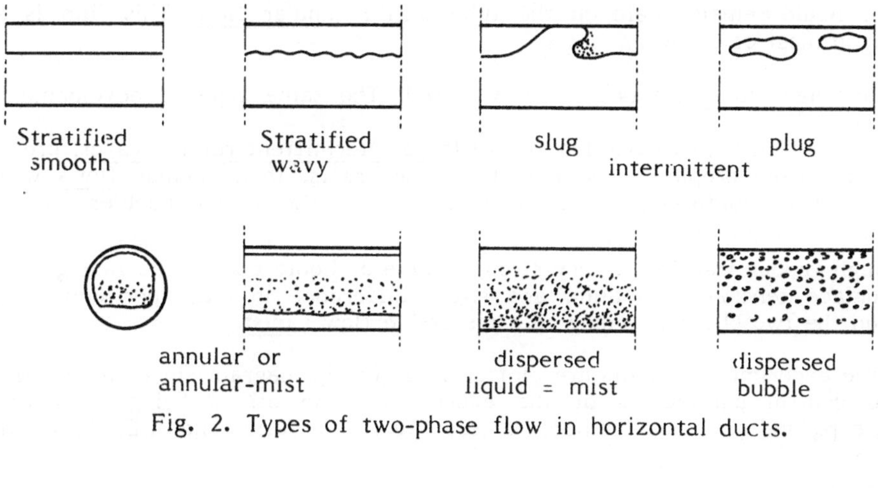

Fig. 2. Types of two-phase flow in horizontal ducts.

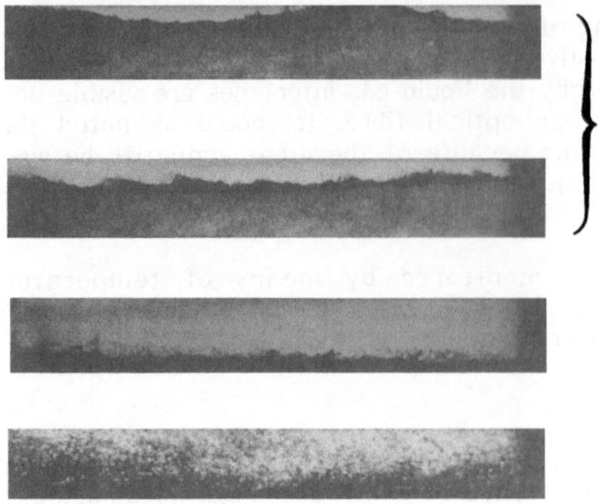

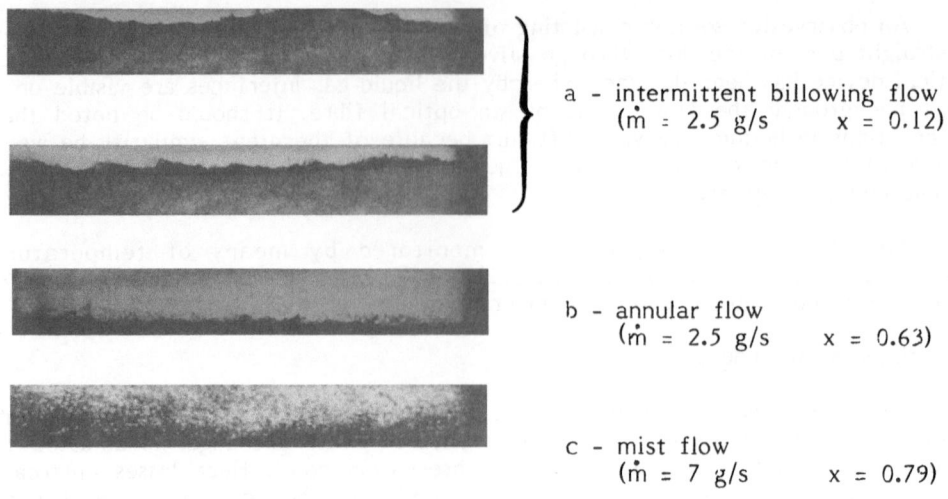

a - intermittent billowing flow
($\dot{m}$ = 2.5 g/s x = 0.12)

b - annular flow
($\dot{m}$ = 2.5 g/s x = 0.63)

c - mist flow
($\dot{m}$ = 7 g/s x = 0.79)

Fig. 3. Typical photos and classification adopted for the flows observed.

Under the working conditions adopted the following flow types were distinguished : stratified wavy, stratified with intermittent billowing waves, annular, mist. Typical photographs are shown on Fig.3. In the stratified and billowing types many small bubbles, due to low surface tension of helium and similar mass per unit volume of the two phases, are present in the liquid under established flow conditions. High fluxes per unit heated surface require great lengths for the establishment of the regime; production of high vapor quality by heater (H1) leads to an overheated gas in the observation section. For that reason most of the results reported here were obtained with the heater (H2). The annular flow regime is hard to identify with our optical system; the presence of a liquid film on the tube wall is shown by streaks due to the index variation between the liquid and the gas and the non constant thickness of the liquid film.

Description of the flow structure change with increasing heat power:

For medium flow rate ($2 < \dot{m} < 4.5$ g/s). At very low vapour contents the liquid level in the tube is too high for the flow to be discernable as stratified wavy or billowing. Once the liquid level falls ($h_L/d < 0.90$) large waves are formed intermittently, spreading at the liquid surface and sometimes breaking on the top of the tube: slug. As the vapour fraction rises the average level of the liquid falls and the frequency (irregular) of wave appearance decreases, cancelling out completely below a liquid level ($h_L/d \simeq 0.25$) at which the faster-flowing gas takes up liquid from the interface to form droplets in the gas phase; an almost static film of liquid remains held on the tube wall : annular flow. This film is also present in most of billowing flows.

For high flow rate ($4.5 < \dot{m} < 7$ g/s). The same type of development is observed :

- for low and medium vapour fractions : large intermittent rolling waves, breaking and losing many droplets picked up from the crests, then annular flows with a very disturbed liquid-gas interface made up of a mixture of gas bubbles and liquid drops;

- for high vapour fractions : very strong uptake of liquid due to the fast gas flow, low surface tension of helium, and slight difference in mass per unit volume between the two phases, leading to a dispersive flow : mist.

The experimental points are plotted in an (x, $\dot{m}$) diagram since these are two of the control parameters of the experiment, the others being the working pressure p_2 and the tube diameter d (see Fig.4). Y. Taitel and A.E. Dukler have developed theoretical models from physical mechanisms to determine the transitions between the different flow regimes[6]. Fig. 4 shows the corresponding boundaries calculated from these models for flow in a horizontal tube of diameter d = 10 mm with the thermodynamic properties of helium under p_2 = 1.1 atm :

$$\rho_G = 18.6\ kg/m^3 \qquad \nu_G = \mu_G/\rho_G = 6.9\ 10^{-8}\ m^2/s$$
$$\rho_L = 123\ kg/m^3 \qquad \nu_L = \mu_L/\rho_L = 2.6\ 10^{-8}\ m^2/s$$

with ρ and ν the mass per unit volume and dynamic viscosity respectively, index G referring to the gas and L to the liquid.

The aspect of the various regions is verified in a first approximation. The intermittent-annular transition in particular takes place for a constant mass vapour fraction corresponding to a constant liquid height (according to equation (1) below). No stratified flow, which according to the theory can only exist at low flow rate (< 1 g/s) and consequently low gas velocity, and no dispersed bubble flow, which according to the theory can only appear at high flow rate (> 30 g/s) when the liquid flow is fast, has been observed.

On the other hand the intermittent-annular transition line value, the only one we are able to check with the first results reported here, is not in good agreement with the theory. It is true that a transition between two flow types is hard to define since the passage from one to another is continuous, and this disagreement is also due partly to inaccuracies in the definition of the various

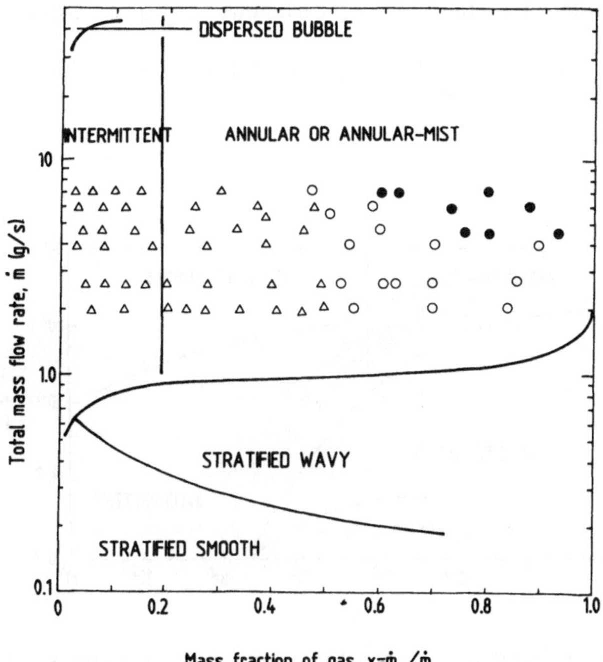

Fig. 4. Flow regime chart for two-phase helium flowing horizontally under 1.1 atm in a 10 mm-diameter pipe.
△ intermittent flow
○ annular or annular-mist flow
● mist flow

flows. Y. Taitel and A.E. Dukler model the intermittent annular flow transition on the basis that in intermittent wave flow the waves touch the top of the tube to form a slug and that consequently the transition corresponds to an average height of liquid $h_L/d = 0.5$. We have chosen to distinguish intermittent from annular flow by the presence or absence of large waves.

For comparison with other fluids or other tube diameters it is useful to work in dimensionless variables. Y. Taitel and A.E. Dukler show that for a given tube inclination each transition is described by a relation dependent on two dimensionless groups and can hence by represented on a two-dimensional chart which accounts for all the working parameters (mass flow rate of each phase, tube diameter, physical properties of fluids ...). Fig. 5 gives the theoretical boundaries and the experimental points expressed as a function of the dimensionless groups X, F, T and K defined by Y. Taitel and A.E. Dukler and calculated with the total mass flow rate $\dot{m}$ and the mass fraction of gas x for turbulent flow of both phases which is the case of the flows studied here when $0.01 < x < 0,97$:

$$X = ((1 - x)/x)^{0.9} (\nu_L/\nu_G)^{0.1} (\rho_G/\rho_L)^{0.4} \tag{1}$$

$$T^2 = (\dot{m}/A)^{1.8}(1-x)^{1.8}\, 0.092\, \nu_L^{0.2}/(d^{1.2}\rho_L^{0.8}(\rho_L - \rho_G)g) \tag{2}$$

$$F^2 = (x\dot{m}/A)^2/(\rho_G(\rho_L - \rho_G)dg) \tag{3}$$

$$K^2 = (\dot{m}/A)^3 x^2(1 - x)/(\rho_L\rho_G(\rho_L - \rho_G)g\nu_L) \tag{4}$$

where A is the pipe cross-section and g the gravity acceleration,
X is the Lockhart-Martinelli parameter,
T can be considered as the ratio of turbulent to gravity forces acting on the gas,
F is a modified Froude number which represents the ratio of inertia to gravity forces on the gas.

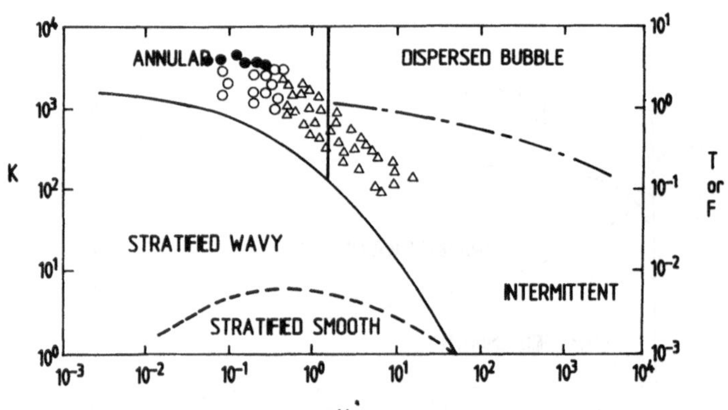

Fig. 5. Generalized flow chart for a horizontal two-phase flow.
- - - - - K vs X
———— F vs X
- ——— - T vs X
} after Ref.6
Δ intermittent flow
o annular or annular-mist flow
• mist flow
} our experiments

CONCLUSION

The first results obtained on helium flows in a horizontal duct cover part of the three regions in which intermittent, annular and mist flows are observed. It should be noted that the curves demarcating the regions are idealised, since actually the different flow types pass continously from one to another. Observation of the development of each regime within each region is the only way to obtain information on the phase distribution inside a tube section and in particular on the fraction of wetting of the tube perimeter by the liquid phase.

Additional studies are under way to define intermittent flows more accurately and to identify the liquid film in intermittent and annular flows by fast cinematography and measurement of the volume fraction of gas, which will enable the gas and liquid velocities to be measured also. In addition the working pressure will be brought close to the critical pressure, where the two phases have similar properties and the flow becomes dispersed bubble, and reduced to below atmospheric pressure until saturated superfluid helium is obtained and the gas and liquid properties differ significantly, at which point stratified flows prevail within the flow rate range adopted.

Moreover in collaboration with the KfK carrying out pressure drop experiments[7] we have been able to identify the same regimes as those described here. A correlation between pressure drops and flow types is being studied.

REFERENCES

1. J.M. Delhaye, Two-phase flow patterns, in : "Two-phase flow and heat transfer in the power and process industries", Hemisphere publishing corporation (1981), p.1.

2. J.L.L. Baker, Simultaneous flow of oil and gas, Oil Gas J. 53:185 (1954).
3. D.S. Scott, Properties of co-current gas-liquid flow, in : "Adv. in Chemical Eng." Vol.4, Academic, New-York (1963), p. 199.

4. J.M. Mandhane, G.A. Gregory and K. Aziz, A flow pattern map for gas-liquid flow in horizontal pipes, Int. J. Multiphase Flow 1: 537 (1974).

5. H.K. Zust and W.B. Bald, Experiment observations of flow boiling of liquid helium I in vertical channels, Cryogenics 21 : 657 (1981).

6. Y. Taitel and A.E. Dukler, A model for predicting flow regime transitions in horizontal and near horizontal gas-liquid flow, AICHE J. 22 : 47 (1976).

7. H. Katheder and M. Susser, Pressure drop of adiabatic and nonadiabatic horizontal two-phase helium flow, in : "Proc. of the 1986 Cryo. Eng. Conf.", Berlin (1986).

PRESSURE DROP AND TEMPERATURE RISE IN He II FLOW IN ROUND TUBES, VENTURI FLOWMETERS AND VALVES

P. L. Walstrom and J. R. Maddocks

Applied Superconductivity Center
University of Wisconsin-Madison
Madison, Wisconsin

ABSTRACT

Pressure drops in highly turbulent He II flow were measured in round tubes, valves, and venturi flowmeters. Results are in good agreement with single-phase flow correlations for classical fluids. The temperature rise in flow in a round tube was measured, and found to agree well with predictions for isenthalpic expansion. Cavitation was observed in the venturis under conditions of low back pressure and high flow rate. Metastable superheating of the helium at the venturi throat was observed before the helium made a transition to saturation pressure.

INTRODUCTION

Most of the experimental work to date with He II has been concerned with phenomena related to superfluid properties. For the present work, the behavior of turbulently flowing He II in various flow circuit elements was experimentally determined and is compared with pressure drop correlations used for classical fluids. The observed behavior, within experimental error, is describable by classic turbulent flow correlations using the normal component viscosity of the helium. The experiments were performed with highly turbulent helium in the temperature range 1.6K up to the lambda point, supplied by a single stroke bellows pump that also served as a mass flow calibration standard.

BELLOWS FLOW CRYOSTAT

All of the experiments were performed by means of a single stroke bellows pump with a capacity of approximately 1.7 liters. By compressing the bellows at constant speed, a steady mass flow can be maintained at the outlet for a time that varies inversely with mass flow rate. The volume-displacement curve of the bellows was measured by filling it with water and measuring the volume of water expelled as the bellows was slowly compressed. These data, together with the instantaneous slope of the bellows displacement traces, provide a volume flow rate calibration. The maximum achievable flow rate is determined by the total pressure drop in the flow circuit; for total flow pressure drops above about 100 torr, gear drive torque limits are exceeded and bellows squirming becomes excessive. Helium outlet temperature from the bellows is varied by varying the vapor pressure over the bath. The bellows is filled with helium from a bath at saturation pressure surrounding it. At the beginning of a stroke, the helium in the bellows is adiabatically compressed; the resultant temperature drop in all cases is less than 1mK. Temperature changes in the loop itself are determined by a particular loop configuration, supply temperature, and mass flow rate. The test loop is in vacuum; outlet flow from the test loop is returned to the bath. A back-pressure valve at the outlet is provided to allow variation of absolute pressure in the test loop.

PRESSURE DROP AND TEMPERATURE RISE IN ROUND TUBES

A flow loop was constructed from a 24 cm dia. helical coil of drawn nominal quarter inch O.D. copper tubing. After the tubing was wound into a coil, the cross section of the inside of the tube was slightly elliptical with a mean diameter of 0.476 cm and ratio of major diameter to minor diameter of 1.03. Pressure taps were installed 4.6 meters apart. The first pressure tap was approximately 20 cm downstream of the start of the first turn. Absolute pressures were measured with Siemens KPY-12 semiconductor pressure sensors[1] that had been precalibrated in a separate apparatus. The sensors were excited with 3 Vdc, and output voltages were amplified with differential amplifiers, digitized, and averaged. In determining pressure drop, baseline output voltages were first subtracted from the amplified sensor output voltage traces. The voltage increments were then converted to pressure increments from the baseline (no flow) pressures. The coil was also instrumented with germanium resistance thermometers installed next to the pressure transducers. Flow data were taken at 1.6K, 1.7K, 1.8K and 2.09K with mass flows ranging from zero to 10 gm/s. Pressure drop data were converted to Fanning friction factor values according to the standard formula

$$f_c = \frac{\pi^2}{32} \frac{D^5 \rho}{\dot{m}^2} \frac{\Delta P}{L} \tag{1}$$

where f_c is the Fanning friction factor, D the hydraulic diameter of the tubing, ρ the fluid density, $\dot{m}$ the mass flow rate, ΔP the pressure drop, and L the flow length. (In the above and following equations, ΔP is taken to be positive for pressure drops.) The Reynolds number was calculated using the expression

$$Re = \frac{\rho v D}{\eta_n} = \frac{4\dot{m}}{\pi D \eta_n} \tag{2}$$

where v is the average fluid velocity and η_n is the normal fluid viscosity.[2] The friction factors from the curved tube measurements were then converted to the values predicted for straight tubing by use of Ito's formula[3]

$$f_c = f_s \left[\left(\frac{D}{2R}\right)^2 Re\right]^{1/20} \tag{3}$$

where f_s is the straight tube friction factor for the same flow conditions and 2R is the diameter of the coil. The correction turned out to be quite large and ranged from 30 to 40 percent for the Reynolds numbers of the experiment. The validity of the derived straight tubing friction factors depends therefore on the applicability of the Ito formula to the flow regime studied.

The results for 1.8K He II flow are shown in Fig. 1, along with earlier data from work by Kashani et. al.[4] for smaller diameter tubing

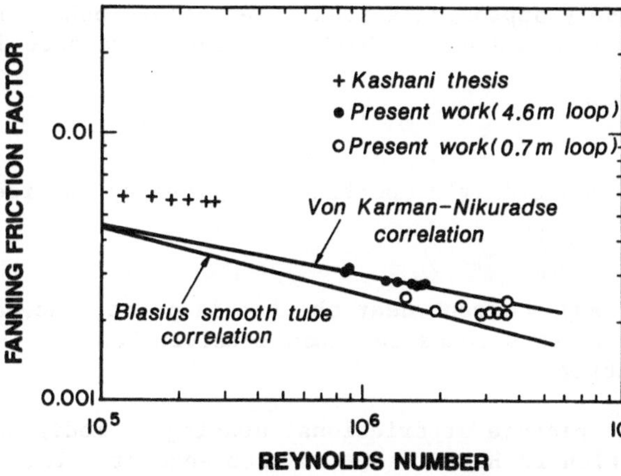

Fig. 1 Dependence of Fanning friction factor on Reynolds number for 0.46 cm. diameter copper tubing at 1.8K.

with greater surface roughness. Our data are represented well by the Von Karman-Nikuradse smooth tube correlation

$$\frac{1}{f^{1/2}} = 1.737 \ln\left(\text{Re } f^{1/2}\right) - 0.396 \tag{4}$$

A tendency for the friction factor points to level out to a constant value at the very highest Reynolds numbers can be seen in the figure. This behavior is observed in the rough pipe limit in classical turbulent fluid flow when the thickness of the laminar sublayer approaches the magnitude of the height of the surface irregularities. To check this, one can insert the friction factor value (2.7×10^{-3}) for the highest Reynolds number data point into Von Karman's rough pipe limit expression for f,

$$\frac{1}{2\sqrt{f}} = 2 \log_{10} \frac{D}{\varepsilon} + 1.14 \tag{5}$$

where ε is the height of the surface irregularities. The resultant value for ε is 2×10^{-5} cm. This is in reasonable agreement with profilometer scans of the surface. Longitudinal grooves with a depth of about $50\text{x}10^{-5}$ cm were found. Longitudinal variations along these grooves were about 10^{-5} cm. Results for the remaining He II temperatures (not shown) were essentially the same as those in Fig. 1 showing good agreement with the Von Karman-Nikaradse smooth tube formula with a tendency to level off to a constant value at the highest Reynolds numbers.

Also shown in Fig. 1 are data for higher mass flow rates taken with a portion of the same coil after it was cut down to an instrumented length of 0.7 m. Here, the agreement with the Von Karman-Nikuradse relation is not as good; the data lie below the line. This may be partly due to entrance length effects, for the loop was less than a complete turn in length and the Ito formula overestimates the curved tube effect in this case. The discrepancy for the short tube data is greater at the higher temperatures and smaller for the lower temperatures. The reason for this systematic trend is not known.

Temperature data were analyzed and compared to the predictions of thermodynamics. Since the flow loop was located in vacuum and insulated with multilayer reflective insulation, heat input by radiation was small. Therefore, if heat conduction in the helium could be ignored, the flow would closely approximate ideal isenthalpic behavior. In isenthalpic flow, the temperature change for small pressure drops is given by the expression

$$\Delta T = -\mu \Delta P \tag{6}$$

where μ is the Joule-Kelvin coefficient, given by the identity

$$\mu = \left(\frac{\partial T}{\partial P}\right)_h = \frac{\alpha T - 1}{\rho C p} \tag{7}$$

For the He II range, except near the lambda point, the magnitude of αT is much less than 1; this means that Joule-Kelvin heating is almost entirely frictional heating.

The simple picture of frictional heating is modified when Gorter-Mellink conduction in He II is taken into account. For comparison with our data, we used a simple constant parameter model in which Gorter-Mellink conduction is superimposed upon convective heat transfer, as in Ref. 4. In the model, the temperature is fixed at the bath temperature at the ends of the tube. The resultant temperature profiles can be expressed as a family of curves parameterized by a dimensionless quantity β, given by the expression

$$\beta = fL\ (\rho v C p)^3\ (\mu \Delta P)^2 \tag{8}$$

where f is the inverse of the Gorter-Mellink conduction parameter, v is the flow velocity and L is the flow length. Our data are shown in Fig. 2 along with straight line fits with slopes given by values for μ from Ref. 5. Offsets have been adjusted for best fits to the data; they are consistent with typical calibration shifts for germanium thermometers. The slopes are generally in good agreement with the experimental data. The points with pressure drops near 20 torr have β values below 20 and would be expected on the basis of the model to lie below the lines, with ΔT

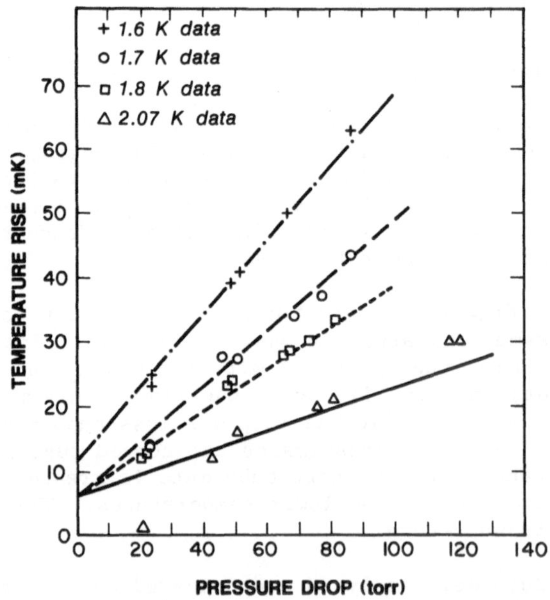

Fig. 2. Temperature rise in He II flow in tubing for various temperatures.

values near those of the intercepts. Further experiments are required to resolve this discrepancy.

PRESSURE DROP IN VALVES

Turbulent flow pressure drop measurements in the He II temperature range were performed for two valves: a nominal 4.3 mm orifice valve with in-line ports, and a nominal 7.1 mm orifice valve with inlet and outlet ports at right angles. The smaller valve was a Hoke, Inc. no. 411M2B brass bellows valve with a quoted flow factor C_v of 0.35. The larger valve was a Nupro Co. no. SS-6UAKTW stainless steel bellows valve with a quoted C_v of 1.36. The flow pressure drop ΔP is related to flow factor C_v by the relation

$$\Delta P = \frac{\dot{m}^2}{C_v^{\,2}} \frac{\rho(\text{water})}{\rho} \tag{9}$$

where $\dot{m}$ is in gallons per minute or by the formula

$$\Delta P = 1.412 \times 10^{-2} \frac{\dot{m}^2}{\rho C_v^{\,2}} \tag{10}$$

where Δp is in torr, $\dot{m}$ in gm/s and ρ in gm/cm^3 . Pressure drops were measured for the two valves in the fully open position during calibration runs for two venturi flowmeters. The valves were located downstream of the venturis in these tests. Pressure sensors were installed approximately 2 cm from the entrance ports of the valves. Inlet and outlet connections were made by soldering tubing into the sockets provided in the inlet and outlet ports with an inner diameter that matched the inner diameter of the ports, so that there was no diameter change from the inlet tubing to the ports. Pressure drops were measured as before by subtracting the absolute pressures measured upstream and downstream of the valve. Data were taken at temperatures of 1.6, 1.7, 1.8, and 2.1K for the Nupro valve and at 1.8K for the Hoke valve. Measured pressure drops at 1.8K for both valves are plotted in Fig. 3 against mass flow squared; the straight

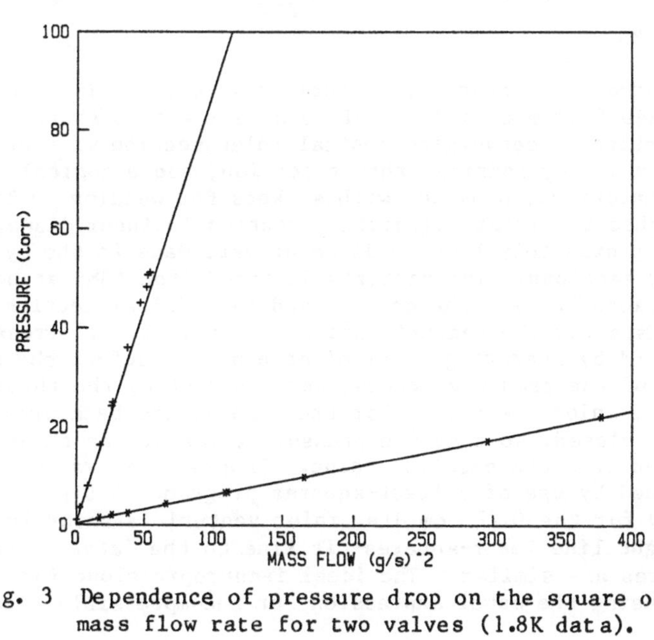

Fig. 3 Dependence of pressure drop on the square of the mass flow rate for two valves (1.8K data).

lines shown are least-squares fits to the data. In the case of the Hoke valve, the highest mass flow data deviate noticeably from the straight line fit to the lower mass flow data shown. For these high mass flow rates ($\dot{m} > 5$ gm/s), cavitation behavior was apparently taking place in the venturi upstream of the valve, in spite of the fact that the back pressure valve was adjusted to give positive throat pressure increments after flow began, and the pressure traces for taps above and below the valve exhibited instabilities. For this reason we restricted the data to mass flow values below 5 gm/s for this valve in determining the slope of the ΔP vs $\dot{m}^2$ curve for comparison with the manufacturer's C_v valve. In the case of the larger Nupro valve, the valve pressure drop behavior when the loop was back pressured was stable and repeatable for all points in the mass flow range (0-20 gm/s), and the points all lie on a straight line.

Least squares slopes from the Nupro valve data at the four temperatures were converted to C_v values according to Eq. 10. The average of the values for the four temperatures was 1.295 with a deviation of ± 0.3%, compared to the manufacturer's quoted value of 1.36. The data for the Hoke valve yielded a value of 0.334 for C_v, compared to the manufacturer's value of 0.35. in both cases, agreement is better than 5% which implies that pressure drops calculated with Eq. 10 and the manufacturer's C_v values will be accurate to 10%. Moreover, since the deviation of both valves from Eq. 9 was nearly the same and of the same sign, a correlation for He II of the same form as Eq. 9, but with a modified C_v can very tentatively be taken to be

$$\Delta P = 1.412 \times 10^{-2} \frac{\dot{m}^2}{\rho C_v'^2} \qquad (11)$$

with $C_v' = 0.95\ C_v$, for Reynolds numbers (calculated on the basis of nominal orifice diameter and normal component viscosity) in the range $1 \times 10^5 - 3 \times 10^6$. Clearly, a much larger data set is required to confirm or modify Eq. 10. Furthermore, Eqs. 9 and 10 become invalid if cavitation takes place in the valve. Some evidence for such behavior was observed at very high flow rates, but since no pressure signal was measured at the valve orifice, the conclusion must be regarded as tentative.

VENTURI FLOWMETERS

We have tested two venturi flowmeters to date in He II. The venturi elements were made from machined stainless steel, with a short cylindrical inlet section, a converging conical inlet section with an included angle of 20°, a short cylindrical throat section, and a conical divergent section. The venturi was provided with sockets for welding in line with tubing that matched the inlet cylindrical section in inner diameter. Pressure taps approximately 1 mm in diameter were made in the cylindrical inlet and throat sections. The venturis deviated from ASME standards in that the transitions between the conical and cylindrical sections were not radiused. Data for the venturis are given in Table I. Pressure drop data were obtained by averaging a set of data points before the end of the flat region of the pressure traces, and subtracting the throat pressure from the inlet pressure. For these data, the back pressure valve was partly closed, so that the pressure trace at the throat of the venturi increased from the baseline value. Slopes of ΔP vs. $\dot{m}^2$ plots were then obtained by use of a least-squares program. A typical data plot (1.8K data) for the 0.775 cm dia. inlet venturi is shown in Fig. 4 along the straight line least-squares-fit line to the data. Data for the other temperatures are similar. The ideal isentropic slope for the venturis was calculated using the usual expression for incompressible flow:

Table I. Venturi Data

Inlet dia. (cm.)	β	T (K)	Calc. Isentropic $\Delta P/\dot{m}^2$	Meas. $\Delta P/\dot{m}^2$	Discharge coeff.	Unrecovered Pressure Drop Ratio
			(torr-s^2/gm^2)			
0.775	0.497	1.6	0.1795	0.1883	0.976	0.159
		1.7	0.1794	0.1880	0.977	0.161
		1.8	0.1792	0.1881	0.976	0.164
		2.09	0.1787	0.1882	0.974	0.162
0.457	0.495	1.8	1.506	1.522	0.995	0.194

$$\Delta P/\dot{m}^2 = \frac{8}{\rho\pi^2 D^4}\left[1/\beta^4 - 1\right] \tag{12}$$

where D is the inlet diameter and β is the throat-to-inlet diameter ratio. The discharge coefficient F_D given in Table I is defined by the relation

$$F_D = \left[(\Delta P/\dot{m}^2)_{isentropic}/(\Delta P/\dot{m}^2)_{measured}\right]^{1/2} \tag{13}$$

Unrecovered pressure drop for the two venturis was also measured by means of pressure taps located downstream; the unrecovered pressure drop ratio is defined as the ratio of the difference of inlet and downstream pressures to the difference of inlet and throat pressures. The values

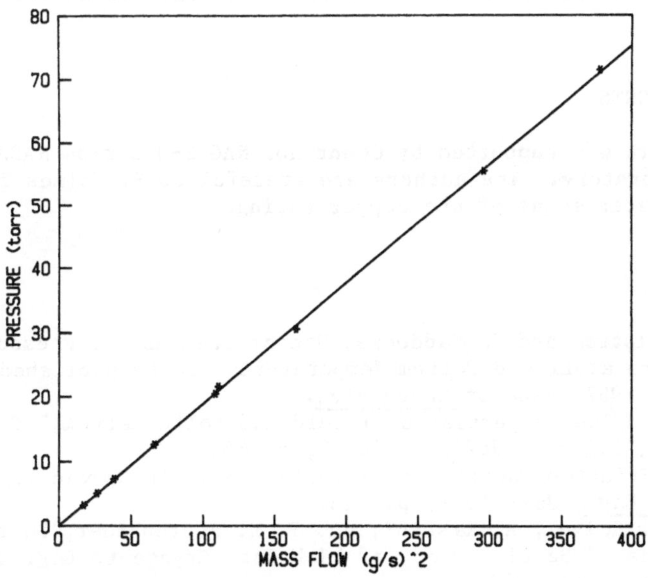

Fig. 4 Dependence of differential inlet-throat pressure upon mass flow rate squared for the 0.775 cm. inlet dia. venturi at 1.8K.

are larger than expected for standard venturis at our high Reynolds numbers and may be the result of the sharp transitions between cylindrical and conical surfaces in the venturis. No systematic dependence on temperature, within experimental error, can be seen in the parameters in Table I.

Pressure data were also taken for helium flow in the 0.775 cm inlet venturi at 1.6K and 2.09K with the back-pressure valve open. The pressures at the venturi throat decreased from the baseline value in these traces. The initial ΔP points lie on a straight line with the same slope as for the back-pressured data when plotted against $\dot{m}^2$. The absolute throat pressure was below saturation pressure before cavitation showing that the helium was in a metastable superheated state. Cavitation was also observed with the smaller venturi, when the back-pressure valve was open. In fact, at very high flow rates, cavitation may have been occuring in the smaller venturi even with the back-pressure valve partly closed, so that the throat pressure increased from the baseline value. A more complete description of the cavitation data for these venturis is included in a paper to be presented at the Space Cryogenics Workshop, Madison, Wisconsin, June 21-23, 1987 (proceedings to be published in Cryogenics).

CONCLUSIONS

Data from the experiments described in this paper, together with results of other investigators, show that the pressure drop behavior of He II in tubing, valves and venturis in highly turbulent flow is described well by classical flow correlations, and that peculiarly superfluid flow phenomena do not manifest themselves, provided that the flow is everywhere single phase. Measurement of the temperature rise in constant diameter tubing is in good agreement with the predictions of thermodynamics for isenthalpic flow. Data for He II flow in venturis indicate that they are an effective mass flow measurement approach for He II, provided the back pressure at the venturi is high enough to prevent cavitation. Accurately machined venturis can be used without calibration for mass flow measurements in He II to an accuracy of at least two percent.

ACKNOWLEDGEMENTS

This work was supported by Grant no. NAG 2-358 from NASA Ames Research Laboratory. The authors are grateful to S. Holmes for making the profilometer scans of the copper tubing.

References

1. P. L. Walstrom and J. Maddocks, Use of Siemens KPY Pressure Sensors at Liquid Helium Temperatures, to be published in the Sept. 1987 issue of Cryogenics.
2. J. Wilks, "The Properties of Liquid and Solid Helium," Clarendon Press, Oxford (1967), Table A5, p. 669.
3. H. Ito, Friction Factors for Turbulent Flow in Curved Pipes, J. Basic Eng., June 1959, p. 123.
4. S. W. Van Sciver, A. Kashani, and J. M. Pfotenhauer,"Technical Applications of He II," Proc. of 11th Int. Cryogenic Eng. Conf., Berlin, April 1986.
5. B. J. Huang, Joule-Thomson Effect in Liquid He II, Cryogenics, Aug./Sept. 1986, vol. 26, p. 475.

THE DESIGN OF FOUNTAIN EFFECT PUMPS

S.W.K. Yuan and T.C. Nast

Lockheed Research & Development Division
Palo Alto, California

ABSTRACT

The potential for in-flight replenishment of liquid helium for future systems such as the Space Infrared Telescope Facility has triggered vast interests in the fountain effect pumps (FEP's). The purpose of this paper is to correlate the FEP data in literatures under a common frame of reference. An empirical equation is proposed that can predict the performance of both the heater activated pumps and the cooler activated devices. According to the present findings, there is a throughput factor associated with each porous material. This throughput factor is inversely proportional to the square root of the Darcy permeability (κ). The proposed equation and the correlation of the throughput factor is very useful for the design of FEP's. For a given porous medium used for FEP, one can easily predict its performance if the permeability of that material is measured.

INTRODUCTION

The transfer of liquid helium in superleaks has been studied extensively[1-4]. Unfortunately, most of these early studies emphasized on the transport of He II in capillaries but not the superleak itself. Valuable information like the temperature difference across the superleak and/or the permeability of such porous media was often omitted in these literature.

One of the earliest investigations on the FEP was by Elsner et al.[5-8]. They were able to describe the transport of He II in a ceramic plug by a quarter law proposed by Vote et al.[9] for the isothermal gravity flow of He II in capillaries (see Ref. 6). The same approach was adopted by Yuan et al.[10] later on.

Recently, Kasthurirengan and Schotte et al.[11] have classified the porous media used for FEP into three groups, namely "Superleaks", "Good Transfer Filters" and normal "Porous Plugs". They were successful in predicting the "Good Transfer Filters" by the Gorter-Mellink equation.

An important issue in the prediction of the FEP results is to note the difference between the finite mass flow mode (e.g FEP's) and zero net mass flow mode (e.g. original Gorter-Mellink equation[12]). Van der Heijden et al.[13] have reported entirely different temperature dependency of the parameter A_{GM} for superfluid flow and zero net mass transport. Moreover, it is important to realize the influence of pore size on throughput transfer. Recent VLPS results show that the transport rate is strongly dependent on the Darcy permeability[14]. It is the purpose of this paper to compare the FEP data in a common frame of reference and to investigate their pore size dependency.

EQUATIONS FOR FOUNTAIN EFFECT PUMPS

Vote et al.[9] proposed the use of power law fitting to describe the isothermal gravity flow of He II through capillaries and narrow slits. The same method was used by Frederking et al.[6] and Yuan et al.[10] to correlate FEP data using a quarter law. According to them, the volumetric flow rate across an FEP can be written as

$$V=vA=A[(\rho_s/\rho)\eta_n/(c_4\rho\kappa^{1/2})]\{\rho\nabla P_T\kappa^{3/2}/\eta_n\}^{1/4} \qquad (1)$$

where ρ and η are the density and viscosity of He II, subscripts n and s stand for normal and superfluid respectively. κ is the Darcy permeability and ∇P_T is the fountain effect pressure ($\rho S\nabla T$). A is the total cross-sectional area of the FEP. The constant c_4 is a function of temperature.

It is found in the present work that the temperature dependency of c_4 can be eliminated by the introduction of a density ratio $(\rho_s/\rho_n)^{1/4}$ in Equation (1). In doing so, the effect of pore size on the FEP transport can be observed clearly. The resulted equation is

$$V=A[(\rho_s/\rho)K_4/\kappa^{1/2}]\{(\rho_s/\rho_n)\eta_n{}^2 S\nabla T\kappa^{3/2}/\rho^2\}^{1/4} \qquad (2)$$

K_4 in the preceding equation is a pure function of the pore size or Darcy permeability. Equation (2) can also be rewritten in the following dimensionless form

$$N_{Re}=K_4(\rho_s/\rho)[(\rho_s/\rho_n)N_{\nabla T}]^{1/4} \qquad (3)$$

with the modified Reynolds number $N_{Re}=\rho v\kappa^{1/2}/\eta_n$ and the dimensionless driving force number $N_{\nabla T}=\rho\nabla P_T\kappa^{3/2}/\eta_n{}^2$

The energy which needs to be supplied to a heater activated FEP or removed from a cooler activated pump is[15]

$$\dot{Q} = \rho V S_H T_H \qquad (4)$$

Subscript H stands for the downstream or hot end of the pump. Experimental results[4,16] indicate that Equation (4) is only valid at small velocities. For large flow rates substantial deviation was observed due to vortex shedding.

The appropriate surface area of the heater (for heater activated pumps) can be estimated through the Kapitza equation

$$A_k = \dot{Q}/\alpha\sigma_{BB}[(T+\Delta T_k)^4 - T^4] \qquad (5)$$

where α and σ_{BB} are the overall empirical transmission coefficient and the black body phonon emissivity respectively. The Kapitza temperature difference is denoted by ΔT_k.

RESULTS AND DISCUSSION

Figure 1 shows the correlation of the cooler activated FEP data of Frederking et al.[6] by Equation (2). The porous material used for this pump was a 1 μm ceramic plug (Al-Si). A single throughput factor of K_4=2.15 was used to fit all the data at various bath temperatures.

Figures 2-5 compare the proposed equation with the data of Kasthurirengan and Schotte et al.[11]. The materials used for the FEP's include Al_2O_3, CeO_2, Ni and Cr. These data also represent cooler activated FEP results. At large ΔT's, integration of Equation (2) might be necessary as indicated by the deviation of the experimental data from the power law.

The heater activated FEP results of Yuan et al.[10] are shown in Figure 6. This pump was made of a 2 μm Mott stainless steel plug. The downstream heater was constructed of a teflon insulated constantan wire (3 mil diameter) with a resistance of about 100 ohms. Various runs correspond to different heater powers.

The throughput factor K_4 is plotted as a function of the Darcy permeability (of the porous media) in Figure 7. It is quite obvious that K_4 is a strong function of the pore size, with a finer plug being a more efficient pump. However this

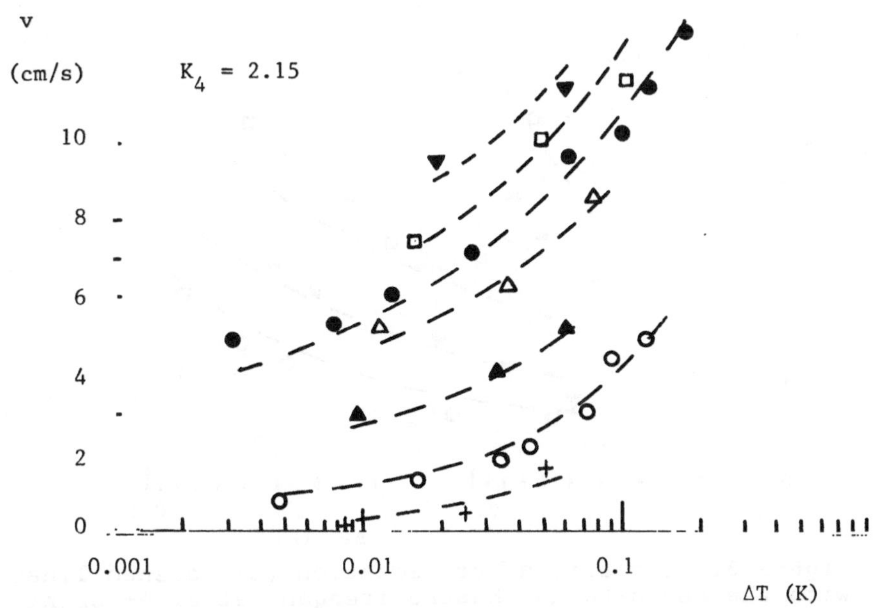

Figure 1. Correlation of Equation (2) (dashed lines) with the cooler activated FEP data of Frederking et al.[6]. Downstream vapor pressure, P_1(mb): ▼,15.2; □,19.9; ●,25.4; △,30.0; ▲,39.4; ○,46.7; +,50.0.

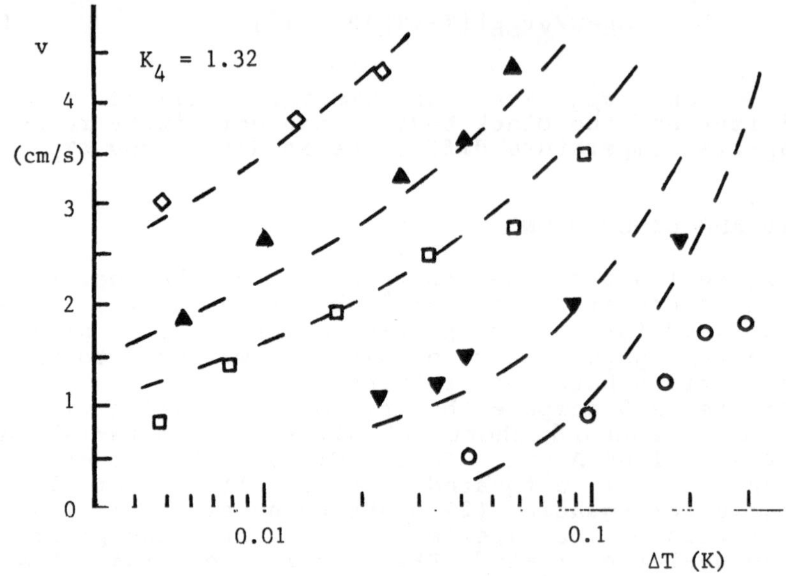

Figure 2. Comparison of Equation (2) (dashed lines) with the FEP data of Kasthurirengan et al.[11] using an Al_2O_3 (O(2)) plug. Downstream vapor pressure, P_1(mb): ◇,11.5; ▲,24.5; □,31.5; ▼,43.4; ○,49.5.

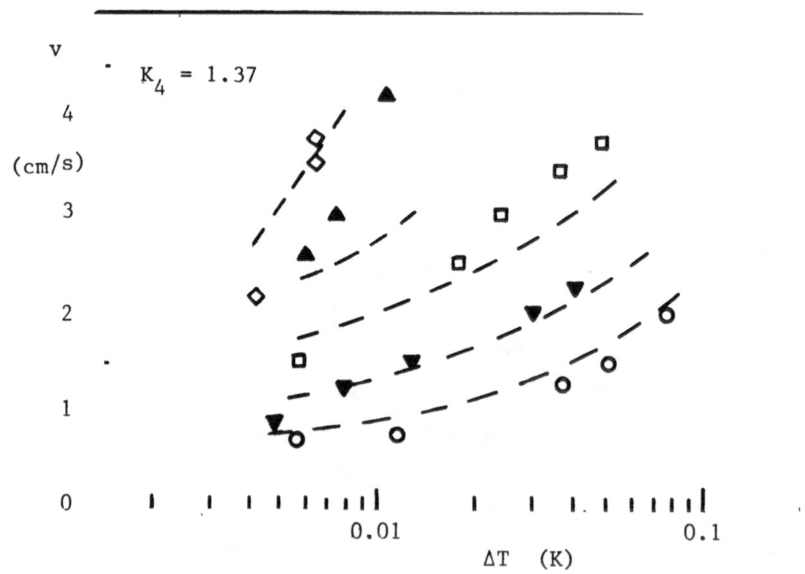

Figure 3. Comparison of Equation (2) (dashed lines) with the FEP data of Kasthurirengan et al.[11] using a CeO_2 plug. Downstream vapor pressure, P_1(mb): ◇,12.5; ▲,20.5; □,29.5; ▼,37.5; ○,42.5.

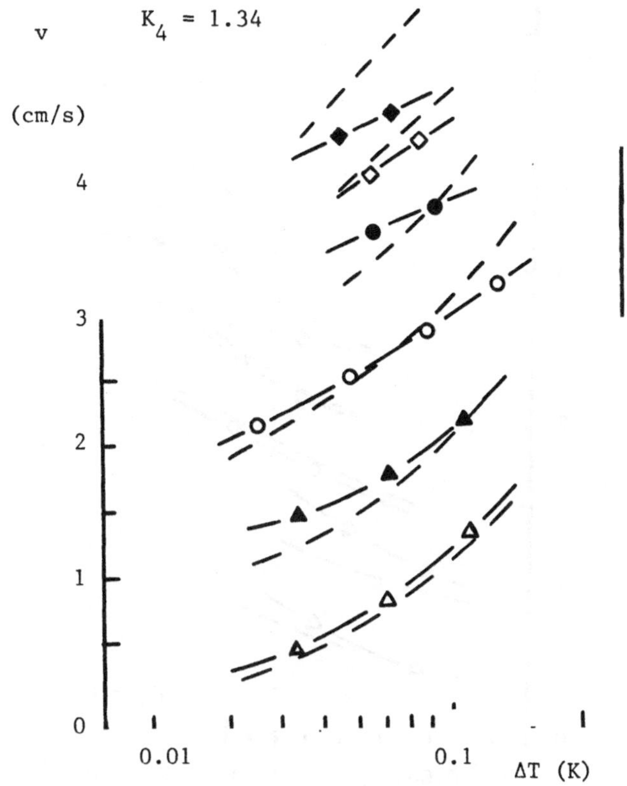

Figure 4. Comparison of Equation (2) (dashed lines) with the FEP data of Kathurirengan et al.[11] using a Ni plug. Downstream vapor pressure, P_1(mb): -◆-,11.5; -◇-,17.5;-●-,24.5;-○-,31.5;-▲-,40.5;-△-,47.5.

is only true up to a certain limit. If the permeability is too low for the transport of He II, the throughput factor should drop sharply. The so-called "Superleak" plug (Al_2O_3, average pore size 0.28 μm, κ=6.46x10^{-12}cm^2) in Ref. 11 have this characteristic. Looking at Figure 7, one can see that plugs with $\kappa \sim 1\times10^{-11}$cm^2 probably have the optimal pore size for the maximum throughput of He II.

It should be mentioned that attempts have been made in fitting the above FEP data by a 1/3 power law also, with

$$V=A(\rho_s/\rho)K_3[(\rho_s/\rho_n)S\nabla T\eta_n/\rho]^{1/3} \quad (6)$$

The throughput factor K_3 shows similar pore-size dependency as K_4. While the quarter law fit the original data better (in v vs ΔT), the 1/3 power equation tends to result in less data scatter when the throughput factor (K_3) is plotted against the Darcy permeability.

Lab size fountain effect pumps have demonstrated flow rates of ~100 litre/hr. Physically, one should be able to transport 1000 litre/hr of liquid helium if the appropriate surface area of the porous media (as calculated from Equation (2)) is used and the heater is efficient enough to sustain the required temperature difference across the pump.

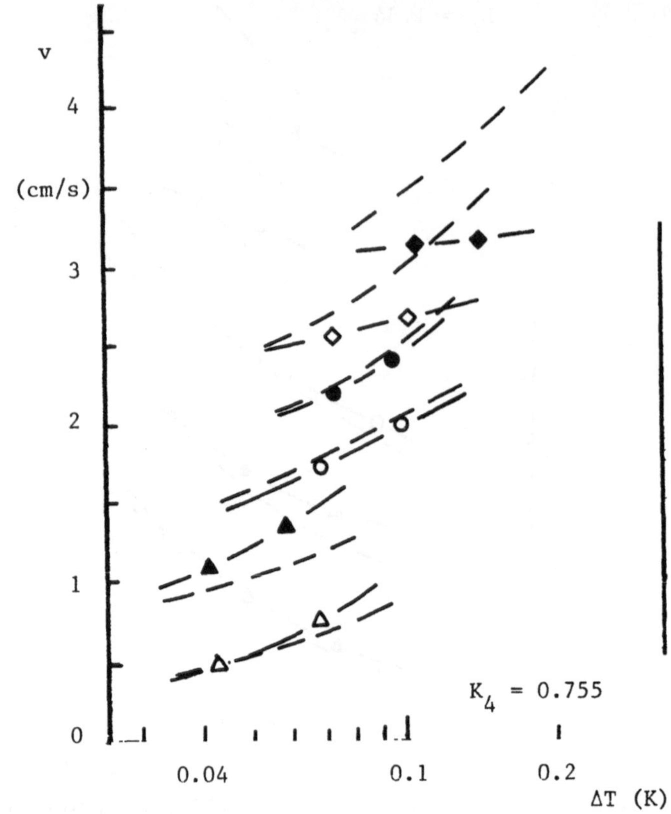

Figure 5. Comparison of Equation (2) (dashed lines) with the FEP data of Kasthurirengan et al.[11] using a Cr plug. Downstream vapor pressure, P_1(mb): -◆-, 11.5; -◇-, 17.5; -●-, 24.5; -O-, 31.5; -▲-, 40.5; -△-, 47.5.

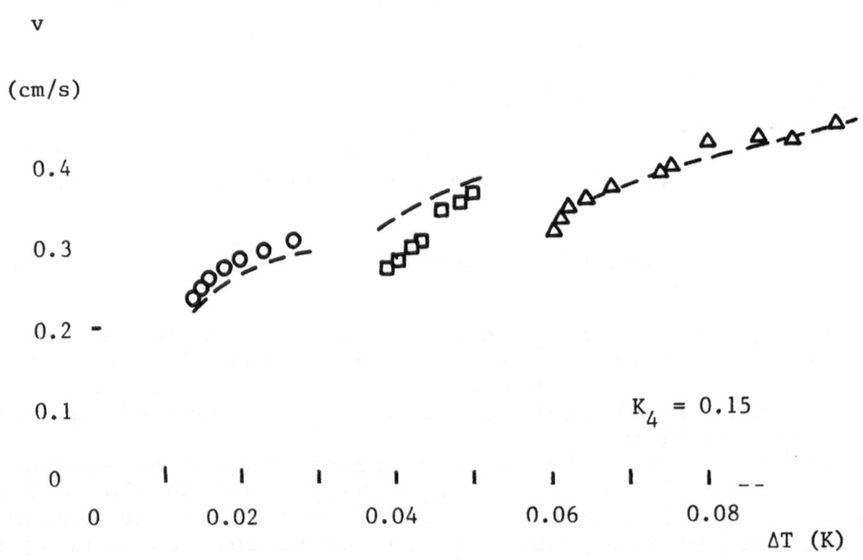

Figure 6. Correlation of Equation (2) (dashed lines) with the FEP data of Yuan et al.[10] using a 2 μm Mott stainless steel plug: O, Run 1; □, Run 2; △, Run 3.

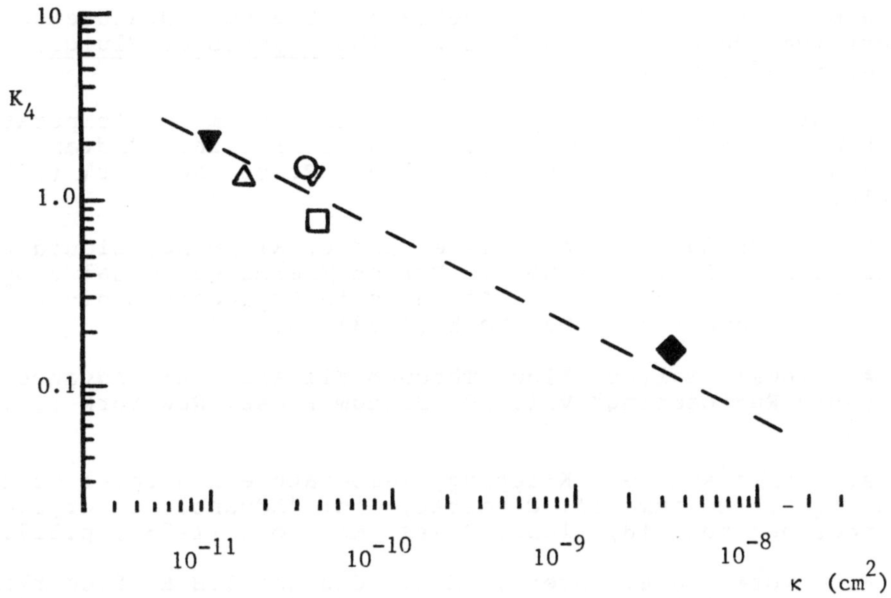

Figure 7. Throughput factor as a function of Darcy permeability. Frederking et al.[6]: ▼,(Al-Si); Kasthurirengan et al.[11]: △,(Al_2O_3); ○,(CeO_2); ▽,(Ni); □,(Cr); Yuan et al.[10]: ◆,(Stainless Steel).

CONCLUSIONS

Most of the FEP data in the literature to date can be predicted by the quarter law (Equation (2)) proposed in this paper. The throughput factor of the above equation was found to be a unique function of the Darcy permeability,

$$K_4 \sim 6.32 \times 10^{-6} / \kappa^{1/2}$$

The present findings are extremely useful for the design of fountain effect pumps. For a given porous plug, its performance as FEP can be predicted if the permeability of the material is measured. For example, if one uses a 1 μm Al-Si ceramic plug (with $\kappa = 1 \times 10^{-11} cm^2$) instead of a 2 μm Mott stainless steel plug, one can reduce the total cross-sectional area of the pump by about 14 times for the same flow rate!

REFERENCES

1. F.A. Staas, K.W. Taconis and W.M. Van Alphen, Experiments on Laminar and Turbulent Flow of He II in Wide Capillaries, Physica 27:893 (1961).

2. F.A. Staas and K.W. Taconis, Superfluid Flow in Wide Capillaries, Physica 27:924 (1961).

3. J.F. Olijhoek, W.M. Van Alphen, R. De Bruyn Ouboter and K.W. Taconis, Adiabatic Superfluid Flow at Supercritical Velocities Through Capillaries, Physica 35:483 (1967).

4. H.E. Corke and A.F. Hildebrandt, Pressure Measurement in Subcritical Pure Superfluid Flow, The Physics of Fluids, Vol. 11, No.3:465 (1968).

5. A. Elsner and G. Klipping, Control System for Temperature and Liquid Level Between 4.2 and 1 K, in: "Advances in Cryogenic Engineering" Vol. 14, Plenum Press, New York (1969), p. 416.

6. T.H.K. Frederking, A. Elsner and G. Klipping, Liquid Flow Rates of Superfluid Helium II During Thermomechanical Pumping Through Porous Media, in: "Advances in Cryogenic Engineering" Vol. 18, Plenum Press, New York (1973), p. 132.

7. A. Elsner, Helium Flow Through Filters, in: "Advances in Cryogenic Engineering" Vol. 18, Plenum Press, New York (1973), p. 141.

8. A. Elsner and G. Klipping, Temperature and Level Control in a Liquid Helium II Cryostat, in: "Advances in Cryogenic Engineering" Vol. 18, Plenum Press, New York (1973), p.317.

9. F.C. Vote, J.E. Myers, H.B. Chu and T.H.K. Frederking, Near-Isothermal Dissipative of Liquid He II at Supercritical Velocities, in: "Advances in Cryogenic Engineering" Vol. 16, Plenum Press, New York (1971), p. 393.

10. S.W.K. Yuan, T.H.K. Frederking and R.M. Carandang, Thermodynamic Performance Evaluation of Liquid Pumps Based on the Fountain Effect of Superfluid Liquid Helium II, in: "Proc, Tenth Intl. Cryo. Engr. Conf." Butterworth, Guildford, UK (1984), p. 301.

11. S. Kasthurirengan, U. Schotte, H.D. Denner, Z. Szucs and G. Klipping, Use of Narrow-Pore Filters in He II Transfer Devices-New Results, Cryogenics, 25:518 (1985).

12. C.J. Gorter and J.H. Mellink, On the Irreversible Process in Liquid Helium II, Physica XV, No.3-4:285 (1949).

13. G. Van der Heijden, J.J. Giezen and H.C. Kramers, Forces in the Flow of Liquid He II (II. Superfluid Flow), Physica, 62:566 (1972).

14. S.W.K. Yuan, The Non-Newtonian Heat and Mass Transport of He II in Porous Media Used for Vapor-Liquid Phase Separation, Ph.D. Thesis, U.C.L.A., (1985); S.W.K. Yuan and T.H.K. Frederking, Non-Linear Vapor-Liquid Phase Separation Including Microgravity Effects, Cryogenics, 27:27 (1987).

15. P. Kapitza, Heat Transfer and Superfluidity of Helium II, Journal of Physics (U.S.S.R.), 5:59 (1941).

16. R. Srinivasan and A. Hoffmann, Investigations on Cooling With Forced Flow of He II. Part 1, Cryogenics, 25:642 (1985).

OPERATING CHARACTERISTICS OF ISOCALORIC FOUNTAIN-EFFECT PUMPS

Peter Kittel

NASA Ames Research Center
Moffett Field, California

ABSTRACT

The governing equations of thermomechanical (fountain-effect) pumps are usually given for pumps operating at a constant temperature difference. These are the thermomechanical and mechanocaloric effects in which the pressure head and mass flow are independent of each other. Here, these equations are recast for a pump operating at a constant heat input (isocaloric). This form more closely represents how such pumps are likely to be used. Under these conditions, the pressure head and mass flow are shown to be related. For ideal pumps, the head and flow are related by a universal curve. For real pumps (those that have normal fluid leakage), a family of curves is developed. These curves approach the curve for an ideal pump at high flow rates. The isocaloric equations are also extended to multistage pumps.

INTRODUCTION

Discussions on fountain-effect pumps (FEPs) usually make the implicit assumption that the pumps operate at a constant differential temperature; i.e., that the temperature at the inlet and outlet are held constant. This assumption greatly simplifies the analyses of pump performance, since under these conditions the pressure head developed by the pump and the flow through the pump are independent of each other. The basic equations governing an ideal FEP (a constant chemical potential device) are the thermomechanical effect[1]:

$$P_1 - P_0 = \Delta P = \int_{T_0}^{T_1} \rho S \, dT \tag{1}$$

where P is the pressure, T is the temperature, ρ is the density, S is the entropy, and the subscripts 0 and 1 refer to the inlet and outlet, respectively; and the mechanocaloric effect[1]:

$$\dot{Q}_1 = \dot{m} S_1 T_1 \tag{2}$$

where $\dot{Q}$ is the heat input and $\dot{m}$ is the mass flow. These equations were derived on the assumption that only the superfluid component is flowing. When T_0 and T_1 are held constant, then ΔP and $\dot{m}$ are clearly independent of each other. This is also true if effects of the thermoconductivity through the plug are included:

$$\dot{Q}_1 = \dot{m}S_1T_1 + \int_{T_0}^{T_1} k\xi(A/\ell)dT \tag{3}$$

where k is the effective thermoconductivity, ξ is the porosity, A is the cross-sectional area, and ℓ is the effective flow path through the pump. Only when the critical velocity is exceeded within the pump does ΔP depend on $\dot{m}$.[2]

FEPs typically operate with temperature differences from a few milliKelvin to several hundred milliKelvin. While such temperatures can be routinely controlled, it is far simpler to operate at constant heat input (isocaloric, $\dot{Q}_1$ = constant). This paper develops the relationships between ΔP and $\dot{m}$ for pumps operating at constant $\dot{Q}$. The derivation will assume that the flow through the pump is always subcritical and that the bath temperature (T_0) is either constant or varies so slowly that its derivatives are insignificant. In addition it is assumed that ρ (the density), η (the viscosity), and C_μ/S (where C_μ is the specific heat at constant chemical potential, μ) are independent of temperature. These are all good assumptions for superfluid helium except for the viscosity near the lambda line.

IDEAL PUMPS

An ideal FEP is shown diagrammatically in Fig. 1. In ideal pumps only the superfluid component flows through the pump. Thus equations 1 and 2 govern its behavior. Heat is applied on the downstream side of a porous plug. This results in both a pressure increase and a mass flow. The heat required is given by (2). Taking the total derivative of (2) results in

$$d\dot{Q} = (\dot{m}S_1 + \dot{m}T_1\partial S/\partial T|_\mu)dT_1 + S_1T_1\, d\dot{m} \tag{4}$$

For the isocaloric case, $d\dot{Q} = 0$; thus

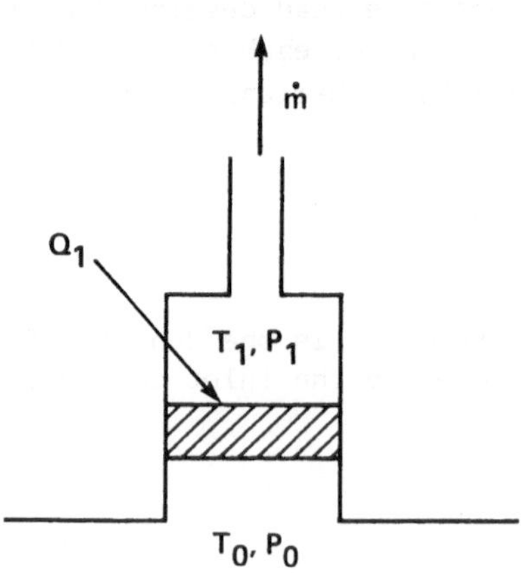

Fig. 1. Schematic of an FEP. (The nomenclature is explained in the text.)

$$(d\dot{m}/dT_1)_Q = -\dot{m}^2(S + C_\mu)/\dot{Q}_1 \tag{5}$$

The relationship between $\dot{m}$ and P is

$$(d\dot{m}/dP) = (d\dot{m}/dT_1)(dT_1/dP) \tag{6}$$

which can be evaluated with the aid of (1) and (5), resulting in

$$\Delta P(1 + C_\mu/S)/\rho\dot{Q}_1 = (1 - S_0T_0/S_1T_1)/\dot{m}$$

$$= 1/\dot{m} - 1/\dot{m}_0 \tag{7}$$

where $\dot{m}_0 = \dot{Q}_1/S_0T_0$ is the maximum possible flow rate. Equation 7 can be further simplified because $C_\mu/S \approx 5.6$ for superfluid helium[3]:

$$6.6\Delta P/\rho\dot{Q}_1 \approx 1/\dot{m} - 1/\dot{m}_0 \tag{7'}$$

This relationship between ΔP and $\dot{m}$ is shown as the $\dot{m}_0^* = \infty$ line in Fig. 2.

TWO-STAGE IDEAL PUMPS

A diagram of a two-stage pump is shown in Fig. 3. The operation of such a pump is discussed in Ref. 4. If both pumps are ideal, then (7) holds for both stages and we may write:

$$(P_1 - P_0)(1 + C_\mu/S)/\rho\dot{Q}_1 = (1 - S_0T_0/S_1T_1)/\dot{m} \tag{8a}$$

and

$$(P_2 - P_1)(1 + C_\mu/S)/\rho\dot{Q}_2 = (1 - S_0'T_0'/S_2T_2)/\dot{m} \tag{8b}$$

These may be combined to yield:

$$(P_2 - P_0)(1 + C_\mu/S) = (\rho/\dot{m})(\dot{Q}_1 + \dot{Q}_2) - \rho(S_0T_0 + S_0'T_0') \tag{9}$$

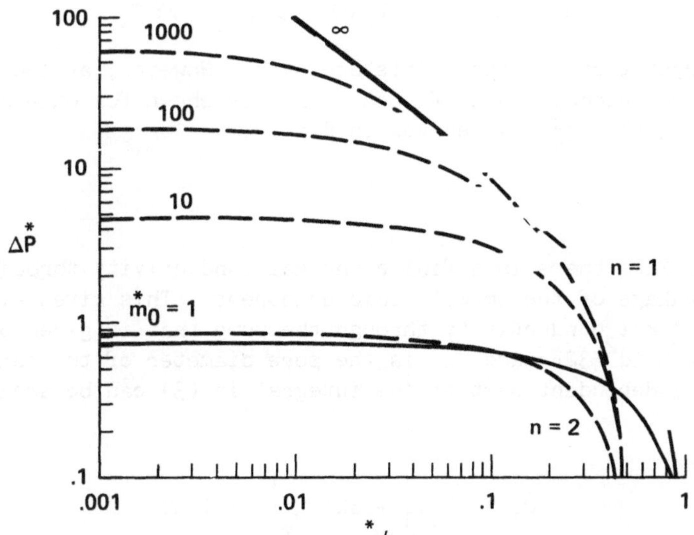

Fig. 2. ΔP vs $\dot{m}$ for one- and two-stage pumps.

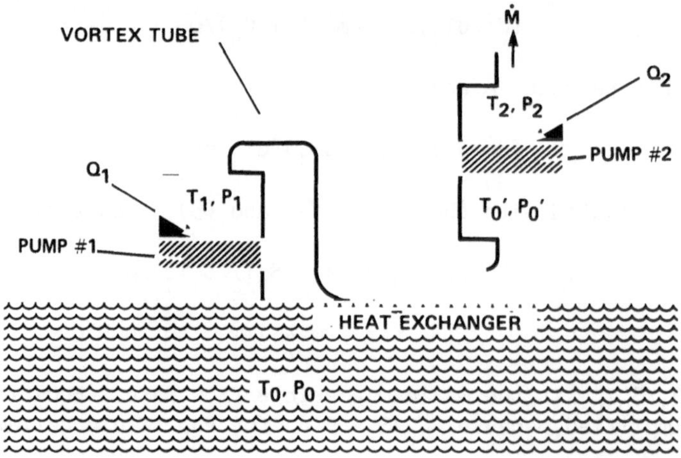

Fig. 3. Schematic of a two-stage FEP. The flow within the vortex tube is highly turbulent, effectively eliminating heat conduction between the heater, $\dot{Q}_1$ and the heat exchanger.

If the heat exchanger between the two stages has an effectiveness of 1, then $S_0'T_0' = S_0T_0$. Furthermore, (9) can be extended to the case of n stages and if the heat input into each stage is identical, then

$$(P_n - P_0)(1 + C_\mu/S) = \rho n\dot{Q}/\dot{m} - \rho n S_0 T_0 \tag{10}$$

Here the total heat input is $n\dot{Q}$. If a similar amount of heat were applied to a single-stage pump, then by (7) the pressure rise for the same flow rate would be

$$(P_n - P_0)(1 + C_\mu/S) = \rho n\dot{Q}/\dot{m} - \rho S_0 T_0 \tag{11}$$

which is larger than for the multistage pump. However, at low flow rates the difference becomes insignificant. This is shown for one- and two-stage pumps as the $m_0^* = \infty$ curves in Fig. 2.

REAL PUMPS

In real FEPs there is a finite thermal conductivity through the pump caused by leakage of the normal fluid component. This gives rise to an effective thermal conductivity through the pump that is given by (3), where $k = \rho^2 S^2 T d^2/32\eta$ and d is the pore diameter of the pump.[5] The temperature-independent part of the integral in (3) can be separated, yielding:

$$\dot{Q}_1 = \dot{m} S_1 T_1 + \rho k' \int_{T_0}^{T_1} S^2 T \, dT \tag{12}$$

where $k' = \rho \xi A d^2/32\eta\ell$. Equations (1) and (12) can be integrated and solved simultaneously to yield:

$$\Delta P^* = -m^* - 1 + (m^{*2} = 2m_0^* + 1)^{1/2} \tag{13}$$

where $\Delta P^* = \Delta P(1 + C_\mu/S)/\rho S_0 T_0$, $m^* = \dot{m}/\dot{m}'$, $m_0^* = \dot{m}_0/\dot{m}'$, and $\dot{m}' = \rho k' S_0 T_0/(1 + C_\mu/S)$. The quantity $\dot{m}'$ is the flow rate characteristic of the leakage. It has been used here to form the dimensionless flow m^* and dimensionless peak flow m_0^*. The quantity m_0^* is a constant that characterizes the pump. In deriving (13) we also made use of the relations $SdT = d(ST)/(1 + C_\mu/S)$ and $2S^2TdT = d(S^2T^2)/(1 + C_\mu/S)$ that are the result of C_μ/S being constant. For a multistage pump of identical stages, the total pressure can be shown to be

$$\Delta P_T^* = n\Delta P^* = n\{-m^* - 1 + [m^{*2} + (2m_0^*/n) + 1]^{1/2}\} \tag{14}$$

DISCUSSION

Equations 13 and 14 are plotted in Fig. 2 for one- and two-stage pumps for a variety of values of m_0^*. For ideal pumps, $\dot{m}' = 0$ and $m_0^* = \infty$. In this limit (13) and (14) reduce to (7) and (10), respectively. From Fig. 2 we see for ideal pumps at low flow rates that the pressure rise for multistage pumps is the same as for a single-stage pump. For real pumps, the multistage pump produces a greater head at low flow rates. In the $m^* = 0$ limit, $n\Delta P^* \approx (2nm_0^*)^{1/2}$. This is a result of the head per stage being smaller, making the total leakage less. At high flow rates the single-stage pump is always superior, always producing a greater head. In the limit of $\Delta P = 0$, the flow goes as $\dot{m} = \dot{m}_0/n$. This is the result of all of the heat going into the mechanocaloric effect.

In a previous work,[2] the leakage was discussed in terms of an attenuation coefficient:

$$\alpha = \dot{m}S_1T_1/\dot{Q}_1 \tag{15}$$

where $0 \le \alpha \le 1$. For a multistage pump, the corresponding definition would be

$$\alpha_n = \dot{m}S_1T_1/n\dot{Q}_1 \tag{15'}$$

Evaluating this results in

$$\alpha_n = (\Delta P^* + 1)\dot{m}/\dot{m}_0 \tag{16}$$

This expression reduces to $\alpha_n = 1/n$ for a case of no leakage ($\dot{m}' = 0$). This is the same result found above in the high flow limit ($\Delta P^* = 0$).

The hydrothermodynamic efficiency of an FEP is given by[4,6]

$$\varepsilon \approx \Delta P_T/\rho\dot{Q}_T \tag{17}$$

This can be evaluated using (14) and (15) to yield:

$$\begin{aligned}\varepsilon &= \alpha_n[n\Delta P^*/(\Delta P^* + 1)]/(1 + C_\mu/S)\\ &= n(\dot{m}/\dot{m}_0)\Delta P^*/(1 + C_\mu/S)\end{aligned} \tag{18}$$

Thus for a real pump, the efficiency goes to 0 for $\dot{m} = 0$ and for $\Delta P = 0$. For an ideal pump, (18) reduces to

$$\varepsilon = (1 - n\dot{m}/\dot{m}_0)/(1 + C_\mu/S) \tag{19}$$

In this case, $\varepsilon \leq 0.15$, as found earlier.[6] Also note from (18) and the definition of ΔP^* that $\varepsilon \propto T^{-6.6}$ and thus the efficiency increases dramatically with decreasing temperature.

The transfer effectiveness η_t is given by[4]

$$(1 - \eta_t) = \Delta m/\dot{m} \tag{20}$$

where $\Delta m = \dot{Q}/L$ is the mass that must be evaporated to keep the process at constant temperature and L is the latent heat. Evaluating (20) results in

$$(1 - \eta_t) = (1/\alpha_n)S_1T_1/L = (\dot{m}/\dot{m}_0)S_0T_0/L \tag{21}$$

As with the efficiency, the effectiveness increases dramatically with decreasing temperature.

SUMMARY

A simple model of the pump characteristics of a fountain-effect pump operating at constant power input has been developed. This model includes the effects of leakage as a single parameter, m_0^*. The model does not include the effects of exceeding the critical velocity (the onset of quantized vorticity). As a result of this model, several conclusions can be drawn:

1. The pressure head decreases monotonically with increasing flow rate.
2. The performance of a real pump is always less (lower pressure head, lower flow) than for an ideal pump.
3. Multistage pumps perform better (greater pressure head) at low flow rates; while single-stage pumps work better (greater flow) at high flow rates.
4. The performance of FEPs improves dramatically with decreasing temperature.

REFERENCES

1. J. Wilks, "The Properties of Liquid and Solid Helium," Clarendon Press, Oxford (1967).
2. P. Kittel, Losses in fountain effect pumps, in: "Proc. Eleventh Intl. Cryo. Engr. Conf.," Butterworth, Guildford, UK (1986), p. 317.
3. F. London, "Superfluids, Vol. 2," Wiley, New York (1954).
4. P. Kittel, Liquid helium pumps for in-orbit transfer, Cryogenics. 27:81 (1987).
5. V. Arp, Heat transport through helium II, Cryogenics. 10:96 (1970).
6. P. Kittel, Orbital resupply of liquid helium, J. Spacecraft Rockets. 23:391 (1986).

OPERATIONAL CHARACTERISTICS OF LOOPS WITH HELIUM II FLOW DRIVEN BY FOUNTAIN EFFECT PUMPS

A. Hofmann, A. Khalil* and H.P. Krämer

Kernforschungszentrum Karlsruhe
Institut für Technische Physik
D-7500 Karlsruhe, FRG

ABSTRACT

Forced flow cooling with helium II is being considered with respect to the design of superconducting magnets with internally cooled conductors. An experimental loop to test such coolant circuits has been investigated. Here, the flow of pressurized helium II is driven by so called self-sustained fountain effect pumps. Such pumps consist of porous plug superfilters where the temperature difference necessary to cause sufficiently high pressures is maintained by the flow of heated helium through appropriately arranged heat exchangers. The test loop has been operated at up to 8 W thermal load. This causes helium II flow rates of about 3 g/s, and the pressure difference ranges up to 0.3 bar. The measured operational characteristics of such loops are discussed. They compare well with theoretical predictions.

INTRODUCTION

Fountain effect pumps (FEP's) are considered to be useful devices to operate superconducting magnets with forced flow of superfluid helium[1]. Such pumps are most attractive when being operated in a so called self-sustained mode where the heat load removed from the cooling loop is used to drive the pump. An experimental loop of such cooling cycles has been set up and it has been tested at different operational conditions. The theoretical expectations are shown to be well confirmed by the experiments.

EXPERIMENTAL SET-UP

The existing apparatus[2] with a containment for pressurized helium II (He II_p), thermally anchored to a pool of saturated helium II (He II_s), has been modified as shown schematically in Fig. 1. The superfilter plug SF is connected at its cold end to the He II_p reservoir and at the upper end to the so called warm end heat exchanger (WHX) where the heat load Q supplied to the test section TS is transferred to the helium coming from the filter SF. The resultant temperature difference, $T_6 - T_5 > 0$, causes a fountain effect pressure Δp which drives a helium flow via the positions 6-7-1-2-3-4-5 through the loop.

The flow rate m is being measured with the acoustic flow meter FM[3] positioned between two heat exchangers HX1 and HX2. HX1 is to lower the helium temperature before re-entrance into the test section. But HX2 proves to be necessary to provide isothermal conditions at the flow meter. Otherwise, even the rather small conductive heat flow opposite to the mass flow would cause a non-permissible temperature gradient in the flow meter.

---

* On leave from Cairo University, Egypt

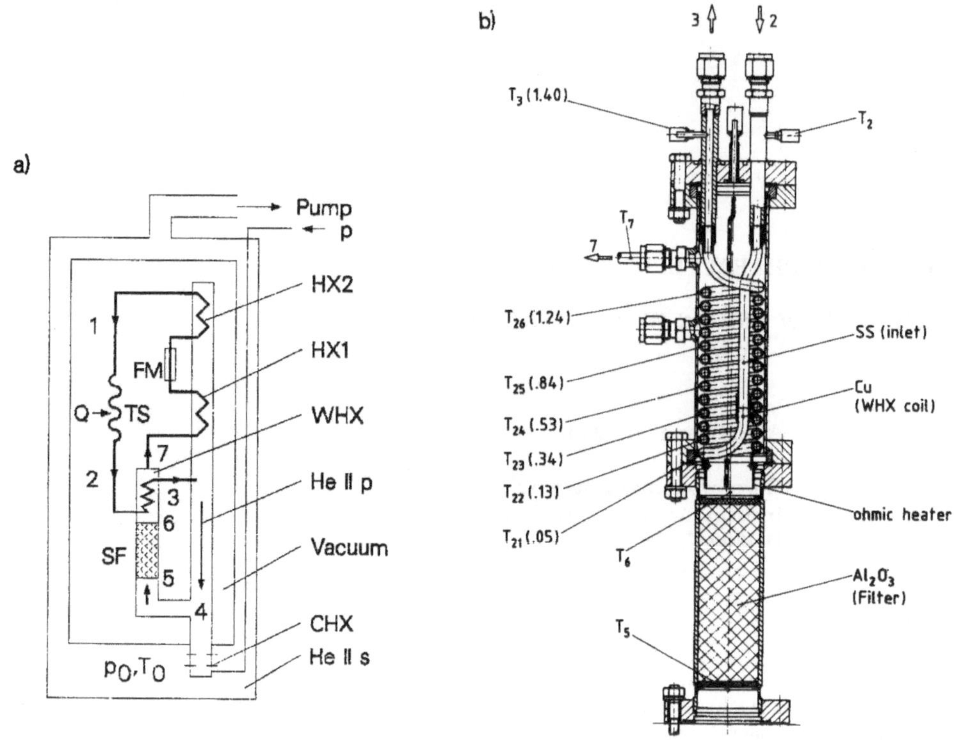

Fig. 1. a) Scheme of the experimental set-up and b) fountain effect pump with superfilter and warm end heat exchanger.

The warm end heat exchanger together with the superfilter is shown in Fig. 1b. This heat exchanger is composed of a 1.4 m long coiled copper tube of 4 and 6 mm inside and outside diameters. The coil with a mean diameter of 30 mm and an axial length of about 85 mm is enclosed in a stainless steel shell connected to the outlet of the filter. The helium heated in the test section TS is fed-in through a stainless steel tube extending close to the filter surface where the highest possible temperature is to be achieved. The numbers 2, 3 and 7 correspond to the respective numbers of Fig. 1a. The heat exchangers HX1 and HX2 are of similar design with about 1 m long coiled copper tubes each. The interconnecting tubes have also 4 mm inner diameter. The distances between positions with different temperature levels are:

WHX outlet (3) to $HeII_p$ reservoir	:	0.2 m
WHX outlet (7) to HX1 inlet	:	1.5 m
TS heater to WHX inlet (2)	:	1.5 m
TS heater to HX2 outlet	:	0.2 m

The overall length of the tubes of this so-called low impedance loop is about 7 m. A high impedance loop with about three times higher pressure drop for the same flow rate is achieved by replacement of a part of the test section by a smaller diameter tube[4]. The filter is made of Al_2O_3 powder with 1.5 μm nominal grain size filled at a length of 100 mm into a 34 mm i.d. stainless steel tube with 1 mm wall thickness. The fabrication procedure is essentially the same as for the 10 mm diameter filters used priviously[2,4]. The volumetric void factor is 50 % and the Darcy permeability measured with helium gas at room temperature is $K = 7.7 \cdot 10^{-15}$ m^2. Both surfaces are protected with a thin layer of cotton attached with a stainless steel grid. Temperature sensors are at both sides of that protection layer. The fact that they do not indicate a detectable temperature difference (less than 1 mK) even at the highest flow of about 3 g/s shows that the protection layers do not deteriorate the FEP performance. The fountain pressure is measured by two piezo resistive transducers (Siemens KPY 10) at both ends of the filter. The temperature sensors needed for the present discussions are at the positions 0 through 7. Further temperature readings within both, the filter and the warm end heat exchanger[5], are beyond the scope of the present paper and will be discussed separately.

EXPERIMENTAL RESULTS

During typical test runs the heater power $\dot{Q}$ is increased stepwise whilst the filter inlet temperature T_5 is kept constant as far as it is possible with respect to the constant temperature refrigeration power. At higher heat load, the cryostat is operated at full power of a 1000 m^3/h Roots blower.

The resultant flow rate m as shown in Fig. 2a first increases steeply with power but a saturational effect becomes apparent at about 7 W. One question to be investigated here is whether this saturation is caused by intrinsic losses within the filter or by limitation of the temperatures at its ends. Partially, the temperatures plotted in Fig. 2a can give an answer to

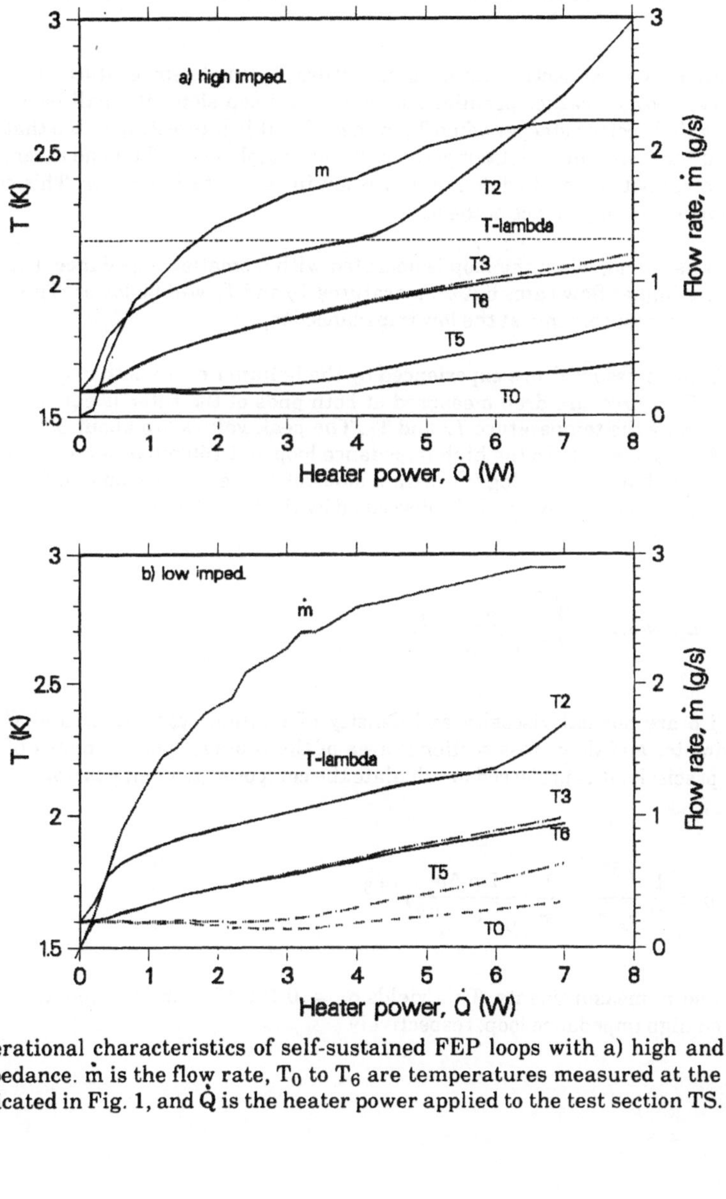

Fig. 2. Operational characteristics of self-sustained FEP loops with a) high and b) low flow impedance. $\dot{m}$ is the flow rate, T_0 to T_6 are temperatures measured at the positions as indicated in Fig. 1, and $\dot{Q}$ is the heater power applied to the test section TS.

that question. The flow rate m is driven by the fountain pressure caused by the temperatures T_5 and T_6 at the cold and warm end of the superfilter, respectively. In an ideal system T_5 should stay constant and the temperature difference T_6 - T_5 should grow with the heat load. In the actual system, T_5 is drifting to higher temperatures, hence the difference T_6 - T_5 does not increase as it should be expected. This may cause the saturational behaviour of the flow rate curve. The increase of T_5 is caused by malfunction of the heat transfer from the pressurized to the saturated pool with temperature T_0. The cryostat which has originally been designed for other applications, is not suited optimally for the installation of this FEP-loop. The cold end heat exchanger, CHX, is undersized for the actual heat load and additionally the filter inlet where most of the applied heat is rejected to the pressurized pool is too far from the heat exchanger. In order to get an analysis of the FEP loop which is independent of this cold end heat exchanger performance, the filter inlet temperature T_5 is taken as reference. It can be kept constant up to a certain level of the heater power, i.e. 2 W at 1.6 K, 7 W at 1.7 K and about 9 W at 1.9 K filter inlet temperature.

The warm end heat exchanger WHX proves to be of better design. The temperature T_7 measured at the outlet of the warm end chamber proves to differ by less than 1 mK from T_6 measured at the filter surface. The coiled tube outlet temperature T_3 (also plotted in Fig. 2) proves to be in general very close to T_6. This terminal temperature difference T_3 - T_6 grows to about 0.03 K at 8 W. A more detailed analysis on this heat exchanger will be given in a separate paper[5].

The curve of the test section outlet temperature T_2 (inlet temperature of the coiled heat exchanger tube) shows a rather peculiar shape with a steep slope at small power, then a flat range which again becomes steeper when T_2 exceeds T_λ. It is interesting to see that the loop can be operated with test section outlet temperatures appreciably above T_λ. (The theoretical limit of $T_{2,max}$ = 3.5 K cannot be veryfied due to limitation in refrigeration power. This could be done with a still higher flow impedance of the loop).

On the other hand, when the loop is operated with a smaller impedance, the same heater power will cause higher flow rates but temperatures T_2 and T_6 will be lower. This is verified by Fig. 2b, the same measurements at the low impedance loop.

In Fig. 3, the pressure drops experienced by the helium flow in both loops are plotted over the flow rates. This pressure drop measured at both ends of the filter is equal to the fountain pressure effected by the temperature T_6 and T_5. The peak values are about 300 mbar pressure difference and 2 g/s flow rate in the high impedance loop and 150 mbar, 3 g/s at low impedance. A straight line fit of log (Δp) vs. log ($\dot{m}$) with a slope of 1.8 seems reasonable for both systems. This means that the pressure drop can be described by the turbulent flow correlation[6]

$$\Delta p = 0.184 \, \frac{\eta^{0.2}}{2\,\rho} \, \frac{L}{D^{1.2} F^{1.8}} \, \dot{m}^{1.8} \tag{1}$$

where η and ρ are normal viscosity and density of helium, resp., L, D and F are length, hydraulic diameter and flow cross sectional area of the tubings. The geometry of the loop not being known precisely, it is preferred to calculate the composed geometry factor

$$\alpha = \frac{1}{F} \, \frac{L^{5/9}}{D^{2/3}} = \frac{1}{\dot{m}} \left(\frac{2\,\rho\,\Delta p}{0.184\,\eta^{0.2}} \right)^{1/1.8} \tag{2}$$

from the Δp and m measurements. This yields $\alpha_1 = 0.190 \cdot 10^8 \ m^{-19/9}$ and $\alpha_2 = 0.345 \ m^{-19/9}$ for the low and high impedance loop, respectively.

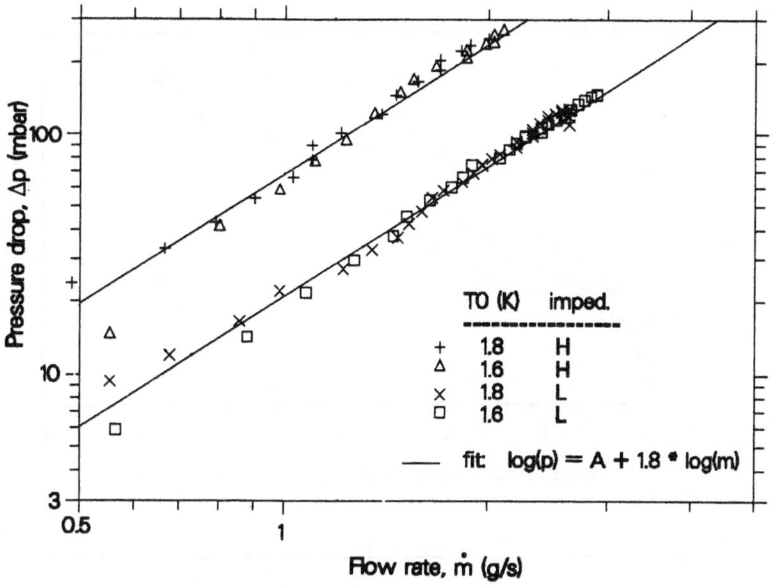

Fig. 3. Pressure drop vs. flow rate in high and low imp. loop.

THEORY AND DISCUSSION OF THE RESULTS

Predictions of the operational characteristics can be obtained as described in a previous paper[7] from the heat balance equations assuming ideal superleaks. In the test section where the heater power $\dot{Q}$ is applied, the internal energy of the helium is changed by

$$u_2 - u_1 = \dot{Q}/\dot{m} \tag{3}$$

In the warm end heat exchanger, WHX, the power

$$\dot{Q}_{23} = \dot{m}\,(h_2 - h_3) \tag{4}$$

is transferred to the helium coming with zero entropy from the (ideal) superfilter. This yields for steady state operation

$$h_2 - h_3 = s_6 T_6 \tag{5}$$

(h is the specific enthalpy). The fountain pressure

$$\Delta p_F = \int_{T_5}^{T_6} \rho\, s\, dT \tag{6}$$

is balanced by the frictional pressure drop given by (2). Together with the state equation of helium

$$p = \phi\,(\rho, T) \tag{7}$$

those equations are sufficient to calculate the operational characteristics such as flow rate, fountain pressure and temperatures at different positions of the loop as functions of the heater power. It is convenient to plot those quantities over the scaled power, $\alpha\dot{Q}$, instead of $\dot{Q}$ (α is the

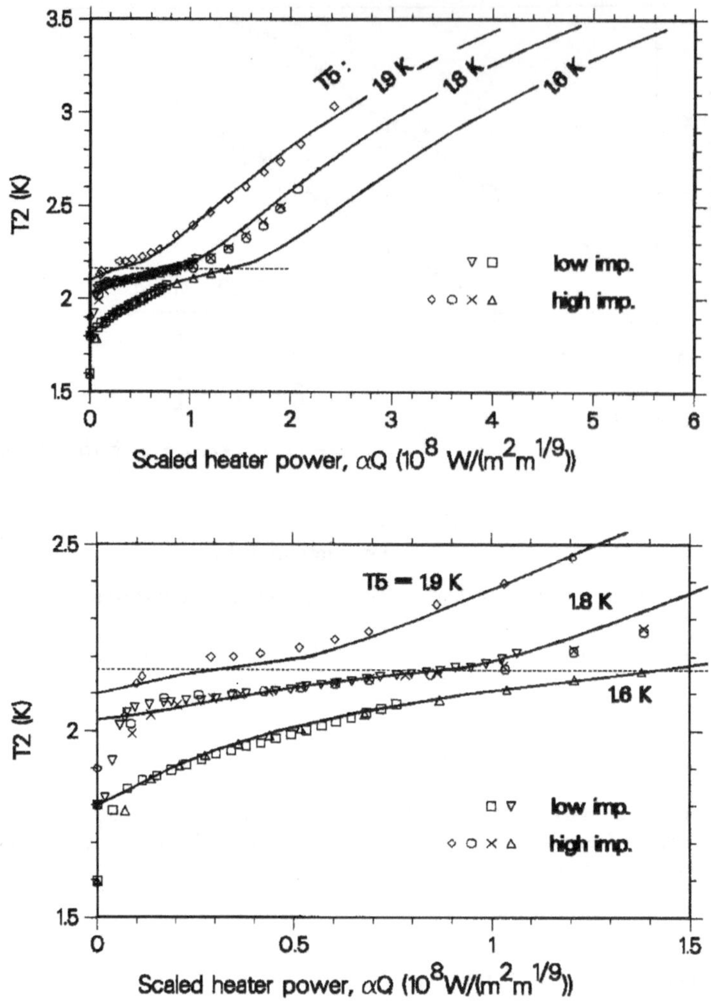

Fig. 4. Calculated and measured test section outlet temperature T_2 for low and high impedance loops plotted vs. scaled heater power. a) Full operational range, b) enlarged low power range.

geometry factor defined by (2)) whereby systems with different flow impedances can be correlated. The present calculations have been done with the helium properties code QEPT 86 of V.D. Arp[8] which allows continuous calculation from He I to He II. The results differ slightly from previous ones[7] where different codes for He I and He II were used. But the calculations with this new code are apparently more consistent. A comparison of measurements and theory is given in Fig. 4, a plot of the test section outlet temperature over the scaled heater power. In general there is a good agreement between measurements and calculations even at very different operational conditions with filter inlet temperatures 1.6 K $\leq T_5 \leq$ 1.9 K, TS outlet temperatures 1.8 $< T_2 \leq$ 2.65 K and different flow impedances. This shows that the assumption of perfect superleak is justified. In this plot only those measurements have been evaluated where T_5 is constant within a 50 mK margin.

There is however a discrepancy at very small heater powers. The theory predicts for T_2 a sudden rise from the initial equilibrium temperature, T_5, to an onset temperatures which is f.i. 2.03 K for T_5 = 1.8 K and 1.8 K for T_5 = 1.6 K. The measured temperature grow steadily but with rather steep slope from T_5 to the flat level range. This behaviour can be explained by the fact that the thermal conductivity of the helium has been neglected in the theoretical model. Obviously, this is allowed for a loop with sufficiently high flow rate where convective energy flow predominates. At small flow rate, however, the major fraction of the power applied to the test section may be removed by Gorter Mellink conduction and there will be a temperature gradient from the heater to the warm end heat exchanger. Thus the fluid temperature measurement depends on the position of the thermometer. The T_2 measurements results from two thermometer positions. For $T_2 < T_\lambda$ it is measured close to the heater (12 cm down stream). At high heater power temperatures close to the heater prove to fluctuate. Therefore, at temperatures $T_2 > T_\lambda$ where the conductivity is very small, T_2 is measured close to WHX at the position indicated in Fig. 1b. Another check of superfilter performance is done by comparison of the measured pressure drop Δp_F calculated from eq. 6 with the measured end temperatures T_5 and T_6 as inputs. Both quantities prove to agree within about 2 mbar even for the highest flow rate of 2.8 g/s. This shows again that within an accuray reasonable for technical applications the porous plug can be treated as an ideal superleak.

Apart from this low power range the agreement is surprisingly good. This means that this self-sustained fountain effect pump is being operated close to its theoretical optimum. There is no limitation of flow by intrinsic dissipation within the filter, the power applied to the test section is transferred with good efficiency to the filter outlet and obviously the filter has a sufficiently high impedance for the normal fluid component. Its flow rate from the high to the low pressure side of the filter is apparently much smaller than the superfluid flow $\dot{m}$ in opposite direction.

For more quantitative assessment the normal fluid backward flow, $\dot{m}_R$, is estimated from the permeability K of the filter. This yields

$$\dot{m}_R = \rho_n A \frac{K \Delta p}{\eta_n L_F}$$

where η_n is the normal viscosity, A and L_F are the helium cross sectional area and the length of the filter. The resultant value for Δp = 300 mbar is less than $\dot{m}_R$ = 0.1 g/s. This is indeed much smaller than the forward flow $\dot{m}$ = 2 g/s.

For the highest flow rate of 3 g/s the mass flux is 8.8 kg/(s m^2) and the mean corresponding flow velocity of v = 0.06 m/s is obviously below its critical value. In earlier experiments[4] the same type of filter but with 1 cm diameter was used. Here, a mass flux of about 21 kg/(sm^2) was achieved. This shows that the present filter has sufficient reserves. Both, its cross section and its length could be reduced if it is to be used for flow rates not greater than 3 g/s and for pressure heads of up to 300 mbar.

CONCLUSION

The experiments have shown that the coolant loop with a self-sustained fountain effect pump behaves largely as expected from calculations based on the assumption that the filter is an ideal superleak. The earlier studies[1,7] where such pumps are considered to be used advantageously in large magnet systems are supported by those experiments.

ACKNOWLEDGMENTS

The authors are much indebted V.D. Arp from NBS for the very useful Fortran code of helium II properties. They would also like to thank B. Vogeley and M. Süpfle for their skilled assistance in the experiments.

This work has been done within the frame of the European Fusion Energy Programme.

REFERENCES

1. A. Hofmann and A. Khalil, Considerations on magnet design based on forced flow of He II in internally cooled cables, in "Fusion Technology 1986", Vol. 2, Published for CEC by Pergamon Press Oxford, p. 1811.

2. R. Srinivasan and A. Hofmann, Investigations on cooling with forced flow of He II, Cryogenics 25 (1985), 641.

3. A. Hofmann and B. Vogeley, Acoustic flow meter for He I and He II application, Proc. ICEC 10, Butterworths, Guildford (1984), 448.

4. A. Hofmann, A. Khalil, H.P. Krämer, J.G. Weisend, R. Srinivasan and B. Vogeley, Investigations on fountain effect pumps for circulating pressurized helium II, in: "Proc. ICEC 11", Butterworths, Guildford, UK (1986), p. 312.

5. To be published.

6. W. Johnson and M.C. Jones, Adv. Cryog. Engineering 23 (1978), 363.

7. A. Hofmann, Thermomechanically driven helium II flow. An option for fusion magnets with internally cooled conductors?, Proc. ICEC 11, Butterworths, Guildford, UK (1986), p. 306.

8. V.D. Arp and K. Agatsuma, Equation of state for liquid helium from 1.4 to 4 K and asymptotic limits at the lamda line, Int. J. Thermophysics (1985), 6, 63.

PERFORMANCE TEST OF A LABORATORY PUMP FOR LIQUID TRANSFER BASED ON THE FOUNTAIN EFFECT

W.E.W. Chen, T.H.K. Frederking and W.A. Hepler

University of California, Los Angeles
Los Angeles, Callifornia

ABSTRACT

A laboratory scale pump has been tested in detail in order to determine the flow characteristics of a heater-activated all-metal thermomechanical pump (fountain effect pump FEP). The emphasis is on the functional dependence of the fountain pressure difference versus mass throughput. A modified Vote et al. power law approximation is employed for a simplified description of the flow rate as a function of the driving force. Flow rates of up to 10 Litres/(hr cm^2) have been obtained despite a large nominal pore size of the porous plug of 2 μm (filtration rating) used for the FEP.

INTRODUCTION

At the discovery of the thermomechanical effect (fountain effect) by Allen and Jones[1], manned space flight was not known. Neither were recent extensions of cryosystems anticipated with operation in the superfluid liquid He II range. The impact of superfluidity, found simultaneously by Kapitza and Allen and Misener, stimulated fountain effect pump (FEP) development for laboratory studies. Examples are references 2 to 5. Tasks in this area include liquid transfer systems based on cooler activation of FEP operation, e.g. references 6-8. In contrast, more recent large scale pump developments have emphasized heater activation, both for space applications and superconducting magnet technology[9]. Space transfer pumps have been documented in two workshops[10,11]. Our present studies have the purpose of determining the functional dependence of the mass flow rate, handled by a pump used previously[12] as a function of the driving temperature difference. The frame of reference of the phenomenological equations of Vote et al.[13] is used. First, we discuss the underlying theory, turn subsequently to experiments, and finally present results and conclusions.

PUMP THEORY

The ideal FEP has been modeled using the change of thermodynamic state at constant chemical potential ($d\mu = 0$). This system is called ideal superleak (ISL). The ISL model of Zemansky[14] is presented in modified form in Figure 1. The ISL model postulates zero entropy convection through the ideal superleak which is the core of the idealized pump. We make use of the two-fluid model noting that there is normal fluid carrying entropy and superfluid capable of flowing through the ISL without flow resistance. Immobilization of the viscous normal fluid in the ISL is a simplifying assumption permitting prediction of the heat submitted (right hand side of Figure 1) and of the heat rejected (left hand side of Figure 1). In each case the ideal amount, TS, taken at the particular temperature, has to be handled by the heat reservoirs of Figure 1. Constant chemical potential implies an enthalpy change of $\Delta(TS)$ across the pump. The two-fluid equations of motion which are useful for the present FEP are summarized as follows.

Superfluid. The ideal superfluid equation is essential to the ideal superleak (ISL) model: there is no flow resistance in the low-speed Landau regime up to the critical superfluid velocity (v_s). The kinetic energy may become noticeable though near v_{sc}, causing a lowering of the static pressure rise achieved by the ISL. The equation may be written as

$$\nabla(\mu + v_s^2/2) = 0 \tag{1}$$

(μ = chemical potential per unit mass). When the kinetic energy term in equation (1) is negligible, the ideal London pressure increment is obtained from $d\mu = 0$ as $dP_T = \rho\ S\ dT$. For a finite temperature difference, we have a thermomechanic P-difference

$$\Delta P_T = \int_T^{T+\Delta T} \rho S(T,P)dT \tag{2}$$

To first order, the pressure increase for a small ΔT is

$$\Delta P = \rho S \Delta T[1+2.8\Delta T/T]; \quad \Delta T/T \ll 1 \tag{3}$$

In the limit of a very small ΔT, the constant property integration of $d\mu = 0$ is obtained: $\Delta P_T = \rho S \Delta T$. For moderate ΔT values, the entropy derivative $(\partial S/\partial T)$ is converted to specific heat information. Adopting a power law for S with d log S/d log T = 5.6, we arrive at Eq. (3).

Normal Fluid. In contrast to the ISL model, the viscous normal fluid component, carrying entropy, may be characterized by its ability to flow through the main FEP component. A porous plug is adopted. Its throughput capability is uniquely determined by the Darcy permeability. The latter is designated as κ, in agreement with the nomenclature of Bird et al. (Ref. 16). The permeability is accessible through forced flow tests. An example is liquid He II flow in contact with a constant temperature bath ($j_n = 0$), once "normal" resistance has been restored. Another method is the zero net mass flow technique with a mass flux density $j = j_s + j_n = 0$, Ref. 15.

Concerning FEP operation above v_{sc}, it is noted that phase separation experiments, Ref. 12, have shown the smallness of $j_n << j$ in porous plugs with pores below 5 μm. Thus, the approximation $j_s = j$ appears to be reasonable in the temperature range considered. Velocities are related to each other by

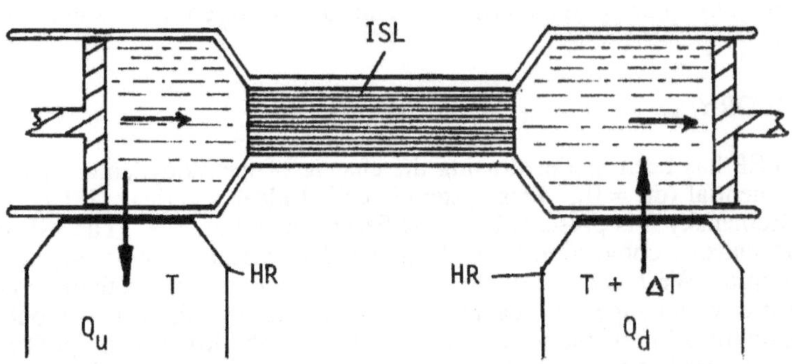

Fig. 1. Ideal superleak (ISL) system[14]; HR heat reservoir; ISL = capillary bundle or porous plug.

$$\overline{j} = \rho\overline{v} = \rho_s\overline{v}_s = \overline{j}_s \tag{4}$$

The transition from zero resistance to full "normal" resistance has been described by the phenomenological equations of Ref. 13. They constitute tangent functions to the flow resistance transition. In contrast to Ref. 13 however, the Darcy permeability is conveniently taken as reference quantity. In particular, $L_c = \kappa^{1/2}$ is a suitable reference length. The flow equations of Ref. 13 have been evaluated for near-isothermal flow primarily in the data range for normal fluid depletion ($\rho_n \ll \rho$; $\rho_s \approx \rho$). Therefore it is necessary to take into account the thermomechanical forces affected by dissipative vortex shedding processes. The latter interact finally, via normal fluid, with the solid grain walls.

The work in recent years suggests that $(\rho_s/\rho_n)\,|\,\mathrm{grad}\,P\,|_T$ may be regarded as an effective potential gradient ($|\,\mathrm{grad}\,P\,|_{eff}$) in the simplified equations. They are given for the porous media FEP as superficial speed $\overline{v}_{so}(\rho_s/\rho) = \overline{v}_o = (\dot{m}/\rho)/A_{tot}$ ($\dot{m}$ mass flow rate, A_{tot} total plug cross section). The resulting modified version of the power law approximations for the low resistance region is

$$\overline{v}_{so} = \overline{v}_o(\rho/\rho_s) = C_4[(\,|\nabla P|_{eff}/L_c)\eta_n^2\rho^{-3}]^{1/4} \tag{5}$$

(C_4 = constant). The relatively high flow resistance regime is characterized by a size-independent mass flux density:

$$\overline{v}_{so} = \overline{v}_o(\rho/\rho_s) = C_3[\,|\nabla P|_{eff}\cdot\eta_n\rho^{-2}]^{1/3} \tag{6}$$

(C_3 = constant). Two examples of FEP use are illustrated in Figures 2 and 3. Figure 2 shows a transfer pump operating in the earth's gravity field. Figure 3 depicts a simplified version of a space transfer system at microgravity.

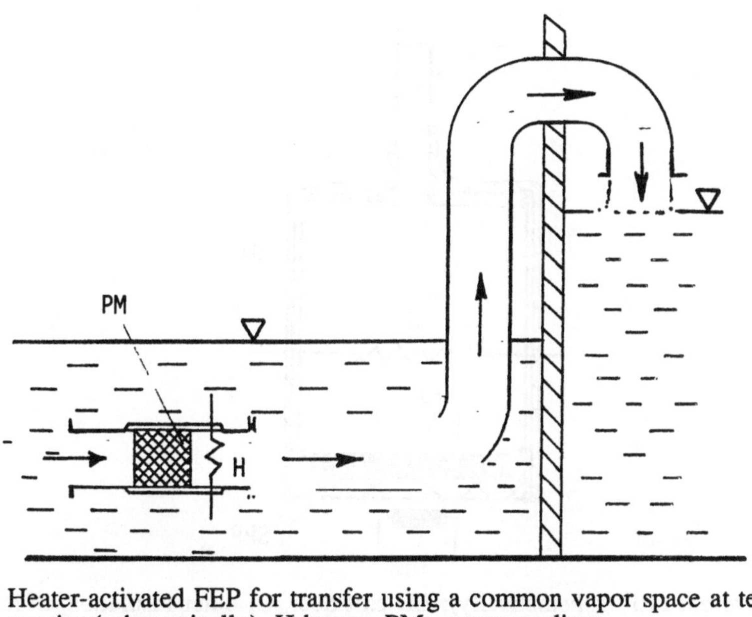

Fig. 2. Heater-activated FEP for transfer using a common vapor space at terrestrial gravity (schematically); H heater, PM porous medium.

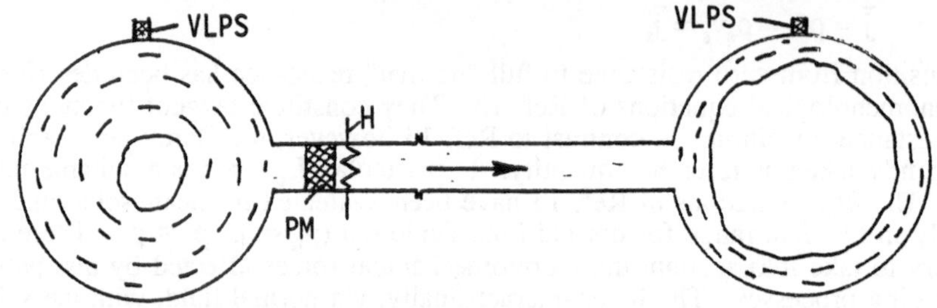

Fig. 3. Liquid transfer system based on FEP use in space at microgravity (schematically); H heater, PM porous medium, VLPS vapor-liquid phase separator.

EXPERIMENTS

The operation of the present FEP as liquid transfer system (Fig. 2) has been outlined in Ref. 12. For the present pump performance tests the transfer section needed for the liquid delivery toward a higher elevation has been removed. Thus the pump is defined as an assembly of components including outer pump walls with inlet section, major pump body consisting of a porous plug, heater and heater chamber, and outlet section. The porous medium is a sintered stainless steel plug ($\kappa = 3.4 \times 10^{-9}$ cm^2, length 0.3175 cm, diameter 1.27 cm, 2 μm nominal particle retention of solid-fluid filtration). The exit duct has circular cross section with a diameter of 0.19 cm (length 20 cm). The pump is shown schematically in Fig. 4. The heater with 100 ohm resistance is located in the downstream chamber. Thermometers upstream and downstream measure temperatures.

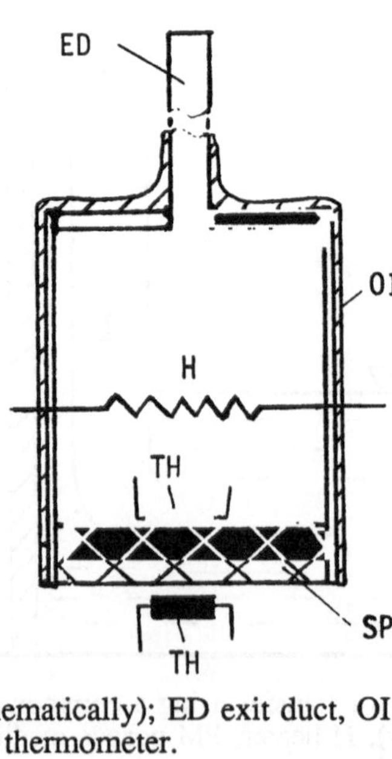

Fig. 4. Present FEP (schematically); ED exit duct, OI outer insulation, SPP sintered porous plug, TH thermometer.

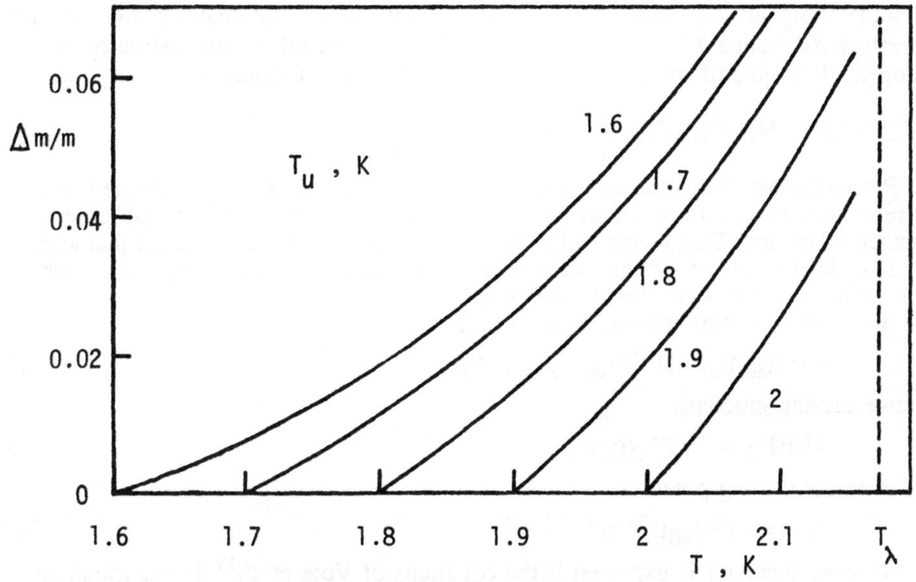

Fig. 5. Relative mass loss for FEP transferring liquid to another reservoir (Fig. 2), using a common vapor space.

The mass flow rate is deduced from the fountain height (Δz) produced at the exit section. From a fountain cross section (A_F) and exit speed (v_F), continuity in liquid flow relates $\overline{v}_F$ to $\overline{v}_o$ by $\dot{m} = \rho A_{tot}\overline{v}_o = \rho A_F\overline{v}_F$. The kinetic flow energy per unit volume is $(v_F^2/2)\rho = g\rho\Delta z$. There is a small vaporization loss as the FEP issues liquid into the common vapor space on top of the liquid bath (Fig. 2). The relative loss is shown in Fig. 5 for different bath temperatures (= upstream temperatures T_u).

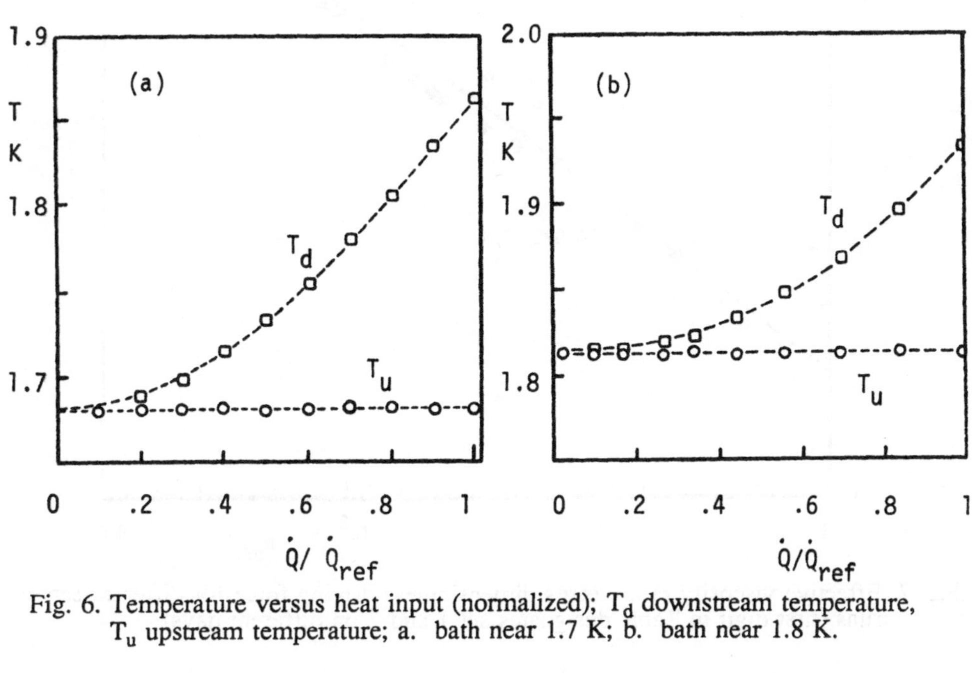

Fig. 6. Temperature versus heat input (normalized); T_d downstream temperature, T_u upstream temperature; a. bath near 1.7 K; b. bath near 1.8 K.

Thermometer signals are displayed in Figs. 6a and 6b versus the heater power. As the temperature difference $\Delta T = T_d - T_u$ is increased, more and more flow resistance is developed. It is quantified subsequently as resistance ratio (effective viscosity ratio).

DISCUSSION AND CONCLUSIONS

Data Presentation. The data are presented in a generalized coordinate frame using a flow resistance ratio, introduced in Ref. 13, in modified form. The "normal" reference resistance of laminar Darcy convection is adopted, instead of the hydraulic diameter (Ref. 13). In the present context, the reference resistance is $(\eta_n/\kappa) = (\eta_n/L_c^2)$. Our dimensionless variables are based on Eqs. (5) and (6).

Effective shear viscosity ratio (resistance ratio):

$$y = \eta_{eff}/\eta_n = (|\nabla P|_{eff}/\bar{v}_o)(\rho/\rho_n)(\rho/\rho_s) \quad (7)$$

Effective driving gradient:

$$|\nabla P|_{eff} = |\nabla P|_T(\rho_s/\rho_n) \quad (8)$$

Dimensionless driving potential:

$$N_{eff} = |\nabla P|_{eff} L_c^3 \rho/\eta_n^2 \quad (9)$$

The following behavior is expected if the equations of Vote et al.[13], in the modified form, are applicable to the present FEP operation. At low resistance ratios the derivative $n = \log d\, y/d \log N_{eff}$ is expected to tend toward a value of $n = 3/4$. At increased ratios, we expect that the derivative changes to $n = 2/3$ or even less. Finally, as the lambda temperature is approached, the derivative n ought to drop to low values because of the decrease of (ρ_s/ρ) toward zero.

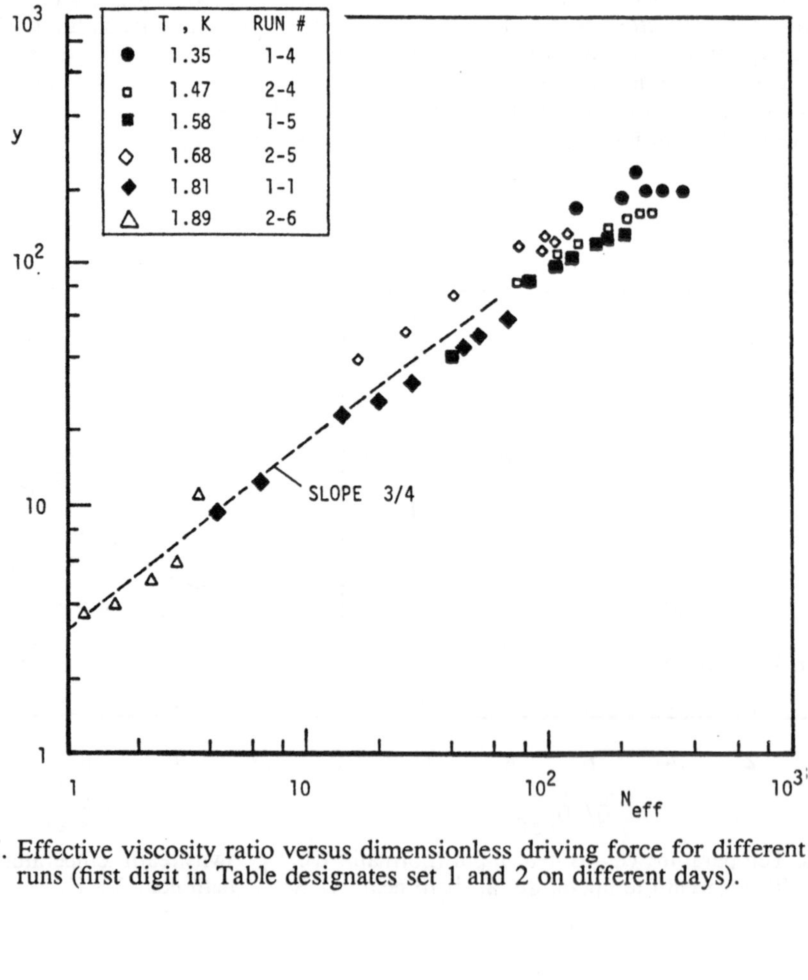

Fig. 7. Effective viscosity ratio versus dimensionless driving force for different sets of runs (first digit in Table designates set 1 and 2 on different days).

Figure 7 shows the FEP data in the frame $y = y(N_{eff})$. It is seen that the data do indeed show trends in agreement with the Vote et al. resistance transition. There is a significant difference: the ratio y is no longer restricted to the range $0 < y < 1$, as in the near-isothermal transition[13]. Instead, rather large values of y may be attained in FEP when a large temperature difference is imposed across the pump plug. The constant C_4 for the low resistance region is evaluated as 3.2.

Pump Performance Characterization. The pump performance functions of interest include not only the mass flow rate as a function of the driving force, but also the heater-to-flow power conversion energetics. The latter has been discussed as energetic effectiveness (ε_e) for space transfer tasks (Fig. 2)[17]. The ε_e- values have been calculated for the ISL system. In most pumps tested so far, there are losses due to vortex shedding during the resistance increase. This behavior is taken into account by a pump efficiency with respect to the ISL case (η_{ISL}) and constant chemical potential respectively. An example of $\eta_{ISL} = \dot{Q}_{ISL}/\dot{Q}$ is shown in Fig. 8 for a circulation pump reported by Hofmann et al.[18] It is seen that η_{ISL} decreases monotonically as the heater power is increased. From the definition of $\varepsilon_e = (W_F/Q_d)_{ISL}$, the real heater power is obtained as

$$\dot{Q} = \dot{Q}_d = (\dot{W}_F/\varepsilon_e)/\eta_{ISL} \qquad (10)$$

The flow power is $\dot{W}_F = \dot{m}(\Delta P_T/\rho)$, and the function $\dot{m}$ versus ΔT complete the FEP characterization including heat input required for a specified ΔT.

Conclusions. It is concluded that the phenomenological equations of Vote et al.[13] provide useful guidance also for FEP operation with inclusion of the effective driving term. Further, our heater-activated FEP appears to have pump characteristics quite consistent with the cooler-activated transfer system, e.g. Ref. 19. Despite a large pore size, flow rates of up to 10 Litres/(hr cm^2) have been reached.

NOMENCLATURE

A	cross section
A_{tot}	total cross section of porous medium
C_i	constant (i=3,4)
j	mass flux density (j_i with i=n,s)
L_c	characteristic length of plug = $\kappa^{1/2}$
$\dot{m}$	mass flow rate
N_{eff}	dimensionless driving force
n	power law exponent
P	pressure
$\dot{Q}$	heater power
q	heat flux density
S	entropy per unit mass
T	temperature
v	velocity (v_i with i=F,n,s)

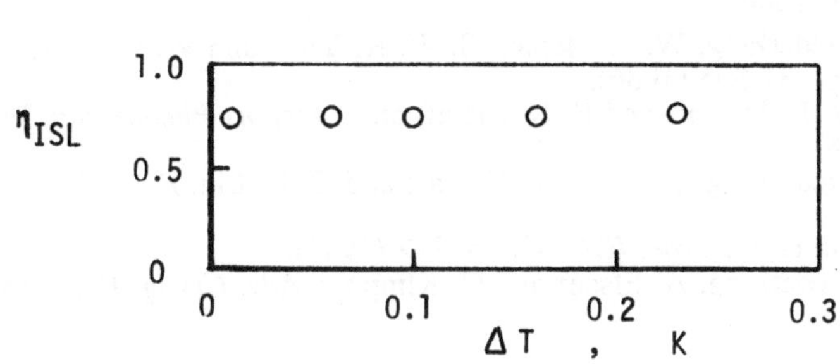

Fig. 8. Efficiency (η_{ISL}) with respect to the ideal superleak system.

W_F	flow power
y	resistance ratio (effective viscosity ratio)
z	vertical position coordinate
ε_e	energetic effectiveness
η	shear viscosity
η_{ISL}	efficiency with respect to ideal superleak
κ	Darcy permeability
μ	chemical potential
ρ	density (ρ_i with i=n,s)

Subscripts

d	downstream
eff	effective
F	fountain
ISL	ideal superleak
n	normal fluid
o	superficial value
ref	reference
s	superfluid
T	thermomechanical
u	upstream

ACKNOWLEDGEMENTS

The work has been supported by NASA Ames Research Center (Grant NAG 2-412). We acknowledge Gary T. Eckwortzel's machining of components for the project.

REFERENCES

1. J. F. Allen and H. Jones, *Nature* 141:243 (1938).
2. J. R. Pellam and P. P. Craig, *Phys. Rev.* 108:1109 (1957).
3. J. N. Kidder and W. M. Fairbank, *Phys. Rev.* 127:987 (1962).
4. F. A. Staas, K. W. Taconis and W. M. Van Alphen, *Physica* 27:893 (1966).
5. B. M. Broulik and G. B. Hess, *Physica* 94B-169 (1978).
6. A. Elsner and G. Klipping, *Adv. Cryog. Eng.* 18:317 (1973).
7. H. D. Denner et al., *Cryogenics* 18:166 (1978).
8. S. Kasthurirengan et al., *Cryogenics* 25:518 (1985).
9. R. Srinivasan and A. Hofmann, *Cryogenics* 25:641 (1985).
10. P. Kittel, *Cryogenics* 26:59 (1986).
11. G. Klipping, *Cryogenics* 27:3 (1987).
12. S. W. K. Yuan and T. H. K. Frederking, *Cryogenics* 27:27 (1987).
13. F. C. Vote, J. E. Myers, H. B. Chu and T. H. K. Frederking, *Adv. Cryog. Eng.* 16:393 (1971).
14. W. M. Zemansky, *Heat and Thermodynamics,* McGraw-Hill, New York, 5th Ed., p. 509 (1968).
15. T. H. K. Frederking, W. A. Hepler, S. W. K. Yuan and W. F. Feng, *Adv. Cryog. Eng.* 31:505 (1986).
16. R. Bird, W. E. Stewart and E. N. Lightfoot, *Transport Phenomena,* Wiley, New York (1968).
17. T. H. K. Frederking, P. Kittel, T. C. Nast and C. K. Liu, *Proc. ICEC-11,* p. 323 (1986).
18. A. Hofmann et al., *Proc. ICEC-11,* p. 312 (1986).
19. T. H. K. Frederking, A. Elsner and G. Klipping, *Adv. Cryog. Eng.* 18:132 (1973).

LAB TESTS OF A THERMOMECHANICAL PUMP FOR SHOOT

Michael J. DiPirro and Robert F. Boyle

NASA/Goddard Space Flight Center
Greenbelt, Maryland

ABSTRACT

Laboratory tests of a thermomechanical (TM) pump utilizing a commercially available porous disk have been conducted. Various size disks, heater configurations and outlet flow impedances have been used to characterize scale models of the pump proposed for the Superfluid Helium On-Orbit Transfer (SHOOT) Flight Experiment. The results yield the scalability of the TM pump to larger diameters and hence larger pumping rates, the dependance of flow rate on back pressure and heater power, and the limits of pumping speed due to internal losses within the porous disk due to mutual and superfluid friction. Analysis indicates that for low back pressures the flow rate is limited by the superfluid friction rather than the mutual friction. For the porous plug used in the early tests this amounts to a practical limit of 4.4 liters per hour per square centimeter. For our baselined flight plug area of 180 cm^2 this yields 790 liters per hour.

INTRODUCTION

A number of future space facilities require liquid helium for cooling of detectors, instruments, or even an entire telescope. Exhaustion of the liquid helium limits the useful life of these facilities. Superfluid helium (SFHe) is the desired state for most users.[1] To refill these helium users on orbit a large SFHe resupply tanker (SFHT) is being developed by NASA. A number of critical components must be developed to enable the transfer of SFHe on orbit. The Superfluid Helium On-Orbit Transfer (SHOOT) Flight Experiment project is developing these components in preparation for a shuttle-attached demonstration flight in mid 1991.[2,3] One of the critical components is the SFHe pump. SHOOT is investigating two approaches; one using a mechanical centrifugal pump and one using a thermomechanical (TM), or fountain-effect pump.

The thermomechanical effect in SFHe is a direct conversion of heat to fluid pressure. It can most easily be understood in terms of the two fluid model of SFHe due to Tisza and London. Briefly, the model assumes that SFHe is made up of two interpenetrating fluids; a "normal" fluid with a non-zero viscosity and entropy, and a "super" fluid with effectively zero viscosity and entropy. Applying heat to a container of SFHe produces a relative abundance of the normal component creating a chemical potential difference, which causes a counter flow of normal and super component. If, in the presence of the applied heat, one restricts the flow of the normal component with a fine porous medium for example, one can create a large chemical potential imbalance without an excessive amount of heat. The super component will still flow from the cooler to the warmer side of the porous medium. However, the normal component, due to its non-zero viscosity, will be impeded from flowing. Indeed, if the pores are fine enough (a superleak), no normal component, and hence no heat, will flow through the medium at all. By adding enough heat to maintain the temperature difference, one can continuously pump the SFHe. The inlet and outlet temperature determine the pressure across the plug by London's equation:

$$\Delta P = \int_{T_i}^{T_o} \rho S dT \tag{1}$$

The helium reaching the warm side of the superleak has no entropy. Thus, to prevent cooling, heat must be supplied to the helium which flows through the plug at a rate

$$\dot{Q}_h = \dot{m}\, S_o\, T_o \tag{2}$$

At the inlet of the pump, the reverse process is occuring; the super component is being drained leaving the remaining fluid entropy rich, or hotter. This process is known as the mechanocaloric effect, given by

$$\dot{Q}_{mc} = \dot{m}\, S_i\, T_i \tag{3}$$

To maintain the inlet temperature, T_i, this heat must be removed, commonly by pumping the vapor over the bath on the inlet side. An ideal TM pump must therefore provide heat $\dot{Q}_h$ at the pump outlet and remove heat $\dot{Q}_{mc}$ at the pump inlet to operate in the steady state mode. Note that the part of $\dot{Q}_h$ which represents an ideal TM pump loss is that part in excess of the mechanocaloric power, $\dot{Q}_{mc}$.

This paper will explore these effects and the effects due to the non-ideality of a real TM pump, and compare the expected losses with those observed in the testing of TM pump using a wire mesh heater and porous plugs made of sintered stainless steel.

ANALYSIS OF PUMP LOSSES

Two types of losses are of interest in SFHe transfer, loss of helium due to heat inputs, and loss of pump pressure due to non-ideal behavior of the TM pump. Most of these losses are covered in the paper by Kittel[4]. Total heat input to an ideal TM pump versus the output pressure is shown in figure 1. As shown, the amount of heat which is due to the mechanocaloric effect is given by the P=0 line and the losses due to the pressure required are the remainder. For a given flow rate the heat required to produce a given pressure is approximately linear with pressure, and is independent of the inlet temperature. The mechanocaloric effect, whichshows up entirely at the pump inlet, is very dependent on the inlet temperature due to the rapid variation of S with temperature ($S \approx cT^{5.6}$).

One of the largest potential heat losses in a less than ideal TM pump is the heat which is conducted back through the pump to the inlet side. This conducted heat flow, $\dot{Q}_c$, is due to the non-zero flow of the normal component from the hot side to the cooler side of the pump. The heat flow per unit area depends on the normal fluid velocity through

$$q = \rho S T v_n \tag{4}$$

Depending on the magnitude of q and the geometry of the flow path, the flow of the normal fluid may be laminar, Gorter-Mellink, or fully turbulent[5]. For smaller diameter flow paths

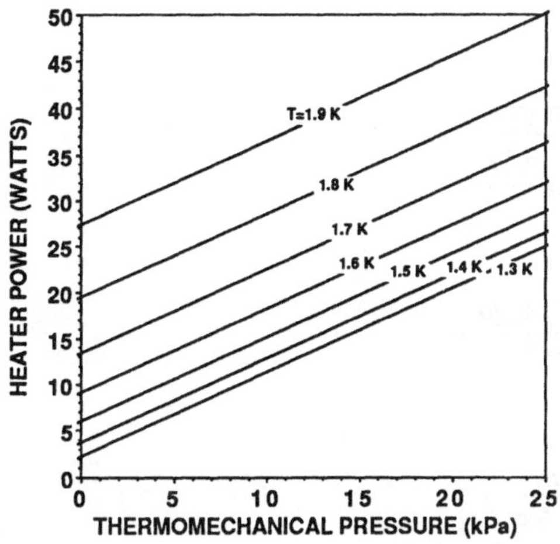

Figure 1. Heater power required to pump 500 L/h back pressure at various inlet temperatures.

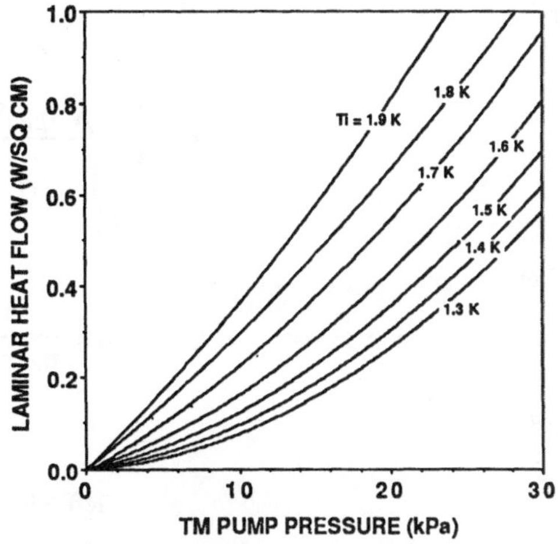

Figure 2. Laminar heat flow per unit area conducted through porous medium with permeability 1.33×10^{-10} cm^2 and thickness 6.4 mm.

and lower heat flows, the flow is likely to be laminar. In this case the heat flow is given by

$$q_l = \frac{K_p}{t} \int_{T_i}^{T_o} \frac{(\rho S)^2 \, T \, dT}{\eta_n} \tag{5}$$

where q_l is the heat flow per unit area of porous disk, K_p is the Darcy permeability, t is the thickness of the plug, and η_n is the viscosity of the normal component. An example using the permeability of our porous disks is given in Figure 2. Note that q in equation (4) is the heat flow per unit pore area. To compare to q_l, q must be multiplied by the porosity, ε. Under laminar flow conditions, one does not expect the ideal TM pressure as given by equation (1) to be diminished. For larger heat flows and pore sizes mutual friction between the normal component and the superfluid component develops. This is known as the Gorter-Mellink mutual friction and the flow in this regime is called Gorter-Mellink flow. A term is added to the usual equations of motion to produce a pressure gradient opposing the superfluid component flow

$$\nabla \mathbf{P} = \rho \, S \, \nabla T - \nabla \mathbf{P}_{GM} - \mathbf{F}_s \tag{6}$$

$$\nabla \mathbf{P}_{GM} = A_{GM}(T) \, \rho_s \, \rho_n \, (v_r)^2 \, \mathbf{v}_r$$

where $\mathbf{F_s}$ is pure superfluid friction and $\mathbf{v_r}$ is the relative velocity of the super and normal components, $\mathbf{v_s} - \mathbf{v_n}$. This mutual friction term presumably only becomes operative when v_r exceeds the superfluid critical velocity, v_{sc}. For negligible v_n, the superfluid friction, when v_{sc} is exceeded, dominates the mutual friction term.[6] Equation (6) may be put in an integral form and solved for either the pressure or heat flow. Following Keller[7] and neglecting $\mathbf{F_s}$ we write these as

$$\Delta P = \int_{T_i}^{T_o} \frac{\rho \, S \, dT}{1 + a \dfrac{q_{eff}^3}{q}} \tag{7}$$

and

$$q = \frac{K_p}{t} \int_{T_i}^{T_o} \frac{(\rho S)^2\, T\, dT}{\eta_n \left(1 + a \dfrac{q_{eff}^3}{q}\right)} \tag{8}$$

where

$$a = \frac{A(T)\ \rho_n/\rho}{\eta_n\ (\rho_s/\rho)^3\ S^2\ T^2}$$

and

$$q_{eff} = q + \frac{\dot{m}}{\rho\, A\, \varepsilon}$$

respectively. When the second term in the integrand denominator becomes small compared to 1 the integrals become those for the ideal TM effect and laminar heat flow. If we take an example based on actual flow measurements we can evaluate the size of the Gorter-Mellink term. In our case, we substitute the values:

$$T_i = 1.667\ \text{K}$$
$$T_o = 1.847\ \text{K}$$
$$q = q_c = .1592\ \text{W/cm}^2$$
$$K_p = 1.32\text{x}10^{-10}\ \text{cm}^2$$
$$\dot{m}/A = .062\ \text{g/s cm}^2$$

and result with the term aq_{eff}^3/q equal to 0.025 at its largest ($T = T_h$). This also our worst case example, so the typical Gorter-Mellink correction to the TM pressure and heat flow is neglible in this case. We also note that in order for the mutual friction term to be fully operative, the vortex line spacing within the superfluid should be smaller than the pore size. This is not achieved with the small pores and low heat flows that we had.

Finally, we note that the superfluid friction term in equation (6) depends on the quantity $(v_s - v_{sc})^\alpha$, where α is in the range of 1 to 2, with perhaps two distinct regions, one for $(v_s - v_{sc}) \sim 1$ or smaller and the other at higher values of v_s.

TEST APPARATUS

Four different pumps (#1-#4) were tested in three different sets of apparatus (A-C). All pumps were based on 6.4 mm thick, 0.1 micron filtration grade (nominal 1 micron pore size) porous disks from Mott Metallurgical. Pumps #1 and #2 were 2.5 cm diameter disks used in A and B. Pump #3 was a 7.6 cm diameter disk used in C. Pump #4 was composed of two disk sections arranged in parallel totalling 73.5 cm^2. Each pump had a heater made from 0.13 mm diameter constantan wire. Pump #1 had a heater wound on a bobbin around its upper circumference. Pump #2 had a cross-hatched wound heater suspended approximately 2.5 mm above its top surface. The wires were spaced approximately 2.5 mm apart across the top surface of the porous disk. Pump #3 had two heaters woven in one direction only, approximately 3 mm above its top surface. The heaters were interleaved with a spacing of 3 mm, such that each heater had a 6 mm space between its wires. Pump #4 had two independent heaters, one over each disk, with spacing of 3mm and a height over the disk of 2.5 mm. The heaters could be operated individually, in series, or in parallel during each run. The permeability of the material was measured by room temperature liquid flow and by pressurized normal liquid helium flow at low temperatures. The value was found to be $1.32\text{x}10^{-10}$ cm^2. This is consistent with values presented elsewhere for other size porous disks made by Mott.

Germanium resistance thermometers (GRT's) were located on the top and bottom of each porous disk, fastened with Stycast 2850 GT epoxy. The epoxy covered less than 0.2 cm^2 of the porous disk. The GRT's were calibrated in our lab against National Bureau of Standards (NBS) calibrated GRT's using standard DC readout techniques. Their relative calibration accuracy was better than 0.1 millikelvin (mK) over the temperature range from 1.3 to 1.9 K. Their absolute accuracy was approximately that stated in the NBS calibration, one mK. The relative calibration was checked *in situ* as well. The heaters were terminated in a four lead arrangement so that checks of voltage and resistance (or voltage and current) could be made to accurately determine the heat dissipation. Outlet pressure was determined by Teledyne Taber model 2215 strain gauge pressure transducers with a readout precision of 4 Pa and arelative accuracy of 14 Pa after *in situ* calibration. Inlet pressure was determined by measuring the temperature and correcting the saturated vapor pressure by the head of helium present.

Apparatus A was used for low flow rate measurements using a 30 L supply dewar and a 15 L receiver. The supply and receiver were connected by a standard transfer line 3 m in length with a 3.76 mm inner diameter flow tube. Bath levels and flow rates were determined by using American Magnetics superconducting level sensors (1% precision). Apparatus B was used for low flow rate measurements within a single dewar[8]. The back pressure on the pump was varied by using different exit tubes. Flow rates were measured by timing the fill of a graduated beaker (2% accuracy after averaging). A GRT was placed at the bottom of the beaker to check for any temperature difference between the SFHe in the beaker and the bath (less than 1 mK was seen). The beaker was periodically emptied by using a plunger. The apparatus also had a provision to attach fluid acquisition devices to the pump inlet. Apparatus C (Figure 3) could accommodate higher flow rates, approaching those to be demonstrated on orbit. The supply and receiver dewars were 100 L capacity connected by a 4 m length, 7.75 mm inner diameter transfer line. A 614 L/s vacuum pump was used to pump on the supply dewar, while a 47 L/s pump was used on the receiver dewar. During several of the transfers the 47 L/s pump was valved off and the receiver dewar was pressurized by an external source of gaseous helium to pressures above the lambda transition. Here the receiver pressure was monitored with an MKS Baratron.

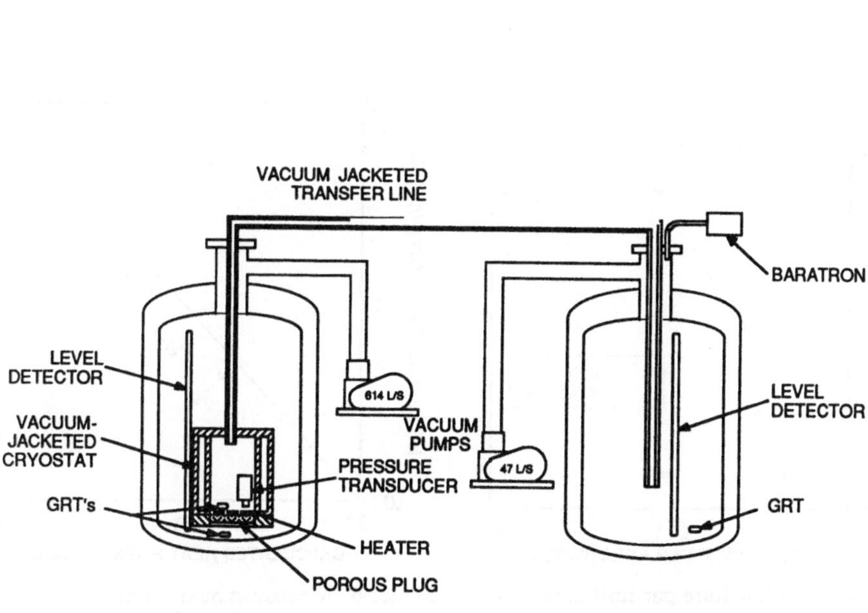

Figure 3. Apparatus C for testing thermomechanical pumps.

Each apparatus was calibrated for heat leak through the pump housing into the surrounding bath. For example, Apparatus C has a conductance of 1.2 W/K at 1.6 K. These parasitics were close to those calculated for the three systems using the conductance of the stainless steel alone, and were used to calculate the parasitic heat conducted through the TM pumps.

TEST RESULTS AND ANALYSIS

A typical run consisted of pumping the SFHe down to operating temperature, turning on the heater to the pump, waiting for thermal equilibrium in the supply and receiver dewars, and noting the temperature, pressure, flow rates, etc. At the higher heater powers and flow rates, the temperature within the supply dewar rose throughout most of the run. Here, a "snapshot" of the data was taken. In these runs the flow rate was found to be relatively constant at the beginning and end of each run, hence these are the times when the indicated flow rates and other parameters were measured.

To determine the effect of heater placement over the porous disks, we repeated measurements using one of the two heaters in pumps #3 and #4 and found no observable change in temperature, pressure, or flow rate. This is an especially drastic situation in pump #4 where one heater is over each porous disk. Since our results indicate that each pump must participate in the flow to achieve high flow rates, this indicates that the heat can be applied a fairly large distance from the pump with no loss in performance. It should be carefully noted, however, that the fluid space within the housing for pumps #3 and #4 had a cross-sectional area of about 1.8 times the area of the pumps, hence the flow velocity was much smaller in the housing, allowing easier heat flow in the opposite direction to the mass flow.

The flow rates per unit area versus the heater power applied is shown in Figure 4. These data represent the raw data for pumps #3 and #4. We then subtract the ideal TM pump heat as shown in Figure 1, and subtract the heat conducted through the pump housing from the raw heater power. Since there is no apparent temperature gradient in the SFHe above the pump, we assume all of the remaining heat applied to be conducted through the SFHe within the pump back into the supply dewar. This heat, normalized to the pump area, is then compared in

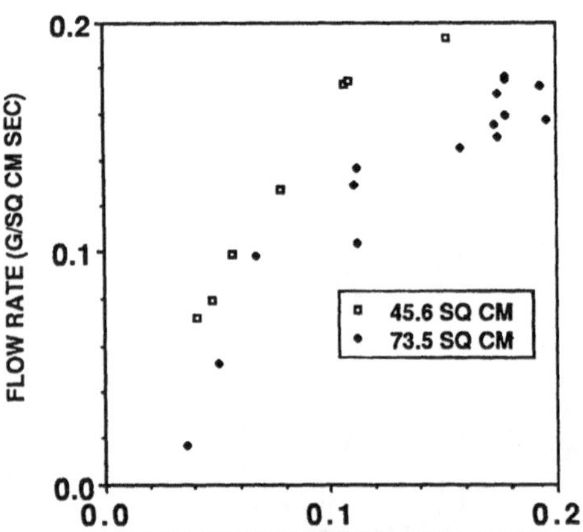

Figure 4. Flow Rate per unit area of TM heater pump power per area.

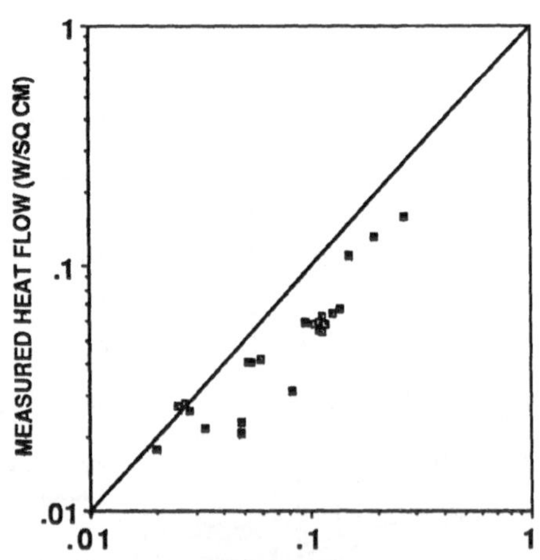

Figure 5. Measured heat conducted through pump vs. laminar heat flow calculated.

Figure 5 to the laminar heat flow calculated from equation (5). We see that good agreement exists up to heat flows of the order of 2.7×10^{-2} W/cm^2.

The overall pressure vs. flow performance for pump #4 is indicated in Figure 6. Since heater powers were not optimized for each pressure, the curve represents the "as achieved" and not the ultimate capability of the TM pump.

The transfer efficiency is defined as the mass of helium delivered to the receiver at the temperature of the supply dewar divided by the mass lost from the supply dewar (after cooldown of receiver and transfer line). Losses due to transfer line heat leaks (about 3 W) and dewar parasitics are subtracted from this number to arrive at the pump efficiency. The overall system losses expected on SHOOT are presented elsewhere.[9] The efficiencies measured lay in the range of 94.4 to 98.0 % for all transfers except those in which the receiver dewar was overpressured with gaseous helium, or when the flow rate was less than 90 liters per hour. In this case the residual heat necessary to hold a column of helium against gravity was substantial compared to that actually used to move the helium. In comparison with the efficiencies given, equivalent flows through an ideal TM pump would have efficiencies in the range of 98 to 99% when operated at the same temperature.

At first the data was analyzed presuming the Gorter-Mellink mutual friction to be the dominant mechanism for the deviation of the pressure and heat conducted from their ideal values. However, as shown earlier, the Gorter-Mellink correction proved to be small in the region of temperature ($T_c = 1.3$ to 1.68 K, $T_h = 1.45$ to 1.85 K), heat flow and mass flow in which we operated. We then approach the problem as one of superfluid dissipation. We calculate the superfluid velocity by the continuity equation, measured mass flow rate, and equation (4).

$$v_s = (\rho_n/\rho_s)q_c/(\varepsilon\rho ST) + \dot{m}/\rho\varepsilon A \quad (9)$$

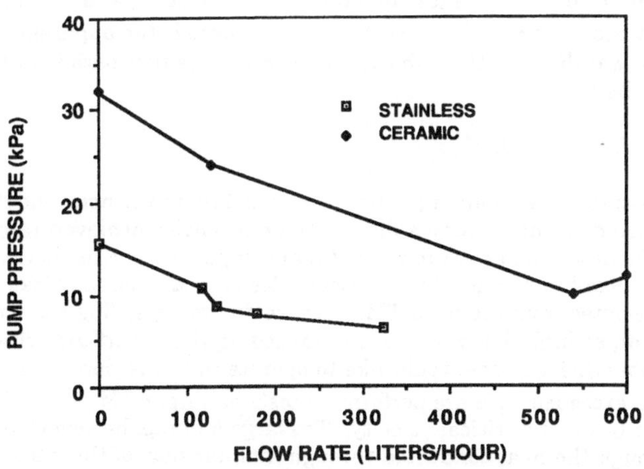

Figure 6. Demonstrated performance of presure vs. flow in 73.5 cm^2 stainless steel TM pump. Also shown is the early data for a 180 cm^2 alumina ceramic TM pump.

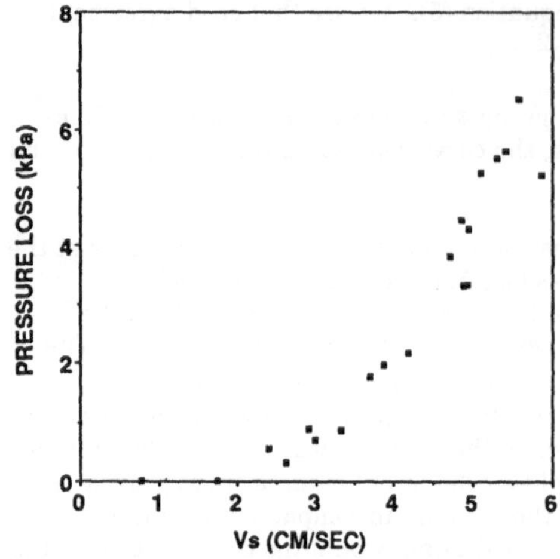

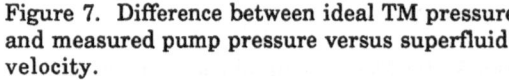

Figure 7. Difference between ideal TM pressure and measured pump pressure versus superfluid velocity.

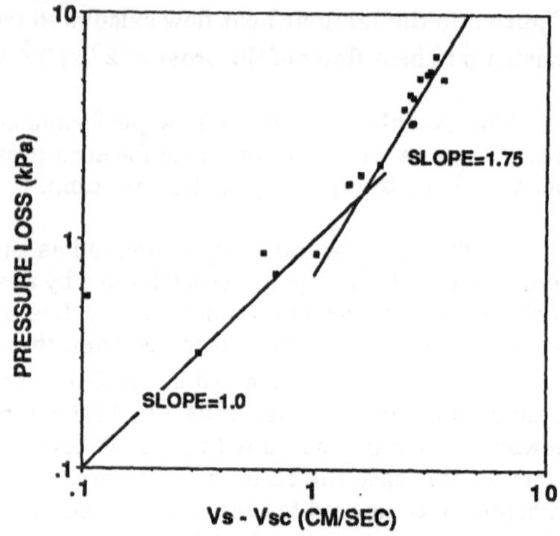

Figure 8. Pressure difference vs. superfluid velocity exceeding the critical velocity, v_{sc}.

The mass flow term represents the majority (56% to 88%) of v_s. The difference between the measured and ideal TM pressure is then plotted against v_s in Figure 7. Note the near zero difference in pressures up to velocities of 2 to 3 cm/sec. A similar effect is seen when the difference between the calculated laminar heat flow and the measured heat conducted is plotted against v_s. Figure 8 is a logarithmic plot of the same pressure difference vs. (v_s - v_{sc}), where v_{sc} , the superfluid critical velocity, is determined to be 2.3 cm/second by Figure 7. From this plot we can see that the pressure decrease from the ideal TM pressure varies as $(v_s - v_{sc})^{1.75}$ at high flow rates, consistent with the Blasius relation for ordinary turbulent fluids, and as reported by Keller[10] for superfluid helium in small channels. Further, below 1 to 2 cm/sec. above the critical velocity, the data isconsistent with a linear fit, as Keller also reported. In contrast to Keller's data, the pressure drop (gradient) is much larger, perhaps by a factor of 10 to 15. It does seem clear, however, that the decrease from the ideal TM effect is caused by the superfluid component exceeding the critical velocity, and not the mutual friction as previously assumed.[11] Note that the v_s given here is the apparent velocity of the superfluid if the flow was directly through the pump. Since the flow path is tortuous the actual critical velocity will be higher.

SUMMARY AND FUTURE WORK

We have demonstrated TM pumping at flows up to 325 liters per hour and at pressures up to 16 kPa using sintered stainless porous disks. The efficiencies achieved ranged from 90% to 98% depending on the flow rate, pressure required, and supply dewar temperature. The flow rate appears to be limited by the superfluid critical velocity in the porous disks used. The pressure achieved deviates from the ideal TM pressure due to exceeding the critical heat flow in the normal fluid, or, at high flow rates, due to exceeding the superfluid critical velocity. To achieve the maximum efficiency one would like to operate in the region where the critical velocity has not been exceeded. We are performing tests on larger TM pumps (180 cm^2) to keep the flow near or below the critical velocity. Too large a pump, however, would lead to excessive losses through the heat conducted through laminar flow of the normal component.

a much higher (32 kPa) ideal pressure achieved and a flow rate per unit pore area equivalent to that achieved in our lab. Beside the higher pressures achieved the material should have less conducted heat and hence higher efficiency due to the smaller pore size. We are in the process of testing this material in a flight-like pump arrangement. Flows as high as 600 L/h have already been achieved. The results of these tests will be reported in a future publication.

REFERENCES

1. S. R. Breon, Liquid helium servicing from the space station, 1987 Space Cryogenics Workshop, Madison, to be published in Cryogenics.
2. M. J. DiPirro and S. H. Castles, Superfluid helium flight demonstration using the thermomechanical effect, Cryogenics 26: 84 (1986).
3. M. J. DiPirro and P. Kittel, The superfluid helium on orbit transfer flight demonstration, Paper No. DA-3, 1987 Cryogenic Engineering Conference.
4. P. Kittel, Losses in fountain effect pumps, Proc.11th International Cryogenic Engineering Conference , Butterworth, London, (1986) p. 317.
5. see for instance, J. Wilks, "The Properties of Liquid and Solid Helium", Clarendon Press, Oxford, (1967).
6. J.T. Tough, Prog. Low Temp. Phys 8: 135 (1982).
7. W. E. Keller, "Helium-3 and Helium-4", Plenum, NY, (1969) p. 321.
8. J. E. Anderson, D. A. Fester, and M. J. DiPirro, Acquisition system testing with superfluid helium, 1987 CEC, Paper FC-2.
9. M. J. DiPirro, The Superfluid Helium On-Orbit Transfer flight experiment: performance estimates, 1987 Space Cryogenics Workshop, Madison, to be published in Cryogenics.
10. W. E. Keller and E. F. Hammel, Physics 2: 221 (1966).
11. U. Schotte and S. Kasthuriringan, Proc. Space Helium Dewar Conference, Huntsville (1983), p. 109; R. Srinivasan and A. Hofmann, Cryogenics 25: 641 (1985).
12. G. L. Mills, et al., Experiments on transferring helium II with a thermomechanical pump, 1987 CEC, Paper BC-3.

EXPERIMENTS ON TRANSFERRING HELIUM II WITH A THERMOMECHANICAL PUMP

G. L. Mills, A. R. Urbach, A. J. Mord, B. H. Brandreth,
L. A. Hermanson, H. A. Snyder

Ball Aerospace Systems Division
Boulder, Colorado

ABSTRACT

Porous plugs were used as thermomechanical pumps to transfer helium II from one dewar to another through a transfer line and two valves. Two different sizes of porous plugs were used, and a scanning electron microscope image of the plugs is presented. Eighty litres were transferred at a maximum flow rate of 90 L/hr. Temperature, pressure, flow rate, and heater power data are presented and compared to a theoretical model.

INTRODUCTION

Thermomechanical (fountain effect) pumps have long been suggested as a means for pumping large quantities of helium II.[1] More recently, the unique properties of helium II have made it useful for cooling space instruments. Several space science missions are now being planned which would benefit greatly from on-orbit resupply of helium II.[2] The use of a thermomechanical pump in accomplishing this resupply has been examined at a system level.[3] The characterization of thermomechanical and centrifugal pumps as components has been investigated.[4,5]

We have performed a series of experiments to demonstrate that large volumes of helium II can be transferred at high flow rates with a thermomechanical pump from one dewar to another through valves and lines which are similar to the plumbing arrangement which would be necessary to accomplish such a transfer on-orbit. In one experiment, the use of a thermomechanical pump to fill an empty, warm transfer line and dewar was demonstrated. We also demonstrated that a thermomechanical pump could fill the receiving dewar completely (zero ullage). In addition, temperature, pressure, and flow rate data taken during the test were used to verify a computer model which we have developed.[6]

TEST APPARATUS

A simplified flow schematic of the test apparatus is shown in Figure 1. The large vessel has a volume of 570 L. It is surrounded by three vapor cooled shields, multilayer insulation, and an outer shell.

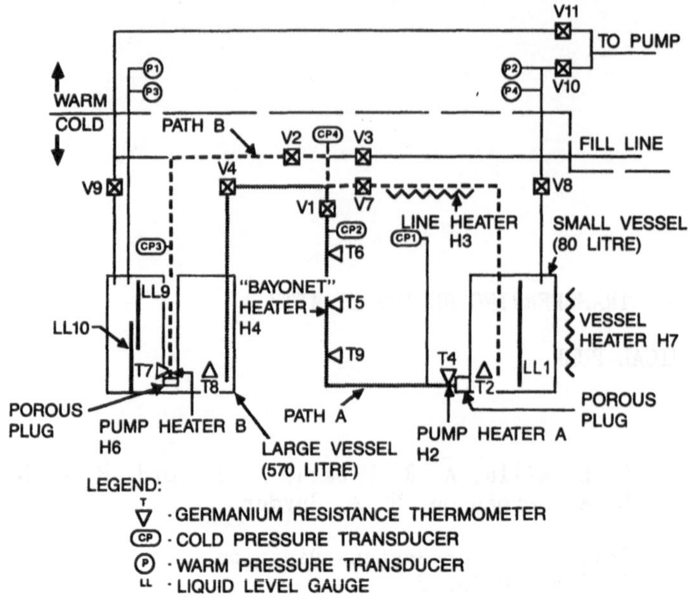

Figure 1. Flow schematic of helium transfer experiment

The large vessel is annular with a 45.7 cm diameter instrument cavity in the middle. The small vessel is an 80 L welded stainless steel cylinder with dished heads. It is located in the large vessel instrument cavity but is thermally isolated from the large vessel by fiberglass supports, 20 layers of multilayer insulation, and vacuum.

The plumbing and valve arrangement and the two thermomechanical pumps allowed helium II to be pumped from either the small vessel to the large vessel (path "A") or from the large vessel to the small vessel (path "B"). The path A transfer line was 695 cm long, 447 cm of which was corrugated stainless steel flexible tubing which had a minimum ID of 1.25 cm and a maximum ID of 1.88 cm. The balance of the transfer line was rigid stainless steel tubing with an ID of 1.13 cm. The path A transfer path had an electric heater wrapped around a section 7.0 cm long and two globe valves, downstream of the porous plug. A germanium resistance thermometer was placed on each side of the porous plug and at three locations downstream of the porous plug and in the large vessel which acted as a receiver dewar. Three Siemens KPY-12 pressure transducers were placed downstream of the porous plug.

Transfer path B was 350 cm long and was almost entirely corrugated tubing like that used in path A and included globe valves downstream of the plug. A section of path B 92 cm long downstream of the plug was wrapped with heater wire. There was a GRT mounted on the downstream side of the plug and a pressure transducer downstream of the plug.

The vent lines from the large and small vessels were routed through high conductance valves and a heater hose to a vacuum pumping system with a capacity of 700 L/S down to one torr.

The porous plugs used in both flow paths were made of a porous ceramic similar in composition to mullite ceramic. Bubble tests of these plugs showed them to have maximum surface pore diameter of 0.5 micron. By weighing the plugs dry and then saturated with water, we determined

Figure 2. Scanning electron microscope photograph of porous plug interior

that the plugs had a 30 percent void volume. Several plugs from the same batch were broken across the thickness and scanning electron microscope (SEM) photographs were taken of the plug microstructure. Figure 2 and other photographs show that the microstructure is quite rough, with many void dimensions much larger than 0.5 microns. SEM photographs of the surface of the plugs indicated less voids on the surface than inside the plug. The plug manufacturer, Coors Ceramics, has reported a "skin effect" such as this in other applications.

The porous plug used in path A was 1.27 cm thick and 3.81 cm in diameter. It was bonded into a stainless steel retainer and a GRT was bonded to each side with Stycast epoxy. The resulting exposed surface area was 9.7 cm^2. The retainer was sealed between the heater housing and a flange on the side of the small vessel using bolts and indium wire. The heater was mounted approximately 0.5 cm from the surface of the plug.

The porous plug used in path B was 1.27 cm thick and 2.54 cm in diameter. It was bonded into a stainless steel retainer with a resulting exposed surface area of 4.10 cm^2. The retainer was sealed to the heater housing using indium wire. The heater housing was welded on to the end of the vacuum jacketed transfer line which hung inside the large vessel.

The heaters used in both flow paths were precision wire wound resistors which had very low resistance change with temperature. They were mounted in the fluid and sized so that the heat flux stayed substantially below the critical heat flux which would cause boiling, approximately 3 W/cm^2.

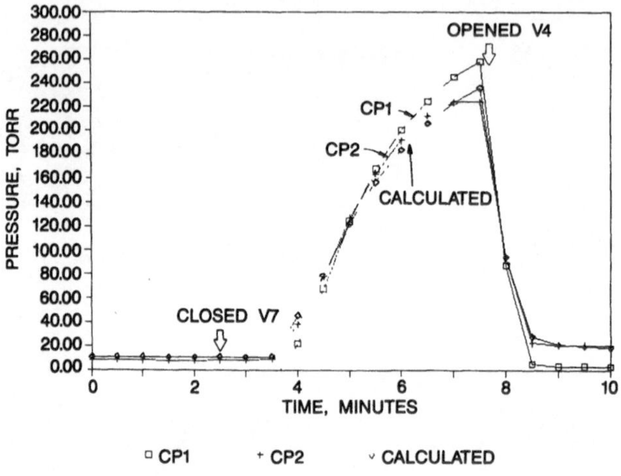

Figure 3. Pressure measured and calculated during no-flow experiment

The small vessel was wrapped with phosphor-bronze wire to provide a capability for warming it up. This heater wire was varnished down and the small vessel was covered with 10 layers of multilayer insulation.

The test apparatus was prepared for the test by pumping a high vacuum (better than 10^{-6} torr) on the guard vacuum and vacuum pumping and helium purging the vessels and all plumbing several times. The two vessels were filled with normal liquid helium and then converted to helium II by vacuum pumping. The GRTs, which had been calibrated by the manufacturer, were recalibrated by measuring the helium vapor pressure with pressure transducers that have an accuracy of 0.1 torr. The GRTs were repeatable to within 10 mK.

TESTS AND RESULTS

The first experiment was to demonstrate the "London pressure" that could be generated by having the thermomechanical pump pressurize a closed line. This was done by closing valves V4, V2, V3, V7, and opening valves V8 and V10. No heat was supplied by the heater, but there was a temperature gradient across the plug, with the downstream side of the plug being warmer, because the small vessel was being pumped by the vacuum system. Figure 3 shows the resultant pressure rise with time for the three cold pressure transducers that were connected into the downstream side of the plug. The pressure transducers were accurate to within 10 torr and cold pressure transducer 2 (CP2) could read a maximum of 220 torr. Figure 3 also shows the pressure on downstream side of the plug calculated from the temperature on either side of the plug and the "London relation": $\Delta P = \int \rho s dT$ where the temperature, T; entropy, S; and density, ρ are integrated across the plug.

The next experiment was to transfer approximately 40 L from the small dewar to the large dewar along path A. This was done by opening valve V4 and allowing helium to flow with no input from the heater. The flow rate began to drop off and after 12.5 min, heater H2 was supplied 3 W of power. At a time of 18.0 min the H2 power was increased to 6 W and at 23.5 min, it was increased to 12 W. At 30 min the power to H2 was reduced to zero and the transfer was allowed to continue for another 14 min.

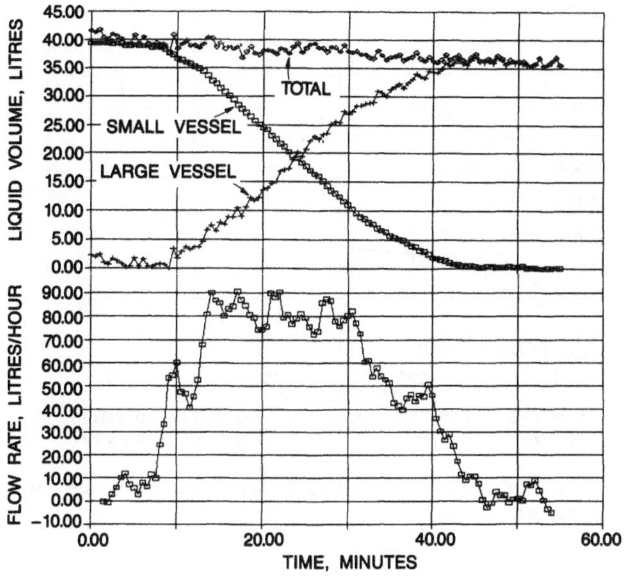

Figure 4. Liquid volume and flow rate during transfer experiment

Figure 4 shows the liquid volume in both vessels, the total liquid volume and the flow rate between the two vessels. The liquid volume shown for the small vessel is approximately the total volume while the large vessel volume has had an offset factor subtracted from it so that the volume is zero at the start of the flow test. The flow rate was determined by taking the derivative with respect to time of the small vessel liquid volume. This data has a noticeable amount of noise on it which has been traced to the liquid level probe power supplies.

It can be seen in Figure 4 that the total liquid volume decreased 3.5 ±0.5 L during the transfer of 36 L or a 9.7 ±1.4 percent loss. This compares to a loss of 2.8 L which would be expected due to the total electrical heat input and the small vessel heat leak. The difference is probably due to the heat leak into the large vessel, which is not known. In an ideal transfer, the total helium loss has been calculated[1] to be 1.3 percent at the supply temperature in the experiment of 1.5 K. The experiment was not an ideal transfer in that the power levels over 3 W caused additional helium boiloff, but no additional flowrate. If the transfer had been run with a continuous 3 W input, the resulting helium loss would have been 0.97 L or 2.4 percent. It appears that the most efficient heater power would have been somewhat less than 3 W.

Figure 5 shows the temperature profile as a function of distance from the supply dewar to the receiver dewar along the transfer line for various heater inputs and flow rates. These profiles show that the flow appears to be restricted by the porous plug at high heater powers.

A similar experiment was performed in which helium was transferred from the small vessel to the large vessel using path A, but additional heat was added at heater H4 to simulate the heat leak from a bayonet coupling. Figure 6 shows the temperature profile as a function of distance from the supply dewar to the receiver dewar along the transfer line for various heater inputs and flow rates. Note that the flow rate and temperature profile are similar for the case where 6 W are put in at H2 and where 4 W are put in at H2 and 2 W are put in at H4.

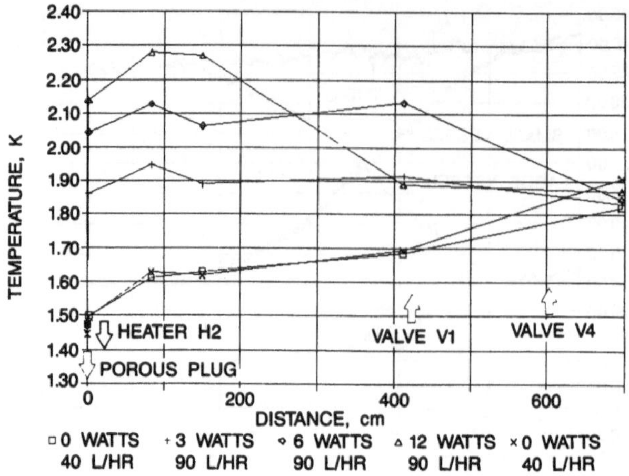

Figure 5. Temperature profile along path A at various heater powers and flow rates

The final experiment consisted of cooling and filling an initially warm vessel. The small vessel was heated above 60 K. Valves V7, V8, V9, V10, and V11 were initially opened and the rest of the valves were closed. Valve V2 was opened in an attempt to start the transfer. No transfer started, even when 1.1 W was applied to heater H6. It was noted that the pressure in the receiving (small) vessel and line was 3 torr while the pressure in the supply vessel was 16 torr as shown on Figure 7. Valve V10 was then closed and the pressure in the small vessel and transfer line increased until it was equal to the pressure in the supply dewar. Within a few minutes the pressure in the small vessel went up and the temperature came down rapidly, indicating that liquid helium was being transferred. The small vessel was then filled at a rate of 30 L/hr until it was completely full.

It is apparent that a necessary condition for liquid to break through a porous plug is that the pressure on the downstream side of the plug be equal or greater than the supply side.

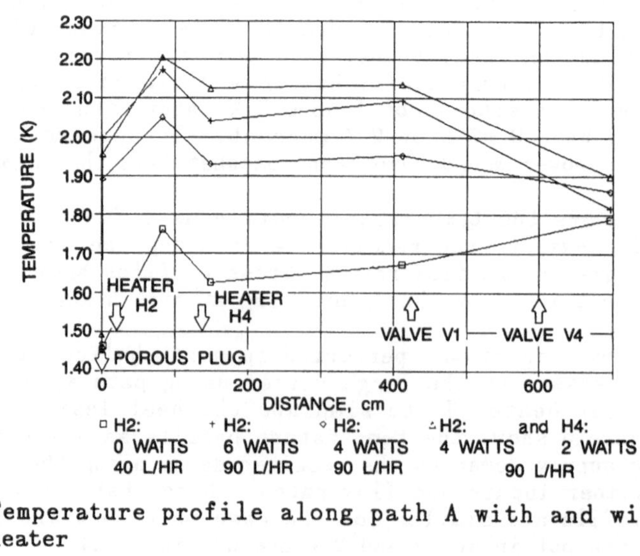

Figure 6. Temperature profile along path A with and without second heater

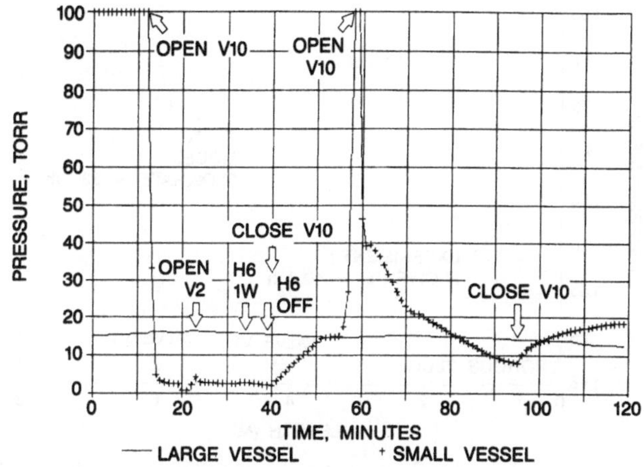

Figure 7. Pressure in vessels during fill of an initially warm small vessel

COMPARISON TO THEORY

Our theoretical understanding of helium II flow through porous plugs and transfer lines has been incorporated in a computer model.[6,7] The current model is running on an IBM PC-AT using the Lotus 1-2-3 software. In the model, the porous plug and transfer line are divided into many one-dimensional finite elements. The number of elements and the length, diameter, and heat input of each element can be easily changed. Mass flow and energy flux are conserved between the elements. The superfluid component flow is driven by the gradient of the Gibbs free energy and dissipated by the Gorter-Mellink interaction with the normal component. The normal component flow is driven by the pressure gradient and is dissipated by viscous friction. The friction factor is calculated from the Reynolds number.

Since the model is one dimensional, it does not account for the pressure drop due to the turning of the fluid as it goes through bends and valves. This is accounted for by using the concept of equivalent length; that is, geometries which cause the fluid to turn are modeled by making the flow path longer. However, the amount by which the flow is increased for a given geometry, can only be determined by empirical data.

Another parameter which must be determined empirically is the tortuosity of the plug. The tortuosity is the ratio of the average flow path length to the thickness of the plug. A tortuosity greater than one affects the flow by increasing the flow path length and decreasing flow area due to conservation of plug volume. The tortuosity used in the model, 2.6, was chosen to match the observed temperature rise across the plug. The equivalent length was then chosen to match the observed flow rate and transfer line temperature profile.

Figure 8, 9, and 10 show the observed and calculated temperature profiles and flowrates from the supply dewar to the receiving dewar for three different cases, using the same model. Only the boundary conditions of supply and receiver dewar temperature and heater power were changed. Note that the model temperature profiles and flow rates are in good agreement with the experiment.

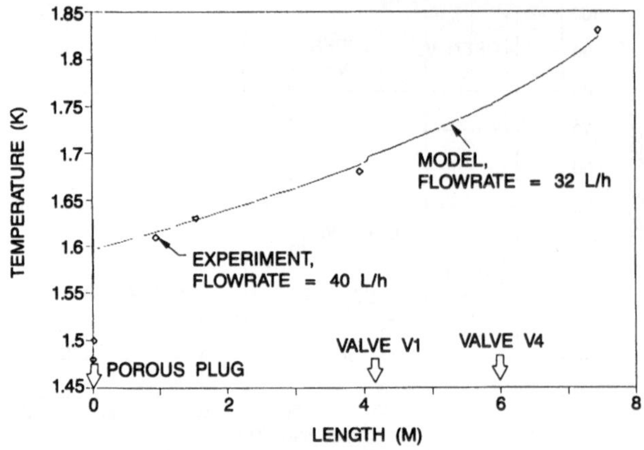

Figure 8. Observed and calculated temperature profiles with no heater power

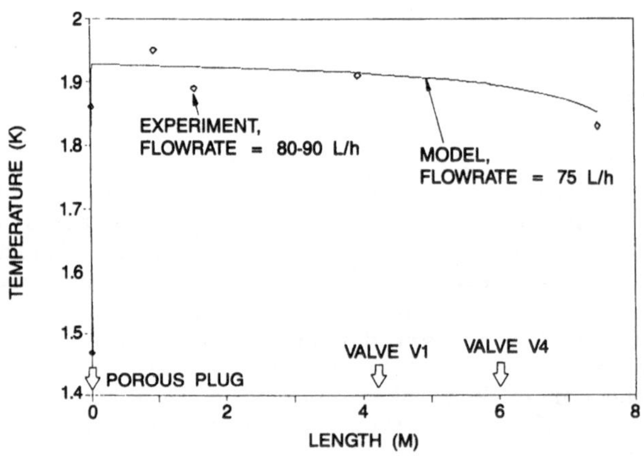

Figure 9. Observed and calculated temperature profiles with 3 W heater power

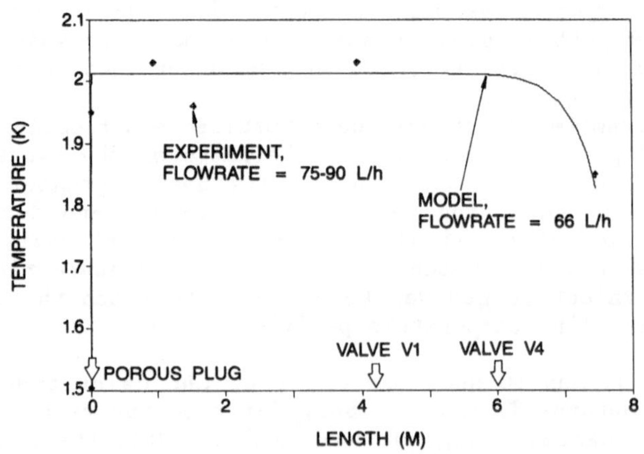

Figure 10. Observed and calculated temperature profiles with 6 W heater power

CONCLUSIONS

1. Thermomechanical pumps have been demonstrated to move large volumes of liquid at high flow rates through realistic lines and valves.

2. Thermomechanical pumps have been demonstrated to be able to fill a warm, empty tank and line.

3. The thermomechanical pump used in this experiment can generate pressures over 240 torr.

4. The BASD computer model can predict the flow rate and temperature profiles of thermomechanical pump helium II transfer systems.

ACKNOWLEDGEMENTS

This work was supported, in part, by NASA Ames Research Center. The authors wish to acknowledge Lloyd Davis, William Deshler, Dale Durbin, Ralph Einertson, Joe Kennedy, Diane Rossi and Jerry Wells for their assistance in obtaining the experimental data.

REFERENCES

1. C. M. Lyneis, M. S. McAshan, H. A. Schwettman, Applications of the Fountain Effect in Superfluid Helium, Department of Physics, Stanford University, September 1968. NTIS AD678719.
2. W. F. Brooks, Helium Transfer for the space station era, Cryogenics 26:61 (1986).
3. A. J. Mord, et. al., Concepts for on-orbit replenishment of liquid helium for SIRTF, Cryogenics 26:68 (1986).
4. P. Kittel, Liquid helium pumps for in-orbit transfer, Cryogenics 27:81 (1987).
5. V. Arp, Comparison of centrifugal and fountain effect pumps, Cryogenics 26:103 (1986).
6. H. A. Snyder, Dewar to dewar transfer model for superfluid helium transfer, Space Cryogenics Workshop, University of Wisconsin - Madison, 21-23 June 1987.
7. L. A. Hermanson, A. J. Mord, H. A. Snyder, Modeling of superfluid helium transfer, Cryogenics 26:107 (1986).

CHARACTERIZATION OF A CENTRIFUGAL PUMP IN He II

J. G. Weisend II and S. W. Van Sciver

Applied Superconductivity Center
University of Wisconsin-Madison
Madison, Wisconsin

ABSTRACT

As part of an effort to determine the feasibility of helium transfer in space a centrifugal pump was tested in He II at a variety of flow rates, pump speeds and fluid temperatures. The pump which has a straight bladed impeller 6.86 cm in diameter generated a maximum pressure rise of 15 kPa and a maximum flow rate of 22 g/s for the conditions of the test. Pump performance seems independent of fluid temperature and is good agreement with the values predicted by the manufacturer. Over the range of flow coefficients, measured maximum efficiency is around 50%. Cavitation is observed in the pump and is thought to be highly dependent on the local heating of the helium in the pump. Preliminary measurements of the noise spectra of the pump suggest a possible mechanism to predict the onset of cavitation.

INTRODUCTION

Centrifugal pumps have important applications in superfluid helium (He II) systems. Proposed uses of these pumps including transferring helium to satellites[1] and forced flow cooling of superconducting magnets and other devices.[2] Successful employment of centrifugal pumps in practical systems requires a thorough understanding of pump behavior in He II. Previous researchers[3,4,5] have shown that centrifugal pumps do function in He II. A basic question pertains to whether the superfluidity of He II results in any unique behavior in centrifugal pumps. More practical questions pertain to pump performance characteristics including: the produced head, flow rate and efficiency in He II; the amount of heat deposited in the helium by the pump, the conditions under which the pump cavitates and the amount of hydraulic noise generated by the pump.

DESCRIPTION OF EXPERIMENT

The pump tested in these experiments was manufactured by Barber-Nichols Engineering Co. it consists of a 6.86 cm diameter impeller with six straight blades housed in a volute casing, which has a diffuser throat diameter of 1.15 cm. Figure 1 is a schematic of the pump impeller

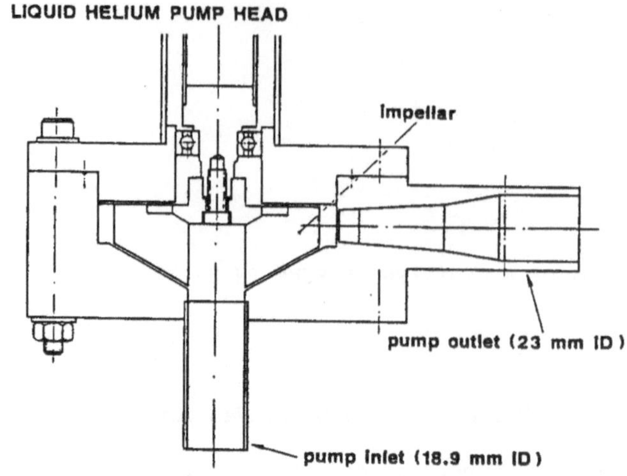

Fig. 1. Schematic of pump impeller housing.

housing. There is no inducer in the pump. A 44.5 cm long shaft connects the impeller with a room temperature motor. The impeller shaft, which rotates at moderate frequencies (f ~ 100 h_z) is supported with gas lubricated ball bearings. Original design criteria required that the pump have a maximum mass flow of 100 g/s, generate a maximum head of 100 kPa and dissipate less than 10 watts into the helium. The pump is an experimental unit designed to work under a variety of coolant conditions including He I, He II and two-phase helium. It is not optimized for He II transfer in space.

Figure 2 is a schematic of the flow loop in which the pump was tested. The loop consists of a section and a discharge line connecting the pump to a saturated helium bath. The suction line contains a 4.12 m long, 1.99 cm O.D. straight stainless steel tube in line with a 2.54 cm O.D., 52 cm long bellows. Two discharge lines have been studied. The first consists of a 4.1 m long, 0.95 cm O.D. straight copper tube and a 15 cm long 0.95 cm O.D. bellows. At the end of this line is a cryogenic valve permitting adjustment of the flow resistance in the test loop. The second discharge line contains a 61 cm long, 0.95 cm O.D. bellows and a 1.27 cm O.D., 3.07 m long straight copper tube. The cryogenic valve was removed from the second discharge line.

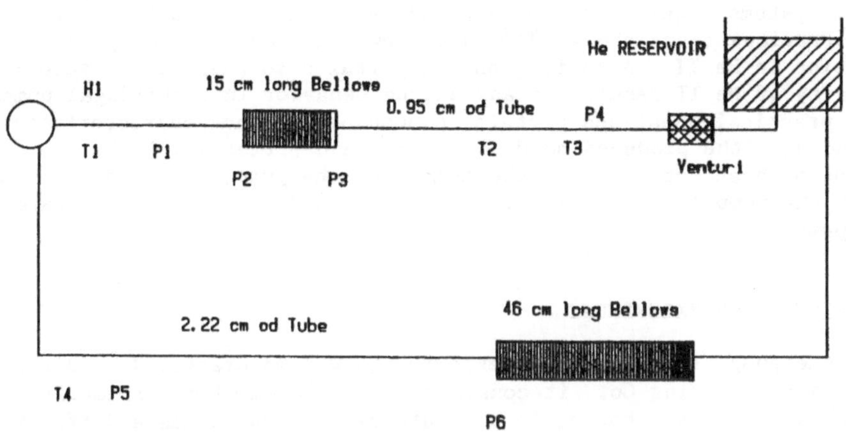

Fig. 2. Helium pump test loop schematic.

As shown in Figure 2, the test loop is instrumented to measure temperature at various points in the loop, pressure rise across the pump and pressure drop through the bellows and straight tubing. The pressure sensors are Siemens KPY 12 transducers which have a nominal range of 0 to 0.2 MPa. These transducers are calibrated _in situ_ at the start of each experiment and are accurate to ± 60 pascals. Thermometers T1, T2 and T3 are Allen Bradley nominal 75 Ω carbon resistors. Thermometer T4 is a Scientific Instruments Ge Resistor. The thermometers are calibrate _ex situ_ and inserted into the loop. Temperature measurements are accurate to within ± 10 mK absolute, but relative changes of 1 mK are detectable. The flow rate was measured by a venturi flow meter installed in the discharge line. The venturi was calibrated previously in a positive displacement flow system by Walstrom and Maddocks.[6] Unfortunately, the venturi meter cavitates at flow rates greater than 10 g/s. To allow for higher flow rate measurements the venturi was used to measure an average friction factor for the straight copper tube in the discharge line. For flow rates above venturi cavitation the pressure drop in the straight section of the discharge line is used to determine the mass flow rate. The mass flow are accurate to ± 0.2 g/s.

Measurement of heat deposition with the flow loop assembly is required to determine the pump thermodynamic efficiency. To accomplish this task, the rate of helium evaporation from the saturated reservoir was measured at room temperature and atmospheric pressure by means of a Hasting laminar flow meter located on the vent line of the helium vacuum pumps. This method allowed for measurement of the total heat applied to the flow loop to within ± 0.1 watt.

The test loop is operated inside the recently constructed Liquid Helium Flow Facility (LHFF) consisting of a 5 m long horizontal cryostat with a vertical cryostat at one end containing the saturated helium bath. The impeller housing and test loop are vacuum insulated and surrounded by an actively cooled 4.5 K radiation shield inside the LHFF.

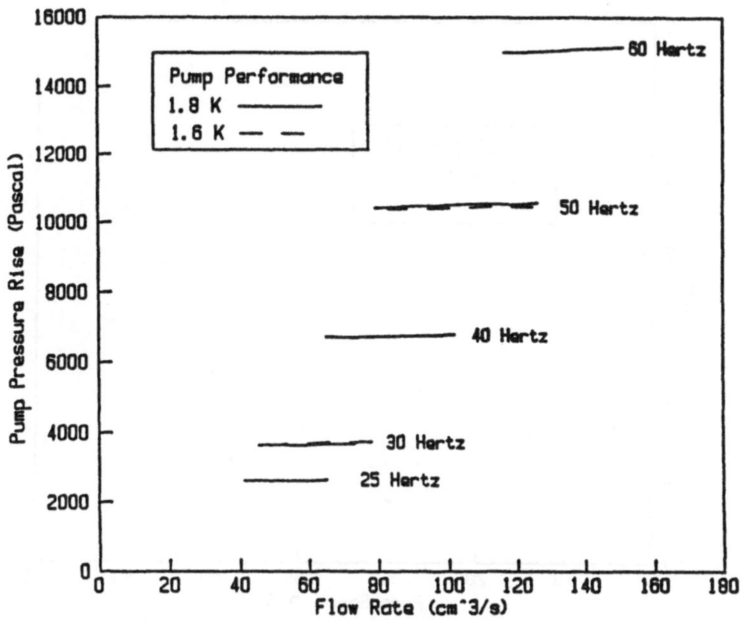

Fig. 3. Pump pressure rise as a function of flow rate.

PUMP PERFORMANCE

Figure 3 shows the total pressure rise and flow rate generated by the pump for a series of constant pump speeds. For each speed the flow rate was adjusted by changing the position of the valve in the discharge line from fully open to almost fully closed. The maximum flow rate observed was 150 cm^3/sec (22 g/s), the maximum pressure rise across the pump was 15 kPa. Due to cavitation in the pump, data were not taken at pump speeds much higher than 60 Hz. The flatness of the curves is explained by the fact that the pump is tested over a small section at the lower end of its range. Flow impedences in the test loop prevent the pump from reaching higher flow rates at the pump speeds tested. It should be noted that there is no appreciable difference in pump performance between 1.8 K and 1.6 K, suggesting that the behavior is independent of superfluid properties.

Figure 4 shows the total heat deposited in the helium by the pump and the moving fluid. This heat was determined by measuring the boiloff rate in the helium reservoir. The 1.37 watt heat load at zero pump speed results from the heat leak into the test loop most probably dominated by the thermal conduction down the pump shaft. The heat dissipated in the helium increases approximately quadratically with pump speed. Since mass flow rate increases proportionally to pump speed the quadratic dependence of the heat generation suggest that frictional processes in the pump and test loop are the principal loss mechanisms. The heat deposited is constant at a given pump speed.

The data in Fig. 3 can be cast in non-dimensional form by defining two parameters: the head coefficient, Ψ, and the flow coefficient, Φ.

$$\Psi = \frac{\Delta P}{\rho N^2 \pi^2 D^2} \tag{1}$$

$$\Phi = \frac{\dot{V}}{\pi D N A} \tag{2}$$

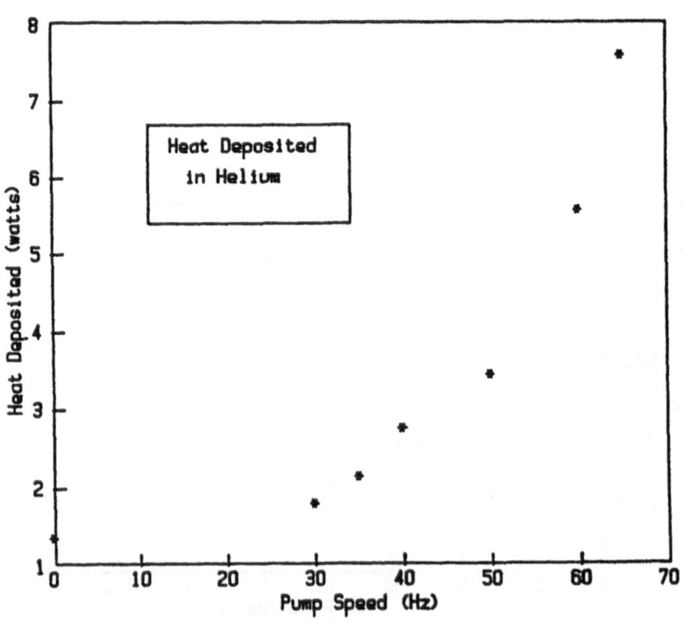

Fig. 4. Heat deposited in helium as a function of pump speed.

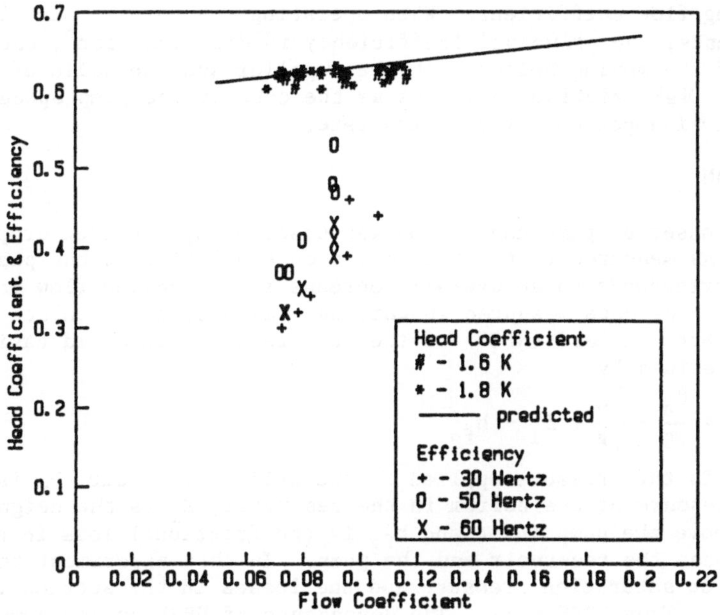

Fig. 5. Head coefficient and process efficiency.

where ΔP is pressure rise across the pump
N is pump speed (Hz)
D is impeller diameter
A is area of the diffuser
$\dot{V}$ is volumetric flow rate

These non-dimensional terms permit the generation of a universal curve for the pump. Plotted in the upper portion of Fig. 5 are the non-dimensional pump performance data. Again note the results are independent of temperature. The solid line in the figure is the head and flow coefficients of the pump predicted by Barber-Nichols. This prediction was based upon treating He II as an ordinary fluid. There is close agreement between the observed data and predictions again supporting the hypothesis that the pump performance does not depend on superfluid properties.

The measured efficiency of the pump as a function of flow coefficient and pump speed is also displayed in Fig. 5. The high effective thermal conductivity of He II results in sufficient heat transfer through the pump inlet to raise the temperature of the helium at the pump inlet above the bath temperature. It is not possible therefore to define the efficiency in the conventional way, in terms of the enthalpy difference across the pump. We have chosen to define the efficiency η as

$$\eta = \frac{\Delta P \dot{V}}{Q} \tag{3}$$

where ΔP is the total pressure rise generated by the pump, $\dot{V}$ is the volumetric flow rate and Q is the total heat dissipated by the pump and the moving fluid. The value of Q is determined by subtracting the measured heat leak with the pump off.

In Fig. 5, the efficiency of the pump is seen to depend on the pump speed as much as it does on flow coefficient. Points with the same flow coefficient but at different pump speeds have widely different efficiencies. For a given pump speed, however, the efficiency drops with decreasing flow coefficient. When operating the pump at such low flow coefficients, the principal inefficiency is disk function[7], the interaction of the moving helium with the impeller and the walls of the pump housing. Disk friction increases as the cube of the pump speed and is relatively independent of the flow rate.

CAVITATION

The onset of pump cavitation was observed by a sudden drop in the pressure as measured by the transducer on the outlet of the pump. This effect corresponds to an overall decrease in the helium flow rate through the loop. The data measured in this way was used to calculate the Net Positive Suction Head (NPSH) of the pump required to avoid cavitation. NPSH is defined by

$$NPSH = \frac{P_a}{\rho g} - \frac{P_v}{\rho g} + Z_i - h_{fg} \qquad (4)$$

where P_a is the pressure applied to the helium reservoir, P_v is the saturation pressure of the helium in the reservoir, Z_i is the height of the helium above the pump inlet and h_{fg} is the frictional loss in the suction line between the reservoir and the pump. In this experiment the reservoir was at saturation pressure and the losses in the suction line may be neglected. Thus NPSH = Z_i. The dependence of NPSH on the pump speed is shown in Fig. 6.

The key to understanding cavitation in this pump appears to be related to the heat deposited locally by the pump into the helium in the impeller housing. The heat raises the temperature and saturation pressure of the helium in the pump increasing the possibility that the helium reaches its saturation pressure at some point in the pump, thus causing cavitation. An illustration of this effect is seen when comparing the required NPSH at a given pump speed and different flow rates. As previously discussed, the pump efficiency drops with decreasing flow rate at a constant pump speed. The combination of this decreased efficiency and the reduced convection heat transfer at lower

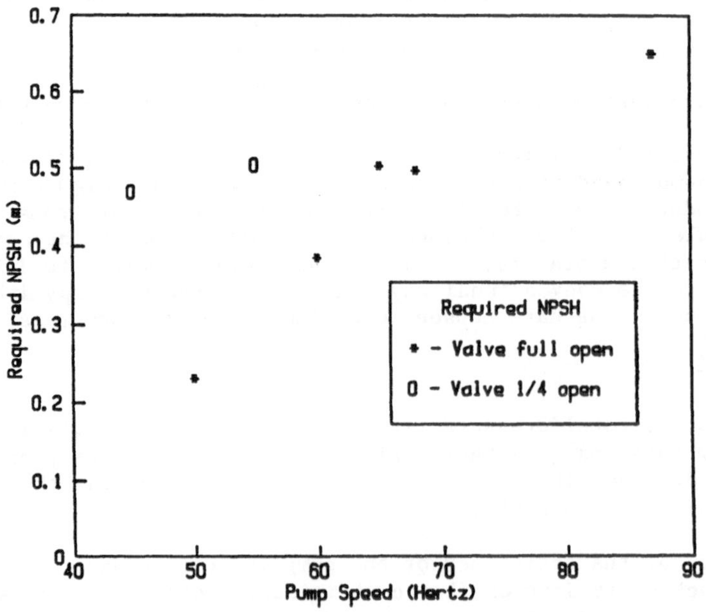

Fig. 6. Required net positive suction head as a function of pump speed.

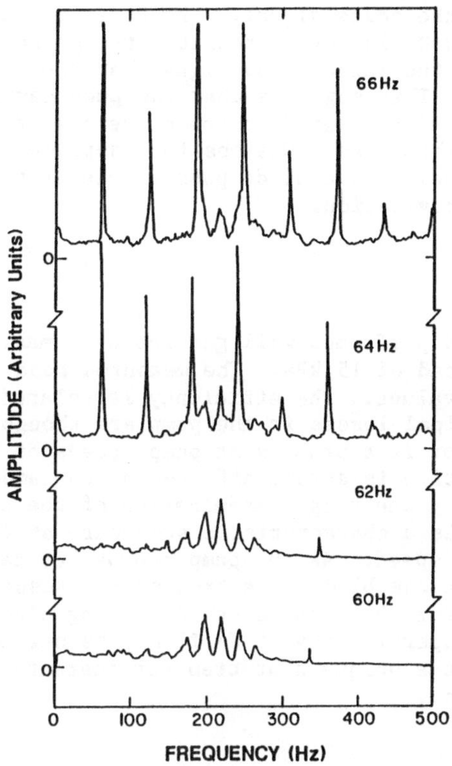

Fig. 7. Pump noise spectrum near cavitation.

flow rates results in an increase of the temperature of the helium in the pump as seen by the thermometer at the pump outlet. Thus, a higher NPSH is required to avoid cavitation. Looking at Fig. 6, it is seen that the two cases where the back pressure value is only 1/4 open (and the flow rate is correspondly reduced) require a NPSH approximately 0.3 meters higher than corresponding speeds with the value full open.

NOISE GENERATION

In order to examine the noise produced by the pump, the frequency spectrum of the pressure transducers signals were studied. Figure 7 shows the frequency response of the transducer on the pump outlet as the pump nears cavitation. The bottom most trace shows a broad 5 peaked signal around 220 Hz. This signal is first observed at lower pump speeds growing out of the background noise spectrum. The peaks location appear independent of pump rotation speed. The next trace from the bottom shows the pump at a slightly higher speed. Here the broad multipeaked signal has grown and an additional signal is seen at 349 Hz which is approximately the blade pass frequency (6 x pump speed) of the pump.* Previous work[8] done on centrifugal pump noise suggests that the primary noise should be seen at the blade pass frequency. However, in this experiment we only see significant noise at this frequency as the pump nears cavitation. Perhaps this noise signal is a precursor to cavitation. It is possible that small bubbles may be generated in the pump at this point and are too small to noticeably affect pump performance but are large

*Pump speeds quoted are nominal and may vary by a few Hertz.

enough to increase the noise signal. The upper two traces show the pump cavitating. The 220 Hz is still present although somewhat suppressed. More noticeable are the large noise signals at each of the harmonics of the pump frequency. This suggests that the pump cavitation results in noise at all frequencies related to the blade pass frequency. Further work is planned in this area. One possible application would be to use the increase in signal at the blade pass frequency to more precisely study the onset of cavitation.

CONCLUSION

The pump tested performed well generating a maximum flow rate of 22 g/s and a maximum head of 15 kPa. The measured head coefficient agrees with the predicted values. The efficiency is related closely to pump speed and the principal losses in the pump are thought to be due to disk friction. Cavitation is a problem at pump speeds of 60 Hz or higher and the onset of cavitation is strong affected by the amount of heat deposited in the helium by the pump. Examination of the noise spectra at the pump outlet indicates a characteristic structure at 220 Hz that seems independent of pump speed. As the pump approaches cavitation there is an increase in noise at the blade pass frequency. Results from this experiment suggest that for the flow rates and tubing sizes of most practical He II systems the superfluidity of He II may be neglected in fluid mechanics modeling but the unique heat transfer characteristics of He II must be taken into account.

ACKNOWLEDGEMENTS

The authors gratefully acknowledge the technical assistance of J. Babson and E. Dreier. This work was funded by NASA Ames Research Laboratory under grant NAG2-358.

REFERENCES

1. P. Kittel, Liquid helium pumps for in orbit transfer, Cryogenics, 27:81 (1987).
2. A. Hofmann and A. Khalil, "Considerations of Magnet Design Based on Forced Flow of He II in Internally Cooled Cables," presented at the 14th SOFT Conference, 1986.
3. P. M. McConnell, "Liquid Helium Pumps," National Bureau of Standards, Boulder, CO (1973).
4. P. R. Ludtke, "Performance Characteristics of a Liquid Helium Pump," National Bureau of Standards, Boulder, CO (1975).
5. W. G. Steward, Centrifugal pump for superfluid helium, Cryogenics, 26:97 (1986).
6. P. L. Walstrom and J. R. Maddocks, "Pressure Drop and Temperature Rise in He II Flow in Round Tubes, Venturi Flow Meters and Valves," Paper DD-2, ICMC/CEC Conference, St. Charles, IL, June 1987.
7. Private Communication, J. Hunt, Barber-Nichols Engineering.
8. J. Tourret, A. Badie-Cassagnet, G. Bernard, J. Foucault and J. Kermarec, "Experimental Studies of Noise Emission and Noise Generation from a Centrifugal Pump," Proc. ASME WAM (1985).

PERFORMANCE OF A SMALL CENTRIFUGAL PUMP

IN He I AND He II

P. R. Ludtke, D. E. Daney, and W. G. Steward

Chemical Engineering Science Division
National Bureau of Standards
Boulder, Colorado

ABSTRACT

The performance and NPSH requirements for a small centrifugal pump were measured over the temperature range of 1.6 to 4.2 K. A close-coupled cryogenic induction motor powers the single stage pump which has a 50 mm diameter impeller. The pump performance (head and capacity) was the same for both He I and He II, in the absence of cavitation. Developed heads up to 16 m and capacities up to 900 L/h are achieved at 7000 rpm. Both a six-blade propeller inducer and a three-blade screw inducer were tested. The screw inducer requires significantly less suction head than the propeller design; the NPSH requirements are less than -100 mm for He I, and dependening on flow rate, range between 35 and 165 mm for He II.

INTRODUCTION

NASA plans several future missions wherein He II cooled satellites will be replenished with He II in space. One option for transferring the He II is to use a simple centrifugal pump. He I is routinely transferred and circulated with centrifugal pumps at superconducting magnet facilities, but there has been little previous experience pumping He II. The results reported herein are from a program to study the feasibility of pumping He II in a milli-gravity environment using a small centrifugal pump; specifically, to compare the pump performance in He II with He I, and to investigate the cavitation characteristics of the pump in both He I and He II with near-zero section head. This pump was used in an earlier helium pump study [1,2] involving mostly He I.

EXPERIMENTAL APPARATUS

The experimental pump apparatus shown in figure 1 resides in the lower half of a 15 cm I.D. x 91 cm deep strip-silvered dewar which has a pipe flange seal on top. This dewar is surrounded by a 25 cm I.D. strip-sil

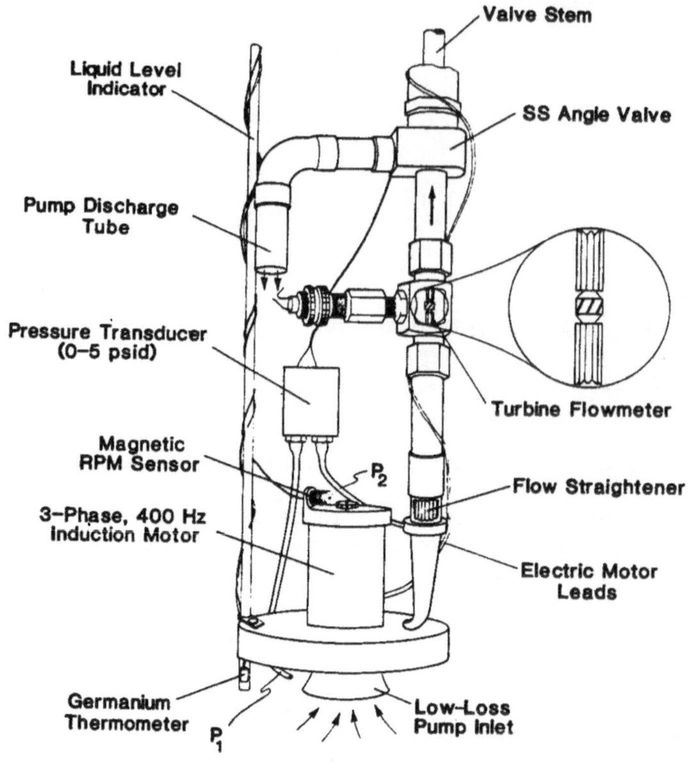

Figure 1, pump test apparatus

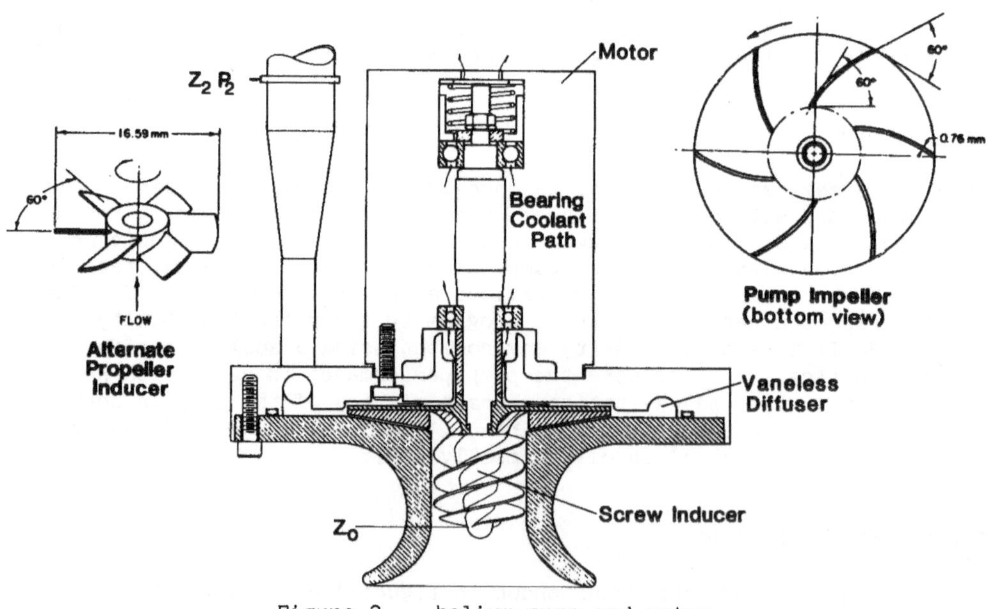

Figure 2, helium pump and motor

vered glass dewar filled with liquid nitrogen for thermal shielding. The two strip-silvered dewars provide visual, video, and photographic capability. Liquid helium is pumped from and discharged right back into the bath. The flow rate is controlled by an extended stem valve with a micrometer adjustment.

The close-coupled pump and motor are shown in figure 2. The propeller inducer shown separately is 16.6 mm diameter and has six blades inclined at 29 degrees (61 degrees from the axis of rotation). Because the propeller inducer gave unsatisfactory performance at low suction head in He II, a new state-of-the-art screw inducer was designed and installed on the pump as shown. This inducer is 18.2 mm diameter, 18.4 mm long, and has three blades with a 12 degree inlet tip angle and a 21 degree exit tip angle. The blade pitch and the root diameter increase from inlet to exit, and the leading edge of the blades are faired back and honed sharp. A special adapter mates the screw inducer to the impeller.

A bottom view of the impeller is shown separately in figure 2; the 50 mm diameter impeller has six vanes which curve back 60 degrees from the tangent. There is a vaneless diffuser on the periphery of the impeller. There is a bearing coolant flow path wherein part of the flow from the high pressure side of the pump is diverted between the back hub of the impeller and the pump housing, up through the motor ball bearings, and out through the motor housing. A removable cover plate (impeller shroud) on the bottom of the pump is sealed with a Teflon o-ring.

The motor and pump have a design speed of 6000 rpm. The motor specifications are given below.

Type:	squirrel cage induction
Number of poles:	eight
Electrical input:	3-phase, 400 Hz, 208V L-L
Power rating:	35 Watts, maximum
Ball bearings:	440C SS balls and races, filled Teflon cages
Motor housing:	All stainless steel
Efficiency at 4 K:	62 to 72 percent[3]

INSTRUMENTATION

A variable reluctance type pressure transducer of all welded construction measures the static pressure across the pump at points P_2 and P_1, as shown in figure 1.

A turbine type flowmeter with a 9.33 mm bore measures the flow through the test loop. There are flow straighteners on both sides of the turbine wheel, and a non-magnetic low-drag pickoff senses the rotating turbine blades. Two partially enclosed ball bearings support the turbine wheel; the balls, races, and cages are all constructed of 440C stainless steel. A thin-wall tube bundle five diameters upstream pre-conditions the flow to the meter.

The pump and motor speed are measured with a magnetic sensor and a ferromagnetic spur gear attached to the motor shaft. A three-phase electrical power supply with variable voltage and variable output frequency powers the motor, and a polyphase wattmeter measures electrical power to the motor.

The pressure transducer was calibrated at 4.2 and 1.8 K, and there was a slight difference in the calibration. The standard deviation of the calibration points about the curve fit was 133 and 67 Pa at 4.2 and 1.8 K, respectively. The turbine flowmeter was calibrated in a calibration apparatus especially fabricated for this purpose. Daney[4] conducted the calibration at 1.25 through 4.2 K and found no discernable difference in the calibration for He I and He II. The standard deviation for the turbine flowmeter calibration was 0.8 percent.

TEST PROCEDURE

The test procedure for measuring H-vs-Q performance is as follows. The dewar is filled to a level of 500-600 mm with liquid helium. The electrical input parameters are set to run the motor at 5000, 6000, or 7000 rpm with 8% motor slip, or else the motor is run with 400 Hz input frequency and 115 volts L-N. If the test is run at 4.2 K, the ullage pressure is controlled by a check valve. If the test is run at temperatures below 4.2 K, the bath is maintained at the desired pressure using a throttle valve and two 104 L/s vacuum pumps. The pumping capacity of the two pumps allow us to maintain a bath temperature of 1.8 K with a power input up to 10 watts, and at 1.6 K with lower power levels.

Once the electrical input parameters are checked with the pump running, the flow valve is opened in incremental steps. During the test, the liquid helium circulates within the dewar in equilibrium with the ullage pressure with no thermal stratification.

PERFORMANCE TESTS

All of the pump performance data reported in figures 3 through 7 were measured with the six-blade propeller inducer shown separately in figure 2. Moreover, the helium liquid level in the dewar was maintained at levels >350 mm in order to keep the flowmeter and pressure transducer immersed in liquid helium.

Figure 3 shows the pump performance (H-vs-Q) data for normal helium at 4.2 K for pump speeds of 5000, 6000, and 7000 rpm. Figure 4 shows a comparable plot for He II at 1.8 K. This is a typical falling curve type plot for a centrifugal pump with no significant differences from the normal helium data.

Figure 5 shows a slightly different treatment of the same 1.8 K data, plus a line representing a curve fit to the earlier[2] normal helium data. The ordinate on this plot is the pseudo-head (H*), and the abcissa is the pseudo-flow (Q*), as defined on the plot.

Figure 6 shows the H-vs-Q data for four different temperatures (1.6 to 4.2 K), indicating the data is consistent even though the measurements were conducted on He I and He II. For this set of data, the input frequency to the motor was 400 Hz and the voltage was 115 volts L-N (pump speed about 5750 rpm). The preceding data indicates that there is no apparent difficulty, nor anything unique about pumping He II with a centrifugal pump. He II obeys the pump affinity laws much the same as more conventional fluids.

Figure 7 shows the complete pump and motor performance data for a speed of 7000 rpm with 8% slip. The motor has best efficiency at about 10% slip. The best efficiency occurs at a flow rate of 1.5×10^{-4} m^3/s. The motor efficiency is comparatively low for electric motors, however this is typical for small electric motors operating in liquid helium.[5]

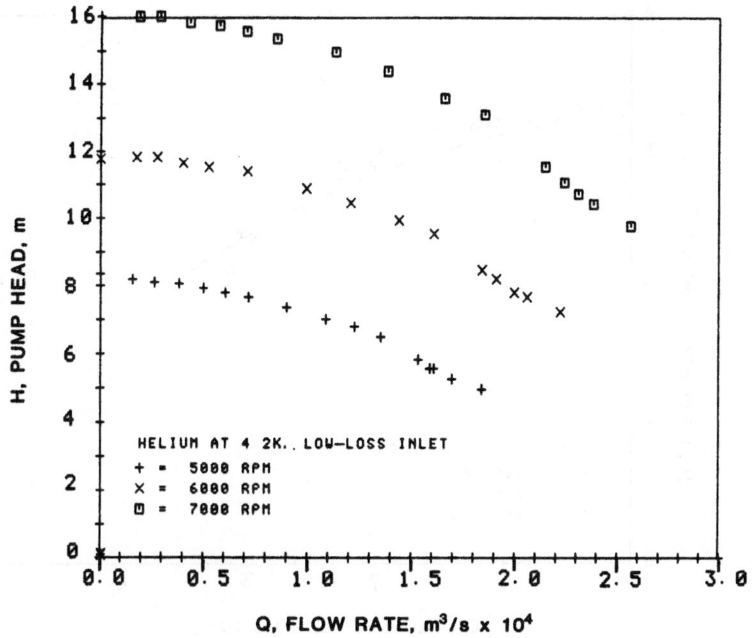

Figure 3, H vs Q data @ 4.2K

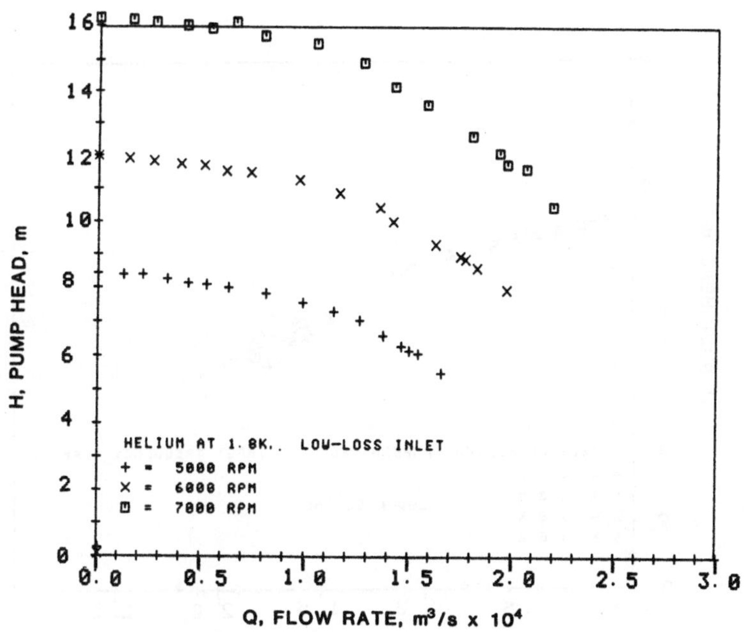

Figure 4, H vs Q data @ 1.8K

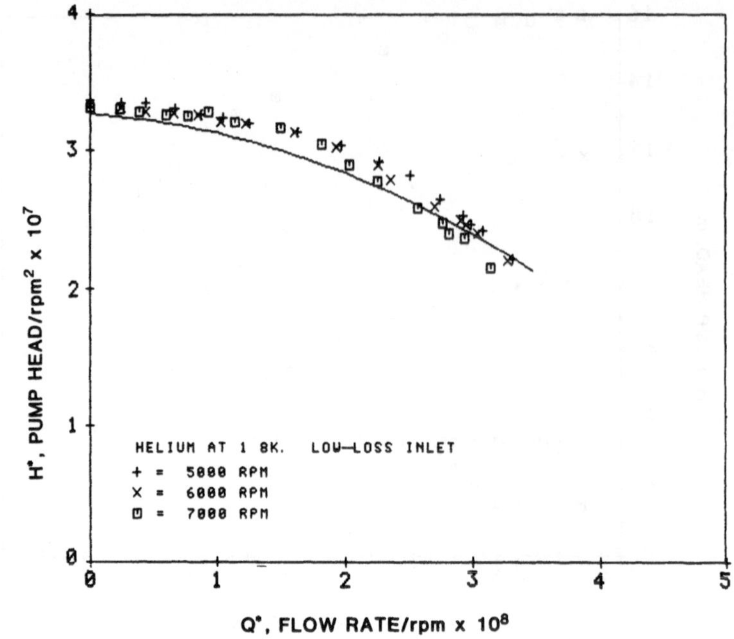

Figure 5, H* vs Q* data @ 1.8K

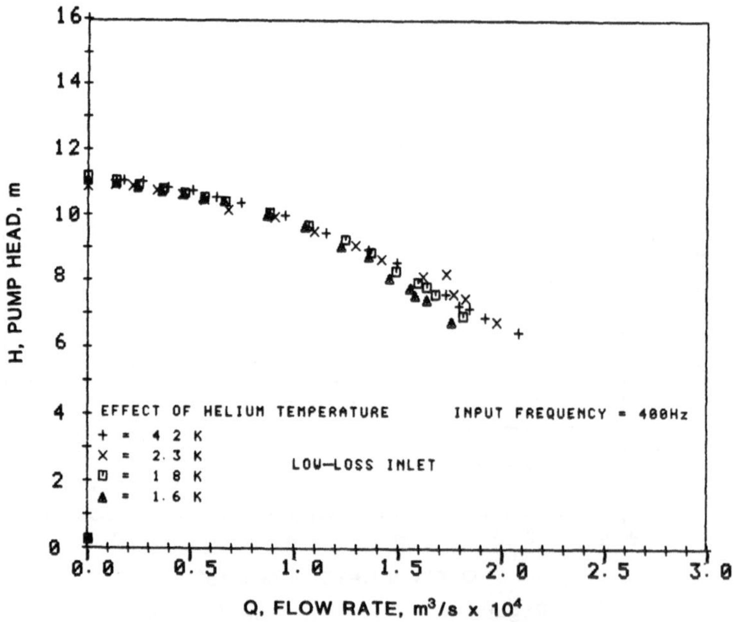

Figure 6, H vs Q @ 1.6 to 4.2K

The efficiency of the pump is comparatively low; however, one must take into consideration that the helium being pumped through the bearing coolant path is not trivial. The area of the annular space between the motor shaft and the pump housing is 0.125 cm^2. This bearing coolant flow has not been measured, and is not taken into account when computing the fluid power, which is used to calculate the total efficiency and the resulting pump efficiency.

CAVITATION TESTS

The next phase of our work led us to investigate the pump performance at low suction head (near zero NPSH) similar to the pump inlet parameters one would incur during transfer of He II in the near-zero gravity of space. From the earlier work,[2] we found that the NPSH requirement for normal helium was zero or less.

We conducted suction head tests wherein the pump parmeters were set and the test was started at a high liquid level (400-500 mm). As the electrical power into the motor slowly vaporized the liquid helium thus decreasing the suction head, the pump ΔP, flow rate, power input, and motor/pump speed were recorded. The suction head decreased until cavitation occurred, or the liquid helium reached the bottom of the pump inlet. The zero reference elevation (Z_0) for all of these tests was the leading edge of the inducer, as shown in figure 2.

Early in the test program we found that the pump with the propeller inducer started to cavitate in He II at NPSH values of about 300 mm. At this point we decided to install a state-of-the-art screw inducer on the pump to improve the performance at low suction head.

The new screw inducer dramatically improved the pump performance at low suction head; the pump starts to cavitate in He II at 150 mm, instead of at 300 mm using the propeller inducer. The new screw inducer had a slight effect on the pump performance at higher suction head; in both He I and He II we noted a slight decrease in head and capacity at high flow rates.

Figure 8 shows the results of the suction head tests in He I and He II. The pump performance in He II is dependent upon the NPSH to the pump. The screw inducer significantly improved the performance at low NPSH but the performance decreases drastically once cavitation starts, and the developed head decreases to zero as the NPSH goes to zero.

The pump performance in He I is an entirely different story. Early in the test program we found that the pump head suffered no degradation until the liquid level reached the bottom of the pump inlet bell (NPSH =-10 mm), where the pump began to ingest vapor along with the liquid. At this point we decided to extend the pump inlet to a level 100 mm below the leading edge of the screw inducer, and to make the pump inlet from polycarbonate so we could see and record any cavitation occurring in the pump inlet. With the new pump inlet, we can pump He I at 4.2 and 2.3 K down to a liquid level of -100 mm, with no degradation in head and negligible decrease in capacity until the pump starts to ingest vapor. Moreover, we observe no vapor formation inside the inlet tube.

The inlet head loss in the bell is negligible, and the head loss in the long entry tube is about one percent of the discharge velocity head. Since the vapor pressure of the helium at the pump inlet is the same as the ullage pressure, the liquid helium level we observe in the suction head tests is essentially the NPSH of the pump.

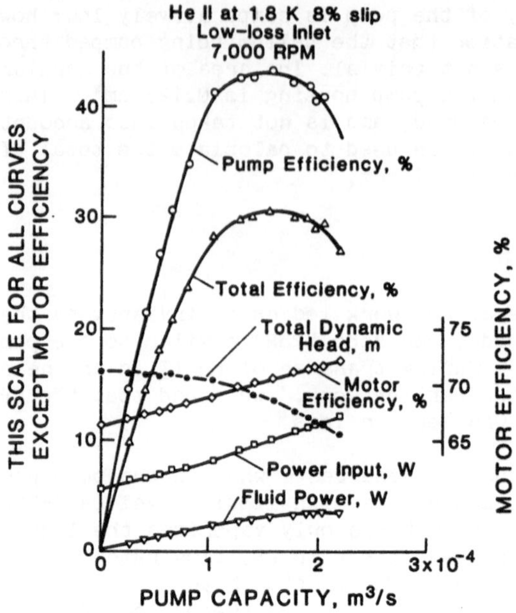

Figure 7, pump performance @ 1.8K

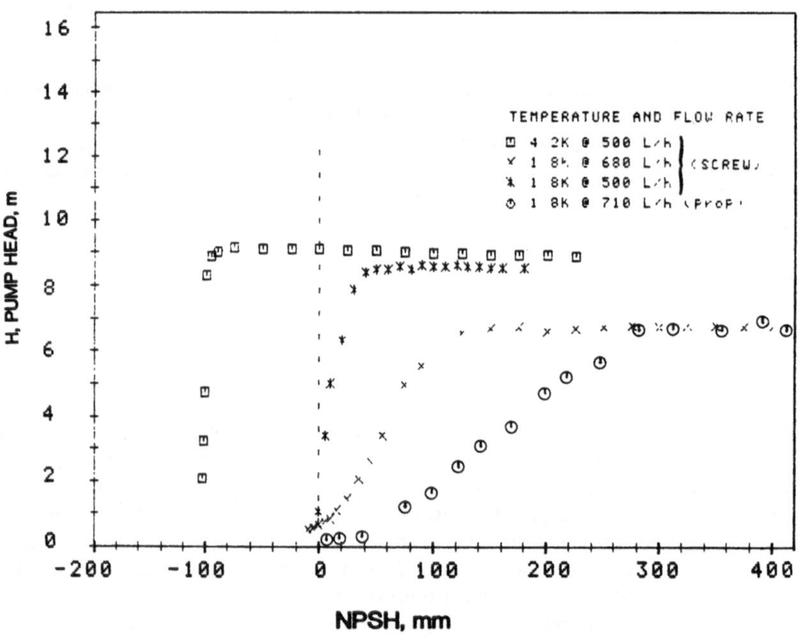

Figure 8, performance vs suction head

CONCLUSIONS

This study indicates that there is nothing difficult or unique about pumping He II with a pre-induced centrifugal pump, provided there is sufficient NPSH for the pump. He II obeys the pump scaling laws in spite of the unique characteristics of the superfluid component.

He II is prone to cavitation in a centrifugal pump, whereas conversely, He I appears to be moderately immune to cavitation. If we define an NPSH requirement (NPSHR) based on the average NPSH where cavitation begins, plus a 10% increase, then for this pump with the screw inducer, the NPSHR is less than -100 mm for He I, and depending on flow rate, ranges between 35 and 165 mm for He II.

The state-of-the-art screw inducer significantly decreased the NPSH requirements of this pump in He II, but had minimal effect on the head and capacity.

ACKNOWLEDGEMENTS

We wish to thank Adel Saad, Paul Hermann, and Glen Smith of Sunstrand Corporation for contributing the screw inducer design; without it the study would have been far less complete.

This work was sponsored and funded by NASA Ames Research Center in support of the SIRTF Program.

NOMENCLATURE

H = pump head = $\frac{P_2 - P_1}{\rho g} + \frac{V_2^2 - V_1^2}{2g} + Z_2 - Z_o$, m

H^* = pseudo-head = $\frac{\text{head}}{\text{pump speed}^2}$, $\frac{\text{m}}{\text{rpm}^2}$

LHe = liquid helium

NPSH = net positive suction head, m

P_2 = static pressure at the pump discharge (static pressure tap at Z_2)

P_1 = static pressure just outside pump inlet bell at elevation Z_o

Q = pump capacity, m^3/s

Q^* = pseudo-pump capacity = $\frac{\text{pump capacity}}{\text{pump speed}}$, $\frac{m^3}{\text{s-rpm}}$

V_1 = fluid velocity just upstream of the inducer at elevation Zo

V_2 = average fluid velocity within the pump discharge tube at the static pressure tap P_2

Zo = pump reference datum plane (leading edge of the inducer)

Z_2 = elevation of pressure tap P_2

REFERENCES

1. P.M. McConnell, "Liquid Helium Pumps," NBSIR 73-316, National Bureau of Standards, Boulder, Colorado (1973)

2. P.R. Ludtke, "Performance characteristics of a liquid helium pump," NBSIR 75-816, National Bureau of Standards, Boulder, Colorado (1975).

3. W.G. Steward, Centrifugal Pump For Superfluid Helium, Cryogenics, Vol. 26, 1986

4. D.E. Daney, Behavior of turbine and Venturi flowmeters in superfluid helium. Proceedings Cryogenic Engineering Conference, 1987.

5. Design Study of ac Motors to Operate in Helium from Liquid to Gaseous State, Redmond, J.H. and F.W. Bott, Journal of Spacecraft and Rockets, Vol. 5, No. 9, 1968.

PUMP PERFORMACE REQUIREMENT FOR THE

LIQUID HELIUM ORBITAL RESUPPLY TANKER

J. H. Lee

NASA Ames Research Center
Moffett Field, California

Y. S. Ng

Sterling Federal Systems, Inc.
Palo Alto, California

ABSTRACT

The Liquid Helium Orbital Resupply Tanker (currently renamed to Superfluid Helium Tanker) will greatly enhance the lifetime of the space missions which require superfluid helium. The Superfluid Helium Tanker pump performance requirement is driven by the superfluid helium replenishment needs of the Space Infrared Telescope Facility (SIRTF). SIRTF is one of the space missions which will require on-orbit superfluid helium resupply in the 1990s. The Superfluid Helium Tanker will carry at least 10000 L of superfluid helium and provide a minimum pump head of 170 torr (0 to 200 L/h) to cool SIRTF from 150 K to 2 K. When the SIRTF tank starts to collect liquid, a minimum flow rate of 300 L/h with a pump head of 60 torr is required to fill the 4000-liter tank.

INTRODUCTION

The system concept and design of the Superfluid Helium Tanker (SFHT) is currently being studied at the NASA Johnson Space Center. Primary users of the SFHT, such as the Space Infrared Telescope Facility (SIRTF), Advanced X-Ray Astrophysics Facility (AXAF), Gravity Probe-B (GP-B), Large Deployable Reflector (LDR), and ASTROMAG will benefit from on-orbit superfluid helium replenishment at two to three-year intervals to extend their mission lifetimes up to about 10 years.

SIRTF is chosen to be the design driver for the SFHT system design concept because it represents one of the larger near term astrophysics missions which require superfluid helium resupply. The liquid helium pump is one of the critical elements of the SFHT. Its ability to meet the needs of the users, in particular SIRTF, will be demonstrated on the Shuttle-based Superfluid Helium On-Orbit Transfer (SHOOT)[1] experiment in mid-1991. The purpose of this paper is to estimate the pump performance characteristics (pump head, flow rates) required to cool down a warm SIRTF dewar and then fill the SIRTF dewar within the constraint of the servicing mission.

SIRTF TELESCOPE SYSTEM DESCRIPTION

SIRTF is a one-meter class cryogenically cooled infrared observatory currently under study by NASA and is planned to be launched in the mid-1990's[2]. Its 4000-liter superfluid helium cryogenic system provides an initial cryogen lifetime of 2 years with 50% margin; thereafter, on-orbit replenishment of the SIRTF cryogenic system will be required every 2 years. Major components of the SIRTF telescope are shown in Fig. 1. The focal plane science instruments and the optical system are cooled by conduction to a 4000-L toroidal superfluid helium tank. The helium tank is suspended off the dewar main shell by twelve low thermal conductance supports. The insulation system consists of three vapor-cooled shields (VCS) with double aluminized

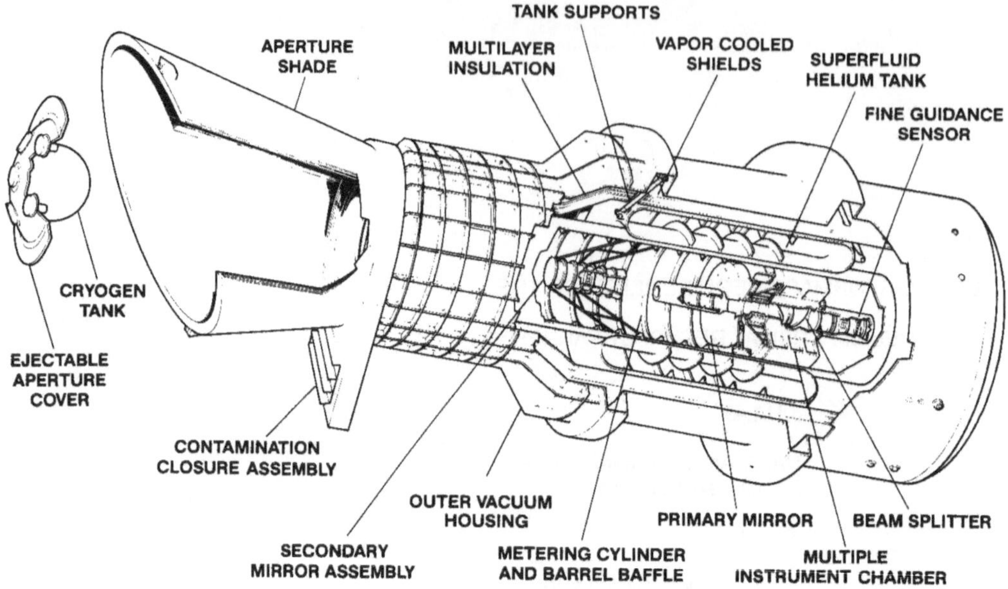

Fig. 1. SIRTF telescope configuration.

mylar/two silk-net spacers multilayer insulation. The contamination closure assembly consists of a "gate-valve" which is commandable to close during on-orbit safe-hold mode operations and also during servicing operations.

Fluid Management System

A concept of the cryogenic fluid management system is shown in Fig. 2. Vapor exiting from the helium tank is routed through the normal operation vent line to five cooling stations: the nominal "7 K" multiple instrument chamber instrument station, the forward barrel baffle, the inner vapor cooled shield (IVCS), the middle vapor cooled shield (MVCS), and the outer vapor cooled shield (OVCS). The six stepper motor- driven cryogenic valves used in the SIRTF fluid management system are similar to those which are currently under development for the SHOOT program by the Utah State University. Two types of the standard NUPRO™ valves were considered in the SIRTF cooldown and cold transfer performance analysis: the 1/2-inch and the 3/4-inch bellows valves. The nominal 1/2-inch bellows valve uses a copper stem insert while the 3/4-inch bellows valve uses a Solid Stellite #6B stem insert. The warm valves (V-7, V-8, V-9) are the nominal 1-inch ball type solenoid valves made by Whittaker. A similar type of warm valve was flown on the Infrared Telescope (IRT) dewar in the SpaceLab II mission, and will be flown on the SHOOT dewars. Two porous plug phase separators are required for the fluid management system. One phase separator which accommodates a vent rate of 6 to 10 mg/s will be used to contain superfluid helium during on-orbit science observations. The diameters of the operational vent line and the high flow vent line are, respectively, 1.19 cm and 1.83 cm diameter. The phase separator on the high flow vent line can accommodate a high vent flow rate during initial collection of liquid in the tank. Thermal lag from telescope components during cooldown imposes substantial parasitic heat on the helium tank; the temperature of the liquid helium collected initially may be higher than the lambda point temperature. Other fluid components such as the pressure relief valves and burst discs are omitted from Fig. 2 for clarity.

Cryogenic Servicing Operations

During an on-orbit cooldown or cold transfer operation, valve (V-1) and the internal fill valve (V-2) together with the poppet valve in the EVA compatible fluid coupling form a two failure-tolerant system. The transfer line between the SFHT and SIRTF is assumed to be a nominal 1.27 cm diameter, 4 m long flexible bellows section. The transfer line is precooled before helium enters the cryogen tank; bypass valve (V-3) is opened to allow the warm gas to leave through the high flow vent line. During on-orbit cooldown of SIRTF, the tank is cooled primarily by liquid helium

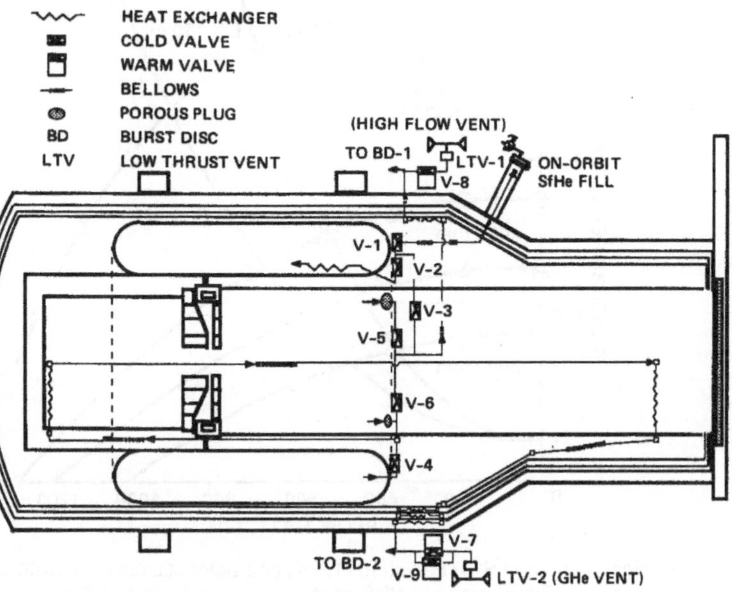

Fig. 2. SIRTF fluid management system.

flowing through a 4-m long tank heat exchanger. Opening the normal vent bypass valve (V-4) and the crossover valve (V-6) allows the warm gas to bypass the porous plugs and cool the VCS and parts of the telescope structure by way of the normal operation vent line and the high flow vent line heat exchangers. Owing to its shorter line and larger diameter, the high flow vent line can accomodate about 3 times as much vent flow as the normal vent line. When the tank starts to collect liquid, V-4 and V-6 are closed and the high flow vent valve (V-5) is opened. The helium bath temperature can be reduced to the SIRTF operating temperature of 1.8 K by venting to space. During ground-fill operation, either the high flow vent line or the on-orbit superfluid helium (SfHe) fill line can be used as a fill line.

REFERENCE SERVICING MISSION

A Space Station-based maintenance and refurbishment operation is currently baselined for SIRTF[3]. However, as a contingency, it is expected that the SIRTF and the SFHT will be designed to provide servicing compatibility with the Space Shuttle. Accessibility of SIRTF and the servicing duration limit for SIRTF at the Space Station (SS) are determined by the Orbital Maneuvering Vehicle (OMV) orbital plane change capability and the differential orbit regression rate between SIRTF and the Space Station[4]. For a nominal 900 km SIRTF orbit and an assumed constant SS altitude of 450 km, SIRTF can be accessible every 9 months when the two orbit planes become coplanar. The servicing duration for SIRTF at the SS or at the Space Shuttle is about 5 days. The roundtrip SIRTF retrieval mission by the OMV takes about 16 hours. The cryogenic servicing logistics for the EVA crewmembers are outlined in Reference 4. The estimated duration for SIRTF cooldown and/or cold transfer, thermal stabilization, and final top-off is about 3 days. With a planned 18-month cryogenic servicing interval, SIRTF is expected to have its helium supply replenished when the tank is still partially full. However, if the SIRTF retrieval opportunity is missed, the SIRTF helium tank may become empty after 9 months and the dewar as well as the telescope may warm to 150 K. So, for this contingency operation, the SFHT has to provide sufficient cryogen to cool the warm SIRTF tank to liquid helium temperatures prior to filling the helium tank. A vacuum shell temperature of 300 K was assumed for SIRTF during cryogenic servicing at the Space Station.

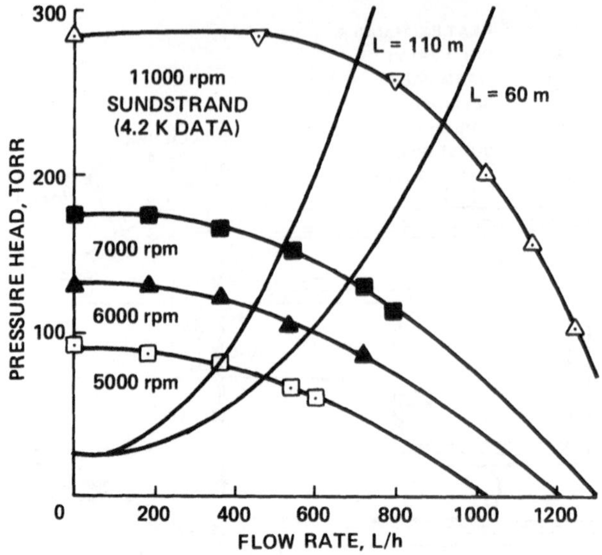

Fig. 3. Mechanical pump and SIRTF system (1.8 K) characteristic curves. Data for 5000 rpm, 6000 rpm, and 7000 rpm from NBS centrifugal pump tests at 1.8 K.

SIRTF COOLDOWN AND COLD TRANSFER ANALYSIS

Superfluid Helium Pumps

In order to assess the pump performance (pump head, flow rates) required to cool a warm SIRTF from 150 K to 2 K, we determined the flow impedance (equivalent lengths) of the fluid management system and applied this to some known pump characteristic curves to determine the operating condition at the pump outlet. Two types of pumps are currently being considered for the transfer of superfluid helium in space, the centrigufal pump (CP) and the fountain effect pump (FEP). The National Bureau of Standards (NBS) has been evaluating the performance of a centrifugal pump operated with Helium II[5], the characteristic curves for pump speeds of 5000 to 7000 rpm are reproduced in Fig. 3. Also shown in Fig. 3 is the pump characteristic curve for the Sundstrand centrifugal pump (Model 145610) operated with Helium I. One major concern in using the CP with He-II is that a net positive suction head is required; therefore, more tests are required to determine the suitability of the CP for low gravity operations. The FEP is simpler and more reliable than the CP because it has no moving parts. Recent laboratory tests of the FEP have demonstrated maximum flow rates of 325 L/h [6], and a maximum static pressure head of 290 torr [7]. Our present analysis is based on the performance characteristics of both the NBS and the Sundstrand centrifugal pumps, with a constant pressure head of, respectively, 170 torr and 284 torr for the range of liquid flow rates (15 to 180 L/h) applicable during the cooldown of a 150 K SIRTF.

System Equivalent Lengths

The system flow impedance was estimated in terms of equivalent lengths and diameters. Loss coefficients for fluid components such as the Nupro cold valves, bends and elbows, bellows, tank inlet (expansion), and tank outlet (contraction) were transformed into equivalent lengths. If the 1/2-inch valves were used, flow impedance across these valves dominate the system pressure drop during cooldown. If the 3/4-inch valves were used, half of the system pressure drop was contributed by the 3/4-inch valves.

The system characteristic curve (pressure drop vs. flow rate) for cold transfer was obtained using a computer program which solved the one-dimensional two-fluid model of He-II flow in a transfer line[8]. The system pressure was determined by the line pressure drop and the saturated vapor pressure at the receiver tank (SIRTF) required to avoid helium boiling in the transfer line. Figure 3 shows the characteristic curves for the 3/4-inch valve (L=60 m) and 1/2-inch valve (L=110 m) systems at a temperature of 1.8 K. An equivalent diameter of 1.19 cm was assumed. The 4-m flexible transfer line contributed 60% of the line pressure drop for the 3/4-inch valve case and 35% for the 1/2-inch valve case. The intersections of the pump and system characteristic curves defines several possible operating conditions of the cold transfer system.

For the cooldown analysis, the equivalent lengths, diameters, and system temperatures were used as inputs for a computer program which modeled a one-dimensional, compressible, ideal gas flow with heat transfer and friction. During system cooldown, two-phase flow may prevail[9]. Two- phase flow provides greater cooling capacity than pure gas flow; therefore, the cooldown time determined from the analysis could be conservative. The cooldown analysis assumed that the entire liquid column in the flow tube was converted into gas upon entering the tank heat exchanger. Using a constant pressure head for the 284-torr and for the 170-torr pumps, the vent flow rate was determined for each system (tank and VCS) temperature. Figure 4 shows a plot of the liquid flow rate versus system temperature for the 3/4-inch valve system.

Thermal Design Parameters

Several SIRTF design parameters have significant impact on the cooldown time. Since the optical system is cooled by conduction to the helium tank, the joint conductance between the tank and the optical system should be carefully controlled. For SIRTF, a total area-to-length (A/L) ratio of .63 cm was assumed for this interface augmented by copper straps. Another critical design parameter is the conductance required to cool the fused silica primary mirror. Several methods to bond copper braids to glass have been pursued by various groups[10,11]. So far, none of these methods have proved to be reliable in space flight environment; however, the copper-filled rubber cements which bonded the copper braids to the glass mirror survived repeated thermal cycling in laboratory tests without spalling the fused silica mirror[12]. In the analysis, the SIRTF primary mirror was assumed to be cooled by 300 copper braids (16 strands of .13 mm diameter wires per braid) attached to the fused silica mirror by copper filled rubber cement bonds. Another important design parameter is the temperature of the contamination closure assembly during cooldown and cold transfer operation. A preliminary design resulted in a temperature of 250 K for the contamination closure assembly which radiated 7 W onto the forward barrel baffle after the helium tank was filled; this resulted in a 30 mg/s boiloff. If the contamination closure assembly temperature can be reduced to 150 K, then the boiloff will be reduced to about 10 mg/s.

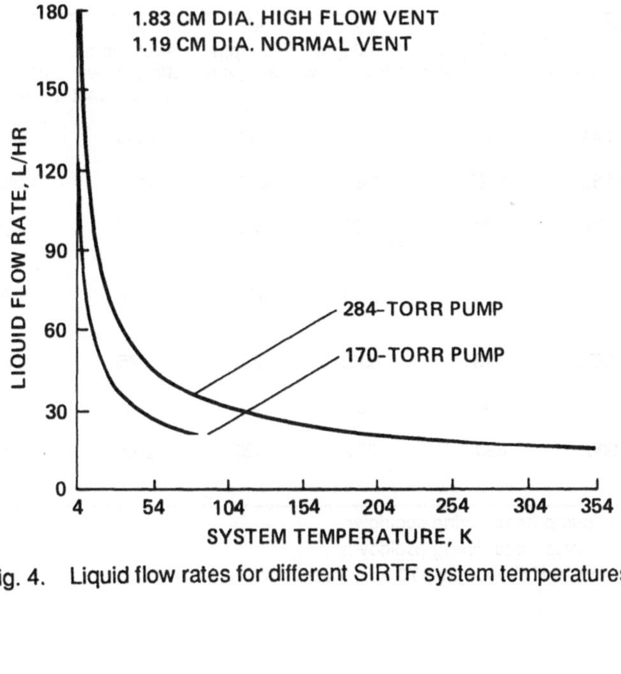

Fig. 4. Liquid flow rates for different SIRTF system temperatures.

Thermal Analysis Results

The transient cooldown behavior of the SIRTF dewar and telescope system was analyzed using the Ames Research Center Cryogenic/Thermal Analyzer Program (ACAP). During cooldown, heat was removed from the telescope components, tank, and dewar insulation system weighing a total of 2000 kg. The transfer line was cooled from 150 K to 4 K in about 15 minutes. Figures 5 and 6 show the temperature profiles of the major SIRTF components during cooldown and cold transfer operations. In Fig. 5, the pump speed was set at a constant 7000 rpm during cooldown and cold transfer operations. The primary mirror cooling rate was about 5 K per hour once it was cooled past 120 K. One possible way to reduce helium consumption was to throttle the flow rate so that the tank and the primary mirror have similar cooling rates (Fig. 6).

Results from all the analysis cases are summarized in Table 1. Cryogen consumption can be reduced if longer cooldown time is permitted. This can be accomplished by reducing the pump head (Case b), varying the pump head during cooldown (Case g), or increasing the system flow impedance (Case c & e). Comparing cases (b) and (g), varying the pump head during cooldown reduced helium consumption by 550 L with only marginal increase in cooldown time. Results also show that if a 170 torr pump head is used during cooldown, a SIRTF fluid management system using the1/2-inch valves (Case d), or the 3/4-inch valves but without a high flow vent path (Case f) is not recommended. A previous study[13] shows that if SIRTF is cooled from 150 K to 2 K within 30 hours using a constant flow rate of 200 L/h, helium consumed during cooldown is 800 L more than Case (a) which assumed a constant pump head with variable flow rates. Assuming a 72-hours cooldown (from 150 K) and fill operation, a SFHT with a 10000 L capacity would be adequate for SIRTF. This amount includes losses from transfer inefficiency and evaporation cooling of the SIRTF helium bath to 1.8 K, and a .1% per day helium loss assuming a hold time of 10 months (on the SS and on ground). As a reference, SIRTF cooldown and fill operation starting at 300 K would require 17000 L; but science instrument changeout which would necessitate warming the SIRTF telescope to 300 K is currently not required by the SIRTF Program Office[3].

PUMP PERFORMANCE REQUIREMENTS

For system cooldown, the SFHT pump should deliver a pressure head of at least 170 torr between 0 to 200 L/h; the drop-off in pressure head within this flow range should probably be less than 5%. If SIRTF cooldown time is limited to 45 hours and the dewar thermal stabilization and final top-off after cold transfer is limited to 10 hours, then a minimum flow rate of 300 L/h is required to fill the 4000-L SIRTF tank within the 5-days SS servicing mission constraint. At the 300 L/h transfer rate, the transfer line pressure loss is 25 torr, and helium is exiting the transfer line

Table 1. Helium Consumption

Cases	(a) 11000 rpm* 3/4-in. valve	(b) 7000 rpm# 3/4-in. valve	(c) 11000 rpm 1/2-in. valve	(d) 7000 rpm 1/2-in. valve	(e) 11000 rpm 3/4-in. valve no high vent	(f) 7000 rpm 3/4-in. valve no high vent	(g) variable+ speed 3/4-in. valve
Cooldown (L)	2722	1540	1372	875	1554	897	1011
Fill (L)	4185	4093	4051	4070	4100	4080	4063
Conversion Loss (L)	400	400	400	400	400	400	400
Transfer Inefficiency (L)	963	718	680	582	721	588	609
Total (L)	8270	6751	6503	5297	6775	5965	6083
Tank cooldown (h)	30	41	46	71	43	65	45
Cold transfer rate (L/h)	900	680	720	525	900	680	680

* 284 torr constant pump head during cooldown
# 170 torr constant pump head during cooldown
+ Between 150 K and 20 K (7000 rpm/170 torr), between 20 K and 4 K (5000 rpm/90 torr)

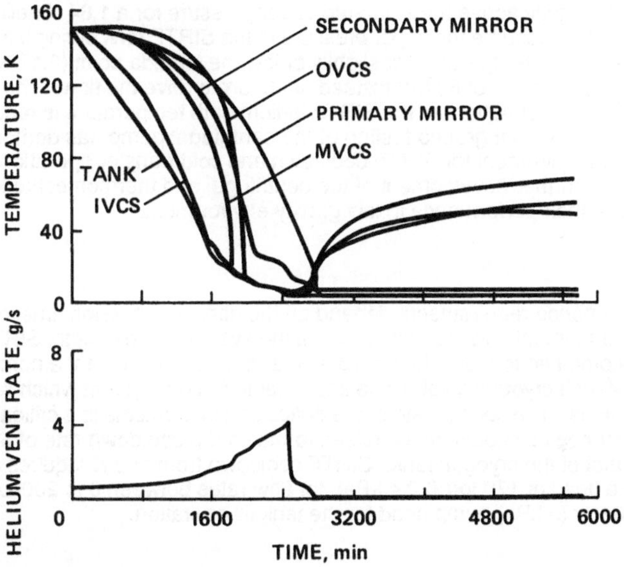

Fig. 5. SIRTF cooldown temperature profiles. Constant170 torr pump head, 3/4-inch valve system.

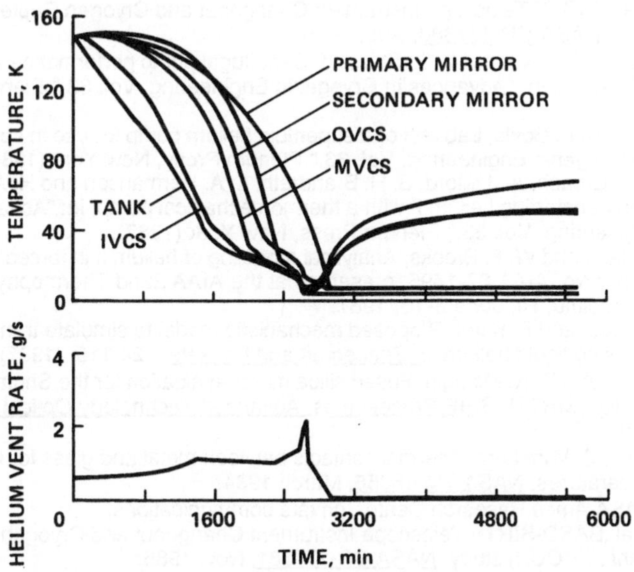

Fig. 6. SIRTF cooldown temperature profiles. Constant170 torr pump head reduced to 90 torr at 20 K. 3/4-inch valve system.

at 14.8 torr which is slightly above the saturated vapor pressure for a 1.84 K helium bath in the receiver tank. But, If we assume the vapor pressure in the SIRTF tank during initial fill operation is 35 torr which gives a bath temperature at slightly below the lambda point (2.17 K), then the pump has to supply a pressure head of 60 torr instead of 40 torr to drive the flow. The required pump head will decrease to about 40 torr as the SIRTF helium bath temperature is reduced to 1.84 K by evaporative cooling. Recent ground testing of the centrifugal pump has demonstrated that the pump performance requirement for SIRTF cooldown and cold transfer operations can be achieved. However, further development of the centrifugal and thermomechanical pumps is required to optimize their performance in low gravity environment.

CONCLUSION

The pump performance requirements depend on the servicing mission timeline constraint as well as the fluid managment and thermal design of the system (for example, SIRTF) to be cooled and filled. It is recommended that a 1.83 cm diameter high flow vent path, a normal helium phase separator, and 3/4-inch cryogenic valves be implemented into a system which requires cooldown and cold fill operations. The cooling rate of the optics and instruments is a critical parameter. Pump operation strategies should be exercised to match the cooldown rate of the optics and instruments with that of the cryogen tank. SIRTF cooldown from 150 K requires a pump with a minimum pressure head of 170 torr (22.4 kPa) for flow rates between 0 to 200 L/h and a flow rate of 300 L/h with 60 torr (8 kPa) pump head for the tank fill operation.

REFERENCES

1. M. J. DiPirro and P. Kittel, The Superfluid Helium On-Orbit Transfer (SHOOT) flight demonstration, in:"Advances in Cryogenic Engineering, Vol. 33," Plenum Press, New York (1987).
2. W. F. Brooks, R. K. Melugin, J. H. Lee, and L. Lemke, Space Infrared Telescope Facility (SIRTF) observatory design, SPIE Proceedings, 619:1 (1986).
3. C. B. Wiltsee and L. A. Manning, Servicing operations for the SIRTF Observatory at the Space Station, Paper No. AIAA-87-0504, presented at the AIAA 25th Aerospace Sciences Meeting, Reno, NV, Jan. 12-15, 1987.
4. T. C. Nast, et al., SIRTF Telescope Instrument Changeout and Cryogen Replenishment (STICCR) study, NASA CR-177380, Aug. 1985.
5. P. R. Ludtke, D. E. Daney, and W. G. Steward, Centrifugal pump performance with normal and superfluid helium, in: "Advances in Cryogenic Engineering, Vol. 33," Plenum Press, New York (1987).
6. M. J. DiPirro and R. F. Boyle, Lab test of a superfluid helium pump for use in space, in: "Advances in Cryogenic Engineering, Vol. 33," Plenum Press, New York (1987).
7. G. L. Mills, A. R. Urbach, A. J. Mord, B. H. Brandreth, L. A. Hermanson and H. A. Snyder, Experiments on transferring helium II with a thermomechanical pump, in: "Advances in Cryogenic Engineering, Vol. 33," Plenum Press, New York (1987).
8. Y. S. Ng, J. H. Lee, and W. F. Brooks, Analytical modeling of helium II in forced flow conditions, Paper No. AIAA-87-1495, presented at the AIAA 22nd Thermophysics Conference, Honolulu, HI, June 8-10, 1987.
9. Y. S. Ng, J. H. Lee, and P. Kittel, Proposed mechanistic model to simulate transfer line cool-down process using liquid helium, J. Spacecraft and Rockets, 24:115 (1987).
10. W. P. Barnes, Jr. and R. K. Melugin, Fused silica mirror evaluation for the Shuttle Infrared Telescope Facility (SIRTF), SPIE Proceedings, Advanced Technology Optical Telescopes II, 444:200 (1983).
11. K. O'Brien and F. C. Witteborn, Thermal contacts between metal and glass for use at cryogenic temperatures, NASA TM 85856, March 1984.
12. J. H. Miller, NASA Ames Research Center, private communications.
13. A. J. Mord, et al.,BASD-SIRTF Telescope Instrument Changeout and Cryogen Replenishment (STICCR) study, NASA CR-177381, Nov. 1985.

CRYOGENIC CHARACTERISTICS OF THE TOPAZ THIN SUPERCONDUCTING SOLENOID

A. Yamamoto, T. Mito, H. Kimura, T. Haruyama, H. Inoue, H. Yamaoka, E. Shiba,* H. Kichimi and Y. Doi

National Laboratory for High Energy Physics (KEK), Oho-machi, Tsukuba-gun, Ibaraki-ken, 305, JAPAN

T. Ohba and T. Ohmori

Ishikawajima-Harima Heavy Industries, Co. Ltd., Marunouchi, Chiyoda-ku, Tokyo, 100, JAPAN

ABSTRACT

The TOPAZ thin superconducting solenoid was designed and built with an "internal winding method", and has been successfully operated since 1984. Cryogenic characteristics of the TOPAZ solenoid were measured in various operational conditons and evaluated. Their results are described.

INTRODUCTION

The TOPAZ thin wall superconducting solenoid, 2.9 m in diameter and 5.1 m in length, was designed and built as a major component of the high energy particle detector "TOPAZ" used at the TRISTAN e^+e^- colliding accelerator.[1,2] The design magnetic field is 1.2 tesla in an iron return yoke at the current of 3650 A, and the total wall thickness in the radial direction is 0.7 radiation lengths (X_0). Figure 1 gives a bird's eye view of the TOPAZ solenoid and a cross section at one end of the solenoid is shown in Fig. 2. The main parameters of the magnet are given in Table I.

The magnet is expected to be as thin as possible. Safety against quenches is, however, another important aspect. The design principle is, therefore, that the magnet should be as thin as possible with the constraint that the maximum coil temperature rise is less than 80 K to avoid large thermal stress, even if the entire stored energy was dumped in the coil.

Essential to the successful fabrication of the TOPAZ thin wall solenoid was the "internal winding method".[2,3] An aluminum stabilized superconductor was directly wound onto the inner surface of a support cylinder with compression stress along the conductor. This applies a similar pressure upon the support cylinder as a magnetic internal pressure and, therefore allows the coil to be tightly pressed against the inner wall of the support cylinder with an adequate preload. This realizes better mechanical and thermal contact of the coil to the support cylinder with less

* Visiting researcher from The Furukawa Electric Co. Ltd.

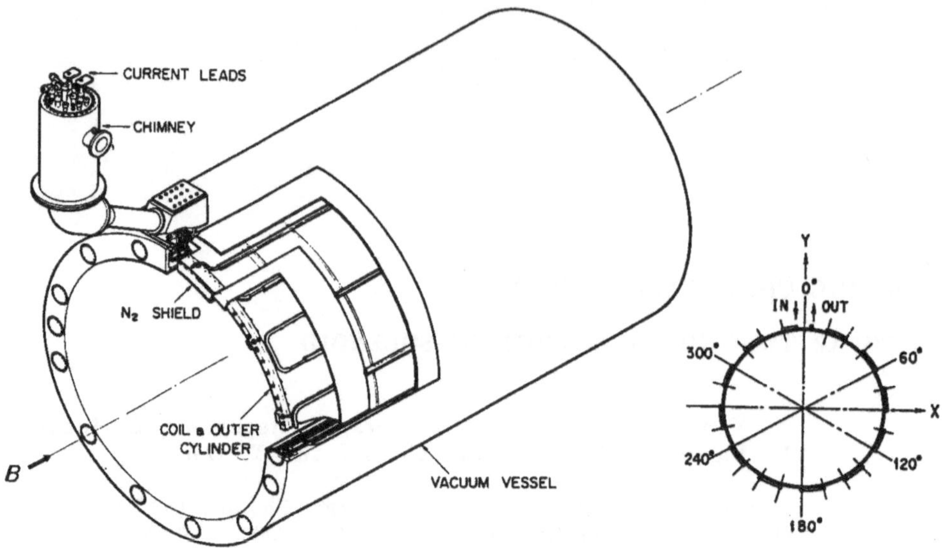

Fig. 1. Bird's eye view of the TOPAZ thin superconducting solenoid.

epoxy resin and, therefore, high reliability in supporting and cooling.

Two phase helium was forced to flow through cooling pipes welded on the outer surface of the outer support cylinder. These cooling pipes (inner diameter = 18 mm and length = 135 m) were placed in parallel to the coil azis at 15 degrees intervals and connected in series. The coil was supported from the vacuum vessel by 12 compression rods made of glass fiber reinforced plastic (GFRP) at each coil end against R − Z (radial and axial) direction, and supported by 4 rods at each end to prevent coil rotation.

COOL−DOWN

The solenoid was indirectly cooled by forced flow of gas and two−phase helium with a 300 W (or 100 l/h) helium refrigerator. The system was operated automatically by the DDC (direct digital control) system in

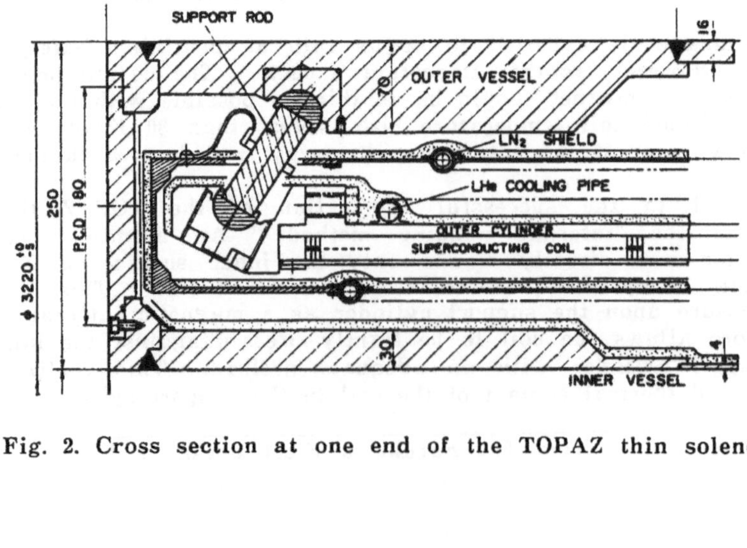

Fig. 2. Cross section at one end of the TOPAZ thin solenoid.

Table 1. Main parameters of the TOPAZ thin superconducting solenoid.

Dimensions	
Coil	**2.9 mφ x 5.1 m**
Cryostat	**(2.72–3.22)mφx5.4m**
Maximum magnetic field	**1.21 T**
Total radiation thickness	**0.70 Xo**
Current	**3650 A**
Number of effective turns	**1332**
Inductance	**2.92 H**
Stored Energy	**19.5 MJ**
Conductor	
Cross section (over all)	**3.6 mm x 18 mm**
Materials	**Nb·Ti/Cu/Al(1:1:20)**
RRR of pure Al (@0 T)	**2500**
Ic (2.4 T, 4,2 K)	**>9000 A**
Thermal load into the coil	**<30 W (+ 20l/h)**
Typical helium mass flow	**5 - 8 g/s**
Cold mass	**4.5 tons**
Total weight	**10 tons**

order to cool the thin and large solenoid carefully, under well controlled conditions[4,5]. The data–logging system was also equipped to monitor the cool–down by means of a number of temperature sensors, strain gauges, voltage taps etc, in amount of about 200 monitoring points in the coil and cryostat. The coil and radiation shield were cooled down with the following constraints in the cool–down process.

dT/dt (300K -> 100 K)	2.5 – 3.0 K/h
(100 K -> 4.5 K)	3.0 – 4.0 K/h
ΔT (Coil: mid – low)	<30 K
ΔT (GHe–in – Coil)	<60 K
ΔT (N2–Shield)	<40 K
ΔT (GN2–in – Shield)	<60 K
ΔT (Coil – Shield)	<40 K

The cool–down was carried out according to a programmed sequence. Helium mass flow rate and inlet gas temperature were automatically controlled to ensure a constant cooling speed within present limits Figure 3 shows trends during the cool–down of (a) the helium gas, coil and thermal radiation shield temperatures, (b) the inlet and outlet helium gas pressures and (c) the typical strain in a support rod at $\theta = 315°$. The temperatures of 'coil–high' and 'coil–low' are located at $\theta = 300°$ and $\theta = 60°$, respectively.

The compressive (negative) strain induced in the support rod became maximum at a coil temperature of about 220 K, as shown in Fig. 3 (c). It is due to to the saggita induced by rotation of the support rod according to the thermal contraction of the coil. Figure 4 shows strain ballances in the 12 support rods at various temperatures. The oval shape of the strain ballance at 220 K is reflected by the temperature distribution in the coil. Typical cooling characteristics in two cooling conditions at the nominal coil temperature (@ $\theta = 300°$) of 220 K are summarized in Table 2. The cool–down with a cooling speed of 3 K/h fully satisfies constraints mentioned above . In this way, the magnet was cooled to 4.5 K within 5 days under safe conditions.

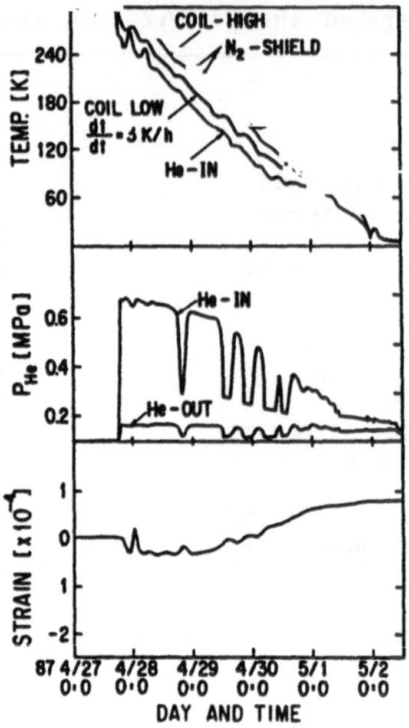

Fig. 3. Trends of (a) coil temperature, (b) He pressure and (c) support strain during cool-down.

CRYOGENIC CHARACTERISTICS

In steady state operation at a two phase helium mass flow of 5 – 10 g/s and at an inlet pressure of 0.2 kgf/cm2 (20 kPa), the temperature distribution in the coil was measured. Figure 5 (a) shows a typical result in which the coil temperature is 4.5 K +/– 0.1 K in most areas except for both ends of the coil where it was locally 5.7 +/– 0.1 K due to localized conductive heat-in-leakage of 0.4 W at each port of the suppport rod. Figure 5 (b) shows a 3D caluculation of the temperature distribution at the coil end, which well agrees with the measured result. We have understood that an additional cooling path at the coil end along the circumference should be helpful to cool the coil end below 5 K.

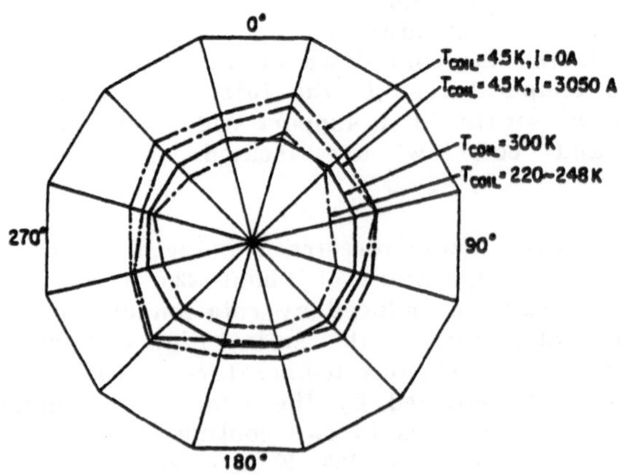

Fig. 4. Ballance of strains in the support rods during cool-down and excitation process.

Table 2. Cooling characteristics at T_{coil} = 220 K

dT/dt	2.5 K/h	3.0 K/h
Coil		
He Pressure (inlet)	0.64 MPa	0.64 MPa
He Pressure (outlet)	0.17 MPa	0.17 MPa
Mass flow rate	~16 g/s	~16 g/s
Temp (He inlet: θ=345°)	199 K	202 K
Temp (He outlet: θ=5°)	238 K	242 K
Temp (coil low: θ=300°)	222 K	222 K
Temp (coil high: θ=60°)	245 K	250 K
ΔT (coil)	23 K	28 K
Strain (support: θ=300°)	−290 μstr	−291μstr
Stress (support: θ=300°)	0.1 MPa	0.1 Mpa
N2 shield		
Temp (N2: inlet)	211 K	210 K
Temp (N2: outlet)	251 K	253 K
Temp (shield low: θ=300°)	225 K	224 K
Temp (shield high :θ=60°)	249 K	251 K
ΔT (shield)	24 K	27 K

The temperature difference between the coil and two phase helium was measured to be less than 0.1 K and it agreed with a previous calculation in which we assumed a heat transfer coeffient of 0.1 W/cm2 between two phase helium and the cooling pipe, a thermal conductivity of 0.06 W/m.K in the ground insulation of the coil and a thermal load of 0.1 W/m2 into the coil due to thermal radiation.

The thermal load of the magnet was measured in three independent ways and they were compared with each other. The minimum mass flow rate to keep the liquid helium level (quality factor: X = 0) constant in the liquid helium pot for the current leads, located at the downstream end of the cooling path, was measured to be 2.2 g/s at T_{coil} = 4.4 K. Subtracting the mass flow of 0.2 g/s for the current leads measured at that time and the estimated thermal loads of liquid helium transfer lines (10 − 15 W) coming into the magnet, we have obtained a thermal load of 25 − 30 W into the lquid helium temperature.

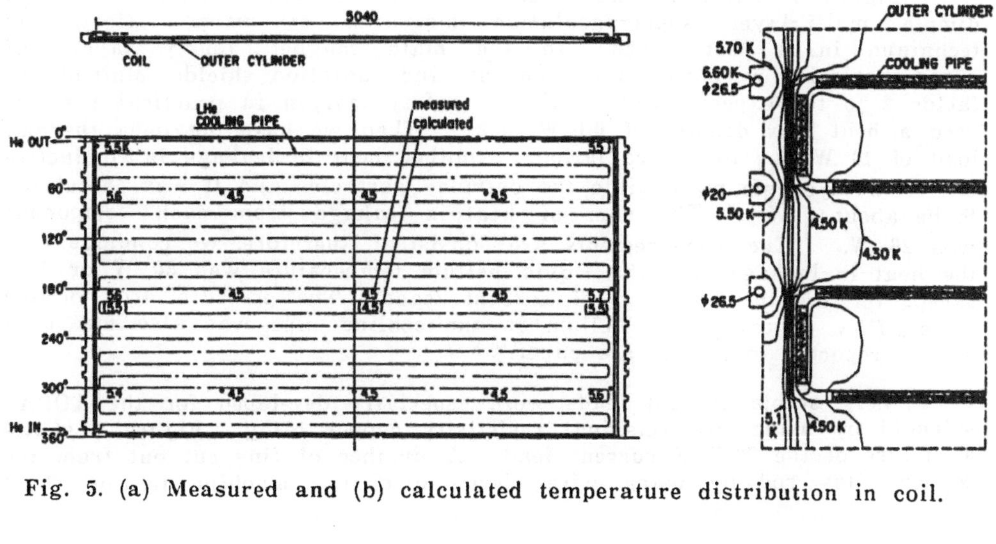

Fig. 5. (a) Measured and (b) calculated temperature distribution in coil.

Table 3. Cryogenic characteristics of the TOPAZ solenoid in the steady state operation.

Coil	
Typical He mass flow	5 g/s – 10 g/s
Quality at downstream end	< 0.5
Pressure (inlet)	0.02 – 0.025 MPa
Δp (@I.D.=18mm, l=135 m)	0.001 – 0.0025 MPa
Thermal load(@ I = 0 A)	< 30 W + 14 l/h (C.L.)
(@ I = 3050 A)	< 30 W + 17 l/h (C.L.)
Temperature(LHe)	4.4 K
Temperature(O.C.)	4.45 K
Temperature(Coil)	4.5 K (5.7 K)
N2 Shield	
Typical N2 mass flow*	8 g s (36 l/h)
Thermal load (radiation shield)	< 330 W
Temperature (Shield)	85 – 90 K

* The radiation shield of the control dewar is connected in series.

The second measurement was made by using a calorimetric method, in which the temperature rise was measured at $\dot{m}_{He} = 0$. The heat in leakage into the coil was obtained from the following equation,

$$Q = \sum M_i \cdot \int_{T_1}^{T_2} C_i \quad dT \; / \; (t1-t2) \qquad (1)$$

where M is the mass, C is the specific heat, T is the temperature and t is the time. The term i indicates each material component. The heat-in-leakage was measured to be 28 W or less in the measurement at $T_1 = 5.5$ K and $T_2 = 7.0$ K except for the chemney.

These gross measurements were compared with the sum of thermal load in each component measured separately. The heat-in-leakage from the support rod of 25 mm diameter and 95 mm length was measured in a test cryostat by using a calorimetric method. The measured result was 0.36 W with LN2 thermal anchor. By using this experimental data, we estimated heat-in-leakage from the support rods in the main magnet. It was calculated to be 12 W or less in total for 32 support rods.

Figure 6 shows an experimental results of heat-in-leakage through dimple multi-layer super-insulation measured by using calorimetric technique in a test bench.[6] In the main magnet, fourty layers of super-insulation were wound on the coil and radiation shield. Multiplying factor 2 to the experimental result for safety margin in practical use, we used a heat flux density of 0.1 W /m2 . Then we have obtain a thermal load of 10 W due to the radiation. Another heat-in-leakage is conductive one from cryogenic piping into the chimney port etc., and it was calculated to be about 4 W. The sum of heat-in-leakage from each component was 26 W. The measurements are consistent, therefore, we conclude that the heat-in-leakage into the liquid helium temperature was 30 W or less except for the thermal load of current leads. The pressure drop of the forced flow of two phase helium in the cooling pipe was also measured and is reported in a separate paper.[7]

Very stable current leads were especially developed for the TOPAZ solenoid by using low reisidual resistivity copper alloy. Figure 7 shows a picture of the TOPAZ current lead. A number of fins cut out from the copper alloy rod by using wire electro-discharge machine improve heat

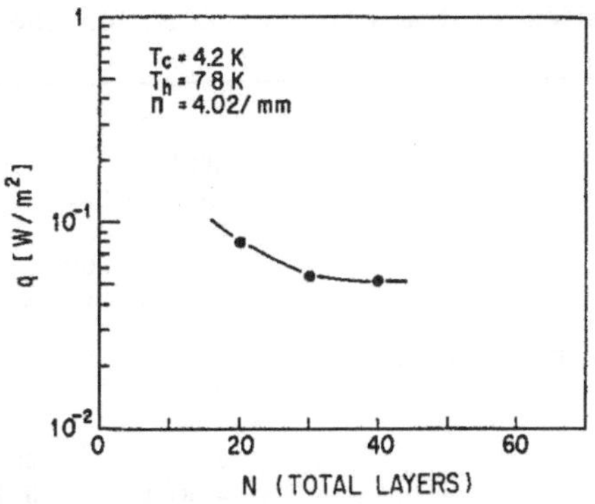

Fig. 6. Measured heat flux density through multi-layer super-insulation.

exchange between leads and gas and provide additional heat capacity. The main parameters and characteristics measured in a test cryostat are given in Table 4. The tranjent stability against an emergency stop of helium mass flow was also tested. The helium mass flow was forced to stop at I = 4000 A. The current was kept constant during 16 min and measured the temperature profiles. As shown in Fig. 8, The growth of the peak temperature was less than 100 K, and no damage was found after the test.

A summary of the cryogenic characteristics of the TOPAZ solenoid is given in Table 3.

QUENCH CHARACTERISTICS

The solenoid was successfully excited up to the maximum current of 3650 A corresponding a magnetic field of 1.2 T with no training behaviour. The operational safety and stability against a quench and other failure were investigated. Results of the tests and their evaluation were reported in a separate paper.[8] Cryogenic characteristics of a forced heater–quench (HT–Q) and a switch–off (SW–OFF) test are summarized in Table 5. During the test, the helium mass flow was forced to stop and safety valves, located at both inlet and outlet ports of the liquid helium line, were forced to open just after the power supply was turned–off. Quench back was initiated within 1.2 sec and the entire coil immediately turned into the normal state.

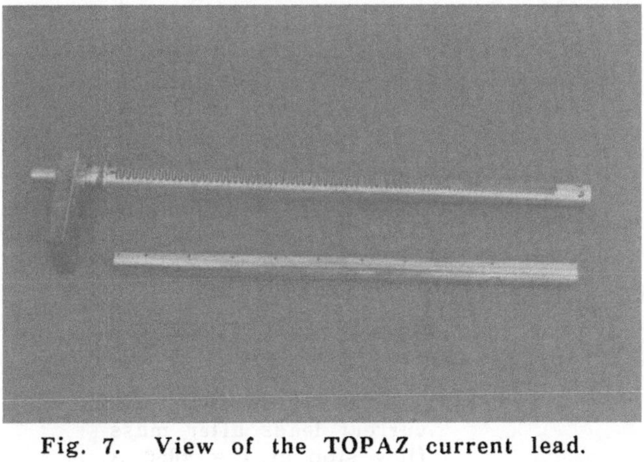

Fig. 7. View of the TOPAZ current lead.

Table 4. Characteristics of the current leads for the TOPAZ solenoid.

Parameters			
I	5000 A		
Materical	Cu·Ag(0.2%)		
RRR	32		
(I·l)/A	1.4E5 A/cm		
Outer dimension	O.D. 2.6cmφ x 50 cm		
Performance(@ T−top=260 K)			
Current	0 A	3000 A	5000 A
Voltage drop	0	23 mV	50 mV
Δp	2 kPa	3 kPa	5 kPa
LHe per pair	14 l/h	17 l/h	21 l/h

This was clearly observed from the immediate temperature rise at the coil ends. The measured temperature growth in the coil agreed with calculations by using the quench simulation code "QUENCH"[9] under an assumption of pure aluminum resistivity of 2.4 x 10^{-11} Ω·m in the stabilizer, taking the magnetoresistivity of aluminum into account. Pressure rise in the current lead pot was monitored during the quench. The maximum pressure was kept safe enough under a design value of 1.1 MPa.

CONCLUSION

Cryogenic characteristics of the TOPAZ solenoid were measured in various operational conditions and were evaluated. Their results reasonably agree with the calculations. We conclude that the TOPAZ thin superconducting solenoid fabricated by using an internal winding technique contains enough safety margin for long term operation.

ACKNOWLEDGEMENTS

We would like to thank Professors S. Ozaki, K. Takahashi H. Hirabayashi and S. Iwata of KEK for their continuous encouragements during this work. We deeply appreciate Professor Y. Watanabe and other member of the TOPAZ collaboration of KEK for their helpful discussions and cooperation.

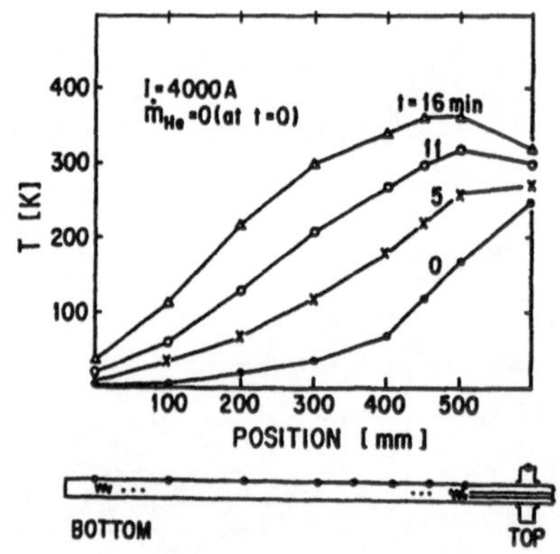

Fig. 8. Temperature growth of the current leads after mass flow stop at I = 4000 A.

Table 5. Characteristics of forced heater-quench and switch-off.

Test item	HT-Q	SW-OFF
Current	3050 A	3650 A
Magnetic field	1.0 T	1.2 T
Dump Resistor	0.1 Ω	0.1 Ω
Input power	35 W	
Input time	0.5 sec	
Time duration until SW-off	10 sec	
Max. temp. after SW-off (meas)	41 K	37 K
Max. temp. after SW-off (cal)*	37 K	34 K
Final aver. temp. in coil	31 K	35 K
Energy dump in coil	1.5 MJ	2.3 MJ
Energy recovery ratio	89 %	88 %

* calculated value with program "QUENCH" (see text)

REFERENCES

1 TOPAZ collabaration: Tristan Proposal, TRISTAN-EXP-002, KEK(1983).
2 A. Yamamoto, H. Inoue and H. Hirabayashi, A thin superconducting solenoid wound with the internal winding method for colliding beam experiments, Journal de Physics, C1 (1984) p. 337.
3 A. Yamamoto, N. Kimura, H. Kichimi, H. Inoue, H. Yamaoka, T. Mito, H. Hirabayashi, M. Ikeda, K. Ohishi, E. Shiba, T. Sato, I. Okada and S. Shiozawa, Test results of the TOPAZ thin superconducting solenoid wound with the internal winding method, Proc. of 9th Int. Conf. on Magnet Technology, Zurich (1984) p. 167.
4 Y. Doi, T. Mito, T. Haruyama, N. Kimura, O. Araoka, M. Tadano, S. Suzuki, Y. Kondo, M. Kawai, H. Yamaoka, A. Yamamoto and Y. Aw Cryogenic system of the TOPAZ thin superconducting solenoid, Proc. ICEC-11, Berlin, (1986) p.424.
5 T. Mito et al., Control system for helium refrigerators of TRISTAN detector magnets, presented in this conference (CF-6).
6 T. Ohmori, M. TsuchiyaT. Tair, M. Takahashi, A. Yamamoto and H. Hirabayashi, Multilayer insulation withaluminized dimpled polyester film, Proc. of ICEC-11, Berlin-West (1986) p.567.
7 T. Haruyama et al., Pressure drop in forced two phase cooling of the large thin superconducting solenoid, presented in this conference (DD-1).
8 A. Yamamoto, T. Mito, N. Kimura, T. Haruyama, H. Yamaoka, O. Ara M. Tadano, S. Suzuki, Y. Kondo, M. Kawai, H. Kichimi and Y. Doi: Quench characteristics and operational stability of the TOPAZ thin superconducting solenoid, Jpn. J. Appl. Phys. Letters, Vol. 25, No. 6 (1986) L443.
9 N. M. Wilson, "Superconducting magnet" Oxford Univ. Press., Oxford (1983) p. 218.

PRESSURE DROP IN FORCED TWO PHASE COOLING OF

THE LARGE THIN SUPERCONDUCTING SOLENOID

T. Haruyama, T. Mito, Y. Doi and A. Yamamoto

National Laboratory for High Energy Physics, KEK
Tsukuba, Ibaraki, Japan

ABSTRACT

The TOPAZ superconducting solenoid is cooled with forced two phase helium since it is large in size (2.9 m in diameter, 5.1 m in length) and a thin configuration. A cooling pipe, 18 mm in diameter and 150 m in total length, is wound around the superconducting coil. The relation between the helium mass flow rate and pressure drop through the magnet was obtained by a liquid helium pump which has a capacity of 300 l/h. The quality of helium in the cooling pipe was calculated from the mass flow rate and heat input to the solenoid. The pressure drop of the two phase helium flow was estimated for the case of (1) a homogeneous flow model and (2) a separated flow model. Experimental results agreed with the theoretical results of the homogeneous model. Also, the flow pattern in the cooling pipe was estimated as "stratified flow" on the Baker-diagram of helium.

INTRODUCTION

Recently, in the field of high energy physics, large thin superconducting solenoid magnets have been developed for particle detectors[1-3]. The forced two phase helium cooling method has been applied for a number of reasons. It does not need as much helium as the pool boiling method and is easy to control. However, it is necessary to estimate the relation between the mass flow rate and pressure drop across the magnet in the case of two phase helium. Only a few papers have reported on these items. At the moment, experimental data is sparse.

In this paper, we describe the experiment and some analysis about the pressure drop of forced two phase helium flow across the large superconducting solenoid magnet. The magnet has been constructed and tested for the high energy particle detector TOPAZ at TRISTAN. The estimate of the two phase helium flow pattern in the cooling pipe is included in this report.

MAGNET COOLING SYSTEM

Magnet

To minimize the radiation length (X_o) of the magnet, the TOPAZ magnet was designed to be a thin structure[4]. The main parameters of the magnet are shown in Table 1. Also a bird's eye view of the magnet is shown in Figure 1. The cooling pipe on the superconducting coil has a configuration that runs parallel to the axis of the magnet. It is not wound spirally on the magnet in order to avoid pressure ocsillations in the pipe caused by a "garden hose effect". The cooling pipe is made up of 24 sections. Each section has two sets of 5 m long straight pipe and four bends.

Cryogenic system

The main parameters of the cryogenic system are listed in Table 2. An experimental determination for the relation between the helium mass flow rate and pressure drop was carried out using the liquid helium pump[5]. A schematic drawing of the flow diagram is shown in Figure 2. The liquid helium discharged from the helium pump has a low quality and may be zero quality after passing through the heat exchanger in the control Dewar. This condition is important to estimate the quality of helium in the transfer tube and magnet. The mass flow rate of liquid helium was measured by a previously calibrated venturi-tube flowmeter . The inlet and outlet pressure of the magnet cooling pipe were measured by electronic pressure transducers. The flowmeter and pressure transducer have an accuracy of 1 % for industrial use.

FORCED TWO PHASE HELIUM PRESSURE DROP

Experiments and Results

Figure 3 shows a schematic drawing of the experiments displaying typical parameters when the mass flow rate is 5 g/sec. The mass flow rate of helium was changed from 8 to 20 g/sec by controlling the helium pump operating speed. It was assumed that zero quality helium (completely liquid) may be discharged from the control Dewar. Part of liquid may be evaporated by the heat load of the transfer tube and magnet cooling pipe.

The test results are shown in Figure 4. Theoretically estimated values of the pressure drop are also shown in the figure. The dashed-line, (1), shows the pressure drop in the case of liquid single phase flow. The solid and chained lines, (2) and (3), show estimates for the case of homogeneous and separated flow model, respectively. Details of these estimates are discussed below.

In our experiments, the garden hose effect was not observed. It may be considered that the helium flow in the cooling pipe was regarded as a holizontal flow. In the steady state, the magnet was cooled by 5-7 g/sec

Table 1. Main parameters of the TOPAZ magnet

Coil Length	(m)	5.1
Coil Inner Diameter	(m)	2.9
Current	(A)	3650
Central Field	(T)	1.21
Stored Energy	(MJ)	19.5
Cold Mass (Al)	(ton)	4.5
Cooling Pipe Length	(m)	150
Cooling Pipe Inner Diameter	(mm)	18
Heat Input (at 4.4 K)		< 30 W + 15 l/h

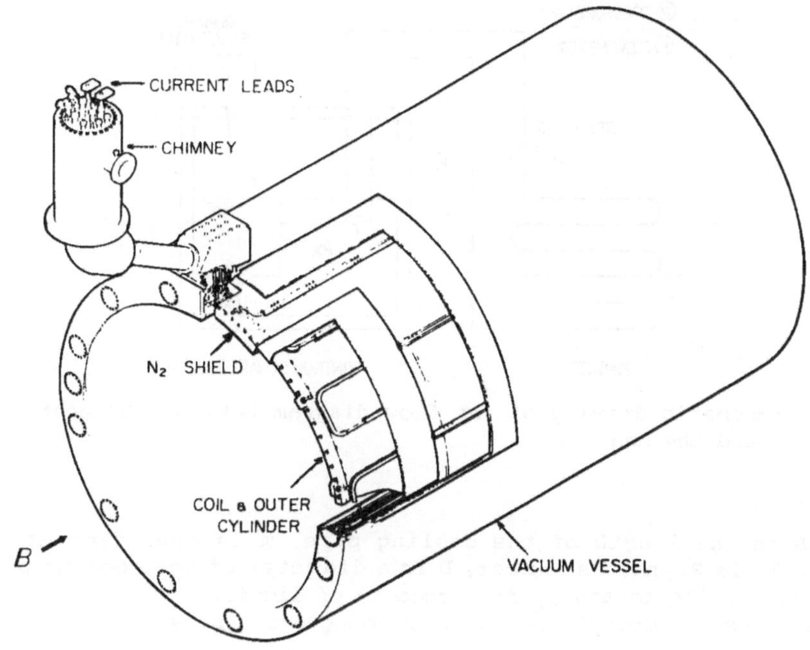

Fig. 1. Bird's eye view of the TOPAZ magnet. LHe cooling pipe is seen inside of LN_2 shield cooling pipe.

of helium mass flow rate. So, the magnet temperature was about 4.5 K because the pressure drop across the magnet was estimated as 0.01-0.02 kg/cm^2.

Analysis

The pressure drop for two models in Figure 4 was estimated as follows:

At first, the pressure drop for the liquid single phase was calculated as,

$$\Delta P_L = (0.184/Re^{0.2})(8L\dot{m}^2/\pi^2 D^5 \rho_L) \tag{1}$$

$$Re = 4\dot{m}/\pi D \mu_L \tag{2}$$

Table 2. Main parameters of the cryogenic system

Compressor	1.1 → 16 kg/cm^2 abs., 1800 Nm^3/h
Cold box	300 W at 4.4 K (100 l/h)
Turbine	230000 rpm (First), 163000 rpm (Second)
Control Dewar	800 liters of LHe
LHe Pump	1.0 → 1.3 kg/cm^2 abs., 300 l/h, 20000 rpm

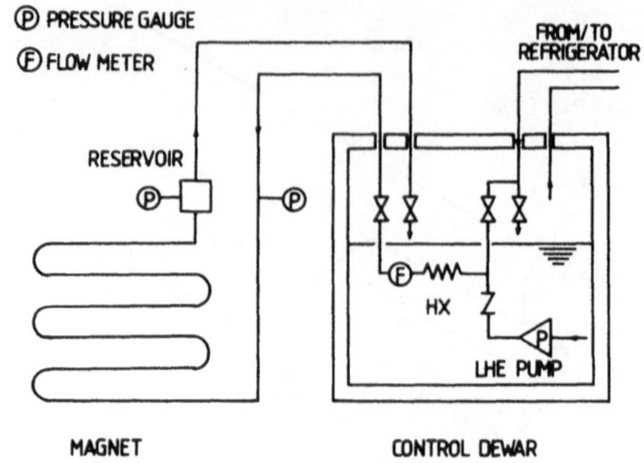

Fig. 2. Schematic drawing of the flow diagram between the control Dewar and the magnet.

where L is the length of the cooling pipe, $\dot{m}$ is mass flow of liquid helium, Re is Reynold's number, D is a diameter of the cooling pipe, ρ_L is density of liquid and μ_L is viscocity of liquid.

The pressure drop in the case of homogeneous flow model[6] is written as

$$\Delta P_{ho} = f_{ho}(x)\Delta P_L \tag{3}$$

$$f_{ho}(x) = [1/(x_2-x_1)]\int_{x1}^{x2}[1+x(\rho_L/\rho_G-1)][1+x(\mu_L/\mu_G-1)]^{-0.2}dx \tag{4}$$

$$x_1 = Q_1/\dot{m}\ell_v \tag{5}$$

$$x_2 = (Q_1+Q_2)/\dot{m}\ell_v \tag{6}$$

where x_1 and x_2 are the quality at inlet and outlet of the magnet respectively, Q_1 and Q_2 are heat input to the transfer tube and the magnet, and ℓ_v is the latent heat of the liquid. They are estimated as Q_1 = 7.5 W, Q_2 = 25 W and ℓ_v = 20.9 J/g, respectively. Density and viscosity of saturated gaseous helium are expressed as ρ_G and μ_G.

Also, for the separated flow model[7], the pressure drop can be written as

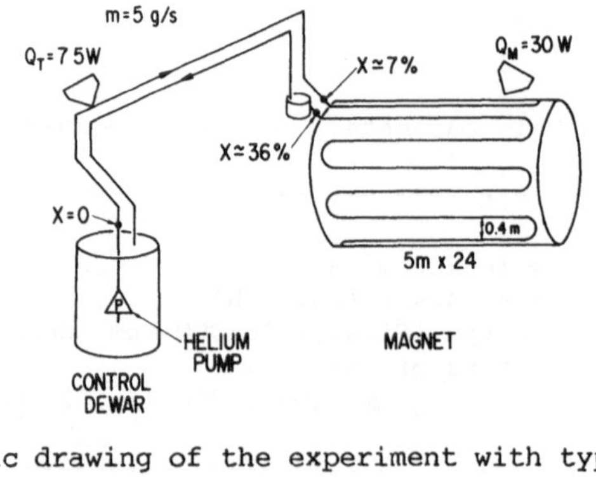

Fig. 3. Schematic drawing of the experiment with typical parameters.

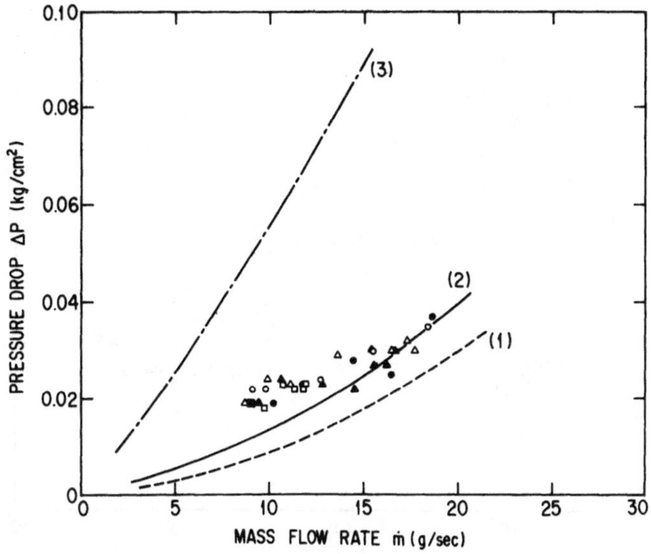

Fig. 4. Experimental results of the relation between the mass flow rate and pressure drop. Theoretically estimated lines are also shown. (1) ΔP_L (Liquid single phase flow), (2) ΔP_{ho} (Homogeneous flow model), (3) ΔP_{se} (Separated flow model)
Pump speed... ; 20000 rpm, ; 18000rpm, ;17000 rpm, ;16000 rpm ; 12000 rpm.

$$\Delta P_{se} = f_{se}(x)\Delta P_L \tag{7}$$

$$f_{se}(x) = [1/(x_2-x_1)]\int_{x1}^{x2}(1-x)^{1.8}\phi_L dx \tag{8}$$

$$\phi_L = 1 + 20/X + 1/X^2 \tag{9}$$

$$X = (\mu_L/\mu_G)^{0.1}(\rho_G/\rho_L)^{0.5}[(1-x)/x]^{0.9} \tag{10}$$

The pressure drop at the bends of the cooling pipe was also estimated to be negligible. The flow pattern in the cooling pipe was estimated on the Baker-diagram of helium[8,9,10]. It is shown in Figure 5 as a parameter of mass flow rate. The two parameters of the Baker-diagram are $(L/G)\lambda\psi$ for the abscissa, and G/λ for the ordinate. L is the liquid's mass velocity and G is the gas's mass velocity, $\lambda \propto (\rho_G\rho_L)^{1/2}$ and $\psi \propto (\mu_L/\rho_L{}^2)^{1/3}/\sigma_L$ where σ_L is the surface tension of the liquid. The flow pattern in the cooling pipe may be regarded as "stratified flow" on the Baker-diagram. However, it is difficult to define the boundary between the stratified and bubble flow regime with any certainty.

In Figure 4, the deviation of the experimental data from the calculated homogeneous model becomes larger as the mass flow rate decreases. This may correspond to the shift of the flow pattern from the bubble to the stratified flow regime in Figure 5. It is considered that

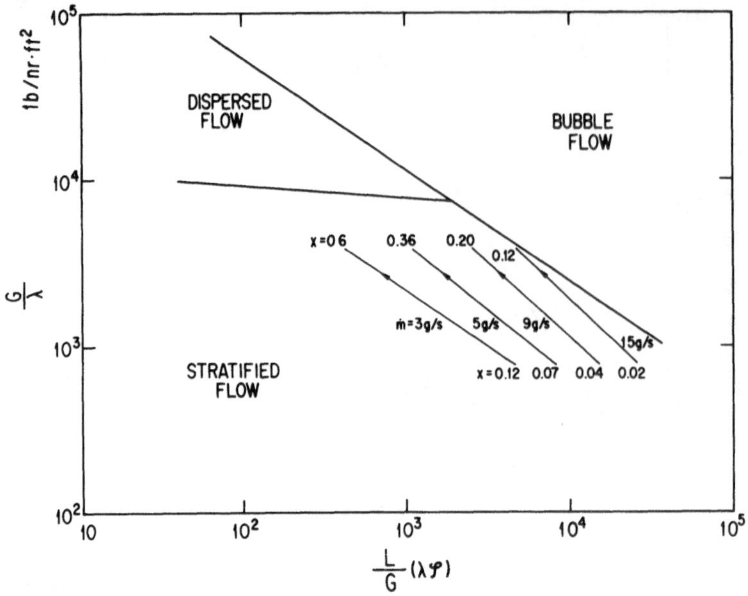

Fig. 5. Flow pattern estimates on the Baker-diagram of helium.

as the mass flow rate becomes small, the flow pattern in the cooling pipe may become more completely separated flow and the pressure drop becomes larger.

SUMMERY

The relation between the mass flow rate and pressure drop of forced two phase helium for the TOPAZ thin superconducting magnet was measured. In typical operation, the pressure drop was 0.01 kg/cm^2 with a mass flow rate of 5 g/sec. The experimental data agreed with the homogeneous flow model of two phase helium at large mass flow rate. The Baker-diagram of helium was used to estimate the deviation of the experimental data from the homogeneous flow model and to predict the flow pattern in the cooling pipe.

ACKNOWLEDGEMENTS

The authors wish to thank Messers Y. Kondo, H. Suzuki, N. Kimura H. Yamaoka of KEK, Y. Ibaraki of TEISAN Co, Ltd., and I. Kawamura of Hitachi Co, Ltd., for their contribution to these experiments.

REFERENCES

1. M. A. Green et al, Forced Two-Phase Helium Cooling of Large Superconducting Magnets, in: "Advances in Cryogenics Engineering," Vol. 25, Plenum Press, New York (1980), p. 420.
2. Y. Doi et al, Cryogenic System of the TOPAZ thin superconducting solenoid, in:"Proc. Eleventh Intl. Cryo. Engr. Conf.,"Butterworth, Guildford, UK (1986), p. 424.
3. M. Wake et al, The first excitation of VENUS thin solenoid magnet, in: "Proc. Eleventh.Intl. Cryo. Engr. Conf.," Butterworth, Guildford, UK (1986), p. 454.

4. A. Yamamoto et al, Performance of the TOPAZ Thin Superconducting Solenoid Wound with Internal Winding Method, J.J.A.P. 25:L440 (1986)
5. Y. Doi et al, Helium Liquifier and Refrigerator for e^+e^- Colliding Beam Detector TOPAZ, Hitachi Review 34,(1985), p.127
6,7 D. Butterworth, Empirical Method for Pressure Drop, in: "Two-Phase Flow and Heat Transfer," D. Butterworth and G. F. Hewitt, eds., Oxford University Press, Oxford (1977), p. 58.
8 C.H.Rode and J.Theilacker,"Preliminary Results of FermilabTwo Phase Helium Tests", FNAL SSC CRYO 85-9, Fermi National Accelerator Laboratory, Illinois (1985)
9 K. Haraguchi et al, "Flow Pattern of the Cryogenic Two-phase Fluid (in Japanese)", Preliminary for presentation on Teion Kougaku 34, B1-5
10 P.H. Eberhard et al, Two Phase Cooling for Superconducting Magnets, in: "Advances in Cryogenics Engineering," Vol. 31, Plenum Press, New York (1986), p. 709.

CRYOGENIC SYSTEM OF A 3 TESLA SUPERCONDUCTING SOLENOID FOR THE AMY PARTICLE DETECTOR AT TRISTAN

Y. Doi, T. Haruyama, T. Mito, K. Tsuchiya, S. Terada,
T. Ohmori, M. Maki, O. Araoka, M. Tadano, S. Suzuki,
H. Suzuki, Y. Kondo and M. Kawai

National Laboratory for High Energy Physics, KEK
Tsukuba, Ibaraki, Japan

N. Yamashita and Y. Ibaraki

Teisan Co, Ltd.
Harima, Hyogo, Japan

ABSTRACT

The cryogenic system of the AMY superconducting solenoid magnet was designed, constructed and tested. The coil has a diameter of 2.5 m, a length of 1.5 m and a cold mass of 17 tons (CU/SUS). It is cooled with a pool-boiling method by immersion in a liquid helium bath of 500 liters. In the initial cooling test, the magnet was successfully cooled down within 7 days by a helium refrigerator with a rated cooling capacity of 100 l/h (300 W at 4.4 K). The refrigerator and magnet were fully controlled by an industrial-use computer system. The coefficient of performance (COP) of the refrigerator (Q_r/W_c) was also measured as 0.00135 with liquid nitrogen consumption of 70 l/h.

INTRODUCTION

Three superconducting magnets have been built for high energy particle detectors at TRISTAN. Two of them, TOPAZ and VENUS magnets[1,2], have a thin solenoid configuration in order to reduce the radiation length for high energy particles. Those magnet are cooled with forced two-phase helium.

The AMY magnet does not have a thin configuration because it must produce a high magnetic field of 3 Tesla. The stored energy is two or three times larger than the other magnets and thus it is desirable to have it completely cryostable. It is cooled with the pool-boiling method.

In this paper, we describe the design, construction and the results of the initial cooling test of the cryogenic system. Initial tests were carried out on December 1986 and when it was operated for a month without trouble.

SYSTEM DESIGN

Magnet

The main parameters of the AMY superconducting magnet are listed in Table 1. It must be noted that the cold mass consists mainly of SUS and Cu. Also, the magnet is cooled in the cryostat which contains 500 liters of liquid helium, so it will be essentially cryostable for magnet quench. The detailed characteristics of the magnet have been reported[3].

Heat Load

To design an adequate refrigerator, it is necessary to estimate the heat load correctly. The heat load of major components are listed in Table 2. The value of heat load for the magnet and Dewar were obtained experimentally from the magnet cooling test. However, the value of the heat load due to the transfer tube was calculated. A hundred watts at 4.4 K and 25 L/H of liquid helium was estimated as the total steady state heat load of this cryogenic system.

Cryogenic system

The parameters of the major components of the cryogenic system were listed in Table 3. Also the flow diagram of the cryogenic system and the photograph of the system are shown in Figures 1 and 2. The main compressor unit consists of a two-stage screw compressor with a four-stage oil removal device. The cold box has two C-3 type turbines of L'AIR LIQUID with static gas bearings. There are two tube trailers for make-up and recovery gas, each with a storage capacity of 1050 Nm^3 at 150 kg/cm^2. The system gas can be purified with a purifier that has a capacity of 100 Nm^3/h. Almost all the valves in the figure 1 are electro-pneumatic valves

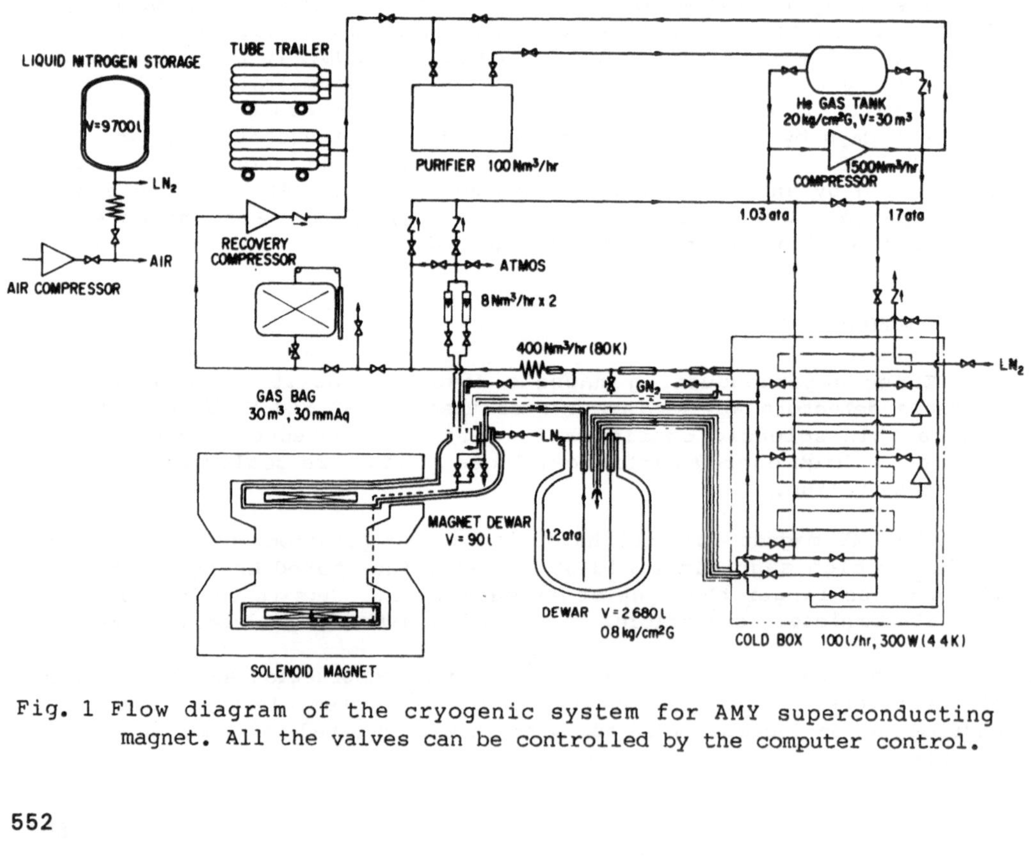

Fig. 1 Flow diagram of the cryogenic system for AMY superconducting magnet. All the valves can be controlled by the computer control.

Table 1. Parameters of AMY magnet

Solenoid	Current	(A)	5000
	Central Field	(T)	3.0
	Coil Inner Diameter	(m)	2.386
	Coil Outer Diameter	(m)	2.584
	Stored Energy	(MJ)	40
Cryostat	He Vessel Inner Diameter	(m)	2.326
	He Vessel Outer Diameter	(m)	2.680
Cold Mass	Cu/SUS	(kg)	1.7×10^4

Table 2. Estimated Heat Load of Component at 4.4 K

Magnet (including cryostat)	< 35 W + 25 l/h
Liquid He dewar	< 5 W
Transfer Line	60 W
Total	100 W + 25 l/h

Table 3. Main components of the cryogenic system

Compressor	1.03 → 19.0 kg/cm^2, 1500 Nm^3/h
Cold box	100 l/h (300 W at 4.4 K) with LN_2 2 turbines with static gas bearing
Liquid He Dewar	2680 liters
Liquid Nitrogen Tank	9700 liters

Fig. 2. Photograph of the cryogenic system for AMY superconducting magnet. The Dewar, cold box and a part of the magnet can be seen.

that can be controlled by the computer control system. The instrumental air for valves is supplied by two air compressors of 36 Nm^3/h and backed up by the evaporated gas from the liquid nitrogen tank. In the steady state the system contains about 500 liters of liquid helium in the magnet cryostat and 400 liters in the Dewar.

COOLING CHARACTERISTICS

Cool-down time estimate

A cool-down time of the AMY magnet can be estimated with a simple calculation. The relation between the mass flow rate and cool-down time may be written as follows:

$$\dot{m} = \Delta Q/\Delta H \Delta t \tag{1}$$

$$\Delta Q = M_m(C_{m1}T_1 - C_{m2}T_2) + M_g(C_{g1}T_1 - C_{g2}T_2) \tag{2}$$

$$\Delta H = 5.2\Delta T \tag{3}$$

where $\dot{m}$ is mass flow rate for magnet precooling, ΔQ is removed energy from the magnet for small interval Δt, ΔH is enthalpy differential between inlet and outlet gas. M_m is a cold mass of the magnet; C_{mi} is the specific heat of the magnet at temperature T_i; M_g is the mass of helium in the cryostat; C_{gi} is the specific heat of helium gas at T_i.

Figure 3 shows the results of simulation for three temperature ranges. For example, if the mass flow rate is kept at 10 g/sec, it takes 5 days to cool down the magnet from 300 to 75 K. Also, it will take 1.5 days from 75 K to 30 K and 0.5 days from 30 K to 5 K.

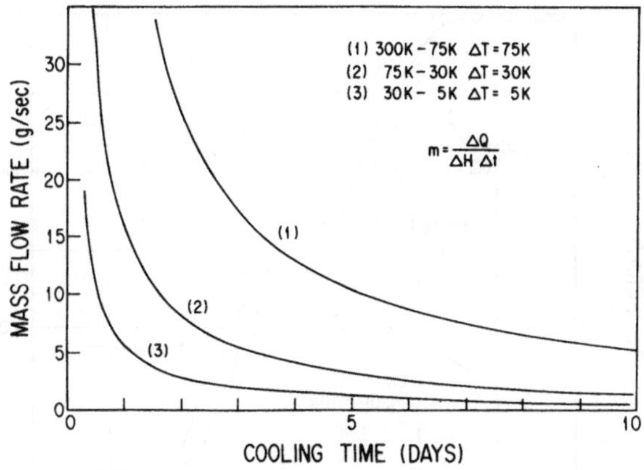

Fig. 3. Simulation result of required mass flow rate and cool down time (Equation (1)).

Contamination

One of the most important considerations for reliable long term operation is maintainance of high purity of the helium in the system[4]. A gas chromatograph is used to monitor the contaminations in helium gas, those are, oxygen, nitrogen, carbon-oxide, carbon-dioxide and so on. The water contamination is also monitored by an electronic dew point meter which uses a thin film aluminium oxide as a sensor. The oil mist in the discharged gas from the compressor was also measured by an "oil check kit" of BALSTON, an aerosol monitor of CGA and an infrared-adsorption method.

In the main compressor, a final adsorber is installed after three stage coalescers in order to remove the oil mist effectively. It consists of charcoal and molecular-sieves.

The oil contamination in the discharged gas was measured as < 0.03 vol. ppm by the "oil check kit" and infrared-adsorption method. The aerosol monitor did not have an adequate resolution for this level of oil contamination. The contamination variation of the system after the compressor start-up was shown in Figure 4. It is seen that the gas became pure gradually as the purifier was operated. The criteria of the impurity for the turbine operation is 10 ppm.

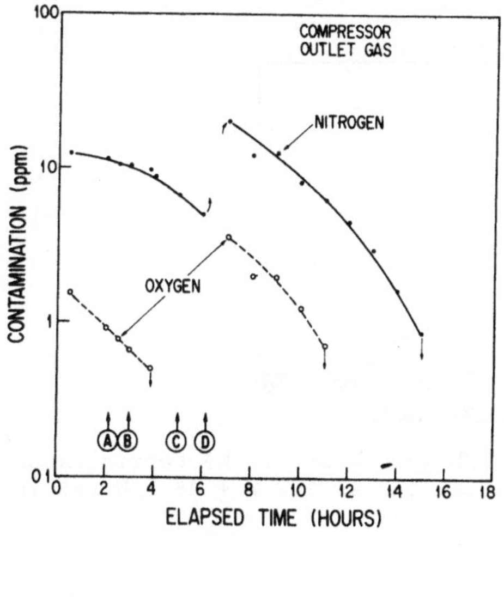

Fig. 4
Contamination variation after the compressor start up.
A: Compressor start
B: Purification start
C: Flow make to cold box
D: Turbine line cleaning

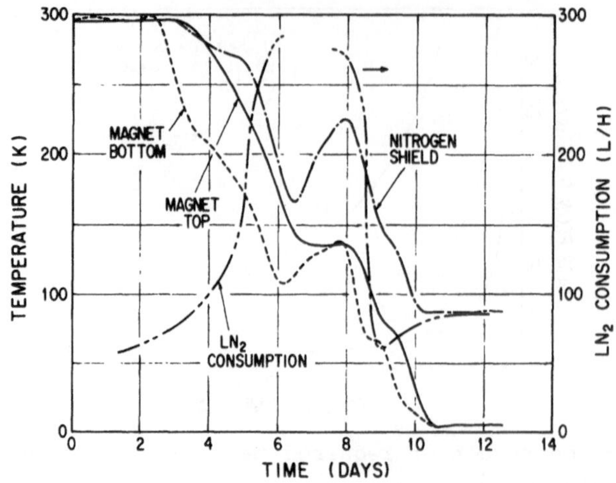

Fig. 5. Cool-down characteristics of AMY magnet. The magnet, shield temperature and LN_2 consumption rates are also plotted.

Magnet cool-down and excitation

The cool-down characteristics of the magnet are shown in Figure 5. The temperature profile of the magnet and liquid nitrogen shield are plotted for typical operation. In order to avoid an excess thermal shrinkage of the magnet, the cool-down speed was fully controlled by the computer control system. The magnet was cooled at a speed of 3-5 K/h by mixing cold and warm gas under the control of the computer program.

The disturbance of the cool-down curve at 6-8th days was due to the turbine trouble in this initial cooling test. Without such trouble, the magnet might have been cooled down within 6 days. The maximum consumption of liquid nitrogen in precooling was about 300 l/h and which reduced to 70 l/h in the steady state.

After the magnet was excited up to full current of 5000 A, a power shut-off test was carried out. Almost all the stored energy of 40 MJ was dumped in the dump resister and the magnet did not quench. The sequential program for the emergency dumping of the system worked well, providing protection for the magnet and the refrigerator. Even when

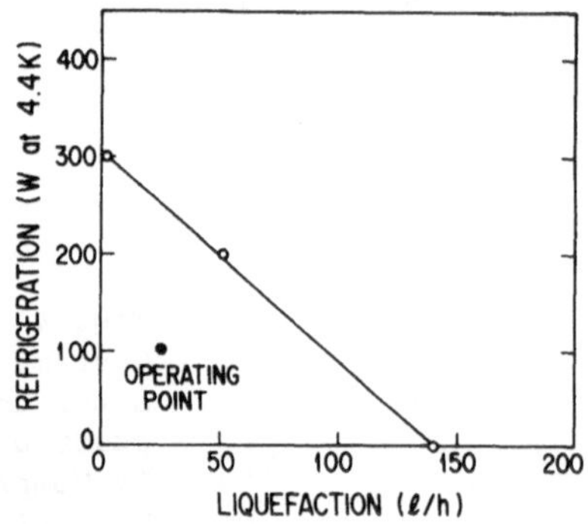

Fig. 6. Performance curve of 100 l/h (300 W at 4.4 K) refrigerator for AMY. Operating point is estimated as 100 W + 25 l/h and is also shown in the figure.

the power was shut down at the full current, the cryogenic system recovered to its steady state condition within 40 minutes.

Performance test of refrigerator

The cooling capacity of the refrigerator was also measured. Tests were carried out to see how much liquid helium could be liquefied in the Dewar and how much power could be added to an electric heater in the Dewar without changing the liquid level. The results are shown in Figure 6. The normal operating point is also shown. The coefficient of performance (COP) was defined as Q_r/W_c, where Q_r is cooling capacity at low temperature and W_c is a compressor work. The value of COP was calculated as 0.00135 with the liquid nitrogen consumption of 70 l/h. For the comparison of cryogenic refrigerators, it is valuable to know the figure of merit (FOM) which is defined as COP/COP_i where COP_i is the COP for the ideal refrigerator. The FOM is calculated as 0.09.

SUMMARY

The cryogenic system for the AMY superconducting magnet was designed constructed and tested. The helium refrigerator has a cooling capacity of 100 l/h or 300 W at 4.4 K and it took 7 days to cool down the 17 tons cold mass of the magnet. The magnet was completely cryostable and the system recovered within 40 minutes after the power was shut down at full current of 5000 A. The coefficient of performance (COP) and the figure of merit (FOM) were also measured as 0.00135 and 0.09 respectively when the liquid nitrogen was consumed at a rate of 70 l/h.

ACKNOWLEDGEMENTS

The authors would like to express our gratitude to Professors S. Ozaki, K. Takahashi and H. Hirabayashi for their continuous support and encouragement in this work.

REFERENCES

1. A. Yamamoto et al, Performance of the TOPAZ Thin Superconducting Solenoid Wound with Internal Winding Method, J.J.A.P. 25: L440 (1986).
2. R. Arai et al, Test Operations of the Venus Superconducting Magnet at KEK, Nucl. Instr. and Meth. A254: 317 (1987).
3. K. Tsuchiya et al, "A 3 Tesla Superconducting Solenoid for the AMY Particle Detector at TRISTAN," KEK Preprint 86-63, National Laboratory for High Energy Physics, JAPAN (1986)
4. R. K. Barger et al, Operational History of Fermilab's 1500 W Refrigerator Used for Energy Saver Magnet Production Testing, in: "Advancesin Cryogenic Engineering," Vol.31,Plenum Press,New York (1985), p. 657.

PERFORMANCE TEST OF THE HERA 3 x 6500 W

HELIUM REFRIGERATION PLANT

M. Clausen, C. Gerke, G. Horlitz,
H. Lierl, K.-D. Nowakowski, S. Rettig,
W. Stahlschmidt

Deutsches Elektronen-Synchrotron (DESY)
Hamburg, Germany

P. Beurer, H. Egolf, H. Herzog,
F. Langenecker, B. Ziegler

Sulzer, Winterthur, Switzerland

ABSTRACT

The HERA-superconducting proton ring magnet system, part of a 820 GeV proton/30 GeV electron colliding beam facility at DESY, Germany, will be cooled by a helium liquefier/refrigerator as described previously. The refrigerator has been specified by DESY, detailed design, fabrication and assembly have been executed by industry. All assembly work was finished by April 1987. A short description of the final design of the plant is given. Some first results of performance tests as values for refrigeration power at 4.4 K and between 40 K and 80 K as well as liquefaction rates at 4.4 K with one coldbox system out of three will be presented. Preliminary values for efficiencies of cold-boxes and compressor groups are calculated.

INTRODUCTION

A storage and colliding facility called HERA (Hadron-Electron-Ring-Accelerator) for 820 GeV protons and 30 GeV electrons consisting of two magnet rings within a tunnel of 6366 m circumference is in construction at and around the site of DESY in Hamburg/Germany. The superconducting proton ring is equipped with dipole magnets of 4.6 T magnetic induction, quadrupoles and different correction elements: dipoles, quadrupoles, sextupoles and octupoles. The cryogenic system for this ring has been described in detail several times[1,2,7] as well as individual papers on dipoles and quadrupoles do exist[3,4,5].

CRYOGENIC SYSTEM

In the final state the cryogenic system (fig. 1: letters and numbers in paranthesis refer to the captions of the

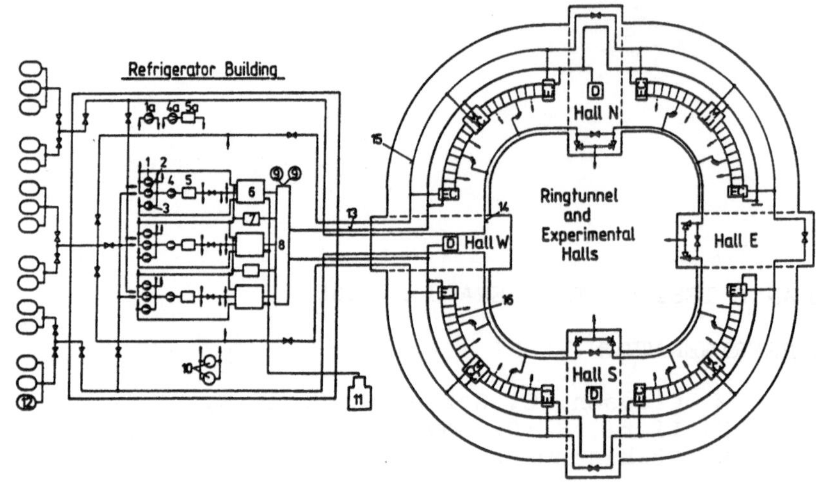

Fig. 1. The Cryogenic System of HERA - compressors for J-T-flow (1, 2), turbine flow (3) and high pressure (4), oil removal (5), cold boxes (6), gas heaters (7), valve box (8), He-dewars (9), cryog. purifiers (10), LN_2-dewar (11), gas tanks (12), tunnel transferlines (13), quench gas header (14), warm gas supply (15)

figures) is composed of 422 superconducting dipoles with cryostat lengths of 9.764 m and 224 quadrupoles with cryostates of 3.978 m. The magnets are arranged in strings within the tunnel.

The whole ring is subdivided into 4 quadrants of 90° bend, separated by 4 experimental halls containing partially (at present in halls N and S only) superconducting detector solenoids (D). Two octants (16) which are the smallest individual cryogenic units in the ring are part of one quadrant.

Between adjacent octants and at their ends are middle boxes (MJC, MCC, MJJ) and endboxes (EC, EJ) which contain precoolers (C) or JT-valves (J) or combinations of them. All boxes are connected to a cryogenic transferline (13), supplying the northern and the southern half ring without being through connected in the experimental hall east. The system has been described in [2,7].

The requirements for refrigeration are listed in tab. 1. In addition to the 4.5 K refrigeration the refrigerators have to supply cold helium gas of 40 K for heat shield cooling which is returned at 80 K.

Table 1. Heat Losses of Components and System Summary

Component	q (4.4 K) [W]	m_c (4.4 K) [g/s]	q (40/80 K) [W]
octant (calculated)	813.6	4.15	2562
octant (design)	1220	4.15	2818
8 octants	9760	33.2	22545
4 detectors	2000	4.4	6000
ring transfer line	1100		11300
design values	12860	37.6	39845
specif. values	13550	41.0	40000

CENTRAL REFRIGERATION PLANT

The whole refrigeration capacity is concentrated in a central refrigerator building. The refrigeration power being necessary for the whole ring is provided by two independent refrigerators each being capable of supplying one half ring. A third one of identical capacity was added thus giving 100% redundancy for each of the two refrigerators in charge or supplying 50% more capacity for the whole ring when all three cold boxes are running at the same time. The main components are listed in figs. 1 and 2.

The arrangement of the refrigerator components in the refrigerator hall is shown in fig. 3. Fig. 4 is a photograph of the hall with the coldbox group in the background.

Some more details of the cryogenic system are demonstrated in fig. 2: The JT flow requires two pre-stage compressors (1, 2) operating in parallel. The turbine flow is handled at higher pressure ($\sim 2 \times 10^5$ Pa) by means of the third pre-stage compressor (3). The total flow rate of the pre-stages is

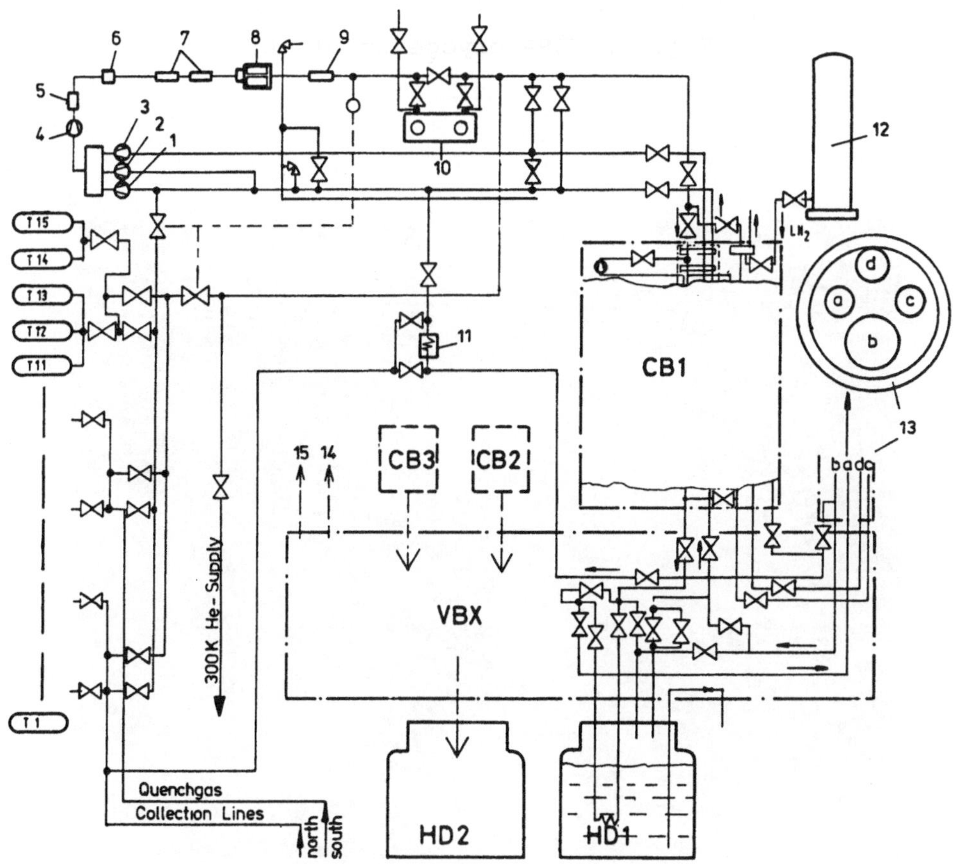

Fig. 2. Principal Function Scheme of a Refrigerator Path. cold boxes (CB1, CB2, CB3), compressors (1, 2, 3, 4,), oil removal (5, 6, 7, 8, 9), cryog. purification (10), gas heater (11), LN_2-dewar (12), tunnel, transferlines (13, 14), magnet measurements supply line (15), gas storage tanks (T1...T15), LHe-dewars (HD1, HD2)

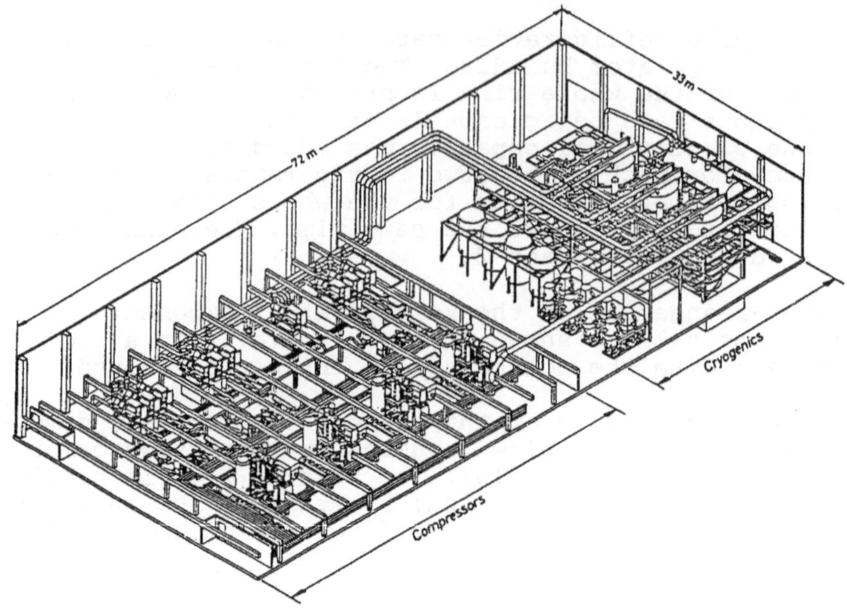

Fig. 3. HERA cryogenic plant.

Fig. 4. Photograph of the Cryohall

compressed to the final pressure of about 18 x 10 ^{5} Pa by means of the HP-stage (4) followed by filter and drying devices (5, 6, 7, 8, 9).

The plant is completed by warm gas storage tanks (T1 ... T15), He-dewars HD1, HD2, a large LN2 tank (liquid N2 is foreseen to be used for cool downs only, not permanently) and the valve box VBX mentioned above.

Fig. 2 shows only one refrigeration path out of three. Each of these can be operated individually or together by means of the valve box (VBx). The plant is controlled by a process control system which is described separately in this conference[8].

The requirements for a single cold box system are listed in tables 2 and 3. A simplified flow scheme of the box is shown in fig. 5. A thermodynamical analysis of the circuit has been given elsewhere[6].

Table 2. Main Data per Cold Box

cooling capacity			
isothermal	at T = 4.4 K	6500	W
return vapor overheating	at T = 4.4 K	275	W
lead cooling gas rate	at T = 4.4 K	20.5	g/s
shield cooling	at 40 < T < 80 K	20000	W
He mass flow rates			
JT circuit		461	g/s
turbine circuit		410	g/s
total		871	g/s
efficiencies			
cold boxes		0.512	
compressors		0.494	
whole system		0.253	
Carnot (T_o = 310 K, T = 4.3 K)		0.0141	
specific power consumption		281	W/W
flow rates for cool down (warm gas return only)			
T = 80 K, LN_2 cooling, turbines off		1.000	g/s
T = 40 K, precooling, turbines on		120	g/s
liquefaction rates			
without LN_2 precooling		45	g/s
with LN_2 precooling		90	g/s
turbines			
number of turbines		7	
design		dynamic gas bearings	
fabricator		Sulzer	

Table 3. Compressors

number of compressors per coldbox	4
design	oil lubricated screw compressors, slide valve controlled
fabricator	Aerzen

stage	type	nominal mass flow	electrical input power
JT-low pressure 1	VMY 536 M	264 g/s	492 kW
JT-low pressure 2	VMY 536 H	197 g/s	366 kW
turbine low pressure	VMY 536 H	410 g/s	397 kW
high pressure	VMY 536 H	871 g/s	1590 kW
total			2845 kW

THE PRESENT EXPERIMENT

The test period of the system started early in 1987, in different steps, first compressors separately. A large fraction of the compressors has already passed a running time of more than 1000 hours with satisfying performance.

A first cool down test with one coldbox was initiated by the end of May (run 1) and a second one in the first days of June 1987 (run 2). In both cases the refrigerator was connected to one of the large dewars.

For realizing the condition for pure refrigeration, an electrical heater in the dewar was powered with constant dissipation and the system was controlled to maintain a constant liquid level in the dewar. It was possible to main-

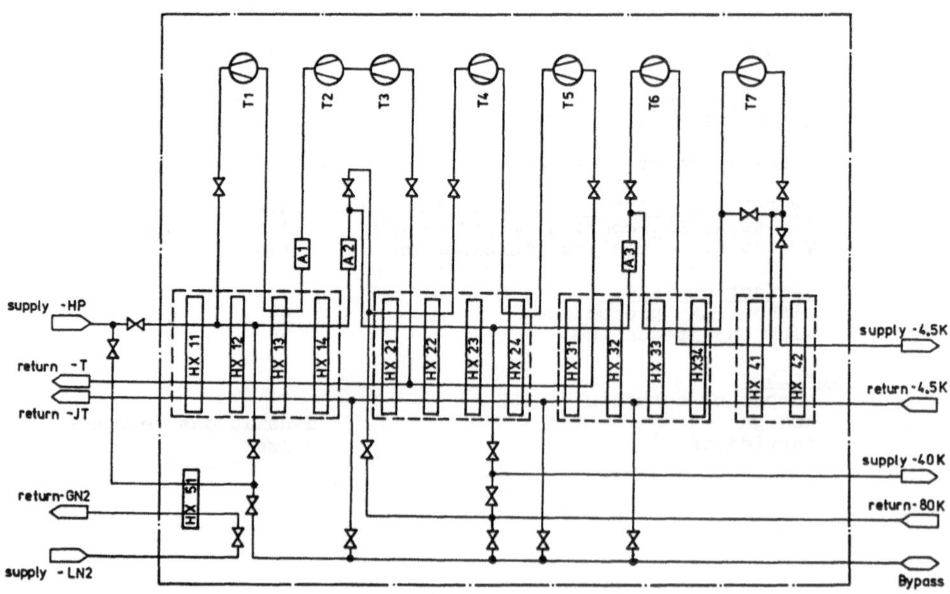

Fig. 5. The HERA Coldbox.

tain constant operating conditions over several hours with turbines T1, T2, T3 switched off. The results are listed in table 4.

During run 2 the dewar was also connected as well as the heater was powered.In addition a second heater of 22.5 kW was used to warm up a simulated shield cooling gas flow from 40 K to 80 K. Furthermore the liquid level was allowed to increase at a constant rate. The liquefaction rate was calculated from the pressure variation in one of the warm gas storage tank sub-groups. The results of this experiment are listed in Table 5 and displayed in a T-S-diagram (Fig. 6), where averaged data have been used.

FINAL REMARKS

Although an accurate optimization of all thermodynamic conditions was not yet possible the first results show good running performance and efficiencies as expected. The tests will be continued in the near future.

ACKNOWLEDGEMENT

The authors wish to express their gratitude to the operating crews of Sulzer and DESY.

Table 4. Compressor Running Data.

		run 1	run 2
JT circuit			
low pressure	[10^5Pa]	1.02	1.05
inlet temperature	[K]	291.0	292.4
mass flow rate	[g/s]	442.8	490.0
electrical power	[kW]	640.5	815.3
turbine circuit			
low pressure	[10^5Pa]	2.0	2.24
inlet temperature	[K]	286.67	289.0
mass flow rate	[g/s]	214.08	380.0
electrical power	[kW]	156.8	273.6
interstage pressure	[10^5Pa]	3.05	3.93
HP circuit			
inlet pressure	[10^5Pa]	3.04	3.76
inlet temperature	[K]	302.37	307.8
mass flow rate	[g/s]	658.22	870.0
high pressure	[10^5Pa]	15.83	18.35
electrical power	[kW]	1291	1489
total electrical power	[kW]	2088	2578
efficiency JT-stage		0.46	0.482
turbine stage		0.34	0.469
HP stage		0.53	0.593
compressor group efficiency		0.48	0.517

Table 5. Coldbox Running Data

		run 1	run 2
inlet pressure	[10^5Pa]	14.94	17.13
inlet temperature	[K]	290.0	291.2
inlet mass flow rate	[g/s]	658.2	870.0
JT return pressure	[10^5Pa]	1.04	1.07
JT return temperature	[K]	288.2	289.8
JT return flow rate	[g/s]	442.8	476.5
turbine return pressure	[10^5Pa]	2.02	2.29
turbine return temperature	[K]	279.3	285.1
turbine return flow rate	[g/s]	214.1	380
effective compressor efficiency at inlet*)	E_{com}	0.457	0.485
refrigeration at 4.4 K	[W]	7540	7330
refrigeration at 40/80 K	[W]	-	22500
liquefaction rate	[g/s]	-	13.5
system total efficiency	E	0.00361	
carnot efficiency	E_c	0.0154	
carnot fractional efficiency	E/E_c	0.234	
cold box efficiency	E_{box}	0.513	
spec. power consumption	W(300K)/W(4.4K)	277	

*) This value is lower than the compressor efficiency in tab.4 because of the pressure and temperature losses in the filters and cryogenic purifiers.

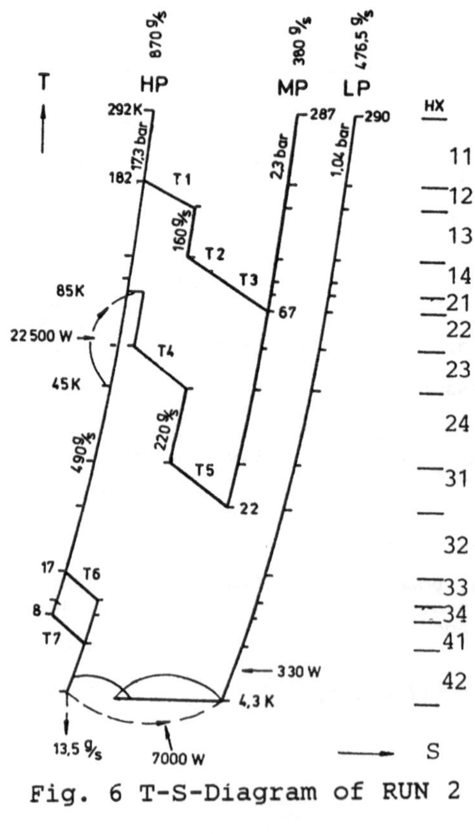

Fig. 6 T-S-Diagram of RUN 2

REFERENCES

1. G. Horlitz, Proc. of the Intern. Cryogenic Engineering Conf., ICEC 10, Helsinki, p. 377
2. H. R. Barton Jr., M. Clausen, G. Horlitz, G. Knust, H. Lierl, Advances in Cryog. Engineering, Vol. 31, Plenum Press New York (1986) p. 635
3. S. Wolff, Proc. of the IX Intern. Conf. on Magnet Technology, Zürich 1985, p. 62-67
4. R. Auzolle, P. LeMarrec, A. Patoux, J. Perot, J. M. Rifflet, Proc. of the ICFA Workshop on Superconducting Magnets and Cryogenics, Brookhaven Nat. Lab., Upton N.Y., 1986
5. C. Daum, J. Geerinck, H. Möller, R. Heller, P. Schmüser, P. Bracké, Proc. of the ICFA Workshop on Superconducting Magnets and Cryogenics, Brookhaven National Lab., Upton N.Y. 1986
6. B. O. Ziegler, Advances in Cryogenic Engineering, Vol. 31, Plenum Press, New York (1986), p. 693
7. H. R. Barton Jr., M. Clausen, G. Horlitz, H. Lierl, ICFA Workshop on Superconducting Magnets and Cryogenics, Brookhaven Nat. Lab.,Upton N.Y.,May 12-16, 1986
8. M. Clausen, K. H. Mess, Chr. Gerke, S. Rettig Cryogenic Engineering Conference 1987, St. Charles, USA, to be published

CRYOGENIC EXPERIENCE AT THE HERA MAGNET MEASUREMENT FACILITY

H. R. Barton, Jr., M. Clausen,
G. Keßler and S. Rettig

Deutsches Elektronen-Synchrotron DESY
D-2000 Hamburg 52, Germany

ABSTRACT

The proton storage ring of HERA will contain 422 superconducting dipole magnets[1] and 224 superconducting quadrupole magnets.[2] At the newly constructed HERA magnet measurement facility, each of the superconducting magnets will be individually tested before being installed in the 6-km ring tunnel. This paper describes the design and initial operation of the helium cryogenic distribution system that allows independent testing of the HERA cold-iron magnet. Operation of each magnet at liquid helium temperature insures the quality of the magnetic characteristics and the cryogenic performance. Six test stands are completely instrumented to monitor the cryogenic conditions of the magnet under test. A process-control computer system is used to record the measurements and automate the cryogenic operation of the measurement facility. Helium for magnet testing is supplied by one of the 6.5-kW helium refrigerators for HERA. A large, helium subcooler is included in the measurement facility to produce the correct operating temperature at the test stands. The performance of the cryogenic system and the schedule for magnet testing are discussed.

DESCRIPTION OF THE CRYOGENIC SYSTEM

The cryogenic system of the HERA magnet measurement facility at DESY contains the following major components:

a. Helium refrigerator/liquefier
b. Storage capacity for liquid and gaseous helium
c. Cryogenic purifier
d. Helium subcooler
e. Six test stands
f. Process-control computer

These components are installed in two building located on the DESY site. The Cryo-Hall contains three identical helium refrigerators. Each has a capacity of 6.5 kW of refrigeration at 4.2 K or 1200 L/hr of liquid helium without consumption of liquid nitrogen. These plants contain gas-bearing turboexpanders and are described elsewhere.[3] The adjacent Magnet Test Hall contains a helium subcooler and six magnet test stands. A block diagram for these components is shown in Fig. 1. Helium is transferred between the two buildings using a fourfold transfer line.

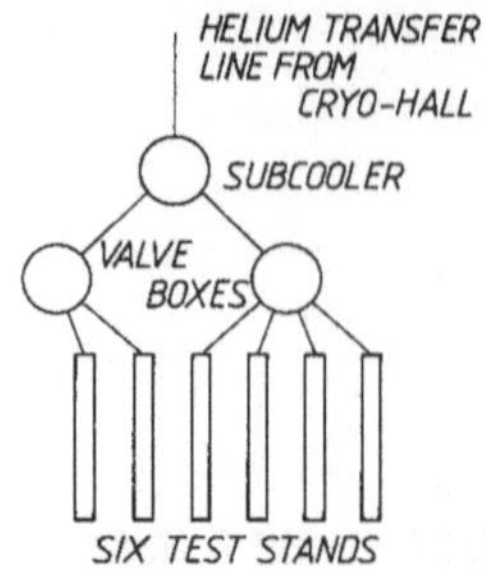

Fig. 1. Block diagram of the components in the Magnet Test Hall.

Fig. 2 shows the arrangement of the components inside the Magnet Test Hall. A large helium subcooler is located above the test stands on the balcony of the building. The subcooler has a liquid volume of 200 L and contains a hear exchanger with the following design data:

		High-Pressure Stream	Low-Pressure Stream
Inlet Temp.	(K)	5.10	4.30
Inlet Press.	(b)	4.1	1.1
Outlet Temp.	(K)	4.35	4.30
Outlet Press.	(b)	4.0	1.1
Flow	(g/s)	480	480

The subcooler is connected symmetrically to two valve boxes also located on the balcony. The valve boxes contain isolation valves which remain open during normal operation but can be closed in case of system repair.

Fig. 2. View of cryogenic components located inside Magnet Test Hall.

When a magnet is installed on the test stand, the ends of the magnet cryostat are closed by two end boxes. These contain the helium supply and return lines to the magnet as well as the cryogenic valves used for shut off and control. The supply boxes also contain the helium-cooled power leads needed to supply the coil of the magnet with current. The main, 6500-A current leads are made from electrolytic, tough pitch copper rod with a single, cooling spiral cut into the surface for helium flow. The end box on the other end of the magnet contains a JT-valve and is connected to the Cryo-Hall by a vacuum-insulated, bypass line. The bypass is used during cool-down and warm-up of the magnet to return gas to the refrigerator. The helium flowing in the shield circuit of the magnet also is returned to the refrigerator using the bypass.

Figs. 3 and 4 show the flow diagram for the cryogenic system. This flow diagram contains the subcooler, one of the valve boxes and one test stand. The control and isolation valves and all connections in the helium circuits are included. Check valves (omitted in the figure) are included in the system wherever a warm line connects to a 4-K circuit. Relief valves installed in the facility are not shown in Figs. 3 and 4. Instrumentation is represented by P for a pressure sensor, T for temperature diodes and F for a flow meter.

INSTRUMENTATION

The single-phase, two-phase and shield circuits of the magnet are completely instrumented in both end boxes. Individually calibrated silicon diode thermometers are installed directly in the helium stream and the electrical leads to these diodes are taken out through cold feedthroughs which are sealed with Conoseal gaskets. The calibration curves for the diodes are stored in a microprocessor responsible for temperature conversion. Pressure is measured at each end of the magnet using Rosemount pressure transducers. Helium flow in the 4-K and the shield circuits is measured using orifice flowmeters with Rosemount differential pressure transducers.

PROCESS-CONTROL COMPUTER

A process-control computer system is installed to automate the operation of the magnet measurement facility and to assist the operators. This computer system contains multiple, distributed computers in the Cryo-Hall and the Magnet Test Hall connected by a redundant communication link. Separate computers are assigned to operate the cryogenic plant, to control the magnet measurement facility, to perform data analysis and to generate displays on the operators' consoles (see Fig. 5). The operators are aided by the results of a simulation running on one of the computers during testing. The signals from all transducers are sent to the process-control computer system which stores this data and is capable of displaying process measurements on the operators' consoles. The computer system performs all interlock functions.

HELIUM TRANSFER LINES

The magnet measurement facility contains a fourfold transfer line to the Cryo-Hall which is 85 m in length. This consists of a 4-K supply and return, a 40-K supply and a bypass return contained inside two separate vacuum jackets. Fig. 6 shows the section of this transfer line between the two buildings.

Another 300 m of vacuum insulated transfer line are installed within the Magnet Test Hall to connect the components of the system. These are primarily individually insulated sections approximately 10 m long with

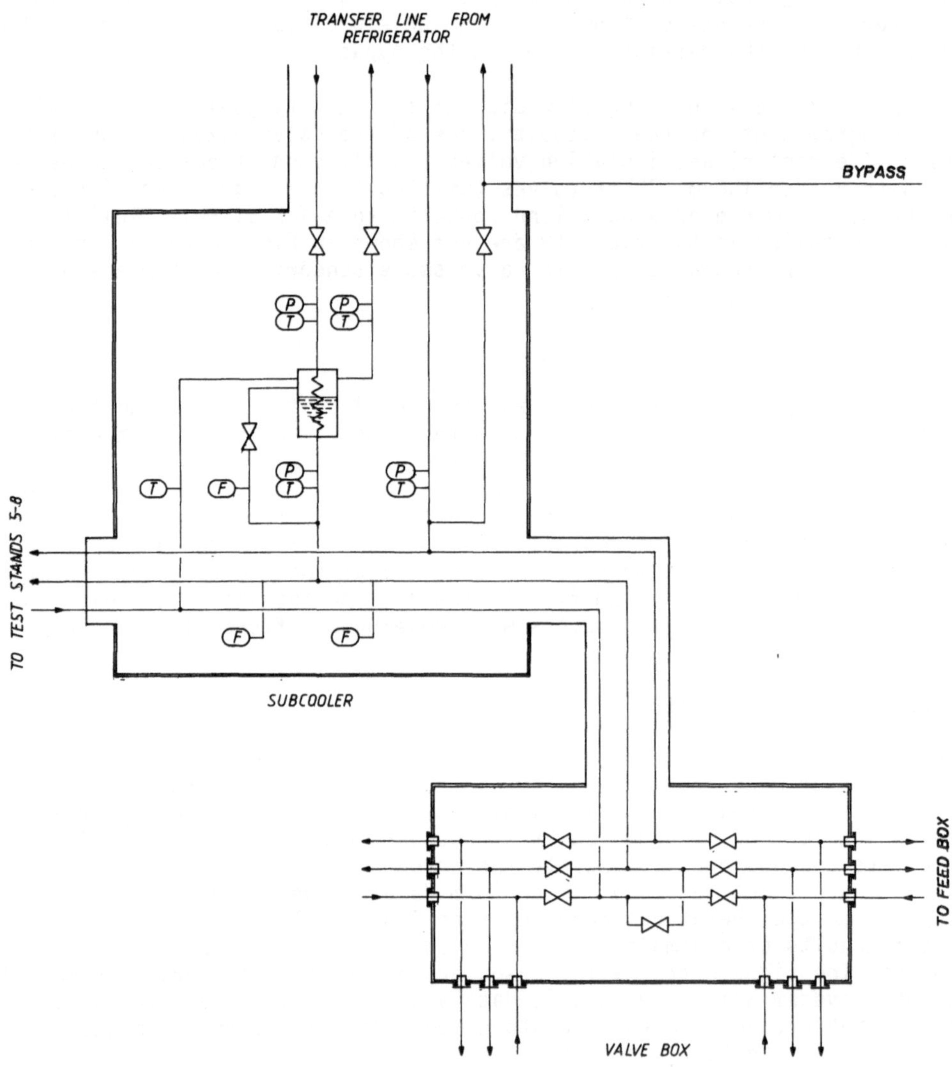

Fig. 3. Process and instrumentation diagram for subcooler and one of the valve boxes.

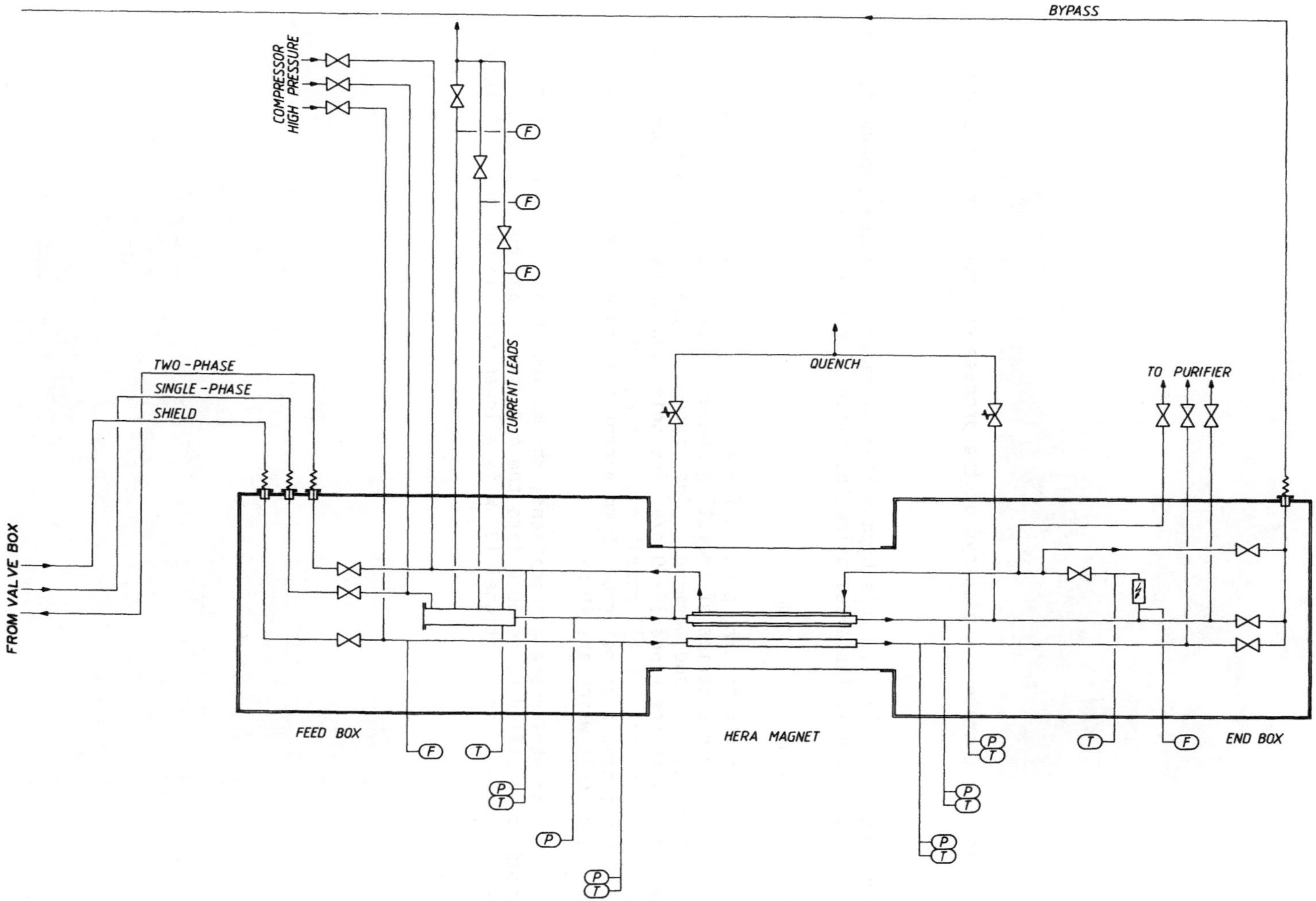

Fig. 4. Process and instrumentation diagram for one test stand.

Fig. 5. Operators's displays of the process-control computer system.

bayonnet ends (Fig. 7). This bypass line is vacuum insulated, connects all test stands and is installed below the floor of the Test Hall.

SYSTEM FUNCTIONS

The complete testing program for a superconducting magnet requires that it undergo a series of operating phases: installation, purging, cool-down, testing and warm-up.[4] The cryogenic system was designed for helium flow in the range 10-70 g/s. The selection of valves was based on calculations done for all possible operating temperatures. The majority of the valves have a diameter of 25 mm except for those in the two-phase return circuit which are 40 mm.

During steady-state conditions, the maximum heat leak for each test stand is specified to be 50 W into the 4-K circuit and 600 W into the shield circuit. An additional 100 W is allowed for the subcooler. Although

Fig. 6. Fourfold helium transfer line between the Magnet Test Hall and the Cryo-Hall.

Fig. 7. Individual transfer lines connecting the valve boxes to the end boxes on the test stands.

the cryogenic plant has excess refrigeration capacity, the liquefaction load on the plant during cool-down of a magnet is large. We estimate that 7000 L of liquid helium will be required to cool a HERA dipole magnet (with a cold mass of approximately 9000 kg) to operating temperature in a 12-hour period.

TESTING SCHEDULE

Six test stands have been installed; an additional two are under construction. With eight test stands in full operation, 1344 testing hours per week are available. If a magnet under test remains on the test stand no longer than 60 hours, production testing could be as high as 22 completed magnets per week. In reality, refrigeration is limited and the weekend days will not be fully utilized. An example of a proposes testing schedule is shown in Fig. 8. A detailed analysis of the schedule shows that the capacity of the cryogenic plant is not exceeded at a measuring rate of 16 magnets per week. The arrival of magnets at DESY from the

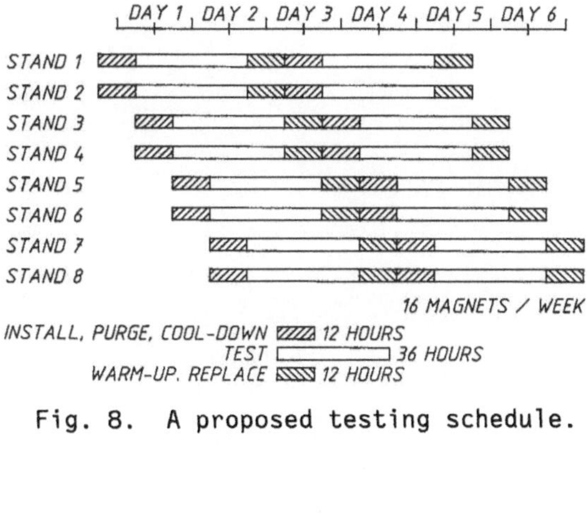

Fig. 8. A proposed testing schedule.

that the capacity of the cryogenic plant is not exceeded at a measuring rate of 16 magnets per week. The arrival of magnets at DESY from the manufacturers will begin late in 1987 and reach in delivery rate of 16 magnets per week in 1988.

REFERENCES

1. H. Kaiser, "Design of Superconducting Dipole for HERA," DESY HERA 1986-14, Deutsches Elektronen-Synchrotron DESY, Hamburg (1986).

2. R. Anzolla, P. Le Marrec, A. Patoux, J. Perot and J. M. Rifflet, Superconducting Quadrupoles for HERA, in "Proceedings of the Workshop on Superconducting Magnets and Cryogenics," Brookhaven National Laboratory BNL 52006 (1986), pp. 195-198.

3. P. Beurer, et al., Performance Test of the HERA 3 X 6500 W Helium Refrigeration Plant, paper CB-4 at this conference.

4. H. R. Baron, Jr., M. Clausen and G. Keßler, Cryogenic System for the HERA Magnet Measurement Facility, in: "Proc. Eleventh Int. Cryo. Engr. Conf.," Butterworth, Guildford, UK (1986), pp.173-177.

CRYOGENIC TESTING AT THE SSC STRING TEST FACILITY

J.C. Theilacker

Fermi National Accelerator Laboratory*
Batavia, Illinois

ABSTRACT

A multipurpose cryogenic testing facility was constructed and became operational in 1986 at Fermilab. It consists of a standard Tevatron satellite refrigerator and 450 liter subcooling dewar housed within a building adjacent to a Tevatron refrigerator. A 11 m by 15 m high bay contains power supplies necessary to power Superconducting Super Collider (SSC) magnets, a cryogenic feedcan to interface the refrigerator to the magnet string, and an area for testing Tevatron components. Currently, a 120 m, 4 m diameter tunnel has been constructed off the high bay to house a half cell of SSC Design D magnets. Capabilities have made to extend the tunnel up to 1 km in length.

Testing at the facility includes: SSC long string test, heat leak measurement, cold compressor tests, turbine test, and cold leak checking.

INTRODUCTION

The proposed SSC is a 20 TeV proton synchrotron[1]. It utilizes nearly 10,000 superconducting dipole and quadrupole magnets to steer the protons around its 83 km circumference. To cool the magnets to 4.3 K, there will be ten refrigeration stations spaced around the ring. Each of these refrigerators will cool four 4 km long magnet strings.

The primary collider components are the 7680 dipole magnets for bending the proton beam to form the closed orbit path. Each dipole is 17 m long, utilizes cold iron to achieve 6.6 T field, and is housed in a low heat leak cryostat[2]. To date, three full length prototype dipoles have been manufactured. Cryogenic, electrical, and magnetic testing of individual magnets are being done at the Magnet Test Facility (MTF) at Fermilab. Results of these tests have been presented by Strait[3,4].

Following individual magnet testing at MTF the magnets are sent to a new facility where they are being assembled into a magnet string. Incorporation of a magnet string up to 1% of the SSC length will be possible. Testing of a long string of magnets is critical in assuring

* Operated by Universities Research Association, Inc., under contract with the U.S. Department of Energy.

that the magnet design is electrically and cryogenically, compatible with the SSC project. Also, it will allow operational feedback to be incorporated into the final component design.

This paper will describe the string test facility built at Fermilab, and present the cryogenic testing plans through FY88.

SYSTEM DESCRIPTION

The string test facility is located adjacent to the berm of the main ring tunnel, at the E4 Tevatron refrigerator. It consists of four main enclosures; a refrigerator building, a 11 m by 15 m utility building, a "tunnel" section, and a control room (Fig. 1). The refrigerator building is built directly off the E4 Tevatron satellite refrigerator. This allows direct access to several Tevatron utilities such as; subcooled liquid nitrogen and 4.5K supercritical helium from the transfer line, low conductivity water for power supplies and bus work cooling, and connection to the ACNET controls system.

Directly between the refrigerator building and mock tunnel section is the utility building. It has a high bay area for off-loading and staging equipment. Power supplies for the string test have been installed here along with the feedcan to interface the refrigerator to the magnet string.

Extending from the utility building is a 120 m long mock tunnel section, enough for a SSC half cell. The tunnel is made up of a concrete slab and 4 m used tunnel sections from the main ring. In order to follow the terrain, the slab (and thus the magnets) is installed with a 0.4% downward grade. This allowed the slab to match with Kautz Road. Beyond a half cell, the "tunnel" will follow the road bed and have a maximum length of 1.2 km.

The refrigeration system consists of a standard Tevatron satellite refrigerator (1000 W at 4.5 K) with two dedicated screw compressors for steady state operation (Fig. 2). The compressors will be common with the

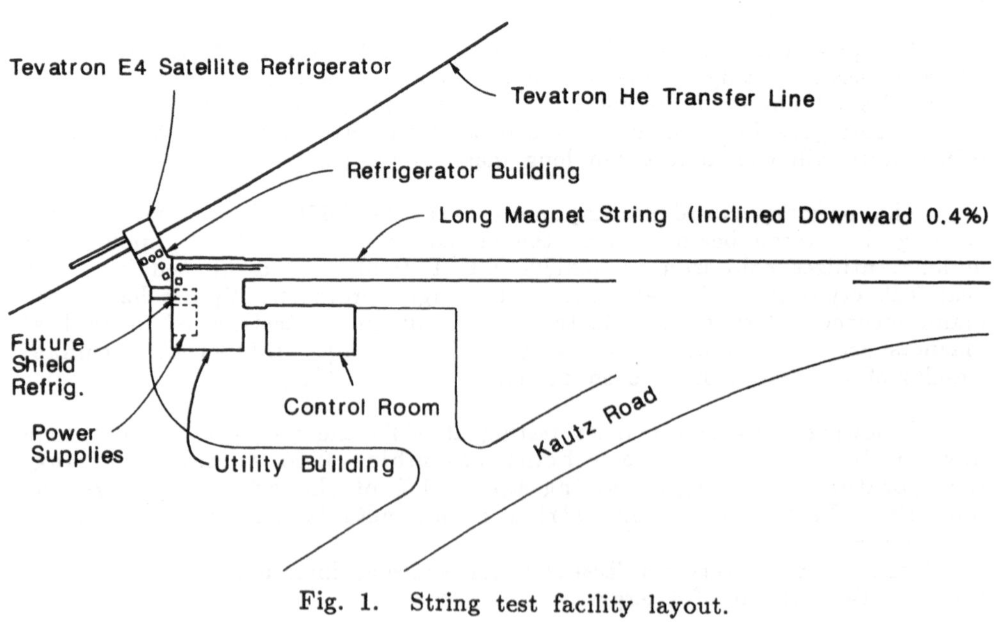

Fig. 1. String test facility layout.

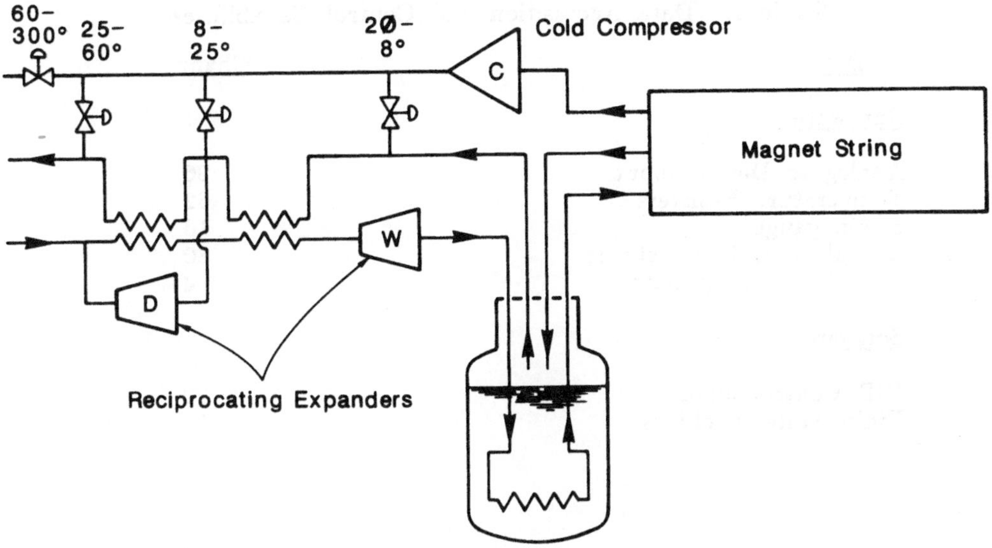

Fig. 2. String test refrigeration system.

Tevatron system during system cooldown and heat leak measurements. During power testing, the compressors will be isolated from the Tevatron to prevent quench pressure excursions from effecting the physics program.

Transient conditions (quench recovery, fast cooldown) will be assisted using partial production of the 4500 liter/hour Central Helium Liquefier (CHL) via the Tevatron transfer line. Small refrigerator transients due to controls or "breathing" of inventory within the magnets will be buffered from the magnet string by a 450 liter subcooling dewar. As the string test approaches one percent of the SSC length, a second refrigerator (a prototype Tevatron Satellite) will be added to provide the 20 K shield refrigeration.

Capabilities have been provided to operate at lower temperatures (down to 2.5 K) using a cold compressor. Depending on the configuration, it will pump on vapor between 2.5 K and 20 K. Three bayonets have been added to the coldbox to accept return gas in the temperature ranges (two- phase to 8 K), (8 K to 25 K), and (25 K to 60 K).

Control of the cryogenic system is accomplished by a in-house Z80 based microprocessor system. It is an expanded version of the Tevatron satellite refrigerator control system. Data acquisition hardware has been added for analog to digital input, temperature resisters, strain gauge, digital control and status, and digital to analog output as summarized in Table 1. Builtin software capabilities include independent proportional, integral, and derivative (PID) control loops and finite state machines. Finite state machines can be tailored for automatic system cooldown and quench recovery.

Information is transferred to and from the microprocessor to the Tevatron central control system through a high speed link. This allows for central alarm monitoring, data logging, and remote control of the system.

Table 1. Data Acquisition and Control Capabilities

Device	Channels
Hardware	
Analog to Digital input	96
Temperature Resisters	32
Strain gauge	16
Digital control and status	80
Digital to analog output	4
Software	
PID Control loops	20
Finite state machines	15

TESTING

Testing and development programs for the string test can be separated into three major categories; Installation, Cryogenic, and Power supply/Quench Protection. Although this paper emphasizes the cryogenic aspects, an outline of all three categories of testing is provided below.

Test and Development

Installation

Development:
A. Installation equipment
B. Alignment techniques
C. Leak checking techniques

Cryogenics

Testing:
A. Heat load measurement 20K and 80K shields
B. Flow impedance measurement
C. Fast magnet warmup and cooldown
D. Ramping effects
E. Quench effects
F. Failure mode

Development:
A. Quench recovery techniques
B. Control strategies

Power Supply/Quench Protection

Testing:
A. Peak field capabilities
B. Cyclic operation
C. Quench behavior

Development
A. Quench detection requirement
B. Energy extraction
C. Bypass Circuit
D. Heater requirements

The initial string testing will be with two 17 m Design D dipoles. Although many of the cryogenic and electrical tests are more interesting with longer strings (several cells), there will be much to learn with the first two magnets. The primary cryogenic test will be a heat leak measurement of the 20 K and 80 K magnet shields. Single phase and shield heat leaks for Design D magnets have been measured previously[6]. The unique feature of this measurement will be to include the heat leak of the magnet interconnection region (Fig. 3). Heat leak will be found by measuring the temperature rise from the mid point in one magnet, through the interconnection region, to the midpoint of the second magnet. Heaters at the shield inlets will allow heat leak to be measured as a function of shield temperature. Measurement of the 20 K shield heat leak is particularly important, since earlier measurements showed higher values than expected.

Flow impedance measurement for the five cryogenic circuits (reference magnet cross-section Fig. 4) is particularly important to insure proper system operation during off design conditions (fast cooldown/warmup and refrigeration transport during refrigerator repair). Preliminary measurements will be made at elevated temperature until several cells are installed.

Development of techniques for fast warmup and cooldown of these cold iron magnets is necessary to meet the design criteria of 24 hours warmup or cooldown time. Two parallel paths of heater cables have been installed in the single phase bypass holes of the iron in the first magnets (final design will have separate penetrations in the iron for the heater cables). Time and temperature safety circuits will assure that the coil is not overheated.

The final design of the SSC requires that liquid helium (or liquid nitrogen for the 80 K shield) be used as a cooldown source for a room temperature string of magnets. Studies will be done to test for yielding on problem areas, bowing of the shields, and sticking of the slide mechanisms as cooldown parameters approach design.

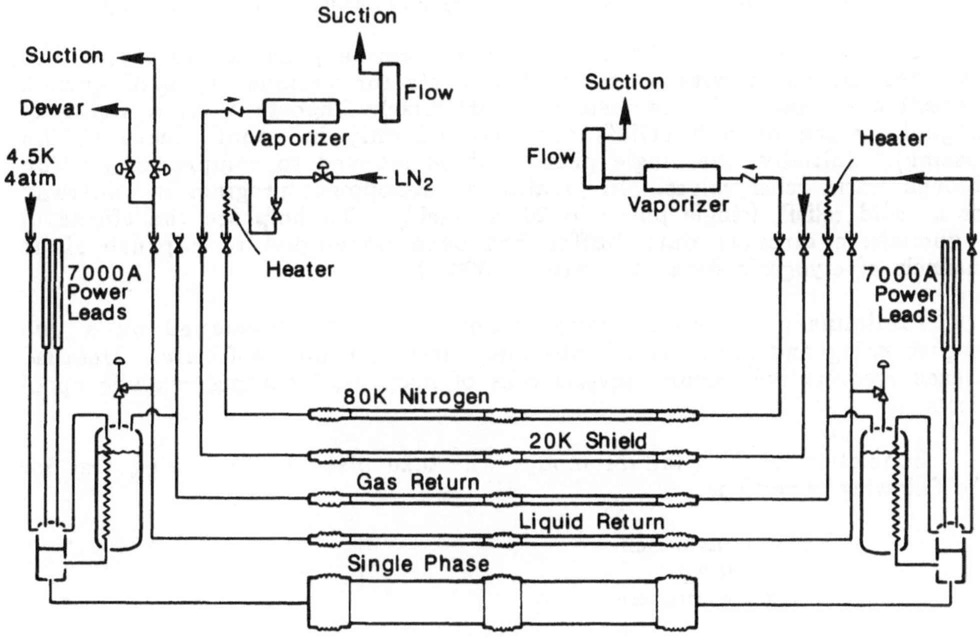

Fig. 3. Magnet string schematic.

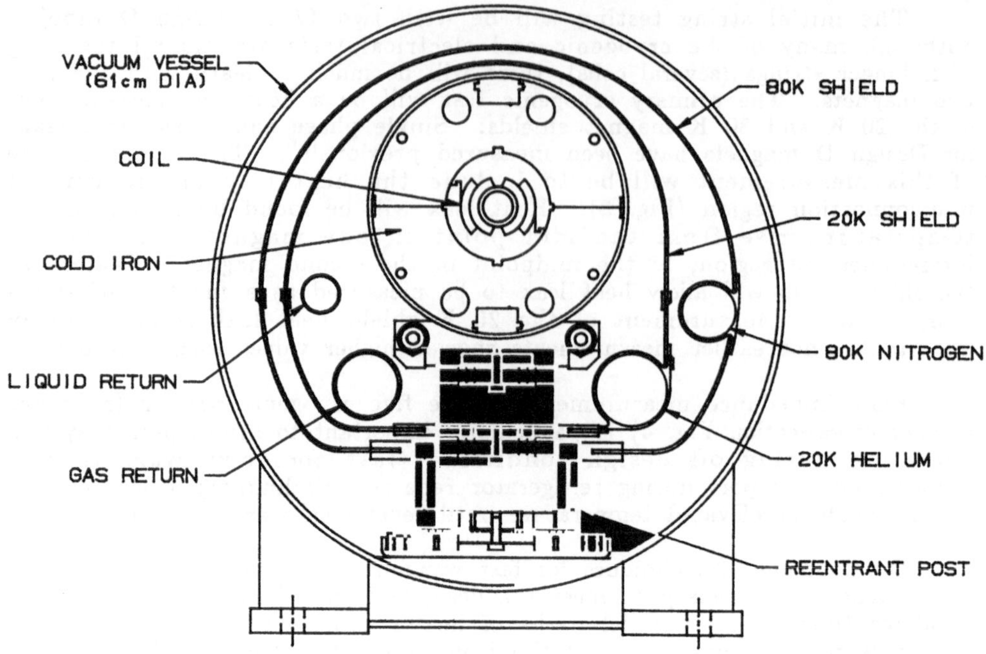

Fig. 4. Cryostat cross section at a suspension point.

During the power supply/quench protection testing, the magnets will be ramped (repeated cycle powering the magnets from injection energy to a given magnet field). The AC losses (hysterisis and eddy current) which result will be used to study the "breathing" effects of the magnet string. These AC losses can be used to simulate synchrotron radiation loading.

Quench effects will be studied to determine peak temperatures and pressures associated with a string of magnets for various types of quench protection schemes. These tests are particularly important on a string of magnets, since quench reliefs are planned only in spool pieces (120m spacing). Initially, the single phase will be relieved to compressor suction through warm relief valves. In parallel, a development program is underway for a "cold relief" (single phase to 20 K shield). To help see the effects of a quench, a circular data buffer has been developed to capture eight channels of cryogenic data at a rate of 300 Hz.

Preliminary quench recovery technique will be developed on a two magnet string and incorporated into finite state machine software. Detailed quench recovery will require several cells of magnets, including realistic spool pieces.

Simulation of SSC failure modes will take place at the string test for the following conditions.

- Vacuum avalanche
- Power outage
- Cryogenic rupture

Vacuum avalanche following quenches can be difficult to recover from, particularly on very long strings. During power outages inventory is stored in the magnets by allowing all cryogenic circuits to pressurize to 20 atm. Design heat loads will prevent inventory loss for up to 30 hours. Controlled simulation of cryogenic ruptures could be valuable in addressing safety issues such as: vacuum space relieving, use of carbon steel vacuum vessels, or release rates for ODH calculations

CONCLUSIONS

A long string test is an important verification of the SSC systems design. Test results will allow early feedback to systems design for deficiencies or to improve operations and reliability.

REFERENCES

1. M.S. McAshan, "Refrigeration Plants for the SSC", SSC-129, SSC Central Design Group, Lawrence Berkeley Laboratory, Berkeley, California (1987)
2. T. Nicol et al, SSC magnet cryostat suspension system design and performance, presented at the 1987 Cryogenic Engineering Conference, St. Charles, Illinois.
3. J. Strait et al, "Tests of Prototype SSC Magnets," T-1451 (SSC-N-321), Fermi National Accelerator Laboratory, Batavia, Illinois (1987).
4. J. Strait et al, "Full Length Prototype SSC Dipole Test Results," T-1450 (SSC-N-320), Fermi National Accelerator Laboratory, Batavia, Illinois (1987).
5. T.J. Peterson and J.D. Fuerst, Tests of cold helium compressors at Fermilab, presented at the 1987 Cryogenic Engineering Conference, St. Charles, Illinois.
6. N.H. Engler et al, SSC long dipole magnet model construction experience, presented at the 1987 Cryogenic Engineering Conference, St. Charles, Illinois.

THE MIRROR FUSION TEST FACILITY CRYOGENIC SYSTEM - PERFORMANCE, MANAGEMENT APPROACH, AND PRESENT EQUIPMENT STATUS

D. S. Slack

Lawrence Livermore National Laboratory
University of California
Livermore, California

W. C. Chronis

Continuous Electron Beam Accelerator Facility
Newport News, Virginia

ABSTRACT

The cryogenic system for the Mirror Fusion Test Facility (MFTF) is a 14-kW, 4.35-K helium refrigeration system that proved to be highly successful and cost-effective. We were able to meet all operating objectives, while remaining within a few percent of our initial cost and schedule plans. The management approach used in MFTF allowed us to make decisions quickly and effectively, and it helped keep our costs down. Manpower levels, extent and type of industrial participation, key aspects of subcontractor specifications, and subcontractor interactions are reviewed, as well as highlights of the system tests, operation, and present equipment status. Organizations planning large, high-technology systems may benefit from our experience with the MFTF cryogenic system.

INTRODUCTION

In this paper, we discuss the management approach used to design and operate the cryogenic system for MFTF. Equipment and technical aspects are discussed in Slack et al., and Chronis and Slack.[1,2] In addition to highlighting management and operational aspects of the MFTF cryogenic system, we discuss system documentation and present equipment status.

MANAGEMENT APPROACH

When the MFTF project was approved in 1982, LLNL decided to subcontract most of the cryogenic system hardware to private enterprise. At that time, the MFTF cryogenic group consisted of two engineers and two mechanical technicians. We wanted to keep the group small to permit rapid decision making and to keep costs low. The group remained this size until near the start of operations. In early 1985, a controls engineer was added to help program the equipment controllers and do controls analysis.

CRYOGENIC SYSTEM SPECIFICATIONS

The cryogenics group wrote a request for proposal (RFP) in 1982. The RFP was primarily a performance specification requiring delivery of hardware, installation, and testing. The RFP stated, for example, that a liquid helium refrigerator of 8 kW plus 30.8 G/s at 4.35 K was required. We also specified those areas where the subcontractor had to interface hardware with other equipment. For example, the subcontractor was assigned required stay in zones within the MFTF building. We also specified helium distribution piping sizes and relief devices where magnet natural convection flows or quench vent pressures were affected. This was necessary since the cryogenic system subcontractor did not have magnet design details. The specification could be termed a hybrid because it required the subcontractor to deliver a certain performance but under the hardware restrictions imposed by LLNL.

In the interest of economy, we tried to minimize the analysis and paperwork required by the subcontractor. We did ask the subcontractor to check the sizing of pipes that provide natural convection flow to the magnets, to provide a backup to calculations done at LLNL. We also required a seismic analysis. In addition, the specification required a 30 to 50% reserve capacity above calculated loads in the cryogenic system. The reserve proved necessary for several reasons. The MFTF plant and capital equipment (PACE) acceptance tests could not have been completed without the reserve, as a result of problems encountered with poor insulating vacuum in the vessel and incomplete flooding of some nitrogen shields. Of course, given more time, these problems would have been corrected.

Three areas were considered important in obtaining the best system at minimum cost. They were to:

- o Write a specification that required minimum changes after the contract was awarded.
- o Encourage effective competitive bidding.
- o Provide maximum cooperation and assistance to the subcontractor throughout the project.

The first item simply required adequate analysis prior to releasing the RFP. In regard to the second item, bidding for the MFTF cryogenic system was limited primarily to Koch, Inc., and CVI Corp. Both companies were serious enough to establish a true competitive bidding situation. Providing maximum cooperation was made easy by the attitude and diligence with which CVI Corp. took on the job once awarded the contract.

Under the contract, CVI provided the following:

- o An 8-kW plus 30-g/s helium refrigerator,
- o A 400-kW nitrogen subcooler,
- o A 500-kW nitrogen reliquefier,
- o An extensive helium distribution system including 13 supply dewars for magnets and cryopanels and a forced-flow cooling system,
- o A nitrogen distribution system.

Under a previous contract for the first phase of MFTF (MFTF-A), CVI Corp. supplied a 3-kW helium refrigerator, a 100-kW nitrogen subcooler, a distribution system, and parts of a helium recovery system. LLNL supplied high-pressure helium gas storage totaling 343 m^3 at 12 MPa, and a 12-MPa, 600-hp helium recovery compressor. All of these components make up the MFTF cryogenic system.

TRAINING AND OPERATION

System operating instructions and manuals for this complex a system could not be economically or effectively completed before actually running the system. Operational procedures written prior to start-up of an untried, complex system are generally inaccurate and serve only as an outline or overview. Consequently, our RFP required only minimal procedure write-up prior to system operation. Start-up and initial operation--first of system components, then the entire system--were directly controlled by engineers who designed and built the system. The engineers used system drawings, particularly piping and instrumentation drawings, with various check lists. The system had five operator stations over a distance of 200 m, so several people needed to be involved in operating the system. Walkie-talkies provided the most effective communication method during early phases of operation. Modern industrial-grade walkie-talkies operated well, even through the steel building used to house MFTF.

During start-up operations, we found it important to be at the equipment where we could observe sights and sounds that would be unobservable to even the best instrumented remote operator. Sights, sounds, and vibration changes in machinery and piping provided valuable diagnostic tools. We also found that the most effective and safest way to start a new system of this type is to ensure that personnel most involved with designing the system are in operational control. Instant communication must be maintained between operators.

Prior to start-up operations, operators need to be trained so they can become familiar with equipment layouts, location of instrumentation, purpose of the equipment, and operational philosophies. In the interest of economy, the number of technicians associated with the cryogenic group remained at two until just prior to start-up for the PACE acceptance tests. In hindsight, more should have been involved earlier. As a result of our decision, we needed to train 24 new people on the job (6 people for each of 4 shifts), while debugging the new system. During the cryogenic system start-up and PACE acceptance tests, which lasted about 6 months, a few key personnel worked up to 100 h/week, which at times tested the limits of their endurance.

SYSTEM OPERATION AND PROBLEMS

In general, the system performed very well. Preliminary and acceptance tests by CVI Corp. demonstrated performance to specification. Subsequently, the system cooled the 1.05 x 10^6 kg of magnets in MFTF to 4.5 K as required. It is noteworthy that this was done within a very tight time schedule. The PACE acceptance tests were completed without a full-system test of the cryogenic system. During cooldown and operation of the 22 MFTF magnets, failure of the cryogenics system for more than a few days would have forced us to abort PACE because of schedule and budget constraints and the time required for to recool the system. Consequently, problems were dealt with on-line.[2]

Oil Carry-Over

After a month of operation, the effectiveness of heat exchanger 1 (room temperature to 80 K) in the 8-kW cold box decreased. This change resulted in loss of capacity and in icing of compressor return lines, which forced a shutdown. About 3 L of an odorous oil substance was removed from the heat exchanger. After cleaning and restart, the system operated another month and the problem reoccurred. Analysis of oil from the heat exchanger showed light ends of the compressor oil. Analysis of gas entering the cold box during operation showed an oil content of about 1 ppm, which would result in the 3 L/month carryover of oil actually found in the cold box.

Propylene glycol, the lightest end of the polypropylene glycol oil used in the compressors, has a vapor pressure that would more or less account for 1 ppm carry-over as a vapor. We concluded that the compressor oil either had excessive light ends or the light ends were being generated during operation. In MFTF, the second stage operates at an 8:1 compression ratio, which gives an adiabatic compression temperature of about 690 K. This temperature may be causing the carryover problem. At a 4:1 compression ratio, the adiabatic compressor temperature is about 520 K. Thermal degradation of the compressor oil starts at 480 K. The presence of abundant oil in the compressor reduces temperatures well below adiabatic; however, the evenness of oil distribution and its effect to cool the helium during compression is not well understood. Consequently, we believe compression ratios in the future systems should be limited to 4:1. To solve this problem, we plan to install a freeze-out heat exchanger upstream of the cold box.

SYSTEM DOCUMENTATION

A mixture of published and unpublished data exists relating to the MFTF cryogenic system. Published documents incluue Slack et al., Chronis and Slack, and Krause and Kozman.[1-3] Additional data include:

- o Complete, well-documented drawings maintained in LLNL central drawing files.
- o A two-volume design data book covering numerous calculations and data accumulated during the design and construction phases of MFTF.
- o A 27-volume operations and maintenance manual compiled by CVI Corp.
- o Disks containing software for equipment programmable controllers.

EQUIPMENT STATUS

Presently, some repairs are needed to return the MFTF cryogenic system to full operational status. Specifically, we need to:

- o Add a freeze-out systems for oil carryover problems,
- o Replace an impeller in the nitrogen reliquefier compressor.
- o Repair an absorber bed in the helium recovery system purifier.

An estimated cost for these repairs as well as other minor repairs would be $100,000 to $200,000. The system can be operated in its present condition but not for extended duration. We plan to operate the system every 6 to 12 months to maintain operational capability while mothballed.

SUMMARY

In designing and developing the cryogenic system for MFTF, our management approach at LLNL allowed us to make decisions quickly and effectively. It also helped keep costs down while allowing us to meet our operating objectives.

ACKNOWLEDGMENTS

Work performed under the auspices of the U.S. Department of Energy by the Lawrence Livermore National Laboratory under Contract W-7405-Eng-48.

We acknowledge CVI Corp. of Columbus, Ohio, and its leaders. As subcontractor for most of the cryogenic equipment, they ensured that their equipment met the needs of MFTF. We also thank Victor Karpenko, program manager, for a get-the-job-done atmosphere, and Robert Nelson, cryogenic system project manager, for his effective leadership.

REFERENCES

1. D. S. Slack, R. Nelson, and W. Chronis, "Cryogenic System for the Mirror Fusion Test Facility," Cryogenics 31: 583 (1985).

2. W. Chronis, D. S. Slack, "Operation of the Cryogenic System for the Mirror Fusion Test Facility," Lawrence Livermore National Laboratory Report UCRL-96002, Livermore, California (1987).

3. K. H. Krause, T. A. Kozman, "MFTF-B PACE Test," Lawrence Livermore National Laboratory Report UCID-20819 (October 1986).

THE TORE SUPRA 300 W - 1.75 K REFRIGERATOR REPORT

Guy M. Gistau, Michel Bonneton, and John W. Mart

L'Air Liquide Advanced Technology Division
Sassenage, France

ABSTRACT

The TORE SUPRA refrigerator is now completely installed and all the acceptance tests have been completed. All equipment has performed above design levels and, as it was planned, totally automatically. The novel equipment such as liquid ring pumps and notably the cold centrifugal compressors have run perfectly for more than 1000 hours. Commissioning and acceptance tests are reported.

INTRODUCTION

The TORE SUPRA refrigerator is used to cool the toroidal field coils of the superconducting tokomak[1] The various operating modes are obtained entirely automatically. The following simultaneous operating parameters correspond to "normal operation": (see Table 1 of reference 2)
10 700 to 32 000 W at 80 K
745 W at 4.0 K
0.003 kg/s liquid helium
300 W at 1.75 K
A description of this refrigerator is given in reference 2.

THE PRELIMINARY TESTING

We chose to carry out the machine trials and the debugging of the corresponding automatic control (simultaneously). In this way we were sure to be able to repeat, in automatic operation, all tests covered ; this enabled us to concentrate our efforts on the testing in hand. We have thus also amassed more than 3000 hours in automatic operation. The testing sequence was classical: the warm machines (compressors, liquid-ring pumps), then the cold box. We first carried out the cooldown to the operating mode called "normal operation", this procedure requiring all the machines to be in operation.

THE TEST CRYOSTAT

In order to be able to measure the performance and observe the behaviour of the refrigerator, the French Atomic Energy Commission constructed a test cryostat which was connected to the refrigerator coldbox as shown in Fig. 1.

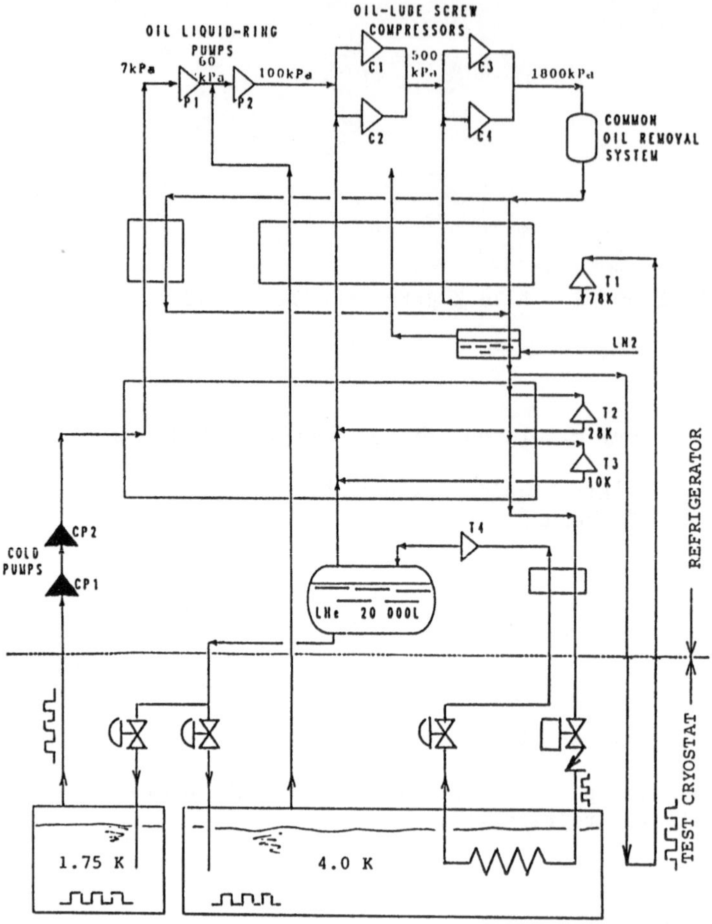

Fig. 1. Test cryostat connected to the refrigerator.

A circuit equipped with up to 40 kW of electrical heating simulated the radiation shields operating at the 80 K level. A circuit with 36 kW of electrical heating simulated the heat load of the toroidal field coils in their thick-walled enclosure. Part of this circuit is immersed in a bath of liquid helium at 4.0 K which also contains an immersed electrical heater of 1500 W. The bath of liquid helium at 1.75 K is similarly equipped with 600 W of electrical load. However, since there is no heat exchanger upstream of the expansion valve, the available refrigeration is shared between the heater immersed in the liquid and another heater element which warms up the helium to 3.46 K. It is pointed out that the volumes of the test circuits and liquid baths are very small compared to those in the tokomak.

THE TEST PROGRAMME

The major points of the test programme were:

- cooldown simulation
- the measurement of constant cryogenic powers at the different temperatures requested by the contract
- behaviour of the refrigerator when submitted to variable heat loads
- behaviour of the refrigerator when some machine (a compressor, a liquid ring pump, a turbine, a cold compressor or the two-phase expander) is stopped

- behaviour of the refrigerator when supply of electricity or cooling water or instrument air is cut, and automaticaly restarted.
- exchanging a turbine or a cold compressor without stopping the refrigerator
- 400 h running time under automatic control with regular change of operating mode

In this brief description of the testing we will only linger on important aspects. To carry out the various performance measurements in the different operating modes, we decided as a general rule, to apply the contractual heat loads at the defined temperatures (80 K, 4.0 K, 1.75 K) and to measure the consequent variation in the level of liquid helium ; the differences in the performance are thus quantified. By calculating the power input to the equivalent Carnot cycle, one can thus compare the different results, albeit that the refrigeration is produced at different temperatures.

The test in normal operating mode

All the machines are in operation simultaneously in this mode. The screw compressors' flow rate was automatically controlled according to the requirements of the cold box. This presented only one difficulty which is due to the large variation of the compressor throughput resulting from a small change in slide valve position when operating near to the maximum. The work involved in alleviating this problem was more time-consuming than we had foreseen. We found that the liquid-ring pumps' characteristics are superior to those expected. In fact, the suction pressure of the 1st stage machine P1, corresponding to 300 W of cryogenic heat load, establishes itself at 4.0 kPa instead of 7.6 kPa. As a result the suction pressure of the cold pumps is lower (1.08 kPa instead of 1.14 kPa) and the total compression ratio is reduced (4.04:1 instead of 6.94:1). Nevertheless we have verified that the calculated cold pumps compression ratio can be attained, or even exceeded, by throttling the discharge of the 2nd stage cold pump, (Fig.2). We noted an excellent flexibility in operation of the cold pumps since, with a constant rotational speed, the compression ratio changes in a ratio of 1.84:1. More detailed information on cold pumps is given in "The Tore Supra refrigerator cold centrifrigal compressors report" in the '87 CEC conference.

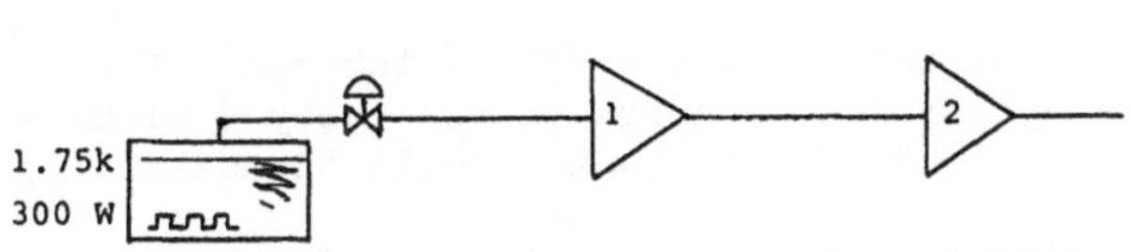

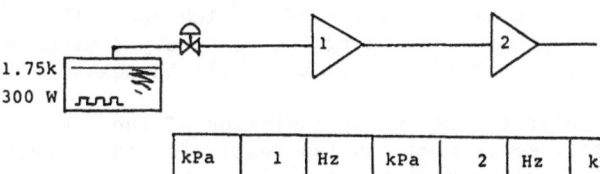

	kPa	1	Hz	kPa	2	Hz	kPa
- Process conditions (theoretical)	1.14	3.10	410	3.53	2.24	460	7.92
- Naturel operation	1.08	2.62	411	2.83	1.54	457	4.37
- Operation at process conditions	1.14	3.18	410	3.63	2.25	459	8.15
- Other operation	1.21	3.29	410	3.98	2.27	459	9.02

Fig. 2. Operating conditions of cold pumps.

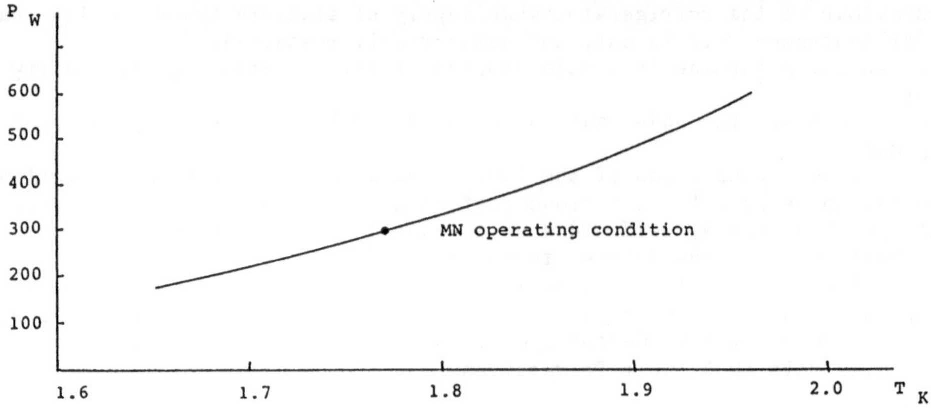

Fig. 3. Cold power versus operating temperature.

The liquefaction rate exceeded the contractual value. We were able to observe its variation as a function of the temperature of the cryogenic turbine T3. The contractual refrigeration load of 300 W in the bath of liquid Helium II was achieved at 1.25 kPa being equivalent to 1.72 K ; higher values were obtained with variation of this temperature (Fig.3). The measured performances exceed the contractual values by 10 %. Carnot efficiency is 13.6 %. We also determined that certain pipe runs in the cold box which were poorly insulated were the cause of abnormal thermal inleak. After correction, the refrigerator performance should improve.

The test in idle mode at 2.1 K

This operating mode reduces the overall energy consumption of the system during those periods when the tokomak is not in use. The turbine T2 and the cold pumps are stopped, the latter no longer injecting energy into the cycle. The liquid ring pumps alone must ensure a pressure of 3.6 kPa above the bath of Helium II. Due to the fact that their characteristic is better than that expected, we could inject a greater power than the contractual value into the Helium II bath; 220 W instead of 133 W. In spite of that, the pressure over the liquid was 3.34 kPa. When the Tokomak is connected to the refrigerator and the real heat loads are known, the rotational speed of the liquid-ring pump P1 will be adjusted as necessary giving an expected saving of about 100 kW. The measured heat loads are in excess of contractual requirements by 32 %. The other operating modes have demonstrated operating margins of between 10 and 30 %.

The test with variable heat loads

During normal operation of the tokomak, the variation of toroidal field strength induces energy pulses every 4 minutes in the thick-walled coil housing (260 kJ) and in the coils themselves (30 kJ).

In order to have an appreciation of the likely behaviour of the refrigerator, we carried out two tests. In the first we injected energy pulses (155 kJ every 4 minutes) into the Joule Thomson circuit. This simulated the energy coming from the thick-walled housing. We observed an overpressure of 0.2 MPa, and verified the correct operation of the devices

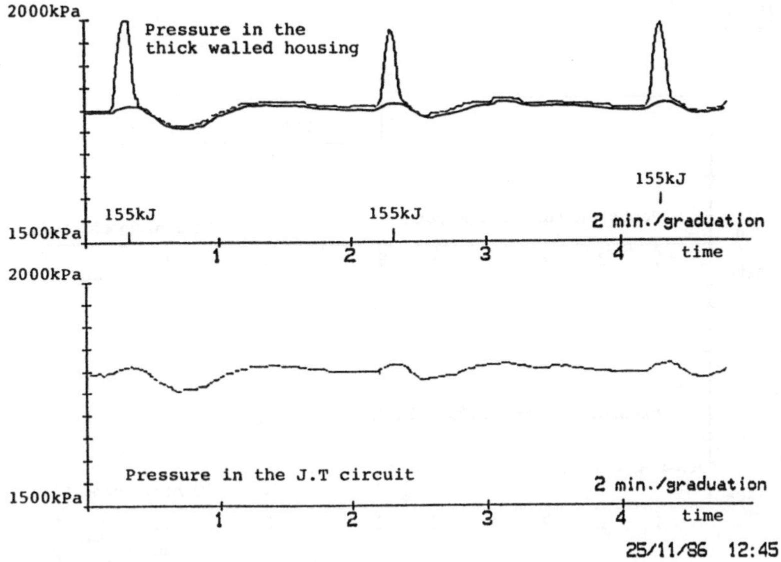

Fig. 4. Overpressure in the JT circuit.

to prevent reverse flow in the Joule-Thomson exchanger and overpressure at the inlet of the wet expander (Fig. 4). It will be necessary to carry out fine adjustments when the tokomak is connected. In the second test, power was cyclically injected into the baths of liquid helium at 4.0 K and 1.75 K. This simulates the energy coming from the thick-walled housing carried by the gas in the Joule-Thomson circuit flowing into the bath at 4.0 K, and the energy received by the coils in the bath at 1.75 K.

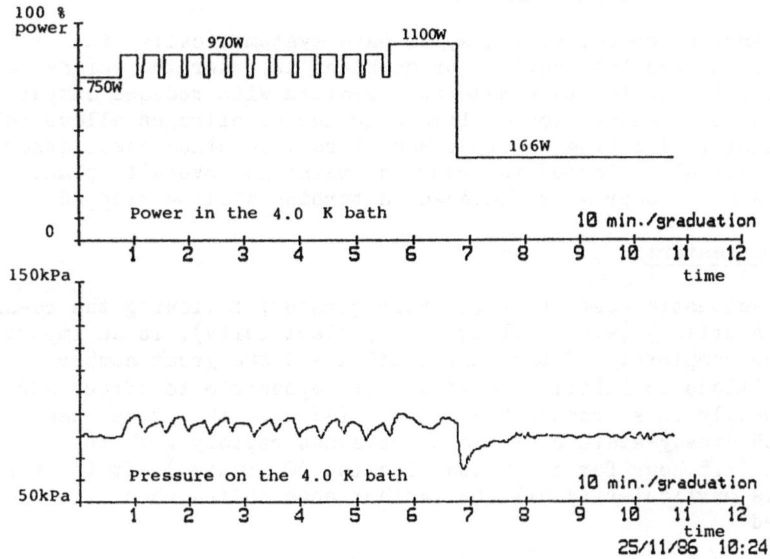

Fig. 5. Energy pulses in the 4.0 K bath.

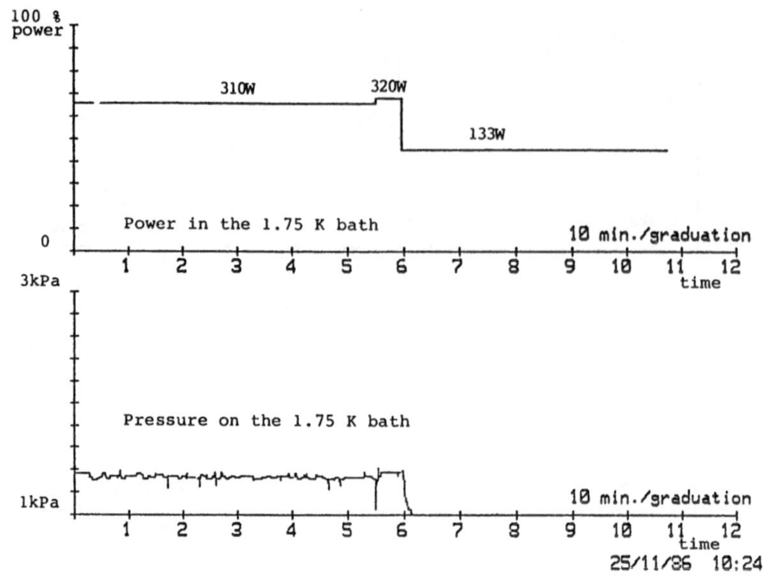

Fig. 6. Energy pulses in the 1.75 K bath

Figures 5 and 6 show that the variation of pressure in the baths is acceptable. Since, as stated earlier, the volume of the test cryostat is relatively small, we believe that the behaviour of the refrigerator will be satisfactory under real operating conditions.

Unscheduled shutdown of machines

In various operating modes, we have systematically investigated the effect of unscheduled shutdown of machines in order to verify a smooth, automatic, transition to a mode of operation with reduced output. In certain cases the evaporation of liquid helium or nitrogen allows full power to be maintained during an experimental run. In other cases other machines are automatically stopped in order to maintain overall plant stability e.g in case of compressor shutdown, a turbine will be stopped.

Automatic restart

An automatic restart of the refrigerator, following the re-establishment of a utility (air, cooling water, electricity), is an important function. The complexity of the installation and the great number of actions and conditions to fulfil make it almost impossible to effect such a procedure manually in a reasonable time. Indeed, the time needed to re-establish steady-state conditions increases rapidly with the duration of shutdown (1.5 hour for a shutdown lasting 20 minutes). It is very likely that this particularity will be further accentuated when the tokomak is connected.

400 hour test

During the 400 hour test running period we simulated the likely operation of the refrigerator in different operating modes, associated with the operating pattern of the Tokomak. Although each mode had already been checked, the important aspect of this test was to observe the automatic change of operating mode according to the diagram, Fig.7. We were able to

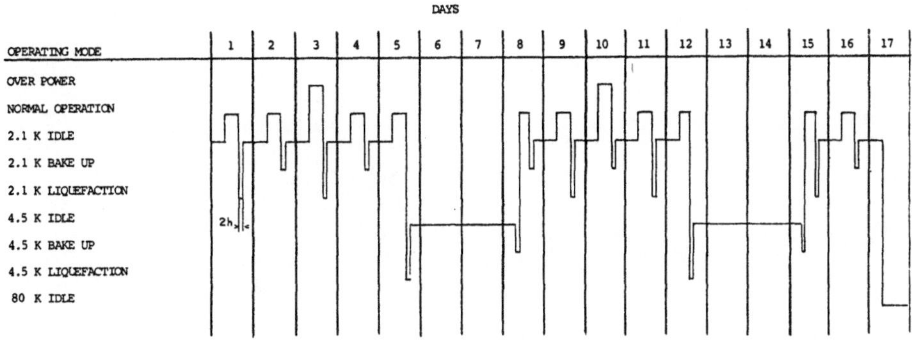

Fig. 7. 400 hour test programme

verify that the operation of the refrigerator is really totally automatic, three re-starts took place following shutdown due to various external faults.

CONCLUSIONS

During the tests, we have checked that : (1) the garanteed performances have been reached, (2) the liquid ring pumps operate correctly, (3) the cold pumps have a very flexible behaviour, (4) the energy pulses are correctly accepted, and (5) the automatic operation from stop to the different modes is not only possible but necessary. We have gained experience in automatically setting the screw compressor flow rate. The experience we had on operating a series of cold pumps on a real refrigerator encourages us to envisage increasing the number of stages in order to eliminate the room-temperature pumping system.

REFERENCES

1. R. Aymar et al, Conceptual design of superconducting Tokomak : TORUS II SUPRA, IEEE Transactions on Magnetics MAG-15 : 542 (1979).
2. G. M. Gistau, G. Claudet, The TORE SUPRA 300 W - 1.75 K refrigerator, in : "Advances in Cryogenic Engineering", Vol 31, Plenum Press, New York (1986), p. 607.

Fig. [illegible] 400 [illegible] test program.

verify that the operation of the cel[illegible] [illegible] three re-start[illegible] [illegible] place following [illegible] due to various external faults.

CONCLUSION

During the tests we have checked that (1) the [illegible] have been reached; (2) the [illegible] [illegible] [illegible] [illegible] [illegible] [illegible]

REFERENCES

1. [illegible] et al. [illegible]
2. [illegible]

SYSTEM DESIGN AND VERIFICATION TEST RESULTS OF CRYOGENIC SYSTEM FOR DEMONSTRATION POLOIDAL COIL

E. Tada, T. Kato, T. Hiyama, K. Kawano, M. Hoshino, and S. Shimamoto

Japan Atomic Energy Research Institute
Naka-machi, Naka-gun, Ibaraki-ken, Japan

ABSTRACT

The Demonstration Poloidal Coil (DPC) program for investigation of large superconducting pulse-coil technology required for Fusion Experimental Reactor (FER) has been conducted. In this program, there are three superconducting pulse coils (DPC-U1, U2, and EX) so as to simulate mechanical interaction of real pulse coils in fusion reactor. These coils are forced-flow pulse coils with a cable-in-conduit conductor and are designed to be cooled by supercritical helium at 350 g/sec under conditions of 10 bar and below 4 K. For this purpose, a new cryogenic pump system, which is composed of a circulation pump and a cold compressor, is developed. In addition, a newly developed current lead with a cable-in-conduit shape has been selected for the current leads and its rated current and insulation testing voltage are specified to be 30 kA and AC 16 kV, respectively. The latest design status of the whole cryogenic system and the verification test results of the cryogenic devices such as the cryogenic pump system and current lead are described in this paper.

INTRODUCTION

The Japan Atomic Energy Research Institute (JAERI) has been developing large superconducting magnet for the Fusion Experimental Reactor (FER), which is the next tokamak after JT-60 and aims to demonstrate the engineering feasibility of a large tokamak reactor in Japan. Since application of superconductivity for FER's poloidal coils is indispensable for a long pulse operation, JAERI has been carrying out a series of projects for the development of superconducting poloidal coils[1] as well as the toroidal coil development[2]. In order to investigate the pulse-coil technology in real coil arrangement, JAERI started the program of the Demonstration Poloidal Coil (DPC) in 1985 [3]. In this program, three 1.0-m ID pulse coils, two of which (DPC-U1 and DPC-U2) utilize Nb-Ti with rated current of 30 kA and the other (DPC-EX) Nb_3Sn with 10 kA, are being developed. These three coils can be tested in the Superconducting Engineering Test Facility (SETF) of JAERI in the arrangement shown in Fig. 1.

In parallel with the coil technology development, JAERI has been making efforts to develop the cryogenic technology[4] for a large helium cryogenic system with a refrigeration capacity of 100 kW at 4 K required for FER. In this cryogenic technology development, some key cryogenic components such as a supercritical helium circulation pump, a cold compressor, a turbo-expander, and a vapor-cooled current lead are being developed in collaboration with several Japanese industries. Since DPC requires a large amount of supercritical helium of 350 g/sec below 4 K in nominal operation, a cryogenic circulation pump and a cold compressor, which have been developed through the cryogenic technology development, are newly installed as key components of the DPC test facility. Furthermore, a newly developed current lead [5] with a rated current of 30 kA is selected as the vapor-cooled current leads for DPC. This paper describes the detail design of the whole cryogenic system and the verification test results of cryogenic components installed for cooling DPC.

SYSTEM DESIGN

Three forced-flow superconducting coils, whose requirements on cooling condition are listed in Table 1, are being developed in the DPC program. In order to satisfy these requirements, the existing 350-ℓ/h, 1.2-kW helium liquefier/refrigerator [6] is used as a main system of the test facility and cryogenic components such as a cryogenic pump system, which can produce and supply supercritical helium with 350 g/sec at 10 bar and below 4 K, are newly installed and connected to the refrigerator. The existing refrigerator has a cool-down capacity of 10 kW at 100 K; this permits to cool the coil system with a total cooling weight of 25 ton including the supporting structures from 300 K to 4 K within 120 hr. In addition to the cryogenic pump system, a liquid helium storage tank, a cold recovery tank, a warm recovery tank, and three pairs of vapor-cooled current leads are installed. The coil system is connected through cryogenic pipings both to the refrigerator and to the pump system as shown in Fig. 2. This piping arrangement is selected in accordance with the following procedure specifications.

(COOL-DOWN)
- The coil system and cryogenic pump system is cooled by the refrigerator from 300 K to 4 K.

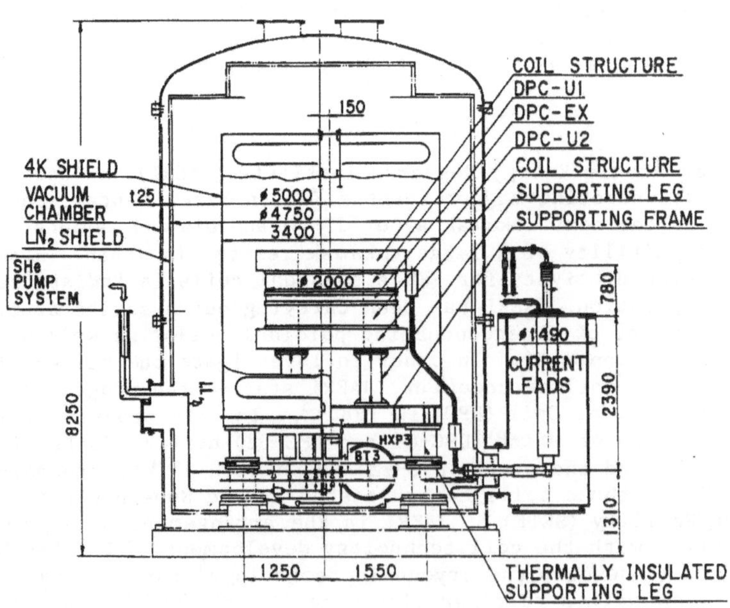

Fig. 1 Schematic view of coil system inside vacuum chamber

Table 1 Requirements on cooling condition

ITEMS		REQUIREMENTS	
(COOL-DOWN)			
cool-down time		120 hr	
total cooling weight		25 ton	
(NOMINAL OPERATION)		DPC-U1&U2	DPC-EX
inlet temperature		4 K	4 K
inlet pressure		10 bar	10 bar
mass flow rate		300 g/sec	50 g/sec
pressure drop		1 bar	1 bar
(LIQUEFACTION LOAD)			
DPC-U1&U2	(30 kA x 4)	6.0 g/sec	
DPC-EX	(10 kA x 2)	1.0 g/sec	

(STEADY-STATE REFRIGERATION)

- The coil system is cooled by supercritical helium supplied from a cryogenic pump system connected to the refrigerator.
- In case of the pump system failure, supercritical helium with 60 g/sec is supplied from the refrigerator to the coil system.
- In case of coil quench, helium gas inside coils is stored into a cold recovery tank at lower temperature.
- In case of refrigerator failure, the pump system is cooled by liquid helium from a liquid helium storage tank for keeping supply of supercritical helium to the coil system for a short period.

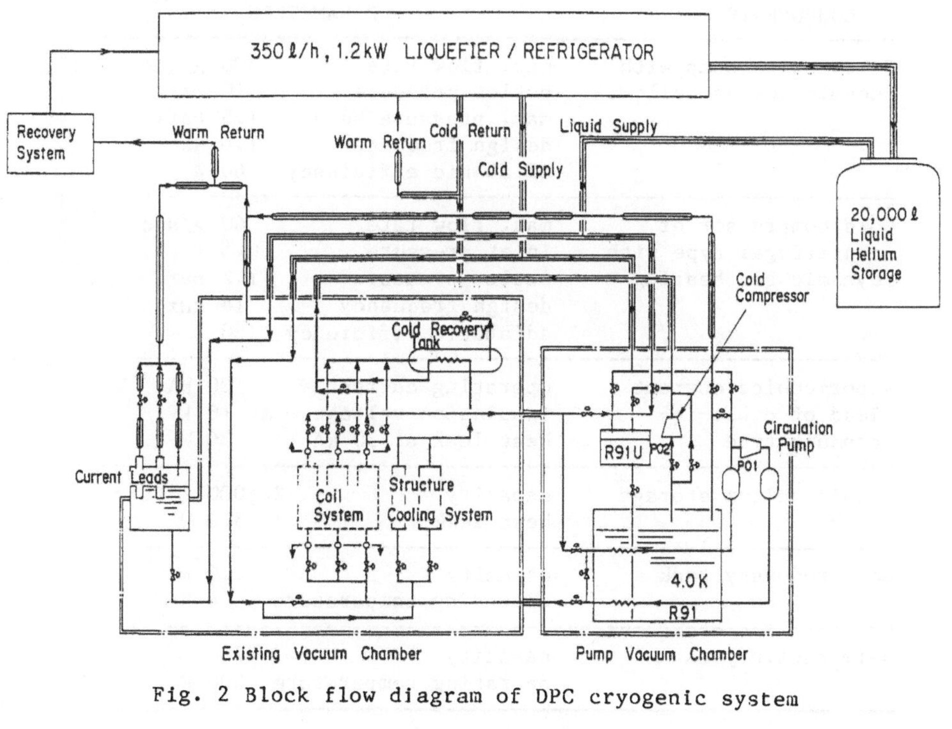

Fig. 2 Block flow diagram of DPC cryogenic system

(CURRENT LEAD COOLING)

- Vapor-cooled current leads is cooled by liquid helium from a liquid helium storage tank.

The coil system is thermally insulated by a 4-K shield plate in order to avoid an excessive heat leak from the outside (see Fig. 1). Furthermore, a cold recovery tank is installed to recover helium gas from the coil system at a low temperature so as to keep the maximum pressure in the coil system within 30 bar during a quench. The cold recovery tank and the supporting structures including a 4-K shield plate have a cooling channel connected to the refrigerator and are cooled below 10 K during nominal operation.

COMPONENT DESIGN

The cryogenic system of DPC is composed of the existing liquefier/ refrigerator and several cryogenic devices which are newly installed. The major parameters of these components are specified as listed in Table 2.

The maximum capacity of the cryogenic pump system, which is composed of a circulation pump and a cold compressor, is specified to produce and supply supercritical helium with mass flow rate of 500 g/sec at 10 bar and below 4 K in accordance with the cooling requirements of the coil system. A reciprocating pump with double acting bellows driven by oil pressure is selected as the circulation pump for generating supercritical helium. A cold compressor, which is a centrifugal compressor with dynamic gas bearing, is to make lower operating temperature than 4 K by evacuating at saturated vapor pressure under cryogenic conditions. The adiabatic efficiency of the pump system is specified to be 60 %.

A three pairs of current leads with a cable-in-conduit shape are installed for the DPC coil system. The rated current and insulation

Table 2 Key parameters of major cryogenic components

COMPONENTS	PARAMETERS	
Circulation pump with double acting bellows	max. flow rate	500 g/sec
	outlet pressure	10 bar
	max. pressure head	1.5 bar
	design frequency	1.0 Hz
	adiabatic efficiency	60 %
Cold compressor of centrifugal type with dynamic gas bearing	max. flow rate	60 g/sec
	inlet pressure	0.5 bar
	outlet pressure	1.2 bar
	design frequency	14 kHz
	adiabatic efficiency	60 %
Vapor-cooled current lead of cable-in-conduit type	operating current	30 kA
	insulation voltage	AC 16 kV
	heat leak at 30 kA	30 W
Liquid helium storage tank	capacity	20,000 liters
	heat leak	5.0 W
Cold recovery tank	capacity	1.0 m^3
	operating temperature	4 K
Warm recovery tank	capacity	700 m^3
	operating temperature	300 K

testing voltage of the current leads are specified to be 30 kA and AC 16 kV, respectively. The thermal performance of the current lead is designed to have an ideal self-cooling condition with heat leak of 1.0 W/kA.

A liquid helium storage tank is designed to have a capacity of 20,000 liters in order to both cool the current leads and initially fill the liquid helium bath of the cryogenic pump system. The capacity of the cold recovery tank is specified to be 1.0 m^3 so as to keep the maximum pressure of the coil within its proof pressure of 30 bar during a coil quench. A warm recovery tank, whose capacity is around 700 m^3, is also newly installed in order to recover and store helium gas evaporated from the current leads and a liquid storage tank.

THERMAL DESIGN

The total static heat load including the conduction, radiation and joule heat is estimated to be around 270 W. In addition to the static heat load, a pumping load of the cryogenic pump system for circulating supercritical helium below 4 K is applied as a refrigeration load. Figure 3 shows calculation results of the thermal balance in the nominal operation in which supercritical helium with mass flow rate of 350 g/sec at 10 bar and 4 K is supplied to the coil system and total pressure drop is 1.0 bar. In this calculation, temperature and pressure at JT heat exchanger of the refrigerator is fixed to the design condition at which a refrigeration capacity of 1.2 kW can be achieved with JT flow rate of 64 g/sec. It is clear from Fig. 3 that the refrigerator has enough refrigeration capacity to cool the coil system and cryogenic pump system since the required JT flow rate is 47.6 g/sec corresponding to around 75 % of the maximum capacity.

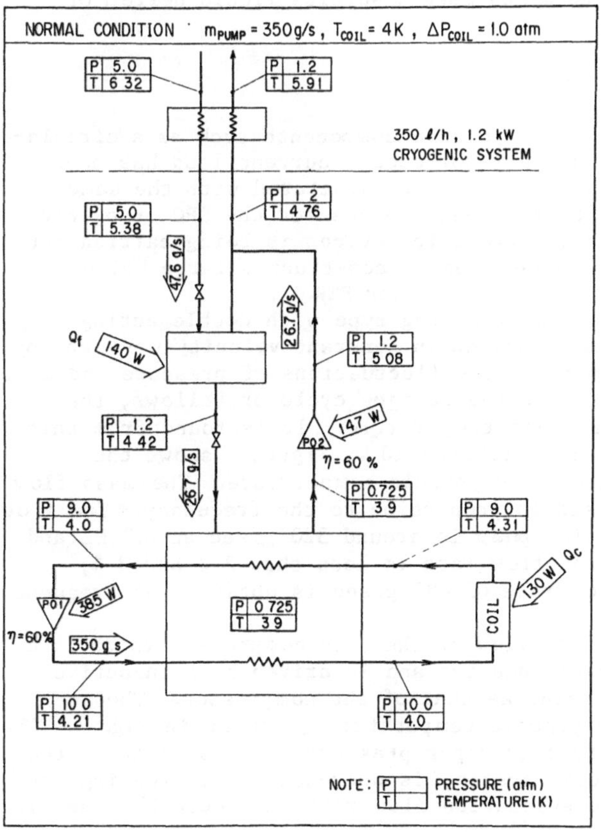

Fig. 3 Calculation result of thermal balance in nominal operation

Table 3 Thermal balance of DPC

ITEMS		NOMINAL	STAND-BY
OPERATING CONDITIONS ;			
Inlet temperature	(k)	4.0	4.4
Inlet pressure	(bar)	10	10
Mass flow rate	(g/s)	350	100
Operating time	(hr)	8	14
HEAT LOAD ;			
Circulation pump	(W)	385	130
Cold compressor	(W)	147	0
Coil system	(W)	60	0
Facility	(W)	210	210
Current leads	(ℓ/h)	210	130
Surplus capacity	(ℓ/h)	-90	+120
balance/day	(ℓ/h)	+ 40	

As a result of this calculation, thermal balance of the whole cryogenic system including the coil system is summarized in Table 3. The total liquefaction and refrigeration capacities required for the nominal operation in day time are estimated to be 210 ℓ/h and 800 W, respectively. In this case, the total heat load exceeds the maximum capacity of the existing liquefier/refrigerator, resulting in a decrease of the stored liquid helium at a rate of around 90 ℓ/h. In case of stand-by mode in night time, however, the existing liquefier/refrigerator can produce an extra liquid helium of around 120 ℓ/h into the liquid helium storage tank. Accordingly, the existing liquefier/refrigerator has sufficient capacity for cooling the whole system so as to make a surplus liquid helium of around 1,000 liters in a day.

VERIFICATION TEST

The verification test of key cryogenic components such as a circulation pump, a cold compressor, and a vapor-cooled current lead has been carried out by using the prototype devices manufactured with the same design specifications as that of the real devices for the DPC cryogenic system. The development of these cryogenic devices is being carried out in collaboration with Japanese industries to construct a large helium cryogenic system with a capacity of 100 kW for FER.

The circulation pump is a reciprocating type with double acting bellows as shown in Fig. 4 and is driven in constant velocity movement by oil pressure so as to avoid an excessive fluctuations of pressure and mass flow rate. In order to investigate the fatigue cycle of bellows, the fatigue test was carried out at room temperature. It is found from this test that the allowable cycle is more than 10^7. Figure 5 shows the typical characteristics measured at cryogenic temperature. The mass flow rate and pump head are increased in proportion to the frequency. The flow rate generated by the circulation pump is around 320 g/sec at 0.8 Hz and the fluctuations of pressure and flow rate is less than 2 % and 3 %, respectively. The maximum flow rate of 480 g/sec is obtained at frequency of 1.0 Hz.

Figure 6 shows the schematic view of the cold compressor which is a centrifugal type with dynamic gas bearing and is driven by a inductive motor installed in the same casing as that of the compressor. The measured characteristics at cryogenic temperature is shown in Fig. 7. The inlet pressure, which is a saturated vapor pressure corresponding to the operating temperature of the coil system, is decreased by increasing the frequency. Since the required evacuation flow rate is around 27 g/sec in

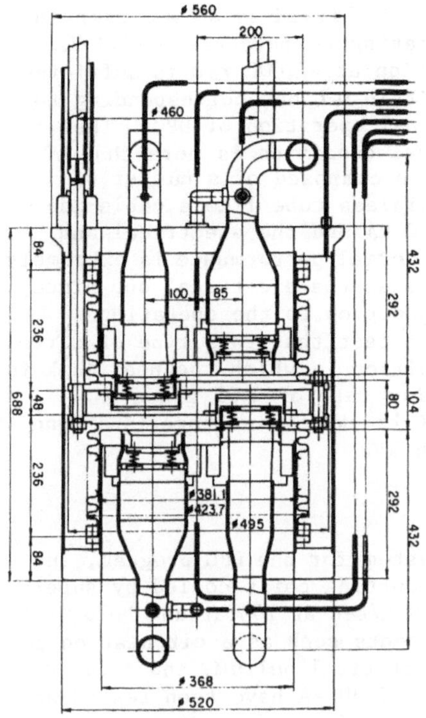

Fig. 4 Schematic view of circulation pump

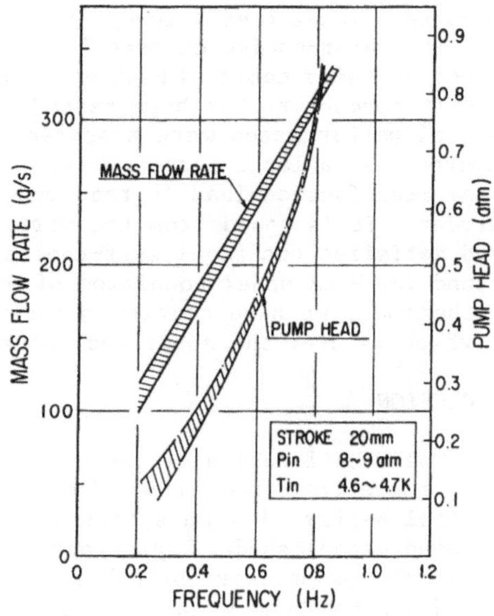

Fig. 5 Measured characteristics of circulation pump

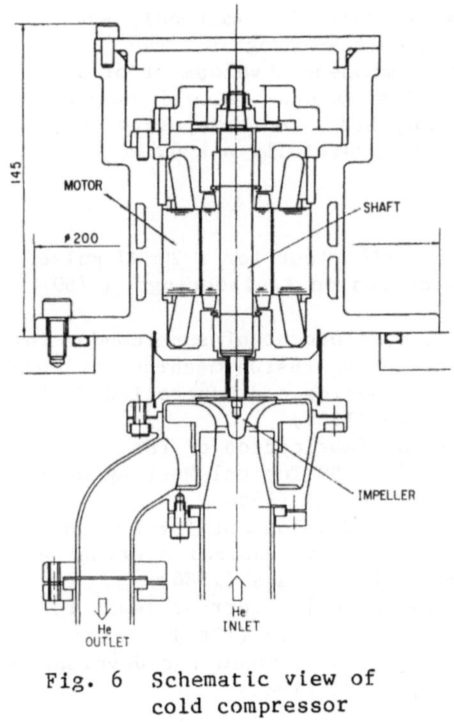

Fig. 6 Schematic view of cold compressor

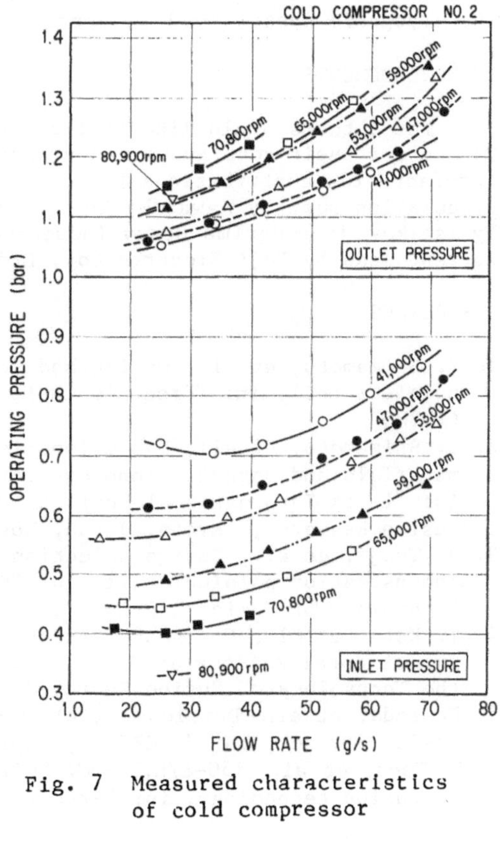

Fig. 7 Measured characteristics of cold compressor

the nominal operation of DPC, it is found from Fig. 7 that the minimum operating pressure of around 0.33 bar can be achieved at 80,900 rpm; this saturated pressure corresponds to the operating temperature of 3.3 K. Furthermore, we have found that the operation at 47,000 rpm is sufficient in order to keep the operating pressure within 0.7 bar corresponding to operating temperature of 4.0 K in the nominal operation of DPC. The measured adiabatic efficiency at the nominal condition is more than 60 %.

The vapor-cooled current lead, which is composed of a number of copper stranded cables inserted into a stainless tube like a cable-in-conduit conductor, has been tested up to 30 kA and the electrical and thermal performances were measured. The thermal performance is extremely important in a large current system as well as the electrical one since a large liquefaction load is required in proportion to the operating current. It is found from the verification test that this type of current lead satisfies the ideal self-cooling condition in which the heat leak is around 1.0 W/kA under condition of mass flow rate of 0.05 g/sec-kA. Furthermore, we have confirmed that the AC break down voltage is around 45 kV which is over the specified value of 16 kV.

CONCLUSION

The detail design of the cryogenic system for the DPC program, in which there are three forced-flow superconducting coils cooled by supercritical helium with mass flow rate of 350 g/sec at 10 bar and below 4 K, has been established. Key cryogenic components such as a circulation pump and a cold compressor for generating supercritical helium, and a vapor-cooled current lead with the rated current of 30 kA have been tested and it is found that the measured characteristics of these components satisfy the specifications of DPC. The construction work of the whole cryogenic system is being carried out since April, 1987 and will be completed at the end of 1987. Thereafter, the first performance test of the cryogenic system will be carried out during early 1988.

ACKNOWLEDGMENTS

The authors would like to thank Drs. S. Mori, K. Tomabechi, and M. Tanaka for their continuous encouragements on this program. The manufacturing contributions for cryogenic component development of a circulation pump by Kawasaki Heavy Industries Co.,Ltd., a cold compressor by Ishikawajima-Harima Heavy Industries Co., Ltd., and a vapor-cooled current lead by Fuji Electric Co., Ltd., are gratefully acknowledged.

REFERENCES

1. S. Shimamoto, et al., Design and verification test for a 20-MJ pulsed poloidal coil, in: "Proc. 10 th Symp. on Fusion Engineering," p.760 (1980)
2. S. Shimamoto, et al., Evolutions in the development of superconducting materials and magnet technology for the coming fusion machine in Japan, in: "11 th International Conf. on Plasma Physics and Controlled Nuclear Fusion Research," Kyoto, Japan, Nov. 13-20 (1986)
3. H. Tsuji, et al., Design selection for the fabrication of the demonstration poloidal coil, in: "Proc. 7th ANS Topical Meeting on the Technology of Fusion Energy," June 15-19, Nevada (1986)
4. T. Kato, et al., Cryogenic system component development for Fusion Experimental Reactor at JAERI, in: "Proc. 7th ANS Topical Meeting on the Technology of Fusion Energy," June 15-19, Nevada (1986)
5. E. Tada, et al., Development of 30-kA vapor-cooled current leads for fusion devices, in: " ICEC-11," April 22-25, Berlin (1986)
6. E. Tada, et al., 350-ℓ/h,1.2-kW helium cryogenic system for development of fusion technology, in: "Proc. ICEC-9," p.93 (1982)

CRYOPUMP OF THE NEUTRAL BEAM INJECTOR FOR JIPPT-IIU

Y. Ohtu, S. Kataoka, T. Ohi

Mechanical Engineering Research Laboratory
Kobe Steel Ltd., Kobe 651, Japan

S. Kitagawa, Y. Oka, O. Kaneko, T. Kuroda

Institute of Plasma Physics, Nagoya University
Nagoya 464, Japan

K. Sakurai

Faculty of Engineering, Nagoya University
Nagoya 464, Japan

Abstract

A high-powered neutral beam injector (NBI) with an electrical output of 3 MW was constructed to heat the tokamak plasma up to a few keV in the JIPPT-IIU torus at the Institute of Plasma Physics, Nagoya University. A pumping speed of about 300 m^3/sec was required in the cryopump to reduce reionization of the neutral beam particles and cold particle flux into the plasma. Several configurations of a Litton-array-type cryopump were studied by Monte Carlo calculation prior to building the actual cryopump.
A cylindrical Litton-array-type cryopump stacked on the neutral beam line was designed to meet the space limitations of the NBI. A small Litton-array-type cryopump composed of a single cryopanel and radiation shield was first built for experimental testing which eventually led to the design of the full-scale cryopump. The cryopump in the NBI has successfully operated since May, 1986. The measured pumping speed of 300 m^3/sec at 2.3×10^{-2} Pa and 4.2 K working temperature is close to the design value of 310 m^3/sec.

Introduction

NBI technology has made remarkable progress following tokamak experiments which have demonstrated the effectiveness of NBI heating. An NBI requires pumps capable of handling large hydrogen and deuterium gas loads while also maintaining pressure below 10^{-2} Pa in the presence of energized neutral beam particles. With existing technology, the only economical way to provide the necessary pumping speed is by means of a liquid helium condensation cryopump [1]. A major problem in the design of cryopumps is to reduce the thermal

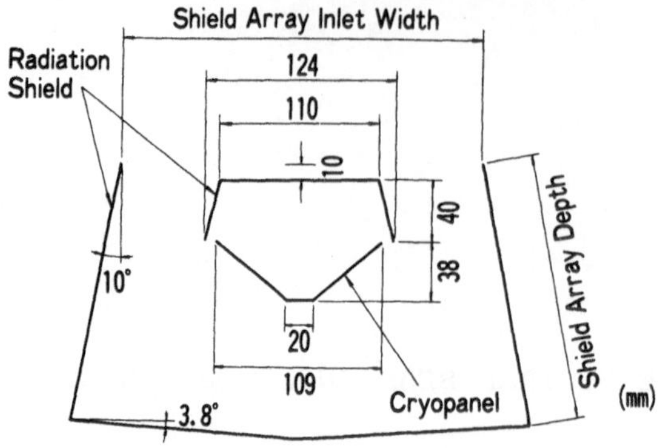

Fig. 1. Two-dimensional model used for the Monte Carlo calculation.

load on the cryopanels while maintaining adequate transmission probability for gas molecules.

The JIPPT-IIU torus utilizes an NBI to heat the plasma. In order to achieve the required pumping speed within the specified space limitation, a Litton-array-type cryopump was designed and manufactured. The results of the design and operation of this cryopump are now described.

Monte Carlo Calculation of Transmission Probability

In order to meet the space limitations for the NBI, the cylindrical cryopump was stacked on top of the neutral beam line. There are several different examples of cryopump design. In the case of a cylindrical cryopump, a Litton-array-type is especially attractive in terms of its compactness, structural simplicity, and ease of manufacturing.

Prior to building the actual cryopump, the effect on installation area of a two-dimensional Litton-array configuration was examined. This configuration is

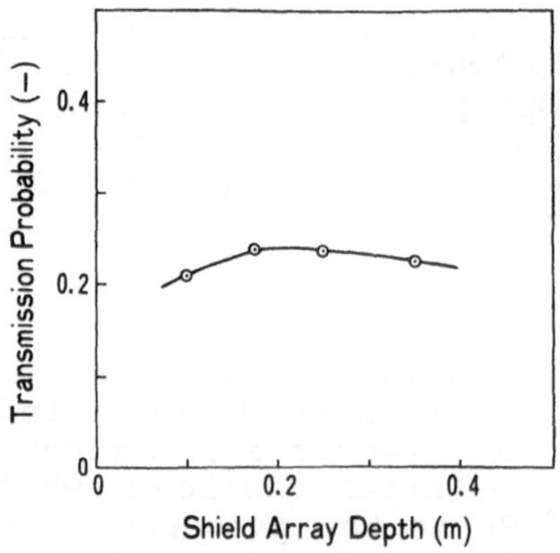

Fig. 2. Variation of transmission probability with shield array depth. Shield array inlet width is 0.24 m.

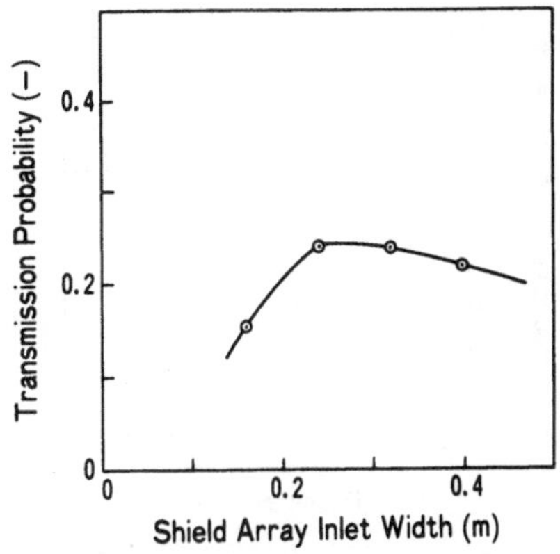

Fig. 3. Variation of transmission probability with shield array inlet width. Shield array depth is 0.175 m.

illustrated in Fig. 1. Specifically, the particle transmission probability was studied by Monte Carlo calculation.

Figure 2 shows the relationship between the shield array depth and transmission probability when the shield array inlet width is 0.24 m. Transmission probability increases with shield array depth, up to a depth of 0.175 m. At this depth or greater, transmission probability stays at a constant value of 0.24.

Figure 3 shows the relationship between shield array inlet width and transmission probability when the shield array depth is 0.175 m. Transmission probability reaches a maximum value of 0.24 at an inlet width of 0.24 m.

These results indicated an optimum shield array depth of 0.175 m and an element inlet width of 0.24 m in order to minimize the cryopump's outside diameter.

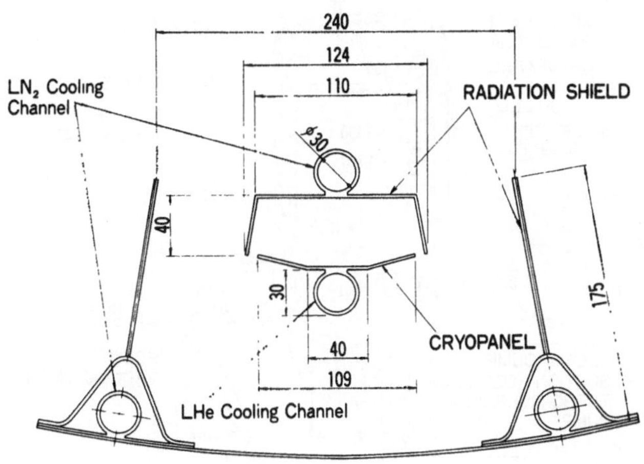

Fig. 4. Cross-sectional view of cryopump element.

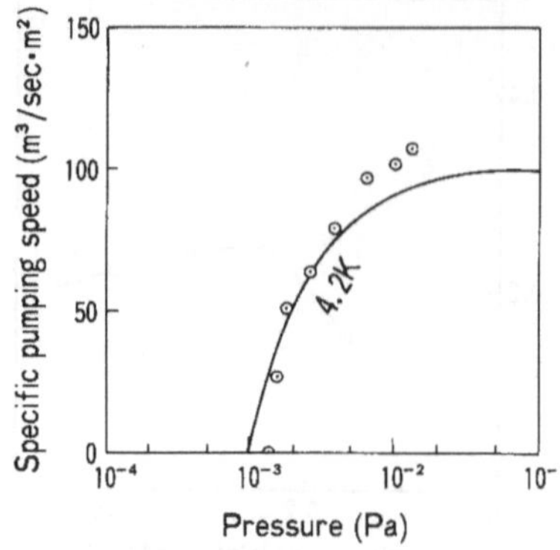

Fig. 5. Specific pumping speed as a function of pressure in the small cryopump.

Measurement of pumping speed in a small cryopump

Figure 4 illustrates a cross-sectional view of the small cryopump that was built to verify the results of the Monte Carlo calculations. The gas particle inlet is 0.3 m high and 0.24 m wide.

Pumping speed was measured in a test vacuum vessel, based on the JIS B 8317 test code [2]. The vacuum gauge was calibrated with a McLeod gauge. The cryopump's pumping speed was first measured without a radiation shield to block gas particle transmission in front of the cryopanel. The measured pumping speed corresponded to a theoretical transmission probability of 1, demonstrating the reliability of the measuring method.

Figure 5 shows the results of pumping speed measurements for the small cryopump. The solid line indicates the calculated pumping speed based on vapor pressure of the

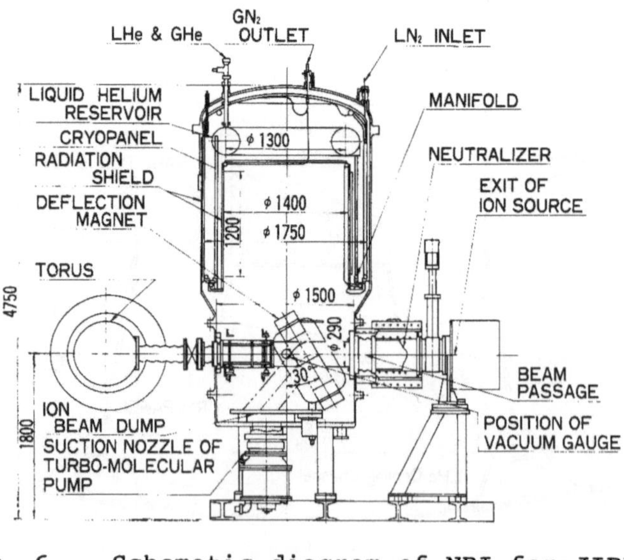

Fig. 6. Schematic diagram of NBI for JIPPT-IIU.

condensed hydrogen that has been affected by thermal radiation from the shield [3]. At low pressure, the measured pumping speed agrees well with the calculated value.

However at high pressure, the flow of hydrogen gas molecules changes from molecular flow to viscous flow. Thus, the jet discharge of hydrogen gas molecules against the cryopanel causes a higher pumping speed compared with that under molecular flow. This is illustrated by the upper three values plotted in Fig. 5.

The measured pumping speed at lower pressure corresponds to a transmission probability of 0.24. This experimental value agrees well with the Monte Carlo calculation.

NBI Cryopump

The size of the element for a full-scale cylindrical cryopump was then determined based on these results from the smaller experimental cryopump. In order to meet the space limitations of the NBI, the cylindrical cryopump was stacked on top of the neutral beam line. This type of cryopump has an optimum inside height and gas molecule inlet diameter which determines a particular transmission probability. Given a particle transmission probability of 0.24 for the cylinder's inner surface, the inside height and inlet diameter of the cylindrical cryopump was determined through Monte Carlo calculation for an axially symmetrical shape.

The cryopump layout is shown in Fig. 6 and its cross-sectional view is shown in Fig. 7. Hydrogen gas is pumped out from the inlet area which measures 1.4 m in diameter. The full-scale version has an outside diameter of 1.75 m. The length of the cryopanel is 1.2 m and the design value for pumping speed is 310 m^3/sec.

The interior of the cryopump is divided into 18 sections. Elements of the cryopump, illustrated in Fig. 4, occupy 17 of these sections. The liquid helium supply tube, located in the remaining section, is S-shaped to compensate for thermal contraction during cool-down. Above the supply tube and 17 cryopanels is a torus-shaped helium reservoir, connected to them by welding. Below them is a manifold, also connected by welding.

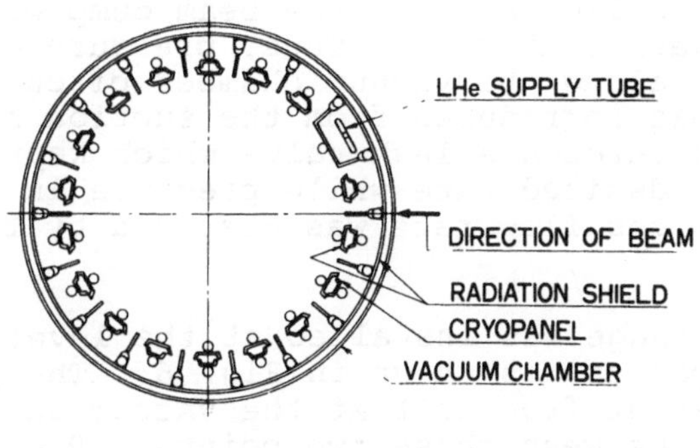

Fig. 7. Cross-sectional view of the cryopump.

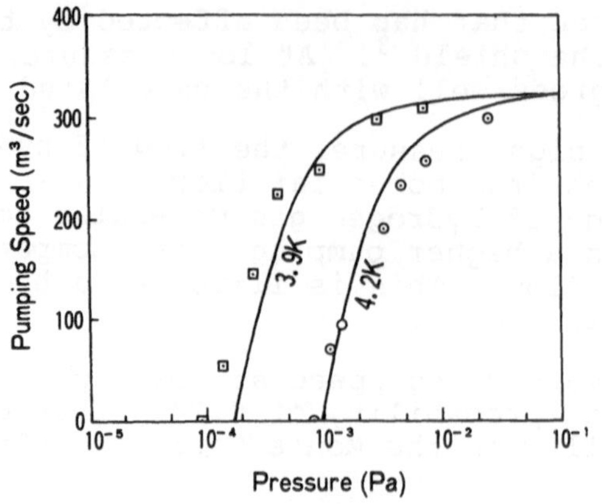

Fig. 8. Pumping speed as a function of pressure in the NBI cryopump.

Liquid helium circulates by gravity through the cooling channel which is integrated with the cryopanels. The liquid helium reservoir functions as a phase separator while vaporized helium gas returns to the recovery system of the helium liquefier.

Aluminum was chosen as the basic structural material so that the complicated cryopanel shapes with integrated cooling channels could be made at low cost by the extrusion technique. Other advantages of aluminum include good thermal conductivity, low-temperature mechanical properties, and short radioactive lifetime.

The cryopanels, radiation shields, and liquid helium reservoir are all made of aluminum while the pipes connecting the top flange of the vacuum chamber to the cryopump are made of stainless steel. The surface of the cryopanel is mechanically polished in order to obtain an emissivity value of 0.05. The surface of the radiation shield which faces the cryopanel is coated with black organic paint.

Before performance testing of the cryopump, the problem of jet flow of hydrogen gas molecules was first studied. Jet flow causes the distribution of pressure along the circumference of the vacuum chamber. During testing, hydrogen gas molecules were introduced from the suction nozzle of the turbo-molecular pump. In this type of flow, the molecules collide against the beam dump without jetting out into the vacuum chamber. Thus, pressure distribution does not occur along the vacuum chamber circumference. The hydrogen gas was introduced from the suction nozzle into the vacuum chamber through a leak valve which kept the gas flowing at the desired rate while pressure in the chamber was measured. The gas flow rate was measured by the burette method.

A vacuum gauge was installed at the level where the neutral beam passes, as shown in Fig. 6. The pressure of the inlet is different from that at the vacuum gauge position due to conductance between those two points. Pressure at the inlet of the cryopump was therefore calculated from the measured pressure, the hydrogen gas flow rate, and this conductance. To calculate the pumping speed, the gas flow rate was divided by the cryopump's inlet pressure.

Figure 8 shows the results of measured pumping speed at both 4.2 K and 3.9 K working temperature in the NBI. The measured pumping speed of 300 m^3/sec at 2.3×10^{-2} Pa and 4.2 K working temperature is close to the design value of 310 m^3/sec.

The rate of liquid helium boil-off from the cryopanel was measured by the change of liquid level in the helium reservoir. The heat load to the cryopanel was estimated at 0.28 W/m^2, a typical value for a chevron-type cryopump.

Conclusion

The design for the Litton-array-type cryopump of the NBI in the JIPPT-IIU torus was based on the given space limitations in addition to experimental results from a small cryopump. The full-scale cryopump has been successfully operated since May, 1986. Its measured pumping speed of 300 m^3/sec at 2.3×10^{-2} Pa and 4.2 K working temperature is close to the design value of 310 m^3/sec.

References

1. C. B. Hood, The development of large cryopumps from space chambers to the fusion program, J. Vac. Sci. Technol., A, Vol.3, No.3, May/Jun (1985), P.1684.
2. Japanese Industrial Standard: Test code for Steam Jet Pump, JIS B 8317, (1977).
3. C. Benvenuti, R. S. Calder, and G. Passardi, Influence of thermal radiation of the vapor pressure of condensed hydrogen (and isotope) between 2 and 4.5 K, J. Vac. Sci. Technol., Vol.13, No.6, Nov./Dec. (1976), P.1172.

CRYOGENIC SYSTEM FOR TRISTAN SUPERCONDUCTING RF CAVITY

K. Hara, K. Hosoyama, Y. Kimura, Yuuji Kojima, Yuzo Kojima
S. Mitsunobu, M. Morimoto, H. Nakai, S. Noguchi
T. Ogitsu and Y. Sakamoto

KEK, National Laboratory for High Energy Physics
Tsukuba, Ibaraki, Japan

T. Nakazato

Laboratory of Nuclear Science, Tohoku University
Mikamine, Sendai, Japan

S. Kawamura, K. Matsumoto and S. Saito

Hitachi Ltd. Kasado Works
Kudamatsu, Yamaguchi, Japan

ABSTRACT

The superconducting RF cavities (32 × 5 cell) will be installed in TRISTAN electron-positron collider at KEK to upgrade the electron-positron beam energy up to 33 GeV × 33 GeV. Two 5-cell cavities are coupled together, enclosed in a cryostat and cooled by liquid helium pool boiling. Cryogenic system for the superconducting RF cavities was designed. The capacity of helium refrigerator is 4 kW at 4.4 K with liquid nitrogen precooling and two expansion turbines. The present system is designed to be easily upgraded to 6.5 kW without liquid nitrogen with an addition of expansion turbines and compressors.

INTRODUCTION

TRISTAN electron-positron collider at KEK was commissioned successfully November 1986[1,2]. The electron-positron beam was accelerated up to 25 GeV × 25 GeV by conventional cavities. The beam energy will be upgrade up to 27 GeV × 27 GeV by the installation of additional conventional cavities at summer shutdown in 1987. For further upgrading of beam energy, installation of 32 × 5-cell superconducting RF cavities has been proposed. Following the authorization of the proposal as a two-year project in April 1986, the construction of the cavities and cryogenic system has been started in cooperation with industries. It is expected that the accelerating field of about 5 MV/m will be achieved[3,4] and the beam energy will be upgraded up to 33 GeV × 33 GeV.

A prototype of the 5-cell cavity[3] and cryogenic system[5] with a 300 W refrigerator (BOC Turbocool 100) were constructed and beam tested successfully in TRISTAN accumulation ring in February 1986.

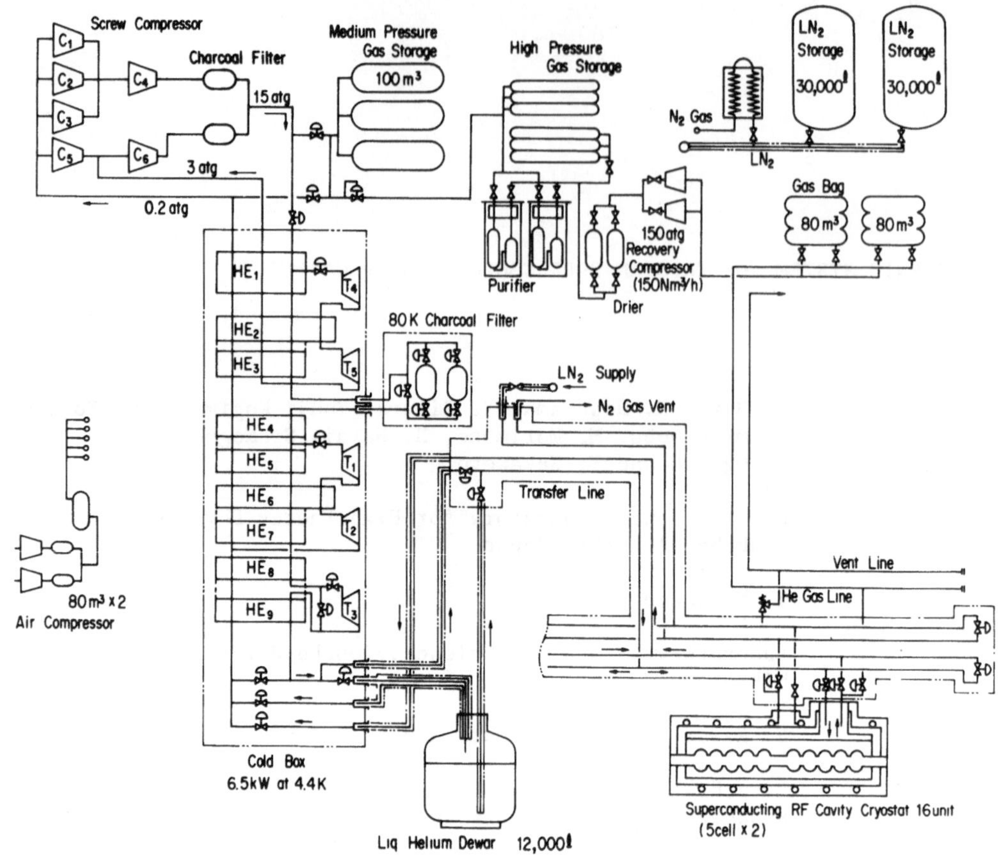

Fig. 1 System flow diagram.

THE CRYOGENIC SYSTEM OVERVIEW

The flow diagram of the cryogenic system for TRISTAN superconducting RF cavities (SCC) is shown on Fig. 1. Figure 2 is an isometric showing the general layout of the cryogenic equipments.

A helium refrigerator cold box and compressor unit will be installed in the building on the ground level. Liquid helium produced by the helium refrigerator and stored in 12,000 L dewar and liquid nitrogen stored in liquid nitrogen tanks will be distributed to the 16 SCC cryostats in the underground RF straight section (about 200 m) of the TRISTAN ring through large size transfer line.

THE SUPERCONDUCTING RF CAVITIES

Main components of superconducting RF acceleration section are 32 × 5-cell 508 MHz superconducting cavities (SCC) installed in TRISTAN RF straight section.

Figure 3 shows the SCC and the cryostat. Two 5-cell 508 MHz cavities made of about 2 mm pure niobium sheet are coupled together, enclosed in a liquid helium vessel with an inner diameter of 700 mm. The amount of liquid helium stored in a cryostat is about 900 L. Each cavity is equipped with antena type RF coupling port on the beam pipe of 180 mmϕ at cavity end. The resonance frequency change due to the helium bath pressure fluctuation in the cryostat is compensated by adjusting the cavity length with piezo electric tuners.

A radiation shield with a temperature of about 80 K is cooled by liquid nitrogen pipe cooling.

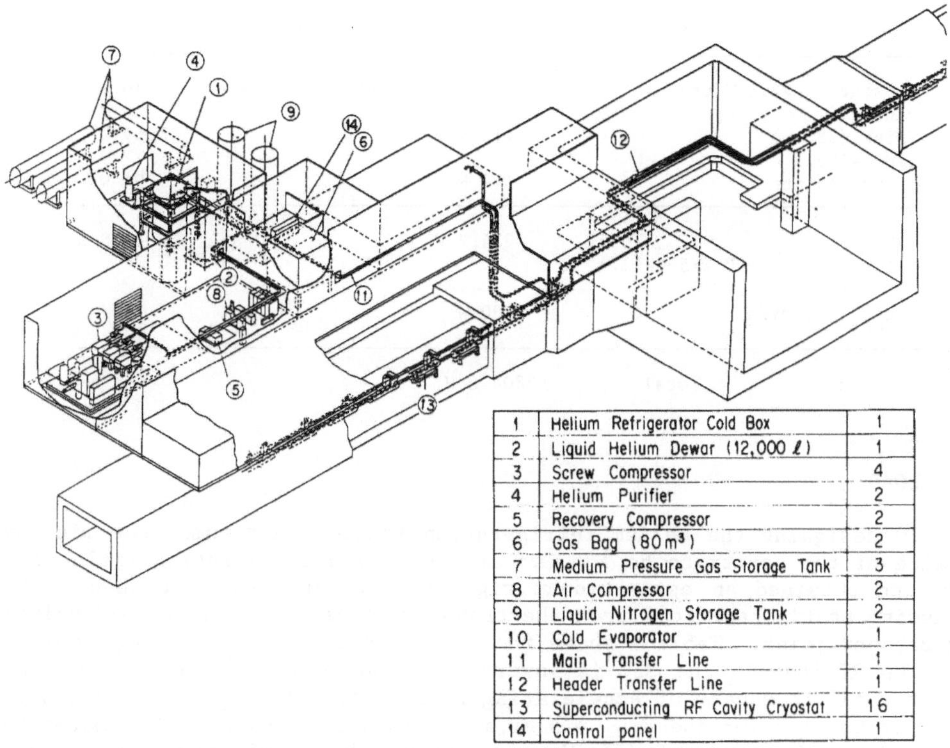

1	Helium Refrigerator Cold Box	1
2	Liquid Helium Dewar (12,000 ℓ)	1
3	Screw Compressor	4
4	Helium Purifier	2
5	Recovery Compressor	2
6	Gas Bag (80 m³)	2
7	Medium Pressure Gas Storage Tank	3
8	Air Compressor	2
9	Liquid Nitrogen Storage Tank	2
10	Cold Evaporator	1
11	Main Transfer Line	1
12	Header Transfer Line	1
13	Superconducting RF Cavity Cryostat	16
14	Control panel	1

Fig. 2 Cryogenic equipments layout.

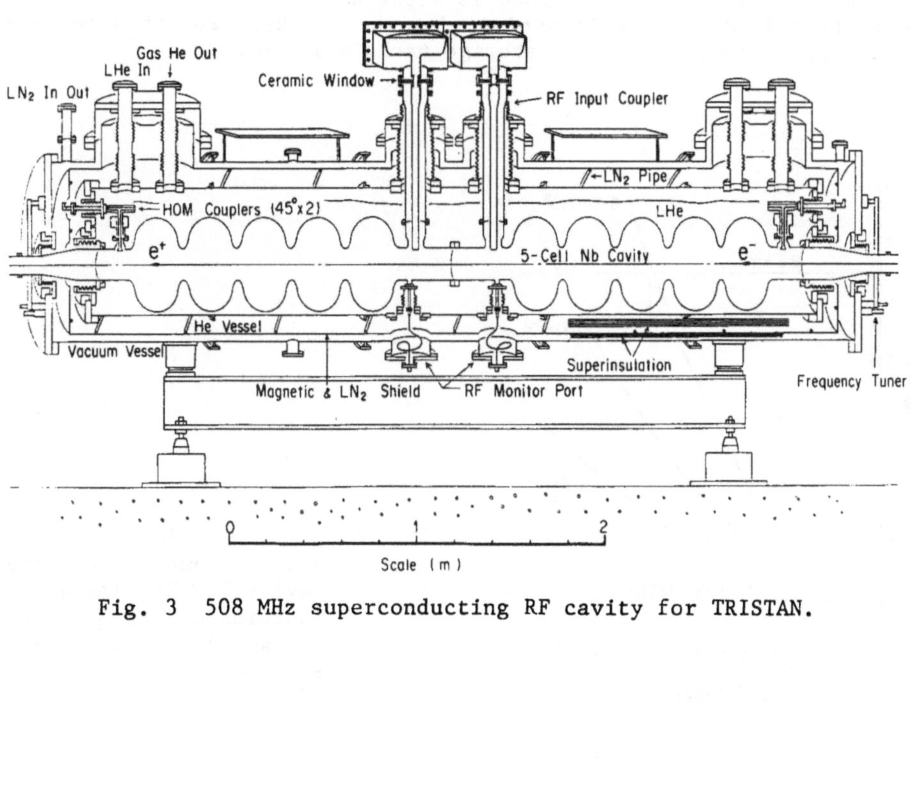

Fig. 3 508 MHz superconducting RF cavity for TRISTAN.

Table 1. Estimated heat loads of the components

		4.4 K		80 K
cryostat	22.8 W/cryostat × 16	364.8 W	48.8 W/cryostat	780.8 W
transfer line (380 m)		412.4 W		5569 W
cold valve & bayonet joint		147 W		294 W
	Total	924.2 W		6643.8 W
RF loss $Q = 1 \times 10^9$ $E_{acc} = 5$ MV/m	90 W/5-cell × 32	2880 W		
	Total	3804.2 W		

HEAT LOADS

In designing the helium refrigeration system, we assumed the minimum Q-value of the cavity to be 1×10^9 at the design accelerating field of 5 MV/m and obtained an estimated refrigerator cooling power of about 4 kW including static heat load from the transfer lines, cryostats, cold valves and Bayonet joint. Table 1 shows the estimated heat loads of components.

The RF loss in the SSC is inversely proportional to the Q-value and proportional to square of accelerating field gradient E_{acc}. At the design value of $Q = 1 \times 10^9$ and $E_{acc} = 5$ MV/m, RF loss in the SCC is estimated to be 90 W per 5-cell cavity. Figure 4 shows the total heat load of the cryogenic system, including about 900 W static heat loss, as a function of E_{acc} at $Q = 0.5 \times 10^9$, 1×10^9, 2×10^9. Drop of Q-value and improvement of E_{acc} causes the heavier heat load to the refrigerator. In the case of $Q = 2 \times 10^9$, total heat load is about 2.4 kW and we get safety factor 1.7. There is a possibility of improving the E_{acc} up to 7 MV/m in the near future. In this case total heat load is about 6.5 kW at $Q = 1 \times 10^9$.

Total heat load at 80 K is estimated to be 6.6 kW. For this cooling liquid nitrogen stored in outdoor storage tanks will be used.

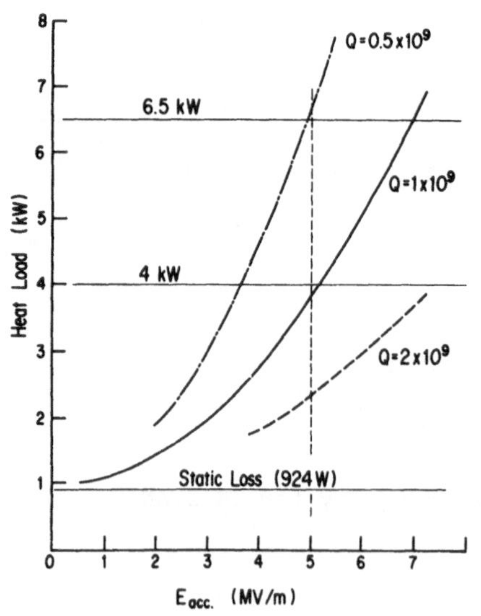

Fig. 4 Total heat load of the 32 × 5 cell superconducting cavity for TRISTAN as a function of the Q-value and accelerating field E_{acc}.

Table 2. Main parameters of turbo-expanders

	T_1	T_2	T_3	T_4	T_5
flow rate (g/s)	135.7	135.7	135.7	89.6	89.6
	152.4*	152.4*			
inlet pressure (MPa)	1.58	0.57	1.60	1.60	1.0
	1.60*				
outlet pressure (kPa)	570	120	400	1000	400
inlet temperature (K)	50.0	25.0	8.0	150.0	95.0
	43.6*	19.7*			
isentropic efficiency (%)	75	75	60	60	55
output power (W)	8,774	6,099	3,020	7,270	7,571
	8,839*	5,305*			
gas bearing					
thrust			static		
journal			dynamic		
roter speed (rpm)	140,000	74,000	50,000	150,000	150,000
		70,600*			
rotor diameter (mm)	35	55	35	35	35
fabricator			Hitachi		

* 4 kW version

COLD BOX AND COMPRESSORS

The flow diagram of the cold box is given in Fig. 1. The cold box is 4 mϕ × 6 m vertical type. At the first stage of the program, 4 kW at 4.4 K refrigeration power is required. In this case instead of using 80 K turbo-expanders T_4, T_5 liquid nitrogen precooling is adopted and 8 K J-T turbo-expander T_3, compressors C_5, C_6 are not used. For the future upgrading of refrigeration power from 4 kW to 6.5 kW, 8 K J-T turbo-expanders T_3 will be required. The main parameters of turbo-expander are listed in Table 2. The T-S diagrams for 4 kW version and 6.5 kW version are shown in Fig. 5, 6 respectively. A dual type 80 K charcoil filter will be able to install in the future if necessary.

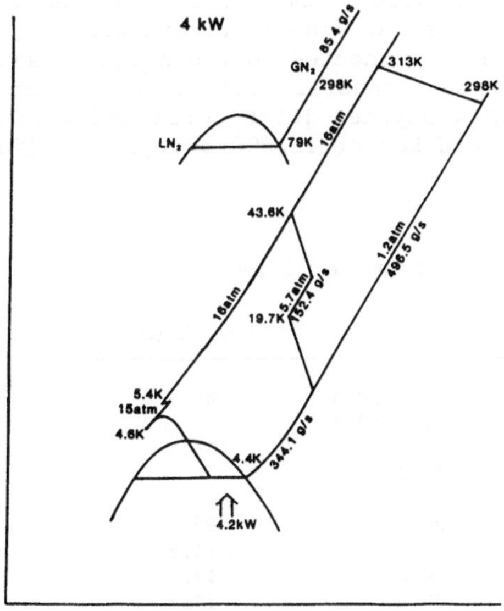

Fig. 5 T-S diagram for 4 kW.

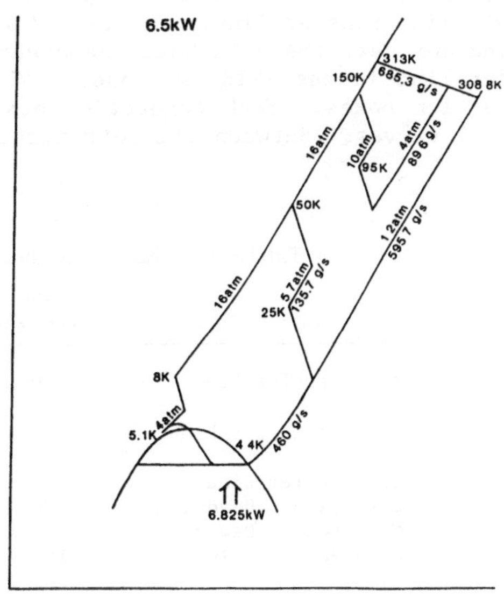

Fig. 6 T-S diagram for 6.5 kW.

Table 3. Main parameters of compressors

	1st Stage C1,C2,C3	2nd Stage C4	1st Stage C5	2nd Stage C6
number of compressor	3	1	1	1
flow rate (Nm^3/h)	10,820	10,820	1397	3738
inlet pressure (kPa)	100	517	100	400
outlet pressure (MPa)	0.517	1.90	0.4	1.97
roter diameter (mm)	320	320	250	250
roter length (mm)	530	355	352	280
roter speed (rpm)	2950	2950	2950	2950
isothermal efficiency (%)	49.9	51.0	47.7	46.5
in put power (kw)	373 × 3	913	127	423
fabricator	Mayekawa			
type	oil flooded screw compressor			
	320L	320S	250L	250S

Oil flooded screw compressors were chosen because they have the highest reliability of all types of helium compressors currently arrailable. For the 4 kW version, three parallel first stage and one second stage is provided. For 6.5 kW version additional two compressors in tandem will be provided. Table 3 lists the main compressor data.

TRANSFER LINE AND DISTRIBUTION SYSTEM

The transfer system consists of two helium lines (supply and return) and two liquid nitrogen lines (supply and return). The main part of transfer system has these four lines in a common vacuum jacket. The helium lines are shielded by an 80 K thin aluminum pipe which is thermally connected to the liquid nitrogen lines. Figure 7 shows the cross section of main transfer line. One end of main transfer line is terminated by the end box near the cold box, the other end is connected to the header transfer line in the TRISTAN tunnel. The header transfer line has the 16 connection boxes. Each connection box has 4 bayonet joint ports and 4 control valves. Between the refrigerator cold box or 12,000 L liquid helium

Table 4. Main parameters of transfer line

	vacuum pipe size	pipe size	Length (m)
Main Transfer Line	450 A	He 50 A/80 A N_2 50 A/50 A	68
Header Tramsfer Line	400 A	He 40 A/100 A N_2 32 A/32 A	174.3
Sub-transfer Line			
Cryostat Header	100 A	He 20 A/32 A	96
Cold Box Dewar	300 A	He 50 A/100 A	11.3
Cold Box Main Tr.	150 A	He 50 A	17.2
	200 A	80 A	17.2
Dewar Main Tr.	200 A	He 80 A	8.6

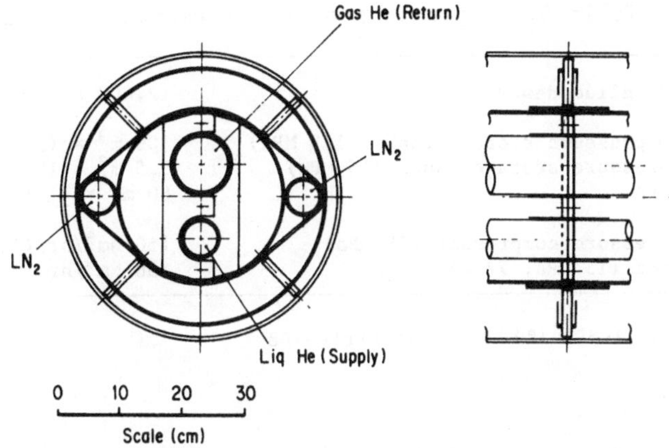

Fig. 7 Cryogenic transfer line.

dewar and end box on the ground level, and also between the connection boxes and SCC cryostats in the TRISTAN tunnel, single channel type transfer lines with bayonet joints are used for easy to handle. Table 4 gives the main parameters of the transfer lines.

COOLDOWN

The superconducting RF cavities will be cooled in parallel by the cold helium gas from ambient temperature. The mass flow rate are controlled by control valves at the connection boxes of header transfer line to keep the inner pressure of the cryostat below 130 kPa. The total cold mass of 16 cryostats is about 16,000 kg. It takes about 3 days to cooldown the SCC from room temperature to 4.4 K and about one day for filling up the liquid helium in the SCC cryostat helium vessels (the total volume is about 14.4 m^3). The cooldown time from ambient temperature to 4.4 K will be restricted by pressure limitation (below 130 kPa) due to the thin structure of pure niobium 5-cell cavity.

OPERATION OF SCC

During the operation of SCC, the liquid helium level in the helium vessels will be controlled by the control valves at the header transfer line to keep the design level. To keep the pressure fluctuation due to the RF power change in the helium vessel, as small as possible, heater in the helium vessel will be used to compensate the RF loss.

Continuous operating cycles of about 3 months without maintenance will be required to meet the TRISTAN operation schedule.

HELIUM GAS RECOVERY AND STORAGE SYSTEM

After the operation of SCC, the liquid helium in the SCC cryostats will be recovered to the liquid helium dewar with capacity of 12,000 L and medium pressure storage tanks (total volume 500 m^3) by firing the heaters in the SCC cryostats. The warm up gas from the SCC cryostats will be stored in gas bag through the helium gas recovery line. The recovery compressor then pumps the gas from the gas bags to high pressure storage

Table 5. Major components of recovery system

liquid helium dewar	12,000 L
mediam pressure storage tanks (1.9 MPa)	100 m^3 × (3 + 2*)
high pressure storage tanks (15 MPa)	0.5 m^3 × 18 × (2 + 2*)
gas bags	80 m^3 × (1 + 1*)
high pressure compressor (15 MPa)	150 Nm^3/hr (1 + 1*)
purifier (15 MPa, 77 K)	150 Nm^3/hr (1 + 1*)

* will be installed in future upgrading

vessels. The high pressure helium gas recovery system consists of two 5 stage air-cooled oil lubricated reciprocating compressors (15 MPa, 150 Nm^3/h × 2) and high pressure storage vessles (15 MPa, 36 m^3). The impure recovery helium gas will be purified by high pressure low temperature (80 K) purifiers and stored in high pressure storage vessels. Table 5 shows the major components of the recovery system.

CONTROL SYSTEM

The TRISTAN cryogenic system will be controlled by means of process control computer system composed of commercially available and reliable components. This process control computer system will also provide data logging and analysis functions and be connected to TRISTAN control system. All controls will be handled with coloured graphic operator consoles.

ACKNOWLEDGEMENT

The authors wish to thank Professors T. Nishikawa and S. Ozaki for the continuous support and encouragement and R. Byrns for many helpful discussions.

REFERENCES

1. Y. Kimura, TRISTAN project and KEK activities in "Proc. XIII the International conference on high energy accelerators", Novosibirsk, U.S.S.R., (1986).
2. G. Horikoshi and Y. Kimura, Status of TRISTAN, in "Particle accelerator conference", Washington, U.S.A., (1987).
3. T. Furuya et al., A prototype superconducting cavity for TRISTAN, in "Proc. XIIIth Internal conference on high energy accelerators", Novosibirsk, U.S.S.R., (1986).
4. Y. Kojima, Research on superconducting rf cavities at KEK, in "Proc. XIth Internal conference on cyclotrons and their applications", Tokyo, (1986).
5. K. Hosoyama et al., A liquid helium transfer line system for superconducting RF cavity, in "Advances Cryogenic Engineering", Vol.31, Plenum Press, New York, (1986), p.1027.

CEBAF'S CRYOGENIC SYSTEM

Paul Brindza, William Chronis, Claus Rode,
J. Patrick Kelley, Marie Shea

Continuous Electron Beam Accelerator Facility

Newport News, Virginia

ABSTRACT

The CEBAF superconducting linear accelerator[1] requires a 4.8KW cryogenic system to operate 418 superconducting r.f. cavities[2] at 2.0K. The cavities are assembled into 53 cryomodules which are stand alone cryostats. The cryomodules are made up of four cryounits, each containing two r.f. cavities. CEBAF's three experimental halls will be equipped with superconducting magnet particle spectrometers, and a superconducting 4π toroidal spectrometer. The refrigeration loads, capacities and design will be discussed and considerations for the 2K operation temperature will be developed. The operating scenario, the distribution system that is integrated with the cavity cryostat design, and the experimental hall cryogenic system will be presented.

INTRODUCTION

The CEBAF cryogenic system[3] will provide refrigeration at 4.5K and 2.0K and will also supply gaseous helium at 40K for shield refrigeration in the cryogenic portion of the accelerator. A block diagram of the system is presented in Figure 1.

The system will include two cold boxes, a "standard" 4.5K system, and another cold box which will incorporate cold compressors and an additional heat exchanger to produce 2.0K refrigeration. A minimum of three stages of cold compressors have been specified; however, preliminary designs of up to five stages have been investigated.

The system will utilize six oil injected screw compressors with slide valves for efficient capacity control. Four machines, two first stage and two second stage, will operate and supply at least 85 percent of the required cycle flow capacity. In essence, one first stage and one second stage compressor will be spare. The oil removal system will be composed of a bulk oil removal stage on each compressor skid to attain a level of 10 ppm(w) of oil in the helium. The final oil removal will be a minimum of three stages in separate vessels. The gas velocity, based upon the filter inlet area, will be less than 5 cm/sec. and the maximum tangential component of velocity in the annular space shall be less than 20 cm/sec.

CEBAF CENTRAL HELIUM LIQUIFIER (CHL)

The Central Helium Liquifier required at CEBAF will simultaneously produce a minimum of 12KW of shield refrigeration load at 40K, 4.8KW of refrigeration at 2.0K, and liquid production of 10 g/sec. at 4.5K. The 10 g/sec. of liquid production will be returned to the interstage of the main compressor system by way of a cryogenic purifier. Figure 2 presents a cycle that will satisfy the CEBAF requirements.

There are two types of resistive losses in a superconducting rf cavity: residual resistance and BCS (Bardeen, Cooper, and Schrieffer) resistance. The residual resistance is caused by localized resistive areas where defects, impurities, or surface dirt disturbs the superconductive properties. The BCS resistance increases with increasing frequency, and decreases as the operating temperature decreases. Other sources of 2K heat include static heat leaks, conduction of heat dissipated in the input waveguides[4], and absorption of higher-order-mode power generated by the beam currents.

The choice of operating temperature affects the BCS component of the cavity Q, and thereby the rf heat load, as well as the refrigeration costs (both capital and operating). The BCS losses vary inversely with the cavity Q, approximately doubling every 0.2K. The refrigeration costs vary inversely with the temperature; in addition, capital costs increase with the 0.7 power of heat load, while operating costs increase to the 0.85 power.

We have chosen 2.0K as the operating temperature. The BCS losses, while an exponential function of temperature, are still a small fraction of the total heat load at 2.0K. The refrigeration capital cost is flat to 0.5% between 2.0 and 2.2K, below 2.0K not only is it not cost-effective, but it also becomes technically difficult due to the very low vapor pressures, less than 0.031 atm. Above 2.5K (0.1 atm) we could delete one stage of vacuum pumping, but the BCS losses are so large that it would not be economical.

This leaves us with an operating range of 2.0 to 2.5K. We have chosen to size the distribution system to be optimized for 2K operation with a flow safety factor of two times the calculated heat load. Since the possibility of higher cavity gradients in the future will tend to shift the optimum toward lower temperatures, this will permit future beam energy increases without requiring replacement of the relatively expensive distribution system. The project loads and capacities are presented in Table 1[8].

Table 1
Refrigeration Requirements

	He Temp (K)	Calculated Load (W)	Design Load (W)	Pressure (atm)	Safety Factor (%)	Flow (g/sec)
Linac cavities	2.0	3310	4800	0.031	145	240
Linac heat shields	38-52*	8000	12,000	3.5	150	160*
End station liquefaction	4.4	165 L/hr	288 L/hr	2.8	175	10

*Lower flows with colder inlets are acceptable; i.e., 150 g/sec with 36-52 K, or 120 g/sec with 32-52K.

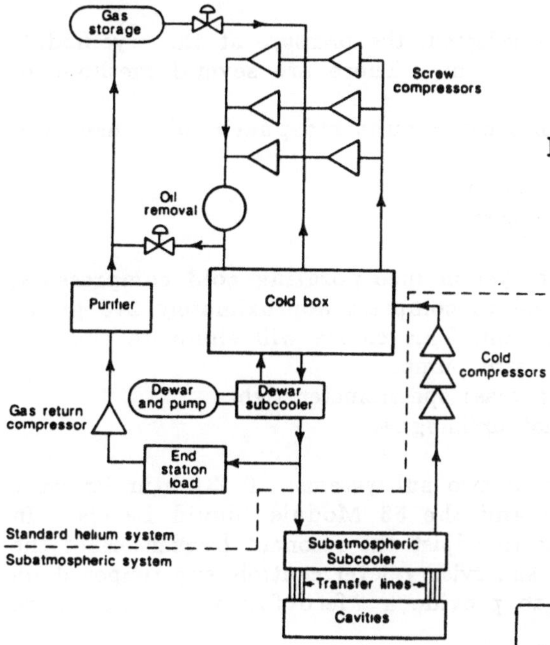

Figure 1 Block diagram of CEBAF refrigerator.

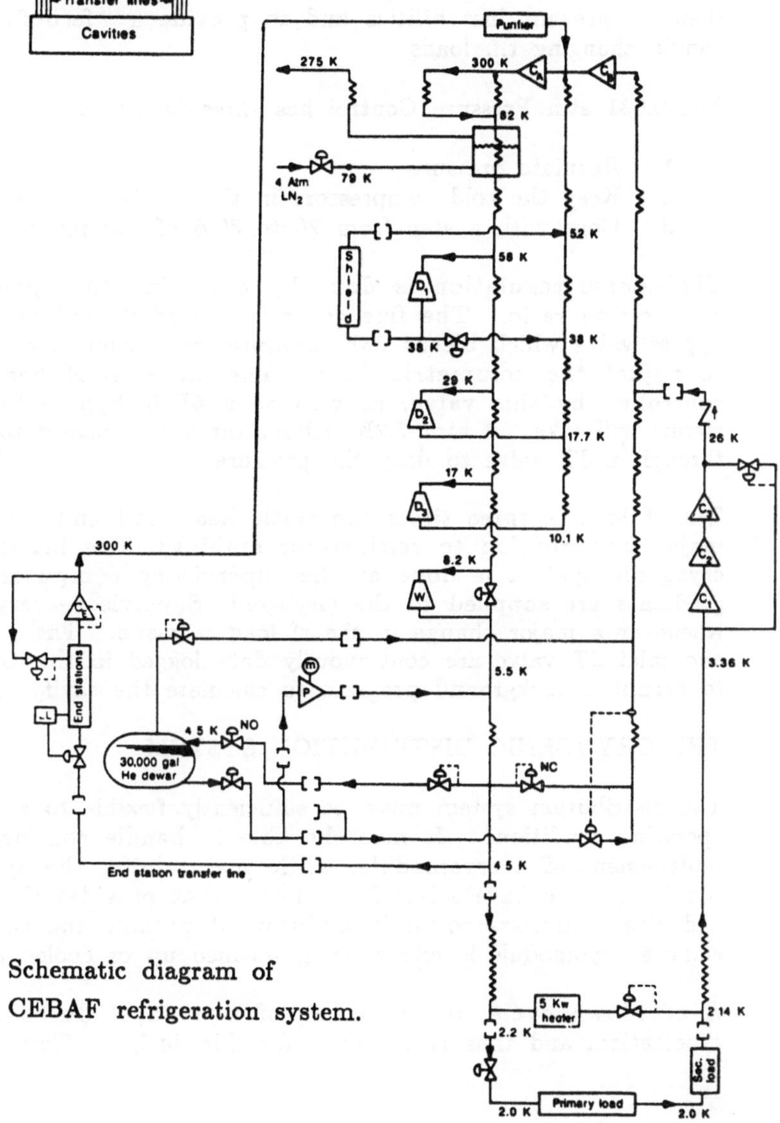

Figure 2 Schematic diagram of CEBAF refrigeration system.

Because of the 2K load temperature selected, the pressure at the cryomodule must be closely maintained at 0.031 atm. There are several methods of maintaining this pressure:

1. Operate with main compressor subatmospheric plus use cold compressors
2. Operate with warm vacuum pumps
3. Operate with cold compressors.

CEBAF has chosen to operate the system incorporating cold compressors. The cold compressor will be designed to compress approximately 240 grams per second from 0.030 atm to 1.2 atm. This choice will eliminate the two major problems of subatmospheric warm pumps:

1. Keeping the warm piping mass spectrometer tight
2. Gigantic low pressure heat exchangers.

The 2.0K Control System consists of two subsystems: 0.031 atm Pressure Control (vapor pressure at 2.0K) and the 53 Module Liquid Levels. In addition, there may be a fast tuner to adjust the resonant frequencies of the cavities. These control loops have supervisory level controls superimposed on them to prevent instabilities and to provide a "feed forward" function to handle changing r.f. loads.

The 0.031 atm Pressure Control has three functions:

1. Regulate pressure
2. Keep the cold compressors in the stable operating regions
3. Cope with a step from 20 to 80% of full refrigeration capacity.

The coarse regulation is done by adjusting the speed which sets the compression ratio. The fine control, if needed, will be accomplished by a bypass valve which bleeds 25K Compressor Output Gas back into the inlet to adjust the volumetric flow. The steps in rf heating loads will be controlled by this valve, as well as a 5KW bypass heater. The heater circuit will take 2.8 atm 2.2K helium through a heater to warm it and then through a JT valve to drop the pressure.

The rf load is three times the static heat load and therefore represents a major perturbation to refrigerator stability. The interface of the rf and cryogenic system is done at the supervisory computer level; the 418 rf gradients are supplied to the Cryogenic Supervisor every 1000 seconds and whenever a major change in the rf load is made. The gradient, liquid level, and inlet JT valve are continuously data logged in the Cryogenic Supervisor to permit a background program to calculate the cavity Q.

THE CRYOGENIC DISTRIBUTION SYSTEM

The distribution system must be sufficiently flexible to allow a wide range of operating conditions. It must be able to handle contingencies, such as the replacement of a cryomodule, while maintaining the system in a standby condition. We have selected a solution that provides the required flexibility and also minimizes costs; in addition, it permits the accelerator to operate while a cryomodule is either being warmed up or cooled down.

A cooldown cycle, for CEBAF, of two days' duration is a reasonable expectation, and thus is the basis for this design. Tests at Cornell at very high cooldown rates have shown no degradation of performance; therefore there are no constraints on the system imposed by thermal stresses at the cooldown rates that we will achieve. For single cryomodules we plan a 30-minute cooldown for the shields and a 6-hour cooldown for cavities.

The system depicted in Figure 3 is based upon using the string of cryomodules[5] as part of the supply transfer line, and a transfer line for the return flow. The cryomodules are series-connected in an "H" pattern utilizing U-tubes and internal flow to distribute 2.2K helium at 2.8 atm and 38K helium at 3.5 atm. Each cryomodule is connected to the return cold vacuum line to maintain its 0.031-atm internal pressure, and the shield flow is returned to the transfer line at four places, one at the end of each arm of the "H". This series-parallel system minimizes the cost of the distribution system.

If a unit must be removed, the cryomodule containing it would be isolated from the supply by removing the U-tubes at each end of the module. These U-tubes are replaced by a U-tube which spans the gap created by the cryomodule and allows the helium supply to the remaining modules to be resumed in a short time. The cryomodules in the other three arms of the "H" are completely unaffected by this operation. Those in the affected arm, downstream of the break, must rely upon the large helium inventory (1500L) in each module to maintain the temperature during the very short transition time. The modules upstream of the isolated module may still be supplied with 2K helium. The removal and replacement of the U-tubes will not take more than 10 or 15 minutes, which is much less than the several-hour stand-alone capacity of each cryomodule.

In this system, the transfer lines are of a simple design[8], which can be mass-produced easily and economically (Figure 4). The system will be easy to control, because it has few control valves, each with a well-defined function. A control valve at the end of each branch of the "H" will maintain the shield at a temperature between 38K at the inlet and no more than 50K at the outlet. A control valve at each cryomodule will maintain the liquid level in each module in the full state, while the parallel connection to the cold vacuum line will keep the pressure in each module at 0.031 atmosphere for 2K operation.

END STATION CRYOGENIC DISTRIBUTION SYSTEM

The design for the CEBAF end stations includes several large superconducting dipoles, quadrupoles, and an 8 coil toroid. (Table 2).

The toroid and quadrupole magnets will be force-cooled and the dipoles will be refrigerated poolboiling. The magnets will be powered and protected individually thus avoiding the problems of quench propagation possible with systems that are series connected. The end station configuration is shown in Figure 5. There are only two interface points to the CHL.

<u>End Station Transfer Line:</u> Supercritical helium is supplied from the CHL dewar subcooler which can operate either from the wet expander output or the dewar liquid pump output. The end station satellite refrigerator will receive this helium from the CHL and convert it to refrigeration power for all three end stations. Subcooled liquid nitrogen is used to cool the transfer line shield, arriving at the end stations only slightly subcooled. It is then used for the magnet shields and released as 80K gas.

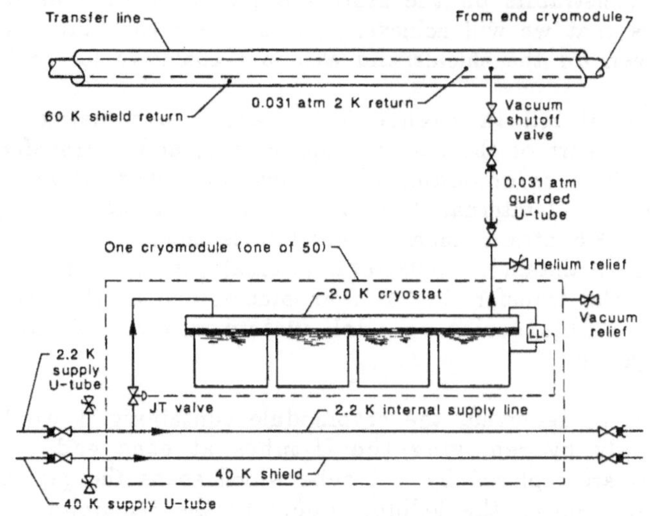

Figure 3 Cryomodule flow schematic.

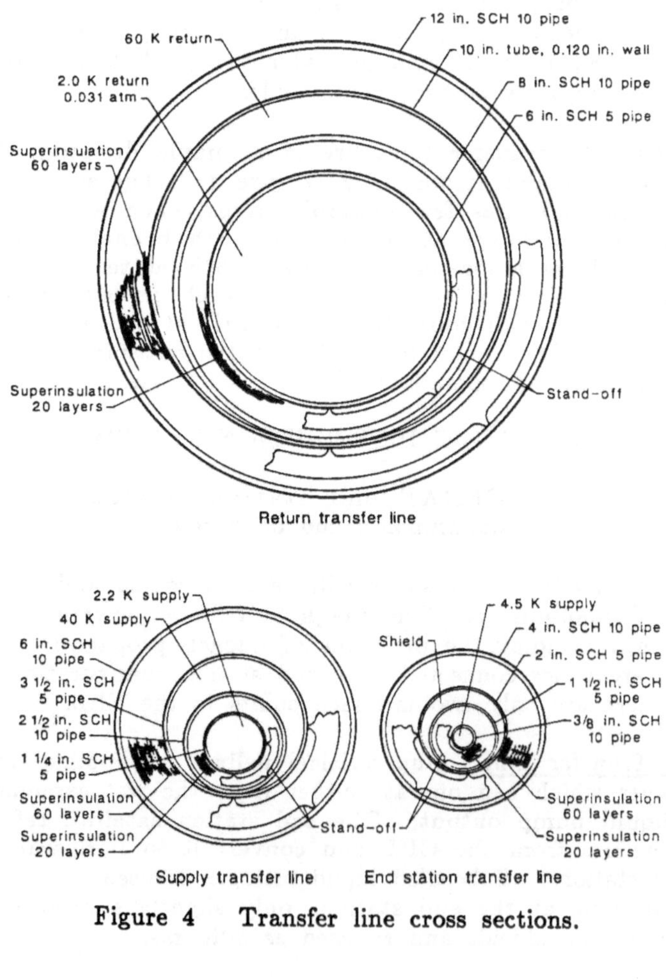

Figure 4 Transfer line cross sections.

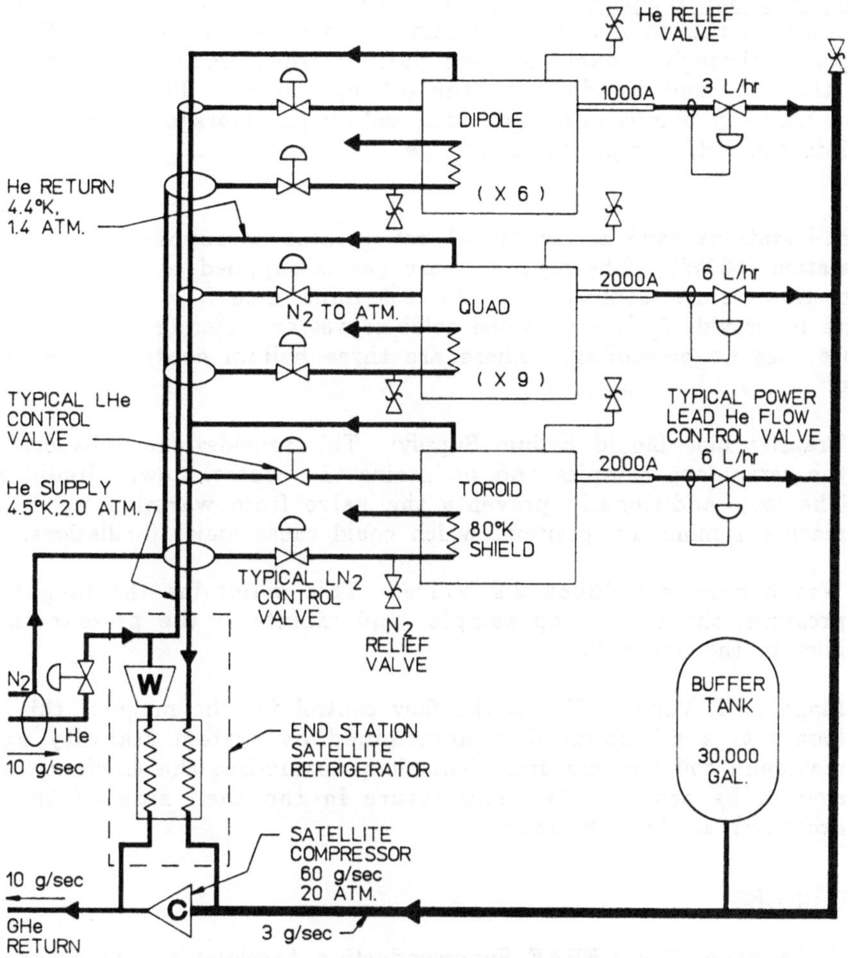

Figure 5 **End station configuration.**

Table 2
End Station Loads

	Lead flow liquid required (L/hr)	5K heat load (Watts)	80K heat load (L/hr)
Magnets	78	550	43
Satellite & Trans. Line	87	25	5
Total Load	165	575	48

Refrigerator capacity 288 L/hr (175%) 996 W (173%) 240 (500%)

End Station Gas Header--Return: The warm helium is manifolded into a low pressure buffer tank and an 18 atm return screw compressor. The CHL flow passes through a dual 10 g/sec purifier which removes contaminants other than H_2 and Ne down to the 0.1 ppm level. The output of the purifier normally dumps into room temperature gas storage (tank farm) and then into the CHL compressor interstage.

The end stations have a "satellite" refrigerator to supply the force flow refrigeration (600W). The high pressure gas is supplied by the end station helium flow return compressor. An 8% imbalance flow is used by the satellite to provide 5.5K gas to the coldbox wet expander or JT valve. The satellite uses no precooling. There are three helium control loops in the system:

A. Transfer Line Liquid Helium Supply: This provides the imbalance flow; the servo loop samples and maintains the control dewar liquid level. The loop additionally prevents the valve from warming up when it reaches a minimum position, which could cause major oscillations.

B. Wet Expander Coldbox JT Valve: This maintains the magnet coil pressure; the servo loop samples and maintains the pressure at the inlet to the first coil.

C. Magnet JT Valve: This is the flow control for the magnet; this servo loop sets a minimum flow determined by system stability and/or maximum coil temperature. This loop maximizes the efficiency of the system by sensing the temperature in the shell side of the heat exchanger at the 15K point.

REFERENCES

1) C. Leemann, The CEBAF Superconducting Accelerator - An Overview, in: "Proc. Linac '86", TV2-3, P. 194.

2) G. Biallas et al, The CEBAF Cavity Cryostat, in: "Proc. Linac '86", MO3-14, P. 73.

3) P. Brindza and C. Rode, The CEBAF Cryogenic System, in: "Proc. Linac '86", MO3-15, P. 76.

4) P. Brindza et al, An Optimized Input Waveguide for the CEBAF Superconducting Linac Cavity, in: "Proc. 1986 Applied Superconductivity Conference", IEEE Trans Mag-23, P. 615.

5) G. Biallas et al, The CEBAF Superconducting Accelerator Cryomodule, in: "Proc. 1986 Applied Superconductivity Conference", IEEE Trans Mag-23, P. 619.

DESIGN CONCEPTS FOR THE ASTROMAG CRYOGENIC SYSTEM

M. A. Green

Lawrence Berkeley Laboratory
Berkeley, California

S. Castles

NASA/Goddard Space Flight Center
Greenbelt, Maryland

ABSTRACT

This paper describes the proposed cryogenic system used to cool the superconducting magnet for the Space Station based ASTROMAG Particle Astrophysics Facility. This 2-meter diameter superconducting magnet will be cooled using stored helium II. The paper presents a liquid helium storage concept which would permit cryogenic lifetimes of up to 3 years between refills. It is proposed that the superconducting coil be cooled using superfluid helium pumped by the thermomechanical effect. It is also proposed that the storage tank be resupplied with helium in orbit. A method for charging and discharging the magnet with minimum helium loss using split gas-cooled leads is discussed. A proposal to use a stirling cycle cryocooler to extend the storage life of the cryostat will also be presented.

THE ASTROMAG SUPERCONDUCTING MAGNET

The ASTROMAG experiment requires a magnetic field for charged particle detection, momentum resolution and energy resolution. The scientific capabilities of the ASTROMAG experiment depend on the size, shape and placement of the magnet coils. The coil configuration in turn strongly influences the cost and complexity of the facility. The magnet and cryogenic design presented in this paper represents a relatively simple magnet design concept which can be carried to the Space Station using the shuttle.

The design concept for the strawman magnet configuration for the ASTROMAG Particle Astrophysics Facility has the following characteristics:[1,2]

1) The magnet coils are designed as relatively high-current coils which operate at stored magnetic energies approaching 20 MJ. The coils are designed so that the ratio of stored energy to active coil mass is about 15 J per gram of active coil mass. About 1/3 of

the overall magnet system mass is expected to be active coil mass. The coils are arranged so that the net magnetic dipole magnet is zero. The field drops to earth's field about 20 meters from the center of the magnet.

2) The magnet coils are to be located outside of the helium storage tank. This permits the physics experiment to be brought close to the coil package. (The goal is to put the closest detector about 10 centimeters from the coil package.) In addition, the stored magnetic energy which is dumped into the coils during a quench can be decoupled from the liquid helium in the storage tank allowing it to be cooled back down with helium vapor after it quenches.

3) The strawman design calls for spherical or near spherical tankage for the liquid helium. This will minimize the tank mass for a given stored helium volume and tank pressure rating; also, the surface to volume ratio is favorable. The difficulty with such a tank is the added complexity of the superinsulation system.

4) The magnet cryostat vessel will have dished, inward facing heads at the ends to permit the detectors to be brought close to the coil ends without the weight penalty associated with flat or near flat heads. The outside surface of the cryostat would be cylindrical near the coil and conical over the central helium tank. This increases the room inside the cryostat for cold mass supports, vapor-cooled shields and intercepts and a possible helium refrigerator thermal strap for the outermost shield and intercept.

5) The magnet, which will have a persistent switch, will be powered through vapor-cooled retractable leads. These leads will be part of the vapor-cooled shield and intercept circuit. This concept permits one to maximize the charging efficiency of the magnet. The lead current will be high enough to permit the coil to charge and discharge at a reasonable rate, and the voltages within the coil will be acceptably low during a quench. (See Table 1)

Table 1 presents some basic parameters for a high-current density version of the 2-coil HEAO-type magnet proposed as a strawman configuration for ASTROMAG. The location of the magnet coils with respect to the helium storage tank and the outer vessel is shown in Figure 1.

It is proposed that the magnet coils be cooled from the liquid helium storage tank by circulating the helium through tubes built into the coil packages. Helium II from the storage tank at 1.6 to 1.8 K will be circulated through the coils using a thermomechanical (fountain effect) pump.

THE MAGNET HELIUM II CRYOGENIC SYSTEM

The helium storage tank will be spherical with attached support rings. The magnet coil packages are attached to these support rings which also mate with the cold mass support system. A schematic representation of the cold mass is shown in Figure 1. As shown, the coil packages have 8 tubular struts between them to carry the 4.405×10^5 N (44.9 metric tons) tensile force between the coils. These struts connect the coils to the helium tank. It is proposed that the struts be made from low thermal conductivity MP-35 N steel to thermally isolate the coils from the helium tank. This isolation allows a thermally efficient recovery from a magnet quench.

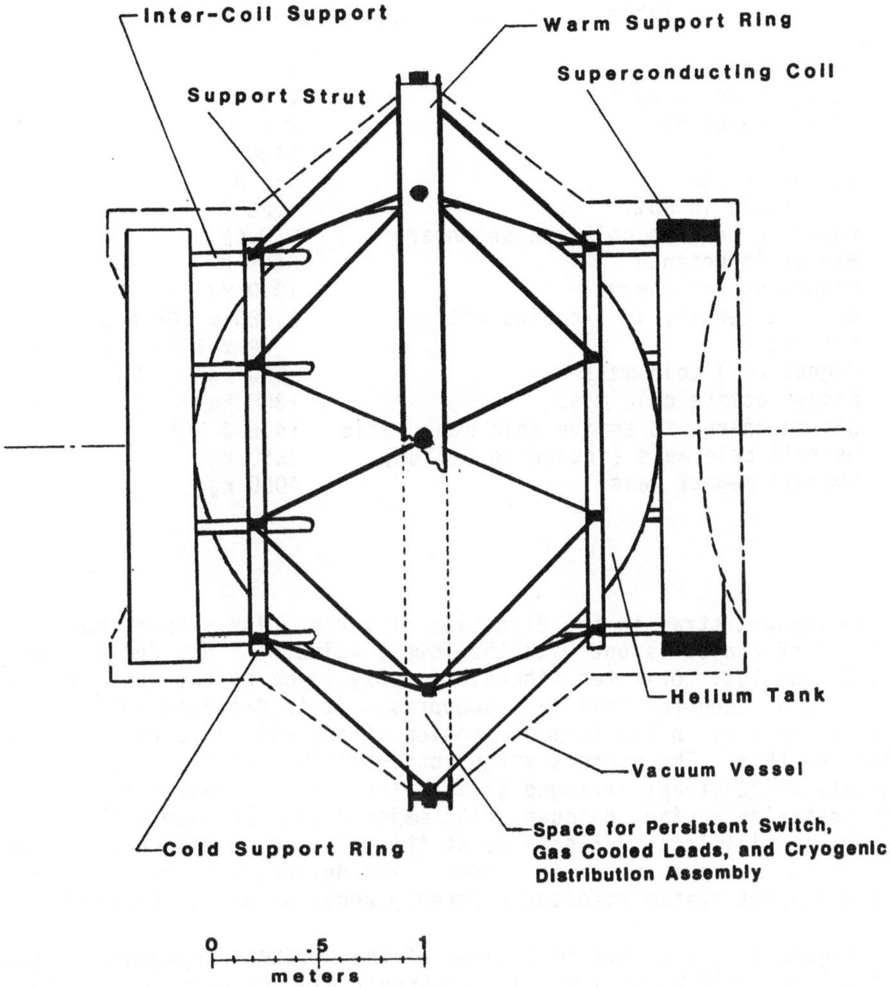

Figure 1. A schematic representation of the ASTROMAG cryostat showing the coils, the helium tank, the support struts and the vacuum vessel.

The largest single heat leak into the helium vessel is through the cold mass support system. The heat leak into the helium vessel through an ideal cold mass support system (where there is enough room for the support rods and their end terminations) is proportional to the cold mass, the design resonant frequency squared and the integrated temperature function Γ (Γ is defined as the integral of thermal conductivity with temperature from the lowest intercept temperature to the helium vessel temperature). The heat leak through the support straps into the vessel is inversely proportional to the elastic modulus

The ASTROMAG spherical cryogen tank has an outside diameter of 2.2 meters. The thickness of the tank would normally be determined by the magnet stored energy. If the magnet coils are thermally isolated from the helium tank, the helium tank thickness is determined by the 1 atm external pressure which is applied during leak testing. The helium tank can contain 4,600 liters of helium (almost 700 kg of helium at 1.8 K) with an additional 15 percent of the tank volume as ullage.

Table 1. Strawman Magnet Parameters

Number of coils	2
Coil outside diameter	2.0 m
Coil separation	2.2 m
Number turns	6400
Design current	754 A
Peak field in coil	6.25 T
Coupling between coil and secondary	> 0.92
Magnet inductance	66.9 H
Magnet stored energy	19.0 MJ
Current density in s/c plus matrix	2.985×10^8 A m^{-2}
EJ^2 limit	1.69×10^{24} J A^2 m^{-4}
Magnet coil cold mass	1500 kg
Magnet active cold mass	1300 kg
Stored energy to active cold mass ratio	14.6 J g^{-1}
Overall cold mass (including helium)	2650 kg
Overall magnet mass	4000 kg

of the support strap in the direction of force. The support rod material of choice is one with the lowest value of Γ divided by the elastic modulus. Oriented fiberglass epoxy is among the best materials which can be used for cold mass supports. It is proposed that this material be used in the form of tension straps with 2 thermal intercept points on them. The support strap configuration shown in Figure 1 consists of 32 straps arranged so that the magnet cryostat is self-centering during cooldown. The major design problem with support systems is minimizing deflections at the tank and on the outside support ring as the support system is loaded. The deflections are kept small so that a support system resonant frequency above 35 Hz is obtained.

Figure 2 is a schematic diagram of the ASTROMAG cryogenic system. Cold valves, cold burst discs and crossover pluming associated with ground operations and shuttle safety requirements have been omitted to provide a clear exposition of the cryogenic system associated with the magnets and magnet leads. Figure 2 shows one of the 2 methods being investigated for circulating helium through the coils and the persistent switch using a thermomechanical effect pump.[3,4] Figure 2 also shows how the boil-off gas is used to cool the retractable gas-cooled 750 A electrical leads, the shields and heat intercepts on the cold mass supports. The shield-intercept flow design for the ASTROMAG magnet cryostat is similar to the 2-shield HEAO long-life cryostat built and tested during the early 1970's.[5] This cryostat demonstrated that helium could be stored within a tank with a magnet for up to 2 years.

The HEAO design intercepted heat from the 100 A current leads on the pipe leading from the tank to the first shield. The ASTROMAG cryostat design calls for enhanced heat transfer leads split into 2 parts.[6] The warm end can be disconnected from the cold end. The cold end will be permanently connected to the magnet, helium vessel and persistent switch. After the gas leaves the cold part of the retractable lead, it passes through the 2 gas-cooled shields and support intercepts to the warm retractable part of the gas-cooled lead system. This configuration will result in increased efficiency during charging and discharging because shields and intercepts will be cooled while there is increased flow through the leads. As a result, the boil-off from the helium tank will be reduced for a considerable time after the magnet has been charged or discharged.

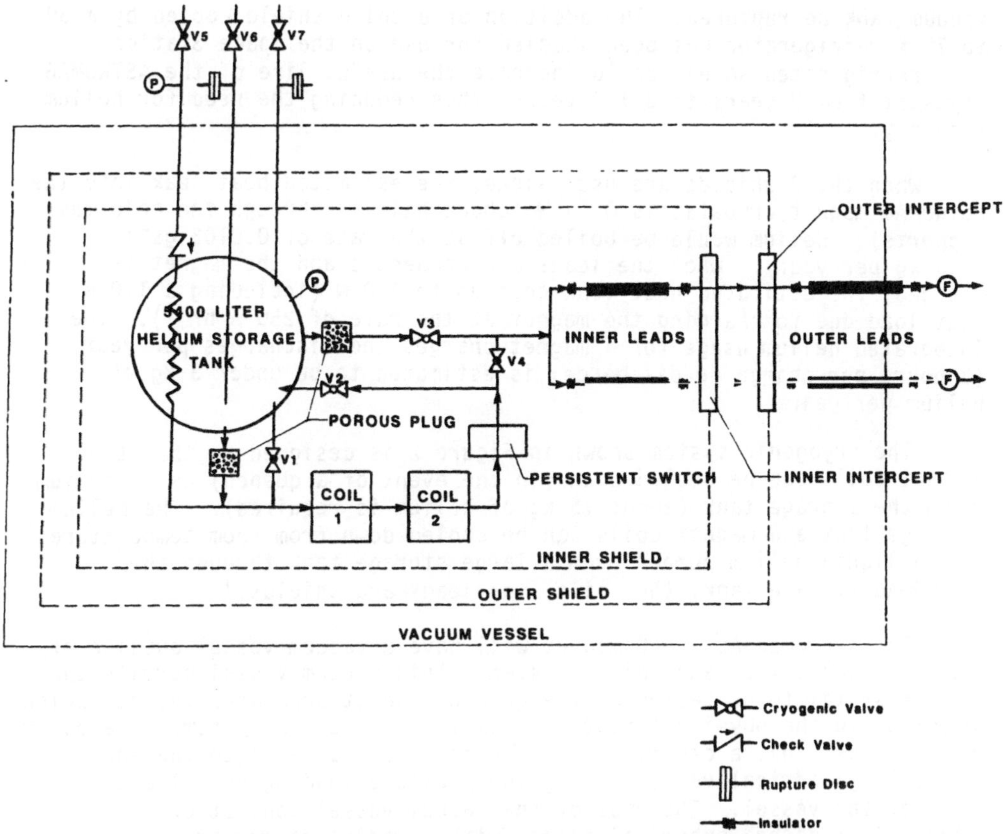

Figure 2. A schematic diagram of the ASTROMAG cryogenic system. (The arrows show the flow through the coil driven by a simple thermomechanical pump. The arrows also show gas flow through the leads and intercepts. The shields are cooled by conduction.)

The 750 A gas-cooled leads proposed for the ASTROMAG magnet can operate within the vacuum vessel in any orientation.[6] The leads would be connected to the gas side of the porous plug. The leads can carry the full cooldown gas flow with a relatively low pressure drop. The leads are designed to withstand high temperature so that they can operate for a long time without gas flow.

In a 2-shield cryogenic system, the most important part of the insulation lies outside the outer shield. It is proposed that 100 to 120 layers of multi-layer insulation be used outside the warm shield. This insulation must be applied very carefully, because it is important that the heat load to the first shield be minimized. The insulation between the outer and inner shield can be reduced to 40 or 50 layers. Inside the inner shield only 2 layers of heavy foil are required outside the foam layer on the helium tank. The layer of foam is a safety feature to reduce the influx of heat to the helium vessel should the vacuum tank be ruptured. The addition of a third shield cooled by a 50 to 70 K refrigerator has been studied for use on the Space Station. This refrigerated shield could increase the useful life of the ASTROMAG cryostat from 2 years to 3 1/2 years, thus reducing the need for helium resupply.

When the 2 shields are used alone, the estimated heat leak into the ASTROMAG magnet cryostat is 0.24 W (about half is through the cold mass supports). Helium would be boiled off at the rate of 0.0103 gs^{-1} (325 kg per year). When the leads are connected and the magnet is charged, the estimated heat leak goes up to 3.0 W (including a 1.0 W heat load due to charging the magnet at the rate of 250 A hr^{-1}). The integrated helium usage for 4 magnet charges and discharges per year (3 hours per charge or discharge) is estimated to be under 3 kg of helium per year.

The cryogenic system shown in Figure 2 is designed so that the magnet coils can be cooled down (in the event of a quench) using helium from the storage tank (about 75 kg of helium is required). The helium storage tank and magnet coils can be cooled down from room temperature using liquid helium pumped from a large storage tank through the ASTROMAG storage tank, the coils, the leads and shields.[7]

The strawman ASTROMAG magnet will have a vacuum vessel outside of the superinsulation and shield system. This vacuum vessel permits the magnet system to be tested on the ground, and it separates the gas which might be in the physics detectors from the insulation system. The outer vacuum vessel has a central support ring which connects to the shuttle. Conical-cylindrical vessels carry the pressure loading from 1 atm outside the vessel. The ends of the vacuum vessel consist of lightweight dished spherical heads (with a radius of curvature of 3 meters). These heads (dished inward) permit the physics detectors to be brought to a distance of 10 centimeters from the coil package. The central support ring transmits forces from the cold mass supports to the shuttle.

The projected magnet system cold mass including a pair of 750 kg coils, a persistent switch, the tank and 700 kg of helium is 2650 kg. An aluminum vacuum vessel plus the shields is estimated to have a mass of 1350 kg. (The use of composite structures could reduce the vacuum vessel mass further.)

CRYOGENIC SAFETY

The ASTROMAG cryogenic system will be designed to meet the shuttle and Space Station safety requirements. Specifically, the requirements of Mil Standard 1522 will be satisfied. The major safety requirements to be met are as follows:

1. The dewar will be two-failure tolerant. To meet this requirement, the cryogen tank will incorporate 3 independent, high-flow rate vent lines.

2. An engineering model cryogen tank will be proof pressure tested.

3. Fracture critical parts will be dye penetrant inspected.

4. The cryogen tank and plumbing system will be capable of withstanding a catastrophic loss of dewar guard vacuum.

The dewars being flown on the Superfluid Helium On Orbit Transfer Flight Demonstration[8] meet these requirements and provide heritage for the design of the ASTROMAG cryogenic system.

SUMMARY

The ASTROMAG cryogenic system for the superconducting magnet has two features not found in ballon magnet cryostats or in previous designs for space-borne cryostats. First, the superconducting magnet coils are thermally isolated from the helium tank so that the coil can be recooled after a magnet quench. Secondly, circulation through the magnet coil is a thermomechanical effect pump which has no moving parts, and it should be very reliable. The projected life for the helium system is over 2 years. This life can be extended to 3 1/2-years by adding a third shield and intercept which is cooled by a small stirling cycle refrigerator.

ACKNOWLEDGMENTS

This research is supported by the Office of Astrophysics, NASA, and the Office of Basic Energy Science, U.S. Department of Energy, under Contract Number DE-AC03-76SF00098.

REFERENCES

1. Interim Report of the ASTROMAG Definition Team on the Particle Astrophysics Magnet Facility ASTROMAG (1986).

2. M.A. Green, et al, ASTROMAG: A Superconducting Particle Astrophysics Magnet Facility for the Space Station, to be published in the IEEE Transactions on Magnetics MAG-23 (1987).

3. M.D. Pirro and S. Castles, "Superfluid Helium Transfer Flight Demonstration Using the Thermomechanical Effect", Cryogenics 26 (1986), p. 84.

4. A. Hofmann, et al, "Considerations on Magnet Design Based on Forced Flow of Helium II in Internally Cooled Cables", Intor-Report (1986).

5. W.L. Pope, et al, "Superconducting Magnet and Cryostat for a Space Application", Advances in Cryogenic Engineering 20 (1974), p. 47.

6. R.G. Smits, et al, "Gas-Cooled Electrical Lead for Use on Forced Cooled Superconducting Magnets", Advances in Cryogenic Engineering 27 (1981).

7. M.A. Green, et al, "The Operation of a Forced Two-Phase Cooling System on a Large Superconducting Magnet", Proceedings of ICEC-8, Genova, p. 72 (1980).

8. M. DiPirro and P. Kittel, "The Superfluid Helium On Orbit Transfer (SHOOT) Flight Demonstration", to be published in Advances in Cryogenic Engineering 33 (1987).

A CRYOGENIC SYSTEM FOR FUSION NEUTRON IRRADIATION AND

IN-SITU MEASUREMENTS ON SUPERCONDUCTORS*

P. A. Hahn and M. W. Guinan

Lawrence Livermore National Laboratory
Livermore, California

ABSTRACT

A specially designed liquid helium flow cryostat for irradiating Nb_3Sn specimens below 20K with fusion (14.6 MeV) neutrons at the Rotating Target Neutron Sorce (RTNS-II) in Livermore and subsequent in-situ measurements of the critical parameters T_c and j_c in magnetic fields up to 15.6 T has been employed in recent experiments. After each irradiation cycle, the flow cryostat was removed from the target without warmup and inserted into a split superconducting solenoid, field enhanced by holmium flux concentrators. This work emphasizes engineering aspects of the cryogenic system, the technical constraints encountered during design and, finally, its performance under operating conditions.

INTRODUCTION

In previous experiments the specimens to be irradiated were mounted at the tip of a cold finger machined from OFHC-(Oxygen Free High Conductivity) copper. The whole setup was then positioned close to the beam spot of the rotating target and irradiated until the desired fluence has been accumulated. After irradiation the HELITRAN (R.G. Hansen & Associates) flow cryostat was inserted into a 3.5 cm diameter room-temperature bore of a 13 T split solenoid superconducting magnet for measuring T_c and j_c as a function of the magnetic field as is illustrated in Fig. 1a.

A practical approach to enhance the maximum field beyond 13 T is the use of holmium as a flux concentrator. In the actual upgrade flat cylinders were inserted into the clear "cold" bore of the magnet with access to the field region between the two adjacent cylinders through a radial gap (1.1 cm x 2.5 cm) of the split solenoid mounted in a bath cryostat. The reduced space would require an unacceptable small diameter of the room temperature bore of 0.6 cm due to insulation, and an even

*This work was performed under the auspices of the United States Department of Energy by Lawrence Livermore National Laboratory under Contract No. W-7405-Eng-48.

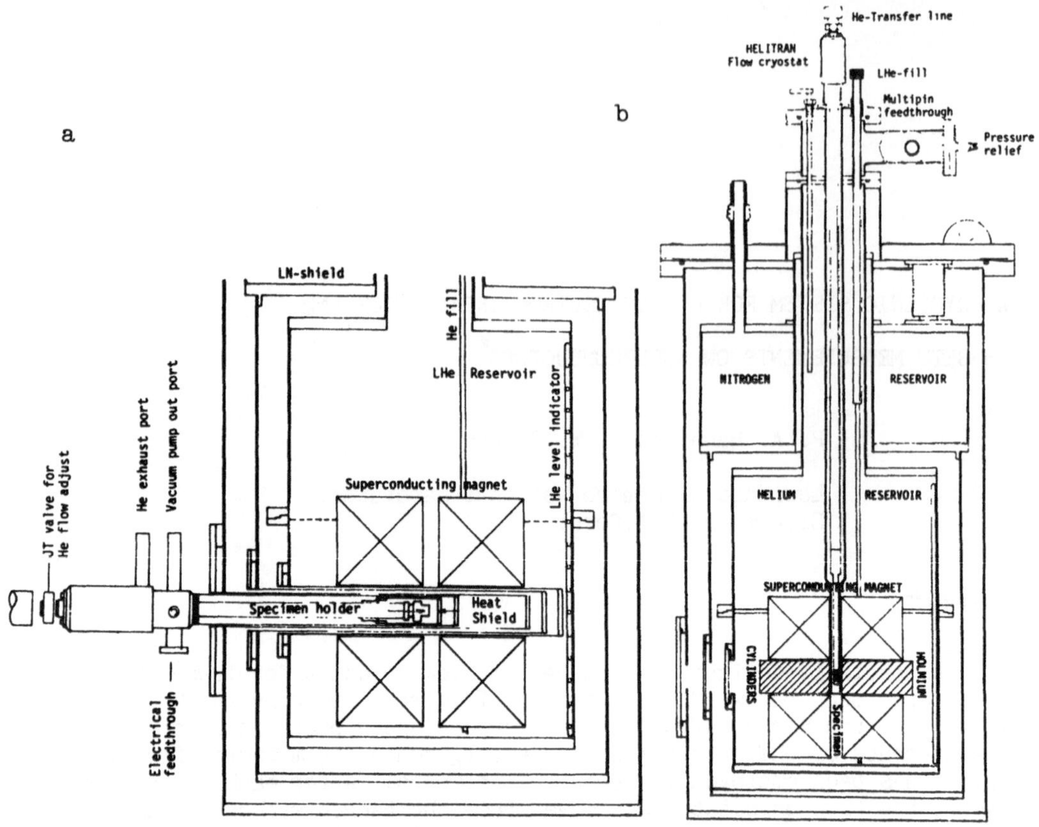

Fig. 1a. Previous experimental setup with room temperature access to the 13 T magnet

b. Design with Ho-flux concentrators in the bore of the superconducting magnet; access to the field is obtained through the LHe-bath and the gap of the split solenoid

smaller diameter cold finger which itself must be maintained in a vacuum chamber during irradiation. For that reason the concept of room temperature access to the high field region of the magnet had to be abandoned with the crucial implication that the field could no longer be accessed over a relatively short distance of 26 cm from the "sideflange" of the bath cryostat. The remaining alternative was to insert the cold finger of the flow cryostat (with surrounding vacuum chamber) directly from the top plate of the bath cryostat as displayed in Fig. 1b. Thus, part of the flow cryostat will be immersed in liquid helium (LHe) which in fact helps maintaining a temperature of 4.2 K and below during j_c-measurements. On the other hand its unusual length (120 cm) and, as a result, increased heat load might impose serious limitations upon the obtainable temperature at the sample location during irradiation when the vacuum chamber is at room temperature.

FIELD ENHANCEMENT OF THE SUPERCONDUCTING MAGNET

Design Considerations

The design and test of holium flux concentrators for an identical superconducting solenoid has been reported earlier[1]. For flat faced poles the net gain in field at the intersecting point of the midplane with the axis has been calculated to 3.0 T (± 0.1 T) if an external field of 13 T is applied. A rather small improvement could be obtained by tapering the adjacent ends of each cylinder. A computer code[2] for tabulating the central field while varying different parameters such as tapering angle, pole diameter and - most important - gap size predicts an additional 0.2 T field increase for a pole diameter of 3.0 cm, a tapering angle of 60° and the actual gap size of 1.1 cm. Special attention should be given to the effect of tapered poles on the shape of the magnetic field. Since the maximum field of the magnet heavily depends on the magnitude of the field at the innermost turns located near the gap where the field values are highest during operation tapering of the holmium cylinders translates into a slight field increase in this section provoking a premature quench. It has been estimated that the theoretical gain in field by tapering will be more than compensated by the lower quench current[3,4]. As an alternative lowering the bath temperature from 4.2 K to 2.5 K (by pumping) has been considered; it would enhance the parameter H_{c2} of the superconducting magnet material by about 5% according to the parabolic law

$$H_{c_2}(T) = H_{c_2}(0)\left(1 - \frac{T^2}{T_c^{\,2}}\right) \tag{1}$$

and assuming 22.0 T for $H_{c2}(0)$ and 16.5 K for T_c which is typical for commercial Nb_3Sn[5]. However, besides j_c the magnet performance is servely limited by the internal stresses between individual turns and layers. Any attempt to exceed the manufacturers specifications, especially the quench current, may result in a destructive quench of the magnet. Therefore no effort was made to test the magnet below 4.2 K.

Magnet Calibration

The use of Ho flux concentrators required recalibrating the magneto resistance probe built into the magnet versus an external reference. A small calibrated search coil (1.1 cm o.d., nA = 136.5 cm^2 at 77 K) in conjunction with a flux standard and an electronic integrator[6] (R_{inp} = 51.1 kΩ) was positioned on the bore axis, but 3 mm off the intersecting point with the midplane in accordance with the sample location of the superconductors in the actual experiment. In four runs data were recorded in ascending and descending fields, i.e. the time (in order to correct for any drift in the integrator later on), the integrator output voltage (0.5 V/T), the reading of the magnetoresistive probe and the magnet current. After each run the coil orientation has been reversed 180° in order to correct for possible field variations along the axis. With the exception of the 4th run where the electronic drift was twice as high as in previous ones the interpolated field values agreed all within 0.02 T accuracy. For establishing the calibration chart (field versus reading of the magnetoresistive probe) we included the averaged results of the first three cycles in descending field where the holmium is completely saturated. A summary of the conducted tests is listed below:

Maximum field obtained with Ho-insert	:	15.6 T
Magnet current at 15.6 T	:	160 A
Gain by Ho-insert	:	2.6 T

HELITRAN FLOW CRYOSTAT

Design Considerations

The design considerations are based on the requirements during the 14 MeV neutron irradiation and the subsequent j_c-measurements. A schematic diagram of the experimental setup with the flow cryostat, the

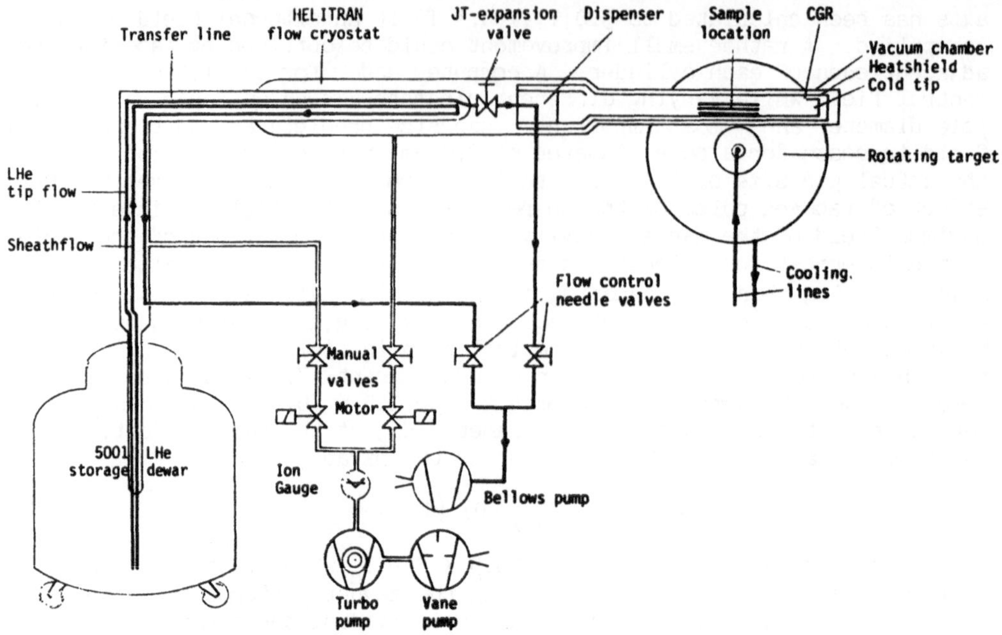

Fig. 2. Schematic view of the HELITRAN flow cryostat, inserted transferline and supporting environment

removable insert tube (HELITRAN-bayonet, Air Products, Inc.) and the supporting environment is shown in Fig. 2. Due to the unusual length of the flow cryostat there will be an increased heat load from thermal radiation of the surrounding vacuum chamber and from neutron heating. The experimental outline requires that the temperature at the tip of the cold finger must be maintained below 20 K during neutron irradiation and below 14 K when the beam is shut off in order to measure T_c as a function of neutron fluence. The main reason for irradiating the superconducting specimens at temperatures between 14 K and 20 K is related to the significant amount of hydrogen released by the involved electrical insulation (epoxy, kapton). At conditions below the triple point (13.8 K at 0.07 atm) the produced hydrogen remains frozen and would seriously disturb temperature control during the first isochronal anneal after irradiation when all of the accumulated hydrogen evaporates causing a huge increase in pressure and temperature. When irradiating above 13.8 K the gaseous hydrogen is continuously removed by the vacuum pumps. However, the question remains whether the mechanisms of defect production at this slightly elevated temperature are the same as at 4.2 K, the actual condition in future applications; fortunately there is strong evidence that this is the case since neither interstitials nor vacancies are mobile below 40 K in Nb_3Sn.

The only feasible design for the required "oversize" flow cryostat implies an extended HELITRAN-bayonet (86.4 cm) with reduced cooling capacity (~1.0 W); in this case the cold finger illustrated in Fig. 3 is a copper rod 20 cm long and 5 mm in diameter. A preliminary estimate for the obtainable temperature has been derived from solving the heat equation in one dimension

$$\frac{\partial u}{\partial t} - \frac{\kappa}{\rho . s} \Delta u = \frac{A}{\rho . s} \qquad (2)$$

where

A ... heat load, W cm^{-1}
κ ... thermal conductivity, W cm^{-1} K^{-1}
s ... specific heat, J g^{-1} K^{-1}
ρ ... (mass) density, g cm^{-3}

where $u = u(x,t)$ is the temperature as a function of distance x from the cold spot (x = 0, location of LHe dispenser) and time t. The heat load A(x,t) from the surrounding vacuum chamber is assumed to be constant over the entire length and does not change with time, i.e. A = const. According to Boltzmann's law

$$P = \sigma \cdot T^4 \qquad \sigma = 5.67 \cdot 10^{-8} \text{ W m}^{-2} \text{ K}^{-4} \tag{3}$$

the thermal radiation emitted per unit area increases with the fourth power of the absolute temperature. As an example the heat load on the cold finger has been calculated based on the following geometry:

Length of vacuum chamber	:	L = 20 cm
Diameter (vacuum chamber)	:	d = 1.1 cm
Temperature (vacuum chamber)	:	T = 300 K

Hence, the surface area not including the top and bottom (which is small compared to the jacket) is found to be 69 cm^2 ($d \cdot \pi \cdot L$). At 300 K the total black body radiation amounts to 3.2 W or, equivalently, 0.16 W cm^{-1} for the external heat function A. After an initial period of cool down the system will enter the state of equilibrium, i.e. $\partial u/\partial t = 0$. The function u(x), now solely depending on the distance x from the cold spot, has to satisfy

$$\Delta u(x) = -\frac{A}{\kappa} = \text{const.} \tag{4}$$

which has the form of Poisson's equation. Under the given conditions with the external heat load exceeding the maximum cooling capacity at the cold spot the imposed boundary condition, u(0) = 4.2 K, cannot be established. However, a coaxial heat shield placed between the cold

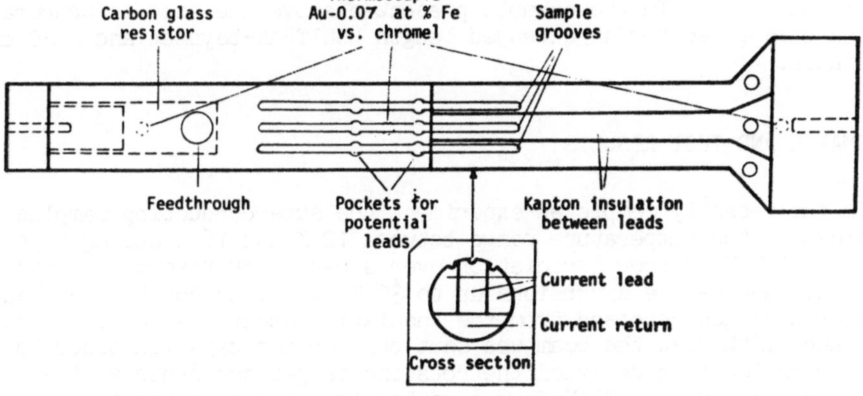

Fig. 3: Display of the cold tip indicating the locations of samples, temperature sensors and leads

finger and the vacuum chamber can improve this situation significantly. Since the emission coefficient of a gold plated and polished surface is only 3% of a black body, the heat load can be reduced from 3.2 W to approximately 0.1 W. An immediate solution of equ. 4 can be obtained by two subsequent integrations while obeying the boundary condition stated above,

$$u(x) = 4.2 + A/2 \;.\; \kappa x^2, \qquad 0 < x < L \tag{5}$$

which exhibits a parabolic dependence of temperature versus x. An uncertainity in the obtainable temperature at the end of the cold finger (x = L) arises from the variation of the thermal conductivity κ with temperature. According to the specifications for OFHC-copper κ drops from 10 W cm^{-1} K^{-1} at 20 K to 6 W cm^{-1} K^{-1} at 100 K. In a conservative estimate (κ = 6 W cm^{-1} K^{-1}) the temperature at the tip is found to rise from 4.2 K to only 4.4 K, which clearly lies well within the requirements for the neutron irradiation (T = u(L) < 20 K). In order to account for the effect of neutron heating (0.5 W) which is restricted to the tip of the cold finger (x = L) one can estimate the rise in equilirium temperature by appending a δ -function on the right side of equ. 4, i.e.

$$\Delta u(x) = - \frac{A}{\kappa} - \frac{B}{\kappa} \; \delta(x-L) \tag{6}$$

where B stands for the neutron generated heat.

The corresponding solution,

$$u(x) = 4.2 + A/2 \;.\; \kappa \, x^2 + B/\kappa \, x \tag{7}$$

would add another 2.5 K to the result of eq. 4. Thus, for 1.0 W cooling capacity the lowest obtainable temperature during irradiation is expected near 7 K.

The limiting factors in the design are the thermal conductivity depending on the purity of the OFHC-copper and the shielding against thermal radiation. In the example presented above the rise in temperature has been calculated for an extended length HELITRAN-bayonet and a 20 cm cold finger.

PERFORMANCE AND TEST RESULTS

In the recently completed experiment the superconducting samples were maintained in the temperature range between 12 K and 16 K during high insensity 14 MeV neutron irradiation over a period of five weeks with occasional temperature excursions up to 30 K (< 1 min) due to beam induced bursts of hydrogen released from the insulating epoxy. After the first, third, and fifth week the beam was shut off for two days (in order to allow for radioactive decay of the rotating target and other active components) and the HELITRAN flow cryostat inserted into the 15.5 T superconducting magnet at which point the liquid helium flow to the cold tip has been increased sevenfold to provide the necessary cooling during j_c-measurements. The actual performance as summarized in Table 1 has

Table 1. System performance parameters observed during the experiment

	Operating Mode	
Parameter	During Neutron Irradiation	Inserted in Magnet
Temperature range at sample location	12 K < T < 16 K	4.1 K < T < 4.6 K for j_c-measurement
LHe flow rate*	0.9 l/h	6.25 l/h
Approx. cooling capacity	0.75 W	4 W
Lowest temperature (no current in samples)	-	3.3 K
Max. useful current carrying capacity	-	25 A

*) A flow rate of 1l LHe per hour (0.125 kg/h) yields a cooling capacity of 0.83 W; a small percentage thereof is used for cooling the transferline between the HELITRAN flow cryostat and a 500 l LHe storage dewar

exceeded design based expectations in some points, particularly the cooling capacity when the Joule-Thompson needle valve was partially closed (expansion cooling); on the other hand the maximum acceptable current for measuring j_c (B) had to be limited to 25 A in order to insure a temperature at the sample location below 4.6 K. Applying a higher current would inevitably lead to gradients in the cold tip, uncertainties in the sample temperature and the necessity to extrapolate j_c-values taken above 4.6 K.

ACKNOWLEDGEMENTS

The authors gratefully acknowledge the advice on the Ho-flux concentrators by R. W. Hoard of MFE and the assistance in calibrating the magnet by D. H. Nelson and D. A. Van Dyke of LBL and R. A. Van Konynenburg (LLNL). We highly appreciate the professional expertise of R.G. Hansen & Associates in the design and manufacture of the HELITRAN flow cryostat.

REFERENCES

1. R. W. Hoard, S. C. Mance, R. L. Leber, E. N. Dalder, M. R. Chaplin, K. Blair, D. H. Nelson, and D. A. Van Dyke, UCRL-90638 Preprint (1984).
2. P. A. Hahn, "Tapert Holmium Evaluation - The Program TAHOE", code available from Nuclear Chemistry Division, Lawrence Livermore National Laboratory (1985).

3. S. Sackett, "JASON - A Code for Solving General Electrostatic Problems - Users Guide", UCID-17814 Report (1978).
4. S. Sackett and R. Healey, "JASON - A Digital Computer Program for the Numerical Solution of the Linear Poisson Equation", UCRL-18721 Report (1969).
5. N. Wilson in: "Superconducting Magnets", pp.3, Clarendon Press Oxford (1983).
6. D. H. Nelson, D. A. Van Dyke, R. W. Hoard, and S. C. Mance, "Magnetic Measurements of a LLNL Superconducting Dipole with Holmium Flux Concentrators" LBID-926 (LBL) Report (1984).

COLD COMPRESSION OF HELIUM FOR REFRIGERATION BELOW 4 K

Hans Quack

Sulzer Brothers Ltd
Winterthur, Switzerland

ABSTRACT

In normal helium refrigeration cycles the refrigerant helium is compressed at or above room temperature. But under certain circumstances a partial compression at lower than ambient temperature is advantageous. The most common use of cold compression is in case of operation at temperatures below 4.4 K, where the vapor pressure of helium is below 1.2 bar. Here the influence of the cold compression on the cycle efficiency is discussed. A review of the available types of compressors is given and the prototype of a supersonic radial turbocompressor is shown with active magnetic bearings and direct electrical drive.

INTRODUCTION

In normal helium refrigeration cycles the refrigerant helium is compressed at or above room temperature. But under certain circumstances a partial compression at lower than ambient temperature is advantageous. The most common use of cold compression is in case of operation at temperatures below 4.4 K, where the vapor pressure of helium is below 1.2 bar. The purpose of the cold compression is

-to reduce the volumetric flow rate of the vapor to make the refrigerator more compact,
-to reduce the parts of the plant, which operate below àtmosheric pressure, and so to keep the chance for impurity inleak as small as possible,
- if possible to increase the overall plant efficiency.

CYCLE EFFICIENCY

The influence of the cold compression on the cycle efficiency is rather complex. This can be demonstrated by the two examples shown in Fig. 1. The first cycle consists of

- isothermal compression at room temperature
- isobaric cooling at high pressure
- nearly isentropic expansion in a cryogenic expander
- nearly isentropc compression in a cold compressor
- isobaric warming up at low pressure.

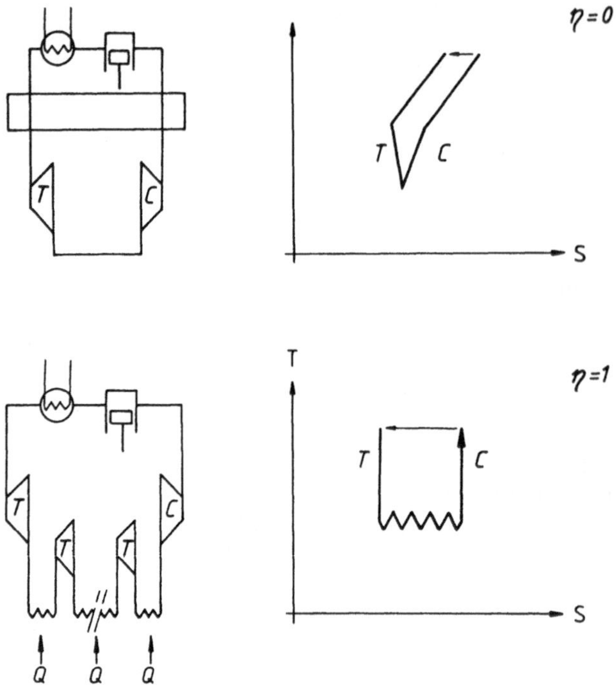

Fig. 1
Two cycles with cold compression

Even though all steps of the cycle are ideal or nearly ideal, the cold compressor has destroyed all the refrigeration produced by the expander. The resultant refrigeration output is zero and the cycle efficiency is zero. Some people use this exaple as proof, that the cold compression is generally disadvantageous. The purpose of a refrigerator is to produce refrigeration, but the cold compressor destroys refrigeration. Therefore it is bad.

The other example in Fig. 1 shows the well known Carnot cycle with

- isothermal compression at room temperature
- isentropic expansion
- isothermal expansion at low temperature with addition of heat
- isentropic compression with a cold compressor to room temperature.

The Carnot cycle is the most efficient cycle. So this is the 'proof', that the cold compressor is the prerequisite for the ideal cycle.

The conclusion one should draw from the two examples in Fig. 1 is that it is probably not the cold compression which makes the difference between a good and a bad cycle. It is of much more importance what else is happening in the cycle.

To obtain more insight into the problem, I would like to introduce a cycle, which takes into account the real thermal properties of helium, i.e. the fact that in some temeprature regions the specific heats between low and high pressure streams are different. Fig. 2 shows a cycle, where helium is compressed isothermally at room temperature to 20 bar. The high pressure stream is cooled at constant pressure to 4.2 K. Then follows an isothermal expansion with heat addition to the saturated liquid line and afterwards the evaporation of the liquid. At this temperature level the specific heat of the low pressure gas is larger than the specific heat of the high pressure gas. Thus there is a surplus of cold capacity, when the low pressure gas enters the countercurrent heat exchanger. So we can leave the heat exchanger after a small warmup and make a compression in a cold compressor, before going back into the heat exchanger.

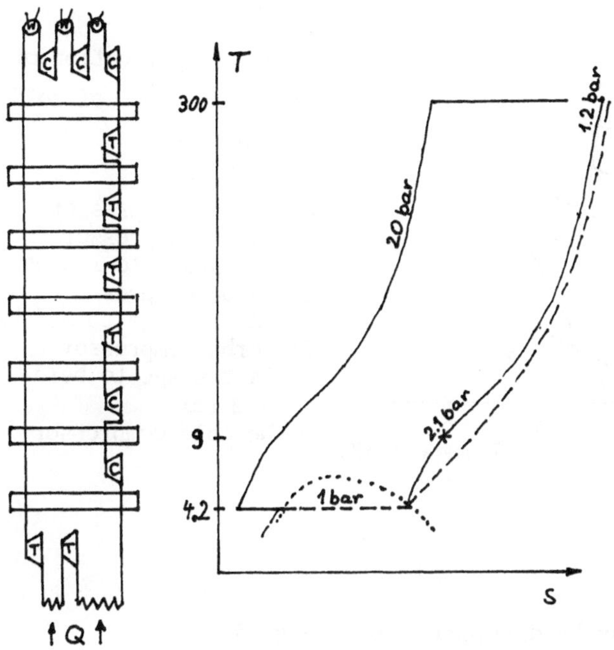

Fig. 2
Ideal refrigeration cycle with helium as working fluid

One can repeat these two steps several times as long as the specific heat of the low pressure stream is larger than the one of the high pressure stream, i.e. up to a temperature level of about 9 K, and it is possible to increase the pressure from 1 bar to about 2.1 bar.

When one continues to higher temperatures, there is a deficiency of cold capacity in the low pressure stream. Therefore one has to insert, now and then, a little expansion to make up for it. So one can continue until room temperature is reached. If all this is done in many small steps with ideal machines and ideal heat exchangers, one ends up at room temperature with a pressure of about 1.2 bar. The cold compression between 4.2 and 9 K provided a larger pressure rise, than the expansions between 9 and 300 K consumed.

There exists another scenario, where one can show that cold compression is thermodynamically advantageous. This is when one has a very low pressure flow with relatively large pressure drop in the heat exchanger and in addition a rather inefficient room temperature compressor.

TYPES OF COLD COMPRESSORS

Once one has convinced oneself that the use of a cold compressor is thermodynamically advantageous, or at least not too detrimental to the system, there comes the question of the availability of the machine itself. Fig.3 shows the author's view of the state of the art. Coordinates of Fig.3 are the refrigeration rate in Watts and the working temperature of the refrigerator in Kelvin. This temperature is not necessarily identical to the suction temperature of the cold compressor, because it is sometimes of advantage to superheat the vapor before compression.

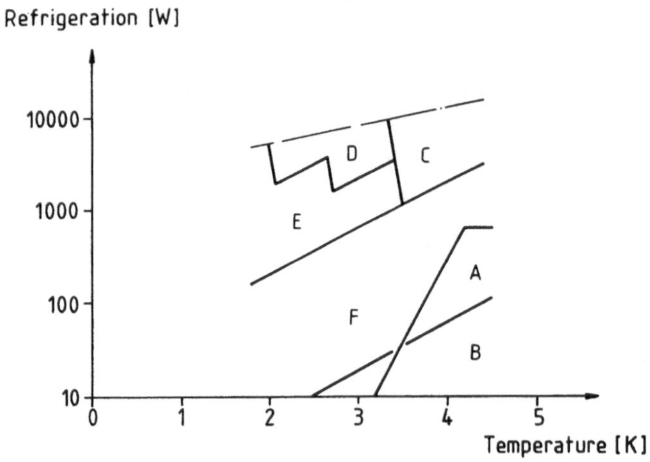

Fig. 3
Range of application of different types of cold compressors
A: Ejector
B: Piston compressor
C: Turbo compressor one stage to 1.2 bar
D: Turbo compressor multi stage to 1.2 bar
E: Turbo compressor last stage to below 1.2 bar
F: No cold compressor

The following fields can be distinguished in Fig. 3:

A: The ejector is a very simple device without moving parts. Its poor thermodynamic efficiency limits its use to the smaller capacity range.

B: Piston compressors can be used for small volumetric throughputs. The lowest temperature will be determined by the possible reduction of the dead volume.

C, D and E describe the range of cold turbo compressors.

C: Here the exhaust pressure of 1.2 bar can be reached in one stage of compression.

D: Several stages of compression are needed to raise the final exhaust pressure to 1.2 bar.

E: Some pressure increase can be obtained from one or several stages of turbo compression, but it is not practical to try to reach 1.2 bar by cold compression only.

F: There remains quite a large field, where no technical solution for a cold compression exists nowadays. The volumetric flow rate is too large for a piston compressor and too small for a turbo compressor. So here is a wide open field for inventors! The only technical solution available today is warming up to room temperature and compression by room temperature vacuum pumps.

TURBOCOMPRESSORS AS COLD COMPRESSORS

Compressor Wheel

There are two different effects which limit the pressure ratio which one can obtain with one turbo compressor stage: Sonic velocity and circumferential speed. Fig. 4 shows that for helium the crossover point for these two limiting effects is at a suction temperature of about 10 K. Below this temperature the sonic velocity at inlet and/or outlet of the compressor wheel limits the possible pressure ratio of a single stage to about 4 to 5. Above 10 K it is the mechanical properties of the wheel or other parts of the rotor which limit the circumferential speed. The warmer it gets, the smaller is the possible pressure ratio. At ambient temperature the limit is about at a temperature ratio of 1.15 to 1.2.

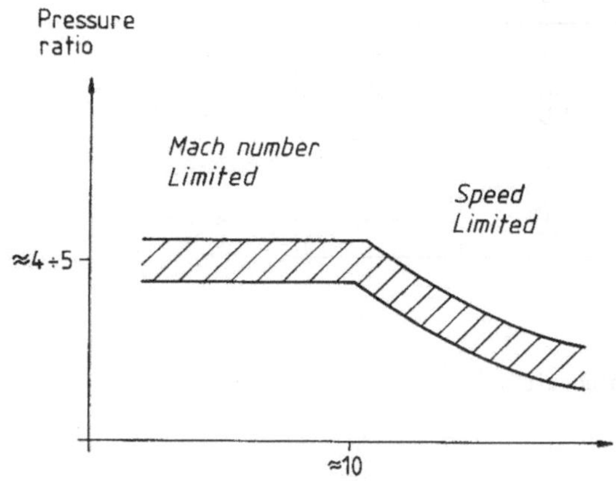

Fig. 4 Possible pressure ratio in one turbo compressor stage

Fig. 5 Wheel of supersonic compressor

Considerable effort has been spent over the last twenty years to improve the performance of compressor stages which are Mach-number limited. Fig. 5 shows the result of this development work. The special shape of the wheel allows efficient and stable operation with supersonic velocity at inlet and outlet.

We are of the opinion that for this size of wheel and for the cold compressor application the open wheel has certain advantages over the shrouded wheel, because the shroud would add additional weight to the overhanging wheel and the bonding of the shroud to the vanes is just an additional risk item.

Bearings and Drive

Fig. 6 shows a matrix of possible bearing and drive systems for cold compressors. The table distinguishes also between the compression to 1.2 bar, i.e. for the field C in Fig. 3 and on the other hand the first stage of compressors for fields E and D of Fig. 3, where the exhaust pressure is below 1.2 bar. The difference between these two categories is the pressure in the bearing room. If this pressure is above atmospheric pressure, one can easily vent oil and helium out into a recovery or recirculating system. If the bearing system is below atmosperic pressure, oil may still be pumped out, whereasfor helium one would need a room temperature vacuum pump, i.e. exactly that component, which one wanted to replace with the cold compressor.

		Gas turbine	Oil turbine	El. Motor w.gear	El. Motor direct	exhaust pressure
Gas bearings		++	⊕	o	+	>1 bar
		o	⊕	o	⊕	<1 bar
Oil bearings		+	+	+	+	>1 bar
		o	⊕	⊕	⊕	<1 bar
Magnetic bearings		+	⊕	+	+	>1 bar
		o	⊕	⊕	++	<1 bar

o not feasable

⊕ feasable, but problematic

\+ feasable

++ optimum choice

Fig. 6 Bearings and drives for cold compressors

Fig 7 Prototype cold compressor with magnetic bearings

One can see from Fig. 6 that for the system with exhaust pressure 1.2 bar there are several feasable solutions to chose from. The choice of the system with gas bearings and turbine drive as optimum comes from the personal preference of the author.

For the systems with exhaust pressure below 1.2 bar there seems to be only one feasable combination: Active magnetic bearings with direct motor drive. Since this technical approach is of course also usable for field C in Fig. 3, our company decided already in 1981 to persue this way for the cold compressor development. Fig. 7 shows the prototype machine: On the right side the cartridge, i.e. the replacable unit with motor, bearings, rotor and stator. On the left is the electronic supply system system for the medium frequency motor and the active magnetic bearings.

TESTS OF COLD HELIUM COMPRESSORS AT FERMILAB

T.J. Peterson and J.D. Fuerst

Fermi National Accelerator Laboratory*
Batavia, Illinois

ABSTRACT

Fermilab has tested two cold helium compressors for possible installation in the satellite refrigerator buildings of the Tevatron cryogenic system. Operating conditions required to obtain an overall Tevatron energy upgrade from 900 to 1000 GeV are (for each of 24 machines): 52 g/s mass flow rate, 0.7 atm inlet pressure, 1.4 atm exhaust pressure. Acceptable efficiency is in the 60% range. Both Creare, Inc. and Cryogenic Consultants, Inc. (CCI) have supplied units for evaluation. The Creare machine is a high speed centrifugal pump/compressor which yielded 60% adiabatic efficiency (including an approximately 20 watt heat leak) with a 1.0 atm inlet pressure and 55 g/s flow rate. Certain mechanical difficulties were present, chiefly the device's inability to withstand two-phase flow. CCI supplied a reciprocating unit which, after initial testing and modification, achieved 59% efficiency with an approximate 35 watt heat leak at a 0.7 atm inlet pressure and 48 g/s flow rate. Although the device lacks the smooth, quiet operating characteristics of a turbomachine, it has endured mechanically throughout testing and is entirely insensitive to two-phase flow.

INTRODUCTION

The beam energy of Fermilab's Tevatron is presently limited to about 900 GeV by dipole magnetic field strength. Operational data and dipole magnet quench data indicate that, in order to reliably increase the energy of the Tevatron from 900 GeV to 1000 GeV, superconducting magnet temperatures have to be lowered by about 0.5K. This requires a pressure decrease of the boiling helium in the "two-phase" channel of the magnet strings. Calculations and computer simulations performed by Fermilab and by Air Products, Inc., have indicated that a cold compressor located on the satellite refrigerator two-phase return line and pumping either saturated helium vapor or two-phase helium may be the most cost-effective means to attain lower temperatures (see Fig. 1). Twenty-four cold compressors (one in each satellite refrigerator) each pumping approximately 52 g/s of helium from 0.7 atm (saturated vapor or two-phase) to about 1.4 atm would be required. A program of testing cold compressors was begun to determine

*Operated by Universities Research Association, Inc. under contract with the U.S. Department of Energy.

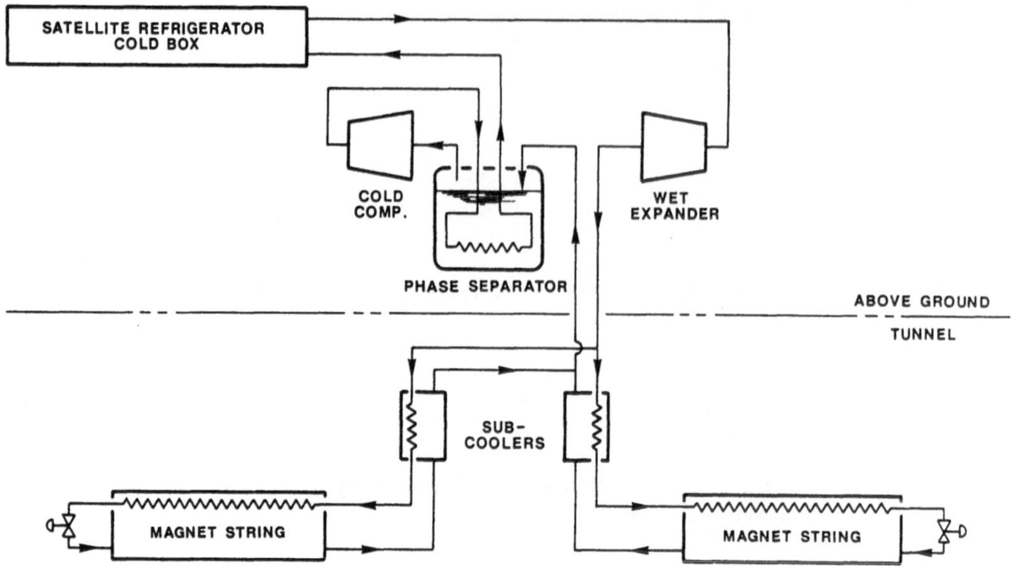

Fig. 1. Satellite refrigerator with cold compressor.

which, if any, might be suitable for use in our satellite refrigerators. Requirements include 60% minimum efficiency under design conditions and extreme reliability. The Tevatron typically runs for six months without scheduled maintenance, and all 24 compressors must be operational to provide 1000 GeV capability.

Two cold compressors have been tested at Fermilab over the past 18 months: a centrifugal unit designed and manufactured by Creare, Inc. and a reciprocating unit designed and manufactured by Cryogenic Consultants, Inc. (CCI). Each was loaned to Fermilab for testing at helium temperatures. A brief test description follows.

TEST METHOD

Figure 2 illustrates the test configuration for the cold compressor tests. Mass flow was controlled by varying wet expander speed and was determined via a flat plate orifice meter (FI4). Instrumentation consisted of pressure taps (PI's), vapor pressure thermometers (TI's), and carbon resistors (TR's). A differential pressure type liquid level gauge, superconducting liquid level indicator, and resistance type dewar heater were also included. In addition to the standard refrigerator instrumentation, each compressor was fitted with a pressure tap, carbon resistor, and vapor pressure thermometer at both inlet and exhaust locations. Zero offsets were periodically obtained for these devices as testing progressed to ensure accuracy. Instrumentation readbacks were processed by Fermilab's existing instrumentation and control system[1,2] which permitted real-time graphical displays and hardcopies of relevant parameters.

Methodical accumulation of data was obtained from each machine through selective manipulation of one parameter (for instance, mass flow) while maintaining control over other parameters such as compressor speed. In this manner a data map blanketing the expected range of operating conditions was created for each compressor.

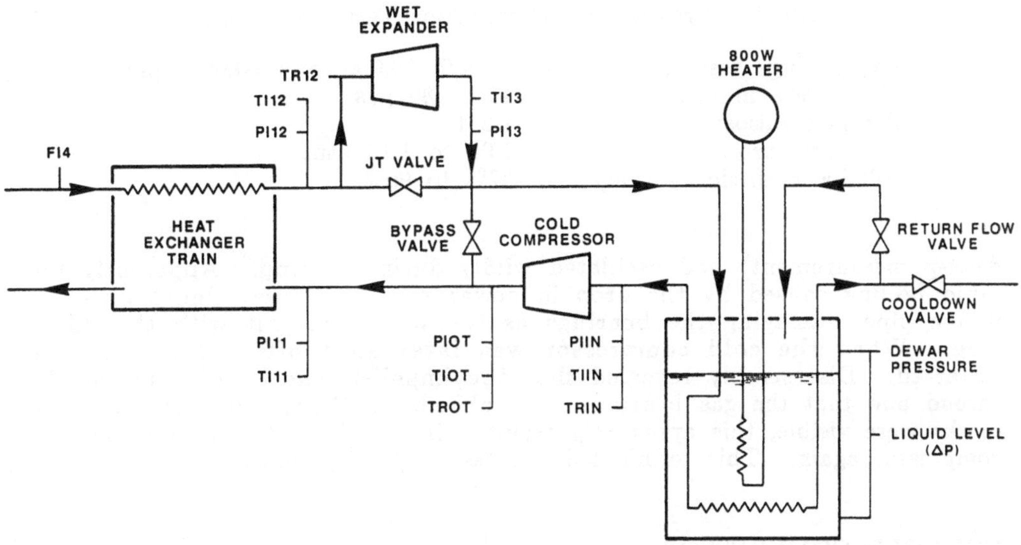

Fig. 2. Cold compressor test configuration.

CREARE COLD COMPRESSOR

Description

A generic pump/compressor has been developed by Creare and was described at the previous Cryogenic Engineering Conference.[3] It is a high-speed (up to 100,000 rpm) electric motor driven centrifugal machine with dynamic gas radial bearings and a permanent magnetic thrust bearing. This cold compressor was tested at Fermilab in January 1986 and again in September 1986 with a different impeller. The results of the January 1986 test are described in a paper by W.D. Stacy.[4] Since in January the cold compressor was operating with an impeller designed for Brookhaven's application,[5] our operating conditions put it thermodynamically far from design conditions. Therefore efficiencies were expectedly low. For the tests in September (described below) the impeller was designed for a 1.0 atm, saturated vapor inlet, 1.4 atm exhaust, and 55 g/s flow. These conditions were given to Creare before a subatmospheric (0.7 atm) inlet condition was considered feasible.

Test Results

Table 1 summarizes the range of test conditions while Table 2 lists some typical data points. Efficiency at design conditions was 60% (with the addition of a tentative heat leak estimate of 20 W ±5 W). Figure 3 illustrates useful head (isentropic enthalpy change) as a function of flow rate and efficiency vs. pressure ratio. The unit performed well during the two days of data accumulation (the unit was shut off each night). However, during startup on the third day the machine "buzzed" twice for a fraction of a second each time. Compressor speed momentarily decreased but operation continued. An ongoing computer plot of dewar liquid level showed that, although the dewar was about 2/3 full, the liquid level readback (a

Table 1. Creare Cold Compressor Test Conditions

Speed range tested:	50,000; 66,000; and 80,000 rpm
Mass flows in data:	41 to 80 g/s
Pressure ratios:	1.2 to 1.5
Inlet pressures:	1.00 to 1.15 atm
Adiabatic efficiency range:	37% to 60%

delta-p measurement) had oscillated wildly during startup. Apparently the rapid boiling caused by the drop in dewar pressure threw liquid into the intake pipe, damaging the bearings as the wheel was hit with the liquid slug. When the cold compressor was later shut off it could not be restarted. Disassembly revealed that the impeller wheel had touched the shroud and that the gas bearings had rubbed. Although only slight polish marks were visible, this apparently resulted in our inability to run the cold compressor again. This terminated our testing in September.

CCI COLD COMPRESSOR

Description

Cryogenic Consultants, Inc. has developed a reciprocating pump/compressor which was delivered to Fermilab in September 1986 for tests. This cold compressor has not been described previously in the literature; therefore a brief description follows. The machine consists of a single stainless steel piston/cylinder arrangement with all-metal spring loaded poppet-style inlet and exhaust valves. The piston shaft extends through a shaft seal in the cover plate to a connecting rod, flywheel, and electric variable speed drive. There are no cams or pull- or push-rods; the valves act as check valves. Some dimensions and specifications are given in Table 3, with Table 4 summarizing the range of test conditions and Table 5 listing some typical data points.

Table 2. Creare Cold Compressor: Four Representative Test Results

Date	9.25.86	9.24.86	9.24.86	9.25.86
Time	14:55	17:30	15:51	12:00
Inlet temp., K (TRIN)	4.464	4.426	4.418	4.373
Inlet temp., K (TIIN)	4.489	4.439	4.381	4.375
Inlet press., atm. abs.	1.103	1.089	1.056	1.042
Inlet enthalpy, J/g	31.52	31.18	31.29	31.16
Inlet entropy, J/gK	8.48	8.41	8.48	8.46
Exhaust temp., K (TROT)	5.469	4.970	5.507	5.263
Exhaust temp., K (TIOT)	~5.4[a]	4.965	~5.4[a]	~5.3[a]
Exhaust press., atm. abs.	1.610	1.292	1.458	1.451
Ideal exhaust enthalpy, J/g	34.28	32.22	33.54	33.37
Exhaust enthalpy, J/g	36.02	34.22	37.27	35.54
Mass flow, g/s	54.7	50.0	74.2	60.4
Flow, liters/s	3.24	2.94	4.45	3.68
Useful head, J/g	2.76	1.04	2.25	2.21
Comp. speed, krpm	66	50	80	66
Adiabatic efficiency, %	61.3	34.2	37.6	50.5

[a] Gas bulb region

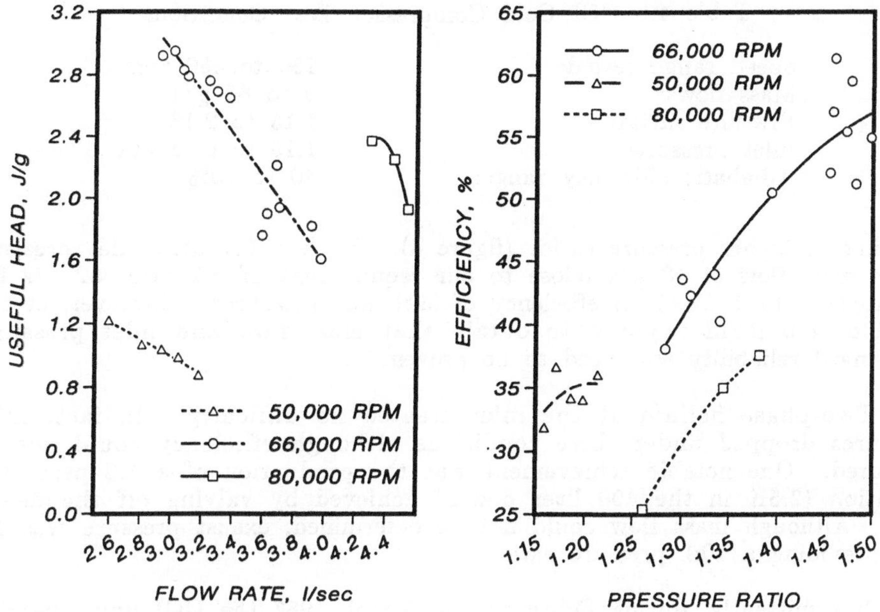

Fig. 3. Creare cold compressor performance.

Test Results

The CCI cold compressor was run almost continuously from October 1, 1986 until mid-December 1986. Adiabatic efficiencies ranged from 15% to 30% although mechanically the valve seals, piston seal and piston shaft seal all performed well. CCI concluded that the valves were undersized and that the low efficiency was caused by extra work done in overcoming valve pressure drops.

The compressor was taken off line, warmed up, and given a new cylinder, piston, and head assembly with larger valves (provided by CCI). The reconditioned unit was tested and exhibited efficiencies between 40 and 70 percent (including a tentative heat leak estimate of 35 W ±10 W). Efficiency at 1 atm inlet, 1.35 atm exhaust with 49 g/s of flow (approximately the Creare design point) was 51%, with higher efficiencies

Table 3. CCI Cold Gas Pump, Model No. CCI-CGP406X300

Bore:	4.375 in. (11.113 cm)
Stroke:	3.000 in. (7.62 cm)
Cylinders:	One (1)
Piston area:	15.0 $in.^2$ (97 cm^2)
Displacement:	45.1 $in.^3$ (739 cc)
Pump speed:	100 - 470 rpm
Motor:	1-1/2 hp DC TEFC Motor, 180 VDC Armature, 1750 rpm
Speed controller:	240 V, 60 Hz, 1 phase (Input) for 1-1/2 hp DC Motor, with Run, Stop, Variable Speed, Local Control

Table 4. CCI Cold Compressor Test Conditions

Speed range tested:	250 to 450 rpm
Mass flows:	0 to 65 g/s
Pressure ratios:	1.15 to 2.16
Inlet pressures:	1.19 to 0.55 ata
Adiabatic efficiency range:	40 to 70%

occuring at higher pressure ratios (figure 4). With a 0.7 atm inlet pressure and a mass flow of 48 g/s (close to our requirement of 0.7 atm, 52 g/s for the upgrade to 1 TeV) an efficiency of 59% was measured. However, at the speed of 450 RPM required to obtain that mass flow and inlet pressure, mechanical reliability will need to be proven.

Two-phase helium at the inlet created no difficulty. In fact, inlet pressures dropped under these conditions although efficiency could not be measured. One notable achievement was the production of a 1.6 psia inlet condition (2.5K in the 400 liter dewar) achieved by valving off the dewar inlet. Although mass flow could not be determined, exaust pressure was 1.2 atm providing a 10:1 pressure ratio.

Intermittently, during February and March 1987 the CCI unit operated at the Magnet Test Facility to serve as a cold compressor for some 3.0 to 3.5K SSC magnet tests. At MTF the unit demonstrated consistent efficiency with high pressure ratios and low mass flow rates (<10 g/s).

CONCLUSIONS AND PLANS

Two fundamentally different cold compressors have been tested under similar operating conditions at Fermilab. The CCI reciprocating machine was tested over a larger range of conditions, including subatmospheric inlet pressures.

The Creare unit performed as predicted at its design point (60% efficiency, 55 g/s flow for 1.0 atm saturated vapor intake and 1.4 atm exhaust). Its gas bearings and lack of valves or mechanical seals give it the advantages of other turbomachinery: potentially long life and low maintenance requirements. Disadvantages include the inability to handle liquid-gas mixtures and a more limited range of flow rates and pressure ratios. Further plans for the Creare unit consist of additional testing with the inclusion of either an integral phase separator/cryostat or a separate phase separator dewar in series upstream of the machine. We are anxious to see how the Creare performs under subatmospheric inlet conditions and

Table 5. CCI Cold Compressor: Five Representative Test Results

Inlet press., atm. abs.	0.55	0.92	1.04	1.19	0.71
Inlet temp., K (TIIN)	3.94	4.20	4.32	4.46	4.12
Inlet enthalpy, J/g	31.85	30.72	30.63	30.45	31.89
Inlet entropy, J/gK	9.654	8.548	8.348	8.117	9.245
Exhaust press., atm. abs.	1.19	1.33	1.35	1.37	1.34
Exhaust temp. K (TROT)	6.00	5.10	4.99	4.86	5.86
Ideal exhaust enthalpy, J/g	38.00	33.21	32.34	31.32	36.86
Exhaust enthalpy, J/g	41.80	34.98	33.96	32.67	40.29
Mass flow, g/s	28	43	49	49	48
Compressor speed, rpm	376	300	297	250	450
Pressure ratio	2.16	1.45	1.30	1.15	1.90
Adiabatic efficiency %	61.8	58.6	51.4	39.2	59.1

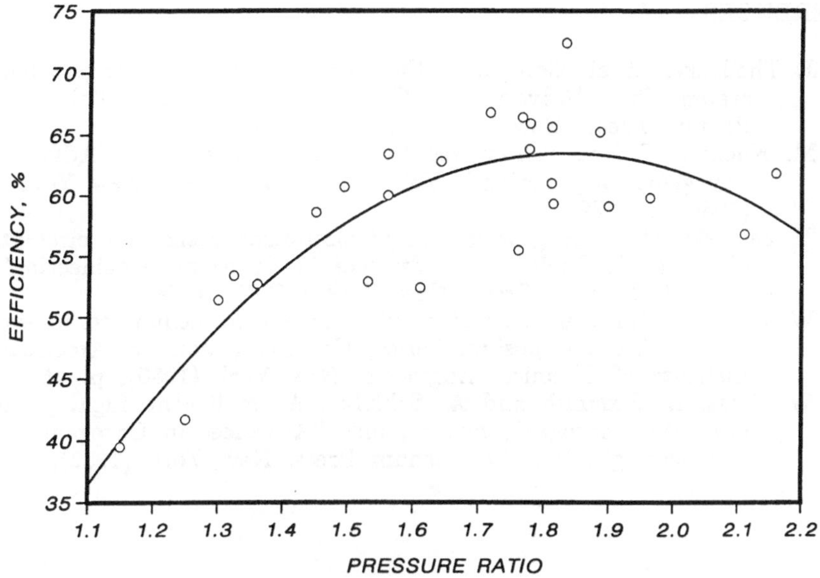

Figure 4. CCI cold compressor performance.

higher pressure ratios; these data points are crucial to our evaluation of this unit and should not be difficult to obtain provided there exists adequate phase separation capability.

The CCI cold compressor is still only marginally acceptable from an efficiency standpoint (51% for 1.0 atm intake, 1.35 atm exhaust with 49g/s flow, 59% with the 0.7 atm intake and 48 g/s for our proposed upgrade). However, the machine functions well under two-phase inlet conditions and its versatility permitted operation almost as a cold "vacuum pump" for service at very low mass flows (less than 10 g/s) and high pressure ratios. So far it has demonstrated excellent reliability for its 2500 hours of operation, with valves and seals remaining remarkably leak tight. Future plans call for monitoring mechanical performance over many hours of continuous operation at speeds of over 400 RPM. It would also be desirable to better understand the sources of inefficiency and make further improvements.

Our long-term goal is to install prototype cold compressors of both designs in a few selected satellite refrigerator buildings in order to gain real operating experience with them. Since virtually all operation is remotely overseen from Fermilab's main control room, these machines must prove their controlability and reliability under "instrument only" supervision. Such a scenario will also provide magnet temperature data relevant to a cold compressor's actual ability to increase maximum Tevatron energy.

ACKNOWLEDGEMENTS

Wayne Stuber, of Air Products, Inc., performed valuable computer simulations while working at Fermilab as a visiting engineer during the summer of 1986. Dodd Stacy and John White of Creare, Inc., provided valuable assistance in testing the Creare cold compressor, and Clyde Harmes of CCI, Inc., has been especially prompt first in delivering the machine to us and then in delivering spare and improved parts.

REFERENCES

1. J. Theilacker et al, Control of the Tevatron satellite refrigeration system, in: "Advances in Cryogenic Engineering" Vol. 29, Plenum Press, New York (1984), p. 437.
2. M. Kuchnir, Pulsed current resistance thermometry, in: "Advances in Cryogenic Engineering", Vol. 29, Plenum Press, New York (1984), p. 879.
3. T. Jasinski, et al, A generic pump/compressor design for circulation of cryogenic fluids, in: "Advances in Cryogenic Engineering", Vol. 31, Plenum Press, New York (1986), p. 991.
4. W.D. Stacy, Performance test results for a cold helium compressor, in: "AIChE Symposium Series, No. 251", Vol. 82, American Institute of Chemical Engineers, New York (1986), p. 45.
5. W. Swift, H. Sixsmith and A. Schlafke, A small centrifugal pump for circulating cryogenic helium, in: "Advances in Cryogenic Engineering", Vol. 27, Plenum Press, New York (1982), p. 777.

OPERATING EXPERIENCES AND TEST RESULTS OF SIX COLD HELIUM COMPRESSORS

D.P. Brown, R.J. Gibbs, A.P. Schlafke, J.H. Sondericker, K.C. Wu

Brookhaven National Laboratory*
Associated Universities, Inc.
Upton, New York

ABSTRACT

Three small and three large cold helium centrifugal compressors have been operated at Brookhaven National Laboratory between 1981 and 1986. The three small cold compressors have been installed on a 1000 W refrigerator for testing a string of superconducting magnets and for R&D purposes. The three large units are components of the BNL 24.8 KW refrigerator to be used to provide cooling for the RHIC project. These compressors are used either to circulate a large amount of supercritical helium through a group of magnets or to pump on the helium bath to reduce temperature in the system. One small circulating compressor tested employs tilting-pad gas bearings and is driven by a DC motor. The two small cold vacuum pumps tested use oil bearings and are driven by oil turbines. The three large oil-bearing cold compressors are driven by DC motors through a gear box. A unique feature of the large vacuum pump is the combination of two pumps with a total of four stages on the same shaft. The adiabatic efficiencies are found to be 57% for the large vacuum pumps and close to 50% for the large circulating compressor. Good overall reliability has been experienced.

INTRODUCTION

Over the past few years, there has been great interest in using cold compressors in superconducting projects. Two major applications of using a cold compressor are to enhance flow rate through superconducting magnets designed with supercritical force flow cooling and to serve as a vacuum pump to reduce the system temperature. These applications provide the cryogenic system control flexibility, enhance heat transfer and reduce the size of the heat exchangers and compressors. While advantages of using cold compressors are appreciable, their efficiency and reliability need to be proved in practice.

Both the reciprocating and the centrifugal type cold compressors of different construction have been reported. In this report, BNL presents the operating experience and test results of one small gas-bearing circulating compressor, two small oil-bearing cold vacuum compressors and three large oil-bearing cold compressors. Machine configuration and system function of the auxiliary equipment are described. Test results for head, work and efficiency versus volume flow rate are given. The three small units have been previously reported and emphasis will be on the three large units.

*Work performed under the auspices of the U.S. Department of Energy.

SMALL GAS-BEARING CIRCULATING COMPRESSOR

The first cold compressor BNL used is a small gas bearing circulator built by Creare Inc., Hanover, N.H., to circulate 156 g/s of supercritical helium at 5 atm and 4 K with a head rise of 0.33 atm. The contract began in 1978 and the unit was delivered in 1980. The specification for this unit is given in Table 1.

This unit uses a pair of tilting-pad gas journal bearings and a magnetic thrust bearing. A brushless DC motor with speed control is used to drive the system. The shaft is designed for low heat leak. A shrouded wheel is used. The configuration is shown in Fig. 1. With small overall dimensions and light weight, this unit is designed to be mounted on a small penetration of the refrigerator vacuum casing. Installation and removal of the compressor is easily accomplished.

The starting sequence is initiated by a single push button. The bearing gas film flow is fully established in a few seconds of shaft rotation. Speed control is accomplished by adjustments in the DC supply voltage. When the drive motor is turned off the shaft coasts down borne by the gas film until about 7000 rpm, whereupon, the shaft slides to a stop on metal to metal contact.

The controller also contains a remagnetizing feature, in the event the rotor becomes demagnetized. In actual operation, difficulty has been found during startup due to uneven torque occurring in the two-pole driving motor. A good practice is to turn on the unit each time the the cooldown begins. This does not effect the cooldown capacity of the refrigerator as the compressor introduces negligible work into the process at warm conditions.

A few failures occurred in the early stages of the test and were corrected by the supplier. Our last experiment was performed in the spring of 1983 and this unit was run essentially continuously for over three months.

As far as performance is concerned, this unit operates with little noise and vibration through the entire speed range. The heat leak is low and the adiabatic efficiency is about 50% near design point. Overall reliability is good. The compressor is rugged and insensitive to process disturbances. It is easily disassembled for inspection or replacement of parts. The manufacturer provided training for some BNL employees so that they are able to do the basic maintenance.

The quality of the gas bearing surfaces appears to deteriorate with the number of starts and stops the bearings have experienced and also with the presence of particulate matter in the gas stream.

Table 1. Specification for the small circulating compressor

Inlet pressure	4.67 atm
Inlet temperature	3.925 K
Inlet density	143.5 g/L
Outlet pressure	5.00 atm
Outlet temperature	3.989 K
Mass flow rate	156.35g/s
Shaft work	55.8 W
Volume flow rate	1.09 L/s
Speed	16800 rpm
Impeller O.D.	2.70 cm

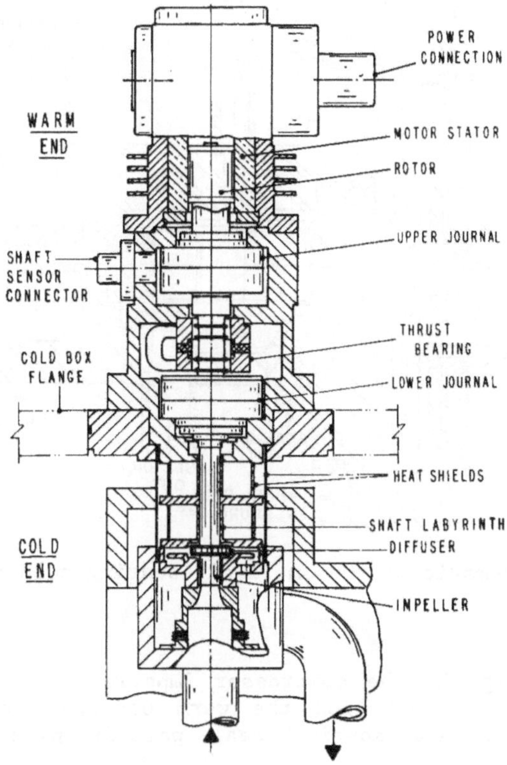

Fig. 1. Schematic of the gas-bearing circulating compressor

A detailed description of this unit is given in reference 1 and 2. In reference 3, the supplier has presented a new generic motor to overcome the starting problem of this unit.

SMALL OIL-BEARING COLD VACUUM COMPRESSOR

Two identical small oil-bearing units have been supplied by the Rotoflow Corporation, Los Angeles, Ca. in 1982. These two units are single stage centrifugal compressors each driven by an oil turbine. They serve as prototype cold compressors to the 24.8 KW BNL refrigerator and were used in series with an ejector to reduce temperature in the refrigeration system. These units are designed to pump a helium bath from 1.4 atm to 0.85 atm. One of the design conditions is given in Table 2. Flow rates are controlled through the compressor speed and can be regulated to match the refrigerator with the load.

Table 2. Specification for the cold compressor

Inlet pressure	0.85 atm
Inlet temperature	4.04 K
Outlet pressure	1.4 atm
Outlet temperature	5.1 K
Mass flow rate	53.6 g/s
Shaft work	291.5 W
Volume flow rate	3.77 L/s
Speed	41000 rpm
Impeller O.D.	4 cm

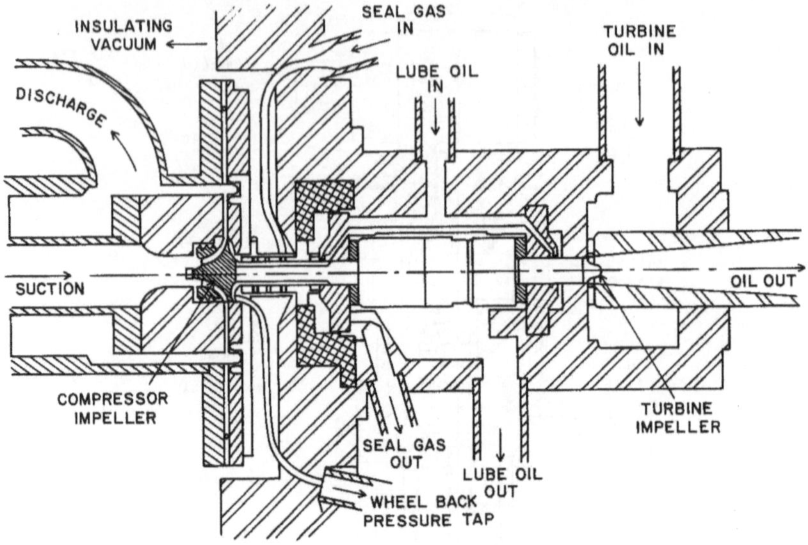

Fig. 2. Schematic of the small oil-bearing cold compressor

As shown in Fig. 2, the compressor impeller is mounted on a hollow section of shaft extension from the warm bearing section. The shaft extension passes through a labyrinth seal, part of which is tapered.

Ambient temperature helium, known as seal gas, is introduced at an intermediate point along the labyrinth to prevent loss of process gas to the warm end and oil migration to the process. The driving turbine is located on the warm end of the shaft. Lubricating oil is supplied to both the journal and the thrust bearings. A skid mounted oil system is used to provide temperature and pressure controlled oil to the turbine and the bearings.

The seal gas pressure within the labyrinth is regulated to be slightly higher in pressure than the compressor impeller back pressure and to maintain this differential as the back pressure varies. The seal gas exits the compressor casing along with the return lubricating oil at a regulated pressure slightly less than the impeller back pressure. This control is to maintain the balance between the process pressure back of the impeller, the seal gas within the labyrinth and the lubricating oil within the adjacent bearing. The bulk of the seal gas is stripped of oil and aerosols and is returned to the refrigerator main compressor suction together with oil vapor. Since the main compressors are of the oil flooded screw type, final cleanup occurs within the compressor station oil removal equipment.

A second helium recovery system, at the 24.8 KW refrigerator, is used to collect the helium gas within the low pressure oil lines and sump. This gas is returned as well. The normal seal gas requirement for a small compressor is about 0.3 g/s (3.5SCFM) and 1.7 g/s for the large machines described below.

In addition to an interlock for low labyrinth seal gas temperature, the compressor will be shutdown for low seal gas differential pressure, low oil pressure, high oil temperature and overspeed. The large turbo compressors are also monitored for shaft vibration. Warning is given at 1.5 mil radial vibration and shutdown occurs when the vibration amplitude reaches 2 mil.

Table 3. Specification for cold compressors C1 and C2

	C1	C2	
Inlet pressure	0.095	0.34	atm
Inlet temperature	2.46	4.58	K
Outlet pressure	0.35	1.389	atm
Outlet temperature	4.72	9.18	K
Mass flow rate	324.0	945.9	g/s
Shaft work	3561.0	21737.0	W
Volume flow rate	153	236	L/s
	325	500	ACFM

Table 4. Specification for cold compressor C3

Inlet pressure	4.15	atm
Inlet temperature	3.47	K
Outlet pressure	5.45	atm
Outlet temperature	3.76	K
Mass flow rate	4054	g/s
Shaft work	5812	W
Volume flow rate	27.4	L/s
	58	ACFM

As can be seen, the oil-bearing system requires considerably more ancillary equipment than the gas bearing system. However, the former provides a stiff bearing system which also is used for large systems. Also the machine themselves are easily and quickly repairable using spare impellers and stock replacement bearing parts.

The cold compressor has been run on the refrigerator for approximately one month. Generally speaking, the unit performs well mechanically even with liquid helium carried into the suction and produces a good head. The adiabatic efficiency is estimated at 50% when there is no liquid present in the suction. Detailed description is given in Ref. 4.

LARGE OIL-BEARING COLD COMPRESSORS

Three large oil-bearing cold compressors were supplied by the Rotoflow Corp. They are major components in the cold end of the 24.8 KW BNL refrigerator and are shown in Fig.3. Cold compressors C1 and C2 are designed to pump the helium pots to 0.1 and 0.35 atm corresponding to 2.49 and 3.27 K respectively. C3 is designed to circulate 4054 g/s of supercritical helium through the load. In Table 3 and 4, the specifications for C1, C2 and C3 are given.

The requirements for the C1 and C2 compressors were combined in a single machine, Figs. 4a and 4b, with the low pot suction entering axially and the intermediate pot suction introduced radially interstage. A total of four stages, two per compressor, were employed so that low Mach numbers are maintained throughout the entire compressor. This design was employed to eliminate the need for a subatmospheric shaft seal at the C1 discharge. Furthermore, the thermal losses caused by seal gas leakages and heat conduction are minimized.

The design and construction of this complex compressor was very difficult and a number of modifications were required. Great emphasis was placed on designing a shaft and impeller system with as high a critical speed as could be achieved. Especially so, because of the overhung wheel

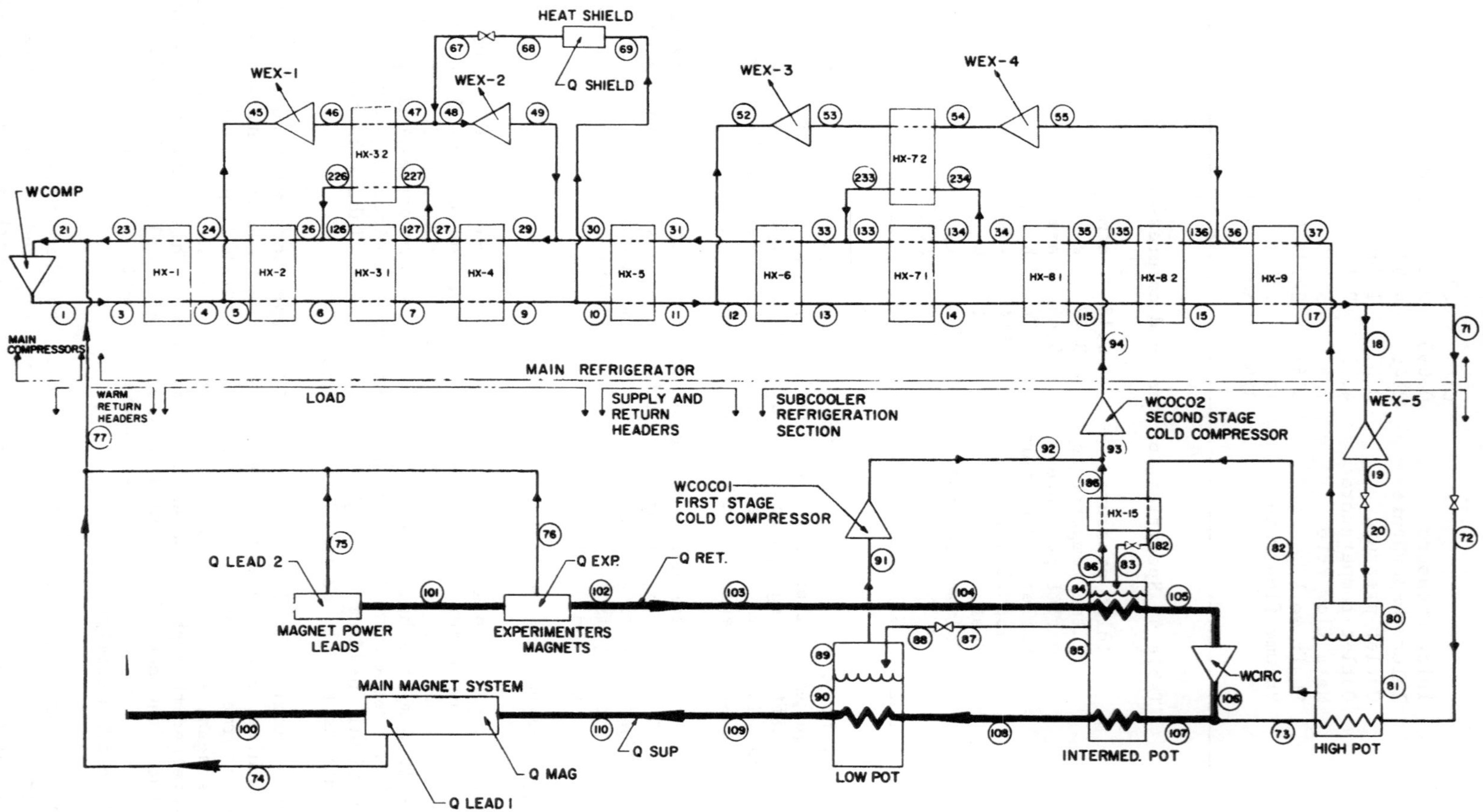

Fig. 3. Flow schematic of the 24.8 KW BNL refrigerator

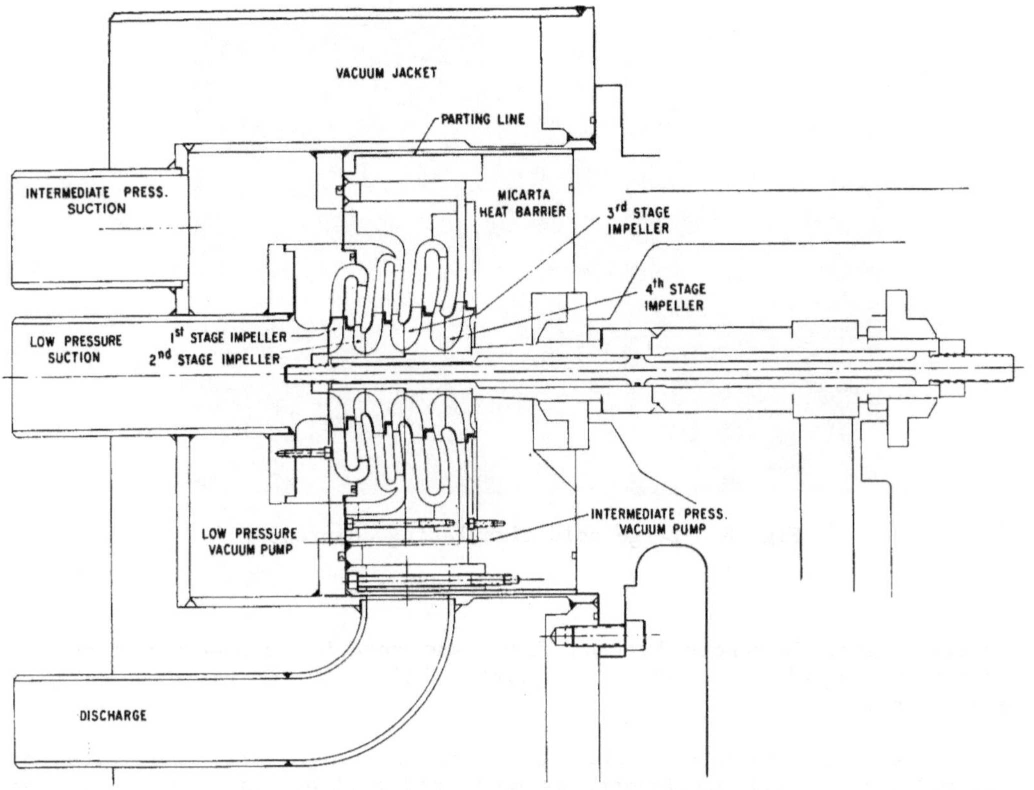

Fig. 4a. Schematic of the four-stage C1/C2

Fig. 4b. Photograph of C1/C2

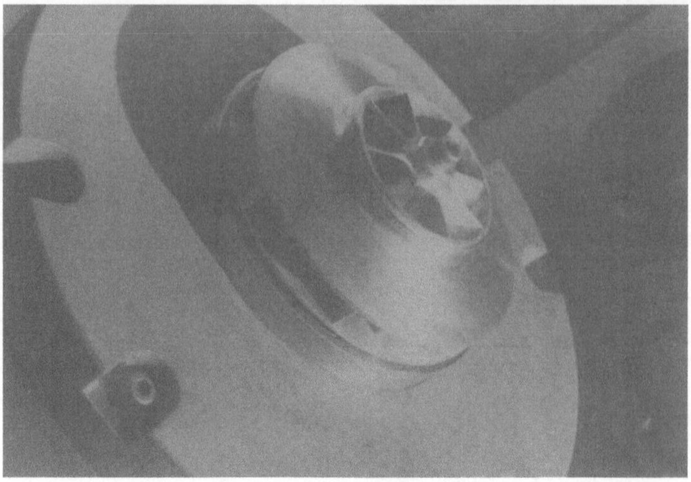

Fig. 5. Large cold circulating compressor

design. The clearances between fixed and rotating components must, of necessity, be close. Also a labyrinth shaft seal is employed at each interstage.

Vibration sensors are adjacent to each bearing. The same values of 1.5 mil and 2.0 mil amplitudes, as mentioned earlier, are used to signal warning and shutdown.

The design speed for this unit is 16000 rpm and the impeller diameters vary from 10.9 to 17.5 cm. This unit has been specified with +10 and -40 percent range to cover other modes of refrigerator operation. Valved bypasses have been installed to prevent surge. The pressure in the helium pots is controlled by a combination of compressor speed and flow bypassing.

C3, as shown in Fig.5, is a single-stage centrifugal unit of conventional design with a 10.4 cm diameter wheel operating at 11000 rpm. It is designed with a safe working pressure of 20 atm. All components of the helium magnet loop are designed to this pressure to allow for process excursions above the normal 5 atm operating pressure without venting. Magnet quenches usually produce pressures in excess of 20 atm and the quantity of gas associated with the excess is vented.

The C3 compressor was never subjected to this system behavior during the performance test. The small circulating compressors, however, were involved with many such events without developing mechanical difficulty.

Both C1/C2 and C3 use the spiral-groove combined journal and thrust bearing described previously. Each unit is driven with a variable speed DC motor with an integral gear drive with a ratio about 10:1. Neither C1/C2 or C3 has been heat stationed.

The cold compressors have been operated near 4K for a total running time of approximately one month during 1985 and 1986 without mechanical failure. Both the C1/C2 and the C3 compressors were disassembled and inspected at the conclusion of the tests. No sign of wear of labyrinth seals and bearings nor contact marks were apparent in either machine with the exception of signs of minor heating and wear of the ambient temperature labyrinth shaft seal.

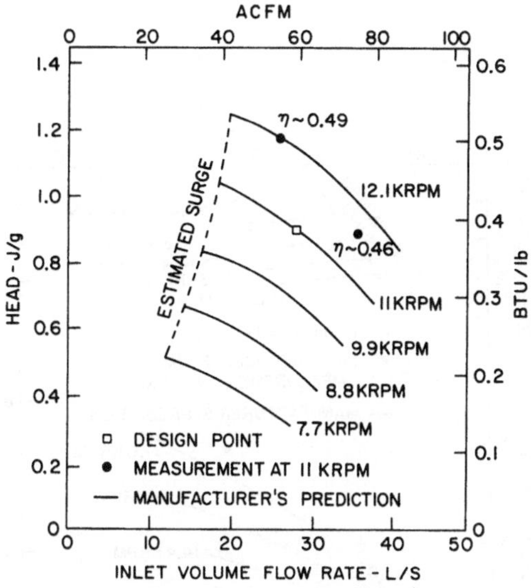

Fig. 6. Head versus volume flow rate for C3

The adiabatic efficiency of these cold compressors, defined as the ideal enthalpy rise over the actual enthalpy rise, is of importance and an effort was made to evaluate the efficiency of these compressors.

Variation of speed from 5000 rpm to 11000 rpm over a few sets of system resistance were performed for the C3 unit. Based on straight calculation from the measured inlet and outlet pressure and temperature, C3 has an average efficiency of 46% at 25.5 L/s flow rate and 1.19 J/g head at 11000 rpm. For 35.4 L/s flow rate and .88 J/g head rise, the efficiency is 43%. Energy balance over the cold end of the refrigerator were performed to check the compressor input work from the straight calculation, it is found approximate 3% should be added to values given above. The design efficiency for C3 is 0.65 at 27.4 L/s with a head rise of 0.88 J/g. A head versus volume flow rate map with average efficiency is shown in Fig.6. Two sets of data are shown in Fig. 7.

The efficiency for C1 and C2 units are hard to evaluate as the two units are mounted on the same shaft and there are no temperature sensors

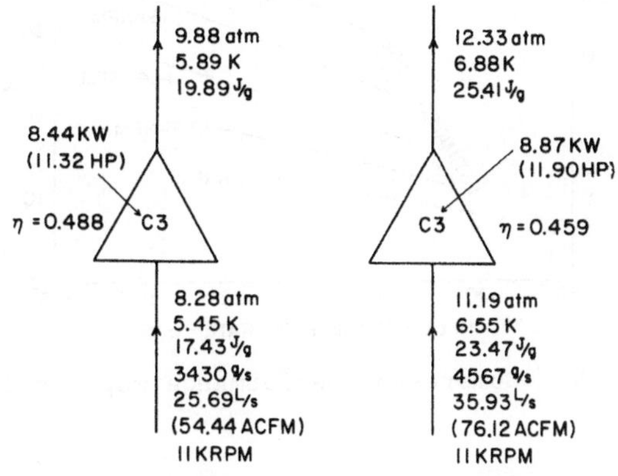

Fig. 7. Two operating conditions for C3

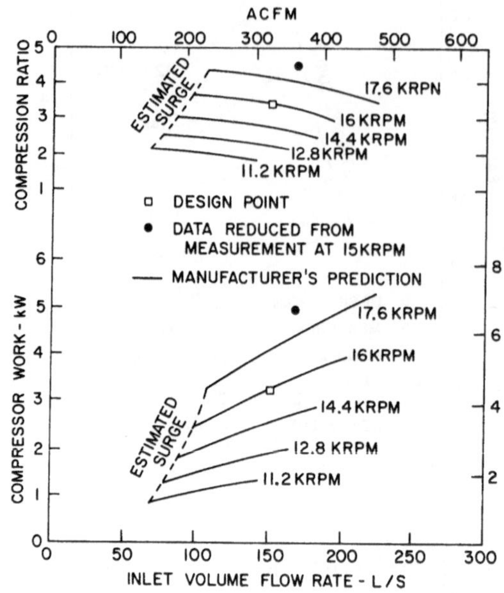

Fig. 8. Compressor performance map for C1

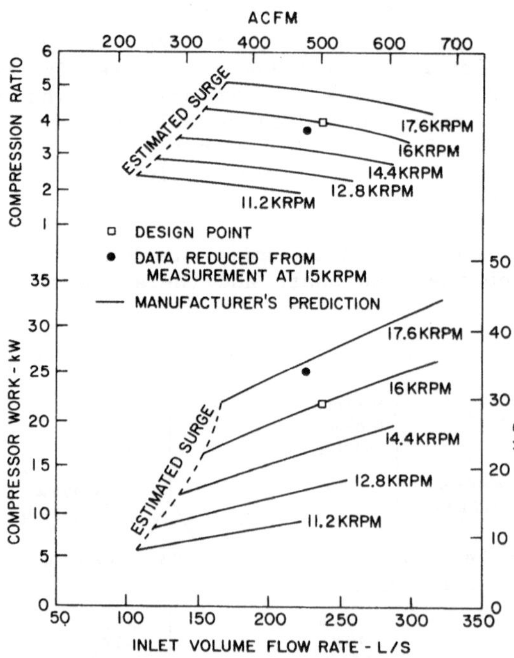

Fig. 9. Compressor performance map for C2

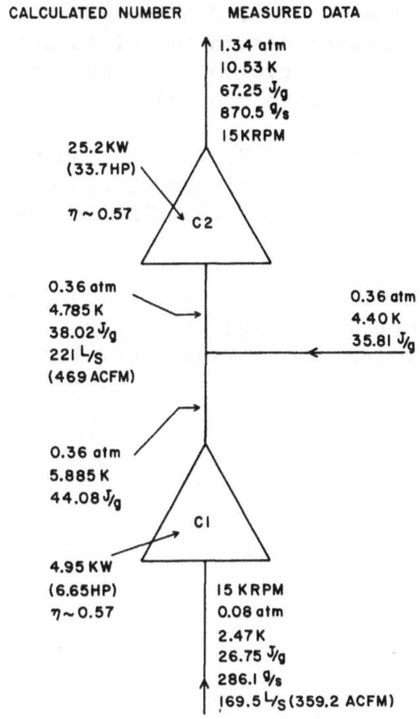

Fig. 10. One operating condition for the C1/C2 unit

available in the compressor interstages. There is, however, a temperature sensor located at the suction inlet which allows one to calculate the compressor efficiencies assuming the same efficiency for C1 and C2. A set of data taken at 15000 rpm with the inlet pressure for C1 at 0.08 atm (2.48 K) suggests that the efficiencies for C1 and C2 are approximately 57%. The design efficiency for C1 and C2 are 0.65 and 0.67 at 16000 rpm. The compression ratio and input work versus inlet volume flow rate for C1 and C2 are given in Fig.8 and 9. A typical set of data is shown in Fig. 10.

CONCLUSION

Cold compressors are important components in the helium system for superconducting application today. BNL has successfully operated and tested six cold compressors. It is not really difficult to operate these machines, but good training for operating personnel and sensible engineering practice is needed. In our experience, the efficiency of these units may not be as high as the suppliers predict, suitable margin should be taken in the system calculation.

REFERENCES

1. W. Swift, H. Sixsmith and A. Schlafke, A small centrifugal pump for circulating cryogenic helium, in: "Advances in Cryogenic Engineering," Vol. 27, Plenum Press, New York (1982), p. 777.

2. K. C. Wu, D. P. Brown, A. P. Schlafke and J. H. Sondericker, Helium refrigerator with features for operation at supercritical pressure, in: "Advances in Cryogenic Engineering," Vol. 29, Plenum Press, New York (1984), p. 495.

3. T. Jasinski, W. Dodd Stacy, S. C. Honkonen and H. Sixsmith, A generic pump/compressor design for circulation of cryogenic fluids, in: "Advances in Cryogenic Engineering, " Vol. 31, Plenum Press, New York (1986), p. 991.

4. A. P. Schlafke, D. P. Brown and K. C. Wu, Combined cold compressor/ejector helium refrigerator cycle, in: "Advances in Cryogenic Engineering," Vol. 29, Plenum Press, New York (1984), p. 487.

5. D. P. Brown, A. P. Schlafke, K. C. Wu and R. W. Moore, Cycle design for the Isabelle helium refrigerator, in: "Advances in Cryogenic Engineering," Vol. 27, Plenum Press, New York (1982), p. 501.

THE 300 W - 1.75 K TORE SUPRA REFRIGERATOR

COLD CENTRIFUGAL COMPRESSORS REPORT

Guy M. Gistau, Yves Pecoud, and Alain E. Ravex

L'Air Liquide Advanced Technology Division
Sassenage, France

ABSTRACT

The cold centrifugal compressors which are part of the TORE SUPRA refrigerator were designed and manufactured by L'AIR LIQUIDE. They were tested at nominal conditions on a special test rig. The refrigerator is now installed and all acceptance tests have been completed. Other tests were carried out off design conditions with the machines installed in the TORE SUPRA refrigerator.

INTRODUCTION

Obtaining superfluid liquid helium, (1.75 K), requires a pumping system, since it is necessary to create and constantly maintain a pressure of around 1.1 kPa above the liquid bath, whilst evaporation of the liquid helium produces a gas flow of 15 to 20 g/s (TORE SUPRA, CEN-CADARACHE, FRANCE).

Technical and economics considerations indicate an advantage if the pressure is increased at low temperature as opposed to pumping at room temperature. This is why we conceived, created and tested compression stages functioning at low temperature and pressure. These stages can operate in series, the inlet conditions of the second stage being equal to the discharge conditions of the first stage.

The conception of each stage was carried out with the aid of a mathematical model which was perfected with the O.N.E.R.A. (Office National d'Etudes et de Recherches Aerospatiales) - National Bureau of Aerospace Research and Design - with intermediate experimental testing. These tests were carried out in a test loop operating at design conditions. Further testing was undertaken in the TORE SUPRA refrigerator installation again under design conditions in order to extend our knowledge concerning cold compressor characteristics.

REALIZATION OF THE COLD PUMPS

Initially, a theoretical study was carried out not only to calculate the two cold pumping stages to be integrated into the TORE SUPRA installation, but to enable us to carry out ourselves a calculation of com-

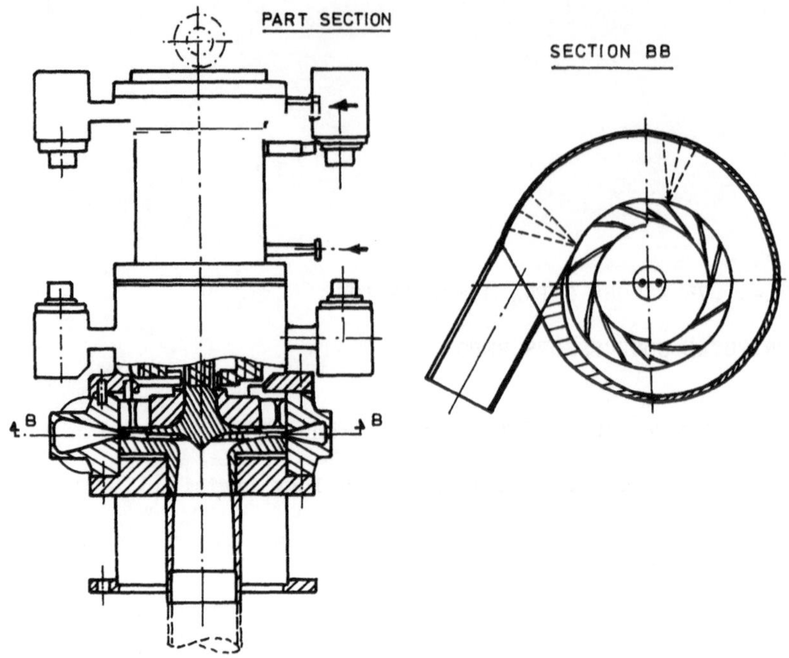

Fig. 1. Cold compressor and wheel.

plete compression systems operating at conditons different from those in the TORE SUPRA project. Secondly, a design was carried out leading to the manufacture of one type of cold pump.

The resulting pumping stages can be considered under two headings:

- compressor (parts in contact with the fluid pumped)
- bearings and drive system

Compressor

This sub-assembly comprises : a shaft, a wheel, a diffuser and a volute casing, all these components being designed for optimum performance at the required operating conditions. The compressor is shown in Fig. 1.

The model used allows us to calculate the variation of conditions of the gas traversing the wheel and diffuser. Thus, assuming certain values of gas velocity at the inlet, the model allows us to predict pump performance (efficiency and compression ratio) corresponding to the inlet conditions (temperature, pressure, flow rate) which we define. The model takes into account the different aerodynamic losses encountered in the rotor and the diffuser. Furthermore, it determines the profile to be given to the fixed and rotating vanes. Thus the coordinates of the blade shape are calculated directly, permitting fabrication by numerically-controlled machines.

Drive shaft. The material chosen for the construction of the shaft had to have a low thermal contraction. For reliable operation the natural critical frequency of the shaft had to be greater than the maximum operating rotational speed of the machine. A calculating program allowed us to determine the natural vibration and to visualize the deformation of the shaft.

The wheel. The open wheel is constructed in an alloy having good mechanical and thermal properties at low temperature. The profile of the blading is calculated to give a desired compression ratio under nominal flow conditions whilst preserving a maximum isentropic efficiency. The wheel and its associated bearings are precision-balanced.

Diffuser. The diffuser is concentric to the wheel and contains fixed blading. Its role is to convert the kinetic energy of the gas leaving the wheel into pressure by means of the blading which is designed to minimise irreversible losses (turbulence etc...)

Volute casing. The volute casing is the duct carrying the gas from the outlet of the diffuser to the compressor discharge pipe. Its role is to smoothly reduce the velocity and direction of the gas leaving the diffuser.

Magnetic bearings and drive motor/bearing assembly

This consistes of (1-) a central stator fixed with respect to the pump housing. It contains the motor windings, detector coils, the journal bearing coils and the magnetic thrust bearing. It also houses standby roller bearings for radial and axial support in case of total magnetic bearing failure. These roller bearings can support three or four touchdowns at full speed without intermediate inspection. The bearing is shown in Fig. 2. and (2-) a rotor, containing the magnetic circuit for the journal and thrust bearings and (3-) a shaft joins the motor to the wheel of the cold pump.

This mobile assembly (rotor, shaft, wheel) shown in Fig. 3. has a different dynamic behaviour from that of a machine with mechanical bearings ; in the case of magnetic bearings the electronic surveillance gives greater precision with respect to vibration control.

Frequency converter. The stator is fed by a frequency converter ; the speed is defined by the refrigerator's automatic control system.

Electronic controller. The position of the rotor with respect to the detector coils is adjusted by the electronic controller acting on the current flowing through the electro magnets. The power supply to the frequen-

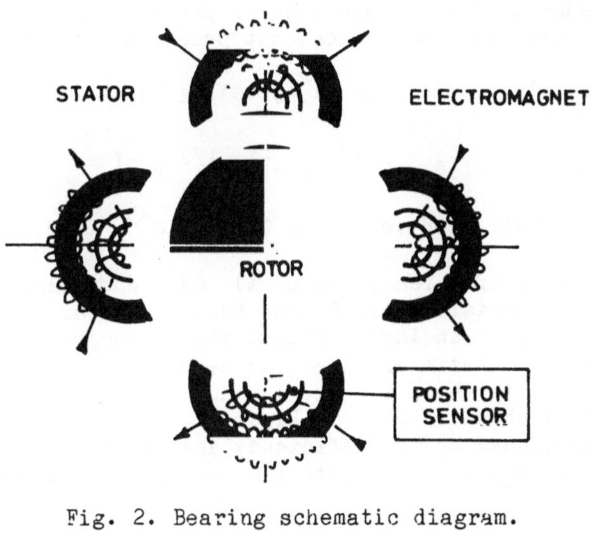

Fig. 2. Bearing schematic diagram.

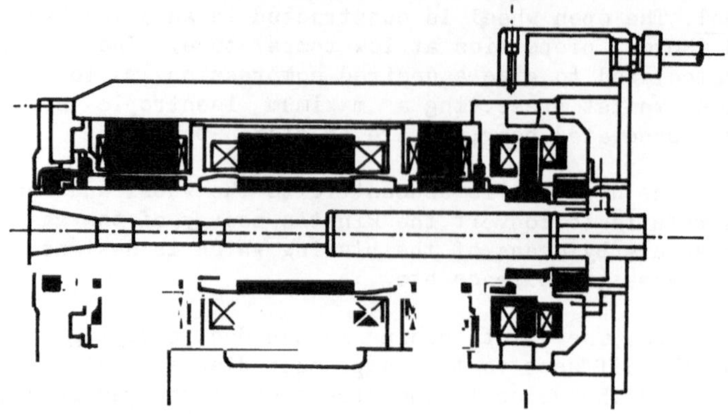

Fig. 3. Electrospindle.

cy converter is also controlled by the electronic system. The system includes a backup battery in case of main supply failure ; the machine can be brought to a halt without incident. In addition, a rotating braking field can, if required, increase the deceleration of the motor in the event of power supply failure. An axial control allows for the clearance between the compressor wheel and the housing to be adjusted to give optimum performance.

RESULTS

Test bench

For the acceptance of the machine for the TORE SUPRA project, the objective was to verify the following points : (1-) Calculated efficiency and compression ratios and (2-) Validity of irreversible loss calculations (thermal losses).

The tests allowed us to plot the compressor characteristics, to improve our knowledge of the optimisation of parameters and thus to have a reference frame, not only theoretical but practical for future projects.

Low pressure stage. The nominal operating point for TORE SUPRA is when :

$$Q\sqrt{T}/P = 2.56 \times 10^{-5} \text{ kg s}^{-1} \text{ K}^{1/2} \text{ Pa}^{-1}.$$

The values investigated vary from 2.24×10^{-5} to 2.78×10^{-5}.

The maximum values of efficiency (52.8 %) and compression ratio (4.1) attained were quite satisfactory. Since these values are greater than theoretical predictions, an investigation of the compressors' characteristics with respect to rotational speed was carried out. The efficiency versus rotational speed is shown in Fig. 4. and the compression ratio versus rotational speed is shown in Fig. 5. Figure 1. indicates an optimum speed but even at much-reduced speed, efficiency and compression ratio are acceptable and the energy injected into the cycle is correspondingly low (318 W)

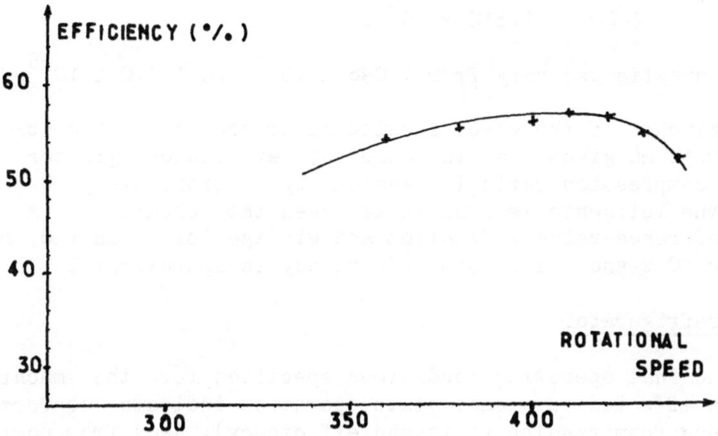

Fig. 4. Cold compressor efficiency as a function of rotational speed.

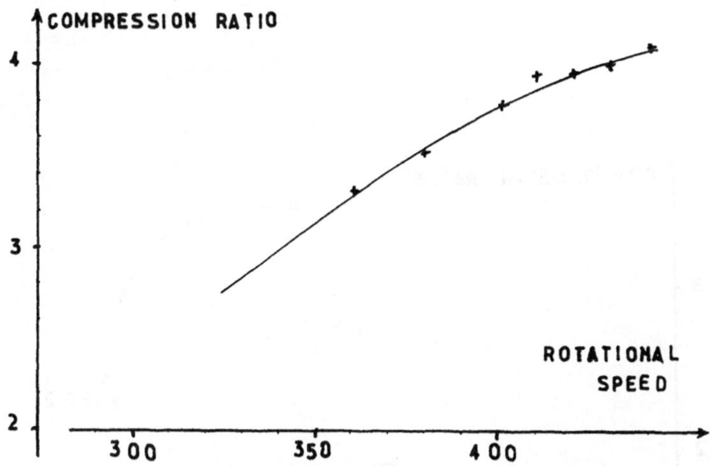

Fig. 5. Compression ratio of a cold compressor as a function of rotational speed.

The discrepancy between theoretical and measured performances arises from two principal sources : (1-) the reduction of running clearance between the wheel and its housing which can be adjusted for optimum efficiency and (2-) a high motor efficiency (close to 0.8)

High pressure stage. The nominal operating point for TORE SUPRA is when

$$Q\sqrt{T}/P = 1.212 \times 10^{-5}.$$

The values investigated vary from 1.096×10^{-5} to 1.410×10^{-5}.

The influence of the wheel running clearance is considerable ; a reduction of 0.12 mm gives an increase of efficiency greater than 3.5 points, the compression ratio increasing by approximately 0.2. On the other hand the influence is less marked when the clearance is increased above the reference value : friction and windage losses in the motor are around 60 to 80 W and drive motor efficiency is approximately 78 to 80 %.

TORE SUPRA refrigerator

Under nominal operating conditions specified for the machines, the measurements effected in steady state operation indicate conformity with specifications (compression ratio and efficiency). Away from nominal conditions (flowrate, rotational speed) it is difficult to apply this theory. In order to understand more clearly, complementary trials were undertaken at TORE SUPRA with the aim of plotting compressor characteristics (compression ratio as a function of $Q\sqrt{T}/P$ at various speeds). The experimental results form the basis of a mathematical model currently being studied. These trials allowed us to draw graphs showing for the compressors tested their compression ratio as a function of $Q\sqrt{T}/P$ for diverse values of ξ (ξ is the ratio $ND/\sqrt{T}$, N is the rotational speed and D is the mean diameter of the wheel). These graphs are shown in Fig. 6. and Fig. 7.

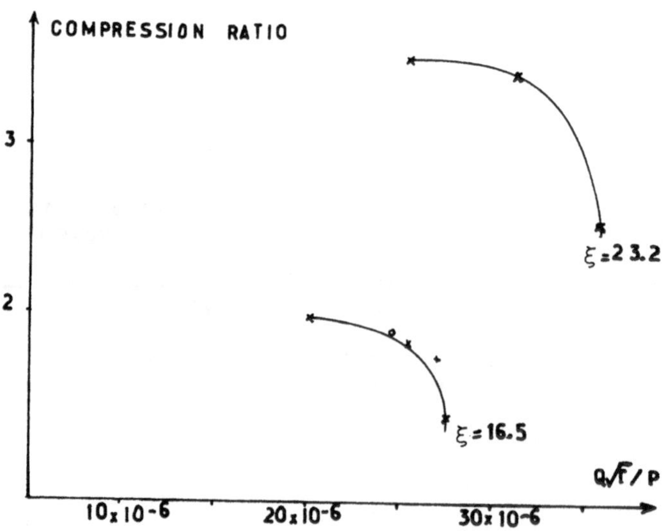

Fig. 6. Compression ratio as a function of $Q\sqrt{T}/P$ for the first compression stage in Tore Supra.

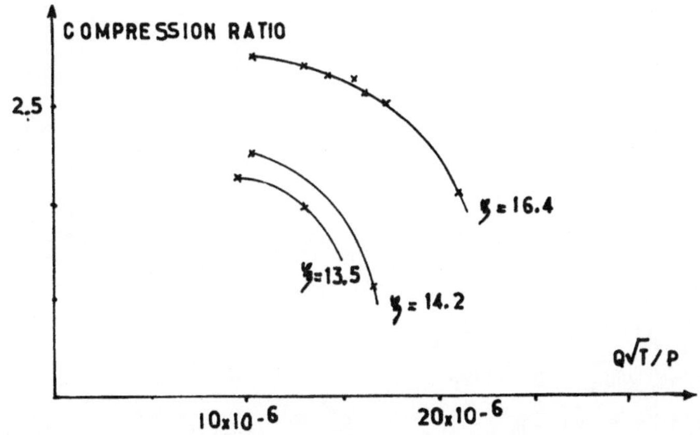

Fig. 7. Compression ratio as a function of $Q\sqrt{T}/P$ for the first compression stage in Tore Supra.

CONCLUSIONS

The results obtained both on the test bench and in the TORE SUPRA refrigerator are very encouraging. They have allowed us to acquire the knowledge necessary to produce cold compressors for future applications (in low temperature helium technology). The eventual aim would be to eliminate the need for a pumping system operating at room temperature.

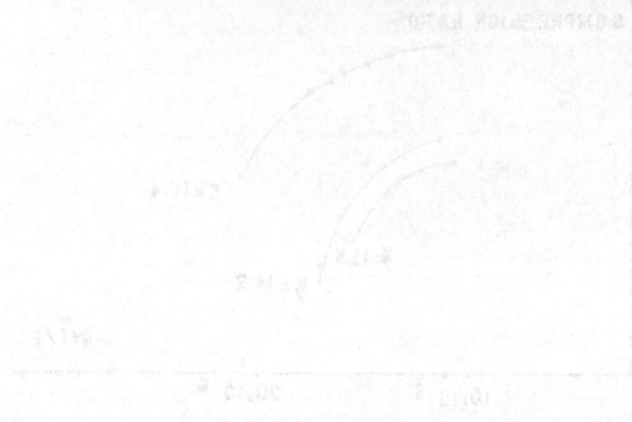

REFRIGERATION FOR ELECTRONICS: SUMMARY OF A PANEL SESSION

Randall K. Kirschman
Physicist in Private Practice
Mountain View, California

Peter J. Kerney
CTI-CRYOGENICS
Waltham, Massachusetts

Ralph C. Longsworth
APD Cryogenics Inc.
Allentown, Pennsylvania

Jon R. Olson
ETA Systems, Inc.
St. Paul, Minnesota

Robert V. Schurter
Koch Process Systems
Westborough, Massachusetts

ABSTRACT

The use of cryogenic temperatures is an option being considered more frequently as a means to enhance the performance of electronic systems. A basic issue in obtaining the benefits of lower operating temperature is integrating the cooling system with the electronic system——what does the electronics community need and what can the cryogenics community provide? This panel will address the engineering issues involved, with emphasis on refrigeration for computer systems operating in the liquid-nitrogen temperature range.

INTRODUCTION

This is a summary of a panel session whose topic was cryogenic refrigeration for large semiconductor-based electronic systems, specifically computers. The coordinator, R. Kirschman, gave a brief overview of the reasons for interest in operating computer systems at low temperatures. Panel members P. Kerney, R. Longsworth, and R. Schurter gave presentations on refrigeration issues related to cooling electronic systems; J. Olson described the new liquid-nitrogen cooled ETA[10] supercomputer.

WHY LOW TEMPERATURE?

Low-temperature operation has been successfully applied to a variety of electronic devices and systems for a variety of reasons. For computers the primary motivation is increased performance, mainly faster operation. The usual concept is a computer system based on CMOS integrated circuits operated at liquid-nitrogen temperature (77 K).

Increasing the overall speed of a digital integrated circuit involves two ingredients: faster individual transistors, and faster interconnections. Lowering a MOS transistor's temperature to the LN_2 range increases its switching speed by about a factor of two, because thermal disturbances to the flow of electrons or holes are reduced, and the carrier mobility is increased. Fast communication between transistors depends on short, low-resistivity interconnections. Lowering the temperature can be beneficial in several ways: resistivity of the metallizations used for interconnections can be lowered, and better materials may be used since electromigration is greatly reduced. The length of interconnections can be decreased because limitations on packing density of transistors, imposed by power dissipation and heat removal, can be alleviated. Thermal conductivities of materials used in integrated circuits and their packages increase at LN_2 temperature; furthermore, digital circuits can, in theory, be operated at lower voltage and thus lower power.

There are other advantages (as well as disadvantages) to operating electronic circuits at low temperatures, but those described above are the primary ones proposed. To obtain the benefits of lower-temperature operation, careful engineering is needed to interface the electronic system to a refrigeration system.

REFRIGERATION FOR COMPUTERS

A wide range of cooling schemes and proven technology is available to the electronic design engineer. The selection among existing alternatives depends on a number of factors and tradoffs; the primary ones are presented below.

Method of Refrigeration

Possible cooling approaches include: open-cycle immersion, closed-cycle mechanical refrigeration (recuperative and regenerative expansion systems), adiabatic demagnetization, and thermoelectric cooling.

The most commonly proposed approach is some type of closed-cycle mechanical refrigerator, perhaps sustaining a liquid-cryogen bath in which the electronic system is immersed. For small loads, less than a few hundred watts (at 77 K), the Gifford-McMahon cycle is considered suitable; for larger loads, the Brayton or Stirling cycles. One would probably want to locate the compressor, and possibly the expander, remote from the electronic system. The system designer must consider the expander's size, noise level, vibration, and electromagnetic interactions. Other factors to consider are the heat-rejection scheme and the support utility requirements.

In designing the cooling system it may be desirable to include sufficient storage capacity to allow continued operation of the computer (perhaps in a reduced capability mode) during refrigerator repair or maintenance.

Temperature

Electronic systems with low power dissipation can be conductively cooled. The system can be interfaced directly to a refrigerator in a vacuum and operated at any temperature within the refrigerator's range.

Immersion cooling, although at a fixed temperature determined by the cryogen, has several advantages: high heat transfer, uniform temperature, and thermal inertia. Cryogens which are non-toxic and non-combustible are helium (4 K), neon (27 K), nitrogen (77 K), and argon (87 K). In addition to their temperatures, their heat transfer characteristics and ability to store refrigeration are also important.

In general, the higher the temperature, the better in terms of complexity, reliability, and efficiency.

Capacity

In determining the required capacity of the refrigerator, it is necessary to evaluate parasitic heat loads in addition to that dissipated by the electronic circuits.

One may want a "standard" refrigerator that has a range of capacities to accomodate a computer product line that has a range of capabilities (and thus a range of power dissipations). Also, the power dissipation of a given computer (and the required refrigerator cooling capacity) may change if it is turned down at night or over weekends. For these situations one can choose between a large refrigerator that can be turned down, or small modular units that can be added and turned on or off incrementally to match the heat load.

Reliability and Maintenance

This is an area of central concern. To the computer user the important question is how frequently and for how long the computer system is "down" as a result of cooling system malfunction or maintenance. Relevant factors are the refrigerator reliability, time to perform servicing, and thermal storage of the cooling system allowing the computer to continue operation when the refrigerator is off. Service strategy is important and includes designs that permit rapid replacement of parts, an organized preventive maintenance program, monitoring for degradation, spare parts inventory, and a responsive service organization.

Cost

In addition to the initial capital cost, there are the costs of utilities, maintenance and service. Capital costs include the refrigerator, cryogen container or vacuum housing, interfacing the refrigerator and electronic system, installation of equipment and utilities, and spare components. Utilities include electricity and possibly cooling water or air conditioning. Service costs include replacement parts, labor, and travel.

In summary, cooled electronic systems are commercially viable and the basic elements exist for developing suitable cryogenic systems for such equipment. The optimum design for a given application requires careful engineering and tradeoffs among the factors discussed above.

ETA COMPUTER SYSTEM

An example of a commercial computer system incorporating cryogenic cooling is the ETA[10] supercomputer developed and manufactured by ETA Systems, Inc. The central processor boards (from 2 to 8), each with about 240 CMOS integrated-circuit chips, are cooled by direct immersion in liquid nitrogen. Cryogenic cooling of the CMOS logic chips provides two benefits: the primary one being increased switching speed (twice the clock speed of an air-cooled system), the second being increased reliability associated with lower semiconductor-device junction temperatures. The integrated circuits are based on room-temperature technology and the system will operate without cooling to LN_2 temperature, at reduced speed.

The development of the cryogenic cooling system began in 1983, with two approaches being investigated: forced convection using a low-temperature gas, and immersion in a cryogenic fluid. Since performance is a primary goal, it is desired to maintain the semiconductor-device junction temperatures as close to 77 K as is reasonable. Because of the heat transfer advantages associated with nucleate boiling and other system-level considerations, the immersion approach was chosen.

The cooling system is made up of two sub-systems: a LN_2 distribution system and a cryogenic refrigeration system. The distribution system comprises a storage dewar (nominal capacity 380 liters), a LN_2 pump as the prime mover, vacuum-jacketed lines for distribution, and a cryostat for immersion of the central processors. The heat from the electronics is exchanged with the LN_2 in the cryostat and the resulting two-phase nitrogen is returned and separated in the storage dewar. The GN_2 is then recondensed by the refrigeration system and returned to the storage dewar as liquid. Each processor board and associated system losses requires about 1 kW of refrigeration at LN_2 temperature.

Several alternative refrigeration systems have been under development at ETA Systems in cooperation with Koch Process Systems and CVI, Inc. One is based on the Brayton cycle using a turbine expander, another is based on the Stirling cycle. Each system has its own set of advantages and disadvantages. The Stirling cycle design, that being the Philips cryogenerator, is presently being used in the ETA[10].

It has been a three-and-one-half year development cycle for the ETA[10] from concept to product. ETA installed its first cryogenically cooled supercomputer in December 1986 at Florida State University; the next delivery is slated for mid-1987 at the German Weather Bureau.

QUESTIONS AND DISCUSSION

During the question and discussion periods, there was much interest in the ETA system, as well as questions on low-temperature electronics in general and the possible impact of the new high transition-temperature superconductors. There was also considerable discussion of the characteristics of various refrigeration approaches and commercial equipment, particularly relating to reliability and maintenance.

BIBLIOGRAPHY

"Low-Temperature Electronics," R. K. Kirschman, ed. IEEE Press, New York (1986).

R. C. Longsworth and W. A. Steyert, Technology for liquid nitrogen cooled computers, IEEE Tr. on Electron Devices ED-34:4 (1987).

"Building the new standard in supercomputers," I/O (ETA Systems, St. Paul, Minnesota), 4:2 (May 1987).

R. C. Longworth and W. A. Sievert, Technology for liquid nitrogen cooled computers, IEEE Tr. on Electron Devices ED-34:4 (1987).

"Building the new standard in supercomputers," ETA ETA Systems, St. Paul, Minnesota), 1:2 (May 1987).

4 K GIFFORD MC MAHON/JOULE-THOMSON CYCLE REFRIGERATORS

Ralph C. Longsworth

APD Cryogenics Inc
Subsidiary of Intermagnetics General Corporation
Allentown, Pennsylvania

ABSTRACT

4 K Gifford McMahon (GM)/Joule-Thomson (J-T) cycle refrigerators are the most commonly used type of closed cycle refrigerator for applications requiring 1 to 4 watts of refrigeration. This paper describes the evolution of this type of refrigerator, its characteristics, application considerations, and gives examples of current applications with emphasis on recondensing helium in Magnetic Resonance Imaging (MRI) cryostats.

INTRODUCTION

Most of the 4 K refrigerators that have been built to date in the 1 to 4 watt capacity range have consisted of a Joule-Thomson cycle 4 K cold stage precooled by a two stage Gifford McMahon cycle refrigerator. Present applications include maser cooling, general laboratory research, and MRI helium cryostat recondensers. Possible future applications include cooling of superconducting electronic devices for computers and magnetoencephalography.

SMALL 4 K REFRIGERATOR COMPARISON

Present day 4 K GM/J-T refrigerators had their origin in the late '50's and early '60's in the development of refrigerators to cool antenna mounted maser-receiver amplifiers. At the 1962 Cryogenic Engineering Conference, three papers were presented that described different closed cycle approaches to producing a few watts of refrigeration at 4 K: cascade J-T,[1] reciprocating Brayton/J-T,[2] and GM/J-T.[3]

The first two units were developed at Air Products and Chemicals, Inc. where S. C. Collins had served as a consultant and passed along his experience in the fabrication of finned tube heat exchangers and construction of the close clearance seal reciprocating pistons used in the two small Brayton expanders. Blockage of the small passages in the heat exchangers with contaminants was perceived to be the major obstacle to achieving long term operation, so diaphragm compressors were used for the cascade J-T system and dry lubricated reciprocating compressors (sealed bearings and filled Teflon piston rings) were developed for the Brayton/J-T system.

The first GM/J-T unit, built at Arthur D. Little (ADL), combined an oil lubricated refrigeration type compressor with a Gifford "gas balancing" type expander. The essential keys to the success of this unit were the technology used to remove oil from the compressed helium and the low expander speed with small pressure loading on the main seal which resulted in long seal life. The refrigeration type compressor used in this and subsequent systems enabled the production of lower cost systems than the other two types of refrigerators.

Only one of the 4 K J-T refrigerators was built, because the diaphragm compressors were large and heavy and efficiency was poor, approximately 10 kW input for 1 W of refrigeration. A number of dry lubricated Brayton cycle units were built; the first operated at 2.5 K and later units were designed for operation at 4.2 K. These units were lighter, more efficient, and better adjusted to outdoor operation than the other two systems. In later units, maintenance intervals of 3,000 hours were achieved. Early maser system reliability was achieved by having two refrigerators in each unit (one backup) which could be serviced without warming the maser. In the end, the lower cost of the GM/J-T system won the competitive battle.

The Jet Propulsion Lab (JPL) purchased a second generation design of a GM/J-T system from ADL, made some modifications to it and installed approximately 50 units in their deep space tracking network. The operation of these units was closely monitored, problems resolved, and detailed service procedures established that enabled them to report in 1980,[4] a maintenance interval of once a year with a total of over one million hours of operation.

In the mid-'70's Air Products introduced a competitive 2 W GM/J-T refrigerator which saw application in research programs. It was envisioned as being suitable for cooling the superconducting Josephson Junction device computers being developed by IBM.[5,6]

The large scale use of 4 K computers did not materialize. However, Magnetic Resonance Imaging has come into widespread use. Most of the MRI systems incorporate superconducting magnets which require between 0.5 W and 4.0 W of refrigeration at 4 K plus additional cooling at higher temperatures to cool radiation shields. MRI has spurred more refrigerator development.[7,8,9] Today there are more than six manufacturers of 4 K refrigerators in the 0.5 W to 4 W capacity range for nonaerospace applications, and they all operate on the GM/J-T cycle.

If applications develop which require 4 K refrigeration in the mW range of capacity, we may see nonmetallic Stirling cycle refrigerators pioneered by Zimmerman[10] or miniature cascade J-T coolers such as those of Little[11] being used. For refrigerant requirements in the 10 W to 100 W range, Collins type refrigerators are most commonly used.[12] This type is typified by the Koch Process Systems Model 1400 helium liquefier/refrigerator. Recent work on Claude cycle helium refrigerators which use small gas bearing expanders[13] is aimed at building a unit in the 4 W to 5 W capacity range which will be competitive with GM/J-T units. At the present time, GM/J-T units are more efficient than the miniature turbine system and less expensive than either the Collins or Claude cycle units.

GM/J-T CHARACTERISTICS

Cycle

Figure 1 is a flow schematic of the GM/J-T cycle. The basic GM refrigerator consists of the compressor and a two stage expander which

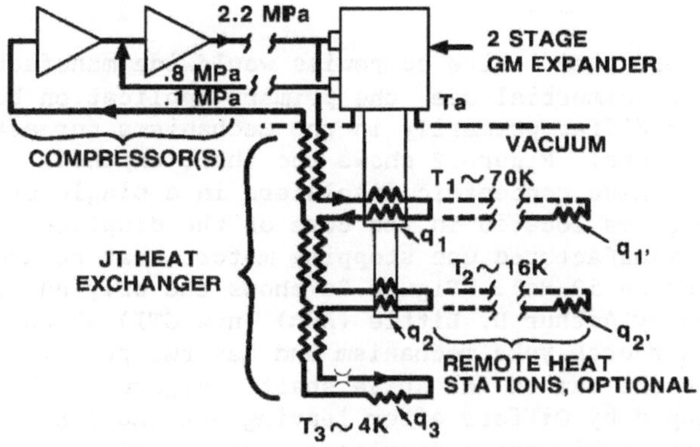

Fig. 1. GM/J-T system schematic

operates with helium at pressures in the range of 2.1 and 0.8 MPa. The expander is used to precool helium in the J-T loop which consists of a counter flow heat exchanger and another stage of compression to bring the return helium pressure at about 0.1 MPa up to the return pressure from the expander. Flow through the J-T loop is typically 5 to 10% of the second stage compressor flow. Precooling temperatures are typically in the range of 50 to 80 K at the first stage and 15 to 20 K at the second stage. It is possible to use the circulating flow of helium through the J-T loop to transport first and second stage refrigeration to remote heat stations.

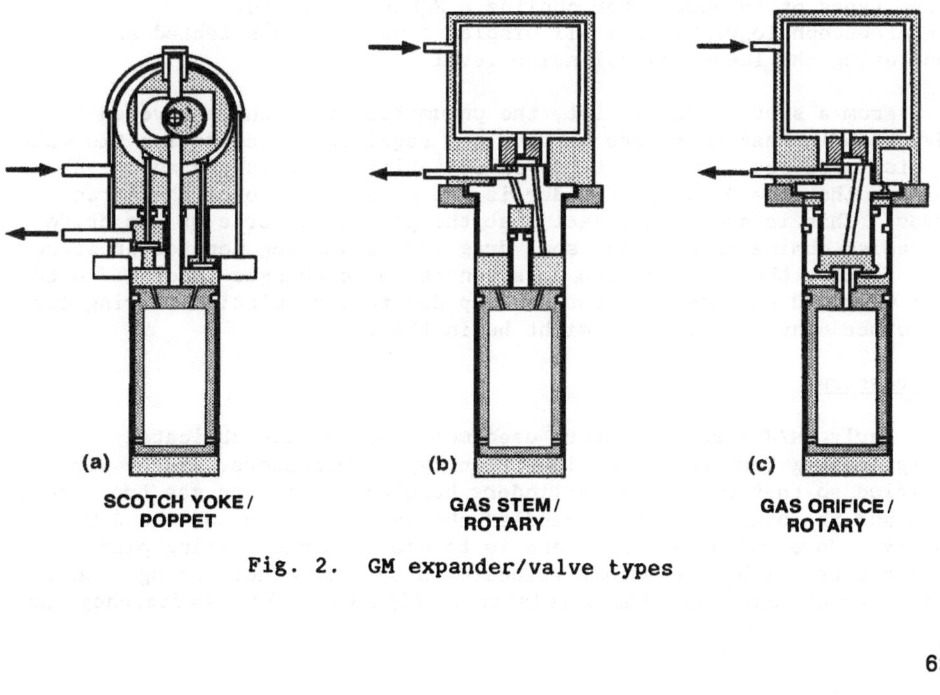

Fig. 2. GM expander/valve types

GM Expanders

There are more than twelve companies worldwide manufacturing GM refrigerators for commercial use, the primary application being cryopumps. These differ primarily in the mechanisms for actuating the displacers and valves. Figure 2 shows the three expander types that are most common. All have concentric displacers in a single stepped cylinder with the regenerators located in the core of the displacer. All units presently being manufactured use stepping motors that rotate at 1.2 Hz on 60 Hz power (1 Hz on 50 Hz). Figure 2a shows the original Cryodyne® design developed by Arthur D. Little (ADL) (now CTI) which drives the displacer with a scotch yoke mechanism and has two poppet-type valves actuated by a second cam on the drive shaft. Figure 2b shows the drive mechanism developed by Gifford after leaving ADL and starting his own company, Cryomech. This has a displacer with a small diameter drive stem that is driven by gas through a rotary valve which switches gas pressure to the drive stem out of phase with the gas flowing to and from the displacer. Figure 2c shows the drive mechanism developed by the author in the Advanced Products Department of Air Products and Chemicals, Inc. (now APD Cryogenics Inc) after having worked with Gifford at Syracuse University and Cryomech on the previous design. This design which is sold under the trade name, Displex®, also has a rotary valve, but it only switches gas flowing to and from the displacer. Displacer actuation is achieved by trapping gas above the displacer and forcing it back and forth through an orifice to a surge volume which is at an intermediate pressure. A slack cap on the top of the displacer helps to optimize the phase relation between the pressure cycle and the displacer motion.

These three types of expanders are generally competitive in cost performance and quiet operation. Differences exist primarily in areas of vibration, ease of service, and reliability. The pneumatically driven units use symmetrical valve discs which give two cycles of refrigeration per revolution of the drive motor. They also rely on proper valve timing to use gas pressure to decelerate the displacer at the ends of the stroke in combination with bumpers to cushion the residual impact. The scotch yoke drive provides essentially sinusoidal motion with no impact and thus has lower inertial vibration. Torque on the drive motor is quite cyclic for the scotch yoke mechanism, and the stepping motor imparts a higher frequency vibration that varies in amplitude. In recent tests of these three types of expanders for cooling a SQUID system for magnetoencephalography, a small Displex expander was selected as generating the lowest signal noise level.

From a service standpoint, the pneumatic drive units have an advantage in that there are no settings required for adjusting the valve or displacer position. In terms of reliability, it is the author's opinion that the Displex expander is the most reliable of the three types. This is due to the fact that the pneumatic force of the drive mechanism dominates over the seal drag forces and regenerator pressure drop forces; thus, it is much less sensitive to changes in drag due to seal wear and changes in pressure drop due to accumulation of ring dust and other contaminants that might be in the gas.

Compressors

Early GM/J-T refrigerators used multipiston oil-lubricated reciprocating commercial refrigeration type compressors. These were modified so that one or two cylinders handled the return gas from the J-T loop and the other cylinders handled the expander flow and the J-T supply. More recently, the trend is to use multiple rolling piston compressors which have become standard in the air conditioning industry. This type of compressor has a relatively high volumetric efficiency and

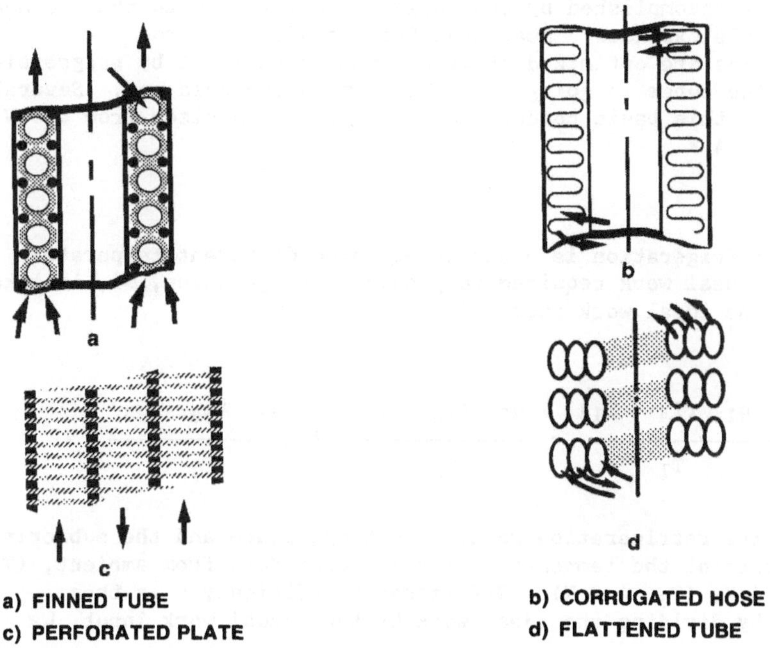

Fig. 3. J-T heat exchanger types

lends itself well to oil injection. Both of these factors make it practical to have a high pressure ratio in the J-T compressor, from 0.1 to 0.8 MPa (1 to 8 atm), which enables the GM expander to operate at a good return pressure.

Heat Exchangers

Figure 3 shows four types of heat exchangers that are in use in small 4 K refrigerators. Finned tube type heat exchangers, Fig. 3a, have historically been most commonly used in the J-T circuit. High effectiveness can be achieved with modest pressure drop loss in the return side. JPL developed a heat exchanger for their 1 W 4 K GM/J-T refrigerators, Fig. 3b, which consists of a helical corrugated tube in a sleeve with a central plug.[4] High pressure helium flows in a spiral pattern on the inside of the corrugation and low pressure helium flows in the outside spiral. This design was an attempt to reduce the cost of the heat exchangers, but we found in building 24 of these systems for JPL that each heat exchanger had to be custom machined to avoid bypassing around the convolutions. Both the finned tube and corrugated tube type heat exchangers have to be relatively long to achieve high efficiency.

Perforated plate type heat exchangers,[13, 14] Fig. 3c, have been developed to obtain high efficiency in a short length with low pressure drop. These were originally developed for systems which used turboexpanders or rotary reciprocating expanders which operate at relatively low pressure ratios and require the high efficiency low pressure drop characteristics of this heat exchanger.

In 1983 we were motivated to develop a new type of heat exchanger, Fig. 3d, which could fit in the annular space around the GM cylinder and match the temperature profile in the neck tube of a helium dewar into which it could be plugged.[7] The concept we developed was to use multiple flattened tubes bonded together and coiled around the expander cylinder. The tubes are continuous from room temperature to 4 K.

Precooling is accomplished by thermally attaching them to the two heat stations of the expander. Heat transfer and pressure drop characteristics are optimized at each temperature level by progressively flattening the tubes in going from the warm to the cold end. Several variations of this basic concept have been tried in sizes from 0.3 W to 4 W capacity at 4 K.

Efficiency

Since refrigeration is produced at three different temperature levels, the ideal work required to produce refrigeration, Wi, is taken as the sum of the ideal work required for each stage.

$$Wi = \frac{q_1 \; (T_a - T_1)}{T_1} + \frac{q_2 \; (T_a - T_2)}{T_2} + \frac{q_3 \; (T_a - T_3)}{T_3} \qquad (1)$$

where q is the refrigeration rate, T is temperature and the subscripts a, 1, 2, 3 represent the temperature levels going down from ambient, ($T_a \approx 300$ K) to 4 K, ($T_3 \approx 4.2$ K). Refrigerator efficiency η is then calculated by dividing the ideal work by the actual work input, Wa,

$$\eta = Wi/Wa \qquad (2)$$

Most GM expanders in use today reach a peak efficiency of 5 to 8% when operating at about 70 K and 20 K. This isn't much different than reported in 1966.[15] Expander efficiency drops off as the precooling temperature is reduced below 20 K, but the efficiency of the J-T loop increases. A precooling temperature of about 16 K results in a better efficiency for a GM/J-T system. The J-T heat exchanger puts a load on the expander which is proportional to the flow rate through the J-T circuit. If a 4 K GM/J-T refrigerator has been designed to maximize the refrigeration produced at 4 K, then the refrigerator efficiency will typically be in the range of 2 to 4% relative to Carnot.

Reliability

In 1975 it was reported[16] that the GM cycle refrigerators in use on parametric amplifier receivers had accumulated over three million hours of operating time and had a meantime between failure (MTBF) in excess of 20,000 hours. These refrigerators were operated by Comsat which, like JPL, had an organization that monitored the refrigerators carefully. In the early years of use, they resolved systematic type problems and instituted maintenance programs that eliminated outages due to wear and contaminant accumulation. Refrigerators were also installed in such a way that they could be serviced easily and spares were kept on hand to minimize down time. In the case of Comsat, they reported less than 1 hour per year off time due to refrigerator problems.

The author is not aware of any more recent reports of reliability. In general, the quality of equipment being built today is better than fifteen years ago. However, reliability of specific units can vary appreciably depending on product maturity, adjustments to supplier changes (especially with regards to compressors), and user care.

The additional J-T compressor in a 4 K GM refrigerator is rather lightly loaded, so its MTBF should be appreciably greater than the expander compressor. The J-T loop is more sensitive to contaminants than a GM expander so more stringent cleanup is required. Two primary areas of concern are air leaking in through fittings in the J-T return line if the pressure is subatmospheric and moisture which becomes adsorbed on interior surfaces if they are exposed to the atmosphere during service. Both of these problems can be avoided by proper precautions. In our laboratory, for example, we ran a system for over 3,000 hours at 0.5 atm absolute return pressure with no warm adsorber in the J-T loop.

One last word about reliability is to note that should a failure occur, the majority of failure modes are soft, that is performance degrades over a period of tens or hundreds of hours, so that repairs can frequently be scheduled and down time can be minimal.

APPLICATION CONSIDERATIONS

Cooldown

Flow rate through the J-T loop for a fixed J-T orifice is approximately proportional to T_a/T_3. When the refrigerator is warm, the flow rate is only about 10% of the cold flow rate, and the J-T effect for helium is positive; that is the gas exits the J-T orifice warmer than it enters. Cooldown does occur after the second stage of the expander cools down because of inefficiencies in the J-T heat exchanger, but it may be slow, i.e., several hours for a small 4 K mass.

Techniques that have been employed to decrease cooldown time include:

- Opening up a variable J-T valve.
- Adding a hydrogen-filled heat switch.
- Adding a bypass line to bring gas from the 4 K heat station direct to the first stage section of the heat exchanger.
- Adding a LN_2 cooldown circuit.

We have found in our 4 K lab systems that a variable J-T valve is preferred, and in units which are inserted into helium cryostat neck tubes, helium boil-off cools the refrigerator quickly.

Temperature Control

Over the limited temperature range of about 3 to 5 K, temperature can be set by adjusting the pressure either by means of a variable J-T valve or a pressure regulator on the return line. Temperature stability of ±0.05 K was reported[6] for a system with a pressure regulator that had an evacuated bellows to make it insensitive to changes in barometric pressure.

Conventional temperature controllers are used for maintaining higher temperatures. Consideration has to be given to the decreasing rate at which refrigeration is produced as the temperature is increased.

Thermal Coupling

The most obvious way to cool something is to attach it directly to the refrigerator. This requires that the refrigerator be warmed to service it. Hoffman[3] describes a system with redundant refrigerators which permits one to be isolated from the load, warmed, and serviced while the other keeps the load cold. We have built a system[5] which employs helium filled thermal couplings that permit passive thermal isolation from the load so the refrigerator can be warmed and serviced, but the refrigerator can't be removed.

More recently, we have demonstrated that a refrigerator which is plugged into the neck of a helium dewar can be removed, warmed, serviced, and reinstalled in about an hour with a loss of only 1.5 L of liquid helium.[8] Other means of thermal coupling are described in reference 17. One important point made in that paper is that the helium in the refrigerator should be kept separate from the helium in the dewar to avoid having contaminants block the J-T loop.

Service

Present GM/J-T systems are typically designed for service at 10,000-hour intervals. Service usually consists of replacing the piston rings (and valve disc) in the expander, replacing two adsorbers in the compressor system, and purging the J-T heat exchanger. This work can be done by trained people in the field but is more frequently done in a service center.

For applications such as MRI where it is desirable to minimize down time, service is facilitated by a) having interchangeable standard components; b) stocking spares in regional facilities, and c) having a thermal coupling mechanism that permits rapid change-out of the expander.

Magnetic Effects

A number of studies have been made to find out the effect of a magnetic field on the refrigerator[18] and the effect of the refrigerator on the homogeneity of the magnetic field in an MRI system. It was found that the standard drive motor used in most GM refrigerators can tolerate external fields of 700 to 800 gauss and with modification can operate in fields up to about 1,800 gauss. We have seen no compressor problems in fields up to 200 gauss.

Tests of GM/J-T refrigerators operating in MRI cryostats have revealed no measurable increase in field inhomogeneity.

APPLICATIONS

General

The use of an appreciable number of 4 K refrigerators, for a specific application, such as those built for Maser receivers, has been isolated. Perhaps an equal number have been built in the last decade for low temperature research and as prototypes of units for projects which may require large numbers of refrigerators in the future. The advent of high temperature superconductors has raised the question of whether or not 4 K refrigeration will be needed for some major applications in the future. Given the fact that superconducting materials always carry more current as the temperature is reduced, and the Josephson effect is greater at lower temperatures, the question about the future of 4 K refrigerators seems to come back to whether systems that require refrigerators will have better performance at lower cost at 4 K than some higher temperature.

MRI

Much work has been done in the last three years to reduce the cost and improve the reliability of 4 K refrigerators for use in recondensing helium in MRI magnets. Several different sizes and types have been developed by our group.

The first unit was designed to plug into the neck tube of a stationary magnet where it provides 3 watts of refrigeration at <20 K and 0.5 watts at 4.5 K. In the magnet cryostat, the warm shield is cooled by liquid nitrogen which in turn may be recondensed by a single stage refrigerator.

The second unit is also designed to plug into the neck tube of a stationary MRI cryostat. It provides all the refrigeration required, 50 watts at <70 K plus 3 watts at <20 K plus 0.5 watts at 4.5 K.[8] Compressor power input is 7.5 kW.

During the past year a mobile MRI magnet was fitted with a single stage GM refrigerator for recondensing the nitrogen and a GM/J-T refrigerator which plugs into the neck tube and where it provides 3 watts at <20 K to cool a cold shield and 2.0 watts at 4.3 K to recondense the helium.[19] The helium vessel is sealed so that when the magnet is charged and discharged, the helium is contained. Pressure rise in the 1-T magnet cryostat is measured to be 24.8 kPa, and recovery to within 19% of steady state pressure is measured to take 5.5 hours.

Since there are several hundred MRI magnets in service today which are refilled with helium through a standard fill port, APD Cryogenics has built two helium recondensers with 25.4 mm diameter bayonets that plug into the helium fill port. One has a net capacity of 1.5 watts suitable for most stationary MRI magnets, and the second has a net capacity of 4.0 watts, which is suitable for most mobile MRI magnets. Input power is 4.5 kW and 8 kW, respectively.

SUMMARY

The successful commercialization of superconducting magnets for MRI has spurred much work on the development of 4 K refrigerators to recondense the liquid helium. The GM/J-T cycle is becoming selected as being most appropriate for the capacity range of 0.5 watts to 4 watts which is primarily because of its cost effectiveness. Reliability and versatility in the way it can be applied are also important factors.

Recent technological advances have been made in the introduction of rolling piston compressors, development of a new compact heat exchanger, and understanding gas purification. We are optimistic about the future of these new 4 K GM/J-T refrigerators.

REFERENCES

1. P. K. Lashmet and J. M. Geist, A closed-cycle cascade helium refrigerator, in: "Advances in Cryogenic Engineering," Vol. 8, Plenum Press, New York (1963) pp. 199-205.
2. K. Zeitz and B. K. Woolfenden, A closed-cycle helium refrigerator for 2.5 K, in: "Advances in Cryogenic Engineering," Vol. 8, Plenum Press, New York (1963), pp. 206-212.
3. T. E. Hoffman, Reliable, continuous, closed-circuit 4 K refrigerator for a Maser application, in: "Advances in Cryogenic Engineering," Vol. 8, Plenum Press, New York (1963), pp. 213-220.
4. W. H. Higa and E. Wiebe, "One million hours at 4.5 K," NBS Special Publication 508, U.S. Government Printing Office, Washington, D.C. (April 1978), pp. 99-108.
5. R. C. Longsworth, "Serviceable refrigerator system for small superconducting devices," NBS Special Publication 607, U.S. Government Printing Office, Washington, D.C. (May 1981), pp. 82-92.

6. E. B. Flint et al., "Performance of a 1 W 4 K cryosystem suitable for a superconducting computer," NBS Special Publication 607, U.S. Government Printing Office, Washington, D.C. (May 1981) pp. 93-102
7. R. C. Longsworth and W. A. Steyert, "4 K refrigerators with a new compact heat exchanger," NBS Special Publication 698, U.S. Government Printing Office, Washington, D.C. (May 1985), pp. 240-249. (Also, U.S. patents 4,484,458; 4,567,943; 4,643,001; others pending)
8. R. C. Longsworth, 4 K refrigerator and interface for MRI cryostats, in: "Advances in Cryogenic Engineering," Vol. 31, Plenum Press, New York (1986), pp. 517-524.
9. T. Koizermi et al., Recondensing refrigerator for superconducting NMR-CT, in: "International Cryocooler Conference," Easton, Maryland, (1986).
10. J. E. Zimmerman and R. Radebaugh, "Operation of a SQUID in a very low power cryocooler," NBS Special Publication 508, U.S. Government Printing Office, Washington, D.C. (April 1978), pp. 59-66.
11. R. Hollman and W. A. Little, "Progress in the development of microminiature refrigerators using photolithographic fabrication techniques," NBS Special Publication 607, U.S. Government Printing Office, Washington, D.C. (May 1981), pp. 160-163.
12. J. L. Smith and G. Y. Robinson, A tribute to S. C. Collins, in: "Advances in Cryogenic Engineering," Vol. 31, Plenum Press, New York (1986) pp. 1-11.
13. H. Izumi et al., Development of small size Claude cycle helium refrigerator with micro turbo-expander, in: "Advances in Cryogenic Engineering," Vol. 31, Plenum Press, New York (1986), pp. 811-818.
14. R. B. Fleming, A compact perforated plate heat exchanger, in: "Advances in Cryogenic Engineering," Vol. 14, Plenum Press, New York (1969), pp. 197-204.
15. T. R. Strobridge and D. B. Chelton, Size and power requirements of 4.2 K refrigerators, in: "Advances in Cryogenic Engineering," Vol. 12, Plenum Press, New York (1967), pp. 576-584.
16. W. H. Hogan, Reliability aspects of cryogenic refrigeration, in: "Advances in Cryogenic Engineering," Vol. 21, Plenum Press, New York (1976), pp. 187-189.
17. R. C. Longsworth, Interfacing small closed-cycle refrigerators to liquid helium cryostats, Cryogenics, 24:253 (1984).
18. R. C. Longsworth and W. A. Steyert, The use of small cryogenic refrigerators near high homogeneity magnets, Cryogenics, 24:243 (1984).
19. F. S. Murray et al., Total cryogen recondensation for mobile MRI, in: "The 6th Intersociety Cryogenics Symposium," Miami Beach, Florida, (1986).

AN INVESTIGATION INTO THE MECHANICS OF JOULE-THOMSON VALVE PLUG FORMATION*

L. Wade

Aerojet ElectroSystems Co.
Azusa, California

C. Donnelly, E. Joham, K. Johnson, R. Phillips,
E. Ryba, B. Self, R. Stanton

Harvey Mudd College
Claremont, California

ABSTRACT

A study was undertaken to examine the phenomenology of plug formation via contaminant condensation in sonic flow Joule-Thomson (J-T) orifices. An apparatus was constructed which allowed plug formation to be visually observed. The cold end of the apparatus consists of a precooler, a counterflow heat exchanger and the J-T expander. Because this is the normal configuration of a cascaded J-T refrigerator cold end, the knowledge gained is directly applicable to existing systems. The test apparatus uses nitrogen gas as the refrigerant and water vapor as the contaminant. Plug formation in a glass J-T expansion valve was observed through a microscope and photographically documented. It was discovered that, for the straight sonic orifices used during these tests, plug formation occurred only in the orifice itself. No contamination condensation was observed on the orifice faces. Mechanisms are proposed which describe the observed characteristics of the plug nucleation and growth.

INTRODUCTION

Space missions which require continuous cryogenic refrigeration for five to ten years are being planned. In support of these requirements several efforts are underway to develop long life refrigerators utilizing sorption compressors and J-T expansion valves.[1,2] While sorption refrigerators are not susceptible to wear related mechanical failure, contaminants in the refrigerant can plug the small orifice used in conventional J-T expansion valves, thereby causing the refrigerator to fail. Solid phase contaminants are easily filtered out, but the condensation of gas-phase contaminants in the J-T orifice presents a much more serious challenge to refrigerator reliability.

---

*Research sponsored by the Aerojet ElectroSystems Co. under internal research and development funding.

The reliability of J-T expanders can be improved by designing the valve so that it is not susceptible to condensation plug formation and/or so that it can be cleaned in situ once a plug has formed. J-T valve designs have been reported which attempt to improve reliability.[3,4,5] A J-T design capable of removing the plug by defrosting should heat the plug directly rather than the entire orifice area. This requires knowledge of precisely where the plug forms. Designing a valve which is inherently resistant to plug formation requires an understanding of plug nucleation and growth mechanics. Since previous investigators have not published detailed information on the phenomenology of plug formation, Aerojet and Harvey Mudd College launched an effort to study the condensation of gas phase contaminants in a J-T valve.

TEST PROCEDURE AND APPARATUS DESIGN

The design goals for the test apparatus were that it should : (1) closely model a typical J-T refrigeration system; (2) allow the formation of the plug in the J-T valve to be monitored; and (3) be versatile enough to be utilized in future studies. The test program examined only straight sonic orifices. Pure nitrogen gas (99.997%) was used as the refrigerant and water vapor was the contaminant.

The apparatus, shown schematically in figure 1 consists of a high pressure gas bottle, an impurity reservoir with a bypass line, a precooler, a tube-in-tube counterflow heat exchanger and a J-T expansion valve. The first two components supply contaminated high pressure gas. The precooler, counterflow heat exchanger and J-T valve are typical of conventional J-T refrigerator cold end designs.

The precooler consists of a heat exchanger in an ethylene-glycol and dry ice bath. A water trap was incorporated into the design of the heat exchanger in the precooling bath to remove water which was in

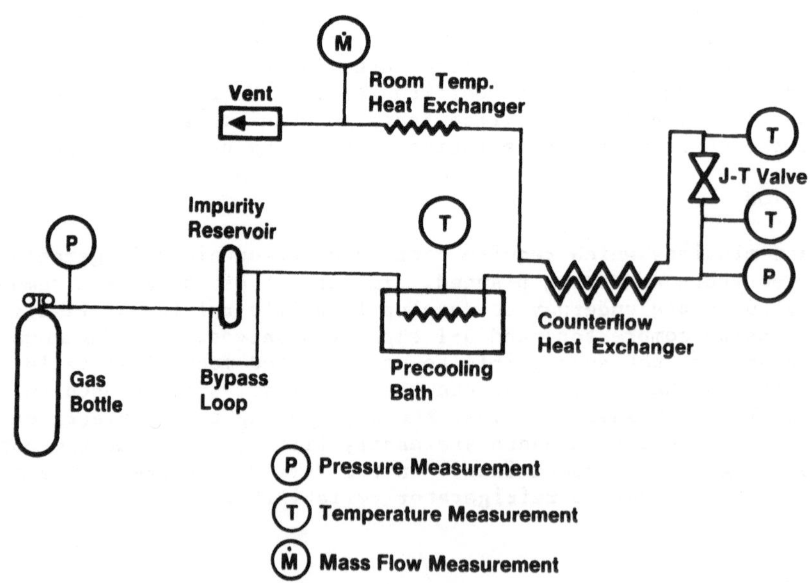

Fig. 1. Schematic of the test apparatus.

excess of the partial pressure of a saturated vapor at the bath temperature. The contamination concentration was therefore a function of precooler temperature. No measurements were made of the concentration. The contamination concentration values reported in this paper are calculated from the saturated equilibrium pressure of water at the precooler bath temperature. The contamination concentration of the refrigerant as it entered the J-T value is probably lower that the estimated values as some of the water froze out in the counterflow heat exchanger.

Three brass J-T valves with copper orifices and three glass J-T valves were used during the test program. Valves of both types were fabricated with orifice diameters of $2*10^{-4}$, $4*10^{-4}$ and $6*10^{-4}$ m. The orifices in the metal J-T valves were made by drilling through the copper with very fine bits. The orifices in the glass J-T valves were made by shrinking glass around a rod of known diameter and then ring sealing the orifice onto the inner glass tube. All six of the valves have an orifice length to diameter ratio of ten.

Testing was conducted in two phases. First, verification testing was conducted with the brass J-T valves to refine the test procedures and to determine the sensitivity of the J-T valve to plug formation. During the verification testing the temperature on both sides of the J-T valve, the temperature of the precooling bath, the pressure upstream of the J-T valve, and the mass flow rate were recorded. Prior to the onset of plug formation, the mass flow rate showed no measurable variation with time.

During the second phase of testing, glass J-T valves were installed and the plug formation was observed through a microscope. A Nikon zoom (10 to 60x) stereoscopic microscope was used for both visual and photographic observations. The stereoscopic microscope allowed the entire orifice and both of its faces to be observed. Photographs were taken of the evolution of the plug. A resolution of $2*10^{-6}$ m during visual observation was estimated by comparing the size of the smallest discernible masses to the orifice diameter. The resolution of the photographs is estimated to be $2*10^{-5}$ m.

The independent variables during experimentation were the high pressure, the precooler temperature and the orifice diameter. The outlet pressure was fixed at approximately $2*10^{5}$ Pa. Mass flow rate is a function of the orifice diameter, the high pressure, and the J-T entrance temperature. J-T exit and entrance temperatures are a function of precooler temperature, exit pressure, and mass flow rate. Once the

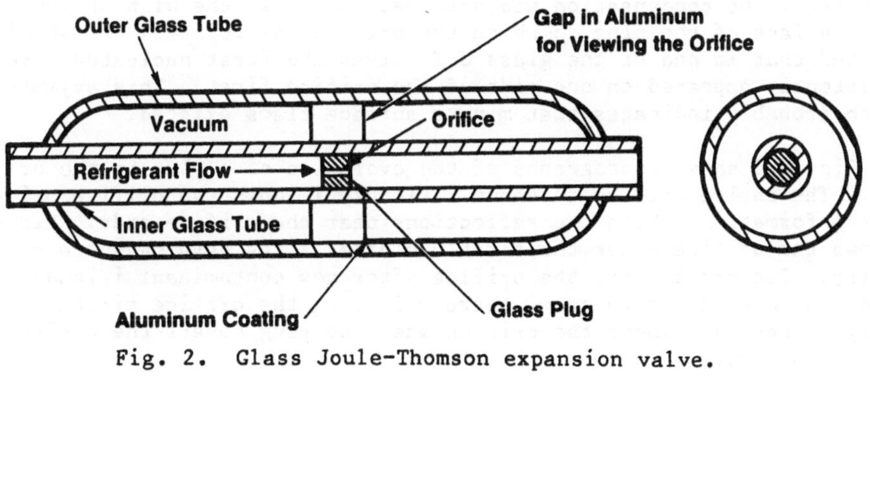

Fig. 2. Glass Joule-Thomson expansion valve.

precooler temperature, high pressure and orifice diameter were set, the dependent variables remained constant. Test conditions were not varied during individual tests.

During the course of the test program, plug formation was observed under a variety of conditions. Precooler temperatures were varied between 285 K and 255 K. The inlet pressures were varied between $2.5*10^6$ and $4.5*10^6$ Pa. Mass flow rates were varied from $3*10^{-4}$ to $2*10^{-3}$ kg/s. The gas temperature on the high pressure side of the J-T valve varied between 255 K and 230 K. The temperature on the low pressure side of the expansion valve was between 3 K and 8 K lower than that of the high pressure side. Calculations of the contamination concentration indicate that the contamination concentration was varied from 30 to 500 parts per million (ppm). Most of the tests were run with an estimated contamination concentration of 70 ppm.

TEST RESULTS

The test apparatus was cooled down with dry nitrogen until it reached steady state conditions. The bypass line was then closed and the nitrogen was run through the impurity reservoir. It normally took about 30 minutes after the establishment of steady state conditions for the plug to begin to form. Forty to fifty minutes later the refrigerant mass flow rate would reach a minimum value of between forty and eighty percent of the initial flow rate. The percent change in mass flow rate was found to be inversely proportional to the orifice cross-sectional area. When the plug reached its maximum size, a portion of it would break off and the mass flow would return to nearly its initial value. After approximately thirty minutes, the mass flow would return to its previous minimum value. This cycle would repeat for four to six hours until the heat exchanger froze up and testing was stopped. The variation in mass flow rate as the plug formed was always very slow. While the rate of formation varied with the test conditions, the same pattern of development was consistently observed.

The first observable indication of plug formation was the appearance of a few, very small islands of frozen contaminant in the orifice near the low pressure exit. The frozen contaminant would then begin to grow axially, against the gas flow. As the first condensed islands began to extend up the orifice, more nucleated sites would appear near the low pressure exit. The frozen contaminant had a white porous appearance. As the plug developed, the newly nucleated sites would also grow towards the high pressure side of the orifice. The plug grew circumferentially primarily through the formation of new nucleation sites. The primary direction of growth was always towards the high pressure side of the orifice. When the plug had covered the surface of the orifice, it would thicken radially until a portion of the plug would break free. No condensation was observed on either the high or low pressure face of the plug in which the orifice was located. It should be noted that in one of the glass J-T valves the first nucleated sites consistently appeared on one side of the orifice first. This asymmetric growth probably indicates that minute surface flaws existed.

Figure 3 shows photographs of the evolution of a plug in the orifice. The entire orifice is shown. Picture 1 shows the orifice prior to plug formation. Note the reflections near the orifice ends. Picture 2 shows the orifice after the first nucleated sites have begun to grow axially. Picture 3 shows the orifice after new contaminant islands have formed and have begun to grow, thereby filling the orifice circumferentially. Picture 4 shows the orifice when the plug covers the entire orifice surface.

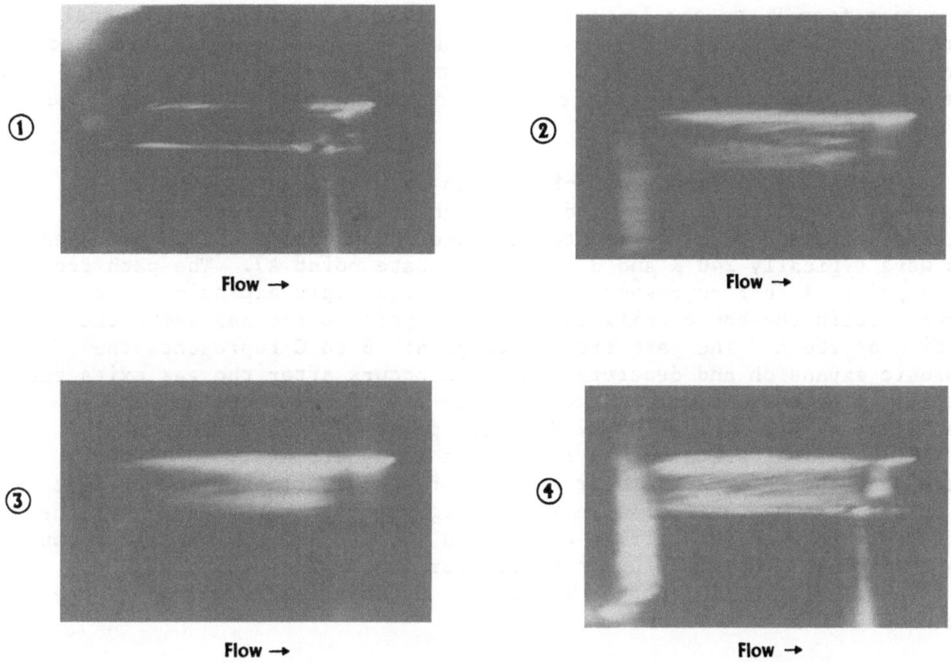

Fig. 3. Photographs of the evolution of a plug in the J-T orifice.

DISCUSSION OF RESULTS

This discussion relates only to straight sonic orifices as studied experimentally. For knife edge orifices, or subsonic expanders, different factors than those discussed may play an important role in plug formation. Some of the additional factors which are of interest when viewing the problem from a broader perspective include: orifice material surface energy and thermal conductivity, orifice surface morphology and shape, contaminant gas and particulate contamination.

Our hypothesis on why the plug formed in the orifice is that the thermodynamic path taken as gas expands through a sonic J-T orifice is not isenthalpic although the inlet and outlet states do lay on the same enthalpy curve. It is postulated that the thermodynamic path actually followed can be broken down into two portions. The first part is the sonic expansion within the orifice. The second part is the subsonic expansion and deceleration which occurs immediately after the gas exits the orifice.

Because frictional effects are comparatively small in a short orifice, the flow within the orifice may be considered to be undergoing an isentropic expansion.[6] The pressure of the refrigerant stream as it exits the orifice (P_m) is approximately 0.53 times the pressure at the entrance of the orifice (P_h) and the temperature of the gas at the exit of the orifice (T_m) is approximately 0.83 times the entrance temperature (T_h). During the sonic portion of the expansion, the change in potential energy due to the pressure change from P_h to P_m and the temperature change from T_h to T_m is converted into directed kinetic energy as the gas velocity is accelerated to the speed of sound.

Nitrogen gas leaves the J-T orifice at sonic velocity. During the second portion of the proposed thermodynamic path which begins when the gas exits the orifice, the gas velocity is substantially reduced and the expansion from P_m to the low pressure P_l occurs. Because no external work is done the potential energy from this expansion and the directed kinetic energy of the gas are converted to random kinetic (thermal) energy, thereby increasing the temperature of the gas to a value close to T_h.

Figure 4 which shows the T-S diagram for nitrogen and the state points as the nitrogen is expanded through the J-T orifice. At the entrance of the J-T valve, the temperature and pressure of the nitrogen gas were typically 240 K and $3.0*10^6$ Pa (state point A). The path from state points A to B represents the nearly isentropic expansion which occurs within the sonic orifice. At state point B the gas exits the orifice at 200 K The path from state points B to C represents the subsonic expansion and deceleration which occurs after the gas exits the orifice. The temperature of the nitrogen gas after expansion through the J-T valve was typically 232 K (state point C). Note that the initial and final state points lie on the isenthalp $H=3.9*10^5$ J/kg. Therefore, the temperature near the low pressure outlet of the orifice is colder than anywhere else in the refrigerator. The large temperature difference between the orifice outlet and the rest of the refrigeration system explains why the plug initially forms near the outlet of the orifice rather than in another section of the orifice or on one of the two faces.

Only a small portion of the contaminant actually condenses in the orifice. Given a nitrogen mass flow rate of $1.5*10^{-3}$ kg/s and a contamination concentration of 70 ppm, approximately $3.2*10^{-4}$ kg of water will pass through the orifice before the plug reaches its maximum size.

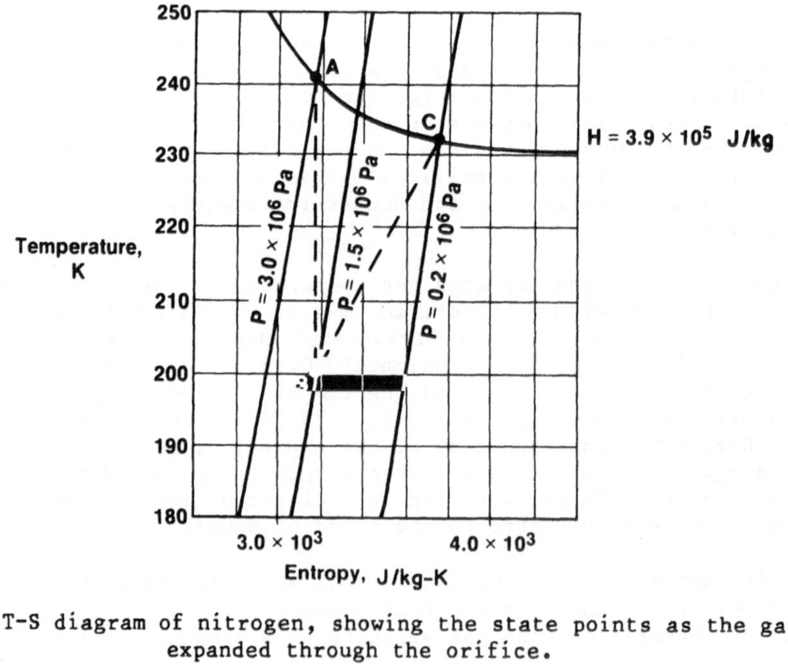

Fig. 4. T-S diagram of nitrogen, showing the state points as the gas is expanded through the orifice.

Assuming that the mass flow rate is linearly proportional to the cross-sectional area of the orifice and that the plug density is 500 kg/m^3, the maximum mass of water in a typical plug was $1.6*10^{-7}$ kg. Therefore only 0.05 percent of the available water condensed in the orifice.

During the initial nucleation stage the rate of mass condensation was extremely slow. Once the first nucleated sites grew large enough to be visible ($2*10^{-6}$ m) the rate of condensation increased considerably. The velocity distribution of gas molecules undergoing sonic expansion is highly unidirectional. Hence, the probability is much stronger that a water molecule will strike the high pressure side rather than any other side of a nucleated site. As a result, the plug tends to grow axially from the low pressure side of the orifice towards the high pressure side.

The water vapor is not in equilibrium when it enters the orifice. Its concentration is thought to be close to that of its vapor pressure at the precooler bath temperature which is typically 70 K higher than the temperature in the orifice. Because the water vapor is in a supercooled state when in the orifice, the heat of sublimation is quickly dissipated upon condensation. Therefore, newly condensed molecules do not have the time necessary to reorder into a crystalline structure once they condense. The resulting condensed water structure is probably a mixture of porous amorphous and polycrystalline structures. It is hypothesized that the probability of particles of water condensed in the heat exchanger sticking to a nucleated site is low due to their large momentum relative to their Van der Waals force; and that the axial growth therefore occurs molecule by molecule. The observed condensation in the orifice is thought to be porous because it is white (it resembles snow on a window). The index of refraction of solid ice is very close to that of glass. The plug, if solid, would have been difficult to see. Researchers examining the depositions of gases on cryopump surfaces have noted porous amorphous and polycrystalline structures when the gas is supercooled.[7,8]

SUMMARY

The phenomenology of plug formation via contaminant condensation in straight sonic flow J-T orifices was observed. A test apparatus which closely modelled a typical J-T refrigeration system was constructed. Nitrogen gas, contaminated with water vapor, was used as the refrigerant. The nucleation and growth of plugs in the J-T valve was successfully observed and photographically documented. Plugs were observed to form only in the orifice, forming initially near the low pressure side of the orifice. The nucleated sites grew axially against the direction of flow, forming a porous structure.

It is hypothesized that the thermodynamic path taken as refrigerant expands through a sonic J-T orifice is only isenthalpic at the end points. The low pressure side of the orifice is thought to be the coldest part of the refrigerator as the expansion within the orifice is nearly isentropic. As a result plug growth occurred only inside the orifice. The plug grew axially against the direction of flow because water molecules primarily struck the high pressure side of the nucleated sites due to the unidirectional nature of sonic flows. It is proposed that the contaminant was supercooled and did not have enough energy to reorder once it condensed which resulted in a porous and probably amorphous structure. While this test program examined only straight sonic orifices, the apparatus is capable of measuring the effects of factors including: orifice shape, surface quality, and surface material on the rate of plug formation.

REFERENCES

1. K. Karperos, Operating Characteristics of a Hydrogen Sorption Refrigerator, Part I: Experiment Design and Results, Proceedings of the 1986 Cryocooler Conference, Easton, Maryland, Sept. 25-26, 1986.

2. J. A. Jones and P. M. Golben, Design, Lifetesting, and Future Designs of Cryogenic Hydride Refrigeration Systems, Cryogenics 25:212 (1985).

3. K. Hedegard, G. Walker, and S. Zylstra, Temperature Sensitive Variable Area Flow Regulator for Joule-Thomson Nozzles, Proceedings of the 1986 Cryocooler Conference, Easton, Maryland, Sept. 25-26, 1986.

4. J. M. Lester and B. Benedict, Joule-Thomson Valves for Long Term Service in Space Cryocoolers, Spec. Pub. 698, National Bureau of Standards, Boulder, Colorado (1984), p. 257.

5. C. K. Chan, private communication, April 1987.

6. A. H. Shapiro, "The Dynamics and Thermodynamics of Compressible Fluid Flow," Ronald Press Co., New York (1953).

7. L. Ni and L. Jun, Research on Mechanisms of Cryofrost Condensation and Cryotrapping, Cryogenics 27:156 (1987).

8. G. Davey, Cryopumping in the Transition and Continuum Pressure Regions, Vacuum 26:17 (1976).

LOW-COST, COMPACT DILUTION REFRIGERATOR:

OPERATION FROM 200 TO 20 mK*

P. R. Roach and K. E. Gray

Materials Science Division
Argonne National Laboratory
Argonne, Illinois

ABSTRACT

We have developed a self-contained (portable) ^{3}He-^{4}He dilution refrigerator which has been operated below 0.020 K without the use of external pumps or gas handling system. Using charcoal pumps for all pumping, the system requires, besides the usual liquid helium and nitrogen baths, only modest electronics for heaters and thermometers. This device has the advantages of low cost and simple operation and does not require extensive maintenance or operator expertise. The refrigerator can operate below 0.020 K for more than 10 hours before the charcoal pumps need to be recycled.

CONTINUOUS REFRIGERATORS

^{3}He-^{4}He dilution refrigerators of the continuously operating type have been used for many years to provide temperatures from 1.0 K down to as low as 0.002 K. These refrigerators operate on the principle of cooling that is achieved in a mixing chamber when ^{3}He crosses a phase boundary separating almost pure ^{3}He from a dilute mixture of 6% ^{3}He in ^{4}He. The flow of ^{3}He across the phase boundary is driven by the removal of ^{3}He from the dilute phase in a separate container, the still, which is thermally isolated from the mixing chamber but connected to the liquid He in it by a thin tube. The temperature of the still is maintained by a heater at a value such that the vapor above the dilute liquid is almost pure ^{3}He. An external pump then removes this ^{3}He from the still, leaving the ^{4}He behind as a stationary background component. In a continuously operating system, the ^{3}He is re-compressed, cooled to 1 K, whereupon it re-liquefies, and returned to the mixing chamber. In order to minimize the heat load on the mixing chamber from this returning ^{3}He, it is precooled by the cold, dilute ^{3}He exiting the mixing chamber in a series of sophisticated heat exchangers.

Such dilution refrigerators include 1 K pots in which liquid ^{4}He is cooled using an external pump. Outside the cryostat will be, in addition to the pumps mentioned, storage tanks for the ^{3}He-^{4}He gas mixture, various traps for cleaning the ^{3}He before it is returned to the refrigerator, large diameter pumping lines to the cryostat, and numerous valves and gauges for operating the system. Such systems require appreciable training or experience to run and to keep them running. Although they can operate relatively unattended except for replenishment of liquid ^{4}He, the external mechanical pumps and gas handling equipment must

---

* Work supported by the U.S. Department of Energy, BES- Materials Sciences, under contract # W-31-109-ENG-38.

be periodically monitored and maintained to avoid contamination of the returning gas. Contamination with either air or pump oil vapor can cause catastrophic blockages of low temperature capillaries. Such blockages may only require warming the whole system to room temperature or, in more serious cases, surgical intervention into the sealed ^{3}He-^{4}He mixture system. Given operators with sufficient low temperature expertise and dilution refrigerator experience, these systems can run for many days at their lowest temperature and some have been kept cold for many months at a stretch.

ONE-SHOT DILUTION REFRIGERATOR

Principle of Operation

The one-shot refrigerator we have developed operates on the same principle as the continuously operating refrigerator except that ^{3}He is not returned to the mixing chamber after it leaves the still. In addition, the external pumps are replaced by charcoal adsorption pumps operating in the 4.2 K bath of the cryostat. Basically, ^{3}He is removed from the still by a charcoal pump until the concentrated ^{3}He above the phase boundary in the mixing chamber is depleted. Then cooling stops and the system must be recycled by heating the charcoal to desorb the gas and to recondense the ^{3}He into the refrigerator.

Figure 1 shows a schematic of our refrigerator. The insert that is the heart of the refrigerator resides in a dewar containing liquid helium at 4.2 K. The charcoal pumps are enclosed in vacuum jackets that can have He gas admitted to them to thermally anchor the charcoal pumps to the bath for pumping. The charcoal is simply lightly packed into its container; the pumping line extends into an empty region above the charcoal. When the gas in the charcoal is to be regenerated, the charcoal is isolated from the He bath by pumping the gas from the vacuum jacket and applying electrical heat to the outside of the charcoal container. The refrigerator itself is contained in its own vacuum jacket that is kept evacuated.

There are 3 systems in our design: a ^{4}He system for the 1 K pot which is used to condense ^{3}He into the refrigerator and to provide thermal shielding; a ^{3}He system which cools a pot attached to the mixing chamber in order to set up the initial phase boundary in the mixing chamber; and the ^{3}He-^{4}He mixture system for the dilution refrigerator itself. At room temperature the amount of ^{3}He for the ^{3}He pot is small enough that it is stored in its charcoal pump at elevated pressure (~1.6 MPa). The ^{3}He-^{4}He mixture is stored in internal storage tanks just above the bath at ~0.6 MPa. The amount of ^{4}He used is large enough that it must be stored in an external tank (at ~0.6 MPa). The sintered copper powder in the bottom of the mixing chamber is to provide a large surface area for good thermal contact between the liquid helium mixture and the copper walls.

Operating the System

The refrigerator is started (after all components are at 4.2 K) by heating the charcoal pump containing the ^{4}He. When the pressure of the ^{4}He reaches 1 atm, it condenses where it runs through the 4.2 K bath and drops into and fills the 1 K pot. The heating is stopped and exchange gas is admitted to the vacuum space of the pump whereupon it begins to pump and the 1 K pot cools to below 1 K. Then the ^{3}He and mixture charcoal pumps are heated and these gases condense where the tubes pass through the 1 K pot. Although this condensing operation causes the 1 K pot to warm somewhat and causes the liquids to fill their containers at temperatures greater than 1 K, the 1 K pot soon cools below 1 K again and the other systems slowly follow by a refluxing (heat-pipe) mechanism whereby warm gas evaporating from the lower systems condenses at the 1 K pot and cold liquid runs back down the walls of the connecting tubes. When all the systems have reached 1 K, the ^{3}He pot is pumped to below 0.3 K. This sets up a phase boundary in the mixing chamber and draws the maximum amount of concentrated ^{3}He into the mixing chamber above the boundary. Without this ^{3}He pot, all cooling would have to be done by pumping on the still; this would simply cause a phase boundary to form in the still and most of the ^{3}He would be drawn to the colder still and pumped uselessly away. When the ^{3}He pot has cooled the mixing chamber to 0.3 K the still is pumped and heat is supplied to it to keep it at ~0.6 K. The temperature of the mixing chamber is controlled by varying the circulation rate of the ^{3}He (determined by the still temperature) or by adding heat to the mixing chamber.

When the concentrated ^{3}He in the mixing chamber is depleted and cooling stops, it is necessary to recycle the system. The sequence of condensing ^{4}He into the 1 K pot, pumping it to 1.0 K, etc., is then repeated until the mixing chamber is again at 0.3 K with the phase-separated mixture in it. This recycling operation currently takes ~5 hours but we believe this can be reduced to ~3 hours by improvements in the charcoal pumps. Then the mixture is again pumped to achieve the lowest temperatures.

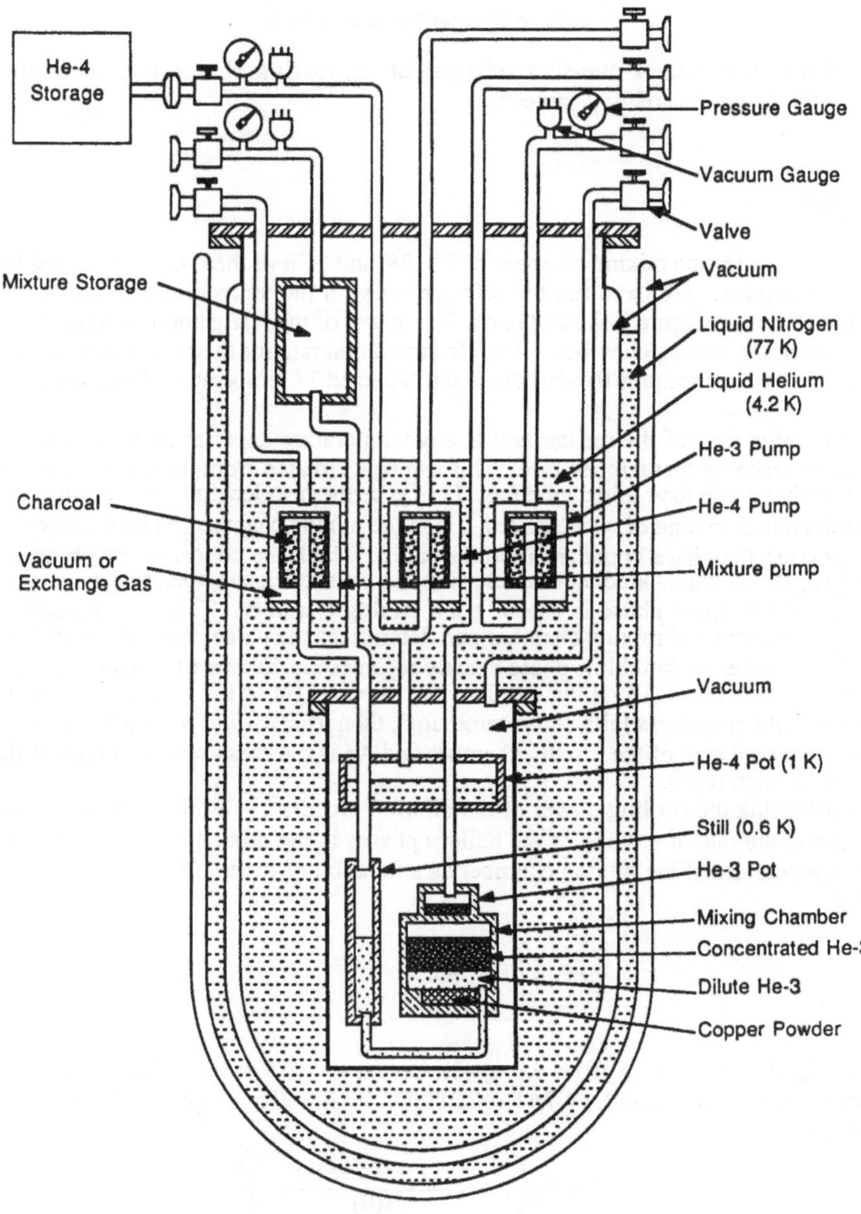

Fig. 1. Schematic of one-shot dilution refrigerator.

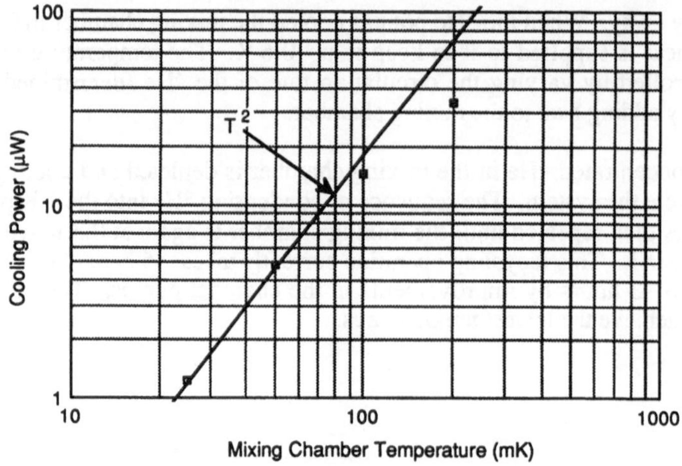

Fig. 2. Cooling power of one-shot refrigerator vs. mixing chamber temperature for a circulation rate of 20 μ moles/s.

Performance

We have cooled the mixing chamber to 0.015K and believe there is the potential for even lower temperatures. We have run the refrigerator with the temperature below 0.020K for more than 10 hours. Figure 2 shows the cooling power of the refrigerator as a function of the temperature of the mixing chamber. The ^{3}He circulation rate for this measurement was 20 μ moles/s. At low temperatures the data show the expected T^2 temperature dependence.

The running time of the refrigerator at any temperature depends on the amount of ^{3}He expended in reaching that temperature. This can be estimated from the known properties of the helium phases and their initial amounts. Let $\dot{Q}_0$ be the heat leak into the mixing chamber, $n_3(0)$ be the initial amount of concentrated ^{3}He in the mixing chamber, $\dot{n}_3$ be the rate at which ^{3}He is crossing the phase boundary (and the rate at which the concentrated phase is being depleted), $n_D(0)$ the initial amount of dilute phase in the mixing chamber, and $\dot{n}_D = \alpha\dot{n}_3$ is the the rate at which dilute phase is entering the mixing chamber. This rate depends on the geometry of the still and mixing chamber and determines the value of α. If the still is above the mixing chamber so that dilute phase exactly replaces the volume of concentrated phase as it is depleted, then $\alpha = 1.32$. If the still is at the same height as the mixing chamber so that the level in both containers falls at the same time, then $\alpha = 0.75/(1 + A_M/A_S)$ where A_M is the cross-sectional area of the mixing chamber and A_S is the cross-sectional area of the still. If $A_M >> A_S$ then $\alpha \sim 0$.

By balancing the cooling power of the dilution process[1], -- $84\,\dot{n}_3\,T^2$, with the external heat leak and the rate of cooling of the helium phases in the mixing chamber one can show that the temperature of the mixing chamber as a function of time, t, during the cool down is given by

$$\frac{T^2-c}{T_0^2-c} = \left(1 - \frac{b}{a}t\right)^{\gamma}, \qquad (1)$$

where $c = \dot{Q}_0/84\,\dot{n}_3$, $a = 24\,n_3(0) + 6.9\,n_D(0)$, $b = 24\,\dot{n}_3 - 6.9\,\alpha\dot{n}_3$, T_0 is the temperature at the start of the cool down and $\gamma = 7/(1 - 0.288\alpha)$. If $\dot{Q}_0 = 0$ and $n_D(0) = 0$ this expression simplifies to

$$T = T_0\left(1 - \frac{\dot{n}_3(1 - 0.288\alpha)}{n_3(0)}\,t\right)^{\frac{\gamma}{2}}. \qquad (2)$$

The cool down time can be obtained from the above expression:

$$t_C = \frac{n_3(0)}{\dot{n}_3(1 - 0.288\alpha)}\left(1 - \left(\frac{T}{T_0}\right)^{\frac{2}{\gamma}}\right). \tag{3}$$

The worst case is for $\alpha = 1.32$. Then Eq. 3 becomes

$$t_C = 1.61\,\frac{n_3(0)}{\dot{n}_3}\left(1 - \left(\frac{T}{T_0}\right)^{0.177}\right). \tag{4}$$

The best case is for $\alpha = 0$, in which case

$$t_C = \frac{n_3(0)}{\dot{n}_3}\left(1 - \left(\frac{T}{T_0}\right)^{0.286}\right). \tag{5}$$

The amount of ^{3}He left after the initial cool down to temperature T is then $n_3(0) - \dot{n}_3\, t_C$. This number divided by the circulation rate, $\dot{n}_3'$, used for the final operation at constant temperature gives the running time at this temperature. This gives, for $\alpha = 0$,

$$t_R = \frac{n_3(0)}{\dot{n}_3'}\left(\frac{T}{T_0}\right)^{0.286}. \tag{6}$$

When running at constant temperature, the cooling power from the circulation just balances the heat load on the mixing chamber, so that $\dot{n}_3'$ in Eq. 6 can be replaced by $\dot{Q}_0/84\,T^2$, where $\dot{Q}_0$ is the heat load. Figure 3 shows this running time available as calculated assuming typical initial conditions for our system: $n_3(0) = 0.25$ moles, and $T_0 = 300$ mK. Not all areas of this plot are accessible in our current refrigerator because they require circulation rates above those that the system can achieve. Clearly, a wide variety of trade-offs can be considered between running time, operating temperature and cooling power.

Design Advantages

One of the important features of our design is that the still is at the same level as the mixing chamber and is of much smaller cross-sectional area. This means there will be a free

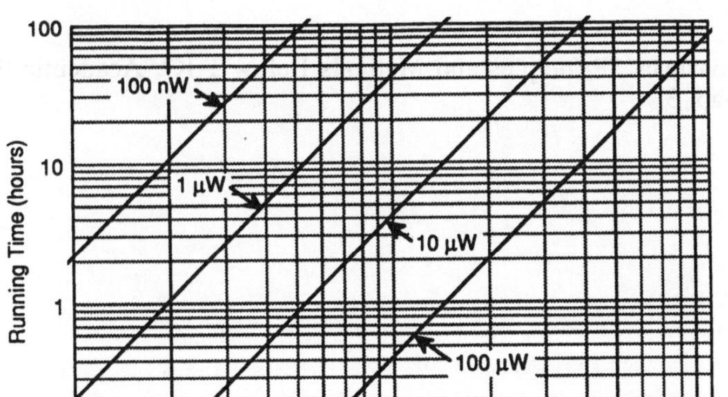

Fig. 3. Calculated running time of system as a function of mixing chamber temperature for a number of heat inputs.

liquid surface in the mixing chamber at approximately the same level as the one in the still. Then, as the ^{3}He in the mixing chamber is depleted, the level in the mixing chamber and still will both drop until the ^{3}He is completely gone from the mixing chamber at which point the mixing chamber will be nearly empty of liquid and the level in the still will be rather low. Since the total amount of dilute phase in the system is approximately constant, the large drop in level of dilute phase in the still can be accommodated by a rather small rise in the level of dilute phase in the mixing chamber because they have very different cross-sectional areas. The advantage of this feature is that the limited cooling power of the refrigerator is not wasted cooling a large amount of dilute phase. (The cool down time for this configuration is described by Eq. 5.) In the standard configuration with the still above and of comparable volume to the mixing chamber, dilute phase would enter the mixing chamber to replace the ^{3}He leaving it and the end point of the operation would be with the mixing chamber full of dilute phase. This greatly reduces the available cooling capacity and running time. (The longer cool down time for this situation is described by Eq. 4.)

Because the ^{3}He is not returned to the mixing chamber, the need for complex heat exchangers is eliminated. Because the external pumps are replaced by internal charcoal pumps, the external gas handling system is eliminated. These simplifications mean the present system is much smaller and, because everything is contained with the cryostat insert, rather portable. The low-temperature part of the system is much simpler without the heat exchangers. Because the heat load on the mixing chamber from returning ^{3}He is eliminated, this design should easily reach temperatures of 0.010K or below. Because the He gas mixture never leaves the low-temperature region during operation, the chance of contamination of the gas is eliminated and the reliability of the system is much improved. Reliability is further enhanced because there are no mechanical pumps needed for the operation of the refrigerator. (An additional benefit of this is that the operation is completely silent and vibrationless.) Because the operation of the system can be entirely controlled with heaters, it is possible to completely automate the running of the system. In that case it is likely that a typical user would make whatever measurements were desired during the 10-hour running time at low temperature and then let the system recycle automatically overnight. Such a control system is currently under development.

ACKNOWLEDGEMENTS

We would like to thank Charles Schaeffer, Paul Morrison, Zenon Sungaila, Yong Jia Qian and Russ Hilleke for their assistance in the construction and testing of various stages of this refrigerator.

REFERENCES

1. O. V. Lounasmaa, "Principles and Methods Below 1 K", Academic Press, London (1974), Chap. 3.

CONTROL OF THE INTERFACE BETWEEN ^{3}He-RICH AND ^{4}He-RICH PHASES USING ELECTRIC FIELDS

Ulf E. Israelsson, H.W. Jackson, and D. Petrac

Applied Sciences and Microgravity Experiments Section
Jet Propulsion Laboratory
California Institute of Technology
Pasadena, California

ABSTRACT

A program to experimentally and theoretically study the feasibility of operating a dilution refrigerator in a microgravity environment has been started at the Jet Propulsion Laboratory. One of the major obstacles to be overcome before the ultimate goal can be realized is to find a replacement for gravity in orienting the boundary interface between the ^{3}He-rich and ^{4}He-rich phases of the liquid in the mixing chamber of the refrigerator. We present preliminary measurements showing that moderate electric fields in a suitable geometry can be used to generate the required orienting forces.

INTRODUCTION

A few space missions being considered by NASA, such as the Space Infrared Telescope Fascility (SIRTF), the Advanced X-ray Fascility (AXAF), and the Large Deployable Reflector (LDR), will need cooling well into the millikelvin temperature region to best utilize the capability of their instruments. Previous work in trying to develop cooling capability into the 0.1 kelvin temperature range for space applications has focussed on adiabatic demagnetization refrigeration (ADR) techniques[1]. The main drawback with the ADR is that it is not a continuous refrigerator. Present designs aim towards a duty cycle (how much of the total time the ADR stays cold) of 90 % at the required operating temperature of 0.1 K[2]. A substantial fraction of a mission lifetime would thus not be available for taking data if an ADR is used to cool the instrument.

The only means identified so far for reaching lower subkelvin temperatures continuously is by the dilution refrigerator[1]. Here cooling takes place in the so called mixing chamber by diluting ^{3}He atoms from a ^{3}He-rich phase into a ^{4}He-rich phase. On earth gravity plays the fundamental role of separating the lighter ^{3}He-rich phase from the heavier ^{4}He-rich phase, thereby enabling the successful operation in a suitable geometry. Gravity does not exist in earth-orbit, so a replacement force must be found if the advantages of the dilution refrigerator are to be realized for space applications[3].

In this paper we present measurements which show that electric field generated forces can be used as a replacement for the gravitational force in moving the two phases into separate regions of space, thus creating and controlling the necessary interface line. The ultimate goal of developing a flight dilution refrigerator (FDR) now appears feasible.

EXPERIMENTAL

The cell used to show control of the $^3He/^4He$ interface line is shown schematically in Fig. 1. The V-shaped geometrical design for the pair of electrodes was selected to conform with a prediction[4] of how to best affect the interface line. The calculation showed that a minimum separation of 0.5 mm, an electrode length of 3.8 cm, and a slope of the lower plate of 0.1 would allow all relevant features of the experiment to be displayed. Inaccuracies in making the spacers that separate the electrodes and define part of our geometry caused the minimum plate separation to be about 0.7 mm, and the slope to be about 0.09. The electrodes were 3.8 cm long and 1.8 cm wide. The plate separation was 4 mm on the wide end of the cell, thus producing a cell volume of 1.26 cm^3. The electrical ground plate is the upper plate shown in the figure, and ground connections are made via copper springs that are fed through the lucite holder and attached to the ^3He pot. Connection to the lower high-voltage electrode is also made by a spring so that the assembly is completely demountable. The high voltage lead is sealed into the lower part of the lucite cell by means of Stycast 2850 epoxy[5]. The lower end of the lead shown in the figure connects to the lead coming through the surrounding exchange gas can from the exterior ^4He bath. The electrodes are made out of copper, and they are carefully electropolished to remove any sharp spots where arcing could get started prematurely. Force is kept on the springs holding the electrodes flush to the lucite spacers by means of a lucite housing as shown in the figure. The housing is sealed by an indium o-ring, and is secured by stainless steel screws to ensure a leaktight seal even at superfluid temperatures. The cell is thermally connected to the ^3He pot by a few copper strips, as well as by the ground leads. Additional thermal contact to the mixture cell is provided by passing the liquid fill line capillary through the inside of the ^3He pot. Both the mixture gas for the cell and the ^3He gas for cooling the pot are passed through a charcoal trap at nitrogen temperature to eliminate contaminants that might interfere with the experiment.

To isolate the ^3He pot and the mixture cell from the main bath, while at the same time allowing visual access into the experimental region, the

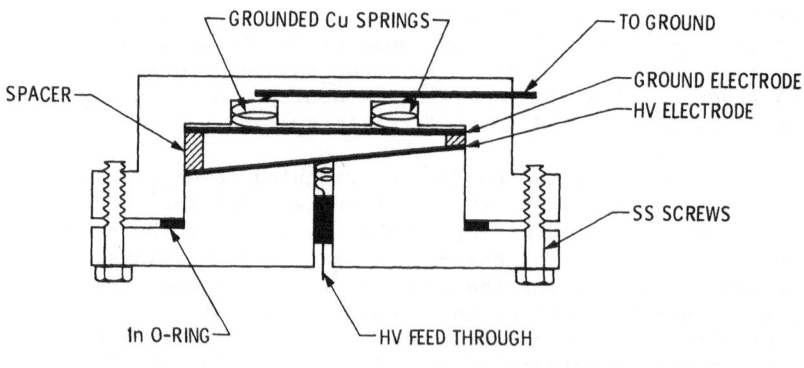

Fig. 1. Schematic drawing of the experimental $^3He/^4He$ mixture cell. It is made out of lucite to allow visual access and to minimize electrical breakdown problems.

exchange gas can is made from lucite. This can is sealed to a lucite top plate by indium o-rings secured by stainless steel screws. Pumping lines for the exchange gas can and for the ^3He pot are made out of thin-walled (0.2 mm) stainless steel tubing, and leaktight feed throughs at the lucite top plate are made with Stycast 2850 epoxy. This method of sealing lucite to lucite, and lucite to stainless, at cryogenic temperatures has proven to work better than expected. The system stayed leak tight even after repeated cycling to helium temperatures.

The temperature of the experimental cell is determined both by measuring the vapour pressure and by a germanium resistor (not shown in Fig. 1.) inside the cell. The ^3He concentration was determined by measuring the vapour pressure at higher temperatures where tabulated values exist[6]. Our experiment utilized a molar ^3He concentration of 60 %, and the temperature was not lowered below 0.6 K.

RESULTS

Using our geometry we are able to apply voltages of up to 10 kV (the power supply limit) to our electrodes without experiencing breakdown through the liquid. This voltage corresponds to a maximum electric field of about 13 MV/m at our closest plate separation. Fig 2. shows qualitatively how the interface line is changing when the electric field is increased from zero to our maximum value. As can be seen, the effect of increasing the field is to pull the ^3He poor phase into the high voltage region, thereby shifting the interface line at a steeper angle from the horizontal. The data shown in the figure are obtained at a temperature of 0.8 K, corresponding to a molar ^3He concentration of 0.72 in the ^3He-rich phase, and 0.52 in the ^3He-poor phase[7]. The dotted circle indicated in the figure shows the limited visual access into the experimental region. The slope of the interface line outside the visual region is therefore an extrapolation from where it can be seen. Quantitative measurements are obtained by measuring the width of the plates where the interface line meets the upper plate, and also by recording how wide the ^3He rich phase is at the left-most point where we have visual access. All measurements are recorded with a level telescope and we estimate our resolution to be about 0.15 mm.

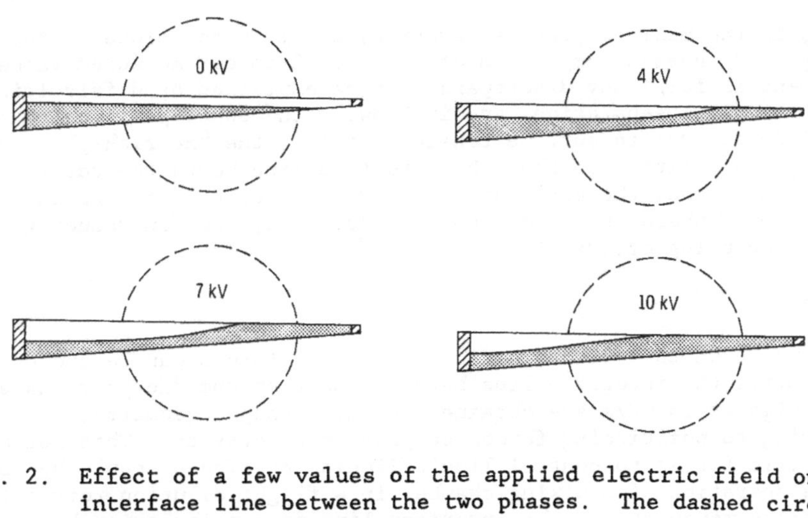

Fig. 2. Effect of a few values of the applied electric field on the interface line between the two phases. The dashed circle indicate our limited visual access.

We can readily conclude from the figures that this method of controlling the interface line using electric fields will work for a flight dilution refrigerator (FDR). Consider if the outlet capillary line from the mixing chamber of a dilution refrigerator leading up to the still is hooked up at the high field region of our geometry. Then in the gravitational field of the earth, cooling in the mixing chamber would not be possible in zero applied electric field since the capillary outlet is covered with ^{3}He-rich phase. However, at an applied electric field of 4000 V, the capillary outlet is covered with ^{3}He-poor phase, and the cooling process will commence. In the absence of gravity the forces required to properly position the phases are orders of magnitude lower, and we see no problem in obtaining the necessary fields.

A theoretical calculation has recently been performed[3] using a slightly different geometry that shows that an electric field of 10 MV/m produces a positioning force that is just as effective as gravity in forcing the two phases into distinct regions of space. In our measurements we have reached 13 MV/m and have thus produced a positioning force nearly twice that of gravity.

The calculation[4] we used to aid in the design of our mixture cell utilizes the momentum conservation theorem including electric fields for ^{3}He/^{4}He mixtures, the Clausius-Mossotti relation for the dielectric constant, and the assumption that the electric field lines produced in our cell are parallel everywhere (the gradual field approximation). An equation is then obtained for the interface line:

$$z_s(x) = 0.5V^2\epsilon_0\Delta\epsilon/\Delta\rho g[\ z_b^{-2}(x) - z_{b0}^{-2}\ ] \qquad [1]$$

Here x is the horizontal coordinate, and z is the vertical coordinate, z_{b0} is the width of the cell where the interface line touches the top plate, V is the applied voltage, g is the acceleration due to gravity, $\Delta\epsilon$ is the difference in dielectric constant between the ^{3}He-poor phase and the ^{3}He-rich phase, and $\Delta\rho$ is the similar difference for the density.

By integrating the interface line across the length of the cell and equating the obtained area with the measured zero field area A of the ^{3}He rich phase, we can solve for the separation of the plates where the interface line touches the top plate. The result is:

$$z_{b0} = z_{max}/\{\ 1 + [2mz_{max}gAV^{-2}\Delta\rho/\epsilon_0\Delta\epsilon]^{1/2}\ \} \qquad [2]$$

where z_{max} is the maximum plate separation, and m is the slope of the lower plate. Figure 3 shows a comparison of equation 2 to our measured values. The agreement is fair; any discrepancy can be explained by difficulties in measuring the plate separation, and in knowing the exact position of our viewing windows. Due to surface tension effects, the ^{4}He rich-phase coats the upper plate a little so that the interface line bends towards the left in fig. 2 and touches the plate at a zero degree angle. We make an extrapolation of where the exact touching point is, thus introducing additional measuring errors.

CONCLUSIONS

We have shown that electric field generated forces can be successfully used to control the interface line between ^{3}He-rich and ^{3}He-poor phases. Fields as high as 13 MV/m are obtained in our v-shaped geometry, corresponding to positioning forces comparable to gravity. This method of using electric fields to control liquid/liquid interfaces should be able to ensure the operation of a mixing chamber in a flight dilution refrigerator. We compare the movements of the interface line to a calculation based on

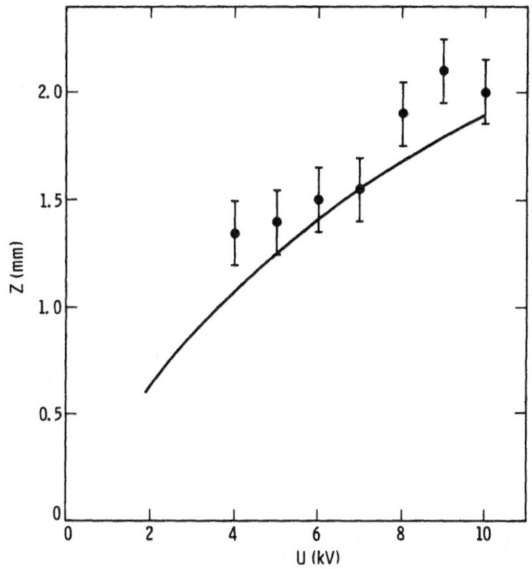

Fig. 3. Width of the experimental cell where the interface line reaches the upper plate as a function of applied voltage. The solid line is a theoretical calculation based on the two-fluid model.

the two-fluid model including electrostriction, and find good agreement with our experimental results.

ACKNOWLEDGEMENTS

The research described in this publication was carried out by the Jet Propulsion Laboratory, California Institute of Technology, under a contract with the National Aeronautics and Space Administration. One of us (U.E.I) acknowledges support from a National Research Council Resident Research Associateship. Helpful comments, and a critical reading of the manuscript by D.M. Strayer is gratefully acknowledged.

REFERENCES

1. O. V. Lounasmaa, "Experimental Principles and Methods Below 1 K," Academic Press, New York (1974)
2. S. Castles, private communication (1986)
3. H. W. Jackson, Can ^3He-^4He dilution refrigerators operate aboard spacecraft?, Cryogenics 22, 59 (1982)
4. H. W. Jackson, unpublished
5. Emerson & Cuming, Inc., Gardena, CA
6. K. W. Taconis, and R. De Bruyn Ouboter, Equilibrium properties of liquid and solid mixtures of helium three and four, in "Progress in Low Temperature Physics," Vol. 4, C. J. Gorter, ed., (1964), p. 38.
7. J. Wilks, "The properties of Liquid and Solid Helium," Oxford UP, London (1967)

MAGNETIC REFRIGERATION: A REVIEW OF A DEVELOPING TECHNOLOGY

John A. Barclay

Astronautics Corporation of America
Astronautics Technology Center
5800 Cottage Grove Road
Madison, Wisconsin

ABSTRACT

Magnetic refrigeration has been used for over 50 years as a technique to achieve temperatures well below 1 K. However, in the last ten years, the technology has been developing for refrigeration applications above 1 K up to, and including heat pumps above room temperature. The work has been multinational in scope and has focused on analysis of magnetic thermodynamic systems, investigation of magnetic materials suitable for refrigerants, and development of prototype refrigerators. Devices providing refrigeration between 1.5 K and 4.2 K, 4.2 K and 20 K, and near room temperature have been emphasized. Recently development efforts have been extended into the 20 K to 80 K temperature range. The understanding of the fundamental limits on magnetic refrigeration is slowly evolving as is a data base for design. The thermodynamic performance of the prototypes has been steadily improving but still has not reached the theoretical limits predicted by analysis.

INTRODUCTION

It is a circumstance of nature that interconversion from one energy form to another ultimately produces thermal energy. Once in the thermal form, it is impossible to completely transform the energy back into another form. It is not surprising, therefore, that enormous interest in the transformation of thermal energy exists. The "thermodynamics" of heat engines, heat pumps, and refrigerators has occupied many scientists and engineers since the scientific revolution. It is also not surprising that an understanding of thermodynamics has played an intricate role in the advancement of industrialized countries. The importance of refrigeration is vividly illustrated by the recent excitement over the discovery of "high temperature" superconductors.[1] The key difference between previously existing superconductors and the new materials is the ease of providing the refrigeration for the superconducting state.

In reflecting upon where refrigeration is necessary; many, many applications immediately come to mind. These range from infrared-sensor cooling to low temperature physics research; from industrial processing to use of cryogenic fuels; and from food preservation to fundamental curiosity about thermodynamics. After some thought about why these applications need refrigeration, five reasons summarize many of the answers:

- enhancement of the ratio of a coherent signal to the random noise of the background in which the signal exists;

- production of a transformation in properties of matter, e.g., superconductivity or superfluidity;

- a need or desire to increase the energy density of a system, e.g. liquefy hydrogen and oxygen for high specific impulse during combustion;

- environmental control such as air conditioning and food preservation; and

- a fundamental study of the refrigeration process itself, i.e. thermodynamics and associated engineering implementation.

These reasons have motivated over a 100 years of refrigeration development since Cailletet first liquefied air by rapid expansion of the compressed gas.[2] Heat engines were studied well in advance of refrigerators and many of the thermodynamic principles were already discovered before refrigeration studies began.[3] Of special importance was the Carnot factor which determines the work, W required to transfer energy Q from a cold temperature T_C to a hot temperature T_H, i.e. $W/Q = (T_H/T_C - 1)$. This relationship shows how rapidly the minimum required work per unit energy transfer increases as the temperature span between T_H and T_C increases.

The "energy intensive" nature of low temperature refrigeration makes investigations of refrigeration methods at these temperatures very important. The efficiency of refrigeration is usually discussed relative to the Carnot factor, i.e.

$$\eta = (W/Q)\,\mathrm{ideal}/(W/Q)\mathrm{real} = W_{ideal}/W_{real} \qquad (1)$$

While efficiency is critically important for many applications, reliability and lifetime along with cost also have significant impact on refrigeration development. For long-life applications, closed-cycle refrigerators are the best prospect. There are several kinds of gas-cycle mechanical coolers that have been developed over many years. These refrigerators operate on a variety of cycles including Stirling cycle, Vuilleumier cycle, Brayton cycle, and Gifford McMahon cycle with several types of expanders and compressors. The type of refrigerator chosen for a particular application depends on the specifications such as cooling power, cold (load) temperature, and hot (sink) temperature. A variety of choices are available to cryogenic system designers.[4]

During the "energy crisis" of the mid 1970's, many energy intensive processes were reviewed to assess whether higher efficiency could be attained by new methods. Magnetic refrigeration, i.e. refrigeration based on the magnetocaloric effect, was one of the techniques uncovered. In the early 1980's the potential for high reliability and long lifetime from a magnetic refrigerator also created significant interest for space applications.

The purpose of this paper is to review the present state of magnetic refrigeration development. It is a shortened version of a more extensive review being published elsewhere. Earlier reviews on magnetic refrigeration have been written by several authors.[5-8]

THEORY OF THERMO-MAGNETIC EFFECTS

Magnetic Thermodynamics

When matter is subjected to a magnetic field, it generally develops a magnetic moment. The description of the magnetic moment and its interaction with thermal and mechanical properties of the system starts by adding the magnetic work term to the internal energy equation; i.e.[9,10]

$$dU = TdS - PdV + \Sigma\mu_j\, dN_j + \mu_o V_M\, HdM \qquad (2)$$

Here the first term on the right is the heat, the second term is the mechanical work, the third is the chemical work, and the fourth and last term is the magnetic work.

The complete magnetic work term associated with a non-hysteretic magnetic material in a solenoid includes a term $(\mu/2)\int H_e^2\, d\tau$ where H_e is the external magnetic field strength. This term does not involve the thermodynamic system itself but comes from the magnetostatic energy of the empty solenoid. The energy stored in the magnet can be absorbed into the internal energy if desired to give a redefined internal energy. If the reference energy point is the charged magnet and the physical volume and molar quantities are held constant, Eqn (2) becomes

$$dU = TdS + \mu_o V_M\, H_e dM \qquad (3)$$

which is the normal starting point for magnetic thermodynamics. Other thermodynamic functions such as the magnetic enthalpy, Helmholtz free energy and Gibbs free energy can be readily obtained.[11]

Magnetocaloric Effect

The entropy change associated with the temperature and magnetic field dependence of the magnetic material can be expressed in terms of independent variables H and T as

$$dS = (\delta S/\delta T)_H dT + (\delta S/\delta H)_T dH \qquad (4)$$

It is convenient to use H and T as independent variables because the magnetization is usually taken as a function of H and T and most magnetic cycles have isothermal or isofield paths. By use of a Maxwell relationship and the heat capacity at constant field intensity, C_H, Eqn. (4) becomes

$$dS = (C_H/T)\,dT + \mu_o V_M\,(\delta M/\delta T)_H dH \qquad (5)$$

Another useful parameter for discussion of magnetocaloric effect is the adiabatic temperature change upon a field change.

It is obtained from Eqn. (5) as

$$dT = -T/C_H \mu_o V_M (\delta M/\delta T)_H dH \qquad (6)$$

These equations illustrate that the two measurable properties of magnetic materials required for analysis of the magnetocaloric effect are the heat capacity and the magnetization as a function of magnetic field and temperature.

Historical Development

The magnetocaloric effect apparently was demonstrated by P. Langevin[12] with liquid oxygen sometime after 1895 when P. Curie studied the magnetic behaviour of paramagnets and before 1918 when "a new magnetocaloric effect" was reported in nickel by P. Weiss and others.[13,14] It was not long before cooling below 1 K was predicted using the magnetocaloric effect in paramagnets.[15,16] The experimental demonstration[17] of the use of this effect in 1933 to achieve temperature below 1 K lead to a nobel prize in chemistry for W.F. Giauque. Since 1933, the magnetocaloric effect from paramagnetism of electrons and nuclei has been used in laboratories to cool to temperatures as low as 10^{-8} K. Many authors have described the sophisticated techniques for magnetic cooling from near 1 K to milli-, micro-, and nano-Kelvin temperatures.[18-21]

In 1954 Heer et al[22] produced the first semi-continuous refrigerator using the magnetocaloric effect that reached approximately 0.2 K. This device cycled every few minutes and hence its refrigeration power was very low. A superconducting solenoid and better heat switches improved the design.[23] Rosenblum et al. used a similar approach for much lower temperature.[24] All of these refrigerators were low-power devices for operation below 1 K.

In 1966 Van Geuns[25] first proposed a magnetic refrigerator utilizing the magnetocaloric effect above 1 K. He noted that two key problems needed to be solved; the transfer of heat from and to the paramagnetic material in a short time; and the handling of the increasing lattice heat capacity as the temperature increases above 1 K. A regenerative cycle using stablized helium gas as the regenerator and a paramagnetic working material was proposed. It consisted of two isothermal stages and two isomagnetization stages which is a magnetic Stirling cycle.[26] In 1976

G.V. Brown[27,28] demonstrated the use of the magnetocaloric effect in a ferromagnetic material near its Curie temperature for pumping heat. Since 1976, a steady development in the magnetic refrigeration design data base and prototype development has occurred.

DEVELOPMENT OF THERMOMAGNETIC DEVICES

The ultimate objective of the study of magnetic refrigeration is to produce an alternative method of refrigeration which fulfills cooling needs more effectively than existing systems at a lower cost. The minimum output from the efforts underway should be a clear understanding of the fundamental scientific and practical engineering limits that define its characteristics. To discover these limits, it is essential to analyze the concepts, produce the component data base, and to design, build and test prototype devices. These three areas form a basis to organize the many publications on magnetic refrigeration.

Analysis of Magnetic Refrigeration Concepts

The theory suggests that magnetic refrigerators should be simple devices. However, there are many parameters in the system and the final designs involve a selection among several coupled variables. The first considerations in the design of a device are the constraints provided by the specification of the cooling application, e.g., the cooling power, the load temperature, the sink temperature, cool-down time, reliability, lifetime, vibration, etc. Given the specifications, the selection process proceeds by making the following selections among several possibilities in each category:

- magnetic material;
- magnetic cycle;
- magnetic field strength and profile;
- frequency of operation;
- material motion;
- magnet configuration;
- heat transfer mode;
- heat transfer geometry;
- structural geometry;
- refrigerator integration components such as pumps, external heat exchangers, drive mechanisms, etc.; and
- instrumentation.

Obviously there are many combinations among these variables. Because efficiency is one of the strong constraints on refrigerator designs, the logical selection among the combination of variables can be approached by the minimization of entropy according to the second law of thermodynamics[29]. This technique is widely used in the design of all types of thermal equipment but has been especially developed by Bejan and Gaggioli[30-34]. Most specific irreversible entropy producing mechanisms can be identified and used to analyze and eventually optimize design. The technique has been applied to magnetic refrigeration[35-40], but more generally only a simple energy balance analysis is used to model the performance.[41-46] There is much more work to be done in this area, especially given the expense of building carefully engineered refrigerators.

Component Data Base Development

The essential features of all magnetic refrigerators have several types of components in common as shown below:

- magnetic material;
- magnet;
- heat transfer medium;
- magnetic work mechanism;
- heat exchangers for sink and source and coupling;
- structural support; and
- instrumentation/power supplies/control mechanisms.

It is necessary to have sufficient information about all of these components before a detailed design can be completely analyzed. For example, the magnetic refrigerant that can be used depends on the temperature range, i.e. below about 20 K, paramagnetic (antiferromagnetic) materials can be used but above about 20 K, ferromagnetic or ferrimagnetic materials near their Curie point must be used. The magnetic material temperature-entropy diagram is the essential property required for cycle choice. Ten years ago, very little information was available on suitable paramagnetic and ferromagnetic refrigerants. Fortunately, this situation has changed and is still changing rapidly largely due to the efforts of Hashimoto and his colleagues in Japan.

Gadolinium gallium garnet, $Gd_3Ga_5O_{12}$, has been the paramagnetic material predominantly used for refrigerators operating between about 1.5 K and 20 K. Other materials such as $Dy_3Al_5O_{12}$[51], $Dy_3Ga_5O_{12}$[52], and mixtures of similar materials[53] have been investigated.

Recent exciting developments on ferromagnetic refrigerants include investigation of materials suitable for refrigeration between 15 K and 77 K. The RAl_2 mixtures and $RAl_{2.2}$ sintered materials are especially promising.[54-56] Work has also been done on ferromagnetic refrigerants for refrigerators above and just below room temperature.[57-59] Once a material has been identified as a suitable refrigerant, there are many more measurements that need to be performed to provide designers with essential data for calculation of thermal expansions, heat leaks, eddy currents, strain, thermal addenda, forces, etc. The number of references in this area is now well over a 100 and would be the subject of a complete review in itself. Several reviews on magnetic materials have been written as a starting point for those who are interested.[47-50]

Magnetic Refrigerator Development

The development of useful devices generally proceeds through stages such as experimental prototype, engineering prototype, preproduction prototype or flight-qualified prototype. The first stage of development proves that the concept is viable, the second proves the feasability under real conditions, and the third stage either proves it is cost effective or mission compatible. Magnetic refrigerators reported to date are primarily in the first stage of development although some of the units being built now are in the second stage.

A useful way to classify the magnetic refrigerators built is by temperature span between T_C and T_H as given below:

- 0.1 K to 1.5 K;
- 1.5 K to 4.2 K;
- 4.2 K to 20 K;
- 20 K to 77 K; and
- near 300 K

Devices in these temperature ranges will be discussed.

0.1 to 1.5 K

The present application of magnetic refrigerators in this temperature range is to cool bolometers for astrophysical measurements in space.[60-66] The cooling powers required are tens of microwatts but temperature stability is extremely important. The devices that have been built and tested are "single-shot" designs with duty factors of 90-95%. Gas-gap heat switches and extremely large coil constants for the superconducting magnets (T/A rather than the more normal 0.1T/A) are two of the innovations developed for these refrigerators. By temperature controlled feedback to the magnet power supply, extremely stable temperatures have been demonstrated.[65] Presently, a flight qualified model of this type of magnetic refrigerator is being built at NASA-Goddard Space Flight Center.[67] Related development is underway at Ball Brothers in Boulder and at the University of California, Berkeley.

1.5 K to 4.2 K

The biggest application of magnetic refrigerators in this temperature range is to produce superfluid liquid helium. Superfluid helium-cooled superconducting magnets of NbTi wire carry about 25% higher current density at 1.8 K compared to 4.2 K. Most of the devices built[68-79] have used GGG as the working material with the exception of early work where $Gd_2(SO_4)_3 \cdot 8H_2O$ or similar materials[80] were used. Direct contact heat exchange between liquid helium and the paramagnetic working material has been used in most of the designs. The difference in condensation and evaporative conductance coefficients has been cleverly used as well.[81] The seal between superfluid helium and normal helium is accomplished either by an expoxy joint (in the static material designs) or by very close tolerance clearance seals. The cooling powers range from tens of milliwatts to several watts with efficiencies as high as a very impressive 79% of Carnot efficiency.[75]

4.2 K to 20 K

Magnetic refrigerators in this temperature range have been built primarily for attachment to a two-stage regenerative or recuperative cryocooler that operates from 15-20 K up to 300 K. This hybrid refrigerator produces temperatures at or below 4.2 K for helium recondensation or general refrigeration. It is an alternative to a two-stage gas cryocooler with a Joule-Thomson expansion loop attached. Several substantially different units[82-89] have been built with cooling powers in the 0.5 to several watt range. The working material has been almost exclusively GGG either static or in reciprocating or rotary motion. The heat transfer modes used are condensing, evaporative, conduction of the gas in a gap, and convective. The efficiencies have been measured in different ways; if the thermodynamic efficiency of the magnetic material is isolated, efficiencies of 40-70% of Carnot have been achieved. However, if all of the component efficiencies such as drive motors, magnet charging and discharging, etc. are included, efficiencies are much lower. Most of the devices are in the experimental prototype stage but several engineering prototypes are being built such as at Astronautics Corporation of America and Hughes Aircraft Company.[90] Most of the devices operate on the Carnot cycle but two new developments are pursuing the regenerative magnetic cycle.[91-93]

20 K to 77 K

The primary application of a magnetic refrigerator in this temperature range is as an upper stage for a 4 to 20 K or 1.5 to 20 K magnetic refrigerator or as a hydrogen liquefier. The use of liquid nitrogen as a cheap and readily available heat sink also explains why 77 K is the upper temperature. (This type of magnetic refrigerator may have special significance now that high-temperature superconductors can operate well above 77 K.) The key technical significance of development of magnetic refrigerators in this temperature range is that the regenerative or recuperative heat transfer problem must be effectively solved before high efficiencies can be achieved. Although much analysis and preliminary design calculations[94] have been done, only one experimental prototype has been built so far.[95] It is a regenerative device that uses a composite of several sintered $RAl_{2.2}$ compounds as the working magnetic material. The regenerator is about 10 times the mass of the magnetic material and is made of a lead-antimony alloy. The heat transfer is accomplished by conduction of the gas in a very small gap between the magnetic material and regenerator.

Near Room Temperature

The applications near room temperature are heat pumps for waste-energy conservation, refrigerators for air conditioning and general cooling; and in some cases the upper stages of magnetic refrigerators for much lower temperatures. All of the devices built[96-102] in this range use gadolinium as the working material. The designs have reciprocating or rotary motion and some have static magnetic material with charging and discharging magnets. Some are recuperative and others are regenerative but all use convective heat transfer between the working material and heat transfer fluid. All units are experimental prototypes designed to prove the concepts. Cooling powers are in the watt to kilowatt range and reported efficiencies do not include all factors so mean little at this experimental stage. Fully engineered units in this temperature range need to be developed.

SUMMARY

This review is a snapshot of the progress in a rapidly developing technology. The studies of magnetic materials, heat transfer and other components in the data base are expanding to complement the analysis and numerical modeling of potential magnetic refrigerators. Devices are being built and tested to validate the models and increase our understanding of the fundamental limits and engineering difficulties of implementation. It is not easy to build a highly efficient refrigerator of any type, but development to date shows steady improvement toward that goal with magnetic refrigerators.

ACKNOWLEDGEMENTS

It is a pleasure to acknowledge many discussions with members of the Thermo Magnetic Devices Department at Astronautics Technology Center and with many colleagues around the world who share in the excitement of working on magnetic refrigerators. I also thank several key individuals in Government funding agencies and private industry who share a vision of what might be and have had the courage to support the development of magnetic refrigeration because of the promise it offers.

REFERENCES

1. J.G. Bednorz and K.A. Muller Z. Phys. B64: 189 (1986).

2. K. Mendelssohn, "The Quest for Absolute Zero; the Meaning of Low Temperature Physics", Taylor and Francis, Ltd., London, (1977).

3. See, for example, I. Kolin, "The Evolution of the Heat Engine", Longman Group Ltd., London (1972).

4. J.A. Cunningham and R.A. Mollicone, S.P.I.E., V. 253: 170 (1980).

5. W.A. Steyert; J. de Physique 39: C6-1598 (1978).

6. T. Hashimoto, Adv. in Cryog. Eng. 32: 261 (1986).

7. A.F. Lacaze, G. Claudet, A.A. Lacaze, adn P. Seyfert; Proc. ICEC-10, pg. 23; Helsinki, (Butterworth & Co., Guildford, U.K. 1984).

8. J.A. Barclay, Proc. of Cryocooler Conf., Boulder NBS-SP-698, May, 1985.

9. H.B. Callen, "Thermodynamics", J. Wiley & Sons, Inc., New York (1959).

10. J.A. Stratton, "Electromagnetic Theory", McGraw Hill Co., Inc., New York (1941); Chp II.

11. W.C. Overton, Jr., Los Alamos National Laboratory internal report, (1985) unpublished, see also Y. Iwasa, J.L. Smith, Jr., C.P. Taussig - AFWAL-TR-86-3113.

12. L. Goldstein, Emeritus Staff Member T-Division, Los Alamos National Laboratory; private communication.

13. P. Weiss and A. Piccard; C.R. Acad. Sci (Paris) 166: 352 (1918).

14. P. Weiss and R. Forrer, Am. Phys. 5: 153 (1926).

15. P. Debye, Ann. Physik 81: 1154 (1926).

16. W.F. Giauque, J. Am. Chem. Soc. 49: 1870 (1927).

17. W.F. Giauque and D.P. MacDougall, Phys. Rev., 43: 768 (1933).

18. N. Kurti and F. Simon, Proc. Roy. Soc., A149: 152 (1935).

19. C.G.B. Garret, "Magnetic Cooling" (Harvard University Press, J. Wiley & Sons, Inc. (1954).

20. R.R. Hudson, Principles and Applications of Magnetic Cooling, (North Holland, 1972).

21. O.V. Lounasmaa, Experimental Principles and Methods Below 1 K, (Academic Press, 1974).

22. C.V. Heer, C.B. Barnes, and J.B. Daunt, Rev. Sci. Inst. 25: 1088 (1954)

23. J.E. Zimmerman, J.D. McNutt, and H.V. Bohm; Cryogenics 2: 153 (1962).

24. S.S. Rosenblum, H.A. Sheinberg, and W.A. Steyert; Cryogenics 16: 245 (1976).

25. J.R. van Geuns; Phillips Res. Rep. Suppl. 6: (1966) "A Study of a New Magnetic Refrigerating Cycle"

26. There are inconsistencies in the early literature on magnetic refrigeration about names of various cycles. According to first law of thermodynamics in this paper (Eqn 3), the cooresponding gas-magnetic intensive variables are P and .H and the extensive variables are -V and M. For I. Kolin's description of the gas cycles, this definition leads to the following definitions of magnetic cycles: Carnot equals 2 isothermal and 2 isentropic steps; Stirling equals 2 isothermal and 2 isomagnetization steps; Brayton (Ericsson 1833) equals 2 isothermals and 2 isofield steps. All real cycle are probably polytropic but a common nomenclature helps.

27. G.V. Brown, J. Appl. Phys. 47: 3673 (1976).

28. G.V. Brown, IEEE Trans. Magn. MAG-13, 1146 (1977).

29. R.C. Tolman and P.C. Fine, Rev. Mod. Phys. 20: 51 (1948).

30. A. Bejan and J.L. Smith, Jr., Adv. Cryog. Eng. 21: 247 (1975)

31. A. Bejan; J. of Heat Transfer 99: 374 (1977).

32. A. Bejan, in Advances in Heat Transfer, J.P. Hartnett and T.F. Irvine, Jr., (Eds.), Vol. 15, 1-58 (1982).

33. A. Bejan; Entropy Generation Through Heat and Fluid Flow; (J. Wiley & Sons, New York 1982).

34. R.A. Gaggioli and W.J. Wepfer; Energy 5, 823 (1980) (see also several other references by Gaggioli et al).

35. J.A. Barclay, NBS-SP-698, May '85 (Eds. R. Radebaugh, B. Louie and S. McCarthy).

36. F.N. Mastrup; Lecture notes presented at workshop on Magnetic Refrigeration; Los Alamos National Laboratory, September, 1984.

37. J.A. Barclay; "A Comparison of Gas and Magnetic Refrigeration", Proc. 22nd Heat Transfer Conf.; Niagara Falls, NY, Aug. 1984

38. M.E. Wood and W.H. Potter; Cryogenics 25: 667 (1985)

39. J.A. Barclay and S. Sarangi; Cryog. Processes and Equipment, Pg. 51 (1984) - Proc. of ASME Cryog. Symposium Dec. 1984.

40. A.F. Lacaze, R. Beranger, G. Bon Mardion, G. Claudet and A.A. Lacaze, Cryogenics 23: 427 (1983).

41. C.P. Taussig, G.R. Gallagher, J.L. Smith, Jr. and Y. Iwasa; Proc. 4th Int'l Cryocooler Conf., Easton, MD, Sept. 1986.

42. T. Hashimoto, T. Numazawa, and T. Maro; Adv. Cryo. Eng. 29: 597 (1984)

43. N. Yamada, Y. Iwasa, Y. Watanabe, and J.L. Smith, Jr., Proc. ICEC10, Helsinki, Eds. H. Collan, P. Berglund, M. Krusius, pg. 109 (1984).

44. S. Castles, NASA Rept. X-732-80-9, Feb. 1980 "Design of an adiabatic demagnetization refrigerator for studies in astrophysics".

45. J.A. Barclay; Cryogenics 20: 467 (1980).

46. W.A. Stewart; J. Appl. Phys. 49: 1216 and 1227 (1978).

47. T. Hashimoto; Adv. Cryog. Eng. 32: 261 (1986)

48. J.A. Barclay and W.A. Steyert; Cryogenics 22: 73 (1982)

49. T. Hashimoto, Cryog. Eng. (Japan) 20: 10 (1985) (In Japanese)

50. T. Hashimoto, T. Numazawa, M. Shino, and T. Okada; Cryogenics 21: 647 (1981)

51. T. Yazawa, T. Numazawa, T. Hashimoto, T. Kurijama, H. Nakajome, and H. Ogiwara; Proc. ICEC11, Berlin (Butterworth, Guildford, U.K. 1986) pg. 275, also: R. Li et al. Adv. Cryog. Eng. 32: 287 (1986).

52. A. Tomokiyo, H. Tayama, T. Hashimoto, T. Aomine, M. Nishida, and S. Sakaguchi; Cryogenics 25: 271 (1985).

53. B. Daudin, A.A. Lacaze, and B. Salce; Cryogenics 22: 439 (1982).

54. C.B. Zimm, J.A. Barclay, and W.R. Johnson; J. Appl. Phys. 55: 2609 (1984).

55. T. Kuzuhara, et.al.; Proc. ICEC11, Berlin (Butterworth, Guildford, U.K. 1986) pg. 280.

56. M. Sahashi, et.al.; Proc. INTERMAG '87, Tokyo, April 1987; paper DH-09 (to be published).

57. H. Osterreicher and F.T. Parker; J. Appl. Phys. 55: 4334 (1984)

58. P.K. Ghosh and S.K. Dutta Roy; Indian J. of Pure and Applied Phys. 23: 362 (1985).

59. G.F. Green, W.G. Patton, and J. Stevens; The magnetocaloric effect of some rare earth metals, 1987 Crogenic Engineering Conference, St. Charles, IL.

60. P. Kittel, Cryogenics 20: 599 (1980).

61. P. Kittel, J. Energy 4: 266 (1980).

62. S. Castles, NASA Rept X-732-80-9, Feb. 1980.

63. P. Kittel, Physica 108B: 1115 (1981).

64. R.D. Britt and P.L. Richards; Int'l. J. Infrared and Millimeter Waves 2: 1083 (1981).

65. P. Kittel, Cryogenics 23: 477 (1983).

66. P. Kittel, Adv. Cryog. Eng. 27: 745 (1982).

67. S. Castles, private communication.

68. W.P. Pratt, Jr., S.S. Rosenblum, W.A. Steyert, and J.A. Barclay, Cryogenics 17: 689 (1977).

69. J.A. Barclay, O. Moze, and L. Paterson; J. Appl. Phys. 50: 5870 (1979).

70. C. Delpuech, et al.; Cryogenics 21: 579 (1981).

71. A.F. Lacaze, A.A. Lacaze, R. Beranger, and G. Bon Mardion; Proc. ICEC-9, Kobe, Japan (Butterworth and Co., Guildford, UK 1982), pg. 14.

72. A. Lacaze, Doc. Ing. Thesis, Grenoble, Oct. 1982 (in French).

73. A.F. Lacaze, et al.; Adv. Cryog. Eng. 27: 703 (1982).

74. A.F. Lacaze, et al.; Cryogenics 23: 427 (1983).

75. A.F. Lacaze, et al.; Adv. Cryog. Eng. 29: 573 (1984).

76. Y. Hakuraku and H. Ogata; Jap. J. Appl. Phys. 25: 140 (1986).

77. Y. Hakuraku and H. Ogata; J. Appl. Phys. 60: 3266 (1986).

78. Y. Hakuraku and H. Ogata; Jap. J. Appl. Phys. 24: 1111 (1985).

79. Y. Hakuraku and H. Ogata; Teion Kohgaku 19: 311 (1984), (in Japanese).

80. C. Delpuech; Doc. Ing. Thesis, Grenoble, Nov. (1980) (in French).

81. Y. Hakuraku and H. Ogata; Cryogenics 26: 171 (1986).

82. D.D. Deardorf and D.L. Johnson; TAD Progress Rept 42-78, Jet Propulsion Laboratory, April-June 1984.

83. H. Nakagome, et al.; Adv. Crog. Eng. 29: 581 (1984).

84. H. Nakagome, et al.; Proc. ICEC-11, Berlin (Plenum Press, 1986) pg. 246

85. T. Kuriyama, et al.; Proc. ICEC-11, Berlin (Plenum Press, 1986) pg. 251

86. T. Numazawa, T. Hashimoto, and H. Nakagome; Adv. Cryog. Eng. 31: 771 (1986).

87. J.A. Barclay, et.al.; Adv. Cryog. Eng. 31: 743 (1986).

88. H. Nakagome, et al.; Adv. Cryog. Eng. 31: 753 (1986).

89. T.F. Fujita, et al.; Adv. Cryog. Eng. 31: 763 (1986).

90. F.N. Mastrup, Hughes Aircraft Company; private communication.

91. C.P. Taussig, G.R. Gallagher, J.L. Smith, Jr., and Y. Iwasa Proc. 4th Int'l Cryocooler Conf., Easton, MD., Sept. 1986.

92. G.M. Claudet; Adv. Cryog. Eng. 31: 733 (1986).

93. P. Seyfert; Grenoble; private communication.

94. J.A. Barclay, et.al.; Los Alamos National Laboratory Progress Reports; 1983-1984-1985; (unpublished).

95. T. Hashimoto et al., private communication; to be reported at the 1987 Cryogenic Engineering Conference.

96. G.V. Brown; J. Appl. Phys. 47: 3673 (1976).

97. G.V. Brown; Am. Soc. of Heating, Refrig., and Air Cond., Engineers Trans. 87: 783 (1981).

98. J.A. Barclay and W.A. Steyert; Electric Power Research Institute Final Report EL-1757, April 1981.

99. S.S. Rosenblum, W.A. Steyert, and W.P. Pratt, Jr., Los Alamos National Laboratory Report, LA-6581 (May 1977).

100. G. Patton, G. Green, J. Stevens, and J. Humphrey; Proc. 4th Int'l Cryocooler Conf., Easton, MD, Sept. 1986.

101. L.D. Kirol; Idaho National Engineering Laboratory; private communication; the 1987 Cryogenic Engineering Conference.

102. W. Peschka; DFVLR; Stuttgart; private communication.

RECENT PROGRESS IN MAGNETIC REFRIGERATION STUDIES

T. Hashimoto, T. Yazawa, R. Li, T. Kuzuhara
and K. Matsumoto

Department of Applied Physics,
Tokyo Institute of Technology,
Tokyo, Japan

H. Nakagome and M. Takahashi,

Energy Science & Technology Lab., Toshiba R & D Center,
Kawasaki, Japan

M. Sahashi, and K. Inomata

Metals & Ceramics Lab., Toshiba R & D Center,
Kawasaki, Japan

A. Tomokiyo and H. Yayama

College of General Education, Kyushu University,
Fukuoka, Japan

ABSTRACT

After the 1985 Cryogenic Engineering Conference, two directions for the fundamental investigations on the magnetic refrigeration to expand the refrigeration range above ∿15 K have been developed by our group. One is the improvement of the refrigeration characteristics able to refrigerate from ∿20 K for the Carnot magnetic refrigerator and the other is the fundamental study of the Ericsson magnetic refrigerator. As for the former purpose, we used a new magnetic material, $Dy_3Al_5O_{12}$, as the refrigerant in a reciprocating Carnot magnetic refrigerator instead of $Gd_3Ga_5O_{12}$. Consequentially, we succeeded in expanding the refrigeration range. As for the latter, we have established the method to make the refrigerant suitable for the ideal Ericsson cycle including two kinds of iso-magnetic field processes. Now, the investigation of the Ericsson magnetic refrigeration cycle and refrigerator is starting.

INTRODUCTION

Recently, we have developed several kinds of Carnot magnetic refrigerators for He liquefaction[1-3] and verified the excellent refrigeration characteristics, especially the high refrigeration efficiency of these[4,5]. The temperature ranges of the operating cycle in these machines are, however, limited to only the range lower than 15 K by the magneto-caloric characteristics of the magnetic refrigerant $Gd_3Ga_5O_{12}$(GGG) under the condition that applicable magnetic field strength for the refrigeration operation is equal to or less than 6 T.

To expand the refrigeration range above ~15 K, two directions of the fundamental investigations are thought to be necessary. One direction is to utilize a new paramagnetic refrigerant instead of GGG for the Carnot magnetic refrigerator, whose magnetic entropy change is sufficient to refrigerate still near 20 K. The other is the development of the Ericsson magnetic refrigerator.

In the former work, one of the authors discussed at the 1985 CEC-ICMC joint conference at Boston[6] the anisotropic spin system having the large effective g-value is a more promising refrigerant compared to the isotropic spin system whose g-value is equal to 2. As an example, in Fig.1[7] are shown the magnetic entropy change ΔS_J of $Dy_3Al_5O_{12}$(DAG) in the case of applying the magnetic field parallel to the <111> crystalline field axis and that in GGG. As the effective spin value J is equal to 1/2 in DAG in contrast to the largest value J = 7/2 in GGG, the absolute value of entropy S(T,B) in DAG is ln(1/2)/ln(7/2) times smaller than that in GGG. However, since the effective g-value is 5 times that in GGG, ΔS_J of DAG is larger than that of GGG above 15 K range.

In the latter case we have to clarify two subjects. The first is how to make the refrigerant suitable for the Ericsson magnetic refrigeration cycle including two kinds of iso-magnetic field processes. The second is how to make the regenerator applicable to an actual Ericsson magnetic refrigerator.

In the present paper, therefore, we will select and report on two topics from the results of our recent investigations corresponding to the two subjects mentioned above. First, we will show the experimental results which verify the effective improvement of the refrigeration characteristics obtained by using the DAG single crystal as the refrigerant instead of GGG in the reciprocating Carnot magnetic refrigeration. Then,

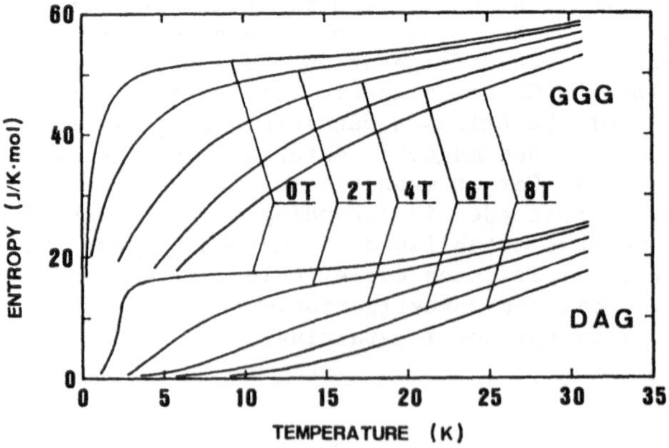

Fig. 1. The variation of the entropy S(T,B) in $Dy_3Al_5O_{12}$(DAG) as a function of temperature in the various magnetic fields B. For the comparison that in $Gd_3Ga_5O_{12}$(GGG) is also shown.

we will show the experimental investigation of the method to make the magnetic refrigerant in order to realize of the ideal Ericsson magnetic refrigeration cycle. The experimental results on the characteristics of the Ericsson magnetic refrigerator and its regenerator will be reported in another paper at this conference[8].

IMPROVEMENT OF REFRIGERATION CHARACTERISTICS IN THE CARNOT MAGNETIC REFRIGERATOR

In the temperature range above ∿15 K the thermal excitation energy of spin ∿kT becomes comparable to or more than the Zeeman splitting energy of spin levels, $g\mu_B B$ induced by the applied magnetic field ∿6 T. Moreover, in the magnetic substances whose Debye temperature is ∿600 K, the lattice entropy becomes not negligible compared with the magnetic entropy, and, as a result, it is difficult to liquefy He gas from the temperature above ∿15 K when using GGG as a refrigerant and a magnetic field lower than ∿6T. Therefore, to obtain the large splitting width of energy level, we have to use a paramagnetic material including magnetic ions that have the large g-factor as shown previously[6].

From this point of view, one of the most promising refrigerants for the Carnot magnetic refrigerator is $Dy_3Al_5O_{12}$(DAG) in which the effective g-value of Dy ion is 10.6 as the applied magnetic field is parallel to the <111> direction[7] and, moreover, whose thermal conductivity is comparable to that of GGG. For the purpose of clarifying the superiority of DAG over to that of GGG, a comparison between the refrigeration characteristics of DAG and GGG in the demagnetization process of the static magnetic refrigerator[1] is shown as a function of the initial temperature of demagnetiza-

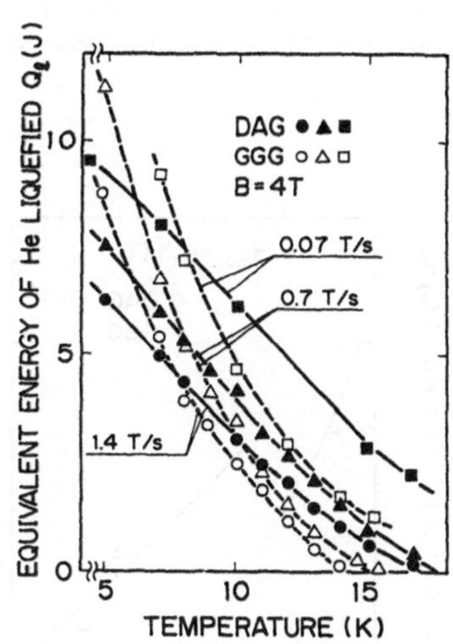

Fig. 2. The He liquefaction power of the static magnetic refrigerator using DAG and GGG in the demagnetization process given as a function of the initial temperature of the demagnetization.

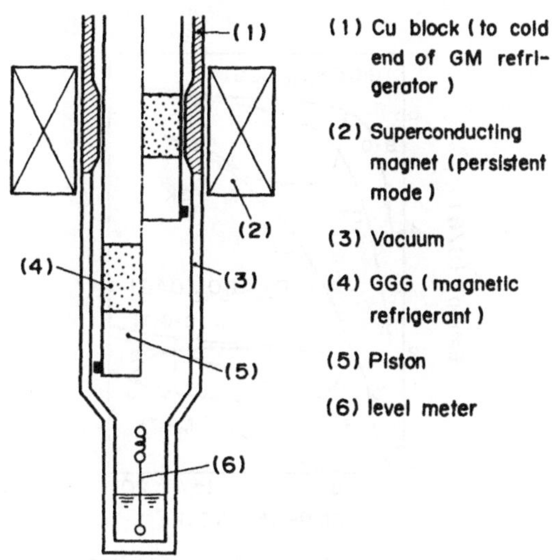

Fig. 3. A schema of the important parts of the reciprocating magnetic refrigerator used for the experiment in refrigeration characteristics in the Carnot cycle.

tion in Fig. 2 of Ref. 9. This result clearly shows that the refrigeration power of DAG is larger than that in GGG above ∿15 K range.

We have investigated the refrigeration characteristics of several kinds of the Carnot magnetic refrigerators, whose refrigerants are DAG single crystals instead of GGG. As an example of those investigations, we will present an experimental result of the refrigeration characteristics of the reciprocating Carnot magnetic refrigerator using DAG as refrigerant compared with GGG.

In this study, we used a test machine similar to that in the previous paper[5], the most imporant part of which is shown in Fig. 3. The actual operating refrigeration cycle in this machine on the entropy plan was estimated both from the temperature variation of the refrigerant caused by the piston motion and from the magnetic field gradient of the machine. Therefore, experimental error in the above estimation is thought to be less than a few per cent in the case of slow cycle frequency less than ∿0.1 H_Z from the measurement of the relaxation time of the heat transfer between the refrigerant and the thermometer.

In the temperature range from 20 K to 4.2 K, we operated this machine using both $Dy_3Al_5O_{12}$(DAG) and $Gd_3Ga_5O_{12}$(GGG) as a refrigerant.

Figure 4 shows a typical example of the experimental results on the refrigeration cycle in DAG and GGG, where the cycle frequencies of those are 0.15 Hz and the refrigeration powers Q of those two are equal to each other and 0.6 J/cycle. It is a notable thing that in the DAG case the heat reservoir of high temperature side at ∿17 K is higher than that in GGG(= 13 K) and is usable to obtain the same refrigeration power as GGG.

In order to clarify the difference between the refrigeration characteristics in DAG and GGG the experimental results of the amount of He liquefied in operating under the identical optimum condition with DAG and

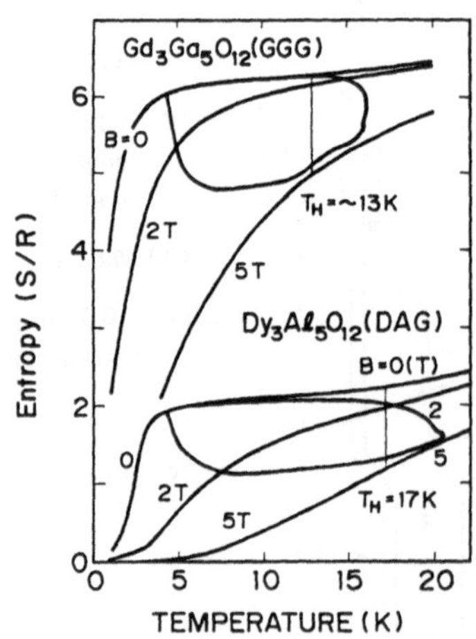

Fig. 4. An example of the experimental results on the comparison of the refrigeration cycles of GGG and DAG whose refrigeration powers are equal to each other.

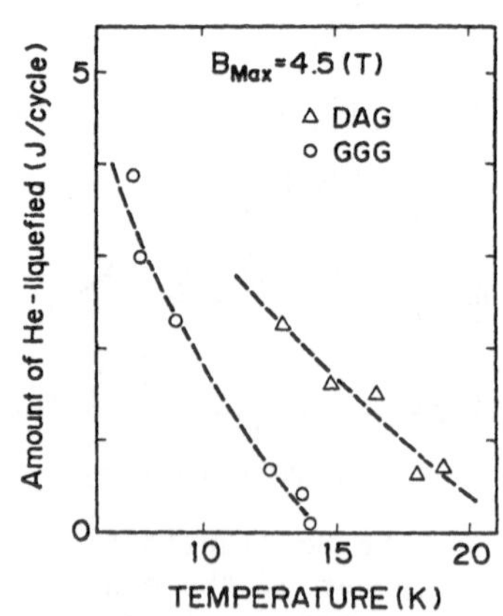

Fig. 5. The refrigeration power for He liquefaction in the reciprocating machine using DAG and GGG. The horizontal scale shows the temperature of the heat reservoir.

GGG are shown in Fig. 5. The horizontal scale shows the temperature of the heat reservoir. This result clearly shows that in the case of the maximum magnetic field B_{max} equal to 4.5 T the ability of DAG in He liquefaction fairly exceeds that of GGG in the temperature range above ∿10 K, and, moreover, it is notable that DAG is able to liquefy even from ∿20 K in spite of the existence of large refrigeration losses in this test machine[5] compared with the previous prototype[2].

Conclusively, the refrigeration characteristics of DAG are superior to those of GGG for the purpose of expanding the initial temperature of refrigeration up to ∿20 K. This experimental result is very important in showing the possibility that a high efficiency cascade refrigerator can be constructed by connecting the Carnot magnetic refrigerator with the gas refrigerator near 20 K. Therefore, for the refrigerator in this temperature range, the most important subject remaining in the future is to develop the magnetic refrigerant of a Dy compound in which the principal axis of the anisotropic crysalline field applied to Dy ions is unique, such as in $DyPO_4$.

FUNDAMENTAL INVESTIGATION FOR THE ERICSSON MAGNETIC REFRIGERATOR

After Brown[10], there is no notable advance in the regenerative magnetic refrigerator in comparison with that in the Carnot. One of the most important reason about this situation is thought to be that the Carnot magnetic refrigerator is suitable for the feasibility study of the magnetic refrigeration because of its simplicity of structure and operation in comparison with the Ericsson magnetic refrigerator. However, if we intend to make high power refrigeration having large refrigeration range, for instance from ∿25 K to 4.2 K, we have to select the Ericsson cycle instead of the Carnot cycle because the starting temperature of refrigeration is limited to less than ∿20 K as long as the paramagnetic substance is used as the refrigerant, as shown in the previous section. Moreover, in the light of the advances of investigations of high Tc superconductors, the possibility of the magnetic refrigerator operating in a higher temperature range seems to be greater. Therefore, our fundamental investigation of the Ericsson magnetic refrigeration had started in the temperature range from 77 K to 4.2 K which was selected since we have a easily usable heat source, that is, liquid N_2 and liquid He.

The most probable Ericsson magnetic refrigeration cycle includes two kinds of iso-magnetic field processes, which satisfy the Carnot condition, is shown in Fig. 6. The figure shows that this cycle is composed of four processes: iso-thermal magnetization (A → B), iso-magnetic (strong) field (B → C), iso-thermal demagnetization (C → D) and iso-magnetic (weak) field (D → A). In order to realize this cycle we have to configure the magnetic refrigerant and regenerator to produce the B → C (absorbing heat from refrigerant) and D → A processes (giving the heat to refrigerant). We will show the investigation of the refrigerant[11].

Since the thermal energy increases considerably, above 20 K a ferromagnetic substance having enough internal magnetic field to arrange the spin system must be used for the refrigerant in the temperature range near the Curie temperature. However, its magnetic entropy change ΔS_J varies as a function of temperature in the constant magnetic field as shown in Fig. 7[6,12,13], and this temperature variation of ΔS_J is not suitable for the Ericsson cycle. So, we tried to make the complex type of magnetic refrigerant necessary to satisfy the condition of the constant ΔS_J in the refrigeration range.

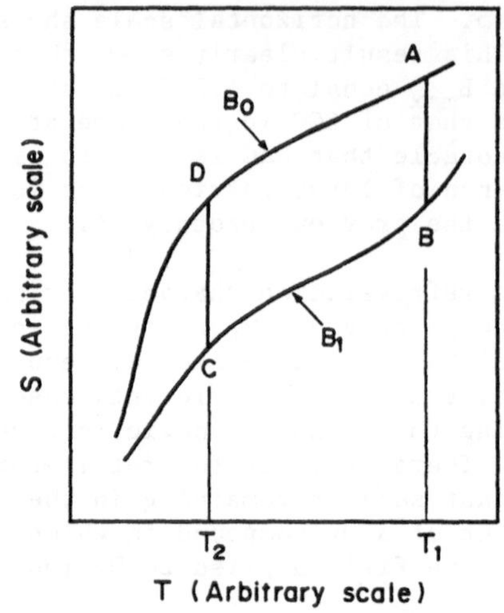

Fig. 6. The ideal Ericsson cycle satisfying the Carnot condition: A→B is the iso-thermal magnetization process, B→C the iso-magnetic (strong) field process, C→D the iso-thermal demagnetization process and D→A the iso-magnetic (weak) field process.

Figure 8 shows the layer structural complex magnetic refrigerant[13]. These materials are made by the press and sintering method. Since the refrigerant must have the high thermal conductivity, it is necessary to make the material dense. Therefore, we used the liquid phase sintering method. In the R-Al system there are two kinds of homogeneous compounds RAl_2 and RAl_3, and the melting point of RAl_3 is lower than that of RAl_2. Therefore, in selecting the mixed compound $RAl_{2+\delta}$ of those two kinds of compounds, those powdered materials (~few μm) were pressed and annealed in Ar atomsphere at 1100°C for 2 hours which is almost the average of the melting

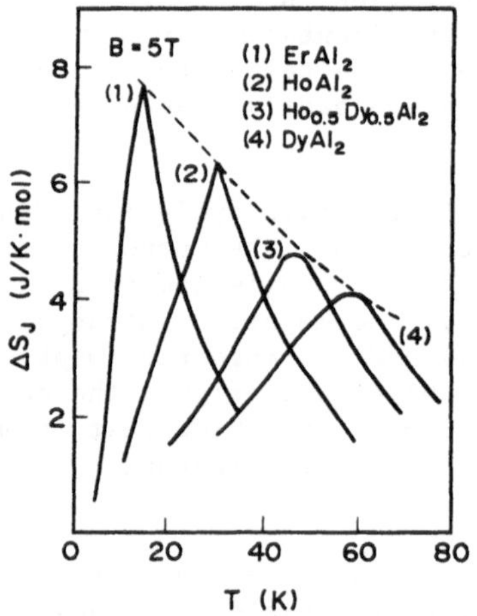

Fig. 7. The magnetic entropy change ΔS of the ferromagnetic RAl_2 series given as a function of temperature, induced by the magnetic field change from 5 T to 0 T.

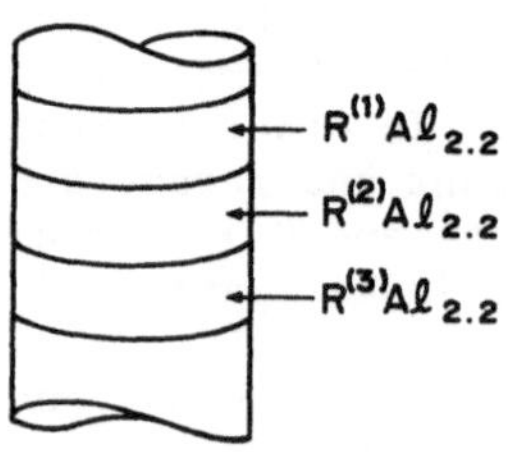

Fig. 8. Schematic figure of the layer structural complex magnetic refrigerant made by the press and sintering method.

temperatures of those two. From the measurement of the density in the resultant material obtained as a function of excess amount δ of Al in the $RAl_{2+\delta}$ series, the final value of δ in the constituent material used for the layer structural complex material, as shown in Fig. 8, is defined as $\delta = 0.2$[14].

In this investigation, using four kinds of $RAl_{2.2}$ compounds, $ErAl_{2.2}$, $HoAl_{2.2}$, $(Ho_{0.5}Dy_{0.5})Al_{2.2}$ and $DyAl_{2.2}$, we made the layer a structurally complex material[14]. In the sintering process at high temperature, the new solid solutions different from the composite materials of the four kinds of layers at the interlayer region of those by the atomic diffusion of R ions were probably produced. It is confirmed that the thickness of the new layer created by the atomic diffusion is less than ∿200 μm[14], and, therefore, these new layers produced in the sintering process gives little effect on the character of the resultant refrigerant.

Based on the investigation of the magnetic entropy of the $RAl_{2.2}$ system[10], we selected the composite material and its ratio to be $[ErAl_{2.2}]_{0.306}[HoAl_{2.2}]_{0.153}[(Ho_{0.5}Dy_{0.5})Al_{2.2}]_{0.025}[DyAl_{2.2}]_{0.516}$. Using the static caloric measurement technique, we observed the specific heat C(T,B) of the resultant layer structural material in the temperature range from 4.2 K up to 77 K in the various magnetic fields B. The experimental error is less than 5 %[13].

The experimental results are shown in Fig. 9. In B = 0 the four kinds of peaks correspond to the Curie temperatures of composite materials in each layer, although each peak is not so sharp in comparison with that of an usual homogeneous magnetic material. It is thought that the large local fluctuation of the composition exists in the two phase solid solution $RAl_{2.2}$.

Using the thermodynamic relation, the entropy S(T,B) of this resultant material is given by the following formula:

$$S(T,B) = \int_0^T \{C(T,B)/T\}dT. \qquad (1)$$

Using eq. (1) and the experimental results shown in Fig. 9, we

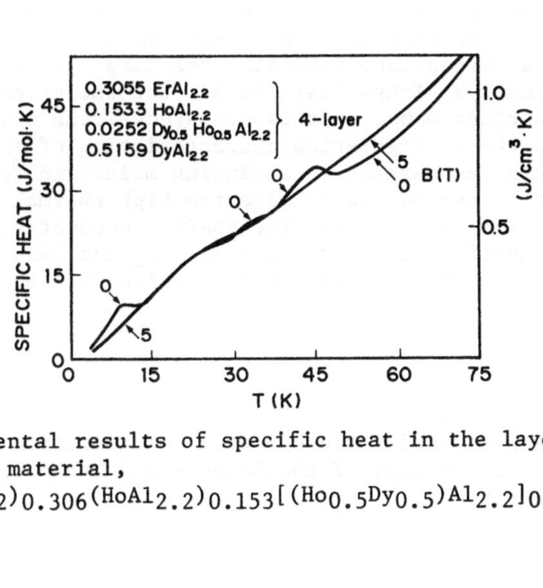

Fig. 9. Experimental results of specific heat in the layer structural complex material, $(ErAl_{2.2})_{0.306}(HoAl_{2.2})_{0.153}[(Ho_{0.5}Dy_{0.5})Al_{2.2}]_{0.025}(DyAl_{2.2})_{0.516}$.

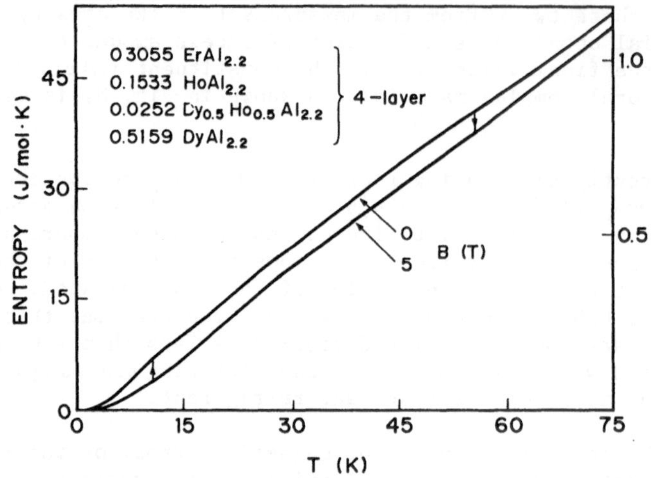

Fig. 10. Entropy of the layer structural complex material calculated from the experimental results shown in Fig.9 by use of Eq.1. ΔS_J of this complex material is constant from ∿50 K to ∿15 K.

obtained the entropy of the layer structural complex refrigerant as shown in Fig. 10. This result clearly shows that in the wide temperature range from 15 K up to 55 K the magnetic entropy change produced by the change of the magnetic field from 0 T to 5 T, $|\Delta S_J(T,5)| = S_J(T,0) - S_J(T,5)$, is almost constant and, moreover, that the absolute value of $\Delta S_J(T,5)$ is 5 J/K·mol large enough to refrigerate, which is comparable to that of DAG at ∿18 K.

Conclusively, from the above experimental results it is clear that it is the only possible way to make the complex magnetic material using the ferromagnetic material series, whose Curie temperature is controllable, for the production of the refrigerant suitable for the ideal Ericsson magnetic refrigerator.

Now our group is investigating additional areas. For the Carnot refrigerator, a compact refrigerator to liquefy the evaporated He gas in the cooling system of the superconducting magnet is going to be developed by Toshiba Co. A new refrigerant including the large g-value magnetic ions is being developed by Tokyo Inst. Tech. For the Ericsson refrigerator, as shown in another paper in this conference, the development of the regenerator and the refrigeration characteristics of the Ericsson refrigerator using this regenerator is advancing mainly in Tokyo Inst. Tech. For the refrigerant, a new method to make the high thermal conductive refrigerant has been developed because the thermal conductivity of the layer structural magnetic compounds, such as RAl_2 and RNi_2 is 10^{-2} times lower than that of Au. These experimental results will be published in the near future.

ACKNOWLEDGEMENTS

This study was performed through Special Coordination Funds for Promoting Science and Technology of the Science and Technology Agency of the Japanese Government.

REFERENCES

1. H. Nakagome et al., The Helium magnetic refrigerator I - Development and experimental results -. in: "Advances in Cryogenic Engineering", Vol.29, Plenum Press, New York, (1984), p.581.
2. H. Nakagome et al., Reciprocating magnetic refrigerator for helium liquefaction, in: "Advances in Cryogenic Engineering", Vol.31, Plenum Press, New York (1986), p.753.
3. H. Nakagome et al., A rotating magnetic refrigerator for helium liquefaction, in: "Proc. Eleventh Intl. Cryo. Engr. Conf. ICEC11", Butterworth, (1986), p.246.
4. T. Numazawa, T. Hashimoto and H. Nakagome, Improvement of liquefaction efficiency of the heat pipe type magnetic refrigerator, in: "Advances in Cryogenic Engineering", Vol.31, Plenum Press, New York (1986), p.771.
5. T. Fujita et al., An experimental study simulating the helium liquefaction process in a reciprocating magnetic refrigerator, in: "Advances in Cryogenic Engineering", Vol.31, Plenum Press, New York (1986), p.763.
6. T. Hashimoto, Recent investigations on refrigerants for magnetic refrigerator, in: "Advances in Cryogenic Engineering", Vol.32, Plenum Press, New York (1986), p.261.
7. R. Li et al., Magnetic and thermal properties of $Dy_3Al_5O_{12}$ as a magnetic refrigerant, in: "Advances in Cryogenic Engineering", Vol.32, Plenum Press, New York (1986), p.287.
8. K. Matsumoto, T. Ito and T. Hashimoto, An Ericsson magnetic refrigerator for low temperature, to be published in Advances in Cryogenic Engineering Vol.33.
9. T. Yazawa et al., The characteristics of $Dy_3Al_5O_{12}$ as a refrigerant for a Carnot magnetic refrigerator, in: "Proc. Eleventh Intl. Cryo. Engr. Conf.", Butterworths, (1986), p.275.
10. G.V. Brown, Magnetic heat pump near room temperature, J. Appl. Phys. 47 : 3673 (1976).
11. T. Hashimoto et al., A new method for producing the magnetic refrigerant suitable for the Ericsson magnetic refrigerator. in: "Proc. Intermag. Conf." (1987) to be published IEEE.
12. T. Hashimoto et al., Investigations on the possibility of the RAl_2 system as a refrigerant in an Ericsson type magnetic refrigerator, in: "Advances in Cryogenics Engineering - Material", Vol.32, Plenum Press, New York (1986), p.279.
13. A. Tomokiyo et al., Specific heat and entropy of RNi_2 (R: rare earth heavy metals) in magnetic field, in: "Advances in Cryogenics Engineering - Material", Vol.32, Plenum Press, New York (1986) p.295.
14. T. Hashimoto et al., New application of complex magnetic materials to the magnetic refrigerant in an Ericsson magnetic refrigerator, to be published in J. Appl. phys., Vol.62, 1987.

AN ERICSSON MAGNETIC REFRIGERATOR FOR LOW TEMPERATURE

Koichi Matsumoto, Takatoshi Ito and Takasu Hashimoto

Department of Applied Physics, Faculty of Science
Tokyo Institute of Technology
Tokyo, Japan

ABSTRACT

An Ericsson magnetic refrigerator has been designed and built for research in magnetic refrigeration in the temperature range of 20 - 77 K. In this temperature range, both magnetic refrigerant and refrigeration cycle are different from those below 20 K. The main features of our magnetic refrigerator were as follows: magnetic refrigerant and regenerator material were made of ferromagnetic material and lead respectively, with the thermal conductivity of gaseous helium used to transfer heat between the magnetic refrigerant and the regenerator material. The initial experimental results are described in this paper.

In our experiment, sintered $DyAl_{2.2}$ was the magnetic refrigerant. As an initial experimental result, at the cycle of 300 s and a field of 5 T, regenerator produced a temperature of 50.3 and 58.7 K at the cold and hot end, respectively, while the magnetic refrigerant changed in temperature between 48.3 and 59.1 K. This temperature change of $DyAl_{2.2}$ was about 1.7 times as large as the maximum adiabatic temperature change with the changing field from 0 to 5 T. The key problems in this design are discussed.

INTRODUCTION

Magnetic refrigeration makes use of the magnetocaloric effect where magnetic material changes in temperature, absorbs, and exhausts heat with changes in magnetic field. Magnetic refrigeration has following advantages compared with gas refrigeration: high efficiency in principle, compactness, high reliability, etc. Magnetic refrigeration has been studied by some investigators in various temperature range[1], especially in the range of 1.7 - 4.2 K [2,3] and 4.2 - 15 K [4,5,6] .

Magnetic refrigerators are composed of two parts: magnetic material and a heat transfer system. Our investigation[7] has shown that the RAl_2 series which R is rare earth atom is a promising candidate for a magnetic refrigerant between 20 - 77 K. As for the thermal cycle, it is necessary to apply regenerative cycles instead of a Carnot cycle, because the lattice entropy grows almost as large as the magnetic entropy above 20 K. Moreover, the high entropy density of the magnetic material causes many difficulties in solving the heat transfer problem.

Lead was used in our apparatus. The thermal conductivity of a gas is expected to be useful for the heat transfer between the two solids, magnetic substance and regenerator material. We previously reported the experimental results on testing a regenerator based on this heat transfer mechanism[8]. In this paper, we describe the initial experimental results of the Ericsson magnetic refrigerator.

DESIGN SUMMARY

The ideal Ericsson cycle is illustrated in the entropy-temperature diagram of a ferromagnetic material, Fig. 1, compared with the Carnot cycle. It is clear from Fig. 1 that the Ericsson cycle (1-2-3-4) has a temperature span much greater than the Carnot cycle (1'-2'-3-4'), and utilizes the entropy change more effectively. The Ericsson cycle consists of the following processes:

- 2-3 : isothermal magnetization process, $B = B_1$ to B_2,
- 3-4 : isofield cooling process, $B = B_2$,
- 4-1 : isothermal demagnetization process, $B = B_2$ to B_1,
- 1-2 : isofield heating process, $B = B_1$.

Since the slopes of 1 to 2 and 4 to 3 aren't generally the same, this ideal cycle can't be operated without magnetic field control in these processes. In fact, an appropriate magnetic field control is very complicated. Without field control, a certain modified Ericsson cycle must be the steady cycle in which the heat transferred into (Q'_R) and out of (Q_R) the regenerator are equal.

The magnetic substance is cooled and heated when it exchanges heat with the regenerator which has temperature gradient. Our device operated on the Ericsson cycle with fixed magnetic material and moving regenerator, because our superconducting magnet had the short field length of 12 cm. As a result, magnetization and demagnetization processes were obtained by charging and discharging the magnet, and the regenerator was driven through a rod which was, in turn, driven by a motor at room temperature.

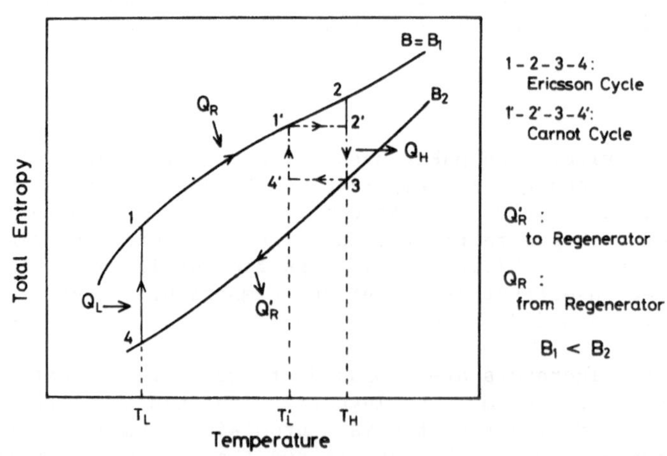

Fig. 1. The Entropy-Temperature diagram for a ferromagnetic material. The solid lines are the curves of constant magnetic field, B_1, B_2. The cycle (1-2-3-4) is the Ericsson cycle and the cycle (1'-2'-3-4') is the Carnot cycle.

The operated cycle is schematically shown in Fig. 2:

(A) charging the magnet from B_1 to B_2, then magnetic substance increased in temperature and rejected heat,
(B) moving the regenerator upward, where the magnetic substance was cooled down in the high field, B_2,
(C) discharging the magnet to B_1, then the magnetic substance decreased in temperature and absorbed heat,
(D) moving the regenerator downward, where the magnetic substance was heated in the field of B_1.

Then, the Ericsson cycle was complete. The temperature of the high end was controlled with gaseous helium and heater in order to maintain it constant.

Figure 3 shows the schematic of the experimental apparatus without the driving motor. In addition, an enlarged cut away view shows the most important part, the magnetic substance and the regenerator. Table 1 summarizes the design parameters of the apparatus.

When thermal conductivity of a gas is the heat transfer mechanism, heat transfer coefficient is obtained by

$$h = \kappa / \delta , \tag{1}$$

where κ is thermal conductivity of the gas and δ is the gap between the surfaces of the magnetic substance and the regenerator matrix. A precision planer surface was machined on the heat transfer surfaces of the regenerator and magnetic material. Hard lead, which is an alloy of lead and antimony, was used as the regenerator material, since pure lead is too

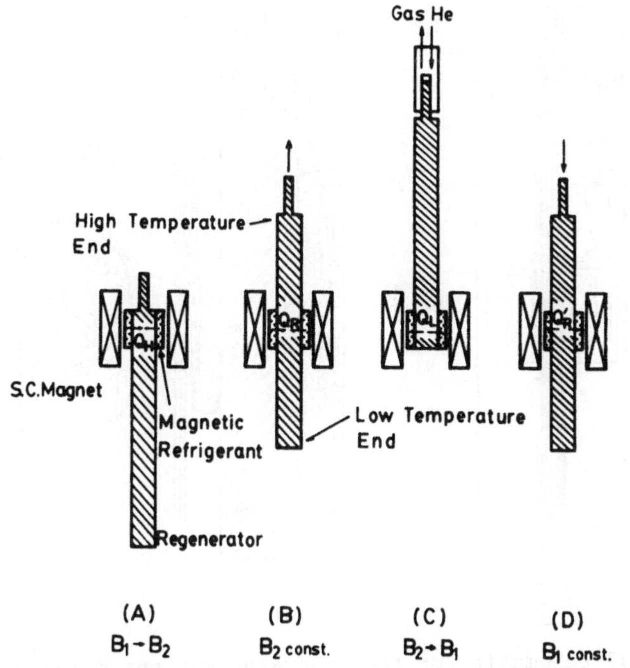

Fig. 2. Operation principle of the apparatus: (A) magnetization from B_1 to B_2 , (B) isofield (B_2) cooling, regenerator is driven up, (C) demagnetization from B_2 to B_1, (D) isofield (B_1) heating, regenerator is driven down.

Table 1. Testing Apparatus Design Parameters

Working substance		
material	-	$DyAl_{2.2}$,sintered (0.55 mol., 20 x 80 x 7 mm, 2 pieces)
Regenerator		
material	-	hard lead
volume, length	-	377 cm^3 , 49 cm
Volume ratio of regenerator to magnetic material	-	16.8
Regenerator moving stroke	-	40 cm
Heat transfer area of magnetic material	-	30.4 cm^2
Heat transfer fluid	-	gaseous helium at 1 atm

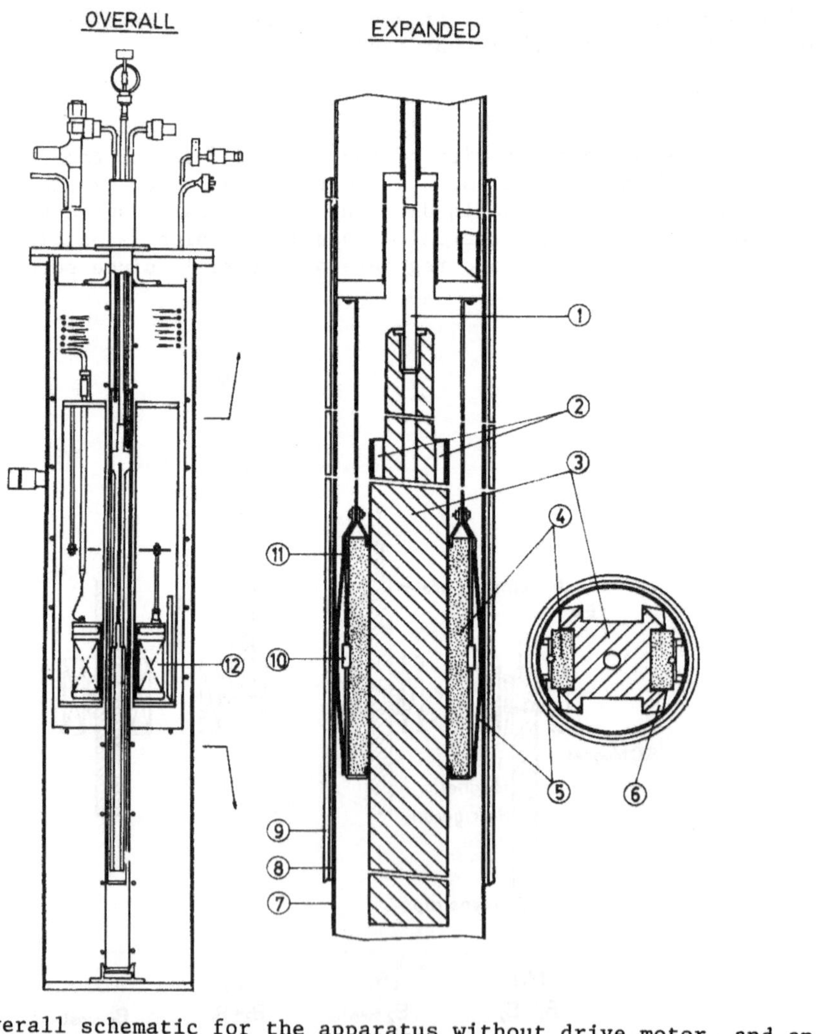

Fig. 3. Overall schematic for the apparatus without drive motor, and an expanded diagram of the main portion. (1. drive rod, 2. heater, 3. regenerator material, 4. magnetic substance, 5. holder spring, 6. regenerator guide, 7. can, 8. vacuum, 9. 77 K shield, 10. carbon glass resistance thermometer, 11. holder, 12. magnet).

soft for machining. Moreover, so as to reduce the gap, the magnetic substance holder had a spring, which was sufficiently soft to prevent frictional heating. Helium gas which was the heat transfer fluid had volumetric heat capacity much smaller than magnetic substance and regenerator material in this temperature range so that mixing effect of helium was ignored.

RESULTS AND DISCUSSION

The experimental arrangement, shown in Fig. 3, was operated without any heat load. The regenerator was oscillated up and down, and the 5 T field turned on and off at appropriate times during the cycle, as previously described. Accordingly, B_1 was equal to 0 T and B_2 was equal to 5 T in Fig. 1 and 2. Since there were too many cycle parameters, such as operating temperature, magnetic field, cycle time, etc., we will describe here only one typical cycle operation.

The regenerator was initially at about 51 K which was near the Curie temperature of sintered $DyAl_{2.2}$. After 11 cycles, the temperature of the top of the regenerator reached 58.7 K and the temperature at the bottom was 50.3 K. This operation was not completely stable, because of the difficulty of keeping the temperature at the high end constant. The cycle consists of four parts: 65 s both in isofield heating and in isofield cooling, 90 s in heat absorption, and 80 s in heat rejection. The magnet was charged and discharged lineally in 45 s.

Figure 4 indicates the operation which was obtained in the manner described above. Figure 4-(A) shows demagnetization process, Fig. 4-(B) represents the regenerative process, and Fig. 4-(C) shows the magnetization process. In all three figures, solid lines represent the temperature of the magnetic substance at its center, and broken lines represent those of the regenerator. In Fig. 4-(A) and (C), the temperature change is represented as a function of time and the broken lines represent the temperature of the regenerator material corresponding to the magnetic material's position. The magnetic material decreased in temperature with the decrease of field, then cooled the regenerator, as shown in Fig. 4-(A). As shown in Fig. 4-(C), the magnetic material increased in temperature on charging the magnet, then heated the regenerator.

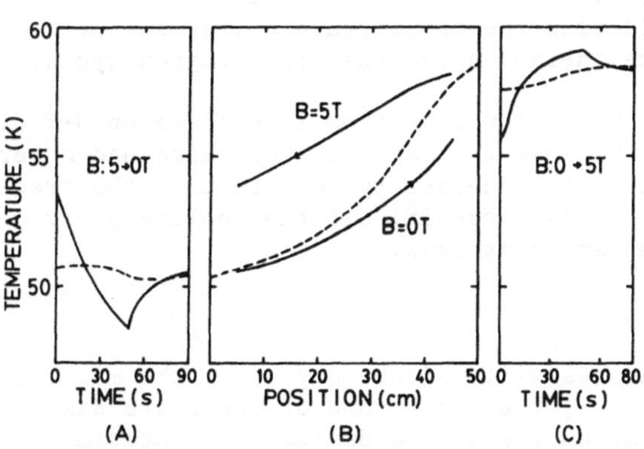

Fig. 4. The operation of the refrigerator: (A),(B) and (C) show demagnetization, regenerative and magnetization process, respectively. The solid lines represent the temperature of magnetic material and the broken lines show the temperature of the regenerator. They are explained in the text.

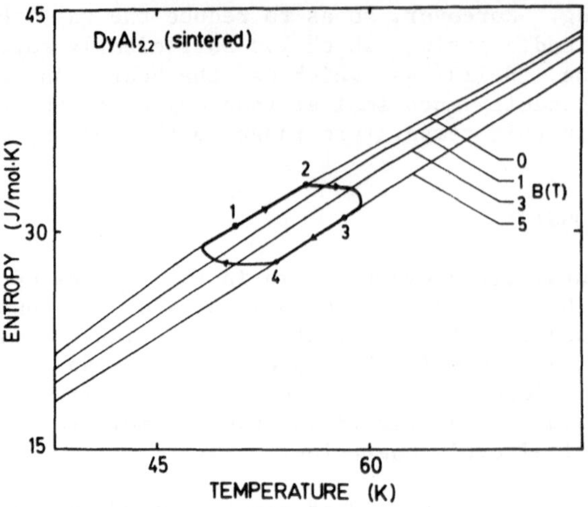

Fig. 5. Thermodynamic cycle is represented on the Entropy-Temperature diagram of sintered $DyAl_{2.2}$. The points, 1, 2, 3 and 4 correspond to those in Fig. 1.

Figure 4-(B) which represents the regenerative process shows the temperature change of the magnetic substance and the average temperature distribution of the regenerator as a function of regenerator position from the bottom. The temperature of the regenerator varied a little around the average value in regenerative processes. The magnetic substance was cooled and heated in a constant magnetic field as the regenerator moved up and down. As a result, a temperature span of 8.5 K has been achieved in the regenerator, and magnetic substance changed in temperature a span of 10.8 K.

The result of this operation is mapped on the entropy-temperature diagram for the sintered $DyAl_{2.2}$ as shown in Fig. 5. The points, 1, 2, 3, 4 in Fig. 5 correspond to those in Fig. 1. In Fig. 5, the process from 2 to 3 and 4 to 1 were similar to an adiabatic change. Particularly in the process from 4 to 1, almost all the heat was absorbed in zero field. In conclusion, considering the temperature difference between 2 and 3, similarly 4 and 1, the thermal cycle was like Brayton cycle.

The cooling power calculated from this trace on S-T diagram was about 0.25 W. However, because the magnetic substance had a certain temperature distribution in it, it is necessary to note that the trace on S-T diagram varied in position. As a result, all the cooling power was wasted on regenerator losses and heat leak.

ORIGIN OF LOSSES

Regenerator losses will undoubtedly limit the maximum temperature span as well as efficiency; therefore some of these are discussed below. In general, refrigeration losses are divided into static losses such as thermal conductivity of the regenerator material, and dynamical losses such as heat transfer loss and frictional heating. In all events, it is necessary to decrease cycle time as well as reduce losses in order to expand the temperature span, because the maximum heat absorbed per cycle is limited by the entropy change of magnetic material as shown in the S-T diagram.

The Brayton-like cycle as shown in Fig. 5 and the very slow cycle of our experiment were obviously caused by insufficient heat transfer between the magnetic substance and the regenerator material. Regenerator loss due to insufficient heat transfer implies additional heat load, because it gives rise to enthalpy flux. In order to study heat transfer between the magnetic substance and the regenerator material, we observed the temperature change of the magnetic substance when the regenerator having a temperature gradient oscillated in zero magnetic field. The heat transfer coefficient calculated from this temperature change was ∿50 mW/cm^2 K at 63 K, then the gap between the heat transfer surfaces was estimated at ∿100 μm using equation (1). The value ∿100 μm was unexpectedly large. Considering that our system had heat transfer area only 30.4 cm^2, some improvements are essential such as high precision surfaces so that they make the gap less than ∿50 μm, large heat transfer area per volume, for example thin magnetic substance.

The longitudinal thermal conductivity of the regenerator and the heat leak from the can wall caused the static losses. This longitudinal thermal conductivity loss, in particular, increases with the expansion of temperature span. This countermeasure ought to be important in expanding the maximum span.

Frictional heating was thought to be as important as dynamic loss. However, this effect was not very serious, owing to the low pressure of the holder spring. However, it was observed that the temperature of the can wall at the magnetic substance oscillated as the regenerator moved up and down. This temperature oscillation was attributed to a heat leak through the magnetic substance holder and, of course, gave rise to additional heat load.

We were not able to analyze these losses quantitatively, since there was no clear distinction among some losses.

CONCLUSION

The design and initial results of an Ericsson magnetic refrigerator in the temperature range about 50 K have been presented. The temperature span of the regenerator was 8.5 K at 300 s operation using 5 T field and $DyAl_{2.2}$. Considering the Brayton-like cycle and long cycle time, heat transfer between the magnetic material and the regenerator material was insufficient. Nevertheless, this experiment demonstrates that the regenerative cycle can produce a temperature span larger than the Carnot cycle.

ACKNOWLEDGMENTS

This study was performed through Special Coordination Funds for Promoting Science and Technology of the Science and Technology Agency of the Japanese Government.

REFERENCES

1. G. V. Brown, Magnetic heat pumping near room temperature, J. Appl. Phy. 47: 3673 (1976)
2. A. F. Lacaze et al, Double acting reciprocating magnetic refrigerator: Recent improvements, in: "Advances in Cryogenic Engineering", Vol.29, Plenum Press, New York (1984), p. 573-579.
3. Y. Hakuraku and H. Ogata, Thermodynamic analysis of a magnetic refrigerator with static heat switches, Cryogenics 26:171 (1986)
4. T. Numazawa et al, The helium magnetic refrigerator II: Liquefaction process and efficiency, in: "Advances in Cryogenic Engineering", Vol.29, Plenum Press, New York (1984), p.589-596.

5. H. Nakagome et al, Reciprocating magnetic refrigerator for helium liquefaction, in: "Advances in Cryogenic Engineering ", Vol.31, Plenum Press, New York (1986), p. 753-762.
6. J. A. Barclay et al, Experimental results on a low-temperature magnetic refrigerator, in: "Advances in Cryogenic Engineering", Vol.31, Plenum Press, New York (1986), p.743-752.
7. T. Hashimoto et al,Investigations on the possibility of the RAl_2 system as a refrigerant in an Ericsson type magnetic refrigerator, in: "Advances in Cryogenic Engineering - Materials", Vol.32, Plenum Press, New York (1986), p.279-286.
8. K. Matsumoto et al, A fundamental study of a regenerator for an Ericsson magnetic refrigerator, in: "Proc. Eleventh Intl. Cryo. Engr. Conf.," Butterworth, Guildford, UK (1986), p.256-260.

ANALYSIS OF MAGNETIC REFRIGERATION WITH EXTERNAL REGENERATION

Steven R. Jaeger and John A. Barclay

Astronautics Corporation of America
Astronautics Technology Center
Madison, Wisconsin

William C. Overton, Jr.

Los Alamos National Laboratory
Los Alamos, New Mexico

ABSTRACT

Regenerative magnetic refrigerators have the potential for improved cooling capacity and temperature span over the more common Carnot devices. Here, two of expected major contributing factors to loss of performance for refrigerators using external regeneration are studied: non-ideal temperature-entropy properties of the magnetic material and non-ideal regeneration. Both factors were found to contribute significantly to reducing the useful range and efficiency of these refrigerators.

INTRODUCTION

Most magnetic refrigerators built to date are based on the non-regenerative, non-recuperative Carnot cycle.[1] The non-regenerative refrigerators are easier to design and build than regenerative refrigerators; however, the temperature difference that they can span is limited to about 15 degrees kelvin for temperatures below 20 K and less at higher temperatures. To increase the operating temperature span, it is necessary to link refrigerators in series or to use recuperative, external regenerative, internal regenerative, or active magnetic regeneration designs. Here, external regeneration means using a fluid to transport heat between the magnetic material and the regenerator so the regenerative material can remain outside of the high magnetic field region. Internal regeneration means moving the regenerator so that it is in direct thermal contact with the magnetic material. This requires moving the regenerative material with respect to the magnetic field. Active magnetic regeneration means using the magnetic material itself as the regenerator.

This paper examines a simplified externally regenerative refrigerator in order to better understand the factors that affect their operating temperature range and performance. The computer model developed for this analysis can be used for preliminary screening of magnetic materials and regenerative refrigerator designs for their use in various applications.

ANALYSIS

The approach here is similar to that used by Overton.[2] The specific factors that were assessed were how the performance was affected by real magnetic materials and non-ideal regeneration. Cross et al[3] provides a more in depth discussion of what is the ideal shape of material temperature-entropy diagram for different magnetic cycles.

This analysis neglects many additional causes of inefficiency that would be present in real regenerative magnetic refrigerators including: non-ideal heat transfer at the hot and cold ends, heat leaks, friction, and flow pressure drop in the fluid used to transfer heat to and from the regenerative material. More detailed modeling of specific refrigerator designs will be required to quantify these effects once specific designs are chosen.

Computer Model

Figure 1 shows a schematic of a magnetic refrigerator with external regeneration from Overton.[2] Heat transfer between the magnetic material and the hot and cold heat exchangers as well as the regenerator stages occurs via a circulating fluid that is controlled by the manifold and valve system shown. The following is a description of how the magnetic refrigerator would operate using a Brayton cycle. An Ericsson cycle could be used by making slight adjustments as mentioned at the end of this description.

During normal operation only one of the valves would be open at a time. After the magnetic material is moved into the high field region, the valve V_H is opened to allow heat from the magnetic material to be rejected to the hot heat exchanger. When the material approaches the temperature of the hot sink, V_H is closed. The magnetic material is then cooled by opening and then closing valves V_1 through V_5 in series and allowing heat to be transferred to the regenerator stages. At the end of regeneration, all the valves are closed and the magnetic material is removed from the magnetic field. Adiabatic demagnetization causes the temperature in the magnetic material to decrease further. Valve V_C is then opened to allow the magnetic material to absorb heat from the cold heat exchanger. When the magnetic material approaches the temperature of the cold heat exchanger, V_C is closed. The magnetic material is then heated by opening and then closing valves V_5 through V_1 in series and allowing heat to transfer from the regenerator stages to the magnetic material. The regenerator valves are open in reverse order during this portion of the cycle in order to produce a stratified temperature profile in the regenerator which allows heat transfer occurring between the magnetic material and the regenerator to take place across a minimum temperature difference. After the regenerative heat transfer occurs, all valves are again closed and the magnetic material is moved back into the high magnetic field. Adiabatic magnetization causes the temperature of the magnetic material to increase above the temperature of the hot sink and the cycle starts over.

An Ericsson cycle would require opening valve V_H before the magnetic material was fully magnetized and opening valve V_C before the magnetic material was fully demagnetized. V_H would need to be opened once the temperature of the magnetic material reached the sink temperature and then the magnetic material would have to be further magnetized at a slow rate so that, in theory at least, the heat transfer at the hot end would

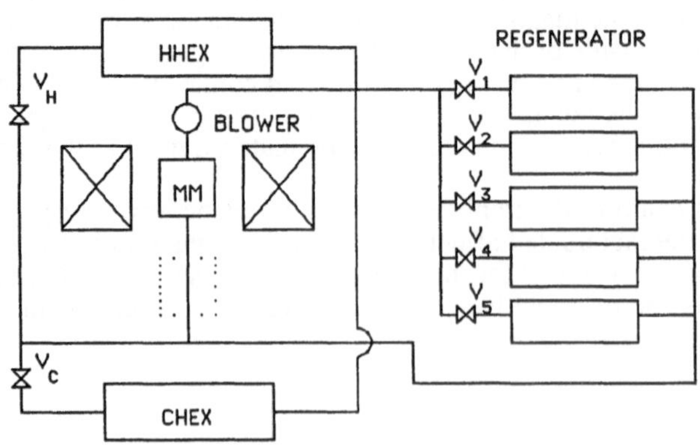

Figure 1. Schematic of external regenerative magnetic refrigerator

take place across an infinitely small temperature difference. Similarly, T_C would be opened once the temperature of the magnetic material reached the cold source temperature.

The following assumptions were made in developing the computer model of the regenerative refrigerator:

- The magnetic material and the individual regenerator stages are isothermal. There is no temperature gradient within them.
- Each regenerator stage is thermally isolated from the others. There is no conductive heat transfer between the regenerative stages.
- The regenerative heat transfer takes place while the magnetic material is in constant magnetic field.
- The heat transfer at the hot and cold ends takes place while the magnetic material is in constant magnetic field for the Brayton cycle and at constant temperature for the Ericsson cycle.

Output of the computer program consists of contour plots of the heat absorbed at the cold end, refrigerator efficiency with respect to Carnot, and refrigerant capacity[4] all as a function of the hot and cold temperatures spanned by the refrigerator.

<u>Results</u>

Figure 2 shows the linear temperature-entropy model used in to approximate GdNi.[2] The eight Tesla curve is approximated by a straight line throughout the entire region of the graph. The zero field curve is approximated by two straight lines of different slopes joined at the Curie temperature. The model matches GdNi reasonably well near the Curie temperature but is less accurate at lower temperatures. An advantage of using a simple model, however, is that it is easy to calculate the performance of some ideal cases by hand to check the accuracy of the computer model.

Figure 2 also shows the the path the magnetic material would take undergoing an Ericsson cycle with perfect regeneration operating between 40 K and 75 K. The cycle is distorted slightly from the ideal Ericsson cycle due to the non-parallel temperature-entropy curves.[3] This causes an decrease in efficiency from the ideal Carnot efficiency. Imperfect regeneration would reduce the cooling capacity at the cold temperature which further reduces the efficiency of the magnetic refrigerator.

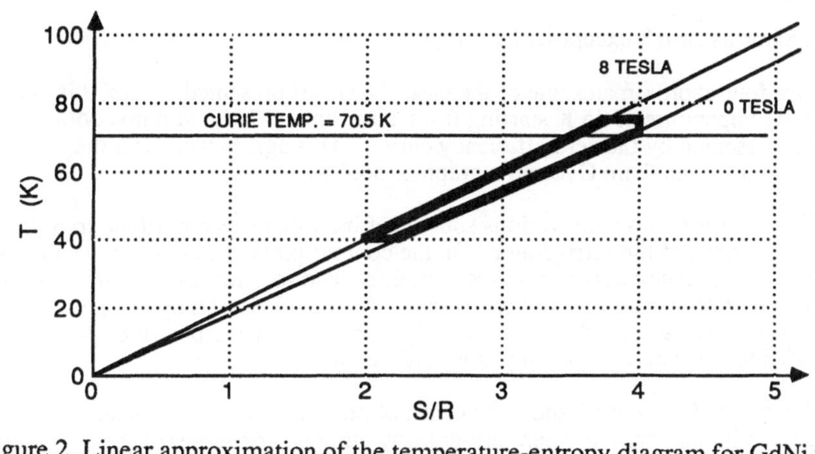

Figure 2. Linear approximation of the temperature-entropy diagram for GdNi with example magnetic refrigeration cycle shown.

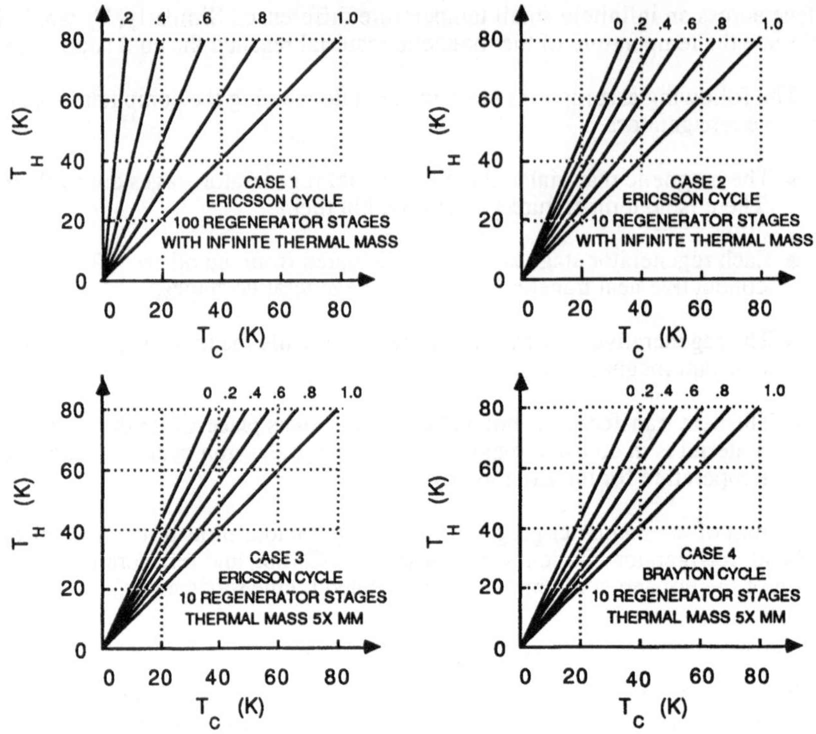

Figure 3. Refrigerator efficiency with respect to Carnot (COP/COP_{IDEAL}) as a function of the source and sink temperatures (T_C and T_H) for four cases.

Figure 3 shows plots of efficiency with respect to Carnot as a function of the sink and source temperatures (T_H and T_C) of the magnetic refrigerator for four cases using the above model for GdNi:

1. Ericsson cycle with 100 regenerator stages with infinite thermal mass.

2. Ericsson cycle with 10 regenerator stages with infinite thermal mass.

3. Ericsson cycle with 10 regenerator stages with the thermal mass of each regenerator stage equal to approximately five times the thermal mass of the magnetic material.

4. Same as case 3 except with a Brayton cycle.

Case four approximates one of the cases Overton[2] presented. Here, it is possible to reach a cold temperature of 35 K starting from a hot sink of 75 K with no cooling load (which is represented by the zero efficiency curve). This agrees well with the maximum cold temperature taken from Overton's graph of 36.4 K.

Figure 3 shows how the various sources of inefficiencies contribute to reducing the overall performance of the refrigerator. On the contour plots, the areas above and to the left of the zero efficiency curve for cases two, three, and four represent regions where the given magnetic refrigerators are unable to span. An ideal Ericsson cycle operating with an ideal magnetic material would be able to refrigerate over any temperature span desired with an efficiency equal to the ideal Carnot efficiency.

The results from case 1 show the effect of non-ideal magnetic material. A magnetic material that has a temperature-entropy plot that converges at the cold end of the temperature span of the refrigerator causes a decrease in efficiency with respect to the Carnot efficiency even when there is ideal regeneration. Theoretically it is still possible to provide cooling all the way to absolute zero, but the ratio of efficiency with respect to the Carnot efficiency approaches zero as the cold temperature of the refrigerator approaches absolute zero.

Case 2 shows the added effect of not having an infinitely stratified regenerator. Regenerative heat transfer across finite temperature differences reduces the maximum temperature span of the refrigerator and also reduces the efficiency within the regions where the refrigerator works.

Case 3 shows the added effect of finite thermal mass of the regenerator. The oscillating temperatures of the regenerative stages during the cycle further decreases the regenerator effectiveness which increases both of the trends noted for case 2.

Case 4 shows the added effect of going from the reversible isothermal heat transfer of the Ericsson cycle at the hot and cold ends to the non-ideal constant field heat transfer of the Brayton cycle. However, the case modeled is an ideal Brayton cycle in that it is assumed the magnetic material stays in thermal contact at the hot and cold ends long enough so that its temperature becomes equal to the equilibrium temperature of the sink and source. Because of this, the Brayton cycle does not have any reduction in the temperature range it can span over the Ericsson cycle but the heat transfer at the hot and cold ends across a finite temperature difference does cause a decrease in efficiency.

CONCLUSIONS

A simplified computer model of a typical magnetic refrigerator using external regeneration has been developed to screen different magnetic materials and refrigerator designs. It does not predict the performance of actual refrigerators because many real sources of inefficiencies were left out. It does , however, provide a quick means of evaluating whether inherent limitations as a result of the choice of magnetic material and general regenerator design will make it impossible for a specific design to satisfy the requirements for given applications. Preliminary results from using this model with an approximate model for GdNi as the magnetic material yield the following conclusions:

- If regeneration takes place at constant magnetic field then the shape of the temperature-entropy diagram will limit the performance of the refrigerator.
- In order to span large temperature differences, it is necessary to have very close to ideal regeneration. This requires a well stratified regenerator with sufficient thermal mass to keep its temperature fluctuations small.

REFERENCES

1. J. A. Barclay, Magnetic refrigeration: a review of a developing technology, paper presented at this conference.

2. W. C. Overton Jr., "Analysis of Magnetic Refrigeration Systems with Staged Regenerators Based on an Idealized Magnetic Entropy Model", LA-10676-MS, Los Alamos National Laboratory, Los Alamos, New Mexico (1986).

3. C. R. Cross et al, Optimal temperature-entropy curves for magnetic refrigeration, paper presented at this conference.

4. M. E. Wood and W. H. Potter, General analysis of magnetic refrigeration and its optimization using a new concept: maximization of refrigerant capacity, Cryogenics, 25:667, (1985).

ROTARY RECUPERATIVE MAGNETIC HEAT PUMP

Lance D. Kirol and Michael W. Dacus

Idaho National Engineering Laboratory
EG&G Idaho, Inc.
Idaho Falls, Idaho

ABSTRACT

A bench scale rotary magnetic heat pump now being built is described. The unique design feature of this heat pump is the method for achieving recuperator fluid flow, which relies simply on parallel flow paths; the primary flow leg allows heat transfer between external load and sink and magnetic working material, while parallel flow accomplishes recuperation. The bench scale test is intended to demonstrate feasibility of the concept and to verify that all significant loss mechanisms are identified and treated properly in performance models, but is not a scaled down version of a practical heat pump. Working material is gadolinium foil 76 μm thick with 127 μm spaces for fluid flow. Magnetic fields are created by neodymium-iron-boron permanent magnets with an air gap field of about 0.9 Tesla. Due to the low field (practical heat pumps will use superconducting magnets with field strength around 9 T), temperature lift is limited to 11 K.

BACKGROUND

Many configurations of continuous magnetic heat pumps are possible; the three major categories are reciprocating, rotary, and those in which working material is stationary and the magnetic field is switched on and off.[1] Previous studies have shown rotary recuperative devices to have significant advantage over the others.[2] Reciprocating regenerative magnetic heat pumps perform less efficiently because temperature rise of regenerator material necessary for thermal energy storage and mixing of the column of regenerator fluid reduce regenerator effectiveness. Heat pumps relying on switching of superconducting magnets require significant refrigeration work to remove heat generated from ac losses in the magnet. The ac losses themselves are small, but refrigeration work required to keep the magnets at 4 K is about 400 times the heat actually removed. (As new high temperature superconductors become available, refrigeration work will be significantly reduced and field switching heat pumps may become the preferred configuration.)

Although rotary recuperative magnetic heat pumps now appear to be the most promising, there exist engineering problems in developing a recuperative heat pump with solid working material, specifically in devising suitable schemes for pumping recuperator fluid. Difficulty is increased because the preferred forms of working material are thin closely spaced plates, small particles, or fine wire screens, all of which reduce the feasibility of inserting any kind of flow seal in the rotor. At least one rotary recuperative heat pump has been built, this being a recuperative Brayton cycle magnetic refrigerator in which recuperator fluid is in thermal contact with the working material except during magnetization and demagnetization.[3] This machine employed a clever scheme of segmenting the rotor and using peripheral seals to force recuperator fluid to flow from segment to segment around the rotor. The refrigerator was tested at room temperature with gadolinium as the refrigerant, and actual performance was significantly below predictions.[4] Our analysis of this refrigerator indicated that performance was degraded primarily by friction in dynamic mechanical seals and imperfect magnetic field profile. In general, heat generation due to mechanical friction reduces recuperator effectiveness and significantly degrades performance, and poor magnetic field profiles make T-S curves of magnetized and demagnetized refrigerant less parallel, which also reduces the amount of recuperation possible. The heat pump design presented here is intended to eliminate problems with dynamic seals and frictional heat generation, and reduce complexity of the overall design. Also, considerable care was taken in obtaining acceptable magnetic field profiles.

CONCEPT DESCRIPTION

The heat pump consists of a rotor of magnetic working material with flow passages to allow heat transfer fluid to move through the rotor in good thermal contact with the magnetic material. The rotor moves within a housing with ports for fluid to enter and exit the rotor, with the ports being in a specific orientation relative to the magnetic field, as shown in Figure 1. Fluid (water, in this case) is pumped into the housing at point 1 where the rotor enters the region of increasing magnetic field. Fluid entering at 1 is free to flow through the rotor in either direction, to outlet ports 2 or 4. Flow resistance is less toward port 2 because rotor motion is concurrent with flow in this direction and because the distance is less; most fluid entering at 1 flows to 2, but some flows against rotation to 4. Likewise, fluid entering at 3 flows to 4 and 2, with most of the flow going to 4. As the rotor moves between 1 and 2, it becomes magnetized and heats up slightly, but fluid flowing between 1 and 2 with the rotor removes most of the heat of magnetization. Magnetization of the working material is nearly isothermal as shown on the T-S diagram in Figure 1b, with heat--equal to the integral of TdS from 1 to 2--being transferred from working material to fluid and ultimately to the load. In the constant field region from 2 to 3, working material is cooled against colder fluid entering at 3, such that at 3 working material and fluid are nearly equal in temperature. From 3 to 4 working material demagnetizes while absorbing heat from the fluid, so demagnetization is also nearly isothermal. The rotor, in traveling from 4 to 1, is heated by fluid flowing from 1 to 4. Recuperative heating of the working material from 4 to 1 and recuperative cooling from 2 to 3--accomplished by fluid flowing counter to rotation in these regions--are essential for obtaining large temperature lifts with an efficient magnetic cycle. The cycle executed by this heat pump approximates an Ericsson cycle.

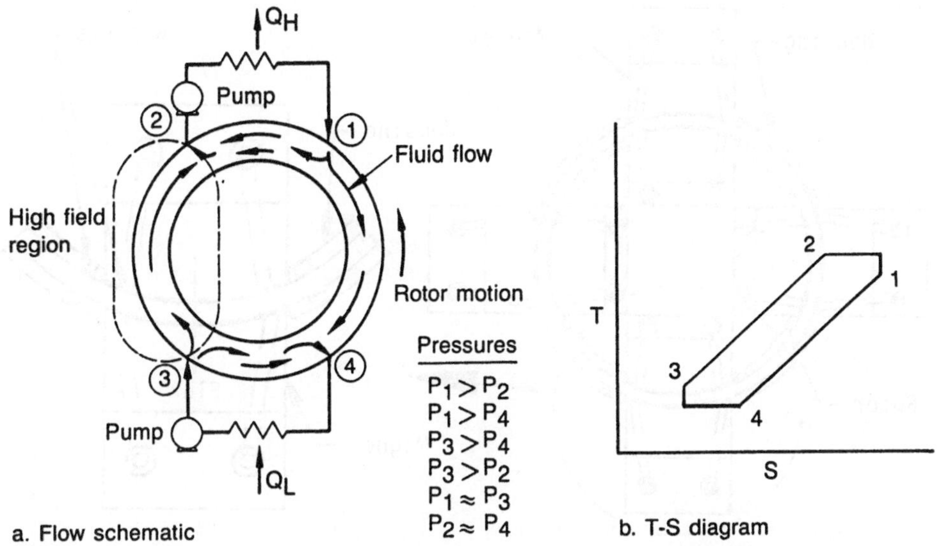

Figure 1. Rotary recuperative magnetic heat pump with the pressure drop through field change regions driving recuperator fluid flow.

BENCH SCALE TEST OBJECTIVES

The bench scale magnetic heat pump is intended to:

- Demonstrate relatively high efficiency (greater than 50% of Carnot) in a continuous magnetic heat pump
- Verify that the design concept is physically realizable and is operable
- Verify that all significant loss mechanisms have been identified and treated properly in analyses.

DESIGN

Design parameters were selected through studies performed using a computer model. Each heat pump configuration was modeled over a range of speeds and an operating curve--efficiency (percent of Carnot COP) versus normalized heat pumping capacity--was generated. Optimization was performed by comparing operating curves rather than specific operating conditions.

Working material is gadolinium. Gadolinium is used for this first demonstration because its room temperature Curie point (293 K) is convenient and because gadolinium is malleable and well characterized thermodynamically. Materials identified for applications above and below room temperature are typically rare earth transition element alloys;[5,6,7] such alloys are very brittle and methods to form them into thin plates have not been developed. Also, it is desirable to use a material for which experimentally verified entropy data with various applied fields are available. Performance issues related to heat pump configuration will not be confounded by questions about working material behavior.

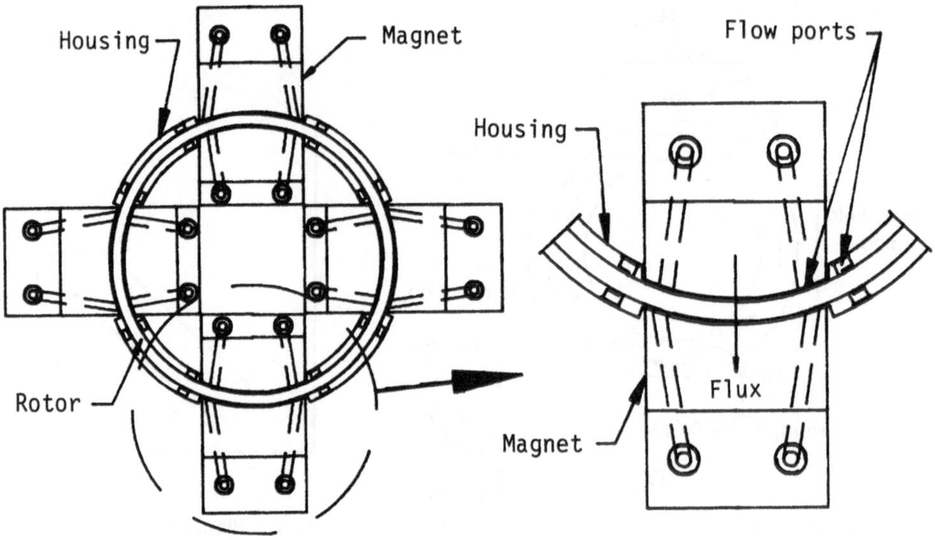

Figure 2. Bench scale heat pump layout.

The heat pump consists of a rotor turning through four high field areas created by permanent magnets, as shown in Figure 2. Working material executes four complete thermodynamic cycles each revolution (instead of one, as illustrated in Figure 1). Four cycles per revolution are used because use of permanent magnets limits lift (at reasonable efficiency) to about 11 K, and optimum length of the recuperative section with 11 K lift is 50 mm. This equates to a rotor diameter of 40 mm when two recuperative sections use 40% of the circumference each, and each field change region is 10% of the circumference. A diameter of 40 mm is too small to allow necessary machining on the inner radius of the housing. Four field changes per cycle allow a practical diameter.

<u>Magnets</u>

Permanent magnets are used to provide regions of high magnetic field because demonstration of the heat pump with permanent magnets is considerably less expensive than with superconducting magnets, and because all objectives of a bench scale test can be met with relatively low fields available from permanent magnets. Peak fields are about 0.9 T.

One permanent magnet assembly is illustrated in Figure 3. Each magnet consists of two blocks of permanent magnet material: one adjacent to the rotor on the outer radius and and one on the inner radius. Each block of magnetic material is 59.5 mm wide by 30.8 mm high and 28.6 mm long in the direction of magnetization. Permanent magnets are epoxy bonded to a steel flux return assembly. Magnet pole faces are machined to match the radius of the rotor, except a flat is machined on the inner pole face to aid in shaping the field. Figure 4 shows desired field profile, measured profile without the flat, and calculated profile with the flat. Actual profile with the flat has not yet been measured. (Measured field profile will differ from the calculated field because finite element analyses of the field did not include flow ports in the magnets.) Field profile without the flat differs significantly from desired profile and, as discussed later, performance is severely degraded.

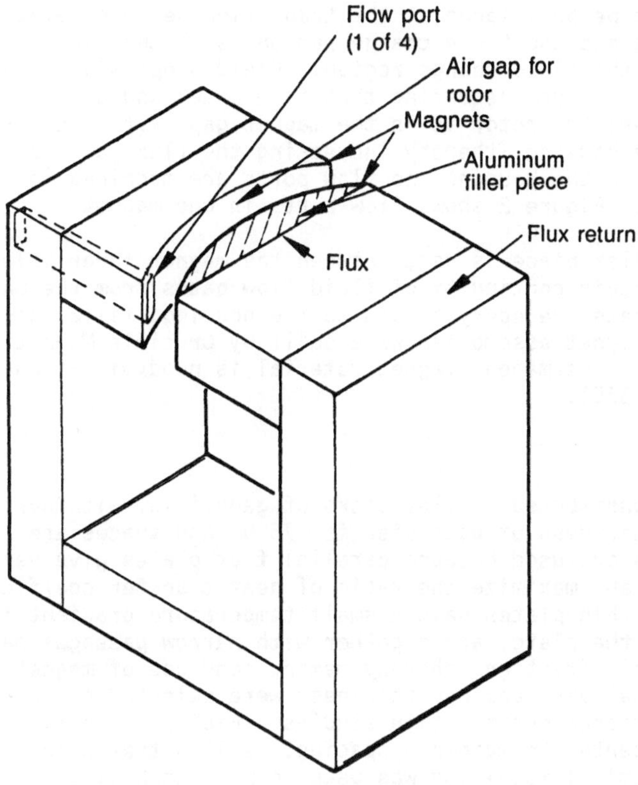

Figure 3. Permanent magnet assembly

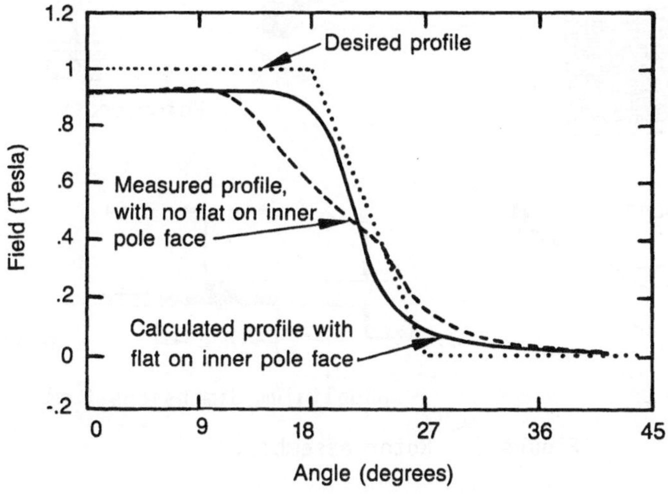

Figure 4. Magnetic field profiles

Flow ports must be positioned at the transition between relatively flat magnetic field and the field change region, such that inlet and outlet ports bound the field change region. Field drops significantly inboard of the magnet edge, requiring that flow enter and exit the rotor at points where the rotor is in the magnet gap. It is not desirable to thicken the housing--thereby increasing the flux gap--to provide flow paths in the housing, so flow ports are machined in the magnets themselves. Figure 2 shows flow ports in one magnet.

An aluminum filler piece is used between the magnet inner pole face and housing to maintain continuity of fluid flow paths from the magnets to the rotor. Magnets are epoxy bonded to the housing, filler piece, and flux return. Magnet assemblies were built by Crucible Magnetics (Elizabethtown, KY). Permanent magnet material is neodymium-iron-boron (Crucible's Crumax 355).

Rotor

The rotor is constructed of flat disks of gadolinium with narrow spaces between. Thickness of each disk is 76 μm and spaces are 127 μm wide. Disks are used because parallel flat plates give very good heat transfer and maximize the ratio of heat transfer coefficient to pressure drop. Thin plates have a small temperature gradient for conduction through the plate, and together with narrow passages have a relatively small void fraction, thereby making good use of magnetic field volume. Actual disk and gap thickness were selected based on computerized performance optimization studies. Each disk is machined with a rib in the center to maintain spacing, as illustrated in Figure 5. Photochemical machining was used to mill each disk from a

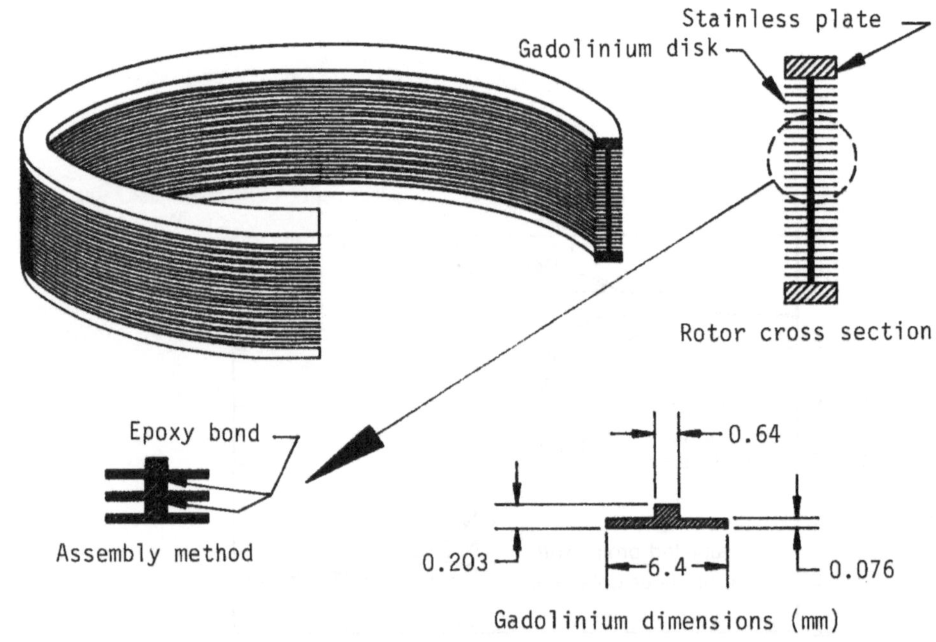

Figure 5. Rotor assembly.

sheet of gadolinium 206 μm thick. Accuracy of chemical machining was not as good as is possible with stainless steel, probably because of oxide inclusions in the gadolinium and lack of experience with this material, but pieces approaching tolerance limits of 12.7 μm on thickness and 127 μm on diameter have been produced. Disks are epoxy bonded together to form a rotor, with thicker (2.5 mm) stainless steel top and bottom plates. Ball bearing races are machined in end plates, and one plate has gear teeth for the drive mechanism.

Rotor plates are parallel to the plane of rotation (that is, disks rather than cylinders are used) because they are simpler to manufacture and assemble into a rotor with accurate spacing. The magnetic field is applied parallel to plates to minimize the demagnetization field of the magnetic working material. Radial build (outer radius minus inner radius) of the rotor is only 6.4 mm. Radial build is small for three reasons: (1) to minimize the permanent magnet flux gap thereby making relatively large fields possible, (2) to minimize fringe flux to give sharp field drop at the magnet edge, and (3) to minimize radial flow variations. Performance is sensitive to maintenance of heat capacity balance between working material and recuperator fluid in the recuperative sections of the heat pump, and large radial build makes such a balance difficult to achieve. Mass rate of working material is maximum at the outside circumference, while fluid flow is highest at the inner circumference because the flow path is shorter. In the current design, the rib in the center of rotor disks serves as a flow divider as well as disk spacer; the rotor is divided radially with separate flow control on the inner and outer diameters, making the radial distance over which acceptable heat balance must be maintained only 3.2 mm.

The assembled rotor has a total of 125 gadolinium disks with a working material stack height of 2.54 cm. Total mass of gadolinium is 270 g.

<u>Housing</u>

The rotor turns in an aluminum housing that provides bearing races and flow ports, and positions the magnets. Aluminum is used because it is nonmagnetic and easily machined. The housing is in two parts: the containment for the rotor and a top closure with flow connections. Flow ports in the housing are at the transition between field change and low flow regions. Rectangular flow passages are milled from the exterior surfaces and closed with covers epoxy cemented in place. Wall thickness of the housing between magnets and the rotor is 760 μm. Thin walls are used to minimize width of the magnet flux gap. Magnet separation is maintained by housing material above and below the rotor cavity, and by the flux return assembly.

PERFORMANCE

A computer model was utilized to predict performance of the bench scale heat pump, and aid in design. The model is steady-state, predominately focusing on conditions at each state point with appropriate iterations to ensure that the cycle closes and the first and second laws of thermodynamics are not violated. Heat transfer rate limitations are included, using published correlations for heat transfer coefficients. Published correlations are also used for pressure drop. Gadolinium temperature-entropy-magnetic field data are from mean field calculations, which have been shown to agree very well with experi-

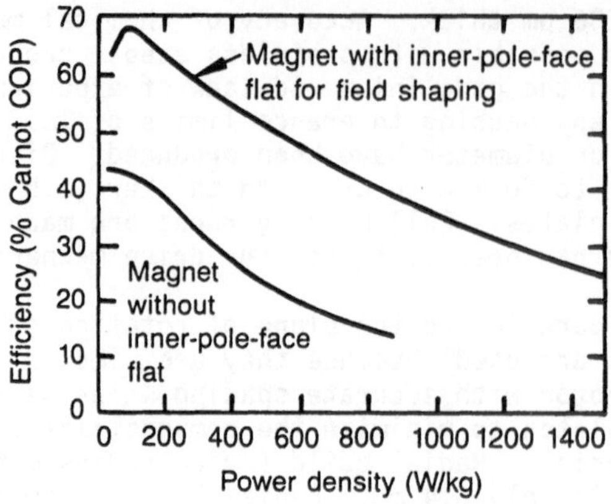

Figure 6. Bench scale magnetic heat pump operating curve.

mental data.[8] Figure 6 shows a computer generated operating curve for the bench scale heat pump. Results are presented as efficiency (percent of Carnot COP) versus heat pumping capacity normalized for working material mass. The curve represents the heat pump operating at increasing speeds, from 60 s/revolution to 1.6 s/revolution. Right of the efficiency peak, increasing speed gives increasing capacity but decreasing efficiency because performance is limited by heat transfer in the recuperator sections. Efficiency shows a maximum because at very low speeds thermal conduction in the rotor and recuperator fluid predominate, and a significant fraction of the heat pumping capacity is lost to internal conduction.

Also shown in Figure 6 is an operating curve for the heat pump with magnetic field profile from a magnet with no flat machined on the inner pole face (Figure 4). Importance of obtaining the desired field profile is evident.

Most significant factors are included in the model, but there are some factors that will affect performance and are not modeled. Among these are:

- Eddy current heating in the magnetic working material
- Mechanical friction (efforts have been made to minimize friction, however)
- Thermal conduction through the housing
- Fluid flow between the housing and rotor
- Magnetic hysteresis (near the Curie temperature, there is very little magnetic hysteresis)
- Epoxy cement overflow onto heat transfer surfaces
- Surface roughness of disks

- Radial and axial variation in magnetic field
- Radial flow variations (these have been scoped for the radial build involved and are not large)
- Uneven plate spacing
- Demagnetization field (small with applied field parallel to disk surface)
- Oxide coating on disks (measured to be only a few hundred angstroms thick).

These factors taken together will cause performance to be below that illustrated in Figure 6. The magnitude of the cumulative effect is unknown. Verification of our computer model against other experimental apparatus indicates that predicted performance will be approached fairly closely.

ACKNOWLEDGMENTS

This work was supported by the U. S. Department of Energy, Office of Industrial Programs, under contract no. DE-AC07-67ID01570.

REFERENCES

1. G. V. Brown, Basic Principles and Possible Configurations of Magnetic Heat Pumps, ASHRAE Transactions, 87, 2, 1981.

2. L. D. Kirol et al., "Magnetic Heat Pump Feasibility Assessment," EGG-2343, Oct. 1984. (Available from National Technical Information Service, Springfield, VA.)

3. W. A. Steyert, Stirling-Cycle Rotating Magnetic Refrigerators and Heat Engines for Use Near Room Temperature, Journal App. Physics 49:1219 (1978).

4. J. A. Barclay and W. A. Steyert, "Magnetic Refrigerator Development, EPRI Report EL-1757," Electric Power Research Institute, Palo Alto, CA (1981).

5. T. Hashimoto, T. Numasawa, M. Shino, and T. Okada, Magnetic Refrigeration in the Temperature Range from 10 K to Room Temperature: the Ferromagnetic Refrigerants, Cryogenics 21:647 (1981).

6. C. B. Zimm, J. A. Barclay, and W. R. Johanson, "Low-Hysteresis Materials for Magnetic Refrigeration: $Gd_{1-x}Er_xAl_2$, Los Alamos Report LA-UR-83-2890," Los Alamos National Laboratory, Los Alamos, New Mexico (1983).

7. H. Oesterreicher and F. T. Parker, Magnetic Cooling Near Curie Temperatures Above 300 K," Journal App. Physics 55:4334 (1984).

8. S. M. Benford and G. V. Brown, T-S diagram for Gadolinium near the Curie Temperature, Journal App. Physics 52:2110 (1981).

OPTIMAL TEMPERATURE -ENTROPY CURVES FOR MAGNETIC REFRIGERATION

C. R. Cross, J. A. Barclay, A. J. DeGregoria, S. R. Jaeger, J. W. Johnson

Astronautics Technology Center
Madison, Wisconsin

ABSTRACT

Magnetic refrigeration utilizes the temperature dependent entropy change produced in a ferromagnetic or paramagnetic material when subjected to a change in magnetic field. By blending different materials together the temperature-entropy (T-S) behavior of the refrigerant may be tailored to maximize system performance. Optimal T-S diagrams for magnetic refrigerators using Carnot, Ericsson and Brayton cycles were determined and compared to results of thermodynamic numerical models.

INTRODUCTION

Magnetic refrigeration utilizes the entropy change produced in a ferromagnetic or paramagnetic material when subjected to a change in magnetic field. In these refrigerants the entropy change with magnetic field is maximum near the magnetic ordering temperature. The temperature-entropy (T-S) diagram of the magnetic material is the principle factor in determining the refrigerator maximum operating range and performance, and is generally not ideal for a given magnetic thermodynamic cycle. However, the shape of the T-S diagram can be modified by blending a number of magnetic materials with different ordering temperatures together, as has been recently demonstrated by Hashimoto et al[1]. The object of this paper is to examine the behavior of several classic thermodynamic cycles operating with differently shaped T-S diagrams and determine the major factors that control refrigerator performance.

THERMODYNAMIC REQUIREMENTS

The refrigerator cooling capacity, Q_C for an ideal reversible cycle is related to the ideal work and heat rejection, Q_H by the energy balance

$$Q_C = Q_H - W_{IDEAL} \quad (1)$$

and limited by the maximum amount of heat that can be rejected at the hot temperature. The second law of thermodynamics further requires the magnitude of the entropy changes produced at the hot and cold temperatures, ΔS_H and ΔS_C respectively, to obey the relation

$$\Delta S_H \geq \Delta S_C \quad (2)$$

For reversible, isothermal heat transfer, $Q_C = \Delta S_C T_C$ and $Q_H = \Delta S_H T_H$. Thus, for a specified hot temperature, the heat rejection and thereby the cooling power, is limited by the maximum ΔS_H.

An important thermodynamic property intimately related to the T-S curves is the material specific heat defined as

$$C_B = T\, dS/dT \quad (3)$$

Thus, the specific heat of a material at a constant applied field decreases with increasing slope of the T-S curve. This is of particular significance when discussing regenerative/recuperative cycles, as we shall see.

The optimal T-S diagram is determined by the thermodynamic requirements of the cycle used to produce refrigeration, and is also influenced by non-ideal processes. Irreversible processes occurring as part of the cycle result in entropy production, thereby reducing cycle efficiency. This additional entropy that must be rejected at T_H modifies the T-S curves required for optimal performance. Typical entropy generating processes are non-isothermal heat transfer, heat leaks, and friction losses. The first and second laws for real cycles can then be written

$$W_{REAL} = Q_H - Q_C + \Delta S_{IRR} T_H \tag{4}$$

$$\Delta S_H = \Delta S_C + \Delta S_{IRR} \tag{5}$$

where ΔS_{IRR} is the irreversible entropy generated through the cycle execution.

The influence of the magnetic material T-S behavior on a number of refrigeration cycles will be examined in the following sections. First, the effect of the magnetic refrigerant material T-S diagram will be discussed with respect to the ideal thermodynamic cycle; i.e. ignoring irreversible effects. Then, the ideal cycle will be contrasted with a real cycle to point out additional phenomena revelant to determining optimal T-S curves.

Ideal Carnot Cycle

For an ideal Carnot cycle, the effects of the refrigerant temperature-entropy behavior is relatively straightforward. As shown in Figure 1(a), the definition of the Carnot cycle requires the extreme values of entropy at the high and low temperature isotherms to be equal, thus implicitly satisfying the second law constraint that the magnitudes of the entropy changes be equal, as expressed in eq. 2. Therefore, the optimum T-S curves for an ideal Carnot cycle are simply those that maximize the quantity S_3-S_1.

Non-Ideal Carnot Cycle

By examining the T-S diagram of an ideal Carnot cycle and superimposing non-ideal processes, their effects can be readily seen. Figure 1(b) shows a Carnot cycle and the results of heat leaks and/or frictional losses. The entropy produced during the partial demagnetization process from points 3 to 4' reduces the available cooling power and the required work input. During the partial magnetization process, points 1 to 2', the entropy generation results in an increased work requirement and heat rejection. Both of these deviations from isentropic processes result in decreased efficiency.

The other principal irreversibility, non-isothermal heat transfer, is illustrated in Figure 1(c). For a real process some finite temperature difference is required to drive the heat transfer at the hot and cold sinks. Rejecting energy at T_H non-isothermally, the process occurring from points 2' to 3', results in an increase in cycle work W and Q_H, and depending on the location of 3', possibly a significant reduction in $\Delta S_{BMAX-BMIN}$. Non-isothermal heat sbsorbtion, points 4' to 1', results in a reduced Q_C, an increased W, and another possible reduction in cycle ΔS.

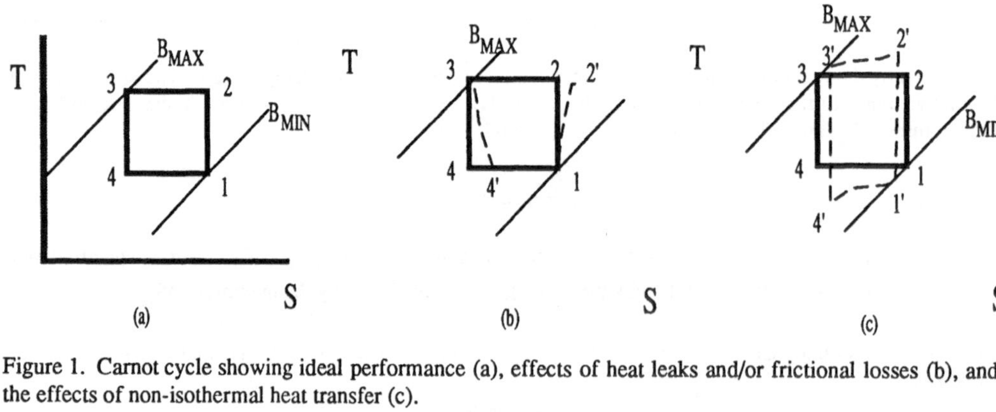

Figure 1. Carnot cycle showing ideal performance (a), effects of heat leaks and/or frictional losses (b), and the effects of non-isothermal heat transfer (c).

The implications for the T-S curve shapes in these irreversible processes is primarily in the heat transfer. Therefore, the steeper the slope of the T-S curve at point 1, the closer points 1 and 1' are at the end of the heat transfer process. This in turn reduces the overall cycle degradation produced by a finite temperature difference remaining at the end of the heat rejection process. Increasing the slope of the T-S curve at point 3 produces the same result.

Ideal Ericsson

The principal difference between Carnot and Ericsson cycles is that regeneration is employed in the Ericsson cycle to enable a greater temperature span from a specified low temperature heat transfer isotherm, T_C, to a specified high temperature heat transfer isotherm, T_H. Energy is absorbed by the regenerator through the temperature range T_C to T_H during the high field, hot-to-cold flow period and returned to the fluid during the low field, reverse flow part of the cycle. An alternate regeneration method utilizes the magnetic refrigerant itself as the heat storage medium. However, this refrigeration cycle, based on a Brayton cycle is disucssed in a later section of this paper.

In order for this exchange to comply with conservation of energy, the energy transferred to and from the magnetic material in the hot and cold portions of the cycle must balance. Graphically, this energy balance is shown in Figure 2 by the equal areas A12B and C43D. This constraint in turn requires that ΔS between the low and high-field curves remains constant, or for the case of straight line T-S curves, that the slopes of the high field and low field lines be equal. When the high and low field lines are parallel, the second law constraint that $\Delta S_H = \Delta S_C$ is automatically met, and the cycle performs at full Carnot efficiency. Thus, for an ideal Ericsson cycle, "parallel" T-S curves (constant $\Delta S_{BMIN\text{-}BMAX}$) are optimal with no other constraints. Of course, the greater the ΔS between the low and high field curves the greater the refrigeration obtainable for a single circuit of the T-S diagram.

Non-Ideal Ericsson

As the Ericsson cycle is dependent on regeneration, which in turn requires "parallel" T-S curves, the effect on refrigerator performance when the T-S lines are *not* parallel is of great interest. For clarity, the T-S curves used in the following discussion are straight lines, but the conclusions apply to general curves as well.

The principal factor in analyzing the effect of non-parallel T-S lines is how ΔS from low to high field varies with temperature. Figure 3(a) shows the T-S plot of an Ericsson cycle attempted on a magnetic material where ΔS increases with increasing temperature, typical of a refrigerator operating below the Curie temperature of a ferromagnetic material. In order for the regenerative heat transfer to balance, the area under the curve from points 3 to 4 must equal the area under the curve from points 1 to 2'. This implies that point 2' has to be at a lower temperature than T_H as shown in the figure. Another way of looking at this is the magnetic material in the high field has less heat capacity than the material in the low field, which implies that the regenerator cannot heat the low field material all the way to T_H. In order to reject heat at T_H it is necessary to include part of a Carnot cycle and perform isentropic partial magnetization to go from point 2'

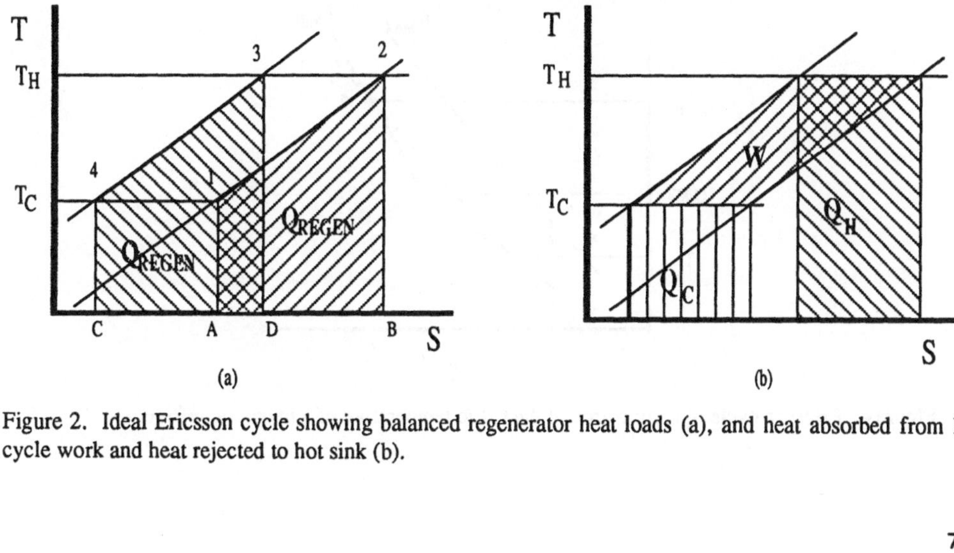

Figure 2. Ideal Ericsson cycle showing balanced regenerator heat loads (a), and heat absorbed from load, cycle work and heat rejected to hot sink (b).

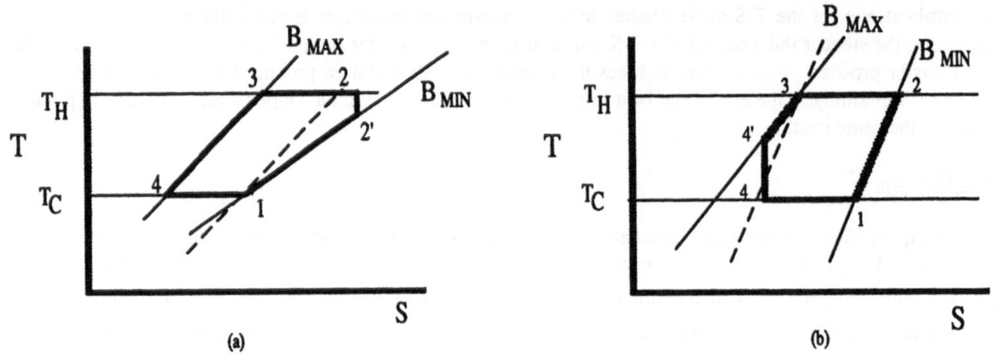

Figure 3. Ericsson cycle executed using a magnetic material with ΔS increasing with increasing temperature (a), and a material with ΔS decreasing with increasing temperature (b).

to the T_H line and then complete the magnetization isothermally at T_H. The dashed line in the figure illustrates the cycle that could be followed if the B_{MIN} line was parallel to the B_{MAX} line. The cycle operating with diverging constant field lines will require more work and will reject more heat than the cycle performed with parallel constant field lines.

Figure 3(b) shows the T-S diagram for operating an Ericsson cycle using a magnetic material with constant field lines converging at high temperature, representative of a refrigerator above the ordering temperature of the magnetic refrigerant. For this case, it is not possible to cool the material in the high field completely to T_C because it has a greater heat capacity than the material in the low field. As shown in the figure, it is necessary to use some of the demagnetization to lower the temperature to T_C before heat can be absorbed. The result of operating this cycle compared to a cycle where the high field line parallels the low field line (shown as a dashed line in the figure) is that while Q_H remains the same for the two cycles, the work done on the magnetic material is increased thus decreasing the available cooling power, as dictated by eq. 1. This also results in a reduction of the refrigerator efficiency.

Figure 4 illustrates the case of operating a magnetic refrigerator around the Curie temperature of a ferromagnetic material. The net result is that extra work is required to travel around the bulge in the cycle compared to a cycle with parallel constant field lines. Depending on the relative sizes of ΔS at the hot and cold temperatures, this will cause either a decrease in the cooling capacity or an increase in the amount of heat rejected. For the case shown where ΔS_H equals ΔS_C, the cooling power is reduced from the ideal parallel line case because the heat capacity of the magnetic material in the high field is greater than the material in the cold field.

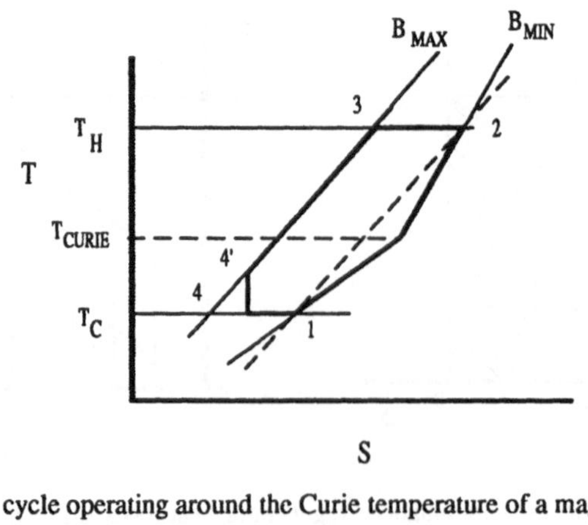

FIgure 4. Ericsson cycle operating around the Curie temperature of a magnetic refrigerant.

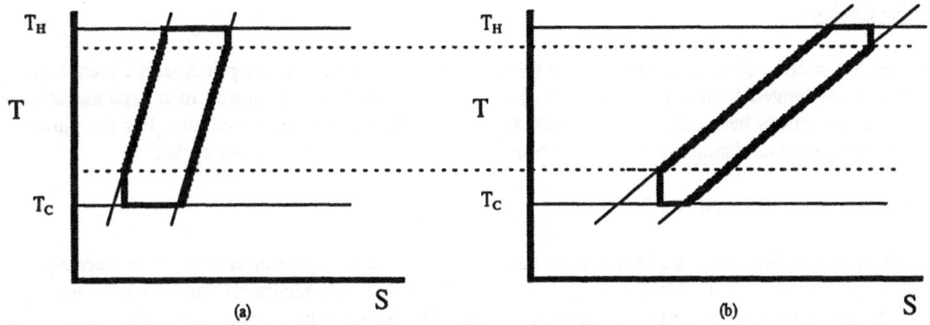

Figure 5. Comparison of Ericsson cycles operating using magnetic materials with different specific heats.

Cycles with Non-Ideal T-S curves and Non-Ideal Heat Transfer

Figure 5 shows two cases of ideal Ericsson cycles operating on ideal magnetic materials with parallel constant field lines. The ΔS between the constant field curves is the same in both cases, but in case (a) the curves are steeper than in case (b), indicating that the magnetic material in case (b) has the higher heat capacity. The cycles shown are based on a regenerator effectiveness of 0.9 with isothermal heat transfer at the hot and cold ends. It is evident from this figure that steep curves are desirable for real refrigerators with non-ideal regeneration.

Brayton Cycle

Brayton cycle shown in Figure 6 is similar to the Ericsson cycle except for the manner in which the field change is accomplished. In the Ericsson cycle the field change processes occur at constant temperature, while in a Brayton cycle the field change is a constant entropy process. By comparing Figures 2 and 6, it can be seen that the Brayton cycle has less cooling capacity and greater heat rejection than the Ericsson cycle.

For real cycles the distinction between Ericsson and Brayton cycles is less well defined. Non-isothermal heat transfer in an Ericsson cycle shifts the shape of the T-S diagram closer to that of a Brayton cycle. Also, Brayton cycles typically do not undergo completely isentropic magnetization or demagnetization. Heat transfer during the field change is performed whenever possible to increase efficiency nearer to that of an Ericsson cycle. With respect to the above discussions however, the Brayton and Ericsson cycles are equivalent in their responses to varying material T-S diagrams.

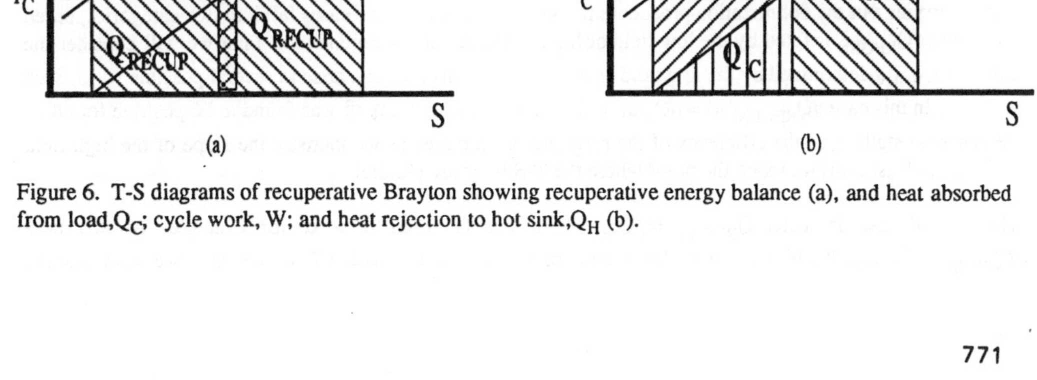

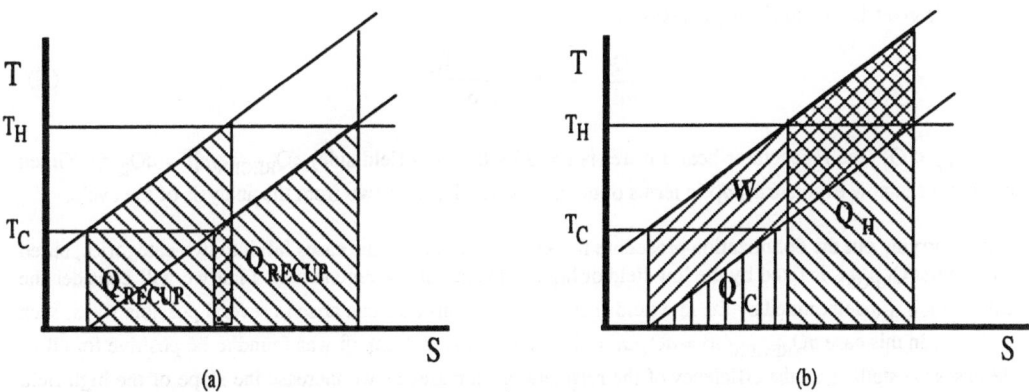

Figure 6. T-S diagrams of recuperative Brayton showing recuperative energy balance (a), and heat absorbed from load,Q_C; cycle work, W; and heat rejection to hot sink,Q_H (b).

Recuperative Cycles

The operation of a recuperative cycle is much the same as a regenerative cycle, the principal difference being that the energy exchanged between the hot and cold legs of the cycle is not stored in a separate storage material, i.e. a regenerator bed; rather the heat transfer fluid itself is the storage medium. For the purposes of this paper, no functional distinction is made between regenerative and recuperative cycles.

NUMERICAL ANALYSIS AND RESULTS

In order to decide which traits are more important for a specific refrigerator design, it is necessary to develop detailed thermodynamic models of the individual designs which accurately model inefficiencies to investigate the performance with different magnetic materials. Some results produced in this manner are discussed in the following sections.

Recuperative Brayton Cycle

To gain some insight into the effect of the shape of the T-S curves of the magnetic refrigerant on the efficiency of a Brayton cycle refrigerator, we have used a "linear" model of the Brayton cycle: both the high field and low field T-S curves are assumed to be straight lines as shown in Figure 7(a). The slope α and intercept a of the low field line are input data, along with T_C and T_H of the cycle, the low field and high field regenerator effectivenesses ε_1 and ε_2, and the entropy S_H corresponding to T_H on the high field line. We will allow the slope, β, of the high field line to vary; the intercept, b, is computed from T_H, S_H and β. The effect of varying β is that the high field line "pivots" around the point T_H, S_H.

Given the input data and the current values for β and b, the cycle is completely determined. The heat transfer to the regenerator (or in the recuperator) plays an important role in this analysis. The maximum possible heat transfer on both the low field and high field sides of the regenerator is easily computed from the T-S diagram. These quantities are multiplied by ε_1 and ε_2 to form Q_1 and Q_2, the actual low field and high field heat transfer. Assuming that the regenerator is balanced, Q_{REGEN} will be the minimum of Q_1 and Q_2. Given Q_{REGEN}, all of the points on the cycle are easily determined.

We define an efficiency, η, as the ratio of Q_C to Q_H. Differentiating η with respect to β and simplifying, we find:

$$\frac{\partial \eta}{\partial \beta} = \frac{1}{Q_H}\left(\frac{\partial Q_C}{\partial \beta} - \eta \frac{\partial Q_H}{\partial \beta}\right) \qquad (6)$$

Both terms in eq. 6 will depend on $dQ_{REGEN}/d\beta$ as well as other cycle parameters. Recall that Q_{REGEN} is the minimum of Q_1 and Q_2. As Q_1 does not depend on the slope of the high field line, $dQ_1/d\beta$ is identically zero. Thus if $Q_1 < Q_2$, i.e. the regenerator heat transfer is "fixed" by the low field side, then $dQ_{REGEN}/d\beta=0$. $dQ_2/d\beta$ is given by the following expression:

$$\frac{\partial Q_2}{\partial \beta} = -\varepsilon_2 \frac{T_H^2 - T_C^2}{2\beta^2} \qquad (7)$$

So if $Q_2 < Q_1$ (the regenerator heat transfer is fixed by the high field side) $dQ_{REGEN}/d\beta = dQ_2/d\beta$. Given eq. 7, we can now evaluate $d\eta/d\beta$ in terms of quantities which are known from the analysis of the cycle.

When carrying out the optimization procedure for a particular model, the cases fall into two categories, based on whether Q_{REGEN} is fixed by the low field or high field side of the regenerator. First we will consider the case of Q_{REGEN} determined by the low field characteristics. This case can arise in several possible ways, such as $\beta < \alpha$. In this case $dQ_{REGEN}/d\beta = dQ_1/d\beta = 0$. The result is that $d\eta/d\beta$ was found to be positive for all of the cases we studied, so the efficiency of the refrigerator increases as we increase the slope of the high field T-S line, at least until we reach the point where the T-S lines are parallel.

The second case is when $Q_{REGEN} = Q_2$, which occurs when $\beta > \alpha$ for example. In this case $dQ_{REGEN}/d\beta=dQ_2/d\beta$ which is always less than zero as seen from Equation 2. In the cases we have studied,

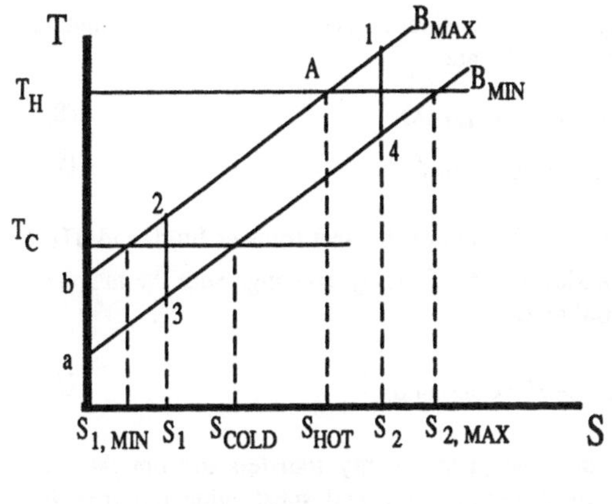

(a)

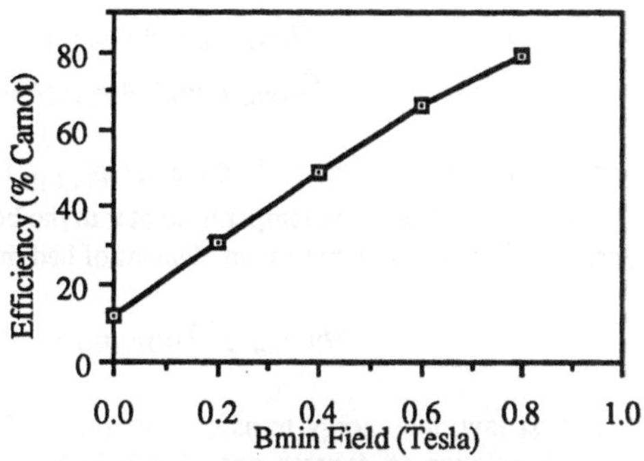

(b)

Figure 7. Diagram of parameters used in "linear" model of a Brayton cycle (a). Efficiency as a function of low B field value for lumped parameter model of recuperative Brayton cycle (b).

this term dominates the expression for $d\eta/d\beta$. If we begin with parallel lines, $d\eta/d\beta$ can be either positive or negative. If it is negative, a decrease in β should cause an increase in efficiency. However, decreasing β will lead to the regenerator heat transfer being determined by the low field side, which as we have seen above gives $d\eta/d\beta > 0$. Thus the optimum efficiency occurs for $\beta = \alpha$ even though $d\eta/d\beta$ is not zero.

If the parallel T-S lines lead to a positive $d\eta/d\beta$, then increasing β will decrease $d\eta/d\beta$. This is a consequence of the expression for $dQ_{REGEN}/d\beta$; note that the magnitude of $dQ_2/d\beta$ decreases as β increases. Eventually we reach the point where $d\eta/d\beta = 0$ for some $\beta > \alpha$. In this case, by increasing the slope of the high field line, we are decreasing both Q_C and Q_H from their parallel line values. The efficiency improves because Q_H is decreasing more rapidly than Q_C, or equivalently, the cycle takes less work to execute, per unit of Q_C.

To further investigate the effect of the T-S curves on refrigerator efficiency, a simple, lumped parameter thermal network model of a recuperative Brayton cycle refrigerator has been developed. The refrigerant used is GGG (gadolinium gallium garnet), T_H is fixed at 20 K and the difference in magnetic field between the low field and high field sides is 5.2 T. The T-S curves for GGG clearly show that in the range of 0 to 6 Tesla the slope decreases as the field strength increases. Therefore, the ratio β / α (in the temperature range 10 to 20 K) is larger as we increase the fields used from 0 T (low field value) to 0.8 T, although it is still less than 1. Thus we would expect greater efficiency for a low field value of 0.8 T. Using T_C = 13.5 K, several runs were made with different values of the low field; results are given in Figure 7(b). The efficiency increases monotonically with the low field value as expected.

Thus the results of the simple lumped parameter model of a recuperative Brayton cycle refrigerator are in good qualitative agreement with our simple linear model of the T-S curves and the conclusions drawn from examination of T-S cycle diagrams in the previous sections. The linear model predicts that if the slope of the high field line is less than that of the lower field line, efficiency can be increased by increasing the slope of the high field line, at least until the lines are parallel. If the lines are parallel, it is possible for the efficiency to be increased further by increasing β, depending on such things as the temperature span of the refrigerator and the change in temperature due to adiabatic magnetization.

Brayton Cycle with Active Magnetic Regeneration

When the thermal mass of the magnetic refrigerant material is used as an internal regenerator we refer to the device as an Active Magnetic Regenerative Refrigerator (AMRR)[2]. The behavior and requirements of an AMRR are quite different from that of the standard regenerative or recuperative cycle. This is due to the coupling of the temperature profile requirements imposed for efficient regenerator with the temperature requirements required by the refrigeration cycle.

Starting with the same analysis approach used by Taussig, et al [3], consider an element of magnetic material forming the regenerative bed of the refrigerator. The energy flows are

$$Q_{REJ,FLUID} = m_{FLUID}Cp_{FLUID}dT_{HOT,FLUID}(x) \tag{8}$$

$$Q_{ABS,FLUID} = m_{FLUID}Cp_{FLUID}dT_{COLD,FLUID}(x) \tag{9}$$

where m_{FLUID} is an element of fluid mass, Cp_{FLUID} is the specific heat of the heat transfer fluid, and dT(x) is the incremental change in temperature at x in the bed produced in the fluid by exchange with the magnetic material. An entropy balance on an element of bed material gives

$$Q_{REJ,BED} / T_{HOT,BED}(x) = Q_{ABS,BED} / T_{COLD,BED}(x) \tag{10}$$

By the first law, the energy transfers from/to the fluid must equal the energy transfers to/from the bed. Using this relation to express eqs. (9-10) in terms of fluid termperature, and substituting the resulting equation into eq. (10), we obtain the expression

$$dT_{HOT,FLUID}(x) / T_{HOT,BED}(x) = dT_{COLD,FLUID}(x) / T_{COLD,BED}(x) \tag{11}$$

Now, in order to minimize the irreversible entropy generation produced by non-isothermal heat transfer, the fluid temperature must everywhere equal the bed temperature. Therfore, substituting bed temperatures for fluid temperatures in eq. (11) we integrate to produce

$$T_{HOT,BED}(x) / T_{COLD,BED}(x) = K \tag{12}$$

Eq. (12) gives the relationship of bed temperature ratio throughout the length of the bed as a positive constant greater than unity.

Thus, the required temperature profile for minimum irreversible entropy generated by non-isothermal heat transfer between the magnetic material comprising the regenerator bed and the heat tranfer fluid is given by eq. (12). Combining eq. (12) with the definition of specific heat given in eq. (2) results in

$$\Delta S_{Bmax\text{-}Bmin} = (K-1)\, T\, dS/dT \tag{13}$$

For straight line T-S curves, eq. (13) is satisfied by a temperature entropy diagram as shown in Figure 8(a).

The performance of AMRRs using T-S curves as given by eq. (13) were investigated using two different analysis techniques. A computer code, using a finite difference technique based on the computational methodology of Schmidt and Willmott[4], was developed to solve the four coupled partial differential equations describing the time dependent performance of an AMRR. Longitudinal heat conduction through the heat transfer fluid and along the bed are neglected, and the bed is treated as one-dimensional. In the limit of infinite regenerator thermal mass, the four coupled partial differential equations in space and time reduce

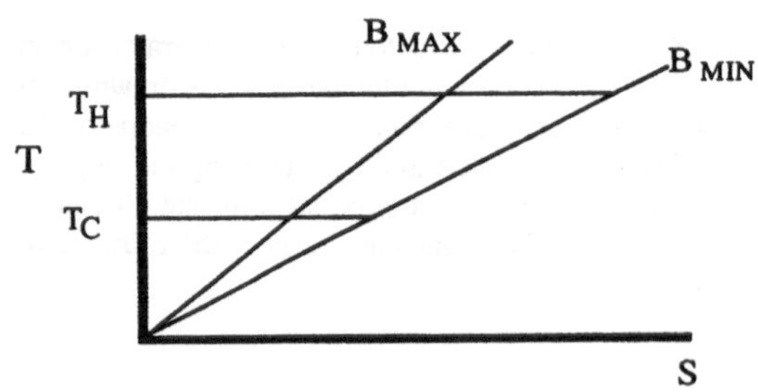

(a)

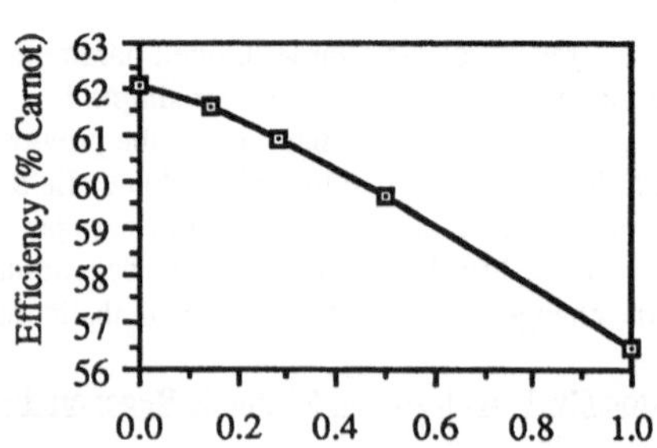

(b)

Figure 8. T-S diagram of magnetic refrigerant predicted as optimum for an Active Magnetic Regenerative Refrigerator (a). Efficiency of an AMRR vs. γ showing effect of varying respective slopes of magentic refrigerant T-S high and low field lines (b).

to two ordinary differential equations in space. This simplified system of equations can be solved quickly and accurately and was used to give an independent check on the finite bed mass code. (The details of the analysis are not discussed in this paper due to space limitations.)

Some results from this analysis are shown in Figure 8(b). The parameter $\gamma = \Delta S_{T=0\ K} / \Delta S_{T=100\ K}$ plotted along the horizontal axis is a measure of the relative slopes of the high and low field T-S lines. When γ = 1.0 the lines are parallel, when γ = 0.0 the lines intersect at T = 0 K. As predicted by the preceding analysis, the maximum efficiency occurs at a value for γ of 0.0.

CONCLUSIONS

It is interesting to note that some of the traits of an ideal magnetic material for a regenerative/recuperative cycle may be conflicting. Specifically, constant field curves can be made more parallel by the careful blending of magnetic materials with different ordering temperatures. But, this will cause the constant field curves to become less steep than the curves for the individual materials, thus increasing the effectiveness demanded of the regenerator in order to maintain performance. Also, blending materials will reduce the peak ΔS for a given field change. Depending on the temperature range being spanned with a single stage, there is a point of diminishing return at which the increase in curve "parallelism" is offset by a decrease in overall ΔS.

Magnetic refrigerators utilizing Carnot, Ericsson and Brayton cycles perform most efficiently with low specific heat magnetic materials, i.e. materials that have a large isentropic temperature change with change in magnetic field. This is true for non-ideal as well as ideal cycles, as the effects of non-ideal processes are minimized with low specific heat materials.

Ericsson/Brayton cycles utilizing separate materials for regeneration exhibit optimal performance with magnetic refrigerants having parallel T-S curves ($\Delta S_{BMIN\text{-}BMAX}$ = constant). As the magnitude of the heat transfer inefficiencies increase, i.e. as the cycle tends closer to a true Brayton, the energy balances in the system require T-S lines with ΔS increasing with increasing temperature. This corresponds to a larger ΔS required at T_H to reject the sum of the ΔS_{COLD} and ΔS_{IRR} as expressed in eq. 5.

For an AMRR, a Brayton cycle utilizing the thermal mass of the magnetic refrigerant as a regenerator, the need to minimize irreversible losses in the regenerator leads to T-S curves with ΔS decreasing to zero at 0 K. This is in contrast to Brayton cycles utilizing external regenerators or recuperators where the temperature profile through the regenerator is not so intimately linked with the temperature profile of the magnetic refrigerant.

No where in these analyses was a case encountered where T-S lines converging with increasing temperature was advantageous.

ACKNOWLEDGMENT

Work sponsored by Flight Dynamics Laboratory, Air Force Wright Aeronautical Laboratories, Aeronautical Systems Division (AFSC) United States Air Force, Wright-Patterson AFB, OH 45433-6533, under contract F33615-86-C-3431.

REFERENCES

1. T. Hasimoto et al, New application of complex type of magnetic materials to the magnetic refrigerant in the Ericsson magnetic refrigerator, preprint, to be published in J. of Appl. Physics (1987)
2. J. A. Barclay, The theory of an active magnetic regenerative refrigerator, in: NASA report NASA-CP-2287 (1983).
3. Taussig et al, Magnetic refrigeration based on magnetically active regeneration, in: "Proc. 4th Intl. Cryocooler Conf., Easton, Maryland (1986).
4. Frank W. Schmidt and A. John Willmott, chapter 8, in: "Thermal Energy Storage and Regeneration", Hemisphere Publishing Corporation, Washington D.C. (1981).

THE MAGNETOCALORIC EFFECT OF SOME RARE EARTH METALS

Geoffrey Green, William Patton, and John Stevens

David Taylor Naval Ship Research and Development Center
Electrical Machinery Technology Branch
Code 2712
Bethesda, Maryland

ABSTRACT

The Navy is investigating a reciprocating magnetic refrigeration concept that uses an active regenerator. The geometry of this active regenerator is an embossed ribbon configured of a formable material (generally a rare earth metal) having a reasonably large magnetocaloric effect at its Neel or Curie temperature. Several rare earths have shown large increases in heat capacity at their respective transition temperatures. Samples of two rare earth metals, terbium and holmium, were subjected to a 7-tesla change in magnetic field over a range of temperatures and the adiabatic temperature change measured. Results indicated a 10.3-°C temperature change in terbium at 237 K and a 6.1-°C change in holmium at 136 K. In addition, no thermal hysteresis effect was observed in these materials near their Neel temperatures.

INTRODUCTION

Recently the U.S. Navy at the David Taylor Naval Ship Research Center (DTNSRDC) in Annapolis, Maryland, has been investigating the development of a magnetic refrigerator. The magnetic refrigerator under study is a reciprocating type using an active regenerator as the working material.[1,2] In its initial stage, it uses a ferromagnetic material that is formed in an embossed ribbon and helically wound into small disks, or "pancakes." These pancakes are than stacked one on top of the other to form the active regenerator. Unfortunately, a regenerator made of a single material having a temperature-dependent magnetocaloric effect will have a limited operating temperature, and therefore several materials are required, each having a magnetocaloric effect at a lower temperature than the last. In addition, a refrigerator using an embossed ribbon type regenerator requires not only that the ferromagnetic regenerator material have a reasonably large magnetocaloric effect but also that it have metallic properties allowing it to be formed into a ribbon configuration. These required characteristics are quite different from the characteristics of the intermetallic compounds and alloys that have been investigated by Barclay and others.[3]

The magnetocaloric effect for an adiabatic temperature change in a material can be defined by the equation:

$$dT = -\ (T/C_H)(\partial m/\partial T)_H dH \qquad (1)$$

where:

- T = temperature, kelvin;
- C_H = constant field heat capacity, joules per kilogram per kelvin;
- m = magnetization, tesla per kilogram; and
- H = internal magnetic field, tesla.

This equation indicates that increasing the magnetic field will increase the temperature, and decreasing the field will decrease the temperature.

Benford and Brown[4,5] have measured a large magnetocaloric effect in two rare earth metals, gadolinium and dysprosium, near each material's transition temperature. Measurements of these two rare earths also indicated that no significant thermal hysteresis , i.e. internal thermal loss, was associated with a change in field; in other words, the measured effects were adiabatically reversible. This reversibility is critically important if the material is to become an effective component of a magnetic refrigerator, because thermal hysteresis could otherwise significantly reduce or eliminate the cooling effect of the refrigeration cycle.

Two rare earth metals that seem to have a large change in zero field heat capacity near their Neel temperature (Tn) are terbium[6] (Tn=235 K) and holmium[7] (Tn=135 K). The heat capacity characteristics of these two materials are illustrated in Figs. 1 and 2. These materials have the potential to produce a large magnetocaloric effect. If there is no thermal hysteresis, as was noted with dysprosium near its Neel temperature, terbium and holmium would be of great importance. These materials would fill the temperature gap left between gadolinium, with a Curie temperature of 293 K, and dysprosium, with Tn=178 K, allowing the development of an active regenerator spanning a temperature range from 300 K down to 100 K in a single stage.

To determine if terbium and holmium might indeed fill this gap, an experimental test apparatus was fabricated to measure the adiabatic magnetization temperature change and the thermal hysteresis effects on these two rare earth metals. Results of the tests, which were quite promising, are described below.

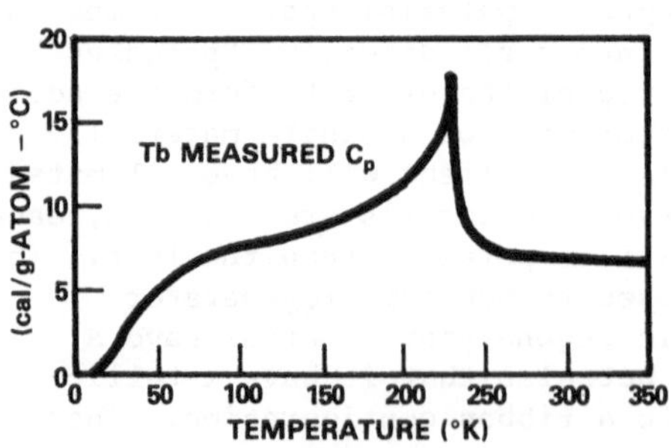

Fig. 1. Gram atomic heat capacity (cp) of terbium.

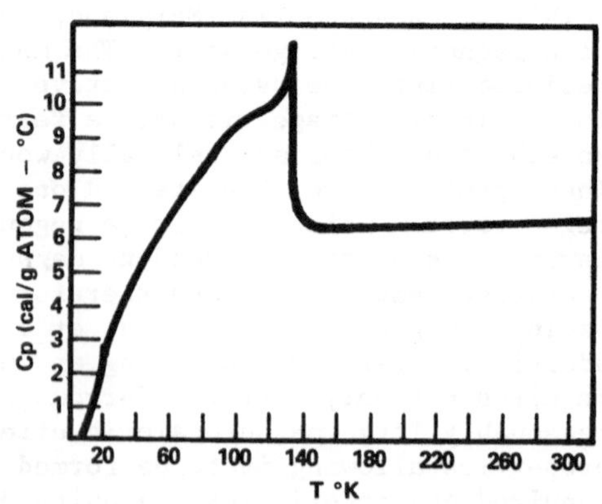

Fig. 2. Gram atomic heat capacity of holmium.

EXPERIMENTAL APPARATUS AND PROCEDURES

The two rare earth specimens tested (one of terbium and one of holmium) were in the form of 100-g ingots, as cast, having a minimum purity of 99.9%. The specimens were machined to a cylindrical shape with a diameter of approximately 1.9 cm and a length of 3.8 cm.

To measure the specimens' temperature response to magnetic field changes, we used a superconducting magnet fabricated at DTNSRDC. The magnet was a simple solenoid, 12.13 cm in diameter and 25.4 cm long, with an 8.26-cm bore. The magnet used niobium-titanium superconductor and contained a vacuum-insulated dewar with vapor-cooled shields and a warm bore. Figure 3 shows the field distribution in the magnet bore. With the magnet on, there was a nearly constant 7-tesla field over a 5-cm length of the bore.

A vacuum-insulated specimen container was designed to fit into the warm hole portion of the magnet dewar, as shown in Fig. 4. The rare earth specimens were suspended and held in place by a stainless steel wire 0.317 cm in diameter. This wire was tensioned with a series of 10 Belleville spring washers. A temperature control arrangement was used that consisted of a liquid nitrogen cooling loop attached to a copper cylinder to which an electrical heater was bonded. In addition, the bottom flange of the specimen container had several connections, one for the thermocouple and heater lead feedthrough, a second for the evacuation of the inner volume, and a third for the inlet and exhaust of the liquid nitrogen cooling loop. Five thermocouples were placed on the specimen at the following locations:

1. Top end, on the surface
2. Bottom end, on the surface
3. Middle, on the surface
4. Middle, half the depth to the center
5. Middle, in the center

A sixth thermocouple was placed on the copper cylinder to which the heater was bonded. This provided a feedback loop to control the operating temperature of the specimen and minimized the radiation heat leak. Figure 5 is a photograph of the specimen assembly.

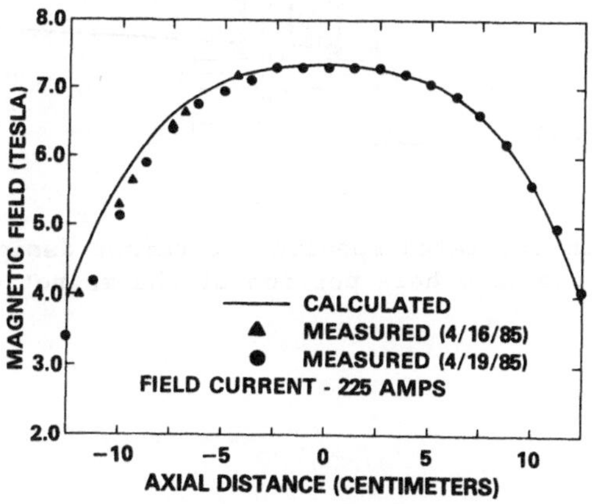

Fig. 3. Field distribution in the bore of the superconducting magnet.

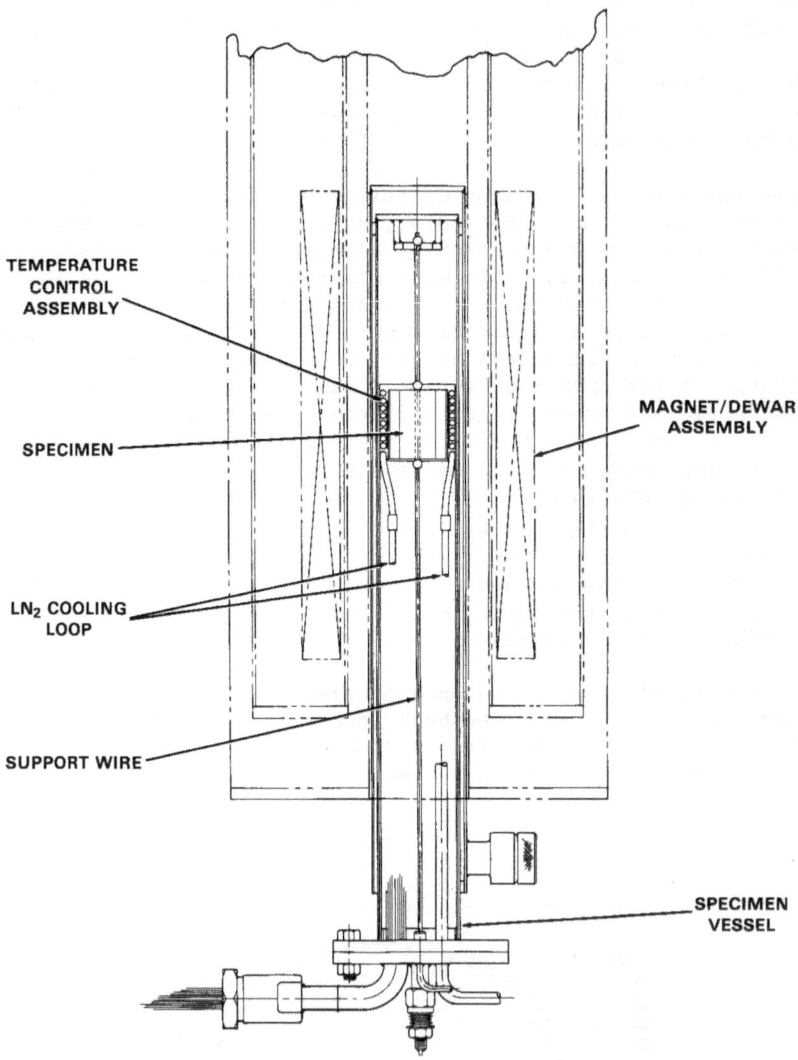

Fig. 4. Vacuum-insulated specimen container designed to fit into the warm hole portion of the magnet dewar.

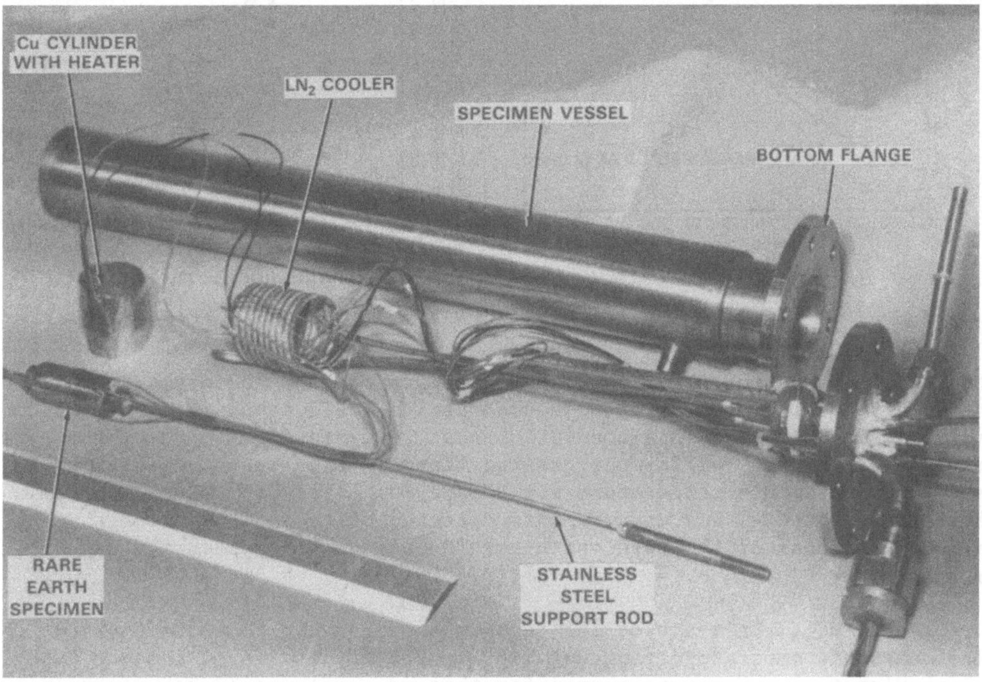

Fig. 5. Rare earth specimen assembly.

The experimental apparatus was used to measure adiabatic magnetization temperature change and thermal hysteresis effects of the two rare earth metals, terbium and holmium. Throughout the tests, two temperature readings were taken at each temperature point to make sure that the data were repeatable. The tests were conducted as follows: Each rare earth sample was cooled to the lowest desired temperature point and the volume inside the specimen vessel was evacuated to a pressure between 50 and 10 μmHg. The copper cylinder surrounding the specimen was then set at the same temperature as the specimen. A set of temperatures was recorded at zero magnetic field, and then the field was increased to the maximum, 7 T. Another set of temperature readings was recorded, and the field was removed. When the field had returned to zero, a third set of temperature readings was taken. The temperature of the specimen was taken as the average of the three thermocouples located in the middle of the specimen at different depths and the temperature change (ΔT) was plotted against the initial temperature of the specimen before the field was applied.

The heater was then used to raise the sample temperature to the next set point, and another series of two sets of temperatures recorded. This process was continued over the full temperature range desired for each specimen.

RESULTS

The magnet cycle time was about 80 seconds: 30 s to ramp up to full field, 10 s at full field to record temperature measurements from the thermocouple output, 30 s to ramp down, and another 10 s to record temperature at zero field. The only significant heat leak in the system was along the stainless steel wire that attached the specimen to the container. Assuming a maximum temperature difference of 200 K between

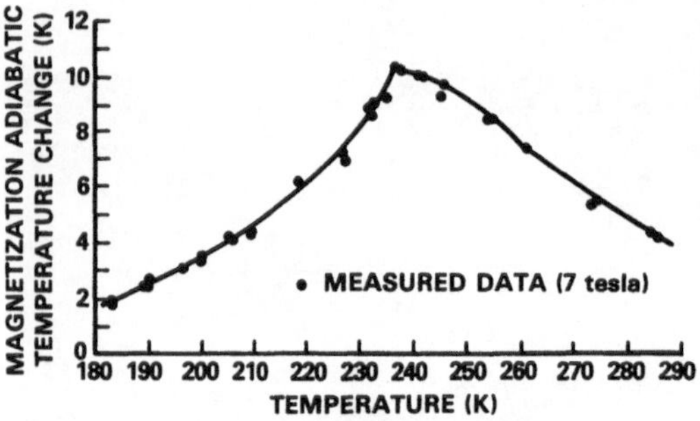

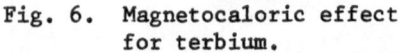

Fig. 6. Magnetocaloric effect for terbium.

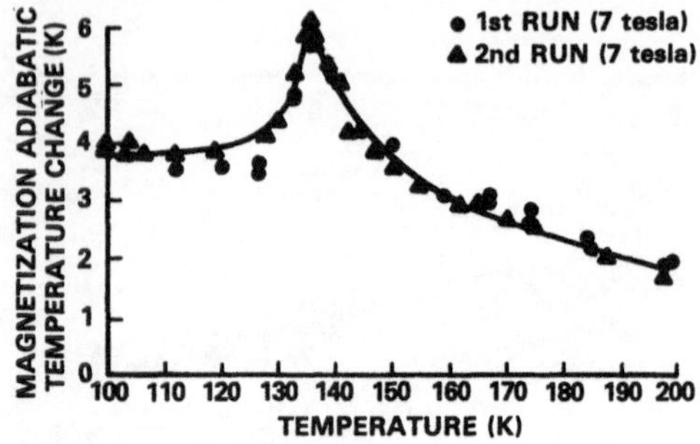

Fig. 7. Magnetocaloric effect for holmium.

the specimen and the outer container, conduction heat loss through the stainless steel wire over an 80-second period would cause the 100-g specimen to warm up 0.4°C. This conduction heat leak was assumed to be the maximum error. Any difference greater than 0.4°C between the starting and ending specimen temperature would have to be assumed to result from hysteresis effects. Our measurements detected no thermal hysteresis effect in either of the rare earth metals based on the predicted experimental error.

Results of the tests on the terbium specimen are shown in Fig. 6. The measured temperature range was 180 to 285 K. The largest magnetization adiabatic temperature change for this material was 10.3°C at 237 K. The results for the holmium specimen (Fig. 7) show that, in the measured temperature range (100 to 200 K), the peak magnetization adiabatic temperature change was lower, 6.1°C at 136 K.

The terbium results indicated a high adiabatic temperature change over a reasonably large temperature span. In fact, a 4°C change or larger occurred over an 80-°C temperature span, whereas the holmium had a much narrower span (20°C) with an adiabatic temperature change of 4°C or higher.

CONCLUSIONS

These experimental measurements indicated that magnetocaloric effects in both terbium and holmium are significant. In addition, no measurable

Table 1. Heat capacity (Cp) characteristics of several rare earth metals.

Material	Temperature of peak Cp (K)	Cp/Cp_{low}
Gadolinium (Gd)	291	86
Terbium (Tb)	230	153
Dysprosium (Dy)	175	106
Holmium (Ho)	131	66
Europium (Eu)[a]	88.8	2445
Erbium (Er)[b]	83	33
Thulium (Tm)[c]	55	62

[a]From Gerstein et al.[8]
[b]From Skochdopole et al.[9]
[c]From Jennings et al.[10]

thermal hysteresis effects were found in either of the two rare earth metals over a wide temperature span near their respective Neel temperatures. Although terbium showed a larger magnetocaloric effect than holmium over a wider temperature range, this effect in holmium was still large enough to produce some significant cooling in an effective magnetic refrigerator.

Several other rare earth metals have also been found to have large spikes in heat capacity below 100 K, and these materials are also presumed to have reasonably large magnetocaloric effects. Table 1 compares these materials with materials for which our measurements have shown significant magnetocaloric effects. Some of these rare earths have a very narrow heat capacity spike, especially europium, and therefore might have a limited effective temperature range. However, all of these rare earths have the desired mechanical characteristic that they can be formed into a ribbon configuration. If, in addition, they are found to exhibit a reasonably large magnetocaloric effect, one could use a combination of these materials in a magnetic refrigerator for cooling from room temperature to the range of 20 K or below in a few stages.

REFERENCES

1. G. Patton, G. Green, J. Stevens, and J. Humphrey, "Reciprocating Magnetic Refrigerator," Proc., 4th Intl. Cryocooler Conf., Easton, Md., Sep. 1986, David Taylor Naval Ship Research and Development Center (in press).

2. G. Green, G. Patton, J. Stevens, and J. Humphrey, "Magnetocaloric Refrigeration," David Taylor Naval Ship Research and Development Center, DTNSRDC report (in preparation).

3. Barclay, J.D., "An Analysis of Liquefaction of Helium Using Magnetic Refrigerators," Los Alamos National Laboratory, Report LA-8991 (1981).

4. S.M. Benford and G.V. Brown, "T-S Diagram for Gadolinium Near the Curie Temperature," J. Appl. Phys., Vol. 52, p. 2110 (1981).

5. S.M. Benford, "The Magnetocaloric Effect in Dysprosium," J. Appl. Phys., Vol. 50, p. 1868, (1979).

6. B.C. Gerstein, M. Griffel, L.D. Jennings, R.E. Miller, R.E. Skochdopole, and F.H. Spedding, "Heat Capacity of Holmium from 15 to 300 K," J. Chem. Phys., Vol. 27, p. 394 (1957).

7. L.D. Jennings, R.M. Stanton, and F.H. Spedding, "Heat Capacity of Terbium from 15 to 300 K," J. Chem. Phys., . 27, p. 909 (1957).

8. R.E. Skochdopole, M. Griffel, and F.H. Spedding, "Heat Capacity of Erbium from 15 to 320 K," J. Chem. Phys., Vol. 23, p. 2258 (1955).

9. B.C. Gerstein, F.J. Jelinek, J.R. Mullaby, W.D. Shickell, and F.H. Spedding, "Heat Capacity of Europium from 5 to 300 K," J. Chem. Phys., Vol. 47, p. 5194 (1967).

10. L.D. Jennings, E. Hill, and F.H. Spedding, "Heat Capacity of Thulium from 15 to 360 K," J. Chem. Phys., Vol. 34, p. 2082 (1961).

PREPARATION AND FABRICATION OF RARE EARTH MAGNETIC MATERIALS

B.J. Beaudry

Materials Preparation Center, Ames Laboratory
Iowa State University
Ames, Iowa

ABSTRACT

Binary alloys between the rare earth metals (elements) and metals from the first or second transition series elements exhibit a large range of magnetic properties. Their application in the field of permanent magnets has grown rapidly. Many alloys of the rare earths with transition series elements have properties suitable for magnetic refrigeration. Preparation of these alloys in high purity and in forms suitable for refrigeration devices presents many problems. Trade offs between purity and magnetic properties must be made to obtain structurally sound refrigeration materials at a reasonable cost. Laboratory methods used to prepare and study these alloys will be presented. Large scale methods to produce permanent magnet materials have been developed. These methods and their possible application to magnetic refrigeration alloys will be presented.

INTRODUCTION

Rare earth intermetallic compounds have been shown to have suitable properties for magnetic refrigeration.[1-4] Intermetallic compounds usually are brittle and this presents problems in their preparation and fabrication. Only a little work has been reported on fabrication of compounds which have properties suitable for magnetic refrigeration. However, the use of rare earth intermetallic compounds (such as $SmCo_5$, Sm_2Co_{17} and $Nd_2Fe_{14}B$) as permanent magnetic materials has achieved industrial prominence.

Experiences of researchers in the field of permanent magnets are of interest not only for methods of fabricating these brittle intermetallic compounds, but also for lower cost processes. In the search for intermetallic compounds with the best properties for magnetic re-

*Operated for the U.S. Department of Energy by Iowa State University under contract no. W-7405-ENG-82. This work was supported by the Office of Basic Energy Sciences.

frigeration, little attention should be paid to the cost of research materials. The permanent magnet industry has shown cost reductions of 100 times and more are possible. The philosophy that is proposed here is to (1) establish intrinsic properties of high purity, well characterized samples, and (2) develop methods to make them cheaply and strong.

PREPARATION OF PURE INTERMETALLIC COMPOUNDS

Determination of the intrinsic properties of a particular intermetallic compound requires well characterized, pure materials. Many papers have been published which were based on phenomena caused by impurities - not just foreign matter, but other phases from the same binary system. Intermetallic compounds should be prepared from pure components and the number of phases present determined metallographically and confirmed by x-ray diffraction. It is much easier to see a small quantity of second phase by optical microscopy (metallography) than by x-ray diffraction. The use of pure components does not guarantee pure compounds. Contamination by the crucibles, the atmosphere or simply by handling of the samples in air sometimes accounts for the introduction of impurities.

The use of an arc melter for the preparation of intermetallic compounds is popular because it is an easy way to eliminate crucible and atmosphere contamination. For experimental quantities, generally it is the easiest and most practical method of melting.

Preparation of alloys between elements of the first transition series and the rare earth elements can be done in a ceramic crucible if the non-rare earth element is melted first and the rare earth metal is added to the molten metal. As the rare earth is added, the melting temperature of the alloy decreases. By following the liquidus temperature of the binary phase diagram, alloys in the central portion of the diagram will be molten at a lower temperature where the attack on the ceramic crucible by the molten rare earth is greatly reduced. If the rare earth metal is impure, the second phase impurities will float to the top of the molten alloy. Since the solubility of the interstitial impurities is decreased by these lower temperatures, some of the interstitial impurities which were present in the original rare earth will be removed as interstitial compounds in the slag. This method is especially useful for large scale production where impure metals can be purified by this slagging effect.

Congruent melting compounds can be formed directly from the melt. If the compound is formed peritectically, lengthy heat treatments are required to achieve a one phase sample. It is imperative that the alloy be examined to confirm the phase diagram since not all published phase diagrams are accurate. Recently we attempted to prepare $CeIr_5$. Based on the published phase diagram[5] which showed $CeIr_5$ was a congruent melting compound, a one phase sample was expected for the as-arc-cast sample. An examination of the microstructure revealed a classical example of peritectic rimming, see Fig. 1. Heat treatment at 1650°C for 200 hrs gave an equilibrated 2-phase structure (see Fig. 2). Microprobe studies showed the second phase to be Ir indicating that the

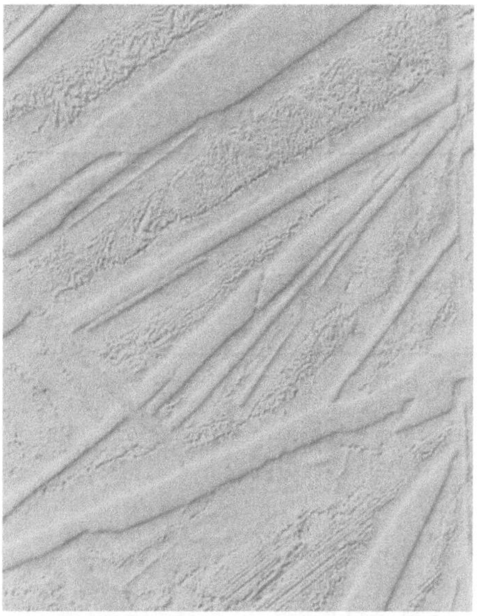

Fig. 1. $CeIr_5$ composition alloy in the as-arc-cast condition. Shows peritectic rimming. 400X

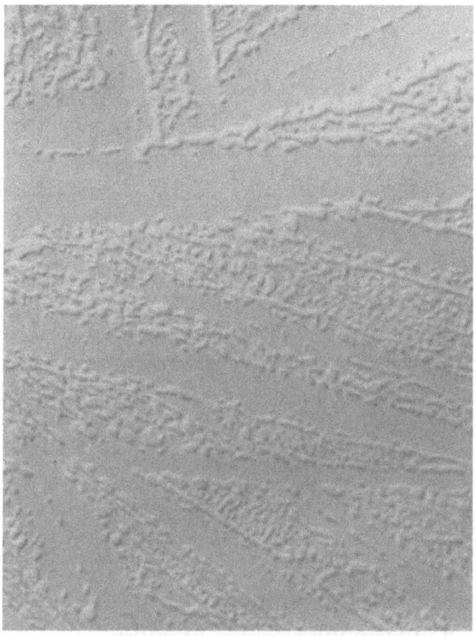

Fig. 2. $CeIr_5$ composition alloy after 200 hrs at 1650°C. Shows two phase. Ir is minor phase in equilibri with $CeIr_x$. 310X.

compound in this region is on the Ce-rich side of $CeIr_5$ and actually forms peritectically. The published phase diagram needs to be corrected.

The preparation of one phase intermetallic compounds can also be difficult if it is assumed that the 3-9's pure rare earth metals available commercially are 99.9% pure with respect to all impurities. Gschneidner[6,7] reported on the purity of commercially available rare earth metals. His conclusion was that advertised purities of rare earth metals with claimed purities of 99.9% usually contained one to five at.% total interstitial impurities. An example of Gd prepared in our Laboratory compared to commercial purity Gd is given in Table I. Fortunately, once the quantity of these second phase materials is known, correction can be made to the composition. It has been our experiences that a large portion of these impurities are not soluble in the intermetallic compound and will either be present as second phase particles mixed in the alloy or they will float to the top if the samples are cooled slowly in a crucible as discussed above. Obtaining a pure binary intermetallic phase when one of the components is impure requires that additional weight of the impure component be added. This slagging that occurs is especially effective at removing impurities when low melting eutectics are present in the binary system such as those between the rare earth metals and elements of the first transition series.

Interstitial impurities in the rare earth metal, in the alloying element, and in the atmosphere used in the melting furnace all will

Table I. Typical purity of Ames Laboratory and commercial purity Gd in weight and atomic ppm.

Impurity	Ames Laboratory Gd Wt.ppm	Ames Laboratory Gd At.ppm	Commercial Gd Wt.ppm	Commercial Gd At.ppm	Gd-rich Compound	At. ppm Gd in Compound
H	2	310	72	11,090	GdH_2	5,545
C	6	80	213	2,790	Gd_3C	8,370
N	1	11	90	1,010	GdN	1,010
O	22	216	1,027	10,090	Gd_2O_3	6,727
F	<3	<25	250	2,100	GdF_3	700
Total ppm	34	672	1,652	25,080	-	22,352/2.24%
	At.% free Gd 99.94				At.% free Gd = 97.7%	

form stable compounds with the rare earth metal. As mentioned above these compounds will frequently "slag out" and will not have a great effect on the intermetallic compound. Since the rare earth metal which is tied up as a compound with the interstitial element is lost, the final composition of the intermetallic alloy is low in rare earth content. The importance of this fact can be seen in the case of $PrNi_5$ which is used as an adiabatic demagnetization refrigeration material.[8] The resistance ratio (RR) of $PrNi_5$ had been shown to be related to its effectiveness as a refrigerant material.[9] In preparing $PrNi_5$ for Pobell,[8] a dramatic change in RR was noted when the composition was varied 0.6 at.% Pr in Ni. The $PrNi_{5.23}$ (-0.6 at.% Pr) had a resistance ratio of 2 while the Pr-rich alloy $PrNi_{4.78}$ (+0.6 at.% Pr) had a RR of 49. Subsequently, in a systematic study of the effect of the ratio of Pr to Ni on the RR and resistivity versus temperature, Cho[10] showed that Ni contents greater than $Ni_{5.0}$ gave a higher residual resistivity and thus a low RR. As a result of his study, the $PrNi_5$ we prepared for other laboratories[11] was prepared to a stoichiometry of $PrNi_{4.985}$ with 99.9 at.% pure Pr. The effect of using 98 at.% pure Pr can be shown by simply substituting $Pr_{0.98}$ for $Pr_{1.0}$. $Pr_{0.98}Ni_{4.985}$ is identical to $PrNi_{5.087}$ which is unacceptable for a refrigerant material. Thus it is important to determine intrinsic properties of intermetallic compounds prepared from pure materials and in some cases like $PrNi_5$ it is essential to know the true purity of the rare earth metal to achieve the desired stoichiometry. Correcting for the presence of impurities in computing the nominal composition is a viable, but poor second choice.

COST CONSIDERATIONS

The attitude or question "Why fool around with pure materials when they are too costly to make a product?" must be addressed. Accurate property measurement is time consuming and therefore costly. The cost of pure, well characterized samples is a small portion of the overall cost of the study. In the permanent magnet industry, the properties of $SmCo_5$[12] and $Nd_2Fe_{14}B$[13] alloys were determined with pure materials. When the commercial nature of these alloys became obvious, many new methods to prepare $SmCo_5$ were devised, including a method in which $SmCo_5$ was prepared directly from the oxide.[14]

Another example is the $Nd_2Fe_{14}B$-type permanent magnets. At least two methods that promise to result in low cost materials have already been devised. The first was that of General Motors[15] in which Na And $CaCl_2$ are used to reduce Nd_2O_3 directly in the presence of Fe. A second method was patented by F. A. Schmidt and D. T. Peterson of the Ames Laboratory.[16] In their method FeF_3 and NdF_3 are co-reduced with calcium. The reaction is initiated by a hot filament in a sealed crucible (bomb) and relies only on chemical heat to achieve separation of the alloy and slag phase. Both of these methods should greatly reduce the cost of $Nd_2Fe_{14}B$.

The important point is that researchers in magnetic refrigeration alloys should continue to use high purity, well characterized samples to determine the intrinsic properties of potentially useful materials. When a good material has been found, large scale production methods can be devised which will lower the cost.

FABRICATION OF BRITTLE INTERMETALLIC COMPOUNDS

Proposed designs[2-4] of magnetic refrigerators which use rare earth intermetallic compounds require thin sheets of these materials. Due to their brittle nature the as-cast materials do not have sufficient strength to withstand the repeated cycle through the magnetic field and the thermal stresses created in the cycle.

The compound GdNi is one of these brittle materials with good refrigeration potential. Wan[17] hot pressed 100 mesh powders of GdNi at 925°C into a cylinder one inch diam. by one inch long. Slices 0.025" thick cut from the cylinder had sufficient strength to be used in a refrigeration device. Metallographic examination of this material showed a dense equi-axed grain structure with few intergranular cracks. Cutting the hot pressed billets into thin slices results in a large percentage of waste. An alternate method which would not require slicing is needed. A method which is currently being developed in our laboratory is hot rolling. A great deal of work remains to be done.

The preparation of powders used in hot pressing is also a tedious task. Rapid solidification to obtain starting materials for hot pressing could possibly reduce or even eliminate the grinding operation, as has been shown by Lee[18] who prepared thin sheets of Nd-Fe-B for permanent magnet materials. However, rapid solidification increased the coercivity of the permanent magnet material which would be undesirable for refrigeration materials.

SUMMARY

Many intermetallic compounds exist that are potentially excellent refrigeration materials. Their intrinsic properties based on pure materials must be established. Methods to prepare the compound found to have the optimum properties must then be established as well as suitable fabrication methods.

REFERENCES

1. J. A. Barclay, W. C. Overton, Jr., and C. B. Zimm, Thermomagnetic properties of GdPd, "U. Eckern", A. Schmid, W. Weber, and H. Wuhl, eds., Elsevier Science Publishers B.V., 157 (1984).
2. A.A. Azhar, D. D. Mitescu, W. R. Johanson, C. B. Zimm, and J. A. Barclay, Specific heat of GdRh, J. Appl. Phys. 57 (1) 3235-7 (1985).
3. K. H. J. Buschow, J. R. Olijhoek, and A. R. Miedema, Extremely large heat capacities between 4 and 10 K, Cryogenics, 15 261-4 (1975).
4. T. Hashimoto, T. Numasawa, M. Shino and T. Okada, Magnetic refrigeration in the temperature range from 10 K to room temperature: the ferromagnetic refrigerants, Cryogenics, 21 647-53 (1981).
5. V. Dmitrieva, et al., "Phase diagram and characteristics of alloys of the Ir-Ce system", Redkzemel nye Metally Splarey I Soedineniya. Izdatel'stro Nauka, Moscow, 185 (1973).
6. K. A. Gschneidner, Jr., "Preparation and purification of rare earth metals and effect of impurities on their properties", Science and Technology of Rare Earth Materials, Academic Press, Inc., New York, 29 (1980).
7. K. A. Gschneidner, Jr., "Ultra pure rare earth metals and compounds: their preparation, properties and effects of impurities". Paper to be presented at 31st IUPAC Congress, Section 5. - Inorganic Chemistry, Sofia, Bulgaria, July 13-18, 1987.
8. R. M. Mueller, Chr. Buchal, H. R. Folle, M. Kubota, and F. Pobell, A double-stage nuclear demagnetization refrigerator, Cryogenics, 20 395-406 (1980).
9. K. Andres and S. Darack, Cooling of ^3He to 1 mK by nuclear demagnetization of $PrNi_5$, Physica 86-88 B&C 1071 (1977).
10. Weol Dong Cho, "A study of the anomalous behavior of the electrical resistivity in alloys near the $PrNi_{5.0}$ composition." M. S. Thesis, Iowa State University, Ames, Iowa 50011 (1982).
11. Based on the successes of Pobell (Ref. 8), 35 laboratories from all over the free-world have used $PrNi_5$ prepared by the Material Preparation Center, Ames Laboratory at Iowa State University in their adiabatic demagnetization refrigerators to obtain temperatures below 0.5 mK.
12. KSVL Narasimhan, Low oxygen processing of $SmCo_5$ magnets, J. Appl. Phys. 52, 2512 (1981).
13. J. F. Herbst, J. J. Croat and W. B. Yelon, Structural and magnetic properties of $Nd_2Fe_{14}B$, J. Appl. Phys. 57 (1) 15 (1985).
14. R. E. Cech, Cobalt rare earth intermetallic compounds produced by calcium hydride reduction of oxides, J. Metals 26 (2) 32 (1974).
15. Ram A. Sharma, Neodymium production processes, J. Metals p. 33 (February 1987).
16. F. A. Schmidt, D. T. Peterson and J. T. Wheelock, "Preparation of rare earth-iron alloys by thermite reduction", U.S. Patent 4,612,047.
17. Chung-Chu Wan, Aerospace Corp. Private communication (1987).
18. R. W. Lee, Hot pressed Nd-Fe-B magnets, Appl. Phys. Lett., 46 (8) 790 (1985).

MEASURED PROPERTIES OF GdNi FOR MAGNETIC REFRIGERATION APPLICATIONS

C.B. Zimm, W.F. Stewart, J.A. Barclay, C.K. Campenni

Astronautics Technology Center
Madison, Wisconsin

W. Overton, C. Olsen, D. Harding, R. Chesebrough

Los Alamos National Laboratory
Los Alamos, New Mexico

W. Johanson

Pomona College
Claremont, California

ABSTRACT

The crystalline ferromagnetic intermetallic compound GdNi, Curie point 70 K, shows potential as a working material for magnetic refrigeration around its Curie point. We report on cryogenic measurements of heat capacity, adiabatic temperature change upon application of a magnetic field, and magnetization for 0 to 9 T magnetic fields; also zero field magnetic susceptibility, electrical resistivity and thermal expansion. The resistivity measurements on as-cooled polycrystalline specimens show the formation of cracks upon thermal cycling; this effect is reduced in hot-pressed specimens. Several of the properties were remeasured on hot-pressed samples and indicated little or no change in the thermomagnetic characteristics compared to the data on bulk crystalline samples. The experimental data are compared to the results of mean-field theory calculations.

INTRODUCTION

Ferromagnetic materials are the usual choice for magnetic refrigeration (MR) above 20 K because of the enhanced magnetocaloric effect near their Curie temperature.[1] Gadolinium-containing materials are a particularly good choice because the spin only configuration of Gd gives a large entropy associated with the ordering of the spins and a near absence of crystal field effects due to the zero orbital angular momentum, L. The crystal-field-induced effects of magneto-striction, magnetic anisotropy, hysteresis, and enhanced background (non-field dependent) heat capacity are generally undesirable for MR. GdNi, a relatively dense ferromagnetic compound with a Curie point of 70 K, shows promise as a magnetic refrigerant in a temperature range of great technical interest. The physical properties of GdNi relevant to MR applications form the subject of this paper.

SAMPLE PREPARATION

GdNi is a congruently melting intermetallic compound crystallizing in the orthorhombic CrB structure.[2,3] The samples used for heat capacity, susceptibility and magnetization measurements were produced via arc-melting by Ames Laboratory[4] using ingredients of at least 99.9% true atomic purity. The resistivity samples and material used to make hot-pressed specimens were produced by arc-melting commercial nominally 99.9% pure Gd and 99.99% pure Ni. The concentration of non-metallic impurities was not included in the purity rating of the commercial Gd sample; the O and N content of approximately 1% tended to segregate out as non-metallic inclusions, making adjustment of the Gd to Ni ratio necessary in order to get phase purity. The phase purity of all specimens was checked by metallography and was at least 99%. The density was measured to be 8705 kg/m^3.

The hot-pressed specimen reported here was prepared by pressing GdNi powder at 800°C in an inert atmosphere without any binder. Attempts were made to press GdNi at a lower temperature using Cu binder, but reaction between the Cu and Gd occurred, producing a sample exhibiting a depressed Curie point and additional magnetic transitions.

MEASUREMENT TECHNIQUES

The apparatus used to make the heat capacity and adiabatic temperature change measurements was described earlier[5]. Several modifications were made to obtain the data reported here, including using a much longer (1 to 10 sec) heat pulse due to the low thermal diffusivity of GdNi near the Curie point, and mounting the sample on fine threads for better thermal isolation. The technique used was to put a known heat pulse into the sample, and then to measure the total temperature change of the sample. Corrections were made for heat losses during the equilibration time of the sample, for heat dissipation in the heater leads, and for thermal addenda. The magnetic field was applied parallel to the long axis of the cylindrical sample, reducing demagnetization effects. The adiabatic temperature change upon magnetization or demagnetization was measured by monitoring the temperature of the sample upon inserting or removing it from the bore of a superconducting solenoid. Corrections were made for the small magnetic field dependence of the carbon glass thermometer used.

Static magnetization, M, was measured by a vibrating sample magnetometer with a superconducting solenoid having a maximum field of 8 T. The static field was applied parallel to the long axis of the samples producing a relatively small correction for demagnetization.

The differential magnetic susceptibility, $\delta M/\delta H/_{H=0}$, was measured using a mutual inductance bridge. The small AC field applied to the sample was 30 A/m rms at 17 Hz. The applied field and the detected component of the susceptibility were parallel to the long axis of the cylindrical samples.

The integrated thermal expansion, $\Delta L/L$, was measured directly by a dilatometer with a mechanical dial indicator. The sample was immersed in LN_2 or ambient air to measure the expansion between the fixed temperature points. Calibration was checked using a sample of OFHC copper.

Resistivity, ρ, was measured by conventional 4 wire AC technique using a lock-in amplifier to detect the voltage across the sample. The voltage leads were spring-loaded copper wires; the current leads were attached with silver-filled conductive epoxy.

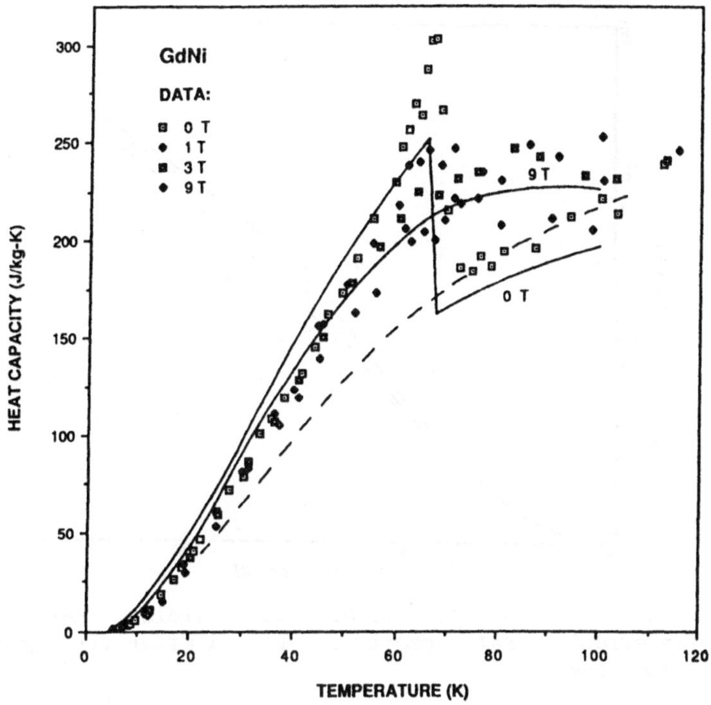

Fig. 1. Heat capacity for GdNi for several applied magnetic fields. The points measured; the solid curves are from a mean field calculation; the dashed line is an estimate of the non-magnetic (lattice) heat capacity.

Figure 1 shows the measured heat capacity, C_B. Note that the zero field heat capacity shows a strong λ-type feature at the Curie point due to the second-order ferromagnetic phase transition. Only a single sharp transition is seen, indicating a uniform sample and verifying the absence of crystal field quenching or other complications. The application of a magnetic field rapidly reduces the height of the peak and moves the maximum to higher temperature, verifying that the applied magnetic field assists the ferromagnetic exchange interaction in removing entropy from the system, even above the zero-field Curie point.

The mean field-calculated heat capacity, also shown in Fig. 1, uses no adjustable parameters; it uses the known values of the Curie temperature (70 K), the spin (7/2), the g factor (2), the Debye temperature (190 K), and the density. The fit to the measured data is only fair, especially at the Curie temperature. This is not surprising in light of the well known inadequacy of mean field theory near the critical point.[6] The problem is especially bad for the heat capacity, the calculation of which involves taking derivatives of the magnetization. Also sketched in Fig. 1 is an estimated curve for the non-magnetic component of the heat capacity, drawn assuming that the magnetic field suppresses the magnetic heat capacity well below the Curie point and that, in the absence of crystal field effects, the magnetic heat capacity disappears a few degrees above the Curie point.

The heat capacity may be integrated to give the entropy using the relation:

$$S(T,B) = S(0,B) + \int_0^T \frac{C_B}{T}\, dT$$

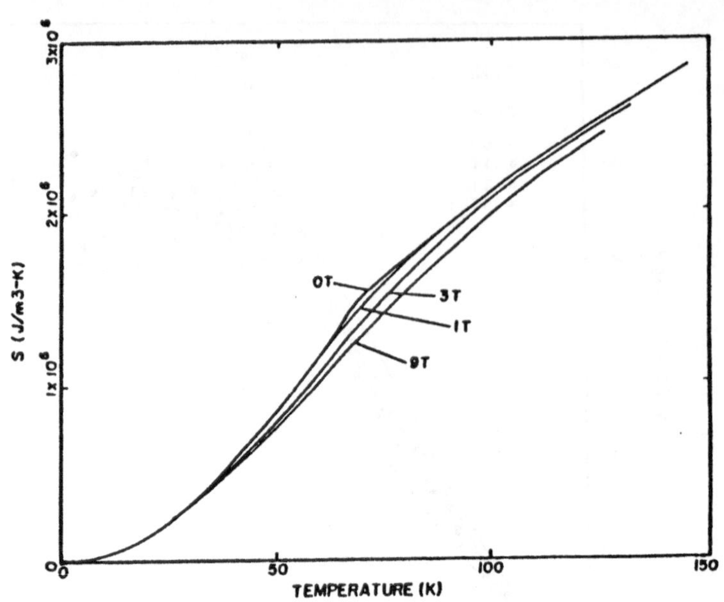

Fig. 2. Entropy of GdNi for several applied fields. Results of integrating heat capacity for Fig. 1 after fitting and smoothing.

The result using smoothed experimental heat capacity data is shown in Fig. 2. Note that most of the field dependence of the entropy of GdNi occurs within 25 K of the Curie point.

The measured changes in temperature of GdNi upon adiabatically changing the magnetic field (ΔT_S) from zero to 1 T, or from 0.2 T to 5 T, are shown in Fig. 3. Note that the temperature change is strongly peaked near the Curie point, a fact with major implications for MR. If the MR is to span more than 40 K, either a series of materials must be used, each with an optimal Curie point, as is possible in an active magnetic regenerator, or a material blended from several different ferromagnets[7] must be used. The reduced ΔT_S resulting from blending puts strong requirements on the quality of the regenerator in a MR.

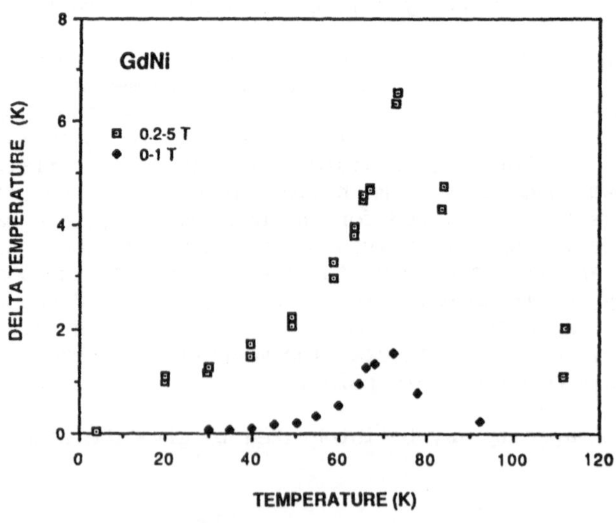

Fig. 3. Measured adiabatic temperature change of GdNi upon magnetization or demagnetization, plotted according to the average of the hot and cold temperatures.

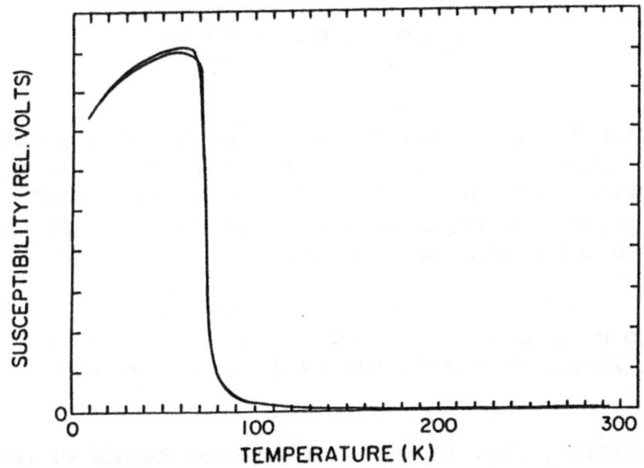

Fig. 4. Differential susceptibility of GdNi measured with a 30 A/M rms applied 17 Hz field. Note marked change at the Curie temperature of 70 K.

A sensitive way to detect a ferromagnetic phase transition and to verify the absence of magnetic hysteresis is to measure the differential susceptibility χ, shown in Fig. 4. Note that χ changes by an order of magnitude within 5 K of the Curie point, verifying that the coercive force is less than 100 A/m. Such a low value is consistent with the L = 0 character of Gd and the relatively low observed magnetic anisotropy.[8] The large value below the Curie point is demagnetization limited; its slight decline at very low temperature is mostly due to the skin effect.

Fig. 5 shows the static magnetization below, above and near the Curie point. The magnetization saturates quickly at 4.2 K, shows curvature at the Curie point, and is almost linear well above the Curie point. The saturated moment corresponds to a magnetic moment gJ = 6.96 $\pm$ 0.04 Bohr magnetons, showing full alignment of the Gd spins. The curvature in M near the Curie point is consistent with the large observed field dependence of the heat capacity, coupled by the equation

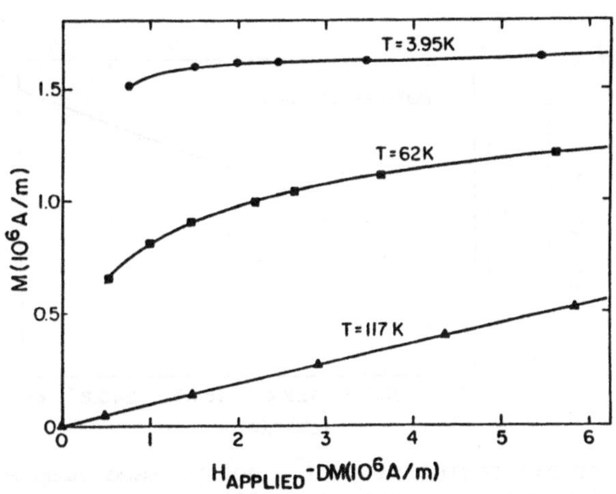

Fig. 5. Static magnetization of GdNi. Data was taken at constant temperature while the applied field was slowly changed.

$$C_B(B,T) - C(0,T) = T\int_0^B \left(\frac{\partial^2 M}{\partial T^2}\right)_B dB$$

The magnetization is needed for designing an MR, because the forces exerted on the magnetic material are proportional to the magnetization times the magnetic field gradient. The temperature dependence of the magnetization shows that magnetic forces are larger when the magnetic material is used below its Curie point.

The total thermal expansion of GdNi between 77 K and 300 K, $\Delta L/L$, was measured to be 1.86×10^{-3}. This value is much less than commonly used cryogenic stainless steels but is close to the values for titanium alloys.

The resistivity, Fig. 6, of a hot-pressed sample of GdNi for superimposed cooling and warming curves, is consistent with data published elsewhere.[3,9] The magnetic ordering transition has a conspicuous effect. Above the Curie point, the magnetic scattering is large and temperature independent, exposing a linear temperature dependence of typical of metallic electron-phonon scattering similar to that observed in the non-magnetic compound LaNi.[9] Below the Curie point, the ordering of the spins causes the magnetic scattering to drop, eventually reaching a sample-dependent residual value. The relatively high value of ρ near the Curie point has two consequences for MR: eddy currents will probably not be a problem but thermal conductivity will be low. The latter prediction comes from the approximate correlation between the thermal conductivity and the electrical conductivity given by the Wiedemann-Franz relation. Hence use of large chunks of ferromagnetic magnetic material should be avoided, a requirement also imposed by the necessity of efficient heat transfer to a fluid medium.

The resistivity of an unannealed arc-melted sample of GdNi, shown in Fig. 7, shows dramatic thermal cycling effects. The changes are due to cracks, also observed by a previous author in a single crystal of GdNi occurring preferentially along the crystallographic b direction.[3]

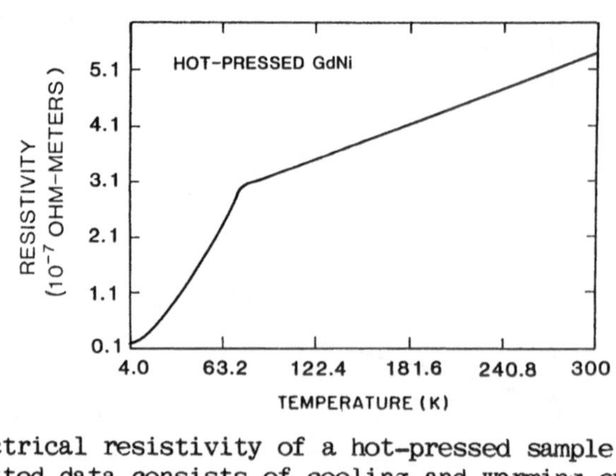

Fig. 6. Electrical resistivity of a hot-pressed sample of GdNi. The plotted data consists of cooling and warming curves that superimpose.

We observed no thermal cracking in any hot pressed samples, with or without binder. The arc melting process involved rapid (10 sec) cooling of the sample; the grain structure for pure specimens is also fairly coarse. Apparently either the slow cooling of the hot pressed samples or their fine grain structure suppresses the thermal cracking. The hot-pressed specimens were also much less brittle than bulk-melted specimens.

SUMMARY

The measured properties of GdNi show it is a very promising magnetic refrigerant. The magnetic transition, observed from heat capacity and susceptibility, is sharp, producing a relatively large ΔT_S. The measured magnetization, resistivity and heat capacity, and the derived entropy curves should be useful for designing magnetic refrigerators. Some of the limitations of GdNi, pointed out by the limited range of large ΔT_S, the potentially low thermal conductivity, and the occurrence of thermally-induced cracking, point out the desirability of additional materials development work connected with MR.

ACKNOWLEDGEMENTS

Funding for this work was received from DOE, DARPA and NASA/KSC.

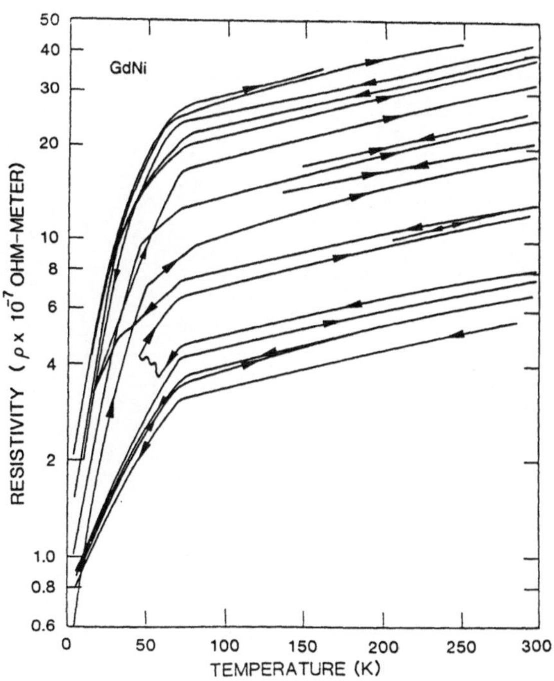

Fig. 7. Electrical resistivity of an unannealed arc-melted sample of GdNi. Note severe thermal hysteresis.

REFERENCES

1. T. Hashimoto et al, Magnetic refrigeration in the temperature range from 10 K to room temperature: the ferromagnetic refrigerants, Cryogenics 21: 647 (1981).
2. R.E. Walline and W.E. Wallace, Magnetic and structural characteristics of lanthanide-nickel compounds, J. Chem. Phys., 41: 1587 (1964).
3. K. Mori and K. Sato, Temperature dependence of electrical resistivity in ferromagnetic GdNi single crystal, J. Phys. Soc. Japan, 49: 246 (1980).
4. Center for Material Preparation, Ames Laboratory, Ames, Iowa.
5. J.A. Barclay et al, An apparatus to determine the heat capacity and thermal conductivity of a material from 1 to 300 K in magnetic fields up to 9T, in: Advances in Cryogenic Engineering, Vol. 30, Plenum Press, New York (1984) p. 425.
6. N.W. Ashcroft and N.D. Mermin, Solid State Physics, Holt Rinehart and Winston, New York (1976) p. 715.
7. T. Hashimoto, et al, New application of complex type of magnetic materials to the magnetic refrigerant in the Ericsson magnetic refrigerator, J. Appl. Phys. to be published.
8. K. Sato, K. Kozachi and K. Mori, Temperature dependence of magnetocrystalline anisotropy energy and saturation magnetization of GdNi single crystal. J. Appl. Phys., 52: 2084 (1981).
9. E. Gratz, et al, Magnetic Properties and Electrical Resistivity of (Gd_xLa_{1-x}) Ni, $(0 \leq x \leq 1)$, J. Mag. Mag. Mat. 54-57: 459 (1986).

CURRENT DEVELOPMENTS IN NASA CRYOGENIC COOLER TECHNOLOGY

Stephen H. Castles

NASA/Goddard Space Flight Center
Greenbelt, Maryland

ABSTRACT

NASA's two main areas of observational scientific endeavor, astrophysics and earth science, have scientific requirements that dictate the use of sensors with more energy resolution than the sensors used in earlier space flight instruments. With present sensor technology these scientific goals can often only be met with the use of sensors and instruments operating at cryogenic temperatures. In response to this need, NASA is developing a broad range of cryogenic coolers and supporting technology. These coolers will be capable of providing sensors and instruments with stable operating temperatures from room temperature down to 0.1 K or less. In addition to providing the desired operating temperature and cooling power, these coolers must have working lifetimes commensurate with the 10 to 15 year lifetime expected for major future NASA facilities. To meet this lifetime requirement NASA is developing long lifetime mechanical coolers and the capability to service stored cryogen coolers on-orbit. The types of coolers currently being developed by NASA include radiative coolers, solid cryogen coolers, surface tension confined liquid cryogen coolers, mechanical coolers, liquid helium dewars, He^3 adsorption coolers, adiabatic demagnetization refrigerators and dilution refrigerators.

INTRODUCTION

NASA astrophysics and earth sciences missions are requiring increasing numbers and types of cryogenic coolers. To date these coolers have been used to increase the sensitivity and signal to noise ratio of detectors by providing the required cryogenic operating temperatures and by cooling associated optical components to decrease the detector background radiation. A list of planned NASA missions using cryogenic coolers that are either in the flight hardware phase or in the development phase is presented in Table 1. For those missions that include both cryogenically cooled and ambient temperature instruments, the cryogenically cooled instruments are listed below each mission.

In this paper "cryogenic cooler" will refer to any system capable of producing temperatures below 160 K. A list of the various types of space flight coolers either being used or being developed by NASA is presented in Table 2. Excellent reviews have previously been written that adequately describe the history of cryogenic cooler technologies[1]. Therefore the emphasis of this paper will be on the current status of the technology for each type of cooler. On-orbit servicing of stored cryogen coolers will also be discussed. Major NASA scientific facilities such as AXAF, ASTROMAG, EOS, LDR, Space Telescope, and SIRTF may have lifetimes of up to 10 or 15 years. To meet these lifetime goals, long lifetime mechanical coolers and the servicing of stored cryogen coolers will be required.

Table 1. NASA Missions Using Cryogenic Coolers

HARDWARE PHASE	DEVELOPMENT PHASE
• Upper Atm. Research Satellite (UARS) - Cryogenic Limb Array Etalon Spectrometer (CLAES) - Improved Stratospheric and Mesospheric Sounder (ISAMS) • Shuttle High Energy Astrophysics Lab. (SHEAL II) - Broad Band X-Ray Telescope (BBXRT) • Cosmic Background Explorer (COBE) - Far Infrared Spectrometer (FIRAS) - Diffuse Infrared Background Experiment (DIRBE) • Gravity Probe - B (GP - B) • Lamda Point Experiment • Superfluid Helium On Orbit Transfer Flight Demonstration (SHOOT)	• Adv'd X-Ray Astrophysics Facility (AXAF) - X-Ray Spectrometer (XRS) • ASTROMAG • Earth Observing System (EOS) - Moderate Resolution Imaging Scanner-Nadir (MODIS - N) - High Resolution Infrared Spectrometer (HIRIS) • Second Generation Space Telescope Instruments - Near Infrared Camera and Multiple Object Spectrometer (NICMOS) - Hubble Imaging Michelson Spectrometer (HIMS) • Space Infrared Telescope Facility (SIRTF) • Superconducting Gravity Gradiometer • Superfluid Helium Tanker (SFHT) • Thermal Transport Critical Point Exp.

Many people are now considering what effect the new high temperature superconductors may have on aerospace cryogenics. Two points are noteworthy. First, the requirement for cryogenic coolers will not be eliminated. For example, of 14 planned missions using liquid helium, only 2 are using it to cool superconductors alone. The remaining payloads are cooling detectors or, in a few cases, doing research on liquid helium itself. In fact, new technologies using the high temperature superconductors may increase the requirements for coolers. Also, it now appears that if the new superconducting materials fulfill their promise, valuable new aerospace cooler technology will become viable.

RADIATIVE COOLERS

Radiative coolers operate by providing a thermally isolated cold patch with a high emissivity surface which radiates to space. Radiative coolers are widely used. They are passive and therefore reliable, have low weight and are relatively inexpensive. Therefore, they are the cryogenic cooler of choice when they can meet both the temperature requirement and the cooling load requirement. Unfortunately, radiation from the sun, the earth albedo and the payload itself severely limits the performance of radiative coolers, making them sensitive to the payload orbit. Multistage radiative coolers have reached 71 K in

Table 2. NASA Cooler Technology

- Radiative coolers: 60 - 300 K
- Solid cryogen coolers: 10 - 160 K
- Surface tension confined liquid cryogen coolers: 10-160K
- Single stage mechanical coolers: 40 - 160 K
  - Stirling cycle, Reverse Brayton cycle, Adsorption, Joule-Thomson, Pulse tube
- Multistage mechanical cooler: 2 - 80 K
  - Stirling cycle, Reverse Brayton cycle, Adsorption, Joule-Thomson, Pulse tube, Magnetic
- Long lifetime liquid helium dewars: 1.5 - 5 K
- Helium-3 adsorption refrigerators: 0.3 - 0.6 K
- Adiabatic demagnetization refrigerators: 2 - 300 mK
- Dilution refrigerators: 2 - 300 mK
- On-orbit liquid helium transfer (12 Payloads, including AXAF, SIRTF, ASTROMAG, LDR)

geosynchronous orbit. But because of radiation from the sun and earth, radiative coolers have limited capability as a cryogenic cooler in the 28 degree inclination, low earth orbit envisioned for the Space Station and typical of many Shuttle launches. Also, at cryogenic temperatures the heat load that can be radiated per unit cold patch surface area becomes small.

Although radiative coolers are an established technology, the Jet Propulsion Laboratory (JPL) is developing a more efficient, light weight radiative cooler, the V- groove isolation radiator[2]. Particular attention has been paid to good thermal isolation of the cold patch from the warm payload and effective shading of the cold patch. To date, engineering models have been built and tested both thermally and structurally. As expected, the test results show a strong interaction between radiator temperature, cooling load and shade temperature. As a typical example, for an orbit that will allow a shade temperature of 160 K, a 0.18 m^2 cold plate will radiate 200 mW at 75 K.

SOLID CRYOGEN COOLERS

Solid cryogen coolers store solid cryogens that sublime to space. The various cryogens[1,3] can provide cooling down to almost 8 K. However, hydrogen, which is the only cryogen available in the temperature range between 8 K and 13.5 K, is presently banned from use for payloads flying on the Shuttle. Solid cryogen coolers presently being fabricated for NASA missions include a two-stage neon/CO_2 cooler for the CLAES instrument on UARS and two argon coolers for BBXRT. The CLAES cooler, which is being produced by Lockheed Missiles and Space Company (LMSC), was a solid hydrogen cooler prior to the new Shuttle ruling on hydrogen. Ball Aerospace Systems Division (BASD) is fabricating the BBXRT coolers.

Solid cryogen coolers are the most commonly used stored cryogen cooler[3]. Two-stage methane/ammonia coolers that have flown to date have obtained lifetimes in the range of 1 year. LMSC is now producing a long lifetime methane/ammonia cooler for the Air Force that is estimated to have a lifetime of approximately 3 years while providing 0.48 W of cooling power at 65 K, as well as providing cooling at higher temperature stages. Cryogen launch weight will become prohibitively large for appreciably longer lifetimes or higher heat loads, favoring the use of either mechanical coolers or serviceable stored cryogen systems.

SURFACE TENSION CONFINED LIQUID CRYOGEN COOLERS

Solid cryogen coolers are difficult to service on-orbit since they must first be filled with liquid cryogen, which is difficult to phase separate from its vapor, and then the liquid must be frozen in place. The coolers are also difficult to change out as a unit for most payloads. The Goddard Space Flight Center (GSFC) is presently investigating a surface tension confined liquid cryogen system[4] that is intended to be serviceable with liquid cryogen on-orbit. The cryogen tank of this stored liquid cryogen cooler is filled with a low density, small pore size, open cell foam. Surface tension holds the liquid cryogen in the tank while allowing the vapor to vent to space. To date, laboratory tests have been performed on samples of foam in small containers of cryogen. The tests have provided preliminary data on the thermal, structural and thermodynamic properties of the foam/liquid cryogen system. However, much research and development will be required before this technology can be relied upon for a major space flight mission requiring a high reliability cooler. Even so, a small surface tension confined liquid cryogen cooler is being developed for a low cost NASA sounding rocket experiment to be launched in November 1987, to look at Supernova 1987a.

MECHANICAL COOLERS

Mechanical coolers can be classified as regenerative or recuperative. Regenerative coolers contain reciprocating parts that force the working fluid back and forth through a regenerator. The regenerator stores heat energy, exchanging it between the fluid entering the cold end of the cooler and the same fluid as it exits from the cold end. The reciprocating parts of regenerative cycle coolers must be dynamically balanced or they will produce levels of vibration that are typically unacceptable for flight experiments. Examples of regenerative cycle coolers are the Stirling, Vuilleumier, Gifford-McMahon and Solvay cycles.

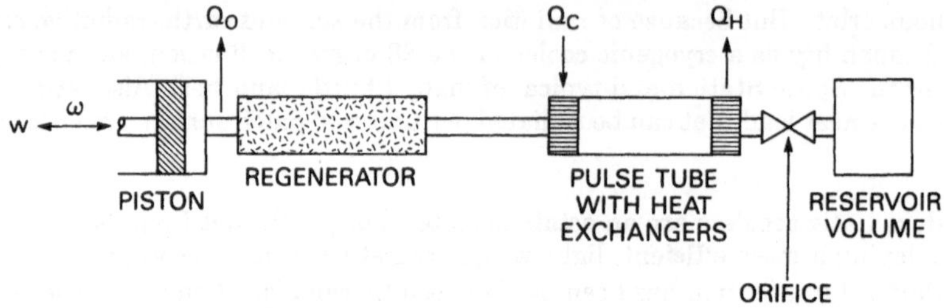

Figure 1. Schematic of pulse tube refrigerator

Notes. Work W moves piston at frequency ω. Heat Q_0 and Q_H are rejected to the heat sink and heat load Q_c is absorbed at the cold end of the pulse tube.

Recuperative cycle mechanical coolers use counterflow heat exchangers (recuperators) to exchange the heat between the working fluid leaving the cold end and the working fluid approaching the cold end. The Brayton and Joule-Thomson cycle are the most common examples. The recuperative cycle coolers do not require reciprocating parts and therefore have the potential to use rotating parts that do not produce appreciable levels of vibration.

NASA is presently pursuing the development of several potential flight mechanical coolers (see Reference 5 for a more general treatment of mechanical coolers). The single stage Stirling cycle cooler being developed by Philips Laboratories for the GSFC is in the most mature stage of development[6]. This cooler incorporates linear motors and magnetically levitated moving parts (piston, displacer and counterbalancer) with clearance seals. An engineering model, which is designed to produce 5 W of cooling at 65 K, is presently being life tested at Philips Labs. It has over 3 years of operating time with no detectable degradation. A protoflight cooler is now being fabricated. The active balancer is designed with sufficient bandwidth to counterbalance the primary operating frequency (19 Hz) and the first 2 harmonics. The Philips Stirling cycle cooler is baselined to fly on the XRS instrument on AXAF.

Stirling cycle coolers are presently the most thermodynamically efficient mechanical coolers for temperatures down to about 15 K. However, the Philips cooler requires extensive electronics to control the radial magnetic bearings and to operate the counterbalancer. A small Stirling cooler that uses a spring to levitate the piston and displacer, instead of magnetic bearings, is under development at Oxford University for the ISAMS instrument on UARS[7]. This cooler produces 0.875 W of cooling at 80 K with an input power to the compressor of less than 30 W.

A cooler that is thermodynamically similar to the Stirling cooler is the pulse tube cooler. This cooler, shown schematically in Figure 1, has a reciprocating piston but replaces the Stirling's displacer with a resonant cavity, the pulse tube. Recently, an orifice and reservoir have been added to the pulse tube to allow control of the phase shift between the piston and the pulse tube resonance. The Ames Research Center (ARC) is presently funding the development of this cooler at the National Bureau of Standards[8]. With the use of an orifice, it may be possible to improve the efficiency of the pulse tube cooler until it approaches that of the Stirling cycle cooler, at least at temperatures above about 75 K.

JPL is developing the adsorption cooler[9], a Joule-Thomson cooler that replaces the usual mechanical compressors with adsorption compressors. The adsorption compressors are inefficient relative to normal compressors but can use heat at high temperatures instead of electrical power. On interplanetary missions, a radioactive thermoelectric generator could provide the necessary heat to power the cooler. JPL has designed, fabricated and tested

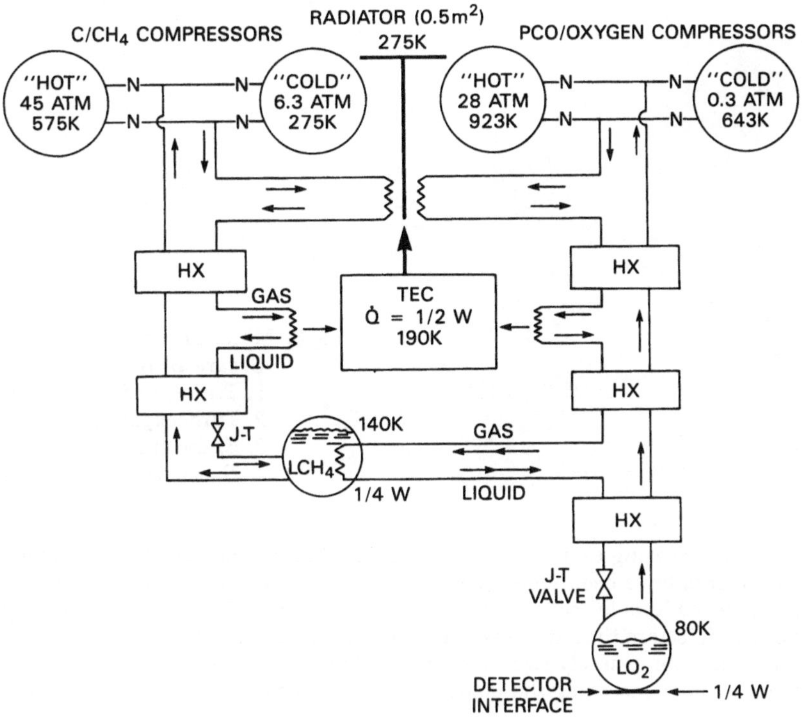

Fig. 2. Two stage adsorption cooler.

engineering models of several stages that operate over various temperature ranges. The only moving part, the valves, have been life tested and the compressors have been thermally cycled to demonstrate their reliability. These tests indicate that the adsorption cooler has the potential of being a reliable, long lifetime cooler.

Because of its low efficiency the adsorption cooler generally does not compete directly with efficient coolers such as Stirling or Brayton cycle coolers. However, unlike many mechanical coolers, small cooling power adsorption coolers are feasible . As an example, the 2 stage adsorption cooler shown in Figure 2, which also uses a Thermoelectric Cooler (TEC), has been designed to produce 1/4 W of cooling at 80 K with an input power of 75 W.

The GSFC is funding Creare, Inc. to develop a reverse Brayton cycle cooler[10]. This cooler uses miniature turbo-machinery which produces no appreciable vibration. A 3.5 mm diameter expander turbine shaft is shown in Figure 3. By using miniature components the input power required by coolers with the Creare technology should be consistent with the relatively small input power available on NASA missions. The GSFC/Creare cooler is designed to produce 5 W of cooling at 65K with an input power of 160 W. The compressor and expander for GSFC's engineering model cooler are presently being fabricated. Creare is also life testing the bearings and shaft of an expander. As of June 1987 these critical parts have been operating for 40 months without detectable degradation. Another bearing and shaft have experienced more than 10,000 start-stop cycles.

Several developments are underway to improve the efficiency of the Creare technology and to produce a cooler requiring less input power. The GSFC is funding Creare to develop an efficient, compact, all metal heat exchanger. The Air Force is funding Creare to develop an even smaller expander and to develop an expander with cold bearings. Cold bearings will significantly reduce the parasitic heat loads on the cooler. If successful, these developments should result in a more efficient, smaller cooler. For example, a cooler producing 1 W at 65 K would require input power of 95 W.

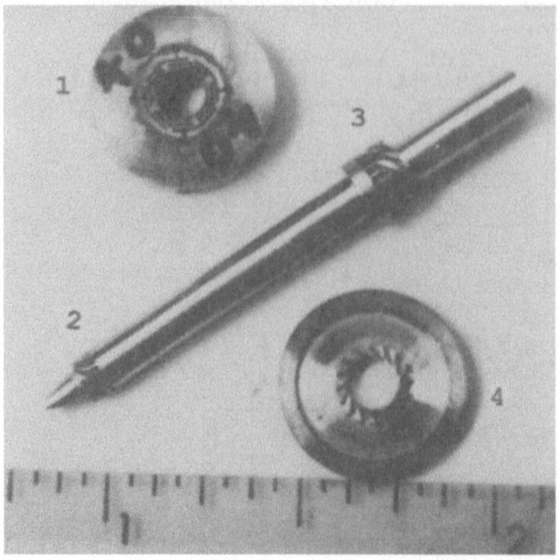

Fig 3. Miniature turbine parts for Creare Brayton cycle cooler:
1. Tilting pad gas bearings,
2. Turboexpander end,
3. Brake end,
4. Nozzle ring for the brake.

NASA in now preparing to develop multistage mechanical coolers. Two distinct sets of cooling requirements have been identified. The first requirement, for approximately 1/2 W of cooling power at 30 K, is to cool optical elements for atmospheric science experiments and to cool new types of detectors that are presently under development. This cooling requirement could be met by an appropriately sized 2 stage Stirling cycle cooler, using essentially present technology and without exceeding the input power available on platforms such as EOS.

The second cooling requirement is for refrigeration at 2 to 4 K for the many future, long duration payloads requiring sensors operating at or below liquid helium temperatures. To fulfill this requirement, new technology must be developed to significantly improve the efficiency of multistage mechanical coolers. In fiscal year 1988, the GSFC and ARC will fund advanced low temperature regenerator concepts; Marshall Space Flight Center (MSFC) and GSFC will pursue new recuperator designs; and the ARC will begin the development of an efficient, low temperature cooling stage. It may be possible that an efficient, low temperature cooler such as a 2 to 15 K magnetic refrigerator can be used as part of a hybrid cooler to provide a more efficient cooler than can be produced with any single cooler technology. While there are encouraging early indications that major advances in multistage cooler technology may be possible, it will be many years before efficient mechanical coolers operating at 2 to 4 K will be available for flight.

LIQUID HELIUM DEWARS

Aerospace liquid helium dewars operating at 1.5 to 2 K are now an established technology. The Infrared Astronomy Satellite (IRAS) dewar, launched in 1983, successfully completed its mission with a lifetime of 10 months. The COBE dewar[11], shown in Figure 4, has been through an extensive acceptance test[12] and has been delivered to the GSFC. The COBE dewar should have a 14 month on-orbit lifetime. BASD built both the IRAS and COBE dewars. In addition to these 2 long lifetime helium dewars, 2 smaller dewars were flown on Space Lab - 2[13,14].

Technology to improve the hold time of liquid helium dewars is now being developed. LMSC, with funding from ARC, has fabricated and tested passive orbital disconnect struts for supporting the cryogen tank[15]. Both ARC and BASD are investigating the use of new materials such as alumina composites for supporting the cryogen tank[16,17].

GSFC is investigating the use of a hybrid cooling system, namely the use of a mechanical cooler to intercept the parasitic heat load on the cryogen tank. Thermal models indicate that for a helium dewar with 3 vapor cooled shields, cooling the outer vapor cooled shield to 50 - 70 K with a mechanical cooler can lengthen the dewar lifetime by 50 to 100 %. Depending on the

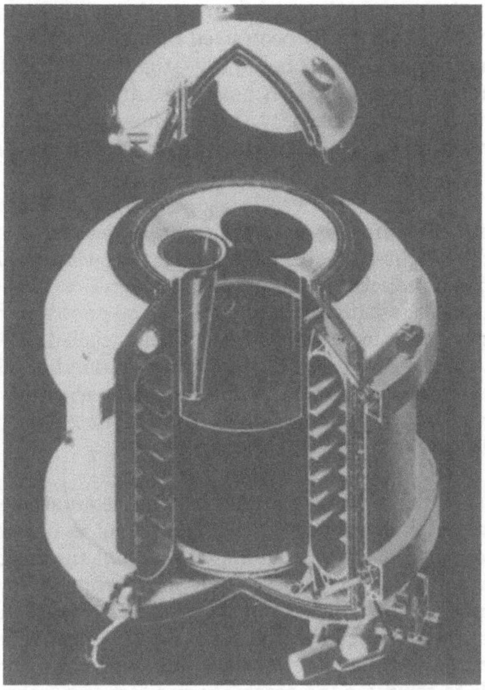
Fig. 4. COBE dewar isometric.

size of the dewar, this will require cooling power of between 1 and 5 watts at the vapor cooled shield, which, in turn, will require approximately 100 to 250 W of input power. As mentioned earlier, providing on-orbit cooling power with mechanical coolers at 2 K is presently beyond the state-of-the-art because such a mechanical cooler requires excessive input power.

NASA is developing on-orbit liquid helium servicing to further increase the working lifetime of payloads requiring liquid helium temperatures. To date, 12 future payloads have requested liquid helium servicing. The technology required for liquid helium servicing is now being tested at the GSFC, at ARC and at a number of companies. The SHOOT flight demonstration of this technology is scheduled for 1991[18,19].

REFRIGERATION BELOW 1 K

Three potential refrigeration techniques exist for cooling detectors below 1 K: He^3 refrigerators, adiabatic demagnetization refrigerators (ADR's) and dilution refrigerators. The He^3 refrigerator can cool detectors to 0.3 K while ADR's and dilution refrigerators can cool to a few mK. He^3 refrigerators and ADR's operating at 0.1 K are in an advanced state of development while the dilution refrigerator will require much development before it can be proposed for flight.

In a low-g He^3 refrigerator, the He^3 is condensed into a copper sponge which is heat sunk to the liquid helium bath. This cold stage is thermally isolated and then an adsorption pump is used to pump on the He^3 to obtain the lowest possible temperature, about 0.3 K. When the He^3 runs out, the cycle is repeated. ARC has been developing He^3 refrigerators for a number of years[20]. Recently, MSFC has funded Alabama Cryogenic Engineering (ACE) to investigate new He^3 refrigerator configurations, including the use of a vortex pump in the cycle.

An ADR uses a superconducting magnet to magnetize and demagnetize a paramagnetic material. Paramagnetic, hydrated salts have traditionally been used but more stable sapphire and glass based paramagnetic materials are now being investigated. When the cold stage is thermally isolated from the helium bath (the ADR heat sink) and the paramagnetic salt is demagnetized, cooling power produced by the paramagnetic salt cools the cold stage to the required operating temperature. The temperature can then be regulated by regulating the residual magnetic field applied to the paramagnetic salt.

The GSFC[21], ARC[22] and the University of California at Berkeley have been developing a flight ADR. Two problems associated with a flight ADR are particularly noteworthy. The high current leads typically used with superconducting magnets would significantly decrease the lifetime of the stored helium; and the fringing field from the magnet would be sufficiently large to interfere with the proper operation of the detectors being cooled. The GSFC appears to have solved each of these problems. A low current superconducting magnet that requires only 1 Amp per Tesla has been procured and successfully tested. With this type of magnet a flight ADR, optimized for minimum weight and designed to cool detectors to 0.1 K, will require a magnet that draws less than 2 Amps.

Recently, GSFC has completed the design of a shielded magnet which has an active bucking coil to provide 3-dimensional magnetic shielding. This bucking coil is designed to decrease the magnetic field at a cylindrical surface around the magnet so that the residual field will be less than the critical field of a type 1 superconductor. An appropriate superconductor placed at the cylindrical surface will then completely contain the residual field. The procurement of such a low current, shielded magnet from Cryomagnetics is in progress. The outer, passive superconducting shield will be 20 cm in diameter and 30 cm long. Using this magnet the GSFC intends to complete the assembly and test of a technology demonstration model ADR by the spring of 1988. This ADR will be a full scale model of the ADR to be flown on the XRS instrument on AXAF.

MSFC, JPL and ARC are investigating the use of a dilution refrigerator in space. The main difficulty to be overcome is that standard dilution refrigerators use gravity for fluid management. JPL is investigating the use of an electrostatic field for fluid management while MSFC and ARC are considering different cycles to eliminate phase separation problems. As part of this effort, MSFC is funding ACE to investigate incorporating vortex coolers into a dilution refrigerator. While much interesting work has been accomplished, even more still remains to be done before dilution refrigerators will be ready for flight.

CONCLUSION

The number of NASA instruments using cryogenic coolers is continuing to grow. With their more sensitive, cooled detectors these instruments are capable of performing such excellent science that they often embody the key scientific goals within their respective disciplines. In response, NASA is increasing its support for the development of improved aerospace cryogenic coolers.

A wide variety of cooler technologies are being developed to meet the varying temperature and cooling power requirements of NASA instruments. Several new cooler technologies are now being proposed for flight. Many of us are looking forward to the day when NASA investigators can choose from a variety of flightworthy cryogenic coolers to meet their cooling requirements.

REFERENCES

1. A. Sherman, History, Status and future applications of spaceborne cryogenic systems, in: "Advances in Cryogenic Engineering," Vol. 27, Plenum Press, New York (1981), p. 1007.

2. S. Bard, Development of a high-performance cryogenic V-groove radiator, Journal of Spacecraft and Rockets, in press (1987).
3. T. C. Nast, Status of solid cryogen coolers, in: "Advances in Cryogenic Engineering," Vol. 31, Plenum Press, New York (1986), p. 835.
4. S. H. Castles and M. E. Schein, Development of a space qualified surface tension confined liquid cryogen cooler, to be published in: "Advances in Cryogenic Engineering," Vol. 33, Plenum Press, New York.
5. O. C. Ledford Jr., Mechanical cryorefrigerators for spacecraft, presented at the 1987 Cryogenic Engineering Conf., St. Charles, Illinois (1987).
6. F. Stolfi and M. Goldowsky, A magnetically suspended linearly driven cryogenic refrigerator, in: "Proc. Second Biennial Conf. on Refrigeration for Cryogenic Sensors", NASA Conference Publication 2287 (1983), p. 263.
7. T. W. Bradshaw and J. Delderfield, Performance of the Oxford miniature Stirling cycle refrigerator, in: "Advances in Cryogenic Engineering," Vol. 31, Plenum Press, New York (1986), p. 801.
8. R. Radebaugh and S. Herrmann, Refrigeration efficiency of pulse-tube refrigerators, presented at Fourth Intl. Cryocooler Conf., Easton, Maryland (1986).
9. J. Jones, Sorption cryogenic refrigeration, status and future, to be published in: "Advances in Cryogenic Engineering," Vol. 33, Plenum Press, New York.
10. J. Valenzuela, H. Sixsmith, and W. Swift, Long-lifetime, closed-cycle cryocooler for space, presented at Fourth Intl. Cryocooler Conf., Easton, Maryland (1986).
11. R. A. Hopkins and S. H. Castles, Design of the superfluid helium dewar for the Cosmic Background Explorer (COBE), Proceedings of SPIE 509:207 (1985)
12. R. A. Hopkins and M. G. Ryschkewitsch, Measured ground performance and predicted orbital performance of the superfluid helium dewar for the Cosmic Background Explorer, in: "Cryogenic Optical Systems and Instruments 2: Proc. SPIE", Los Angeles, California (1987), p. 134.
13. E. W. Urban and D. R. Ladner, Comparison of I. R. telescope cryogenic performance - laboratory versus space, presented at Space Cryogenics Workshop, Madison, Wisconsin (1987).
14. P. V. Mason et al, Preliminary results of the Spacelab 2 superfluid helium experiment, in: "Advances in Cryogenic Engineering," Vol. 31, Plenum Press, New York (1986), p.869.
15. R. T. Parmley and T. C. Nast, System structural test results: six PODS III supports, in: "Advances in Cryogenic Engineering," Vol. 31, Plenum Press, New York (1986), p.941.
16. R. A. Hopkins And D. A. Payne, Optimized support systems for spaceborne dewars, Cryogenics 27:209 (1987).
17. R. A. Hopkins et. al., Next-generation tension strap supports for spaceborne dewars, presented at AIAA Thermophysics Conf., Honolulu, Hawaii (1987).
18. M. DiPirro and S. Castles, Superfluid helium transfer flight demonstration using the thermomechanical effect, Cryogenics 26, p. 84 (1986).
19. M. DiPirro and P. Kittel, The superfluid helium on-orbit transfer (SHOOT) flight demonstration, to be published in: "Advances in Cryogenic Engineering," Vol. 33, Plenum Press, New York.
20. P. Kittel, ^{3}He cooling systems for space - status report, "Proc. Space Cryogenics Workshop," Berlin (1984), p.116.
21. S. H. Castles, Design of an adiabatic demagnetization refrigerator for studies in astrophysics, in: "Proc. Second Biennial Conf. on Refrigeration for Cryogenic Sensors", NASA Conference Publication 2287 (1983), p.389.
22. P. Kittel, Temperature stabilized adiabatic demagnetization for space applications, Cryogenics 20: 599 (1980).

HIGH PRESSURE RATIO CRYOCOOLER WITH INTEGRAL EXPANDER AND HEAT EXCHANGER

J.A. Crunkleton, J.L. Smith, Jr., Y. Iwasa

MIT Cryogenic Engineering Laboratory
Cambridge, Massachusetts

ABSTRACT

A new 1 W, 4.2 K cryocooler is under development that is intended to miniaturize helium temperature refrigeration systems using a high-pressure-ratio Collins-type cycle. The configuration resulted from optimization studies of a saturated vapor compression (SVC) cycle[1,2] that employs miniature parallel-plate heat exchangers. The basic configuration is a long displacer in a close-fitting, thin-walled cylinder. The displacer-to-cylinder gap is the high-pressure passage of the heat exchanger, and the low-pressure passage is formed by a thin tube over the OD of the cylinder. A solenoid-operated inlet valve admits 40 atm helium to the displacer-to-cylinder gap at room temperature, while the solenoid-operated exhaust valve operates at 4 atm. The single-stage cryocooler produces 1 W of refrigeration at 40 K without precooling and at 20 K with liquid nitrogen precooling.

INTRODUCTION

Miniature liquid-helium-temperature cryocoolers that are efficient, compact, and reliable are not generally available. The primary efforts in liquid helium temperature cryocooler development have been focused on the modification of two-stage, 10 K to 15 K units intended for cryopumping applications. The Stirling and Gifford-McMahon cycles have been modified by using a Joule-Thomson loop to reach 4.2 K. Such efforts have focused on reliability rather than thermodynamic efficiency.

Efforts at this laboratory to find a better cryocooler configuration began with a computerized parametric optimization study of the SVC cycle that included component losses. The SVC cycle was chosen over the purely regenerative cycles operating with a Joule-Thomson loop because the SVC cycle requires much less power input per watt of refrigeration at liquid helium temperatures.[3] The optimization study was performed by examining SVC cycle performance with different heat exchanger and expansion engine configurations. Component losses included were expander and compressor inefficiencies, heat exchange temperature difference, fluid friction, heat conduction, and radiation heat leak. The optimization study identified a cryocooler design that incorporates the integration of a heat exchanger and expansion engine in a concentric tube configuration.

Design of the integral heat exchanger and expander component was intended to combine the advantages of regenerative heat exchange and counterflow heat exchange in a single package that would use a displacer type expansion engine. In totally regenerative cycles, the pressure ratio is effectively limited by the gas volume in the regenerator, which must be large enough so that the low-pressure flow pressure drop through the regenerator matrix is not excessive. In the integral expander and exchanger configuration, the displacer-to-cylinder gap-regenerator passes high-pressure flow, allowing the pressure ratio to be chosen for greater work extraction rather than minimal pressure drop. Because the low-pressure return flow effectively adds heat capacity to the regenerator walls, counterflow heat exchange significantly reduces vulnerability to diminishing regenerator wall heat capacity at low temperatures, adding yet another advantage to the integral configuration.

DESCRIPTION OF THE SINGLE-STAGE CRYOCOOLER

A schematic of the 53 cm long integral concentric tube heat exchanger and expansion engine is shown in Figure 1. Most parts are made of 304 stainless steel; the room-temperature parts for the exhaust valve are made of brass. The displacer is coated with S-400 Teflon to reduce sliding friction. Displacer support disks and outer shell round-out-rings maintain shape during pressure cycling. The outer shell is 29.4 mm diameter. The room-temperature inlet valve is solenoid operated. The exhaust valve operates at the refrigeration temperature and is operated by a solenoid located at room temperature. A Teflon coated stainless steel tube located inside a stainless steel sheath provides the mechanical connection between the warm solenoid and the cold valve. Aluminized superinsulation covers all internal parts. The primary component of the stroking mechanism is a hydraulic system used to absorb the expander power output during the expansion stroke and to supply power during the exhaust and recompression stroke. An electronic linear position transducer is connected to the hydraulic cylinder shaft. The position transducer output is supplied to an electronic control circuit which is then connected to the inlet valve, exhaust valve, and hydraulic directional control valve solenoids. The control circuit is designed to make actuation timings for the three solenoid valves independent. Also shown in Figure 1 is a liquid nitrogen cooled radiation shield and heat exchanger. The experiment was run with and without liquid nitrogen precooling. The radiation shield was present in both cases, but was liquid nitrogen cooled only during cryocooler operation with liquid nitrogen precooling.

Performance Algorithm

Design of the integral exchanger and expander required development of a comprehensive performance algorithm which consists of two parts: one part to model heat exchange and another to model an adiabatic expansion process followed by a lumped heat input. The model for heat exchange includes regenerative heat transfer between the reciprocating displacer and cylinder with counterflow heat exchange between the high- and low-pressure flow streams. The result of this heat exchange model is an average enthalpy flow at a cross section perpendicular to the engine axis consisting of enthalpy flows due to heat exchange and displacer motion. Results from the performance algorithm are used to obtain information such as the magnitude of heat load sustained by the cryocooler at a specified temperature and the temperature versus time data for the displacer, cylinder, and outer shell at the cold end. The information obtained can be compared with experimental results to better understand the various loss mechanisms.

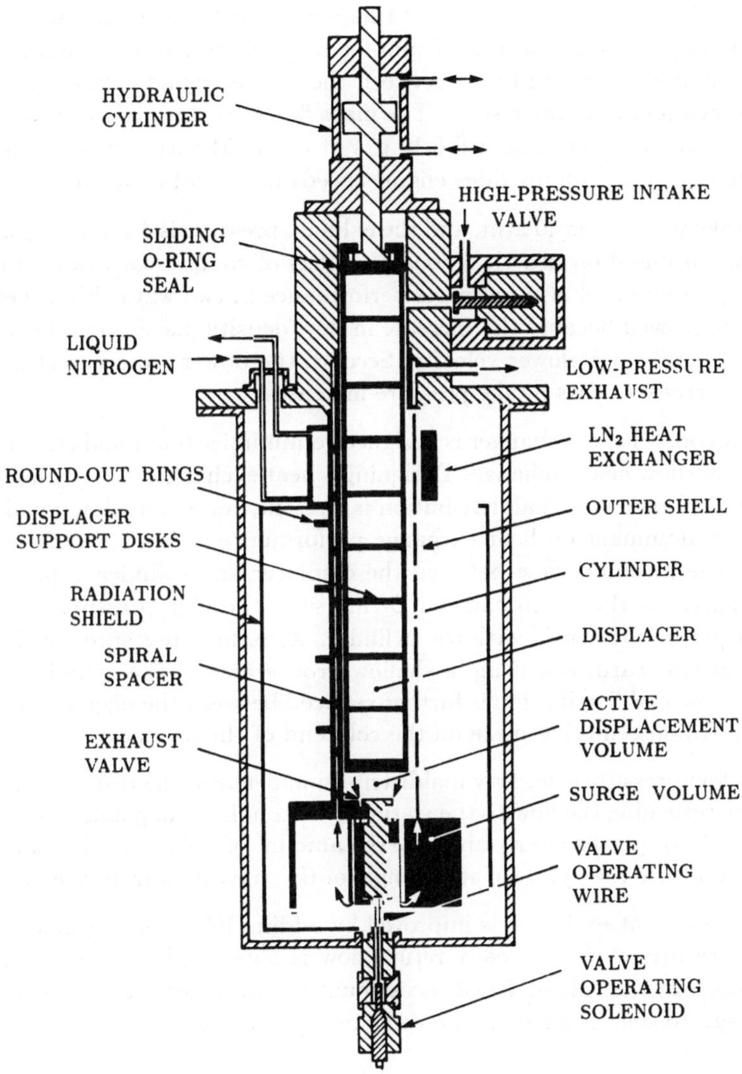

Fig. 1 Schematic of integral heat exchanger and expansion engine.

Performance Specifications

Designing the cryocooler to operate both with and without liquid nitrogen auxiliary cooling requires compromise in performance specifications and component dimensions. For example, expansion engine bore and stroke must be selected to yield mass flows large enough to withstand the effect on performance of displacer motion heat leak when auxiliary cooling is not used, yet small enough to keep heat exchange inefficiencies from dominating performance with auxiliary cooling. A bore of 26.6 mm and stroke of 26.7 mm was selected for this experiment. The tube wall thicknesses measure 0.9 mm for the displacer, 0.7 mm for the cylinder, and 0.5 mm for the outer shell. The radial gaps are 0.13 mm between the displacer and cylinder and 0.25 mm between the cylinder and outer shell. The mass flows are approximately 0.1 g/s at 90 RPM without auxiliary cooling and 0.15 g/s at 65 RPM with auxiliary cooling. The hydraulic drive mechanism provides engine speeds of 30 to 140 RPM.

The intake pressure is 40 atm, and the exhaust pressure is 4 atm. The intake pressure was chosen based on practical considerations of compressor availability. Having the exhaust pressure at 4 atm increases performance in two ways. First, heat transfer efficiency is improved because the increase in flow density places more helium close to the exchanger walls at a slower velocity. Second, the power requirement at the warm compressor decreases as its intake pressure increases.

The cryocooler heat exchanger is designed to minimize flow maldistribution and to enhance counterflow heat exchange. Designing a heat exchanger for very high efficiency is a frivolous effort if flow maldistribution is present; moreover, flow maldistribution becomes more dominant on heat exchange performance as heat transfer efficiency is increased.[4] The incoming flow between the displacer and cylinder is periodic and is directed axially. At the warm end, an O-ring seals the high-pressure gas and maintains the displacer concentric with the cylinder. Also, high-pressure flow is uniformly distributed at the warm end using a shallow groove around the cylinder at the inlet valve port. Flow maldistribution is further reduced between the displacer and cylinder by placing a Rulon centering piece on the cold end of the displacer.

On the low-pressure side, flow maldistribution between the displacer and cylinder is reduced by returning the flow to the warm end through spiral passages which encircle the cylinder. Two spiral spacers minimize the amount of leakage in the axial direction. The spacers encircle the cylinder at an angle of 68° measured from the vertical axis.

Counterflow heat exchange is improved by adding RC filter characteristics to the low-pressure return. A more steady return flow is obtained by adjusting the amount of surge volume at the cold end and the amount of flow resistance in the low-pressure return passage to obtain an RC time constant approximately equal to the period of one cycle.

Hydroforming of the Concentric Tubes

A precise fabrication technique had to be developed for the three tubes comprising the displacer, cylinder, and outer shell. Minimizing flow maldistribution required a technique that would maintain constant gap widths between tubes. Eliminating displacer motion friction and easing assembly of the tubes required that the tubes be very straight. Also, for the cryocooler to be viable for further development, the fabrication technique had to be simple and relatively inexpensive. Using these criteria, a hydraulic tube forming technique was implemented.

All three tubes were hydroformed within the same forming cylinder. This cylinder was gun drilled using 1117 free-cutting steel to very close tolerances. The most critical specifications are 0.08 mm straightness on the 76 cm length and 0.01 mm tolerance on the 29.46 mm inner diameter.

The outer shell was formed first. End plugs were welded into the thin-walled tube to be formed. This tube was then filled with hydraulic fluid and placed inside the forming cylinder where it was pressurized until it pressed against the forming cylinder wall. After the hydraulic pressure was relieved the tube was removed from the forming cylinder and the end plugs were removed. To obtain a 0.25 mm radial gap between the cylinder and outer shell. two layers of Mylar, each 0.13 mm thick, were placed inside the outer shell. Care was taken to fit each Mylar sheet tightly into the outer shell without overlap. The cylinder tube was then formed against the Mylar spacers. A single layer of Mylar was next used to obtain the 0.13 mm gap between the displacer and cylinder. After the Mylar spacer was in place, the Teflon coated displacer tube was formed. Separation of the tubes required use of a threaded shaft screwed into the end cap of one tube and pushing against the opposite end cap of another tube.

The hydroforming process resulted in very straight tubes, with straightness improving with increasing wall thickness. The thin-walled outer shell was bowed by 0.05 mm over a 56 cm length. Straightness measurements indicated that all three tubes were in a continuous bow over their entire length.

Fabrication of the Spiral Spacer

The two spiral spacers that encircle the cylinder at an angle of 68° measured from the vertical axis are 0.27 mm thick and approximately 0.5 mm wide. The spacers are made of Stycast 2850 epoxy from Emerson Cummings. The spacer thickness was selected to provide a slight interference fit between the cylinder and the outer shell. A brief description of the fabrication procedure follows.

The hydroformed cylinder tube was mounted on a case-hardened carbon steel mandrel that had been centerless ground for a tight fit and for straightness. The mandrel was then mounted in a lathe. Much care was taken to center both ends of the mandrel so that the spiral spacers would be of the same thickness over the entire length. To provide the 68° helix angle, the lathe was modified so that the carriage moved at a pitch of 36 mm per spindle revolution.

Two casting grooves, approximately 0.5 mm wide and 0.5 mm deep, were constructed using three layers of Kapton tape (0.13 mm Kapton with 0.04 mm adhesive). A tape applicator mounted to the lathe carriage was used to accomodate both rolls of tape used to make a single groove. Using the tape applicator with the lathe allowed the grooves to be equally spaced.

The resulting grooves were then filled with epoxy and allowed to harden. A carbide tool was used to turn the spacers to a diameter required to give 0.13 mm of interference with the outer shell. This interference was necessary to ensure minimal helium flow leakage in the axial direction.

After the end pieces were welded to the cylinder and outer shell tubes, assembly required raising the temperature of the outer shell using heating tape and lowering the cylinder temperature using liquid nitrogen. An assembly jig was used for this procedure. Lowering of the outer shell onto the cylinder was done manually, which provided adequate speed control and axial alignment.

Inlet and Exhaust Valves

Valving considerations play a major role in the cryocooler development. For adequate cryocooler performance, the inlet and exhaust valves must provide:

1) sufficient valve sealing to prevent gas leakage from high to low pressure when the valve is closed,
2) minimum resistance to gas flow when the valve is open, and
3) quick response to input signals to open and close.

Sealing for both valves is achieved by using Kel-F valve washers, which provide compliance between the valve plunger and the stainless steel valve seats. Also to provide sealing, the pressure on the valve seat is maintained at twice the magnitude of the pressure difference between the high- and low-pressure sides.

Flow passages for both valves are designed large enough to make pressure drops through the valves very small. This design is possible since only one valve is located at the cold end, leaving room for a large exhaust valve passage.

Quick valve response to input signals is essential for the inlet and exhaust valves in this experiment. Valve timing is controlled using a linear displacement transducer attached to the hydraulic piston in conjunction with an electronic control circuit. Power to move the valves is provided by commercial solenoid plungers.

Hydraulic Drive Mechanism

The drive mechanism for the expander must absorb the energy transferred to room temperature from the expansion work and maintain displacer motion. During the expansion portion of the cycle, the hydraulic system is used to dissipate work. The system then supplies power during the exhaust and recompression portions of the cycle. Directional control valve timing is controlled similar to the inlet and exhaust valves.

To achieve constant displacer motion, the hydraulic pump was operated as a pressure source and two pressure compensated flow control valves were used to control the displacer speed. One pressure compensated flow control valve was used for the expansion stroke, and the other was used for the exhaust and recompression stroke.

EXPERIMENTAL RESULTS

Heat leaks due to conduction and displacer motion were measured individually before pressure cycling the cryocooler for refrigeration. The experiments consisted of cooling the surge volume to near liquid nitrogen temperature, and then measuring the amount of energy that must be removed from the surge volume to maintain a constant temperature. Axial conduction along the heat exchanger and radiation heat leak from the vacuum jacket comprise the static heat leak. The measured heat leak was 1.2 W, which corresponds well with predicted values for conduction and radiation. The heat leak due to displacer motion was measured by stroking the displacer while keeping the exhaust valve open to maintain constant pressure in the engine and heat exchanger. The magnitude of displacer motion heat leak is the difference between the energy removed from the surge volume to maintain constant temperature with displacer motion and the static heat leak measurement. The displacer motion heat leak was 1.5 W. The magnitude of this heat leak varied very little between engine speeds of 35 and 110 RPM, indicating very little displacer-to-cylinder friction.

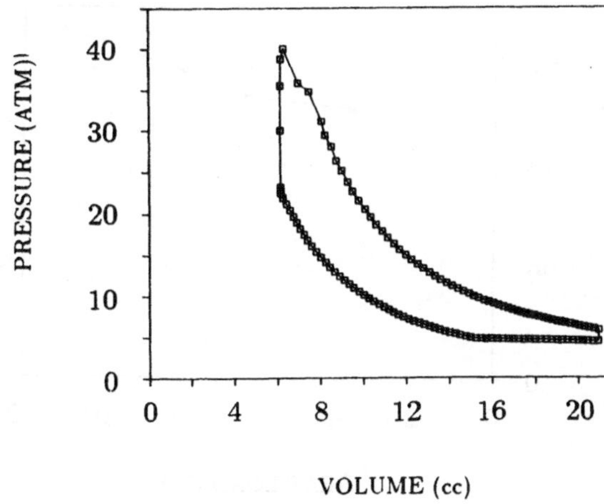

Fig. 2 Pressure versus volume for the case of no precool and no heat load.

During cryocooler operation without auxiliary cooling, the minimum exhaust temperature recorded was 35 K at an engine speed of 90 RPM and an average mass flow rate of 0.11 g/s. Diode sensors were used to measure temperature. A pressure versus volume plot is shown in Figure 2. The indicated work output is 18 W. Pressures were measured using a piezoelectric transducer operating at the cold end. The plot indicates that optimal operation requires substantial recompression. The plot also indicates erratic behavior when the intake valve closes. Data points taken at 5 ms intervals are shown to help explain this behavior and the resulting consequences. When the inlet valve opens, the pressure begins to build in the dead volume. After the dead volume is charged, the electronic controls switch the hydraulic directional control valve to begin the expansion stroke. The initial motion of the displacer is very fast because the pressure compensated flow control valve intended to steady the displacer motion reacts slowly. Fast displacer motion is indicated by the large space between the data points during the initial portion of the expansion stroke. A consequence of the fast displacer motion is a prohibitively large mass flow rate into the cold working volume during the initial portion of the expansion stroke. Further examination of the p-v plot reveals that over 50 % of the mass flow per cycle occurs during the initial 5 % of the cycle time. In numbers, approximately 0.04 g of the 0.071 g per cycle flows into the cold working volume during the first 30 ms of expansion. The resulting high mass flow rate on the high-pressure side during such a small portion of the total cycle results in poor counterflow heat exchange. One would surmise that the overall cryocooler performance would be increased appreciably with an improved drive mechanism, which would limit the displacer speed at the start of the expansion stroke.

The data discussed so far are for operation with no heat load. To measure the performance with heat loads, two Minco thermofoil heaters were placed inside the surge volume. Starting from the no-load condition, several heat loads were added to the cold end while maintaining the engine speed at 90 RPM and the mass flow rate at 0.11 g/s. Constant mass flow rate was maintained by adjusting the amount of recompression and the amount of exhaust blowdown. Figure 3 is a plot of the minimum exhaust temperature reached for each load. At the largest applied load of 11 W, very little recompression was used to maintain the 0.11 g/s mass flow.

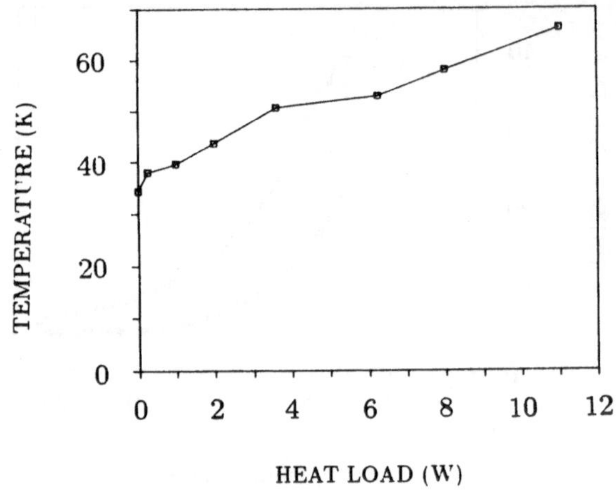

Fig. 3 Exhaust temperature versus heat load with no precooling.

The single-stage unit was also run using liquid nitrogen in a precooling heat exchanger and in a radiation shield. The minimum exhaust temperature reached was 19.5 K at an average flow rate of 0.15 g/s and an engine speed of 65 RPM. The p-v plot for this condition is very similar to the one obtained without auxiliary cooling. The p-v plot indicates a power output of 13.7 W. Almost identical amounts of recompression were required for both conditions. Heat loads of up to 6.3 W (where little recompression could be used) were applied while maintaining the mass flow rate at 0.15 g/s and the engine speed at 65 RPM. Figure 4 shows the minimum exhaust temperatures versus heat load.

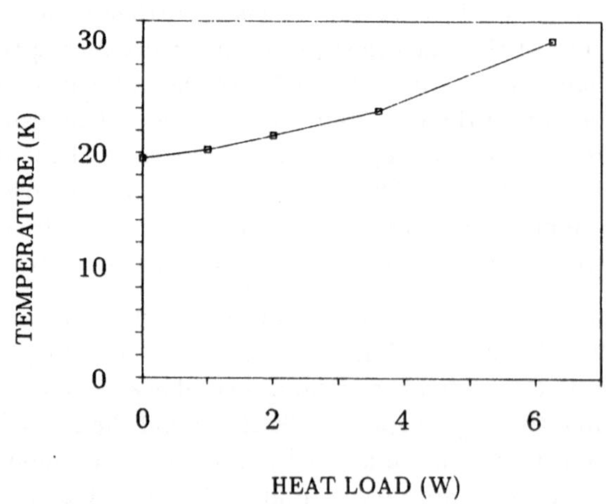

Fig. 4 Exhaust temperature versus applied heat load for the case of liquid nitrogen precooling.

DISCUSSION OF RESULTS

Theoretical results from the performance algorithm can be correlated to experimental results to get a better idea of the magnitude of the various loss mechanisms. Performance specifications that are inputs to the performance algorithm were adjusted until the magnitudes of work extracted, heat load, mass flow rate and cold end temperatures matched the values measured from the experiment. Correlation between the performance algorithm outputs and the experimentally measured values was excellent for data points obtained both without and with auxiliary cooling. The correlation suggests that, without auxiliary cooling, the adiabatic engine efficiency is 0.8 and that the heat exchange temperature difference at the cold end, for no heat load, varied between 18 K and 39 K. Also, the algorithm suggests that for 18 W of power extracted, 15 W is lost to counterflow heat exchange, with the balance of the work being lost to 1 W of static heat leak and 2 W of displacer motion heat leak. With auxiliary cooling, the engine efficiency is 0.8 and the no-load heat exchange temperature difference at the cold end varied between 12 K and 19 K. The 13.7 W of power extracted is consumed by 12.2 W of enthalpy flow at the cold end due to heat exchange, 1 W of static heat leak, and 0.5 W of displacer motion heat leak.

These correlations between the performance algorithm and the experiment data are not expected to be of utmost accuracy. The correlations can be used, however, to indicate areas for improvement for future cryocoolers utilizing this design. The salient observation is that counterflow heat exchange performance limited the cryocooler performance. As discussed earlier, this poor heat exchange performance is primarily due to most of the high-pressure side mass flow occuring during the first 30 ms of the cycle, which is the result of slow response of the hydraulic components. Another observation is that cryocooler performance was not severly affected by diminishing heat capacity of the exchanger walls. This conclusion is supported in two ways. First, the performance algorithm, which does not include the effect of diminishing heat capacity, predicts the observed cryocooler performance quite reasonably. Second, heat capacity is never lost at the cold end because of the low-pressure counterflow. A plot of the low-pressure side flow is shown in Figure 5 for the case of auxiliary cooling. The minimum mass flow is approximately 0.09 g/s.

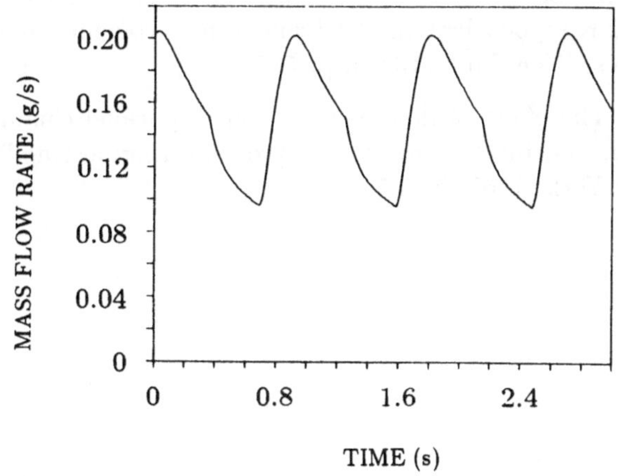

Fig. 5 Low-pressure side mass flow rate for the case of liquid nitrogen precooling and no applied heat load.

CONCLUSIONS

A new cryocooler configuration has been designed, assembled, and tested. The prototype machine produced refrigeration of 1 W at 40 K without precooling and at 20 K with liquid nitrogen precooling. The results indicate that a high pressure ratio single-stage cryocooler, using a combination of regenerative and counterflow heat exchange, is a viable component in a multi-stage miniature liquid helium temperature refrigerator.

Very high mass flow rates on the high-pressure side during a small portion of the cycle period resulted in poor counterflow heat exchange. A performance algorithm indicates that poor heat exchange accounts for most of the thermodynamic losses. Best performance would result if the high-pressure mass flow were of the same order of magnitude as the low-pressure flow throughout the cycle. One way to help average the flow rate would be to have slow displacer motion during the intake stroke. Additional averaging of the flow rate might be obtained by throttling at the inlet valve during the initial pressurization of the dead volume; this would, however, require complex valve action to prevent throttling during the intake stroke.

ACKNOWLEDGEMENTS

This work is supported by the Toshiba Corporation of Japan. This publication is based on a thesis to be submitted in September, 1987 by J.A. Crunkleton in partial fulfillment of requirements for the degree of Doctor of Philosophy at the Massachusetts Institute of Technology.

REFERENCES

1. M. Minta, J.L. Smith, Jr., Helium liquifier cycles with saturated vapor compression, in: "Advances in Cryogenic Engineering," Vol. 27, Plenum Press, New York (1982), p. 603.

2. M. Minta, J.L. Smith, Jr., An optimum cold end configuration for helium liquefaction cycles, in: "Advances in Cryogenic Engineering," Vol. 29, Plenum Press, New York (1984), p. 479.

3. F.W. Pirtle, P.A. Lessard, J.M. Kaufman, P.J. Kerney, Thermodynamic aspects of small 4.2 K cryocoolers, in: "Advances in Cryogenic Engineering," Vol. 27, Plenum Press, New York (1982), p. 595.

4. F.B. Fleming, The effect of flow distribution in parallel channels of counterflow heat exchangers, in: "Advances in Cryogenic Engineering," Vol. 12, Plenum Press, New York (1967), p. 352.

DEVELOPMENT OF A SPACE QUALIFIED SURFACE TENSION CONFINED LIQUID CRYOGEN COOLER (STCLCC)

Stephen H. Castles and Michael E. Schein

NASA-Goddard Space Flight Center
Greenbelt, Maryland

ABSTRACT

The NASA-GSFC is developing a new type of cryogenic cooler for use with spaceflight payloads. This cooler will be capable maintaining instrumentation within the temperature range of 10-120 K. Known as the Surface Tension Confined Liquid Cryogen Cooler (STCLCC), it will allow liquid cryogens to be flown in space without the risk of liquid being entrained in the vent gas. The cooler contains a low density (85-95% free volume) open cell material. This acts as a "sponge" with surface tension trapping the liquid cryogen within its pores. The surface tension effect keeps liquid away from the cooler's vent during launch, zero-g operations, and landing. It also prevents any liquid slosh which can produce vibrations and microphonic noise in sensitive instrumentation. Other benefits of the STCLCC include the potential to be serviced on-orbit and an inherent simplicity of operation which enhances system safety. It also promises a reduction in cost, complexity, and pre-launch servicing requirements over the present state of the art, the solid cryogen cooler. The STCLCC program is in the preliminary design and verification stage, with early test results confirming acceptable thermodynamic, mechanical, and cryogen retention properties of several first generation sponge materials.

INTRODUCTION

The demand for on-orbit cryogenic cooling is expected to rise sharply through the next decade and beyond. The primary near-term cause for this is the increasing sensitivity requirements of astrophysics and earth observing instruments. In the longer term, a variety of cooling demands are anticipated for superconducting instrumentation, biological and microgravity research, and materials processing.

Existing spaceflight cryogenic coolers have been adequate for free-flying spacecraft launched to date using expendable launch vehicles. With the advent of the shuttle, a number of new design constraints have been placed upon cryogenic cooler design, including:

a] Longer hold times between final ground servicing and launch
b] Enhanced and expanded safety requirements
c] Capability to be safely returned to Earth in the event of a launch abort

A new series of long duration scientific facilities in the Space Telescope class are being prepared for service in the near future. Their introduction and the inherent flexibility of the shuttle (and eventually the space station) logically suggests some further capabilities which would be desirable in a cryogenic cooler. These include:

d] The capability of being serviced on-orbit
e] Compatibility with manned operations, either Extra-Vehicular Activity (EVA) or activities inside a shuttle cabin or station module.

The remainder of this paper describes the STCLCC and contrasts it to available cooler technology. Presently under investigation at the NASA-Goddard Space Flight Center, this new technology is designed to provide all of the above capabilities. To date, preliminary laboratory tests have investigated STCLCC thermodynamic and mechanical properties. While promising results have been obtained, much more research and development is required before the full potential of this concept is known or realized.

PRESENT STATE OF THE ART

The present state-of-the-art in space qualified cryogenic coolers are mechanical coolers, liquid (superfluid) helium dewars, and Solid Cryogen Coolers (SCC)[1,2]. Mechanical coolers are the system of choice for payloads with large heat loads or long lifetimes. In this application, the weight penalty of a stored cryogen is usually excessive. Presently, mechanical coolers also require large amounts of power (>100 W) with an associated heat dissipation capability and sophisticated control electronics to suppress vibration.

Liquid helium dewars are baselined for many future on-orbit cooling duties where there is a requirement for extremely low temperatures (~2 K). Examples are the Cosmic Background Explorer (COBE), the Advanced X-Ray Astrophysics Facility (AXAF), and the Space Infra-Red Telescope Facility (SIRTF). These dewars are now an established technology. The capability of reservicing them in space is presently being developed by NASA and will be proven by the Superfluid Helium On-Orbit Transfer (SHOOT) shuttle flight demonstration. The low heat capacity of liquid helium requires that the total heat load be small (< 200 mW), the amount of helium on-board be relatively large (>200 liters), and that the dewars be protected by a sophisticated thermal insulation system.

The STCLCC is being developed for two distinct classes of payloads which are not suitable for either of the above types of coolers. The first are payloads with a long duration mission but without a large power generation capability. These include second and third generation Space Telescope instruments and the Large Deployable Reflector. Capabilities d] and e] listed above are particularly important for these payloads. The second class of payloads are inexpensive instruments for shuttle sortie missions. Capabilities a], b], and c] are of primary concern for these payloads.

The needs of these two categories are presently being met by the solid cryogen cooler (SCC). A cross-section of a generic SCC is given in Figure 1a. These coolers use cryogens such as methane, carbon dioxide, nitrogen, argon, or neon loaded in a tank as a liquid or a gas. The cryogen is then frozen in place by running a coolant (typically liquid helium) through a heat exchanger coil wrapped around the tank. An alternative method of freezing a liquid is lowering its vapor pressure by pumping on the vent line. This type of system has been used by the SCC's made by Lockheed Missiles and Space Company for the NIMBUS series of satellites. SCC's are presently baselined for a number of upcoming shuttle payloads, including the Broad Band X-Ray Telescope (BBXRT) and the Cryogenic Limb Array Etalon Spectrometer (CLAES) experiment aboard the Upper Atmosphere Research Satellite (UARS).

In practice, a number of factors have made development of a shuttle-qualified SCC difficult. In particular, the shuttle's relatively long ground hold between final servicing and launch (8 to 18 days) makes it difficult to assure that the cryogen in the tank will not melt, even if on-board pumps are used. If liquid is in the tank at launch, it is hard to prove that the liquid will not find its way into the system plumbing. This creates the possibility of liquid venting in the payload bay, causing a loss of cryogen and creating cold temperature hazards to equipment and EVA personnel.

Besides the possibility of venting liquid, jeopardizing mission success, and creating safety concerns, the SCC has other drawbacks. One is that it would be difficult or impossible to service on-orbit since the cryogen needs to be loaded into the SCC on-orbit as a liquid or gas and frozen in place. A quantity of coolant such as liquid helium would have to be provided to freeze the cryogen, increasing the launch weight and complexity of the servicing system. The liquid venting problem also must be solved before an SCC can be used inside manned areas of the shuttle or space station.

STCLCC SYSTEM DESCRIPTION

The STCLCC addresses the free liquid problems inherent in the SCC by eliminating the requirement to freeze the cryogen. Instead the liquid cryogen is bound in place by trapping it in an open cell micropore "sponge". The surface tension of the liquid inside this sponge precludes it from migrating inside the tank or entering the vent line.

A cross section of a conceptual STCLCC cooler is shown in Figure 1b. In many ways it is similar to common liquid cryogen storage dewars. In this drawing, the inner cryogen tank is shown suspended within an outer vacuum shell by a series of support straps. Multilayer insulation is used to block radiation heat transfer to the inner tank.

To these standard features the STCLCC dewar adds some modifications. The most obvious of these is the surface tension effect (STE) sponge filling most of the volume of the inner tank. The sponge design concept shown in Figure 1b has the sponge material filling the tank except for stress relief cutouts at the fill and vent lines. Some mission scenarios may require only a partial filling of the tank with the STE sponge. An example would be a one week shuttle sortie mission where much of the cryogen will boil off during the planned pre-launch ground hold.

In a zero-g environment the liquid phase of the cryogen may be expected to migrate away from sources of heat that convert it to vapor. The sponge materials examined so far are all excellent thermal insulators, so they will do little to counteract this tendency. It is possible that a large fraction of the cryogen in the tank could eventually come to rest in an area of the tank where there are no conduction heat sources, away from the tank walls and the cold finger . Protected by the low thermal conductivity of the sponge and the surrounding vapor, this liquid will lose its usefulness as a coolant. In order to reduce or eliminate this effect, the conceptual STCLCC design includes a number of heat transfer fins. These fins are thin metal plates thermally anchored to the cold finger, which provide an efficient heat conduction path from the cryogen to the cold finger. Spacing between the fins is a design variable dependent on the cryogen used, the sponge material, and the geometry of the tank. A similar fin system was flown in an SCC design on the High Energy Astrophysical Observatory (HEAO) series of spacecraft.

While the sponge materials are remarkably durable, they are composed of brittle microscopic fibers which could be a source of particulate contaminates in the vent gas. A number of approaches to counteract this problem are being considered. One is to bond the surface of the sponge facing the vent line with a porous binder to keep free particles from breaking off. Another is to cover this same surface with a sheet of micropore material such as Gore-Tex™ to filter these particles out before they get to the vent line. A variation of this idea is to add a conventional filter to the vent line as shown in Figure 1b. All of these approaches will be tested during the STCLCC development program.

Since the object of the STCLCC is to allow the use of a liquid cryogen both on the ground and in space, some mechanism has to be provided to ensure that the inner tank on-orbit pressure is above the cryogen's triple point, below which the cryogen could freeze and possibly damage the STE sponge. The easiest way to accomplish this is to include a pressure regulator on the vent line. This unit may be as complicated as a controllable absolute pressure regulator for instruments requiring a precise operating temperature, or as simple as a check valve when operating temperature requirements are not so severe. An example of the latter would be a one atmosphere gauge pressure poppet valve. On the ground the tank pressure would be two atmospheres absolute, while in space the tank pressure would drop to one atmosphere absolute. For liquid nitrogen this translates to cryogen temperatures of 84 K on the ground and 77.4 K on-orbit.

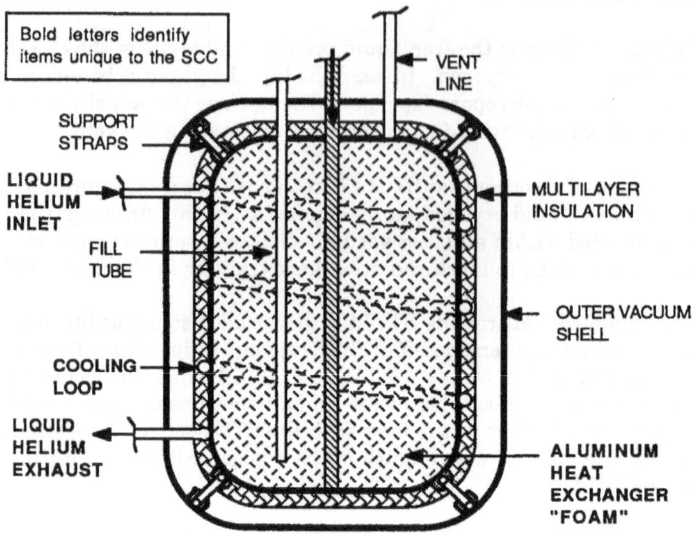

Figure 1a - Solid Cryogen Cooler Cross-Section

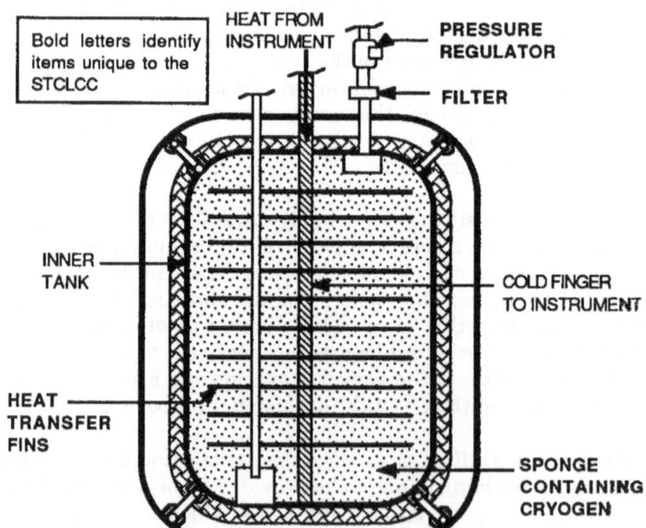

Figure 1b - STCLCC Concept

Procedures for filling the STCLCC are similar to those for filling any liquid cryogen dewar. During tests all candidate sponge materials have shown excellent resistance to erosion from the relatively high speed vent gases generated during system cooldown. Good design practice would allow the incoming liquid stream to have its energy dissipated by the tank wall as shown in Figure 1b. Any liquid cryogen coupler may be used for ground fill purposes. An EVA compatible liquid cryogen coupler capable of handling temperatures down to that of liquid helium is being developed as part of the SHOOT flight demonstration.

THE STCLCC CONCEPT - ADVANTAGES AND DISADVANTAGES

Besides eliminating the free liquid problem, the STCLCC concept has a number of advantages over the SCC. It is much more accommodating of prelaunch ground delays. The ground hold of the cooler is theoretically limited only by its tankage capacity since the cryogen is always in a liquid state. Servicing operations on the launch pad are also simplified since no freezing and subcooling operations are required. Refilling the STCLCC on-orbit would also be a relatively simple operation, assuming an on-orbit liquid supply dewar is developed. Also, any concern about liquid slosh producing undesirable vibration induced microphonic noise in sensitive detectors is eliminated by the sponge.

The STCLCC does have some disadvantages. The first is that some tank volume is lost to the sponge. Part of this volume loss is also experienced by the SCC, which generally contains a 3 - 5% by volume aluminum "foam" to enhance the heat transfer inside the cooler. The aluminum foam does not contribute any surface tension effect since it has very large pores (> 1 mm). Another STCLCC concern is that the sponge may be a particulate contamination source although, as mentioned above, this may be addressed by a filter on the vent line. Finally, the STCLCC suffers a loss in cooling power per unit weight and volume compared to the SCC. The latent heat of evaporation is about 10% less than the latent heat of sublimation and the liquid density is about 10-15% lower than the solid density for common cryogens. However, in operation the SCC rarely realizes the advantage of a higher solid cryogen density. If the cryogen is loaded as a liquid, the fill capacity of the SCC is exactly the same as an equivalent STCLCC. Topping off the SCC after freezing is usually not attempted due to launch pad servicing time and safety constraints.

THE SURFACE TENSION EFFECT SPONGE

Three types of sponge material have been investigated to date. Some relevant properties for each of these are given below in Table 1.

Table 1. Sponge Material Properties

Fiber Material	Silicon	Carbon/Carbon	Alumina
Product Name and Manufacturer	H.T.P.-6, Lockheed Missiles & Space Co, Sunnyvale, CA.	"NucFil", Nuclear Filter Technology, Golden, CO.	ZAL-15, Zircar Products, Inc., Florida, NY.
Intended Purpose	Space Shuttle thermal tile	Nuclear waste gas filtration	Furnace and crucible insulation
Density	0.083 g/cm^3	0.248 g/cm^3	0.234 g/cm^3
Actual Free Volume	96%	89%	94%
Volume Filled by LN_2	81 %	76%	77%
Effective Pore Size	11 μm	7 μm	9 μm
Wicking Height (LN_2)	4.1 cm	6.4 cm	5.1 cm
Crushing Pressure	310 kPa	207 kPa across fibers >1000 kPa with fibers	345 kPa

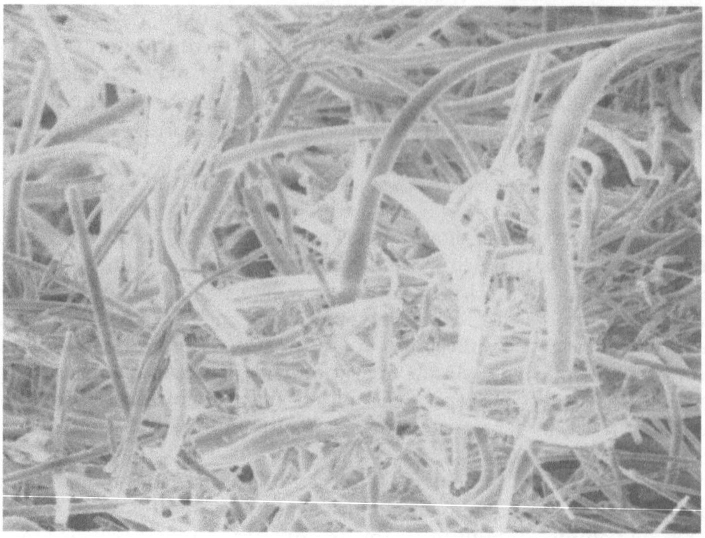

Figure 2 - Silicon STE Sponge Internal Structure (Magnification: 640X)

The internal structure of the silicon sponge is a matted mass of fibers oriented in random directions, as shown in Figure 2. The internal structure of the alumina material is similar. The structure of the carbon/carbon fibrous composite is slightly different, with fibers aligned preferentially within a thin layer. These layers are then built up to form the bulk material.

Low density, high free volume, and small pore size are the main sponge design goals. The requirement for the first two properties are obvious for spaceflight systems where weight and volume constraints are critical. A small pore size translates into a large wicking effect, which is synonymous with the force the sponge exerts on the cryogen to keep it in place. A simple example of the meaning of wicking height is: the silicon sponge will support a 4.1 cm high column of liquid nitrogen in 1 g, or a 41 cm column in 0.1 g. The maximum adverse on-orbit acceleration a STCLCC tank might expect to see would be from a firing of the shuttle's Orbital Maneuvering System (OMS). This acceleration is typically 0.02 g, which means that a full STCLCC dewar using the silicon sponge could have a maximum dimension of about 2 meters in the direction of the OMS induced acceleration vector without the cryogen migrating out of the sponge. Larger accelerations due to gravity or launch follow known directions, and the design orientation of the tank will ensure that any free liquid migrates away from the vent.

Using sponge density and available free volume as a selection criteria, it would appear that the silicon shuttle thermal tile material is presently the best candidate for use as a STCLCC sponge. However, this material has the largest pore size and therefore the weakest cryogen retention characteristics. The STCLCC's presently planned are for shuttle sortie and sounding rocket missions. They all have relatively small cryogen tanks (about one cubic foot), so the pore size difference is not a problem for these systems.

Tests to date have resulted in fill levels about 15% less than the actual free volume. This lost volume may be due to trapped gas and/or incomplete wetting. One of the goals of the on-going STCLCC development program is to find or produce a sponge and develop fill techniques which will allow a 95% fill using a cryogen with a surface tension equal to that of liquid nitrogen.

STCLCC DEVELOPMENT - PAST EFFORTS

Efforts at creating spaceflight coolers similar to the STCLCC have been made previously. A small dewar containing a zeolite or activated charcoal was designed by Dr. John Hendricks for the Space Astronomy Lab at the University of Florida[3]. Another spaceflight dewar containing a carbon material was developed by Kadel Engineering Corporation for the Air Force. It was tested with both liquid nitrogen and liquid helium. At that point the intention was to pump on the liquid nitrogen in the dewar to convert it to a solid. Unfortunately, the surface tension effect prevented the nitrogen from freezing. The carbon was removed and the tank was recently completed as a conventional solid cryogen cooler. Similar attempts to use the sponge with liquid helium were ended after the incorporation of a porous plug phase separator[4]. It should be pointed out that this technique is only applicable to superfluid helium dewars.

Another program at NASA has used a surface tension effect sponge to contain a liquid cryogen under zero-g conditions. At the MSFC Mr. Orville Weaver has developed an apparatus to create a constant cryogenic environment using liquid nitrogen for a series of micro-g experiments aboard the NASA KC-135 Zero-g Test Aircraft. He reports excellent results for both cryogen retention and thermal stability using the alumina sponge material described in Table 1[5].

STCLCC DEVELOPMENT - PRESENT EFFORTS

The NASA-GSFC STCLCC development program has been organized into two complimentary efforts, The first is the testing and characterization of cryogen behavior within STE sponge materials. The second is the conceptual design and optimization of prototype STCLCC dewars for a variety of applications. A small STCLCC dewar is presently under construction in support of the Supernova X-Ray Spectrometer (SXS) series of sounding rocket launches to study Supernova 1987a. These are scheduled for launch beginning in November, 1987 out of Woomera, Australia.

Another application of the STE sponge has been with the NASA-GSFC SHOOT program. Initial testing has been carried out using the silicon fiber material as one component of a superfluid helium acquisition system feeding a thermomechanical pump. It has demonstrated excellent acquisition properties while allowing flow rates through the sponge in excess of requirements.

TESTING

STCLCC testing to date has concentrated on the STE sponge. First, the thermodynamic behavior of both "empty" and sponge filled tanks were compared to see if adsorption or liquid migration would have any affect on boiloff rates. The boiloff behavior of an empty small scale (7.5 cm inner diameter, 15 cm in length) foam-insulated tank was characterized with liquid nitrogen. The tank was then filled with STE sponge material (first the carbon/carbon composite in one test and then the silicon fiber material in a second test) and again charged with liquid nitrogen. There was no discernable difference between the "empty" (no sponge) and "full" (with sponge) cases. It was therefore concluded that the sponge materials presently under investigation may be thought of as "transparent". This means the boiloff behavior of the STCLCC may be predicted directly using saturated liquid/vapor data.

Preliminary tests have also been performed on sponge wicking, mechanical strength, and thermal transport properties. Positive results have encouraged us to design further tests to make more precise measurements of these properties under a wider variety of pressure, temperature and loading conditions. Parallel efforts will explore various methods of supporting the sponge within the tank. Test equipment is also being designed for a series of static loading, vibration, and acoustic tests of sponge materials within a cryogen tank. These tests will determine if the brittleness of the sponge materials translates to a mechanical sensitivity to launch or landing load environments. Dynamic loads will be applied to sponge materials both with and without cryogens on-board. When these loads tests are run with cryogens, a filter system will be installed on the vent line. A series of in-line filters with various well-defined pore sizes will be examined after each test under a scanning electron microscope to characterize the particulate shedding tendencies of the sponge materials.

STCLCC DESIGN EFFORTS

The STCLCC concept is a simple one. It is amenable to a wide variety of applications. For a particular mission the detailed design of an appropriate STCLCC will include optimizing the following variables:

1. Type and amount of liquid cryogen (determined by operating temperature and mission life requirements)
2. Percentage of tank to be filled with sponge
3. Cold finger placement and heat exchanger fin spacing
4. Heat load to the cryogen (to be minimized)
5. Vent line tubing size, pressure regulator, and filters (if needed)
6. Sponge support system inside the tank.

STCLCC design efforts which include the above are or soon will be underway in support of proposed NASA projects, including SXS and the Near-Infrared Spectrometer, a shuttle Spartan class payload. It is anticipated that this technology will be applicable to future spaceflight projects where passive, low-cost coolers are required, or when on-orbit servicing of a cooler would increase a mission's lifetime.

REFERENCES

1. T. C. Nast, Status of Solid Cryogen Coolers, in: "Advances in Cryogenic Engineering", Vol. 31, Plenum Press, New York (1986), p. 835
2. A. Sherman, History, Status, and Future Applications of Spaceborne Cryogenic Systems, in: "Advances in Cryogenic Engineering", Vol. 27, Plenum Press, New York (1981), p. 1007
3. Personal Communication with Dr. John Hendricks, University of Florida
4. Personal Communication with Mr. Gary Frodsham, Kadel Engineering Corp., Danville, Indiana
5. Personal Communication with Mr. Orville Weaver, NASA-Marshall Space Flight Center

SMALL TURBO-BRAYTON CRYOCOOLERS

Herbert Sixsmith, Javier Valenzuela, and Walter L. Swift

Creare Inc.
Hanover, New Hampshire

ABSTRACT

The development of miniature turbomachinery has made possible the extension of the operating range of turbo-Brayton cryocoolers down to applications requiring as little as 1 - 5 watts of refrigeration at 4.2 to 70 K. This paper reviews the status of this technology and describes two cryocoolers presently under development at Creare: a 5 watt, 70 K cooler for space sensors, and a 0.6 l/hour helium liquefier for Magnetic Resonance Imaging systems. Cycle configuration trades are discussed and bounds on turbo-Brayton cycle cooling temperatures and capacities are presented. The design and performance of prototypical hardware is described.

INTRODUCTION

There has been increased activity in developing cryocooling systems for relatively low cooling loads - from 1 W to 100 W at temperatures from 77 K to below 4.2 K. For at least a large portion of this range of requirements, there has been strong activity in developing turbomachines for use in reverse Brayton cycles to meet these cooling requirements. The other candidate cycles which are relatively mature in their developments in terms of demonstrating successful operation include G-M, Stirling and/or Vuilleumier, and reverse Brayton employing positive displacement machines. In general, the current state of the art with respect to these systems is that each has been shown to operate successfully. Further developments are now aimed at increasing reliability, extending life, extending range of operation, improving performance, reducing weight and system complexity, and lowering cost.

In 1982, Johnson[1] presented a map of cycle applications for temperatures and loads for long lifetime cryocoolers. At that time, coolers for the 1 - 5 w range were dominated by cycles using positive displacement machines. At the same conference, Sixsmith and Swift[2] described the successful demonstration of a miniature self-acting gas bearing which should be useful in extending the range of cryogenic coolers using turbomachines to lower loads and temperatures. Since that time, these bearings have been incorporated in very small machines, and coupled with improved heat exchanger designs, have been integrated into a high performance single stage cryocooler for long-life operation in space[3].

This paper reviews important technical elements of the basic components of a reverse Brayton cycle. A single stage cycle is used to illustrate the major losses in performance and some of the design issues for components. A second, two stage system intended for MRI cooling is briefly discussed. These developments have resulted in producing cycle Carnot efficiencies at relatively low loads which are competitive with those achieved in systems using positive displacement machines. This is important because mechanical features of the hardware used in the turbo - Brayton systems allows for reduced size, weight, and vibration in comparison to positive displacement machines.

THE TURBO BRAYTON CYCLE

A single stage reverse Brayton cycle is illustrated in Figure 1. At the high temperature end, gas is compressed (and heated) from p_1 to p_2. The heat of compression is rejected through an aftercooler (or some heat may be removed in the compression process itself, depending on the compressor design). At the low temperature end, the gas expands through the turbine from p_3 to p_4, where p_3 and p_4 differ from p_1 and p_2 by the

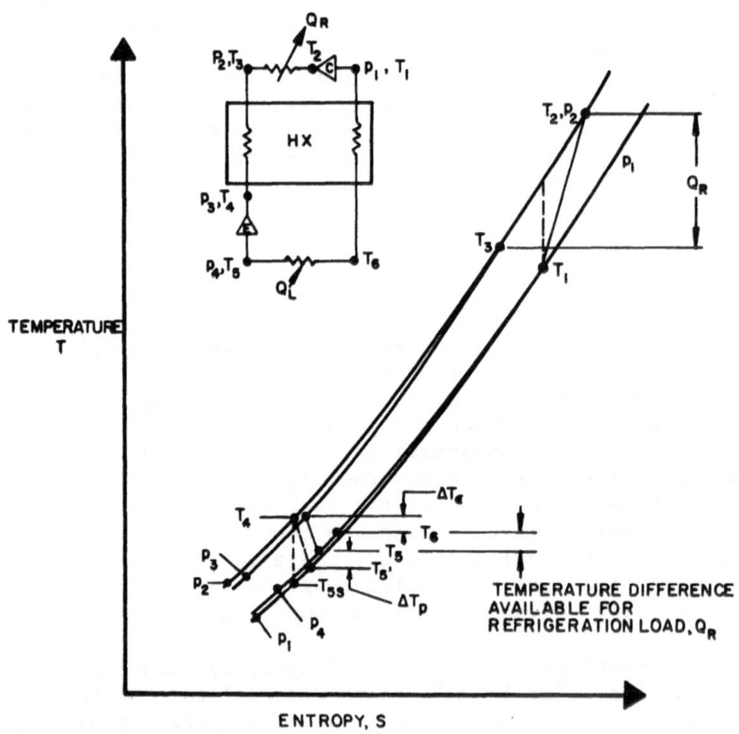

Figure 1. Single Stage Reverse Brayton Cycle. E is the Expander, C is the Compressor, Q_L and Q_R are the Load and Heat Rejected, Respectively.

frictional pressure losses in the heat exchanger. The high pressure stream of gas from the compressor is precooled by the low pressure stream in a counterflow heat exchanger. The useful refrigeration produced by the turbine is used to absorb heat from the load Q_L, and to absorb parasitic heat loads which result from irreversibilities and from conduction, convection and radiation from ambient.

The state points in Figure 1 clearly indicate the major sources of loss in the cycle:

1) The compression process is non-isentropic. The input power to the cycle is proportional to the inverse of the compressor efficiency.

2) Pressure losses and ineffectiveness in the heat exchanger are major sources of loss. In Figure 1, the dashed line from T_4 to $T_{5'}$ represents the expansion in a turbine from p_2 to p_1 (ie. no pressure losses). Because of the pressure losses in the cycle the actual expansion occurs from T_4 to T_5. The loss in refrigeration is represented by ΔT_P. Since this loss originates primarily in the heat exchanger it should be included in the calculation of overall effectiveness in order to compare one heat exchanger design with another[3]. Similarly, the impact of heat exchanger thermal ineffectiveness (ΔT_ε) shown in the Figure adversely affects cycle efficiency.

3) The third major loss occurs in the turboexpander, and is associated with irreversibilities in the expansion process as well as heat in-leak from the warm end of the machine. The effect is illustrated by comparing the temperature drop through the turbine (T_4 - T_5) with the isentropic expansion from T_4 to T_{5s}.

For a given set of cycle sink and source temperatures, T_3 and T_6, decreases in load Q_R will necessitate a decrease in component sizes. In the compressor, reduced size has the effect of increasing the operating speed and increasing losses relative to overall power level. In the turboexpander, size also decreases, speed increases and the relative losses increase. However, because there is heat leak from ambient to the cold end of the turbine, (principally due to conduction through the housing and shaft), the losses increase very rapidly in this component relative to the net power. For a well designed miniature turboexpander, the heat in-leak to the fluid in the turboexpander Q_T is approximately:

$$Q_T \cong 0.05\ D_T\ (T_B - T_5) \qquad (1)$$

where D_T is the turbine tip diameter in cm, T_B and T_5 are the warm and cold end temperatures of the turboexpander in K and 0.05 is an empirical constant which takes into account material conductivity as well as specific features in our designs. Q_T is given in watts. For a 0.3 cm diameter turbine (the smallest we've developed to date), with warm and cold temperatures of 270 K and 70 K, respectively, the heat leak is 3 W. It is clear that as loads are reduced to values near 1 W, most of the useful refrigeration produced by the turbine is used to overcome the internal heat leak producing a substantial reduction in the efficiency of the turboexpander. It is also clear from this expression that for a given net refrigeration output from the turbine it is desirable to reduce the turbine diameter to the smallest practical size and/or reduce the warm end temperatures. While these conclusions seem obvious, there are practical issues which must be addressed in order to reduce heat leak in the machine. It is these issues which generally dominate the final design of the machine.

In the absence of pressure losses and ineffectiveness in the heat exchanger, the coefficient of performance of a single stage reverse Brayton cycle pumping heat from T_6 to a rejection temperature of T_3 is:

$$\begin{aligned} \text{COP} &= Q_L/P \\ &= (T_6/T_3)\eta_c\eta_T/\pi^{(k-1)/k} \end{aligned} \qquad (2)$$

where P is the compressor input power, η_c and η_T are compressor and turboexpander efficiency, respectively, k is the ratio of specific heats and π is the pressure ratio p_2/p_1. The inverse of the COP, P/Q_L is plotted in Figure 2, to show the relationship between input power, pressure ratio and temperature ratio. This figure also demonstrates the strong impact that heat exchanger effectiveness has on performance. The single dashed line on the figure shows the ratio of input power to refrigeration load for T_3/T_6 of 2.5 with assumed compressor and turboexpander isentropic efficiencies of 0.5 each. The two solid curves directly above the dashed line show the effect of heat exchanger performance on overall cycle power. As the pressure ratio is lowered for a given set of temperatures, the temperature drop through the turbine (T_4-T_5) is reduced, until it approaches (T_4-T_6), reflecting the ineffectiveness of the heat exchanger. As a result, there is for each temperature ratio θ, and for each combination of ε, η_c and η_T, an optimum pressure ratio which gives the lowest input power to refrigeration ratio P/Q_L. However, the range of optimum pressure ratios for a given set of component performance parameters is reasonably broad. The figure shows that the optimum pressure ratio increases substantially with decreasing heat exchanger effectiveness. It should be noted that pressure losses have been assumed to be negligibly small in this illustration. In general they become very important at small loads and should be included in the cycle trade studies. This can be done

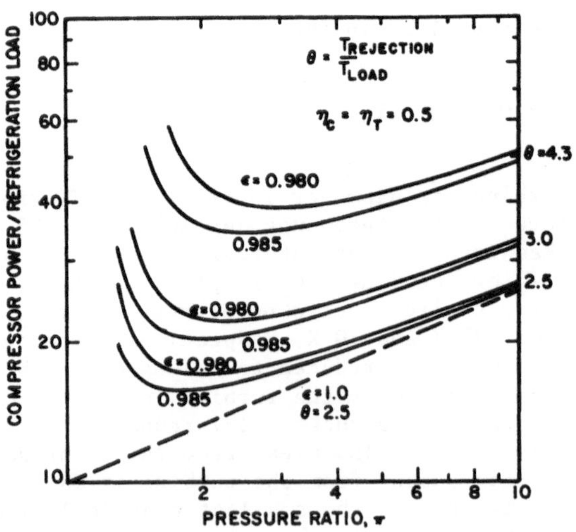

Figure 2. Performance Curves for a Single Stage Reverse Brayton Cryocooler Using Turbomachines.

Table 1. Cycle Specifications for a Single Stage Reverse Brayton Cryocooler

Fluid	Neon
Load Temperature, T_6	70 K
Inlet Temperature, T_1	260K
Pressure Ratio	1.6
Mass Flow Rate	1.0 g/s
Compressor Efficiency*	0.4
Heat Exchanger Effectiveness	0.985
Turboexpander Efficiency	0.5
P/Q_L	30 W/W

* Includes Motor Losses

relatively simply by incorporating the ΔT_p associated with pressure loss in the definition of heat exchanger effectiveness (See Reference 3). It should also be noted that the effect of η_c and η_T is, to first order, to scale the curves in the vertical direction.

The single stage cooler cycle described in the preceding paragraphs has been used to identify the important component performance characteristics and how they affect and limit cycle performance. Detailed trade studies along the lines described above were performed to establish system and component specifications to supply 5 W of cooling at 70 K. Components for this cycle are currently being developed[3]. Table 1 lists the important component and cycle parameters for this system.

The development of this cycle represents state-of-the-art technology in terms of the characteristics of the components. The system is intended for long life cooling in space, an application which has tended to drive the development of most small cryocoolers over the last 20 years. The major technology drivers, however, are also comparable to system requirements for commercial use on earth.

1) Contamination - in space, the absence of contamination within the system is extremely important. Maintenance is not practical, operational life of 5-10 years is required for some applications. For commercial applications (MRI cooling, computer cooling), the gas must be continuously cleaned during operation and the cleaning system must be simple and effective. The mean length of time between maintenance requirements are increasing.

2) Reliability - this is extremely important in space because of the expense of replacement and/or redundancy. It is also important in commercial applications because any down-time of the cooled system can be very expensive.

3) Cycle efficiency - very important in space applications because the power supply and conditioning, and the heat transport and rejection systems impose extreme weight and size penalties on the overall launch budget. In commercial applications ease of manufacture and first cost considerations can reduce the impact of cycle performance.

4) Size and weight - these are very important in space, less so in commercial applications.

5) Robustness - this requirement varies in commercial applications, it is very important in a launch environment for space.

6) Vibration signature - this is modestly important in some commercial applications. It is very important in space applications where sensor and optics structure performance may be affected.

It is because of these technology drivers that gas bearing-based turbomachines were selected for the single stage space application. The low cooling load (5 w) requires small components which in turn results in very high rotational speeds for the turbomachines - the compressor design speed is 480,000 rpm and the turbine speed is 570,000 rpm. These high speeds require gas (or magnetic) bearings, which, in turn, allow for long life operation without contamination and with no perceptible vibration.

The key elements of technology developments to achieve this involve miniaturization of the rotating machines, development of a high effectiveness, all-metal, compact heat exchanger and the development of a small high speed, low voltage ac induction motor with reasonable efficiency. The following sections briefly discuss each of these components.

TURBOEXPANDER

A functional schematic of the turboexpander is shown in Figure 3. The machine consists of a warm end, a thermal isolation section and a cold end. The warm end housing contains the gas bearings which support the shaft and a brake circuit to absorb work produced in the turbine. In the cold end of the machine, the cryogen expands through radial nozzles and the turbine which provides useful refrigeration. The middle housing element provides both a structural connection and thermal isolation between the warm end and cold end of the machine. It is through this thermal isolation section that heat flows from the warm end to the cold end.

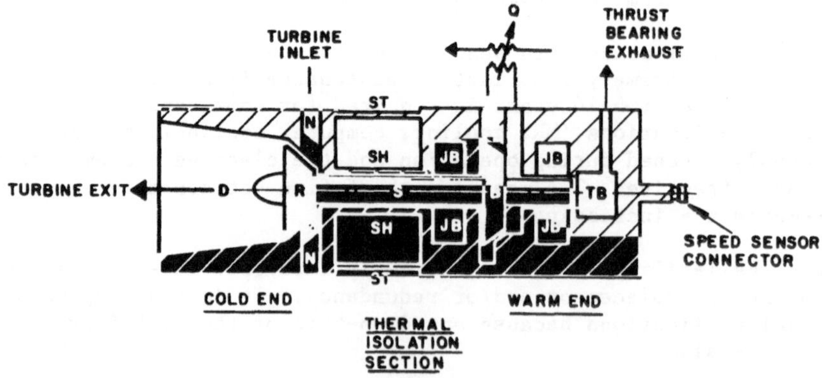

KEY		
R TURBINE ROTOR	B BRAKE COMPRESSOR	ST STRUCTURAL SHELL
N TURBINE NOZZLES	JB JOURNAL BEARINGS	
D TURBINE DIFFUSER	TB THRUST BEARING	
S SHAFT	SH SHAFT SLEEVE	

Figure 3. Turboexpander Schematic

For this single stage cycle, the turbine is 0.318 cm (0.125 in.) diameter and has a design speed of 570,000 rpm. In addition to the normal considerations of achieving good aerodynamic performance in such a small machine, critical elements of the design involve stable and robust operation at the high rotational speeds, and reliable start/stop operation.

The cycle conditions dictate the turbine rotor diameter and rotational speed. Self acting gas bearings described in Reference 2 are used to provide reliable operation. Bearings of this size and type have demonstrated reliable operation at speeds well in excess of these requirements. Several years of continuous operation and thousands of stop/start cycles have been accumulated without degradation in performance. Shaft speed is generally limited only by the first critical speed in bending. This limitation, for a given diameter, determines the shaft length, and the overhang distance into the cold end, which strongly affects the conductive heat leak penalty to the cold end. Development efforts on this machine are fairly mature. Reliable operation of machines of this size has been demonstrated at temperatures well below 20 K. Tests on machines of comparable size indicate that the design goal of 50% in overall efficiency can be achieved.

HEAT EXCHANGER

For the single stage cryocooler described above, a detailed analysis was performed[3] to establish the design of the heat exchanger which had high effectiveness with acceptably small pressure losses. The resulting design employs a concentric arrangement of copper discs and rings which are supported in thin walled stainless steel tubing as shown in Figure 4. The discs are provided with 1060 slots each and the rings are each provided with 1500 slots. The thickness of the discs is about 0.05 cm and the space between the discs is about 0.075 cm.

Slots provide a much better rate of heat transfer than holes for given values of mass flow rate, pressure drop and plate thickness. The minimum practical hole pitch to diameter ratio in a perforated plate is about 1.5. When the rate of heat transfer of this hole spacing is compared with the heat transfer rate for slots, the product of the area and the heat transfer coefficient for holes is about 40 percent lower than for slots.

There are six major sources of loss which tend to reduce to performance of a heat exchanger. These are:

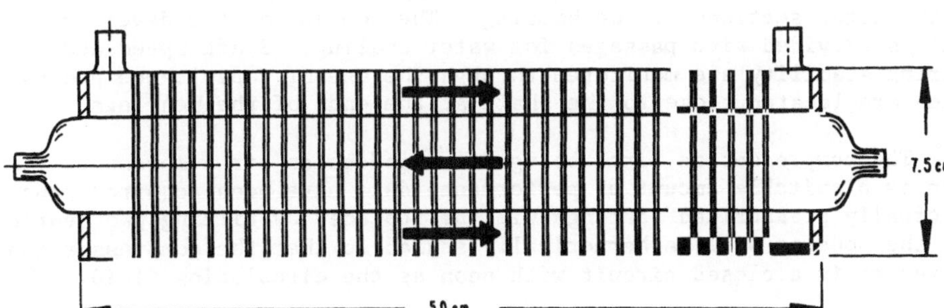

Figure 4. Compact Counterflow Heat Exchanger

1. Gas film thermal resistance.

2. Transverse thermal resistance in the metal between the high pressure and low pressure passages.

3. Axial conduction of heat

4. Step-wise temperature gradient due to the finite number of plates.

5. Flow mal-distribution

6. Pressure drop

The thermal resistance of the gas film can be reduced by reducing the thickness of the film and increasing the number of passages. But this will increase the pressure drops in the high pressure and low pressure sides with a corresponding reduction in the expansion ratio across the turbine. In a good design the loss of refrigeration due to the reduced expansion ratio will be greater than the gain due to the reduced film thickness. The heat exchanger should be designed so that the sum total of the losses approaches a minimum. However, there is a tendency for the volume and the weight to increase rapidly as attempts are made to reduce the sum total of the losses. In general, high effectiveness ratios call for large cross sectional areas and long lengths. The design challenge for the heat exchanger under development is to establish fabrication methods which will accommodate the large number of plates and their conductive attachment to the tubes. The development of this component is just underway.

THE COMPRESSOR

Work is also proceeding on the development of a miniature centrifugal compressor. An assembly layout of a bench test model is shown in Figure 5. The diameter of the impeller is 1.79 cm and the maximum speed is about 8,000 r/s. The impeller is of shrouded type with radial clearance seal surrounding the inlet eye. This seal is the same diameter as the seal at the rear of the impeller to balance the thrusts due to the pressures in the impeller housing. In the bench test configuration, the shaft is supported on externally pressurized journal bearings to facilitate balancing and developmental testing. The axial position is controlled by means of a passive magnetic thrust bearing. The middle portion of the shaft is designed as the rotor of a solid rotor three phase induction motor. The laminations for the stator are fabricated from a low loss magnetic alloy. This material has a very low thermal conductivity. In order to raise the rate of cooling, the laminations are spaced by a number of copper laminations to conduct heat to the outer sections of the housing. The housing of the developmental unit is provided with passages for water cooling. Shaft speed and bearing stability are monitored through the use of capacitance probes which are located close to the shaft at the ends of the bearings.

The compressor is first being developed as an air compressor. As soon as a suitable amount of performance data have been acquired, the externally pressurized bearings will be replaced by tilting pad bearings and the housing will be hermetically sealed so that the compressor can be tested in a closed circuit with neon as the circulating fluid.

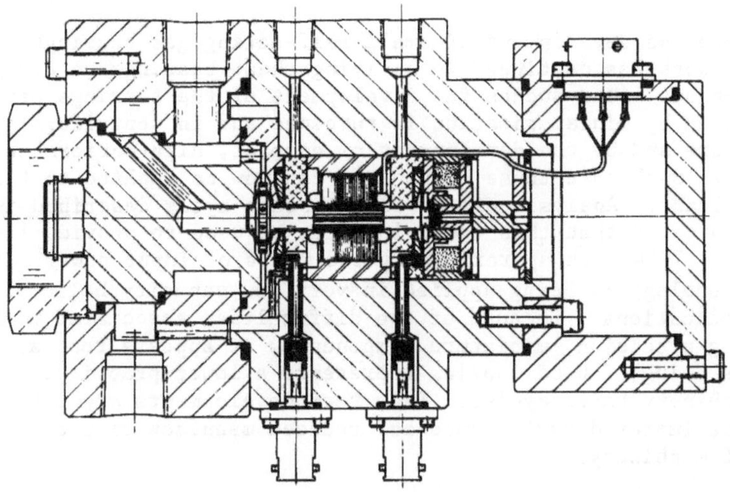

Figure 5. Developmental Centrifugal Compressor

Table 2. Cycle Specifications for a Two Stage MRI Cooler

Fluid	Helium
Shield Cooling Temperature	70 K
Shield Cooling Load	40 w
Liquefaction Rate	0.6 l/hr
Pressure Ratio	10:1
Warm Turbine Size	0.51 cm
Cold Turbine Size	0.51 cm
Warm Turbine Efficiency	0.5
Cold Turbine Efficiency	0.55
Warm Turbine Inlet Temperature	75 K
Cold Turbine Inlet Temperature	15.5 K
Cycle Mass Flow Rate	1.4 g/s
Cycle Input Power	6.4 Kw

TWO STAGE COOLER FOR MRI

The component technology developments described above are being used in several applications. The most significant other cycle being developed at this time will be used to provide shield cooling and liquefaction of helium for a Magnet Resonance Imaging (MRI) superconducting magnet cryostat. Two turboexpanders are incorporated in a Joule-Brayton cycle without external precooling. The turbines are in series to reduce the expansion ratio across each. Table 2 lists important cycle specifications and expected performance. The compressor for this stage is a unique oil flooded positive displacement machine currently under developmental testing. The heat exchangers will be high effectiveness, parallel plate design.

SUMMARY

Since the development of small self-acting gas bearings, announced in 1982, work has continued in applying these bearings to cryogenic turbomachines. The successful development of the bearings has been accompanied by miniaturization in turbines and in centrifugal compressors and by the development of compact, high performance heat exchangers such that these components can now be applied to fairly low cooling loads. Analysis of the single stage cycle described in this paper indicates that this technology can be used to provide 1 watt of cooling at 70 K with approximately 100 watts of input power. Similarly, this technology is being applied in very low capacity helium liquefaction applications. Because of the difficulties associated with producing precision, miniature components, it appears that at least for the time being, 1 W of cooling represents a lower practical bound for turbomachinery based cycles. Near term developments are likely to emphasize improved performance and reduced manufacturing cost for this class of machinery.

ACKNOWLEDGEMENTS

We would like to acknowledge the support of NASA/GSFC in the development of the single stage cooler, DHHS/NIH for support in the development of the MRI cooler and DARPA, AFWAL and AFSTC for their support in some of the underlying technology.

REFERENCES

1. Johnson, A.L., Spacecraft-Borne Long Life Cryogenic Refrigeration Status and Trends, in: "Refrigeration for Cryogenic Sensors", M. Gasser, ed., NASA CP-2287 (1983), p. 47.

2. Sixsmith, H. and Swift, W.L., A Miniature Tilting Pad Gas Lubricated Bearing, "Refrigeration for Cryogenic Sensors", M. Gasser, ed., NASA CP-2287 (1983), p. 189.

3. Valenzuela, J.A., et al., Long-Life, Closed-Cycle Cryocooler for Space, Creare TN-413, presented at 4th International Cryocooler Conference, Easton MD (1986).

SMALL VUILLEUMIER COOLER

Hideto Yoshimura

Mitsubishi Electric Co.
Amagasaki, Hyogo, Japan

Masakuni Kawada

Electrotechnical Laboratory
Tsukuba, Ibaraki, Japan

ABSTRACT

There has recently been much interest in spacecraft borne long lifetime mechanical coolers in Japan, because the cooling loads of IR detectors (IRCCD) for Earth observation instruments have become greater than the cooling capacity of radiant coolers. Therefore active development programs of several types of long lifetime mechanical coolers are in progress. The Vuilleumier cooler is one of the most feasible mechanical coolers for space craft system, because a cooler of this type has the potential advantages of long lifetime operation, small vibration, compactness and low weight. We have designed, built and successfully tested a feasibility model of the spacecraft borne Vuilleumier cooler for the first time in Japan. The major cooler performances are 1.8 watts of cooling capacity at 80 K, 52 K of no-load temperature, 28 minutes of cooldown time, and 130 watts of heater input power.

This paper describes at first the principal design features of the small Vuilleumier cooler (The feasibility model of the spacecraft borne Vuilleumier cooler) and then performance test results. This paper further describes the calculation method of the thermal performance and the comparison between calculations and experimental data.

INTRODUCTION

There has recently been much interest in spacecraft borne long lifetime mechanical coolers in Japan, because the cooling loads of IR detectors (IRCCD) for Earth observation instruments have become greater than the cooling capacity of radiant coolers. Therefore active development programs of several types of long lifetime mechanical coolers are in progress.

The Vuilleumier cooler is a heat-driven machine patented by Rudolph Vuilleumier in the United States in 1918. A cooler of this type has the potential advantages of long lifetime operation, small vibration, compactness and low weight. Therefore much effort has been done in developing reliable calculation method of its performance and in the development of long lifetime design concepts for the spacecraft system in United States[1-4] and in another countries.[5]

However there has been little experience and therefore only limited knowleges about the Vuilleumier cooler in Japan. So we have built a feasibility model of the spacecraft borne Vuilleumier cooler for the first time in Japan. Our objectives are development of reliable calculation method of Vuilleumier cooler's performance, demonstration of advantages of the Vuilleumier cooler and confirmation of feasibility of the Vuilleumier cooler for the spacecraft system.

SYSTEM DESCRIPTION

The useful lifetime of conventional mechanical coolers are limited by two major factors : wear and outgassing. Generally wear is small in the Vuilleumier coolers, because the pressures are nearly equal throughout the system at any moment and mechanical loads on bearings and seals are very small. To prevent outgassing we adopted a clean motor developed for use of high vacuum environment. We also adopted high vacuum grease for ball bearings located in the crankcase and baking procedure before filling with clean helium to eliminate helium gas contamination.

The small Vuilleumier cooler (The feasibility model of a spacecraft borne Vuilleumier cooler) was designed to provide over 1 W of cooling capacity at 80 K. A photograph and a cross section of the cooler are shown in Fig. 1. Heat is supplied by a electric heater and rejected by water. The cooler is driven by a small stepping motor with pullout torque of 46 N*cm at operating speed. Drive mechanism is a simple crank-type.

The hot cylinder is made of SUS 316. A electric heater is wound around the hot cylinder head. A water tube made of copper is also wound and brazed around the bottom of the hot cylinder. The cold displacer is made of a phenolic resin tube filled with 200 mesh stainless screen disks 0.85 cm diameter by 10.7 cm long. The hot displacer is made of a machinable ceramics tube with low thermal conductivity filled with 60 mesh stainless screen disks 1.0 cm diameter by 9.2 cm long. A rider ring made of VESPEL SP-21 [DUPONT] is mounted at lower part of the hot displacer. The crankcase is made of aluminium alloy to achieve low weight. Inside of the crankcase heat exchanger by which helium gas gets cooled is soldered. Kap-seals of 15-percent glass loaded teflon ensure that helium gas flows through regenerators, and prevent excessive leakage between the working volume and the crankcase volume. The thermal insulator of the hot cylinder is made of calcium Silicate heat insulating material.

The thermodynamic design values of the small Vuilleumier cooler are shown in Table 1.

Displacer positions are measured using a rotary encoder that is attached to the drive motor. The alternating pressure in the system is measured using a pressure transducer mounted at the heat exchanger. From both signals of the rotary encoder and the pressure transducer we measured the PV diagram thus the gross cooling capacity produced by the machine. The temperature at the cold head is measured using caliblated thermocouples soldered on the cold head. A heater coil wound round the cold head provides the means of applying an external load to measure the net cooling capacity produced by the cooler. The temperature of the hot cylinder wall is monitored and controlled by a thermocouple brazed on the hot cylinder head.

(a) Photograph

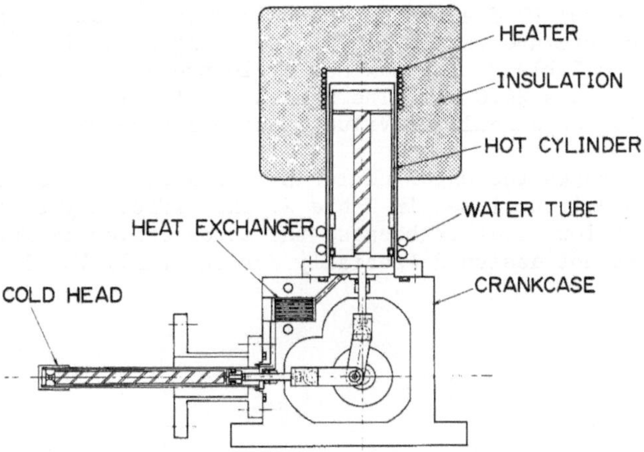

(b) Cross Section

Fig. 1. Photograph and cross section of small Vuilleumier cooler.

Table 1. Thermodynamic Design Values

Hot displacer diameter	, mm	40
Cold displacer diameter	, mm	10
Stroke	, mm	10
Hot displaced volume	, cm^3	12.6
Cold displaced volume	, cm^3	0.79
Reduced dead volume at 80 K	, cm^3	4.45
Phase	,	90

PERFORMANCE

Fig. 2 shows the cooldown performance of the small Vuilleumier cooler. The operating condition is listed in Table 2. The warmup time of the hot cylinder wall was about 9 minutes and the cooldown time of the cold head was about 28 minutes. The no-load temperature of the cold head was 52 K.

Fig. 3 shows the cooling capacity performance. The operating condition is listed in Table 2. The gross cooling capacity and the net cooling capacity at 80 K were 4.3 W and 1.8 W respectively. The difference between the gross cooling capacity and the net cooling capacity (2.5 W) means total heat losses. Dashed lines in the figure are calculated values about which we will discuss later.

Fig. 4 shows PV diagram of the cold volume that was measured when the cold head temperature was 80 K. The operating condition was the same as that listed in Table 2. The maximum cycle pressure was 34.0 atm and minimum cycle pressure was 26.0 atm. The pressure ratio is 1.3. The dashed line in the figure is the calculated value about which we will discuss later.

Table 3 shows the major cooler performance that was measured under the operating condition listed in Table 2. The efficiency of the motor driver was extremely low. That is because the motor driver is for general use and therefore was not designed specially for our small Vuilleumier cooler.

Table 2. Operating Condition

Charge pressure	, atm	30
Cycle speed	, rpm	540
Hot wall temperature	, K	923
Intermediate temperature	, K	300

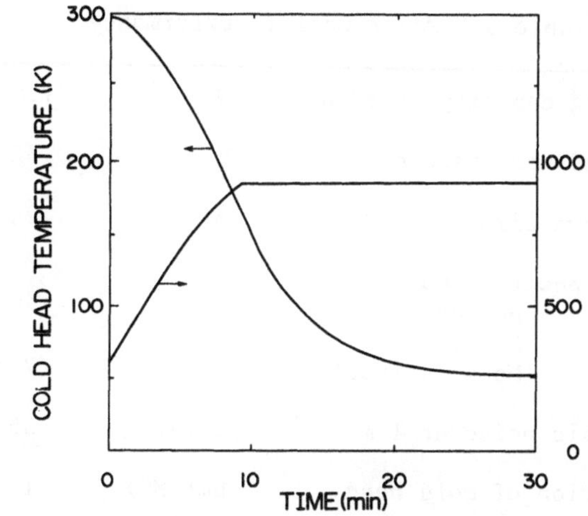

Fig. 2. Cooldown performance.

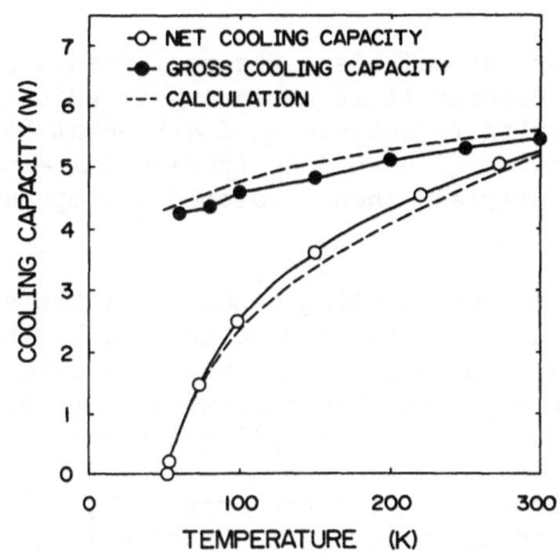

Fig. 3. Cooling capacity performance.

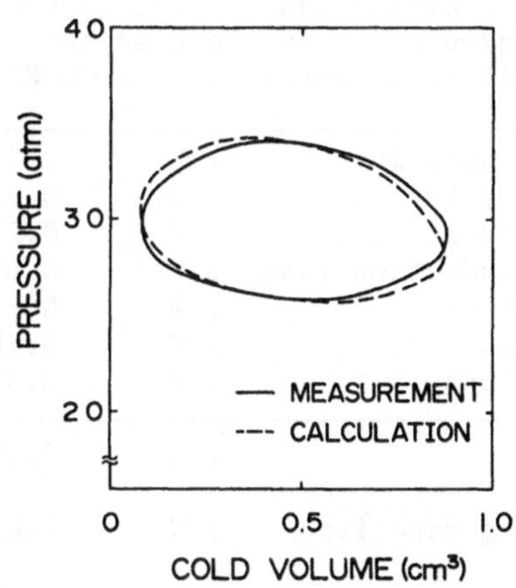

Fig. 4. PV diagram of the cold volume.

Table 3. Major Cooler Performance

Cooling capacity at 80 K	, W	1.8
No-load temperature	, K	52
Cooldown time	, min	28
Input power at 80 K	, W	
Heater		130
Motor		5
(Motor driver)		(75)
Acoustic noise at 1 m	, dB(A)	55
Vibration of cold head	, μm(RMS)	14
Weight	, kg	6

PERFORMANCE CALCULATION

The basic procedure to calculate the performance of the Vuilleuimer cooler is well known[1]. However it is necessary to adjust some parameters using experimental results to achieve good agreement between calculations and measurements. At first we have done fundamental experiments to estimate each heat losses correctly and then developed a computer program using the experimental results.

PV diagrams thus a gross cooling capacity at the cold volume and a indicated work at the hot volume are calculated by the Schmidt theory. A net cooling capacity is calculated by subtracting various heat losses from the gross cooling capacity. The following seven different losses are computed separately based on the decoupling assumption.

(1) Regenerator heat loss, (2) Pressure drop loss, (3) Cylinder conduction loss, (4) Displacer conduction loss, (5) Shuttle loss, (6) Pumping loss and (7) Insulation loss. A total input power is calculated by adding various heat losses to the indicated work at the hot volume. The heat loss items are same as that of the cold head.

Table 4. Heat Losses and Cooling Capacity (Calculation in the case that Cold Head Temperature is 80 K)

Regenerator heat loss	, W	0.55	
Pressure drop loss	, W	0.92	
Cylinder conduction loss	, W	0.27	
Displacer conduction loss	, W	0.01	
Shuttle loss	, W	0.61	
Pumping loss	, W	0.58	
Insulation loss	, W	0.01	
Total loss	, W	2.95	
Gross cooling capacity	, W	4.63	
Net cooling capacity	, W	1.68	(1.80*)

* Measurement

Table 5. Heater Power Requirement (Calculation in the case that Cold Head Temperature is 80 K)

Regenerator heat loss	, W	10.8
Pressure drop loss	, W	10.4
Cylinder conduction loss	, W	27.7
Displacer conduction loss	, W	9.6
Shuttle loss	, W	8.7
Pumping loss	, W	1.8
Insulation loss	, W	27.0
Total loss	, W	96.0
Indicated work	, W	19.0
Total input power	, W	115.0 (130.*)

* Measurement

The dashed lines in Fig. 3 and Fig. 4 are calculated values that were computed by the program. Table 4 shows the calculated values of heat losses and cooling capacity in the case that the operating condition of the cooler is the same as that listed in Table 2 and the cold head temperature is 80 K. The calculated net cooling capacity is 1.68 W and the measured value is 1.8 W. Table 5 shows the calculated values of heat losses, indicated work and total input power in the case that the operating condition of the cooler is the same as that listed in Table 2 and the cold head temperature is 80 K. The calculated total input power is 115 W and the measured value is 130 W .The agreements between measurements and calculations are quite well. This agreement confirmed that the calculation method is effective and applicable to the thermal design of the Vuilleumier cooler.

CONCLUSION

The small Vuilleumier cooler (The feasibility model of the spacecraft borne Vuillmeuier cooler) is designed, fabricated and successfully tested for the first time in Japan. We are going to perform a long term operation of our machine to demonstrate the long lifetime possibility of the Vuilleumier cooler.

The calculation method of the thermal performance has been developed. Calculations agree well with measurements. This agreement confirmed that the calculation method is adequate to guide the design of the Vuilleumier cooler.

At present the small Vuilleumier cooler needs a larger electric input power per unit cooling capacity than the Stirling cycle cooler. However, it is possible to chose another type of power input for the Vuilleumier cooler such as solar and isotope power.

REFERENCES

1. G. Walker, " Cryocoolers Part 1: Fundamentals " , Plenum Press, New York (1983), p.185
2. G. K. Pitcher and F. K. du Pre, Miniature Vuilleumier Cycle Refrigerator, in: "Advances in Cryogenic Engineering," Vol. 15, Plenum Press, New York (1970), p.447

3. A. Sherman, Selected Vuilleumier Refrigerator Performance Characteristics, in: "Advances in Cryogenic Engineering," Vol. 18, Plenum Press, New York (1972), p.352
4. M. S. Crouthamel and B. Shelpuk, A Combustion-Heated Thermally Actuated Vuilleumier Refrigerator, in: "Advances in Cryogenic Engineering," Vol. 18 Plenum Press, New York (1972), p.339
5. J. Guolin, X. Chiqiong and X. Miaogen, A Test and Analysis of a Vuilleumier Refrigerator with Magnetic Linkage, in: "Proc. 11th Intl. Cryo. Engr. Conf.,"Butterworth, Guildford, UK (1986), p.300

SMALL SCALE FREE DISPLACER STIRLING CRYOCOOLER

C. Heiden and G. Reich

Institut für Angewandte Physik
Justus-Liebig-Universität Gießen
Gießen, Federal Republic of Germany

ABSTRACT

A small Stirling cryocooler was built consisting of a three stage plastic displacer unit similar to J. Zimmerman's design and a ceramic compressor. The whole machine is hermetically sealed using dye shaped stainless steel bellows. By counter balancing the weight of the displacer, it was possible to run the machine in the free displacer mode at stroke frequencies of the order of 1 Hz with cooling performance similar or even better (due to non sinusoidal motion) than in the case of the crankshaft driven displacer. This feature allows for simpler design and more versatility in placing the displacer unit with regard to the compressor. Magnetic interference from the latter can be reduced by increasing its distance to the displacer, which is of importance when cooling SQUIDs. The performance under various running conditions was tested.

INTRODUCTION

Coolers for cryoelectronic devices like SQUIDs have to meet rather stringent specifications concerning electromagnetic interference noise and vibrations. One design philosophy, adopted first by Zimmerman and coworkers[1], was to build a low power Stirling cryocooler using plastic materials like glass fiber reinforced epoxy for the displacer unit. Such coolers were then built by several groups[2,3,4] demonstrating the feasibility of this approach but also its inherent draw backs. One handicap so far has been the motion control of the displacer that was coupled either mechanically to the crankshaft of the compressor drive unit, or was driven by a computer controlled servo motor[3], or was moved pneumatically. The two latter methods have the advantage to decouple locally the displacer unit from the compressor thereby reducing greatly interference signals, that are generated by the compressor. Here we report on another approach towards the same goal that has been used for a long time in small Stirling cryocoolers for infrared sensors: The split cycle free displacer machine, in which the displacer is driven by the pressure wave of the working fluid. This leads to a rather simple construction in which the only connection between compressor and displacer unit is the tube for the working fluid, which, depending on the volume of the compressor, can be made comparatively long.

DETAILS OF CONSTRUCTION

The displacer unit is made similar to designs of J. Zimmerman[1]. To implement gap regeneration, a nylon displacer is machined to fit with close tolerance in a surrounding cylinder of glas fiber reinforced plastic of 2 mm wall thickness. To enlarge the surface effective for regeneration, axial slots of 0.1 mm width and a depth of 4 mm were sawed into the displacer. A unit with three steps was made. The radial clearance between displacer and sleeve was of the order of 30 µm. The mass of the displacer is balanced by a counter mass attached to a lever (Fig. 1). The connecting rod between lever and displacer is part of a close tolerance (~10 µm) clearance seal. The whole unit is hermetically sealed against the atmosphere by a dye shaped bellows, that is operated in the bending mode leading to a very long life time (>10^8 strokes). The maximum stroke of the displacer is 10 mm, it can be reduced by setting screws of a stroke limiter at the lever. The dimensions of the displacer are the same as in Ref. 2.

This unit was driven by the pressure wave from a compressor that was designed for a long life with negligible gas contamination. Cylinder and piston of the compressor are made of alumina (Fig. 2), and so are the integrated parts of the clearance seal, piston shaft and associated ceramic bore. The driving force is transmitted via a lever from a crankshaft in order to allow, in similar fashion as with the displacer unit, for a hermetic enclosure of the compressor using a stainless steel bellows seal. The ceramic compressor cylinder itself is encased in a stainless steel sleeve which is anchored in two needle bearings to allow for the necessary pendulum motion during the piston stroke (see Fig. 3).

Compressor and displacer unit are connected by means of copper tubing with 2 mm inner diameter. To monitor the performance of the cooler, platinum resistance thermometers, a silicon pressure transducer and an inductive position sensor for the displacer stroke are used.

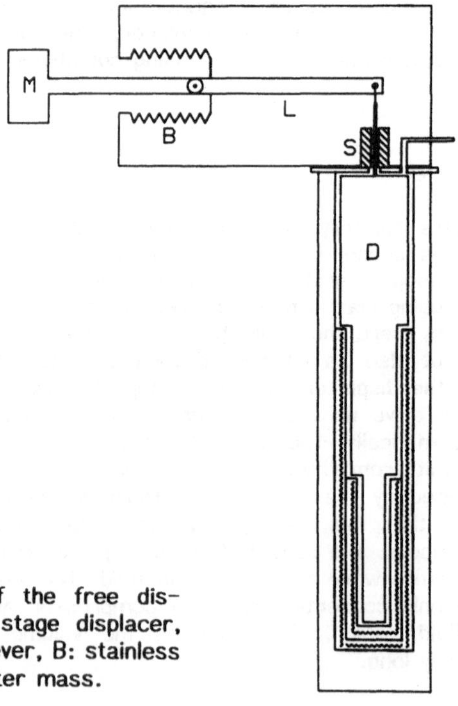

Fig.1. Schematic drawing of the free displacer unit. D: three stage displacer, S: clearance seal, L: lever, B: stainless steel bellows, M: counter mass.

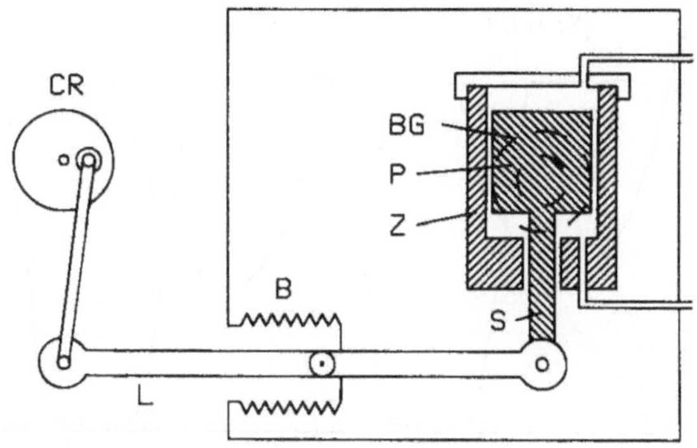

Fig. 2. Sketch of the ceramic compressor unit. CR: crankshaft, L: lever, B: stainless steel bellows, S: ceramic shaft , Z: ceramic cylinder, P: ceramic piston, BG: needle bearings.

Fig. 3. Photo of the compressor unit. The stainless steel case (lower right) is taken off to show the stainless steel sleeve of the ceramic compressor unit.

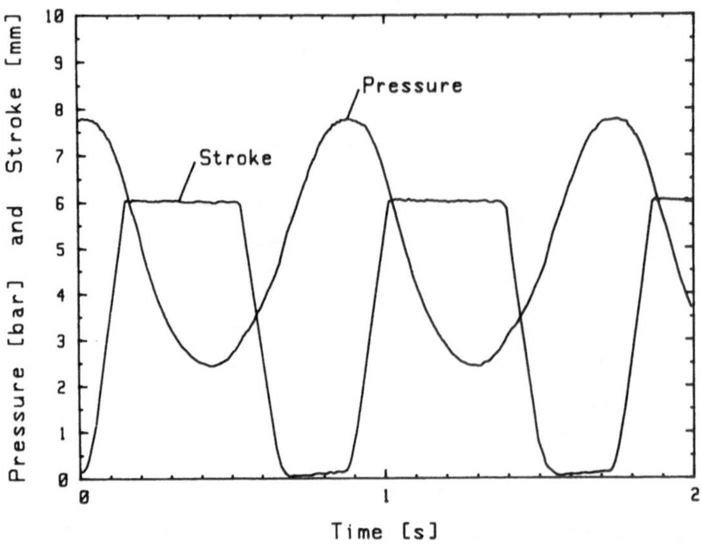

Fig. 4. Stroke and pressure diagram of the free displacer unit.

PERFORMANCE

The cooler was tested, varying among others the parameters of the pressure wave: average pressure $\overline{p}$, pressure ratio $r = p_{max}/p_{min}$ (influenced by the length of the connecting tube) and the stroke frequency. Helium was used as working fluid. The stroke frequency could be varied between 0.6 Hz and 2.2 Hz. With proper balancing of the displacer mass, a pressure wave driven displacer motion could be realized for all stroke frequencies in the above mentioned range that led to cooling. In Fig. 4 the pressure at the gas inlet of the displacer unit and the position of the displacer piston are plotted as function of time for a particular set of parameters. The phase shift between pressure wave and position depends mainly on the masses of piston and counter weight and on the gas friction in the regenerator gap. The cooling behavior, obtained for another set of parameters is shown in Fig. 5. An almost linear cool down behavior with a comparatively high cooling rate at the cold tip of about 100 K/h is observed down to temperatures near 70 K, below which the cooling rate slows down. Lowest end temperatures for this unit were 22 K.

The cooling rate for the free displacer was found to be noticeably higher compared to the case of the crankshaft driven displacer at otherwise similar working conditions. This is attributed to the more effective nonsinusoidal motion of the free displacer (Fig. 4).

Important for good cooling performance appears to be a sufficient light construction of the displacer. If the moment of inertia becomes too high, a more sinusoidal motion is observed leading to a degraded performance. It may be worth mentioning that cooling here is observed for both senses of rotation of the crankshaft, in contrast to the case of the crankshaft driven displacer.

Lower end temperatures may be obtained by increasing the number of stages of the displacer unit. Perhaps it may also be possible to use a conical displacer, or, better, a combination of cylindrical stepped displacer with a conical section.

CONCLUSION

Plastic free displacer split Stirling cryocoolers can be operated satisfactorily at low stroke frequencies of the order of 1 Hz. This facilitates a better local separation from

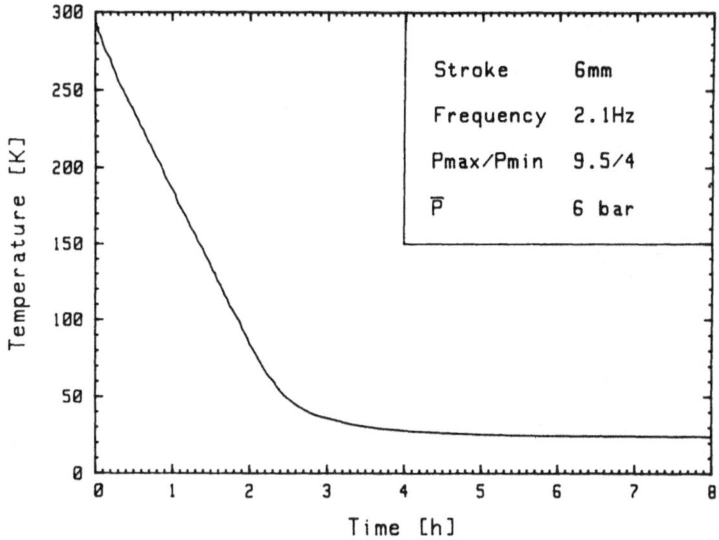

Fig. 5. Typical cool down behavior of the displacer unit.

the compressor thereby reducing significantly magnetic interference noise and gaining flexibility in the placement of the displacer unit, which may add an attractive feature for using such a cooler for SQUIDs.

ACKNOWLEDGEMENT

We gratefully acknowledge valuable assistance of Mr. R. Dittrich during this work.

REFERENCES

1. J. E. Zimmerman, D. B. Sullivan, "A Study of Design Principles for Refrigerators for Low-Power Cryoelectronic Devices", NBS Technical Note 1049, U.S. Department of Commerce, National Bureau of Standards, Washington D.C. (1982).
2. C. Heiden, Miniature refrigerators for cryoelectronic sensors, in: "Advances in Solid State Physics XXIV", P. Grosse ed., Vieweg, Braunschweig (1984).
3. N. Lambert, Simple construction and performance of a conical plastic cryocooler, in: "Proceedings of the Third Cryocooler Conference", NBS Special Publication 698, Ray Radebaugh, Beverley Louie, and Sandy McCarthy eds., National Bureau of Standards, Boulder (1985).
4. Chen Guobang, in: "Proceedings of the 4th International Cryocooler Conference", to be published.

DEVELOPMENT AND EXPERIMENTAL TEST OF AN ANALYTICAL MODEL OF THE ORIFICE PULSE TUBE REFRIGERATOR*

Peter J. Storch and Ray Radebaugh

Chemical Engineering Science Division
National Bureau of Standards
Boulder, Colorado

ABSTRACT

The promise of high reliability and high refrigeration capacity is responsible for a recent surge of interest in pulse tube refrigeration. This work involves the development of an analytical model describing behavior of the orifice pulse tube to gain a better understanding of the refrigeration process. Due to oscillating gas flow, the system is described in terms of average enthalpy flow with such simplifying assumptions as an ideal gas and sinusoidal pressure variation. Phasor analysis is used to represent the temperature, pressure, and mass flow rate waves in vector form. Also discussed in this paper is the verification of the model in which analytical predictions are compared to experimental measurements. The results confirm predictions by the model that refrigeration power is proportional to the average pressure, the pulse frequency, the mass flow ratio, and the square of the dynamic pressure ratio. Also, a temperature probe was devised to measure the average temperature profile and dynamic temperature in the tube. As a result of simplifying assumptions, magnitudes of refrigeration power from the model are between 3 and 5 times greater than experimental values.

INTRODUCTION

The need for reliable cryocoolers is a well known fact today, particularly in space applications. The pulse tube refrigerator has potential for high reliability because it only has one moving part, the compressor, and operates at low pressures and pressure ratios. High refrigeration capacity and good intrinsic efficiency are other advantages which make the pulse tube a desirable alternative to other cryocoolers. At the National Bureau of Standards (NBS), a low temperature of 60 K has been reached using a single stage pulse tube with an orifice.[1] This system produced 12 W of refrigeration power with an efficiency of 40% of Carnot at 80 K and a frequency of 6 Hz.

The pulse tube refrigerator consists of a compressor at room temperature, the pulse tube with a heat exchanger at each end, and a regener-

*Research funded by NASA Ames Research Center under contract no. A-34964C(RCW).

ator between the compressor and the cold end of the tube. These components make up a system called the basic pulse tube, which was studied extensively by Gifford and coworkers.[2] In 1984 Mikulin inserted an orifice at the closed end of the tube which allowed gas to flow into and out of a large reservoir volume.[3] Figure 1 shows the NBS version of this new pulse tube and is referred to as the orifice pulse tube. The refrigeration cycle begins as the piston moves forward and the gas is cooled to the cold end temperature as it flows through the regenerator and into the tube. The gas in the tube is compressed adiabatically and is heated as it travels toward the closed end. During this high pressure period, heat is rejected from the system in the hot end heat exchanger. In the basic pulse tube, gas in the tube is cooled by transferring heat with the tube wall, while in the orifice pulse tube the gas is cooled by adiabatic expansion due to flow through the orifice. The piston then moves back and gas flows out of the tube and back through the regenerator. Gas in the tube is then cooled due to adiabatic expansion and gas flowing through the cold end heat exchanger absorbs heat from the refrigeration load. The cycle ends with a low pressure period during which gas in the tube is heated in the basic pulse tube by heat transfer from the tube wall. In the orifice pulse tube, gas in the tube is heated by adiabatic compression due to flow through the orifice. The cycle results in an average enthalpy flow from the cold end to the hot end which establishes a constant temperature gradient in the tube and provides a continuous refrigeration effect.

Previous experimental work indicates that the orifice creates a much greater enthalpy flow in the tube than heat transfer to the tube wall.[1] Therefore, the subject of this study is the orifice pulse tube and the purpose is to develop an analytical model of the refrigeration power. Such a model will identify the various parameters which influence performance and will be useful in optimization and design.

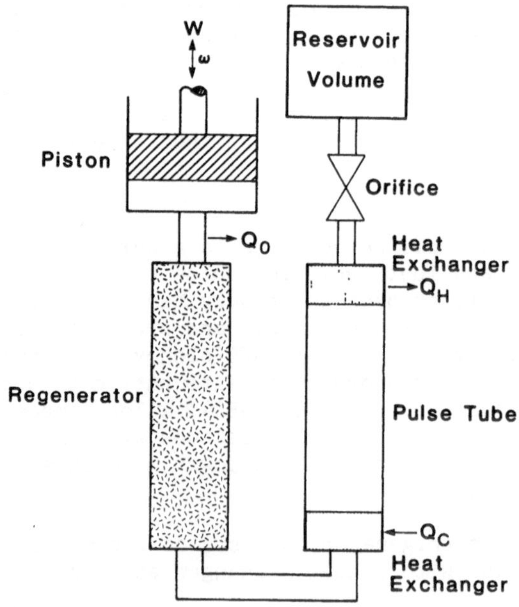

Figure 1. Schematic of the orifice pulse tube refrigerator.

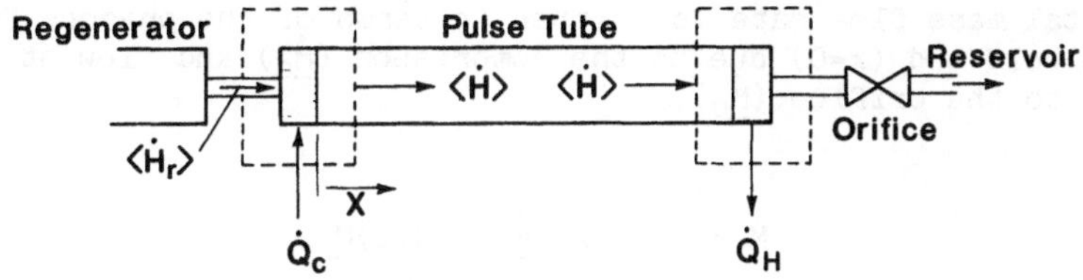

Figure 2. Energy balance system showing time averaged heat and enthalpy flows.

ENTHALPY FLOW ANALYSIS

The system to be analyzed for enthalpy flow is shown in figure 2 and consists of the working fluid in the tube. Due to the oscillating gas flow, enthalpy flows are time-averaged over one cycle. Applying the first law of thermodynamics to the cold end of the tube gives the following expression for the heat absorbed by the system

$$\dot{Q}_c = \langle\dot{H}\rangle - \langle\dot{H}_r\rangle, \tag{1}$$

where $\langle\dot{H}\rangle$ is the average enthalpy flow in the tube and $\langle\dot{H}_r\rangle$ is the average enthalpy flow from the regenerator. Since flow through the orifice is isothermal for an ideal gas, the first law at the hot end says that the heat rejected from the system is equal to the enthalpy flow in the tube,

$$\dot{Q}_h = \langle\dot{H}\rangle. \tag{2}$$

The energy balances reveal two interesting points about the pulse tube refrigerator. First, unlike other refrigerators both the net refrigeration, $\dot{Q}_c$, and the gross refrigeration, $\dot{Q}_h$, can be measured. Second, since there is no addition of heat at any point along the tube, the enthalpy flow must be constant in this region.

The average enthalpy flow over one cycle assuming an ideal gas is given by

$$\langle\dot{H}\rangle = (C_p/\tau) \oint_0^\tau \dot{m}T_d \, dt \tag{3}$$

where τ is the period of the cycle, C_p is the heat capacity, $\dot{m}$ is the mass flow rate, and T_d is the dynamic temperature.

PHASOR ANALYSIS

In order to easily describe the oscillations in the tube, the pressure and mass flow rate waves are assumed to be sinusoidal. This assumption is good for small pressure ratios and greatly simplifies the mathematics involved in the analysis. The sine waves are then represented in the frequency domain by stationary vectors or phasors. The magnitude and the angle of a phasor are equal to the amplitude and phase of the sinusoid respectively. Phasors are represented here by underlined capital letters. For further details on the phasor analysis and any other aspect of the model, see reference 4.

The total mass flow rate is written in terms of the phasors for flow at the cold end (x=0) due to the compressor ($\underline{\dot{M}}_c$) and flow at the hot end due to the orifice ($\underline{\dot{M}}_o$),

$$\underline{\dot{M}} = \{1-X(x)\}\underline{\dot{M}}_c + X(x)\underline{\dot{M}}_o, \tag{4}$$

where X(x) is the ratio of mass in the tube up to point x to the total mass. The term X(x) satisfies the conditions X(0) = 0 and X(1) = 1 and is a function of the temperature profile in the pulse tube. The phasor diagram in figure 3a demonstrates how $\underline{\dot{M}}_o$ lags $\underline{\dot{M}}_c$ by the phase angle θ.

The pressure wave in the tube consists of a static and a dynamic component and is assumed to be only a function of time. The dynamic pressure is written as the sum of two phasors: the pressure due to flow from the compressor ($\underline{P}_c$) and the pressure due to flow through the orifice ($\underline{P}_o$),

$$\underline{P}_d = \underline{P}_c + \underline{P}_o. \tag{5}$$

Figure 3b shows the pressure phasor diagram and in general, pressure will lag mass flow rate by 90 degrees as shown with $\underline{P}_c$ and $\underline{\dot{M}}_c$. In the case of $\underline{P}_o$, there is an additional 180 degree lag since $\underline{\dot{M}}_o$ is decreasing the pressure in the tube when $\underline{\dot{M}}_c$ is increasing the pressure. Flow through the orifice is assumed to be laminar, which means $\underline{\dot{M}}_o$ is proportional to the total pressure and $\underline{P}_d$ has the same phase angle θ as $\underline{\dot{M}}_o$.

The dynamic temperature is derived from the following energy balance on an element of gas in the tube assuming no heat transfer between the gas and the tube wall

$$d(\rho A_g u)/dt + d(\dot{m}h)/dx = 0, \tag{6}$$

where ρ is the gas density, A_g is the cross-sectional area of the tube, u is the specific internal energy, and h is the specific enthalpy. Using the continuity equation, the definition of internal energy, and the definition of enthalpy for an ideal gas, a differential equation in terms of temperature and pressure is obtained,

$$\partial T/\partial t = (RT/C_pP)\ \partial P/\partial t - (RT\dot{m}/A_gP)\ \partial T/\partial x, \tag{7}$$

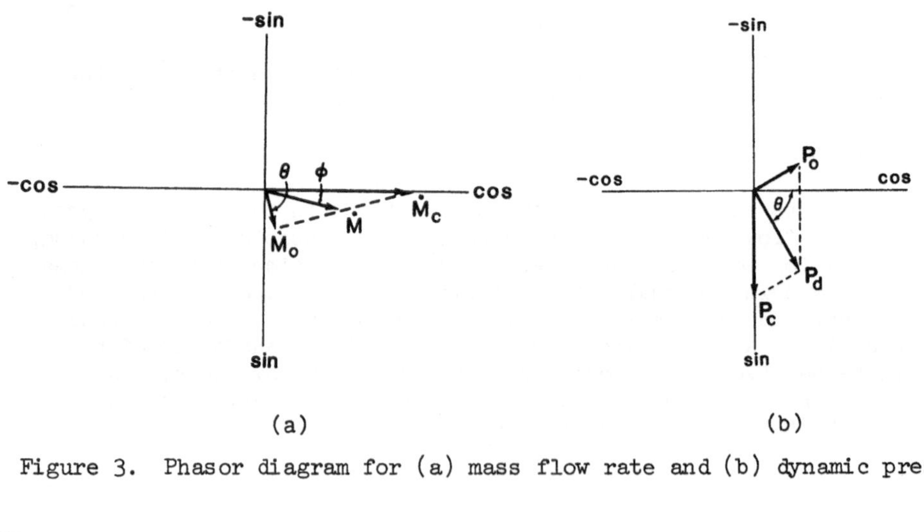

Figure 3. Phasor diagram for (a) mass flow rate and (b) dynamic pressure.

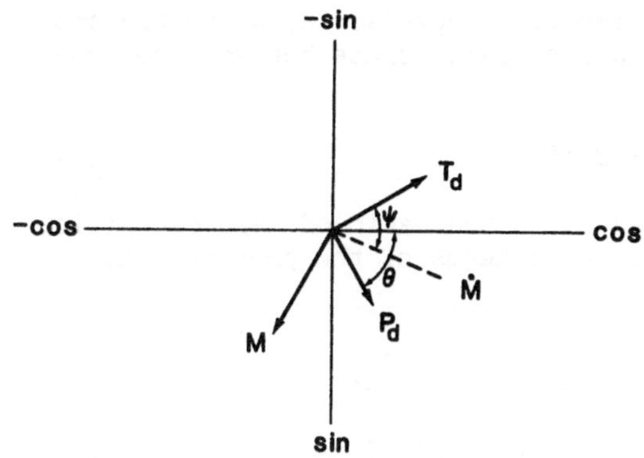

Figure 4. Phasor diagram for dynamic temperature.

where R is the gas constant. For small pressure ratios, $\Delta P/P$ and $\Delta T/T$ are small and T and P are equal to the average values T_a and P_a. Equation (7) is then easily integrated over time to arrive at the phasor representation of the dynamic temperature in terms of $\underline{P}_d$ and $\underline{M}$, the mass moved past point x,

$$\underline{T}_d = (RT_a/C_pP_a)\ \underline{P}_d - (RT_a/A_gP_a)(\partial T_a/\partial x)\underline{M}. \quad (8)$$

Figure 4 shows the phasor diagram for the dynamic temperature.

ENTHALPY FLOW EQUATIONS

The average enthalpy flow in the tube is found according to equation (3) by integrating the product of $\dot{m}(x,t)$ and $T_d(x,t)$ over one cycle in the time domain,

$$\langle\dot{H}\rangle = 1/2\ RT_a\ (P_d/P_a)\ [(1-X)\dot{m}_c\cos\theta + X\dot{m}_o]. \quad (9)$$

An expression for $\cos\theta$ is written in terms of the mass flow ratio using figure 3b where T_{cp} is the compressor temperature and T_c is the cold end temperature,

$$\cos\theta = P_o/P_c = (\dot{m}_o/\dot{m}_c)(T_{cp}/T_c) \quad (10)$$

Then using the following equation for the mass flow rate at the cold end where V_{cp} is the fixed swept (total) volume of the compressor and ν is the frequency,

$$\dot{m}_c = (P_aV_{cp}/RT_{cp})\pi\nu, \quad (11)$$

the enthalpy flow is put in the dimensionless form

$$\langle\dot{H}\rangle/P_aV_{cp}\nu = (\pi/2)(P_d/P_a)(\dot{m}_o/\dot{m}_c)(T_a/T_{cp})\ [(1-X)(T_{cp}/T_c) + X]. \quad (12)$$

The product of the terms involving T_a and X in equation (12) cancel and the model is independent of position,

$$\langle\dot{H}\rangle/P_aV_{cp}\nu = (\pi/2)(P_d/P_a)(\dot{m}_o/\dot{m}_c). \quad (13)$$

Therefore the model is consistent with the first law which requires a constant enthalpy flow in the tube.

The following equation is applied for a reversible, adiabatic process with a small pressure ratio and zero dead volume between the compressor and the tube,

$$(P_d/P_a) = 1/2\ \gamma(V_{cp}/V_t) \tag{14}$$

where V_t is the pulse tube volume and γ is the ratio of heat capacities. Two dimensionless forms for a fixed V_t in terms of the pressure ratio and the volume ratio are

$$\langle\dot{H}\rangle/P_a V_t \nu = \pi(1/\gamma)(P_d/P_a)^2\ (\dot{m}_o/\dot{m}_c), \tag{15}$$

$$\langle\dot{H}\rangle/P_a V_t \nu = (\pi/4)\gamma(V_{cp}/V_t)^2\ (\dot{m}_o/\dot{m}_c). \tag{16}$$

The average enthalpy flow given in equations (13), (15), and (16) represent the analytical model for the gross refrigeration power produced by the orifice pulse tube.

DISCUSSION OF RESULTS

In order to verify the model, various parameters in the model were investigated experimentally to determine their affect on refrigeration power. The test apparatus consisted of a pulse tube of length 130 mm and diameter 19.1 mm with an adjustable orifice. Temperature and mass flow rate were measured at the cold end of the tube while pressure and temperature were measured at the hot end. Equation (15) says that for a constant pulse tube volume the refrigeration power will be proportional to the average pressure P_a, the frequency ν, the mass flow ratio m_o/m_c, and the square of the dynamic pressure ratio P_d/P_a. Experimental results presented in figures 5 and 6 show that the model is successful in predicting the dependence of performance on the important parameters. Also, a comparison of measurements with nitrogen and helium show that

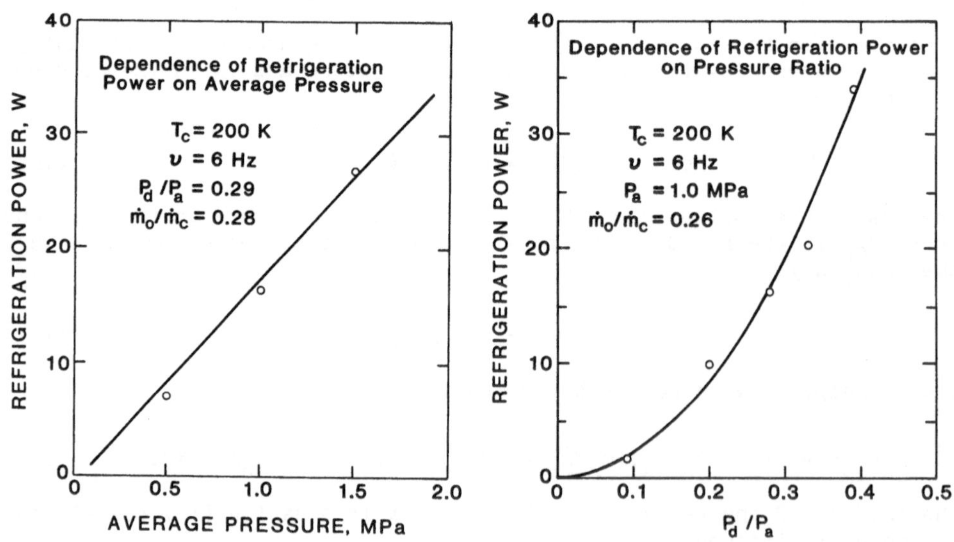

(a) (b)

Figure 5. Experimental measurements of refrigeration power as a function of a) average pressure and b) dynamic pressure ratio.

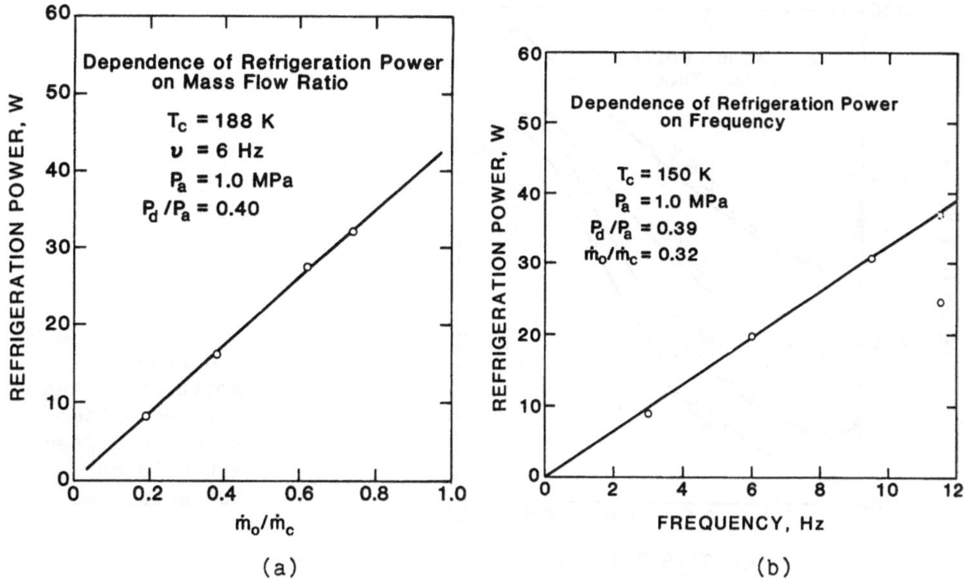

Figure 6. Experimental measurements of refrigeration power as a function of a) mass flow ratio and b) frequency.

equation (15) is correct in predicting that for a fixed volume ratio V_{cp}/V_t, helium with a larger heat capacity ratio is the best working fluid.

The temperature wave was measured at various positions along the tube with a thin foil thermocouple for a fast response. Figure 7 displays the average temperature profiles which have a steep slope with linear behavior in the middle of the tube. The model was found to be thermodynamically consistent by predicting constant enthalpy flow along the tube. However, theoretical values for the refrigeration power were between 3 and 5 times greater than the experimental measurements. This result is attributed to the assumptions of sinusoidal behavior and no end effects. Figure 8 shows the deviation of the actual temperature wave from sinusoidal behavior and that the experimental T_d is much smaller at the two ends than predicted by the sine wave. The model also does not account for dead volume in the system which will decrease the temperature change produced by flow through the orifice. The ratio of theoretical refrigeration power to the experimental value as a function of cold end temperature is shown in figure 9 which can be used to correct for the ideal assumptions in the model.

CONCLUSIONS

The analytical model presented here for the orifice pulse tube refrigerator results in a simple expression for the gross refrigeration power which agrees with experiments regarding the dependence on all the parameters. The magnitude predicted by the model is 3-5 times higher than experiment and is presumably due to the simplifying assumptions used in the model.

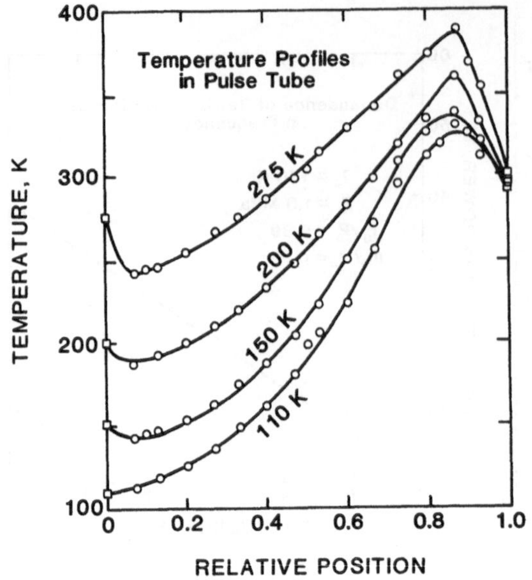

Figure 7. Average gas temperature as a function of position in the orifice pulse tube for different cold end temperatures.

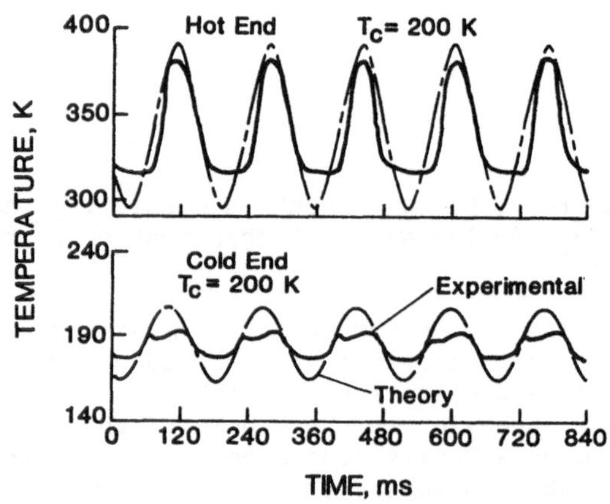

Figure 8. Comparison of experimental temperature wave with theoretical sine wave at the cold and hot ends.

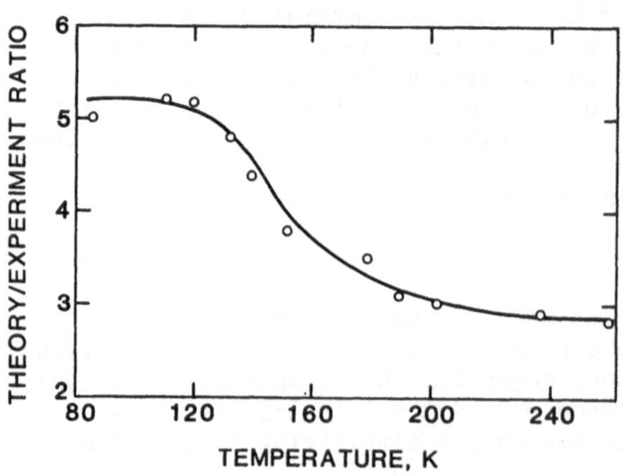

Figure 9. Ratio of theoretical to experimental refrigeration power as a function of cold end temperature.

REFERENCES

1. S. Herrmann and R. Radebaugh, Measurements of the Efficiency and Refrigeration Power of Pulse-Tube Refrigeration, National Bureau of Standards Tech. Note 1301, Sept. (1986).

2. W.E. Gifford and R.C. Longsworth, Pulse Tube Refrigeration, Trans. of the ASME, 63:264 (1964).

3. E.I. Mikulin, A.A. Tarasov, and M.P. Shkrebyonock, Low Temperature Expansion Pulse Tubes, "Advances in Cryogenic Engineering," Plenum Press, New York Vol. 29 (1984) p. 629.

4. P.J. Storch and R. Radebaugh, Analytical Model of Refrigeration Power of the Orifice Pulse Tube Refrigerator, NBS Technical Note, to be published.

PULSE TUBE WITH AXIAL CURVATURE

Yuan Zhou, Wen-xiu Zhu and Yen Sun

Cryogenic Laboratory
Chinese Academy of Sciences
Beijing, People's Republic of China

ABSTRACT

Experiments are made to compare the performance of coiled pulse tubes with those of straight ones having similar cross sections, lengths and operating conditions. The performance degradation of coiled pulse tubes is also reported when the ratio of the axial radius to the radius of tube cross section is reduced. The inflence of flow resistance on refrigeration performance is discussed. It is possible to construct a pulse tube refrigerator within a compact space with good performance.

INTRODUCTION

Since the pulse tube refrigerator was developed by Gifford and Longsworth in 1963[1], investigations on it have been done by many scientists[2-5]. Recently the "orifice" type pulse tube refrigerator has been improved to achieve 60 K in one stage at NBS[6]. Further, the comparison of the intrinsic cooling efficiency of the pulse tube refrigerator with that of other type of refrigerators has been made by Radebaugh and Herrman . The intrinsic efficiency of the device has been shown to be comparable to that of a Stirling refrigerator being greater than that of a Joule-Thomson refrigerator. Moreover, the pulse tube refrigerator has only one moving part, which is at room temperature. Because of its high reliability and simplicity, pulse tube refrigerator shows potential applications in various fields. However, when space is limited, it may be more practical to make the pulse tube coiled, so that the tube volume can be enlarged and refrigeration power increased. In order to improve the flexibility in constructing pulse tubes with good performance within a compact space, the pulse tube with axial curvature is investigated in this paper. The basic arrangement of our pulse tube refrigerator, especially that of the orifice, is the same as that in NBS[6]. We have been able to obtain a low temperature of 96 K using a straight pulse tube 12 mm in diameter by 400 mm long with maximum and minimum pressures of 0.65 MPa and 0.4 MPa respectively at a frequency of 10.5 Hz, when helium is used as the working gas with valveless compressor. A low temperature of 105 K was achieved with a coiled pulse tube, whose radius of curvature is 45 mm, and which has the same cross-section and length, and is operated under the same condition as that of the said straight pulse tube. A low temperature of 77 K was achieved with another coiled pulse tube, which has a radius of curvature of 60 mm, and is 800 mm long wound in two turns, and operated between maximum and minimum pressures of 0.75 MPa and 0.5 MPa respectively at a frequency of 10.5 Hz. The influence of orifice valve setting and frequency on the cold end temperatures has also been investigated experimentally.

EXPERIMENTAL APPARATUS AND PULSE TUBE SPECIMENS

A schematic diagram of the test apparatus is shown in Fig.1. The orifice is a 10 turns needle valve outside a diffusion pump. A vacuum of 2.67×10^{-3} Pa can be obtained with the pump. The reservoir volume is 2 L. The regenerator is a stainless steel tube 24 mm in diameter by 105 mm long filled with 780 discs of 150 mesh bronze screen. The absolute pressure in the pulse tube was measured with a transducer mounted at the hot end. Eight thermocouples were attached to the pulse tube at equal intervals from the cold end to the hot end to detect the temperature distribution along the tube. The flow straightener was inserted between the pulse tube and the regenerator to make the gas flow pattern planar. The pressure oscillation was generated by a valveless compressor which is one meter away from the test set-up. The volume displacement of the above compressor is about 280 cm^3, with the rotating speed of the shaft ranging fom 12 rpm to 700 rpm. A by-pass valve is disposed between the high pressure side and the crankcase to adjust the pressure ratio. The dimensions of the pulse tube specimens used in the experiment are listed in Table 1.

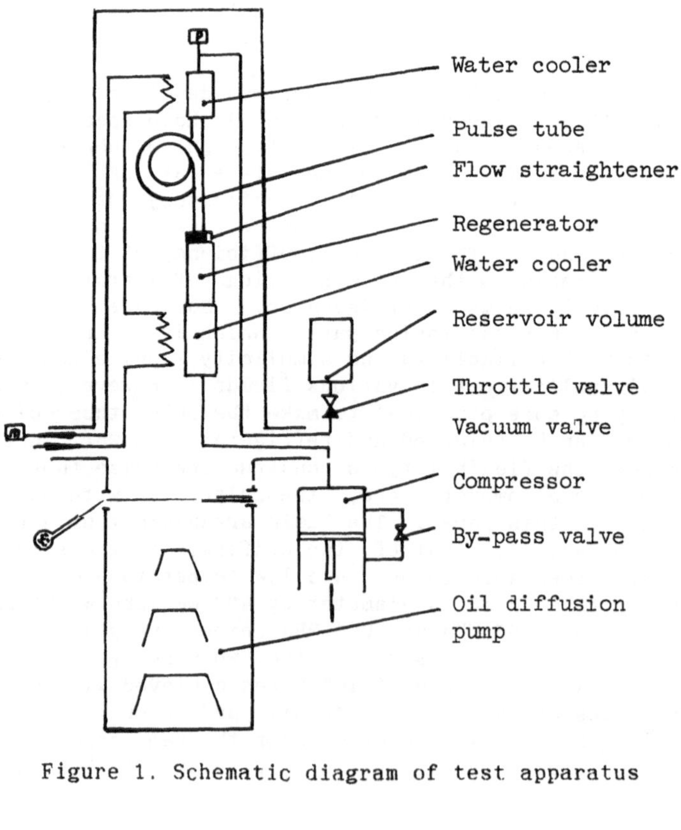

Figure 1. Schematic diagram of test apparatus

Table 1. Test Specimens

specimen No.	length (mm)	I.D. 2r(mm)	thickness of wall(mm)	radius of curvature R(mm)	turn(s)	R/r
I(straight)	400	12.0	1.0			
II (coiled)	400	12.0	1.0	25	1	4.2
III(coiled)	400	12.0	1.0	35	1	5.8
IV (coiled)	400	12.0	1.0	45	1	7.5
V (coiled)	800	12.0	1.0	60	2	10.0

All of the tubes are made of stainless steel. The open hot end is soldered to a copper end piece which contains thirteen punched copper discs in order to enhance the heat transfer from the gas to the copper walls. The volume of this isothermal section is 7% of the pulse tube. Outside the copper wall is a cooling water jacket. Fig.2 shows the five pulse tube specimens I, II, III, IV and V with the regenerator and the hot end heat exchanger removed.

TEST RESULT

The experiments on both straight and coiled pulse tubes were done with the same equipment under the same conditions, so that the results can be compared directly. An approximate mean pressure of 0.55 MPa and a pressure ratio of 1.63 was used for Specimen I, II, III and IV . Helium gas was used for all the runs. The frequency varied from 6 Hz to 10.5 Hz. A mean pressure of 0.63 MPa and a pressure ratio of 1.5 was only for Specimen V. The influence of the pulse frequency and the orifice valve setting on the performance characteristics of the pulse tube refrigerators Specimen I and IV are illustrated in Fig.3 and 4. The influence of the radius of curvature of a helically coiled pulse tube on its refrigeration performance is shown in Fig.5. The cold end temperatures obtained with Specimen I, II, III and IV are 96 K, 122 K, 108 K and 105 K respectively at a frequency of 10.5 Hz with orifice valve

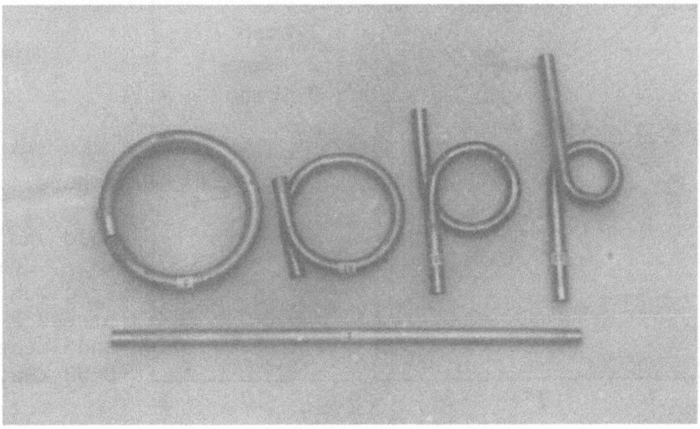

Figure 2. Five specimens

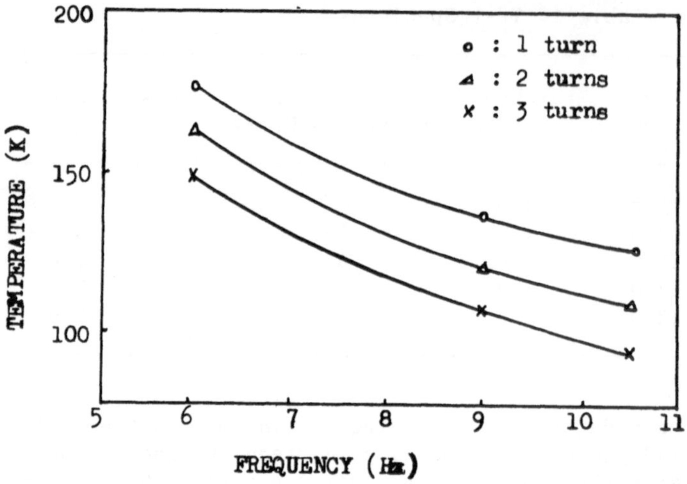

Figure 3.

Cold end temperature vs. frequency for specimen I, with pressure ratio 1.53, mean pressure 0.55 MPa.

setting of 3 turns. Fig. 6 shows the temperature distributions along the four specimen tubes. Comparing the test result of the straight tube with those of the coiled ones we see that their temperature distribution curves have the similar shapes. Fig.7 shows the temperature distribution along the tube of Specimen V. The cold end temperature recorded with Specimen V was 77 K at a frequency of 10.5 Hz and orifice valve setting of 5 turns under a mean pressure of 0.63 MPa and a pressure ratio of 1.5. The cold end temperature of 84 K was achieved with a mean pressure of 0.55 MPa and under otherwise the same conditions as the above.

A increase of the mean pressure and the volume of the pulse tube will definitely lower the cold end temperature (Fig.7). Unfortunately, experiments with higher mean pressures could not be performed due to the limited torque of the compressor motor.

Because the regenerator and the hot end heat exchanger are not well designed, the test results presented above are not optimal.

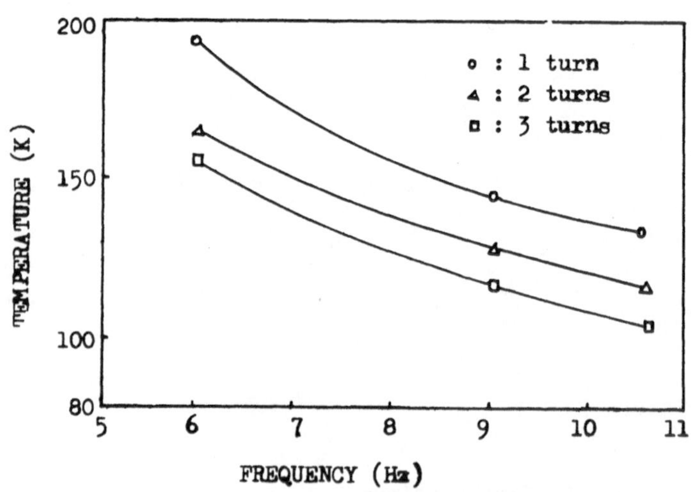

Figure 4.

Cold end temperature vs. frequency for specimen IV, with pressure ratio 1.63, and mean pressure 0.55 MPa.

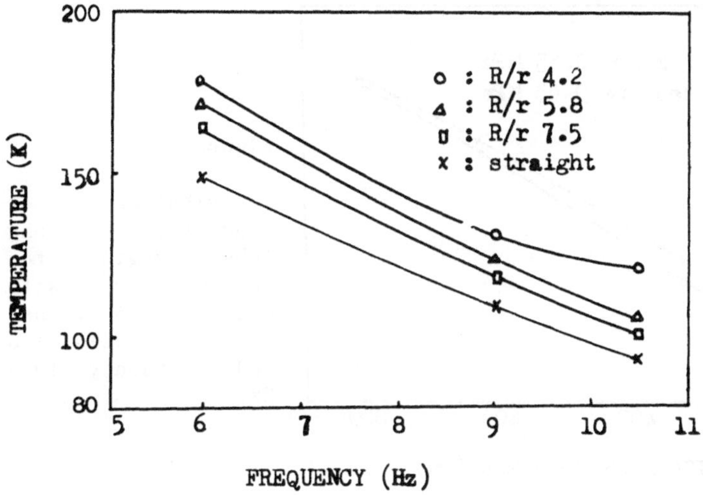

Figure 5.

Cold end temperature vs. frequency for specimen I, II, III and IV, with oriffice 3 turns, pressure ratio 1.63, and mean pressure 0.55 MPa.

ANALYSIS AND DISCUSSION

The variations of the cold end temperature with frequency and orifice valve setting are similar for the specimens I and IV (Fig.3, 4), which demonstrates the similarity in refrigeration mechanism for both straight and coiled pulse tube refrigerators. However, there are some differences between the two patterns.

In a coiled tube, the centrifugal force of a flowing fluid acting at right angles to the main direction of flow produces a pair of secondary flows, which causes higher flow resistance in the tube and also higher heat transfer rate between the working gas and the tube wall than in a straight tube. This results in a larger refrigeration loss in a coiled pulse tube refrigerator. Fig.5 shows that the cold end temperature obtained by Specimen I is lower than that obtained by coiled pulse tubes.

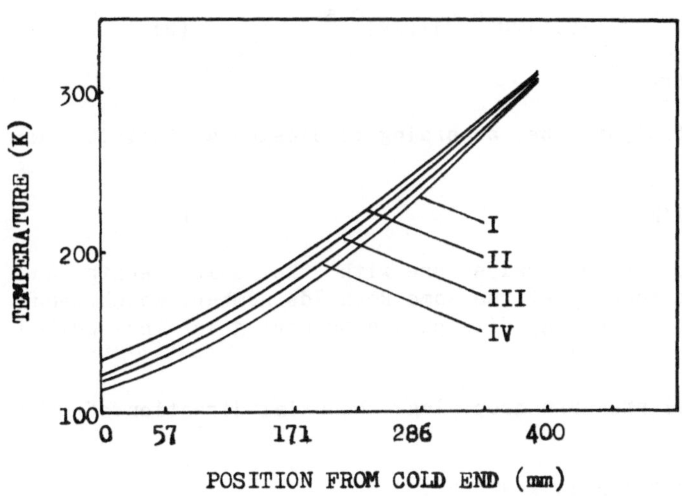

Figure 6.

Distribution of temperature along tube length for specimen I, II, III and IV, with oriffice 3 turns, pressure ratio 1.63, mean pressure 0.55 MPa and frequency 9 Hz.

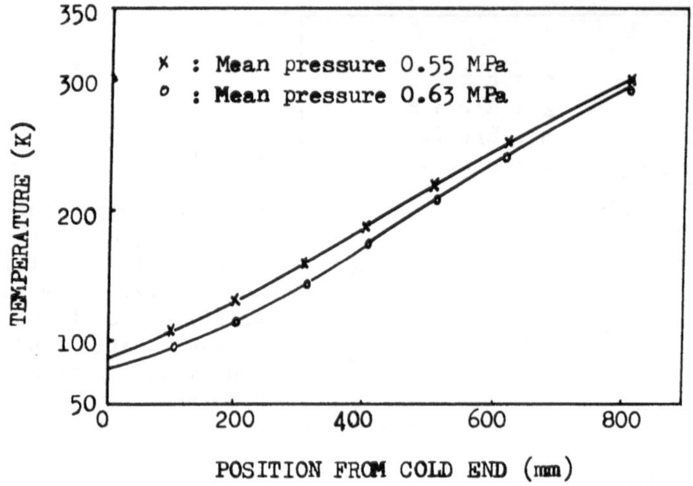

Figure 7.

Distribution of teperature along tube length for specimen V, with oriffice 5 turns, pressure ratio 1.5 and frequency 10.5 Hz.

Because of the complexity in analyzing the alternative and compressable flow process in a pulse tube, a simpler model of steady flow is adopted for qualitative analysis.

In our case, the range of Reynolds number is $7x10^3$-$1.2x10^4$, showing that the flow is turbulent. When the flow is turbulent, the mean velocity V is always large enough to produce the secondary flow which has a strong effect in determining the resistance coefficient a_c.

The definition of the resistance coefficient for a coiled tube a_c is given by the following equation

$$a_c = -\partial p/(R\partial\alpha) \text{ x } 2r/(\rho V^2/2) \qquad (1)$$

where p is the pressure in the tube, R the radius of curvature, α the angle of turn of the coil. From the work of Mori and Nakayama[7], we have

$$a_c = 0.3D^{-0.2}(1+0.112D^{-0.2})(r/R)^{0.5} \qquad (2)$$

where 300>D>0.034, $D=Re(r/R)^2$.

In the case of a straight tube, according to Blasius's formula, we have

$$a_s = 0.314Re^{-0.25} \qquad (3)$$

For a coiled tube and a straight one with the same length and diameter, a_c is greater than a_s at the same Reynolds number, which means a greater pressure drop and exergy dissipation occurring in the coiled tube.

From (1), the total pressure drop along the axial direction of the coiled tube is

$$P_c = 2\pi R/(2r) \text{ x } a_c \text{ x } \rho V^2/2 \qquad (4)$$

So exergy dissipation rate E_c due to the friction loss is

$$E_c = P_c \times 2\pi R \times \pi r^2/(t/2) \quad (5)$$

where t is the time for one cycle. For a straight tube with the same size we have

$$E_s = P_s \times L \times \pi r^2/(t/2) \quad (5')$$

where $P_s = L/(2r) \times a_s \times \rho V^2/2$, L is the length of the straight tube. So

$$E_c/E_s = a_c/a_s \quad (6)$$

In the Reynolds number range considered ($Re=7\times10^3-1.2\times10^4$), a_c/a_s = 1.22-1.28.

From (6) we know that the coiled pulse tubes will generate less refrigeration power or have a higher cold end temperature when no refrigeration power is given.

The influence of R/r on refrigeration power is considered as following. In the range of D considered, $0.112D^{-0.2}<<1$, so formula (2) can be approximately written as

$$a_c = 0.3Re^{-0.2}(R/r)^{-0.1} \quad (8)$$

When R/r increases, a_c decreases, leading to the decrease of T_c.

From (8) we can also have

$$\partial a_c/\partial(R/r) = -0.03Re^{-0.2}(R/r)^{-1.1}<0$$

If $Re_2>Re_1$,

$$[\partial a_c/\partial(R/r)]_{Re_1} > [\partial a_c/\partial(R/r)]_{Re_2}$$

Because T_c increases with a_c, so

$$[\partial T_c/\partial(R/r)]_{Re_1} > [\partial T_c/\partial(R/r)]_{Re_2}$$

This implies that the cold end temperature difference between two different coiled pulse tubes at higher frequency is smaller than that at lower frequency, and coincides with the test results in Fig.5 (Curve II is not reliable because the tube was not well manufactured).

From equation (8) we know, the smaller the ratio R/r is, the more the performance of a coiled pulse tube refrigerator diverges from a corresponding straight one. If K is designated as a factor of reduction, we have

$$K = \Delta T_{c,c}/\Delta T_{c,s}$$

where $\Delta T_{c,c}$ is the temperature difference between the two ends obtained by a coiled pulse tube, $\Delta T_{c,s}$ is the temperature difference between the two ends of a straight pulse tube with the same sizes and operated under the same conditions. It is clear that K is a function of the ratio R/r, i.e. $1-K \propto 1/(R/r)$. In our experiments the hot end temperature of all the pulse tubes are 293 K, thus $\Delta T_{c,s}$=197 K, therefore the value of K for specimen II, III and IV are 0.87, 0.94 and 0.95 respectively, and the corresponding R/r are 4.2, 5.8 and 7.5.

In addition to the above, Patankar et al[8] shows that there is a significant difference in heat transfer rate between a coiled tube and a straight one. Generally speaking, for the "orifice" type pulse tubes, the heat transfer between gas and the tube wall is detrimental to refrigeration performance according to the analysis made at NBS . But in our case, all the tubes are quite long and made of stainless steel, so the influence of different heat transfer rates should be small.

In designing the orifice pulse tube with coiled tube, it is advisable not to make the ratio R/r too low,preferably be 8 or higher, in order to get a satisfactory performance.

CONCLUSIONS

The performance of coiled pulse tube will be degraded by its greater flow resistance. When R/r is high enough, the coiled tube will perform nearly the same as the relevant straight one. It is possible to construct a practical compact cryocooler with coiled pulse tube when space is limited.

ACKNOWLEDGEMENTS

We would like to thank Prof. C. S. Hong and Y. Z. Zhu for their helpful technical discussions concerning this paper.

REFERENCES

1. W. E. Gifford and R. C. Longsworth, "Pulse Tube Refrigeration", ASME paper No. 63-WA-290.
2. R. C. Longsworth, An experimental investigation of pulse tube refrigeration heat pumping rates, in: "Advances in Cryogenic Engineering", Vol. 12, Plenum Press, New York (1967), p. 608.
3. R. A. Lechner and R. A. Ackerman, Concentric pulse tube analysis and design, in: "Advances in Cryogenic Engineering", Vol. 18, Plenum Press, New York (1973), p. 467.
4. E. I. Mikulin et al, Low temperature expansion pulse tubes, in: "Advances in Cryogenic Engineering", Vol. 29, Plenum Press, New York (1984), p. 629.
5. R. Radebaugh et al, A comparison of three types of pulse tube refrigerators: New method for reaching 60 K, in: "Advances in Cryogenic Engineering", Vol. 31, Plenum Press, New York (1986), p. 779.
6. S. Herrmann and R. Radebaugh, "Measurements of the efficiency and refrigeration power of pulse tube refrigerators", NBS Technical Note 1301 (1986).
7. Y. Mori and W. Nakayama, Study on forced convective heat transfer in curved pipes, Intl. J. Heat and Mass Transfer, 10: 37 (1967).
8. S. V. Patankar et al, Prediction of turbulent flow in curved pipes, J. Fluid Mech., 67: 583, (1975).

SORPTION CRYOGENIC REFRIGERATION - STATUS AND FUTURE

Jack A. Jones

Applied Technologies Section
Jet Propulsion Laboratory
Pasadena, California

ABSTRACT

This paper describes sorption refrigeration development, which represents a relatively new breakthrough in cryogenic cooling. Sorption refrigerators have virtually no wear-related moving parts, have negligible vibration, and offer extremely long life (at least ten years). In sorption compressors, low pressure gas is physically adsorbed or chemically absorbed to cooled solids. When heated an additional 100°C to 200°C the gas becomes greatly pressurized and is desorbed, i.e., vented, from the solids. Precooling and expansion of the gas causes partial liquefaction, thus providing net cooling. Recent testing at JPL includes a 1000-hour life test of a hydrogen chemisorption refrigerator (14K-30K), a feasibility test of a nitrogen physisorption refrigerator (100K-200K) and a demonstration test of an oxygen chemisorption compressor (for 55K-90K). Although first stage sorption refrigeration systems require more power than mechanical systems, multiple-stage sorption systems are at least three times more efficient and at least ten times lighter than mechanical refrigerators for 7K-10K cooling (SH_2 vacuum sublimation onto hydrides). Due to the high reliability, long-life, light-weight, low vibration characteristics of sorption refrigeration, it is presently being considered for many spacecraft applications and may eventually have many ground applications as well.

INTRODUCTION

Sorption refrigeration is a method of cooling wherein gas is compressed by means of physical surface a<u>d</u>sorption or chemical internal a<u>b</u>sorption, and then passed through an expansion device, such as a Joule-Thompson (J-T) expansion valve, thus creating net cooling. In the compressor portion of sorption refrigerators, low pressure gas is sorbed to solids at various temperatures. When the gas-solid combinations are heated an additional 100°C-200°C, the gas becomes greatly pressurized and is vented from the solid. The high pressure gas is then precooled and expanded through a J-T valve, where the gas is partially liquefied. The liquid is boiled by absorbing heat from electronics, e.g., infrared (IR) detectors, and the resulting low pressure gas is eventually reabsorbed by the solid sorbent. Sorption refrigeration systems have no moving parts, other than very-long-life check valves, and thus could eventually have a potential lifetime of decades operating with virtually no vibration.

SORPTION REFRIGERATOR STAGES

There are a number of long-life, vibrationless sorption refrigeration systems that can be used at various temperature levels, and the following sections will deal with specific recommended gas-solid combinations for each range. In order to minimize overall power and weight requirements, there are some general recommended rules to follow when considering various system stages:

1. When possible, use chemisorption instead of physisorption, as chemisorption systems generally operate more efficiently and at higher temperature ranges, thus resulting in much lighter heat rejection systems.

2. When a physisorption system is used, select a gas that has a relatively high boiling point, thus higher Van der Wals attractive forces, and that does not require excessive operating pressure ratios. The corresponding sorbent should have both high sorption loading and low void volume in order to minimize total power requirements.

3. Reject the compressor heat at the highest possible temperature level. Always attempt to reject the compressor heat by direct passive radiation or convection rather than by another refrigeration stage. Parallel sorption stages, i.e., all compressor heat rejected independently, result in additive power requirements, whereas series sorption stages, i.e., compressor heats rejected to another sorption stage, result in multiplying power requirements.

4. Precool the gas stream as low as possible, and preferably liquefy it, prior to the J-T heat exchanger in order to maximize overall J-T cooling efficiency.

The following J-T refrigeration stages are recommended for each respective temperature level and are discussed in detail in the following sections:

110K-150K	Charcoal/Methane (C/CH_4) Physisorption
55K-90K	CH_4 + Praseodymium-Cerium-Oxide (PCO) Chemisorption
14K-30K	CH_4 + PCO + LH_2 Hydride Chemisorption
7K-10K	CH_4 + PCO + LH_2 Hydride + SH_2 Hydride Chemisorption
4K-5K	CH_4 + PCO + LH_2 Hydride + Mechanical Helium System

Charcoal/Methane Physisorption (110K-150K)

For cooling detectors or other equipment to temperatures above about 175K, the simplest active refrigeration system is the thermoelectric cooler (TEC). This type of "reverse thermocouple" relies on the Peltier effect in that a temperature difference is created between two ends of dissimilar materials when a voltage is applied. This type of cooling has no moving parts, and thus very long life, but is generally very inefficient at temperatures below 175K.

For long-life, vibrationless cooling in the 110K to 150K region, a charcoal/methane physisorption refrigerator is recommended. A typical cycle is shown in Figure 1. For this system, 6.3 atm (0.636 MPa) methane gas is physisorbed at 275K onto a low void volume, dense saran charcoal[1]. As the system is heated to 575K, most of the gas is desorbed at 45 atm (4.54 MPa). The high pressure gas is precooled by a counterflow heat exchanger, liquefied by a TEC at 190K, followed by another counterflow heat exchanger, and finally expanded to 6.3 atm (0.636 MPa) at 140K. As the low pressure liquid is boiled, the resulting cold vapor then travels back through the heat exchangers, thus precooling the entering high pressure gas. The low pressure gas is finally reabsorbed into a cooled charcoal canister to complete the cycle. The amount of heat that must be removed from the CH_4 stream by the TEC at 190K is slightly less than the amount of heat added at 140K, assuming parasitic heating is minimized. Thus, the specific powers of the TEC and CH_4 stages (as well as later subsequent stage) are approximately additive.

A computer program has been written that can readily size any gas-solid sorptions system (physical or chemical) if the basic properties and isotherm data are known. The program is based primarily on the equations as started by a previous 'SORPTION' computer program developed at JPL for hydrogen and C/N_2 sorption sizing[2]. The new program, 'SORB', is a shortened version which is more versatile and user-friendly, but eliminates pressure drop and thermoelectric cooler efficiency calculations. The 'SORB' program has been used to size the C/CH_4 sorption refrigeration system.

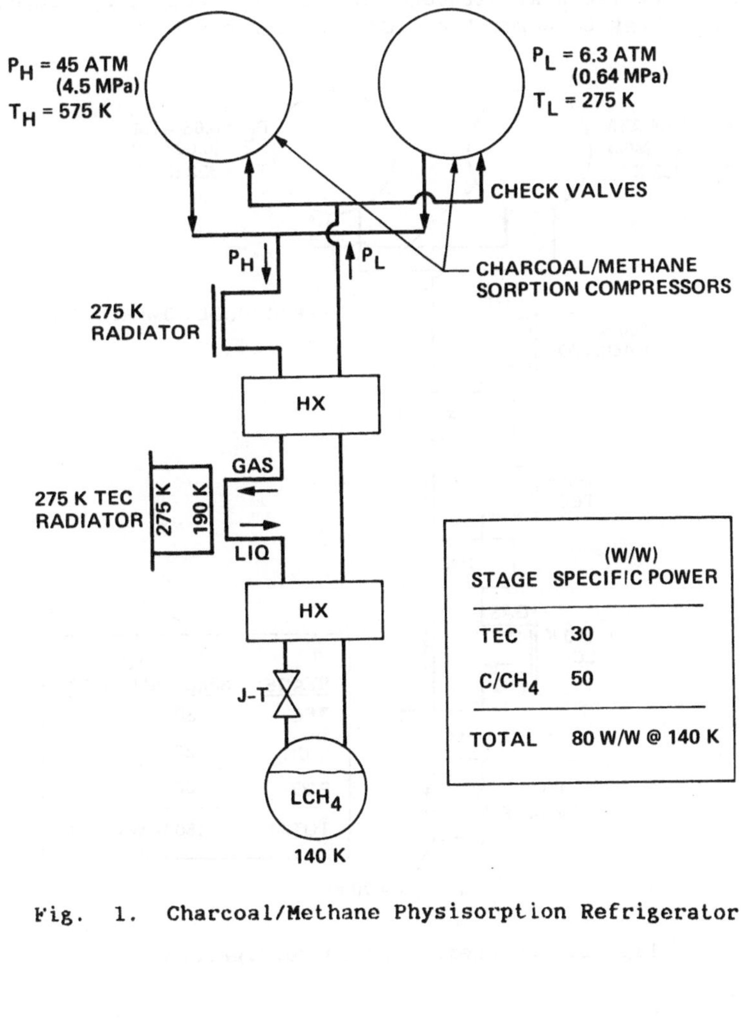

Fig. 1. Charcoal/Methane Physisorption Refrigerator

A great variety of temperature and pressure operating ranges is readily available for the C/CH_4 system. Lower heat rejection temperatures result in substantial power and weight reductions for both the C/CH_4 and TEC compressor systems, while higher heat rejection temperatures result in lowered performance, but smaller radiators per watt of heat. If a 300K heat sink is used instead of a 275K heat sink, the total TEC power increases from 30 W to 45 W and the C/CH_4 power increases from 50 W to 60 W.

Tests on a C/CH_4 physisorption system are presently in progress at JPL. A somewhat similar charcoal/nitrogen (C/N_2) physisorption refrigerator has already been demonstrated at JPL in the 100K-120K region[3].

PCO Chemisorption (55K-90K)

In the past year, over twenty potentially usable oxide chemical reactions were studied at JPL in order to determine potential use as an oxygen chemisorption refrigerator stage[4,5]. Of these, only doped SrO_2 and $Pr_{1-n}Ce_nO_x$ (PCO) were found to be fully reversible in the 1 atm pressure range and at temperatures below 700°C (for materials strength considerations). The SrO_2 compound, which has about a 10% usable oxygen content, sintered at temperatures above about 500°C (10 atm) and could not reabsorb oxygen. Only the PCO (1% usable oxygen content) compound was found to be completely reversible in the full range of pressures (0.1 atm to 100 atm) and temperatures. By using a charcoal/methane (C/CH_4) upper stage and precooling to 140K, a PCO refrigerator should be able to attain temperatures of 90K to 55K (Figure 2). The total power requirements to reach 70K are about 180 W/W, although this can be reduced by recycling some of the PCO waste heat to power the C/CH_4 compressors.

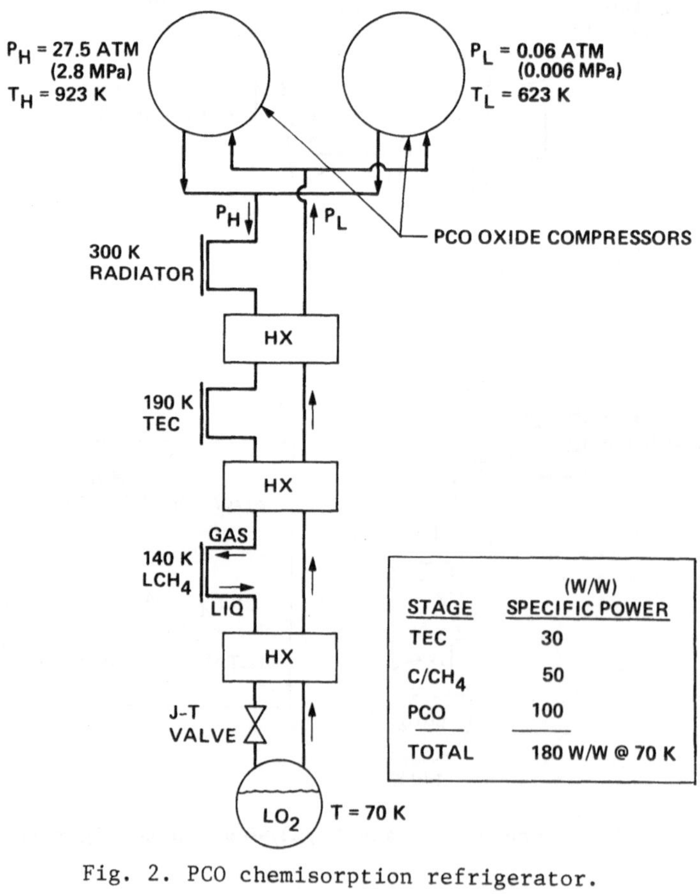

Fig. 2. PCO chemisorption refrigerator.

The original PCO compound was first discovered by Mullhaupt of Union Carbide[6] to be useful as the oxygen carrier in a reversible cyclic oxidation-reduction process for separating oxygen from air. Mullhaupt successfully cycled the PCO compound 7,000 times to prove reversibility, and in fact, actually showed somewhat improved PCO performance over the course of the test. Recent tests on a PCO oxide compressor at JPL have confirmed predicted compressor performance characteristics, and life testing on the system is expected in the near future.

<u>LH_2 Hydride Chemisorption (14K-30K)</u>

The most well-known sorption refrigeration system is the hydride system, and it was tested in the Netherlands in 1972[7]. It was later developed and endurance tested at the Jet Propulsion Laboratory[8]. For this system which is shown in Figure 3, low pressure hydrogen is absorbed into $LaNi_5$ at about room temperature. When heated to 100°C, the hydrogen is pressured from 1 atm to 40 atm (4.04 MPa) and then vented. After being precooled with LO_2 at 70K, the high pressure hydrogen is expanded to 0.10 atm (0.010 MPa) at 14K. The precooling at about 70K is necessary in order for the hydrogen to be significantly below its inversion temperature of about 200K and thus provide significant refrigeration when expanded.

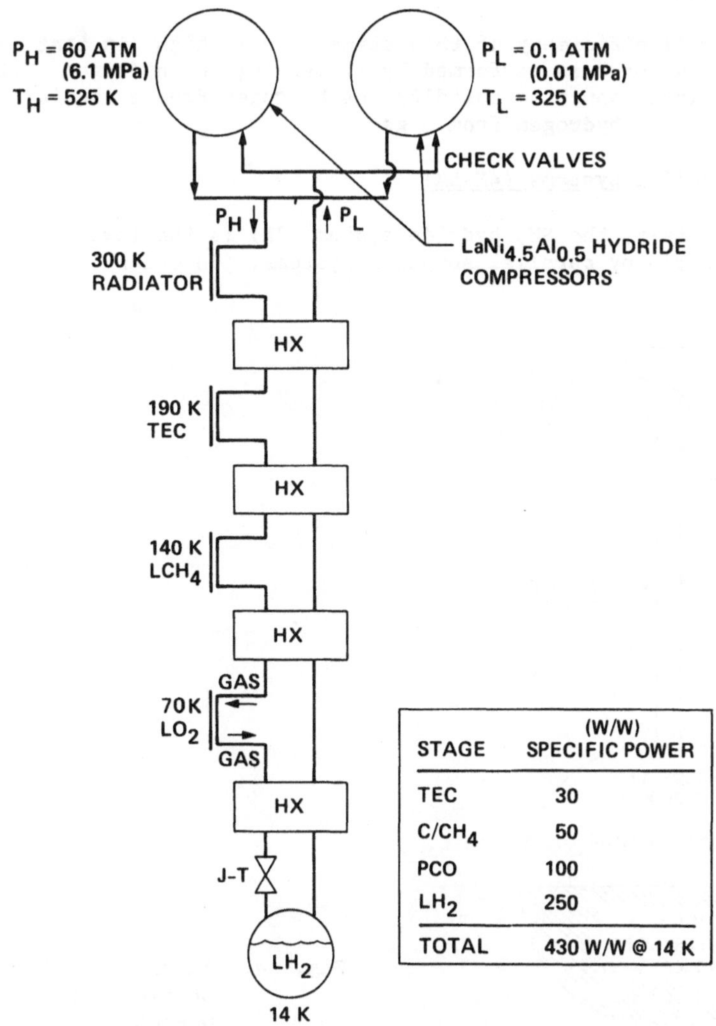

STAGE	(W/W) SPECIFIC POWER
TEC	30
C/CH_4	50
PCO	100
LH_2	250
TOTAL	430 W/W @ 14 K

Fig. 3. LH_2 Hydride chemisorption refrigerator.

A complete hydride refrigeration system has accumulated over 1,000 hours of operation between 14K and 29K at JPL. In a separate test, an ERGENICS hydride compressor, without an attached cryogenics section, has operated successfully for 5,800 continuous hours[8].

SH_2 Hydride Chemisorption (7K-10K)

To produce temperatures as low as 7K-10K, the vapor pressure over collected liquid hydrogen is lowered by means of absorbing the hydrogen vapor onto a very low pressure warm hydride, such as palladium[9]. With a 1.7 torr partial pressure of hydrogen (corresponding to a hydride temperature of about 0^oC), the hydrogen is solidified and sublimes at 10K (Figure 4). A cryogenic sensor can be continuously maintained at 10K by means of providing two Dewars of hydrogen. While one is being liquefied, the other is being sublimated. The cryogenic sensor can be attached to both Dewars by means of a thermal switch, thus being continuously maintained at 10K.

Open-cycle SH_2 systems with mechanical vacuums have operated at Lockheed at temperatures as low as 7K[10]. Since closed-cycle hydride systems are known to operate at equally low absorption pressures[11], temperatures of 7K are thus expected to be attainable with this type of sorption system.

The overall efficiency of this stage is very high, in that most of the "refrigeration work" is performed by liquefying the hydrogen. It takes relatively little power to solidify the hydrogen from a liquid as compared to liquefying the hydrogen from a gas.

Alternate Cooling Systems (4K-5K)

Unfortunately, the SH_2 hydride system (7K) is the lower limit of cooling possible by parallel sorption systems. In order to attain temper-

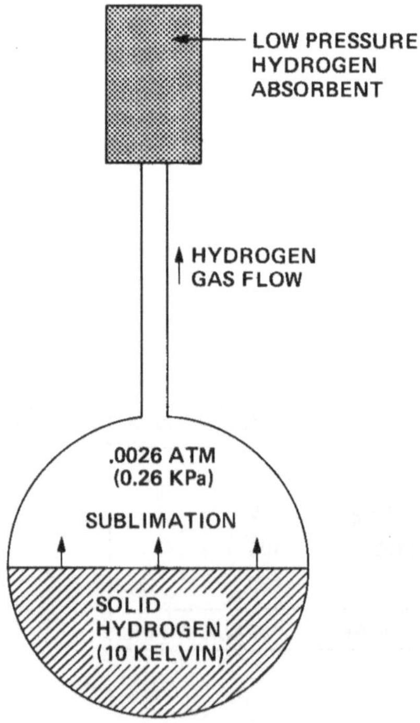

Fig. 4. An SH_2 Hydride Chemisorption Operation

atures below 7K by sorption, it is necessary to use helium gas that is sorbed at temperatures of about 20K or below, thus requiring a hydride stage in series with the helium stage. Since it requires at least 100 W of heat to be rejected at 14K for every watt removed at 5K for the C/He system, it would thus require at least 60 kW of power to provide 1 W at 5K with C/He sorption[12]. Somewhat less power would be required if a solid hydrogen stage was added to further precool the helium, but the overall system complexity would increase. There are basically two types of alternate refrigeration systems, in addition to the C/He sorption system, that can attain 4K-5K temperatures. The first system is the adiabatic demagnetization refrigerator (ADM), which, in theory, is a relatively efficient refrigeration stage which can remove heat at 4K-5K or below and reject it above 14K by means of alternately magnetizing and demagnetizing a cryogenic paramagnetic material. It should be noted, however, that even with a very high 75% Carnot efficiency, it would thus require rejecting over 4 times as much heat at 14K as is removed at 4.2K. With a C/CH_4-PCO-LH_2 hydride upper stage, the total specific power requirements thus become about 2kW for cooling at 4.2K.

It is important to note that with a mechanical helium compressor rejecting heat at room temperature (thus parallel staging as opposed to ADM's series staging), the total heat requirements are only about 1,100 watts (Figure 5). The present problem with the mechanical helium J-T system, however, is that contamination from oil-lubricated compressors eventually clogs the tiny J-T orifices. Substantial progress has been made in recent years regarding non-lubricated helium compressors, and in fact, one such compressor has been demonstrated over 9,000 hours of successful operation[13].

Alternate expansion systems, such as the Gifford-McMahon (G M) are much more "forgiving' in regard to oil contamination due to their greatly increased regenerator expansion flow areas. Unfortunately, present cryogenic regenerators lack specific heat, and thus thermal performance, below 15K. Progress in this area must await improved cryogenic regeneration performance, which is presently in progress at JPL.

The comparison of various long-life 4K-10K refrigeration stages is shown in Table 1, as updated from Reference 12.

SUMMARY AND CONCLUSIONS

The development of sorption refrigeration and the identification of various operating stages has for the first time made it possible to obtain efficient, very long life (at least ten years) and low vibration cooling (as low as 7K) for cryogenic sensors. Sorption refrigerators have virtually no wear-related moving parts other than very long life room-temperature, low-frequency check valves. The sorption systems that are recommended for various cooling ranges are as follows: charcoal/methane physisorption (110K-150K); PCO oxide chemisorption (55K-90K); LH_2 hydride chemisorption (14K-30K); and SH_2 hydride chemisorption (7K-10K). Although first stage sorption refrigeration systems require more power than mechanical systems, multiple stage sorption systems are at least three times more efficient and at least ten times lighter than mechanical refrigerators for 7K-10K cooling (Table 1).

Unfortunately, at present there appears to be no readily available, efficient, long-life refrigeration system, either sorption or mechanical, for cooling in the 4K-5K region. Charcoal/helium sorption systems are likely to be very heavy and require up to 60 kW of power to produce one watt of cooling at 5K. Other more efficient mechanical stages that may be

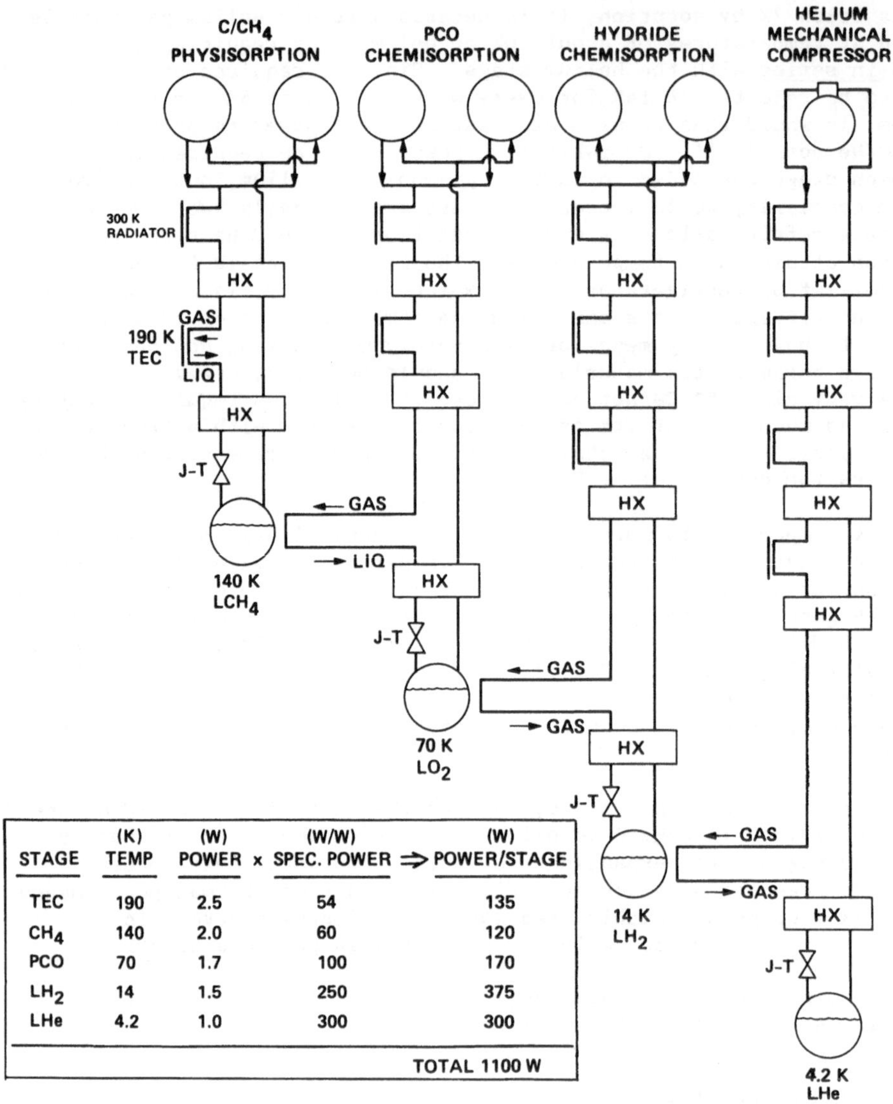

STAGE	(K) TEMP	(W) POWER	x	(W/W) SPEC. POWER	⇒	(W) POWER/STAGE
TEC	190	2.5		54		135
CH_4	140	2.0		60		120
PCO	70	1.7		100		170
LH_2	14	1.5		250		375
LHe	4.2	1.0		300		300
						TOTAL 1100 W

Fig. 5. Sorption/Mechanical J-T Hybrid Refrigerator

developed for long-life cooling in the 4K-5K region are a mechanical compressor helium J-T system (requires development of non-lubricated helium compressors) or mechanical helium G M regenerator systems (can use present oil-lubricated compressors, but requires cryogenic regenerator development).

The sorption stages (methane oxide hydride) as well as the suggested mechanical helium stage (J-T or G M) all reject their heat at room temperature or above, and thus are very convenient to use for ground-based operations. Furthermore, due to the high reliability, long-life, light-weight, low vibration characteristics of sorption refrigeration, it is presently being considered for several future JPL spacecraft applications.

Table 1. Mechanical and sorption multi-stage refrigerator comparisons.

Stages	System Spec Power, W/W[a]	Refrig Weight/ Cooling, kg/W[a]	LEO Rad Weight/ Cooling, kg/W[a,b]	System Weight/ Cooling, kg/W	Focal Plane Vibration
<u>Mechanical</u>:					
1. R^3 Brayton (60K)/R^3 Brayton (10K)	1670	225	33.4	258.4	yes
2. VM (55K)/VM (15K)/VM (10K)	10,000	130	200.0	330.0	yes
3. Turbo (65K)/Turbo (25K)/Turbo (8K)	2200	100	44.0	144.0	yes
4. Stirling (65K)/Stirling (10K)	1700	160	34.0	194.0	yes
5. Stirling (80K)/Stirling (20K)/ He J-T (4.4K)	3000	254[c]	-	-	yes
<u>Sorption</u>:					
6. C-CH_4 (140K)/PCO (70K)/LH_2 Hydride (14K)/SH_2 Hydride (7-10K)	500	13	20.1[d]	33.1	no
7. C-CH_4 (140K)/PCO (70K)/LH_2 Hydride (14K)/C-He (5K)	60,000	70	1760.0	1830.0	no
<u>Sorption/Mechanical Hybrids</u>:					
8. C-CH_4 (140K)/PCO (70K)/LH_2 Hydride (14K)/Mechanical He J-T (4.2K)	1100[e]	40	25.0	65.0	no
9. C-CH_4 (140K)/PCO (70K)/LH_2 Hydride (14K)/Adiabatic Demag (4.2K)	2000[e]	100	35.0	150.0	no

[a]Powers and weights are per watt of cooling at third stage.
[b]All radiators are 10K cooler than corresponding refrigerator "heat sink temperature."
[c]USSR Salyut 6 flight, 127 kg for 1/2 watt cooling at 4.4K.
[d]Radiator weight for 225K C-N_2 is a cascading temperature radiator with a fluid loop.
[e]Total power can be further reduced by 20% to 40% by using waste heat from helium and/or oxide compressors.

ACKNOWLEDGEMENTS

The research described in this paper was carried out by the Jet Propulsion Laboratory, California Institute of Technology, under a contract with the National Aeronautics and Space Administration.

REFERENCES

1. D. F. Quinn et al, Solid adsorbents for storage of CMG for automotive use - saran carbon, "Alternate Energy Conference," Windsor, Ontario, Canada (1985).

2. K. H. Barhydt, "General Computer Model for Predicting the Performance of Gas Sorption Refrigerators," JPL Internal Document Final Report No. D-2600 (1985).

3. S. Bard, Development of an 80-120K charcoal/nitrogen adsorption cryocooler, "International Cryocooler Conference," Annapolis, MD (1986).

4. J. A. Jones, "Oxygen Chemisorption Cryogenic Refrigerator," NASA Patent Application No. MPO-16734-1-CU (1985).

5. J. A. Jones, Oxygen chemisorption compressor study for cryogenic J-T refrigeration, "AIAA Thermophysics Conference," Honolulu, Hawaii (1987).

6. J. T. Mullhaupt, "Process and Composition for Separation of Oxygen from Air Using Pr-Ce Oxides as the Carrier," U. S. Patent No. 3,980,763 (1976).

7. H. H. Van Mal and A. Mijnheer, Hydrogen refrigerator for the 20K region with $LaNi_5$ hydride thermal absorption compressor for hydrogen, in: "Proc. ICEC 4, IPC Science and Technology Press," Guilford, UK (1972).

8. J. A. Jones and P. M. Golben, Design, life testing, and future designs of cryogenic hydride refrigeration systems, Cryogenics Vol. 25 (1985).

9. J. A. Jones, "Ten Degree Kelvin Hydride Refrigerator," U. S. Patent No. 4,641,499 (1987).

10. T. C. Nast, "Study of a Solid Hydrogen Cooler for Spacecraft Instruments and Sensors," Final Report LMSC-D766177, Lockheed, Palo Alto, CA (1980).

11. T. Flanagan, University of Vermont, Private Communication, September (1986).

12. J. A. Jones, Sorption refrigeration comparison study, "11th International Cryogenic Engineering Conference," W. Berlin, W. Germany (1986).

13. R. W. Breckenridge, Refrigerators for cooling spaceborne sensors, in "Proc. Soc. Photo-Optical Instrument Engineers 245 Cryocooled Sensor Technology," San Diego, CA (1980).

NEW DESIGN OF AN ADIABATIC DEMAGNETIZATION CRYOSTAT FOR SPACE APPLICATION

Junya Yamamoto

Low Temperature Center
Osaka University
Suita, Osaka, Japan

Akio Sato and Masashi Sahashi

Toshiba Research and Development Center
Kawasaki, Kanagawa, Japan

ABSTRACT

Developing a cooling method of far-infrared detectors in micro-gravity circumstances in the 0.1 K region, we have designed a new adiabatic demagnetization cryostat and improved a fabrication method of magnetic salt. The cryostat was designed for laboratory use (under 1 G), however most thermal conditions were planned for use in space. An optical window was prepared for the real combination with detector element. The measured heat leak was about 130 mW which enabled us to operate the cryostat more than 50 hours. Low operating current superconducting magnet which was indirectly cooled by liquid helium, was mounted on the cryostat. The magnetic field of 3 T was obtained by current of 11.5 A. The success of the magnet guaranteed the system to work in space where fluid interfaces are not clear. Manganese ammonium alum was grown from the saturated solution and purified.

INTRODUCTION

In recent years there has been an interest in using a far infrared detector at extreme low temperatures, because noise equivalent power (NEP) of a far infrared bolometer decreases as T^n where T is the bath temperature and $3/2<n<5/2$[1]. As obtaining low temperature, pumped ^{4}He which is established technique either at laboratory or space[2] can give us temperature around 1.8 K. A ^{3}He cryostat with an adsorption pump can reach the 0.3 K region,[1,3] which is mostly developed for a balloon-borne detector.

Historically adiabatic demagnetization of paramagnetic salt has been a powerful method to get extreme low temperatures, however others like a dilution refrigerator have took over it for a laboratory refrigerator. As improving space application of far infrared spectroscopy, the adiabatic demagnetization method becomes important to cool detectors because it neither has fluid interface nor requires low temperature valves. Several experiments and theoretical consideration have been done to show the fundamental operation and to stabilize the temperature[4,5].

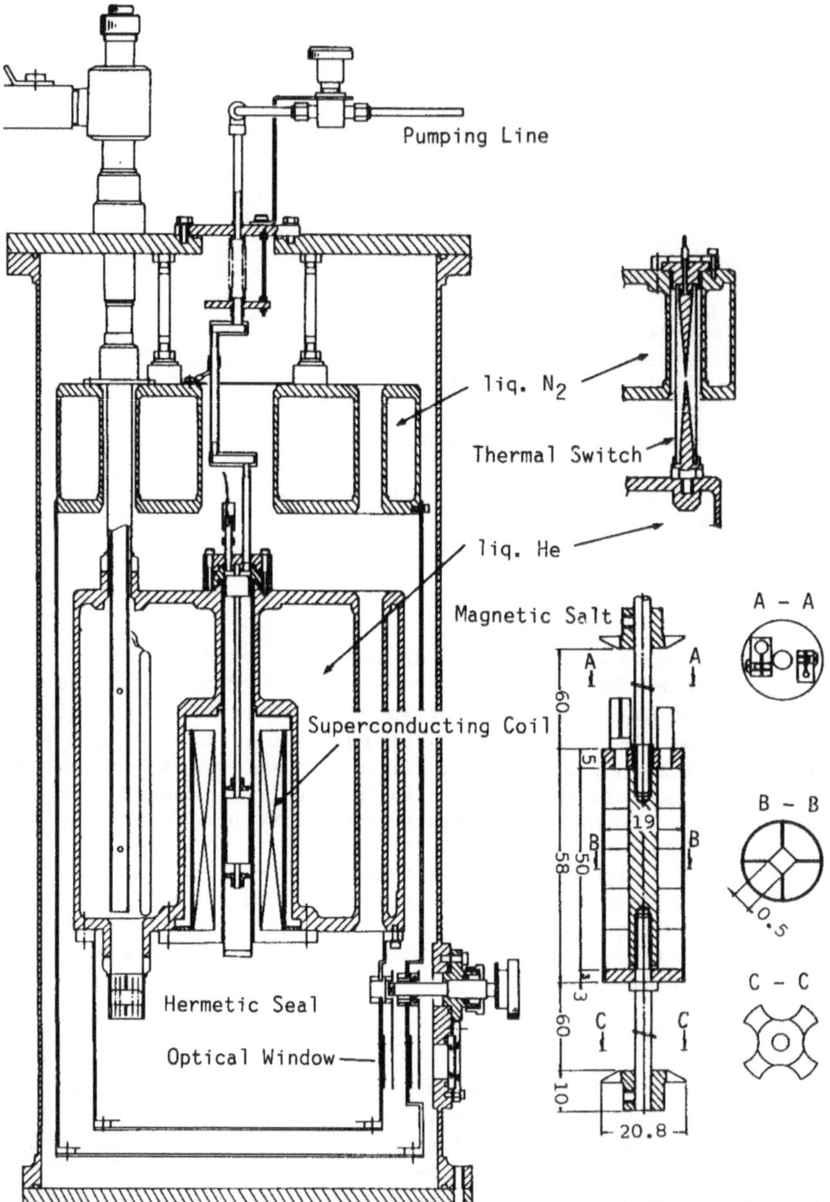

Fig. 1. Schematic Drawing of the adiabatic demagnetization cryostat with an indirect cooled superconducting coil. The liquid nitrogen and helium vessels are connected by gas filled thermal switches for precooling the coil. The adiabatic demagnetization cell which is shown at the right half of the drawing, is set in vacuum at the center of the coil.

To realize the cryostat in orbit, a superconducting magnet must work with low current to produce enough magnetic field regardless of the surface of liquid helium. This paper describes the work on a newly designed cryostat, an indirect cooled superconducting magnet, and preparation of magnetic salt.

CRYOSTAT

Schematic drawing of the test cryostat is shown in Fig. 1. The cryostat is made of aluminum alloy for reduction of weight. The total weight of the cryostat including magnet is 37 kg. The helium and nitrogen baths capacity is 9.5 L and 4.0 L, respectively. Nitrogen-filled thermal switches[6] connect mechanically and thermally between baths to ease intricacies of precooling the magnet. The magnet is set in vacuum surrounded by radiation shields, and connected thermally to the bottom of the helium bath by 24 bolts. The magnet is cooled only by conduction through coil bobbin made of copper. The connection faces of the coil bobbin and the bottom of helium bath are plated with gold for reduction of contact resistance. Gas-cooled power leads are designed for current of 15 A along with other measuring wires, and the expected heat leak through them is only 18 mW. The surface of 4 K and 77 K radiation shields, and the innerface of the room temperature wall are buffing to decrease emissivity.

Temperature decreases of the dewar and the magnet in the precooling stage are shown in Fig. 2. The cryostat was only cooled by liquid nitrogen (LN_2) filled in the nitrogen dewar. The helium dewar was cooled by the thermal switches. Slow temperature decrease was observed even when LN_2 dewar was empty. After transferring liquid helium, the thermal switch acted as a simple stainless steel tube. The measured total heat leak was about 130 mW, which enables us to operate the cryostat more than 50 hours. The lowest temperature after pumping was 1.6 K with a 300 L/min. pump, because of a narrow pumping tube.

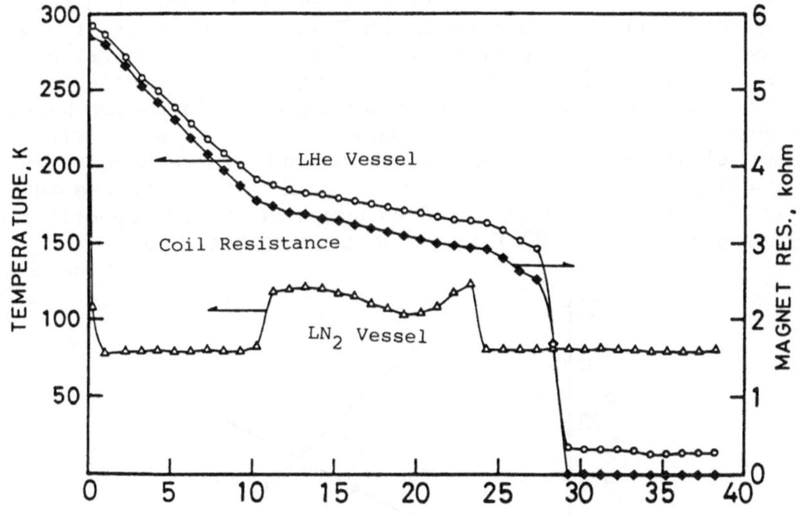

Fig. 2. Precooling characteristics of the cryostat. The helium vessel is cooled by thermal switches from the nitrogen vessel.

Table 1. Specification of indirectly cooled superconducting coil

Conductor		Coil	
superconductor	NbTi (37-63)	clear bore	24.0 mm
copper ratio	0.93	inner diameter	35.2 mm
number of filament	931	outer diameter	68.4 mm
diameter (bare)	0.209 mm	height	152.3 mm
diameter		number of turn	33397 turn
(with insulator)	0.27 mm	field constant	0.264 T/A
critical current		inductance	13.2 H
(6 T, 4.2 K)	25.6 A	maximum field(direct)	5.0 T
length	5435 m	maximum field(indirect)	3.6 T
		filling material	Epoxy

INDIRECT COOLED SUPERCONDUCTING MAGNET

The specification of the coil is shown in Table 1. A 10 mm thick flange, a 5.5 mm thick inner wall of the coil frame, and 24 copper bars outside of the coil winding which are made of high purity copper, provide thermal path in the coil. The gold plated flange is tighten to the gold plated aluminum bottom of the He bath by 24 of 6 mm bolts to get high thermal conductance. The coil was impregnated by epoxy to improve thermal diffusion for the initial cooling and propagation of quench. Designed voltage between terminals is 500 V to suppress the temperature rise, when the coil will quench.

Coil performance was compared with short sample characteristics as shown in Fig. 3. The sweep rate of the current was 0.077 A/s (0.02 T/s). In bath cooling condition, the magnetic field of 5 T was obtained after several quenches. On the other hand, 3.6 T was the maximum field by the indirect cooling. Difference of maximum fields between direct and indirect cooling mainly come from ohmic heating in the hermetic seal and conduction through connecting superconducting wires in vacuum between the coil and hermetic seal, because training effect was observed with bath cooled case, while degradation of quench current was recorded with the indirect cooling. Below 3.6 T of magnetic field, the coil was so stable that the coil current was controlled by computer. Stray magnetic field at the detector position is 0.02 T when the central field is 3 T.

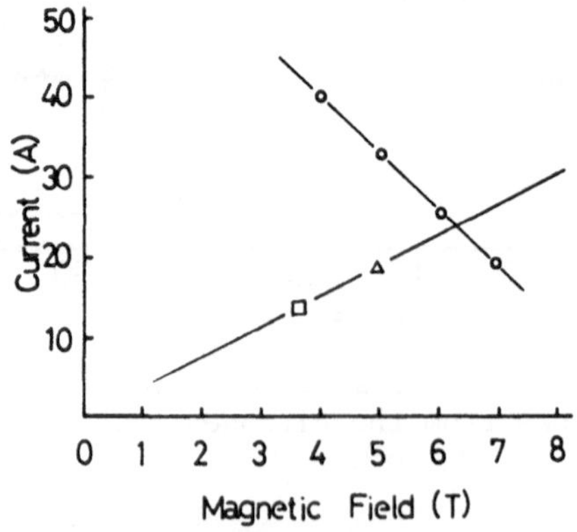

Fig. 3. Load line of the coil and critical current characteristics. △ and □ show the maximum operating points with direct and indirect cooling.

ADIABATIC DEMAGNETIZATION CELL

An adiabatic demagnetization cell is shown in the right half of Fig. 1. The cell is placed in a space where the pressure of helium gas can be controlled by an adsorption pump or an external pump. Section B-B shows the individual magnetic salt shapes. The center pillar and fins prepare the thermal paths from the salt. Estimated maximum heat input from 4.2 K to the cell is 7.4×10^{-7} W without thermal boundary resistance.

MAGNETIC SALT

Single crystals of manganese ammonium alum ($Mn-NH_4$ alum) were prepared by the falling temperature technique for crystal growth[7]. Powder of the alum obtained from a commercial supplier was recrystallized to get single crystals. In this process the temperature of the $Mn-NH_4$ alum was accurately kept at 45°C to solve, where the solubility of alum was 330 g(L solution)$^{-1}$. The solution temperature was controlled by point programme to fall according to the chosen schedule, for example 1 °C/h. A number of single crystals of 1 ∿ 6 cm^3 size was obtained by using the above method. The purified crystals mechanically polished to the shape fitted in the adiabatic demagnetization cell.

CONCLUSION

Low evaporation helium cryostat with an indirect cooled superconducting coil was completed. The success of the indirect cooled coil will develop the adiabatic demagnetization in space. Thermal input can be diminished using much thinner wire which is mostly used for ac electrical machine. The adiabatic demagnetization cell was prepared to get the 0.2 K temperature region with manganese ammonium sulphate salt. The salt (alum) is carefully purified by the falling temperature technique. This work is financially supported by the Institute of Space and Astronautical Science, under contract No. D-4-1.

REFERENCES

1. J. V. Radostitz, I. G. Nolt, P. Kittel, and R. J. Donnelly, Portable ^{3}He detector cryostat for the far infrared, Rev. Sci. Instrum. 49:89 (1978).
2. A. R. Urbach and P. V. Mason, IRAS cryogenic system flight performance report, "Advance in Cryogenic Engineering," Vol. 29, Plenum, New York (1984), p.658.
3. J. Yamamoto, A ^{3}He cryostat using a charcoal adsoption pump for a far-infrared detector, Japan. J. Appl. Phys. 14:1807 (1975).
4. R. D. Britt and P. L. Richards, An adiabatic demagnetization refrigerator for bolometers, Int. J. Infrared and Millimeter Waves 2:1083 (1981).
5. P. Kittel, Temperature stability limits for an isothermal demagnetization refrigerator, "Advance in Cryogenic Engineering," Vol. 29, Plenum, New York (1984), p.613.
6. J. Yamamoto, Improvement in the heat transfer of a gas filled thermal switch, "Advance in Cryogenic Engineering," Vol. 29, Plenum, New York (1984), p.299.
7. "Crystal Growth," B. R. Pamplin, ed., Pergamon Press, London (1975), p.560.

BAYONET FOR SUPERFLUID HELIUM TRANSFER IN SPACE

G.E. McIntosh, D.S. Lombard, D.L. Martindale

Cryolab, Inc.
San Luis Obispo, California

M.J. DiPirro

NASA Goddard Space Flight Center
Greenbelt, Maryland

ABSTRACT

A prototype superfluid helium bayonet for potential space applications has been developed and evaluated with a low heat leak test apparatus. Measured heat leak of the 13 mm (1/2 inch) bayonet pair is 0.21 W at 1.8 K with an uncertainty of +0.09/-0.05 W. Bayonets are fabricated with thin, electron beam (EB) welded tubes which are EB welded to machined nose and flange pieces. Low heat leak structural integrity is provided by a 0.9 mm thickness of filament wound fiberglass-epoxy. Superfluid creep is restricted by KEL-F nose seals which form vacuum-tight extensions to the bayonet cold end pieces.

INTRODUCTION

In order to develop the technology required to efficiently and safely transfer SFHe on orbit, NASA has undertaken a program to develop the components necessary, culminating in an attached shuttle experiment called the Superfluid Helium On-Orbit Transfer (SHOOT) Flight Experiment[1]. It was recognized early in the program that a number of critical components would represent advances in the state-of-the-art. One of these components is the disconnectable coupler that would be used in the transfer line between the supply dewar and the receiver facility. Ordinary transfer line couplers take the shape of a bayonet with a vacuum jacket in each mating half, an elastomer warm end seal and a close-fit type seal at the cold end. For normal fluid helium (temperatures above 2.2 Kelvin), which can support an overpresssure, the close-fit seal is all that is required to keep the liquid away from the warm end of the coupler, and thus keep the heat input to the liquid relatively small (approximately 1 to 2 W). SFHe will not easily support an overpressure in the presence of its vapor, and thus would bypass the close-fit type seal. Moreover, the high effective thermal conductance of SFHe would immediately transport heat into the flowing liquid, thus destroying the efficiency of the system.

If one considers a transfer line to be made up of two couplers and a vacuum jacketed, multilayer insulation (MLI) blanketed transfer tube, it would be reasonable to assign one half of the parasitic heat to the couplers and the other half to the line. Using this formula, for a flow of 200 litres per hour (approximately 8 g/s) through the transfer tube, approximately 4 W per coupler would bring the SFHE to the lambda transition. However, a more severe impact of this parasitic heat load is reflected in the vent requirements on a receiver dewar. As an illustration, assume that SFHe leaving the superfluid pump in a supply dewar is at a temperature of 1.65 K. Also assume that the receiver operating temperature is 1.8 K. If the requirement for a transfer operation is to fill the receiver to 100% at its operating temperature or lower, then all parasitic heat to the transferred SFHe above that which raises the liquid temperature to 1.8 K would need to be removed by the venting system of the receiving dewar during transfer. Otherwise, a series of top-offs and cooldowns of the receiver would have to be performed - a very time consuming operation.

As an example of normal dewar vent capability, the Cosmic Background Explorer (COBE) dewar will vent the flow equivalent of about 0.5 W at a tank temperature of 1.8 K, limited mainly by the porous plug phase separator. With an 8 g/s flow through the transfer line, the liquid absorbs approximately 2.9 W in going from 1.65 to 1.8 K. Thus the total heat leak allowed in the transfer line and couplers in order to accomplish a 100% fill without top-off would be 3.4 W. With 25% of this parasitic heat being attributable to each coupler, that leaves a desired 0.85 W or lower heat load per coupler[2]. Thus, enhanced thermal performance over and above the low temperature seal problem is required.

CRITICAL DESIGN FEATURES

Superfluid helium bayonets combine thermal and mechanical requirements of conventional cryogens with those unique to He II. It is essential that the heat leak be very low to keep temperatures below the lambda point. There must also be an effective cold seal to prevent superfluid creep into the space between the thermal isolator tubes and into any purge/vent tube. These two requirements effectively exclude conventional metal bayonets with flow restrictive nose seals or close-fitting annular spaces. For example, a typical bayonet pair having a mean isolator tube diameter of 2.54 cm with male and female tube thicknesses of 0.406 and 0.254 mm, respectively, and a thermal length of 25.4 cm would have a heat leak of 0.61 W. An allowance for insulation heat leak raises the total to 0.75 W, 50% higher than the 0.5 W goal. This indicates the need for unusually long metal bayonets (with correspondingly higher lateral losses) or the use of low thermal conductivity composite materials. Because composite materials diffuse helium[3], a metal diffusion barrier is required to maintain a long-term static high vacuum in the insulation space. The apparent solution to the heat leak/diffusion problem is to combine metal and composite features into bayonets with thin metal liners for leak tightness and a composite over-wrap for low heat leak mechanical strength.

As shown in Figures 1, 2 and 3, the development bayonet design is based on a 0.051 mm thick (0.002 inch) stainless steel liner and a 0.89 mm (0.035 inch) filament wound over-wrap of epoxy/fiberglass. Thermal lengths of the isolator tubes are 281 mm for the male bayonet and approximately 430 mm for the female. Nominal diameters of the male and female bayonets are 28.7 and 30.7 mm respectively. Conduction heat leak of the bayonet pair is calculated using thermal conductivity integrals from 2 to 295 K. A value of 29.189 W/cm is used for stainless steel and 1.464 W/cm for the composite over-wrap. The calculated total conduction heat leak for the bayonet pair is 0.13 W.

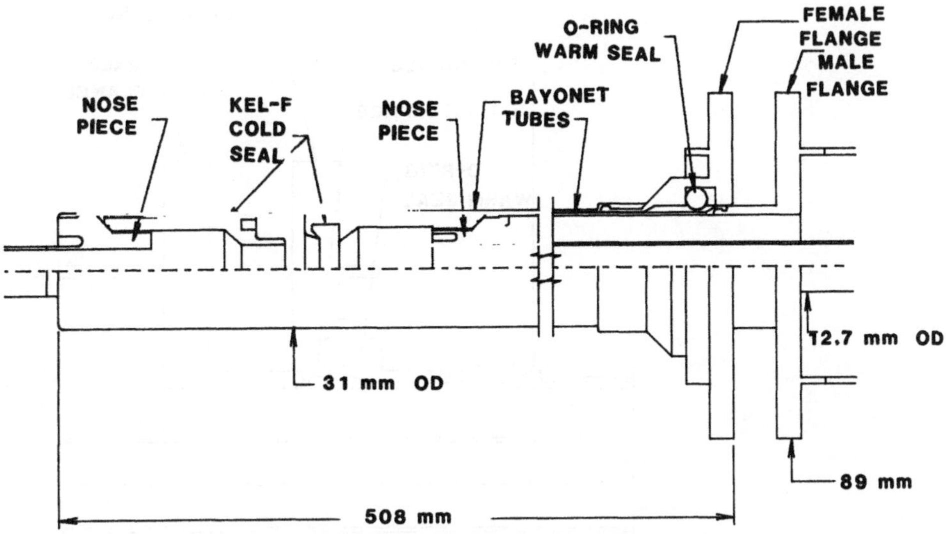

Fig. 1 Superfluid helium bayonet assembly

Initially, lateral MLI insulation heat leak was calculated in several increments for each bayonet using linear temperature gradients on the isolation tubes. Using this approach, calculated insulation heat leaks were 0.12 W for the female bayonet and 0.08 W for the male with a total of 0.20 W. A subsequent analytical solution for the female bayonet combined conduction and insulation into a single, lower, value resulting in an insulation heat leak of 0.08 W by backing out the conduction quantity. Insulation heat leak for the male bayonet was reduced similarly to 0.04 W. Thus, the recalculated heat leak of the bayonet pair was as follows:

Solid Conduction	0.13 W
Insulation	
Female	0.08
Male	0.04
Total Calculated	0.25 W

A nose seal capable of preventing superfluid creep was vital to the design. The original plan was to use a lip seal machined from Vespel grade SP-211. However, KEL-F material was immediately available and a nose seal assembly as shown in Figure 2 was machined and tested in liquid

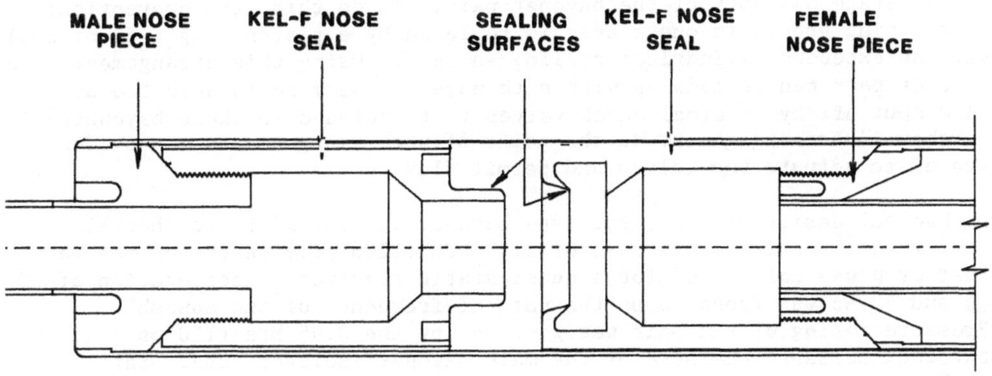

Fig. 2 Cold end details

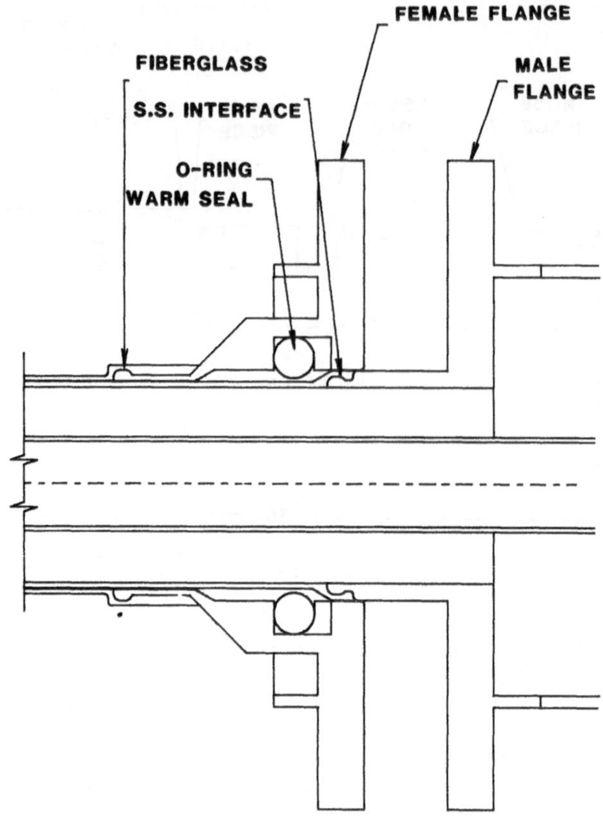

Fig. 3 Warm end flange details

nitrogen. The first pair was vacuum tight at liquid nitrogen temperature and could be mated or separated either warm or cold. Based on this success, a series of KEL-F lip seals was fabricated with diametral interferences ranging from 0.15 to 0.76 mm on a sealing diameter of 17.1 mm. All of these seals were functional but one with 0.86 mm interference was too tight and the lip portion of the joint cracked when mated. The functional seals exhibited warm and cold helium leak rates in the range from 10 E-6 to 10 E-7 atm-cc/s.

The warm end flange assembly shown in Figure 3 reflects the ultimate goal of two stage assembly of the bayonet pair. To do this, the conventional flat o-ring groove in one flange is replaced by a piston ring type of seal with an extended cylindrical sealing surface. Using this arrangement, the bayonet pair can be made up with both nose and warm seals actuated but flow shut off by internal check valves (not included in these bayonets). Further linear engagement to the point of warm flange contact provides travel to actuate the valves and permit flow.

Mechanical design of this prototype bayonet was secondary to thermal performance. A maximum stress of 8.32 MPa (1206 psi) in the composite over-wrap was calculated for a quasi-static transverse acceleration of 10 g, and 52 hz was found to be the natural frequency of the assembly. Pressure rating of the pair was governed by the 0.95 MPa (138 psi) external collapse pressure of the male bayonet isolator tube. Safe working pressure of the assembly is somewhat less than 1/2 of collapse or approximately 0.414 MPa (60 psia).

BAYONET FABRICATION

Because welding 0.051 mm thick stainless steel foil is beyond normal hand welding techniques, the isolator tubes were hand-rolled around a mandrel and placed in a holding fixture for electron beam (EB) welding. Isolator tubes were also EB welded to accurately machined (+0.051/-0.000 mm) stainless steel end pieces. When welding was completed, each bayonet assembly was positioned in a mandrel for filament winding. The 60 degrees (from axial) filament winding was done by a specialty company as was the EB welding. The bayonets were machined to final dimensions after curing the epoxy composite.

The EB welding vendor experienced considerable difficulty in getting sound seam welds in the 0.051 mm foil, probably due to inadequate tooling. Similar problems were encountered in EB welding tubes to the end pieces. Filament winding results were generally unsatisfactory because the machined parts did not have enough tie-in surface, adhesion to the stainless tubes was poor due to inadequate surface preparation and because the winding mandrels did not provide the desired rigidity of support. In particular, the winding vendor was not successful in sizing a mandrel loose enough to be withdrawn and tight enough to prevent some wrinkling of the stainless steel foil caused by winding tension.

TEST APPARATUS

Apparatus to test the superfluid bayonets requires a device which has little or no end effects, substantial volume for extended test periods and low tare heat leak so that the bayonet contribution is clearly distinguishable. The cryostat developed for this purpose consists of a 35 litre copper isothermal reservoir, a copper liquid nitrogen-cooled shield, approximately 3.5 cm of loosely wrapped (10 layers/cm) MLI, two thin stainless steel tubes heat stationed to the nitrogen shield and a thin G-11 valve actuator stem. The stub tube below the end of the female bayonet is also nitrogen shielded to reduce end heat leak. A valve at the bottom of the reservoir separates it from the test bayonet assembly allowing tare boiloff measurements. Nitrogen shield cooling is accomplished by intermittent flow through a continuous tube terminating in a small external "keepfull" reservoir equipped with a float-operated check valve. The male bayonet and cryostat share a common vacuum. The female bayonet and lower shield have their own vacuum. Calculated heat leak of the cryostat is 0.043 W including an allowance of 0.009 W for insulation. This value, based on a vacuum of 1.333 E-3 Pa, is probably too low for the actual test conditions.

The test setup for normal helium is shown in Figure 4. It is a conventional cryogenic boiloff arrangement with a precision gas meter. Ambient temperature and barometric pressure instrumentation were also available for readings during the test program. The apparatus was modified for superfluid testing by adding a 0-200 Torr mechanical vacuum gauge, a 56 CFM two-stage vacuum pump and a flow control device.

TEST PROGRAM

The test sequence consisted of determining the tare heat leak of the cryostat with normal helium, normal heat leak of cryostat and bayonet, superfluid overall heat leak and superfluid tare. Following these tests, the bayonet was subjected to 98 liquid nitrogen cycles with bayonets warmed to room temperature between each cycle. The final element of the

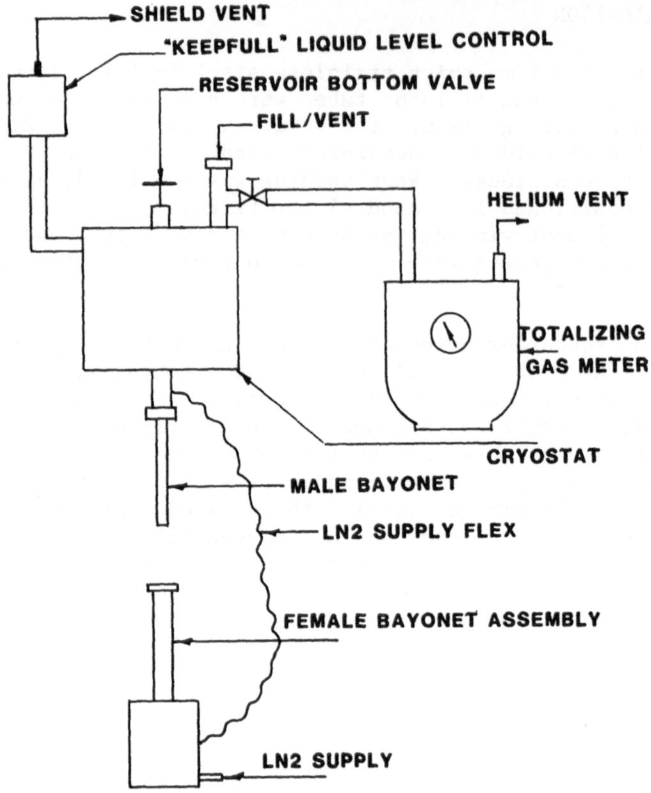

Fig. 4 Liquid helium bayonet test setup

test required a repeat of the normal and superfluid tests to determine performance impact of 98 thermal cycles. Also, at several points during the helium tests, a purge/relief line was used to check helium leakage past the nose seal.

TEST RESULTS

The experimental program was plagued by bayonet leak problems. Because of inadequate fiberglass anchoring and modulus mismatch, all of the stainless steel liner tubes eventually cracked from mechanical loads. These cracks resulted in vacuum leaks between the liner and composite over-wrap in every case. Repairs with Emerson Cumming 2850 FT black epoxy made the bayonets reasonably vacuum tight, but there was steady vacuum deterioration throughout the test sequence.

Despite vacuum problems, representative heat leak values were obtained prior to the liquid nitrogen cycling tests. The most consistent set of normal and superfluid values were taken in a two-day continuous run with the following results:

Normal Helium:	Total heat leak	0.251 W
	Reservoir tare	0.167
	Bayonet assembly	0.084 W
Superfluid Helium:	Total heat leak	0.376 W
	Reservoir tare	0.177
	Bayonet assembly	0.199 W

Since precise pressure control was not realized during the superfluid tests, the two reservoir tare values of 0.167 and 0.177 W were considered equivalent with the normal value more accurate. This had the effect of increasing the bayonet superfluid heat leak to 0.209 W which was the final value concluded from the program.

Nose seal leakage tests with the cryostat in operation did not yield accurate results because the purge/relief capillary line was not designed for efficient vacuum pumping. During the second helium test run after over 100 thermal cycles, the leak rate was estimated to be 10 E-4 atm-cc/s. A warm nose seal leak rate of 3.5 x 10 E-4 atm-cc/s was measured in a follow-up check on the KEL-F units used throughout the test program. Evidently, wear from the extended life test plus other miscellaneous make and break cycles increased the leak rate by two or three orders of magnitude. The final nose seal leak rate may have been great enough to allow some superfluid creep and attendant heat load.

CONCLUSIONS

Superfluid bayonets with heat leaks less than 0.25 W were demonstrated. The bayonets could be improved mechanically by better design and fabrication. Such bayonets are basically fragile and should be protected from mechanical bending loads at all times. KEL-F nose seals were functional with superfluid helium but further development should be aimed at determining leak rate versus surface finish and effective useful life. Cryostats for bayonet testing should be carefully designed and fully instrumented to provided all pertinent data.

ACKNOWLEDGMENT

This work was performed in fulfillment of NASA Goddard Space Flight Center Contract NAS5-29224 under the direction of Dr. Michael J. DiPirro.

REFERENCES

1. M.J. DiPirro and P. Kittel, The Superfluid Helium On-Orbit Transfer flight demonstration, 1987 Cryogenic Engineering Conference, Paper DA-3.
2. M.J. DiPirro, The SHOOT flight demonstration: Performance Estimates, 1987 Space Cryogenics Workshop, Madison, Wisconsin.
3. D. Evans and J.T. Morgan, Cryogen containment in composite vessels, in: "Advances in Cryogenic Engineering - Materials," Vol. 32, Plenum Press, New York (1986), p. 127.

THE SUPERFLUID HELIUM ON-ORBIT TRANSFER (SHOOT) FLIGHT EXPERIMENT

Michael J. DiPirro

NASA/Goddard Space Flight Center
Greenbelt, Maryland

Peter Kittel

NASA/Ames Research Center
Moffett Field, California

ABSTRACT

The SHOOT flight demonstration is being undertaken to verify component and system level technology necessary to resupply large superfluid helium dewars in space. The baseline configuration uses two identical 210 liter dewars connected by a transfer line which contains a quick disconnect coupling. The helium will be transferred back and forth between the dewars under various conditions of flow rate, parasitic heat load, and temperature. An astronaut Extra-vechicular Activity (EVA) is also planned to manually de-mate and mate the coupling. A number of components necessary for the flight are being developed. These components are described here.

THE MISSION

A large number of future orbiting facilities will require liquid helium to cool individual instruments or the entire facility itself. Among the former are the Advanced X-ray Astrophysics Facility (AXAF) containing the superfluid helium (SFHe) cooled X-ray Spectrometer; a second generation infrared instrument aboard the Space Telescope; and the multiple instruments on the Large Deployable Reflector. Among the latter are the Space Infrared Telescope Facility (SIRTF), and the Particle Astrophysics Magnet Facility (Astromag) aboard the space station. All these facilities require SFHe to cool through the vaporization of liquid helium. The gaseous helium is then vented to space. Meeting the lifetime goals of these facilities of from 10 to 20 years, requires the SFHe to be replenished every two to three years. Of the three methods of accomplishing the refill (return to earth, dewar replacement , or on-orbit helium transfer) the transfer of SFHe on orbit is the most technically feasible and cost effective[1]. To develop the technology to transfer SFHe in orbit, a Shuttle based experiment has been devised. Critical components are now being developed through ground based tests. These components will be integrated into the SHOOT flight experiment, and as applicable, will be used in the SuperFluid Helium Tanker (SFHT), which will perform the actual servicing missions. The SHOOT objective is to demonstrate the on orbit performance of the critical components and system level operation of the technology necessary for the successful SFHe servicing of orbiting facilities.

SIRTF has been selected as the model facility for computing the refill requirements[2]. The present SIRTF design has a 4000 liter SFHe tank. The SFHT must accommodate this volume; losses due to pre-servicing and servicing parasitic heat loads; and a cooldown of

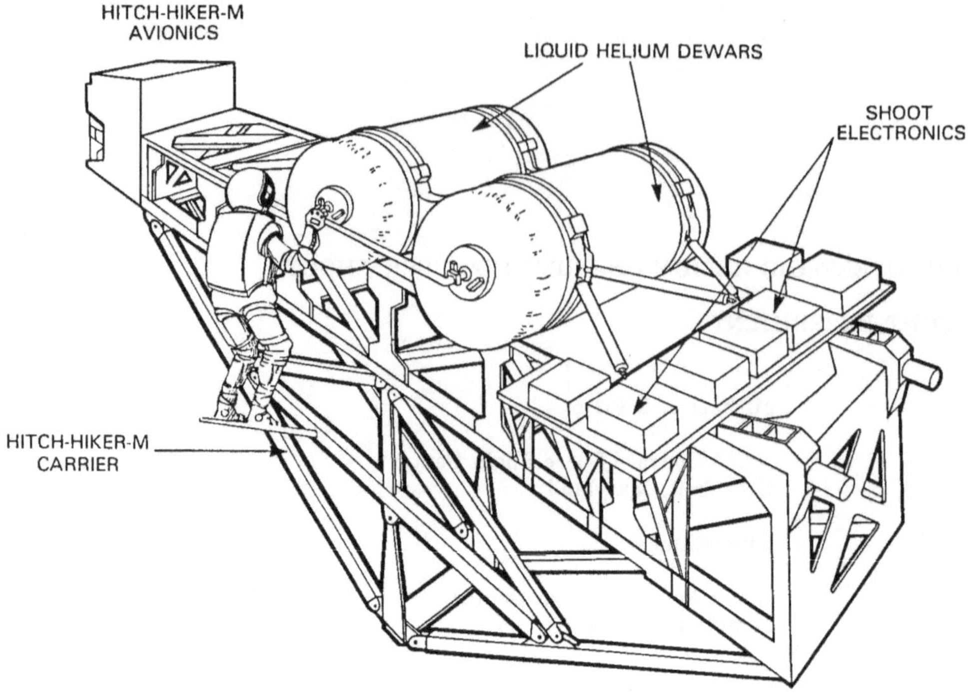

Figure 1. The SHOOT flight experiment on the Hitchhiker-M carrier

SIRTF from a temperature of 150 kelvin (K) in the event that SIRTF cannot be serviced before its SFHe is depleted. Cooldown will start with a flow rate of less than 15 liters per hour for efficient heat transfer, and the rate will increase as SIRTF cools. The cooldown time is estimated to be 45 hours. Present estimates put the tanker capacity at 10,000 liters or more. Shuttle time lines indicate that a transfer rate of at least 300 litres per hour with a goal of 500 liters per hour or more will be required once SIRTF is cold. During the initial part of the cooldown a pressure head of 22 kPa will be required. During cold Dewar transfer a head of less than 8 kPa will be required to refill SIRTF at the desired rate. SHOOT will demonstrate these transfer rates under steady state conditions (constant temperature and pressure) in the source and receiver dewars.

To transfer normal helium on the ground one can simply increase the pressure over the source liquid and use gravity to position the fluid over an outlet. This was demonstrated by Ball Aerospace Systems Division in both the Infrared Astronomical Satellite (IRAS) and the Cosmic Background Explorer (COBE) Dewars. The source Dewar was pressurized to 60 kPa at a temperature of 2.2 K. This allowed the top-off of a partially filled Dewar at a rate of 250 liters per hour with a loss of about 25% of the transferred liquid.[3] Due to its high thermal conductivity it is difficult to significantly overpressurize SFHe in this way. Furthermore, because of its low vapor pressure (< 5 kPa) it is not easy to transfer SFHe by creating a differential pressure between source and receiver Dewars. One may, however, successfully transfer liquid helium just above the lambda transition using overpressurization.
This parasitic heat input is given by

$$\dot{Q} = \dot{m}TS,$$

where $\dot{m}$ is the mass flow rate through the TM pump, T is the temperature of the source dewar (pump inlet) and S is the specific entropy of the SFHe at the pump inlet. Because in the temperature range from 1.4 to 1.9 K (the range of interest for the source dewar) the specific entropy varies as $T^{5.6}$ it is important to maintain as low a temperature as possible in the source dewar to minimize losses. The ideal pump will also require some extra heat to create a TM pressure to drive the SFHe into the receiver dewar. Non-ideal TM pump losses have been estimated.[4] Recent tests indicate real TM pump efficiencies of better than 95% at transfer rates of 325 liters per hour.

He I phase separation is achieved by restricting the flow of fluid out the vent. Liquid that does pass the restriction is vaporized, cooling the remaining helium within the dewar. This is the thermodynamic vent system (TVS) envisioned for venting other cryogenic fluids in orbit. The technique being developed for SHOOT will involve a compact version of a TVS in which the throttling device and the heat exchanger are the same unit. The device will contain either narrow slits in high purity copper or small pores in OFHC copper.[9] The high thermal conductivity of the copper, provides a good thermal path between the liquid vaporizing in the slits (or pores) and the bulk liquid in the tank. The cooldown efficiency is as good as that achieved in cooling a dewar in one g. Tests to date on sample slits indicate a phase separation efficiency of 95 to 100%, limited by our ability to detect very fine droplets visually. A test to measure the phase separation efficiency calorimetrically is being assembled.

An alternative is to pump the liquid between the dewars. This is the method chosen for SHOOT. Two types of pumps are being investigated: a centrifugal pump and a thermomechanical (TM) or fountain effect pump. Other mechanical pumps could also be used - bellows, reciprocating, etc. The pump feeds directly into a transfer line going to the receiver dewar. The transfer line has at least one coupling to allow an astronaut to connect and disconnect the source and receiver Dewars. The transfer line and coupler must be designed for low heat leak for two reasons. First, to conserve helium the parasitic heat leak must be as small as possible. A transfer line heat leak of 20 W for example would result in an 11% loss of helium at 200 liters per hour. Second, the heat which goes into the transferred SFHe must be taken out at the receiver dewar. The larger this heat load, the larger the receiver dewar vent line must be. This has an adverse effect on the low heat leak design of the receiver dewar.

Without gravity, positioning the liquid at the transfer line entrance (or pump inlet) becomes a much more difficult problem. For this reason a fluid acquisition system (FAS) must be installed within the dewar. The FAS must provide SFHe to the pump inlet at all flow rates and at a pressure to prevent cavitation. The system must be self priming and be able to counteract residual accelerations of the order of a milli-g in an any direction. The proposed system uses the surface tension of the SFHe to position the liquid. This system has the advantage that little extra driving pressure is needed, and this does not effect the fluid transfer.

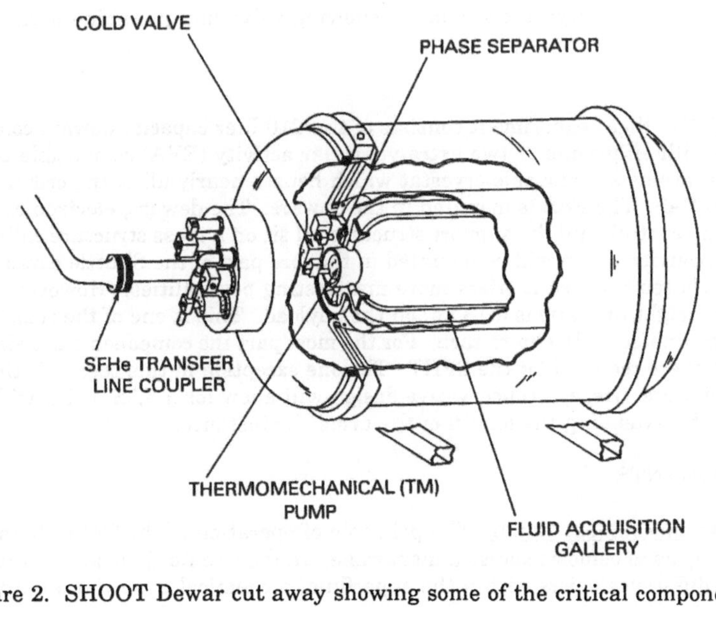

Figure 2. SHOOT Dewar cut away showing some of the critical components

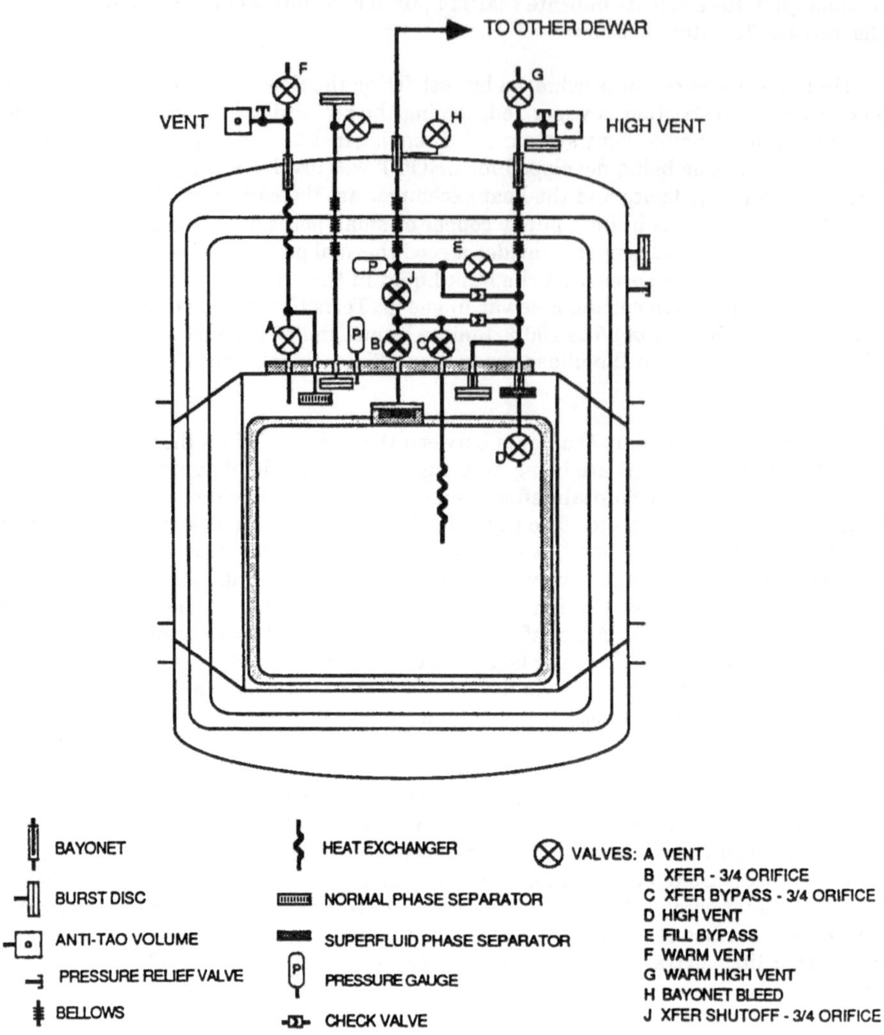

Figure 3. Dewar/cryostat schematic showing valve and plumbing arrangement

The SHOOT flight experiment consists of two 210 liter capacity dewars connected by a transfer line with either one or two extra-vehicular activity (EVA) compatible couplings. Each dewar contains a removable cryostat which houses nearly all of the critical components except for the FAS. The FAS is mounted in the dewars. The dewars, electronics for monitoring and control, and the support structure all sit on a truss structure called the Hitchhiker-M carrier which will be mounted in the fore part of the Shuttle cargo bay. This location was chosen because it offers more manifesting possibilities. However, its location places severe weight limitations (518 kg) on the payload. This is one of the reasons the size of the dewars are limited to 210 liters each. For the most part the components are sized to correspond to those required for the SFHT. The one exception in this regard is the FAS which must fit the dewars. Here, a conservative design will allow for a FAS which will work in the SFHT merely by extending the length of the arms, for instance.

THE COMPONENTS

Thermomechanical Pump. The principle of operation of the TM or fountain effect pump is analogous to osmosis across a membrane. In the two fluid model of SFHe when a temperature difference exists within the superfluid, a chemical potential difference results which drives the two components in opposite directions: the normal component which carries entropy flows toward the colder temperature and the super component flows toward the warmer region. If the normal component's flow is restricted, by its non-zero viscosity in small pores for instance, a flow imbalance exists, with net flow toward the warmer region. The application of heat on the downstream side of a porous medium can then be used as a pump. The advantages of this type of SFHe pump are its simplicity, reliability (no moving parts), and ease of flow control.

Losses occur in an ideal pump by the inverse to the TM effect, called the mechanocaloric effect, which occurs at the pump inlet. Here, since the super component alone flows through the pump, the remaining SFHe is rich in normal (entropy rich) component. This appears as a heating. In order to maintain the low temperature in the source dewar, helium must be boiled.

Centrifugal Pump. The centrifugal pump is being developed with the National Bureau of Standards[5]. It is a single stage centrifugal pump capable of flows up to 800 l/hr and capable of producing a pressure rise of 22 kPa. Three inducers have been tested. The first is a 6-bladed fan. With this inducer the pump shows evidence of cavitation for less than 300 mm of net positive suction head (NPSH) in superfluid and almost no performance at the desired operating point of 0 NPSH. In an effort to improve this a new inducer was tested. This one is a 3-bladed screw type with variable pitch. With this inducer the pump can pull a negative suction head (<-100 mm NPSH) in normal helium. However it still cavitates for < 100 mm NPSH in superfluid. To further reduce the cavitation, a jet inducer has been designed and is currently being tested. Early tests indicate that superfluid can be pumped at rates greater than 300 l/hr at a negative NPSH. For the jet inducer part of the pump's output is injected into the inlet, entraining liquid from the tank. About 20% of the flow is diverted to the jet. Even with this, the centrifugal pump is expected to have a transfer efficiency slightly better than that of a TM pump[4].

Fluid Acquisition System. Maintaining liquid at the pump inlet requires an FAS. Three types are currently under investigation for use in SHOOT. The first type is a screen channel device, also known as a gallery system, which phase separates liquid from gas by surface tension in the fine pores of the screen. For normal liquids the tank is pressurized resulting in liquid being forced into the channels covered with the screen, while the vapor is excluded from the channel because the screen is wet by the liquid and the surface tension of the liquid prevents a breakthrough. SFHe however does not easily support an overpressure in the saturated state, so the fluid in the channel will be slightly under the saturated vapor pressure, perhaps in a metastable state. A positive pressure within the channel could be obtained through the use of the TM effect, however the the effect will not be large due to the relatively large (>5 microns) size of the pores and short path length through the screen. Heat will more easily flow through a screen than through a TM pump, short circuiting the TM effect. An open cell small capillary device has been tested on a small scale as an alternative.[6] The small cells provide capillary action and also provide a small TM effect through the waste heat generated at the pump inlet (the mechanocaloric effect in the TM pump).

Phase Separators. Two types of phase separators are required for the SHOOT program. One separator is needed to provide phase separation and low pressure drop during transfer when relatively large (10 to 30 W) heat inputs will be generated in the source Dewar. This phase separator must also provide complete phase separation at low heat inputs (low flow rates). To accomplish this SHOOT will use a porous cylinder made of sintered stainless steel, similar to the porous plugs used in IRAS and COBE but larger in diameter and pore size. The larger diameter and pores will prevent the well known "knee"[7] in the pressure drop versus cooling power curve from occurring in our operating power range. A candidate small scale plug has been tested at GSFC. The results indicate that a plug of about 200 cm^2 will handle a flow equivalent to a 20 W heat input without an excessive (< 200 Pa) pressure drop.[8]

The second type of phase separator is designed to phase separate normal helium (He I) as well as SFHe. The requirement for this type of phase separator on SHOOT is driven by the last servicing of the dewars occurring 2 to 4 days before launch. During servicing the

Dewars will be filled and pumped to below the superfluid transition, then the vent valves will be closed and all vacuum pumps will be removed. Due to weight constraints, no on board vacuum pump will be used, hence from the time of last servicing until ascent, the helium within the dewars will be warming. Depending on the parasitic heat input and length of time before launch, the liquid may warm to above the superfluid transition. During ascent, a barometrically operated switch will open vent valves. From this point the Dewars will begin cooling. If the temperature of the helium is above 2.2 K, phase separation of He I and its vapor will be needed. (We do not expect the SFHT temperature to rise above the lambda point and require this type of phase separator.)

Transfer Line Coupling. A crucial interface between the source and receiver dewars is the transfer line coupling. This coupling must have a low heat leak to the SFHe being transferred, low flow impedance, and must be astronaut EVA compatible. A typical laboratory transfer line coupling for use with normal liquid helium has an o-ring type seal at the warm end and a close-fit type seal at the cold end of a long thin walled stainless steel tube. The close fit seal is sufficient in this use since escaping liquid vaporizes and its backpressure prevents more liquid from escaping. For SFHe the pressure is small in the first place and an overpressure cannot be maintained in any event, hence the nose seal must be tighter. Cryolab developed a new type of bayonet coupler for use with SFHe using an improved nost seal and lower thermal conductance material to provide a measured heat leak of less than 0.25 W.[10] A contract through the Johnson Space Center to develop a fully EVA compatible coupler for use on SHOOT and the SFHT is being negotiated.

Mass Gauging. Mass gauging in orbit for normal cryogens is a major technology problem with a number of techniques being investigated including RF, gamma ray absorption, and pressure pulse. SFHe, however, because of its high thermal conductivity, lends itself to the very simple technique of heat capacity measurement. SHOOT will use the heat pulse method of determining the heat capacity of the SFHe within the Dewar. The technique works as follows. Germanium resistance thermometers (GRT) are used to make precise temperature measurements within the dewar. A heat pulse of selectable power ($<$ 40 W) and measured duration (between .25 and 10 seconds) is sent through a heater or series of heaters within the dewar. The GRT is monitored during and after the heat pulse. Appropriate extrapolations are made to the pre and post pulse temperature drifts to determine the temperature rise due to the pulse. The mass is the heat input divided by the temperature change divided by the specific heat of SFHe at the average temperature. Note that the heat capacity of the tank and components is small compared to that of the SFHe over the operating temperature range. To achieve high accuracy the GRT's will be read out to a precision of 0.1 mK or better and their accuracy will be 1 mK to precisely identify the specific heat (over the SHOOT temperature range of 1.4 to 1.9 K the specific heat varies - like the entropy - as $T^{5.6}$). Coupled with the accurate knowledge of the voltage and length of the pulse, SHOOT will determine the mass of SFHe within the dewar to 1% of capacity.

Flow Meters. To determine the flow rate at any particular time it is not practical to perform multiple mass gaugings. A SFHe flowmeter is therefore desirable. Three types are under investigation. One type uses the mass and thermal flow properties of SFHe to measure the mass flow rate via heat flow.[11] The temperature upstream and downstream of a heater is measured. The flow rate is computed from the temperature, temperature gradients, heater power and geometry of the flow tube. A second type of flowmeter is a turbine meter[12]. It was found to have repeatable readings in both superfluid and in normal fluid helium. However, it cavitates easily in superfluid if the backpressure on the meter is low. Furthermore, it is difficult to cryo rate the small bearings and two-phase flow can over-spin the rotor. For these reasons we are no longer considering this meter. The final type of meter tested is a venturi meter[12,13]. At high flow rates this meter behaves as if superfluid helium is a classical Newtonian fluid. It gives repeatable results in both normal and superfluid helium. It does cavitate if the backpressure is low. However, this is not expected to be a problem in our application.

Friction Factor. The flow of liquid helium through the transfer line in SHOOT is expected to be at Reynold's numbers of 10^4-10^6. For modelling the transfer process we needed to know the friction factor at these flow rates. Fortunately the University of Wisconsin had just built a helium flow facility in which we have been able to measure the friction factor of superfluid[13]. For smooth tubes it was found that at these high Reynold's numbers that superfluid behaves like a Newtonian fluid, following the von Karman correlation. For bellows the pressure drop appears to be 50% greater than predicted.

Cryogenic Valves. Each dewar on SHOOT alternately acts as a source and receiver for helium transfer. This coupled with the requirement for two fault tolerance for astronaut EVA forces SHOOT to incorporate 3 warm and 6 cryogenic valves in each dewar. Reliability is crucial; the loss of any one of the valves could severely limit the chances of fulfilling the mission. Utah State University's Center for Space Engineering is developing two sizes of stepper motor operated cryogenic valves for SHOOT.[14] Both are based on standard Nupro valves, with a custom gear drive, stepper motor and limit switches attached. Nominal 3/4" Nupro valves are used in the transfer path to provide acceptable pressure drops at high flow rates. Nominal 1/2" valves (the same type used on COBE) are used in vent and bypass lines. Thermal cycling and vibration tests have shown no detectable leakage ($<10^{-9}$ std. cc/sec.).

THE SYSTEM

Safety. The stored energy in a liquid helium dewar can be large. If the guard vacuum, which provides most of the thermal insulation for the liquid, is lost due to main shell failure or cryogen tank failure, the build up of pressure within the cryogen tank can be very rapid. Containment of the expanding gas is out of the question for all but the very smallest dewars, even in ground systems. Hence, to make the payload safe in the event of a rapid pressurization of the cryogen tank, the controlled release of the helium is required. Passive relief devices are desired, since active control (such as valve actuation) is more prone to failure. The passive devices used for SHOOT are burst disks developed by Ametek, Straza Division.[15] These burst disks use a patented plunger mechanism to burst a uniform diaphragm, rather than using a scored diaphragm. Tests have shown the former to be more reliable and repeatable than the latter type. Burst disks are connected to the cold and warm ends of two nominal 1.9 cm diameter vent lines. One line is dedicated to the pressure relief function, the other serves as a low impedance vent during the transfer process. Each of the lines alone are capable of venting the helium evolved for the worst case heat input scenario.[16] In addition, SHOOT is considering a coating of thermally insulating material applied over the outer surface of the cryogen tank to limit the heat input rate to the fluid.

Transfer Efficiency. The overall efficiency of a refill operation in general, from weight and volume at launch to amount of SFHe delivered to a payload, depends on the operating scenario: condition of the helium at launch, delay between launch and transfer operations, etc. This type of system analysis is not within the scope of this paper. During a transfer the efficiency depends on the interactions of many of the components of SHOOT. If we define the transfer efficiency as the mass of helium delivered to the receiving dewar at the temperature and pressure of the source Dewar, divided by the mass decrease of the source dewar, then efficiencies better than 90% are expected at the required flow rates when the receiver starts out cold. Reference 8 contains detailed estimates of the steady state end to end efficiency of SHOOT.

EVA Operations. The purpose of the EVA is to demonstrate the human factors associated with helium resupply. In particular, with handling the transfer line and with operating the coupling. The EVA will also be an opportunity to verify the astronaut interface with the control/data system. In demonstrating the EVA coupling and transfer line the thermal performance and the flow impedance will be measured before and after the coupling operation. The procedure for the EVA follows the these steps: 1) launch with the coupler mated, 2) perform several transfer operations, 3) demate then re-mate the coupler, and 4) perform several more transfer operations. This approach allows all of the EVA functions to be performed during a single EVA.

Data/Command System. Our approach to the data command system is to build it in several stages of increasing complexity but with the system operational as soon as the first stage is complete. The system will start with a process controller as a core. This will consist of real time data acquisition and real time control part. This simple system will allow for manual control using pre-packaged routines and for displaying and archiving data. Later this system will grow to fully automatic control system. In parallel, an expert system shell is being developed. This shell will contain fault diagnosis of valves, sensors, and pumps. Finally this will grow to a full expert system with fault work-arounds. The primary means of experiment control will be in real time from the ground. In addition there will be a somewhat more limited capability for control and data acquisition from the aft flight deck of the orbiter. The latter controller will provide the astronauts with the status of the experiment during the EVA and will allow them to control the transfers.

REFERENCES

1. T.C.Nast, et.al. "SIRTF Telescope Instrument Changeout and Cryogen Replenishment (STICCR) Study" NASA CR 177380 (Aug. 1985); and A.J.Mord, et.al. "BASD- SIRTF Telescope Instrument Changeout and Cryogen Replenishment (STICCR) Study" NASA CR 177381 (Nov. 1985)
2. J.H.Lee and Y.S. Ng; "Pump Performance Requirement for the Liquid Helium Orbital Resupply Tanker" Cryogenic Engineering Conference (St. Charles, 1987), Paper FC-3.
3. M.G. Ryschkewitsch, private communication.
4. P. Kittel: "Orbital Resupply of Liquid Helium" J. Spacecraft and Rockets, 23 (1986), 391; M.J. DiPirro and S.H. Castles, Cryogenics 26 (1986), p. 84; P. Kittel: "Losses in Fountain Effect Pumps" Proc. 11th International Cryogenic Engineering Conference (Butterworth, 1986) p. 317; M.J. DiPirro and R.F. Boyle, Lab tests of a thermomechanical pump for SHOOT, CEC, 1987, Paper BC-2.
5. P.R.Ludtke, D.E.Daney and W.G.Steward; "Centrifugal Pump Performance with Normal and Superfluid Helium" Cryogenic Engineering Conference (St. Charles, 1987) and P.R.Ludtke and D.E.Daney; "Cavitation Characteristics of a Small Centrifugal Pump in He I and He II" Space Cryogenics Workshop (Madison, WI, 1987)
6. J.E. Anderson, D.A. Fester, and M.J. DiPirro, "Acquisition System Testing with Superfluid Helium", CEC,1987, Paper FC-2.
7. M.J. DiPirro, F. Fash, and D.C. McHugh, "Precision Measurements on a Porous Plug for Use in COBE", Proceedings of the 1983 Space Helium Dewar Conference, Huntsville, Alabama, p. 121; G. Karr and E.W. Urban, Cryogenics 20 (1980) 266; J.B. Hendericks and G.R. Karr, Proc. of ICEC 9, (1982).
8. M.J. DiPirro, "The SHOOT Flight Demonstration: Performance Estimates", Proceedings of the Space Cryogenics Workshop, June, 1987, to be published in Cryogenics.
9. M.J. DiPirro and S.H. Castles, Cryogenics 26 (1986), 84
10. G.E. McIntosh, et al., "Bayonet for Superfluid Helium Transfer in Space", CEC,1987, Paper FC-4.
11. S.R. Breon and M.J. DiPirro, "He II Flow Meter", CEC, 1987; the technique makes use of the investigations reported in A. Kashani and S.W. Van Sciver, Adv. in Cryo. Eng. 31, (1985) 489.
12. D.E.Daney; "Behavior of Turbine and Venturi Flowmeters in Superfluid Helium" Cryogenic Engineering Conference (St. Charles, 1987) and D.E.Daney; "Cavitation in Flowing Superfluid Helium" Space Cryogenics Workshop (Madison, WI, 1987)
13. J.R.Maddocks, S.W.van Sciver, P.L.Walstrom and J.G.Weisend; "Pressure Drop and Flow Metering in He II" Space Cryogenics Workshop (Madison, WI, 1987)
14. R.H. Haycock, "Status Report on NASA Contract NAS5-29422", Report number CSE/87-007, to be available from the Clearinghouse for Federal Scientific and Technical Information, Springfield, VA 22151.
15. "Verification Test Report for Cryogenic Burst Disk Assembly", Report No. 8-480454 Rev. A, March, 1987, to be available from the Clearinghouse for Federal Scientific and Technical Information, Springfield, VA 22151.
16. G. Colon, O. Figueroa, and R.A. Callens, private communication.

CRYOGENIC AND THERMAL DESIGN FOR THE SUPERFLUID HELIUM ON-ORBIT TRANSFER (SHOOT) EXPERIMENT

J. H. Lee, S. Maa and W. F. Brooks

Telescope Systems Branch
NASA Ames Research Center
Moffett Field, California

Y. S. Ng

Sterling Federal Systems
Palo Alto, California

ABSTRACT

During the next 10 years several space missions will require superfluid helium replenishment at two or three-year intervals to extend their lifetimes. A shuttle experiment is in hardware development which will demonstrate many of the key components including fluid pumps, acquisition devices, flowmeters, quick disconnect EVA couplers and process controllers. The experiment will consist of a series of transfers between two 200-liter dewars. This paper describes the design, analysis and trades conducted for the external thermal design of the dewars and support electronics as well as the optimization and prediction of the dewar and cryostat assembly performance. Of particular importance in the analysis are the ground hold and standby performance of the dewars as well as the temperature of the helium bath during high flow rate transfers. The Hitchhiker carrier is a flight of opportunity and the exact orbit inclination and shuttle profiles will not be known until approximately one year before launch. As a result, the external thermal design focused on the stability of the dewar, transfer line, and electronics temperatures in a number of representative orbits.

INTRODUCTION

During the 1990's NASA is planning to launch several missions which will require on-orbit replenishment of liquid helium tanks. An example of such a mission is the Space Infrared Telescope Facility (SIRTF)[1] which has a nominal dewar size of 4000 liters and a lifetime of 2-3 years. To extend the mission to five years the helium tank will be serviced on orbit. This will be accomplished at either the Space Shuttle or Space Station when it becomes available in the late 1990's. Previous studies have been conducted to determine the requirements for the on-orbit servicing equipment or tanker[2,3] and indicate that a tanker approximately 8-10,000 liters will be required to insure sufficient cryogen to service a payload that has run out of helium and requires a cooldown. Transfer rates of at least 300 liters/hour and preferably 500 liters/hour will be required to allow servicing from the Space Shuttle during a 7 day mission. Because containing and pumping large quantities of liquid helium in a low-g environment presents several unique problems, NASA is currently developing the Superfluid Helium On-Orbit Transfer (SHOOT) experiment. This experiment will fly aboard the Space Shuttle to obtain key data on critical components such as the pump, liquid acquisition device, and flowmeter. In addition, system configuration and operational approaches to on-orbit transfer including astronaut Extra Vehicular Activity

(EVA) operations will be verified. The experiment can the monitored and controlled from either the Shuttle Aft Flight Deck or a ground Payload Operations Control Center. This paper details the design and analysis of the exterior thermal control for the experiment as well as the thermal performance of the flight dewars and cryostats.

SYSTEM CONFIGURATION AND REQUIREMENTS

The general configuration of the experiment is shown in Fig. 1 which is a 2-D view of the experiment from the Shuttle Aft Flight Deck Window. Two 210 L superfluid helium dewars weighing 177 Kg each are connected by a low heat leak transfer line. The transfer line is mated to the dewars using SIRTF prototype disconnect fluid couplers which are compatible with astronaut EVA. Inside the dewar, a cryostat contains the valve manifolds, pumps and fluid management devices required to transfer the helium between the two tanks. The specific pump to be flown has not yet been selected and prototype thermomechanical and centrifugal helium pump systems are currently undergoing testing. The dewars and associated controllers and sensor electronics are mounted on a cross bay structure called a Hitchhiker Carrier. The total experiment weight on this modified Mission Peculiar Experiment Support Structure (MPESS) is limited to 517 kg. The Hitchhiker avionics supplies experiment power and command and data handling interfaces to the Shuttle but the experiment must supply its mission unique electronics. For the SHOOT experiment this consists of pump and flowmeter drive electronics, temperature and pressure and level sensor circuits, valve drive electronics and backup power. The dewars and electronics will be integrated with the carrier at the launch site.

THERMAL DESIGN

Requirements

The experiment will be flown as a payload of opportunity and will not control the Shuttle attitude or the exact orbit inclination or altitude. This necessitates a versatile temperature control system which can react to a wide range of heat fluxes. The objective of the thermal design is to minimize the outer shell temperature of the dewar and transfer line, to maintain the electronics within their survival temperatures for the worst cases, and to determine the conditions under which the helium transfer and EVA operations can be performed. For survival purposes the electronics must be maintained between 233-343 K.

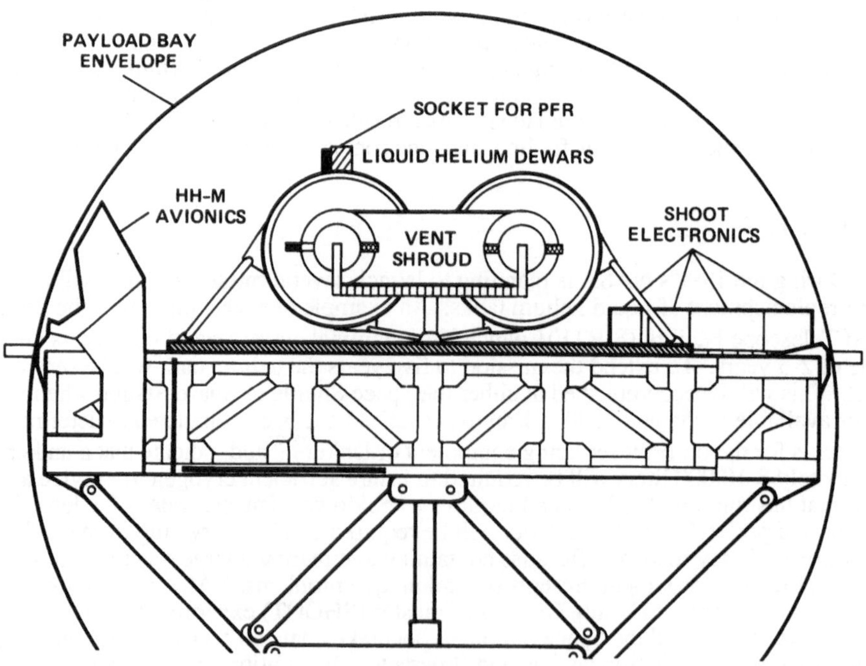

Fig. 1. SHOOT system on Hitchhiker-M carrier

For helium transfer operations the electronics must be maintained between 253-323 K and the EVA associated hardware must be between 155-386 K. The power dissipation of the electronics boxes is a key element in determining the system performance. There are 6 electronics boxes which vary from 0-45 watts standby and 0-50 watts operating power. The box power dissipation is given in Table 1.

Table 1 SHOOT electronic box power dissipation

Electronic Box	Standby	Normal	Weight
Heater Box (Hlvs)	45 W	50W	16KG
Command and Data Handling (C&DH)	15	15	7
Power Distribution Box (PDU)	20	28	9
Aux. Power (APU)	0	0	16
Temp. and Pressure (TPS)	20	20	5
Flow Measurement (FM)	5	5	5

Fig. 2 is a graphical representation of the four orbit scenarios which have been chosen to explore the extremes and expected nominal temperature of the experiment. Case one and case two are two high inclination orbits with the cargo bay facing the sun and space. Case 1 represents the hottest orbit and case 2 the coldest. Case 3 represents a low inclination orbit in which the sun is in the orbit plane and the cargo bay faces the earth for the entire orbit. Case 4 represents a 52 degree inclination orbit in which the bay faces the sun for the entire orbit. Cases 3 and 4 represent a range of expected heat loads. In all cases the Shuttle was assumed to remain in the orbit long enough to reach equilibrium. Other key assumptions were as follows:

Orbit Altitude:	296 km	Solar Constant:	1 400 W/m^2
Albedo Factor:	.3	Earth IR Emission:	242 W/m^2

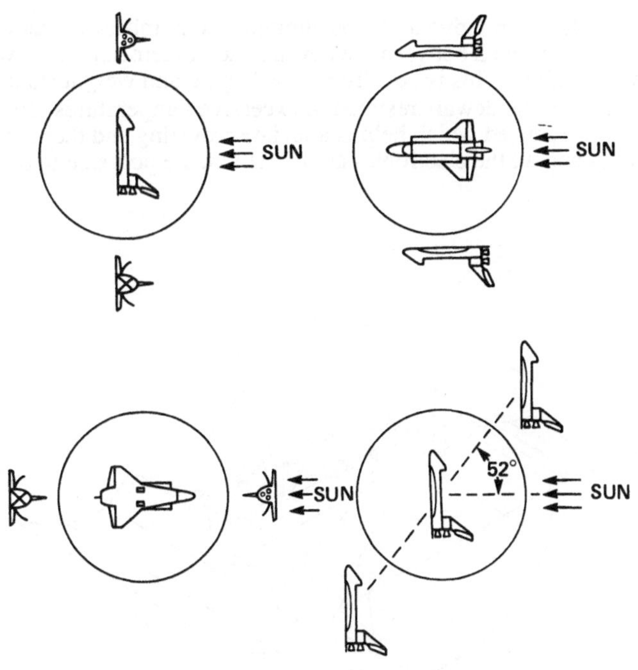

Fig. 2. Mission models for thermal design.
Case 1-Upper left = b = 90°, hot. Case 2-Upper right = b = 90°, cold
Case 3-Lower left = b = 0°, nominal. Case 4-Lower right = b = 52°, hot

The extreme weight limitation required the thermal control system to be as light as possible which precluded the use of heat pipes and louver type radiators. No modifications to the STS or H/H-M surfaces were allowed since the Hitchhker program approach is to minimize integration and interface complexity. Finally the forward portion of the experiment facing the AFD must be kept clear for EVA access.

Thermal Model

Thermal math models of the STS cargo bay, experiment H/H-M carrier the dewar and the flight electronics were constructed using SSPTA and SINDA[4]. SSPTA was used to calculate radiation couplings and absorbed environmental heat fluxes. These results were then entered into SINDA to generate steady state and transient temperature distributions. The temperature distribution and thermal response of the dewars are used as boundary conditions for the cryogenic model to evaluate dewar and cryostat performance. The thermal model, shown in Fig..3, contained the following elements:

1. STS Cargo Bay, forward bulkhead, STS radiators and a blocking surface to simulate nearby payloads. Thermal properties for beta cloth, MLI and silverized Teflon tape radiators used on the STS cargo surfaces and were obtained from STS flight data.[5]

2. The dewar thermal blankets were modeled as 18 layers of double aluminized Kapton MLI, with beta cloth as the outer most layer.

3. The dewar support structure and MPESS painted with white paint (chemglaze properties).

4. The aluminum electronics boxes were covered with beta cloth and/or MLI. Some surfaces of the boxes are used as radiators with bare metal surfaces covered with 5 mil thick silverized teflon tapes. The electronics is thermally isolated from the MPESS support bridge.

Trades

Trades were conducted relative to the positioning and number of electronics boxes. Initially the boxes were positioned between the dewars and underneath the dewars on the outside of the dewar assembly. This is best from a packaging and weight standpoint; however, the shadowing by the dewars resulted in excessive temperatures. In the final design the number of boxes was reduced which helped alleviate crowding and thermal trapping in the gaps between the boxes and all the electronics are located on the port side to obtain a better

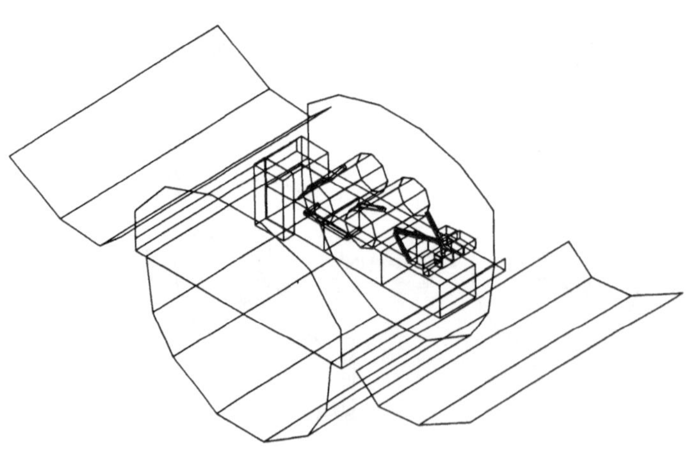

Fig. 3. Thermal model of SHOOT system in cargo bay

viewfactor to space as shown in Fig. 1. All the sides of the boxes which are not radiators are insulated with 18 layers of MLI with a final layer of beta cloth. Each box is assumed to be thermally isolated from the MPESS using a system of titanium bolts an fiberglass spacers. A radiator fin was required on the top of the Power Distribution Box to lower the box temperature. There are a total of 12 radiator surfaces with an effective area of 0.8 m^2.

Table 2. External temperature distribution

COMPONENT	Case1 (hot)	Case2 (cold)	Case 3 (nominal)	Case4 (52 deg/hot)
DEWARS	318-391 K	116-168 K	244-289 K	227-379 K
TRANSFER LINE	347-353	145-147	248-286	255-348
HEATER BOX	342	258	301-308	323-328
POWER DIST	334	262	296-308	2320-325

For the high inclination orbits both the transfer line and dewars are outside the EVA operation temperature limits during the orbit. For the (hot) case 1, the electronics are within the operation limits so that helium transfers can be performed. For the cold case 2 the electronics can be maintained within their operational temperatures limits by adding 10-15 Watts per box resulting in the temperatures given in column 2 of Table 2. For the low inclination orbit case 3 with the orbiter bay facing the earth at all times both the dewar and EVA hardware as well as the electronics are within their operating temperatures for the entire orbit. For case 4, the 52 degree inclination with the bay facing the sun continuously, the MPESS stucture and several of the electronics boxes exceed their operational limits for part of the orbit but all stay within their survival limits.

CRYOGENIC DESIGN

Requirements

The dewars must be sized to provide enough helium on orbit perform to perform 8 helium transfers between the dewars. Six superfluid helium tests ranging from 100-800 L/h and two warm transfers into a dewar that has been raised to 20 K and 150 K respectively are planned. During ground hold, the helium is maintained in the normal condition at 4.2 K. The helium is converted to superfluid during launch and orbit insertion. Ground hold is baselined as 4 days although there is the possibility that the system may have ground holds as long as 12 days. Servicing of the dewars will occur in both the horizontal and vertical orientations. The dewar fill, vent and transfer manifold as well as the pump must provide a minimum flow rate of 300 L/h (500 L/h as a goal) under temperature and flow impedance conditions which simulate the SIRTF manifold. This requires the pump to develop a dynamic head of approximately 170 torr for cooldown of the warm dewar. The manifold must be two failure tolerant and as a result 6 valves and 5 burst disk are required per dewar. The plumbing manifold shown in Fig. 4 must be sized to allow efficient operation for the fountain effect pump. The dewars are constructed to withstand Shuttle launch loads and the support straps are sized based on fatigue.

Model

The dewar was modeled using a software program (ACAP) developed at Ames and the model was subsequently verified on SINDA. A representation of the model is shown in Fig. 5. Each dewar weighs 177 kg with about 98 kg of cryostat and helium suspended on the tensioned strap support system. Each dewar's six tensioned S-glass straps have dimensions of 2.05 cm width, 0.1 cm thickness and 13.34 cm length. The 304 stainless steel fill vent plumbing manifold is modeled as .038 cm wall thickness 1.19 cm diameter lines except for the high vent line which has a diameter of 1.83 cm. The wall thickness has recently been changed to .05cm which has not been incorporated in the model; it is expected to increase the heat leak and increase helium boiloff. The bellows section in each line are 3.37 cm long with .1mm wall thickness except for high flow line which is 2.54 cm in length and 15 mm wall thickness. There are 5 layers of MLI between the tank and inner vapor cooled shields (VCS) 15 layers between the vapor cooled shields and 35 layers between the shield and outer wall.

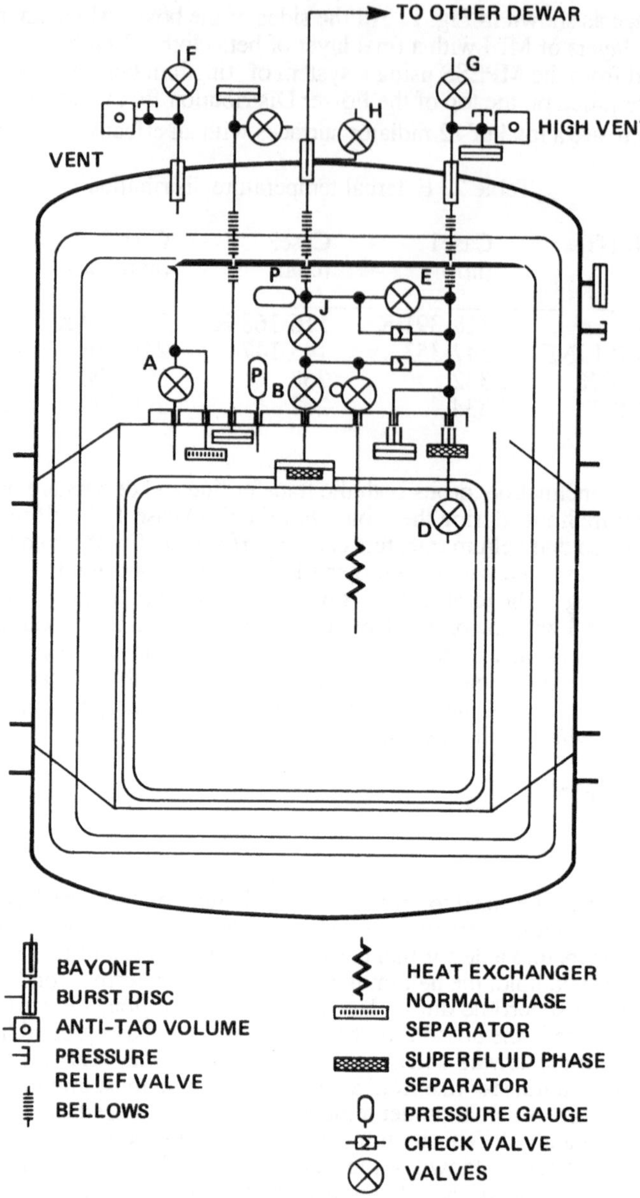

Fig. 4. SHOOT Dewar configuration

The MLI performance is based on the model used for IRAS [6]. The wiring harness is modeled as 128 #36 AWG manganin and 96 #28 AWG manganin wires. Current estimates for the actual wire count for the #36 AWG is 88. The outer shell is modeled as covered with 18 layers of MLI with a beta cloth cover. The orbit average results from the thermal model are used to set the temperature of the beta cloth. Each of the vapor cooled shields is modeled as perfectly connected to the cryostat and straps but a heat exchanger effectiveness is calculated in ACAP. This effectiveness is almost always above 95%. The helium tank itself is modeled as having the specific heat associated with the cold fluid but the actual vessel material is ignored.

Trades

In the initial design the strap thickness was sized for infinite lifetime with a thickness of 0.16 cm. Because of the high helium boiloff during ground hold the strap thickness was reduced to 0.1 cm thickness. The initial configuration also had the transfer line connected directly from the outer shell to the tank. The model indicated that if the transfer line on each cryostat is thermally linked to the inner vapor cooled shield 60 mW is added to the helium during transfer but this design change significantly reduces the ground and on-orbit standby parasitics. The initial configuration had a 175 liter tanks and resulted in a negative margin after a 4 day ground hold, 4 day on-orbit hold and 8 transfer tests. The dewars were therefore increased to 210 liters. An additional trade was performed to determine the ability of the dewar to withstand a 4 day hold in the superfluid state. This requires the helium bath vent to be valved off which stops vapor cooling of the shields. Under these conditions the bath rose above the lambda point at approximately 2 days. As a result the decision was made to maintain the fluid in the normal state during ground hold with the helium venting into the payload bay. Special high conductivity high vent plugs are used to convert the helium to superfluid on orbit.

Performance

The dewar ground hold performance at 4.2 K helium temperature is 0.75%/day boiloff by mass. The dewars are sensitive to the MLI performance and when the MLI is degraded by a factor of 1.6 to account for manufacturing, the helium boiloff rises from 0.75%/day to 1%/day. For comparison the IRAS[6] system was 0.85% per day by mass and the two spacelab 2 dewars were approximately 5.0% per day [7,8]. The key to the SHOOT dewar performance is the low cross section tensioned fiberglass support strap system and the bellows isolators in the fill/vent manifold. After vertical servicing the dewars are assumed to be 95% full at 4.2 K which is equivalent to 26 kg of helium per

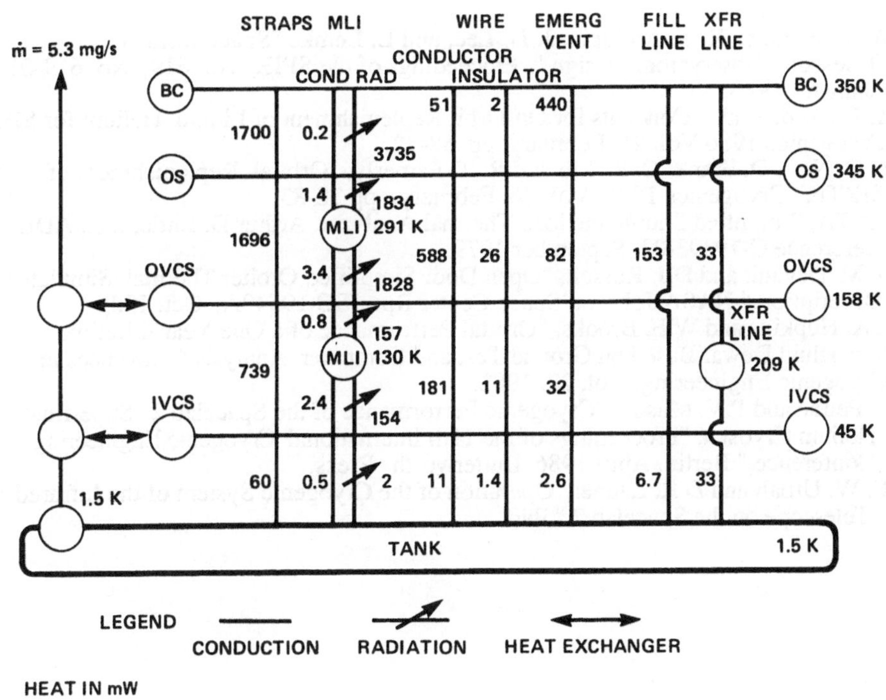

Fig. 5. SHOOT Dewar thermal network model

dewar. The current model predicts that during a 5 day hold the system will boil away 1.3 kg (11 L) of normal fluid per dewar. The conversion to superfluid during launch and insertion will leave 15 kg of superfluid (103 L) per dewar and the on-orbit hold for 4 days will require 1 kg (15 L) per dewar. This results in 95 liters of superfluid or roughly two half full tanks at the start of testing. The six cold tank to tank transfers at superfluid temperature will take a total of 4 hours and will consume 58 liters. The two warm transfer will consume 52 liters which leaves a residual of 40 liters per tank or a reserve of 19 %. As an example of the criticality of the dewar design, if the ground hold was extended to 12 days which is being considered and if the dewar were to consume 5.0%/day, there would be only 5 kg or 30 liters remaining in each dewar at the start of testing and an extremely curtailed test program would have to be constructed.

An analysis of the behavior of the pumps and tanks during a high flow transfer when 30 W is dissipated in the main bath was performed. The model indicated that the 1.19cm diameter fill/vent manifold valves were the principal impedance and were not large enough to maintain the bath temperature below 1.5 K during the helium transfer. To allow the fountain effect pump to operate efficiently the three fill/vent valves were changed to to 1.91 cm inlet diameter. The current requirement is to maintain the bath at or below 1.4 Kelvin with up to 20 W dissipated in the dewar helium tank by the fountain effect pump. Current model indicates that with all vent lines open the dewar temperature wil rise to 1.6-1.8 K during a transfer in which the helium pump dissipates 20 Watts.

DISCUSSION

The overall thermal and cryogenic design meets the performance requirements and can accomodate the anticipated variations in the ground hold and the orbit selection. The dewars are very low boiloff rate dewars even by space standards and close attention to detail in construction will be required although systems with the requreid low heat leaks have been produced for previous space missions.

REFERENCES

1. W. F. Brooks, R. K. Melugin, J. H. Lee, and L. Lemke,"Space Infrared Telescope Observatory Design,", Proceedings of the SPIE, Vol. 619, No. 619-01, January 1986.
2. A. J. Mord, et al., "Concepts for On-Orbit Replenishment of Liquid Helium for SIRTF," Cryogenics 1986 Vol. 26, February, pp 68-72.
3. T. C. Nast, D. Frank, C. K. Liu and R. T. Pamerly, "Orbital Replenishment of SIRTF," Cryogenics 1986, Vol. 26, February, pp 78-83.
4. SSPTA, Simplified Shuttle Payload Thermal Analyzer, Arthur D. Little, Inc., ADL Reference C-79393-03, September 1979.
5. G. M. Devault and D.J. Russell, "Open Door Simplified Orbiter Thermal Simulator Description," NASA Johnson Space Center Rpt. JSC-19540A, Oct. 1985
6. R.A. Hopkins and W.F. Brooks, "Orbital Performance of a One Year Lifetime Superfluid Dewar Based on Ground Test and Computer Analysis," Advances in Cryogenic Engineering, Vol. 27, 1982.
7. D. Petrac and P.V. Mason, "Cryogenic Performance of the Spacelab 2 Superfluid Helium Cryostat, "Proceedings of the 11th International Cryogenic Engineering Conference," Berlin, April 1986, Butterworth Press.
8. E. W. Urban and D.R. Ladner,"Operation of the Cryogenic System of the Infrared Telescope on the Spacelab 2," ibid

ACQUISITION SYSTEM TESTING WITH SUPERFLUID HELIUM

John E. Anderson
Dale A. Fester

Martin Marietta Denver Aerospace
Denver, Colorado

Michael J. DiPirro

NASA Goddard Space Flight Center
Greenbelt, Maryland

ABSTRACT

In support of NASA's Superfluid Helium On-Orbit Transfer (SHOOT) Flight Experiment, two pumping approaches are being evaluated. NASA-GSFC is conducting tests on the thermomechanical (TM) pump while NASA-ARC is sponsoring tests on a centrifugal pump at the NBS Cryogenic Laboratory. For either of these approaches, a common concern is the method that should be employed for fluid acquisition. While technology is available for design, fabrication, and operation of fluid acquisition systems with conventional fluids, performance with superfluid helium (SFHe) is uncertain. As a result, Martin Marietta joined with NASA to investigate the use of capillary acquisition systems. Assurance is needed that the devices will work with superfluid helium under an adverse 10^{-4}g operational environment. Minus one-g outflow tests were conducted with SFHe in conjunction with the TM pump set-up in the Cryogenic Laboratories at GSFC. Both fine mesh screen and porous sponge systems were tested. A screen acquisition device was also tested with the low-NPSH centrifugal pump. Results to date show that the screen and sponge are capable of supplying SFHe to the TM pump inlet against a one-g head up to four cm which is more than sufficient for the SHOOT application. Results with the sponge are reproducible while those with the screen cannot always be repeated. Further tests to characterize these systems are required.

INTRODUCTION

Present and future NASA orbiting facilities will require superfluid helium (SFHe) to cool individual instruments (e.g., the X-ray Spectrometer aboard the Advanced X-ray Astrophysics Facility) or the entire facility (e.g., the Space Infrared Telescope Facility). Most of these SFHe users desire on orbit refilling of their liquid helium dewars to allow the instruments to function over a long lifetime, improving their utility and cost effectiveness. To this end NASA, through the Office of Space Flight and the Office of the Space

Station, has undertaken a program to develop the technology to transfer liquid helium on orbit and to demonstrate that technology on a shuttle-attached payload called the Superfluid Helium On-Orbit Transfer (SHOOT) Flight Experiment[1]. The SHOOT will demonstrate the components and techniques needed to develop a large (10,000+ liter) SFHe resupply tanker. A number of components are required to be developed for the SHOOT program including the SFHe pump and the fluid acquisition system for the dewars, SFHe phase separators for the vent lines, a mass gauging system, low heat leak transfer line and quick disconnect couplings that are astronaut compatible, and a SFHe flow meter. Of these development items, the fluid acquisition system has the greatest requirement for on-orbit testing and hence represents a challenge to design a successful system for flight without the ability to verify its performance completely on the ground. Due to the interaction of the fluid acquisition system with the SFHe pump and other components, one quickly realizes the need for a system level flight experiment to verify the design. The differences in properties and behavior between SFHe and conventional fluids further magnify the acquisition system challenge and underscore the need for thorough ground testing as a precursor to the flight experiment.

ACQUISITION SYSTEM EVALUATION

Operational requirements and guidelines for the SHOOT experiment were used to provide the liquid acquisition system design requirements shown in Table 1. Candidate capillary acquisition system concepts were then defined that might meet the requirements. Both screen and sheet metal vane devices were considered. From a comparative analysis, a screen channel system was selected for further evaluation since it can operate over a wider range of environmental accelerations than vane systems and can provide larger design margins. The small pores of fine-mesh screen potentially maximize capillary forces with SFHe which has a very low surface tension. A preliminary design of a four-channel acquisition system is shown in Fig. 1. The channel is formed by solid sheet metal on three sides with 325 X 2300 mesh stainless steel screen supported by perforated plate on the fourth side facing the tank wall.

A pressure loss analysis was conducted on the acquisition device that included hydrostatic, channel internal flow, fluid acceleration, and screen entrance loss factors. The channel internal flow friction factors were based upon experimental results obtained from previous studies.[2,3] Pressure drop through the screen was based upon the Armour-Cannon relation.[4] It was assumed that the device would

Table 1. SHOOT Liquid Acquisition System Requirements

Steady, Gas-Free Liquid Delivery to the Pump	
Tank Geometry	
- Internal diameter, m	0.57
- Length, m	0.89
- Tank Volume, L	210.00
Flow Rate	
- Transfer Rate, L/h	500.00
- Demonstration Rate, L/h	1000.00
Acceleration Environment	
- Ambient or Steady State, g	10^{-5}
- Crew Movement Perturbations, g	10^{-4}
- Attitude Control Thruster Firings, g	10^{-2}

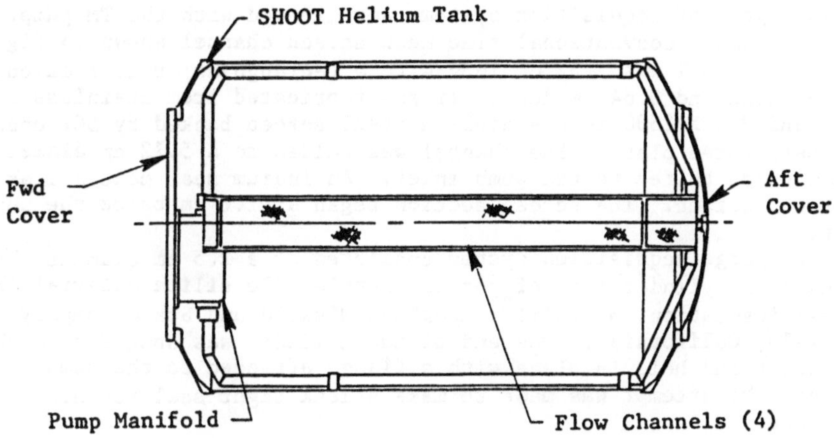

Fig. 1 Preliminary Design - SHOOT Liquid Acquisition System

operate in essentially the same manner as a similar system would with conventional cryogens. The screen bubble point was scaled by the ratio of surface tension; the results showed that helium could be outflowed at the required rate to very low residuals, i.e., high expulsion efficiency, even under the highest adverse acceleration of 10^{-2}g.

The use of a sponge composed of many closely spaced radial vanes or of fine porous material that could provide a reservoir of SFHe at the TM pump inlet was also considered. A porous material was selected because of the smaller pore size requirement dictated by the low surface tension of SFHe. This approach could be used either by itself or in combination with screen channels feeding the sponge.

ACQUISITION SYSTEM TESTING

Both screen and sponge acquisition systems were tested with a thermomechanical pump and a screen system was tested with the centrifugal pump. The systems, tests conducted, and results obtained are discussed.

Thermomechanical Pump Systems

The TM pump is a simple device which converts heat to fluid pressure using the thermomechanical or fountain effect unique to SFHe. The device consists of a resistive wire heater on the downstream side of a porous medium. Such systems have been used in laboratories on a small scale for many years. For SHOOT, consideration is being given to the use of either a sintered stainless steel or ceramic material having pores on the order of one micron. This pore size sacrifices some TM pump efficiency for ease of manufacture. Some of the results of the TM pump tests at GSFC are presented in another paper.[5] Due to the inverse of the TM effect, the mechanocaloric effect, some of the heat supplied to the SFHe on the downstream side of the porous media is transferred to the upstream side of the plug. This, in addition to the parasitic heat thermally conducted through the pump or pump housing, provides a heat source at the inlet of the pump.

Two types of acquisition systems were tested with the TM pump. The first was a conventional fine mesh screen channel shown in Fig. 2. The test channel was 7.62 cm long with a rectangular cross section, 2.54 cm wide and 0.64 cm deep. It was fabricated from stainless steel plate and 325 X 2300 mesh stainless steel screen backed by 50% open area perforated plate. The channel was welded to a 5.72 cm diameter flange which bolted to the pump inlet. An indium seal made a leak tight connection. The screen section began at 1.6 cm below the pump inlet.

The sponge acquisition system consisted of a 2.5 cm diameter by 7.5 cm long cylinder made of porous shuttle tile silica material (HTP thermal insulation, 6 lbm/ft^3, Lockheed Missile and Space Company, Sunnyvale, California). One end of the cylinder was snug fit to the pump inlet and held in place with a flange attached to the pump housing. No attempt was made to make a leak tight seal at this interface.

Both acquisition devices were tested in the system shown schematically in Fig. 3. The acquisition device was suspended below the bottom of the TM pump porous plug inlet. The pump outlet was connected via a capillary tube to a graduated glass beaker that was emptied periodically by using a plunger. Germanium Resistance Thermometers (GRT's) were placed at the pump inlet, pump outlet, helium bath and in the receiving beaker. The relative accuracy of these GRT's for the temperature range of the tests was approximately 0.1 milliKelvin (mK). Teledyne Taber Model 2215 pressure transducers with a resolution of 0.03 torr and a relative accuracy of approximately 0.1 torr were located at the pump inlet and outlet. The dewar included 2.5 cm diameter windows located 180 degrees apart to allow direct observation of the liquid level in the beaker. Two superconducting level detectors were used in the SFHe bath to measure the liquid level when it was above the windows. Flow rate was

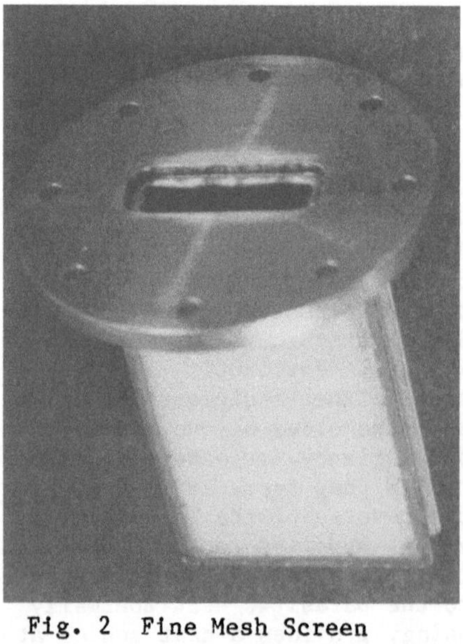

Fig. 2 Fine Mesh Screen Channel Device

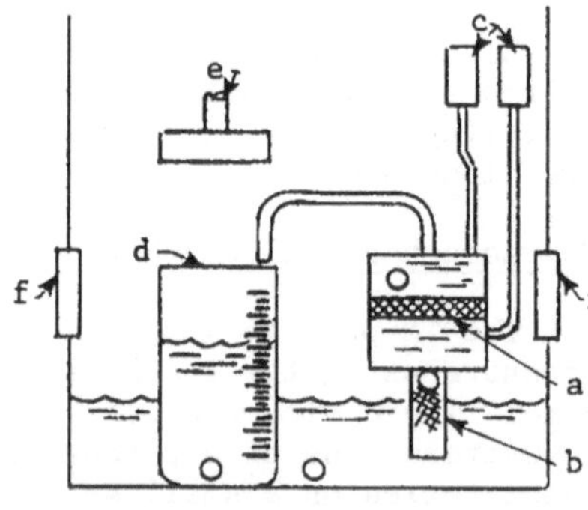

a) TM Pump
b) Acquisition Device
c) Pressure Transducers
d) Beaker
e) Plunger
f) Window
o) GRT

Fig. 3 TM Pump/Acquisition Device Test Apparatus

determined from the fill rate of the beaker. The volume ratio of the bath to the beaker was approximately 6:1 and a typical flow measurement involved beaker level changes of 2 to 3 cm; hence, the bath level (helium head) decreased from 0.3 to 0.5 cm during a flow measurement.

The data acquired during numerous tests with the screen device are summarized in Fig. 4. For clarity, the data for each input power level are represented by straight line segments. To indicate the amount of data scatter within a test, the raw data are included for one run. Different flow rates were achieved in two ways. First the pumping speed of the TM pump could be varied by the amount of heat applied. Second, three different impedance capillaries were used at the pump outlet. Out of 21 runs, eight showed some indication of pumping with a negative head at the pump inlet. Some of the runs which did not show flow below the pump inlet lie on top of the data shown and are not represented in the figure. Of the eight runs showing pumping with a negative head, there seemed to be no correlation with flow rate, outlet capillary size or bath temperature (bath temperatures between 1.35 and 1.65 K were used). At least two of the eight runs showed pumping with negative heads large enough to expose portions of the screen.

During all of the runs, temperature differences across the screen were not detected (i.e., $<$ 0.1mK). Using the saturated vapor pressure, the measured inlet pressure, and correcting for bath hydrostatic head, there was also no pressure difference across the screen (i.e., $<$ 0.1 torr). By the same token, data taken at the same heater power levels without the screen channel attached to the pump inlet showed the same flow rates for non-negative helium heads, indicating that the test screen channel presented negligible impedance to SFHe flow.

It was believed that a thermomechanical effect might exist across the fine mesh screen used in the channel acquisition system. To investigate this possibility, a small 325 X 2300 mesh screen disc was evaluated in a TM static pressure test. A pressure increase across the screen of 0.35 torr was measured at 0.55 w/cm^2. This corresponds to a SFHe head of 3.3 cm. While this pressure difference is relatively large, the heat flow per area required to achieve this head is also large compared to that used in the TM pump tests. These

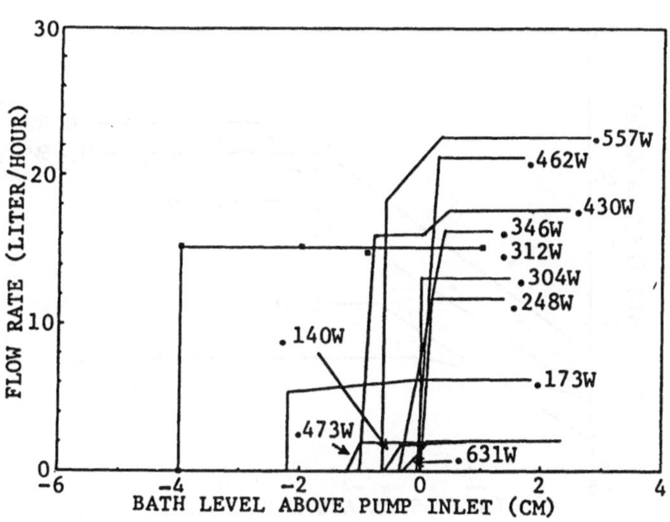

Fig. 4 TM Pump/Screen Acquisition System Test Results

results show that a TM effect across the screen was not the primary reason for the observed pumping with a negative helium head.

Results obtained with the sponge acquisition device are summarized in Fig. 5. For clarity most of the data are represented by straight line segments. All data for one run at 0.46 W heater power are shown to indicate typical data scatter. Note that more data points are shown here than for the screen acquisition tests. In the screen tests, flow rates were comparatively constant, allowing averaging of a few beaker fillings to represent one data point. With the changing flow rate at large negative heads with the sponge system, this averaging or smoothing of data was not possible. In all of the runs, the flow rate did not drop to zero until the bath level dropped below the silica cylinder at the bottom of the dewar. Accurate determination of the bath level near the bottom of the dewar was not possible due to the position of the view port.

All of the sponge system tests demonstrated a pumping capability with a negative helium hydrostatic head. For the operating range of the TM pump involved, 0 to 0.5 W (the flow rate does not increase significantly above this heater power), a negative head of at least one cm was supported with no decrease in flow rate. As in the screen channel tests, the flow rate for a given power for any non-negative head was the same with or without the sponge attached to the pump inlet.

The irreproducibility of the screen channel test results is apparently due to the metastable nature of the SFHe under conditions of slightly lower than saturated vapor pressure. The results of the static screen test indicate that the driving pressure of the TM effect across the screen is far too small to cause a measurable head rise under static conditions with a screen as large as that used in the pump tests. The present screen provides a TM effect only about 1/15 as large as the ideal TM effect at a heat flow of 0.3 W/ cm^2 when one considers the temperature difference across the screen. Under conditions of flow, dissipation across the screen and within the channel will further decrease this head.

The sponge material tested with pores of the order of 20 to 50 microns and an 80+% open geometry appears to have the advantage of performance verification on the ground. Its ability to raise SFHe reproducibly from one to over three cm across the flow range of the

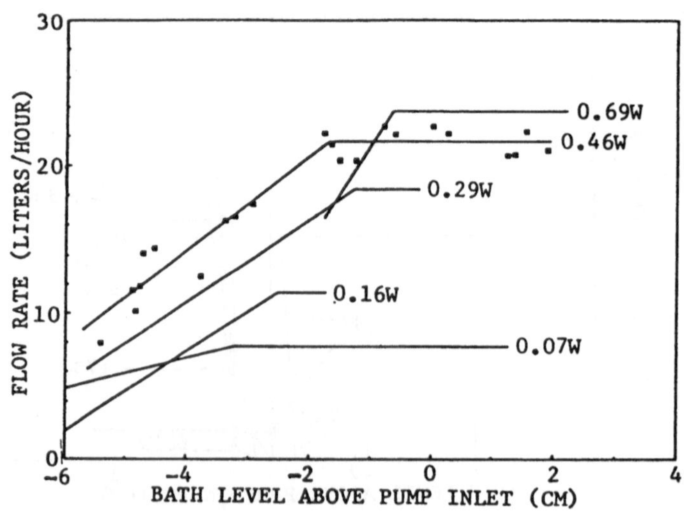

Fig. 5 TM Pump/Sponge Acquisition System Test Results

pump scales to 10 to 30 m at 10^{-3}g, more than enough to boost SFHe in the bottom of a tanker. It remains to be seen whether the cross-sectional area of the sponge can be reduced to less than that of the pump and maintain this flow. It is also unknown how well the sponge material can be drained by the TM pump. The acquisition of SFHe in the sponge material is probably a combination of capillary and thermomechanical forces. In static lab one-g tests, it was shown to raise liquid nitrogen at least 3.2 cm above a liquid bath. Scaling to the surface tension and density of SFHe gives a static head rise of 0.7 cm due to capillary action alone. Further testing of the sponge material is needed to determine SFHe flow impedance over the long distances required to drain the SHOOT dewar or a SFHe resupply tanker.

Centrifugal Pump System

While the TM pump system was being tested at NASA-GSFC, a similar program for a centrifugal pump system was being conducted by the National Bureau of Standards Cryogenic Laboratory under the sponsorship of NASA-ARC.[6] The pump tested was a single stage, electrically driven pump manufactured by Barber-Nichols of Denver, Colorado, and was capable of providing the 500 L/hr flowrate.

The primary purpose of these tests was to investigate pump performance with both normal and SFHe. Test time was provided to investigate operation with a screen device attached to the pump inlet.

The acquisition device was designed and fabricated for attachment to the pump inlet flange. The cylindrical device, 10.22 cm diameter and 8.9 cm high, is shown attached to the pump in Fig. 6. The cylindrical and bottom surfaces were formed from 325 X 2300 mesh stainless steel screen welded to a flange bolted to the pump inlet. Screen support was provided by eight 0.3 cm rods welded to the flange and a bottom ring.

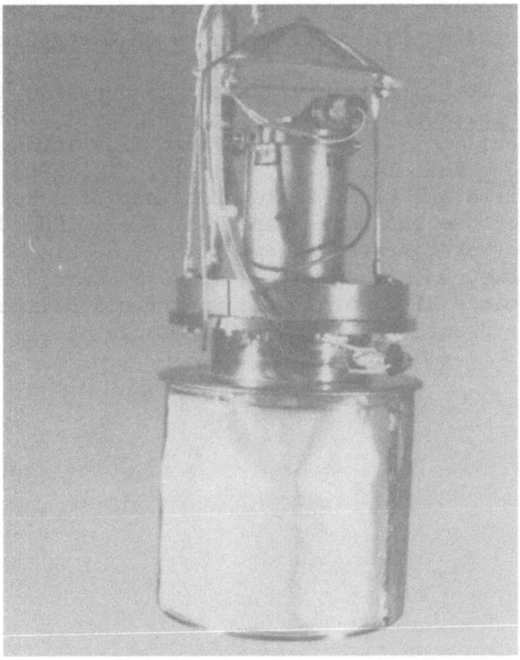

Fig. 6 Centrifugal Pump and Screen Acquisition System Test Assembly

The combined system was flow tested with normal helium at 2.3K and 4.2K. For both temperatures, the liquid level in the test dewar dropped below the pump inlet before the flow through the pump began to decrease rapidly due to cavitation. A level of seven cm below the pump inlet was obtained at 2.3 K and 2.5 cm at 4.2 K. This difference is attributed to the difference in surface tension at the two temperatures. The negative hydrostatic pressure under one-g conditions is significantly greater than would be required on orbit.

Testing of the pump and acquisition device was also attempted with SFHe at 1.8 K. At this temperature, the pump cavitated when the dewar liquid level was about 7.5 cm above the pump inlet. Thus, the screen was covered by liquid and no information regarding the ability of the screen to prevent vapor ingestion is available. Similar cavitation occurred with the pump alone, indicating that the screen presents very low resistance to flow.

SUMMARY

Promising results have been obtained with the fluid acquisition systems tested in combination with the TM pump. Pumping capability with a negative SFHe head much greater than required on orbit has been demonstrated. Some of the results, such as the randomness of the screen system performance, are not completely understood. Further in-depth testing is needed to thoroughly characterize both the screen and sponge acquisition systems in combination with the TM pump and to provide an acceptable design for either system.

Additional testing of the screen acquisition system in combination with the centrifugal pump is also required with SFHe, since no satisfactory evaluation of the screen has been performed to date.

REFERENCES

1. M. J. DiPirro and P. Kittel, The Superfluid Helium On-Orbit Transfer (SHOOT) Flight Demonstration, Paper No. DA-3, 1987 Cryogenic Engineering Conference.
2. A. Kashani and S. W. VanSciver, Steady State Forced Convection Heat Transfer in He II, Advances in Cryogenic Engineering, 31:489(1986).
3. L. C. Yang et al, Characterization of Superfluid Helium Transfer, Cryogenics, Vol. 21, No. 4, pg 207 April, 1981.
4. J. C. Armour and J. N. Cannon, Fluid Flow Through Woven Screens, AICHE J., Vol. 14, No. 3, pg. 415, May, 1968.
5. M. J. DiPirro and R. F. Boyle, Lab Test of A Superfluid Helium Pump for Use In Space, Paper No. BC-2, 1987 Cryogenic Engineering Conference.
6. P. R. Ludtke, D. E. Daney, and W. G. Steward, Centrifugal Pump Performance With Normal and Superfluid Helium, Paper No. BC-8, 1987 Cryogenic Engineering Conference.

DIRECT LIQUID CONTENT MEASUREMENT

APPLICABLE FOR HE II SPACE CRYOSTATS

M. Wanner

Linde TVT
D-8023 Hoellriegelskreuth
W.-Germany

ABSTRACT

The He II cryostat ISO requires a precise measurement of the liquid content throughout the 18 month mission. For that purpose a direct calorimetric method was suggested: a well defined heat pulse into the He II bath causes a small temperature increase which can be measured and which is directly correlated to the liquid mass through the LHe-specific heat. To study this measuring principle especially under the potential zero gravity constraints of disconnected liquid volumes a setup was established which allowed to investigate the heat transfer between separated liquid volumes. The results for different fluid configurations confirm that even for completely disconnected volumes the heat is almost immediately distributed throughout the whole liquid by a process of evaporation and recondensation.

INTRODUCTION

The continuous control of the residual lifetime during the mission of a He II space cryostat such as ISO depends on the precise knowledge of the momentary content of stored cryogen.

In a microgravity environment however the location of the liquid gas interface in the He vessel is undetermined. Hence the standard level probes such as superconductive wires can only be used during pre-launch operations.

Different alternatives have been considered for ISO such as the measurement and numeric integration of the He vent rate (as used for example in IRAS) or the rotation of the whole cryostat in order to produce an artificial gravity[1].

The considerable errors resulting during the evaluation of the LHe-mass from such indirect measurements can be reduced by a direct calorimetic method.

A well defined heat pulse Q applied to a superfluid He bath for a time Δt will immediately distribute throughout the liquid due to the excellent heat conductivity of He II. Hence the bath temperature will closely follow the heat pulse according to

$$\Delta T = Q^* \Delta t / M_{He} {}^* C_{svp}$$

By knowledge of the specific heat C_{svp} the liquid mass M_{He} can be deduced basically from a measurement of the temperature increase ΔT. (For sufficiently large He II volumes the specific heat of the tank itself is negligible).

This latter method will obviously work satifactorily on ground and has also been tested in a small He-experiment during a short shuttle flight [2].

For the much larger He system of ISO it has however to be verified that the heat will be dissipated throughout the whole liquid within a sufficiently short period, even under the severe conditions that liquid volumes in the tank might not be connected or only be connected by a superfluid film along the tank walls.

For that case it was estimated that neither gas heat conduction nor film flow along the walls can achieve a thermal equilibrium within a time scale of 1000 sec.

The only process that theoretically predicts nearly immediate heat exchange is the evaporation from the warmer volume and subsequent recondensation onto the colder volume driven by the vapor pressure difference between the volumes. The main task of the experimental study was therefore to verify that this process is in fact dominating.

EXPERIMENT DESIGN

Although it was not possible to perform the experiment within the actual ISO volume of 2200 l the design parameters have been chosen in such a way to be representative for the actual application.

Table 1 shows the basic requirements and parameters of the experiment as compared to the ISO application.

Table 1

Requirement		Dim.	Experiment	ISO
accuracy	$\Delta M/M$	%	5	5
temperature increase	ΔT	mK	10	5
He II volume (max.)	V	L	8	2200
bath temperature	T_B	K	1,8	1,8
base heat load to bath	Q_C	mW	100	130
maximum heat for ΔT	$\dot{Q}.\Delta t$	Ws	33	4300
power	$\dot{Q}$	W	2	20
heat pulse length	Δt	s	16,5	216
temperature resolution	ΔT_B	mK	< 0.1	< 0.25

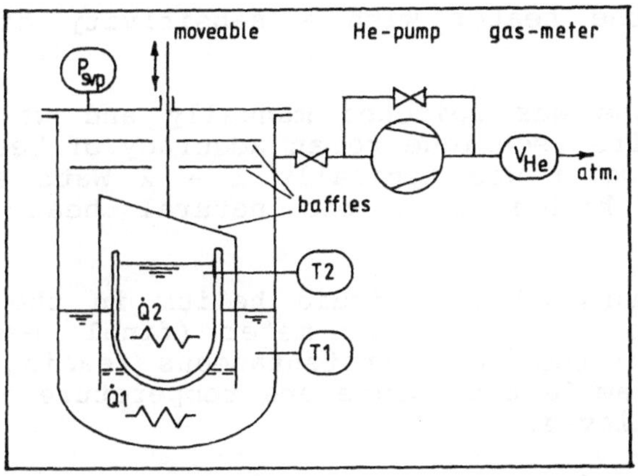

Fig 1. Schematic of the test setup

The small temperature increase was chosen in order not to disturb the cryosystem and to limit the required heat for the 2200 l ISO-cryostat.

The experimental setup as shown in Fig. 1 consists essentially of a conventional glass dewar which was pumped and stabilized to the desired bath temperature.

A temperature stabilization of typically 0,1 mK/min was achieved by a manual control valve in the vent line. An automatic stabilization was considered not suitable in order not to counteract the temperature variations caused by the heat pulse.

Within the main bath a moveable glass beaker serves as a secondary volume. By changing the submersion of the beaker the liquid can be studied as a single volume, as two volumes connected by a film or as two disconnected volumes.

Special attention was given to the beaker design. The double walled and vacuum insulated glass beaker minimizes the heat transport through the walls. A suspension by 3 thin wires restricts the film flow between the two bath and an additional radiation baffle surrounds the beaker completely with the exception of small slits for visual observation of the liquid level.

The two heaters in the main bath and in the beaker are standard wire wound resistors with a nominal resistance of 56 Ω. In the actual ISO application the design will have to provide sufficient heat transfer area and a cavity structure to retend the liquid even during microgravity accelerations. The temperature sensors are 100 Ω Allen Bradley resistors which are driven by 10 μA excitation current to increase the signal. This current induces a small self-heating which however was compensated by calibrating the sensors in situ with the same current.

A simple laboratory electronic was used which consisted of a power supply for the heaters and two parallel Wheatstone bridges for measuring both the bath temperatures with an absolute accuracy of 5 mK and the small temperature changes induced by the heater with a sensitivity of better than 0.3 mK.

The heat pulse was switched manually and was measured by standard 4 wire technique to an accuracy of better than 1 %. The applied power was typically 1 - 2 Watt and hence was considerable higher than the natural heat load to the He-bath.

The actual mass of the liquid helium in the cryostat was evaluated from a volume gas meter (final reading when all LHe has evaporated minus instantaneous reading). By correction to the ambient pressure and temperature an accuracy of 0.3 % was achieved.

After a general checkout of the equipment the following configurations were studied:

- beaker completely submerged (single volume)
- beaker partly submerged (two volumes connected by a superfluid film)
- beaker completely above the LHe level (two volumes but no heat transport via the film)

For each configuration the heat pulse was successively applied to the beaker and to the main bath and the signals of T1 and T2 were monitored and evaluated as a function of time.

RESULTS AND DISCUSSION

A typical recorder output of a test sequence is shown in Fig. 2 where the beaker was raised completely above the main bath and the heat pulse was applied to the bath.

As can be seen, both temperatures increase linearly with time and thus demonstrate that the applied heat is immediately distributed between both volumes.

The small time offsets of the recorder signals are due to the different recorder pen positions. In fact no time delay was detected within the accuracy of our setup (appr. 0.1 sec)

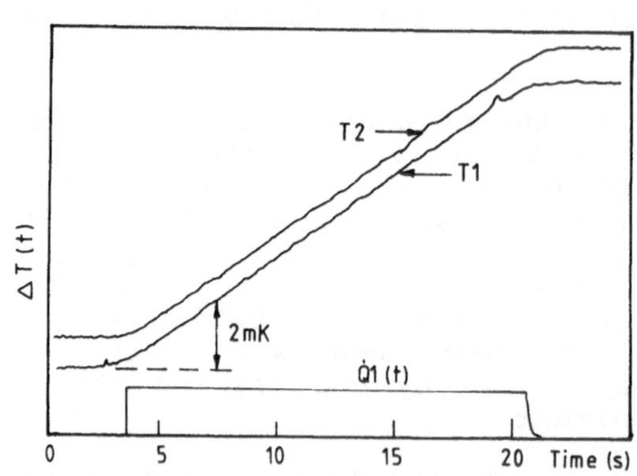

Fig 2. Recorder plot of the temperature variations T1 and T2 following a heat pulse Q1 into the bath

The small differences between the signals of T1 and T2 during the heating period are caused by the slightly different calibration curves of the two thermometers.

The surprising instantaneous heat transfer from the bath to the beaker can be explained only by a process of evaporation and immediate recondensation due to the excellent heat conductivities of He II. Any contribution to the heat transport by means of the superfluid film flow from the upper to the lower volume would be in the opposite direction. The recondensation process was also visualized after prolonged heating by a macroscopic change of the liquid levels in the beaker and in the bath.

Essentially the same features have been observed independant whether the beaker was completely or only partly submerged or whether the heat was applied to the bath or to the beaker.

A more detailed investigation of the driving temperature difference T1-T2 during heating showed that a small but constant temperature difference of 0.3 mK can be detected when the heat is applied to the beaker whereas in the opposite case any temperature gradient between the bath and the beaker is beyond the limits of our measurement (see Fig.3). The slight decrease of the signals with time is again due to the different individual calibration curves of the sensors.

The experiment data have been evaluated according to

$$M_{He} = \frac{U_H * I_H * \Delta t}{C_{svp} * \Delta T1,2}$$

and were compared with the reference mass M_{He} as measured by the gas meter (U_H = heater voltage, I_H = heater current). Specific heat data of He II at saturated vapor pressure have been taken from Refs. 3 and 4. The result of this evaluation is plotted in Fig. 4 in the He mass range 0,25 - 1 kg. The results are independent of the fluid liquid configuration.

As can be seen, the data points follow closely the 45° straight line which indicates the theoretical prediction.

The scatter of the individual data points from a straight fit line is only of the order of 1 %. Nevertheless there appears a systematic deviation of about 5 %.

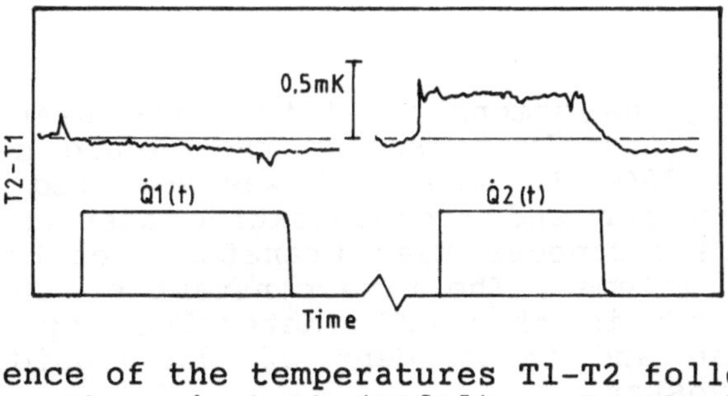

Fig 3. Difference of the temperatures T1-T2 following a heat pulse to the main bath (Q1[t]) or to the beaker(Q2[t])

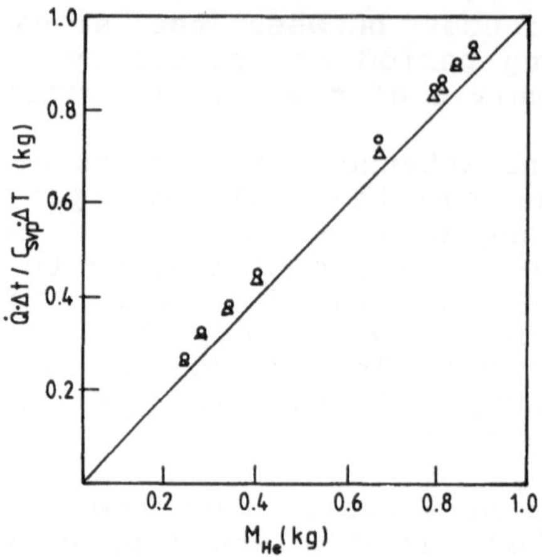

Fig 4. Comparison of the liquid mass evaluated from the heat pulse method as compared to the actual mass M_{He} C_{svp} data from Ref. 3 (o) and Ref. 4 (Δ)

The different contributing error sources shall be shortly addressed:

The mean statistical errors resulting from the stability of the heater supply voltage, the measuring accuracies of the heater voltage, current and the pulse length as well as the accuracy of the ΔT-measurement (signal resolution and slow drifts of the bath temperature) were estimated in total to about 2 %.

On the other hand, the systematic error of appr. 5 % is mainly due to uncertainties in the specific heat data and of the absolute bath temperature. The specific heat data quoted by the two literature sources differ by appr. 3 %. In addition due to the pronounced temperature dependance of C_{svp} systematic errors in the absolute temperature of only 10 mK will lead to an additional error in the specific heat and hence in the calculated mass of 3.6 %. Moreover such absolute temperature errors induce also errors in ΔT because the slopes of the R-T characteristic change with temperature.

Other sources of inaccuracy such as the specific heat of the He vapor and of the solid materials (glass dewar, heater elements, thermometers) are of the order of 0.1 J/K and hence negligible compared to the 2800 J/K of specific heat of the He-bath.

CONCLUSIONS

The experiment has demonstrated that the heat pulse method is capable of directly measuring the liquid content of the He II tank of ISO. In detail it was verified that the process of evaporation and recondensation assures an effective and almost instantaneous heat transfer even between disconnected liquid volumes. The time constant of the liquid content measurement is thus only determined by the length of the heat pulse and independent of the liquid distribution within the cryostat.

The essential requirement is only that the heater element is in contact with a sufficient amount of the liquid phase in order not to "dry out" during the heating period. This requirement can be met by placing the heater element inside the He II tank and provide a capillary structure with characteristic dimensions such that during the maximum expected accelerations the surface tension forces still dominate and keep the liquid close to the heater surface.

On the other hand it became obvious that the absolute temperatures of the bath must be measured to better than 10 mK and more precise specific heat data are required in order to achieve the desired high accuracy of this method.

ACKNOWLEDGEMENT

The author acknowledges stimulating discussions with M. A. Davidson (ESA) and T. Passvogel (MBB) as well as experimental support by H. Rüdiger. This work was done under ESA and MBB contract (MBB contract no. R 3141/3411 R)

REFERENCES

1. "Direct Liquid Content Measurement Study, Phase II, Final Report" ESA contract report, to be published

2. D. Petrac, P. V. Mason Proc. ICEC 11, p. 347 (1986)

3. J. Wilks, "The properties of liquid and solid Helium" Clarendon Press (1967) p. 666

4. "The Thermodynamic Properties of He II from 0 K to the Lambda Transition", NBS-Technical Note 1029 (1980).

THERMAL PERFORMANCE OF THE COSMIC BACKGROUND EXPLORER SUPERFLUID HELIUM DEWAR, AS BUILT AND WITH AN IMPROVED SUPPORT SYSTEM

Richard A. Hopkins and Dan A. Payne

Ball Aerospace Systems Division
Boulder, Colorado

ABSTRACT

Launch of the Cosmic Background Explorer (COBE) is planned for 1990. A critical element of the COBE is the 650 L superfluid helium dewar that will maintain the cryogenic instrument assembly at ~2 K for an estimated 14 months. Life testing of the dewar was completed in the autumn of 1985. The support system heat conduction is a major part of the overall heat load to the cryogen. The COBE dewar uses a tension strap system made of fiberglass/epoxy; this is the most thermally efficient system used to date in flight hardware. However, the thermal efficiency of the COBE support system could be considerably improved by using a more optimized orientation of straps and a newly-developed composite material instead of the fiberglass/epoxy. A thermally optimized tension strap support system is defined based on recent fatigue testing of alumina/epoxy straps. This material has a modulus of elasticity over twice that of fiberglass/epoxy, and therefore displays greater fatigue strength. A support system using alumina/epoxy can have both smaller strap size and greater stiffness than a fiberglass system. The thermal conductivity is somewhat higher than that of fiberglass/epoxy. The more optimized strap orientation provides considerably greater strap length and thermal resistance, and the alumina material provides about 50 percent higher resonance frequency. Thermal performance predictions are compared for the COBE dewar (1) as built, (2) with fiberglass/epoxy straps in an optimized configuration, and (3) with alumina/epoxy straps in an optimized configuration. The optimized support designs provide a predicted cryogen lifetime improvement of about 50 percent.

INTRODUCTION

An artist's rendition of the COBE superfluid helium dewar is shown in Fig. 1. Extensive thermal performance testing of the dewar was completed in the autumn of 1985. The system was designed for a space shuttle launch, but the decision has been made to launch COBE on a Thor-Delta. Launch is planned in 1990. The dewar structural system is adequate for either launch vehicle. The dewar has a 650 L cryogen storage capacity and houses the cryogenic optical assembly, consisting of a far infrared absolute spectrophotometer and a diffuse background experiment. The cryogen tank is surrounded by three vapor-cooled shields (VCS) and supported from the vacuum shell by twelve fiberglass/epoxy tension straps. The design is described in more detail in Reference 1.

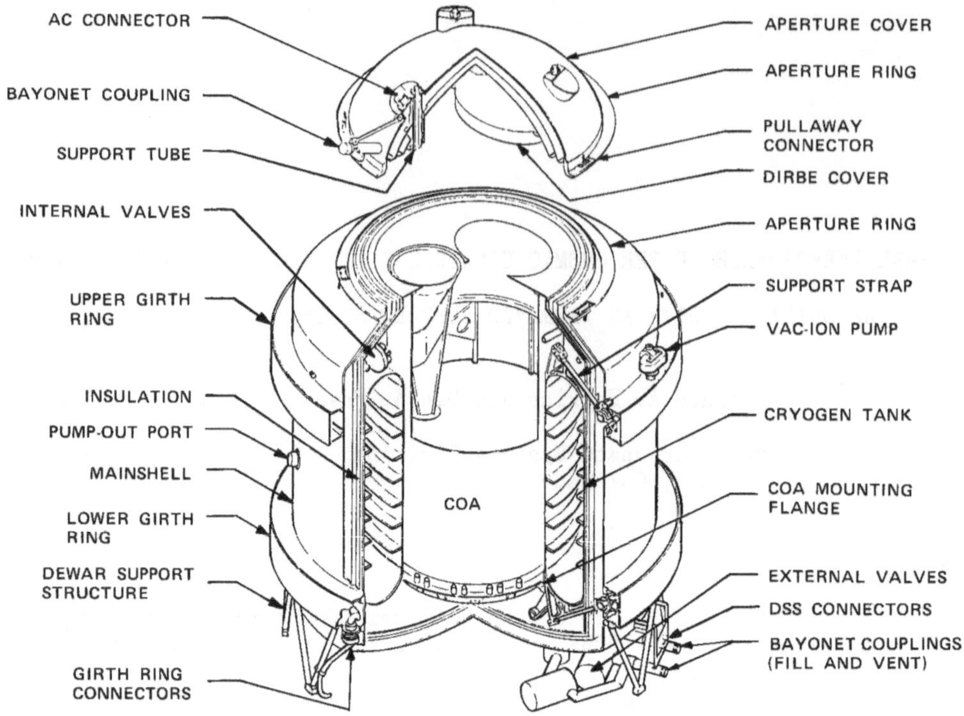

Fig. 1. COBE Dewar.

The COBE dewar tension straps are identical in geometry to those of the Infrared Astronomical Satellite (IRAS) dewar, launched in 1983. The glass and epoxy materials are slightly different. The orientation of straps in these dewars is not thermally optimized. The installed straps lie in radial planes, rather than forming a zig-zag pattern around the circumference of the dewar. A radial plane is any plane that includes the centerline of the dewar. The radial arrangement is easier to assemble, but it results in straps that are stiffer than necessary to provide adequate strength. With the zig-zag pattern straps are longer, providing greater thermal isolation, and the support system has a lower resonant frequency.

A tension strap system using a zig-zag configuration is the most thermally efficient support approach for spaceborne dewars. Efforts have been made over the last 15 to 20 years to develop various types of dual support concepts which would reduce the conducted heat input compared to the tension strap system. In general, the dual support approach uses a primary support to accommodate conditions of large dynamic loading and a smaller secondary support to hold the cryogen tank in place while in a static loading environment. The primary support is to some extent thermally disconnected from the cryogen tank during static conditions (e.g., on-orbit). A careful comparison of dewar performance with an optimized tension strap system and with the currently most advanced dual support concept has been performed[2], considering system level design constraints and requirements. That analysis shows no significant advantage with the dual support approach for a dewar that uses an extensive vapor-cooling system, as needed in the mass-optimized design for a long-life application. This is due primarily to the thermal interaction between the vapor-cooling system and the primary support tube, that must be sized for buckling strength and is considerably larger in cross section than a tension strap.

Sizing of tension straps is dictated by tensile ultimate strength, tensile fatigue strength, and the minimum allowable resonant frequency of the support system. Fatigue strength is generally the design driver. An overall system resonant frequency as low as 10 Hz is acceptable for launch on the space shuttle[3] or an expendable vehicle and is also usually compatible with orbital (attitude control) requirements. Compliances between the dewar and the launch vehicle must be considered, and a coupled loads analysis may be needed to verify the structural integrity of the system. Depending on the specific application, this means that the support system resonant frequency must be somewhat greater than 10 Hz (e.g., 15 Hz).

Several types of fibers are candidates for use as low thermal-conductance structural members. The ideal composite for the dewar tension strap application has high ultimate and fatigue tensile strengths to minimize cross-sectional area, high modulus of elasticity to maximize resonant frequency, and low thermal conductivity to minimize heat conduction. Thermal contraction coefficient may also be an issue in some cases. The straps in the IRAS and COBE dewars are made of S-glass. Graphite fibers have been used extensively in various applications, especially where thermal isolation is not required, because of superior strength and very high modulus of elasticity. The thermal conductivities of glass and graphite composites are similar at very low temperatures, but at room temperature glass has the advantage by a factor of about 5. Consequently, graphite provides better thermal efficiency only if the strap warm end is at a very low temperature (e.g., less than 100 K).

Straps made of alumina (85 percent Al_2O_3 and 15 percent SiO_2) fibers and the same epoxy used in the COBE straps have been tested by the National Bureau of Standards (NBS) to determine fatigue strength, modulus of elasticity, and thermal conductivity from 4 to 300 K.[4] Fatigue strength depends on configuration as well as material properties. To provide a direct fatigue strength comparison with the COBE fiberglass straps, also tested by NBS, the alumina test strap geometry was made identical to the COBE straps. The geometry is shown in Fig. 2. The data, shown in Fig. 3, indicate a considerably greater fatigue strength for alumina/epoxy, especially for applications with high numbers of cycles. The modulus of elasticity is 2.3 times greater with alumina, which provides greater stiffness and resonant frequency. The thermal conductivity of alumina/epoxy is somewhat smaller at very low temperatures, but greater at warm temperatures, as shown in Fig. 4. The current data are higher than the previous measurements reported in Reference 5 because of higher fiber content in the test material (i.e., 70 percent compared to 50 percent).

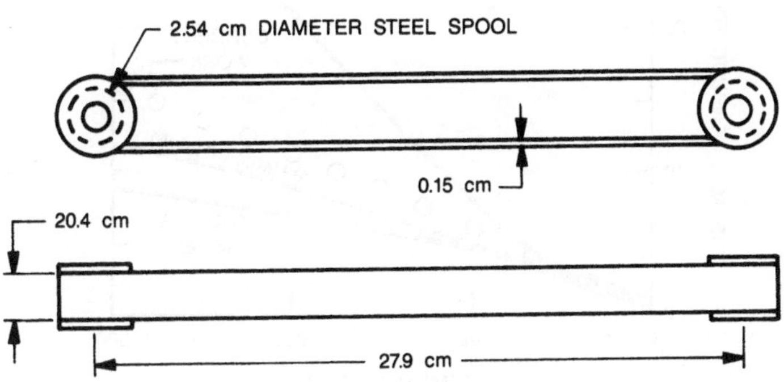

Fig. 2. Test strap geometry.

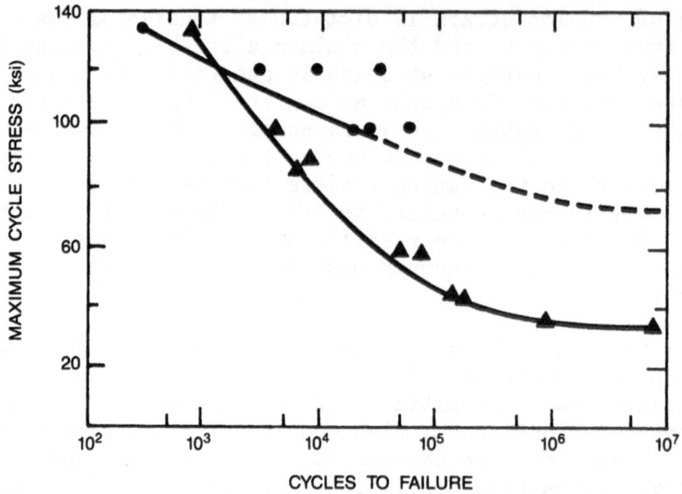

Fig. 3. Strap fatigue test results at R = 0.05 and room temperature, where R is the ratio of minimum-to-maximum cyclic stress (Δ S2-glass/SCI REZ 081 epoxy; • Sumitomo alumina/SCI REZ 081 epoxy).

The comparison of strap heat conduction for a given application depends on the vacuum shell temperature and the interaction with the vapor-cooling system. Thermally-optimized strap systems are sized for the COBE dewar using both fiberglass/epoxy and alumina/epoxy. Cryogen lifetimes are estimated for the COBE dewar design with the optimized strap systems to determine the improvement compared to the actual COBE dewar.

STRUCTURAL ANALYSIS

The philosophy used in defining a thermally-optimized strap system is to push the structural performance of the straps to the practical limit within the confines of the COBE dewar geometry. This is simply a matter of maximizing length and minimizing cross sectional area. The length is maximized by going from radially oriented straps, as in the actual COBE dewar, to a zig-zag orientation. This orientation minimizes the load that each strap must carry and results in less stiffness than

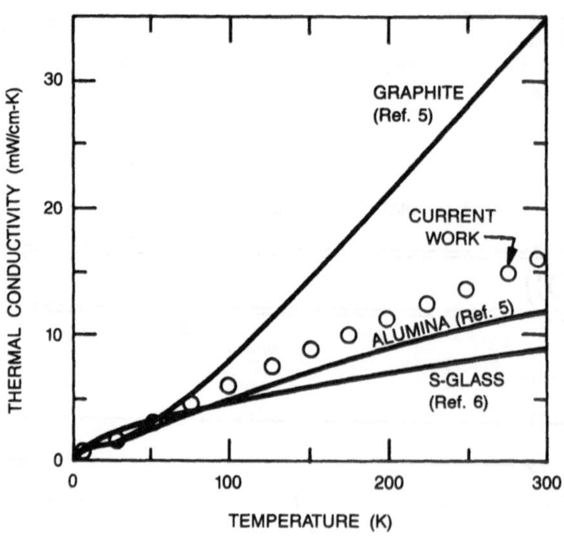

Fig. 4. Thermal conductivity of epoxy composite materials in the fiber direction.

with the radial pattern. The straps zig-zag between the vacuum shell and cryogen tank at an angle of 35 degrees from the plane normal to the dewar axis. This angle provides the same stiffness in both the axial and lateral directions, and the resultant strap length is twice that using a radial pattern. The cross-sectional area is minimized by taking advantage of the fatigue test data, shown in Fig. 3, which became available after the COBE strap sizing was finalized. These fatigue data, along with proper precautions against overloading and fatigue accumulation during test and transportation, allow nearly a 50 percent reduction in strap area compared to COBE. However, using smaller straps results in running higher stresses and higher risks.

A limit load of 13 g in any axis has been found to be reasonable for both the originally planned space shuttle launch and the currently planned Delta launch. Also, COBE experience has shown that 1000 cycles of fully reversing 13 g loads is equivalent to test, transportation, and flight as far as fatigue is concerned, provided that reasonable effort is made to minimize fatigue during test and transportation. A factor of 4 is routine for fatigue analysis to account for uncertainties, so that the design life is 4000 cycles.

Based on the test data, the mechanical properties shown in Table 1 are appropriate for sizing the new strap cases. The modulus of elasticity of the alumina straps is 2.27 times greater than that of the fiberglass straps, and the alumina strap fatigue strength at 4000 cycles is 17 percent greater.

As a result of using the zig-zag orientation and the conservative sizing of the COBE dewar straps, thermally optimized straps have an area-to-length ratio about one-fourth that of the COBE straps. Since heat conduction and stiffness are both proportional to area/length, the reduction in thermal conductance results in an equal reduction in strap stiffness. Thus, the natural frequency of the cryogen tank and instrument on the straps would drop from 34 Hz in the actual COBE dewar to about 18 Hz for the new glass straps and about 25 Hz for the new alumina straps. While an 18 Hz frequency is adequate for most applications, the higher stiffness of the alumina straps would be valuable in reducing the stresses in the dewar plumbing and VCSs due to cryogen tank motion.

THERMAL ANALYSIS

Dewar boil-off rate is computed with the thermal network model developed for the COBE dewar.[7] This model, shown in Fig. 5, has been validated by a carefully planned test program, during which the dewar per-

Table 1. Strap Mechanical Properties

	S-glass	Alumina
Tensile Strength, MPa (ksi)	1370 (199)	1080[a] (157)
Tensile Modulus, GPa (ksi)	60.0 (8,700)	136 (19,700)
Fatigue Strength at 4,000 Cycles, MPa (ksi)	662 (96)	772 (112)

[a] Estimated at 80 percent of the ultimate strength of the raw material, based on test data from the glass straps.

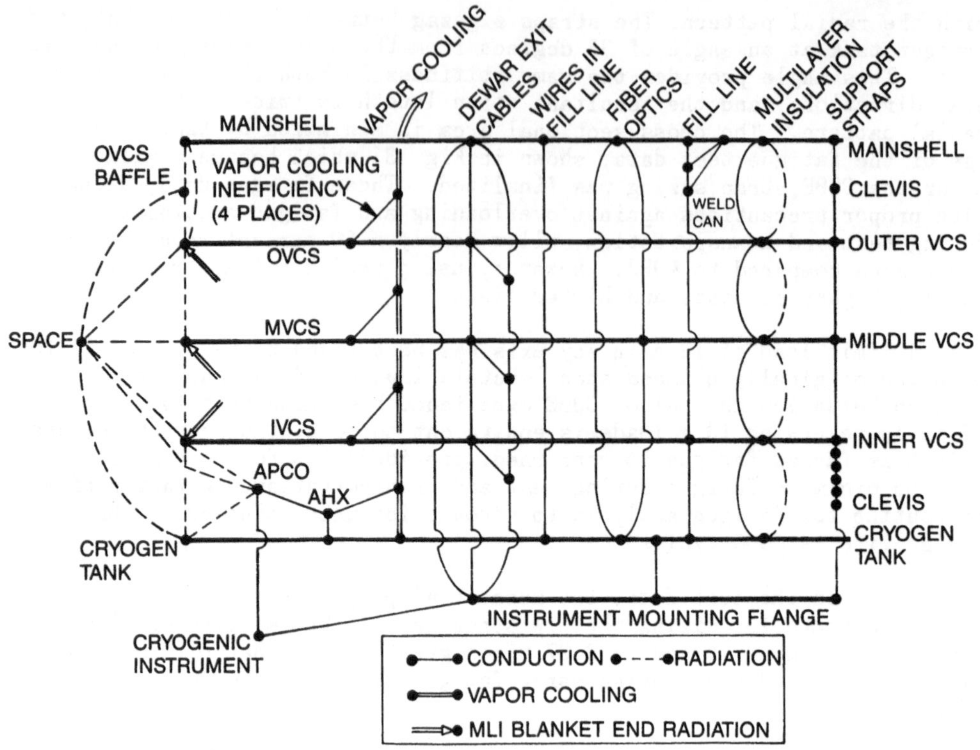

Fig. 5. Dewar thermal network model for survey configuration.

formance was measured at three different stable conditions. Adjusting the model to force agreement with three different sets of test data improves its capability to make valid performance predictions for boundary conditions other than those of the test. This is important since testing was performed with the aperture cover in place and with a vacuum shell temperature of 300 K to contain cost, and the condition on orbit is without the aperture cover and with a predicted vacuum shell temperature of 150 K. If all other boundary conditions are held constant, this temperature difference alone results in a factor of 4 change in boil-off rate. The other boundary conditions are the cryogenic instrument power dissipation (22.4 mW) and the aperture heating (3.6 mW) to the top of the cryogenic instrument package. Launching the COBE on a Delta, rather than on the space shuttle, will not significantly alter the dewar boundary conditions or performance.

Thermal conductivities used for the S-glass and alumina straps are shown in Fig. 4. The alumina data labeled "current work" is used for this analysis.

RESULTS

Support strap configurations and characteristics are given in Table 2 for the as-built COBE dewar and for the zig-zag configuration using either fiberglass or alumina straps. The two thermally optimized systems have area-to-length ratios about 4 times smaller than the COBE dewar straps, and the alumina straps are 14 percent smaller than the fiberglass straps due to superior fatigue strength. Also shown in Table 2 are the support systems resonant frequencies. All are above the required value of about 15 Hz.

Table 2. Support Strap Configurations and Characteristics

	Case 1[a]	Case 2[b]	Case 3[c]
Cross-Sectional Area per Strap (cm^2)	0.613	0.323	0.277
Total length of Strap (cm)	27.9	55.8	55.8
Length from CT to IVCS (cm)[d]	16.0	32.0	32.0
Length from IVCS to MVCS (cm)[d]	1.5	3.0	3.0
Length from MVCS to OVCS (cm)[d]	2.3	4.6	4.6
Length from OVCS to MS (cm)[d]	4.3	8.6	8.6
Support system resonant frequency (Hz)	34.0	17.8	24.9
Limit load (g)	13	13	13

[a]COBE dewar - fiberglass/epoxy straps in radial pattern.
[b]Fiberglass/epoxy straps in zig-zag pattern.
[c]Alumina/epoxy straps in zig-zag pattern.
[d]Thermal lengths include subtraction of 1 cm at each vapor-cooled shield to account for attachment block.

Breakdowns of the cryogen heat loads and VCS temperatures for nominal orbital boundary conditions are given in Table 3. The new strap system, using either alumina or fiberglass, results in about 33 percent lower total heat load, which means a 50 percent greater cryogen lifetime. Use of alumina improves lifetime about 5 percent compared to fiberglass. Instrument power dissipation becomes very dominant for the new systems compared to the as-built dewar, since heat conducted through the straps is much less significant. For example, radiation from the vacuum shell is only 27 percent of the support heat conduction from the vacuum shell for the as-built configuration, and for the fiberglass and alumina zig-zag configurations, this increases to 98 and 78 percent, respectively.

The influence of vacuum shell temperature on lifetime is shown in Fig. 6. In terms of percent of impact, the vacuum shell temperature is still important but has less influence for the new systems. This is expected since parasitic heat input is less significant with a thermally optimized support system.

Table 3. Summary of Heat Loads (mW) and Temperatures (K) for Survey Condition

	Case 1[a]	Case 2	Case 3
Instrument Power	22.4	22.4	22.4
Support Straps	13.2	2.2	0.9
Electrical Cabling	8.2	5.9	5.8
Radiation	4.7	2.1	2.0
Optical Fibers	1.3	1.0	1.0
Fill Line	0.2	0.1	0.1
Total	50.0	33.7	32.2
IVCS Temperature	23.6	17.1	16.7
MVCS Temperature	40.4	33.4	35.0
OVCS Temperature	71.8	67.7	70.5

[a]Cases defined in Table 2.

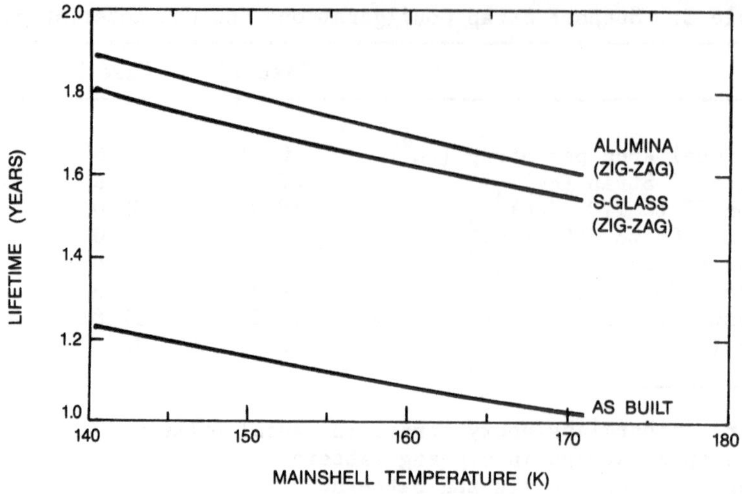

Fig. 6. Effect of vacuum shell (mainshell) temperature on cryogen lifetime.

The influence of cryogenic instrument power dissipation on lifetime is given in Fig. 7. As expected, it has greater influence for the thermally optimized systems, in which parasitic heat input is smaller.

CONCLUSIONS

The support system of the COBE dewar is not thermally optimized, either in terms of configuration or materials. Thermally optimized support systems using either fiberglass/epoxy straps or alumina/epoxy straps have been defined, and the corresponding cryogen lifetime improvements compared to the actual COBE dewar have been calculated. The improvement is about 50 percent in either case, with alumina straps providing about 5 percent longer lifetime than fiberglass straps. Also, alumina straps provide a resonant frequency about 40 percent higher than do fiberglass straps. These advantages justify the continued development of the

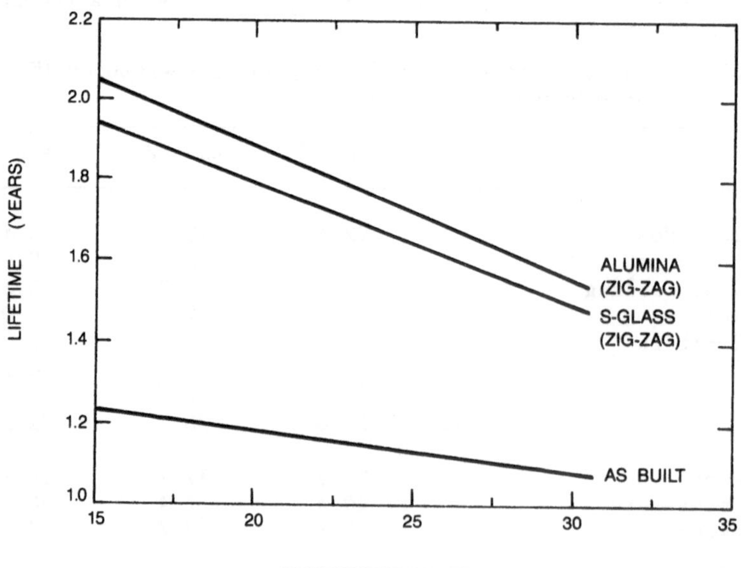

Fig. 7. Effect of instrument power dissipation on cryogen lifetime.

alumina/epoxy composite for flight applications and the use of the zig-zag configuration, which may be somewhat more costly to assemble than the radial configuration, in the design of future long-life dewars.

REFERENCES

1. R. A. Hopkins and S. H. Castles, Design of the superfluid helium dewar for the Cosmic Background Explorer, in: "Proceedings of SPIE," Vol. 509, August 1984.

2. R. A. Hopkins and D. A. Payne, Optimized support systems for space-borne dewars, in: "Cryogenics," Vol. 27, No. 4, April 1987.

3. "Space Shuttle System Payload Accommodations," JSC07700, Vol. XIV, p. 3-24.

4. R. A. Hopkins and D. A. Payne, Next-generation tension strap supports for spaceborne dewars, in: "AIAA 22nd International Thermophysics Conference," June 1987.

5. M. Takeno, S. Nishijima and T. Okada, Thermal and mechanical properties of advanced composite materials at low temperatures, in: "Advances in Cryogenic Engineering," Vol. 30, 1986.

6. J. G. Hust, Thermal conductivity of glass fiber/epoxy composite support bands for cryogenic dewars, Phase II, NBS Report 773.30-83-1, June 1983.

7. R. A. Hopkins and M. G. Ryschkewitsch, Measured ground performance and predicted orbital performance of the superfluid helium dewar for the Cosmic Background Explorer, in: "Proceedings of SPIE," Vol. 619, January 1986.

DESIGN AND TEST OF A MODIFIED PASSIVE ORBITAL DISCONNECT STRUT (PODS-IV)

Iran E. Spradley
Richard T. Parmley

Research & Development Division
Lockheed Missiles & Space Company, Inc.
Palo Alto, California

ABSTRACT

Passive Orbital Disconnect Struts (PODS) can potentially reduce the support conductance a factor of 10 over state-of-the-art tension band nondisconnect supports and can significantly reduce helium dewar weights without reducing lifetime. A series of thermal and structural tests have demonstrated this performance improvement. The major shortcoming of the PODS-III design is lack of side-load resistance to thermal shorting. This paper describes a design modification (PODS-IV) that dramatically increases the side-load resistance to shorting without significantly degrading thermal performance.

INTRODUCTION

The PODS support system has been developed and refined over the past 6 years. These struts were analyzed, designed, fabricated, and tested structurally and thermally.[1,2,3,4] Tests included thermal conductance to helium temperature, thermal expansion data, axial and side-load thermally shorting loads, ultimate loads, evacuation rates, fatigue tests at liquid-nitrogen temperature, effect on strut shorting of simulated asymmetric temperatures of the vacuum shell, model vibration tests at frequencies consistent with Space Transport System (STS) qualification requirements, and determination of principal resonant modes of the strut system.

The PODS-IV supports are modifications of the previously tested PODS-III designed to significantly increase side-load resistance while minimizing thermal conductance degradation.[5] The struts were fabricated and structurally tested. The resulting test data were then compared to previous data and predicted performance to verify the improvement in side-load resistance.

The PODS-IV support is the current selection to support the 1580-L superfluid-helium tank on the Gravity Probe-B program and the 200-L superfluid-helium Independent Research dewar (Fig. 1). Other missions where the PODS-IV supports may be used include SIRTF, AXAF, LDR, OTVs, Space Station, subcritical tankers, and other NASA and DoD missions.

Fig. 1. 200-L superfluid-helium Independent Research dewar supported by PODS-IV.

PODS CONCEPT

Figure 2 presents the PODS-IV support concept. A minimum of six struts (three pairs) is required to support a cryogen tank. (Six struts provide a statically determinate support system in all axes.) As the tank diameter changes due to cooldown or pressurization, the angled pinned-end struts are free to move in and out as the tank moves up or down. A similar adjustment occurs automaticaly as the vacuum shell diameter changes in orbit due to temperature changes.

For purposes of installation, the warm end of the strut provides a length adjustment feature. The threads on the rod-end fitting and length adjustment are a different pitch; consequently, by rotating the adjustment hex, precise length adjustments can be made during strut installation without rotating the strut. This feature allows length adjustments after the vapor-cooled shields (VCSs) are attached to the struts.

The cold end of the strut provides the passive orbital disconnect feature. The cold rod end fitting/stem is connected to the body by a highly thermally resistive, thin-wall graphite/epoxy orbit tube and adjustment bushing. Highly thermally resistive graphite filaments are connected between the body/nut and a collar which is epoxied onto the stem.

The conical load-bearing surfaces are separated from the nut (tension) and body (compression) by an axial gap of 0.0076 cm at operating temperature. At ambient temperature, the gaps are set to take into account the differential shrinkage between the various parts. During 1-g thermal test or orbital flight, the conical surfaces are prevented from touching by this spacing and the filaments. Consequently, heat is transferred from the body to the cold rod end fitting/stem subassembly by radiation and by conduction along the orbit tube and filaments.

During launch, the g-load elastically deforms the orbit tube along its axis; the conical shoulder of the stem rests hard on the body (compression) or nut (tension). The load path bypasses the orbit tube. The major thermal resistance and load path during launch are now the launch tube. Upon achieving orbit, the conical shoulder of the stem passively disconnects from the body or nut. Side loads on the launch tube due to VCS cooldown are transmitted through the PODS-IV mechanism. The filaments elastically deform and strain resist the applied side loads to prevent thermal shorting of the conical shoulder. The major thermal resistance is then the thin-wall orbit tube and filaments.

DESIGN

The PODS launch and orbit tubes are optimized separately to provide the largest possible thermal resistance while meeting the respective structural requirements. This is done by using the support optimization computer program DEWAR developed by David Bushnell.[6,7] The program

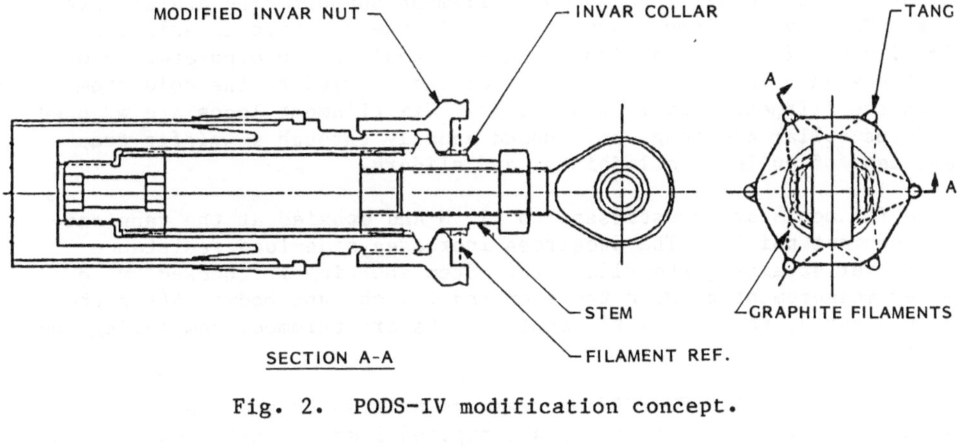

Fig. 2. PODS-IV modification concept.

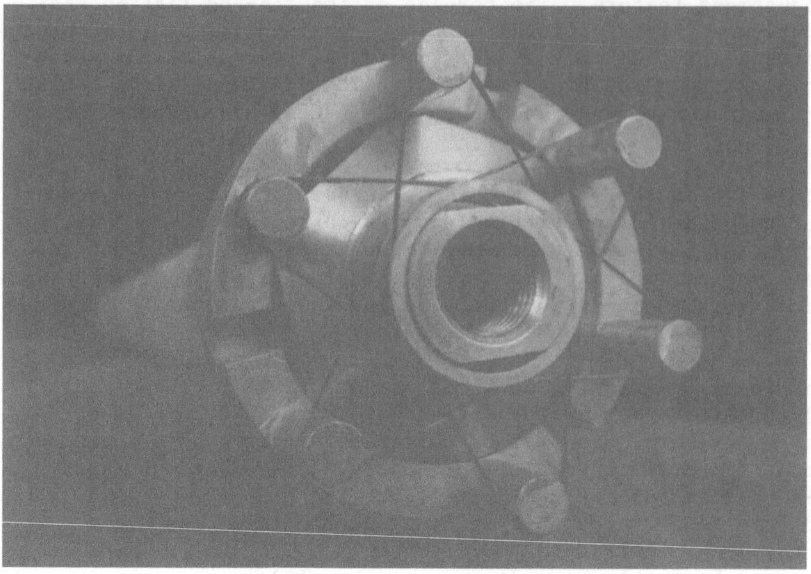

Fig. 3. PODS-IV mechanism.

uses a nonlinear programming algorithm called CONMIN written by Vanderplaats in the early 1970s.[8,9] CONMIN is based on a nonlinear constrained search algorithm developed by Zoutendijk.[10] The DEWAR program accounts for a variety of parameters such as support material structural properties, supported weights, axial and lateral g-loading, resonance frequency requirements, and minimum support thicknesses. Tube lengths, thicknesses, and attachment angles are varied by the program to obtain an optimized design.

The DEWAR program, however, does not account for nonaxial loading (side loads) of the launch tube which can occur due to VCS cooldown and insulation weight. These side loads can cause the semicantilevered launch tube to deflect and "short out" the conical shoulder of the disconnect assembly. The problem created by this condition is how to stiffen the support of the semicantilevered end of the launch tube without significantly increasing support orbital conductance.

The PODS-IV mechanism (Fig. 3) was designed to solve this problem. It consists of a spoked arrangement of 1000-fiber (1K) T300 graphite filaments connected between an invar filament support ring and an invar collar. The invar filament support ring is pressed onto an existing PODS-III nut. In future designs, the ring would be incorporated into the nut design. The collar is slid over and epoxied to the cold stem between the filament support ring tangs. Six filament loops are wrapped around the collar and tangs and loaded equally through a whiffle-tree arrangement of pulley, ball joints, and sliders.

The filaments are prestressed to 111 N and epoxied at the tang and collar contact points. The prestress increases side-load resistance without over straining the filaments. Stem shorting is checked by an ohmmeter measurement between the stem and launch tube body. After the epoxy has cured, the excess filament lengths are trimmed, completing the assembly.

A double precision FORTRAN program was written to analyze the interaction between the filaments and applied load. Static equilibrium is checked by summing the moments around the collar and nut to zero and comparing the resultant filament force angle to the applied displacement angle. From the model, a spoke arrangement of 12 filaments was selected to provide reasonably uniform circumferential side-load resistance.

Filament size is governed by side-load requirements. However, thermal considerations also must be addressed, since the PODS-IV mechanism increases support conductance due the parallel resistive connection of the filaments and the orbit tube. Figure 4 was developed from the following linear conductive relationship used to evaluate thermal conductance

$$q = (T_H - T_C)/R_{EQ}$$

where

$$R_{EQ} = R_{TUBE} * R_F/(R_{TUBE} + R_F)$$

The normal operating temperature of the orbit tube and filaments was assumed to be between a hot temperature (T_H) of 20 to 40 K and a cold temperature (T_C) of 2 K. Conductivities were evaluted at average temperatures. A 15% increase in thermal conductance over the original PODS-III design is predicted for these struts.

TESTING

Initial testing of the PODS supports consisted of circumferentially side loading the struts to check for uniformity. This was done using the PODS-III test setup that uses a rod with pins at each end to support the strut. The rod is supported in a fixture that allows side-load testing horizontally and at inclinations of 15° intervals up to 75°. Testing was performed at 22.2 cm and 30.1 cm from the cold-end support. This corresponds to the VCS #1 and #2 attachment points on the superfluid-helium Independent Research dewar. Loads were applied perpendicular to the launch tube until shorting occurred. Shorting was monitored with an ohmmeter connected between the body and stem. The

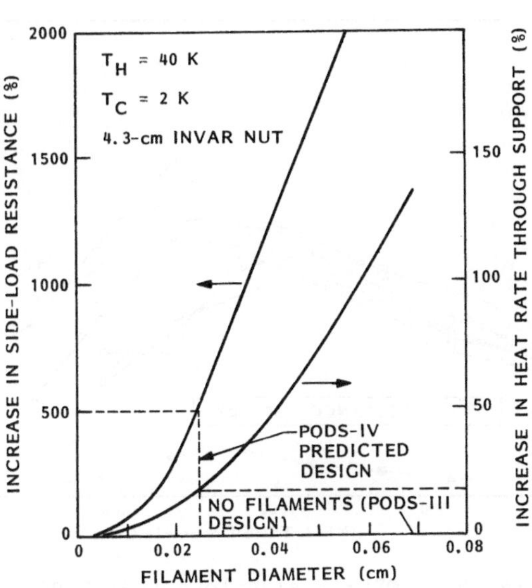

Fig. 4. PODS-IV effect on strut side-load and thermal performance.

standard deviation around the strut circumference was from 5% to 20% of the applied load.

These and previous PODS-III tests were performed at room temperature, which causes the side-load capability to be less than at design conditions. This is due to the gaps between the conical stem surfaces and the nut and body being set to take into account the differential shrinkage between the various parts. To simulate operational performance, the struts were placed vertically in a load cell and tensioned to elastically elongate the orbit tube and change the gap setting. A side load was applied perpendicular to the launch tube with a second load cell.

Six PODS-IV supports were tested. A seventh PODS support was also tested both in the PODS-III and PODS-IV configurations. This support was originally disassembled and inspected after fatigue testing, and then reassembled. Figure 5 presents the results for the two VCS locations. The peak side-load represents the capability of the strut at

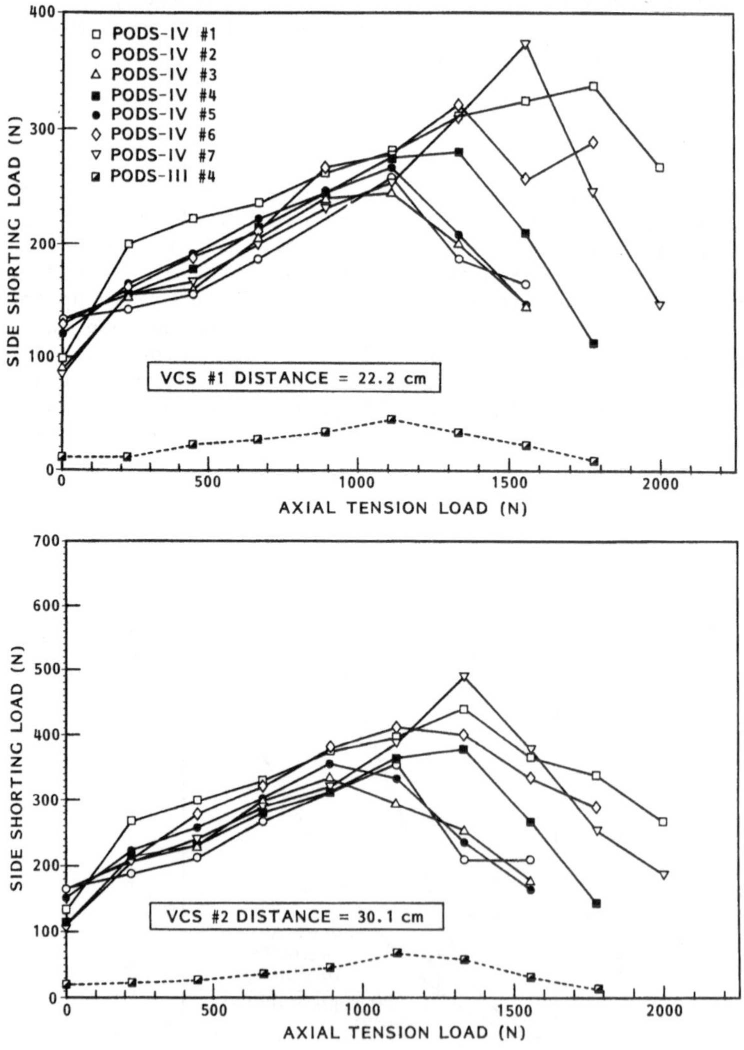

Fig. 5. PODS-IV side-load resistance.

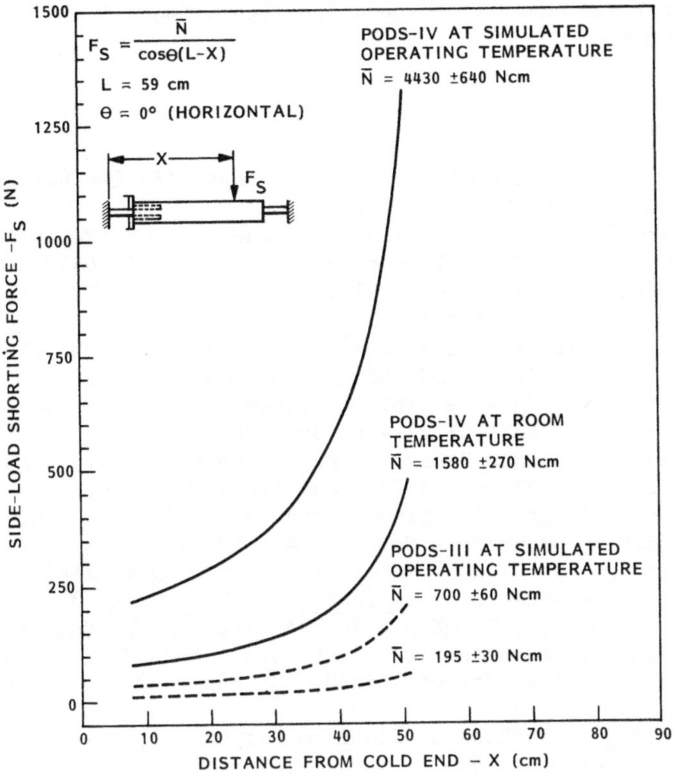

Fig. 6. PODS-IV side-load performance.

operating temperature. The average standard deviation for the PODS-IV struts tested was 17% at room temperature and 14% at simulated operating conditions. The PODS-IV nominal performance represented an increase of 810% at room temperature and 630% at simulated operating conditions over the PODS-III support. Figure 6 shows the results as predictions from the averaged data as a function of side-load placement along the launch tube length.

CONCLUSION

The PODS-IV mechanism represents a straightforward and relatively simple way of increasing PODS-III side-load performance. Minimal modification is necessary to update a PODS-III to the PODS-IV configuration. A nominal increase of 630% to 812% was achieved in side-load resistance over the PODS-III design, with a predicted 15% increase in thermal conductance. The PODS-IV modification represents a major advance over the PODS-III, since the increase in side-load resistance largely removes the limitation on length of the graphic/epoxy orbit tube that was established due to PODS-III side-load limitations. The total conductance of a PODS-IV design can actually be lower than PODS-III because of the high resistance of the longer orbit tube.

ACKNOWLEDGMENTS

This work was sponsored by the National Aeronautics and Space Administration (NASA) under Contract No. NAS2-11946 through the Ames

Research Center and under a Lockheed Missiles & Space Company, Inc. research program.

REFERENCES

1. R. T. Parmley, "Feasibility Study for Long Lifetime Helium Dewar," NASA CR 166254 (Dec 1981)
2. R. T. Parmley, "Passive Orbital Disconnect Strut (PODS-III) Structural and Thermal Test Program," NASA CR 166473 (Mar 1983)
3. R. T. Parmley, "Passive Orbital Disconnect Strut (PODS-III) Structural Test Program," NASA CR 177325 (Jan 1985)
4. R. T. Parmley et al., "Test and Evaluate Passive Orbital Disconnect Struts (PODS-III)," NASA CR 177368 (Aug 1985)
5. I. E. Spradley, "Passive Orbital Disconnect Strut (PODS-IV) Development," NASA CR 177426 (Sep 1986)
6. D. Bushness, "Optimum Design of Dewar Supports," J. Spacecraft and Rockets, Vol. 22, 4:432 (Jul-Aug 1985)
7. D. Bushnell, "Improved Optimum Design of Dewar Supports," AIAA 28th SDM Meeting, Monterey, CA, (6-8 Apr 1987)
8. G. N. Vanderplaats and F. Moses, Structural Optimization by Methods of Feasible Directions, Computers & Structures, 3:739-755 (1973)
9. G. N. Vanderplaats, "COMNIN-A FORTRAN Program for Constrained Function Minimization: User's Manual," NASA TM X-62,282, Ames Research Center, Moffett Field, CA (Aug 1973); version updated Mar 1975
10. G. Zoutendijk, "Methods of Feasible Directions," Elsevier Publishing Co., Amsterdam (1960)

UNIQUE CRYOGENIC FEATURES OF THE GRAVITY PROBE B (GP-B) EXPERIMENT

Richard T. Parmley

Cryogenic Technology Group
Research & Development Division
Lockheed Missiles & Space Company, Inc.
Palo Alto, California

ABSTRACT

Gravity Probe B is an orbital test of Einstein's general theory of relativity using gyroscopes. The precession of the gyroscopes will measure both the geodetic effect (6.6 arc sec/yr) through the curved space-time surrounding the Earth and the motional effect (0.042 arc sec/yr) due to the rotating Earth dragging space-time around with it. To achieve the extraordinary accuracies needed to measure these small precessions, it is necessary to have the gyroscopes operating in the following environments: a vacuum of $<10^{-10}$ torr; an acceleration level of $<10^{-10}$ g; a magnetic field of $<10^{-7}$ gauss; and a temperature near 2 K. This paper discusses the current status of the cryogenic designs required to simplify launch-pad operations while achieving an orbital lifetime of 2 years, as well as the designs that allow the gyroscopes to be installed into or removed from a cold dewar.

INTRODUCTION

The scientific equipment in Gravity Probe B (GP-B) required to measure the geodetic and frame-dragging effects includes four gyroscopes, four superconducting quantum interference devices (SQUIDS) to measure the gyroscope precession, a quartz telescope to provide a frame of reference in space in conjunction with a reference star, and a drag-free mass used to obtain the low acceleration levels required. Boiloff gas from the dewar is used in conjunction with eight proportional gas thrusters to keep the drag-free mass centered in its housing. All this equipment is mounted in a quartz block assembly (QBA), which is mounted in a probe assembly. The probe assembly, in turn, is mounted inside the well of a 1580 L superfluid helium dewar (Fig. 1). The probe must be inserted while the dewar tank and well are filled with 4.2 K liquid helium, which requires the use of a helium-purged air-lock assembly. Once the probe is installed, good structural and thermal contact is required at five heat stations, four located at the vapor-cooled shields and the fifth at the dewar tank neck. An Axial Lok device, described below, provides this structural support between the probe and the dewar, as well as good heat transfer at the five heat stations.

Superfluid dewars that have flown require continuous pumping on the subatmospheric-pressure superfluid helium tank up until launch. For a

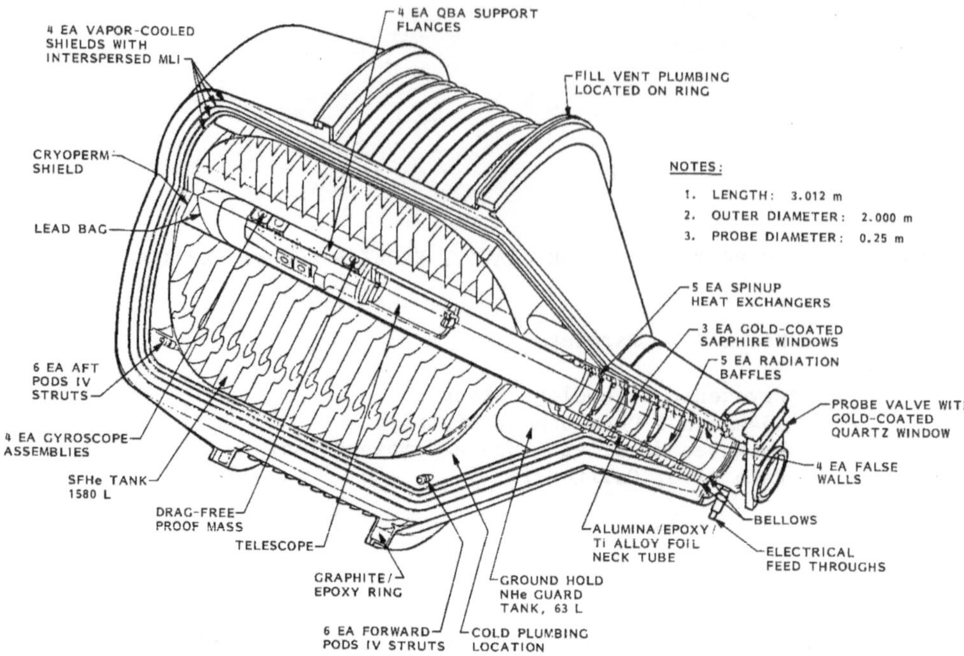

Fig. 1 GP-B Flight Dewar/Probe Concept

shuttle launch, a vacuum pumping system is located in the Orbiter bay; consequently, power is required continuously in the bay for the pumps during ground hold. The pumps have to be flight-qualified even though they operate only on the ground. Their extra weight is taken into orbit, and the potential exists for leaking oil contaminating other payloads. Loss of superfluid helium, which occurs in proportion to the length of the ground-hold period and bay temperature environment, shortens the orbital lifetime. To eliminate this servicing complication, a 63-L guard tank of 4.2 K liquid helium is mounted off the dewar neck tube and shorted to the coldest vapor-cooled shield. The superfluid-helium tank is serviced first followed by the guard tank. The servicing occurs before the assembly is loaded into the Orbiter, when access to the ground support equipment is much easier. A single fill and single vent line are required to service both tanks; only one additional valve, RAV2, and one burst disk, BD2, are required for the guard tank (Fig. 2). No servicing is required onboard the Orbiter for up to 2 weeks. If the launch is delayed beyond 2 weeks, the smaller guard tank is refilled with liquid helium at 1-atm vapor pressure, a much simpler procedure than servicing the superfluid helium tank. The nonvented, subatmospheric, superfluid-helium tank can go for over a month with a temperature rise of only 0.1 K (the superfluid tank is pumped down to 1.7 K). The tank is kept nonvented on the ground and in orbit until it reaches its operating temperature of 1.8 K. This design conserves all of the superfluid helium, maximizing the orbital lifetime.

Almost half the heat leak into the dewar by conduction and radiation is in the probe neck tube region. Consequently, to keep the conduction heat load to a minimum, both the probe and dewar necks are filament wound using a low-conductivity, high-modulus gamma alumina/epoxy composite. The probe neck also has a 0.025-mm titanium alloy liner to achieve the $<10^{-10}$ torr vacuum required inside the probe. Gold-coated windows, walls, and baffles are used along with black

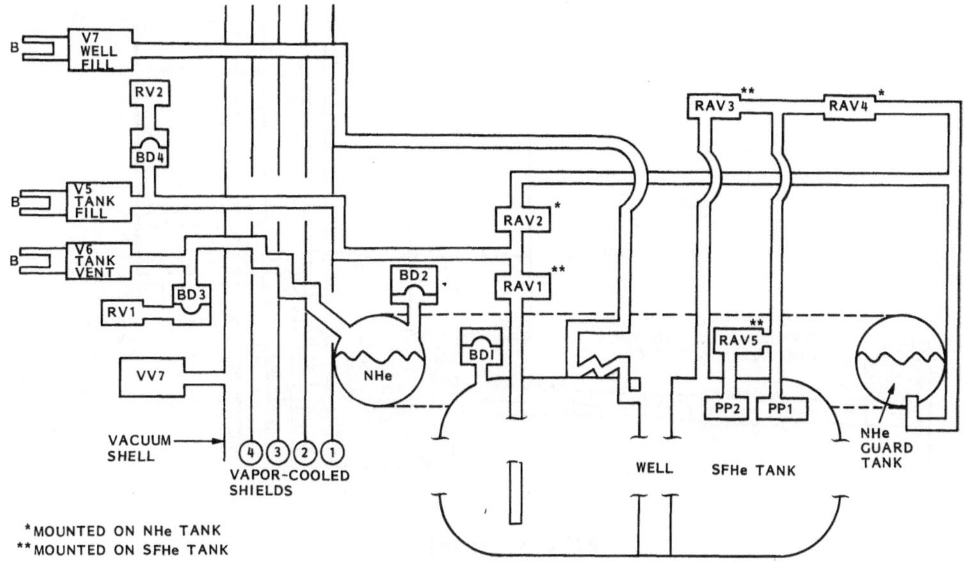

NOTES:

B - BAYONET
BD - BURST DISC
PP - POROUS PLUG
RV - RELIEF VALVE
V - SHUTOFF VALVE
VV - PUMPOUT PORT/RELIEF VALVE
RAV - REMOTELY ACTUATED VALVE

Fig. 2 GP-B Flight Dewar Plumbing

anodized aluminum walls and baffles to reduce the radiant heat load onto the telescope.

Twelve high-efficiency PODS-IV struts are used to support the superfluid helium tank to achieve a 2-year orbital lifetime. Gamma alumina/epoxy and graphite/epoxy composites are used in the strut design.

Details of the Axial Lok, alumina/epoxy neck tube, and PODS-IV supports are provided below.

AXIAL LOK

The Axial Lok provides structural support between the probe and dewar as well as good heat transfer at the five heat stations. It consists of a number of mechanisms, integrated to provide the required structural and thermal functions: (1) Three "dogs" (Fig. 3) hold the probe in the dewar providing the main structural support and heat transfer to the superfluid helium tank. (2) A lead-bag retainer (Fig. 4) presses the lead bag* against the inside of the dewar well as the Teflon-coated probe vacuum shell is inserted. The flattened beryllium copper springs provide the load required for good thermal contact between the lead bag and the well; the retainer also provides lateral support for the probe. (3) Eight thermal shoes (Fig. 5) at each of the four vapor-cooled shield locations (32 total) transfer heat conductively from the probe out to the dewar vent heat exchangers.

The dogs are rotated out of the way during probe insertion. A castellated tool, concentric with a specially designed long allen

*A series of expanded lead bags provides the ultralow magnetic field

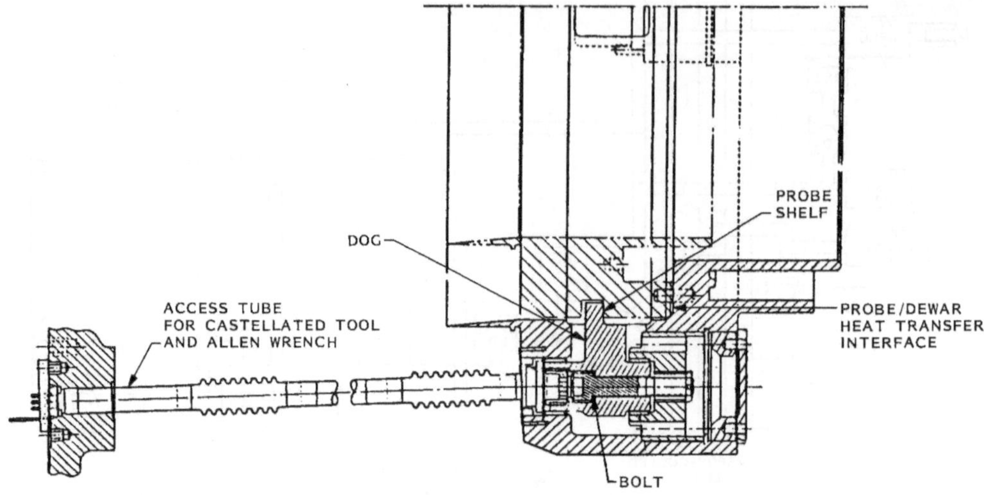

Fig. 3 Axial Lok

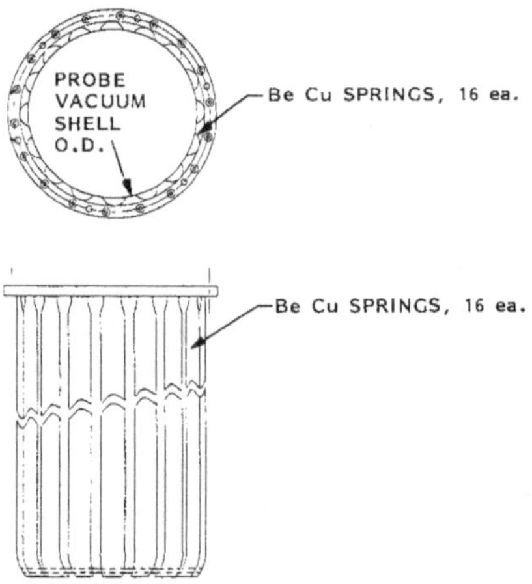

Fig. 4 Lead-Bag Retainer

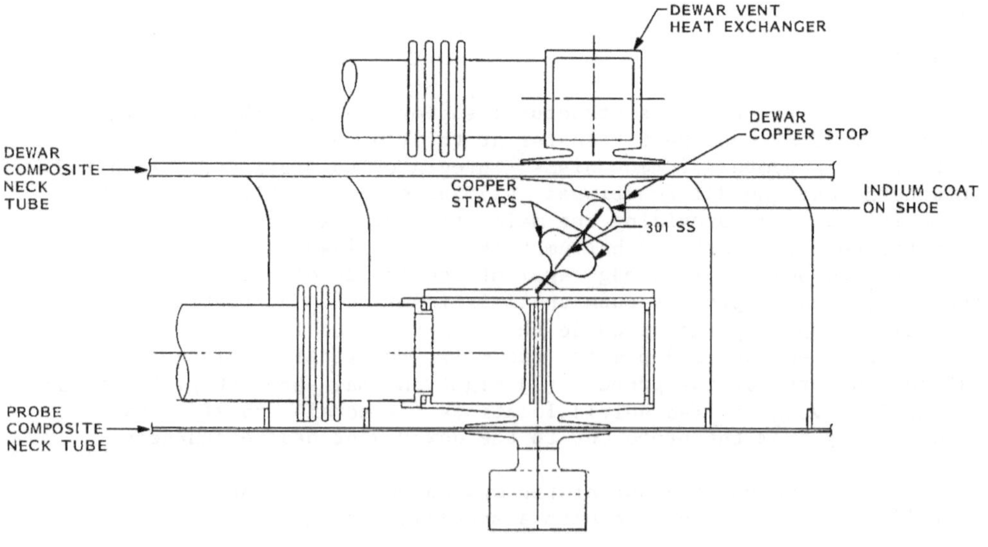

Fig. 5 Thermal Shoe Design

wrench, is inserted into a tube permanently located in the dewar. The castellated tool is used to engage and rotate the dog into the shelf machined into the probe ring. The three bolts on the dogs are then torqued down using the allen wrench. These bolts apply a 26,400 N load through the "dogs" to the probe/dewar-ring interface. The castellated tool and allen wrench are removed and a "shish kabob" of radiation disks mounted on a fiberglass rod is installed in the dewar tubes to reduce radiation tunneling. The probe side of the 6061-T6 Al interface is brush plated with indium (copper strike undercoat) to enhance heat transfer. The installation process is reversed when removing the probe.

As the dogs pull the probe against the dewar interface, each of the 32 thermal shoes engages a separate mating stop on the dewar neck. The thermal shoes are staggered at 11.25° intervals around the circumference so they don't interfere with each other during probe insertion. The 301 stainless steel spring used in the shoe is bent elastically and driven radially outward as the dogs pull the probe into the dewar.

Typical measured radial versus axial loads for each thermal shoe are shown in Fig. 6 as a function of the initial radial gap spacing between the probe shoe and the stop. A large number of these tests have been performed in the test apparatus shown in Fig. 7. Some additional testing remains. Thermal contact is made between an indium brush-plated copper rod brazed to the end of the spring and a concave cylindrical stop located on a copper ring, bonded to the inside of the dewar composite neck. Copper strap loops, 0.125 mm thick, are located on both sides of the spring. The straps are sized to keep the ΔT across the spring to less than several degrees K for the predicted heat loads.

ALUMINA/EPOXY-COMPOSITE NECK TUBES

The probe neck tube provides the aperture for the telescope and allows the probe to be evacuated and inserted or removed from a cold dewar. This design requires a separate neck tube for the dewar.

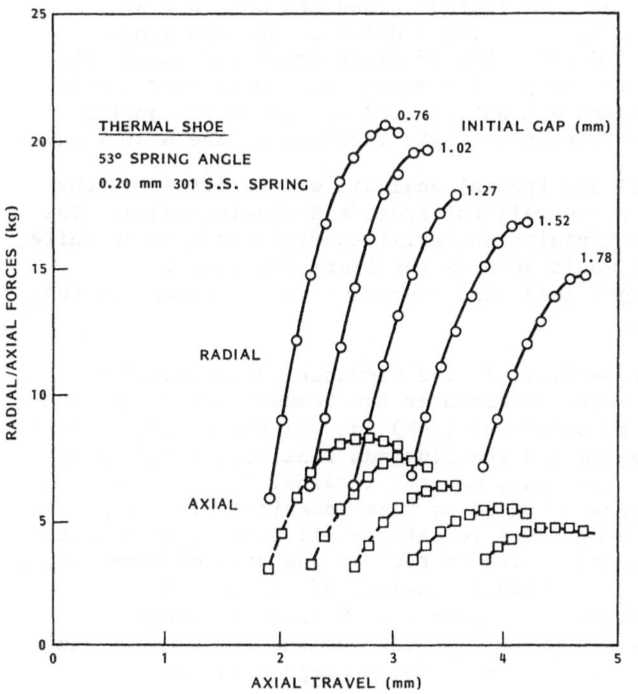

Fig. 6 Measured Thermal Shoe Radial and Axial Loads

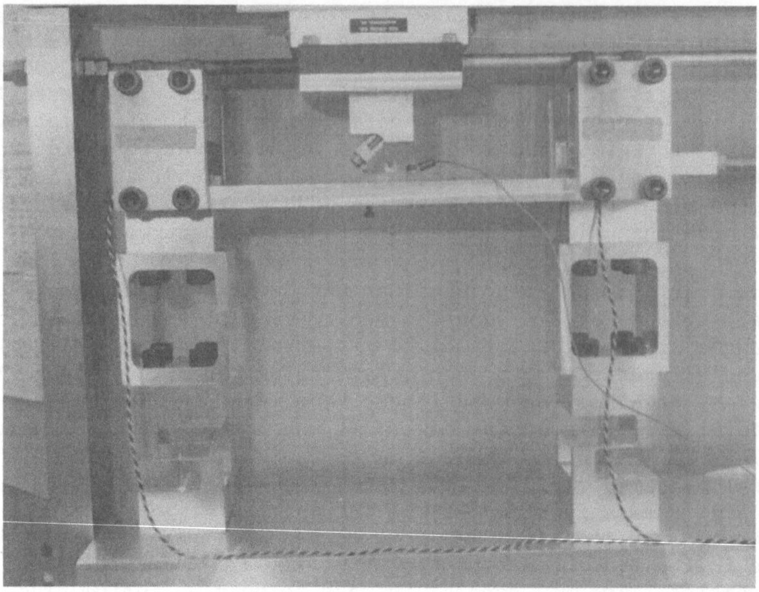

Fig. 7 Thermal Shoe Test Apparatus

The neck tubes for both the probe and the dewar must be designed to withstand a 1-atm compressive load during probe installation as well as 15-g flight loads with vacuum on both sides of the tubes. Because a high proportion of the total parasitic heat load in the dewar is from these tubes, various approaches were investigated to reduce the heat load. The first approach used folded composite neck tubes (to increase the conduction path length), with glass epoxy tubes at the warmer end and graphite epoxy tubes at the colder end (Fig. 8). Subsequently, a new fiber was reported[1] that has a thermal conductivity nearly as low as glass from 300 to 50 K and half that of glass and the same as graphite from 50 K to 0 K (Fig. 9). The modulus of the new gamma alumina/epoxy fiber is over double that of glass/epoxy and nearly the same as that of the graphite/epoxy. Consequently, the simpler single alumina/epoxy neck tube design was substituted for the more complex folded neck tube design, with only a slight increase in the heat load.

An extensive structural and thermal analysis was performed on the probe neck tube to optimize the wall thickness and winding angle. The wind angle was selected to minimize the axial thermal conductance while providing a structure that would survive the 1-atm ground-hold compressive load, 15-g flight load, and thermal strains induced during cooldown.

The thermal strains arise from (1) the mismatch in contraction between the fiber and the epoxy matrix over the temperature range from epoxy curing (414 K) down to operating (2 K) and (2) the contraction mismatch between the composite and the aluminum tank bond (cold end), the stainless steel bond (warm end), and the internal copper rings and external copper heat exchangers bonded at four locations along the composite tube. The best compromise to satisfy all these requirements is a $\pm 40°$ filament-winding angle with respect to the axis of symmetry, a probe wall thickness of 0.46 mm with a diameter of 244 mm, and a measured fiber volume fraction of 55 percent. Stiffening rings are wound at 15.5-mm intervals along the tube to minimize the wall thickness for thermal reasons, yet still provide a 1-atm compressive load

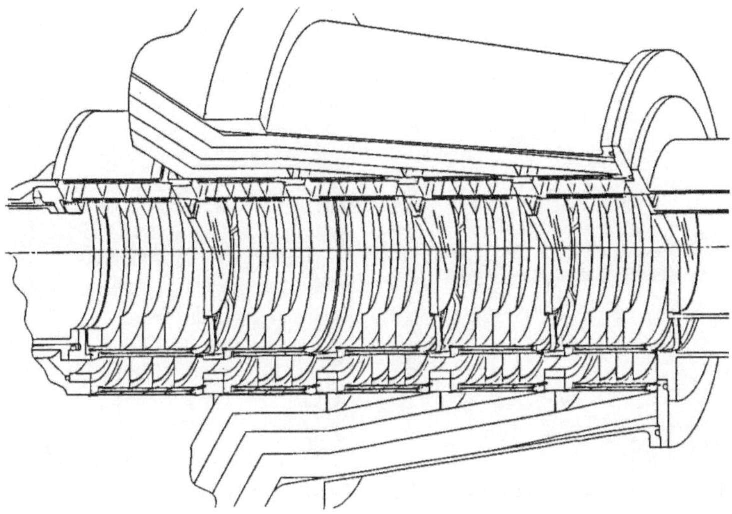

Fig. 8 Folded Neck Tube Design

capability (with a safety margin of 4). The rings also provide a mounting surface for gold-coated Kapton radiation baffles. The baffles reduce the radiation tunneling in the annular space between the probe and dewar necks to < 1 mW.

A full-diameter, one-third-length test section of the composite neck tube was fabricated (Fig. 10) and thermally cycled from room temperature to liquid nitrogen and helium temperatures (Fig. 11). An external helium pressure of 1 atm was placed on the test section and the test article was evacuated. The permeation rate of helium gas was measured with a helium mass spectrometer leak detector. The low permeation rates measured are due to the 0.025-mm titanium alloy liner used. The thermal cycling is continuing. (Following probe insertion, the well is evacuated so there is vacuum on both sides of the probe neck for the remainder of the mission.) The tube has survived the thermal cycling to date with no structural damage. The thermal conductivity of the composite is currently being measured over the temperature range 300 K to 78 K to validate the literature values.[1]

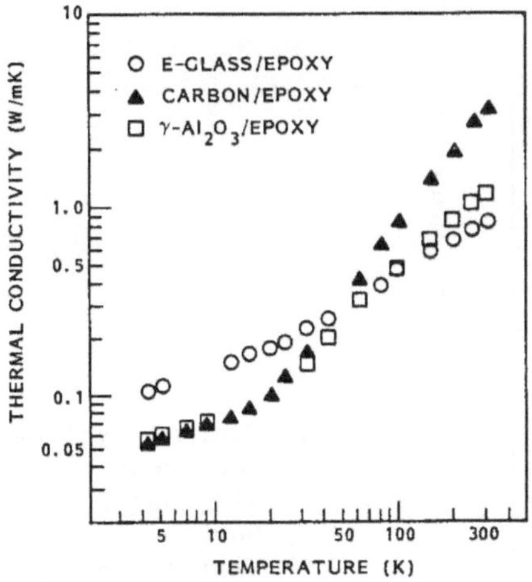

Fig. 9 Composites Thermal Conductivity

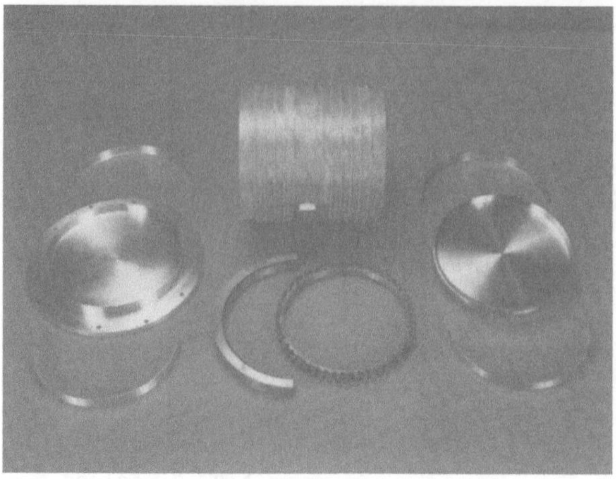

Fig. 10 Alumina/Epoxy Neck Tube Test Article

PASSIVE ORBITAL DISCONNECT SUPPORTS (PODS-IV)

The PODS-IV struts are the most thoroughly ground tested support system to date for cryogenic flight dewars.[2,3,4,5,6] Tests have included thermal performance, static load tests, dynamic load tests, and creep load tests on the struts, both individually and as a system. Six supports can be used on smaller tanks, i.e., on a 200-L superfluid-helium flight dewar (Fig. 12), or 12 supports can be used for longer tanks, i.e., the 1580-L GP-B flight dewar (Fig. 13).

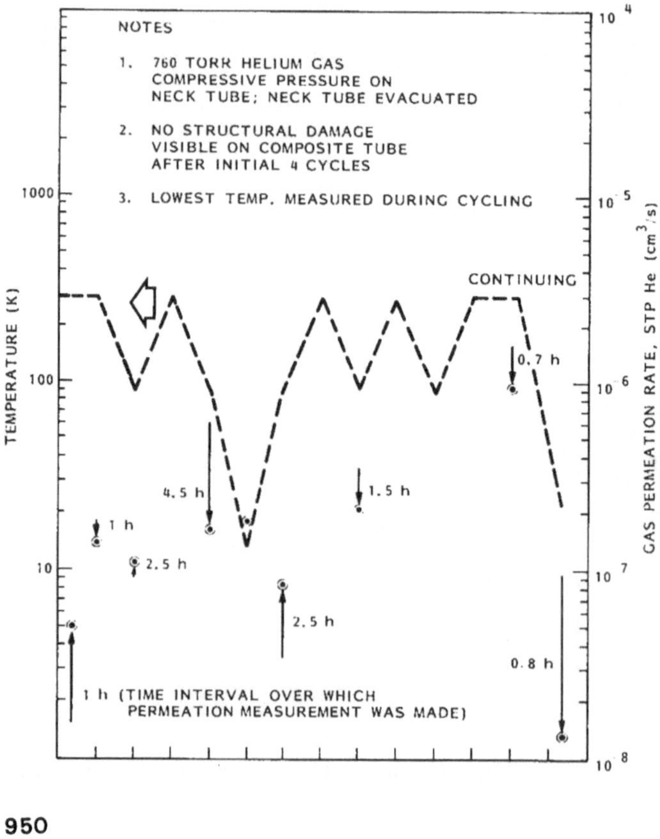

Fig. 11 Thermal Cycling Test Results of Alumina/Epoxy Neck Tube Test Article

Fig. 12 Superfluid Helium Flight Dewar With Six PODS-IV Struts

The only true support-optimization program that has been developed for either PODS struts or tension support straps, DEWAR,[7] was used to size the PODS-IV struts (optimum number of struts, length, diameter, wind angles, number of layers, wall thickness, and mounting angle). The program minimizes support conductance within specified constraints (boundary temperatures, acceleration levels in all axes, minimum allowable frequency during launch or orbit, maximum allowable loads based on tube or strap fatigue strength, column buckling, localized buckling and dimensions of attachment hardware). The program accounts properly for coupling between lateral and tilting motion in the free vibration modes. The flexibilities of the dewar, vacuum shell, and supporting rings are also accounted for. Stress criteria in material coordinates for each lamina of a composite laminated support tube can become active constraints during optimization. Shell buckling is handled via a PANDA-type[7] of analysis for a composite cylindrical shell, rather than through use of the simple classical buckling formula. Thicknesses of individual layers in a laminated composite strut tube can be decision variables for optimization, rather than just the overall thickness of the tube being a decision variable. The conductivity is orthotropic, and the overall conductance now depends on the thicknesses and winding angles of the laminae in the composite strut tube wall.

Note that the 12-support configuration shown in Fig. 13 has the 6 forward supports located almost in a plane and attached to a graphite/epoxy support ring at the warm end of the struts. As the tank cools down, the thermal loads put into the struts are minimal with this arrangement. In orbit, changes in vacuum-shell dimensions due to temperature also put minimal thermal strains in the struts because of the ultralow expansion characteristics of the graphite/epoxy support ring. This ring is also used to mount the star blipper, providing a very stable mounting surface referenced to the telescope inside the probe. The aft six struts are sized for the axial launch loads. All 12 struts provide lateral, tilt and torsional load capability.

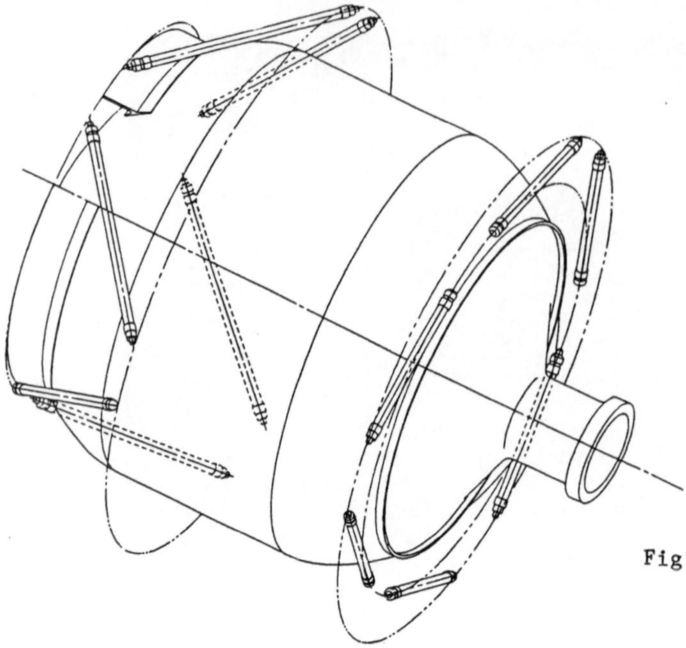

Fig. 13 GP-B Flight Dewar Support System Using 12 PODS-IV Struts

The four vapor-cooled shields are supported off the alumina/epoxy dewar neck described previously. They are thermally grounded and stabilized against side loads by the PODS-IV alumina/epoxy launch tubes. Substituting the alumina/epoxy for the glass/epoxy used previously allows the wall thickness to be cut approximately in half because of the alumina/epoxy's higher modulus. (Column buckling designs the tube.) The side load capability of the strut has been increased 700 percent going from the PODS-III to PODS-IV design; side loads put on the struts due to vapor-cooled shield contractions are well within the capability of the PODS-IV design.

The passive load bypass mechanisms of the strut described in earlier papers[4] does not close on the ground under 1 g (designed for 1.5 g) or in orbit, as demonstrated by ground tests of the complete support system.

A detailed 73-node thermal model of a vapor-cooled helium dewar verifies that the weight of the dewar can be cut in half[2] for the same lifetime by substituting PODS-IV struts for tension straps when both support systems are optimized using the same ground rules (shuttle launch) and the computer program described previously. The above analyses assume no internal instrument heat load.

CONCLUSIONS

A number of unique features have been designed into the probe and superfluid-helium dewar that are driven by Gravity Probe-B program requirements.

These features include: (1) the ability to load the probe into a cold dewar and provide good structural support and thermal contact after insertion, (2) elimination of the need for a pumping system during ground hold, and (3) use of a new composite, gamma alumina/epoxy, to minimize the parasitic heat loads into the dewar through the telescope aperture and the tank supports.

ACKNOWLEDGEMENTS

This work was performed for Stanford University and funded by the NASA/Marshall Space Flight Center. The PODS-IV development was funded by the NASA/Ames Research Center.

Adolf Kratz developed the Axial Lok concept, Gary Reynolds developed the thermal shoe concept, Dave Donegan developed the lead-bag retainer concept, Mark Ferraro and Pat McCormack wound the alumina/epoxy tubes, Mary Wright and Marc Regelbrugge performed the structural analyses, Dave Bushnell developed the support optimization program DEWAR, and Jack Goodman performed the thermal analyses and the temperature cycling of the composite tube.

REFERENCES

1. M. Takeno, S. Nishijima and T. Okada, Thermal and Mechanical Properties of Advanced Composite Materials at Low Temperatures in "Advances in Cryogenic Engineering Materials," Vol. 32, Plenum Press (1985), p. 217.

2. R. T. Parmley, "Feasibility Study for Long Lifetime Helium Dewar," NASA-CR166254, Dec. 1981.

3. R. T. Parmley, "Passive Orbital Disconnect Strut (PODS-III), Structural and Thermal Test Program," NASA CR 166473, Mar 1983.

4. R. T. Parmley, "Passive Orbital Disconnect Strut (PODS-III) Structural Test Program," NASA-CR 177325, Jan 1985.

5. R. T. Parmley et al., "Test and Evaluate Passive Orbital Disconnect Struts (PODS-III)," NASA-CR 177368, Aug 1985.

6. I. Spradley, "Passive Orbital Disconnect Strut (PODS-IV) Development," NASA-CR 177426, Sept 1986.

7. David Bushnell, "Improved Optimum Design of Dewar Supports," 28th SDM meeting, April 6-8, 1987.

AN ADIABATIC DEMAGNETIZATION COOLED BOLOMETER SYSTEM

L. Lesyna, T. Roellig, M. Werner, P. Kittel *

NASA-Ames Research Center
Moffett Field, California

ABSTRACT

We have constructed an adiabatic demagnetization refrigeration (ADR) system suitable for cooling bolometers for use at infrared and submillimeter wavelengths. Because it does not rely on a gravitational field for operation, this refrigeration method has been proposed to cool bolometers intended for NASA's Space Infrared Telescope Facility (SIRTF). To explore the use of ADR in this manner, we have constructed a portable instrument that has been used successfully for ground-based astronomical observations. By employing chromic potassium sulfate as a paramagnetic cooling material, we have successfully cooled bolometers to temperatures of 0.1 K with a cryogenic hold time of greater than nine hours. Our most sensitive detector tested had a measured electrical noise equivalent power of 7×10^{-17} W $Hz^{-1/2}$, which is the lowest value yet recorded for a submillimeter wavelength bolometer. A computer controlled feedback loop was used to regulate the temperature of the bolometer to within 14 μK, which results in a less than one percent change in responsivity for 0.1 K bolometers. We discuss operation of this refrigeration system, and some issues affecting the adaptation of this type of system to the spacecraft environment.

INTRODUCTION

Adiabatic demagnetization refrigeration is of great importance for applications requiring the sensitive detection of millimeter-submillimeter wavelength radiation. The cryogenically cooled bolometer is presently the most sensitive detector at these wavelengths. Bolometers housed in cryogenic refrigerators operating at 0.3 K are in routine use, and the most sensitive have noise equivalent powers (NEP) of about 4×10^{-16} W $Hz^{-1/2}$. Useful gains in sensitivity can be expected by operating optimized bolometers at temperatures of 0.1 K. We have constructed a portable adiabatic demagnetization refrigeration system (ADR) to house bolometers, and have used our system to perform ground-based one millimeter radiometry. NASA's Space Infrared Telescope Facility (SIRTF) will require ADR to cool bolometers to 0.1 K. Many

----

* The present address of L. Lesyna is Grumman Corporate Research Center, Bethpage, New York.

issues concerning successful implementation of ADR aboard spacecraft can be addressed by laboratory and field tests of a portable, compact refrigeration system.

CRYOSTAT

In Fig. 1, we show a cross-section drawing of our cryostat. The cryostat used by us is similar in design to others used for adiabatic demagnetization[1]. It contains both the paramagnetic salt and superconducting solenoid which generates the magnetic field around the salt. The interior of the cryostat is filled with liquid He, which may be pumped on with an external mechanical pump to reach temperatures of less than 2.0 K. The paramagnetic salt is positioned inside the bore of the solenoid in a re-entrant cavity and the bolometer is placed in thermal contact with the salt. The re-entrant cavity may be independently pressurized with He gas, or evacuated to make or break thermal contact between the paramagnetic salt and the liquid He bath. External radiation passes through a vacuum window at ambient temperature and a cryogenic vacuum window a He bath temperature, before striking the detector.

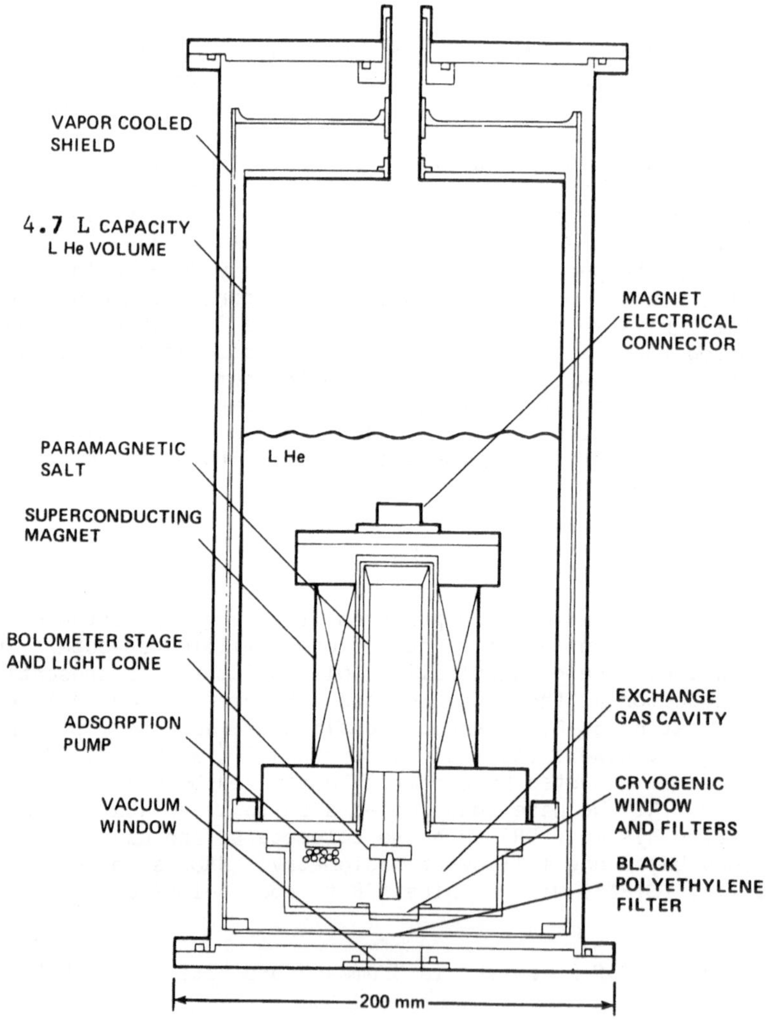

Fig. 1. Adiabatic demagnetization cryostat.

The magnetic field around the paramagnetic salt is generated by passing current through a superconducting solenoid manufactured by American Magnetics, Inc. A field of 6 T can be achieved by passing 30 A through this solenoid. A persistent switch is located in parallel with the solenoid, permitting a zero resistance path for the electrical current. Electrical leads are inserted through the neck of the cryostat to mate with the solenoid electrical connector. These leads are designed to permit He vapor evacuation in order to reduce the bath temperature below 4.2 K. An external power supply delivers a selectable value of current through the solenoid.

In order to make or break thermal contact between the liquid He bath and the paramagnetic salt, a charcoal adsorption pump is employed. It consists of a few grams of activated charcoal cemented with Stycast 2850 FT epoxy onto a small Cu sheet. A 0.5 W Allen-Bradley carbon-composition 1.5 kΩ resistor is imbedded along with the charcoal, and serves as a heater. The sheet is epoxied onto a solid fiberglass post of 10 mm x 10 mm x 10 mm size, and attached to a liquid He cooled surface of the cryostat. Dissipating 0.1 W of power through the resistor drives off a small quantity of He gas, which produces the desired thermal contact. When this power is removed, the He exchange gas is readsorbed, and thermal isolation between the salt and bath is reestablished.

The use of exchange gas to conduct heat to the paramagnet necessitates the use of a cryogenically cooled vacuum window. The window material must be transparent to submillimeter radiation. A design similar to that described by Loose et al.[2] was used. It consists of a quartz window positioned above a teflon spacer, which is in turn inserted into a stainless steel flange. The circumference of the assembly was coated with a flexible epoxy resin (Armstrong A-12 resin, with hardener and resin mixed in a ratio of 2:1). Three layers of Al foil were wrapped tightly around the circumference, resulting in a leak-proof seal. This combination may contract without cracking when cooled, resulting in a useful seal.

We chose the paramagnetic salt chromic potassium sulfate as our magnetic refrigerant. For this substance, the magnetocaloric effect is large enough to reach practical operating temperatures of less than 0.1 K for useful periods of time. The bolometer detector must be placed in thermal contact with the salt. We have achieved this by using the apparatus depicted in Fig. 2. The salt was compressed in a metallic die around one hundred 160 μm diameter Cu wires, and then hermetically sealed within a brass tube with epoxy end caps. The ends of the Cu wires were soldered to a brass bolometer holder, into which the bolometer was placed. A brass cone restricts the field of view of the bolometer to match that of the incoming submillimeter radiation. The composition and geometry of the above materials was chosen to minimize eddy current heating during demagnetization, and to minimize the heat capacity of all addenda to the salt.

The salt pill and addenda were suspended inside the bore of the solenoid by twelve nylon filaments, each approximately 2 cm long (dental floss). The expected heat leak into the salt from a 4.2 K liquid He bath is estimated to be 3×10^{-7} W. A suspension can limit the hold time at 0.1 K, thus it was necessary to choose the suspension geometry to limit the heat leak into the salt. The salt pill is actually suspended within a type 304 stainless steel tube which slips inside the bore of the solenoid. This facilitates removal and installation.

The entropy of the addenda is negligible compared to the entropy of the salt, which is 1.7 J/K at temperatures of 4.2 K. Thus, the addenda does not play an important role in determining whether a salt pill will successfully cool down to temperatures of 0.1 K. The limiting factor is the thermal contact between the Cu wires and the salt. If any heat flows directly into

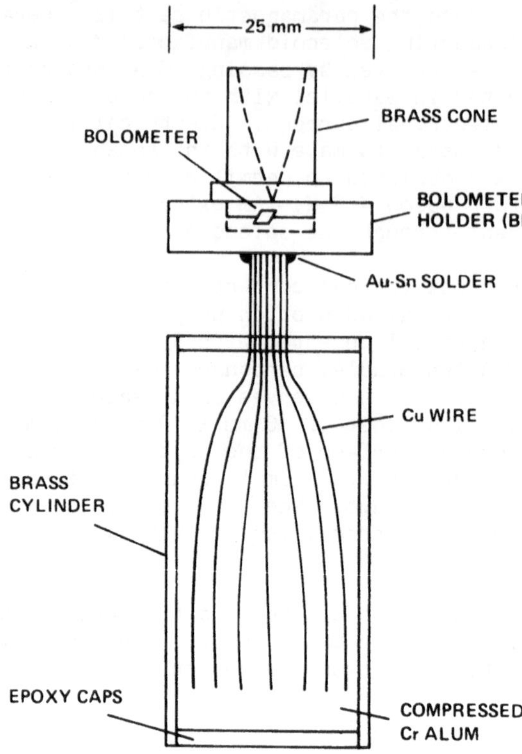

Fig. 2. Paramagnetic salt pill.

the bolometer holder, which is designed to be in good thermal contact with the Cu wires, then a temperature difference is established between the salt and wires. The thermal boundary resistance between the salt and wires is large enough to maintain a significant temperature difference[3].

Because of these considerations, we have found it necessary to take the following precautions. First, all thermal conduction paths lead to the case of the salt pill, whenever possible. This means foregoing direct suspension between the bolometer holder and the liquid He work surface. Second, all wires that lead to the bolometer, such as those for the germanium resistance thermometer and the bolometer itself, are wrapped many times around the case of the salt pill. We employed rubber cement adhesive to bond these wires to the case. By taking these precautions, most of the external heat is directed into the case, and a much smaller amount reaches the bolometer holder.

One method to reduce this troublesome thermal contact problem would be to increase the surface area of the Cu wires. This could be achieved by using many more smaller diameter wires.

With the present configuration of the cryostat, we have reached temperatures as low as 0.05 K, by demagnetizing from paramagnetic salt temperatures of 2.0 K and 6 T magnetic field. We have maintained temperatures of 0.1 K for a time period in excess of 9 hours before recycling the refrigerator. Further details of the cryostat are given by Lesyna[4].

REFRIGERATION CONTROL SYSTEM

In order to perform reliable radiometric detection with a bolometer, it is necessary to limit the range of variations in the temperature of the bolometer heat sink. For the 0.1 K bolometers we have constructed, it is necessary to control the temperature of the refrigerator to within 30 μK to keep variations of the bolometer responsivity to within one percent. In order to achieve this requirement, a closed loop control system was implemented. The temperature of the bolometer heat sink is sensed and compared with the desired heat sink temperature. A change in magnetic field around the paramagnetic salt is governed by a second order control loop algorithm. A block diagram of the electronic instrumentation is presented in Fig. 3. When in use at the observatory, the instruments and cryostat reside in the Cassegrain focus cage which moves with the telescope. The computer controls the instruments by issuing character sequences which are converted to GPIB bus commands.

The temperature of the bolometer heat sink is determined by measuring the electrical resistance of a Lakeshore GR-30A germanium resistance thermometer (GRT) which is calibrated for the temperature ranges 0.05 K to 4.2 K. The resistor is excited by an AC current source within a Linear Research LR-400 resistance bridge set to dissipate a maximum power of 1 pW in the resistor. An analog voltage proportional to the resistance of the calibrated GRT is fed to a multiplexer under GPIB control. A calibrated diode thermometer monitors the temperature of the superconducting solenoid. The diode serves as a warning to the computer to shut down the system as the liquid He runs out of the cryostat. This avoids a possibly damaging solenoid quench. The voltage across the diode is also fed into the multiplexer.

The voltage from the multiplexer is read by a $5\frac{1}{2}$ digit voltmeter. This precision is necessary to insure sensitivity to small temperature variations of the GRT.

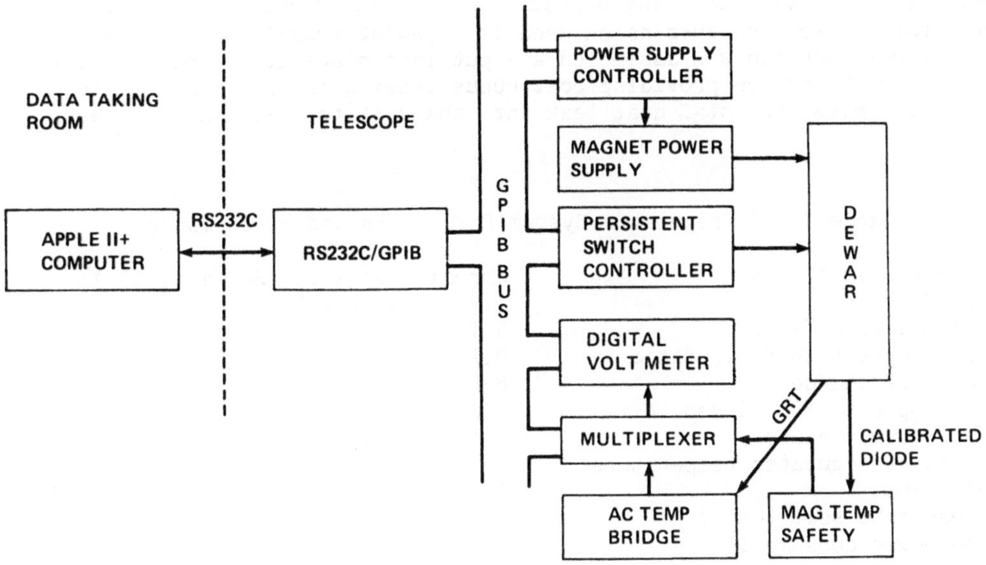

Fig. 3. Refrigeration system electronics.

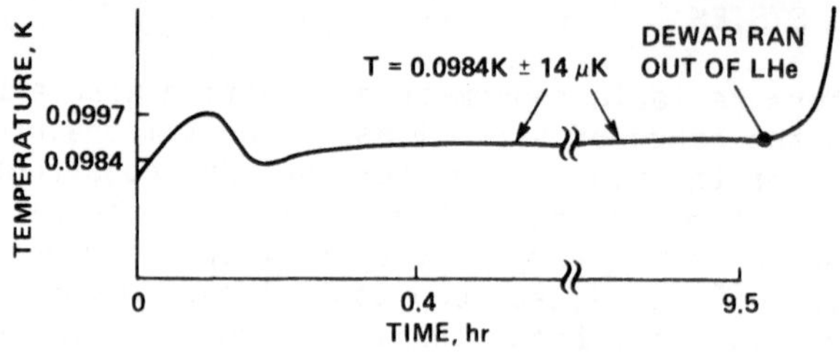

Fig. 4. Sample run of ADR temperature vs. time.

The computer controls a Kepco SN-488 power supply controller, which generates a programmable voltage from a 12 bit DAC. This in turn is fed to a Kepco ATE power supply configured to act as a variable 0 A to 5 A current source. This current is fed through the superconducting solenoid, thus closing the control loop.

For our ADR system, we have realized a temperature stability at 0.1 K of 14 μK. A diagram of temperature vs. time for a sample run is shown in Fig. 4. Adjustment in current through the solenoid was made at intervals of 120 s. This stability is more than adequate to maintain a bolometer responsivity to within one percent over the course of a 9 hour observing period.

Our refrigeration system thus met our goals of achieving and maintaining temperatures of 0.1 K for useful period of time. We have summarized refrigeration system parameters and performance in Table 1.

FUTURE WORK

The change that would make the greatest enhancement to overall performance and reliability is to improve the suspension of the salt pill. The nylon filaments we have used have a tendency to stretch after cycling between ambient and cryogenic temperature. This results in the optical axis of the bolometer field of view assuming unpredictable orientations with respect to the cryostat. Also, the suspension used is somewhat fragile. For these reasons, the suspension was assembled and put into place at the observatory. A more rugged suspension providing continuous tension in the filaments is needed that limits the total heat leak into the salt to less than 10^{-6} W.

Table 1. Refrigerator System Parameters and Performance

Paramagnetic salt, mass	Chromic Potassium Sulfate, 80 g
Entropy S(0,T)	1.7 J/K at T = 2.0 K
Maximum magnetic field	6 T
Solenoid current to field ratio	0.2 T/A
Solenoid inductance	6 H
Lowest temperature achieved (est.)	0.04 K
Regulated temperature	0.10 K ± 14 μK
Hold time at regulated temperature	9 hours
Recycle time	1 hour
Heat leak into salt (est.)	< 5 μW
Average power consumption	0.5 W
Peak power consumption	150 W
Liquid helium consumption (average)	0.7 L/h

The mass and volume of our ADR system is dominated by the electronics, specifically the magnet power supply, resistance bridge and computer. The physical size is thus a burden when moved outside the laboratory. A custom electronics package would be relatively straightforward to build and could occupy considerably less volume.

In order to adapt an ADR of this type to operate in spacecraft, both rugged suspensions capable of surviving a launch, and compact low-mass electronics are essential. In addition, there are other considerations which impose more stringent constraints on the cryogenic system. Liquid helium boiloff must be reduced considerably, as cryogen resupply in space is difficult and costly. Another constraint is set by the power requirement of this system. When the current through the solenoid is changed, the power needed may be two orders of magnitude greater than the average power requirement of 0.5 W. This value could tax the power capabilities of the spacecraft and hence must be made much smaller. Some schemes of avoiding or reducing this power cost have been proposed[1,3]. Other paramagnetic materials should also be considered for spacecraft ADR use. In particular, non-hydrous paramagnetic substances such as cerous metaphosphate[5] and ruby[6] have been used successfully as magnetic refrigerants, and are chemically stable over long periods of time. Unlike many hydrated paramagnetic salts, including chromic potassium sulfate, the above mentioned refrigerants are resistant to deterioration when heated above ambient temperature.

CONCLUSIONS

We have successfully used adiabatic demagnetization refrigeration to cool bolometers to 0.1 K. Our most sensitive submillimeter wavelength bolometer was measured to have a NEP of 7×10^{-17} W $Hz^{-1/2}$, which is the lowest value yet recorded. Our ADR system has been used successfully at Palomar Observatory to perform sensitive millimeter radiometry. We have identified components of the system that are candidates for further study, for the use of ADR on the ground or in space.

ACKNOWLEDGEMENTS

L. Lesyna thanks A. B. C. Walker, Jr., for guidance. We thank D. Lesberg, A. Spivak and R. Zeiger for technical assistance, and E. Kremer for manuscript preparation. This work was supported by a NASA cryogenics RTOP.

REFERENCES

1. R. D. Britt and P. L. Richards, Int. J. Infrared and Millimeter Waves 2:1083 (1981).
2. P. Loose, U. Waas, and M. Wohlecke, Cryogenics 14:470 (1974).
3. P. Kittel, J. Energy, 4:266 (1980).
4. L. Lesyna, Ph.D. Thesis, Stanford Univ. (1987).
5. E. L. Althouse, Cryogenics, 9:177 (1969).
6. J. W. Bakker, H. van Kempem, P. Wyder, Phys. Lett. 31A:290 (1970).

EXPANSION TURBINES AND REFRIGERATION FOR

GAS SEPARATION AND LIQUEFACTION

L. C. Kun

Union Carbide Corporation
Linde Division
Tonawanda, New York

ABSTRACT

A simple method sufficient for the preliminary design of the 50% reaction radial inward flow turbines is outlined. Design charts based on the similarity principles giving guidance to determine the optimum operating speed and the principal dimensions are included. Some conflicting requirements of the disciplines involved; e.g., aerodynamics and manufacturing are discussed.

INTRODUCTION

There are many ways to produce refrigeration; e.g., evaporation/desorption, Joule-Thomson expansion, expansion by work extraction, expansion by doing work on part of the working fluid (such as the Ranque-Hilsch vortex tube, the emptying process, and pulse tube), magnetic cooling, thermoelectric cooling, etc.

Two milestones in industrial air separation occurred in 1895 and 1899 when Carl von Linde used the J-T Process and Georges Claude the combination of the J-T process with a reciprocating expander to generate liquid air, to obtain oxygen for a growing market. A good account of the early pioneering work can be found in references [1,2].

Even today these are the two methods; i.e., the J-T expansion and the expansion by work extraction, which are used almost exclusively in the large scale cryogenic process. While the early successful designs were the impulse type, P. Kapitza, after analyzing the losses in the impulse and the radial inward flow type designs, developed and successfully operated the latter type as he reported in 1939 [3]. Thus by the 1940's the requirements of a thermodynamically efficient air separation or liquefaction cycle were understood to a great extent and the many necessary components were fairly well developed.

CYCLE CHARACTERISTICS

The modern liquefaction cycle is more related to the Claude process except that the refrigeration is generated at two temperature levels (Fig. 1) by turbines. This is a practical compromise between complexity and cycle losses; i.e., investment and operating cost in liquefying nitrogen. The cycle efficiency actually would be higher with three turbines mainly due to the improved matching of the cooling curves in the heat exchangers [4].

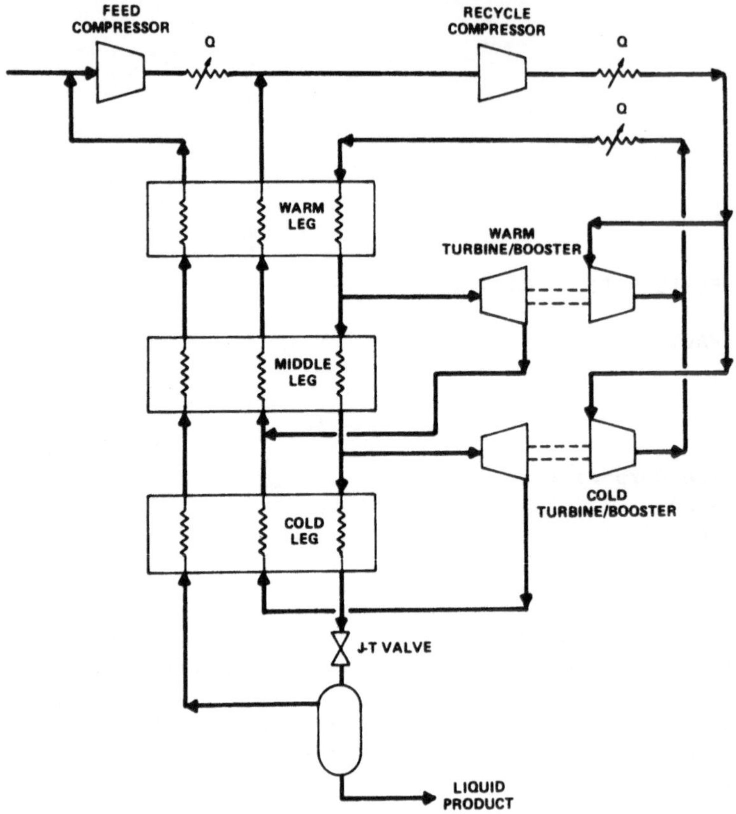

Fig. 1. Modern liquefier cycle.

Finding the economic optimum for such a liquefier is a complex process, as outlined in reference [5]. As can be seen from Fig. 2, the turbine efficiencies exert a large influence on liquefaction power consumption [6] and even in the 90% range there is incentive for further improvement.

TURBINE SIMILARITY PARAMETERS

As a first impression the aerothermodynamic and mechanical design of a turbine appears to be an impossibly complicated undertaking. However the preliminary design task today is greatly facilitated by the publications of the dimensionless design diagrams for many types of turbomachines. Dimensional analysis to develop dimensionless groups governing the characteristics of turbomachines is performed by Shepherd [7]. Design charts for many type of turbines and compressors are published [8].

Perhaps it would be instructional to compare the design input data and the dimensionless groups necessary to perform the preliminary design of a radial inward flow turbine (wheel diameters, blade height at the wheel inlet, operating speed, generated power, etc.), with those which are needed to calculate the pressure drop in a round pipe for the general case. When comparing the needed design inputs in Fig. 3, it is indeed surprising to find that very little added data is needed for the preliminary turbine calculations!

The first two of the design parameters, specific speed and specific diameter, are most useful since starting with cycle design data will result in turbine (wheel) size and operating speed. Their significance is expressed by the fact that, if two geometrically similar machines are operating at the same Reynolds and Mach numbers and, if the ratio of the specific heats of the working fluid is also the same, then the kinematic and dynamic similarity is also assured; i.e., the two machines will have the same efficiency.

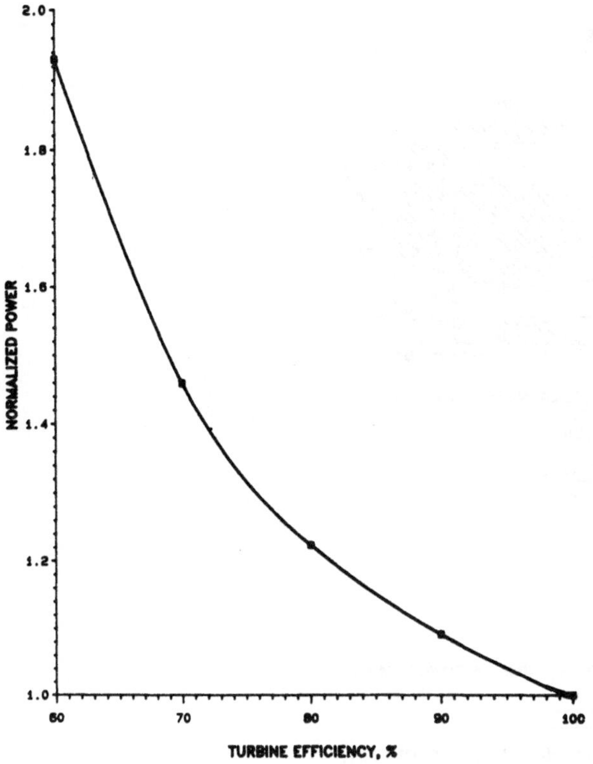

Fig. 2. Normalized liquefaction power as a function of turbine efficiency.

The defined Reynolds number is used to derate the efficiency in that the design charts - n_s d_s diagrams - are given for Re*=$2x10^6$, at and above which no efficiency improvement is expected for radial inward flow turbines but a drop in performance occurs below that value. See Fig. 3.45 [8]. The use of the Laval or Mach numbers are similar in concept, denoting the nature of the flow, design techniques to be used, and the expected losses.

Besides being used in the adiabatic head formula and in the Laval or Mach numbers, the ratio of c_p/c_v= k has an additional role to play in the proportioning of the flow path area schedule; i.e., the "optimum" turbine geometries are similar only for a given k.

The relationship between n_s and the expected efficiency for a turbine designed with the "optimized" geometry and operating at a Re* of $2x10^6$ or above and with a La* of 1 is shown in Fig. 4. A few characteristics of this optimum geometry can be reckoned from Fig. 5. These include the wheel diameter, D (from d_s), the eye diameter d (from ε_{opt}), and the axial clearance s at the wheel tip.

PRELIMINARY DESIGN OF A TURBINE

An illustration of the use of the n_s, d_s numbers and the design charts is given in the Appendix. The calculations for the given design conditions resulted in a preliminary sizing of the turbine wheel with diameters of D=95.44 mm, d=64.9 mm, and a blade height of 3.53 mm at the inlet. We chose 25 mm for the hub diameters and .33 D or approximately 31 mm for the blade axial dimensions of which we set 25% as a constant pitch diameter helix to assure swirl free outlet for the turbine at the design point.

We selected 14 blades for this wheel. Their spacing at the mean outlet radius is 10 mm, providing us with a challenge.

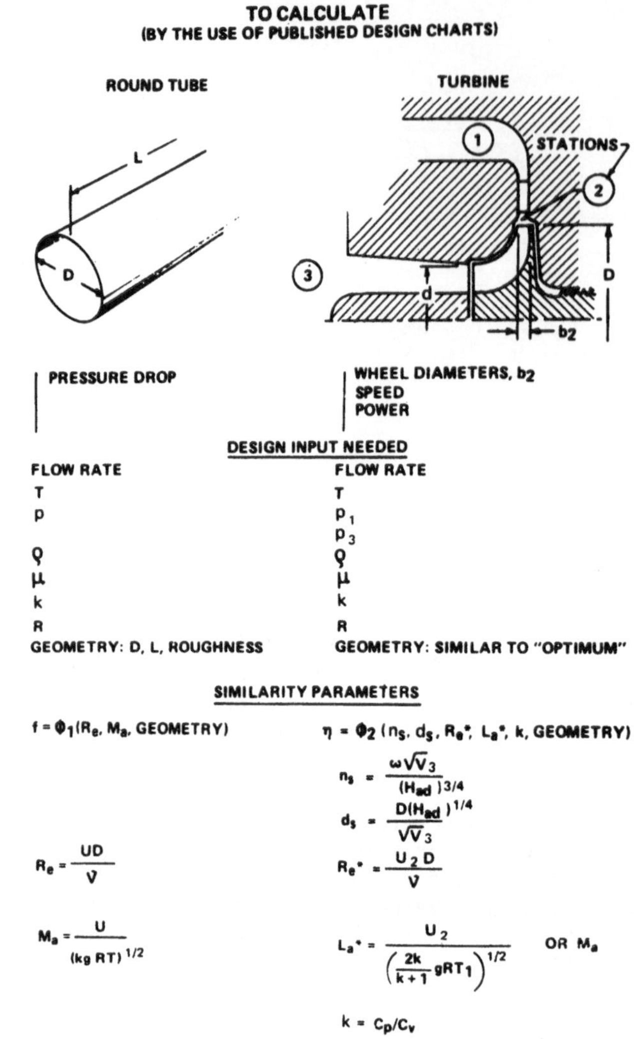

Fig. 3. Comparison of similarity parameters, round tube, turbine.

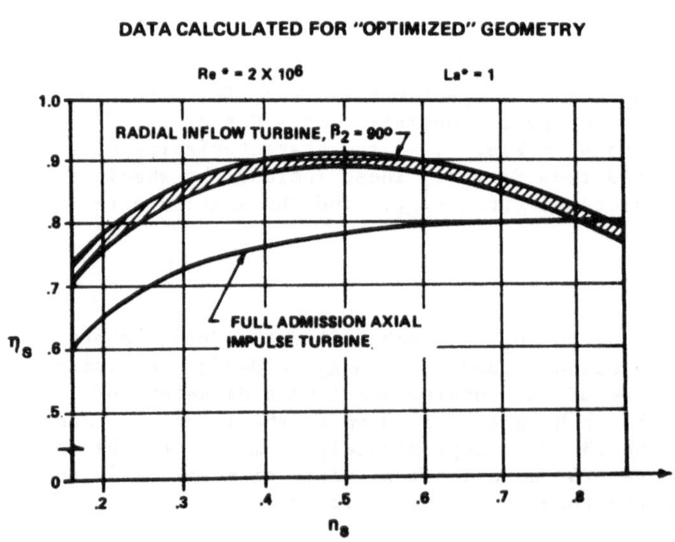

Fig. 4. Efficiency vs. specific speed chart for radial inward flow and axial flow impulse turbines.

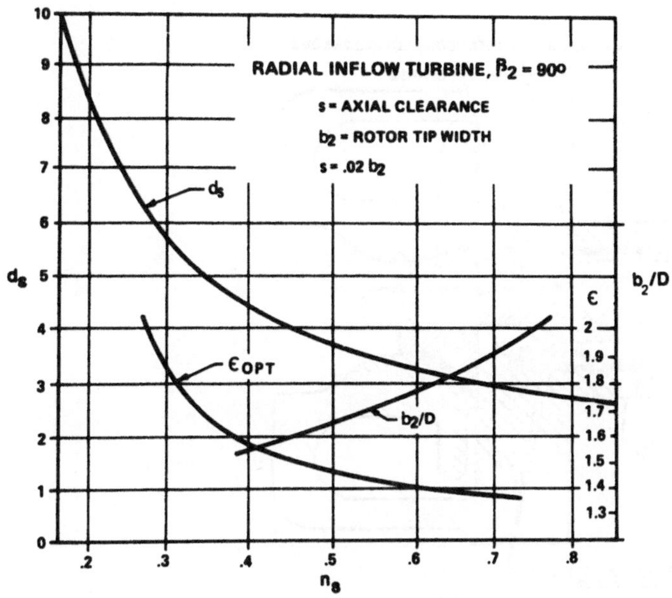

Fig. 5. n_s, d_s chart with geometric parameters.

The η vs n_s chart is developed for 2% trailing edge thickness and 2% clearance between the stationary housing walls and turbine wheel blades. This requirement, in our case, would mean a trailing edge thickness of only 0.2 mm and a clearance of 0.0714 mm between the stationary shroud and the tip of the blades at the periphery of the wheel. A wheel with such thin blades, if it can be fabricated at all, would be exceedingly expensive; therefore we may have to derate our efficiency prediction and/or consider the use of splitter blades. As to the axial clearance of 0.0714 mm, we note that affordable manufacturing tolerances may add up to more than this figure and that the axial clearance of a hydrodynamic thrust bearing, a likely candidate, will also exceed this value. Fortunately we have a solution here. We could cover the wheel with a shroud which is somewhat more tolerant to axial misalignment. This approach however, will require a labyrinth seal at the eye of the wheel. (A good background for an effective seal design is produced in reference [9].) The use of a seal will, however, increase the burden of the designer to assure an acceptable solution considering the bearings and rotodynamics since the large diameter seals tend to act as bearings providing a destabilizing effect [10].

With this example we hoped to achieve the purpose of illustrating some of the series of compromises one has to make to produce a machine without jeopardizing either the optimum efficiency or the reliability of the machine, while keeping the cost within reasonable limits.

ROTORDYNAMICS

The bearing design and rotordynamics have as high an influence on reliability of the machine as aerodynamics has on efficiency. We favor the dual level suspension system for high speed turbines due to the very good field experience with an early version [11,12]. A thorough treatment of the bearing-rotor system stability problems and the benefits provided by the elastic, damped support system is given [13], with experimental results [14] and important design criteria [15].

The schematic of an embodiment, employed by us in a recently developed liquefier turbine, is depicted in Fig. 6. Note that the journal and thrust bearings on the two sides of the shaft are interconnected with a large sleeve and suspended on springs. Such an arrangement would in fact reduce the rotor stability threshold. If damping is introduced, however, the threshold may be

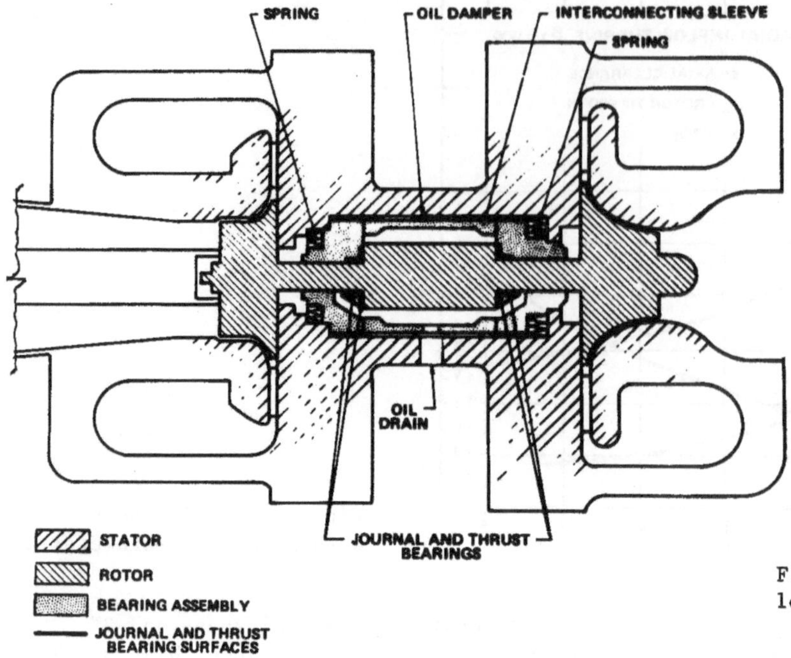

Fig. 6. Dual level suspension.

greatly increased over the original value. There is an optimum damping value which assures the maximum stability limit. Squeeze-film damping is provided in this assembly by introducing the bearing oil into the suitably dimensioned space between the interconnecting sleeve and stationary housing.*

Besides assuring high stability limits with simple journal bearing geometries this design has some added advantages. These include the ability to tune the system criticals, during the design phase, away from the operating speed, and the ability to operate with large unbalance forces or sudden external excitations. This system also allows a reduction of the shaft diameter, resulting in lower seal gas consumption. Computer programs developed by the University of Virgina's Rotating Machinery and Controls (ROMAC) Industrial Research Program were used to analyze the system. References [16,17] provide information on the force coefficients of the squeeze film dampers.

WHEEL STRESSES, STRAINS AND WHEEL DYNAMICS

As a rule the stresses in the turbine wheel will not limit our design. An early glimpse of this can be gained from Fig. 7 reproduced here with the kind permission of its author [15]. This chart shows the stress factors for many wheels where the maximum stresses were reduced to near their lowest level by FEA techniques. This means that in an actual case it might be difficult to design the wheel with lower stresses; however, a careless design may result in much higher figures. An experienced designer would subject the wheel to FEA analysis if the stresses computed by this chart are within 50% of the yield point of the wheel material.

Analysis is also indicated at higher peripheral speeds (over 250 m/sec) to gain information on the deflected shape of the wheel at operating speed. This will help to locate the wheel inlet axially under the inlet nozzles.

The FEA programs are also able to evaluate the dynamic behavior of structures; i.e., to predict the critical frequencies. This is of great importance

----

* US Patent #4,430,011 L.C. Kun: Integral Bearing System 2/7/84.

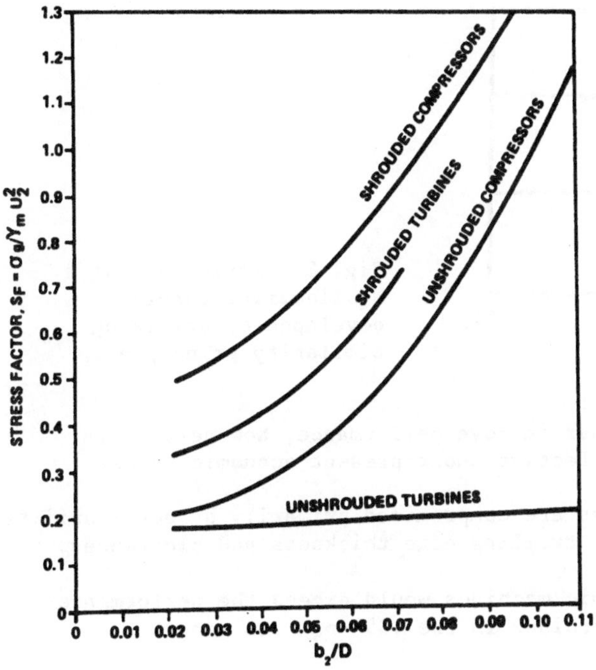

Fig. 7. Stress factors for various turbocompressor and turbine wheel configurations.

for a wheel operating closely to its yield point since it may have little margin left to accommodate any fatigue stresses which may be present.

We have to recognize an important forcing mechanism in evaluating the fatigue conditions at the wheel. As the wheel blades pass under the jets emanating from the stationary inlet nozzles, there will be a periodic excitation proportional to the speed and the number of inlet nozzles. The coincidence of excitation and natural frequencies can be conveniently displayed graphically on the so called Campbell diagram [18]. In so doing, we usually discover numerous frequency interferences and more often than not, the complete lack of operating windows. Fortunately, the actual situation is not as bleak as it appears because only a limited number of modes can be excited with the mentioned mechanism [19]. This imposes the further requirement of establishing the mode shapes for each of the interfering frequencies (See reference [20] for the theory and practice of modal analysis.) The number of inlet guide vanes usually can be finalized only after this information is known.

FIELD EXPERIENCE

Many turbines employing the described basic design principles have been built for air separation plant and nitrogen liquefier applications in the past and have demonstrated over 90% efficiency with high reliability (Fig. 8) [21]. The bearing system has worked without fail. Fatigue of the wheel material, however, caused two failures; one in the development phase due to lack of correct analysis of mode shapes, and the second on a production unit by allowing it to run in a forbidden speed range below the design speed.

SUMMARY

Available state of the art designs and manufacturing techniques should result in efficiencies in the 90% range with high reliability for large (over 150 HP) cryogenic turbines for either air separation plant or liquefier application.

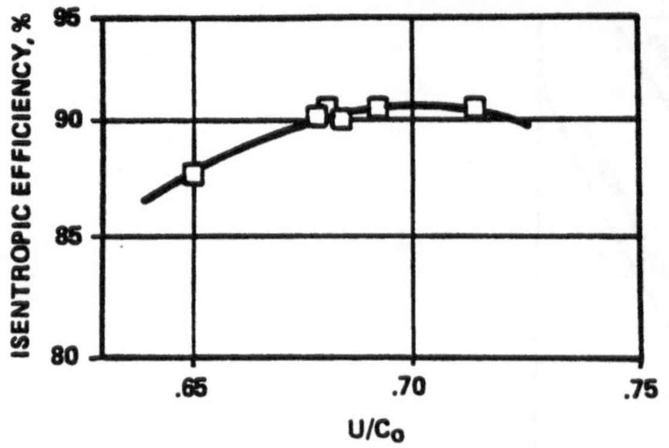

Fig. 8. Efficiency of a N_2 liquefier turbine developed by utilizing similarity principles.

Incentive still exists to further improve performance, however, it is doubtful that this would be cost effective under present economic rules.

The efficiency of small turbines are compromised primarily by our inability to assure optimum dimensions; e.g., trailing edge thickness and clearances.

It is doubtful that other dynamic machines would exceed the performance of the "optimum" radial inward flow turbines in the $.35 < n_s < .65$ range.

ACKNOWLEDGMENT

Appreciation is expressed to the Linde Division of Union Carbide for granting permission to publish this paper. I would also like to acknowledge the team members; contributions from V. E. Bergsten, C. M. Chang, E. P. Eardley, A. P. Evans, N. Nenov, W. J. Owens, R. H. Sentz, G. J. Thrasher, and J. B. Wulf made the described development possible.

APPENDIX: Preliminary Design of a Radial Inward Flow, 50 Reaction Turbine

The design procedure given below is but the most rudimentary. (A more detailed procedure is given in reference [23].) If real gas properties are used, however, it can provide surprisingly good wheel sizes and operating speed.

Working Fluid: N_2, $M = 28.016$, $R = \frac{Ro}{M} = 296.77 \ \frac{Nm}{kg}$, $k = 1.4$ (ideal gas)

$P_1 = 0.6895\text{x}10^6 \ N/m^2$ $\qquad$ $P_3 = 0.1517 \times 10^6 \ N/m^2$

$T_1 = 125$ K $\qquad$ Flow = $\dot{n}$ - 100 kmol/hr

<u>Wheel</u>

Calculate T_3, V_3 and the adiabatic head

$$T_3 = T_1 \left\{ 1 - \eta \left[1 - \left(P_3/P_1 \right)^{\frac{k-1}{k}} \right] \right\} = 86.59 \text{ K}$$

$$V_3 = \frac{\dot{n}}{3600} \ \frac{RoT_3}{P_3} = 0.13183 \ \frac{m^3}{s}$$

$$H_{ad} = \frac{k}{k-1} \ R \ T_1 \left[1 - \left(P_3/P_1 \right)^{\frac{k-1}{k}} \right] = 45{,}494 \frac{Nm}{kg} = \frac{m^2}{s^2}$$

Selecting $n_s = .5$, (Fig. 4) corresponding approximately to the optimum operating speed and since

$$n_s = \frac{\omega \sqrt{V_3}}{H_{ad}}$$

$$N = \frac{60}{2\pi} \frac{n_s H_{ad}^{3/4}}{\sqrt{V_3}} = 41{,}032 \ n_s \text{ RPM}$$

The optimum d_s (Fig. 5) appears to be 3.75, and from

$$d_s = D \frac{H_{ad}^{1/4}}{\sqrt{V_3}}$$

$$D = 0.09544 \text{ m } (95.44 \text{ mm})$$

and since the optimum ε is 1.47 for the chosen n_s

$$d = \frac{D}{\varepsilon} = 64.9 \text{ mm}$$

the blade height at the wheel inlet

$$b_2 = 0.37D = 0.00353 \text{ m } (3.53 \text{ mm})$$

Checking the defined Reynolds and Laval numbers:

$$R_e^* = \frac{V_2 D}{\nu} = 1.93 \times 10^7$$

The kinematic viscosity in this formula is evaluated at turbine outlet conditions. Since the Reynolds number is greater than 2×10^6, the design charts (Figures 4 and 5) are valid.

The Laval number

$$L_a^* = \frac{U_2}{\left(\frac{2k}{k+1} gRT_1\right)^{1/2}} = .9684$$

No shock losses expected.

Since we are designing our turbine with radial blades (i.e., $\beta_2 = 90°$) at the wheel inlet, which means 50% reaction; we can calculate the thermodynamic condition at station No. 2, at the wheel inlet. The pressure

$$P_2 = P_1 \left(1 - \frac{45{,}595}{2} \frac{k-1}{k} \frac{1}{RT_1}\right)^{\frac{k}{k-1}} = .3508 \ \frac{N}{m^2}$$

Nozzles

Knowing the thermodynamic state at the wheel inlet, we can now perform the preliminary design of the nozzles.

Assuming zero incidence angle and knowing the peripheral speed U_2 entrance velocity Cm-2, C_2 and α_2 can be calculated from the velocity triangle. The nozzle angle will be somewhat less than α_2 allowing for the change along the spiral path between the inlet nozzle throat and wheel inlet.

The velocity and pressure at the throat can be calculated by

$$C_{thr} = C_2 \frac{D}{D_{thr}}$$

and compressible Bernoulli. The nozzle height t will be calculated from b_2 allowing for the expansion of the jet in the space between the a walls bordering the nozzles and the wheel. As an approximation

$$t = .95\, b_2$$

The total nozzle area A can now be calculated from the nozzle equation

$$\dot{m} = KA \frac{P_1}{T_1} \sqrt{\frac{2k}{R(k-1)} \left(\frac{P_{thr}}{P_1}^{2k} - \frac{P_{thr}}{P_1}^{\frac{k-1}{k}} \right)}$$

where K is the nozzle flow coefficient.

$$ta = A/Z_n$$

where a is the nozzle spacing, Z_n is the number of nozzles, usually established by wheel dynamics. The relationship between nozzle blade spacing its length and the Zweifel loading coefficient for a low loss design is discussed in [22] (chapter 4, volume 2).

The nozzle shape, location, and throat dimensions will eventually be developed by boardwork graphics or by CAD methods.

Diffuser

A diffuser is an important component of an efficient turbine since between 4-6% of the adiabatic head is present at the wheel discharge as kinetic energy and around 70% of this can be recovered by a well designed diffuser.

NOMENCLATURE

A = total nozzle area
a = nozzle spacing
b = width in axial direction
C = velocity
D = wheel diameter
d = eye tip diameter
d_s = specific diameter
g = gravitational constant
H_{ad} = adiabatic head
k = cp/cv = Ratio of specific heats
L = length
La* = peripheral Laval number
m = total nozzle flow rate
N = shaft speed
$\dot{n}$ = flowrate
n_s = specific speed
P = Pressure
Re* = Machine Reynolds Number
s = clearance
T = temperature
U = linear velocity
u = tangential direction
V = volumetric flow
Z_n = number of nozzle blades

α = nozzle flow angle
β = blade angle
γ_m = density of the metal
ε = D/d - diameter ratio
η = efficiency
μ = viscosity
ρ = density
σ = stress
ω = angular velocity
ν = μ/ρ = kinematic viscosity

SUBSCRIPTS

STATIONS

1 = nozzle cascade
2 = wheel tip
3 = turbine discharge
thr = nozzle throat

GEOMETRIC

m = meridional direction

REFERENCES

1. S.C. Collins and R.L. Cannaday, "Expansion Machines for Low Temperature Processes," Oxford University Press, London (1958).
2. W.B. Gosney, "Principles of Refrigeration," Cambridge University Press, London-New York (1982).
3. J.S. Swearingen, "Turbo-Expanders," Transactions American Institute Chemical Engineers, New York (1947).
4. W.J. Olszewski, Efficient turbine process for liquefaction of cryogens, in: "Advances in Cryogenic Engineering," Vol. 17, Plenum Press, New York (1972).
5. L.C. Kun and T.C. Hanson, High efficiency turboexpander in an N_2 liquefier, Union Carbide Corporation, Linde Division. Paper Presented at the AIChE Spring Meeting (1985).
6. T.C. Hanson, "Liquifier Performance as a Function of Turbine Efficiency," Internal Memorandum, Union Carbide Corp. Linde Division.
7. D.G. Shepherd, "Principles of Turbomachinery," The Macmillan Company, New York (1956).
8. O.E. Balje, "Turbomachines: A Guide to Design, Selection, and Theory," John Wiley & Sons, New York (1981).
9. G. Vermes, A fluid mechanics approach to the labyrinth seal leakage problem, Journal of Engineering for Power Vol. 83, p. 161 (1961).
10. R. Jenny and H.R. Wyssmann, Lateral vibration reduction in high pressure centrifugal compressors, in "9th Turbomachinery Symposium;" Texas A&M, College Station, Texas (1980).
11. L.C. Kun, H.H. Ammann, and H.M. Scofield, Development of novel gas bearing supported cryogenic expansion turbines, in "Advances in Cryogenic Engineering," Vol. 14, Plenum Press, New York (1969).
12. L.C. Kun, H.H. Ammann, and H.M. Scofield, Theoretical analysis of a cryogenic gas bearing with a flexible damped support, in "Advances in Cryogenic Engineering," Vol. 17, Plenum Press, New York (1972).
13. E.J. Gunter, Jr., "Dynamic Stability of Rotor-Bearing Systems," National Aeronautics and Space Administration SP-113 (1966).
14. J. Tonnesen and J.W. Lund, Some experiments on instability of rotors supported in fluid-film bearings, Journal of Mechanical Design, Vol. 100 (1978).
15. D.J. Arnold, Critical stress, dynamics, and bearing problems in rotating machinery, Structural Sciences Research, Inc., Marina Del Ray, California, Paper Presented at AIChE Spring Meeting (1985).
16. L.E. Barrett and E.J. Gunter, "Steady-State and Transient Analysis of a Squeeze Film Damper Bearing for Rotor Stability," NASA CR-254 (1975).
17. A.Z. Szeri, A.A. Raimondi and A. Giron-Duarte, Linear force coefficients for squeeze-film dampers, ASME Journal of Lubrication Technology Vol. 65, p. 326 (1982).
18. W. Campbell, The protection of steam-turbine disk wheels from axial vibration, in "Transactions of the ASME," Vol. 46, ASME, New York (1924), p. 31.
19. F. Kushner, Disc vibration - rotating blade and stationary vane interaction, Journal of Mechanical Design, Vol. 102, p. 579 (1980).
20. D.J. Ewins, "Modal Testing: Theory and Practice," Research Studies Press LTD., Letchworth, & John Wiley & Sons Inc., New York (1984).
21. G.J. Thrasher, "Efficiency of an N_2 Liquefier Turbine," Internal Communication, Union Carbide Corp., Linde Div., (1984)
22. A.J. Glassman, ed "Turbine Design and Application," National Aeronautics and Space Administration, Volume 1-3, NASA SP-290 (1973).
23. L.C. Kun and R. H. Sentz, High efficiency expansion turbines in air separation and liquefaction plants, Paper Presented at the International Conference on Production and Purification of Coal Gas and Separation of Air, Beijing, China (1985).

NITROGEN PRODUCTION FOR EOR

R. F. Pahade and J. H. Ziemer

Union Carbide Corporation
Linde Division
Tonawanda, New York

ABSTRACT

Nitrogen has gained prominence as one of the fluids for enhanced oil and gas recovery (EOR) in the last decade. Large volume and high pressure characteristics of EOR nitrogen application required development of improved air separation cycles. The resulting nitrogen production process cycles for EOR application are described and compared. The nitrogen process cycle utilizes proven technology and equipment. It results in significant capital and energy savings.

INTRODUCTION

Nitrogen injection for enhanced oil and gas recovery has been employed successfully in many domestic oil and gas fields since the 1960's, and has reached a state of maturity. Nitrogen injection has been used in over thirty fields[1] for applications that include immiscible displace ment, pressure maintenance, miscible floods, improved gravity draining, and as a driver gas for carbon dioxide or hydrocarbon miscible floods.

In the early EOR applications, inert-gas generators typically were used to produce nitrogen[2] by burning natural gas in an internal combustion engine or steam system with an oxygen-deficient atmosphere. The exhaust gas is treated to remove corrodents before being compressed to the required injection pressure. During the late 1970's the rising cost of natural gas prompted the use of processes such as cryogenic air separation for EOR nitrogen production. Nitrogen produced cryogenically is free of corrodents which alleviates concerns some oil and gas producers had regarding corrosion problems associated with inert gas generators. A nitrogen purity of less than 10 ppm oxygen typical of conventional industrial gas applications, is specified for cryogenic nitrogen to minimize corrosion.

Most conventional industrial gas applications use relatively small quantities of nitrogen at low pressures compared to petroleum reservoirs that require large volumes of gas at high pressures. A typical large reservoir may require up to 111,000 Nm^3/h NTP (100 MMscfd)* at a pressure

* NTP is at 14.7 psia and 70°F
  scf is at 14.7 psia and 60°F

of 41,000 kPa (6000 psi). Chemical processes normally run continuously, and require an uninterrupted supply of nitrogen. This necessitates having a liquid backup supply large enough to handle planned and emergency plant shutdowns such as those caused by power outages. Conversely, because of the large volumes of reservoirs, ultra-high product availability is not required for EOR nitrogen applications; fairly long interruptions can be tolerated without a measurable impact on the oil and gas production process. Therefore, relatively small quantities of liquid nitrogen normally are provided, and it generally is used to enhance nitrogen plant production, and not for backup. These differences in requirements from traditional nitrogen production facilities spurred development of improved process cycles to improve efficiency and, thus, economics. Cryogenic processes have proven cost-effective for producing large quantities of nitrogen; alternative technologies, adsorption or membranes, economically are best suited for applications requiring much smaller capacities.

EOR NITROGEN PLANT PROCESS CYCLES

Three basic process cycles for producing nitrogen by cryogenically separating air are the single column, double column, and dual column. Variations of the single and double column cycles have long been used for production of industrial gases from air. The dual column cycle represents advanced adaptation of cryogenic technology for EOR applications. This cycle has been successfully reduced to practice by Union Carbide Corporation.

Single Column Cycle

This cycle is widely used for conventional industrial gas nitrogen plants with capacities ranging up to 8,400 Nm^3/h NTP (7 MMscfd). These applications typically require a very high product availability, so relatively large liquid backup systems are common place. The single-column waste expansion cycle has been widely used because it has good liquid-making capability and an intermediate product supply pressure--product compression is not required for many applications.

The single column waste expansion cycle is shown in Figure 1. Air is filtered, compressed, cooled in an aftercooler, and the condensed moisture is removed in a condensate separator. The feed air then enters the reversing heat exchanger which cools the air feed and removes contaminants. Carbon dioxide and water vapor are deposited on the heat exchanger surface by cooling against the product nitrogen and oxygen-rich waste streams. Alternately switching the air feed and waste streams through the exchanger passages allows the deposited contaminants to evaporate into the waste stream which exhausts to the atmosphere. The cooled air stream is fed to a gas phase adsorption trap which removes trace quantities of contaminants that may pass through the reversing heat exchanger. The air stream is then fed to the distillation column where it is separated into an oxygen rich liquid at the bottom and pure nitrogen gas at the top. The air feed itself provides the driving force to accomplish the separation. Part of the nitrogen gas is taken off as product and the balance is condensed in the reboiler/condenser to provide column reflux. The oxygen-rich liquid is subcooled and flashed into the reboiler/condenser where it is vaporized by the condensing nitrogen. The oxygen-rich vapor is heated and then expanded in the expansion turbine to generate refrigeration for the plant. The cold waste stream after expansion is re-warmed and vented to the atmosphere.

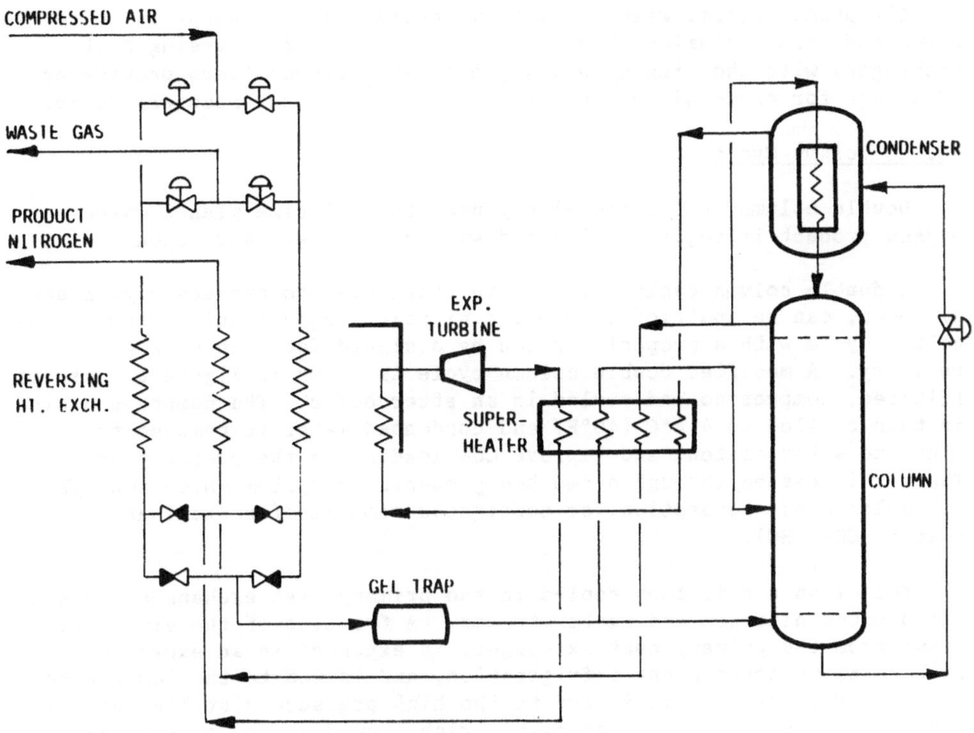

Fig. 1. Single column cycle

For gaseous nitrogen only applications, the single column waste expansion cycle recovers 50-57% of nitrogen in the air feed as nitrogen product. The remaining portion of the feed air, available as waste, is required for self cleaning of the reversing heat exchanger. The optimum nitrogen delivery pressures without product compression range from 520 to 690 kPa (75 to 100 psia). The corresponding air supply pressures range from 620 to 790 kPa (90 to 115 psia). The air fed to the bottom of the column provides the driving force to accomplish separation. Column recovery is limited by equilibrium constraints on the bottom tray. The cycle offers simplicity of operation; however, its relatively low product recovery results in high air feed flow requirements. The high air throughput requirements severely limit the capacity of plants that can be shop packaged and shipped.

Single column cycles with air expansion instead of waste expansion have also been employed for nitrogen production. In this process cooled air is expanded through an expansion turbine to generate plant refrigeration, and the turbine discharge is then fed to the separation column. Air expansion results in reduced column operating pressure which improves product recovery; however, the product pressure is lower (275 to 450 kPa, 40-50 psia) which counterbalances the recovery improvement.

The single column cycle recovery can be improved by the incorporation of a heat pump to enhance product nitrogen recovery. Nitrogen is a logical heat pump fluid since it can be combined with the product nitrogen in the same compression train. The high pressure nitrogen stream is condensed to provide reboiler duty at the column bottom, and is vaporized at column pressure to provide added reflux duty at the top. The savings due to improved product recovery, that is the reduced air flow, is counterbalanced by the additional processing of the nitrogen recycle stream.

Air prepurifiers, which utilize molecular sieve adsorption to remove water and carbon dioxide, have been used instead of reversing heat exchangers with the single column processes. Prepurifiers provide an advantage for cryogenic processes with higher product recovery ratios.

Double Column Cycle

Double column cycles are widely used for all size plants where oxygen product is required with and without nitrogen and argon.

A double column cycle, similar to those used to produce oxygen and nitrogen, can be modified to produce nitrogen product only. The double column cycle with a prepurifier can be designed for high nitrogen recovery. A modified double column cycle is shown in Figure 2. Air is filtered, compressed and cooled in an aftercooler. The compressed air is then chilled to 4.4°C (40°F) and condensed water is removed to minimize water content and improve CO_2 loading on the prepurifier. The air is passed through a two bed prepurifier system which employs molecular sieve adsorption for continuous removal of contaminants (water, CO_2, HC).

The clean air is then cooled in the primary heat exchanger against the product nitrogen and waste streams. A fraction of the air is withdrawn from the primary heat exchanger, is expanded in an expansion turbine to generate plant refrigeration, and is fed to the low pressure column. Most of the air is fed to the high pressure distillation column where it is separated into an oxygen rich liquid at the bottom and pure nitrogen at the top. The nitrogen gas is condensed in the condenser/reboiler providing boil-up for the low pressure column. Part of the liquid nitrogen is used to reflux the high pressure column, and the

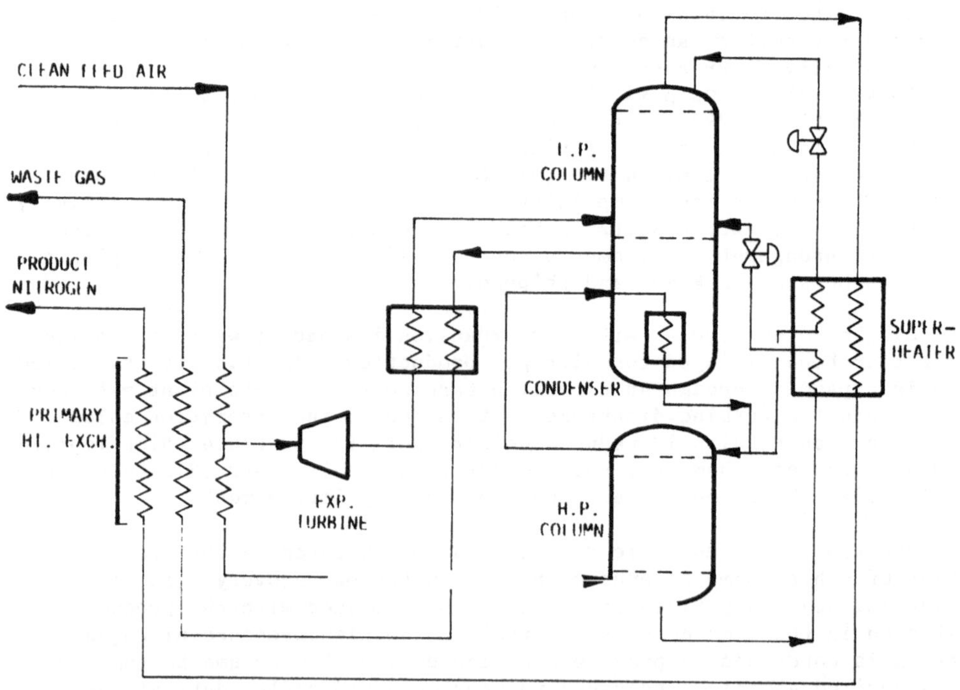

Fig. 2. Double column cycle

balance is subcooled, flashed and used to reflux the low pressure column. The oxygen-rich liquid is subcooled and flashed into the low pressure column. Nitrogen product is produced at the top of the low pressure column and an oxygen rich waste stream is produced at the bottom. The high pressure column provides the low pressure column reboil vapor and low pressure column reflux liquid. At higher operating pressure, the cycle is limited by the high pressure column, capability to generate reflux for the low pressure column. This results in optimum pressure of 830 to 970 kPa (120 to 140 psia) for the high pressure column.

The double column cycle optimized for EOR applications recovers 90 to 95% of the the nitrogen in feed air as nitrogen product. The nitrogen plant product delivery pressures range from 140 to 240 kPa abs (20 to 35 psia). The corresponding air supply pressures range from 690 to 965 kPa abs (100 to 140 psia). The double column cycle has a significantly higher product recovery which results in reduced air flow requirements. The double column cycle separation is more efficient than the single column cycle, but it has a much lower product delivery pressure. Power usage is about 2% lower than the single column cycle.

Other variations of the double column cycle, such as expansion of high pressure nitrogen vapor or expansion of waste oxygen to generate plant refrigeration, could be utilized. These process variations result in about the same performance as the air expansion arrangement described above.

Dual Column Cycle

Several new nitrogen plant cycles were conceptualized that achieved high nitrogen recovery at high pressure to result in improved plant efficiency. The use of multiple single columns in series was identified as the most promising process. This approach also uses the air feed to drive the separation process. Several unique ways of combining multiple single columns were identified, developed, and were, subsequently, patented.[3,4,5]

A dual column cycle [3] is shown in Figure 3. Air is filtered, compressed, and cooled in an aftercooler. The compressed air is then chilled to 4.4°C (40°F) and condensed water is removed. The air is then passed through a two bed prepurifier system which employs molecular sieve adsorption for the continuous removal of contaminants (water, CO_2, HC).

The clean air is then cooled in the primary heat exchanger against an oxygen rich waste stream and two product nitrogen streams. A fraction of the air is withdrawn as a side stream from the primary heat exchanger, expanded in the expansion turbine to generate plant refrigeration, and fed to the low pressure column. The cooled high pressure air is fed to the high pressure column where it is separated into an oxygen rich liquid at the bottom and pure nitrogen at the top. Part of the nitrogen gas is taken off as high pressure product and the balance is condensed in the main condenser to provide reboil vapor for the low pressure column and liquid reflux for the high pressure column. The high pressure oxygen-rich liquid is subcooled, flashed and fed to the low pressure column. The oxygen rich liquid stream and low pressure air from the expander are further separated in the low pressure column. Part of the nitrogen gas is taken off as low pressure product and the balance is condensed in the top condenser to provide liquid reflux for the low pressure column. The low pressure oxygen rich liquid is subcooled and flashed into the top condenser where it is vaporized by condensing nitrogen.

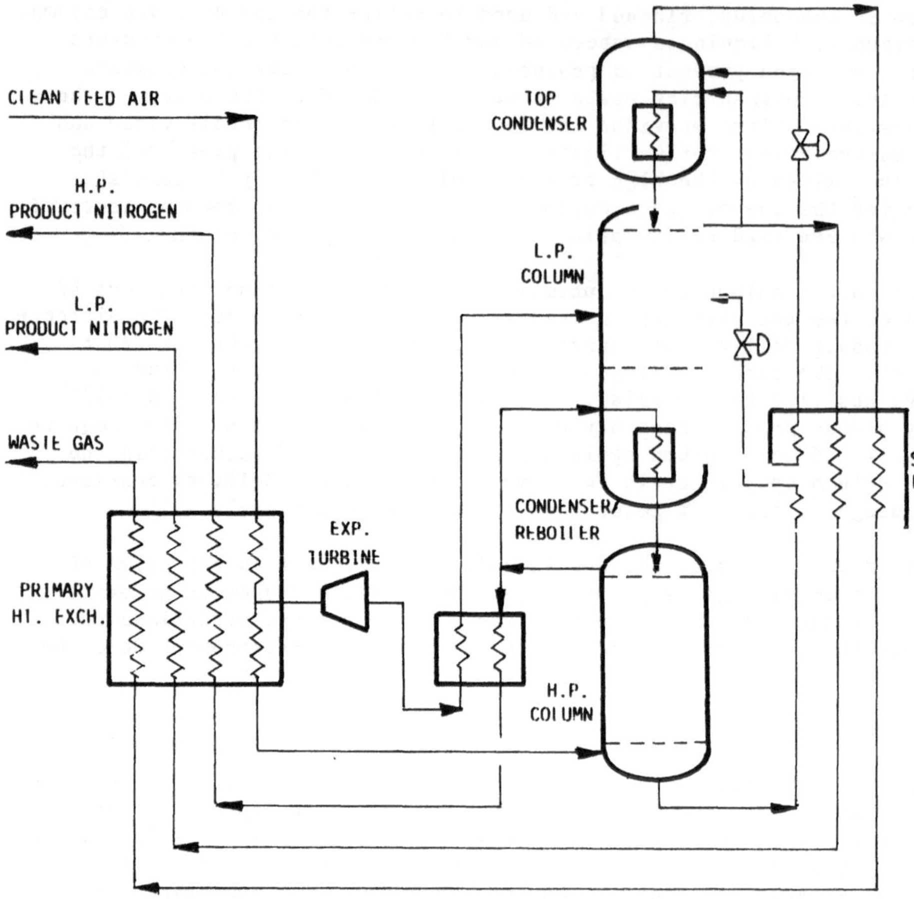

Fig. 3. Dual column cycle

In the dual column cycle the high pressure column provides low pressure column reboil vapor, but, unlike the double column cycle, the low pressure waste stream provides the low pressure column reflux. Thus, operation of the low pressure column is partially decoupled from the high pressure column, resulting in maximum flow of high pressure nitrogen product. The increase in air compression power, compared to the double column cycle, is more than counter balanced by the availability of product at higher pressures. Nitrogen recovery is limited by equilibrium constraints at the bottom of both columns. Integration of the main condenser in the bottom of the low pressure column, with several distillation trays below the high pressure kettle feed, provides an upper column reboil vapor with high oxygen content. This maximizes recovery of low pressure nitrogen.

The dual column cycle recovers 85 to 90% of the feed air as nitrogen product. The nitrogen plant product delivery pressure range from 760 to 830 kPa (110 to 120 psia) and 330 to 380 kPa (48 to 55 psia). The corresponding air supply pressures range from 830 to 1030 kPa (120 to 150 psia). The dual column results in a high product recovery while maintaining energy efficiency. On a separation basis, it results in power savings in excess of 10% over the single column cycle. A cycle

Table 1. Process Cycle Comparison

Cycle	Single Column Waste Expansion	Modified Double Column	Dual Column
Air Cleanup	RHX	Prepurifier	Prepurifier
Product Recovery as % of Nitrogen in Feed Air	56	94	88
% Power Savings (including product compression to 1000 psig)	Base	2	8

comparison is shown in Table 1. The air supply pressure is higher but more than half of the product is available at higher pressures. The technology utilized is proven air separation technology. Compression and air pre-cleanup utilizes existing state-of-the-art equipment. Overall, the dual column cycle results in a low risk, low cost nitrogen plant.

SUMMARY

Three cryogenic nitrogen plant cycles tailored for EOR applications have been discussed. The single column waste expansion cycle is a simple cycle; it is well suited for small nitrogen plant applications. For large plants, the modified double column cycle has the advantage of high product recovery and provides some power and capital savings. The newly developed dual column cycle results in high product recovery and significant power and capital savings. The dual column cycle utilizes proven air plant technology and equipment and offers the best option for large EOR nitrogen applications.

REFERENCES

1. J. P. Clancy et al, Analysis of nitrogen injection projects to develop screening guides and offshore design criteria, Offshore Europe 1983, Society of Petroleum Engineers of AIME, September 1983.
2. W. B. Bleakley, Nitrogen from air and condensate recovery, Petroleum Engineer International, July 1983.
3. R. F. Pahade et al, U.S. Patent No. 4,453,957, "Double Column Multiple Condenser-Reboiler High Pressure Nitrogen Process", June 12, 1984.
4. W. J. Olszewski et al, U.S. Patent No. 4,439,220, "Dual Column High Pressure Nitrogen Process", March 27, 1984.
5. H. Cheung, U.S. Patent No. 4,448,595, "Split Column Multiple Condenser Reboiler High Pressure Nitrogen Process", May 15, 1984.

RECOVERY OF VALUABLE HYDROCARBONS USING DEPHLEGMATOR TECHNOLOGY

D. P. Bernhard and H. C. Rowles

Air Products and Chemicals, Inc.

Allentown, Pennsylvania

ABSTRACT

High efficiency cryogenic processes using dephlegmators have been developed for the recovery of valuable hydrocarbons. These processes have been used in applications for high recovery of heavy components from large flows of light gases. Processes are presented for recovery of ethylene from cracked gas (both grass roots and retrofit), light olefins from FCC unit off-gas, and heavy olefins from dehydrogenation unit off-gas.

The heart of these processes is the dephlegmator, which takes the place of the distillation column or phase separator in conventional processes. Liquid condensed from the feed gas rising in the dephlegmator flows back as reflux. Various refrigeration sources are employed for efficient condensation of the reflux. Brazed aluminum core type heat exchangers are used to provide efficient heat and mass transfer in compact cryogenic units.

INTRODUCTION TO DEPHLEGMATOR TECHNOLOGY

Dephlegmators have been proposed for many years to reduce energy consumption in gas separation processes. Some of the earliest cryogenic gas separation plants designed by Claude used dephlegmators. The full potential of dephlegmation, however, has been hampered by the difficulty of predicting the performance of such equipment. Air Products has solved this problem and has been using dephlegmators commercially in a variety of cryogenic gas separation processes since 1970.

The dephlegmator, while maintaining simplicity of equipment and operation, provides greatly enhanced separation capability over conventional partial condensation processes, which rely on the equilibrium established between vapor and liquid after cooling and phase separation. A dephlegmator can also provide enhanced efficiency in many distillation processes, where it serves the dual function of rectifying column and partial condenser. This is particularly true where high recovery of heavy components is desired from gas mixtures having high concentrations of light components, and where the relative volatility between light and heavy key components is two or more. Integration of a dephlegmator process with conventional distillation, either for preseparation or postseparation, can increase the recovery of valuable hydrocarbon components.

DEPHLEGMATOR OPERATION

Modern dephlegmators are usually brazed aluminum heat exchangers, of special design and construction which endow otherwise ordinary heat exchangers with extraordinary separation capability. Brazed aluminum exchangers, common in cryogenic service, consist of many layers or passages to transfer heat among several streams. Dephlegmators include enhancements to standard brazed aluminum heat exchanger design that permit them also to behave as mass transfer devices. In particular,

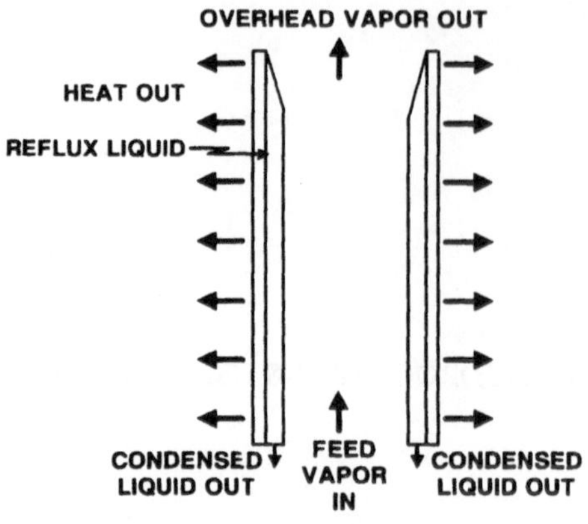

Figure 1
What is a Dephlegmator?

dephlegmator feed stream passages are designed to allow liquid flow downward, countercurrent to the flow of vapor.

Figure 1 illustrates how a dephlegmator functions. The feed vapor stream flows upward within some of the passages, and heat is removed by other process or refrigerant streams in adjacent passages. Some fraction of the feed condenses on the heat transfer surface, forming a liquid which flows downward, refluxing the rising vapor. The reflux becomes enriched with the less volatile components as the more volatile ones are revaporized.

Unlike reflux liquid produced in a conventional distillation column condenser, dephlegmator reflux is generated throughout the mass transfer zone, over a wide temperature range. Conventional distillations have all the refrigeration supplied in the condenser, the coldest temperature point. Since much of the refrigeration can be supplied at warmer temperatures in a dephlegmator, energy savings are obtained.

In comparison to a partial condensation process, a dephlegmator process yields a further benefit. A valuable hydrocarbon liquid is separated from the feed vapor at the coldest temperature in a partial condensation process. By comparison, a dephlegmator process recovers the liquid at a warmer temperature and minimizes the quantity of dissolved light hydrocarbons and inerts. Energy savings, or alternatively higher hydrocarbon recoveries, are reaped in downstream fractionation steps by reducing the concentration of lights.

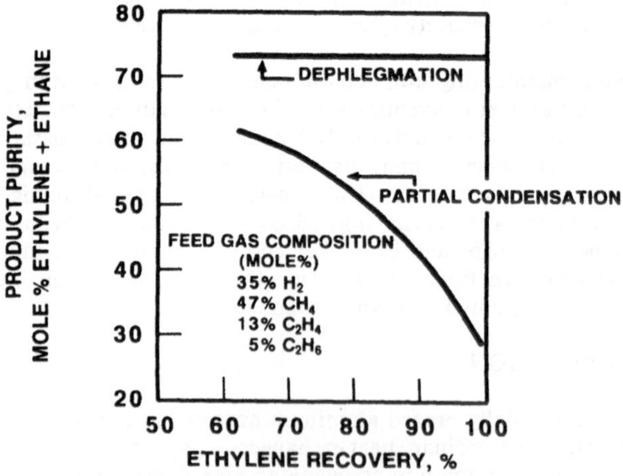

Figure 2
Comparison of Dephlegmator and Partial Condensation Process Performance

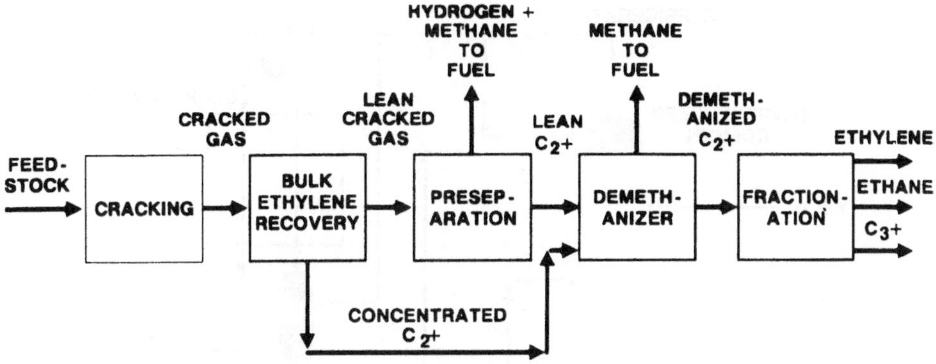

Figure 3
Ethylene Recovery from Cracked Gas

Figure 2 compares the ideal performance of partial condensation to dephlegmation for separating ethylene from a feed gas laden with hydrogen and methane. The partial condensation process achieves ethylene recovery at the expense of purity. The dephlegmator process, on the other hand, achieves high ethylene recovery and purity. The flat dephlegmator curve is calculated for perfect equilibrium between the feed gas and the recovered ethylene-rich product. In practice, heat and mass transfer resistances limit the approach to equilibrium, but the deviation, which depends upon feed and refrigerant stream conditions as well as equipment configuration and size, is very slight.

DEPHLEGMATOR APPLICATIONS

Dephlegmator processes are ideally suited for recovering valuable hydrocarbons such as ethylene from light gas mixtures. Three examples of hydrocarbon recovery processes are presented below to illustrate dephlegmator applications:

(1) ethylene recovery from cracked gas;

(2) olefins recovery from FCC off-gas; and

(3) C_4 hydrocarbons recovery from dehydrogenation unit off-gas.

These are only samples of the broad spectrum of gas separations to which dephlegmator technology can be applied. The same concept, for instance, has been used in processes to separate methane from H_2–N_2 and H_2–CO synthesis gases.

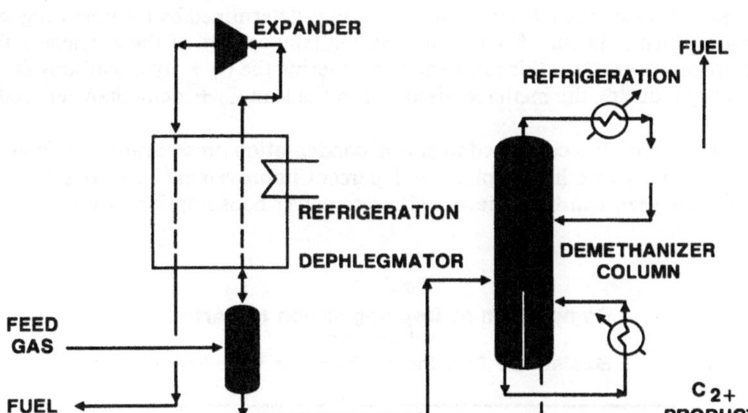

Figure 4
Dephlegmator Preseparation of Ethylene from Cracked Gas

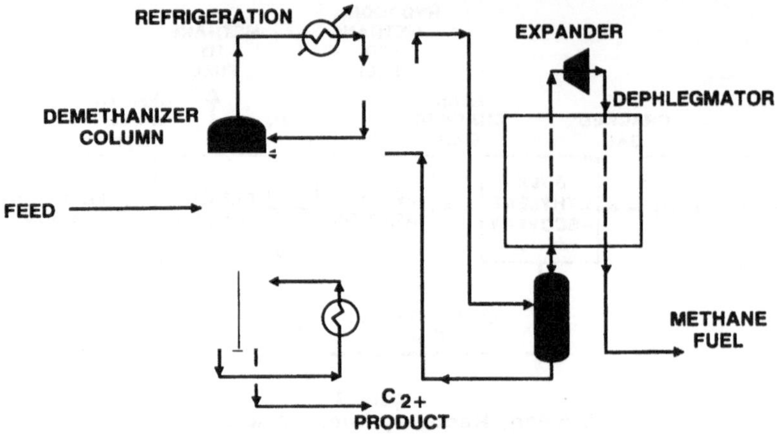

Figure 5
Separation of Ethylene from Demethanizer Overhead Gas

Example 1: Ethylene Recovery from Cracked Gas

Figure 3 shows a block diagram of an ethylene plant. Following bulk ethylene recovery, lean cracked gas is conventionally processed to recover additional ethylene by partially condensing it at − 130°C. A demethanizer, operated at 3.5 MPa, and subsequent fractionation equipment separate ethylene and other hydrocarbon products from the lean and concentrated C_2+ streams. Ethylene is lost to fuel from both the preseparation and demethanization steps.

Dephlegmator processes can prove to be extremely valuable when they are integrated with new or existing ethylene manufacturing facilities in the steps where ethylene losses occur. Dephlegmator preseparation can be substituted for partial condensation preseparation to enhance ethylene recovery, reduce demethanizer condenser refrigeration requirements, and/or increase throughput of an existing plant. An additional processing step, incorporating a dephlegmator, can also be used to recover the ethylene normally lost to fuel from the demethanizer overhead gas. Flowsheets for ethylene recovery from lean cracked gas and from demethanizer overhead gas are shown in Figures 4 and 5, respectively.

For partial condensation preseparation, high methane content in the lean C_2+ feed stream to the demethanizer is the cause of low ethylene recovery. Virtually all the methane supplied to the demethanizer is rejected to fuel from the top of the column. More ethylene is lost as the overhead vapor rate increases because the ethylene concentration is determined by the operating temperature of the condenser, which is in turn fixed by the boiling temperature of the ethylene refrigerant. A dephlegmator process overcomes this problem by recovering the C_2+ hydrocarbons at − 100°C or above, significantly reducing the methane dissolved in the lean C_2+ demethanizer feed.

Dephlegmator preseparation is compared to partial condensation preseparation in Table 1 for a new 450,000 metric ton per year ethylene plant. A 1 percent improvement in recovery, worth almost 1 million dollars per year, can be achieved with less power consumption, lower capital costs for

Table 1
Comparison of Dephlegmation to Partial Condensation
Basis: 450,000 Metric Ton Per Year Plant

ETHYLENE RECOVERY	+1%
POWER SAVING	10%
ANNUAL SAVINGS	$\$1 \times 10^6$ (@20¢/KG)

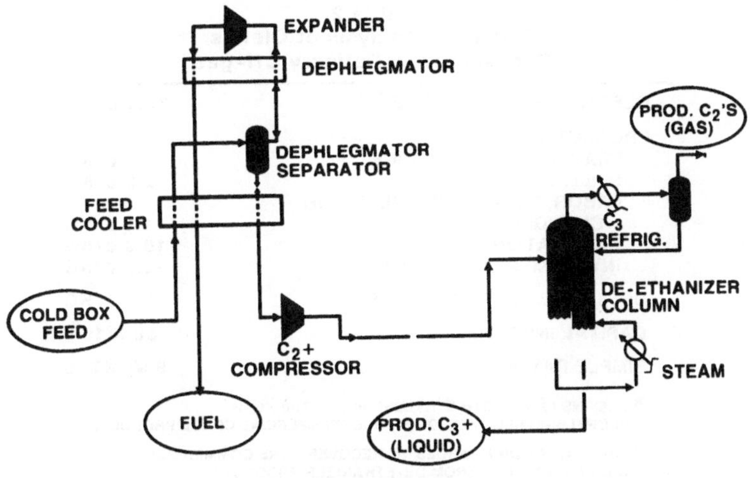

Figure 6
Olefins Recovery from Refinery Off-gas

demethanization and refrigerant compression, and a competitive capital cost for preseparation. A dephlegmator preseparation process, incorporated in an ethylene plant of this size, has continuously demonstrated these marked economic savings since it began operation in 1978.

In an existing facility, a dephlegmator preseparation step can be used to debottleneck the plant. Typically, excess reflux, containing methane and a significant amount of ethylene, is condensed in the demethanizer condenser by evaporating ethylene refrigerant. The excess liquid is then evaporated to refrigerate the partial condensation preseparation process, and is subsequently recycled and compressed by the cracked gas compressors upstream of the bulk ethylene recovery section of the plant. Replacing the partial condensation preseparation step with a dephlegmator preseparation step can reduce or eliminate the recycle and allow increased throughput of cracked gas. To achieve the same fractional recovery of ethylene, the dephlegmator process requires little or none of the refrigeration supplied by the recycle.

The process shown in Figure 5 is specifically tailored to retrofit an existing plant. Refrigeration obtained by work expansion of the methane fuel is utilized in a dephlegmator, which condenses and concentrates the ethylene from the demethanizer overhead vapor. Over 85 percent of the ethylene previously lost, representing almost 1 percent of plant production, can be recovered with no modification to the existing column or auxiliary equipment. Recovered ethylene is returned to the demethanizer. A recent study indicated that the investment in such a retrofit could be repaid in less than a year.

Example 2: Olefins Recovery from FCC Unit Off-gas

The off-gas streams from fluid catalytic cracking (FCC) units are similar to the cracked gas streams produced in ethylene plants. Figure 6 is the flowsheet of a recently installed process in which C_2 and heavier hydrocarbons, previously burned as fuel, are recovered via preseparation and deethanization. The dephlegmator preseparation process recovers a concentrated C_2+ stream containing 12 mole percent methane and inerts, much less than the 36 mole percent lights content achievable by partial condensation. The partial condensation process would require additional capital expense for demethanization to meet product specifications. The deethanizer processes the concentrated C_2+ stream and recovers a C_2 fraction containing less than 2 mole percent propylene and propane and a liquid C_3+ fraction containing 400 molar ppm of ethane. The C_2 fraction is valuable as a supplement to raw feed in an ethylene cracker. The C_3+ product is suitable for further fractionation to recover the individual hydrocarbon constituents, notably propylene.

Concomitant with the excellent purity of the concentrated C_2+ stream recovered by the dephlegmator process is high recovery. The C_2+ streams recovered in the cold box contain over 91 percent of the ethylene, over 99 percent of the ethane, and 100 percent of all the heavier hydrocarbons. High

Table 2
Economic Analysis of Olefins Recovery from Refinery Off-gas

OLEFINS PRODUCT VALUE [(a)]	38.1 ¢/KG
PRODUCTION COSTS:	
FRACTIONATION CHARGE [(b)]	-5.5 ¢/KG
UTILITIES	-2.2 ¢/KG
LABOR, OVERHEAD, AND OTHER CHARGES	-1.1 ¢/KG
RAW MATERIAL	-10.0 ¢/KG
INVESTMENT RECOVERY	-2.2 ¢/KG
NET INCOME BEFORE TAXES	17.1 ¢/KG
CAPITAL INVESTMENT	$8 X 10^6
SIMPLE PAYOUT	9 MONTHS

[a] OLEFINS (3000 KG/H ETHYLENE PLUS 3700 KG/H PROPYLENE) RECOVERED AS PURE COMMERCIAL GRADE PRODUCTS

[b] FRACTIONATION CHARGES TO RECOVER PURE COMMERCIAL GRADE PRODUCTS FROM DE-ETHANIZER PRODUCTS

recoveries and good purities of ethylene in the C_2 product and of propylene in the C_3+ product obtained by the dephlegmator/deethanizer process result in very attractive economics for this project, as Table 2 shows.

Example 3: C_4 Hydrocarbon Recovery from Dehydrogenation Unit Off-gas

Dephlegmator processes for recovery of C_4 hydrocarbons are similar to those for ethylene recovery and find application in the product purification steps associated with C_4 olefins and dienes manufacture from paraffin feedstocks. They provide very high levels of C_4 hydrocarbon recovery while minimizing the concentration of undesirable C_3 hydrocarbons and lighter components in the feed to the depropanizer downstream. The process flow diagram for a C_4 recovery unit currently under construction is shown in Figure 7. The unit recovers 2,900 kg/h of butenes along with other C_4 hydrocarbons from 13,600 kg/h of feed gas containing 91 mole percent C_3 hydrocarbon and lighter impurities. All the refrigeration required for dephlegmator operation is provided by expanding the residual light gas from the top of the dephlegmator. Overall C_4 hydrocarbon recovery is 98.5 percent, and the liquid purity is 86 mole percent C_4 hydrocarbons.

DESIGN METHODS

Dephlegmator performance is determined by a latticework of interrelated transfer processes; accurate predictions of heat, mass, and momentum transfer resistances, as well as fluid thermophysical properties, are essential for equipment design. Extensive tests to simulate a wide range of dephlegmator

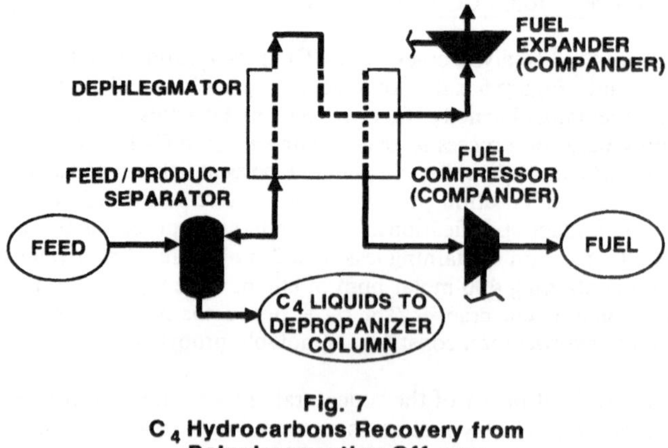

Fig. 7
C_4 Hydrocarbons Recovery from Dehydrogenation Off-gas

operating conditions have been conducted for a variety of test systems. These tests provide data on thermodynamic, hydraulic, and heat and mass transfer characteristics. Together with the results of successful commercial applications and rigorous process simulation techniques, the test data form the basis for highly reliable methods to predict dephlegmator equipment and process performance.

CONCLUSIONS

Dephlegmator processes have been used by Air Products in a wide variety of cryogenic gas separation applications. They are intrinsically simple and reliable, and yet efficient, both theoretically and practically. Their efficiency results from the effective use of available refrigeration to generate internal reflux.

Coupled dephlegmator and fractionation processes and equipment for hydrocarbon separations yield an excellent return on investment. Preseparation or postseparation steps can be added to existing facilities or integrated with new facilities.

Accurate thermodynamic data and valid correlations for heat/mass transfer and flow hydraulics, combined with rigorous process simulation techniques, provide the capability for design of highly efficient cryogenic gas separation systems.

REFERENCES

1. D. P. Bernhard, T. W. Goodwin, and H. C. Rowles, "Recovery of Hydrocarbon Liquids Using Dephlegmator Technology," presented at the NPRA Annual Meeting, Los Angeles (1986).
2. D. P. Bernhard, H. C. Rowles, and R. L. Teichman, "Process for Recovering C_4+ Hydrocarbons Using a Dephlegmator," U.S. Patent 4,519,825 (1985).
3. C. H. Chiu et al., "Dephlegmator Cycle for Ethylene Plants Application," presented at the 1979 International Congress of Refrigeration, Venice (1979).
4. J. A. Pryor and H. C. Rowles, "Recovery of C_2+ Hydrocarbons by Plural Stage Rectification and First Stage Dephlegmation," U.S. Patent 4,002,042 (1977).
5. H. C. Rowles and T. C. Tsao, "Recovery of C_2 Hydrocarbons from Demethanizer Overhead," U.S. Patent 4,270,940 (1981).
6. H. C. Rowles and D. W. Woodward, "Separation of Hydrogen Containing Gas Mixtures," U.S. Patent 4,270,939 (1981).

OPTIMIZATION OF THE OPERATION OF A GAS TERMINAL

C. Fuge, P. Eisele, B. D. Whitehead*

LINDE AG, TVT Division
D-8023 Hoellriegelskreuth
F.R. Germany

ABSTRACT

Optimal operation of complex process plants requires the assistance of computer systems for day-to-day operation.

The application of computers in plant operation comprises:

- Optimization systems to define the global plant set points such as product purities keeping in mind the physical limits of plant equipment and the boundary conditions of the market for feed and products.
- Advanced process control systems which help the operator to keep the plant as close to these global set points as possible.

This paper deals with the various aspects of an optimization system, which was developed for the gas terminal at Kaarstø, Norway. It will be shown that, due to the complex process sequence of this plant, the wide range of feedstock quality and the numerous degrees of freedom for operation, the use of the optimization system is essential to ensure plant operation close to its physical limits at minimum operating cost. The optimization system is designed for off-line usage, but can be extended later for on-line use.

DEFINITION OF OPTIMIZATION

Before discussing optimization in detail, I first want to define the term optimization:

Optimization is the task of finding the values for a set of operating variables which result in an extreme value for an objective function, which can be either maximum or minimum. If for example the objective function is the plant profit, then the maximum value is the optimum. If, as in this case, the objective function is utilities consumption, then the minimum value is the optimum. The objective function is the result of plant operation and could, therefore, be directly measured from production figures, utilities consumption, etc. This, however, is not practic-

* To whom correspondence should be sent

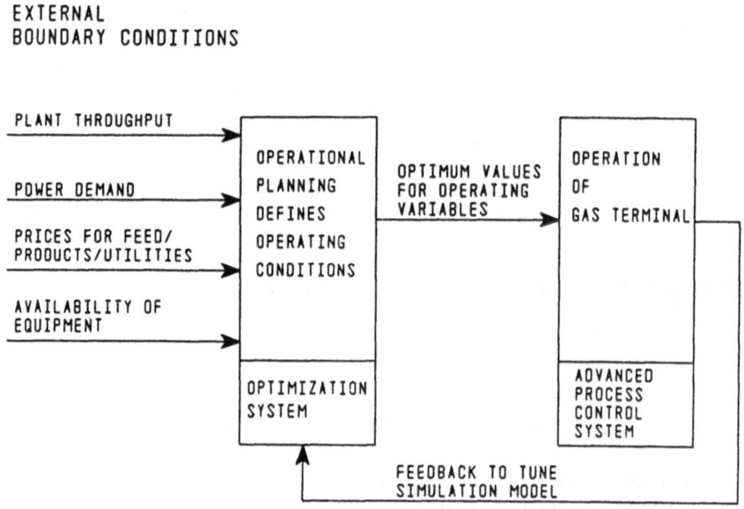

Fig. 1. Flow of information for optimized operation.

able since plant disturbances, inaccuracy of measurements and the time factor make it impossible for the optimizer to rely on real time plant data to find the optimum in a reasonable period of time. The plant performance therefore has to be simulated by a model, which provides the objective function for any set of operating variables with good accuracy and within a short time to the optimizer. Figure 1 illustrates the flow of information between operational planning personnel, who use the plant optimizer, and plant operators, who adjust the plant such that the optimal values for the operating variables are maintained. The feedback of information from the plant is used to fine-tune the simulation model, especially in areas where the result of the simulation can be influenced by time-dependent factors such as fouling of heat exchangers.

OPTIMIZATION ROUTINE

The optimizer itself is the mathematical routine which shifts the operating variables towards the optimum in an iterative way. Various methods have been described ranging from slow statistical methods to more sophisticated fast methods working on the basis of partial differentials for the respective variables. In principle, it can be stated that:

The fast derivative methods are less robust in dealing with local optima than the slow statistical methods when the objective function surface exhibits roughness. For smooth functions the derivative methods are superior because they arrive at the optimum with less iterations. The roughness of the objective function stems from the physical limits of plant equipment which lead to discontinuous relationships between independent and dependent variables. A typical example is the response of a compressor to a load reduction. Above a certain load limit, the relationship between the power consumption and the load is a smooth continuous function. At the load limit, the by-pass controllers become active and the response function shows a discontinuity at this point.

Another example is the fact that the model (and the plant) can choose to drive some pumps either with steam turbines or with electric motors.

Numerous tests have shown, that it is quite likely to arrive at one or more such discontinuities in the course of the optimization. It was, therefore, decided to select a statistical method which starts from a relatively broad pattern of values for the optimized variables and then reduces the range of this pattern based on the results of the previous iterations. Roughness of the objective function is no problem for this method since no derivatives are calculated. The end-of-run criteria can be either the number of iterations or the final dimensions of the pattern, ie the range of variation of the independent variables.

HANDLING PLANT CONSTRAINTS

The optimization has to take the physical limits of plant equipment into consideration. In the real plant the violation of constraints leads to operating disturbances such as loss of product purity or excessive consumption of utilities. The optimizer has, therefore, to respond appropriately when the simulation model indicates a bottleneck situation. For this purpose, penalty functions are used which worsen the objective function by a value which is related to the degree of constraint violation.

This procedure ensures that the operating variables selected at the end of the optimization iterations are in the feasible region and do not lead to a bottleneck situation. The reliable and accurate prediction of equipment constraints is an essential feature of the simulation model. In most cases the optimum mode of operation is either at an equipment capacity bottleneck or at the limits of the market.

FEATURES OF THE GAS TERMINAL

The plant capacity, its product values, the range of operating scenarios and the process sequence complexity determine what potential profit can be gained by the use of an optimization system. From the following brief description of the plant features it is clear, that this gas terminal offers plenty of possibilities to improve plant profit by optimization.

The terminal treats high pressure associated gas from North Sea oil fields. Oil and gas are delivered in a certain ratio which depends on the location of the oil field and changes over the years. The gas is collected from up to four different oil fields, each of which has its typical gas composition. The flow of gas is determined by the oil production and is, therefore, not a free variable. The feed design quantity is 650 t/h. Figure 2 shows the simplified flow sheet of the plant.

The process sequence is:

- Separation of gas from oil in the slug catcher
- Split of gas into two identical trains
- Drying of gas in a fixed bed drier
- Partial fractionation of the gas by extensive use of the refrigeration potential which is set free in the two-step expansion of the gas to the pressure of the demethanizer. No additional refrigeration is necessary for the partial condensation and for the operation of the demethanizer.
- Recompression of the methane from the demethanizers of both trains to the pipeline pressure by parallel gas turbine-driven compressors; recovery of the heat of compression in the interreboilers of the demethanizers.

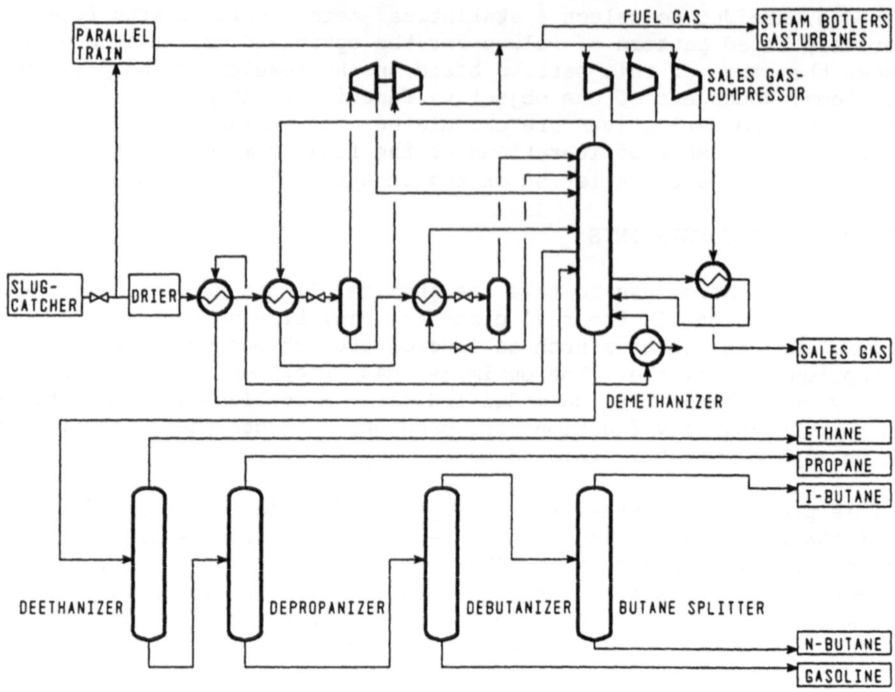

Fig. 2. Simplified flowsheet of gas terminal.

- Fractionation of heavier products into ethane, propane, i-butane, n-butane and gasoline.
- Liquefied products storage in atmospheric tanks including a vapour recovery cycle.
- Refrigeration cycle for the deethanizer and for product chilling.
- Systems for generation of steam and electrical energy including heat recovery from gas turbines.

A high degree of energy recovery as in this plant means an intensive coupling of process steps which makes it impossible for the operator to predict all influences on plant performance when boundary conditions change. The optimization system frees the operator from the need to consider such uncertainties and thereby avoids either a suboptimal conservative mode of operation or operational disturbances.

STEADY-STATE SIMULATION MODEL

The simulation model is the heart of the optimization system in which all details of plant performance for a given set of input data are predicted. From these details the global values for the plant profit and for constraints are calculated and transferred to the optimizer. It is the quality of the simulation which determines the potential of the total optimization package. The model is designed in a modular structure which follows the sequence of the process steps. The coupling of process steps makes it necessary to solve internal iteration loops between groups of modules. Figure 3 shows the overall structure of the simulation model.

The realistic simulation of equipment is based on design data such as heat exchanger surface, performance curves of compressors, efficiency of rotary equipment or heat transfer rates. All these design data are organized in a separate file where they can be easily modified when feedback from plant data indicates that the simulation deviates from reality. This tuning of the model has to be done only once.

The scope of results comprises all technical performance data of equipment. The comprehensive detailed results for mass and energy balances are used to carry out the economic evaluation following the simulation.

Considerable effort was put into the design of the user interface using preformatted input schemes with default values for the input data. A detailed message system helps the user to identify a faulty input value and informs him when in the course of simulation a validity limit has been violated.

The thermodynamic basis is fitted to the ranges for temperature, pressure and composition in the respective plant sections. This approach allows the use of fast flash calculations for the multicomponent heat and mass balances without loss of accuracy.

The single feed columns are simulated with short-cut methods. For the demethanizer a stage-to-stage method is applied to ensure a realistic simulation of the influences of the multiple feed streams including the influence of the expander. Stage efficiency is load dependent. Heat

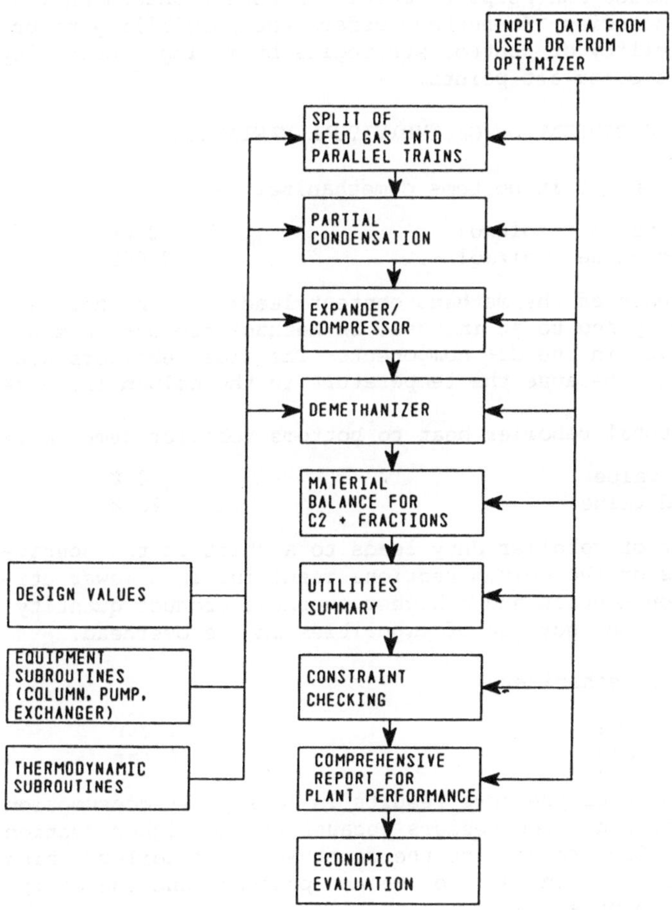

Fig. 3. Structure of rigorous plant simulation model.

transfer coefficients in reboilers and condensers are sensitive to the heat flux rate of the exchanger. Thereby it can for example be simulated whether a too high heat flux leads to a bottleneck in the reboiler because of the change from nucleate to film boiling.

The performance of the compressors for sales gas and refrigerant and of the expansion turbines has an essential influence on the utilities consumption. The simulation uses the characteristic performance curves for each stage (pressure ratio over volume with speed as parameter) and calculates the effect of anti-surge control.

The performance of the gas turbine drivers for the sales gas compressors and the generator is simulated based on manufacturers performance curves for the relationship between power, gas consumption, speed and excess air in the flue gas with the ambient temperature as a parameter.

For each major item of equipment constraints are calculated. Columns are checked for vapour load, compressors for volume and speed limits, turbines for steam load, exchangers for the ratio between heat load and temperature difference.

The decribed features of the simulation model ensure that the plant performance is mirrored for any type of operation in an accurate and transparent way. It is possible to use the simulation model without the optimizer to perform single case study runs. This offers an excellent possibility to investigate the influence of a selected input variable on plant performance and helps to develop a better understanding of complex process dependencies. It further offers the possibility to predict exactly the feasibility of control strategies by making sensitivity tests for selected controller set points.

EXAMPLES FOR OPTIMIZATION OF INDIVIDUAL VARIABLES

1. Methane content in bottoms demethanizer

Starting value	mol/mol	0.04
Optimized value	mol/mol	0.005

The decrease of the methane content leads to a higher reboiler duty and to an increase in methane product caused by an increase in the C2+ components. The side reboilers are constraints because the temperature in the column increases.

2. Part of total reboiler heat to bottoms reboiler demethanizer

Starting value	30 %
Optimized value	20 %

The shift of reboiler duty leads to a shift in the operating lines of the column sections resulting in a lower utilities consumption and a higher overhead product quantity because of an increase of impurities in the overhead.

3. Pressure demethanizer

Starting value	31 bar
Optimized value	36 bar

The increase of pressure results in less power consumption of the sales gas compressors because of the higher suction pressure. Constraints are the demethanizer reboilers which limit a further increase of column pressure and the equipment design pressure.

4. Pressure in partial condensation section

Starting value	72 bar
Optimized value	79 bar

The increase in pressure results in a shift of the partial condensation and in a higher efficiency of the expander which positively influence the operating line in the demethanizer.

These simple examples demonstrate that, even for optimization of single variables, a number of influences have to be taken into account.

In real plant operation many combinations of optimization problems will have to be solved and the boundary conditions will change from day to day. The higher the complexity of the problem, the higher the potential savings from the optimization will be.

Although the incremental change in operating costs is relatively small compared to the total operating costs of a plant of this size, the cost/benefit relation for the optimizer system is favorable, the pay-out time is considerably less than one year. Improvements in plant operation by recognizing plant constraints before they lead to disturbances further add to the justification of this tool, though it is difficult to quantify the effect of such improvements.

The primary boundary conditions for plant operation can be summarized as follows:

- Feed and products prices
- Utilities prices (mainly imported electricity, cooling water, boiler feed water)
- Product stream, flows and impurity limits, set by market conditions
- Fouling status of heat exchangers especially when cooled by sea water
- Expander availability
- Availability of parallel units such as drivers for pumps, sales gas compressors or steam boilers
- Ambient conditions (air and cooling water temperature)

The calculation time for the optimization system depends on the number of optimized variables. Experience has shown that for six variables about 60 iterations are sufficient to identify the optimum. This means, an execution time on a large business computer (with approximately 15 millions instructions/sec computing power) of about three to five minutes. Therefore, the computing time is no obstacle for the frequent use of the system.

CONCLUSIONS

A plant optimization system consisting of an accurate plant simulation model which calculates the objective function and the plant constraints for all types of operation together with a robust optimizer is an efficient tool to improve plant operation in day-to-day operation. The potential savings from its application increase with the magnitude and frequency of changes in the boundary conditions. For the normal fluctuations of the plant operation the investment for the optimization system has a payout time of less than one year.

NEW MEASUREMENTS OF THE TENSILE STRENGTH OF LIQUID ^{4}He

Joel A. Nissen, Erik Bodegom, Laird C. Brodie,
and Jack S. Semura

Department of Physics
Portland State University
Portland, Oregon

ABSTRACT

A piezoelectric hemispherical transducer was used to focus high-intensity ultrasound into a small volume of superfluid helium. The transducer was gated at its resonant frequency of 566 kHz with gate widths less than 1 ms in order to minimize the effects of transducer heating and acoustic streaming. The onset of nucleation was detected with small-angle scattering of laser light from the cavitation zone by microscopic bubbles. The laser probe also was used to confirm calculations of the pressure amplitude at the focus based on the power radiated into the liquid, the geometry of the transducer, and the nonlinear attenuation of the sound. The cavitation pressure observed was significantly greater than previously reported in the literature and in agreement with homogeneous nucleation theory.

INTRODUCTION

It is well known that most liquids exhibit a tensile strength which is smaller than the tensile strengths predicted by homogeneous nucleation theory[1,2,3,4]. This is usually attributed to the difficulties of preparing liquid samples free from dissolved solids and gases which act as sites for heterogeneous nucleation. Liquid ^{4}He holds a unique place in this respect since all foreign gases, with the exception of minute traces of ^{3}He, are frozen at liquid helium temperatures. Furthermore, if the helium is in the superfluid state, all crevices on solid surfaces are filled with superfluid, eliminating the chance of heterogeneous nucleation on helium vapor pockets.

Despite the quantum nature of liquid helium, the Becker-Doring theory of nucleation of the vapor phase from the liquid should be valid down to 0.3 K in ^{4}He and 0.2 K in ^{3}He[5,6]. The Becker-Doring theory as formulated by Blander and Katz[7] relates the rate of critical sized bubble formation per unit volume, J, to curves of temperature and pressure in the metastable region.

$$J = N(2\sigma/\pi mB)^{\frac{1}{2}} \exp\{-16\pi\sigma^3/3kT(P_e-P_L)^2\,\delta^2\}$$

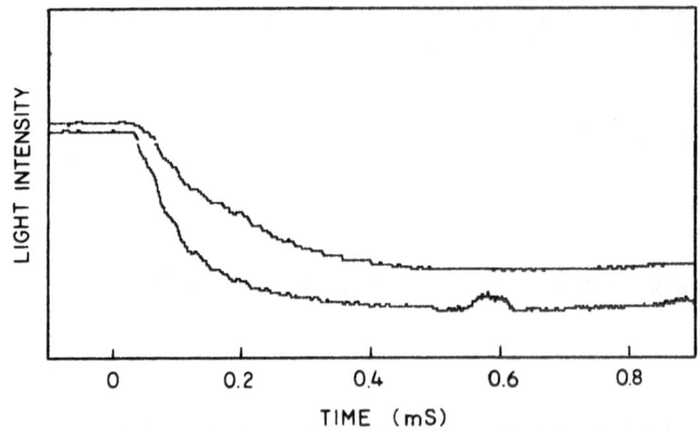

Fig. 1. Light intensity versus time.

At very low temperatures Lifshitz and Kagan[8] have shown that nucleation due to quantum tunneling through the nucleation potential barrier will predominate over nucleation resulting from thermal transitions over the top of the barrier, the mechanism assumed by classical nucleation theories.

Although the Becker-Doring theory of nucleation has been successful in predicting the limit of superheat in positive pressure for both helium isotopes[9,10], previous measurements of the tensile strength (negative pressures) have been in disagreement. Several nucleation mechanisms have been proposed to explain this discrepancy. Akulichev[11] has calculated that negative ions formed by ionization of helium atoms from sources such as cosmic rays would reduce the tensile strength. Quantized vortices have been suggested as a mechanism for premature nucleation as has quantum nucleation.

In analogy with the superheating of cryogens, in which mechanisms for heterogeneous nucleation are suppressed by rapidly heating a small volume of liquid, we have designed an experiment in which the liquid ^{4}He is stressed for less than a millisecond in a volume of 10^{-8} litres. This method greatly reduces the chance of nucleation at the site of an electron bubble or a particle of frozen gas and has led to agreement with classical nucleation theory.

EXPERIMENTAL PROCEDURE

Short bursts of ultrasound were focused by a piezoelectric hemispherical transducer immersed in the liquid helium. The transducer, Channel 5400 Navy I, had an inside radius of 0.00625 m and was operated at its resonant frequency of 566 kHz. Thin wires supported the transducer along its rim at nodal surfaces to minimize ultrasonic radiation into the supports and two grooves were cut into the rim so that the focal zone, which lies half below the rim, would be visible. We used a dewar with transverse optical windows, allowing us to use laser light to probe the cavitation zone with a minimum of distortion.

The He-Ne laser was chosen to detect the onset of nucleation for three reasons. First, according to classical nucleation theory, a void in a liquid must reach a critical radius before it is energetically favorable for the bubble to grow; for ^{4}He, this radius is only 2 nm so it is desirable to have a method of detecting very small bubbles. If there are many bubbles present in the cavitation zone, light scattered out of the laser beam will be detectable when the bubbles have grown to approximately

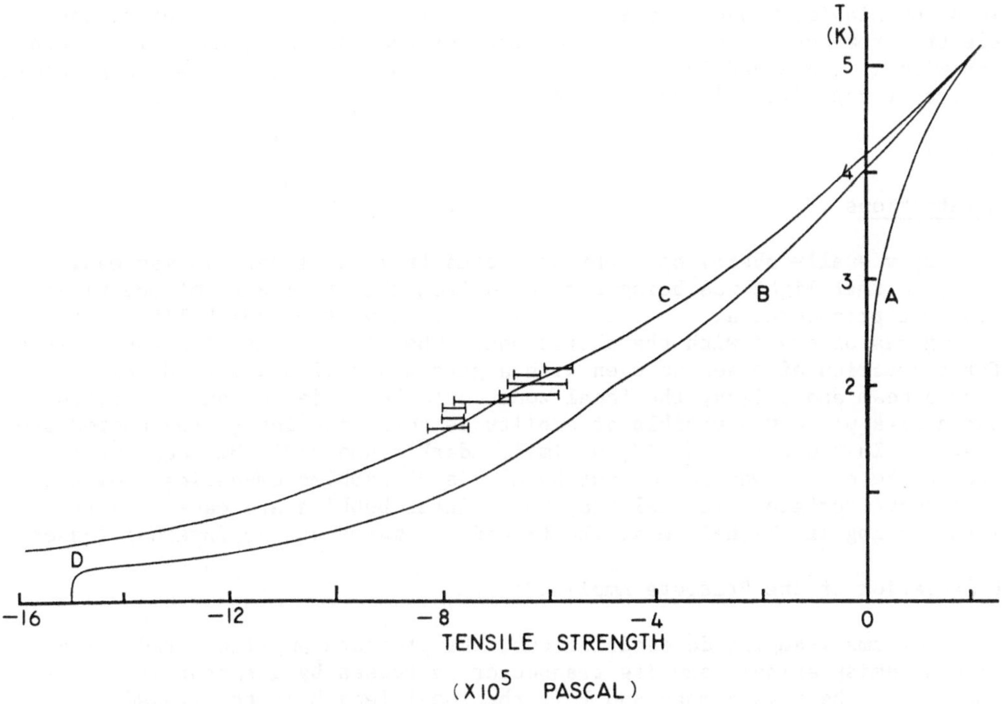

Fig. 2. Tensile strength versus temperature. Curve A is the coexistance curve for ^{4}He; B and C are the predicted tensile strengths according to homogeneous nucleation theory for $J = 10^3$ and $J = 10^{18}$ nuclei per litre per sec, respectively. Point D indicates the onset of quantum nucleation.

600 nm, the wavelength of the light. Bubbles of this radius resonate at a frequency of 50 MHz[11]. At such a high frequency it is very difficult to detect bubbles with sonic transducers. Second, the light probe has clear spatial resolution, enabling the discrimination between cavitation on the transducer's surface and cavitation at the transducer's focus. Third, the laser can be used as a pressure transducer. Briefly, the pressure variations in the liquid cause periodic variations in the index of refraction, giving rise to a diffraction pattern. The square of the acoustic pressure amplitude is proportional to I_1/I_0, where I_1 is the intensity of the first order diffraction and I_0 is the incident light intensity.

Figure 1 shows the output of a photodiode, which detects parallel light passing through the caviation zone. In the upper trace, the piezoelectric transducer is gated on at time zero. In 40 μs, the sound reaches the focal point of the transducer and the light intensity drops as light is diffracted from the beam. For the next 0.4 ms, the piezoelectric transducer builds to full amplitude, after which there is no change in the light intensity. For the lower trace, the driving voltage of the piezoelectric transducer was increased. More light is diffracted from the beam, and a signal appears at 0.5 ms due to cavitation. Not seen in Figure 1 are the large fluctuations in the light intensity that are observable at later times.

Measurements of the acoustic power radiated into the liquid and the light intensity diffracted by the sound were made between 1.6 K and the lambda point along the coexistance curve. To determine the power radiated into the liquid, measurements of the electrical impedance of the piezoelectric transducer were made both with the transducer in the helium bath and with it suspended just above the liquid surface. This was the greatest source of experimental error in the measurements.

RESULTS

Observations

By visually observing light scattered from the laser, it was easily verified that light was being scattered from the focal zone of the piezoelectric transducer at the same pressure amplitudes at which light scattering was observed with the photodiode. The light scattering was visible for a fraction of a second even with a gate width of 0.5 ms. Using a spread beam and a lens, the focal zone could be projected onto a screen, and a dark plume was visible at amplitudes corresponding to the photodiode signal. In the normal fluid, a similar dark plume could be seen for an instant before it was washed out by clouds of bubbles emanating from the transducer surface. We feel that these later bubbles are caused by the superheating of the helium as the transducer warms due to internal losses.

Calculation of the Pressure Amplitude

For small-amplitude oscillations, the pressure amplitude radiating from a hemispherical focusing transducer increases by a factor of $K_p = kF$ where k is the wave number and F is the focal length of the transducer. If we are to increase the amplitude to finite values, however, powerful absorption occurs along the focal path due to the formation of shock waves. The pressure at the focus is given by approximate expression[12]

$$P_F = \pi P_o K_p \{m + (\varepsilon K_p P_o/\rho c_o^2) \ln K_p\}^{-1}$$

where

$$m = \pi - 1 \text{ and } \varepsilon = \tfrac{1}{2}(\partial^2 c/\partial \rho^2)_s \rho_o/c + 1$$

The parameter ε has been shown to be a correct description of the nonlinear attenuation of sound in He II at high static pressures[13].

The light intensity data indicate that the calculated values shown in Figure 2 are not in error by more than 20 percent.

Comparison with Theory

The results graphed in Figure 2 are in agreement with Becker-Doring theory for $J = 10^{18}$ critical nuclei per second per litre. To see if this is reasonable, we can make a crude estimation of the rate that would be necessary to create one critical nucleus in the cavitation zone in one negative pressure excursion. For an aperture angle of 90°, the radius of the focal zone (Airy Circle) is 3/k where k is the wave number. The volume used in this experiment was approximately 10^{-8} litres. In one excursion the liquid is exposed to large negative pressures for a fraction of a microsecond. Therefore, to have a high probability of producing one bubble, J must be on the order of 10^{-15}. It also must be kept in mind that in order to detect breakdown of the liquid there must be either a few large bubbles to grow over many oscillations due to rectified diffusion or thousands of bubbles with sizes on the order of a wavelength of light. It has been assumed in this calculation that the pressure excursion is isothermal; a small temperature change, however, would not substantially alter the results.

CONCLUSION

The tensile strength of liquid ^{4}He presented here are substantially greater than the most negative pressure previously reported of -1.2×10^5 pascal[4]. For the temperature range 1.6 K to 2.15 K, these results are compatible with classical nucleation theory. Extending the observations into the normal fluid is a problem that has still not been solved.

ACKNOWLEDGMENTS

This work was supported by a National Science Foundation grant, DMR-3804249, and by the Environmental Sciences and Resources program at Portland State University (ESR Publication 215).

REFERENCES

1. A. D. Misener and G. R. Hebert, Tensile strength of liquid helium II. Nature 177:946 (1956).
2. J. W. Beams, Tensile strengths of liquid argon, helium, nitrogen, and oxygen, Phys. Fluids 2:1 (1959).
3. R. D. Finch, T. G. Wang, R. Kagiwada, M. Barmatz, and I. Rudnick, Studies of the threshold-of-cavitation noise in liquid helium, J. Acoust. Soc. Am. 40:211 (1966).
4. P. L. Marston, Tensile strength and visible ultrasonic cavitation of superfluid ^{4}He, J. Low Temp. Phys. 25:383 (1976).
5. V. A. Akulichev and V. A. Bulanov, Strength of quantum liquids, Sov. Phys. Acoust. 20:501 (1975).
6. D. N. Sinha, private communication.
7. M. Blander and J. L. Katz, Bubble nucleation in liquids, AIChE J. 21:833 (1975).
8. I. M. Lifshits and Y. Kagan, Quantum kinetics of phase transitions at temperatures close to absolute zero, Sov. Phys.-JETP 35:206 (1972).
9. D. N. Sinha, J. S. Semura, and L. C. Brodie, Homogeneous nucleation in ^{4}He: A corresponding-states analysis, Phys. Rev. A 26:1048 (1982).
10. D. Lezak, L. C. Brodie, J. S. Semura, and E. Bodegom, Homogeneous nucleation temperature of liquid helium three, accepted for publication by Phys. Rev. B (1987).
11. V. A. Akulichev, Acoustic cavitation in low temperature liquids, Ultrasonics 26:8 (1986).
12. L. D. Rozenberg, in: "Sources of High-Intensity Ultrasound, vol. I, Part III, L. D. Rozenberg, ed., Plenum Press, New York (1969), p. 332.
13. A. Hikata, H. Kwun, C. Elbaum, Finite amplitude wave propagation in solid and liquid ^{4}He, Phys. Rev. B 21:3932 (1980).

SIMULTANEOUS PRESSURE AND TEMPERATURE MEASUREMENTS ON HELIUM IN THE HIGH SPEED ROTATING FRAME

R.M. Igra, M.G. Rao, and R.G. Scurlock

Institute of Cryogenics
The University
Southampton SO9 5NH, England

ABSTRACT

As part of continuing thermodynamic and hydrodynamic studies of liquid helium in the high speed rotating frame, we have successfully measured pressures and temperatures simultaneously at the periphery of a 0.5 metre diameter rotor at tip speeds up to 65(m/s). For these measurements a Siemens KPY14 absolute pressure transducer and Southampton miniature diode thermometer have been used. This paper describes the calibration of the pressure transducer and its performance characteristics in centrifugal acceleration fields up to 2200 "g". The change in sensitivity of the transducer after many thermal cycles, between 4.2 K and 300 K, and repeated acceleration to 65(m/s) is less than 0.5%. Pressure and temperature measurements under high speed rotation are discussed by comparing the observed variations, as a function of tip speed, with predicted values of temperature and pressure using available thermodynamic data.

INTRODUCTION

The field windings of superconducting alternators, which are situated at the periphery of a rotor, require refrigeration. Efficient design therefore requires an understanding of the thermodynamic and hydrodynamic states of liquid helium in the high speed rotating frame. At the periphery of a 1 metre diameter rotor with a rotational speed of 3000 rpm (50Hz), centrifugal accelerations of up to 5000 "g" occur. The centrifugal compression causes temperature and pressure changes in liquid helium. Previously the only property of liquid helium that has been measured under these conditions is temperature. To determine the thermodynamic state of the liquid helium at the periphery of a rotor the fluid flow has had to be assumed to be isentropic (adiabatic), Scurlock[1], with an enthalpy change during radial flow of

$$h = \tfrac{1}{2}.w^2.(r_1^2 - r_2^2) \qquad [1]$$

where w=angular velocity of the rotor and r=radial distance from the axis of rotation.

Although calculated and measured temperature changes agree well with these assumptions the actual state of the liquid helium may lie anywhere on the isotherm. One other thermodynamic property that may be measured at the periphery of a rotor is pressure. The pressure gradient dP/dr across an element of fluid at radius r under equilibrium in a non-flowing fluid column is:

$$\frac{dP}{dr} = \rho(r,w).r.w^2 \qquad [2]$$

It should be noted that, although the density ρ is a strong function of radius, it is also a weak function of w.

INSTRUMENTATION

Because of centrifugal compression effects, it is not possible to take pressure gauge lines from the periphery of the rotor to the axis. A pressure transducer was therefore required that could be mounted at the periphery of the rotating cryostat, and operate at tip speeds of up to 65(m/s). Based on the assumptions above, pressure variations of 0-4 bar and temperature changes of 0-0.5 K were expected. The resolution of the Southampton Miniature Diode Thermometers SMDT's[2] we used was ± 1 mK or 0.2%. The voltage resolution of the available DVM's was ± 10 uV d.c. at 100ms integration time. The pressure transducer was therefore required to have a range of at least 0-6 bar and a sensitivity of at least 5(mV/bar), to give the same resolution. Further, because of centrifugal stresses in the thermosyphon arm, the mass of the transducer and its mounting was required to be less than 20 g. An examination of the published data on pressure transducers showed that all the basic types of pressure transducer have been successfully tested at 4.2 K. For the purposes of this work, only integrated circuit semiconductor strain gauge devices were likely to be suitable.

Thornton[3] and Bukovich[4] developed their devices specifically for use in centrifugal field of 5000 "g" at 4K. Bukovich reported in his paper centrifugal effects equivalent to 0.6 bar pressure changes. Neither of these devices met the above requirements. Briemesser[5] reports the measurement of cold gaseous helium pressures at the axis of a rotating cryostat using Siemens AG KPY12 0-2 bar differential transducer. Table 1 contains a summary of the publisher data[3,4,5].

A Siemens AG KPY14 0-10 bar absolute pressure transducer was found to meet all our requirements and was therefore calibrated at 4 K, 77K and 300 K, prior to mounting on the thermosyphon arm. The KPY14 and KPY12 transducers consist of a diaphragm which is etched from a Silicon chip, 2x2x1 mm. Each diaphragm has four strain gauges on it, produced by ion implantation. The chip is mounted in a metal holder.

CALIBRATION OF PRESSURE SENSOR

Calibration was carried out in a small probe, machined out of copper, which could be pressurized from 0 to 6 bar (absolute) with gaseous helium. The probe was immersed directly into a 50 litre liquid helium dewar which acted as a thermal bath. The test transducer was mounted in the probe with a SMDT bonded to its outer surface. When pressurized above atmospheric pressure, the probe became filled with subcooled or supercritical liquid helium by condensation. The hydrostatic pressure head in the probe was less than 2mbar. Pressures within the probe were read with a Wallace and Tiernan Ltd 'Digiread' digital 0 to 2 bar differential pressure gauge (resolution ± 0.05 mbar, accuracy ± 1 mbar). The reference pressure to the 'Digiread' was set to either atmospheric or 2 bar above atmospheric, using a Siemens AG KPY12 differential pressure transducer. The latter was referenced to atmospheric pressure and calibrated by the 'Digiread' during each pressure cycle. Atmospheric pressure was measured by a mercury manometer. The estimated accuracy of

TABLE 1. Pressure Transducer Characteristics

Group	Oxford	MIT	Karlsruhe	Southampton
Reference	Thornton ref(3) 1978	Bukovich ref(4) 1977	Breimesser ref(5) 1983	This paper
Device type	LC-oscillator capacitance gauge. Two types I & II	Bonded resistance strain gauge. Labmade Two types, I & II	Siemens AG KPY12 Semiconductor strain gauge IC device	Siemens AG KPY14 Semiconductor strain gauge IC device
Range	I 1-2bar II 1-20bar	0-1bar Differential	0-2bar Differential	0-10bar Absolute
Sensitivity	I = 48KHz/bar II=466KHz/bar	I =230μV/bar II=420μv/bar at 3V	100mV/bar at 5V 50mv/bar at RT.	11mV/bar at 250.0μA
Sensitivity shift with temperature	-0.09%/K between 4K & 77K for both types I&II		-0.3%/K 4.2K -0.3%/K 77K -0.2%/K 300K	-0.2%/K 4.2K
Linearity	I&II ±0.3% at 4.2K	I =3% FS II=1% FS	±0.5% FS for 240K-400K	±25mbar between 0&5 bar
Heat dissipation	14mW at 5V		5mW at 5V	0.4mW at 250uA
Reproducibility	±0.25mbar after 10 pressure cycles between 1 & 10bar, at 4K			±0.4% FS
Hysteresis		I =1.5% FS II=0.8% FS	<0.5% FS at RT	<1mbar at 4.2K
Size: 1 =length d =diameter	1=30mm dmax=40mm dmin=20mm	1=6mm d=16mm both I & II	1=7-15mm dmax=13mm dmin=6mm	1=7-15mm dmax=13mm dmin=6mm
Mass		<2g est. both I & II	1-2.7g	1-2.7g
Comments	Mechanically complex	Only 2 active components per device.	Response time <1ms at 293K Low cost	Response time <1ms at 293K

the calibration was ± 7 mbar absolute and ± 4 mbar differential. A constant current supply with a stability of 1/2000 was used to energise the transducer.

The reproducibility of the KPY14 was determined by thermally cycling it between 300 K, 77 K and 4.2 K 13 times. During the first 6 cycles the probe was vacuum pumped to 0.1 mbar. Subsequently the device was cycled between 0 and 5 bars at each temperature. The measured pressure change with cycling is shown in Fig. 1. All values are referenced to the 13th cycle.

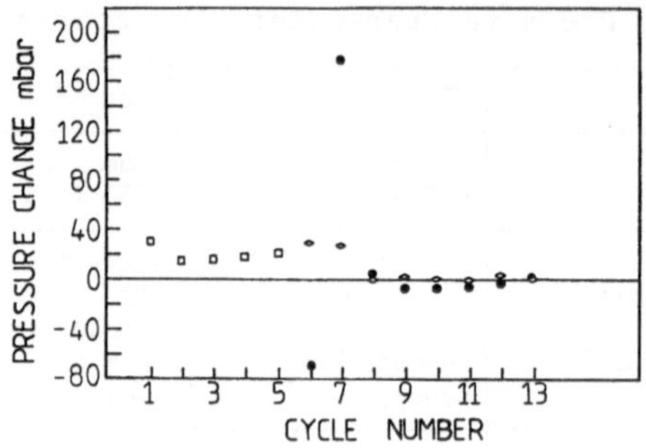

Fig.1 Variation of KPY14 output voltage, at 4.2K, with cycling between 300 K and 4.2 K. The change in output voltage is expressed as a pressure change. □ = 0 bar, ◇ = 0 bar, ● = 5 bar.

The large pressure changes occurring at cycles 6 and 7 are believed to be a stress relieving pheonomena associated with the onset of pressurization and corresponds to a sensitivity change of $\pm$ 2.5%. After this, the sensitivity appears to vary by less than 0.2%. During the 13th cycle to 4.2K the output voltage of the device was measured at 19 pressures between 0 and 5 bar for excitation currents of 100.0, 250.0 and 800.0 μA, see Fig.2. The bridge resistance of the device at 4.2 K was measured as 5261 $\pm$ 6 Ohms, at all 3 currents. Fig.3 is a plot of the sensitivity change with temperature normalized by the sensitivity at 4.2 K. At 4.2 K the temperature sensitivity is -0.2% per K, which equates to a change of 10 (mbar/K) at 5 bar. Using the measured change in the bridge resistance with temperature, the temperature change in the transducer chip, due to self heating, was calculated to be:

0.3 $\pm$ 0.3 K at P below atmospheric pressure
0.0 $\pm$ 0.3 K at P above atmospheric pressure

at 250.0 μA.

This change results from the increase in heat transfer from the chip with increasing pressure in the probe.

After completion of the rotating frame measurements described below, the pressure transducer was recalibrated. Although the zero offset voltage was found to have changed from ~ 4.5 mV to 1.3 mV the overall sensitivity was found to have changed by less than 0.4%, i.e. 16 mbar at 5 bar. This was despite having been cycled to 2200 "g" at least 30 times and cooled 9 times from room temperature to 4.2 K. This value has been taken to indicate the overall reproducibility of the device.

A summary of the KPY14 characteristics is shown in Table 1.

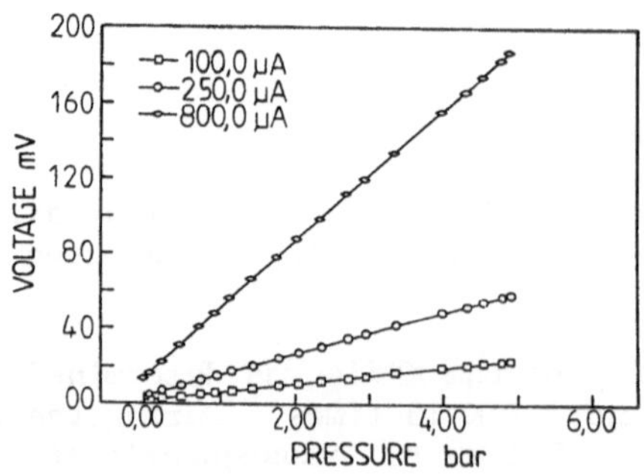

Fig.2 Output voltage of KPY14 against applied pressure, at 4.2 K.

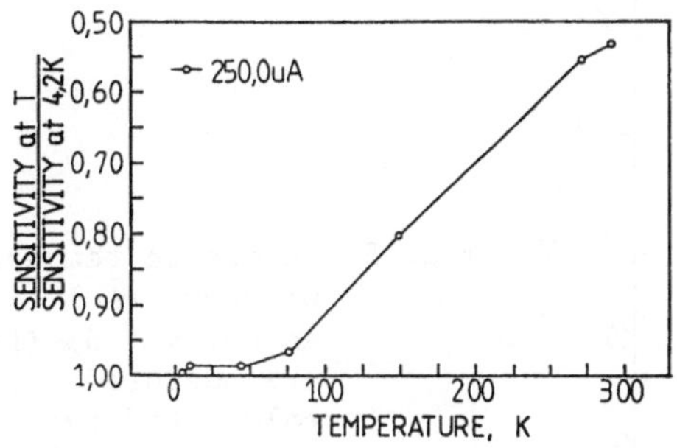

Fig.3 The effect of temperature on the sensitivity of the KPY14, normalized to the sensitivity at 4.2 K.

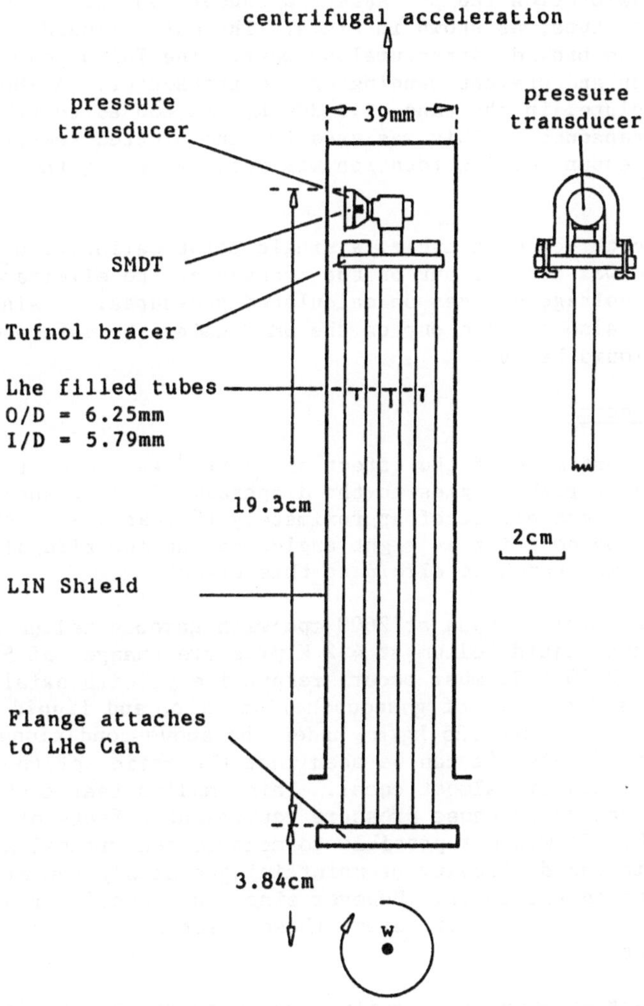

Fig. 4. Liquid helium thermosyphon arm for use in rotor.

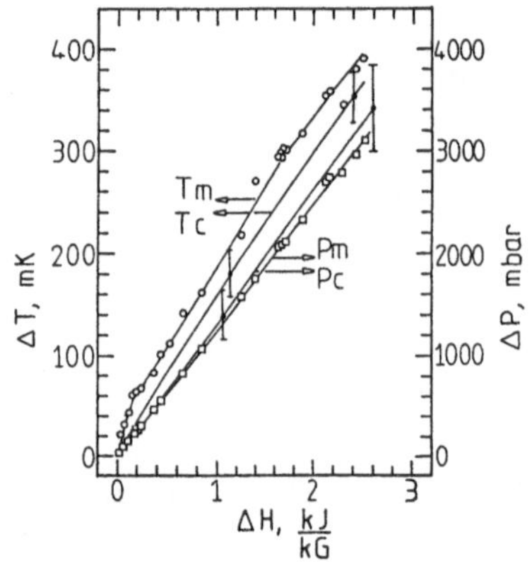

Fig. 5. Change in temperature (ΔT) and pressure (ΔP) vs change in enthalpy (ΔH) (from [1]). T_m,P_m are measured curves. T_c,P_c are calculated curves. Axial pressure on rotor = 0.991 bar, saturated liquid temperature = 4.201 K.

ROTATING FRAME MEASUREMENTS

After calibration, the pressure transducer was soft soldered into the thermosyphon tube, as shown in Fig.4. The outer liquid helium filled tubes are used to provide structural support. The Tufnol cap was used to dampen vibration and prevent bending of the transducer. A SMDT, calibrated to ± 15 mK absolute, in the range 2.7-300 K, was bonded to the radial mid-point of the transducer. This was used for the plotted temperature readings. Axial pressure during rotation was measured using the 'Digiread' and manometer.

During each cooldown cycle, a single point calibration at atmospheric pressure was carried out on the transducer, to eliminate the effects of zero offset voltage changes on calculated pressures. A single point calibration was also carried out on the SMDT using the saturated temperature of the liquid helium.

Centrifugal effects

A simple estimate of the effect of centrifugal acceleration of 2200 "g" acting at right angles to the diaphragm of the transducer suggests a pressure equivalent effect of approximately 15 mbar due to the mass of the diaphragm. By mounting at right angles to the centrifugal acceleration, Fig.4, it was hoped to eliminate this effect.

When the rotor is spun at 2600 rpm with gaseous helium at 294 K, air at 294 K, and liquid helium at 4.2 K pressure changes of 5 ± 3 mbar, 26 ± 5 mbar and 2300 ± 20 mbar occur, respectively, with axial pressure of 1.0 bar. The densities of gaseous helium, air, and liquid helium are 0.16 kg/m^3, 1.2 kg/m^3 and 125 kg/m^3 under the above conditions. Neglecting compressibility effects, it can be seen that the ratios of the pressure changes to densities are almost equal. This implies that centrifugal forces on the transducer cause pressure equivalent effects of less than 3 mbar at 294 K. It was not possible to measure centrifugal stress effects at 4.2 K, due to the difficulty of maintaining a stable temperature with no liquid helium in the rotor. However since the tensile properties of silicon increase at low temperatures, these effects should be less than at room temperature.

The effect of angular velocity, expressed as an enthalpy change, on the measured pressure and temperature changes in liquid helium is shown in Fig. 5. The bulk of the data were obtained during one cooldown cycle. The remaining data were obtained over 6 cooldown cycles and 20 spin-ups. The scatter in data points is less than ± 40 mbar or ± 40 J/kg.

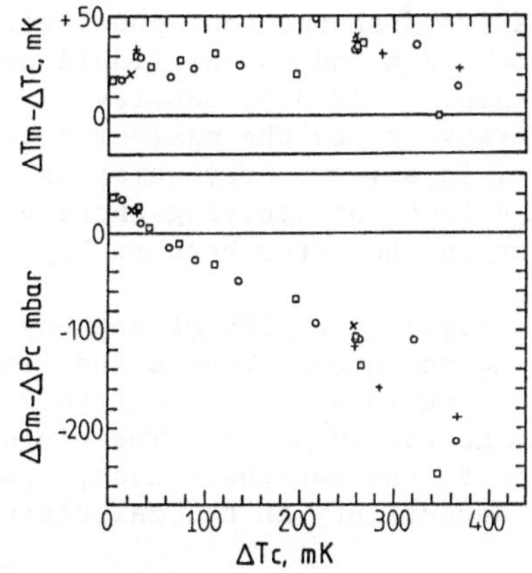

Fig. 6. Deviation between the calculated and measured values shown in Fig. 5.

The experimental uncertainty in the rotational speed measurements equates to about ± 20 J/kg at 3000 rpm. The reason for the sudden rise in temperature, of ~21 mK, between zero speed and the first measurement at 350 rpm is not clear. It is probably associated with an increase in the axial pressure caused by the increased helium boil-off rate that occurs with rotation. This would, however, require that an axial pressure gradient between 4.2 K and 300 K exists. For saturated liquid helium at 1.0 bar, a 20 mK change is equivalent to a 14 mbar change in pressure.

The experimental scatter is believed to arise from three factors. (a) the calibration uncertainties described above. (b) During each run, axial pressure was found to increase by 10-15 mbar as the speed increased. This was due to the helium recovery system. Corrections have been made in the pressure and temperature readings to take account of these changes. (c) The liquid level in the axial reservoir was not controlled or monitored during the course of a rotational run. The gas/liquid interface could therefore have moved outwards radially by up to 4 cm. This would lead to the pressure transducer and SMDT under-reading the change from the stationary state, by a maximum of 80 mbar and 10 mK, respectively.

DISCUSSION

The computer programme HEPROP, Hands[6] has been used to calculate thermodynamic data for comparison with the measured data. HEPROP is based on the standard state equations of McCarty[7]. Also shown in Fig.5 are plots of temperature and pressure change, calculated from the measured enthalpy change assuming isentropic compression. The difference between these calculated values and the measured values are shown in Fig.6.

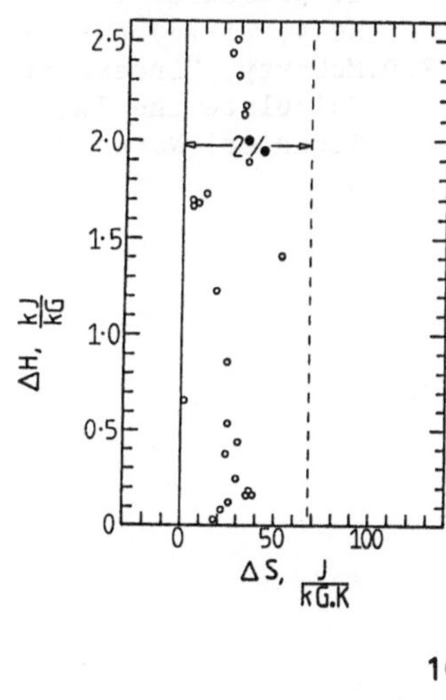

Fig. 7. Entropy change from P,T data under rotation. Axial conditions are P = 0.991 bar, S = 3.430 kJ/kg·K, H = 9.591 kJ/kg.

McCarty[7,8] states that the likely uncertainties in the state equations below 5.2 K and in the liquid are: Pressure $\pm$10%, Temperature $\pm$0.5%, Enthalpy $\pm$2%(5%), Density $\pm$0.5%(1.5%), and Entropy $\pm$2%. The values in brackets are the maximum uncertainties. These values are the uncertainty in any property which is calculated from any two other properties. The effects of applying these values to the absolute fluid properties are shown in the error bars in Figs. 5 and 7.

Fig.7 is a plot of entropy versus enthalpy calculated as a function of the measured pressure and temperature. The deviation between the latter curve and the isentropic state is probably due to the temperature jump mentioned above. The fluid flow condition from the axis of the rotor to the periphery does, however, appear to be isentropic, within the uncertainty in the calculated data.

ACKNOWLEDGEMENTS

We wish to thank Dr.B.Hands for his assistance with the computer programme 'HEPROP' and Siemens AG (UK) for the pressure transducers they gave us. RMI acknowledges US SERC for a research studentship and the British Cryogenic Council for a travel bursary.

REFERENCES

1. R.G.Scurlock, Thermodynamics of helium in high speed rotating machines, in: Proc. of 6th Int. Cryo. Eng. Conf, ICEC6, Eindhoven. Butterworth, Guildford, UK (1976) p.35.
2. R.M. Igra, M.G.Rao, R.G.Scurlock, Thermometric characteristics of the improved miniature Si diodes, in: Proc. of 11th Int. Cryo. Eng. Conf. ICEC11, Berlin. Butterworth (1986) p.617.
3. G.K.Thornton, A high resolution pressure transducer operating at liquid helium temperatures, in: Proc. of 7th Int. Cryo. Eng. Conf. ICEC7,London. Butterworth, (1978) p.608.
4. R.A.Bukovich, J.L.Smith, K.A.Tepper, A bonded-strain-gauge pressure transducer for high-speed liquid helium temperature rotors, in: "Advances in Cryogenic Engineering", vol.23, Plenum Press, New York (1977).
5. F.Breimesser et al, KPY12 - A pressure transducer suitable for low temperature use, in: 8th Int. Conf. on Magnet Technology, Grenoble, (1983).
6. B.A.Hands, "HEPROP - a computer programme for the thermodynamic and thermophysical properties of helium - Third Edition".Dept Report, Cryogenics Laboratory, Dept of Eng.Sci., University of Oxford, Parks Road, Oxford OX1 3PJ, England.
7. R.D.McCarty, Thermodynamic Properties of Helium 4 from 2 to 1500 K at pressures to 10^8 Pa, in: J.Physics.Chem. Ref. Data 2, No.4, (1973) p.923-1042.
8. R.D.McCarty, "Interactive FORTRAN Programs for Micro Computers to Calculate the Thermophysical Propeties of Twelve fluids (MIPROPS)", Technical Note 1097. National Bureau of Standards, USA (1986).

HELIUM ADSORPTION ON ACTIVATED CARBONS AT TEMPERATURES BETWEEN 4 AND 76 K*

Isaura Vázquez, M. Patricia Russell,
David R. Smith and Ray Radebaugh

Chemical Engineering Science Division
National Bureau of Standards
Boulder, Colorado

ABSTRACT

Helium adsorption isotherms have been measured for two activated carbons in the 4-76 K temperature range for pressures from 0.1 to 3000 kPa. Such measurements have not been made previously in this temperature and pressure range, but they are needed for the design of an adsorption compressor for helium gas. This paper describes the measurement and analysis techniques for obtaining the adsorption isotherms. The isosteric heats of adsorption are derived from these isotherms. Adsorption isotherms on various charcoals are correlated using the effective adsorbed film thickness as a fundamental parameter.

INTRODUCTION

Activated carbon can adsorb and desorb large amounts of helium gas at high pressures and low temperatures, which makes this material useful for the design of adsorption compressors. An adsorption compressor could replace the mechanical compressor in refrigeration systems where temperatures as low as 4-15 K and high reliability are desired or required, such as satellite applications of the Joule-Thomson refrigerator. The data at low temperatures and high pressures are also useful for the design of regenerators utilizing adsorption of helium to provide a high effective heat capacity.

The design of adsorption compressors requires a knowledge of the mass adsorbed as a function of temperature and pressure. An extensive literature search has failed to discover data published within the range of pressures and temperatures of this study.

---

* Research sponsored by Air Force Space Technology Center, Kirtland Air Force Base, NM, under contract FY830386601003.

This paper presents measurements of helium adsorption isotherms on two different activated carbons for the temperature range from 4 to 76 K and pressures from 0.1 to 3000 kPa. It also presents the heat of adsorption derived from these isotherms.

EXPERIMENTAL TECHNIQUE

Figure 1 shows a schematic of the experimental adsorption apparatus. The calibrated volume was filled by passing helium gas through a charcoal trap, cooled by liquid nitrogen to remove any impurities. The pressure and temperature of the helium in the calibrated volume were measured using a 5 MPa capacitive pressure transducer and a platinum resistance thermometer, respectively.

Valve V_1 between the calibrated volume and the copper adsorption cell was then opened to allow an increment of the helium gas into the adsorption cell. The adsorption cell contains an internal copper grid structure to enhance thermal equilibrium within the activated carbon, a poor heat conductor. Valve V_1 was closed and the temperature and pressure of the calibrated volume and the adsorption cell were measured. Both the total amount of mass introduced into the adsorption cell from the calibrated volume and the amount of gas in the adsorption cell void space, which has not been adsorbed, were determined by using the density of the gas from the NBS equation of state for helium.[1]

The pressure and temperature of the helium gas in the adsorption cell were measured using another 5 MPa capacitive

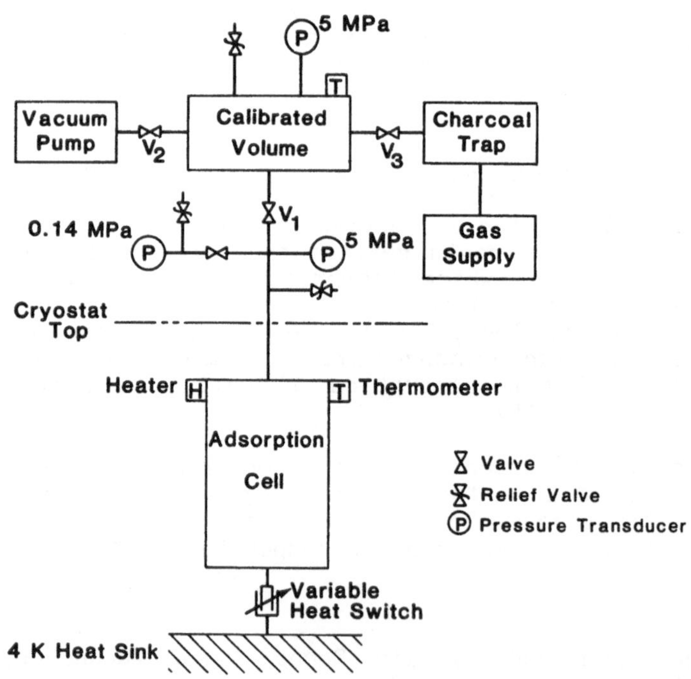

Fig. 1. Experimental apparatus

pressure transducer and a germanium resistance thermometer. The temperature of the adsorption cell was controlled using a resistance heater and a heat switch connected to a liquid helium bath maintained at 4 K.

Two different procedures were used to take adsorption data. The first procedure consisted of filling the adsorption cell with a measured amount of helium gas and changing the conditions in the cell by cooling or heating it, taking adsorption or desorption data along constant mass lines. This process of cooling and heating provides a method to determine any hysteresis in the adsorption (desorption) process. Within the experimental uncertainty, no detectable difference was found between both processes. The second procedure consisted of keeping a constant temperature in the cell and adding known amounts of helium gas to the cell. The higher temperature isotherm data and the 4 K data were taken using this procedure. A detailed description of this experimental technique and apparatus is found elsewhere.[2]

SAMPLE CHARACTERIZATION

The two different activated carbons used in this study were Barnebey-Cheney, type 580-26 (also referred to as carbon A) and Saran A, 100% PVDC (also referred to as carbon B).** Carbon A is a coconut shell charcoal, -18+25 U. S. standard sieve. Carbon B is in the form of 0.5-0.8 mm diameter spheres. Table 1 shows the experimental parameters for each activated carbon. The surface areas of the carbons were estimated by using N_2 with the BET method.[3,4,5]

The apparent density is defined as the mass of adsorbent per unit cell volume. The total cell volume (69.3 cm^3) consisted of the volume occupied by the carbon solid, the macroscopic void volume (macropore volume and interparticle volume) and the micropore volume. Figure 2 shows scanning electron micrographs (SEM) of the carbon A and carbon B. The macropore and interparticle space can be clearly identified. The micropore volume has linear dimensions of the order of nanometers and cannot be identified in these pictures.

Table 1. Sample parameters

Type	Carbon A	Carbon B
Mass, ±0.005 g	26.48	40.51
Specific surface area, ±10 m^2/g	1474	1161
Apparent density, ±0.002 g/cm^3	0.382	0.585
Micropore volume*, ±1.7%	26.6	24.7
Macroscopic void volume*, ±1.7%	58.0	47.0
Solid volume*, ±1.7%	15.4	28.3

* percent of the total cell volume

** Trade names are given here in order to fully characterize the samples, but it does not represent an endorsement by NBS. Other manufacturer's products may work as well or better.

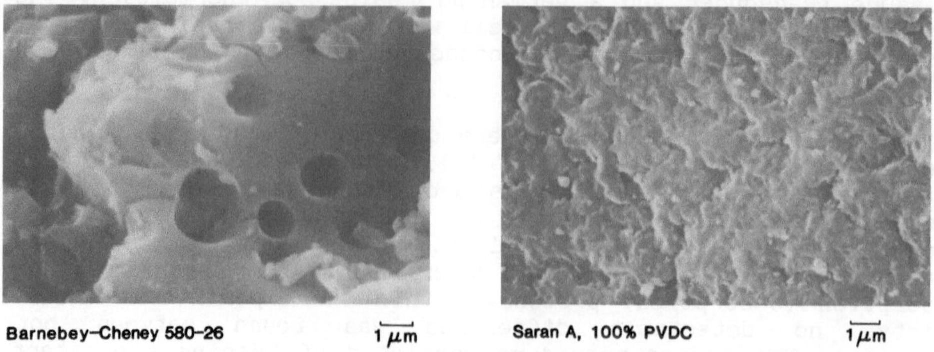

Fig. 2. Scanning electron micrographs for the Barnebey-Cheney 580-26 (carbon A) and the Saran A, 100% PVDC (carbon B).

The total void volume in the cell was measured using helium gas at room temperature.[2] The micropore volume was calculated by dividing the mass of adsorbed nitrogen at saturation by the density of liquid nitrogen at saturation.

RESULTS

Figures 3 and 4 show the helium adsorption isotherms for the carbon A and carbon B, respectively. The specific mass adsorbed (M_a) is plotted as a function of pressure where M_a is calculated by subtracting the mass of He in the void volume of the cell from

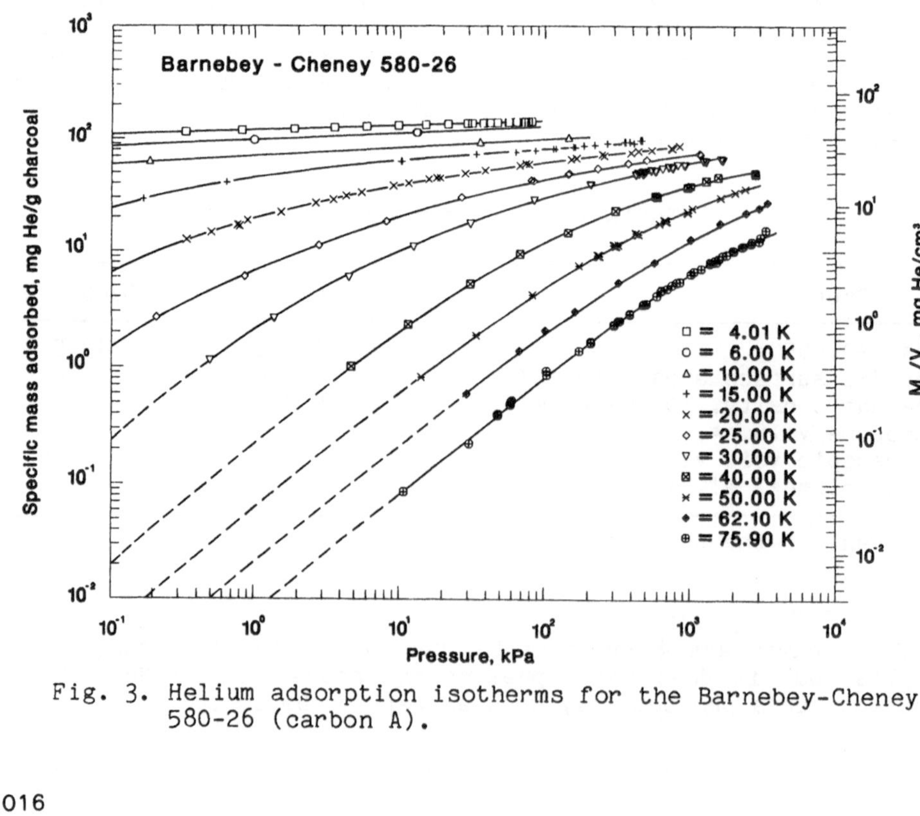

Fig. 3. Helium adsorption isotherms for the Barnebey-Cheney 580-26 (carbon A).

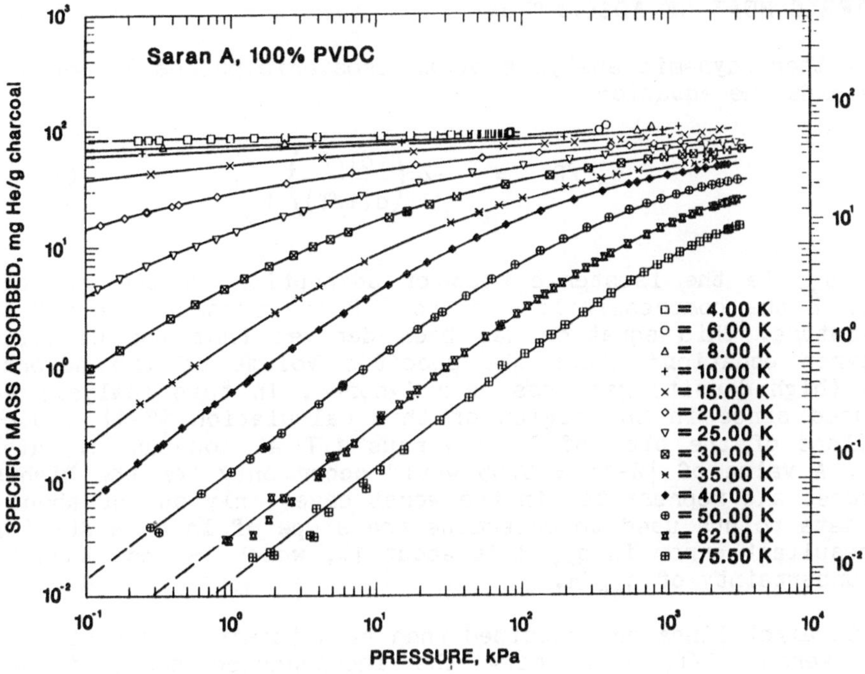

Fig. 4. Helium adsorption isotherms for the Saran A, 100% PVDC (carbon B).

the total amount of He mass in the cell, where the void volume of the cell does not include the volume occupied by the adsorbed film. Even though there are no known measurements of the specific volume of the adsorbed film for this pressure and temperature range, the Van der Waals volume b, (23.7 cm³/mol) should be a good representation of it[7] and was used here to obtain an approximate value of M_a. The value M_a is the mass equivalent of the STP volume (V_e), used by Chan, Tward and Boudaie[8] for H_2, Ne, and N_2 adsorption on activated carbon.

A linear behavior is observed in the 75.9 K isotherm at low coverage for carbon A and in the higher temperature isotherms for carbon B. Such behavior is theoretically predicted at low coverages by Henry's law and is the two dimensional analogue to the ideal gas law.[4] The dashed lines at low coverage for the carbon A represent Henry's law. At low coverage carbon A and B adsorb about the same amount per gram of carbon, but because of its higher density, carbon B adsorbs more per unit volume. Carbon B quickly saturates at a monolayer whereas carbon A continues to adsorb beyond the monolayer. At high coverage the mass adsorbed per unit volume is about the same for both carbons.

The carbon B isotherm data, at high coverages and temperatures from 6 to 15 K, shows a rapid increase in the amount of mass adsorbed with respect to pressure. This behavior is believed to be a consequence of a transition from a dense two-dimensional fluid to a two-dimensional solid.[9] Similar behavior probably exists with the carbon A but that region was not studied.

ISOSTERIC HEAT OF ADSORPTION

A thermodynamic analysis of an isosteric process[2], constant M_a, yields the equation

$$q_{st}/R = -Z \left(\frac{d\ln P}{d(1/T)}\right)_{M_a}, \qquad (1)$$

where q_{st} is the isosteric heat of adsorption, R the gas constant, Z the compressibility factor, P the pressure, and T the temperature. This equation has been derived from the Clausius-Clapeyron equation[2] where the specific volume of the adsorbed phase (high density gas) has been ignored. In this analysis Z=1 was used since in the region of this calculation $|Z-1| < 0.05$. The slope of the plot of ln P versus 1/T at constant M_a gives q_{st}/R. A value of $|Z-1| = 0.05$ would occur only for the highest pressures and represents, in the worst case, only one of about 4 or 5 data points used to determine the slope of ln P versus 1/T. The resultant error in q_{st}/R is about 1%, which is less than the 3-5% uncertainty of q_{st}/R.

Straight lines are obtained when each isostere is plotted as ln P versus 1/T, confirming the independence of q_{st}/R with respect to temperature over the studied range.[2] Figure 5 shows q_{st}/R versus the specific amount of mass adsorbed. A rapid change of q_{st} is observed at the monolayer completion at about 70 mg/g for both carbons.

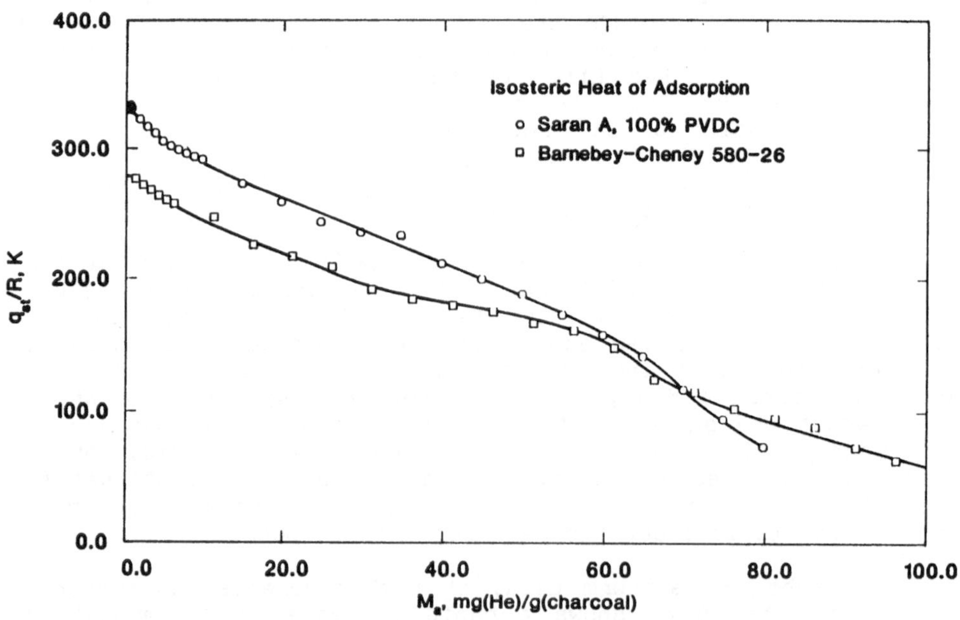

Fig. 5. Isosteric heat of adsorption for the Barnebey-Cheney 580-26 (carbon A) and the Saran A, 100% PVDC (carbon B).

GENERALIZED CORRELATIONS

Some analytic expressions for the mass adsorbed as a function of temperature and pressure have been developed by different investigators. The adsorption isotherm data studied here have been compared to some of these correlations.

Kidnay and Hiza[10] represent their He adsorption data using the Dubinin correlation[6] in which the volume of adsorbed helium, V_t, is a function only of the volumetric adsorption potential (ϵ), in which

$$\epsilon = (RT/V_\ell) \ln(P_s/P), \tag{2}$$

where V_ℓ is the specific volume of the adsorbate in the liquid phase at the normal boiling point, R is the gas constant, T the temperature, P the pressure and $P_s=(T/T_c)^2P_c$ is the equivalent saturation pressure above the critical temperature. V_t is analogous to the adsorbed mass, M_t, which is not the same as the previously discussed M_a. It represents the mass of the adsorbate within the micropore volume (total mass of helium in the cell minus the mass within the macroscopic void volume). This study's 76 K adsorption isotherm data for the carbon A agrees very well with the 76 K isotherm data of Kidnay and Hiza when comparing the mass adsorbed per unit surface area.[2]

For Ne, H_2 and N_2 Chan, Tward and Boudaie[8] used the Maslan correlation[11], which is similar to the Dubinin correlation except that the saturation pressure above the critical point was taken as

$$\ln P_s = A + B/T, \tag{3}$$

where the A and B coefficients are obtained from a linear least squares approximation of the adsorbate saturation values below the critical temperature. They found that the log of mass adsorbed is linear with respect to the adsorption potential, ϵ.

Figure 6 shows that the carbon B isotherm data does not show linear behavior using the Maslan correlation when the log of the effective film thickness, t_a, is plotted versus ϵ. The t_a, analogous to M_a, is equal to M_a divided by the density of the adsorbed film and the adsorbent surface area. It is used instead of M_a with the purpose of later comparison between different adsorbates and adsorbents. Not only do the curves deviate from linear behavior, but there is a temperature dependence as well for temperatures above 20 K. The dashed lines in Figure 6 represent the 4 K and the 76 K isotherms for the carbon A. Again, deviation from linearity and a temperature dependence are observed. The Dubinin correlation also shows a temperature dependence above 20 K and a deviation from straight-line behavior for log t_a versus ϵ.[2]

SUMMARY AND CONCLUSIONS

Extensive data on the He adsorption isotherms of two activated carbons have been presented in the pressure-temperature range proposed for the operation of an adsorption compressor, a range which had not been studied previously. Henry's law is observed at low coverages in both carbons. The derived isosteric

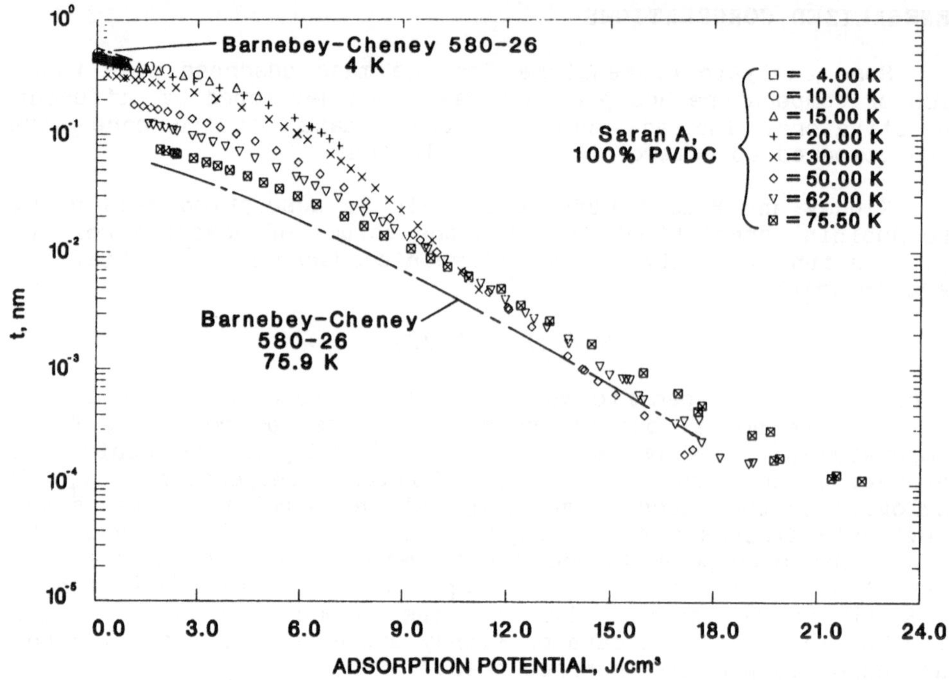

Fig. 6. Helium adsorption isotherms for the Saran (carbon B) using Maslan correlation[11]. The dashed lines are the 4 K and 76 K isotherms for Barnebey-Cheney.

heat of adsorption shows a rapid decrease at the expected monolayer completion. The two carbons behave somewhat differently when comparing the amount adsorbed per unit surface area. The two commonly used correlations do not satisfactorily represent the data over the wide range of pressure and mass adsorbed studied here because of an additional temperature dependence not accounted for by the correlations. Further correlation studies are needed.

ACKNOWLEDGEMENTS

We wish to thank John Stagen of Barnebey-Cheney for supplying us with the sample of 580-26 activated carbon and David Quinn of the Royal Military College of Ontario, Canada for making the Saran Carbon for us.

REFERENCES

1. R. D. McCarty, "Interactive FORTRAN Programs for Micro Computers to Calculate the Thermophysical Properties of Twelve Fluids [MIPROPS]," Tech Note 1097, National Bureau of Standards, Boulder, Colorado (1986).
2. M. P. Russell et al, Helium adsorption on charcoal at low temperatures and high pressures, Cryogenics, to be published.
3. S. Brunauer, P. H. Emmett, and E. Teller, Adsorption of gases in multimolecular layers, J. Am. Chem. Soc. 60:309 (1938).

4. C. Jr. Orr, and J. M. Dallavalle, Gas adsorption for surface area measurement, in: "Fine Particle Measurement, Size, Surface and Pore Volume," Macmillan Co., New York (1959), p. 164-204.
5. Standard test method for specific surface area of carbon or graphite, in: "Annual Book of ASTM Standards," Sect. 15, vol. 15.01:C819-77, ASTM, Pennsylvania. (1977).
6. D. M. Young, and A. D. Crowell, Heats of adsorption, in: "Physical Adsorption of Gases," Butterworths, Washington (1962), p. 70-75.
7. M. M. Dubinin, The potential theory of adsorption of gases and vapors for adsorbents with energetically nonuniform surfaces, Chem. Review 60:235 (1960).
8. C. K. Chan, E. Tward, and K. I. Boudaie, Adsorption of gases and vapors for adsorption of hydrogen, neon and nitrogen on activated charcoal, Cryogenics 24:451 (1984).
9. R. L. Elgin, and D. L. Goodstein, Thermodynamic study of the He monolayer adsorbed on grafoil, Physical Review A 9:2657 (1974).
10. A. J. Kidnay, and M. J. Hiza, High pressure adsorption isotherms of neon, hydrogen and helium at 76 K, in: "Advances in Cryogenic Engineering", Vol. 12, Plenum Press, New York, (1973) p. 730.
11. F. D. Maslan, M. Altman, and E.C.R. Aberth, Prediction of gas-adsorbent equilibria, J. Phys. Chem. 57:106 (1953).

HIGH PRESSURE ADSORPTION ISOTHERMS OF HELIUM ON ACTIVATED CHARCOAL

L. Duband and J. Chaussy

C.N.R.S., Centre de Recherches sur les Très Basses Températures,
BP 166 X, 38042 Grenoble-Cédex, France

A. Ravex

C.E.N.G.-DRF, Service Basses Températures, Laboratoire Cryogénie Technique Grenoble, France

ABSTRACT

Adsorption of helium on pellets and compressed activated charcoal has been measured in the pressure range from 1 to 30 atmospheres and in the temperature range from 10 to 80 K. A Clapeyron-like law ($\ln P_s = A+B/T$) for fictitious saturation vapour pressure P_s above T_{cr} is deduced experimentally from the measurements. This law leads to a fairly good generalized correlation for all experimental data in the framework of Dubinin theory :

$$C = \frac{M\alpha}{b}\exp(-\gamma x) \quad \text{with} \quad x = \frac{RT}{\beta}\ln\left(\frac{P_s}{P}\right)$$

INTRODUCTION

Physical adsorption has been extensively studied, see for example the selective review given by Kidnay and Hiza[1]. However most of the data for helium adsorption isotherms on charcoal are available at low pressures because of the interest for cryopumping or low pressure gas removal process[2]. The present work is mainly extended in the high pressure range (1 atm to 30 atm) for a large scale of temperature (10 K to 80 K) in the framework of thermal compression for Joule-Thomson cryocooling.

EXPERIMENTAL SET UP

The experimental technique used in this study is a volumetric method which needs a relatively simple apparatus described elsewhere[3]. Isotherms of adsorption are determined by measuring the amount of helium adsorbed at various equilibrium pressures and temperatures. The measurements are done using both adsorption and desorption process. Both results agree within experimental accuracy. The helium used for measurements is 4N purity. The amount of adsorbed helium is determined from pressure variation measurements in a calibrated volume at room temperature. The temperature of the adsorption cell is measured with a NbN thermometer and controlled by a regulation.

MEASUREMENTS AND ANALYSIS

Measurements were done with both activated charcoal pellets (m_C = 8.34 g) and compressed activated charcoal (m_C = 12.15 g, compression pressure about $1.5x10^8$ N/m^2) in two different adsorption cells (26.5 cm^3 and 18.25 cm^3). This charcoal supplied by Prolabo (code 22631) has an specific area of 1150 m^2. Before the measurements the adsorbent is outgased for a few hours at a temperature of 150° C by vacuum pumping (the lost in weight for the charcoal is about 15 % : 9.70 g to 8.34 g for pellets and 14.31 g to 12.15 g for compressed). The experimental apparatus allows us to pump the cell through a connecting pipe of 10 mm diameter during this outgasing process and then to reduce the dead volume due to the connecting pipe for the measurements.

The calibrated volume (V_r) at room temperature (T_r) is filled with helium successively at various pressure (P_r) and expanded into the cell. The amount of helium M_{He} in the adsorption cell and plumbing line for each regulated temperature (T_e) is calculated using an ideal gas law approximation in the room temperature calibrated volume.

$$M_{He} = \left(\frac{P_r}{T_r} - \frac{P}{T_f}\right)\frac{V_r M}{R}$$

where P and T_f are the equilibrium pressure and temperature in the calibrated volume after expansion .

Results are to be expressed in terms of true gas storage capacity in the adsorbent without including the capacity effects of the intergranular voids and plumbing volume.

For the gas present in the connecting pipe (about 3 cm^3) we assume a linear temperature gradient and using an ideal gas law approximation, we obtain for the corresponding mass :

$$m_p = \frac{MPV_p}{R(T_r - T_e)} \ln \frac{T_r}{T_e}$$

where M is the molecular weight, P the pressure, V_p the volume of the connecting pipe, T_r the ambient temperature and T_e the temperature of the cell.

The total free volumes of the cells are determined by expanding helium gas at room temperature (adsorption is then negligible) from the calibrated volume into the cells. This total free volume (intergranular voids and pores of the charcoal) has been found to be 23.15 cm^3 in the case of pellets and 13.25 cm^3 in the case of compressed charcoal, corresponding respectively to 87.4 % and 72.5 % of the cell volume. The density of the pellets has been experimentally found to be 0.74 g/cm^3 (before outgasing), allowing in the case of pellets the determination of the intergranular void (13.4 cm^3) which represents 50.5 % of the cell volume. The remaining free volume (9.75 cm^3) accounts for the pores (macropores and micropores) of the charcoal pellets. These pores represent 74.4 % of the volume of a pellet which correspond to a bulk density for the charcoal of 2.5 g/cm^3 close to bulk carbon tabulated value. With these experimental informations we are able to determine the adsorption isotherms using a common definition[2] in which the amount of adsorbed gas is taken equal to the total amount of helium contained in the pore volume (macropore + micropore). The analysis of experimental data in such a way has been described elsewhere[3] and leads to a rather good correlation. However the lack of information on relative proportion between macropore and micropore volumes is quite unsatisfactory because it is known that macropore volume are mostly filled with compressed gas and can be considered to be equivalent to intergranular volume with regard to void volume even though micropores are mostly filled with adsorbed helium.

In the case of compressed activated charcoal we have measured an average density of 0.784 g/cm^3 (before outgasing) close to the pellets density. A tentative of analysis of the experimental data assuming no intergranular void volume, taking the compressed charcoal as a "maxipellet", turned out to be unsuccessful, leading to correlation parameters quite different from those obtained previously[3] with uncompressed pellets. A trial and error method to optimize the lnC versus RT/β lnP fit described in the next section, leads us to assume in the case of compressed charcoal a void volume (either intergranular or macropore) of 4.75 cm^3 and consequently 8.5 cm^3 for micropores. If we assume that compressing the charcoal has little effect on the micropores, identical values for the volume fraction of micropore over bulk charcoal can be assumed for both compressed and uncompressed charcoal. Experimental data are now analyzed considering as adsorbed the amount of helium contained only in the micropore. The isotherms are then determined using the following expression.

$$C = \frac{M_{He} - M_p - \rho V_v}{m_c}$$

where ρ is the density[4] of helium at P and T_e and V_v is the void volume (intergranular + macropores). The experimental results obtained by this way are shown in Fig. 1 and 2 for both compressed and uncompressed charcoal. Volume fractions are reported in Table 1.

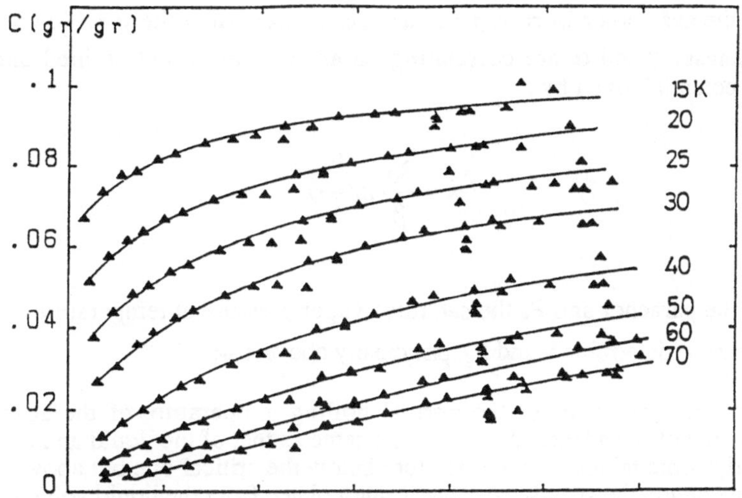

Fig.1. Helium isotherms for pellets

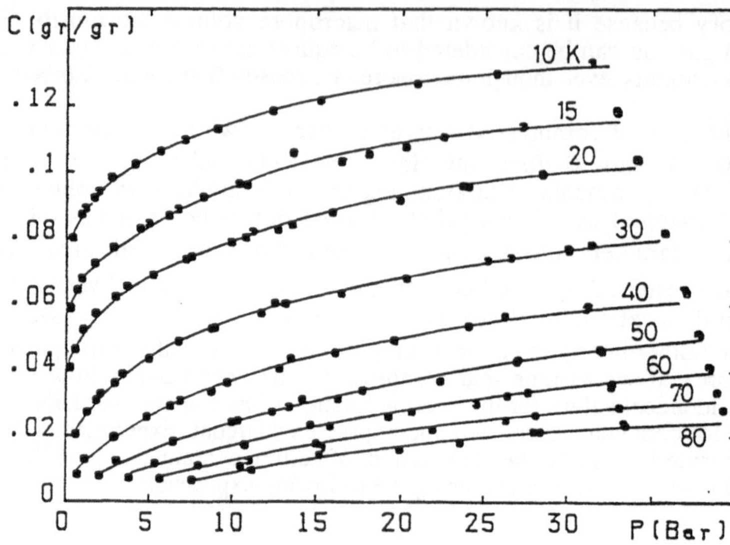

Fig.2. Helium isotherms for compressed charcoal

CORRELATION OF ISOTHERMS

The original theory proposed by Polanyi and later developped by various authors[5] postulates tnat the van der Waals potential of the solid is independent of temperature and is only a function of the distance from the adsorbent surface : the adsorption potential is a unique function of the amount of gas adsorbed. According to Dubinin[6] we used for correlation of isotherms an expression of the form :

$$C = \frac{M\alpha}{v} e^{-\gamma x} \tag{1}$$

where C is the gas mass adsorbed per unit mass of charcoal, v the specific volume of the adsorbed phase, γ and α are correlating parameters to be determined and x is the adsorption potential given by :

$$x = \frac{RT}{\beta} \ln(\frac{P_s}{P}) \tag{2}$$

where β is the parachor and P_S the saturated vapor pressure at temperature T. The main difficulties are to determine v and P_S, principally above T_{CR}.

For temperature under the normal boiling temperature of the adsorbate, the specific volume of the adsorbed phase is the same as that of the liquid and is taken from the thermodynamic tables. For temperature below the critical one but above the normal boiling temperature, the specific volume is interpolated from the liquid volume and above the critical temperature the van der Waals volume b is taken for v. For temperature below the critical temperature of the adsorbate, the vapor pressure is taken from the tables and

Table 1 - Properties of Charcoal

	Cell Volume (cm³) Vc	Mass of dry charcoal (g)	Intergranular void volume cm3/ % of Vc	Macropore volume cm3/ % of Vc	Micropore volume cm3/ % of Vc	Volume fraction of solid charcoal % Vc
Pellets	26.5	8.34	13.4/50.5 %	9.75/36.8 %		12.6 %
				13.2 %**	23.6 %**	
Compressed	18.25	12.15	4.75*26 %		8.5/46.66 %	27.3 %

* Assumed from data analysis
** Calculated assuming no effect of compression on micropores

above a fictitious vapor pressure has to be defined. For helium gas above the critical temperature some authors[2,5,7] took a fictitious vapor pressure $P_s = P_{cr}(T/T_{cr})^2$ where T_{cr} and P_{cr} are respectively the critical temperature and pressure of the gas.

As other authors[5,8], we found this method unsatisfactory (see for example the correlation obtained by this way on Fig. 3 : The dispersion remains quite wide and a systematic shift of the fit with temperature is clearly evident. Moreover it leads to negative values for the adsorption potential x.

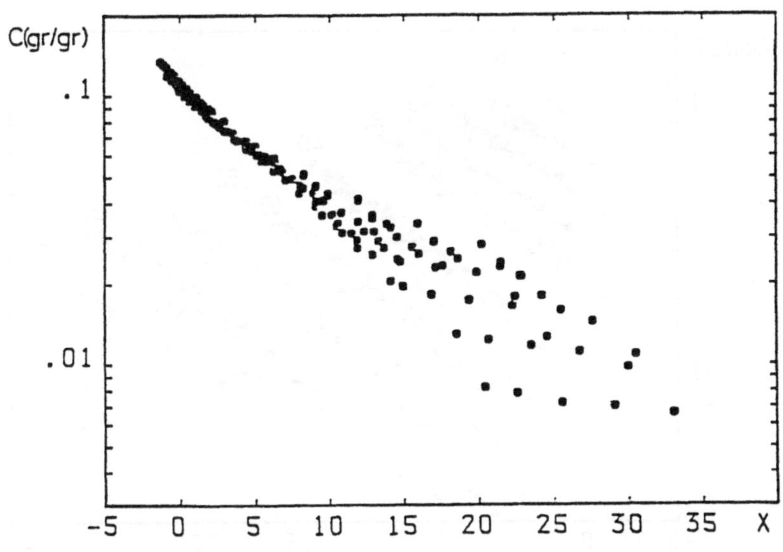

Fig.3. Correlation (LnC versus x) with $Ps=Pcr(T/Tcr)^2$ for compressed charcoal

Table 2

M(g/mole)	b(cm^3/mole)	β(J/mole)	α	γ	A	B
4.003	24	85.27	0.73	0.11	5.8	–26

We used another procedure. The equations (1) and (2) are developed as follows :

$$\ln C = \ln \frac{M\alpha}{b} - \frac{\gamma RT}{\beta} \ln P_s + \frac{\gamma RT}{\beta} \ln P \qquad (3)$$

A plot of lnC versus RT/β lnP is reported in Fig. 4. It leads to a family of parallel lines from the slope of which we get an experimental determination of γ. It is worth noting that for both pellets and compressed charcoal, experimental data lead to the same value for γ. The values of lnC for RT/β lnP = 0 ($\ln C_0(T) = \ln M\alpha/b - \gamma RT/\beta \ln P_s$) for each experimental temperature and both pellet and compressed charcoal are plotted versus T, see Fig. 5. They follow a linear variation, which lead experimentally to assume a $\ln P_s = A+B/T$ law for the fictitious vapor pressure over T_{cr}. Thus we determine experimentally the coefficient A. To get B and consequently α, we equalize at T_{cr} both real[9] and fictitious vapor pressure. The values of different parameters (M,b,β) used for helium are reported in Table 2, as well as the correlating parameters obtained by the way described before.

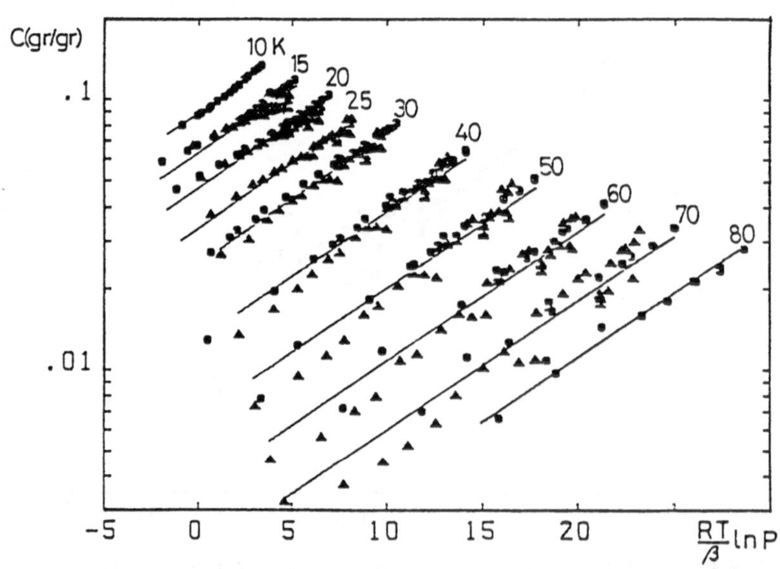

Fig.4. LnC versus (RT/β)LnP▲:pellets;●:compressed charcoal

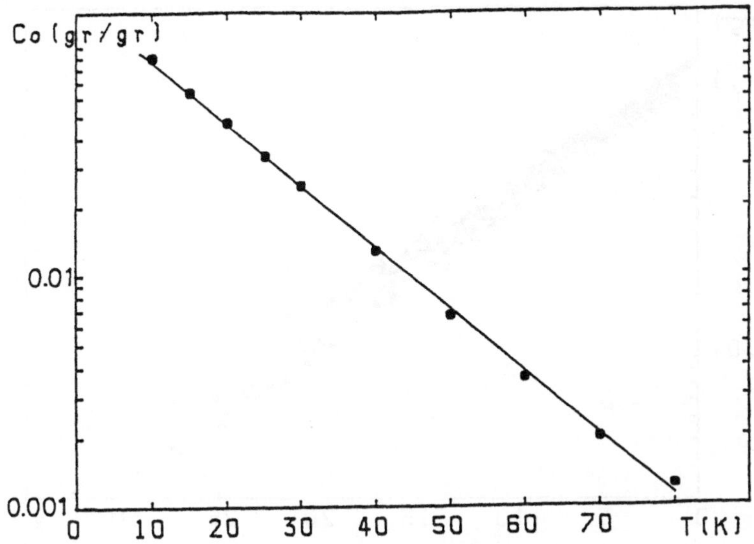

Fig.5.LnC0(T) versus T

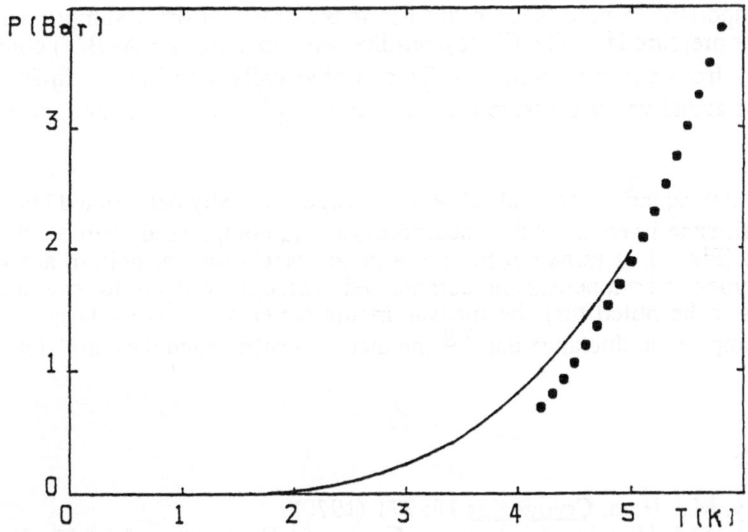

Fig.6. Saturation vapor pressure for helium.—: NBS tabulated values (Ref 9)
: fictitious values obtained from the analysis describe in the text

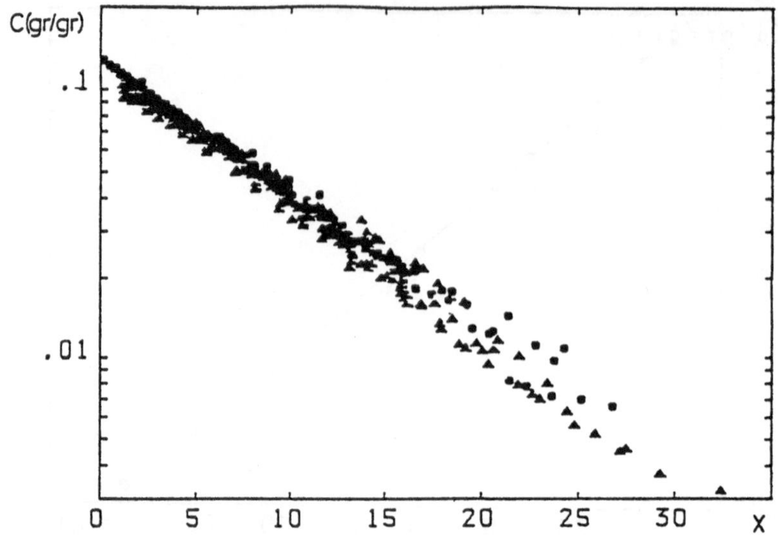

Fig.7. Generalised correlation with fictitious vapor pressure proposed in this work. ▲:pellets;●:compressed charcoal

It is important to note that we did not make any a priori assumption for the fictitious vapor pressure law. The Clapeyron-like behaviour ($\ln P_s = A+B/T$) comes out experimentally from our measurements. This is physically satisfactory. In Fig. 6 we report both the actual vapor pressure for ^{4}He below T_{cr}[9] and the fictitious one above T_{cr}.

Using both equations (1) and (2) with the experimentally determined law for P_s, we made for all experimental results uncompressed and compressed charcoal the fit of logC versus x (Fig. 7). It turns out to give a good correlation for helium adsorption isotherms common to both pellets and compressed charcoal (compare for example with Fig. 3). It has to be noted that the measurements cover a relatively large range of temperature compared to litterature data[7,8] and that the scatter round the correlating line is rather low.

REFERENCES

1. A.J. Kidnay, M.J. Hiza, Cryogenics **10**: 271 (1970).
2. A.J. Kidnay, M.J. Hiza in "Advances in Cryogenic Engineering,"vol **12**, Plenum , New York (1966), p.730
3. L. Duband, A. Ravex, J. Chaussy, Cryogenics **27**: 397 (1987).
4. D. Robert D. McCarty, NBS–Technical Note - 631, Nov. 1972.
5. W.H. Cook, Basmadjian, Canadian J. of Chem.Eng. **42**: 146 (1964).
6. M.M. Dubinin, Chem. Rev. **60** : 235 (1960).
7. C.K. Chan, E. Tward, K.I. Boudaie, Cryogenics **24** : 451 (1984)
8. L.C. Yang, T.D. Vo, H.H. Buriss, Cryogenics **22**: 625 (1982).
9. NBS Monograph **10** (1960).

TEMPERATURE DEPENDENCE OF THE COHESION PARAMETER FOR CALCULATING BINARY VLE VALUES FOR SYSTEMS CONTAINING HELIUM AND NEON

Y. Adachi and H. Sugie

Center of Information Processing Education
Nagoya Institute of Technology
Nagoya, Japan

B.C.-Y. Lu

Department of Chemical Engineering
University of Ottawa
Ottawa, Ontario, Canada

ABSTRACT

The temperature functions for representation of the cohesion parameter "a" of the van der Waals type cubic equations of state have been analyzed in both the subcritical and supercritical regions. The most suitable values of Ω_{ac} $(=a_cP_c/R^2T_c^2)$ for representation of vapour pressures of alkanes are found to be 0.4484 and 0.4254 for the Soave and the exponential functions, respectively. These results indicate that the Soave function is better suited for the Peng-Robinson equation; and the exponential functions, for the van der Waals, Redlich-Kwong and Soave-Redlich-Kwong equations. In the vapor-liquid equilibrium (VLE) calculations for systems containing helium and neon, which have very low critical temperatures, binary data have been used to determine the optimal value of $\Omega_a(=aP_c/R^2T_c^2)$ in the supercritical region. The temperature functions developed for these two gases have been successfully applied to VLE calculations for eight helium-containing, three neon-containing and helium-neon systems.

INTRODUCTION

This investigation concerns the calculation of binary vapor-liquid equilibrium (VLE) values for mixtures, containing a nonpolar compound and a quantum gas (helium or neon), using a simple cubic equation of state (EOS).

One of the important factors for VLE calculations is the capability of the selected EOS in the reproduction of vapor pressures of the components of the concerned mixture. The quality of this reproduction is controlled by the cohesion parameter $a(=\Omega_a R^2T_c^2/P_c)$. When Ω_a is obtained from an exact fit of vapor pressure, its value varies considerably with temperature and increases with the decrease of temperature. Soave[1] observed a linear relationship between $(\Omega_a/\Omega_{ac})^{\frac{1}{2}}$ and $T_r^{\frac{1}{2}}$ in his modification of the Redlich-Kwong (RK)[2] EOS (SRK). The soave form temperature function has been adopted in many EOS subsequently proposed in the literature, such as the Peng-Robinson (PR) equation[3], the Schmidt-Wenzel (SW) equation[4], and the Adachi-Lu-Sugie (ALS) equation[5]. However, the quality of the calculated vapor-pressures varies from equation to equation, indicating that the difference in the Ω_{ac} values of these equations may play a role. The quantity Ω_{ac} is the value of Ω_a at the critical point, and it varies with the equations. For example, Ω_{ac}=0.457236 for the PR equation and Ω_{ac}=0.42748 for the SRK equation. In an earlier study[6,7] on the modification of the van der Waals (VDW) equation, the temperature dependence of the parameter "a" was found to be better represented by an exponential function than the Soave type. On the other hand, the Soave function was found to be more suitable than the exponential function for the PR equation. In order to gain insight of these observations, the temperature dependence of Ω_a has been further analyzed, using the VDW type cubic equations expressed in the form suggested by Schmidt and Wenzel[4],

$$P = RT / (V - b) - a(T) / (V^2 + ubV + wb^2) . \qquad (1)$$

The Ω_a values in the subcritical region can be easily obtained from an exact fit to vapor pressures and the temperature function developed for representing these values is generally used to obtain the Ω_a values in the supercritical region by extrapolation. Although this technique does not present problems in the VLE calculations involving normal fluids, it may not be applicable to systems containing quantum gases due to their low critical temperatures. Graboski and Daubert[8], Elliott and Daubert[9], and Yu and Lu[10] determined specific temperature functions for hydrogen. It is intended in this work to obtain suitable temperature functions for representation of Ω_a values for helium and neon in the supercritical region.

TEMPERATURE FUNCTION FOR Ω_a IN SUBCRITICAL REGION

In the determination of Ω_a from pure-component vapor pressure P^{sat}, the controlling quantity is Ω_{ac}[11]. The calculation procedure has been presented earlier[12].

In order to establish the influence of Ω_{ac} on the selection of temperature functions for representation of Ω_a values, methane and n-decane were selected as typical representations of normal fluids. A total of six Ω_{ac} values were chosen to illustrate its influence on $\alpha(=\Omega_a/\Omega_{ac})$. For consistency, P^{sat} values generated from the correlation of Gomez-Nieto and Thodos[13] were arbitrarily selected as the data base, and used to determine the α values by an exact fit. The temperature dependence of α for these two substances are shown in Fig. 1. For normal fluids, the slope of α vs. T_r curves usually becomes steeper with the increase of the acentric factor of ω. For methane (Fig. 1), α is less than unity for Ω_{ac} = 0.65; and for Ω_{ac} = 0.60, when T_r is higher than 0.7. Thus, problems may occur in the extrapolation procedure. Fortunately, the Ω_{ac} values of all known equations of state are smaller than 0.60.

To determine how accurately the Soave temperature function

$$\alpha^{\frac{1}{2}} = (\Omega_a / \Omega_{ac})^{\frac{1}{2}} = 1 + m(1 - T_r^{\frac{1}{2}}) \qquad (2)$$

can represent the temperature dependence of α and the optimal Ω_{ac} value associated with such a representation, values of $\alpha^{\frac{1}{2}}$ were plotted against $T_r^{\frac{1}{2}}$ for methane in Fig. 2. Equation 2 is represented by a straight line

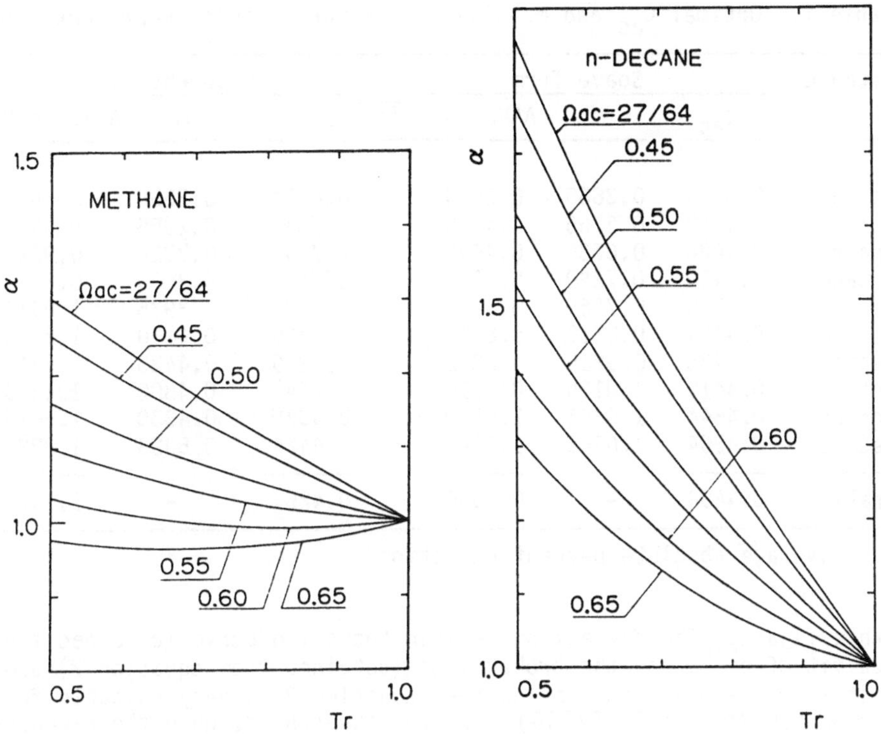

Fig. 1. Temperature dependence of α at several Ω_{ac} values for methane and n-decane.

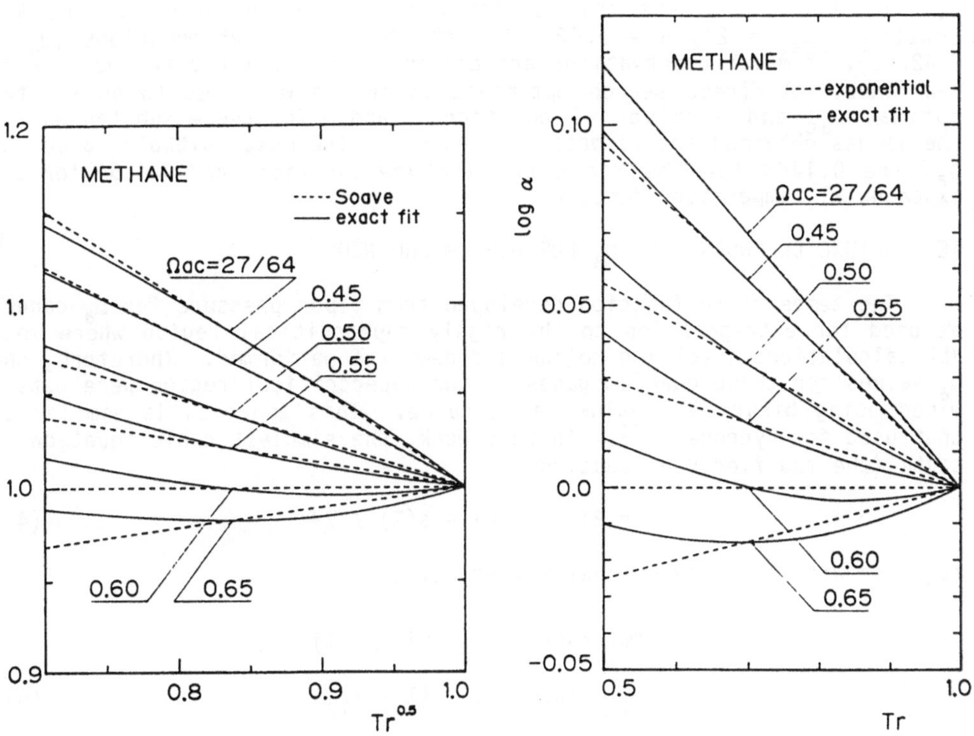

Fig. 2. Plots of Soave and exponential functions for methane.

Table 1. Optimal Ω_{ac} and m Values for Vapor Pressure Representation

Substance	Soave form			Exponential form		
	Ω_{ac}	m	AAPD in P^{sat}	Ω_{ac}	m	AAPD in P^{sat}
methane	0.4589	0.3665	0.3074	0.4224	0.2330	0.5517
ethane	0.4419	0.5765	0.4158	27/64	0.2955	0.8390
propane	0.4424	0.6580	0.4640	27/64	0.3320	0.9280
n-butane	0.4419	0.7290	0.4335	27/64	0.3620	0.8933
n-pentane	0.4528	0.7665	0.6384	0.4224	0.3935	1.0411
n-hexane	0.4484	0.8510	0.8231	0.4269	0.4160	1.2435
n-heptane	0.4484	0.9215	0.8912	0.4279	0.4435	1.3248
n-octane	0.4419	1.0125	0.7216	27/64	0.4800	1.1589
n-nonane	0.4544	1.0355	1.0500	0.4379	0.4830	1.5032
n-decane	0.4564	1.0945	1.3629	0.4424	0.5023	1.8226
overall	0.4484	-	0.7680	0.4254	-	1.1923

AAPD = average absolute percent deviation

in this figure. The deviations between the solid curve (obtained from an exact fit of P^{sat}) and the broken line (obtained from equation 2) are the lowest at Ω_{ac} = 0.45, indicating that equation 2 is very suitable for the PR equation (Ω_{ac} = 0.457236). On the other hand, when the exponential temperature function proposed by Adachi and Lu[6]

$$\log_{10} \alpha = \log_{10} (\Omega_a / \Omega_{ac}) = m (1 - T_r) \qquad (3)$$

was also considered for methane by plotting log α vs. T_r as shown in Fig. 2, the results indicate that equation 3 is very suitable for the VDW equation (Ω_{ac} = 27/64 = 0.421875), the RK and the SRK equations (Ω_{ac} = 0.42748). Similar observations are obtained for the first ten members of n-alkanes. A direct search optimization method was used to obtain the optimal Ω_{ac} and m values of equations 2 and 3 for these substances and the values obtained are reported in Table 1. The most suitable values of Ω_{ac} are 0.4484 for the Soave temperature function, and 0.4254 for the exponential temperature function.

TEMPERATURE DEPENDENCE OF Ω_a FOR HELIUM AND NEON

The temperature function developed from vapor pressure for Ω_a cannot be used for extrapolation to the highly supercritical region where most VLE calculations involving helium and neon are performed. Therefore, the Ω_a values for these quantum gases in the supercritical region were determined using binary VLE values as a guide. This approach is similar to that used for hydrogen[8-10]. In this work, the simplest cubic equation of state (the modified VDW equation[6])

$$P = RT / (V-b) - a(T) / V^2 \qquad (4)$$

together with the conventional mixing rules

$$a_{mixture} = \sum_i \sum_j x_i x_j a_{ij} \qquad (5)$$

$$a_{ij} = (a_{ii} a_{jj})^{\frac{1}{2}} (1 - k_{ij}) \qquad (6)$$

$$b_{mixture} = \sum_i b_i x_i \qquad (7)$$

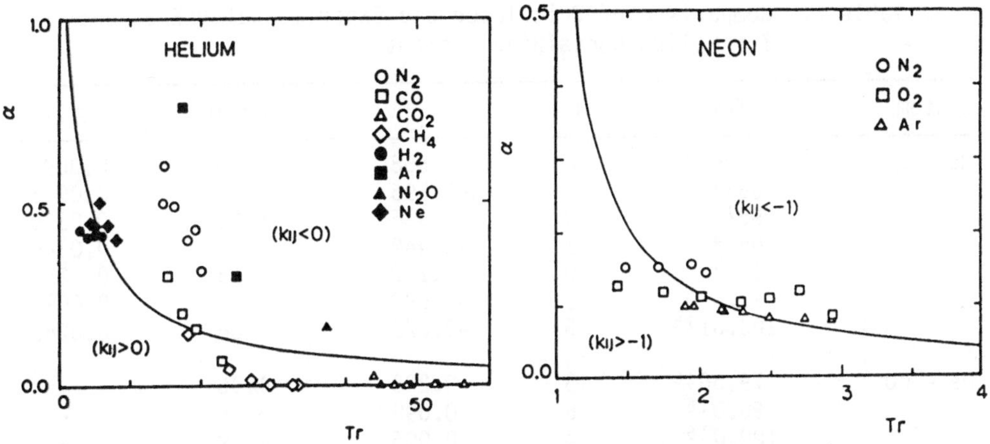

Fig. 3. Locations of correlated optimal α values shown by solid lines.

were used. The Ω_a values were determined directly by minimizing the pressure deviation in bubble-point calculations. These values, presented in terms of α, are plotted against T_r in Fig. 3. Setting the k_{ij} values to be 0 and -1 for helium-containing and neon-containing systems, respectively, the following temperature functions were obtained:

$$\alpha_{He} = (\Omega_a / \Omega_{ac})_{He} = 3.353 / (T_r + 2.353) \tag{8}$$

$$\alpha_{Ne} = (\Omega_a / \Omega_{ac})_{Ne} = 0.1324 / (T_r - 0.8676) \tag{9}$$

The α vs T_r curves represented by these equation are shown in Fig. 3 by the solid lines. The distribution of the points in this figure indicates the validity of equations 8 and 9.

RESULTS OF VLE CALCULATIONS

The temperature functions proposed above are used for calculating binary VLE values for systems containing helium and neon. The results obtained for nine helium-containing binary systems at a total of 42 isotherms are presented in Table 2, including the results obtained for the helium-neon system. The results obtained for three neon-containing binary systems at a total of 15 isotherms are presented in Table 3. The results reported in Tables 2 and 3 are generally satisfactory.

CONCLUSIONS

For a simple VDW type cubic equation, the temperature dependence of Ω_a determined from the fitting of vapor pressures can be classified by the value of Ω_{ac}. In other words, each Ω_{ac} value dictates a specific trend of temperature dependence of Ω_a.

The Soave temperature function is good for equations having Ω_{ac}= 0.4484 (such as the PR equation) and the exponential function is suitable

Table 2. Comparison of Calculated and Experimental VLE Values for Helium Containing Systems

System	T(K)	N	k_{ij}	AAPD in P	$\lvert \Delta y_1 \rvert_{av}$
He - N_2	76.5[14]	28	-0.558	8.99	0.0096
	85[14]	28	-0.596	6.78	0.0096
	95[14]	28	-0.536	5.65	0.0225
	105[14]	31	-0.442	7.27	0.0346
	115[14]	30	0.172	9.65	0.0252
	77.6[15]	9	-0.690	4.46	0.0078
	100.61[15]	5	-0.622	7.06	0.0268
He - CO	79.50[16]	8	-0.238	4.95	-
	90.0[16]	5	-0.098	3.29	-
	100.0[16]	4	-0.006	4.19	-
	120.0[16]	4	0.292	5.66	-
He - CH_4	124.00[17]	5	0.418	4.10	0.0006
	154.00[17]	5	0.856	4.82	0.0072
	174.00[17]	7	1.152	7.32	0.0252
	94.97[18]	8	0.050	5.58	0.0062
	124.85[18]	9	0.402	3.00	0.0081
	139.83[18]	6	0.602	2.35	0.0034
He - C_3H_8	323.15[19]	7	6.326	6.11	0.0373
	298.15[19]	8	5.506	4.59	0.0164
	273.15[19]	8	4.430	1.33	0.0087
	248.15[19]	8	3.812	2.77	0.0065
	223.15[19]	7	3.292	1.86	0.0044
	198.15[19]	7	2.850	1.24	0.0022
	173.15[19]	8	2.318	0.69	0.0004
He - CO_2	229.9[20]	5	0.420	6.79	0.0191
	244.9[20]	4	0.736	1.60	0.0076
	274.9[20]	7	0.994	5.15	0.0224
	253.11[21]	10	1.918	5.44	0.0329
	273.26[21]	12	1.468	5.82	0.0278
	293.13[21]	8	4.330	3.13	0.0337
He - Ne	24.71[22]	6	-0.948	3.30	0.0126
	26.00[22]	8	-0.962	6.13	0.0268
	27.03[22]	8	-0.984	6.57	0.0325
He - H_2	15.50[23]	8	0.162	6.23	0.0012
	20.40[23]	12	0.114	5.42	0.0312
	26.00[23]	9	0.044	4.02	0.0358
He - N_2O	195.0[16]	4	-0.386	1.83	-
	235.0[16]	5	1.014	4.51	-
	255.0[16]	19	1.212	2.15	-
	285.0[16]	8	2.216	3.09	-
He - Ar	91.34[24]	8	-1.060	2.26	0.0068
	130.08[24]	5	-0.526	12.8	0.0183

N = Number of data points. AAPD = Average absolute percent deviation.

Table 3. Comparison of Calculated and Experimental VLE Values for Neon-Containing Systems

System	T(K)	N	k_{ij}	AAPD in P	$\lvert \Delta y_1 \rvert_{av}$
Ne - Ar	84.42[25]	10	-0.782	2.02	0.0073
	87.42[25]	9	-0.828	1.92	0.0037
	95.82[25]	9	-0.906	2.28	0.0148
	101.94[25]	10	-0.984	1.39	0.0080
	110.78[25]	8	-1.052	1.15	0.0076
	121.36[25]	6	-1.120	2.40	0.0158
	129.93[25]	6	-1.196	2.11	0.0127
Ne - O_2	63.35[26]	10	-0.474	4.48	0.0006
	77.69[26]	15	-0.782	11.9	0.0126
	89.44[26]	16	-0.954	6.33	0.0193
	101.46[26]	15	-1.100	2.71	0.0093
	110.39[26]	12	-1.276	7.20	0.0209
	120.03[26]	12	-1.236	2.67	0.0112
	130.00[26]	8	-1.282	3.22	0.0185
Ne - N_2	66.13[27]	5	-0.714	7.61	-

N = Number of data points. AAPD = Average absolute percent deviation.

for equations having Ω_{ac} = 0.4254 (such as the VDW, RK and SRK equations).

For quantum gases helium and neon, it is necessary to develop the temperature function for Ω_a in the supercritical region with the aid of mixture properties. The developed expressions are suitable for VLE calculations.

ACKNOWLEDGEMENT

The authors are indebted to the Natural Sciences and Engineering Research Council of Canada for financial support.

REFERENCES

1. G. Soave, Equilibrium constants from a modified Redlich-Kwong equation of state, Chem. Eng. Sci. 27:1197 (1972).
2. O. Redlich and J.N.S. Kwong, On the thermodynamics of solutions. V. an equation of state. Fugacities of gaseous solutions, Chem. Rev. 44:233 (1949).
3. D.Y. Peng and D.B. Robinson, A new two-constant equation of state, Ind. Eng. Chem. Fundam. 15:59 (1976).
4. G. Schmidt and H. Wenzel, A modified van der Waals type equation of state, Chem. Eng. Sci. 35:1503 (1980).
5. Y. Adachi, B.C.-Y. Lu and H. Sugie, A four-parameter equation of state, Fluid Phase Equilibria 11:29 (1984).
6. Y. Adachi and B.C.-Y. Lu, Simplest equation of state for vapor-liquid equilibrium calculations: A modification of the van der Waals equation, AIChE Journal 30:991 (1984).

7. Y. Adachi and B.C.-Y. Lu, Binary parameter values for VLE calculations, in "Proc. CHEMPOR'85, 4th Intl. Chem. Eng. Conf." Coimbra, Portugal (1985), p. 06/1.
8. M.S. Graboski and T.E. Daubert, A modified Soave equation of state for phase equilibrium calculations. 3. Systems containing hydrogen, Ind. Eng. Chem. Process Des. Dev. 18:300 (1979).
9. J.R. Elliott, Jr. and T.E. Daubert, Revised procedures for phase equilibrium calculations with the Soave equation of state, Ind. Eng. Chem. Process Des. Dev. 24:743 (1985).
10. J.-M. Yu and B.C.-Y. Lu, Prediction of VLE values for systems containing hydrogen using the van der Waals equation, in: "Advances in Cryogenic Engineering," Vol. 31, Plenum Press, New York (1986), p. 1181.
11. Y. Adachi and B.C.-Y. Lu, The controlling quantity in cubic equations of state for phase equilibrium calculations, Can. J. Chem. Eng. 63:497 (1985).
12. Y. Adachi, W.K. Chung and B.C.-Y. Lu, Effect of cohesion parameter on VLE calculations, in: "Advances in Cryogenic Engineering", Vol. 29, Plenum Press, New York (1984), p. 989.
13. M. Gomez-Nieto and G. Thodos, Generalized vapor pressure equation for nonpolar substances, Ind. Eng. Chem. Fundam. 17:45 (1978).
14. W.E. DeVaney, B.J. Dalton and J.C. Meeks, Jr., Vapor-liquid equilibria of the helium-nitrogen system, J. Chem. Eng. Data 8:473 (1963).
15. W.B. Streett, Gas-liquid and fluid-fluid phase separation in the system helium-nitrogen near the critical temperature of nitrogen, Chem. Eng. Progr., Symp. Ser. 63(81):37 (1967).
16. W.R. Parrish and W.G. Steward, Vapor-liquid equilibria data for helium-carbon monoxide and helium-nitrous oxide systems, J. Chem. Eng. Data 20:412 (1975).
17. H.L. Rhodes, W.E. DeVaney and P.C. Tully, Phase equilibria data for helium-methane in the vapor-liquid and fluid-fluid region, J. Chem. Eng. Data 16:19 (1971).
18. C.K. Heck and M.J. Hiza, Liquid-vapor equilibrium in the system helium-methane, AIChE Journal 13:593 (1967).
19. D.L. Schindler, G.W. Swift and F. Kurata, More low temperature V-L design data, Hydrocarbon Processing 45(11): 205 (1966).
20. R.F. MacKendrick, C.K. Heck and P.L. Barrick, Liquid-vapor equilibria of the helium-carbon dioxide system, J. Chem. Eng. Data 13:352 (1968).
21. D.W. Burfield, H.P. Richardson and R.A. Guereca, Vapor-liquid equilibria and dielectric constants for the helium-carbon dioxide system, AIChE Journal 16:97 (1970).
22. M. Knorn, Vapor-liquid equilibria of the neon-helium system, Cryogenics 7:177 (1967).
23. W.B. Streett, R.E. Sonntag and G.J. Van Wylen, Liquid-vapor equilibrium in the system normal hydrogen-helium, J. Phys. Chem. 40:1390 (1964).
24. W.B. Streett, Gas-liquid and fluid-fluid phase separation in the system helium + argon at high pressures, Trans. Faraday Soc. 65:696 (1969).
25. W.B. Streett, Liquid-vapor equilibrium in the system neon-argon, J. Phys. Chem. 42:500 (1965).
26. W.B. Streett and C.H. Jones, Liquid-vapor equilibrum in the system neon-oxygen from 63° to 152°K and at pressures to 5000 psi, in: "Advances in Cryogenic Engineering", Vol. 11, Plenum Press, New York (1966), p. 356.
27. W.B. Streett, Liquid-vapor equilibrium in the system neon-nitrogen, Cryogenics 5:27 (1965).

EVIDENCE OF UNRELIABILITY OF FACTORY ANALYSES

OF ARGON IMPURITY IN OXYGEN

Franco Pavese, Danilo Ferri and Domenico Giraudi

Sezione Termometrica
CNR - Istituto di Metrologia "G.Colonnetti" (IMGC)
Torino, Italy

ABSTRACT

Anomalous values of the triple-point temperature of oxygen samples certified to contain less than 20 ppm of total impurities have been often measured in standard laboratories, with discrepancies of as much as 2 mK - equivalent to an argon contamination as high as 200 ppmv - though reproducible within $\pm$ 0.2 mK. However, the correlation between these wrong values and contamination by argon had never been proven. This paper shows the first qualitative evidence of such a correlation, and also that the analytical data are still inconsistent, from a quantitative point of view, with respect to calorimetric data, by as much as 100 %.

INTRODUCTION

Since the earliest works in cryogenics, the triple point of oxygen is considered a very high-quality reference point for thermometer calibrations as oxygen of high purity is readily available and, when good adiabatic conditions are set, the whole melting of the solid samples takes place in a very narrow temperature range (0.1-0.3 mK).

Nevertheless, with the improvement of the experimental techniques in the last twenty years to an accuracy limit of temperature measurements of a tenth of a millikelvin, the limiting factor for accuracy became the gas purity.

In oxygen, mainly two impurities, argon and nitrogen, affect the triple-point temperature to such a level that 100 ppm of them is enough to change the temperature value by + (0.7-1.0) mK and - (2.0-2.4) mK respectively. Therefore, in order to limit the contribution of these impurities to the total uncertainty of the reference point to within $\pm$ 0.1 mK, their amount should be limited to less than 10 ppmv and 5 ppmv respectively, which is close to the specifications of most research-grade oxygens (99.998 % or better).

With a proper gas handling technique, to avoid contamination - essentially a nitrogen increase - during its transfer from the bottle to the thermal apparatus, everybody should be able in principle to obtain a triple-point temperature value correct within 0.1-0.2 mK. On the contrary, in many different laboratories triple-point temperature values too high by as much as 2 mK have been observed in high-accuracy experiments with oxygen samples of different commercial sources, though the associated impurity analyses never indicated any sizeable contamination.

This lack of understanding the reasons for this temperature error - no reason other than argon contamination is known to be able to raise the triple-point temperature - actually voids the use of the oxygen triple point as an accurate fixed point, as the procedure to comply with to obtain the right thermodynamic state becomes indefinite. Each realization must be "calibrated" against a certified one (or against a calibrated thermometer) to check the correctness of its value .

The aim of this paper is to show, for the first time, evidence of a possible reason for the triple point irreproducibility, and to suggest some possible remedies.

RESULTS AND DISCUSSION

Ancsin[1] has probably been the first in a 1970 paper to record a systematic difference between the triple-point temperature values of oxygen samples taken from different bottles (Table 1).

Table 1 also shows similar differences observed by Pavese[2] on different batches of samples, by Furukawa[3] and during an international intercomparison of realizations of triple points[4,5]. The authors are aware also of several other unpublished occurrences of anomalous values in different laboratories.

These differences are often of the order of few tenths of a millikelvin, so that confidence of their being real systematic errors increased only after the technique of permanently sealing the samples in cells was developed[2]. This because filling of cells is generally done by this technique under better control of the gas manipulation and the technique allows the comparison of different samples sealed at different times

Table 1. Anomalous values observed with oxygen at the triple point

Author	Ref.	Sample (cell N°)	Analysis	Ar (ppmv)	N_2 (ppmv)	ΔT^a (mK)	equivalent Ar (ppmv)
Ancsin (1970)	1	([b])	batch	0.54	5.2	+ 2.3	190
Pavese (1976-86)	2	1-3	batch	<10	8	0^c	--
		4	batch	--	4.4	+ 0.2_5	20
			specific	--	3.2		
			specific	1.2	7.6		
		6-8	specific	3.0	1.8	+ 0.4_5	40
Furukawa (1983)	3	CO-7	batch	<5	<3	+ 1.2	100
Sakurai Pavese (1978-84)	5 4	7801	not available			+ 1.8	150

[a] Difference to the correct value. [b] Open-cell experiment. [c] Reference.

(often years apart) and even in different laboratories, as many times as it is necessary to gain a sufficient confidence level on the observed differences, without involving anymore sample manipulation.

All the values in Table 1 are higher than the temperature value most commonly obtained with oxygen and thus considered the correct one for the triple-point state. The amount of argon required in oxygen to justify the temperature increase ranges from 20 to 300 ppmv, much higher than reported in all the available impurity analyses for these published data.

Therefore, a dilemma arises whether the temperature increase is caused by a physical reason other than argon contamination or the impurity analyses of argon in oxygen are possibly wrong by more than one order of magnitude.

On the other hand, it is not possibile to check purity by thermal analysis with sufficient accuracy, ie, to check if contamination is the reason for the temperature errors, as argon, should it be the reason for the increase, affects very little the melting range of the sample and any broadening could be better attributed to contamination by nitrogen, which, on the contrary, would lower the temperature value.

The correlation between the presence of argon in oxygen and the rise of its triple-point temperature value has been already firmly established by several authors, even for very diluted mixtures [6 7]. On this respect, Ancsin's reported value[6] + 10 μK/ppmv and Compton's value[7] + 15 μK/ppmv appear to be rounded values, the exact ones being + 11 μK/ppmv and + 13 μK/ppmv respectively.

The first step in this research has been to ascertain the ability of the manufacturers to correctly measure the amount of argon, when present in small amounts in oxygen. This was done by asking a few companies to prepare a mixture of 100 ppmv Ar in their research-grade-pure O_2, and then to certify it, with a specific analysis of the Ar content in the lecture bottle.

Two companies (indicated as A and B in Table 2, which contains the data) complied with the request so far: one also supplied a bottle of the pure oxygen from the same batch. Samples from these bottles were sealed in cells at IMGC and then measured with the standard procedure[2].

To compare results, the average value for the triple-point temperature obtained in the recent international intercomparison of fixed point by means of sealed cells has been used as a reference[4]. It agrees to + (0.01 ± 0.15) mK with the value defined by the reference cell IMGC N°1,

Table 2. Data on Ar/O_2 mixtures and results by thermal analysis

Manufacturer	Certified Ar (ppmv)	ΔT^a (mK)	Equivalent Ar contamination (ppmv)[b]
A	107 ± 10	+ 1.1 ± 0.1_5	92 ± 12
B	105 ± 3	+ 1.3 ± 0.1_5	108 ± 12

[a] Measured difference to the correct value. [b] Using the average value + (12 ± 1) μK/ppmv.

used in that intercomparison and in this work.

The pure oxygen of manufacturer A gave a triple point temperature whose value is coincident to the reference one within the reproducibility of IMGC measurements ($\pm$ 0.15 mK).

The mixture of manufacturer A gave a value (1.1 $\pm$ 0.15) mK higher than the former, that of manufacturer B a value (1.3 $\pm$ 0.15) mK higher. The two results are consistent within the combined temperature and composition uncertainties.

Then, manufacturer A was let to analyse the mixture of manufacturer B, toghether with samples taken from the bottles used in Pavese's batches of cells N°1-3 and N°6-8 and in Furukawa's cell CO-7. Results are reported in Table 3: they are consistent in that no argon was found in the reference batch (N°1-3) and that an amount qualitatively proportional to the observed over-temperatures was found in the other bottles, even when former analyses did not (Table 1).

However, the set of argon-impurity values is still not quantitatively consistent with the over-temperatures observed at the triple point: the values appear to be too high by a factor of about 2 with respect to the calorimetric data, but this assumption is not consistent with the data of Table 2.

CONCLUSIONS

For the first time, evidence has been found of a correlation between values of the triple-point temperature of oxygen higher than the correct one and argon contamination, in bottles of commercial oxygen, unreported by their original batch or specific certificates, when available.

This explains some of the peculiar cases formerly reported in the literature, and is probably an explanation also for all the others, as already suspected but never proven yet.

However, quite surprising are the very large errors which affect the analytical determination of the argon content in oxygen. Contrarily to nitrogen, which may be found in a lecture bottle due to poor bottle conditioning or to poor control during filling operation of the bottle from the batch - an occurrence that, too, is experienced far too often -

Table 3. Results of the analyses of manufacturer A

Bottle	Impurities Ar (ppmv)	Impurities N_2 (ppmv)	ΔT equivalent [a] (mK)
from manuf. B	174	3.3	+ 2.1
[b] used for cells 1-3 [2]	<5	14.1	<0.1
[b] used for cells 6-8 [2]	94	8.2	+ 1.1
[b] used by Furukawa [3]	266	8.0	+ 3.2

[a] Using the average value + (12 $\pm$ 1) μK/ppmv; [b] See original analysis in Table 1.

argon content cannot increase on handling or on storage. Therefore, it must be contained in the batch, which is not surprising, should the analytical means for its detection in production be affected by the same degree of "blindness" and unreliability shown in this paper.

The presence of argon in oxygen is probably harmless for most users, but the lack of confidence which results from the present work, on the capability to certify gases may have larger implications than suspected, until the reason for the present unreliability is well understood, and corrected.

This is particularly important with standard reference materials (SRM): the case of gaseous SRM is particularly delicate.

The technique of permanently sealing the gaseous samples, proved again itself in this occurrence, at least for stable molecules and mildly reactive gases, to be a powerful and reliable method for measuring some physical properties and for providing a reference device for non-volatile recording of properties of technical - and commercial - value [8 9].

As far as the triple point of oxygen as a reference temperature point is concerned, the present situation requires the exclusion from use, for absolute realizations of this fixed point, of oxygen obtained from distillation of air, because the argon content could be higher than expected, and required. Only considering commercial sources, oxygen obtained from hydrolysis of water (as in case of manufacturer A) should be preferred.

REFERENCES

1. J.Ancsin, The triple point of oxygen, Metrologia 6:53 (1970).
2. F.Pavese and D.Ferri, Ten years of research on sealed cells for phase transition of gases at IMGC, in: "Temperature, Its Measurement and Control in Science and Industry", J.F.Schooley ed., American Institute of Physics, New York (1982), p. 217.
3. G.T.Furukawa, The triple point of oxygen in sealed transportable cells, Report to the Comité Consultatif de Thermométrie, BIPM, Sèvres, Document CCT/84-31 (1984).
4. F.Pavese, J.Ancsin, D.N.Astrov, J.Bonhoure, G.Bonnier, G.T.Furukawa, R.C.Kemp, H.Maas, R.L.Rusby, H.Sakurai and Ling Shankang, An international intercomparison of fixed points by means of sealed cells in the range 13.81-90.686 K, Metrologia 20:127 (1984).
5. H.Sakurai and O.Tamura, Remeasurements of the triple point of oxygen: comparison between the NRLM sealed cell and an open cell, Report to the Comité Consultatif de Thermométrie, BIPM, Sèvres, Document CCT/84-30 (1984).
6. J.Ancsin, Dew points, boiling points and triple points of pure and impure oxygen, Metrologia 9:26 (1973).
7. J.P.Compton and S.D.Ward, Realization of the boiling points of oxygen, Metrologia 12:101 (1976).
8. F.Pavese, Gases as standard reference materials, in: "Proc. Int. Symposium on Production and Use of SRM", BAM , Berlin, Germany (1980), p. 472.
9. F.Pavese, Triple points of gases in sealed cells: primary temperature fixed point or reference standards ?, in: "Proc. INSYMET '86", Dom Techniky CSVTS, Bratislava (1986), p. 53.

A NEW SQUARE-ROOT-TYPE PSEUDO-CUBIC EQUATION OF STATE

Masahiro Kato and Toshihiro Kiuchi

Department of Industrial Chemistry
Faculty of Engineering, Nihon University
Koriyama, Fukushima, Japan

ABSTRACT

A three-parameter pseudo-cubic equation of state has been previously proposed by the authors. In the present investigation, two sets of a square-root-type pseudo-cubic equation of state are newly proposed. The present equation of state is a sixth power polynomial equation in volume, which may be solved in a manner similar to the conventional cubic equations. The present pseudo-cubic equation of state has in maximum three positive roots in volume. The three-parameter equation newly proposed is successfully applied to the critical isotherms of argon, carbon dioxide, and water. The present equation of state surely satisfies the critical point requirements. To apply the present equation of state for wide range of temperature, the temperature dependency of the one parameter is correlated with temperature and acentric factor. The applicability of the present equation of state is demonstrated for the saturated properties of pure substances including cryogenic fluids. Comparing with the conventional equations of state, the square-root-type pseudo-cubic equation of state gives superior results especially in the prediction of saturated liquid densities.

INTRODUCTION

In the engineering applications, the cubic equations of state are widely used because of their simplicity. Some cubic equations and their modifications have been proposed by many investigators[1-16]. The values of the critical compressibility factor are given as 0.333 and 0.307 for Redlich-Kwong[1,2] and Peng-Robinson[3] equations, respectively. The actual values of the critical compressibility factor are, however, between 0.26 and 0.28 for popular hydrocarbons. Schmidt-Wenzel[4] discussed in detail the cubic equations of state and finally proposed their equation of state. Abbott[16] and Michels-Meijer[17] have pointed out the limited flexibilities on the cubic equations of state. The authors[19-21] have recently proposed a three-parameter pseudo-cubic equation of state and previously applied for the critical isotherms of pure components, saturated properties, and vapor-liquid equilibria of mixtures. In the present investigation, two sets of square-root-type pseudo-cubic equation of state are newly proposed to improve the accuracy for the citical isotherms and saturated liquid volumes of pure components.

SQUARE-ROOT-TYPE PSEUDO-CUBIC EQUATION OF STATE

A square-root-type pseudo-cubic equation of state is newly proposed as follows:

$$P = \frac{R\,T}{V^* - b} - \frac{a}{V^{*2}} \tag{1}$$

where,

$$V^* = \sqrt{(V - c_1)(V + c_2)} \tag{2}$$

$$a = \Omega_a \frac{(R\,Tc)^2}{Pc}, \qquad b = \Omega_b \frac{R\,Tc}{Pc} \tag{3}$$

$$c_1 = \Omega_{c_1} \frac{R\,Tc}{Pc}, \qquad c_2 = \Omega_{c_2} \frac{R\,Tc}{Pc} \tag{4}$$

$$\Omega_a = 27/64, \qquad \Omega_b = 1/8 \tag{5}$$

in which, P, R, T, and V*, denote the pressure, gas constant, temperature, and apparent volume, respectively. As a reference cubic equation, van der Waals equation is simply used as shown in Eq.(1) as shown in a previous paper.[19] For the apparent volume V*, the square-root-type is newly introduced as shown in Eq.(2). As shown in Eq.(5), the dimensionless parameters of a and b, Ω_a and Ω_b, are completely the same with those in van der Waals equation of state. Equation (1) is cubic in the apparent volume V*. Giving the temperature T and pressure P, the apparent volume V* may be analytically evaluated using the Cardano formula. Three real roots of V* are possible. The volume V is then obtained from the following quadratic Eq.(6).

$$V^2 - (c_1 - c_2)\cdot V - (V^{*2} + c_1\cdot c_2) = 0 \tag{6}$$

Giving c_1 and c_2 positive, one positive and one negative value of V is obtained from Eq.(2) for each value of V*. The negative value of V is meaningless, and the present equation of state has therefore a maximum of three positive values of V for a given state point.

The fugacity coefficient ϕ of a pure component for the proposed equation of state is given as follows:

$$\ln\phi = \ln\left(\frac{2\,V}{V + V^* + \delta}\right) + \frac{b}{\lambda}\ln\left(\frac{V + V^* - b + \delta + \lambda}{V + V^* - b + \delta - \lambda}\right) + (Z - 1) - \ln Z + \left(\frac{a/RT}{c_1 + c_2}\right)\ln\left(\frac{V - c_1}{V + c_2}\right) \tag{7}$$

where,

$$Z = \frac{P\,V}{R\,T}, \qquad V^* = \sqrt{(V - c_1)(V + c_2)} \tag{8}$$

$$\delta = \left(\frac{c_2 - c_1}{2}\right), \qquad \lambda = \sqrt{\left(\frac{c_1 + c_2}{2}\right)^2 + b^2} \tag{9}$$

Equations (1) and (2) have four unknown parameters, a, b, c_1, and c_2. These four parameters cannot be evaluated from the critical point data alone. To transform Eq.(1) into a three-parameter form, two sets of square-root-type pseudo-cubic equation are proposed.

Type A

Assuming the limiting value of $(V/Vc)_{P=\infty}$ as (1/3),[19] the dimensionless parameters, Ω_{c_1} and Ω_{c_2}, are given as follows:

$$\Omega_{c_1} = \frac{1}{8}\{ 2 (\theta - \theta^{-1}) + \sqrt{(\theta - \theta^{-1})(\theta - 4\theta^{-1})} \} \quad (10)$$

$$\Omega_{c_2} = \frac{1}{8}\{ 2 (\theta^{-1} - \theta) + \sqrt{(\theta - \theta^{-1})(\theta - 4\theta^{-1})} \} \quad (11)$$

where,

$$\theta = Zc/Zcs, \qquad Zcs = 3/8 \quad (12)$$

in which, Zc and Zcs denote the compressibility factor and the one of van der Waals equation. The actual value of the critical compressibility factor Zc is in general less than (3/8), which makes the value of θ in Eq.(12) positive and less than unity. The values of Ω_{c_1} and Ω_{c_2} are therefore positive which give one positive and one negative value of volume V for each value of apparent volume V*. The temperature dependence of "a" was introduced to satisfy the vapor pressures of n-alkanes, from methane to n-decane.[19]

$$a = K_a \cdot a_c \quad (13)$$

$$\ln K_a = m (1 - \sqrt{Tr}), \quad m = 1.1747 + 3.4583 \cdot \omega \quad (14)$$

where, ω and Tr represent the acentric factor and reduced temperature, respectively. The vapor pressures of n-alkanes were evaluated from Thodos.[23] The critical temperature Tc and critical pressure Pc were taken from Thodos[23] and the critical volumes and acentric factors are from Reid-Prausnitz-Sherwood.[24]

Type B

To satisfy the three critical point requirements strictly and the critical isotherms of pure components, argon, carbon dioxide, and water, as well as possible, the optimum values of the dimensionless parameter were obtained and correlated with the compressibility factor Zc as shown in Fig.1 and Eq.(15).

$$\Omega_{c_2} = 1.1709 - 2.8161 \cdot Zc \quad (15)$$

$$\Omega_{c_1} = Zc - \frac{Zcs^2}{Zc + \Omega_{c_2}}, \qquad Zcs = 3/8 \quad (16)$$

The actual compressibility factor Zc is surely between 0.160 and 0.416, which makes the values of c_1 and c_2 in Eq.(6) positive. Equation (6) therefore gives one positive and one negative value of the volume V for each value of the apparent volume V*. The temperature dependence of "a" was chosen similar to that of Type A:

$$\ln K_a = m (1 - \sqrt{Tr}), \quad m = 1.2294 + 3.3302 \cdot \omega \quad (17)$$

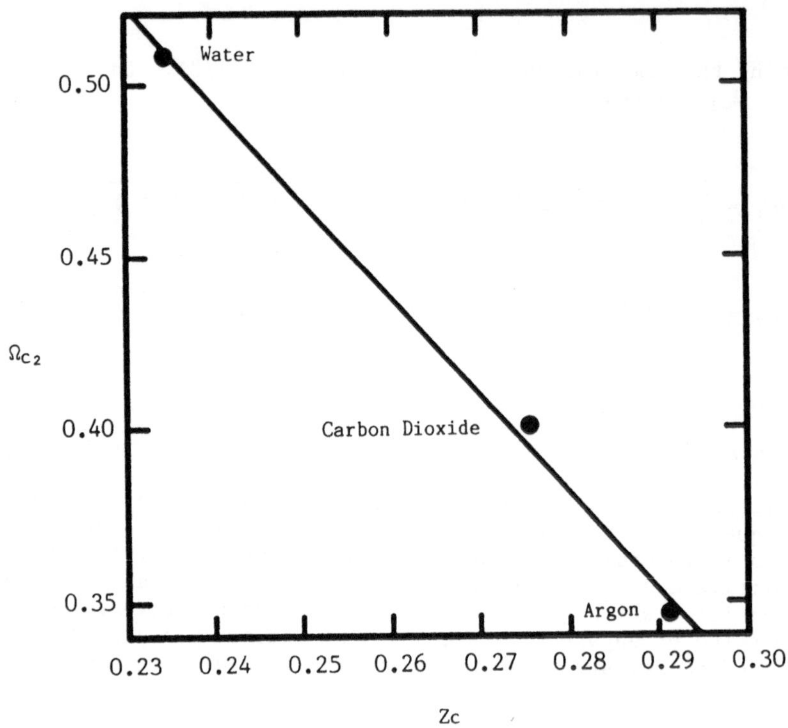

Fig. 1. Correlation diagram of Ω_{c_2} with Zc.

RESULTS

Table 1 gives a comparison between the square-root-type pseudo-cubic equation of state and several others as applied to the critical isotherms of argon, carbon dioxide, and water. In Table 1, the values in "Pure Cubic" represent the limited minimum ones for the universal form of the cubic equations strictly satisfying the three critical point requirements. The RMS(%) values in Table 1 are the deviation in pressure and the "†" on Zc† denotes actual compressibility factors. The pseudo-cubic equation of state strictly satisfies the critical point requirements. As shown in Table 1 in Type B of the present equation of state effectively improved the performance of the critical isotherms. The data for argon and carbon dioxide are those of Michels-Meijer.[17] For water, the interpolated data from the PVT data of JSME[22] shown in the previous paper[19] was used.

Table 2 shows a comparison of the proposed pseudo-cubic equation of state and several other equations of state for saturated properties of 19 pure substances as reported in the data book of Canjar-Manning.[25] The original pseudo-cubic equation of state[19] gave poor results in vapor pressures, comparing with the others in Table 2. The Type B of the present square-root-type equation of state effectively improved the accuracy on vapor pressures as shown in Table 2.

Figure 2 gives the PVT relations of ammonia, showing excellent results of the present square-root-type pseudo-cubic equation of state. Tables 3 to 5 give the calculation results of saturated properties for 12 polar components shown in the data books.[25,26] The present pseudo-cubic equation seems applicable for polar substances.

Table 1. Average Pressure Deviations on Critical Isotherms

Equation	No. of Parameters	Argon[17]		Carbon Dioxide[17]		Water[19,22]	
		RMS(%)	Zc	RMS(%)	Zc	RMS(%)	Zc
Ours-A	3	15.45	0.291†	15.29	0.276†	10.38	0.234†
Ours-B	3	6.28	0.291†	6.54	0.276†	6.10	0.234†
Pure Cubic	5	12.78	0.291†	10.91	0.276†	3.70	0.234†
van der Waals[7]	2	1807	0.375	2109	0.375	3131	0.375
Redlich-Kwong[1]	2	11.91	0.333	31.13	0.333	314.2	0.333
Peng-Robinson[3]	2	20.94	0.307	13.72	0.307	132.0	0.307
Schmidt-Wenzel[4]	3	12.05	0.334	12.57	0.315	128.5	0.307
Patel-Teja[5]	3	11.01	0.329	12.36	0.313	121.7	0.305
Harmens-Knapp[6]	3	13.26	0.321	14.80	0.305	94.34	0.298
Clausius[8]	3	34.53	0.291†	36.40	0.276†	21.93	0.234†
Himpan[9]	3	32.56	0.291†	34.24	0.276†	18.85	0.234†
HCEL[13]	3	11.33	-----	21.47	-----	246.4	-----
CCOR[15]	4	30.05	0.291	32.85	0.273	10.42	0.263
Kubic[18]	3	16.10	0.333	19.99	0.323	69.83	0.318
Ours-P[19]	3	17.23	0.291†	17.71	0.276†	4.79	0.234†

Ours-A:Type(A), Ours-B:Type(B), HCEL:Hamam-Chung-Elshayal-Lu[13]
CCOR:Cubic Chain-of-Rotators,[15] Ours-P:Previous[19]
†:Same with Experimental Values

$$\mathrm{RMS} = \sqrt{\Sigma\{(\mathrm{Pcalc} - \mathrm{Pexpl})/\mathrm{Pexpl}\}^2/\mathrm{Number\ of\ Data\ Points}}$$

Table 2. Accuracies on Saturated Properties of 19 Pure Substances Shown in Canjar-Manning[25] Data Book

Equation of State	$\lvert\Delta P/P\rvert$av,%	$\lvert\Delta V^v/V^v\rvert$av,%	$\lvert\Delta V^l/V^l\rvert$av,%
Ours-A	2.02	5.08	3.63
Ours-B	1.73	4.94	4.68
Soave-Redlich-Kwong[2]	1.62	3.22	14.19
Peng-Robinson[3]	1.13	2.85	7.42
Schmidt-Wenzel[4]	1.08	2.92	7.19
Patel-Teja[5]	1.38	2.77	6.76
Harmens-Knapp[6]	1.62	3.34	6.46
Hamam-Chung-Elshayal-Lu[13]	1.10	7.73	2.63
Cubic Chain-of-Rotators[15]	1.28	7.65	3.45
Kubic[18]	1.18	3.00	7.93
Ours-P[19]	2.03	4.60	3.96

Ours-A:Type(A), Ours-B:Type(B), Ours-P:Previous[19]
P:Vapor Pressure, V^v:Saturated Vapor Volume, V^l:Saturated Liquid Volume
Substances:Acetylene, Ammonia, Benzene, i-Butane, n-Butane, 1-Butene, Carbon Dioxide, Carbon Monoxide, Ethane, Ethylene, n-Hexane, Methane, Nitrogen, Oxygen, n-Pentane, Propane, Propylene, Sulfur Dioxide, Water

Table 3. Accuracies of Vapor Pressures for Polar Substances, $|\Delta P/P|_{av}$ %

Substance	Ours A	B	SRK	PR	SW	PT	HK	HCEL	CCOR	KUB	Ours P
Ammonia[25]	6.09	3.94	0.78	0.52	0.69	0.39	0.73	0.73	1.02	1.05	6.11
Ethanol[26]	2.94	2.58	2.14	1.95	2.88	0.95	2.55	1.34	2.37	0.60	2.80
Heavy Water[26]	12.12	5.89	6.24	2.99	4.96	3.69	4.03	5.73	0.49	3.97	11.35
Methanol[26]	4.34	2.30	2.86	2.44	3.41	0.94	1.82	2.29	0.56	2.23	4.34
R11[26]	2.08	0.94	1.01	1.77	0.53	1.59	2.36	1.91	1.23	4.62	1.90
R12[26]	0.83	0.77	0.88	2.71	1.53	2.52	3.43	1.14	1.04	4.71	0.82
R13[26]	3.07	2.41	1.89	1.40	1.81	1.13	0.86	3.18	2.58	3.26	3.04
R13b1[26]	5.01	2.99	0.84	1.13	0.58	1.07	1.73	1.34	1.27	2.73	4.99
R21[26]	2.21	2.61	1.55	2.46	1.86	2.38	3.24	0.75	1.32	2.67	2.18
R114[26]	4.00	2.42	1.15	1.07	0.51	0.84	1.24	1.68	2.01	2.30	3.97
Sulfur Dioxide[25]	0.97	1.28	2.15	1.74	1.62	1.69	2.09	0.97	1.36	0.68	1.02
Water[25]	6.93	7.07	8.33	4.93	7.25	5.61	5.79	8.68	2.48	4.67	6.94
Average	4.22	2.93	2.49	2.09	2.30	1.90	2.49	2.48	1.48	2.79	4.12

Ours-A:Type(A), Ours-B:Type(B), SRK:Soave-Redlich-Kwong[2], PR:Peng-Robinson[3]
SW:Schmidt-Wenzel[4], PT:Patel-Teja[5], HK:Harmens-Knapp[6]
HCEL:Hamam-Chung-Elshayal-Lu[13], CCOR:Cubic Chain-of-Rotators[15], KUB:Kubic[18]
Ours-P:Previous[19]

Table 4. Accuracies of Saturated Vapor Volumes for Polar Substances, $|\Delta V^V/V^V|_{av}$ %

Substance	Ours A	B	SRK	PR	SW	PT	HK	HCEL	CCOR	KUB	Ours P
Ammonia[25]	6.88	4.47	7.15	6.01	6.64	6.28	5.35	13.21	2.48	7.88	5.87
Ethanol[26]	4.92	4.71	4.44	2.87	3.37	3.62	5.59	16.03	5.55	6.05	4.39
Heavy Water[26]	13.65	8.96	10.31	5.79	8.46	7.00	7.20	16.05	3.43	7.68	12.11
Methanol[26]	4.86	3.38	9.30	6.89	5.45	8.37	10.21	18.36	3.91	8.03	3.80
R11[26]	4.95	3.92	1.88	2.57	1.21	2.20	3.15	7.88	6.67	4.48	4.55
R12[26]	2.53	2.60	1.23	3.06	1.84	2.72	3.74	5.04	4.62	4.60	2.31
R13[26]	4.73	3.92	3.77	2.79	3.39	2.57	2.21	9.43	6.51	4.33	4.57
R13b1[26]	5.88	3.38	1.57	1.63	1.66	1.31	1.52	5.45	4.40	3.29	5.90
R21[26]	4.58	4.98	1.61	2.90	2.36	2.60	3.53	5.19	6.24	3.24	4.18
R114[26]	5.80	4.22	1.99	2.05	1.39	1.82	2.40	5.07	5.97	2.70	5.62
Sulfur Dioxide[25]	4.65	5.16	2.39	1.70	1.57	1.60	2.23	7.53	8.47	1.53	4.05
Water[25]	9.89	11.74	14.03	8.32	12.20	9.48	9.42	22.32	5.66	8.57	9.06
Average	6.11	5.12	4.97	3.88	4.13	4.13	4.71	10.96	5.33	5.20	5.53

Ours-A:Type(A), Ours-B:Type(B), SRK:Soave-Redlich-Kwong[2], PR:Peng-Robinson[3]
SW:Schmidt-Wenzel[4], PT:Patel-Teja[5], HK:Harmens-Knapp[6]
HCEL:Hamam-Chung-Elshayal-Lu[13], CCOR:Cubic Chain-of-Rotators[15], KUB:Kubic[18]
Ours-P:Previous[19]

Table 5. Accuracies of Saturated Liquid Volumes for Polar Substances, $|\Delta V^l/V^l|_{av}$ %

Substance	Ours A	B	SRK	PR	SW	PT	HK	HCEL	CCOR	KUB	Ours P
Ammonia[25]	2.27	2.85	30.72	15.78	19.03	17.89	13.66	13.48	13.79	16.22	3.01
Ethanol[26]	4.39	5.33	24.18	10.23	5.21	5.26	5.77	5.47	4.53	12.39	5.68
Heavy Water[26]	3.21	6.07	41.48	25.54	24.19	23.33	19.28	19.20	19.11	25.39	3.22
Methanol[26]	2.70	9.75	38.75	23.16	14.63	14.32	11.66	10.08	9.28	9.33	4.10
R11[26]	2.68	4.88	11.06	6.26	4.70	5.01	6.25	2.08	2.31	7.85	2.81
R12[26]	2.40	5.47	9.05	5.76	2.95	3.50	5.48	1.54	1.56	5.76	2.57
R13[26]	2.60	2.96	10.13	5.88	4.31	4.58	5.74	2.30	2.31	6.46	3.14
R13b1[26]	4.33	10.57	8.97	5.08	2.37	2.64	4.69	1.14	1.35	4.33	3.72
R21[26]	2.27	3.41	12.55	4.72	4.38	3.69	4.91	0.22	1.15	5.67	2.61
R114[26]	2.15	3.36	9.42	5.55	3.98	4.59	6.87	3.73	3.82	6.66	2.60
Sulfur Dioxide[25]	1.86	2.99	17.48	5.07	7.07	6.09	5.12	0.90	1.72	6.91	2.13
Water[25]	5.94	1.63	40.44	24.81	24.26	23.38	19.29	17.76	18.54	28.51	4.70
Average	3.07	4.94	21.19	11.49	9.76	9.52	9.06	6.49	6.62	11.29	3.36

Ours-A:Type(A), Ours-B:Type(B), SRK:Soave-Redlich-Kwong[2], PR:Peng-Robinson[3]
SW:Schmidt-Wenzel[4], PT:Patel-Teja[5], HK:Harmens-Knapp[6]
HCEL:Hamam-Chung-Elshayal-Lu[13], CCOR:Cubic Chain-of-Rotators[15], KUB:Kubic[18]
Ours-P:Previous[19]

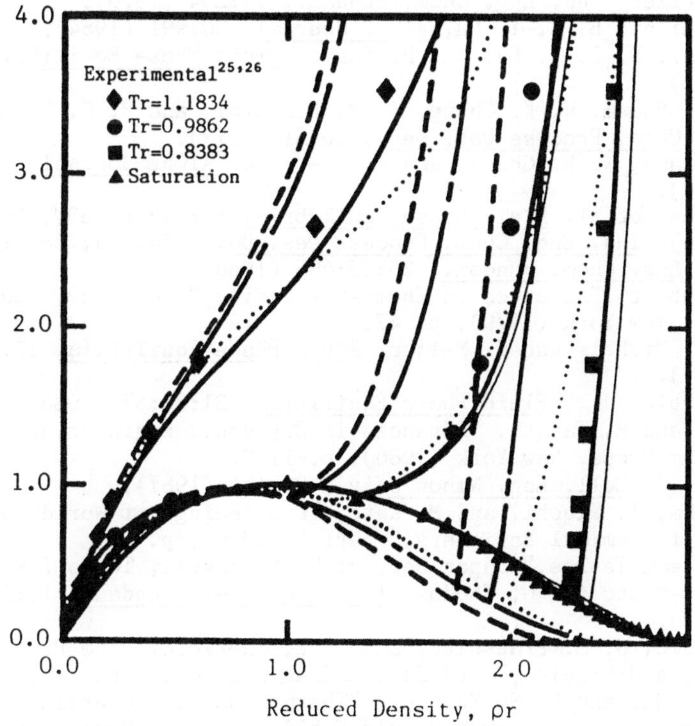

Fig. 2. Calculated and experimental isotherms for ammonia.
——:Ours-A, ——:Ours-B, - - - -:Soave-Redlich-Kwong
— — —:Schmidt-Wenzel, ...:Cubic Chain-of-Rotators

CONCLUSIONS

Two sets of the square-root-type pseudo-cubic equation of state have been proposed in the present investigation. The proposed equation of state can be solved like with the conventional cubic equations of state. The present three-parameter pseudo-cubic equation of state has been successfully applied for describing the critical isotherms of pure components. The temperature dependence of the parameter has been introduced. Consequently superior results have been obtained for the saturated liquid volumes of pure components. The proposed pseudo-cubic equation of state strictly satisfies the critical point requirements. Keeping the simplicity of the cubic equations of state, the accuracies were were effectively improved by use of the pseudo-cubic equation of state. The proposed pseudo-cubic equation of state seems considerably applicable for polar components, especially in the prediction of saturated liquid volumes.

REFERENCES

1. O. Redlich and J. N. S. Kwong, Chem. Rev., 44:233 (1949).
2. G. Soave, Chem. Eng. Sci., 27:1197 (1972).
3. D. Y. Peng and D. B. Robinson, Ind. Eng. Chem. Fundam., 15:59 (1976).
4. G. Schmidt and H. Wenzel, Chem. Eng. Sci., 35:1503 (1980).
5. N. C. Patel and A. S. Teja, Chem. Eng. Sci., 37:463 (1981).
6. A. Harmens and H. Knapp, Ind. Eng. Chem. Fundam., 19:291 (1980).
7. J. D. van der Waals "Over de Continuiteit van den gas-en Vloeistoftoestand," doctoral dissertation, Leiden, Holland (1973).
8. R. Clausius, Ann. Phys. u. Chem., 9:337 (1880).
9. J. Himpan, Z. Phys., 131:17 (1951).
10. G. G. Fuller, Ind. Eng. Chem. Fundam., 15:254 (1976).
11. Y. Adachi and B. C.-Y. Lu, AIChE Journal, 30:991 (1984).
12. Y. Adachi, B. C.-Y. Lu, and H. Sugie, Fluid Phase Equilibria, 11:29 (1983).
13. S. E. M. Hamam, W. K. Chung, I. M. Elshayal, and B. C.-Y. Lu, Ind. Eng. Chem. Process Des. Dev., 16:51 (1977).
14. T. Ishikawa, W. K. Chung, and B. C.-Y. Lu, AIChE Journal, 26:372 (1980).
15. K.-C. Chao et al, Fluid Phase Equilibria, 13:143 (1983), 29:475 (1986), Ind. Eng. Chem. Process Des. Dev., 24:764;768 (1985), Ind. Eng. Chem. Fundam., 25:75;695 (1986).
16. M. M. Abbott, "Advances in Chemistry Series," Vol. 182, Amer. Chem. Soc., New York (1979), p. 47.
17. M. A. J. Michels and H. Meijer, Fluid Phase Equiliblia, 17:57 (1984).
18. W. L. Kubic, Jr., Fluid Phase Equilibria, 31:35;57 (1986).
19. M. Kato and H. Tanaka, "Advances in Cryogenic Engineering," Vol.31, Plenum Press, New York (1986), p. 1169.
20. M. Kato, J. Coll. Eng. Nihon Univ., 28:181 (1987).
21. H. Tanaka, T. Kiuchi, and M. Kato, "Proceedings of World Congress III of Chemical Engineers," part 2 (1986), p. 48.
22. "JSME Steam Tables," Japan Soc. of Mech. Engrs., Tokyo (1980).
23. M. Gomez-Nieto and G. Thodos, Ind. Eng. Chem. Fundam., 16:254 (1977).
24. R. C. Reid, J. M. Prausnitz, and T. K. Sherwood, "The Properties of Gases and Liquids," 3rd Ed., McGraw-Hill Co., New York (1977).
25. L. N. Canjar and F. S. Manning, "Thermodynamic Properties and Reduced Correlations for Gases," Gulf Publishing Co., Houston (1967).
26. "Thermophysical Properties of Fluids," Japan Soc. of Mech. Engrs., Tokyo (1983).

REVIEW OF INSTRUMENTATION FOR SUPERCONDUCTING MAGNETS

J.A. Zichy

Swiss Institute for Nuclear Research (SIN)

Villigen, Switzerland

ABSTRACT

A survey of sensors commonly applied to monitor superconducting magnets is presented. An attempt is made to compile available data on the useful range, accuracy and reliability of these sensors. Next, the control philosophy in some representative superconducting magnet systems is described, including the instrumentation utilized. The representative examples are taken from the broad field of large magnets used in accelerators, fusion reactors and for energy storage. Merits of the different control systems applied are reviewed, including information on operational experience. Finally, desirable developments of magnet instrumentation and control philosophy are discussed.

INTRODUCTION

After fifty heroic years of research on superconductivity, the first high field laboratory magnet demonstrated 10-T field in 1963. The progress in cryogenics and superconducting (SC) magnet technology since then has developed rapidly. Starting from single-core wires of unpredictable performance it progressed to commercially available fully transposed, multistage, copper-stabilized high-current and high-field conductors with hundred thousands of filaments. New materials, NbTi and NbSn, were studied and the technology to fabricate thin wires of a few micrometers diameter is today well mastered. The drive to develop new conductors and to improve magnet engineering came from high energy physics in order to build new powerful accelerators and beam guiding magnets. The biggest success of SC-magnets is the medical application of nuclear Magnetic Resonance Imaging (MRI), routinely used in every major hospital throughout the world. The SC-magnets are either pool-boiling type, i.e., immersed in a helium bath at atmospheric pressure, or forced-flow cooled, utilizing pressurized supercritical helium circulating in the SC-cable. The helium temperature varies between 1.8 K (for superfluid helium) and 4.2 K.

INSTRUMENTATION

The instrumentation of SC-magnets has two purposes; first to control the temperature and strain during the cool-down and warm-up, second to adjust and control the operating conditions (current, field, cooling, strain, etc.) of the magnet.

Thermometry

The temperature sensors should yield data with good accuracy between 1.7 to 300 K in order to cover the whole operating range, including magnets cooled with superfluid helium. Additionally, their reading has to be at the operating temperature independent of the magnetic field. These requirements are too demanding for industrial grade sensors, therefore one usually installs sensors for "high temperatures", say for $T > 30$ K, and another set with overlapping range for the operating temperature.

Thermometers for Cooldown/warmup. Requiring only modest accuracy widens the range of commercially available sensors. If speed in response and small sensor size are critical, the Chromel/Constantan (Type E) thermocouple, recommended by NBS, is a good choice. More accurate are platinum resistance thermometers (PRT).[1] The recently developed Cryogenic Linear Temperature Sensor (CLTS) has good sensitivity, modest accuracy and a linear sensor response.[2] In Table 1 are listed the characteristic properties of these thermometers.

Thermometers for Operation. The sensitivity of the "low-temperature" thermometer to magnetic induction, its accuracy, stability with time and reproducibility after several thermal cycles are of equal importance in choosing a sensor. The $SrTiO_3$ capacitor has the smallest field dependent errors but it drifts with time and thermal cycling. However, it has the unique ability to work reliably in ac-fields. The carbon-glass resistance thermometer (CGRT) with appropriate signal conditioning is commercially available and it has a wider range than other sensors. Its magnetoresistance is orientation dependent, however, the error is also correctable for off-the-shelf sensors, [3,4] see Fig. 1. A disadvantage of this thermometer is that its reading may change several percent after thermal cycling.[1] The carbon resistance thermometer (CRT) is of small size and has a relatively small and reproducible magneto-resistance, although it needs frequent calibration and care in handling.[1,5] Germanium resistors thermometers (GeRT) are commonly used for high accuracy temperature measurements since they have an excellent reproducibility and high sensitivity. Their large orientation dependent magnetoresistance makes it inadvisable to use them in fields higher than 2.5 T.[6] Some properties of these sensors are listed in Table 2.

The combination of commercially available CLTSs with CGRTs provides the best choice of thermometers. Since thermocouples do not need any signal conditioning, they are a cost saving substitution for CLTSs.[7]

Helium Properties

Besides the temperature, the pressure and, for two-phase cooling, the quality factor are measured to characterize the helium. In the bath-cooled coil, liquid level sensors control the filling of the dewar. The forced- flow cooled magnets do as a rule require turbulent flow, hence He-mass flow measurement, in order to maintain a good heat transfer between coolant and conductor. The density, viscosity and other properties of the helium are calculated, from the measured temperature and pressure, utilizing the NBS tables of helium properties.[8]

Table 1. Characteristics of "high-temperature" thermometers

Sensor	Range (K)	Accuracy (K)	Stability (%)	Thermal cycling (%)
Type E	30–1000	1.0–3.0	<0.5	<1.0
PRT	20–>300	0.2–0.5	<0.1	<0.4
CLTS	2.4– 269	1.0–3.0	<0.1	<0.5

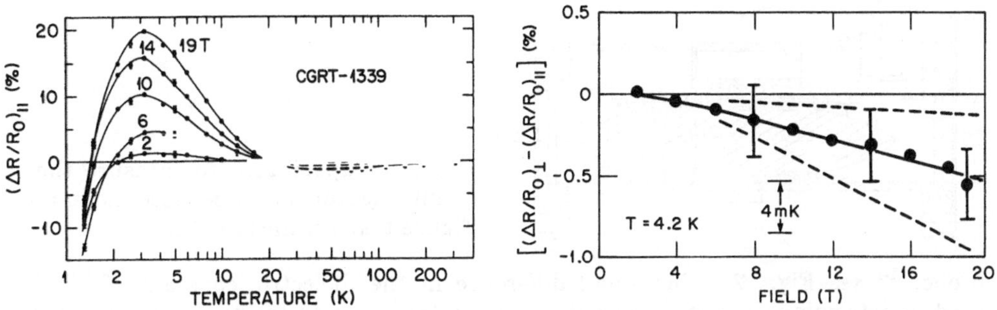

Fig. 1. Magnetoresistance of CGRTs. The graph on the left shows the longitudinal component vs temperature for selected magnetic fields. The graph on the right gives, for 4.2 K, the difference between longitudinal and transversal component vs magnetic field. Figure taken from Ref. 3.

Pressure. Both type of magnets do use pressure gauges, in the bath-cooled coils to indicate pressure excursions in the dewar and in the forced-flow cooled magnets also to evaluate the He-mass flow. The most common gauge, the pressure-to-current transducer, is at ambient temperature and connected by a pipe directly to the helium volume of the magnet. The length of the capillary pipe, made preferably from stainless steel with welded connections, does increase the response time and the risk of helium leaks. Transducers are available for the pressure range from 0 to 40 MPa and if the range is properly selected their accuracy is better than 5%. The transducer is practical since it can be recalibrated or exchanged while the magnet is at liquid helium (LHe) temperature. Different capacitor and strain gauge type cold transducers have been used successfully.[9,10] Their main advantage is a shorter response time and reduced risk of helium leakage since no connecting pipes are needed. Their disadvantage is a zero shift after thermal cycling. Unfortunately, the reading of cold transducers varies with the strength of the magnetic induction.[1]

Liquid Level. From the many types of level sensors only two continuous level indicators are mentioned. The capacitance gauge, consisting of two coaxial tubes immersed in the liquid, changes its capacitance as the liquid level rises. It is an accurate, stable and reliable gauge, however, the most expensive.[12] The resistive gauge utilizes a SC-wire arranged so that the resistance of the wire immersed in the liquid is proportional to the depth of liquid in the dewar. It is a relatively inexpensive device and gives good results. The operating range is limited by the critical temperature of the SC-wire used.[13] Both devices need a current source and their signal can be measured with an ac- or dc-bridge, respectively.

Quality Factor. For two-phase cooling, it is essential to monitor the ratio of evaporated helium to the two-phase helium, the so-called quality factor X. A continuous readout system consists of a concentric cylindrical capacitor, through which the total He-mass flow must pass and of a cold oscillator placed near to the

Table 2. Characteristics of "low-temperature" thermometers

Sensor	Range (K)	Accuracy at 4.2 K (K)	Stability (%)	Thermal cycling (%)	Effect of magnetic fields
$SrTiO_3$	2.2– 60	<0.01	<5.0	large	negligible
CGRT	1.5–300	<0.02	<1.0	<5	correctable
CRT	1.5– 30	<0.05	<1.0	large	correctable
GeRT	4.0–100	<0.01	<0.5	<1	large

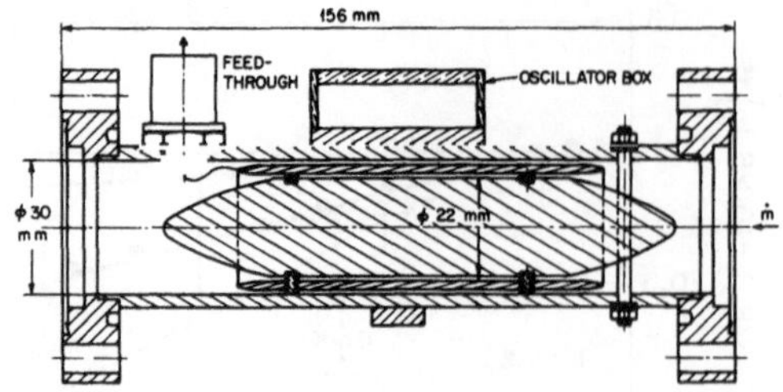

Fig. 2. Capacitance to measure the quality factor of two-phase helium. Figure taken from Ref. 14.

probe, [14] see Fig. 2. The small difference in the dielectric constant of 4% is used to determine the vapor content of the mixture by monitoring the frequency of the oscillator. This gauge measures the quality factor with 1% accuracy, thereby creating a pressure drop of 500 Pa at a He-mass flow of 50 g/s.

He – Mass Flow. This measurement is useful for all magnets during cool-down and warm-up but not indispensible. However, this gauge is necessary to operate a forced-flow coil safely. Usually some type of orifice or a Venturi tube is installed in the helium path and the pressure drop on the gauge is measured. The He-mass flow is then calculated from the pressure drop across the the orifice, using the separately measured pressure and temperature of the coolant in front of the orifice.[15] It has to be kept in mind that these flow meters work reliably only for stationary single-phase flow. The cross-section of the orifice can be chosen such that the He-mass flow meter is accurate in both the laminar and turbulent range of the flow.[16] More expensive in fabrication are Venturi-tubes. Nevertheless, they have the additional advantage of small residual pressure drop. The accuracy of these flow meters is about 5 to 10% in the range of 1 to several hundred grams per second of He-mass flow. A new type of He-mass flow meter has been tested recently utilizing ultrasound.[17] The acoustic signals emitted and received by two piezo ceramics are used to measure directly the velocity of the helium. The signal conditioning is relatively complex; its accuracy at a He-mass flow of 100 g/s is 3%. Further development is in progress to improve the accuracy of the gauge for flow rates lower than 10 g/s.

Strain

The forces acting on the conductor and on the structure are monitored with strain gauges. To compensate the thermal zero shift, two gauges are mounted such that the active sensor experiences the strain and the dummy is only thermally connected to the magnet. The two gauges form one half of a Wheatstone-bridge, with the other half of the bridge at ambient temperature. The calibration factor and the apparent strain do change with temperature and a minimum in the resistivity of the alloys does require the calibration of the gauges versus temperature.[18] The strain of the Karma and modified Karma alloy (nickel-chromium) probes are independent of the magnetic field direction. They display a negligible hysteresis effect in a wide range of strain level (200 to 9000 μm/m). They do have the additional advantage of reproducible magneto-resistance, albeit the parameters of a polynomial equation of fourth order must be determined by calibration if they are used at low magnetic field. In high field ($3 > B > 12$ T) the reading is a linear function of the magnetic induction, see Fig. 3. The accuracy in the above mentioned range is about 10 μm/m.[19] It is possible to evaluate the principal strain by arranging three pairs of strain gauges as a Rosette and combining their reading. Special care must be taken to reduce the dissipated power below about 10 mW/cm^2 to ensure that the substrate and the probe do have the same temperature.

Magnetic Induction

It is often necessary to check the field produced by the magnet, although the field distribution can be calculated reliably.

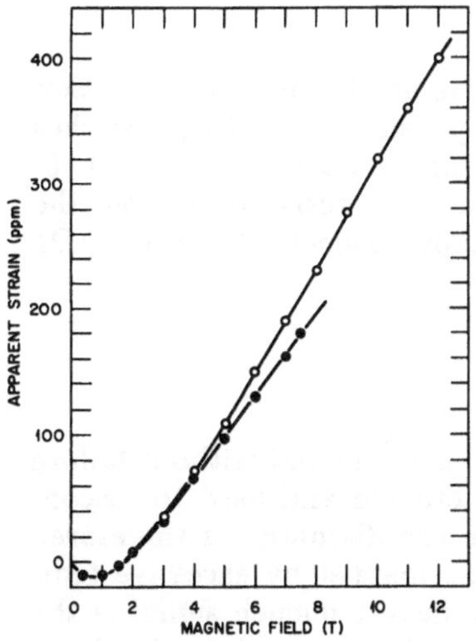

Fig. 3. Apparent strain vs magnetic field in two strain gauges of the same lot. Figure taken from Ref. 31.

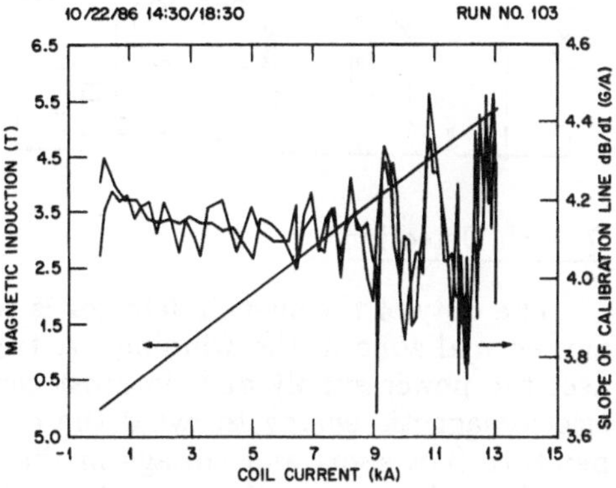

Fig. 4. Magnetic induction B vs current measured in the Swiss LCT coil with a Hall probe. Also shown is dB/dI vs current, displaying wiggles caused by quantum oscillation.

Magnetoresistor. This very popular probe, made from high purity copper wire, has an accuracy of <0.1% in fields larger than 3 T.[20] The probe, usually a coil, must be designed with great care to minimize its size, the temperature sensitivity and to avoid mechanical strain.

Hall Probe. A variety of semiconductor probes together with their associated electronics are commercially available.[21] The signal of the probes is quite linear at low fields; however, at high fields ($B > 5$ T) reproducible wiggles, caused by quantum oscillation, become noticeable,[22] see Fig. 4. The accuracy of the probes is usually better than 0.5%. Thermal cycling may change the average slope of the calibration curve by 1 to 2%.[1,23]

NMR Probe. The performance of this gauge relies solely on the measurement of a frequency, which can be done very accurately. To ensure a sharp resonance the field over the sample volume must be nearly uniform, i.e., 1 in 1000 over several mm. A NMR-probe suitable for LHe-temperatures can be used at any field level and its accuracy is better than 1 ppm.

Pick – up Coil. This device measures the time derivative of the field and therefore it is applied mainly in pulsed fields. Specially shaped coils may be designed to respond preferentially to uniform fields or to higher order components. At LHe-temperatures the thermal contraction of the coil must be taken into account. In steady fields the coil must be moved and is therefore used with advantage in the warm bore of a magnet. The accuracy of a pick-up coil is about 0.1% in fields up to 30 T.[24]

Other Sensors

Acoustic emission probes provide valuable information on the correlation between origin of coil quenching and the mechanical deterioration of the winding insulation.[25,26,27] The displacement sensor is used to study the movement of the winding in its case. Resistive and strain gauge type displacement sensors with warm[28] and cold[29] transducers were successfully employed; they have a range of 25 to 30 mm with a resolution better than 0.1 mm.[30]

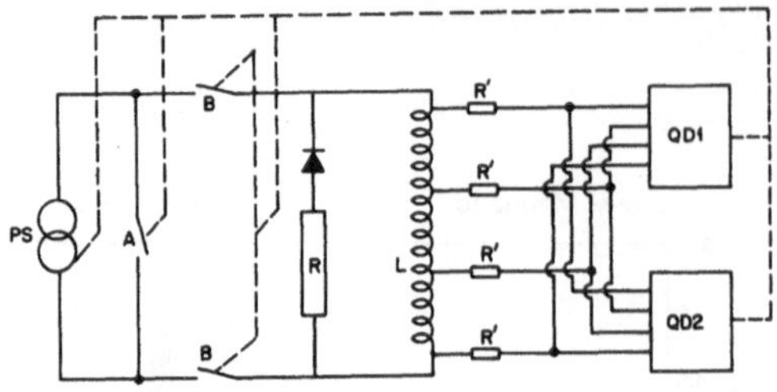

Fig. 5. Schema of a quench protection system: power supply PS; switches A,B; dump resistor R; SC coil L; current-limiting resistors R' on the voltage taps; quench detectors $QD1$ and $QD2$.

Quench Detection

The duty of the quench detector is to indicate immediately and without failure any normal zone in the winding. A trigger signal is created and used to disconnect the power supply and, in most cases, to initiate the discharge of the stored electromagnetic energy to avoid the destruction of the magnet by excessive temperature (hot spot) and voltage in the winding. The fastest quench signal is the dc-voltage drop across the normal zone in the conductor. To access the coil voltage the winding, including the current leads, is equipped with several voltage taps. It is advisable to limit the maximum current in the voltage tap by a series resistance, see Fig. 5. For a single magnet this dc-voltage can be measured either directly with a balanced, resistive-bridge type circuit or indirectly by the voltage induced in a so-called "compensation coil" attached to the conductor or to the magnet.[31] This latter signal must then be amplified and electronically subtracted from the separately measured voltage in the winding. The correctly adjusted circuit eliminates the inductive component of the voltage and responds to the resistive voltage only. When the dc-voltage exceeds the selected threshold for a predetermined time interval, the trigger signal to dump is generated.

In a large multi-coil systems the ramp rate of all coils affects the induced voltage in each magnet if their mutual inductance is not very small. The simple resistive-bridge is not adequate anymore and, therefore, an additional balanced ac-bridge or several compensation coils are used to measure the quench voltages. The SULTAN high field test facility at SIN contains three concentric SC-solenoids for which the induced voltage was compensated for a maximum ramp rate by balancing an ac- and a dc-bridge. A nice example for the second solution is used at the IFSMTF in Oak Ridge to protect six toroidal field (TF) coils arranged in a torus.[32] Each coil is equipped with compensation coils. The quench detector of a TF coil takes signals from both neighbours and from its own compensation coil. For each TF coil a separate hard-wired electronic circuit eliminates the induced voltage, see Fig. 6. The system works reliably and could be extended to include compensation for poloidal field coils too.

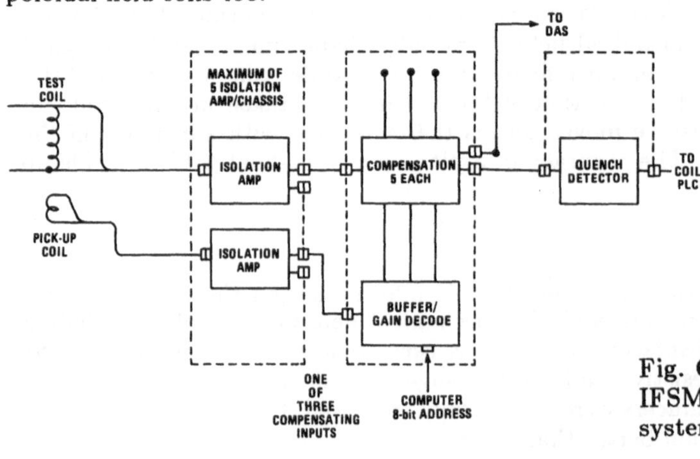

Fig. 6. Block diagram of the IFSMTF quench protection system, taken from Ref. 32.

DATA ACQUISITION

The signal conditioning of the sensors is at ambient temperature because cold electronics is still in development.[33] To register the raw data, convert them to engineering units and to display the data, digital computers of various size and speed are commonly used. The data handling is thereby simplified and results can be reduced to the essential information needed to diagnose any event occurring during a run. The use of digital data handling eliminates several sensor deficiencies like range limitation and nonlinearity and may be used to auto-calibrate the sensors too. A further application is self check and self diagnosis of the sensors and their signal conditioning. To study the history of a quench, an appropriate subset of the signals is registered with a frequency of 100 to 1000 Hz before and during the dump.[34,35] For a review, of the fast growing field of computer application, the reader is referred to the literature.[36,37,38]

CONTROL SYSTEM

The invention of microprocessors and Programmable Logic Controllers (PLC) in the seventies made it possible to design active control systems for the complex, sophisticated SC-magnet systems. In many systems the cool-down and warm-up of the magnets is performed according to a prescribed temperature curve by concurrence with conditions concerning temperature gradients in the magnet.[39] Some control systems do go a step further and include the automatic control of the magnet and the associated refrigerator by a power failure and magnet quench too.[40]

The energy of the SC-magnet is usually discharged by disconnecting the coil from the power supply and forcing the current through a suitable dump resistor. The advantage of this solution is that the helium boil off is reduced, since the energy is discharged externally, albeit it is an active technique which depends on the reliable operation of mechanical switches. A further drawback of this method is the high voltage, of several kVolts, generated in the coil. In a passive protection system the stored energy is transferred either into a short circuited secondary winding or into several shunt resistors. This techniques has the advantage of high reliability and, properly designed, it significantly reduces the internal voltage across the winding. The disadvantage of this technique is the energy dissipation when the magnet is charged and the fact that all the energy is dumped into the cryostat.

A very important aspect of the control system is the redundancy of the protecting circuit. To exclude the possibility of an undetected normal zone, the winding is subdivided into several sections and overlapping segments are connected to separate quench detectors. Another method is to include into the interlock chain the lowest helium pressure, the LHe-level or the minimum He-mass flow.

Magnet Protection

High current density SC-magnets may be protected from local overheating in case of a quench by driving the whole coil normal and distributing instantaneously the stored energy in the whole magnet. This thermal quenchback can be achieved with an auxiliary secondary circuit.[41] Cross-quenching combines thermal and magnetic quenchback by transferring the energy from one part of a coil into another part; thereby eliminating the need for an auxiliary coil.[42]

The coils of large accelerators and the TF coils of a fusion device are connected in series, but there is no practical way to extract a major fraction of the stored magnetic energy from the quenching magnet in such a circuit.[43] At FNL the current is bypassed for groups of magnets from which one or more are quenching by firing thyristors. The energy stored in the quenching magnet is absorbed in the magnet, while the energy of the others is transferred into dump resistors. At BNL, diodes were chosen to bypass the current, because diodes open automatically if the threshold is reached. The HERA quench protection system is based on cold diodes to bypass the current and on subdivision of the circuit to limit the external voltage The current sensing element is a magnetic amplifier which is insensitive to radiation. To distribute the energy in the quenching magnet, heaters induce thermal quenchback.[44]

To protect the TF-coils of the proposed NET fusion reactor, alternate coils are connected to the same power supply. This coil connection imposes the necessary symmetry on the current distribution to ensure vanishing out-of-plane forces while discharging the coils, albeit the current in both circuits must be well balanced. Opening dc-circuit breakers initiates rapid discharge of the stored energy by introducing dump resistors between the coils of the circuit. The failure of one circuit breaker causes a small variation in current which in turn creates an acceptable symmetric variation of the centering forces.[45]

The discharge of the bath-cooled coils of a SC-Magnet Energy Storage (SMES) plant in case of emergency results in excessive consumption or loss of the liquid helium. For a 5500-MWh plant, the proposed protection schema forces a thermal quenchback. Each turn is separately short circuited and by injecting pressurized helium gas on the top into the container the coolant is forced into a LHe-reservoir. The coil and the structure warm up safely, converting the stored energy to heat.[46,47] This procedure must be carried out in 5 to 20 seconds and limits the maximum temperature to about 400 K. The reliability of such a schema depends on minimizing its complexity and reducing the number of active components.

Most MRI magnets built have low fields (<2 T) and employ bath-cooled coils. The main solenoid and several gradient coils operate in persistent mode while the electrical leads are either retracted or disconnected by activating SC-switches. In case of a quench SC-switches are warmed up to force the current into dump resistors. The system usually does not include a refrigerator, since the LHe-consumed in persistent mode is less than 0.3 L/h.[48] Operation is fully automatic to attain maximum availability.[49]

Operating Experience

The information on operational experience with SC-magnet systems is scarce although the reported performance is gratifying. For the last 13 years an 8-m long SCμ-channel was operated at SIN in an irradiated environment.[50] The SC-channel ran without continuous supervision for more then 85,000 hours, although more than then 100 quenches were encountered, caused by technical failures or human errors. The system is shut down once a year for maintenance. The SMES of the Bonneville Power Authority, the only commercial facility realised up to date, also ran without continuous supervision. The operational experience with the SMES itself was rewarding; however, the performance of the cryogenic system was not satisfactory.[51] A small, NbSn tape wound, portable 15-T test magnet, operated by Clarendon Laboratories in Oxford, had only 30 quenches over about 50,000 hours operation during the last 13 years.[52]

CONCLUSIONS

Further research and development is needed to improve the performance of all sensors in magnetic fields. The influence of thermal cycling on the accuracy of the sensor reading must be further studied and reduced. There is little information available on the performance of signal conditioning in ac-fields. It is further important to study the behaviour of all sensors and their electronics in an irradiated environment. Cold transducers are already tested but, to the knowledge of the author, not available commercially. The lack of the following probes must be overcome:

- He-mass flow meter for two-phase helium, and
- Quench detector for coils in variable external fields.

In most applications there are still too many sensors installed.[53] Increasing confidence will surely reduce the number of sensors down to a level which is sufficient for safe operation.

Some basic perceptions should be incorporated in the design of the SC-magnet to enhance safety. High quench propagation velocity enables either earlier quench detection or the use of higher quench thresholds. Improved transversal thermal conductivity reduces the hot spot temperature by distributing the heat rapidly in the winding. High current density cables are promoted in recent applications, clearly favoring forced-flow cooled magnets with short discharge time and good high voltage characteristics.[7] High voltage withstand capability implies winding integrity, i.e., no sensor should be mounted in the winding The same sensors may be used for protection and monitoring, but it is indispensible to provide an independent hard-wired circuitry for coil protection. The protection system should possess built in redundancy. The quench detector must be fail safe to faults in sensor cabling and signal conditioning. The application of passive instead of active components must be further pursued to attain intrinsically save operation. Achieving reliable performance and maximum availability promotes SC-magnet technology in the public and will open new vistas for commercial application too.

ACKNOWLEDGEMENTS

The author wants to express his gratitude to the SC-magnet specialists in several laboratories around the world for the information contributed to prepare this paper. He is much indebted to M. S. Lubell at Oak Ridge National Laboratory (ORNL) for valuable discussions and for the hospitality in the SC-magnet group.

REFERENCES

1. L. G. Rubin et al., **Adv. in Cryog. Eng.**, Vol. 31 (1986), p. 1221.
2. E. P. Balsamo et al., Frascati, Internal Report (1984).
3. H. H. Sample et al., **Rev. Sci. Instrum.** 53:1129 (1982).
4. J. Hua et al., **Cryogenics** 27:90 (1987).
5. Y. Koike et al., **Cryogenics** 25:499 (1985).
6. A. Roy et al., **Rev. Sci. Instrum.** 56:483 (1985).
7. J. A. Zichy et al., **IEEE Trans. Magn.** MAG-21:245 (1985).
8. "Thermophysical Properties of Helium 4 from 2 to 1500 K with Pressures to 1000 Atmospheres," NBS Technical Note 631, National Bureau of Standards, Boulder, Colorado (1972).
9. R. Jacobs, **Adv. in Cryog. Eng.**, Vol. 31 (1986), p. 1277.
10. F. Pavese, **Adv. in Cryog. Eng.**, Vol. 29 (1983), p. 869.
11. G. Kraft et al., **Cryogenics** 20:625 (1980).
12. W. E. Williams et al., **Rev. Sci. Instrum.** 25:111 (1954).
13. R. Ries et al., **Rev. Sci. Instrum.** 35:762 (1964).
14. D. Hagedorn et al., **Adv. in Cryog. Eng.**, Vol. 31 (1986) p. 1299.
15. Deutsche-Norm, DIN 1952, Beuth Verlag GMBH, Berlin (1982).
16. VDI/VDE-Handbuch Messtechnk, Entwurf VDI/VDE 2041 (1975).
17. P. Lavocat, "11th Int. Cryog. Eng. Conf." (1986), p. 577.
18. C. Ferrero et al., **Adv. in Cryog. Eng.**, Vol. 27 (1981) p. 1173.
19. H. S. Freynik et al., **Adv. in Cryog. Eng.**, Vol. 24 (1978), p. 9473.
20. G. B. Scott et al., **J. of Physics E** 1:925 (1968).

21. H. H. Sample et al., Cryogenics 17:597 (1977).
22. J. Cornelis et al., presented at the ASC, Baltimore, MD (1986).
23. L. G. Rubin et al., **Rev. Sci. Instrum.** 46:1624 (1975).
24. G. Kido et al., "9th Int. Conf. on Mag. Tech.," Villigen (1986), p. 821.
25. Y. Iwasa, **Cryogenics** 25:304 (1985).
26. O. O. Ige et al., **Cryogenics** 26:131 (1986).
27. H. Iwasaki et al., "9th Int. conf. on Mag. Tech.," Villigen (1986), p. 830.
28. J. F. Ellis et al., **Rev. Sci. Instrum.** 49:398 (1978).
29. B. Yurke et al., **Cryogenics** 26:436 (1986).
30. H. Zehlein et al., "9th Symposium on Eng. Problems of Fusion Research," IEEE, Vol. 1 (1981), p. 293.
31. P. L. Walstrom, **Cryogenics** 20:509 (1980).
32. S. S. Shen et al., "9th Int. Conf. on Mag. Tech.," Villigen (1986), p. 811.
33. M. G. Rao et al., **Adv. in Cryog. Eng.**, Vol. 31 (1986), p. 1211.
34. K. N. Henrichsen et al., **Adv. in Cryog. Eng.**, Vol. 27 (1981), p. 245.
35. L. R. Baylor et al., "Proc. of the 11th Symp. on Fusion Eng.," Vol. 1 (1985), p. 605.
36. D. Sheats et al., **Adv. in Cryog. Eng.**, Vol. 27 (1981), p. 1183.
37. R. A. Thomas, **Adv. in Cryog. Eng.**, Vol. 29 (1983), p. 911.
38. J. A. Good et al., "9th Int. Conf. on Mag. Tech.," Villigen (1986), p. 824.
39. W. Herz et al., Kernforschungszentrum Karlsruhe, Internal Report (1984).
40. T. Mito et al., "Control System for Refrigerators of TRISTAN Dectector Magnets," this conference, paper no. FC-6.
41. M. A. Green, **Cryogenics** 24:659 (1984).
42. U. Trinks et al., **J. de Phys.** 45:C1-217 (1984).
43. P. F. Smith, **Rev. Sci. Instrum.** 34:368 (1963).
44. K. H. Mess, "Quench Dection at HERA," contribution to the Accelerator Conference in Washington (1987).
45. J. B. Hicks, "Proc. of the 11th Symp. on Fusion Eng.," Vol. 1 (1985), p. 459.
46. Y. M. Eyssa et al., **Adv. in Cryog. Eng.**, Vol. 31 (1986), p. 113.
47. S. M. Schoenung et al., **Adv. in Cryog. Eng.**, Vol. 31 (1986), p. 121.
48. C. Lesmond et al., "9th Int. Conf. on Mag. Tech.," Villigen (1986), p. 255.
49. I. L. McDougall et al., "11th Int. Cryog. Eng. Conf." (1986), p. 36.
50. I. Horvath et al., "9th Int. conf. on Mag. Tech.," Villigen (1986), p. 174.
51. H. J. Boenig et al., IEEE, Vol. PAS-104 (1985), p. 302.
52. H. Jones et al., "9th Int. Conf. on Mag. Tech.," Villigen (1986), p. 211.
53. G. Vecsey, **J. de Phys.** 45:C1-643 (1984).

CRYOGENIC INSTRUMENTATION OF AN SSC MAGNET TEST STAND

K. McGuire, J. Strait, M. Kuchnir, and A. McInturff

Fermi National Accelerator Laboratory
Batavia, Illinois

ABSTRACT

This paper describes the system used to acquire cryogenic data for the testing of SSC magnets at the Fermilab Magnet Test Facility. An array of pressure transducers, resistance thermometers, vapor pressure thermometers, and signal conditioning circuits are used. Readings with time resolution appropriate for quench recording are obtained with a waveform digitizer and steady-state measurements are obtained with higher accuracy using a digital voltmeter. The waveform digitizer is clocked at a 400 Hz sampling rate and these readings are stored in local ring buffers. The system is modular and can be expanded to add more channels. The software for the acquisition, control, logging, and display of cryogenic data consist of two programs which run as separate tasks. These programs (as well as a third program which acquires quench and magnetic data) communicate and pass data using shared global resources. The acquired data are available for analysis via a nationwide DECnet[a] network.

INTRODUCTION

Fermilab is collaborating with Brookhaven National Laboratory, Lawrence Berkeley Laboratory, and the SSC Central Design Group on developing, building, and testing superconducting magnets[1,2] for the proposed Superconducting Supercollider (SSC)[3]. Quench and magnetic field tests[4,5] on these magnets are performed by the Fermilab Magnet Test Facility. Temperature and pressure measurements are necessary to characterize the thermodynamic state of the helium coolant. The temperature of the magnet coil must be known to an accuracy better than 50 mK to understand both the quench performance and the magnetic field quality data. Temperatures and pressures are measured dynamically during quenches to locate the quench origin longitudinally and to understand stresses on the magnet and cryostat.

Figure 1 shows a flow schematic of the test stand and the locations of the sensing devices. Liquid helium enters the supply end box (feed can) in the 'Liquid Supply' line, where some of the flow is diverted to cool the power leads. The inlet temperature is then measured by a carbon resistor before the fluid enters a subcooler. The liquid level in this subcooler is measured both by a differential pressure gauge[b] and by a superconducting wire probe[c]. The temperature of the subcooled liquid is measured with a helium vapor pressure thermometer[d] (VPT) and its pressure is measured by a warm pressure transducer[e] whose tap is after the subcooler. In both magnet interconnect regions, where the transition is made between the end boxes and the magnet, there are a germanium[f], a carbon glass[g], and a platinum[h] resistor for measuring the temperature, and a cold pressure transducer[i]. At the return end the helium is directed back to the supply end box through the 'Liquid Return' line of the magnet cryostat. It then enters the supply end box where its temperature is measured by a VPT, the pressure is measured by a warm pressure transducer, and the mass flow is measured by a venturi tube. The venturi tube was installed at this point due to space requirements inside the feed can. Some of the returning liquid is used to maintain the liquid level in the subcooler. The remaining helium passes through a Joule-Thomson valve and enters the magnet's 'Gas Return' line. Pressure is measured at this point by an inexpensive piezo-resistive bending beam pressure transducer[j] and the temperature is measured by a carbon resistor[k]. The fluid then passes through the magnet and is again turned around to pass through the magnet's '20 K Shield' before returning to the refrigerator. In this mode of cooling, this shield actually operates at about 4.5 K. The instrumentation in the feed can for the '20 K Shield' is the same as that used for the 'Gas Return' line. Instrumentation for the '80 K Shield', which is cooled by liquid nitrogen, consists of platinum resistors at either end, as well as a pressure gauge at the feed end and a flowmeter at the return end, where the nitrogen is warmed up by a heat exchanger and exhausted.

SENSING DEVICES

Pressure measurements were obtained using several different types of pressure transducers. In the locations where high accuracy was required, bonded strain gauge[e] transducers were used. Cold pressure transducers of the sputtered strain gauge[i] type were installed inside the magnet interconnect region to obtain readings with high time resolution. Although these transducers have a large zero shift between 4 and 300 K, their sensitivity is unchanged. They have an excellent time response because they are located directly in the fluid, with no intervening tubing. Delays introduced by the tubing connecting the warm pressure transducers measure about 30 ms at the feed end and 15 ms at the return end of the magnet. Using the cold pressure transducers, the leading edge of pressure pulses caused by quenching can

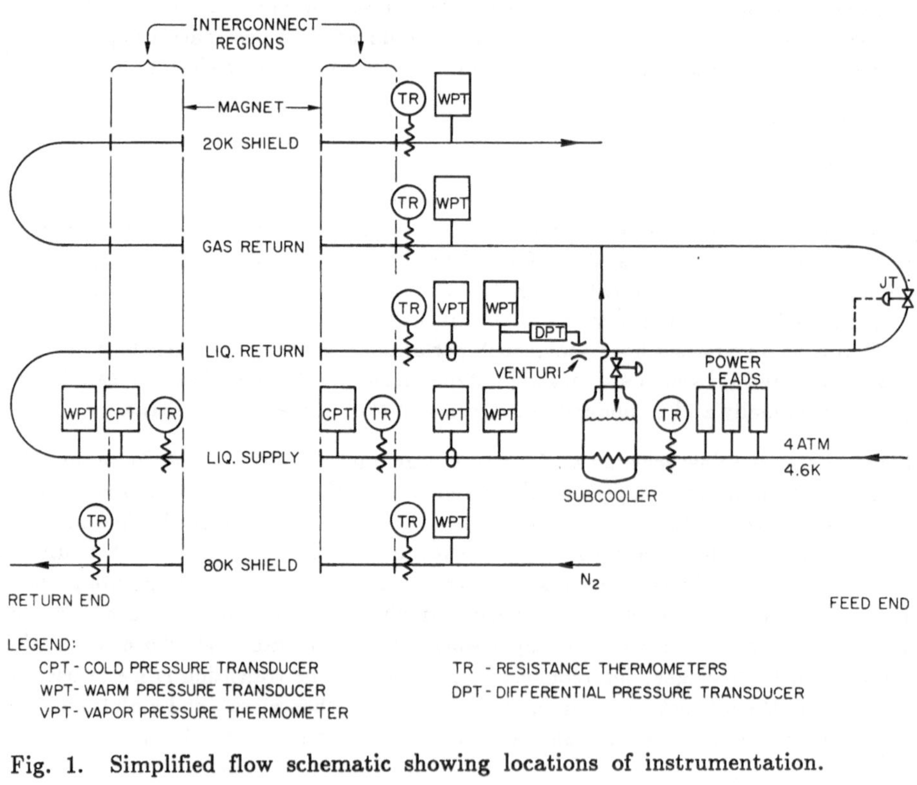

Fig. 1. Simplified flow schematic showing locations of instrumentation.

be observed with resolution of about 2 ms, corresponding to a resolution of 0.3 m in the longitudinal origin of the quench. The signals from the pressure transducers are amplified at the test stand before being cabled into the control room.

Mass flow of helium through the magnet is measured with a venturi tube and differential pressure transducer[l] combination. Turbine flowmeters were ruled out due to planned quench testing of the magnets. Fast pressure rises, such as those expected when the magnet quenches would destroy these devices.

Several different types of resistance thermometers were used for various reasons having to do with range, accuracy, repeatability, and time response. These thermometers, as well as their ranges and reasons for usage are shown in Table 1. In the locations where high accuracy is required, a germanium resistor and a carbon glass resistor are paired. There are two reasons for this: the first reason is redundancy; the other has to do with the accuracy and the effects of magnetic fields on these devices. An attempt is made to combine the best features of these thermometers and to compensate for their weaknesses. The germanium resistors are exceptional in their sensitivity, repeatablity, and time response, but are susceptible to the effects of magnetic fields. The magnitude of this effect is orientation-dependent. While a prototype of the system was being tested, it was possible to see a thermal oscillation with a frequency of 20 Hz with a peak-peak amplitude corresponding to 1 mK (excitation was 10 microamperes). The frequency of this oscillation was made to change by altering the cryogenic configuration of the test stand in which this thermometer was installed. The carbon glass thermometer has good accuracy (approximately as accurate as a silicon diode at 4.2 K) and is not very susceptible to magnetic fields. The carbon glass thermometer can be cross-calibrated against the germanium thermometer in the absence of magnetic fields. Then the carbon glass thermometer can be used as the primary thermometer when there are magnetic fields present.

Some of the resistance thermometers were initially calibrated at the factory from which they were purchased. All of them were mounted, packaged, and calibrated[6,7,8] at the Fermilab Magnet Development Facility. After about a year of service (and five thermal cycles), some germanium and carbon glass thermometers were recalibrated. The new calibration was found to agree with the original calibrations (including independent factory calibration) to within 5 mK.

The resistance thermometers are excited by a DC constant-current source. Depending on the type of thermometer used, the excitation current is 25, 6, or 1.5 microamperes. The current for all thermometers on a single current source runs in series through a resistor at

Table 1. Characteristics of Resistance Thermometers

Thermometer	*Range (kelvins)*	*Reasons for use*	*Disadvantages*
Germanium	1 - 80	Excellent stability - repeatability is ±0.5 mK when thermally shocked between 295 K and 4.2 K Excellent time response	Influenced by magnetic fields Small range
Carbon glass	2 - 200	Good stability - repeatability is ±7.5 mK when thermally shocked between 295 K and 4.2 K High sensitivity below 10 K Good immunity to magnetic fields	
Carbon	2 - 200	Low cost	Unstable with thermal cycling
Platinum	25 - 800	Used to monitor cooldown and warmup	

room temperature (designated R_I) and then through the thermometers. The resistor at room temperature, R_I, has voltage taps on it. By knowing the value of R_I and the voltage drop across it, the current through any thermometer in the chain is known. Four leads (made of twisted pairs of manganin wire) from each thermometer are brought outside the cryostat and the connections to complete the current chain are made externally. This allows thermometers to be bypassed in the event that there is a short circuit or they become open-circuited. It also allows the electronics to be modular. Provisions for changing the direction of the current to measure the offsets due to thermal EMFs and contact potentials is provided by a mechanical relay. This relay can be controlled either locally by a switch on the front panel, or remotely by applying a TTL level signal to a jack on the front panel. To minimize noise pickup, The voltage drops across these thermometers are first amplified locally (see Fig. 2) at the test stand before being cabled into the control room (where they are measured by the readout equipment). These amplifiers have a low impedance single-ended output and the readout devices have high impedance differential inputs.

While the accuracy of the resistance thermometry measurements has not been studied in detail, it is believed that the dominant source of uncertainty comes from the stability of the amplifiers. These problems can be minimized by rebuilding with higher grade components. An error in the gain of 1%, due to thermal or temporal drifts over the approximately 1 month testing time of each magnet, corresponds to an error of about 20 mK for the germanium and carbon-glass thermometers. The internal consistency of the measurements suggest that this is a reasonable estimate of the accuracy. Uncertainties from other sources, in particular the calibration, are considerably smaller than this.

Helium vapor pressure thermometers (VPTs) are used in two locations to check the calibrations of some of the resistance thermometers. The geometry of the VPTs are dictated by the absolute accuracy required and the warm volume of the system (tubing and pressure transducers). Because the absolute accuracy of strain gauge pressure transducers is specified as a percentage of the full scale range, it is desirable to keep this range as small as possible. Since the maximum pressure for helium vapor in equilibrium with liquid is about 227 kPa, the volumes of the system were chosen such that the charging (warm) pressure of the system is slightly higher than this. A large cold volume was used for two reasons - to yield a large dynamic range, and so that errors introduced by uncertainty in the warm volumes would be as small as possible. The volume of the bulb is 0.003 L, while the volume of the cold internal tubing is 0.009 L. The warm volume, consisting of tubing and cavities within pressure gauges and transducers is 0.971 L, including a 0.906 L ballast volume. The final charging pressure is 265 kPa. The calculated time constant is about 0.17 seconds and the operating range (corresponding to the bulb being 10% full to 90% full) is from 2.2 K to 5.1 K. The primary error in

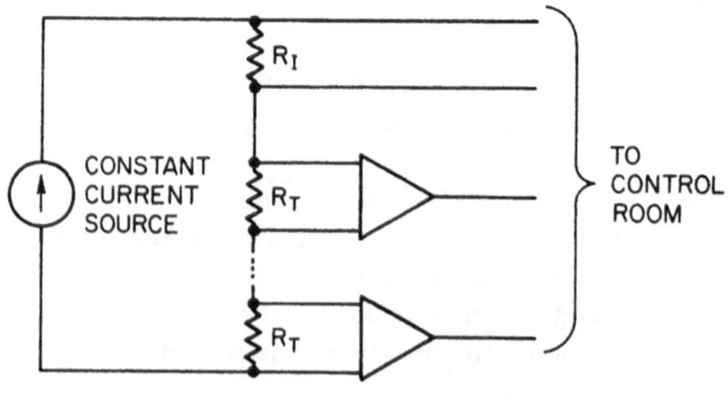

Fig. 2. Implementation of resistance thermometers.

reading the VPTs comes from the accuracy in the pressure transducers[m] used to read them and their associated electronics (0.05% of Full Scale Output). Using a pressure transducer with a range of 50 psia (340 kPa), the accuracy obtained is better than 2 mK.

Table 2 summarizes the accuracy of the measurements.

READOUT DEVICES

The required measurements fall into two categories - those which must be done very fast (where absolute accuracy is not important), and those which must be done with high accuracy (where speed is not required). Since no readout device has the capability of high accuracy and great speed, it was decided to use two readout devices. The fast readings are taken with a multichannel waveform digitizer[n] with 12-bit resolution (10-bit accuracy). The memory in these modules is implemented as sets of ring buffers for each channel. As the waveform digitizer is clocked, a pointer to the next memory location is incremented. When the pointer reaches the end of the ring buffer, it is reset to point to the beginning. This cycle is repeated continuously until a 'stop trigger' is generated when a quench is detected. The waveform digitizer is configured (by a combination of jumpers and software) such that it take a certain number of post-trigger samples when the stop trigger is received. By judiciously choosing the number of post-trigger samples, a number of pre-trigger samples can be saved as well.

A digital voltmeter[o] (DVM) is used for the high accuracy readings. Since most DVMs have at most two inputs, it is impractical to use several of these when there are large numbers of devices to read. Therefore, the signals are multiplexed through a scanner[p]. The scanner uses mechanical relays for switching so as not to degrade or corrupt the signal, and sequentially switches the signal to the DVM.

Table 2. Accuracy of Measurements.

Location	*Measurement*	*Device Type*	*Accuracy*[a]
Power leads	Temperature	Carbon resistor	50 mK
Magnet flow	Flow	Venturi / strain gauge	unknown
Liquid supply	Pressure	Bonded strain gauge	0.375 psi (2.6 kPa)
Liquid supply	Pressure	Sputtered strain gauge	0.625 psi (4.3 kPa)
Liquid supply	Temperature	Germanium resistor	20 mK
Liquid supply	Temperature	Carbon glass resistor	20 mK
Liquid supply	Temperature	Carbon resistor	50 mK
Liquid supply	Temperature	VPT / bonded strain gauge	2 mK
Liquid return	Pressure	Bonded strain gauge	0.375 psi (2.6 kPa)
Liquid return	Pressure	Sputtered strain gauge	0.625 psi (4.3 kPa)
Liquid return	Temperature	Germanium resistor	20 mK
Liquid return	Temperature	Carbon glass resistor	20 mK
Liquid return	Temperature	Carbon resistor	50 mK
Liquid return	Temperature	VPT / bonded strain gauge	2 mK
Gas return	Pressure	Piezo-resistive on bending beam	0.25 psi (1.7 kPa)
Gas return	Temperature	Carbon resistor	50 mK
20K shield	Pressure	Piezo-resistive on bending beam	0.25 psi (1.7 kPa)
20 K shield	Temperature	Carbon resistor	50 mK
80 K shield	Pressure	Piezo-resistive on bending beam	0.25 psi (1.7 kPa)
80 K shield	Temperature	Carbon resistor	50 mK

[a]Accuracy quoted for thermometers are specified at 4.2 K.
Accuracy quoted for pressure transducers are from manufacturer's specifications.

The base of the readout system consists of a CAMAC[9] crate connected to a VAX[q] computer through a CAMAC interface[r]. Both the DVM and the scanner are controlled by the computer through a GPIB[10,11,s] interface which resides in the CAMAC crate. Figure 3 shows a composite of the test stand instrumentation and the associated readout equipment.

OPERATOR INTERFACE

Measurements of pressure and flow are provided to the cryogenic systems operator in two different ways. First there is a gauge panel on the test stand with various pressure gauges and flowmeters which can be read visually. Temperatures at two locations can also be read visually using pressure gauges on the VPTs. All instrumented readings can be obtained from a second source. A cryogenic data acquisition program functions as a data server. It receives data requests from a separate data logging program which is controlled by timers implemented in software. There are separate timers for each of three data logging functions: logging data to a computer file, logging data to a printer, and updating a data display on a computer terminal. This display functions as the operator's primary interface for high accuracy data. When a timer times out, it specifies the appropriate data request by setting bits in a common (global) event flag cluster. It then notifies the cryogenic data acquisition program that it wants data by writing a message to a shared 'mailbox'. The requested data is then acquired and then stored in shared global memory. Completion of the transaction is signalled by a 'Data Ready' event flag. Finally, the data logging program collects the data from the shared global memory and performs the appropriate logging function. Cryogenic data for the program which does the quench and magnetic measurements is obtained the same way.

DATA SHARING

Since the SSC project is a collaboration between Fermilab, Brookhaven National Laboratory, Lawrence Berkeley Laboratory, and the SSC Central Design Group, it is useful to have the data acquired during measurements available on-line to all parties involved. This is easily accomplished with a nationwide network of VAX computers which use DECnet software for

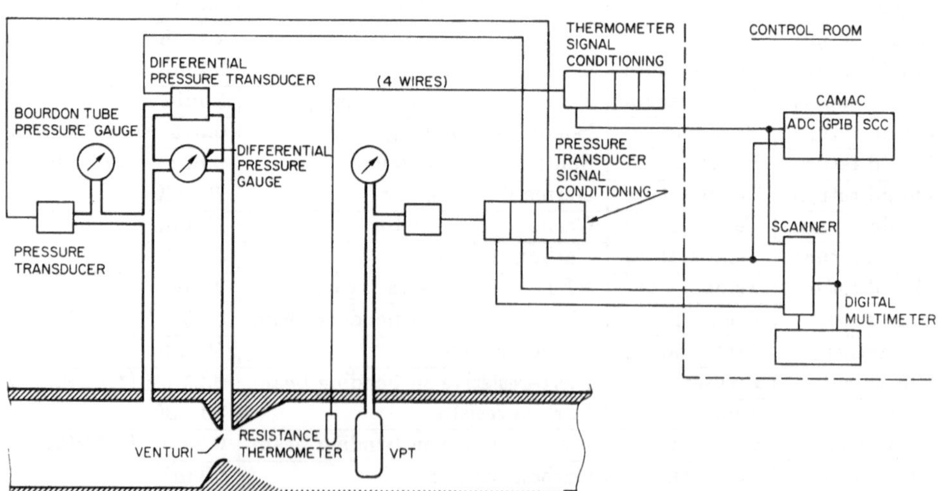

Fig. 3. Composite of test stand instrumentation.

transparent file transfers. Data are initially stored on the VAX at the Fermilab Magnet Test Facility. From there, files can be copied to any computer on the network for special studies and analysis, and are instantly available as the measurements proceed.

ACKNOWLEGEMENTS

Fermilab is operated by the Universities Research Association under contract with the United States Department of Energy. Many people have contributed to the success of this project by providing the necessary support. In particular, we wish to acknowlege the contributions of R. Barger, A. Bianchi, B.C. Brown, J. Garvey, F. Johnson, G. Kirschbaum, D. Lewis, R. Nehring, J. Pachnik, A. Rusy, E. Schmidt, H. Stahl, J. Tague, F. Wilson, and W. Zimmerman. In addition, we wish to acknowlege the work of the Fermilab Accelerator Division Cryogenic Systems group, who designed and built the end boxes, particularly T. Peterson and J. Theilacker; and J. Osterlund and J. Sondericker of Brookhaven National Laboratory, who were consulted in the design stage of the resistance thermometry.

MANUFACTURERS

a. DECnet is a trademark of Digital Equipment Corporation, Massachusetts.
b. Model Capsuhelic, Dwyer Instruments, Michigan City, Indiana.
c. American Magnetics, Oak Ridge, Tennessee.
d. Designed by M. Kuchnir, Fermilab.
e. Model 831 Dynisco, Norwood, Massachusetts.
f. Model N2K, Scientific Instruments, Inc., West Palm Beach, Florida.
g. Model CGR-1-1000, LakeShore Cryotronics, Inc., Westerville, Ohio.
h. Model P1, Scientific Instruments, Inc., West Palm Beach, Florida.
i. Model 1000-05, I.M.O. Delaval, CEC Instruments Division, Pasadena, California.
j. Model PX105, Omega Engineering, Inc., Stamford, Connecticut.
k. Ordinary carbon resistors designed for electronics use, Allen-Bradley, Inc.
l. Model TJE Differential Pressure Transducer, Sensotec, Columbus, Ohio.
m. Model Super TJE Absolute Pressure Transducer, Sensotec, Columbus, Ohio.
n. LeCroy Model 8212A High Accuracy Simultaneous Sampling Data Logger, LeCroy Research Systems Corporation, Spring Valley, New York.
o. Model 3457, Hewlett-Packard, Palo Alto, California.
p. Model 705, Keithley Instruments, Inc., Cleveland, Ohio.
q. Model 11/730, Digital Equipment Corporation, Massachusetts. VAX is a trademark of Digital Equipment Corporation.
r. Model 411 PDP-11 CAMAC Interface, Jorway Corporation, Westbury, New York.
s. Model 3388-G1A GPIB Interface Module, Kinetic Systems Corporation, Lockport, Illinois.

REFERENCES

1. P. Dahl, et al., Construction of cold mass assembly for full length dipoles for the SSC accelerator, "Proceedings of the Applied Superconductivity Conference," Baltimore, Maryland (1986), J. Schooley (Ed.).
2. R.C. Niemann, et al., Design, construction and test of a full scale SSC dipole magnet cryostat thermal model, "Proceedings of the Applied Superconductivity Conference," Baltimore, Maryland (1986), J. Schooley (Ed.).
3. Conceptual Design of the Superconducting Super Collider, SSC Central Design Group, Lawrence Berkeley Laboratory, One Cyclotron Road, Berkeley, California, 94720.

4. J. Strait, et al., Full length prototype SSC dipole test results, "Proceedings of the Applied Superconductivity Conference," Baltimore, Maryland (1986), J. Schooley (Ed.). Also available as Fermilab Technical Memo TM-1450, SSC-N-320, 0102.001.
5. J. Strait, et al., Tests of prototype SSC magnets, "Proceedings of the 12th Particle Accelerator Conference," Washington, D.C. (1987). Also available as Fermilab Technical Memo TM-1451, SSC-N-321, 0102.001.
6. M. Kuchnir, J.L. Tague, Carbon Resistance Thermometers, Fermilab Technical Memo TM-647, 1600.000.
7. M. Kuchnir, J.L. Tague, and H. Cranor, Pulsed Current Resistance Thermometry, Fermilab Technical Memo TM-1099, 1680.00.
8. M. Kuchnir, Pulsed current resistance thermometry, in "Advances in Cryogenic Engineering" Vol. 29, Plenum, New York, (1984).
9. CAMAC stands for Computer Automated Measurement and Control, and is specified by IEEE standards 583, 595, 596, 675, and 683.
10. GPIB stands for General Purpose Interface Bus, and is also known as HPIB (Hewlett-Packard Interface Bus) and ASCII bus. It is specified by IEEE Std. 488 and ANSI Std. MC1.1.
11. Kevin McGuire, Fermilab Magnet Development and Test Facility GPIB Package, Fermilab Technical Memo TM-1287, 2320.000.

BEHAVIOR OF TURBINE AND VENTURI FLOWMETERS IN SUPERFLUID HELIUM

D. E. Daney

National Bureau of Standards

Boulder, Colorado

ABSTRACT

Turbine and Venturi flowmeters were calibrated in both normal and superfluid helium in an apparatus designed specifically for that purpose. Simple in concept, the facility forces liquid helium from a calibrated bellows through the flowmeter. An ambient temperature stepping motor drives the welded stainless steel bellows whose capacity is 10 L. Flow rates range up to 0.35 L/s, and helium temperatures span from 1.25 K to 4.0 K. For the 9.3 mm bore turbine meter, identical meter factors are obtained for normal and superfluid helium. For the Venturi meter, discharge coefficients between 0.98 and 1.0 are obtained with normal and superfluid.

INTRODUCTION

Flow metering is important in most engineering applications of superfluid helium, and for in-orbit transfer of superfluid[1] - the motivation for this study - it is especially important because of the cost and logistics of putting liquid helium in orbit.

Below the critical velocity, neither Venturi or turbine meters should, in theory, operate. In the case of the Venturi, the flow stream lines loop into pressure taps, so that the stagnation pressure is observed at all locations, and the pressure difference is zero.[2] In the case of the turbine meter, the lift on the blades is zero, so there is no torque to overcome friction. Fortunately, flow velocities of engineering interest are well above the superfluid critical velocity, and both Venturi and turbine meters operate in superfluid as if they were in normal helium.

Previous work on superfluid flow metering includes use of an uncalibrated Venturi[3] and calibration of a turbine meter[4] which was reported while our work was in progress.

METER CALIBRATION FACILITY

The meter calibration apparatus (see Figure 1) uses a bellows as the flow source for the test meter. Both the bellows and meter are immersed

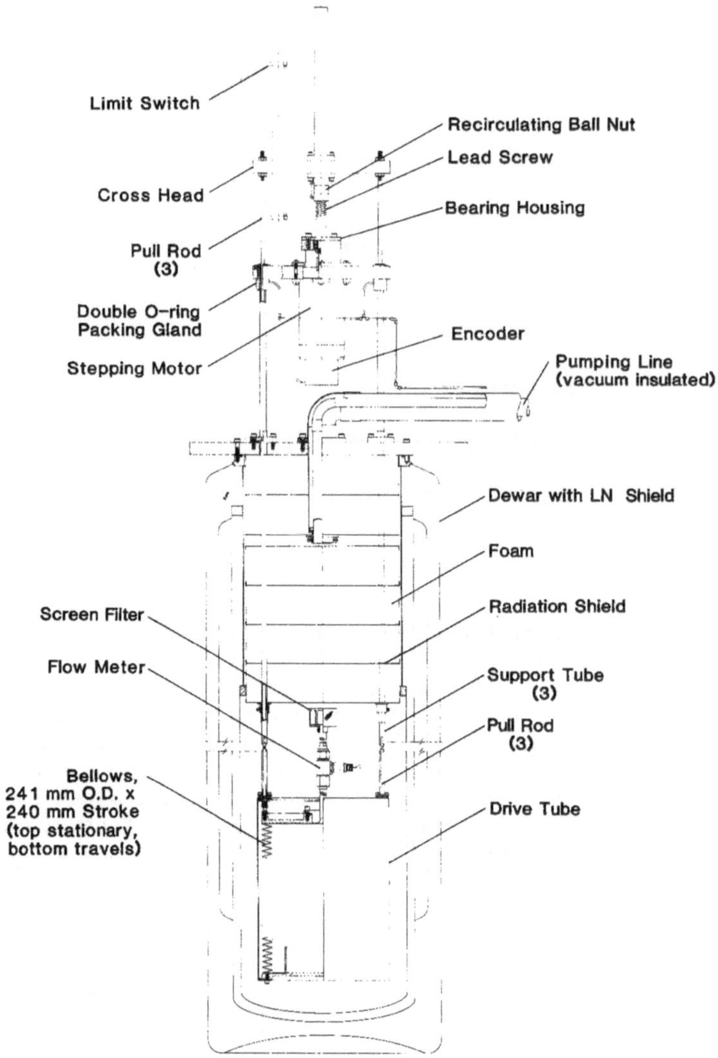

Figure 1. Meter calibration apparatus.

in liquid helium at depths up to 43 cm above the discharge tube. This arrangement is similar to that used by Rivetti, et al.[5] although the details differ. A stepping motor with 25,000 steps per revolution smoothly drives the bellows at flow rates up to 2 L/s, but flowmeter pressure drop generally limits the flow to lower values. For the meter calibrations reported here, the flow rate ranged from 0.007 to 0.35 L/s. The assembly is enclosed within a 30 cm diameter by 152 cm deep liquid nitrogen shielded dewar, and the overall heat leak into the system is 0.4 watts. With the vacuum pumps we normally use, the minimum bath temperature is 1.25 K although larger pumps are available at our laboratory that should allow temperatures down to about 1.0 K.

The bellows, welded from Type 316L stainless steel, has a 241 mm O.D.,a 240 mm stroke, and a 10 L displacement. To facilitate changing meters it has a bolted flange at the discharge sealed by a 50 mm diameter, spring loaded aluminum C-ring. The discharge line is 11.7 mm I.D. with a 6.4 mm inlet radius. We calibrated the bellows with a small bell type gas prover designed specifically for the task. Using gas rather than liquid eliminates problems of entrained bubbles and nonwetting of recesses in the convolutions.

Turbine Meter

The turbine meter is a commercially available one with a six bladed rotor suspended on ball bearings made entirely from 440 C stainless steel. The bore diameter is 9.35 mm, and 4 bladed guide vanes are located on both sides of the rotor. The final 2 mm of the inlet vanes are bent 30° in order to induce swirl and increase the turbine speed. An inductive speed sensor gives low drag over a wide flow range. Although the discharge from the bellows is swirl-free, we installed a flow straightener 6 pipe diameters upstream of the meter to match the way the meter is installed in the NBS pump test facility. The straightener is a bundle of tubes 2.4 mm O.D. by 0.07 mm wall by 20 mm long.

Venturi Meter

The Venturi meter, which is installed without a flow straightener, (see Figure 2) follows the classical or Herschel design, and its dimensions are chosen to give a pressure difference of 3.5 kPa for a 4 K liquid helium flow rate of 800 L/h. The meter pressure difference is measured with a commercial, variable reluctance pressure transducer which is close coupled to the Venturi and operates immersed in the liquid helium bath. The magnetic stainless steel diaphragm of the transducer is welded to the transducer body to give a hermetic seal. The pressure transducer calibration (see Figure 3) shows a 0.2 percent change in sensitivity between 300 K and 4 K and a 3.6 percent change in sensitivity between 4 K and 2.1 K. From 2.1 K to 2.0 K no further significant change was observed. The zero shift, which continues all the way to 1.25 K is accommodated by referencing all measurements to the zero flow condition. The temperature dependence of the transducer output results from both mechanical changes due to thermal stresses and from electrical resistance changes in the coil with temperature. The shift between 300 K and 4 K is surprisingly small, whereas the shift between 4 K and 2 K is unexpectedly large since changes in thermal expansion and electrical resistance below 4 K are small.

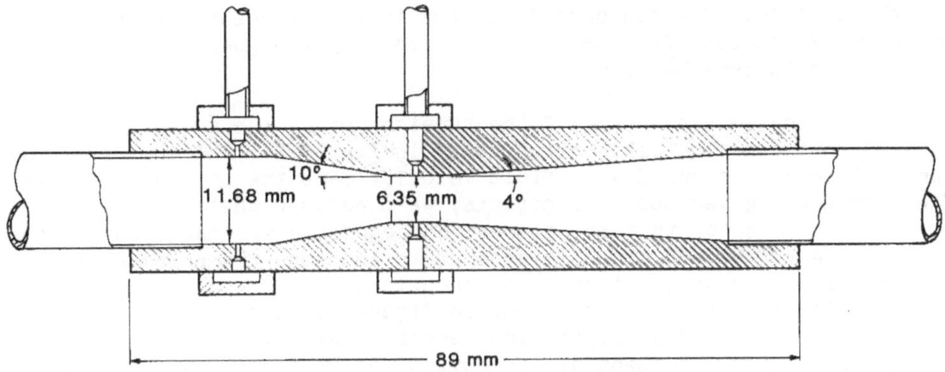

Figure 2. Venturi flowmeter.

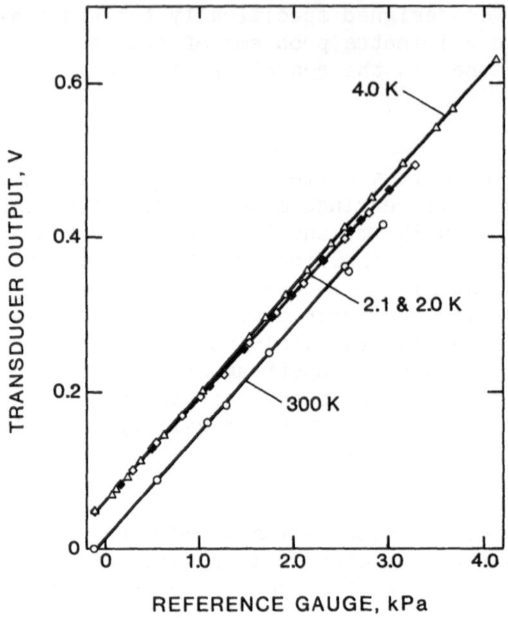

Figure 3. Venturi pressure transducer calibration.

RESULTS AND DISCUSSION

Turbine Meter

The meter factor, which is blade frequency divided by the flow rate, is given in Figure 4 over a flow range of 47:1. Separate curve fits for the normal and superfluid data show a meter output for the superfluid less than 0.5 percent greater than for the normal fluid. This small difference may be caused by the 13 percent greater density of the superfluid or by its lower viscosity. In any case, the overriding feature of the results is the similarity in meter performance with both normal and superfluid helium.

Some scatter in the data at low flow rates is to be expected since the lowest flow rate is just sufficient to overcome bearing friction and turn the turbine. The standard deviation is 0.45 percent for the 4 K data and 0.75 percent for the superfluid data. At higher flow rates the random error is somewhat less.

During the initial tests of the turbine meter, the superfluid gave the jump in the meter factor illustrated in Figure 5. The jump was quite reproducible and occurred at a fluid velocity past the turbine of 3 m/s. After considering various explanations, we concluded that cavitation might be the cause of this phenomenon; consequently we installed a 5.6 mm bore orifice at the end of the discharge to increase the pressure in the turbine. This remedy eliminated the jump in the meter factor and gave the high flow rate results presented in Figure 4. Subsequently, in a variety of circumstances pumping and metering helium, we have observed that He II is highly susceptible to cavitation. We are now studying cavitation in flowing helium in more detail.

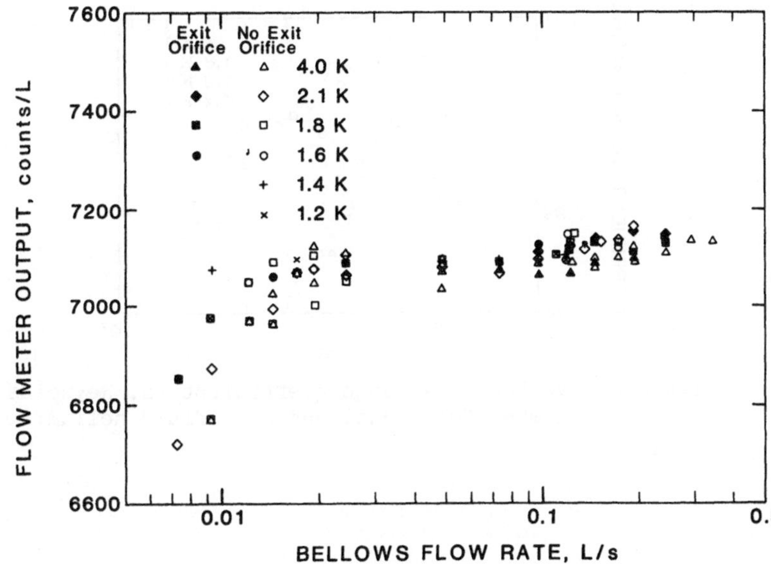

Figure 4. Turbine meter calibration in normal and superfluid helium.

Venturi Meter

The Venturi meter measurements are summarized in Figures 6 and 7 which present the discharge coefficient, C_D, as a function of Reynolds number and throat velocity, respectively. The normal component viscosity is used to evaluate the Reynolds number for the superfluid, and incompressible flow is assumed in computing C_D. Because the Venturi pressure difference at the lowest flow rate is only 3 percent of full scale for the pressure transducer, some scatter in the low flow data is to be expected.

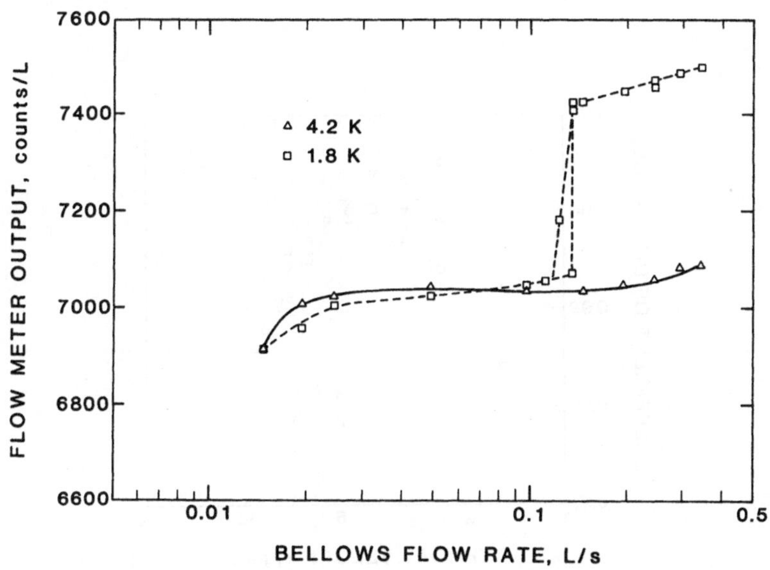

Figure 5. Cavitation effects in turbine meter.

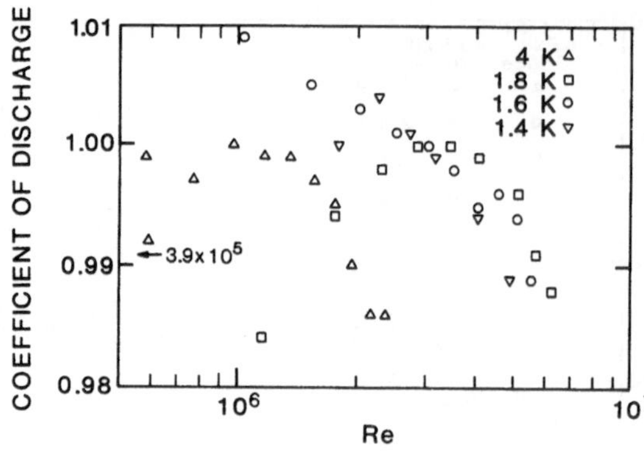

Figure 6. Venturi discharge coefficient vs. Reynolds number for normal and superfluid helium.

There are several possible reasons why the velocity correlates the drop in C_D at high flow rates better than the Reynolds number does. One possibility is that a Reynolds number based on the normal component viscosity may not be appropriate for highly turbulent flow of the superfluid. Another possibility is that the decrease in C_D at high flow rates is caused by fluid compressibility effects, which are proportional to Δp or V^2. Cavitation or a velocity dependent systematic error are yet other possibilities.

We calculate the correction to the discharge coefficient from the following expression for the expansion coefficient

$$Y = \left\{[1-\tfrac{1}{2}\Delta p/\rho_1 c^2(1-\beta^4)]/[1-\beta^4(1+\Delta p/\rho_1 c^2)]\right\}^{\frac{1}{2}} \qquad (1)$$

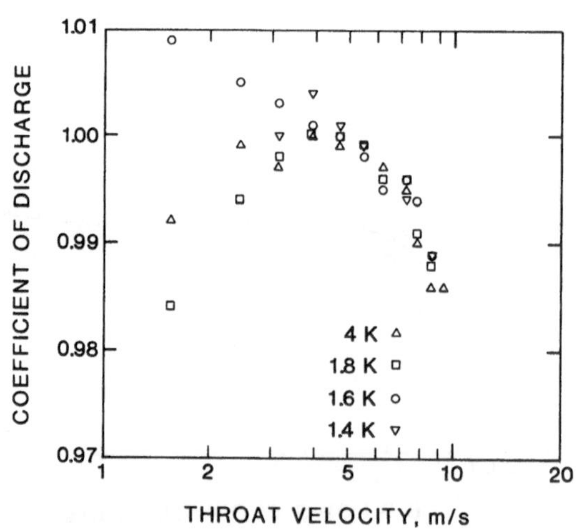

Figure 7. Venturi discharge coefficient vs. throat velocity for normal and superfluid helium.

which is derived in the Appendix. For a pressure difference of 3.5 kPa (0.5 psi), which is near the maximum for these measurements, equation 1 gives a correction to C_D of + 0.01 percent, so fluid compressibility is not responsibile for the tail-off in the discharge coefficient.

Because of our experience with cavitation in the initial turbine meter tests, we installed a 6.4 mm diameter orifice at the exit of the 65 mm long discharge line in order to create back pressure on the Venturi and suppress cavitation. The close agreement of the results for the normal and superfluid, which have radically different cavitation characteristics, indicate that cavitation did not occur. If the tail-off is characteristic of the apparatus, the mechanism is not clear. Stress effects at high flow rates could affect start-up transients but should not effect steady flow results.

CONCLUSIONS

Both turbine and Venturi flow meters accurately measure turbulent flow of superfluid helium, and their meter factors and discharge coefficients are essentially the same for both fluids. Because superfluid is susceptible to cavitation, meters for superfluid systems should be installed so that the pressure at all points within the meter remains above the saturation pressure. In our tests this required an orifice downstream of the meter. In most systems, however, an orifice should not be necessary because of adequate downstream flow resistance.

ACKNOWLEDGMENTS

This work was sponsored by NASA Ames Research Center under Contract No. 21059C(TRK). We thank J. D. Siegwarth for his suggestion that cavitation caused the anomalous turbine meter results.

APPENDIX

The effect of compressibility on the discharge coefficient of a slightly compressible fluid can be calculated by assuming that the local equation of state for an isentropic process is

$$\rho = \rho_o + (\partial\rho/\partial p)_s \cdot p \qquad (1A)$$

where p is the pressure relative to the reference state. Using the thermo-dynamic identity

$$c^2 = (\partial p/\partial\rho/)_s$$

gives

$$\rho = \rho_o + p/c^2 \qquad (2A)$$

as the local equation of state. For liquid helium this is a valid approximation since the sound velocity is only a weak function of pressure.

The enthalpy change for an isentropic process is

$$\Delta h = \int_o^{\Delta p} dp/\rho = 1/\rho\int_o^{\Delta p} dp/(1 + p/\rho c^2) \qquad (3A)$$

Integrating (3A) and using as the first two terms of a power series expansion for log (1 + x) gives

$$\Delta h = \Delta p/\rho_0(1-\tfrac{1}{2}|\Delta p|/\rho_0 c^2) \quad (4A)$$

The change in enthalpy due to acceleration is

$$\Delta h = V_1^2/2 - V_2^2/2 \quad (5A)$$

which combined with the continuity equation and the local equation of state (2A) becomes

$$V_1 = (A_1/A_2)(1+|\Delta p|/\rho_1 c^2) \quad (6A)$$

Combining 4A, 5A and 6A we obtain the Venturi throat velocity

$$V_2^2 = Y(2\Delta p/\rho_1)^{1/2}[1/(1-\beta^4)]^{1/2} \quad (7A)$$

where

$$Y = \{[(1-\tfrac{1}{2}|\Delta p|/\rho_1 c^2)(1-\beta^4)]/[1-\beta^4(1+\Delta p/\rho_1 c^2)]\}^{1/2} \quad (8A)$$

is analogous to the expansion factor for compressible flow of an ideal gas.

NOMENCLATURE

A	Venturi cross-sectional area
c	velocity of sound
C_D	coefficient of discharge = $Q/A_2(1-\beta^4)^{1/2} \cdot (\rho_1/2\Delta p)^{1/2}$
D	Venturi diameter
h	enthalpy
p	pressure
Δp	Venturi pressure difference = p_1-p_2
Q	volume flow rate
Re	Reynolds number = $\rho VD/\mu s$ entropy
s	entropy
V	velocity of fluid
Y	expansion factor defined by equation 8A

Greek

β	Venturi diameter ratio = D_2/D_1
μ	viscosity of normal component
ρ	density

Subscripts

0	reference state
1	Venturi inlet
2	Venturi throat

REFERENCES

1. W. F. Brooks, Helium transfer for the space station era, Cryogenics 26:61 (1986).

2. R. Meservey, Superfluid flow in a Venturi tube, Physics of Fluids 8:1209 (1965).

3. P. M. McConnell, "Liquid Helium Pumps", NBS1R 73-316, National Bureau of Standards, Boulder, Colorado (1973).

4. A. Kashani and S. W. Van Sciver, Steady state forced convection heat transfer in He II, in: "Advances in Cryogenic Engineering", vol. 31, Plenum Press, New York (1986), p. 489.

5. A. Rivetti, G. Martini, and R. Goria, A test circuit for calibration of liquid and supercritical helium flowmeters, in: "Advances in Cryogenic Engineering", vol. 29, Plenum Press, New York (1984), p. 903.

6. P. R. Ludtke, D. E. Daney, and W. G. Steward, Centrifugal pump performance with normal and superfluid helium, proceedings of this conference.

SONIC STANDING WAVE GAS DENSITY MONITOR

Ronald J. Walker

Fermi National Accelerator Laboratory*
Batavia, Illinois

ABSTRACT

Devices described in this document are capable of finding leaks with high sensitivity and reliability and at relatively low cost. Three versions of the basic detector have been developed and tested. One version is a portable instrument which is battery operated and has an output meter which gives a numerical indication of the amount of helium present at the pickup point. The portable version does not have temperature control because the operator can adjust the sensitivity and observe slight differences while searching for a leak location. The second version is mounted in a weather tight enclosure, operates on 110VAC line power, and has temperature control. This version is designed to mount in a fixed strategic position and give an alarm in the event that a helium leak develops. A third device has been developed, which is being used to monitor the density of gas in a large Cerenkov counter.

INTRODUCTION

The operation of Fermi National Accelerator Laboratory involves the use of a large and complex system of superconducting magnets, helium refrigerators and liquefiers, and extensive piping systems. Helium losses are economically significant if leaks are not found and repaired promptly, and the development of devices to automatically locate the leaks is strongly motivated. A large number of units is needed, and they must be very reliable, stable, and inexpensive.

A detector has been developed which meets these needs, based on a sonic resonance in a cylindrical cavity. It is possible to maintain the resonance in a very stable fashion due to changes in the relative phase of the voltage waveform exciting the cavity and the response of a microphone which measures the amplitude of the standing pressure wave at the opposite end of the tube. The pressure amplitude is the resultant of the traveling wave from the speaker and all of the reflections.

---

*Operated by Universities Research Association, Inc., under contract with the U.S. Department of Energy

THE RESONANT TUBE

The basic detector consists of a 15.24 cm long by 2.22 mm diameter cylindrical cell. An earphone type speaker is mounted in one end and a small microphone is mounted in the other end. The electronic circuitry energizes the speaker at a frequency which is equal to the organ pipe resonance at one half wavelength in the cell. This occurs at approximately 1100 Hz if the cell is filled with air at atmospheric pressure and at a temperature of 21 degrees C. The frequency of the resonance will change if the density of the gas changes. The sonic velocity depends on the density and bulk modulus of the gas according to the relation

$$c = \text{const} \sqrt{B/\rho} \tag{1}$$

The definition of the various symbols is included at the end of this text.

In an ideal gas, the bulk modulus is equal to the pressure, and this is approximately true in ambient air. It is evident from equation (1) that a 1% change in density will result in a 1/2% change in sonic velocity and consequently a 1/2% change in resonant frequency in the tube. The resonance we are discussing is a half wave resonance. Since both ends of the tube are closed, there will be a pressure antinode at each end with a pressure node in the center.

The circuitry which drives the speaker is able lock in on the resonant frequency in a very stable manner because of the large change of the relative phase of the signal to the speaker at one end and the output of the microphone at the other end when the system is near the resonance.

This phase difference is exactly 180 degrees at resonance, and varies sharply up or down as the frequency goes off resonance. The phase shift can readily be derived mathematically with a few simplifying assumptions. The pressure in the tube as a function of position and time and can be approximated by the summation of a traveling wave emitted by the speaker and all of the waves generated by reflections at the ends of the tube. It is assumed that the losses of energy in the system can be accounted for by using an effective reflection coefficient (R) which is less than one. The traveling wave emitted by the speaker is traveling to the right with sonic velocity, and the wave is described by the relation

$$y = \sin\left(\omega t - \frac{2\pi x}{\lambda}\right) \tag{2}$$

We are trying to obtain the phase difference between the traveling wave evaluated at the speaker, x = 0 and the total pressure wave at the microphone. We shall evaluate the wave amplitude and phase at the microphone position, x = L. Since there is no phase change on reflection at a closed end, the phase of the multiply reflected waves is determined by the total number of traversals of the tube length, and the amplitude is determined by the number of reflections. The amplitude of the traveling wave at the speaker is arbitrarily chosen to be 1.0.

Consider the contribution to the total amplitude at the microphone position, x = L, by the traveling wave as it is incident on the microphone for the n-th time. The incident wave will have been reflected 2n-2 times, and the reflected wave will have been reflected 2n-1 times. The contribution to the total amplitude of these two waves will therefore be

$$y_n = (R^{2n-2} + R^{2n-1}) \sin\left[\omega t - \frac{2\pi(2n-1)L}{\lambda}\right] \tag{3}$$

Note that the traveling wave has traversed the tube 2n-1 times at this point. If we define

$$A_n = (1 + R)R^{2n-2} \quad ; \quad \phi_n = \frac{2\pi(2n-1)L}{\lambda} \tag{4}$$

we can express the total amplitude at the microphone as

$$y_n = \sin \omega t\, \Sigma_n\, A_n \cos \phi_n - \cos\omega t\, \Sigma_n A_n \sin \phi_n \tag{5}$$

where the summations run from n = 1 to n = ∞. Define

$$C = \Sigma_n\, A_n \cos \phi_n \quad \text{and} \quad S = \Sigma_n\, A_n \sin \phi_n \tag{6}$$

Then the amplitude can be expressed as

$$y = A \sin(\omega t - \phi_n) \tag{7}$$

and the amplitude factor and phase are

$$A = \sqrt{C^2 + S^2} \quad ; \quad \phi = \tan^{-1} S/C \tag{8}$$

In order to obtain a quantitative prediction as to the phase and amplitude in an actual cell, the above sums were evaluated for a series of frequencies and reflection coefficients. The measured phase shift between the speaker coil and the microphone indicated an effective reflection coefficient of 0.96. This coefficient is reasonable considering the irregularities and cavities at the ends of the tube caused by mounting the speaker and microphone. Fig. 1 shows a plot of the calculated amplitude and phase shift for frequencies near the half wave resonance using a reflection coefficient of 0.96.

THE ELECTRONIC CIRCUITS

There exists at the present time a need for two types of instruments based on the principles discussed above. One type needed is a portable unit

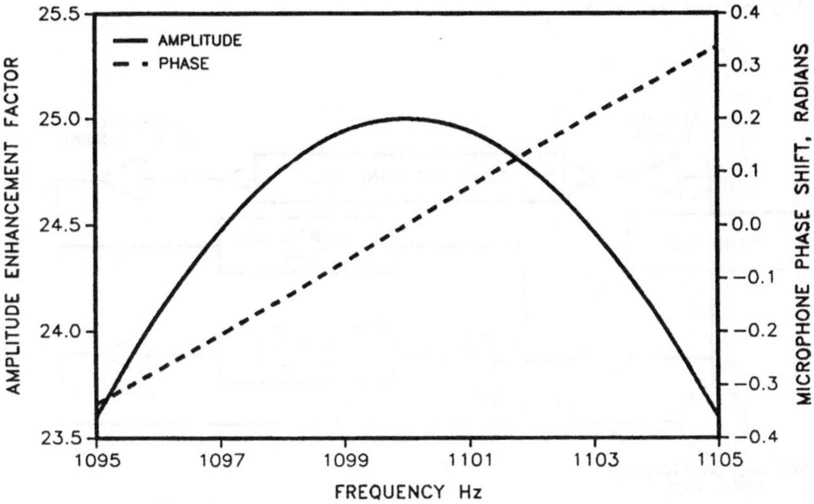

Fig. 1. Calculated amplitude and speaker to microphone phase shift of a typical sonic cell. The assumed reflection coefficient is 0.96.

used by a technician to search for helium leaks. The second type of unit needed is AC powered and it needs to be unattended and automatic in operation. This type will be mounted in a place where helium leaks are likely to appear in operation, such as relief valve locations. Many likely leak sources are outside and the unit needs to be able to stand the extremes of ambient temperature. The fixed unit contains a heater and proportional temperature control circuitry. A third type of unit has been placed in service. This unit uses the same cell and lock in circuit, but operates in the frequency range of 2000 to 3000 Hz which corresponds to the gas mixtures used in a Cerenkov counter used at Fermilab.

How the Phase Lock Works

The operating principle of the phase lock is illustrated in Fig. 2. This circuit is included in each of the three devices in use. The action of the circuit is to adjust the cell to it's resonant frequency and maintain resonance as the conditions in the cell change. The circuit finds the fundamental frequency, which is the half wave mode for a tube with both ends closed. A gas flow is introduced into the cell via small holes in the center, which is at a pressure node and a velocity antinode.

The source of the oscillation is a voltage controlled oscillator (An LM629). The "phase shift" block in Fig. 2 is a simple inductor-capacitor delay that is used to trim the phase of the speaker to optimize the match of phases between the oscillator and the microphone. A small transistor (2N2219A) drives the speaker. The action of the zero crossing elements (two LM311's) is to generate square voltage waves of 50% duty cycle which go positive as the inputs pass through zero.

The polarity of the speaker leads is adjusted so that the microphone and speaker are in phase at resonance for convenience in finding a control algorithm. The phase relationships are such that the the microphone zero crossing moves later relative to the oscillator waveform as the frequency shifts above the resonance. The "10 microsecond pulse" block generates 10 microsecond positive square pulses which rise on the trailing edge of the microphone zero crossing waveform. Therefore, as the frequency shifts below the resonance, the "AND" block (an LM7408) generates a train of pulses at its output. Furthermore, as the frequency shifts above resonance, the output of the AND circuit is zero.

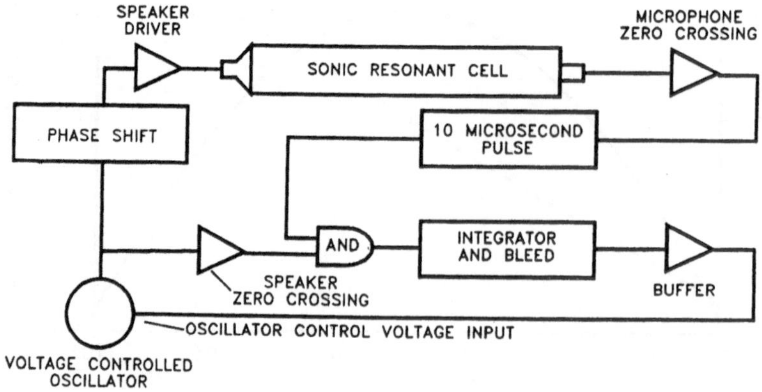

Fig. 2. Block diagram of the circuit which establishes and maintains a half wave sonic resonance in the sonic cell.

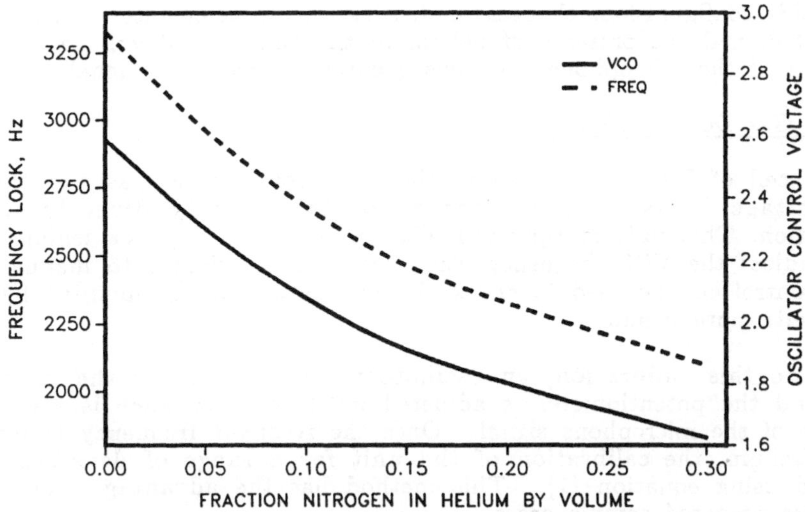

Fig. 3 Frequency lock data and associated voltage controlled oscillator input voltage for a Cerenkov counter gas density monitor using the standard sonic cell as a sensing element.

When a train of pulses is present at the output of the AND circuit, the DC output voltage of the integrator rises causing the frequency of the voltage controlled oscillator to increase. This voltage will continue to increase until the phases of the microphone and coil are matched and resonance is attained. If the frequency of the cell is above resonance, the integrator block will bleed down the control voltage because there are no charging pulses coming from the AND circuit.

This phase lock is very stable because the resonance in the cell is sharp and the phase shift is very sensitive to deviations from resonance.

The Portable Leak Detector

The portable leak detector has a small battery operated pump which draws air from a test point through the cell at the center point. One input of a differential amplifier (LM324) is attached to the control voltage input of the VCO (Voltage controlled oscillator) of the unit described in Fig. 2. The other input of the amplifier is set by an external potentiometer so that the operator can adjust the output of the amplifier to be in range of a digital volt meter. This unit is capable of producing a readable output for a sample pickup of 50 parts per million of helium in air, and gives a good response on a small "soap bubble" leak (about .001 standard cm^3/s)

The Fixed Leak Detector

This unit has twin cells mounted in a common enclosure with the speakers driven in series by the circuit of Fig. 2. A small 110 volt aquarium pump draws air through the two cells in parallel. The inlet of the cell which locks the VCO frequency is taken from a point near the unit and the inlet of the second cell is taken from the point likely to see a helium leak. The microphone of the sample cell is connected to a zero crossing chip in opposite polarity to the reference cell zero crossing which locks the frequency. The outputs of the two zero crossings are connected to the two inputs of a second AND circuit. The output of the second AND is integrated to generate a leak indication.

In this configuration, changes in density due to temperature variations are compensated, and the presence of helium in the sample cell will cause a large phase shift of the microphone and consequently a large error signal.

Cerenkov Gas Density Monitor

The cell of Fig. 2 is placed in the Cerenkov counter gas and the VCO input voltage is used as a measure of the density directly without amplification. This unit is equipped with a dial potentiometer which can be used to adjust the VCO frequency when the unit is switched to manual. The manual control can be used to check the calibration of the monitor by filling the cell with pure helium.

To do this calibration, an oscilloscope is attached to the microphone output and the potentiometer is adjusted until the resonance is seen in the amplitude of the microphone signal. Once the resonant frequency is measured in pure helium, the calibration of the unit for a range of densities can be calculated using equation (1). This method has the advantage that it does not require prepared sample gases.

The calibration curves shown in Fig. 3 were established in the above fashion. Fig. 3 shows a calibration for both VCO input voltage and microphone frequency. After the monitor was installed in the Cerenkov counter, a standard span gas with 20% nitrogen in helium was used to check the calibration curves of Fig. 3. The device responded to the span gas with an error of less than 1% from the calculated calibration curves. The output is very stable and reproducible. The VCO input voltage is linear in frequency.

CONCLUSIONS AND EVALUATIONS

The main motivation for this work was to find a practical leak detector which could operate continuously and be stable and reliable. The fixed leak detector version appears to do this job very well. We plan to eventually deploy approximately 100 of these units in the Fermilab Tevatron cryogenic system. All of our experience at Fermilab indicates that thermal conductivity devices have too much drift due to temperature variations and variations in sample pressure or flow rate to be able to constantly monitor systems without assistance from operators, especially in an outdoor environment.

The portable leak detector was built because it is an easy by-product of the fixed leak detector. The sensitivity of this detector is approximately equal to the soap bubble type of detection, and about a factor of 20 less than the ultimate sensitivity of a good thermal conductivity device. The sonic detector is sufficient to find any economically significant leaks. It is a fact that soap bubbles are more convenient than either the thermal conductivity device or the sonic device in those applications where soap bubbles are acceptable.

The cost factor or possibly reliability are the only reasons to use the portable sonic device instead of the thermal conductivity device. It is the opinion of this author that a major factor in the cost of the thermal conductivity device is the care required in the manufacture of the thermal conductivity cell, whereas no extreme care or close tolerances are required to manufacture the sonic cell.

We plan to manufacture about ten of the portable leak detectors and use them at this laboratory in various locations where the thermal conductivity devices are in use. This program will give some practical experience with the devices and form the basis for judgement on the reliability and usefulness of the portable detector.

The sonic unit installed to monitor the gas density in a 54 cubic meter Cerenkov counter has been very useful and successful. This unit is designed to enable the experimenters to maintain a 20% nitrogen in helium gas mixture in the counter. The popularity of this instrument is due to it's stability and the fact that it does not require a standard gas to perform a calibration or compare with the fill gas. Another useful feature is that it is not sensitive to the sample gas flow used. The experimenters plan to install the single sonic cell inside the counter when the operating schedule permits. After this is done, no sample flow will be required.

Sensitivity of the Sonic Detector

The sensitivity of the sonic detector will be discussed in terms of the fixed unit. If a high sensitivity device is to be used, the dual cell type of detection will be needed to provide sufficient stability. The curves shown in Fig. 1 apply to the phase shift and frequency in the sampling cell of the fixed unit. The output of the AND circuit in the fixed unit is connected to a simple RC integrator whose output in turn is equal to the 5 volt supply voltage multiplied by the duty cycle of the train off pulses coming out of the AND circuit. For example, a phase shift of 0.1 radians corresponds to a frequency shift of 1.3 Hz. and an output voltage of .08 volts. The helium content corresponding to this 0.1 radian phase shift will be 0.236% as calculated from equation (1). An easily readable output is .001 volts which corresponds to about 30 parts per million. This output is obtained directly from the sonic cells without the use of amplifiers and is stable at the 1 millivolt level.

A typical thermal conductivity unit with a maximum sensitivity of 10 parts per million uses a differential amplifier with a gain of 1000 on the output of the dual thermal cell to obtain the maximum sensitivity. This unit appears to have a large amount of drift in the high sensitivity mode.

NOTATION

c	sonic velocity	m/s	L	tube length	m
B	bulk modulus	n/m^2	n	number of tube traversals	1
ρ	density	kg/m^3			
y	relative amplitude	1	R	reflection coefficient	1
ω	angular velocity	Hz	A_n	amplitude coefficient	1
t	elapsed time	s	ϕ_n	phase of nth reflected wave	rad
x	distance down tube	m			
λ	wavelength	m			

C	partial amplitude summation	1
S	partial amplitude summation	1
A	net relative amplitude at microphone	1
ϕ	net relative phase at microphone	rad

AN ELECTRONIC BALANCE FOR WEIGHING FOAMS AT CRYOGENIC TEMPERATURES*

R. O. Voth and J. D. Siegwarth

Chemical Engineering Science Division
National Bureau of Standards
Boulder, Colorado

ABSTRACT

A commercial electronic balance was altered to weigh objects in a cryogenic environment containing combustible fluids. The balance was used to measure the mass gain of open cellular foams touching the surface of liquid hydrogen. The mass gain rate is a function of the foam wicking characteristics (a function of the foam structure). Weighing the empty and liquid filled foam sample in the ullage above the liquid, and the foam submerged in the liquid, gives sufficient information to allow a free volume or porosity to be determined for the foam. These tests are especially important for foams too weak to withstand the surface tension forces of ambient temperature liquids. Details of the design of the cryogenic balance and some results for several types of foams are presented in this paper.

INTRODUCTION

A desired fuel pellet for the inertial confinement fusion (ICF) research experiment is a hollow spherical shell of liquid fuel.[1] An outer hollow spherical shell contains the liquid fuel, but in spheres of the desired size the liquid slumps to the bottom of the shell in 1-g gravity. A method being investigated to support a thin uniform layer within the outer shell consists of lining the container with a rigid foam matrix. Liquid condensed from an initial gas fill or added to the cold target through a tube fills the foam and is held in it by surface tension. Characteristics of the foam important for this use include small pores that will readily wick low surface tension cryogenic fluids, the ability to survive low temperatures, a high liquid capacity compared to the foam volume and easy machinability that allows fabricating to the close tolerances required in this application.

APPARATUS

We investigated the low temperature characteristics of foams in liquid hydrogen or liquid deuterium by suspending them from a modified

*Contribution of the National Bureau of Standards, not subject to copyright. Work sponsored by the Lawrence Livermore Nat'l. Laboratory.

electronic balance. Weighing the foam samples in low temperature environments allowed us to measure the wicking characteristics with the two hydrogen isotopes and determine the mass of liquid held by a known volume of foam.

The electronic balance, shown in figure 1, was constructed to weigh the foam samples in liquid hydrogen. The electronic balance sensing unit was sealed into a chamber maintained at ambient temperature that communicated with the hydrogen in the cryostat ullage volume via a vertical tube. The balance mechanism is from a commercially available balance capable of weighing 200 g to 0.001 g. The electronics associated with the balance were separated from the sensing unit and installed in a purged box to eliminate ignition sources in the experimental area. An extension cord connected the electronics to the sensing unit. The vertical tube penetrates into the inner test Dewar. The electronic balance was suspended from a water actuated cylinder that could move the balance mechanism vertically. Moving the balance allowed us to weigh the foam in the ullage without a liquid fill, with one end of the foam sample touching the liquid surface and wicking liquid and with the sample submerged in the liquid and full in the ullage. From these measurements the wicking rate, the quantity of liquid held by the foam and the foam material density could be calculated. Liquid surface tension influenced the submerged weight unless the sample mechanism D was completely submerged to above the hook at C in figure 1.

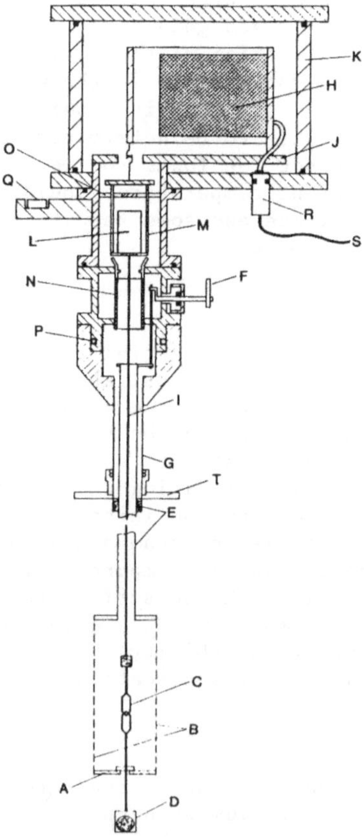

Fig. 1. The electronic balance sensing unit modified for use in liquid hydrogen.

In figure 1, the foam and cage D are shown lifted from the hook C allowing the balance to be zeroed. The foam is raised by a pneumatic cylinder via the bell crank F, the tube E, the chains B, and the washer A. Vertical movement is allowed by the double O-ring slip seal in flange T and the 25 mm O.D. tube G from the electronic balance sensing unit. The sample is suspended from the balance H by a small chain I in the tube E. The balance H is mounted on a cantilevered plate J so that, should gas tight enclosure K flex when pressurized, the leveling of the balance is unaffected. The enclosure is capable of operating with up to one MPa internal pressure. The tungsten reference weight L, used to monitor the balance calibration, is shown placed on its suspension cage M. This cage replaces a section of the suspension I. Some swivel joints in this suspension allows flexing but not rotation,. A pneumatic-cylinder-driven bell crank, similar to F but not shown in the drawing raises N which lifts the reference weight L off the suspension and clamps it against the fixed top support O. A gas tight seal R is provided for the electrical leads S leading to a remote digital readout.

Two adjustments were designed into the balance. The O-ring seal at P, connecting the microbalance assembly to the support tube G, is a gland seal rather than a flange seal so the instrument can be leveled according to the bull's eye level Q while G is held vertical. The flange T is leveled as required to maintain the suspension wire C centered in tube E by moving the mating flange on the cryostat. The mating flange is flexibly connected to the cryostat with a bellows; and is leveled using three threaded standoffs screws.

The lightest foam sample weighed 0.3 grams and there was an inaccuracy of 0.3 % in its mass due to the sensitivity of the balance.

The cryostat used in the foam tests consists of a liquid nitrogen shielded liquid hydrogen container with an internal test container. The internal container can be filled directly with liquid hydrogen or filled with liquid deuterium condensed by the liquid hydrogen in the hydrogen container. Windows in the cryostat permitted lighting, viewing and photographing the foam during the tests.

TEST PROCEDURES

The first foams tested were silica aerogel foams.[2,3] These aerogel foams had pores that ranged in size from 5 to 10 nm and a free volume of about 95% of the total volume of the foam. The combination of small pore size that produced relatively large forces from surface tension and of the weak interstitial material caused this foam to shatter into powder when touched by room temperature liquids.

The initial tests were qualitative. The main interest was whether the foam would wick liquid hydrogen and survive, and whether the foam would hold the liquid when it was suspended in the ullage space. Fortunately, the silica based aerogel was clear enough that with a proper back lighting the wicking liquid interface was visible as it moved up the foam sample. Figure 2 is a picture of the liquid interface in an aerogel foam sample 2 cm long and 2 cm in diameter. The liquid deuterium surface is faintly visible near the bottom of the top foam sample, while the wicking liquid interface is near the top of the sample. The wicking interface was horizontal unless one of the holding cage wires was too close to the foam and the liquid wicked up between the wire and the foam. When the liquid did wick up the side, a complex interface formed because liquid wicked in from the side as well as the bottom.

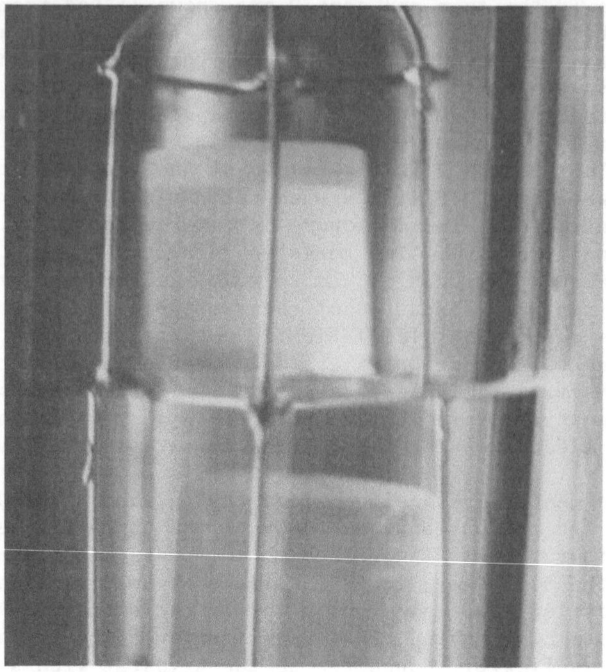

Fig. 2. Silica aerogel foam wicking liquid deuterium. The top sample is nearly full and is opaque. The lower sample is submerged and clear.

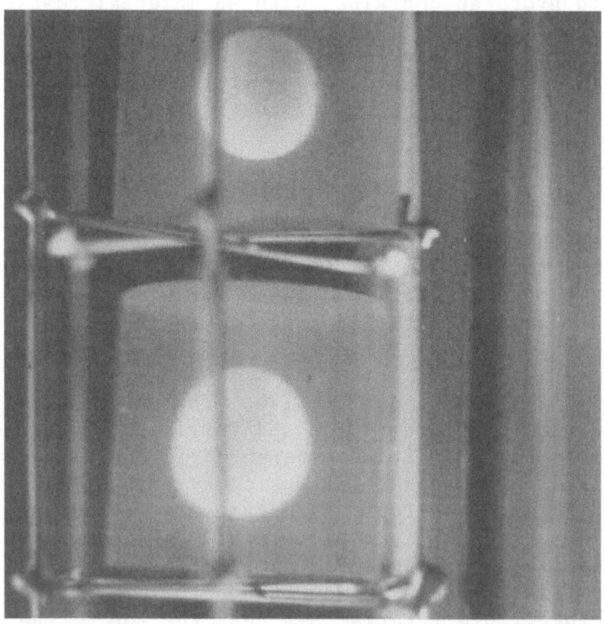

Fig. 3. Silica aerogel foam warming and evaporating the contained liquid deuterium. The white sphere is the remaining deuterium.

When the liquid hydrogen wicking interface reached the top of the sample, just filling with liquid, the sample would remain temporarily opaque. After a short time, the foam would become clear. The top sample in figure 2 shows the opaque quality while the submerged lower sample is clear. The foam remained clear until it warmed and liquid began evaporating. As the liquid evaporates from the sample it first turns white and the edges of the sample become clear while the portion containing liquid remains white and opaque. A photograph of the remaining opaque white spheres is shown in figure 3. We do not believe the remaining sphere is solid deuterium since the temperature and pressure in the test chambers are always above the triple point.

The procedure followed to obtain wicking characteristics of the foams was to fill the cryostat with the test liquid, weigh the sample in the ullage before it picked up any liquid, weigh the sample when the bottom of the sample touched and wicked the liquid, and finally weigh the full sample submerged in the liquid and in the ullage space above the liquid.

RESULTS

Figure 4 shows some typical wicking data for a variety of foams. These curves are based on weight gain-versus-time measurements made as the foam samples wicked full of liquid. The weight gains were converted to volumes and then to interface heights in the foam, based on the liquid density, the foam's free volume, and the assumption that the wicking liquid interface was horizontal. The curves on figure 4 show interface velocity or slope of the filling curve plotted against the height of the interface. The disparities in the starting or minimum wicking height result from variations in how far the foam sample is initially submerged to begin the wicking test. If the sample is dipped very far into the liquid, time is needed before a horizontal interface

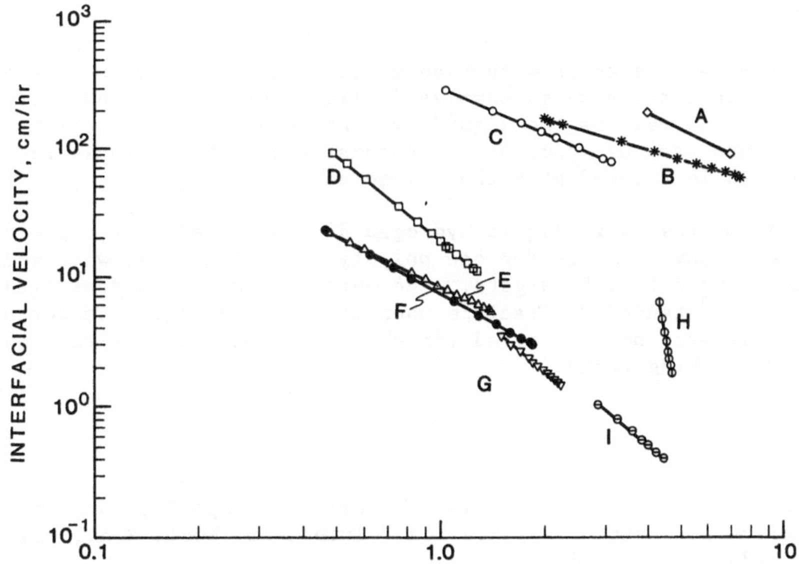

Fig. 4. Liquid interface velocities for various foams.

Table 1. Physical characteristics of the foams shown in figure 4

Foam	Foam Type	Test Fluid	Foam Density mg/cm^3	Free Volume percent
A	TPX	LH_2	40	95
B	Carbon	LH_2	43	90
C	Polystyrene	LH_2	65	95
D	Resorcinol	LH_2	111	89
E	Polyethylene	LH_2	107	89
F	Silica Aerogel	LD_2	101	95
G	Polyethylene	LH_2	107	95
H	Polystyrene	LH_2	65	95
I	Polyethylene	LH_2	107	89

is established and the wicking proceeds from a fixed area. During this time the slope of the wicking curve changes rapidly depending on how far the sample end is submerged into the liquid. These initial data cannot be readily interpreted and are not included on figure 4. The abrupt change in slope when the sample filled is also not shown by the curves.

Most of the foams tested were manufactured specifically for use in ICF experiments. Thus, every attempt had been made to manufacture a foam with small pore sizes and yet a 90% or higher free volume to hold the maximum liquid. Table 1 shows some of the physical characteristics for the foams tested. The free volume is the fraction expressed as a percentage of the foam filled with liquid, based on the weight of a foam sample full of liquid in the ullage. Foams E, G, and I are the same foams but the curves are for three different tests. The difference in slope between foams E and I is due to the short duration of test E. The wicking rate had probably not yet stabilized from the initial submersion. Curve G represents the wicking rate with the wicking liquid subcooled by slightly pressurizing the ullage with helium.

CONCLUSIONS

These measurements show that an electronic balance with a zeroing feature can accurately weigh samples in liquid hydrogen. The balance has been used to measure the liquid hydrogen and liquid deuterium wicking rates into various foams and to demonstrate that the foams retained the liquid while suspended in the ullage above the liquid.

The foams tested in liquid hydrogen display widely varying wicking rates. Most foams, except for one polystyrene sample (H), wick well enough to be used in ICF targets. The aerogel foams with densities less than 100 mg/cm^3 tended to fracture when filled with liquid deuterium but no damage has ever been observed (in the other foams during the low) temperature wicking tests.

REFERENCES

1. D. H. Darling and R. A. Sacks, "Wetted foam capsules for direct drive ICF reactor application," Trans. Am. Nucl. Soc., Vol. 52 (1986) p. 257-258.

2. W. J. Schmitt, R. A. Grieger-Block, and T. W. Chapman, "The preparation of acid catalyzed silica aerogel," Chemical Engineering at Supercritical Fluid Conditions, Ed. Michael E. Paulaitis, Ann Arbor Science (1982), p. 445-460.

3. C. W. Price, W. G. Halsey, and S. C. Sanders, "Structural characterization of silica aerogel using specific surface area measurements," Report, Lawrence Livermore National Laboratory, 1985.

ELECTRO-OPTIC SAMPLER FOR CHARACTERIZATION OF DEVICES IN A CRYOGENIC ENVIRONMENT

Douglas R. Dykaar, Roman Sobolewski,[(a)] James M. Chwalek, Thomas Y. Hsiang, and Gerard A. Mourou

Department of Electrical Engineering, and Ultrafast Science Center at the Laboratory for Laser Energetics
The University of Rochester, Rochester, New York

ABSTRACT

A fully integrated superconducting electro-optic sampler, designed for subpicosecond measurements of the transient response of devices operating in a cryogenic environment, is reviewed in detail. Several different configurations of the sampling head, including a coplanar transmission line geometry with a photoconductive switch fabricated on semi-insulating GaAs and sensor transmission lines fabricated on lithium tantalate, are discussed. The optical part of the system consists of two 80-fs laser pulse trains at a 100 MHz repetition rate, generated by a colliding-pulse mode-locked laser. One pulse train was used to generate a picosecond electrical input pulse, while the second probed the induced change in birefringence of the electro-optic crystal ($LiTaO_3$). By changing the relative delay between the optical excitation and sampling pulses, the temporal evolution of the electrical transient was recorded. The limit of the temporal response of our sampler, as determined by measuring the transient onset of photoconductivity in a GaAs switch, was less than 400 fs. We discuss two applications of this sampler for the study of the picosecond pulse propagation on superconducting transmission lines and for probing the switching of a Josephson tunnel junction.

INTRODUCTION

Electrical sampling systems with sensitivity to resolve events faster than 10 ps are either built with Josephson electronics or are based on femtosecond laser techniques. The Josephson sampler, developed by Faris,[1] was specially designed for transient measurements of superconducting electronics and, as was shown in Ref. 2, is well suited for that purpose. However, it uses a Josephson junction as the sampling gate, and therefore the time resolution (a few picoseconds) is limited to the present state of the art in junction design.[2] Taking advantage of femtosecond lasers, there are two optical systems capable of subpicosecond time resolution that are operated at liquid helium temperatures. These are the Auston sampler,[3] which uses a photoconductive sampling gate, and the electro-optic sampler[4] which is based on the measurement of the change in the birefringence of certain materials (such as lithium tantalate or GaAs) induced by an electric field transient. The former system has recently been used by researchers at IBM[5] to study the picosecond pulse propagation on niobium transmission lines, while the latter has seen a wide range of application and is well-established as the most popular sampling system for ultrafast electronics.[6-8]

In this paper we report the cryogenic implementation of the electro-optic sampler and its applications to testing superconducting devices and circuits. The design represents a natural extension of the room-temperature system into the low temperature environment. All performance parameters of the room-temperature version of the sampler, such as subpicosecond resolution and microvolt sensitivity[4] are either improved or remain unchanged in the cryogenic version. The measured temporal response of less than 400 fs represents the fastest electrical transient ever measured at cryogenic temperatures.

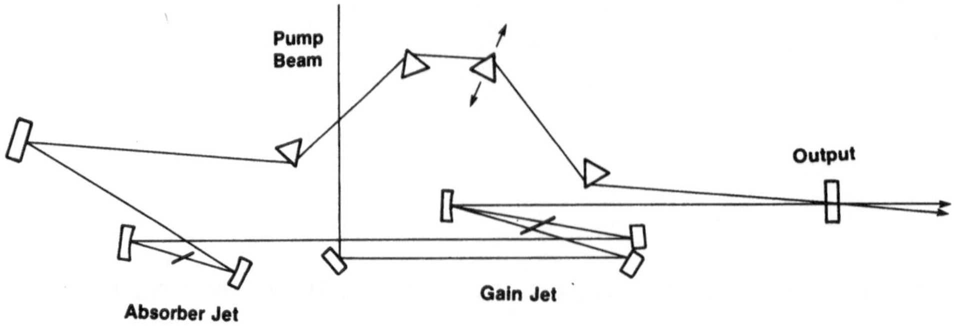

Fig. 1. Colliding pulse mode locked laser system.

The next section discusses the principles of electro-optic sampling. We focus our attention on the femtosecond laser system, and on the electro-optic technique of extracting information about an unknown electrical transient. Section III reviews several designs of the cryogenic, electro-optic sampling system, with the best configuration for our measurements discussed in detail. Examples of applications of the system different measurements are presented in Sec. IV. Final remarks are in the concluding section.

PRINCIPLES OF ELECTRO-OPTIC SAMPLING

Generation of Femtosecond Optical Pulses

The underlying speed of the electro-optic sampling technique is the direct result of the extremely short duration of the laser pulses used to make the measurements. The laser used is a Colliding Pulse Mode Locked (CPM) system which in its present form is capable of generating pulses as short as 27-fs Full Width at Half Maximum (FWHM).[9] The CPM laser has been described in the literature (see e.g. Ref. 10) and only a brief description of the version used in our experiments will be given.

As shown in Fig. 1 the CPM laser is a ring laser using two dye jets, a Rhodamine 6G gain jet and a DODCI (DiethylOxa-DicarboCyanine Iodide) saturable absorber jet. These two dyes are a well-matched pair and produce passively mode-locked pulses. The pulses counter-propagate in the laser cavity and become shortened when they collide in the saturable absorber. This accounts for the two beams shown leaving the cavity in the figure. A commercial cw argon laser pumps the gain jet at about 3 W.

Four Brewster angle prisms are then placed into the cavity. These prisms serve to compensate for any linear optical dispersion which occurs in the propagation paths inside the cavity, through the introduction of negative dispersion. By varying the amount of glass in the beam path, the pulse width is then minimized. The resultant pulses are well below 100 fs FWHM. For stability reasons we chose to work with pulses of 60-80 fs FWHM. The pulse repetition rate is chosen to be 100 MHz and is determined by the round trip propagation time in the cavity. Thus, the output from the cavity is a pair of pulse trains with 10 ns pulse separation. Jitter between individual pulses or between pulse trains is on the order of a few femtoseconds.[11]

Generation and Detection of Ultrafast Electrical Signals

To initiate an electrical transient, a photoconductive switch is excited by the CPM laser pulses. The propagated electrical signal is then detected using the second CPM laser pulse train. Transformation between electrical and optical signals is accomplished through the use of the Pockels effect[10] which alters the state of polarization of an electro-optic medium in proportion to the applied electric field. If the medium is placed between crossed polarizers, the induced birefringence results in a change in the optical intensity. Since this change is an atomic phenomenon, the speed can be as fast as a few femtoseconds.

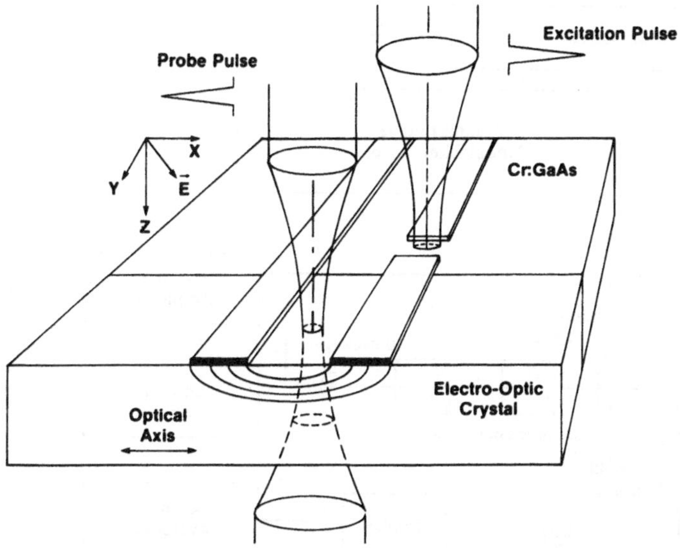

Fig. 2.
Sample configuration.

The configuration of a typical experiment is shown in Fig. 2. Here, a coplanar transmission line is fabricated on a semi-insulating GaAs substrate and continues onto a piece of $LiTaO_3$ crystal butted against the GaAs chip, with the crystal serving as the sensor for the Pockels effect. When the excitation pulse arrives at the switch, photocarriers are generated which effectively close the switch. The rise of the photocarrier population follows the integral of the optical pulse so that the switch is closed on a time scale of 100 fs. Once the electrical pulse begins to propagate on the transmission line, however, it is subject to the dispersion due to signal propagation on transmission lines. By positioning the switch gap very close to the edge of the GaAs, the propagation distance to the first sampling point and the resultant pulse dispersion can be kept very small.

Microvolt Sampling

Figure 3 shows the actual implementation of the electro-optic sampling scheme. It is noted that most electrical signals of interest are in the volt or even millivolt range, whereas the halfwave voltage for the modulators (Pockels cell) used is in the kilovolt range. Significant sensitivity enhancement is therefore needed.

The first step is provided by the laser itself. With a repetition rate of 100 MHz, small signals can be quickly integrated by the photodiodes into large ones. Next, a lock-in amplifier is used to move the integrated diode signal into a frequency regime of low electronic noise.

As shown in Fig. 3, the lock-in amplifies a particular frequency used for modulating either the bias of the photoconductive switch or, via an acousto-optic modulator, the excitation pulse train. Our latest version uses a mixing technique in conjunction with the low-frequency, high-sensitivity lock-in amplifier, which allows detection of signals in the 2–4 MHz region. The advantage of this scheme is that the modulation frequency can now be chosen to be above the 1/f region of the CPM noise spectrum (with the 1/f noise "floor" at about 2 MHz). As in the previous low-frequency measurement scheme,[4] a differential measurement is made using the two polarizations from the final polarizer to increase the signal to noise ratio. The photodiode loads are properly adjusted to maximize diode response at the modulation frequency.

CRYOSAMPLER DEVELOPMENT

In order to characterize superconducting electronics on a picosecond timescale, the electro-optic sampling system used in the room temperature environment had to be modified. First, a dewar had to be designed for such an integrated system. This was accomplished by a special design with a total of four in-line windows (all neither crystalline nor birefringent) to allow laser beams to pass through the dewar undistorted. The liquid helium in the dewar must also be nondistorting, so the temperature in the dewar was reduced below the superfluid transition point. This condition had the added advantage of excellent thermal equilibration and prevented thermal birefringence in the sampling crystal.

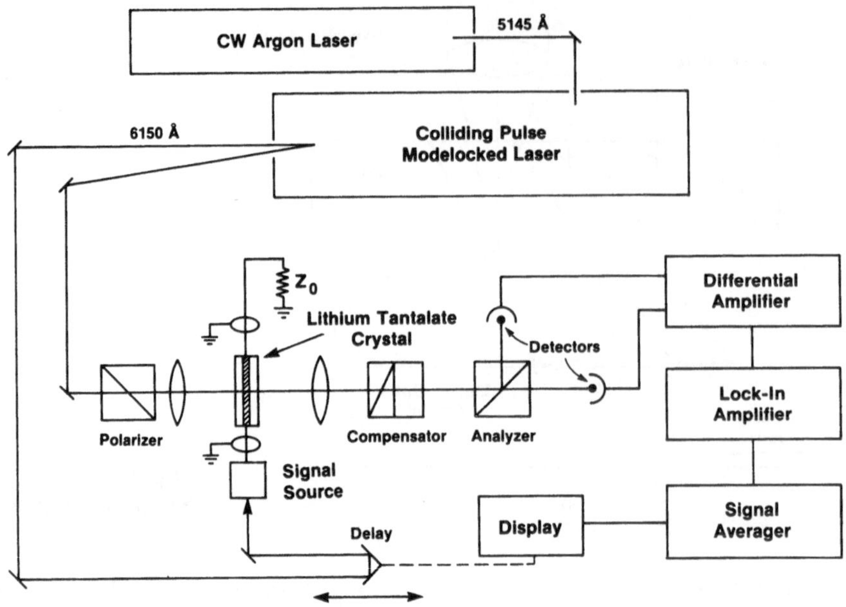

Fig. 3. Sampling system schematic (switch bias modulation is omitted for clarity).

In our first attempt, a simple setup was designed using a coaxial-type geometry with the sampling crystal placed in a cut cable just above the liquid helium level.[12] Using 2-mm (0.085 in.) diameter semirigid coaxial cable and 40 GHz connectors, a system rise time of 16.4 ps was measured. This relatively slow response was a direct result of the long cables (about 1 m) required to route the signal from the room-temperature switch into the dewar.

The next step was to integrate the switch and the sampling crystal.[13,14] Several structures were considered. The reflection mode geometry[13] was tried first. In this geometry the sampling crystal is coated with a dielectric reflective coating. The gapless structure allowed the switch to be defined by the laser excitation beam spot rather than by a predefined gap made into the switch. In addition to allowing very short propagation distance, this structure provided flexibility in changing the propagation path length. A sampler made in this geometry, using lead electrodes, had a measured rise time of 1.8 ps at room temperature. However, it became apparent that it was difficult to properly align the laser beams inside the dewar, and so different geometries were explored. (We note that a similar design has been adopted recently by Gallager et al.[5] who took full advantage of the gapless structure).

The final geometry chosen for most of the experiments is schematically shown in Fig. 2. This geometry has also been used at room temperature, with a measured rise time of 460 fs.[14] In this design, the switch and sampling crystal were edge-polished until a mating surface was produced with micron uniformity. These were then glued down onto a microscope slide, and surface polished. The resultant structure was so uniform that metal electrodes could be evaporated continuously across the interface. The sampling test structure was fabricated using photolithographically defined 20 μm indium lines. The measured time resolution of this test structure was 360 fs, as shown in Fig. 4.

APPLICATIONS

Picosecond Pulse Propagation on Superconducting Transmission Lines

The sampling system described above was used to experimentally characterize propagation of picosecond electrical transients on coplanar striplines. In one experiment,[15] transients with a 10 - 90% rise time of 1.0 ps and a relaxation time on the order of a few

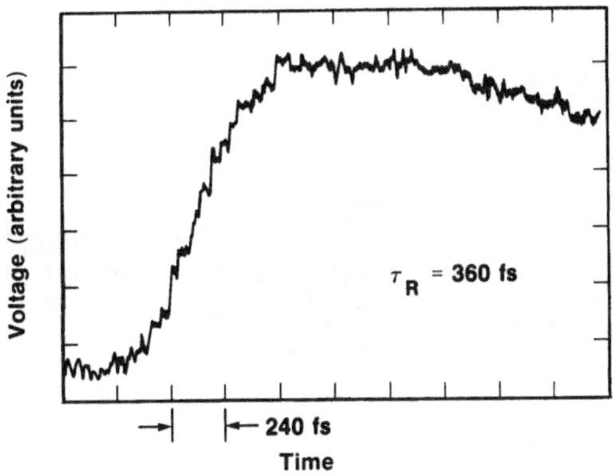

Fig. 4.
Output waveform from the fully integrated cryogenic sampler. Rise time is about 360 fs.

picoseconds were launched onto a superconducting transmission line with coplanar indium striplines. The experimental configuration was identical to that shown schematically in Fig. 2. The output waveforms shown in Fig. 5a were sampled at different points along the lines on $LiTaO_3$, with the first waveform sampled at 75 µm from the switch and designated as 0.0 mm. Next the signals were sampled at 0.15, 0.3, 0.9, and 1.5 mm of propagation. Different characteristics were observed on the output: a ringing was noted on top of the plateau of the transient, and a distinctive pulse sharpening resulted as the slower, high-frequency components of the pulse piled up on the faster moving plateau. The rise time also increased, as the signal propagated along the line.

Figure 5b presents the result of numerical simulations of the experimentally studied indium lines. The calculations are based on a model and its corresponding algorithm developed by Whitaker et al,[16] which is a full-wave model and takes into account both the modal dispersion effects and superconducting energy gap phenomena. One notes the excellent quantitative agreement between the experimental and numerical data. For details, the readers are referred to Refs. 15 and 16.

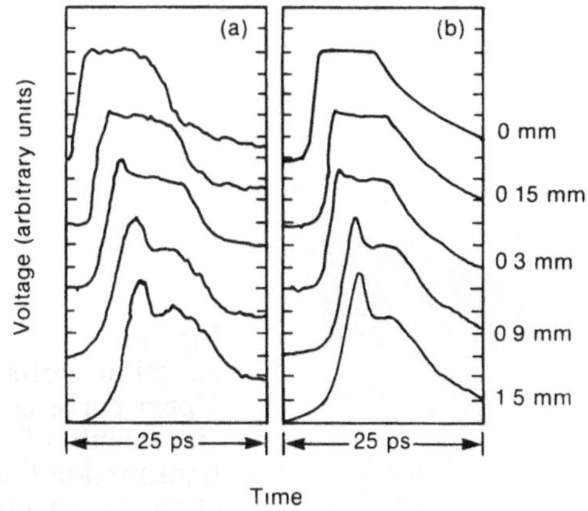

Fig. 5. Sequence of electrical transients propagated on the indium transmission line (T = 1.8 K). (a) Experimental results; (b) numerical simulations.

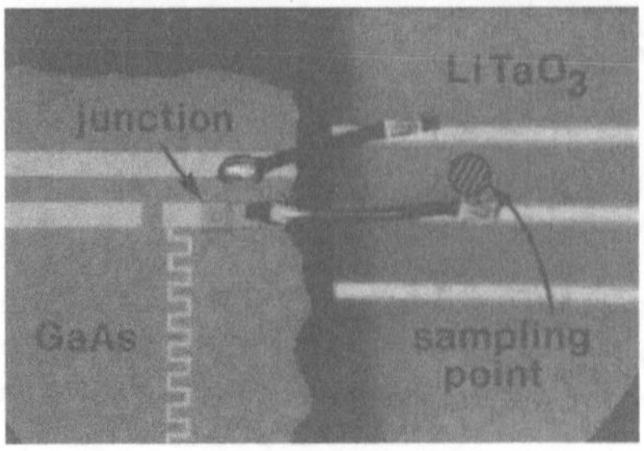

Fig. 6.
Micrograph of the sampling head used in the measurements of the Josephson junction switching process.

Characterization of Josephson Tunnel Junctions

The cryogenic sampler has also been successfully used to study the switching process of an unbiased Josephson junction driven by picosecond electrical pulses.[17] The cryogenic part of the experimental setup is shown in Fig. 6. The circuit, fabricated on GaAs, was made in a standard Pb alloy technology. The junction was integrated with the photoconductive switch (a 50-μm wide gap on the left hand side of the junction) so that the electrical pulses were delivered to the junction virtually undistorted. A 10-μm wide meander line supplied the necessary DC bias to the junction and simultaneously represented a low-pass filter for the high-frequency signals. The junction output was connected to a 50-Ohm Au coplanar waveguide, fabricated on a $LiTaO_3$ crystal, by 18-μm Au wire bonds. The electrical input pulse duration was about 8 ps.

The difficulty of this experiment was in discriminating between the junction response of less than three millivolts and the much larger excitation pulse. In order to overcome this difficulty, the experimental configuration was arranged in the same way as a typical transfer-function measurement in two-port microwave network. For a fixed input signal, we measured the signal propagating through the $LiTaO_3$ transmission line for two different states of the circuit. The top waveform in Fig. 7 was taken with a resistor substituted for the junction (the junction was driven normal, using cw light from the argon laser), while the bottom trace corresponds to the response of the unbiased junction. The difference in the waveforms was attributed to the different propagating conditions caused by the junction switching from the zero-voltage to the finite-voltage state.
The switching process can be visualized in Fig. 8, where we plot the difference between the waveforms presented in Fig. 7. It is seen that the junction started to respond when the input pulse reached its maximum, and the full voltage across the junction was developed on a time scale corresponding to the input pulse duration. After switching, the junction remained in the voltage state. These observations agree well with numerical simulations.[17] For details, the readers are referred to Refs. 17 and 18.

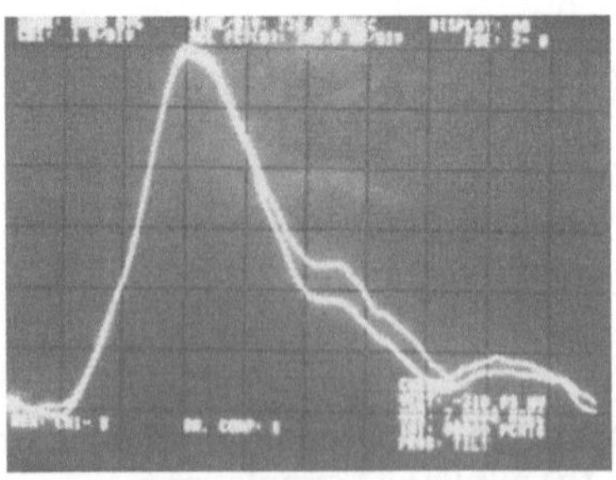

Fig. 7.
Junction voltage waveforms. Upper curve corresponds to the "no junction " condition in the transmission line, and the lower curve is the unbiased junction response.

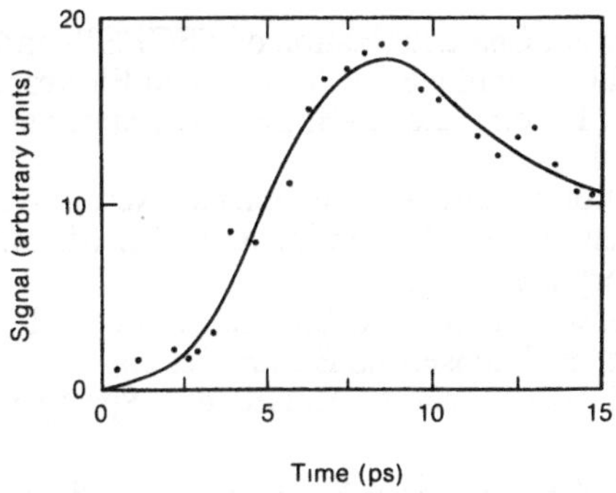

Fig. 8.
Time resolved switching process of an unbiased Josephson junction driven by an 8-ps electrical pulse. The input pulse amplitude was approximately five times the junction critical current.

CONCLUSIONS

In this paper, we reviewed the cryogenic electro-optic sampler for use in picosecond transient studies. The principles of operation, the system configuration, and applications to pulse transmission studies and to Josephson junction switching have been discussed. This sampling system is a versatile tool for studies of any ultrafast low temperature devices (e.g. semiconductor or superconductor-semiconductor hybrid) that have temporal response well beyond the capacity of traditional electronic equipment. In particular, with propagation characteristics on transmission lines well studied and an effective algorithm established, we have at hand capabilities to investigate the transient response of devices in a quatitative and precise way.

ACKNOWLEDGEMENT

Work supported by the National Science Foundation through Grant No DMR-85-06689, by the U.S. Air Force under contract F49620-87-C-0016 to the Ultrafast Optical Electronics Center at the Laboratory for Laser Energetics of the University of Rochester, and by the Laser Fusion Feasibility Project at the Laboratory for Laser Energetics which has the following sponsors: Empire State Energy Research Corporation, General Electric Company, New York State Energy Research and Development Authority, Ontario Hydro, and the University of Rochester. Such support does not imply endorsement of the content by any of the above parties.

REFERENCES

(a) also at the Institute of Physics, Polish Academy of Sciences, PL-02668 Warszawa, Poland.
1. S. M. Faris, Generation and measurement of ultrashort current pulses with Josephson devices, Appl. Phys. Lett. 36:1005 (1980).
2. P. Wolf, B. J. Van Zeghbroeck, and U. Deutsch, A Josephson sampler with 2.1 ps resolution, IEEE Trans. Magn. MAG-21:226 (1985).
3. D. H. Auston, Picosecond optoelectronic switching and gating in silicon, Appl. Phys. Lett. 26:101 (1975).
4. J. A. Valdmanis, G. A. Mourou and C. W. Gabel, Picosecond electro-optic sampling system, Appl. Phys. Lett. 41:211 (1982); see also J. A. Valdmanis and G. A. Mourou, Subpicosecond electro-optic sampling: principles and applications, IEEE J. Quantum Electron. QE-22:69 (1986).
5. W. J. Gallagher et al, Subpicosecond optoelectronic study of resistive and superconductive transmission lines, Appl. Phys. Lett. 50:350 (1987); C. -C. Chi et al, Subpicosecond optoelectronic study of superconducting transmission lines, IEEE Trans. Magn. - in print.

6. K. E. Meyer, D. R. Dykaar, and G. A. Mourou, Characterization of TEGFETs and MESFETs using the electro-optic sampling technique, in: "Picosecond Electronics and Optoelectronics," G. A. Mourou, D. M. Bloom, and C.-H. Lee, eds.,Springer-Verlag, Berlin, Heidelberg (1985), p. 54.
7. D. R. Dykaar et al, Picosecond characterization of ultrafast phenomena: New devices and new techniques, in: "Ultrafast Phenomena V," G. R. Fleming and A. E. Siegman, eds., Springer-Verlag, Berlin, Heidelberg (1986), p. 103.
8. B.H. Kolner, K.J. Weingarten, M.J.W. Rodwell, and D.M. Bloom, Picosecond sampling and harmonic mixing in GaAs, in: "Picosecond Electronics and Optoelectronics," G. A. Mourou, D. M. Bloom, and C.-H. Lee, eds.,Springer-Verlag, Berlin, Heidelberg (1985), p. 50.
9. J.A. Valdmanis, and R.L. Fork, Design consideration for a femtosecond pulse laser balancing self phase modulation, group velocity dispersion, saturable absorption, and saturable gain, IEEE J. Quantum Electron. QE-22:112 (1986).
10. A. E. Siegman, "Lasers," University Science Books, Mill Valley (1986).
11. M. A. Pessot - private communication.
12. D. R. Dykaar, T. Y. Hsiang, and G. A. Mourou, An application of picosecond electro-optic sampling to superconducting electronics, IEEE Trans. Magn.MAG-21:230 (1985).
13. D. R. Dykaar, T. Y. Hsiang, and G. A. Mourou, Development of a picosecond cryo-sampler, in: "Picosecond Electronics and Optoelectronics," G. A. Mourou, D. M. Bloom, and C.-H. Lee, eds., Springer-Verlag, Berlin, Heidelberg (1985), p. 249.
14. G. A. Mourou and K. E. Meyer, Subpicosecond electro-optic sampling using coplanar strip transmission lines, Appl. Phys. Lett. 45:492 (1984).
15. T. Y. Hsiang et al, Propagation characteristics of picosecond electrical transients on coplanar striplines, submitted to Appl. Phys. Lett.
16. J. F. Whitaker et al, Propagation model for ultrafast signals on superconducting dispersive striplines, submitted to IEEE Trans. Microwave Theory Tech. - special issue on microwave CAD.
17. D. R. Dykaar, R. Sobolewski, T. Y. Hsiang, and G. A. Mourou, Response of a Josephson Junction to a stepped voltage pulse, IEEE Trans. Magn. - in print.
18. R. Sobolewski, D. R. Dykaar, T. Y. Hsiang, and G. A. Mourou, Picosecond switching in Josephson tunnel junctions, in: "Picosecond Electronics and Optoelectronics," Springer-Verlag, Berlin, Heidelberg (1987) - in print.

CONTROL SYSTEM FOR HELIUM REFRIGERATORS OF TRISTAN DETECTOR MAGNETS

T. Mito, T. Haruyama, M. Tadano, Y. Kondo, A. Yamamoto
N. Kimura, and Y. Doi

National Laboratory for High Energy Physics
Physics Department
Tsukuba-gun, Ibaraki-ken, Japan

ABSTRACT

A computer control system for the helium refrigerators of the TRISTAN detector magnets (VENUS, TOPAZ, and AMY) was designed and constructed for the purpose of centralized and automatic operation. The control system of these helium refrigerators is programmed to do a series of procedure automatically in order to simplify the operation. The system is also programmed to meet emergencies such as a power failure or a magnet quench, in order to guarantee the long-term operation without trouble. Each part of the system was tested and confirmed to work well after some modifications. All helium refrigerators have been operated together using the centralized control station from January, 1987.

INTRODUCTION

Colliding experiments of the TRISTAN electron-positron collider at KEK started successfully in October, 1986. Three superconducting solenoid magnets and helium refrigerators are distributed over 3 km at the colliding regions of the TRISTAN and are used as fundamental elements of the particle detectors. These magnets and helium refrigerators are different in design and operational conditions and were constructed by different companies. However, the control systems are unified in order to simplify the operation and to control all cryogenic systems together at the central control stations.

The control system was constructed in two stages. In the first stage, the three control systems were tested separately using the local control stations which are located near each refrigerators. The automatic control programs were modified while the systems were in operation. At the second stage, the three control systems were linked by an optical fiber data link so that all cryogenic systems could be operated simultaneously at the central control station.

CRYOGENIC SYSTEM

The main specifications of the three helium refrigerators and magnets are listed in Table 1. The VENUS magnet is cooled indirectly with a forced flow of two phase helium by a 500 W, 4.2 K helium refrigerator[1].

The TOPAZ magnet is also cooled indirectly by a 300 W, 4.4 K helium refrigerator or a backup centrifugal liquid helium pump[2]. The AMY magnet is cooled directly with pool boiling liquid helium by a 300 W, 4.4 K helium refrigerator[3]. Cool down and warm up processes for these magnets are done automatically. The VENUS and TOPAZ magnets are especially very thin so they must be cooled down and warmed up carefully according to the programmed temperature curves to protect the magnets from the mechanical damage due to thermal stresses. The cooling time of the magnet is determined by the temperature and flow rate of the cooling gas. The cooling path of the VENUS magnet is longer than that of the TOPAZ. Therefore, the cooling time of the VENUS magnet is about 3 times longer than that of the TOPAZ due to the large pressure drop and the low flow rate with the same temperature of the cooling gas.

CONTROL SYSTEM

The configuration of the total control system is shown in Fig. 1. The system consists of two parts: control and monitor. The list of input/output signals is shown in Table 2. The distributed process control system (YOKOGAWA, CENTUM) controls all controllable parts of the helium refrigerators and the magnets. It consists of field control stations for process control and operator stations for man-machine interface. The field control station has dual processors to do feedback and sequence control and includes analogue-to-digital or digital-to-analogue converters and signal conditioners to interface the sensors, the control valves, etc. The operator station has a graphics display, a keyboard, a winchester disc, a floppy disc, a printer and a color hard copy unit to do normal operation and system generation. The field control stations and the operator stations are connected by the data highway (1 Mbps). Control programs can be built or modified from the operator station keeping the field control station in operation.

Table 1. Main specifications of cryogenic systems

	VENUS	TOPAZ	AMY
Cold box			
Type	Claude cycle	→	→
Capacity	150 L/H	100 L/H	100 L/H
Refrigeration	500 W at 4.2 K	300 W at 4.4 K	300 W at 4.4 K
Control Dewar			
Capacity	1,000 L	800 L	2,400 L
LHe pump	-	Centrifugal pump	-
Flow rate		300 L/H	
LN_2 storage	15,000 L	10,000 L	9,700 L
Compressor			
Type	Two-stage screw with four-stage oil removal equipment		
Pressure	1.1 → 16.0 kgf/cm^2	→	→
Flow rate	2,300 Nm^3/H	1,800 Nm^3/H	1,500 Nm^3/H
Magnet			
Inner diameter	3.54 m	2.90 m	2.39 m
Length	5.27 m	5.10 m	1.54 m
Central field	0.75 T	1.21 T	3.0 T
Stored energy	11.7 MJ	19.5 MJ	40.0 MJ
Cooling method	Two phase forced flow →		Pool boiling
Cooling path	(11 mmID×4) × 340 m + 14 mmID × 260 m	18 mmID × 150 m	-
LHe capacity	210 L	40 L	500 L
Cold mass	4.3 ton (Al)	4.5 ton (Al)	17 ton (Cu)
Thermal load	40 W + 17 L/H	25 W + 20 L/H	33 W + 23 L/H
Cool down time	14 days	5 days	7 days

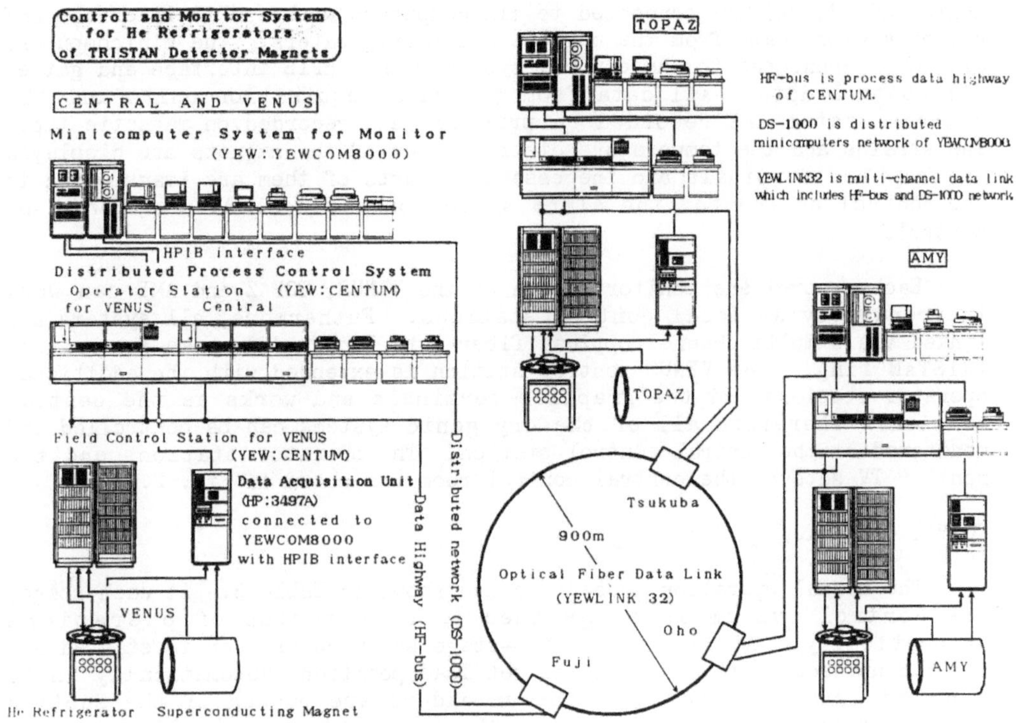

Fig. 1. Schematic diagram of total control system

Table 2. Input/output signals of control and monitor systems

	VENUS	TOPAZ	AMY	TOTAL
Control data	301	298	254	853
Analogue input	85	86	84	255
Temperature	29	33	35	97
Pressure	37	29	27	93
Flow rate	5	7	4	16
Liquid level	6	5	5	16
Speed, etc.	8	12	13	33
Digital input	105	110	101	316
Status	66	60	56	182
Alarm	39	50	45	134
Analogue output	42	38	30	110
Control valve	37	32	28	97
Heater, etc.	5	6	2	13
Digital output	69	64	39	172
ON/OFF valve	20	14	7	41
Switch, etc.	49	50	32	131
Monitor data	170	202	93	465
Refrigerator temp.	11	13	6	30
Magnet temperature	99	103	36	238
strain	56	44	30	130
others	4	42	21	67

The monitor system consists of minicomputers (YOKOGAWA, YEWCOM8000) each with a real time operating system, a graphics terminal, a magnetic tape, a printer, a plotter and a color hard copy unit. Data acquisition units (HP, 3497A) are connected to minicomputers with a HPIB interface and gather monitor data from the magnets and refrigerators. The minicomputers are also connected to the control system with a HPIB interface and gather all control data. All data from the data acquisition units and the control system are recorded on printers and recorded on magnetic tape. The strains and the temperature distributions of the magnets are displayed on graphics terminals and the essential parts of them are transferred to the control system for the alarm system of the cool-down and warm-up control.

Each control and monitor system of the VENUS, TOPAZ and AMY can work independently at local control stations. Futhermore, all systems are linked via a multi-channel optical fiber data link which goes around the TRISTAN ring. The VENUS control station is extended with the additional operator stations and the graphics terminals and works as the central station. Therefore all of the cryogenic systems can be controlled and monitored at the central control station. The operator stations and the monitor TV sets at the central control room are shown in Fig. 2.

NORMAL OPERATION

The normal operation procedures are shown in Table 3. At each stage, the control system is programmed to do a series of operations automatically. At first, the two-stage screw compressor is started and the two unloaders are moved to the set load positions automatically in 10 minutes, during which time the unloaders are paused if the suction pressure becomes too low. The purification of the system is done for about 2 days after the start-up of the compressor. It is now done manually because the impurity monitors are not connected to the control system yet. It is possible, after the connection, to change gas flow patterns automatically while monitoring the impurity.

After the cold box is cooled to liquid nitrogen temperature, the cool down process of the magnet is started. The magnet is cooled down carefully controlling the temperature and the pressure of the cold box output gas. These controls is done using the two valves in the cold box, the cold gas valve and the warm gas valve. The gas is mixed with two

Fig. 2. Operator stations and monitor TV sets in central control room

Table 3. Normal operation of cryogenic system

Main compressor start	(Auto)
↓	
Purification	(Manual)
↓	
Magnet cool down	
Cold box cool down with liquid nitrogen	(Auto)
Magnet cool down start	(Auto)
Liquid nitrogen shields of magnet cool down	(Auto)
Turbine start	(Auto)
Adjust turbine speed	(Manual)
Cold box helium return line change	(Auto)
Store liquid helium in Dewar and magnet pot	(Auto & Manual)
↓	
Steady state	
Current lead cooling gas flow control	(Auto)
Cascade control with magnet current	
Cascade control with current lead temperature	
↓	
Magnet warm up	
Turbine stop	
Cold box and control Dewar warm up to liquid nitrogen temperature	
Magnet warm up	(Auto)
Liquid nitrogen shields of magnet warm up	(Auto)
Cold box and control Dewar warm up to room temperature	
↓	
Main compressor stop	

valves controlling the pressure is done by the cold gas valve and the temperature by the warm gas valve. Futhermore, the flow rate of the magnet input gas is controlled by the valve in the control Dewar so that the magnet is cooled down according to the programmed temperature curve. The cool down program pauses when the temperature differences in the magnet exceed the limit values. The liquid nitrogen shields of the magnet is also cooled down in the same manner as the magnet.

Two turbines are started with the auto-start program when the magnet temperature becomes 140 K. The turbine speed is adjusted manually now, but it will be done automatically after the necessary cooling power is determined at each stage of the cool down process. Then the return line of the cold gas to the cold box is changed depending on the return gas temperature. The magnet is cooled down to liquid helium temperature and liquid helium is stored in the control Dewar and the magnet helium pot. These operations at the finish of the cool down are done manually, because set points of turbine speed and the distribution rate of cold gas to the cold box, to the control Dewar and to the magnet are not fixed yet.

In the steady state, the helium refrigerator is continuously operated in the same condition. Only the flow rates of the current leads are changed depending on the magnet current. The valves of the current leads are feedback controlled by the flow rates of the current lead cooling gas. These set points are cascade controlled by the magnet current (short time operation) or the current lead temperatures (long term operation).

The warm up of the magnet is done using a reverse operation of that used for the cool down. The gas is mixed with two valve controlling the pressure by the warm gas valve and the temperature by the cold gas valve, opposite to that used for the cool down. The pressure control remains with the warm gas valve while the cold gas valve closes in the final stage of the warm up.

Cool down curves of the VENUS magnet are shown in Fig. 3. The magnet was cooled down keeping the temperature differences in the magnet less than 40 K and according to the temperature curve of 0.5 K/H (300 K → 252 K), 0.8 K/H (252 K → 156 K), 1.3 K/H (156 K → 78 K), and 3.1 K/H (78 K → 4.2 K). The cool down program paused when the magnet temperature lagged behind the programmed temperature because of the limit of the magnet inlet pressure. Therefore, it took a longer time than that which was actually programmed.

EMERGENCY OPERATION

The control system is programmed to meet an emergency such as a magnet quench, a compressor emergency stop caused by a power failure or other troubles, and a current lead flow stop as shown in Table 4.

When the magnet quenches, the magnet is isolated from the refrigerator and is connected to the gas bag by the closing or opening of the valves in the instant in order to minimize a disturbance to the refrigerator. The programmed valve set pattern for the quench of the VENUS magnet is shown in Fig. 4. The refrigerator is kept on standby by the emergency program, so the magnet can be cooled down immediately after the pressure in the magnet becomes low. A trend graph of the recovery procedure after the quench is shown in Fig. 5. The magnet quenched due to a noise in the quench detector during the continuous operation in 1986. The magnet was restarted to cool down in 30 minutes after the quench and became standby in 3 hours.

When the compressor stops suddenly due to a power failure or other accidents, the valves are set to the programmed pattern to protect the system from destruction and adulteration of impurities. Actual power failures also occurred during the operation, but there was no trouble, because of the emergency program.

During steady state operation, one the weakest parts of the system is the magnet current leads. The current leads will burn out, when the flows of the cooling gas stop for several minutes. So, they are protected from

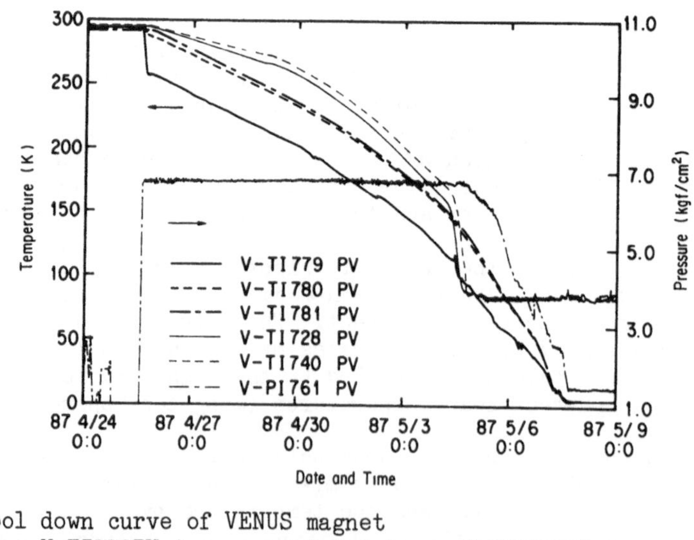

Fig. 3. Cool down curve of VENUS magnet
Here V-TI779PV is magnet inlet temp, V-TI780PV is magnet outlet temp, V-TI781PV is magnet middle temp, V-TI728PV is LN_2 inner shield temp, V-TI740PV is LN_2 outer shield temp, V-PI761PV is magnet inlet pressure.

Table 4. Emergency operation

1. Magnet quench
   - Close or open valves according to Quench-valve-pattern (Auto)
     - Isolate magnet from helium refrigerator (Auto)
     - Connect magnet to recovery system (Auto)
     - Open magnet release valves (Auto)
     - ↓
   - Recovery operation (Auto & Manual)
     - Reset quench sequence (Auto)
       (Close magnet release valves, etc.)
     - Restart magnet cool down (Manual)

2. Compressor emergency stop
   - Set valves according to Compressor-emst-pattern (Auto)
     - ↓
   - Recovery operation (Auto & Manual)

3. Current lead (CL) protection
   - Change return line of current lead cooling gas (Auto)

   Main compressor side
   CL flow stop → ↓ ← CL temperature rise
   Recovery side
   ↓ ← Waiting 1 minute
   Vent side
   ↓

VENUS	TOPAZ	AMY
Shut off power supply after waiting 4 minutes	Slow down current just after vent mode	Slow down current after waiting 4 minutes, and shut off PS, if temperature rise

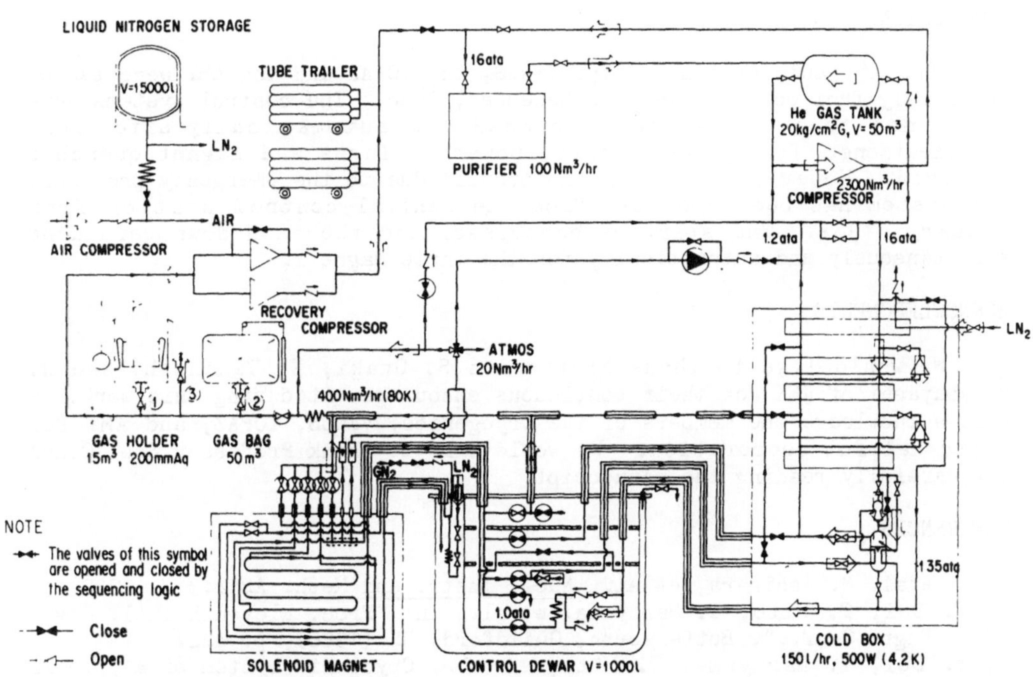

Fig. 4. Programmed valve pattern for VENUS magnet quench. Here dark shaded valves in circle are closed and unshaded valves are opened when magnet quenches.

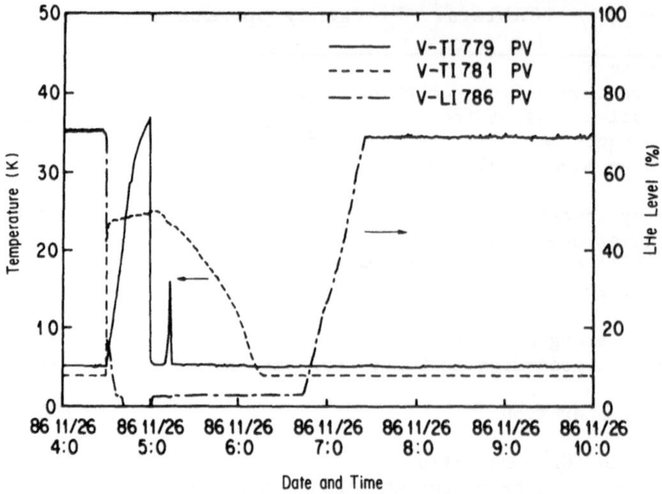

Fig. 5. Trend graph of quench recovery of VENUS magnet. Here V-TI779PV is magnet inlet temperature, V-TI781PV is magnet middle temp, and V-LI78PV is liquid helium level of magnet pot.

burning out after alarm signally flow stop and temperature rise. When an alarm is detected, the return gas line of the current leads is changed to the recovery system from the main compressor suction, and to the vent after waiting 1 minute. If the alarm is not cleared after the return line changes, the current in the magnet is reduced. A flow stop also occurred during the operation in 1987 and the magnet was discharged successfully.

CONCLUSION

The cryogenic systems of the VENUS, the TOPAZ and the AMY were tested separately from October, 1985 to December, 1986. The control systems were confirmed to do a series of procedures automatically after some modifications. During these tests, power failures and magnet quenches occurred. However, there was no trouble due to the emergency programs. The system has been operated from the central control station from January, 1987. The start of compressor and the cool down were done simultaneously and automatically for the three magnets.

ACKNOWLEDGMENTS

We would like to thank Professors S. Ozaki, K. Takahashi and H. Hirabayashi of KEK for their continuous encouragement during this work. We also acknowledge the members of the Cryogenics, VENUS, TOPAZ, and AMY for their helpful cooperation. We would like to thank Professor S.L. Olsen for carefully reading the manuscript.

REFERENCES

1. R. Arai, N. Ishihara, et al., Nucl. Instr. and Meth. A254:317 (1987).
2. Y. Doi, T. Mito, T. Haruyama, et al., in: "Proc. Eleventh Intl. Cryo. Engr. Conf.,", Butterworth, Guildford, UK (1986), p. 424.
3. Y. Doi, T. Haruyama, T. Mito, et al., Cryogenic system of a 3 tesla superconducting solenoid for the AMY particle detector at TRISTAN, in: "Advances in Cryogenic Engineering," Vol. 33, Plenum Press, New York (1987).

EXPERIENCE WITH A PROCESS CONTROL SYSTEM FOR LARGE SCALE CRYOGENIC SYSTEMS

M.Clausen, K. H. Mess, Chr. Gerke, S. Rettig

Deutsches Elektronen-Synchrotron DESY

Hamburg, Germany

ABSTRACT

The superconducting proton storage ring of the HERA project at DESY (Hamburg) will contain 422 dipole - and 224 quadrupole magnets. The magnets will be installed in the 6.3-km ring underground. Three refrigerators produce supercritical helium, which is fed into a transferline that supplies individual sections of the ring. Another transferline provides helium for the magnet test facility. All cryogenic installations will be controlled by a process control computer system. This paper describes some of the experience during specification, installation and the first running period of the system installed at DESY. Individual solutions for distributed process controls, low temperature measurements and some utilities are presented.

INTRODUCTION

Computer controls are one of the main components in a high energy physics laboratory like DESY. Both the large refrigeration system and its process control system are new areas of activity for DESY. In the discussions about the controls of the cryogenic system it was realized early that a cryogenic system also needed an adequate computer system for the controls.

TYPE OF COMPUTER REQUIRED

Obviously only a system of computers could perform all the required features needed to control the refrigerator and the other cryogenic components. An important step towards a common process control system for all cryogenic components was made when DESY decided that the control system for the refrigerator should not be part of the contract with the supplier. In this way, DESY had the freedom to choose the system that DESY would have to work and live with for many years. The decision had the disadvantage that the specification and the final decision for the control system had to be taken ahead of the specification for the refrigerator, because the software engineering was part of the contract with the refrigerator supplier. In this way, the engineers for the computer system came into the uncomfortable situation of trying to define the number of i/o points without a specification for the system that should be controlled. The estimated numbers can be found in table 1, left column.

Table 1. Evolution of computer hardware

cryo component	computer equipment	technical specification	original order	extended order	actually in use
coldboxes	computer	3	1	2	2
	analog inp.	210	240	400	327
	analog outp.	90	96	248	213
	digit. inp.	150	192	720	645
	digit. outp.	150	192	224	137
	an. inp. PT100	-	-	80	62
compressors	computer	1	1	1	1
	analog inp.	100	112	128	104
	analog outp.	36	40	72	62
	digit. inp.	83	96	448	410
	digit. outp.	83	96	240	190
	an. inp. PT100			96	61
magnet test	computer	1	1	1	1
	analog inp.	60	64	128	105
	analog outp.	80	80	40	52
	digit. inp.	20	32	192	187
	digit. outp.	60	64	64	60
ring control	computer	1	1	2	2
	analog inp.	40	48	426 a	b
	analog outp.	40	48	120 a	b
	digit. inp.	80	128	869 a	b
	digit. outp.	80	128	218 a	b

a: these numbers are not finally defined.
b: nothing installed.

THE CRYOGENIC COMPONENTS

The HERA cryosystem consists of three mayor components.:

a. Refrigerator [1,2]
b. Magnet measurement facility [3]
c. Helium distribution system [4]

At least these components should have their own process computer. Nevertheless, all computers should be connected by a communication link. The refrigerator can be split into two logical pieces - the compressors and the coldboxes. It was forseen, that at least these components should have their own computer.

THE BUY OR BUILD CHOICE

Since computer controls are standard tools at DESY, the question was raised, whether it was possible to reduce costs by building a process control system out of the existing computer software and hardware. The main argument to buy such a system was the special software that DESY would have to develop in a very short time. The risk of warranty problems with the supplier of the refrigerator was estimated to be too high. So DESY decided to rely on the software and hardware of a commercial system.

MAIN TECHNICAL CRITERIA

Reliability

Generally cryogenic processes are slow processes. This has the advantage of a low CPU load to the control computer. On the other hand, a single-component failure in the cryogenic system causing a system stop will shutdown the whole accelerator. Therefore the refrigerator has many components in 'hot stand by', that can take over if an error occurs. This leads to the request for the computer system, that the meantime between failure (mtbf) for all computer that control the refrigerator was specified to be 10 years. For the other computer an mtbf of 3 years was specified.

Redundancy. The communication link is the only connection between all the computer in the system. Therefore at least this link - and even better - all components of the communication link should be redundant. In case of redundant computers, a 'hot stand by' was specified.

Uninterruptable power supplies(ups). It was demanded, that also in case of a power failure, the process control system should stay online. This way the operators will have data displays in order to monitore all system functions in every situation. Therefore an "ups" also for at least one operator console was specified.

General Hardware

"State of the art" hardware was desired. At that time, this meant at least a 16-bit processor and more than 128 kbyte of memory. It was preferred to get a system with standard components. All application programs should be downloaded to the microprocessor. All i/o cards should be changeable 'online'. Adresses should be selected by the position of the card and not by jumpers. In addition there should be a standard interface to other programmable logic computers (plc's), to feed the temperature data of the magnets into the control system.

Software

One of the most important criteria in the specification was the software. Because DESY is a research laboratory, one can expect that the final software design will never be reached. Therefore the software must be extremely flexible. All loops and points in the system must be defined by a 'fill in the blanks' method. The sequencially running software must be readable like a BASIC or FORTRAN program. In conclusion, the software should be optimized for short developement times. A set of standard graphics was specified as well as a good graphic editor and debugging tools.

Access to process data, and link to VAX computer. Because the cryogenic controls are only one component of the accelerator controls, it must be possible to get access to all process data from user written programs and to transfer these data to a VAX computer which will be connected to the accelerator control system.

COMPARISON OF DIFFERENT SYSTEMS

It is very difficult to compare computer systems 'on paper'. On the other hand, the differences between the offered systems were big enough to lead to a clear decision. Some systems obviously did not fulfill the specification. For example, there were systems which must be programmed in assembler or even machine language and systems that defined control loops not by software but by means of special hardware cards. Other systems needed many computers to handle the specified i/o channels, did not have enough memory or used one central uninterruptable power supply for their distributed control system.

Reasons to Accept One System

The system that best fulfills all the requirements, and is now running at DESY, is the EMCON D/3 system from Rexnord/USA (Fig 1.). Some technical data are as follows: multibus standard; cpu 8086/87; 1Mbyte memory; completely redundant communication; VAX computer as host system; all system software written in the programming language 'C', and all sources available for the user; access to all process data from the host; remarkable amount of software utilities.

EXPERIENCE

Expansion of the System

Soon after the computer system was ordered, the requirements for the refrigerator were updated. The requested number of i/o channels for the refrigerator was a factor of three to four above the first estimate and nearly a factor of ten above the estimate for the ring controls (Table 1). Fortunately the system is very flexible, so the number of computers increased from one to two for the coldbox and for the ring ring controls only. In addition the numbers of digital channels for the compressors could be reduced to the actual amount, because local plc's were added to the compressors and purifier systems. These plc's are part of the man/machine safety system, enable the operators to locally test individual units and reduce i/o costs.

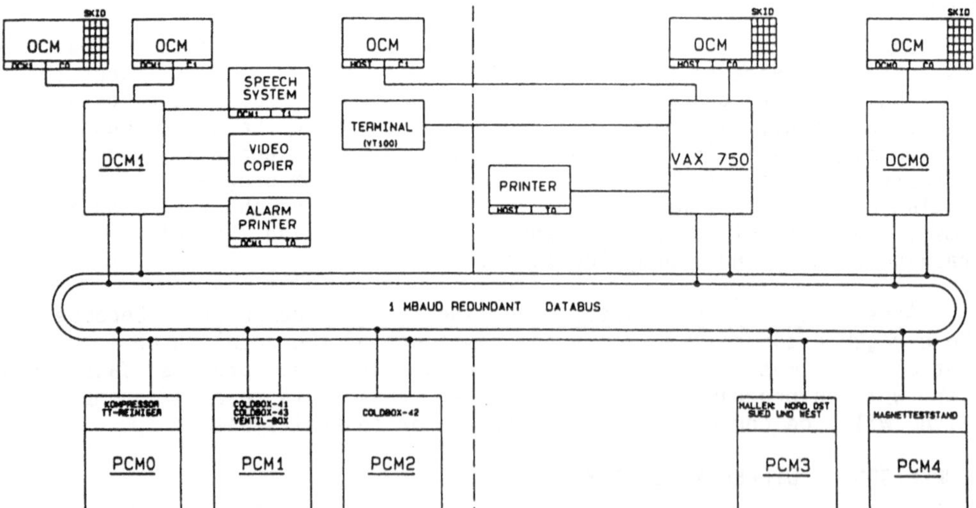

Fig. 1. System layout for the HERA cryogenic system.
OCM : Operator console. PCM : Process computer.
DCM : Display computer. Drucker : Printer.

System Startup

In the first test periods it was found to be very useful to have software that could be easily changed. At this time also the utility to make 'on line' changes was often used. Later the requirements changed and a good documentation seemed to be more important than fast software changes. The experience with the process computers (PCM) was extremly good. Up to now - for more than two years - no redundant computer had to take over, because there was no failure during this period. On the other hand, a memory size of even 1Mbyte is too small for the display computers (DCM). A mass storage system like a winchester should be added to the display computer as the operators tend to create more and more graphics.

The Host System

One item that was not specified, but came along with the D/3 system, is the possibility to use a VAX as host computer. The VAX supports multiple users during program developement . In addition the user-friendly VMS operating system helps to save effort by invoking user defined software utilities. Easy to use database access routines enable the user to calculate efficiencies; program process status screens, or generate data logs (i.e. for acceptance tests). Historical trending simplifies long term error checking.

RING CONTROLS AND LOW TEMPERATURE MEASUREMENT

The control of the cryogenic systems in the 4 experimental halls, the 12 substations in the tunnel and the HERA superconducting magnets has to fulfill special requirements. On one hand, it clearly has to be closely connected to the general cryogenic control system; on the other hand, it has to interact with the accelerator control. Furthermore, the equipment is distributed over a length of 6.5 km, is hardly accessible and almost constantly exposed to ionizing radiation. Moreover, the temperature sensors in discussion show a highly nonlinear behaviour, thus requiring a substantial computing power to yield a readable result.

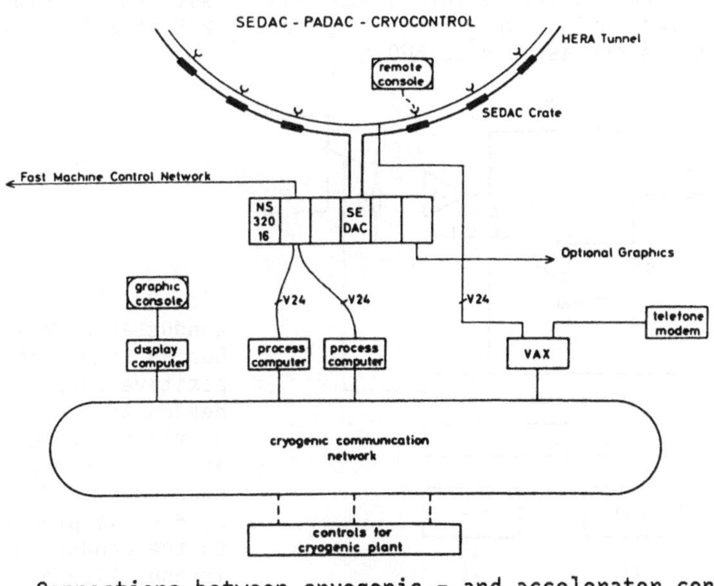

Fig. 2. Connections between cryogenic - and accelerator controls. Remote consoles on the telephone line and in the ring.

The adopted solution is based on the suitably expanded Serial Data Aquisition System (SEDAC) originally designed for PETRA.[5,6] In its latest stage the system consists of a network of NS32016 microprocessors, each controlling a particular task, or part of one. The data flow on the connecting fast packet-switching network between the microprocessors can be substantially reduced by a suitable definition of the tasks. In particular, it is not necessary to have the complete networking operational if one wants to test just a subsystem. Within this network of accelerator control microprocessors, four are set aside to collect and distribute data of the cryogenic installations in the experimental halls and the ring. The communication between the cryogenic control system and these microprocessors will initially be on simple RS 232 lines following a special protocol (Fig. 2). The microprocessors will be downloaded from the accelerator control system and can also be interrogated from the main control room. The microprocessors communicate with up to 32 crates on a single coaxial line up to a length of several kilometer. A crate controller acts on 16 equipment control modules. The connections are transformer coupled; the crate controlers can be build in TTL. Both features are essential to allow operation in radiation hazardous area. The design of the equipment control modules also requires the use of radiation insensitive electronics (at least 10^3 Gy).

TEMPERATURE SENSORS AND - READOUT SYSTEMS

Of particular interest is the aquisition of cryogenic temperature data. In radiation safe areas - like the magnet test facility - silicon diodes will be used[7]. In the tunnel sensors less sensitive to radiation have to be used. This could be carbon glass resistors, carbon composition resistors or newly developed carbon thin film resistors. All of these devices have a negative temperature coefficient. That is to say, for a constant current i the injected power grows $N=i^2R$ with decreasing temperature. Hence it is necessary to measure the current, needed to achieve a constant voltage drop u. In this case, $N=R/u^2$, the injected power decreases with temperature (and heat capacity). The electronics[8] is quite insensitive to offset voltages and hence to radiation damage.Figure 3 illustrates the principle. These units perform quite well. With carbon glass resistors temperature resolutions of $\pm$ 6 mK at 4 K and $\pm$ 2 K at 300 K are obtainable even over distances as long as 800 m.

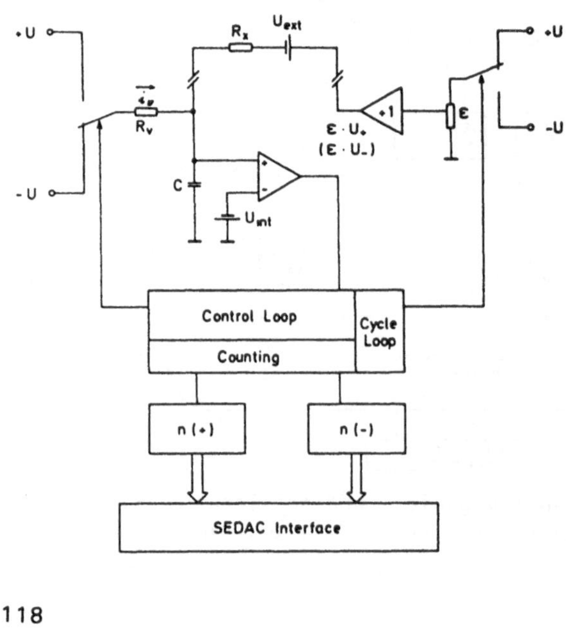

Fig. 3. Principle of the Conductance Measurement During time T the number of positive current pulses n_+ needed to keep the voltage on the capacitor at U_+ or at U_-), respectively, are counted. The difference $n_+ - n_-$ is proportional to the conductance, irrespective of small additional external - internal vol...

In a similar fashion, all sensors that change a resistance can be read out, like pressure transducers or valve positioners. The data handling, necessary to compute the temperatures , will be done in the NS32016 microprocessor. The cryogenic control system can aquire at any time the already calibrated readings.

UTILITIES

EXPERT SYSTEMS

The installation of the control system was an enlightening experience for us on all levels. Not everybody experiences the same difficulties, but everybody should be able to cure a problem that has been solved earlier. For this purpose, an expert system would be very helpful to collect and acces all the expertise.

Due to lack of manpower, we could not build a proper expert system. On the other hand we do not want to loose the valuable expertise gained during the set-up phase. We therefore have decided to store the information in a computer file. To have a primitive means to retrieve the information, we add keywords for every problem. When we want to find out if a specific problem has already been solved once, we perform a sequential search for selected keywords through the file. The keywords for each solution are additionally collected in a separate file which can be examined before the search so that promising keywords can be chosen easily.

REMOTE CONSOLES

As shown in Figure 2. it is possible to connect remote consoles to the host computer. DESY will use personal computer based on the 68000 processor as remote stations. A special software package is written on these computers to display all the graphic functions like on the standard operator consoles shown in Figure 4.

Figure 4. Operator consoles of the process control system with graphics on the screens. In the background, a status screen and two screens of the video system.

CONCLUSIONS

Reliability and flexibility for software and hardware are the two most important things for the specification of a process control system. A powerful host and efficient data access are useful tools. Graphics are intensively used and need mass storage for proper operations. The connection and an open protocol to other process - or data taking systems is a important detail for a flexible system. The choice of which system will be used should be made as early as possible, while the final layout of the system should not be fixed before the process is specified.

The part of the cryogenic control which is in a radiation hazardous area and normally not accessable will be constructed like the accelerator control system. The software will nevertheless run basically in the cryogenic control processor with the exception of the transformation to engineering units. This solution benefits from the advantages of the cryogenic control system, creates economically priced and optimized hardware and enables easy interfacing to the accelerator control.

ACKNOWLEDGEMENTS

With special thanks we would like to mention J. J. Sondericker - BNL for his help on the technical specification.

REFERENCES

1. G. Horlitz, "A Central Refrigeration System for the Superconducting HERA Proton Magnet Ring," DESY-HERA 1984/02, Febr. 1984

2. P. Beurer, et al., "Performance Test of the HERA 3 X 6500 W Helium Refrigeration Plant," paper CB-4 at this conference

3. H. R. Barton Jr, M. Clausen and G. Kessler, "Cryogenic System for the HERA Magnet Measurement Facility," Proc. Eleventh Intl. Cryo. Engr. Conf., Butterworth, Guildford, UK (1986) pp.173-177.

4. G. Horlitz, "Refrigeration of a 6.4-km Circumfence, 4.5 Tesla Superconducting Magnet Ring System for the Electron Protron Collider HERA," Proc. 10th Intl. Cryo. Eng. Conf., Butterworth, Guildford, UK (1984), p.377.

5. H. Freese, G. Hochweller,"The Serial Data Aquisition System at PETRA," IEEE-NS-26, 3, 1979

6. G. Hochweller, et al., "Padac Multi - Microcomputers, Basic Building Blocks for Future Control Systems," IEEE-NS-28, 3, 1981

7. J. Sondericker, "Production and Use of High Grade Silicon Diode Temperature Sensors", Advances in Cryogenic Engineering, Vol 27 Plenum Press, New York and London

8. Designed by M. Swars (DESY), patent pending

CRYOGENIC DESIGN OF THE D-ZERO LIQUID ARGON COLLIDER CALORIMETER

George T. Mulholland, Kurt J. Krempetz, Richard D. Luther,
Robert H. Wands, Katharine J. Weber

Fermi National Accelerator Laboratory*
Batavia, Illinois

ABSTRACT

The D-Zero liquid argon collider calorimeter has been designed for high energy particle physics experimentation as an integral part of the Tevatron accelerator/collider. The high density, hermetic, 125,000 channel calorimeter is an assembly of three axial, cylindrical, $20m^3$ liquid argon cryostats containing 770 tonnes of calorimeter modules comprised of uranium and copper absorber plates and multilayer signal boards. The 228t end calorimeter cryostats can be moved 1.3m axially, while full and operable, to allow access to a Central Tracker mounted in the warm bore of the central calorimeter. The liquid argon is stored 13m below ground at detector-level in a $76m^3$ storage dewar. Refrigeration for the argon in the cryostats and the dewar is provided by liquid nitrogen stored in a $76m^3$ dewar located at ground-level. Special vent and spill sump provisions allow personnel to work within the detector platform below the cryostats and in the accelerator tunnel. After completion of testing in the assembly hall position, the entire 4550t detector is rolled 40m into the collision hall for each running period. Cooldown, warmup, and control and operation of the cryogenic systems of the detector are described.

OVERVIEW

The superconducting Tevatron was added to Fermilab's 400 Gev Proton Accelerator, the main ring, in 1983. An antiproton source was added in 1985, and the system became a p-pbar, 1 Tev/1 Tev, collider in 1987. A Collider Detector surrounding one of the points of the accelerator p-pbar beam crossings can measure virtually all the energy of the colliding interaction (Fig. 1.) The measurement of all the energy is called hermetic calorimetry. Although there are other liquid argon calorimeters and other hermetic collider detectors, the D-Zero (named for the accelerator beam crossing location) liquid argon collider calorimeters will be the first of their kind (Fig. 2). The cryogenic aspects of the liquid argon calorimeter portion of the D-Zero detector are described here.

The liquid argon serves as the particle detector ionizing media in a repetitive cell structure (Fig. 3) of argon, signal board, argon, and Uranium or copper absorber plate, with a superimposed electric field. Local signal board pads indicate location and the electric charge collected is proportional to the ionization and the ratio of the argon to plate absorption lengths. This arrangement provides a dense, intrinsically calibrated, drift-free calorimeter.

*Operated by the Universities Research Association, Inc. for the U.S. Department of Energy.

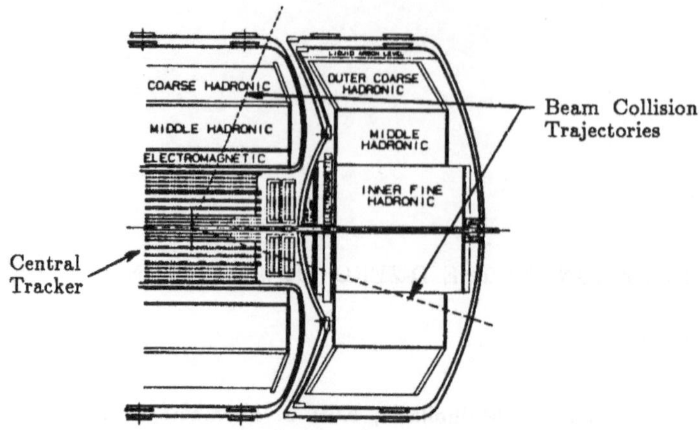

Fig. 1. Calorimeter cross section showing the module layout,the central tracker, and two arbitrary trajectories from the collision point.

CRYOSTATS

The three cryostats serve to hold and insulate the liquid argon, transmit the large module assembly weights to room-temperature supports, and contain all the necessary electrical and cryogenic feedthroughs. The 5m diameter type 304 stainless steel inner vessels are ASME Code-stamped (Section VIII, Div. 1) and are designed for an internal pressure of 1 atmosphere plus insulating vacuum plus liquid head. The inner vessels are also designed to accommodate a full vacuum loading. The vacuum jackets are designed in accordance with CGA-341 with fabrication comparable to ASME Code rules (Div. 1), and are also fabricated of stainless steel to preclude fault brittle fracture.

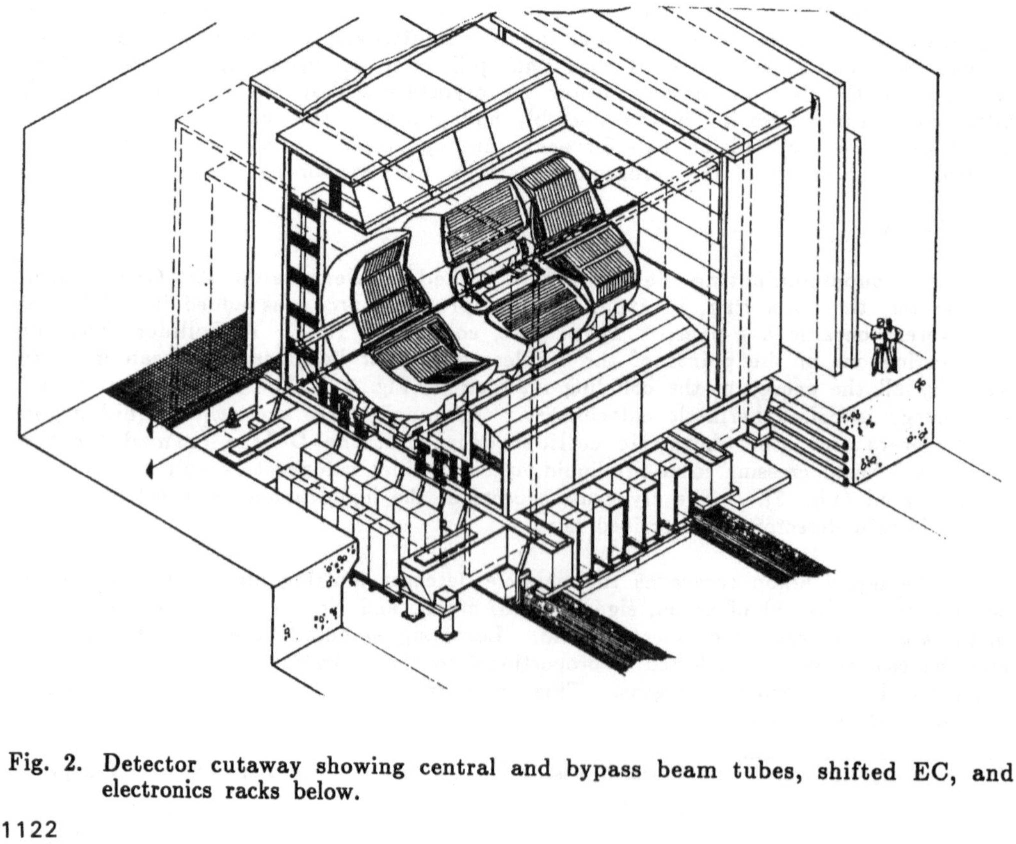

Fig. 2. Detector cutaway showing central and bypass beam tubes, shifted EC, and electronics racks below.

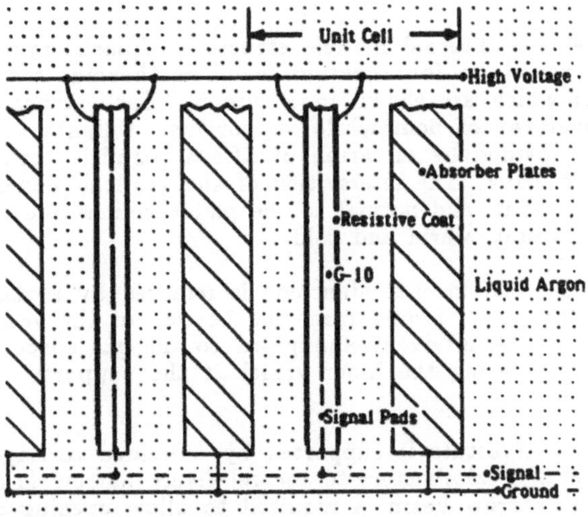

Fig. 3. Cell Structure. 1.5kV is applied, ionization occurs in the Argon gaps, and charge is collected on the copper pads as signal.

The cryostat shapes reflect efforts to maximize the calorimeter hermeticity and absorption depth uniformity as a function of angle, and to minimize the energy absorption of the walls. All the vessel details not covered under the rules the ASME Code (Div. 1) were designed by analysis. Aggressive Code design minimized particle-entry wall thicknesses by use of finite-element (Fig. 4) and ASME Division 2 analysis methods, and independent third party review.

The 1.52m bore of the torus-shaped CC cryostat contains the warm Central Tracker surrounding the interaction point. In contrast each EC vessel contains a 7.5cm center tube in order to bring the detector as close as possible to the Tevatron beam tube. The EC tubes include bellows assemblies to accommodate vessel movements and to minimize deflection-induced stresses. Each vessel also contains a rigid (no bellows) reentrant tube (27cm dia.) near its outer shell to accommodate passage of the Main Ring through the detector.

The calorimeter modules are to be erected inside the fabricated, tested, and stamped vessels. Access requires that the EC heads at one end and the CC heads at both ends be removed. After module assembly, the heads are refitted, welded, and tested as before by an ASME Code stamp holder. The flangeless design enhances the compactness of the vessels and the efficient use of available space made possible by the passive nature of the liquid argon detector.

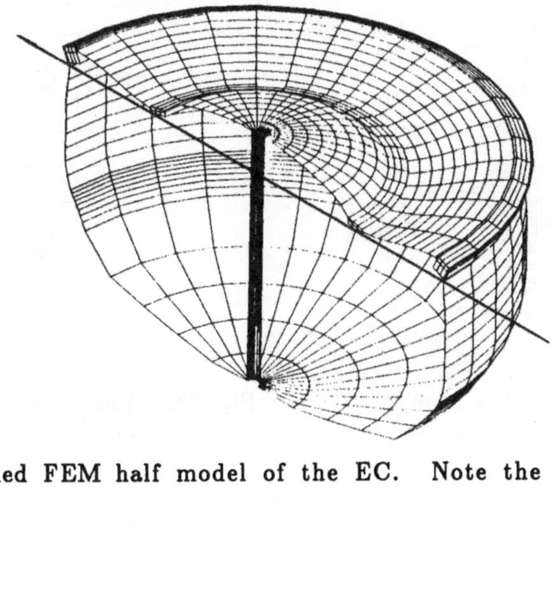

Fig. 4. Fully loaded FEM half model of the EC. Note the exaggerated flange rotations.

SUPPORTS

The inner vessel and internal module structure of each cryostat are supported within the vacuum vessel on four custom-designed thermal contraction support stanchions, details of which are shown in Figure 5. Each stanchion is designed to accommodate a movement of 6cm while supporting weights to 90,900kg. The needed flexibility is provided by an array of fifteen parallel vertical Inconel plates Nicro-brazed to header plates. The intersection of the vertical planes passing through the centers of the stanchions and normal to the plates defines the stationary vertical "centerline" of each cryostat, warm and cold.

The 40W thermal conduction in each stanchion is removed by a thermal siphon heat intercept mounted at the top. Liquid from the bottom of the vessel, driven by the liquid head, is boiled in the intercept and piped to the gas space at the top. The resulting heat load penetrating the inner vessel at the supports is reduced to less than 4 watts.

Each of the three cryostats is supported by a welded carriage structure which transfers the weight of the calorimeters to the relatively narrow (1.5m) central beam of the platform. The narrow beam is required to allow opening of the toroid (see Fig. 6). The heavily loaded carriages are fabricated of 304SS because of the finite possibility of a cryogenic spill. A spherical interface is located between the bottom of the stanchion and the top of the carriage in order to accommodate the distortions of the carriage as the calorimeter is assembled, (i.e., loaded). The radius of the spherical interface is sized to maintain alignment of the stanchions and the spherical surfaces as the carriage arms rotate and translate under load.

A system of hydraulic jacks and drivers and heavy-duty rollers are used to move the 4550 tonne detector system on its platform from the assembly hall into the collision hall (a distance of 25 meters) and into position in the accelerator ring for physics experimentation. Positioning of the calorimeters on the central beam of the platform is accomplished by initial placement. All four toroidal iron pieces and both EC's are roller mounted to provide access to the Central Tracker in the bore of the CC. High-capacity, low-profile, carriage-mounted rollers allow each EC to be moved a distance of 1.3m on hardened steel ways imbedded in the central beam. Jacks are provided on the EC's to unload the rollers and, as required, to trim the vertical position. Each 228 tonne EC is moved while operational and full of liquid argon in order to avoid the time penalty required to empty and refill the liquid (CC=19.9m^3, EC=8.8m^3). Six specially-designed (Cryolab, Inc.) rotary bayonet triplets are required to maintain the cryogenic and vacuum connections during the excursion.

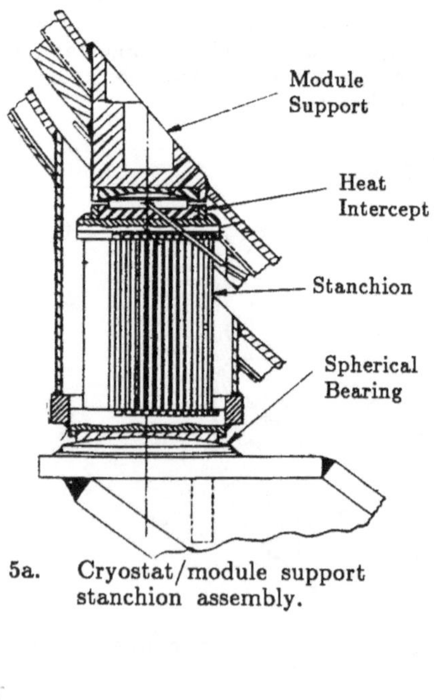

Fig. 5a. Cryostat/module support stanchion assembly.

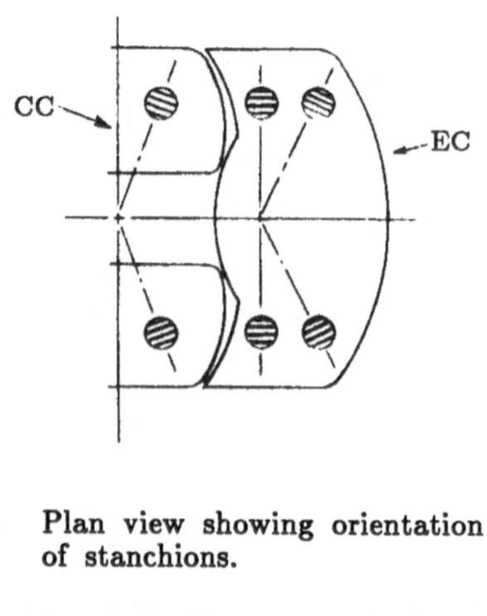

Fig. 5b. Plan view showing orientation of stanchions.

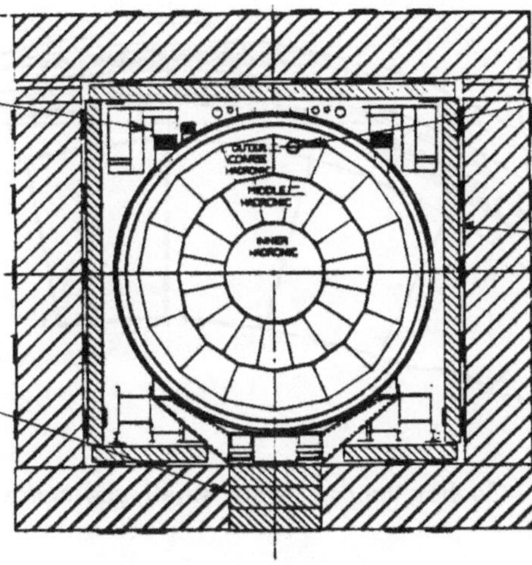

Fig. 6. EC end section.

COOLING SYSTEM

The operating cooling requirement of three cryostats is provided by an inventory-maintained, 76.6m^3, 0.3% per day boiloff, LN2 storage dewar. The ground level dewar location, 8m above the loops, provides 0.64 atmosphere subcooled liquid at the cryostats. Each cryostat is equipped with two sets of cooling loops (cooldown and operating, see Fig. 7). The 10kW max (condensing) operating loop is in the gas space. The two 20kW (condensing) cooldown loops straddle the apex of the calorimeter, operate only during cooldown, and occupy a space beneath the operating liquid level. Loop temperatures from 77.4K to 91.4K are controlled by a common back pressure control valve (Fig. 8). The steady state system refrigeration loads are listed in Table 1. The weights and cooldown loads (by module) are given in Table 2.

Cooldown is provided by (a) natural convection, (b) direct condensation and forced convection, and (c) direct pool boiling, in that order. The cooling loops are set to a temperature above T_c, the two-atmosphere liquid argon equilibrium temperature (a), then just below T_c and heaters boil the condensed liquid (b), condensing is maximized, heaters are off, and liquid level rises (c). The assembly stresses due to the temperature gradients and differential thermal contraction as indicated by a set of module RTD's will limit the rate of cooldown of all calorimeters. Finite Element Models (FEM) of the module assemblies have been developed which provide stress profiles for given temperature gradients on both the inter- and intra-module levels.

Table 1. System Heat Loads

	CC	EC	Totals
Cables	1120	1120	3360
Radiation	150	130	410
Nozzles	83	83	249
Piping	24	24	72
Supports	16	16	48
	1393	1373	4139
Dewars(.3%)			1200
Wet Lines			315
			5654 W

Table 2. Cooldown Loads (10^9 joules)

	CC	EC	System
Copper	7.82	8.41	24.64
G-10	0.50	0.27	1.04
Uranium	3.17	1.20	5.57
304SS	1.00	1.58	4.16
Cryostat	0.90	0.90	2.70
	13.4	12.4	38.11

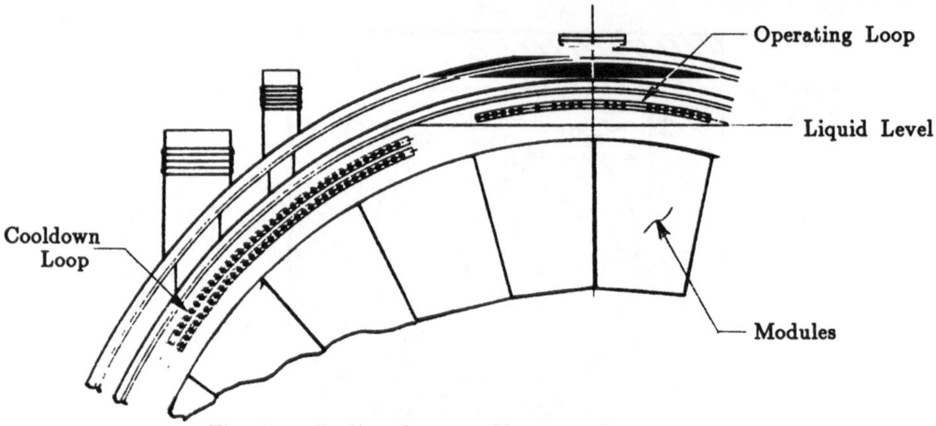

Fig. 7. Cooling loops. Note small gas space.

In tests, isolated and typical ECMH modules have cooled down in 24 hours when limiting the maximum temperature difference to 65K, and warmed-up in 24 hours when limiting the maximum gas temperature to 400K. The communication of the EC vertical absorber plate gaps in the assembly can be easily arranged to support the target two-week cooldown. CC module cooldown testing is not as far advanced, and the cooling suffers from an assembly which aligns the absorber plate gaps in 82 closed polygons concentric about the horizontal axis. Layer to layer communication is by equally spaced, azimuthal, 3cm inter-module and 15cm end plate nominal clearances.

Warm-up is driven by 40kW of redundantly-connected, finned heaters located at the bottom of the cryostat, and may be improved by the pump and purge of the cryostats to He to increase the convection. It is anticipated that a cold pump and purge to He on cooldown can be avoided. The 100C maximum allowable temperature (signal boards permanently degrade at 150C) will control the warm-up heaters. Since warm-ups indicate major repair, it is hoped that very few to no warm-ups will be required.

The cryogenic system has provisions to purify the argon by passing the liquid through molecular sieves. The purifier is used when the electro-negative impurities, predominately oxygen, become greater than one part per million in order to avoid signal loss.

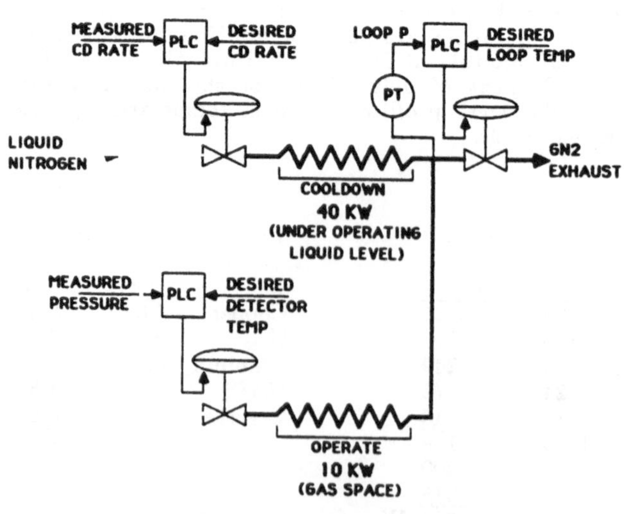

Fig. 8. General Cooling Configuration. Provisions allow setting the cooldown rate, loop temperature, and cryostat pressure independently.

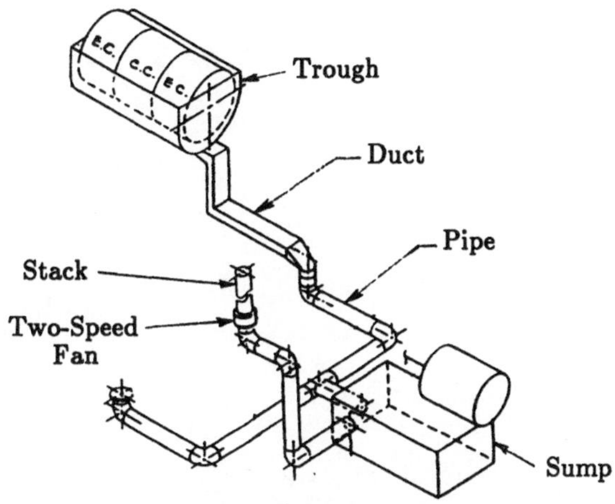

Fig. 9. Spill Containment System

OXYGEN DEFICIENCY HAZARD PREVENTION SYSTEM

Special liquid argon/nitrogen spill provisions have been designed to minimize the possibility of an Oxygen Deficiency Hazard (ODH) to the personnel working in enclosed spaces of the accelerator tunnel or on electronics located below the cryostat elevation in the detector platform. A sheet metal trough on each cryostat is arranged so that all active line connections are above and within it. The trough drains to a flexible duct (to accommodate motion) and then off the platform through a 61 cm diameter pipe to a specially insulated $57m^3$ sump capable of holding all the cryostat capacity. An $85m^3/min$ vent fan continuously draws air in at the top of each cryostat through the trough duct pipe and sump, and exhausts out-of-doors. The vent rate is increased to $280m^3/min$ when temperature changes of the duct are (failsafe) sensed. The system is shown in Figure 9.

These arrangements provide ODH protection against pneumatic depressurization, flashing, and spillage of any single cryostat nozzle from any feedthrough opening less than 5cm in diameter. Note there are nozzles below the liquid level. Active cooling loop and fill lines with greater continuous flow capability will be flow monitored and latched closed at each end on overflow. ODH detectors will sense manned and connector concentration locations, and close all source valves on a fail-safe alarm voting scheme.

CONTROLS

The cryogenic controls are provided by a programmable controller of 300 I/O and 20 PID loops. Temperature (multiplexed), pressure, flow, and vacuum monitoring, control, alarm, interlock, shutdown, and logging support unmanned steady-state operation. Remote communication with the detector VAX computer is by way of an IBM Token Ring and is password-protected.

CURRENT STATUS

The fabricated CC vessel is due to be delivered in September, 1987. It will be cold tested, outfitted and prepared for module loading by the end of the year. The first EC is due to be ordered October 1, 1987 for delivery in eight months.

ACKNOWLEDGEMENTS

The authors would like to note the contributions of Co-op students Leo Engstrom, Iowa State University, Caroline Kurita and Dave Rudland, University of Illinois, and Steve Wintercorn, Marquette University, to this work.

CRYOGENIC PULSED POWER TRANSFORMERS*

John D. Rogers

Los Alamos National Laboratory
Los Alamos, New Mexico

P. W. Eckels, D. T. Hackworth, E. J. Shestak, S. K. Singh

Westinghouse Research Laboratories
Pittsburgh, Pennsylvania

ABSTRACT

Liquid Nitrogen cooled, pulsed power transformers, two each with 14.4 MJ and one with 33.5 MJ energy storage capacity in the primary windings, utilize the low temperature properties of the copper windings for efficient energy transfer. The transformers are to provide currents of 0.95, 0.31, and 0.31 MA to a three stage distributed electromagnetic launcher and are designed to withstand voltages of 200 and 70 kV for the large and small transformers, respectively. The transformers are to be contained in fiberglass reinforced polyester plastic dewars to avoid eddy current coupling and lateral forces that would exist with a metal dewar. The transformers effectively have currents in the primary or secondary windings at different times. Thus, to improve the coupling between the primary and secondary windings, the latter, a one layer winding, is made relatively thin and is supported structurally for magnetic loading against the outer one layer, edge wound, primary winding. Both layers are mounted on an inner G10 support cage by thermal shrinkage. The coils are pool bath cooled. Normal and fault mode analyses indicate safe operations with some precautions for venting nitrogen gas provided.

INTRODUCTION

Three transformers are being built to convert 50 kA current into a high current to drive a distributed rail gun. The energy source is obtained from twenty eight traction motors with flywheels that store 80 MJ of energy. The traction motor bank will charge the transformer primary windings, connected in series, to 50 kA in 3 s. Energy transfer time between the primary and secondary windings is calculated to be 2.5 ms. Energy transfer into the rail gun occurs in about 5 to 6 ms with the transformers operating in parallel in the discharge mode. Energy to be delivered to the rail gun is 30 MJ from the large transformer and 10 MJ from each of the small transformers at peak currents of 0.95 and 0.31 MA, respectively. Residual energy, 20 MJ, is finally dissipated in stainless steel resistors in series with the transformer secondary windings with the secondary circuits shorted after the rail gun has been fired.

*Work performed under the auspices of the U. S. Army, Strategic Defense Command and the Department of Energy.

TRANSFORMER DESIGN

The design involves two layer, air core, solenoid type transformers. The design parameters are given in Tables 1 and 2. Cryogenic aluminum conductors were considered for the maximum gain in reduction of resistive losses at liquid nitrogen temperatures. An overall system energy transfer analysis showed that the very wide primary winding conductors, required to withstand the hoop stresses with aluminum, led to large losses from trapped magnetic fluxes and low coupling between the primary and secondary windings. The overall system analysis showed that copper conductors with a seven fold resistivity reduction at 77 K and smaller radial edge wound conductors improved the overall energy transfer. For this reason, annealed CDA 102 copper was chosen for the primary and secondary windings. The radial edge wound thickness of the primary winding was determined to maintain the hoop stress at about 60 percent of the yield strength. Because the transformers carry current either in the primary or secondary windings in an inverse relationship, the radial force remains constant if losses during energy transfer are neglected. Thus, the secondary winding can be supported by the primary winding. The secondary winding thickness was chosen to withstand the axial force and no hoop load. Use of the thin secondary benefits the coupling.

Table 1. Large Transformer Parameters

<u>Primary Winding</u>	
Height, cm	213
Radius average, cm	125
Number turns	120
Peak current, kA	50
Peak voltage across coil, kV	200
Peak voltage between primary and secondary, kV	110
Charge time, s	3
Energy storage, MJ	33.5
Inductance, mH	26.5
<u>Secondary Winding</u>	
Height, cm	213
Radius average, cm	119.13
Gap between primary and secondary, cm	1.0
Number turns	6
Peak current, MA	0.950
Peak voltage across coil, kV	10
Inductance, mH	0.062
Coupling coefficient between primary and secondary	0.95
Mutual inductance between primary and secondary, mH	1.22
<u>Conductor</u>	
Primary winding edge wound, cm	OFHC copper, 1.71 X 7.2
Secondary winding, cm	OFHC copper, 2.54 thick

Table 2. Small Transformers Parameters

Primary Winding	
Height, cm	152
Radius average, cm	75
Number turns	109
Peak current, kA	50
Peak voltage across coil, kV	70
Peak voltage between primary and secondary, kV	80
Charge time, s	3
Energy storage, MJ	14.4
Inductance, mH	11.5
Secondary Winding	
Height, cm	152
Radius average, cm	69.23
Gap between primary and secondary, cm	1.0
Number turns	16
Peak current, MA	0.314
Peak voltage across coil, kV	10
Inductance, mH	0.222
Coupling coefficient between primary and secondary	0.923
Mutual inductance between primary and secondary, mH	1.47
Conductor	
Primary edge wound, cm	OFHC copper 1.33 X 7.0
Secondary winding, cm	OFHC copper, 2.54 thick

The gap, 1 cm, between the primary and secondary winding was kept as small as possible to allow for electrical insulation to withstand the 200 and 70 kV reflected voltages for the two sizes of transformers and at the same time to obtain as high a coupling coefficient as practical. The reflected voltages on the primary windings result from their being open circuited and the voltage developed across the secondary windings from current driving the rail gun. The primary winding voltages are to be suppressed by varistor stacks across those windings.

The transformers are designed to have reasonable cooling in a bath of liquid nitrogen. The electrical insulation limits the rate of nitrogen vaporization and prevents overpressurization of the dewar. The liquid nitrogen also serves as an electrical insulating material. Energy transfer times are so short that the heating of the windings is essentially adiabatic, and the temperature increases are moderate and do not create thermal stresses.

The secondary winding for the large transformer was formed by machining grooves in a copper cylinder and by flat winding insulated conductor for the small transformers. Turn insulation on the secondary windings is provided by the separation distance in the liquid nitrogen for the large transformer and Kapton wrap for the wound secondaries for the small transformers. Distance between the secondary turns of the large transformer is maintained by using G10-CR spacers which provide compressive strength in the axial direction of the winding. Axial and cylindrical support of the secondary windings is also provided by interlocking axial bars and circular plates to form the secondary support cages which are rigidly attached to the secondary windings.

The primary windings of the transformers are formed by edge winding copper conductors and the conductor segments are joined with butt welds. The conductors are insulated with Kapton wrap, Nomex, and fiberglass wrap. A layer of epoxy fiberglass is attached to the inner bores to allow machining to the appropriate tolerances. The primary windings are direct resistively heated to increase their diameters and then placed over the insulated secondary windings to achieve an interference fit to ensure mechanical integrity during all portions of the duty cycle. A number of Kapton layers provide the electrical insulation between primary and secondary windings. Final assembly of the coils provides the axial support and compression to the wound coils.

The design concept for the large transformer is shown in Fig. 1, and a picture of the secondary winding for the transformer is shown in Fig. 2.

NORMAL AND FAULT MODE PERFORMANCE

For cryogenic purposes, the energy in the transformer during charge, discharge, and under fault conditions represents important design considerations to be accommodated in the leads and transformer coils themselves or to be transferred to an external dump circuit. Careful and conservative analyses were undertaken to bound the conditions of heating and liquid nitrogen vaporization to maintain the integrity of the dewar and transformer system.

Analysis of the differential contraction of the tie bolts and the transformer coils and structure shows that because of the high surface to volume ratio of the tie bolts between the end support plates the bolts will cool preferentially. A long cooldown of seven to ten days will be required to avoid excessive thermal stresses. This will be accomplished by vaporizing liquid nitrogen introduced into the dewar and slowly programming a reduction in temperature to 77 K.

Normal mode operation involves 3 s charging, 2.5 ms energy transfer from primary winding to the secondary, and 6 ms to discharge. The resistive energy deposition during these periods and the trapped flux energy in the primary windings are given in Table 3. Operation of the power supply involves crowbarring the varistor stack across the primary windings to suppress voltage. Thirty percent of the trapped flux energy is diverted into the varistor circuits leaving a residual energy to heat the conductors. The conductor temperature increases are also tabulated.

Three fault mode conditions were evaluated. These conditions considered having the peak energy stored in each of the primary and secondary windings deposited in the respective windings, heating of the leads with a clamped energy dump, and an arc breakdown within the liquid nitrogen bath. The calculated results for the first condition assumes instantaneous adiabatic heating. The expected temperature increases are given in Table 4. Resistive heating during charge adds to these fault temperature increases.

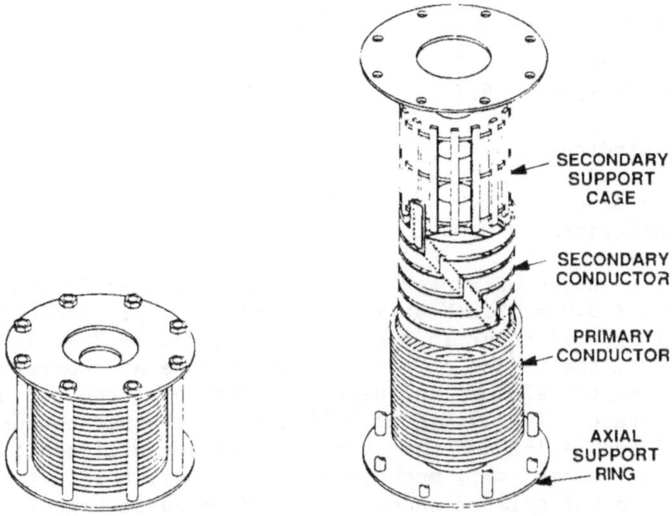

Fig. 1. Expanded assembly schematic of the large transformer.

Fig. 2. Secondary of large transformer mounted on G10-CR support structure.

Table 3. Normal Mode Energy Deposition

Primary Winding	Resistive Energy, MJ	Trapped Flux Energy, MJ	Residual Trapped Flux Energy, MJ	Conductor Temperature Increase, K
Large Transformer	8.5	4.0	2.8	5.0
Small Transformers	6.1	2.5	1.8	8.2
Secondary Winding				
Large Transformer	0.9	NA	NA	<1.0
Small Transformers	0.09	NA	NA	<1.0

For the choice of a peak nucleate boiling heat flux of 30 W/cm^2 for a $\Delta T = 14.8 + 5.0 = 19.8$ K at 1.96 atm from Kosky,[1] chosen conservatively high, cooling will occur in 8 s for a sustained 19.8 K ΔT. Then, at 80 K and 1.22 atm, the volume of gas evolved will be at the rate of 2.5 m^3/s for the large transformer for heat transfer only from the outer primary winding surface. The 19.8 K ΔT will not be attained because of the electrical insulation and the conductor surface will cool; thus the time for cooldown becomes very much longer. However, the initial vaporization rate from the uninsulated secondary could be quite high for a temperature rise of 38 K in which case the gas evolution can be as high as 5 m^3/s before the onset of film boiling. The dewar vent systems are sized to accommodate this fault mode of nitrogen vaporization.

Regardless of the perceived fault mode just discussed, the heating of the secondary leads from the transformers mitigates against such an event. Table 5 gives the temperature increases for the transformer leads for the condition for which the transformers are shorted at the leads just external to the dewars under peak current conditions. Whereas the primary lead temperature increases are tolerable, those for the secondary leads are prohibitively large. To avoid this condition, external dump resistors that reduce the discharge period a factor of five are inserted in the external circuit with crowbar or shorting switches.

The third fault condition for an arc breakdown in the liquid nitrogen examined a range of parameters with energy in either the primary or secondary windings of the transformers. Sustained arcs of 30, 100, and 300 V were examined. Decay times varied from 0.2 with an arc of 300 V with the energy stored in the secondary windings of either the large or small transformers to 12.5 s for a 30 V arc with the energy stored in the primary winding of the large transformer.

Table 4. Fault Mode Energy Dump

Primary Winding	Energy, MJ	Temperature Increase, K
Large Transformer	33.5	14.8
Small Transformer	14.4	11.4
Secondary Winding		
Large Transformer	30	38
Small Transformer	10	27

Table 5. Lead Temperature Increases

	Temperature Increase, K
Large Transformer	
Primary Leads	69
Secondary Leads	181
Small Transformers	
Primary Leads	73
Secondary Leads	217

Assumptions incorporated into the vapor generation and venting calculations were

1. No shock wave is generated.
2. All fluid released is in the gaseous or vapor phase. No liquid is entrained.
3. The gas from the total energy available is generated over two e-folding periods, that is two L/R time constants.
4. The arc is in intimate contact with the fluid to generate vapor during the entire arc period. This assumption implies to some extent that there is no massive local superheating of nitrogen gas in the vicinity of the arc to alter the gas generation or flow.
5. Gas flow through the dewar blow off ports is sonic.
6. The initial gas flow rate is based upon the I^2R in the arc at peak current.

Calculations show that frictional and expansion pressure drop effects are negligible and that the significant pressure drops to drive the gas from the dewar arise from fluid acceleration and contraction effects through the dewar lid blow-off port. Blow-off ports in the lids for the large and small transformer dewars, respectively, were chosen to be 0.213 and 0.118 m. in diameter. For these conditions the pressure for 80 K nitrogen need only to be 1.27 and 1.34 atm, respectively, for the large and small dewars for sonic flow of the gas. Pressure drop calculations show that the corresponding venting pressures will rise to 2.2 and 2.3 atm and the flow will be subsonic.

SUMMARY

Design of the cryogenic transformers has posed several unusual requirements. Besides the usual electromechanical, stress, and electrical insulation features to be considered, transient operation of the transformers has directed design efforts, because of cryogenic and thermal limitations, to impose constraints on the external circuit and on the dewar venting system. Normal mode operation of the transformers have minimal effect on the system. Major energy dumps in the transformers cause vent systems to be oversized and the need for external dump circuits. A submerged arc within the liquid nitrogen bath is a destructive calamity to the transformer affected and demands a major dewar venting capability not to burst the dewar.

REFERENCE

1. P. G. Kosky, dissertation Fig. 18, University of California, Dept. Chem. Eng. (1964), Berkeley, CA.

AEROSPACE GENERATORS USING HIGH-PURITY CRYOGENIC ALUMINUM ROTOR CONDUCTOR

T.A. Keim

General Electric Company
Schenectady, New York

ABSTRACT

It now appears probable that composite electrical conductors consisting of high-purity aluminum strengthened with alloy aluminum will become available. When cooled to liquid hydrogen temperature, such conductor can be operated at current densities previously considered impossible except with superconductor. Using such conductor, it should be possible to design electromechanical apparatus having previously unachievable properties. To obtain a preliminary estimate of the properties of a high-powered electric machine with a hyperconductive aluminum rotor, extrapolations have been made based on existing designs for superconducting generators.[1] This paper explores the limits of this process. Fundamental dimensions are presented for an aerospace generator to produce 50 or 100 MW at 50 kV dc. The layout is derived from the design of the Advanced Superconducting Generator (ASG) previously developed for the U.S. Air Force. The resulting configuration is reasonable from the point of view of centrifugal stress. Results from the ASG, combined with 2-dimensional calculations, are used in lieu of 3-dimensional magnetic calculations to determine the dimensions of the machines. Many parameters of the stator can be selected based on reasonable changes from the ASG. Much detailed mechanical engineering would be required to determine whether the configurations proposed could indeed be achieved. It is particularly problematic to determine whether it would prove possible to design an adequate cooling system for the rotor of this proposed generator. The results can serve as a vehicle for outlining the advances that would need to be achieved to make the machine practical.

INTRODUCTION

In 1985 Schlicher and Oberly[1] published a thought-provoking paper with the central premise that newly developed conductors comprised of a composite of high-purity aluminum and alloy aluminum can, when cooled by liquid hydrogen, be operated at a current density equal to that achieved in helium-cooled superconducting windings. They discuss the direct substitution of such an aluminum winding for the superconducting field winding of the Advanced Superconducting Generator (ASG).[2] While this simplistic approach can only begin to assess the capabilities and limitations of a generator using such a hydrogen-cooled composite field winding, the calculations included in the paper fail to find a compelling reason why the idea cannot be made to work.

The merit of the Schlicher and Oberly paper is the simplicity of the approach. To propose direct substitution of the new conductor for a high-performance superconductor in a demanding application focuses attention on the subject in a manner that would have been unachievable by discussing the use of the same conductor for the same purpose in a less literal manner. The obvious problems with this proposal are clearly not so major that they cannot be reduced or eliminated by obvious measures. The proposal presented as it is therefore forces the reader to recognize the potential merits of the new conductor as a substitute for superconductor in this and other applications.

The prospect of a liquid-hydrogen-cooled generator with the specific weight of a superconducting generator is a very attractive one, in view of the prevalence of proposals to use liquid-hydrogen-fueled turbomachines for low-specific-weight power supplies. The availability of the coolant at no additional effort, and the coincidence of the need for generator coolant and the need to vaporize fuel, are compelling attractions, especially if they can be achieved at no penalty in weight, relative to a superconducting generator. The concept deserves a more detailed evaluation. This paper is an incremental contribution to such an effort.

For superconducting generators, most of what has been determined about the parametric size, weight, and output scaling has been derived from 2-dimensional analyses of a typical cross section. Typically the parameters characterizing a cross section are hypothecated, and properties per unit length computed. One desired property (usually output power) is used to determine the required length for a machine of the assumed cross section, and many derived characteristics of the machine are then determined. The process is repeated as the parameters of the cross section are varied in a methodical manner, and a family of possible designs is thus generated. The resulting family may be evaluated by a number of different measures of merit to determine a desirable configuration for a proposed application.

One important limitation of this procedure is that the resulting family of candidate configurations differ not only in the values of their readily determinable figures of merit, but also in the ease with which (or the extent to which) the hypothesized parameters can be achieved in hardware. To cite a simple example, power increases linearly with length for any set of cross-sectional parameters at constant rotational speed. But the rotor lateral flexibility increases as the cube of length. For a short machine, this may imply nothing more than a reduction in the margin between operating speed and critical speed, but for a long machine, operating speed or rotor cross-section properties may need to be altered to achieve acceptable rotor dynamic behavior.

An important step in a parametric study is therefore the preparation and analysis of design layouts for one or more point designs. The engineering analysis of these layouts may reveal problems that indicate the desirability of selecting cross-section parameters other than those defined by the simple parametric study. The ASG is an example of a point design that has been extensively studied.

By selecting the ASG as a starting point, Schlicher and Oberly were able to discuss the merits of the new conductor without reexamining all the parametric scaling issues. Ultimately, the choice between cryogenic aluminum and superconductor should be made after a study in which parametric scaling is reconsidered. This study should compare the properties of machines that have been optimized for a specific application, with the properties of the different conductors taken into account.

Such a comparison is beyond the scope of this paper. The discussion presented here will focus on a few configurations closely related to the ASG. First, a set of changes to the design of the ASG will be discussed. These changes permit us to describe a 50 or 100 MW machine, apparently more in line with present-day thinking about future needs. Then the impact of a switch to cryogenic aluminum conductor in this design will be discussed.

LARGER RATINGS FROM THE ASG CROSS-SECTION

The specifications of the ASG, including the power rating, were selected to push the technology; the nature of an airborne system that might require such a large amount of power was not clearly defined. Given the subsequent course of events, the selection has proved remarkably prescient. Today proposals are being advanced for systems using hundreds of megawatts of power, with modular generating units in the 50 to 200 MW range. To make this discussion more relevant to today's perceived needs, attention will first be directed to the changes required to convert the 20 MW generator configuration to one capable of 50 or 100 MW. Also, we will, for this discussion, consider machines capable of 50 kV at the output of a direct-connected Graetz bridge (as opposed to 40 kV for the ASG.)

The basic procedure will be to assume that the overall generator is axially longer, but otherwise very similar to, the ASG. This will permit results from the ASG to be used to estimate important end effects. Also, by not straying far from a well-known configuration, we maintain a reasonable degree of assurance that the devices being discussed can be realized. In selecting this procedure, we have implicitly determined that we will consider only 4-pole machines operating at 6000 rpm [100 Hz], and that the machines will be of the conductively shielded type. The resulting configuration is almost certain to be inferior to that which could be achieved for any application with all parameters free, but, as will be seen, the results are not unattractive, so the possibility of still more attractive designs is a bonus.

To implement this procedure, magnetic scaling will be used to determine the required length. The impact of this length change on the achievability of other parameters of the machine will not be considered at length.

[In the following paragraphs, pairs of numbers — the first freestanding and the second enclosed in parentheses — will be used. Wherever this grouping appears, the first number will refer to a machine of 50 MW output; the second, to 100 MW, e.g., 50(100) MVA.]

To relate ac to dc quantities, we will use simple equations for inductive loads. The rated dc current is 1000 (2000) A. The corresponding rms line current is

$$I_b = \sqrt{\frac{2}{3}}\, I_{dc} = 816\ (1633)\ \text{A} \tag{1}$$

These numbers are exactly 2(4) times as large as the armature current of the ASG. This means an armature cross section exactly like that of the ASG can be used by the simple expedient of reconnecting the coils to form 2(4) parallel circuits per phase. Thus we have reasonable assurance that the proposed armature current density can be achieved. The number of series armature turns per circuit is 168 (84).

The output voltage of a 3-phase bridge rectifier is given by

$$V_{dc} = \frac{3}{\pi}\sqrt{6}\, V_{ln} - \frac{3\omega L}{2\pi}\, I \tag{2}$$

Here V_{ln} is the rms voltage (line to neutral), generated internal to the machine, and L is the commutating inductance, which is some transient inductance of the machine, generally close to the subtransient inductance. For purposes of this discussion, we will assume a value of commutating reactance $\omega L = 0.4$ per unit on the ac base, determined by voltage at open circuit and rms current at rated load. This reactance assumption should be conservative; the actual subtransient reactance should be below this value.

For the 50 kV machines being considered here, the base voltage is

$$V_b = \frac{\pi}{3\sqrt{6}}\, V_{dc} = 21.4\ \text{kV} \tag{3}$$

Equation 2 may be rewritten:

$$\frac{V}{V_b} = \frac{3}{\pi}\sqrt{6}\,\frac{V_{ln}}{V_b} - \frac{0.6}{2\pi}\,\frac{I}{I_b} \tag{4}$$

From equation 4, we see that, at rated conditions, the generated voltage is

$$V_{ln} = \frac{\pi}{3\sqrt{6}}\left(1 + \frac{0.6}{2\pi}\right) V_b = 23.4\ \text{kV} \tag{5}$$

for either rating.

To determine the required length of the generator, we need to make the field-to-armature mutual inductance large enough to generate this voltage. Three-dimensional effects are important in this calculation. For the ASG, a relatively accurate 3-dimensional calculation was performed.[3] For this analysis we will not repeat this computationally intensive procedure. Instead, we will use the existing calculation (for a 6-in. straight section) to represent the effect of the end turns and the first 6 inches of the straight section. To estimate the effect of lengthening the generator, we will use a 2-dimensional calculation from Kirtley[4]:

$$\frac{\omega M I_{fb}}{\sqrt{2}\, l} = \frac{dV_{ln}}{dl} = \frac{\sqrt{2}\,\mu_o J_f R_{fo}^2 N_{at}\,\omega}{\pi\,\theta_{wae}(1-x^2)}\left(\frac{R_{fo}}{R_{ao}}\right)^2 \tag{6}$$

$$\times \sin\frac{\theta_{wfe}}{2}\,\sin\frac{\theta_{wae}}{2}\,(1-y^4)\left[-\ln x - \frac{1-x^4}{4}\left(\frac{R_{ao}}{R_s}\right)^4\right]$$

This expression is particularized for the spatial fundamental contribution. The notation follows Kirtley directly. For the cross section of the ASG, this value is 1190 (595) V/in. [46,800 (23,400) V/m].

From the 3-dimensional analysis, we know that the end sections and the first 6 inches generate 66.5 V/turn, or, for the cases of interest, 11170 (5590) V. Note that the end turns contribute approximately the same amount of mutual flux as 3.4 in. [0.086 m] of incremental straight length. The incremental length required to meet the design goals is 10.2 (29.9) in. [0.259 (0.759) m], for a total straight length of 16.2 (35.9) in. [0.411 (0.912) m].

In view of the increase in voltage, it may be prudent to include more space between field and armature than in the ASG to allow for more insulation. From equation 6, the impact ΔM of a change ΔR (with resulting change Δx) on the 3-dimensional contribution to mutual inductance is given by:

$$\frac{M+\Delta M}{M} = \frac{(1-x^2)}{\left[1-(x+\Delta x)^2\right]} \frac{R_{ao}^2}{(R_{ao}+\Delta R)^2} \times \frac{\left[-\ln(x+\Delta x) - \left(\frac{(1-(x+\Delta x)^4}{4}\right)\left(\frac{R_{ao}+\Delta R}{R_s}\right)^4\right]}{\left[-\ln x - \frac{(1-x^4)}{4}\left(\frac{R_{ao}}{R_s}\right)^4\right]} \tag{7}$$

For $\Delta R = 0.5$ in. [0.013 m], $\Delta M/M = -0.108$.

To approximate the impact of this radius change on the length required to generate the required voltage, we will scale the entire straight length by the appropriate factor. Thus for the larger machine radius, the required lengths are 18.2 (40.2) in. [0.462 (1.02) m].

The weight of these revised machines may be estimated by computing the incremental weight per unit axial length of the axially invariant portion, and adding an allowance for the end regions. Table 1 details the areas and densities used to make this computation. The rotor areas are taken closely from the ASG. The stator areas have been adjusted to account for the radius change proposed above.

Table 2 presents the estimated weights for the configurations proposed. The end weight allowance has been made larger for the 100 MW case than for the 50 MW case. The 50 MW end weight allowance is itself larger than the overall weight goal for the 20 MW ASG.

There is some reason to expect larger machines to have larger end weights (higher output torque may require a larger shaft, for example), but much of the increase in end weight allowance with rating may be regarded as a cushion against estimating errors. We see that end-component weight is significant at both power levels. Even with the conservative end weight allowances, the specific weights in Table 2 are quite attractive.

Figure 1 shows conceptually the comparison between the ASG, and the 50 and 100 MW machines discussed here.

There are important differences between the ASG and the larger machines in Figure 1. Several of the more important are:

- Operating and short circuit torques are higher.
- The longer rotors present substantially different rotor dynamics problems.

Table 1 Superconducting Generator Component Masses

Rotor				Stator			
Component	Total Area (m^2)	Density (kg/m^3)	Linear Density (kg/m)	Component	Total Area (m^2)	Density (kg/m^3)	Linear Density (kg/m)
Bore tube	0.0154	8200	126	Bore tube	0.0228	2200	49
Support Structure	0.101	1900	188	Conductor	0.0186	8900	167
Conductor	0.0219	8900	194	Insulation	0.0183	1900	34
Torque Tube	0.0190	8200	155	Banding	0.0008	2200	2
Radiation Shield	0.0039	2700	11	Bulkheads	0.43	470*	203
EM Shield	0.0309	2700	84	Case	0.0394	2700	107
			758		*axial average		562

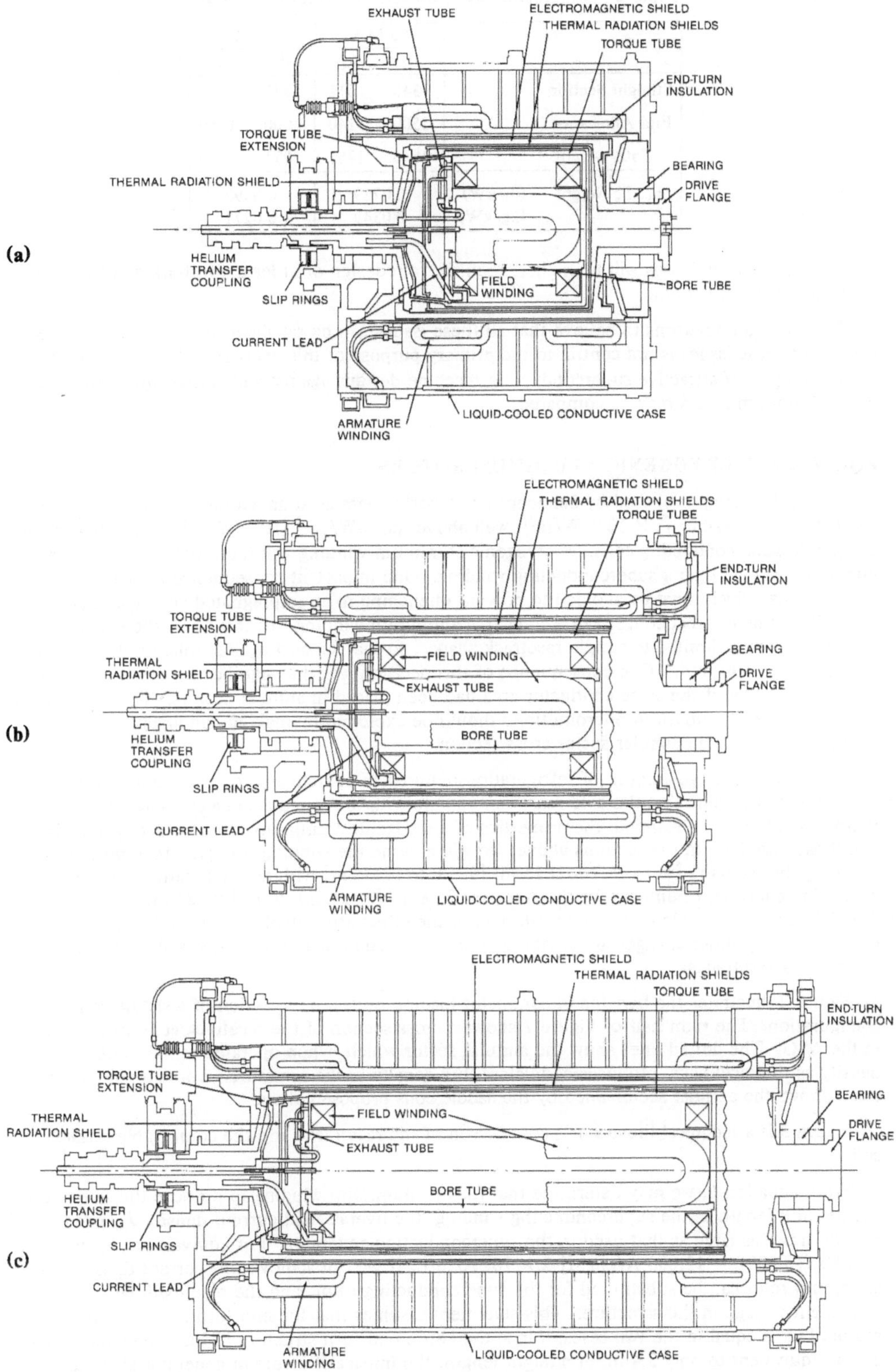

Fig. 1. Comparison of superconducting generators for different ratings but same speed and frequency. (a) 20 MW, 6000 rpm [100 Hz] generator; (b) 50 MW, 6000 rpm [100 Hz] generator; (c) 100 MW, 6000 rpm [100 Hz] generator

Table 2 Superconducting Generator Weight Estimate

		50 MW		100 MW	
		lb	kg	lb	kg
Straight Section		1345	611	2971	1380
End Allowance		2500	1140	3000	1360
Total weight		3845	1751	5971	2710
Specific weight	(lb/kW)	0.0769		0.0597	
	(kg/kW)	0.0349		0.0271	

- Overall armature losses will be higher, requiring a redesign or at least a rethinking of the armature cooling system.

There are good reasons to believe that all these issues can be satisfactorily resolved. But discussion of these issues is not central to the primary purpose of this exercise, which has been to establish a pair of attractive superconducting machine designs against which machines with cryogenic aluminum rotors may be compared.

EQUIVALENT CRYOGENIC ALUMINUM ROTORS

The field winding proposed by Schlicher and Oberly operates at an average volumetric power dissipation of 5 W/cm^2 [5 × 10^{-6} W/m^{3}], well above the mW/cm^3 [10^{-7} W/m^3] levels that are of concern in superconducting coils. A cryogenic aluminum winding will therefore be much more intensively cooled than a superconducting winding. One important consequence is that microscale motions, which must be avoided to prevent quenching in an impregnated superconducting coil, will not be of great concern in a cryogenic aluminum winding. In the ASG, the superconducting coils were limited to simple racetrack shapes, so that support against micromotions could be reasonably achieved.[5] One consequence of this decision was that significantly less mutual flux was obtained than if the same conductor area had been placed in annular sectors. The insensitivity of cryogenic aluminum to micromotions should be exploited by making windings which more nearly approximate an annular sector cross section.

Also, with the large amount of refrigeration required to operate a cryogenic aluminum winding, it seems plausible that the cold mass can operate with much reduced electromagnetic and thermal shielding. To estimate the properties of a cryogenic aluminum rotor for use with stators like those described in the sections above, we will assume the winding occupies an annular sector spanning the radius range from 7 to 9.25 in. [0.178 to 0.235 m] with an arc length of 60° per pole. The equivalent numbers for the ASG are 7.28 in. [0.185 m], 9 in. [0.229 m], and 44°. The difference in outside radius is an estimate of the effect attributable to reduced shielding requirements; the other changes are attributable to substitution of well-designed saddle windings for the existing racetracks.

Figure 2 illustrates the benefits in conductor loading caused by selection of a saddle coil configuration. The right half of Figure 2 shows a cross section of the racetrack coils (solid lines) in the ASG. The dotted lines show the annular sector which, when operated at the same current density as the racetracks, produces equivalent flux density in the armature. The left side of Figure 2 shows the annular sector filled by the saddle coils proposed above.

Figure 3 is a technical illustrator's comparison of the racetracks and the comparable saddle coils.

From equation 6, we may determine that for the aluminum winding to produce the same voltage per unit length as the superconducting winding, the overall field current density J_{fa} needs to be 0.80 times as large as that used in the superconducting generator. For equivalence, then, J_{fa} = 0.80 (15,000) = 12,000 A/cm^2 [1.2 × 10^8 A/m^2]. If we can achieve this current density, the aluminum rotor can be substituted for the superconducting rotor, and the result will be equivalent voltage in the armature. This statement assumes that the end-turn contribution to mutual flux is equal in the two cases. Since the entire end-turn structure in the superconducting case is equivalent to only 3.4 in. of straight length, the impact of errors in generator size caused by this approximation are expected to be small. To achieve generator performance equivalent to a superconducting generator, it therefore may not be necessary to achieve exact current density equivalence with hydrogen-cooled aluminum.

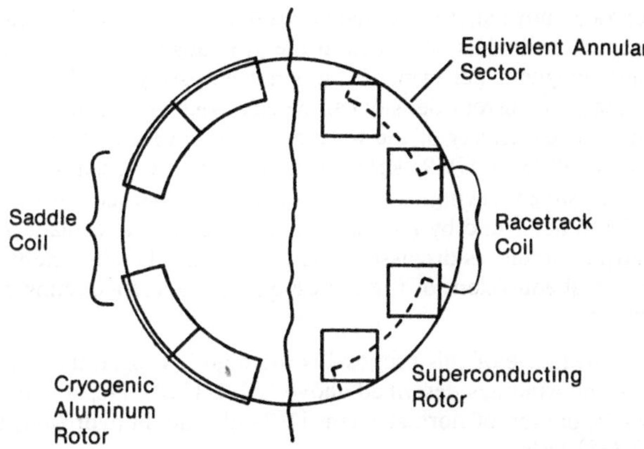

Fig. 2. Cross section of saddle coils compared to racetrack coils.

Consider what might be required to achieve 12,000 A/cm^2 with an aluminum conductor. The armature of the ASG is a good example of a resistive winding in which careful attention has been given to close packing of conductor while allowing adequate room for coolant flow and structure.

Assume that we can achieve slightly better packing in an aluminum rotor. The ASG armature cross section is 40% conductor, 14% cooling passage, and 46% structural material and insulation. Assume 50% conductor, 15% cooling, and 35% insulation for the cryogenic aluminum winding. The current density on the conductor is then 24,000 A/cm^2 [2.4×10^8 A/m^2]. Let us further assume that the conductor is 65% high-purity aluminum. Current density on the high-purity conductor is 36,500 A/cm^2 [3.65×10^8 A/m^2]. If the resistivity of the conductor is 1/500 of room temperature aluminum (ambitious, but possibly achievable), the loss density is 7.63 W/cm^3 [7.63×10^{-6} W/m^3]. in the high-purity conductor or 2.48 W/cm^3 [2.48×10^{-6} W/m^3] on the overall winding. The ASG armature achieves 8 cm^2 of heat transfer area per cm^3 of conductor. If the aluminum is similarly well cooled, the surface heat flux is 0.62 W/cm^2 [6200 W/m^2]. This is clearly an acceptable level, so possibly some heat transfer area can be sacrificed.

The gross volume of the winding is 2510 (4190) in.^{3} [0.0411 (0.0687) m^3], assuming a 14.5 in. [0.368 m] mean length for an end turn. If we assume the insulation density is the same as aluminum (a slight overestimate), the overall winding density is still only 85% that of aluminum. The winding weighs only 213 (356) lb [96.7 (112) kg]. In terms more directly comparable to Table 1, the gross winding area is 76.6 in.^{2} [0.0494 m^2] and the density (as argued above) is 0.085 lb/in.^{3} [2350 kg/m^3], giving a linear density of 6.51 lb/in. [116 kg/m]. This is 4.37 lb/in. [78.1 kg/m] less than the superconducting winding in Table 1.

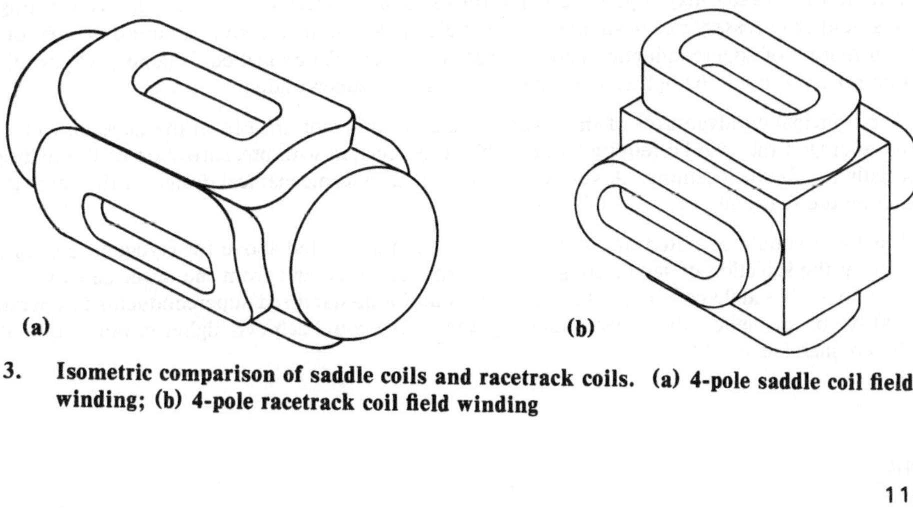

Fig. 3. Isometric comparison of saddle coils and racetrack coils. (a) 4-pole saddle coil field winding; (b) 4-pole racetrack coil field winding

The rotors proposed above should be magnetically equivalent to the rotors proposed for the superconducting generator models described in the previous section. It can be seen that in order to achieve equivalent magnetic performance, it is not necessary to achieve exactly equivalent current density. The required current density can be achieved with reasonable requirements for local heat transfer. The reduced density of the winding material results in some weight saving (approximately 79.5 (176) lb [36.1 (79.9) kg]), even after the increased winding area is taken into account. Some of this weight savings will be offset by increases as a result of the larger rotor diameter (not considered here) and by the need for more elaborate coolant plumbing and possibly a secondary heat transfer circuit, as discussed below. Overall, the aluminum rotors described here should be the functional equivalent of the corresponding superconducting rotors, with equivalent or slightly lower weights.

The principal disadvantage of this proposal is the high hydrogen flow rate required. The total power dissipation in the windings described above is 102 (170) kW. If this level of cooling were to be provided by evaporation of normal 1-atm [101-kPa] liquid hydrogen, the mass flow required would be 0.231 (0.385) kg/s.

Assuming that the coolant picks up an additional 10 kW of heat after leaving the winding cooler, the volume flow of exhaust vapor in such circumstances is 0.2 (0.31) m^3/s. With a 200 ft/s [61 m/s] flow velocity, the duct area required is 32.7 (50.7) cm^2. This is a substantial fraction of the winding gross cross section, certainly more than the fraction allotted to coolant passages. A cold return coupling with adequate area will be a major challenge.

The unutilized refrigeration in the exhaust flow (up to 273 K) is 720 (1200) kW. With refrigerant flows at this level, a very nonstandard cryogenic system design is required. Rather than utilizing return flow to intercept conduction and minimize boiloff, it will be necessary to provide very good isolation of the return stream to prevent excess cooling of other parts of the machine.

The foregoing paragraphs assume that the coolant fluid undergoes a process at 1 atm [101 kPa]. To approximate such a process, it would be necessary to avoid significant radial outflow of high-density fluid. If the flow enters on axis, it must stay on axis. This is not inconceivable, but a compact high-capacity secondary heat transfer system needs to be designed. Coolant could be transferred to the rotor at a radius near the field winding radius. After evaporation, the vapor could be returned on axis. This would require a cryogen transfer coupling with unprecedented properties. If the coolant is brought on board on axis at 1 atm [101 kPa] and then permitted to flow radially to the winding radius, the winding pressure would be 14 atm [1414 kPa]. The fluid would be supercritical, and if the same mass flow were used as in the 1-atm [101-kPa] process, the exhaust temperture would be 35 K. This temperature is inconsistent with the resistance ratio assumption made earlier. This alternative therefore would require mass flows still higher than those proposed above.

CONCLUSIONS

Information drawn from publicly available sources can be used to make reasonable projections of the properties of high-power lightweight cryogenic generators with either superconducting or resistive field windings. It seems plausible, based on the arguments advanced above, that in order to achieve generator performance from a cryogenic aluminum rotor comparable to that achievable with present-day superconducting rotors, direct replacement of the superconducting coils is neither necessary nor desirable. The simple coils described above require only 80% of the current density of superconducting coils in order to achieve functional equivalence. Cryogenic aluminum coils should be lighter than the corresponding superconducting coils.

The principal disadvantages of the cryogenic aluminum rotor arise from the large flow of liquid hydrogen that will be required and problems associated with pressurization of the hydrogen, especially at off-axis locations. A very careful cryogenic and mechanical design of the rotor is required before feasibility can be established.

Finally, it should be noted that several of the advantages cited above for cryogenic aluminum arise from the selection of saddle coils, rather than racetracks, and from the lower density of aluminum coils. If saddle coils could be built with aluminum-stabilized superconductor to operate at 15,000 A/cm^2 or higher, the superconducting alternative could achieve higher power output at nearly comparable weight.

REFERENCES

[1] R.L. Schlicher and C.E. Oberly, Cryogenic Aluminum-Wound Generator Rotor Concept, *Fifth IEEE Pulsed Power Conference*, Paper PTTT-12, June 1985

[2] B.B. Gamble and T.A. Keim, "High Power-Density Superconducting Generator," *AIAA Journal of Energy*, Vol. 6, No. 1, Jan.-Feb. 1982, pp. 38-44

[3] B.B. Gamble, T.A. Keim, and K.F. Schoch, "Advanced Superconducting Generator Program, Phase II Interim Technical Report," Air Force Wright Aeronautical Laboratories, SRD-79-176, December 1979

[4] J.L. Kirtley, Jr., Basic Formulas for Air-Core Synchronous Machines, *IEEE PES Conference*, Paper 71 CP 155-PWR, January 1971.

[5] B.B. Gamble, T.A. Keim, and P.A. Rios, "Superconducting Rotor Research, Phase I Interim Technical Report," Air Force Wright Aeronautical Laboratories, SRD-77-131 (November 1977)

LOW LOSS LIQUID HELIUM TRANSFER SYSTEM, USING A HIGH PERFORMANCE CENTRIFUGAL PUMP AND COLD GAS EXCHANGE

H. Berndt, R. Doll, U. Jahn, and W. Wiedemann

Walther-Meissner-Institut für
Tieftemperaturforschung
Garching, W.-Germany

ABSTRACT

A liquid He transfer system with overall transfer losses < 2 % is described. Energy and running time savings for He-liquefiction up to 30 % are achieved in comparison with conventional transfer using a pressurized liq. He-storage tank. The main components of the system are a high reliable, completely magnetic suspended centrifugal pump submerged in the liquid and a double transfer line allowing cold gas exchange between transport vessel and storage tank. Filling rates up to 1500 L/h and more are possible. Constructional and performance date of the system are reported.

INTRODUCTION

In the past a lot of efforts were made to enhance the performance of He-liquefiers, including expansion engines and turbines, but not too much has been done to diminish the considerable losses, when liquid He is transferred from a large storage tank - mostly connected to a liquifier - into a transport vessel at ambient pressure. Transfer losses of 30 % and more are usual, they must not be explained in detail. During transfer the liquefaction rate of a connected liquefier is diminished when liquid He is withdrawn from the storage tank, since the resulting void volume in the tank must be filled with cold gas from the liquefier. When transferring liquid without operating liquefier the losses are similar drastic, since the transfer pressure must be maintained either by evaporating liquid Helium or by pressurizing the tank with warm He-gas, these losses may be as high as 25 % related to the withdrawn liquid - they result in a prolongated starting time when the liquefier is started again.

Transfer losses are diminished to a few percent, when a reliable, high performance submersible liquid He-pump is available in connection with a double transfer line which allows cold gas exchange between transport vessel and storage tank. Pump transfer is most effective applied, when a liquefier fitted with a cold (or warm) ejector is available, so that liquid at nearly ambient pressure is produced.

A maintenancefree and nearly frictionless operating centrifugal pump and a suitable transfer line was built. A prototype system is in operation since about two months. Though the matching between pump performance characteristic data and transfer system is not optimized, the overall transfer losses could be reduced from more than 30 % to below 2 %. This means savings in energy consumption and running time of the liquefier of about 30 %, compared to the method used before.

Liquid He-centrifugal pump

Since a liquid under its own vapour pressure must be pumped a submersible centrifugal pump seems to be the best solution, when maintenancefree and reliable operation and good performance can be achieved. Fig. 1 shows a schematic drawing of the pump. Liquid enters the pump through the inlet stator (1), is accelerated in the covered rotor (2) (36 mm Ø); enters the guide wheel (3) and flows through 8 vertical channels (4) to the exit connection (5). The rotor with the armature shaft of the driving motor is magnetically suspended. An exact defined axial position of the rotor is achieved by an active magnetic bearing (6). The rotor position is detected by an inductive pick-up system (7). The rf signal of the pick-up system is amplified, processed by an electronic device and fed back to the magnet system (6). During operation the axial position is fixed to better $\pm$ 1 μm, whilst the free distance between rotor and magnet system is $\sim$ 0.1 mm. The sealing between high and low pressure stream is combined with the magnetic bearing at the outside rim of the rotor. This type of axial sealing guarantees two essential advantages: 1. Symmetric sealing avoids axial thrust on the rotor shaft, 2. Sealing at the high pressure end of the rotor assures, that resulting vapour, arising from rotor friction and motor losses is brought back to the supply reservoir, as a consequence a pure one-phase liquid stream is delivered. The radial bearings (8) at both ends of

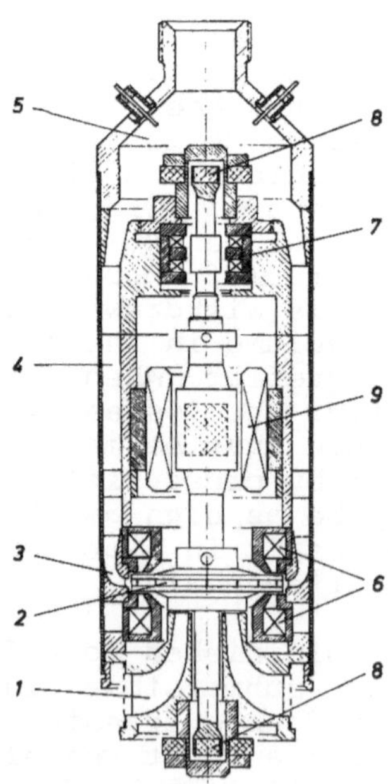

Fig. 1. Longitudinal section of the helium pump.

the rotor shaft are passive bearings consisting of permanent magnets. Eddy current damping of parasitic motions (precession, nutation, rolling) is achieved by suitable means. Revolution speeds up to 10000 rpm have been achieved without instabilities and problems. During transfer operation speeds up to 6000 rpm are used. The rotor is actuated by a self-synchronous motor (9), consisting of a permanent magnet in the rotor shaft, magnetized perpendicular to the rotor axis and two 90 ° displaced coils. One of them gives a generator signal, which is amplified, 90 ° phase shifted and fed back to the second coil. Starting the "frictionless" rotor is no problem. Revolution speeds can be adjusted continiously. In the moment a maximum available driving power of 10 W is enough for transfer operations (higher driving powers can easily be achieved). The motor efficiency is better 95 % and could be further improved by using superconductive windings. (The pump has a maximum outside diameter of 50 mm, total length 180 mm, weight 1.32 kg).

Pump performance data

Performance data were taken in two different test cryostats. Pressure head H(m), flow rate $\dot{V}$(L/h) and power consumption of the motor N(W) were measured simultaneously. Data acqusition was done with a computer. The complete data field is available, hydraulic efficiency $\eta_{hydr} = H_{exper}/H_{theor}$ and total efficiency $\eta_{tot} = p \cdot \dot{V}/N$ have been calculated.

The pump performance characteristic is shown in Fig. 2. The shaded area shows the region where the overall efficiency is greater 50 %. The max. total efficiency including motor losses is 53.3 % at a revolution speed of 115 s^{-1}. At this point a flow rate of 1440 L/h at a pressure head of 9.6 m is available with a total power input of 8.9 W. The onset of the guide wheel efficiency can be seen clearly along the line X, where the pressure head is enhanced by about 28 %. η_{hydr}

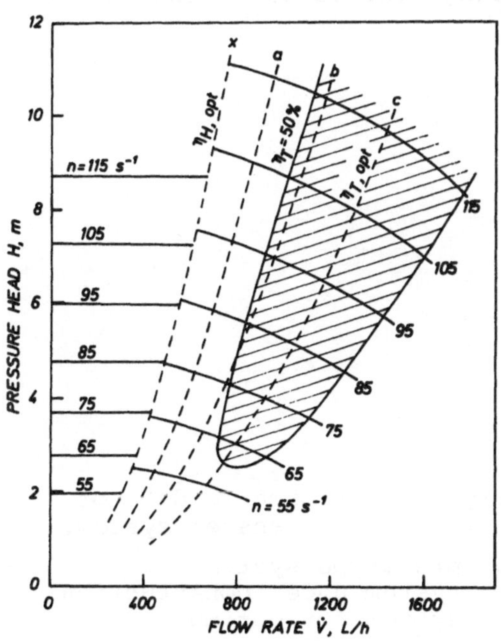

Fig. 2 Pump performance.

$\eta_H = \eta_{hydr} = H_{exper}/H_{theor}$

$\eta_T = \eta_{tot} = p \cdot \dot{V}/N$

a prototype system
b system under construction
c system for optimum efficiency
x onset of guide wheel efficiency

reaches max. values along this line (64.8 % at 115 s^{-1}) and may get still higher at evaluated speeds. Measured data are in very good agreement with theoretical predicitions, that means: $\dot{V}$(L/h) ∿ n(s^{-1}); H(m) ∿ n^2; N(W) ∿ n^3. Line a shows the load line of the present prototype-transfer-system. Matching with pump performance data is not yet optimal (pump works at η_{tot} ≈ 45 %) because there were some constructional restrictions concerning the transfer system. Line b shows the expected load line of the final transfer system which will be installed in a 5000 L storage tank in connection with a Sulzer TCF 100 liquefier. Also with this system the pump does not work at the highest efficiency since we are restricted with the outside diameter of the vacuum tube of the transfer line at the transport vessel end to 25 mm (vessels with 30 mm neck-tube diameter must be filled). Line c shows the load line for optimum efficiency - it can be verified, when only vessels with a neck-tube diameter of ≳ 50 mm are filled. Fig. 3 shows the performance data of the complete system, pump and transfer line: a prototype system, b system under construction, c optimum system. When operating system a with a transfer rate of 833 L/h (n = 100 s^{-1}), the total power input is 5 W, system b needs for the same flow rate only 2.5 W at n = 76 s^{-1} and system c 1.5 W at n = 62 s^{-1}.

Transfer system

The complete transfer device is shown in Fig. 4. Since the pressure drop of the transfer line should be as low as possible flexible transfer lines are not recommended. The vessel to be filled is raised by a mechanical lifting device, combined with a balance. The lifting stage is raised to coupling position, the vessel to be filled is weighed (tare) and further raised to filling position. A sliding Teflon-stuffing box and ball valves at the end of the transfer line and on the head of the transport vessel avoid cold gas exit and gas losses. Coupling is done by a bayonett quick coupling. The receiver end of the

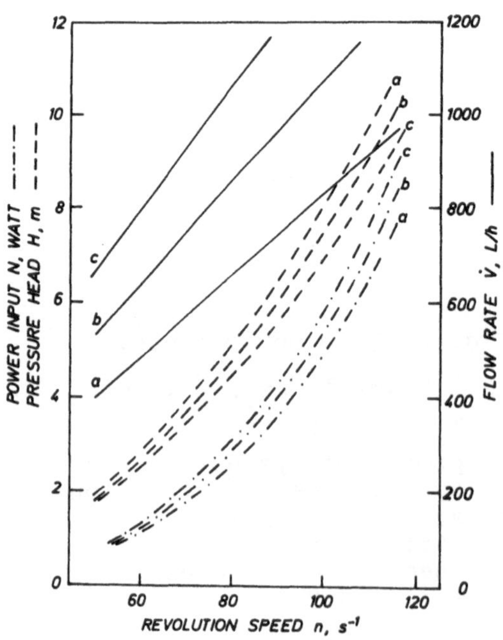

Fig. 3 Performance data of the transfer system.

a prototype system
b system under construction
c system for optimum efficiency

transfer line is protected by a rubber bellow to avoid condensing water on the transfer line surface when the filled vessel is lowered to decoupling position. The rubber bellow-volume is automatically connected to the gas holder when the stage is moving. After decoupling a second weighing procedure is done and the transferred mass of He is registrated by a computer. When starting transfer (pump speed = 70 s^{-1}) the three way valve in the valve box of the syphon is open to the gas holder. As soon as a H_2 vapour pressure thermometer mounted on the line of the back gas stream indicates a temperature $\leq$ 20 K the back gas stream is directed to the storage tank. The end of the filling process is indicated by a sudden change of the pump motor current (change of the pressure drop, when the back gas line is filled with liquid). Onset and end of the filling process can also clearly be seen by observing the balance reading. The complete transfer process besides the operation of the quick coupling and ball valves will finally be controlled by a computer.

EXPERIMENTAL RESULTS

The system was first tested in connection with a 1000 L storage tank, which was filled from another 2000 L storage vessel. During the filling of transport vessel (mostly 100 L content) the total amount of He gas escaping into the gas holder was measured. Initial cool down of the warm transfer system needs 2 - 2.5 L of liquid He, depending on the cooling rate (pump speed = 70 s^{-1}). Recooling of the transfer line at the following filling processes (time between two fillings 10' - 15') needs about 0.5 L of liquid He. During the filling process itself ($\sim$ 7.2' for 100 L at n = 100 s^{-1}) the evaporated amount of liquid from the complete system (transfer line, transport vessel and storage tank) is 0.8 - 1.2 L. For filling a cold 100 L vessel the total loss is 1.3 % to 1.7 %. After successful tests the transfer system was

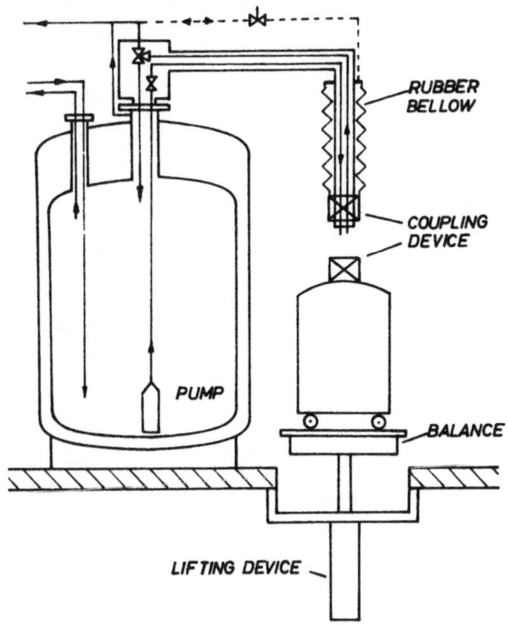

Fig. 4. Transfer device, schematic.

adapted to a 2000 L tank, which is connected to a Sulzer TCF 100 He-liquefier. Operating the system with the liquefier in operation offers no problems. The liquefier is not influenced by the transfer, the energy brought into the tank by the pump (0.6 W-h for 100 L transfer) is absorbed from the liquefier as refrigeration power, the percentage loss for pumping 100 L is further diminished to ∿ 0.3 %. The total loss rate for filling a 100 L transport vessel (cold transfer system and cold vessel) is lowered to about 1 %. With a transfer line, optimally matched to the pump transfer rates > 1500 L/h can be achieved.

CONCLUSION

When transferring liquid He with an adequate reliable pump and cold gas exchange, savings in energy and liquefier running time of about 30 % have been achieved. Filling rates exceeding 1500 L/h can be realized and no cold parts appear during the handling process, an additional safety aspect. Some essential attributes should make the centrifugal pump also suitable for transferring liq. He II in space applications: High reliability, maintenancefree operation, high total performance and pure one-phase liquid flow. We are convinced that total pump efficiencies exceeding 60 % are possible by making some constructional changes in the layout of rotor, guide wheel and sealing gap.

ACKNOWLEDGEMENTS

We would like to thank workshop and technical staff of the Walther-Meissner-Institut, their efforts gave us great help in realizing this project, which was financially sponsored by the BMFT of the Federal Republic of Germany.

A MINIATURE CRYOGENIC HIGH VACUUM VALVE*

J. D. Siegwarth and R. O. Voth

Chemical Engineering Science Division
National Bureau of Standards
Boulder, Colorado

ABSTRACT

A small high vacuum valve has been built for use in a liquid hydrogen handling system. The valve stem tip consists of a polycarbonate resin which closes against a stainless steel seat. Other features of the valve include a closing mechanism that is helium gas operated and a bellows stem seal that allows the valve to open and close without changing the internal volume.

INTRODUCTION

High vacuum valves operating at 20 K are required in a particular liquid hydrogen handling system. Other requirements dictate that the valves must open and close without changing the system volume. The valve must be remotely actuated. For maximum flexibility in locating the valve in the apparatus, the valve actuator has also been located at cryogenic temperature. To minimize the heat introduced by the actuator, the closing force required to seal the valve should be low. One method of reducing the required closing force is to use a valve with a small port, hence a small surface area of seat contact. This was possible since a high valve conductance was not a requirement for this application. The line sizes to be valved are 1.5 to 3 mm O.D.

Though commercially available valves can meet the high vacuum requirements, they are larger than desired and do not meet the zero volume change requirement. A small, commercially available valve fitted with a driver has been demonstrated to produce a high vacuum seal in cryogenic service.[1] This valve has a 7.9 mm diameter orifice, much larger than is needed for this liquid hydrogen system valve. Besides the fact that it was not a zero volume change valve, the 170 kg load required to close it would be expected to introduce a large amount of heat especially if the driver is at cryogenic temperature. Commercially available valves then, are not suitable for this liquid hydrogen handling system.

The need for zero volume change when the valve is actuated requires that the stem seal be at liquid hydrogen temperature regardless of the location of the valve actuator. At the low temperature, the stem seal can only be a metal bellows or diaphragm. A lever passing through a bellows or diaphragm seal near its fulcrum is one means of providing the zero volume change desired. Axial motion through a balanced pair of bellows was chosen for the zero volume change seal for the valves described here.

VALVE DESIGN AND CONSTRUCTION

Several valve designs were tested before arriving at the design shown in figure 1. This valve features a stem tip of polycarbonate resin (Lexan) seating against a polished stainless steel surface. The stem tip is a 60° included cone. The stainless steel seat is a matching cone. The diameter of the large end of the seat cone is 0.75 mm and the port diameter is 0.4 mm. The stem tip is held to the lower stem assembly by a brass cup that also centers the stem tip. The surface of the valve stem behind the stem tip is curved as shown in figure 1 to permit some self aligning of the stem tip into the seat when the valve is closed. The spring under the stem guide disc provides the force required to open the valve since this lower stem assembly is not connected to the portion of the stem above the bonnet.

The seal between the bonnet and the valve body is a 0.13 mm thick teflon gasket. This is a modification of a stepped seal design reported in the literature.[2] The design shown in figure 1 replaces the stepped seal design with a tongue and groove design. This design prevents the gasket from flowing when the joint is assembled. Rather than machining the groove in a single piece, the outer wall of the groove is provided by the close fitting sleeve shown in figure 1. This design makes the joint easier to machine to a close fit since the inner and outer surfaces of the groove may be separately fit to the inner and outer surfaces of the tongue. The gasket is reusable and easy to remove and replace when so desired. This joint has been disassembled an estimated one hundred times during repeated tests of the valve without replacing the Teflon gasket. Six 8-32 screws clamp the flanges together.

Two identical, welded stainless steel bellows form the stem seal. One connects the bonnet to the lower side of the closing yoke. The other connects the upper side of the closing yoke to the top support, figure 1. The top support is fixed to the bonnet. The internal volumes of the bellows communicate with the internal volume of the valve body. When the closing yoke assembly is pressed down, the upper bellows extends by the same amount the lower bellows compresses. The internal volume of the valve remains unchanged as the valve closes because the volume of a bellows is proportional to the length of the bellows over their operating range. This is true provided the internal pressure does not change enough to bulge the convolutions of the bellows.

An air driven piston at ambient temperature, connected via a stainless steel stand-off tube to the closing yoke, closed the valve during earlier tests of the design. The valve was closed during the final tests by a closely-coupled bellows piston at cryogenic temperature. This cryogenic temperature valve actuator is illustrated in figure 2. It consists of a stainless steel welded bellows, with the top end closed, fitted inside a close fitting brass cup. The lower end of the bellows is soldered to the cup so that the space between the cup and the bellows is a closed volume. The closing yoke fits inside the bellows as shown and the actuator is attached to the valve via the two mounting bolts as shown in figure 2. Introduction of pressurized helium gas into

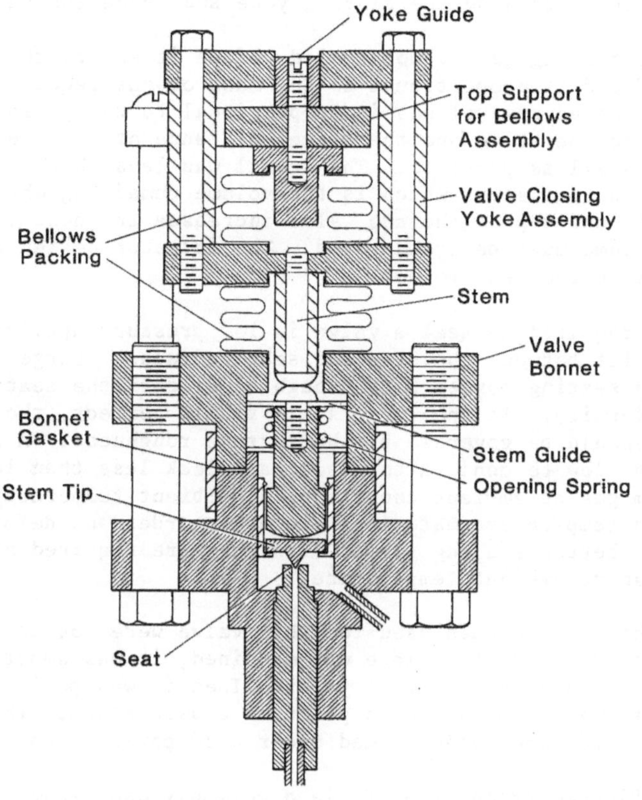

Fig. 1. Cross section of final valve design. The drawing is approximately to scale. The diameter of the bonnet and body flanges are 23 mm.

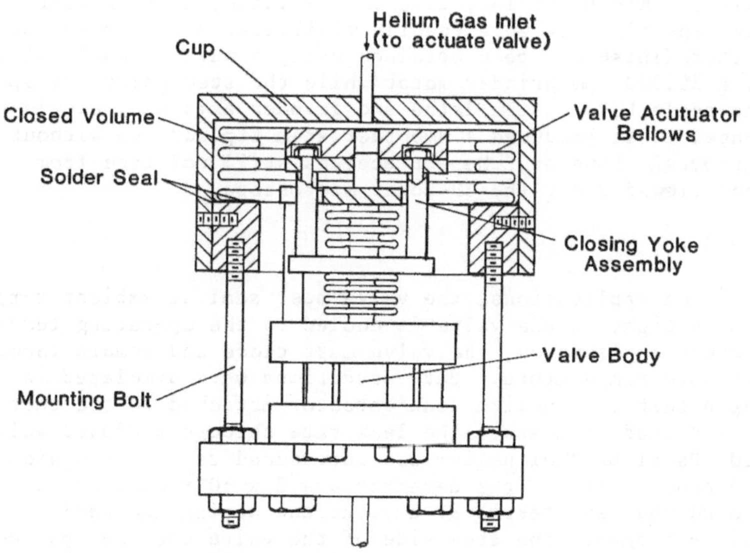

Fig. 2. A cutaway drawing of the cryogenic actuator tested with the valve.

the closed volume compresses the bellows and causes the closed end of the bellows to press down on the closing yoke and force the valve shut.

When the actuating gas is not cooled before it enters the actuator, the heat introduced is proportional to the mass of gas required to close the valve. The added mass of gas is proportional to the pressure required to close the valve and the travel distance of the stem so both should be kept small as possible. The travel was less than 1 mm. The dead volume of the actuator, which is the volume remaining when the bellows contacts the top of the cup, also increases the heat load because that volume must be pressurized. The actuator design of figure 2 minimizes the dead volume.

The force required to seal a valve in low pressure applications depends on the fit between the stem and seat surfaces. Large scale mismatches require seating forces high enough to distort the seating surfaces into conformity. If the large scale fit is perfect, the seating load required should be governed by the surface roughness. A soft stem tip can readily flow to conform to a seat and leak less than 10^{-8} std. cm^3/s of Helium gas at ambient conditions at ambient temperature. At liquid hydrogen temperature materials are much harder and deform less readily. Thus, better sealing surface finishes are required at low temperature than at ambient temperature.

The polishing techniques used for this valve were not at the state of the art. After the seat surface was machined, it was smoothed with 150 grit paper followed by 500 grit paper. Then it was polished with 1 μm Al_2O_3 on a swab or on a wooden point on a swab stick. The finished surface appeared polished when viewed under a 30 power microscope.

Chlorotrifluoroethylene (Kel-F) is a commonly used seat material in cryogenic valves. Polycarbonate resin was, however, chosen for this valve because it was easier to machine and polish by the methods used. Like other plastics used in low temperature applications, polycarbonate at 77 K does not shatter when struck. The polycarbonate stem points were machined, then smoothed with 500 grit paper, and polished with jewelers rouge. The final polishing was done with a toothpaste containing an abrasive. This polishing, done in the lathe, left concentric grooves around the tip that were readily visible with a 30 power microscope. A better finish has been obtained using a water soaked cotton swab spun by a 30,000 rpm grinder motor while the stem point was spun in a lathe. The angle between the lathe and grinder axis was held in the 30 to 60° range. This produced a smoother stem tip surface without concentric grooves. However, the surface was still not free from scratches when viewed under the 30 power microscope.

TESTING METHODS

In one of the applications, the valve must seal at ambient temperature then remain tight as the valve is cooled to the operating temperature. In another application, the valve must close and remain vacuum tight at cryogenic temperature. Both conditions were simulated in liquid nitrogen tests. A helium leak detector attached to the seat side of the valve was used to measure the leak rate through a closed valve when about 10^5 Pa (1 bar) of helium was introduced to the stem side. The estimated sensitivity of the detector was 3×10^{-8} std. cm^3/s. The valve was closed and leak tested at ambient and during the cooling to 77 K in liquid nitrogen. The stem side of the valve was then pumped out and purged with nitrogen gas prior to opening it. The opening was confirmed by the sound of the forepump on the seat side pumping nitrogen gas. The valves were then closed and leak tested again. The valve was generally opened, closed and leak tested several times while cold. If a valve began to leak, the closing load could be increased by increasing the gas pressure in the valve actuator. The closing load was calculated from the piston area of the valve actuator and the gas pressure. The closing load given for various valves in the next section is the maximum value of the load required during the leak testing to seal the valve. The actual load on the seat could be 20% lower. About 20% of the load was required just to compress bellows and springs.

A different procedure was used for the 20 K tests. During the cool-down, the valve remained open with 10^5 Pa (1 bar) of hydrogen filling the system. This pressure was maintained to prevent air from entering and contaminating the valve should leaks be present. When the valve had cooled to 20 K, it was pumped out, closed, filled with 10^5 Pa (1 bar) of helium gas on the side opposite the leak detector and leak tested. The valve was then purged with hydrogen to remove the helium, opened, closed and retested. This was done only once because the helium removal was very slow at 20 K, perhaps because of adsorption of the helium.

The zero volume change feature of the stem seal was tested by filling the valve completely with ethyl alcohol with the fill level visible in a vertical 1/2 mm diameter glass capillary connected to the valve. The change in alcohol level as the valve actuated was measured with a cathetometer with a 0.01 mm resolution.

VALVE TEST RESULTS

The alcohol level in the measurement of the valve volume change drifted over the period of measurement, which limited the accuracy. The internal volume change as the valve was actuated was less than 0.02 mm, the estimated sensitivity of the measurement.

The final valve design is shown to approximate scale in figure 1. The outer diameter of closure bellows assembly, in figure 2, was 41.3 mm. The overall height of the valve was approximately 50 mm.

The valve was tested for 14 closings at liquid nitrogen temperature. The leak rate was less than the minimum sensitivity of the leak detector, 3 x 10^{-9} std. cm^3/s, when the closing load was greater than 4 kg. The detector sensitivity was determined with a standard leak. At lower closing loads, the valve would sometimes show a detectable leak. As the closing load was reduced further, a small but fairly repeatable leak was obtained. The ambient temperature closing load required was always less than that required at cryogenic temperature.

After the 77 K tests, the valve was disassembled and the stem tip examined under the 30 power microscope. There was no evidence of distortion of the stem tip by the seat.

The valve was closed twice and helium leak tested at 20 K with a maximum closing load of 3.4 kg. No leak was detected.

The load per square mm of seat projected in the horizontal plane perpendicular to the valve axis was about 13 kg . Assuming a seat width of 0.5 mm for the valve in reference 1, the seat loads were the same. A better finish on the plastic stem tip on the valve reported here might reduce that load.

The pressure differential on the closing bellows required to provide the 4 kg load with a bellows of 7.7 cm active area was 50 kPa (7.5 psi) in liquid nitrogen. A lower closing pressure of only 43 kPa

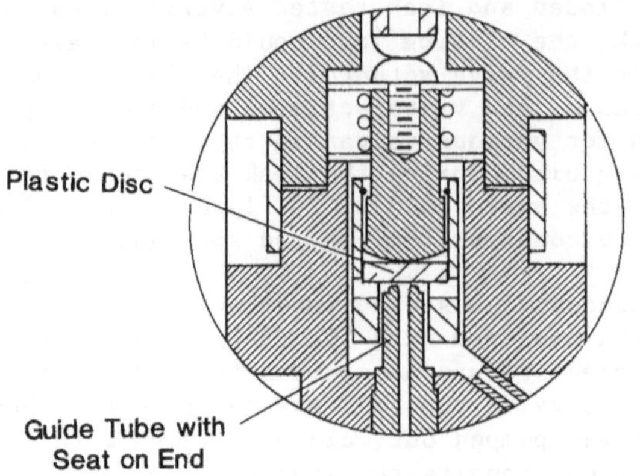

Fig. 3. This shows the alternative seat and stem design tested in the valve shown in figure 1.

(6.2 psi) was sufficient for the 20 K tests. The heat input was detectable when the bellows was activated in liquid hydrogen. Three bubbles of the order of 1 cm^3 formed. This heat load to the valve actuator could be reduced to essentially nothing by supplying the helium closure gas from a cold reservoir at some less critical location in the apparatus.

OTHER DESIGNS TRIED

Some success was obtained with alternate designs, but most were found unsatisfactory. Both coned and flat disc stem tips were tried. Cone shaped stem tips of soft solder, copper, aluminum, filled and unfilled Teflon, and Kel-F were tried. Only the last produced a usable seal. The softness and the poorer machinability of this material, however, caused it to be discarded in favor of polycarbonate resin.

Flat disc stem tips were tested also with some success. Flat discs of some common plastics, Feflon, Kapton and Mylar were tried and failed. In these cases, the stem tip design probably contributed to the failure.

The stem design shown in figure 3 produced some more successful flat disc tests. The extended seat guides and locates the disc to the seat in the design. Some success was achieved with this valve with a chlorotrifluorethylene disc. Some final tests of metal discs were tried with designs shown in figure 3. Indium, 50-50 lead-tin soft solder, unannealed high purity copper and pure gold discs were tried. The first two were soft enough to flow against the seat but would not close leak-tight at 77 K. The copper and gold did not flow significantly. The very slight seat indentation visible was not uniform showing that some misalignment was present in this design. In spite of this misalignment, however, the gold stem disc sealed vacuum tight for eight successive closings at 77 K with a 5 kg closing load. No further effort to determine whether gold might be superior to plastic as a valve stem tip has been carried out.

ACKNOWELDGMENT

This work was supported by Lawrence Livermore National Laboratory.

REFERENCES

1. J. F. Siebert, R. A. Hopkins, and H. A. Chameroy, S. H. Castles, "Development of a Launchworthy Motor-Operated Valve for Containment of Superfluid Helium," Proc. 9th Int. Cry. Engr. Conf., Kobe, Japan, 11-14 May, 1982, pp. 182-183.

2. D. N. Astrov and L. B. Belyanskii, "A High Vacuum Seal for Low Temperatures," Instr. Exptl. Tech. USSR, 2, 506 (1966).

REFERENCES

1. P. Blabert, R. A. Hopkins, and H. A. Chasaroff, [illegible] "Development of a [illegible] Motor-Operated Valve for [illegible] of Superfluid Helium," Proc. 9th Intl. Cry. Eng. Conf., Kobe, Japan, 11-14 May, 1982, pp. 182-185.

2. [illegible] and [illegible], "High Vacuum Seals for Low Temperatures," Instr. Exptl. Tech. [illegible]

TEST AND STUDY FOR A SMALL-SIZED ENVIRONMENTAL SIMULATOR OF SPACE RADIANT COOLERS

Han Jun, Cui Guang-De, and Hong Guo-Tong

Lanzhou Institute of Physics
Lanzhou, China

ABSTRACT

A small-sized space environmental simulator for the testing of radiant coolers has been developed. The chamber is provided with a cryopump system and evacuated to an ultimated pressure of 1.8×10^{-5} Pa. Its cold source is provided by a miniature refrigerator. The cooling capacity of the black cold space target can be kept to 8 W at 20 K and its emissivity is 0.98. The system is easy to operate, clean, and oil-free. The optical specular surfaces of the radiant cooler cannot be contaminated.

INTRODUCTION

Radiant cooling is an effective cooling method for infrared remote sensing systems on spacecraft, and also the ideal cooling source of IR remote sensing equipment and electronic instruments at modern space stations. The radiant cooler (RC) is the critical assembly of an IR system, but there are still some difficulties in the theories and technologies of the development process. Tests and measurements on the ground should be strictly implemented to eliminate risk during space operations. A space environmental simulator (SES) is the key system for such tests and measurements of RCs.

From the 1970s until now, tests of RCs have taken place generally in big- or middle-sized SES systems, and the cold power of the target was supplied by large or middle-sized helium refrigeration or neon liquefaction cycles.[1,2] This equipment has large refrigeration capacity and is suitable for testing large RCs, but the operating cost is high.

Our equipment is a small-sized SES system. It was developed by our institute in 1985. The chamber is evacuated to an ultimate pressure of 1.8×10^{-5} Pa by a cryopump system; the cold source is a miniature refrigerator. The cooling capacity of the cold space target can be kept to 8 W at 20 K, and its hemispherical emissivity is 0.98. The system is accompanied by a multichannel tour-measuring instrument (MCTMI) for automatic data collecting during experiments.

This SES system is suitable for the test and study of medium- and small-sized RCs. It is easy to operate, clean, and oil-free. The specular surfaces of the RC tested in the system cannot be contaminated. The cost of testing RCs in this small-sized SES system is much less than in larger equipment.

STRUCTURE OF THE SES SYSTEM

Figure 1 is the schematic diagram of the small-sized SES system. There are two closed-cycle cryogenic refrigerators in the system: the Solvay (SV) refrigerator used as the cold source of the system and the Gifford-McMahon (GM) refrigerator as the cryopump which operates when the system is baked for outgassing. In order to prevent the thin-walled cylinder of the refrigerator from deforming, the cold space target (or cold background) is "softly" connected through the high-conductivity metal belts (such as copper or silver) to the cold head of the SV refrigerator. The front surfaces (facing the RC) of the cold background are covered with black-painted open-celled aluminum honeycombs. To the back side is bonded some charcoal for a cryogenic sorption pump, all of which is shielded by a liquid nitrogen shroud. When the temperature of the cold target has reached 20~30 K, the cryopump is isolated from the chamber, and the vacuum is maintained by the cryogenic sorption pump.

The temperatures of the cold head of the SV refrigerator and the cold target are measured by gold/iron-chromel thermocouples, while the surface temperatures of the RC are measured by copper/iron-chromel thermocouples. The test data are gathered and printed at adjustable time intervals by the MCTMI apparatus.

ESTIMATE OF SIMULATING ERRORS

Some surfaces of the RC during flight may be illuminated by direct sunlight, earth-reflected sunlight, and earth IR radiation, and may also receive parasitic heat loads from the spacecraft to which the RC is mounted. All the loads received by the RC are are offset by radiating through the radiators of the RC to the cold space and its heat capacity. In order to ensure the operational stability of the IR sensors, the temperature of the cold patch must be controlled within a very narrow range. Therefore, in addition to exactly calculating the heat loads mentioned above, it is very important to test the RC in the SES system on the ground. Comparing the test data and the calculated results, the design of the RC can be finalized. It is evident that the SES system should provide a cold

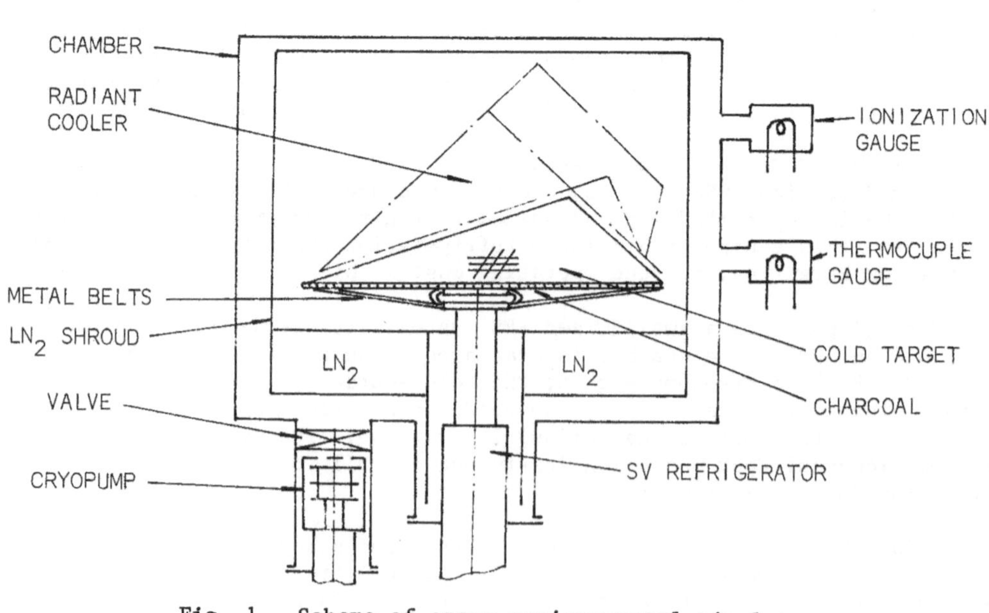

Fig. 1. Scheme of space environmental simulator

black vacuum environment. But the economic efficiency is also an important factor that we should consider. How do the space environmental conditions really appear in the SES system? And how great are the simulation errors?

In Fig. 2 an enclosure consisting of the front surfaces of the RC and the honeycomb surfaces of the cold target is shown. The cone and shield of the RC are high-reflectivity specular surfaces, and others in the enclosure are scattering surfaces. The calculations on the heat exchange between these surfaces and the temperature distribution on these surfaces are very complicated, and are not described here. Based on the theory of analyzing spececraft simulation errors, the error sources and variations of the SES system are only qualitatively discussed and estimated in this report. The results can be used for a design foundation for the system.

Error from Temperature Simulation of the Cold Space Target

The real temperature of the cold space is 3~4 K. To reduce cost, thermal simulation tests of spacecraft are generally carried out by means of a liquid nitrogen heat sink at 80~100 K. The simulating error is less than 1% according to the following equation:

$$\delta_1 = (T_h^4 - T_s^4)/\ T_v^4 \tag{1}$$

where δ_1 is the temperature simulation error, T_v the spacecraft temperature, T_h the temperature of the heat-sink surface, and T_s the real temperature of the cold space.

From Eq. (1), if the temperature of the second-stage cold patch of the RC is steady at 100 K, the temperature of the cold target must be less than 30 K to assure a simulating error of less than 1%.

Error from Limited Size and Emissivity of the Space Target

Real space is a limitless heat sink with an emissivity of $\varepsilon = 1$ which absorbs all the radiating heat from the spacecraft. But the simulated heat sink (or cold background) in the SES system has a finite area with emissivity less than 1. The error resulting from this is given by

$$\delta_2 = A_1/A_2(1/\varepsilon - 1) \tag{2}$$

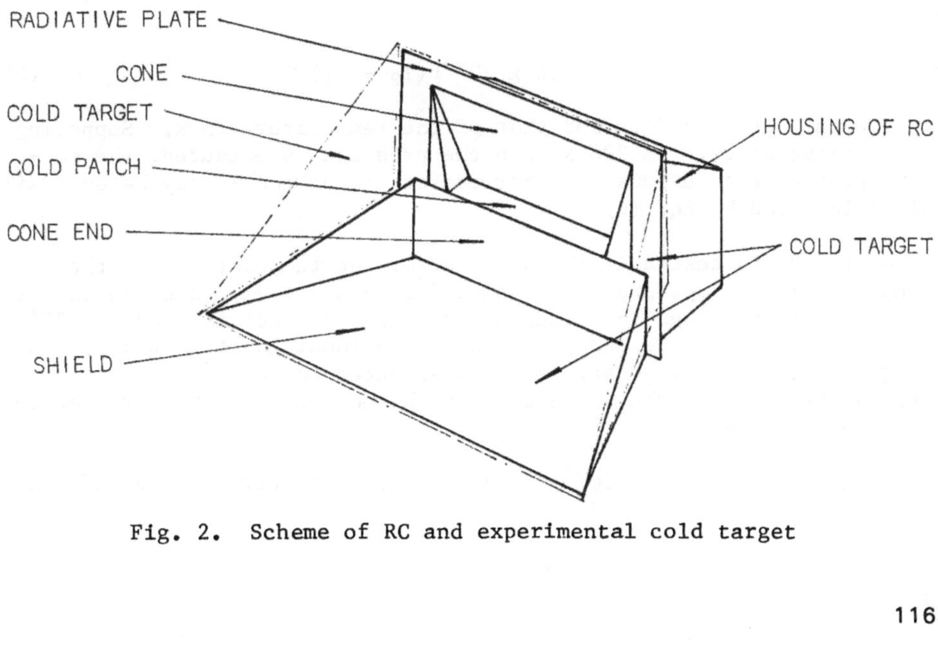

Fig. 2. Scheme of RC and experimental cold target

where δ_2 is the simulating error, A_1, A_2 the areas of the spacecraft outer surface and the surface of the simulated heat sink, respectively, and ε the emissivity of the heat sink.

Extending Eq. (2) to the SES system of RCs, A_1 represents the cold patch area of the RC, and A_2 is the area of the heat sink which is covered with black-painted aluminum honecombs. If $A_2/A_1 \geqslant 10$ and $\varepsilon = 0.95$ (or 0.98), then $\delta_2 \leqslant 0.5\%$ (or 0.2%); if $A_2/A_1 \geqslant 5$ and $\varepsilon = 0.98$, $\delta_2 \leqslant 0.4\%$. So the greater ε and the ratio A_2/A_1, the smaller the simulation error from the limited size and emissivity of the cold background.

Error from Residual Gas Conduction

In order to minimize the conduction of the gas remaining in the vacuum chamber of the SES system, the pressure in the chamber must be reduced to a value at which gas conduction between the heat-transferring surfaces can be neglected. At this low pressure (1.3×10^{-1}Pa), the average free path of a gas molecule is larger than the dimensions of the test chamber. The residual gas conductive heat per unit area between surfaces is

$$\phi = Ka_oP(T_2 - T_1) \tag{3}$$

where K is a constant, a_o the overall accommodation coefficient, P the pressure (Torr), and T_1, T_2 the temperatures of the two heat-transferring surfaces. The equation is suitable for parallel plates, concentric spheres, and coaxial cylinders.

The constant K is calculated by

$$K = (\gamma + 1)/(\gamma - 1)(R/(8\pi MT))^{1/2} \tag{4}$$

where $\gamma = c_p/c_v$ is the ratio of specific heats, R the mole gas constant, M the mass of a mole, and T the absolute temperature where the pressure is measured. K is 0.016, 0.028, and 0.059 for air, helium, and hydrogen respectively.[4] The coefficient a_o depends on a_1, a_2, the overall accommodation coefficients of the two heat-transferring surfaces, and A_1, A_2, the areas of the two surfaces. It is given by

$$a_o = a_1a_2/(a_2 + A_1/A_2(1 - a_2)a_1) \tag{5}$$

Assuming that $A_1/A_2 \leqslant 1$, and $a_o = 1$ for air at a temperature less than 300 K,

$$\phi = 1.6 \times 10^{-2}P(T_2 - T_1) \tag{6}$$

where the unit of P is Torr and that of the temperature is K. Supposing that the temperature T is 300 K when the pressure is measured, the conductive heat of the residual gas between stages of the RC may be approximately calculated by Eq. (6).

When the RC is tested in the SES system, the temperatures of the housing and first stage are 273 K and 159 K respectively, and the radiating power of the first stage (concluding cone and cone end) is 3.438×10^{-3} W/cm^2. If the conductive heat through the residual gas from the ambient housing to the first-stage structure is expected to be less than 1% of that power, the pressure in the space between the housing and the first stage must not be more than

$$P_1 = (3.438 \times 10^{-5})/(1.6 \times 10^{-2} \times 144) = 1.9 \times 10^{-5} \text{ Torr} = 2.3 \times 10^{-3} \text{ Pa}$$

The temperature of the second-stage cold patch is 105 K (Fig. 2), and the radiating power of the patch is 7.0333×10^{-4} W/cm^2. If the conductive heat by residual gas to the patch is expected to be less than 1% of this power, the maximum pressure in the volume between the first- and second-stage structures is

$$P_2 = (7.0333 \times 10^{-4} \times 0.01)/(1.6 \times 10^{-2} \times 54) = 8 \times 10^{-6} \text{ Torr} = 1.67 \times 10^{-3} \text{ Pa}$$

Therefore, from the point of view of minimizing the conductive heat by residual gas, the pressure in the chamber of the SES system must be less than 1.067×10^{-3} Pa, which is more than the limiting pressure (1.8×10^{-3}Pa) of the SES system.

TEST RESULTS

First, the system was tested with a radiation-plate suspended in it. At the beginning of the test, the cryopump was not required to operate, and the vacuum in the chamber was maintained at 1.6×10^{-2} Pa for 12.5 hours (see Fig. 3) by a mechanical pump with a cold trap at liquid nitrogen temperature. After the SV refrigerator had been started, the temperature of the cold background decreased rapidly. It took 4 hours to stabilize at 16.65 K. Following the temperature decrease, the vacuum in the chamber quickly dropped to 1.8×10^{-5} Pa. The temperature of the radiation-plate hung in the chamber reached 120 K only 6.5 hours after turning on the SV refrigerator, and continued to decrease.

A radiant cooler was successfully tested in this SES system in April 1987. The black, cold target cooldown from 196 K to 25 K took 8 hours. The chamber vacuum was 6.67×10^{-4} Pa. The RC was baked for 72 hours at 320 K and cooled down from 202 K to 115 K, which took 62 hours.

CONCLUSIONS

1. A small-sized space environmental simulator (SES system for testing space radiant coolers (RC) has been developed, in which the cold target

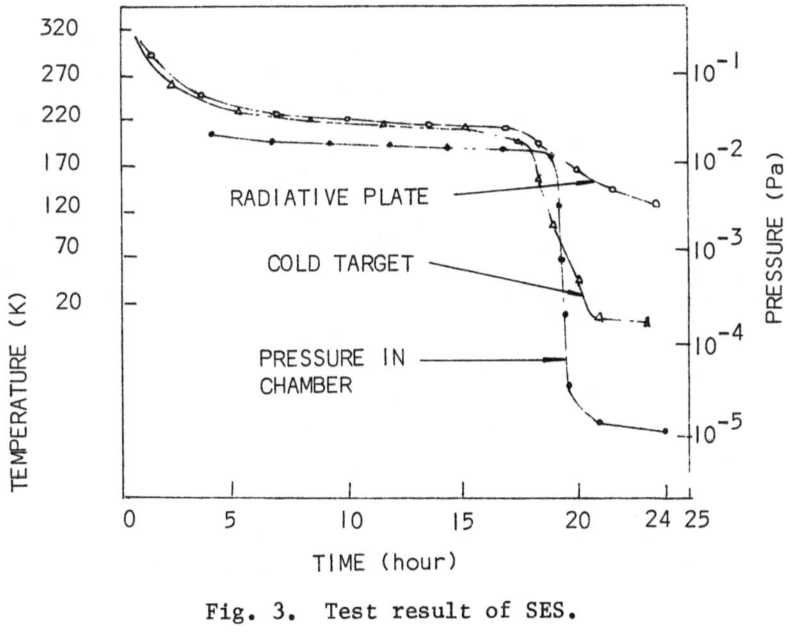

Fig. 3. Test result of SES.

has a mimimum temperature (no load) of 16.65 K and a hemispherical emissivity of 0.98. The vacuum of this system has been lowered to 1.8×10^{-5} Pa by a cryopump.

2. The SES system uses a miniature closed-cycle refrigerator to cool the cold target. It is clean, oil-free, and easy to operate. The critical optical surfaces of the RC tested in this system cannot be contaminated.

3. The test results of the SES showed that the speed of cooling down was fast. Thermal vacuum tests of an RC have been successfully completed in this small-sized SES system.

ACKNOWLEDGMENT

The authors are greatly indebted to Mr. Jiang Ji-Shan, Pan Yan-Ping, Pang Wei, Xu Ze-Yuan, Zhao Tong, Miss Ku Xiu-Ming, Han Mei-Yu, Mr. Gong Tie-Quan, and Miss Zhang Jing-Li for their aid in the preparation and operation of these tests.

REFERENCES

1. R.V. Annable and J.F. Lodder, "Design of a Dual Patch Multi-Element Radiant Cooler," N71-10606, ITT Aerospace/Optical Division, Fort Wayne, Indiana (1970).

2. J.S. Buller, Mercury cadmium telluride space radiation cooler for synchronous orbiting satellites, Laser & Infrared, 5:40 (1981) (Published in Chinese).

3. Huang Ven-Cheng et al., Cryogenic engineering on large-scale space environmental simulator, Cryo. Eng. 1:1 (1979) (Published in Chinese).

4. Yan Shou-Sheng and Lu Guo, Conduction of heat by gas, in: "Experimental Principles and Methods in Low-Temperature Physics," Wang Chang-Tai, ed., Science Press, Peking (1985), p. 211 (Published in Chinese).

A CRYOGENIC SOURCE OF POLARIZED DEUTERONS

A.A. Belushkina, V.P. Ershov, V.V. Fimushkin, G.I. Gaj, L.S. Kotova, Yu.K. Pilipenko, V.B. Shutov, V.V. Smelyansky, A.I. Valevich, and I.V. Zhigulin

High-Energy Laboratory
Joint Institute for Nuclear Research
Dubna, USSR

ABSTRACT

A cryogenic source and ionizer to obtain a polarized deuteron beam at the JINR synchrophasotron are described. A source of this type can also be applied as a jet-polarized target to operate in the internal beam of the accelerator. The dissociation of deuterium and the formation of the atomic beam in the source occur at low temperatures. The gas is pumped out by means of cryopumps.

A superconducting sextupole magnet with a field of 1.0 T is used to separate the hyperfine structure components of atoms. The atomic beam is ionized in a Penning ionizer with a magentic field of up to 7 T. The source operates in a pulsed mode with a pulse of 5-7 s and duration of 500-100 s.

INTRODUCTION

Cryogenic sources of polarized atoms and ions are under development at the JINR High-Energy Laboratory in Dubna. A soruce of ions is used to obtain a high-energy polarized beam at the synchrophasotron.[1-4] A source of atoms can be applied as a jet-polarized target to operate in the internal beam of the accelerator.[5-6] A cryogenic source has a number of distinctive properties which are important:

- gas dissociation and the formation of an atomic beam occur at low temperatures,
- large amounts of gas are pumped out by means of cryopumps,
- magnetic fields are formed by superconducting magnets.

A cryogenic source is compact and requires low power. This is important because it is located in a terminal of the preaccelerator at 700 kV. For two years the "Polaris" source has operated at the synchrophasotron. The accelerated beam of polarized deuterons is used for experiments in high-energy spin physics.

DESIGN AND PRINCIPLES OF OPERATION OF THE SOURCE

A general view of the source of polarized deuterons is shown in Fig. 1. It consists of two parts: a source of atoms and an ionizer.

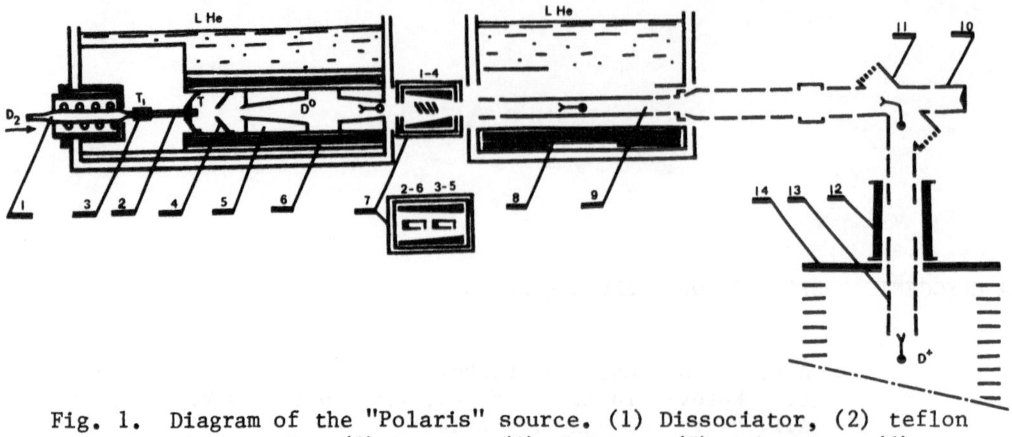

Fig. 1. Diagram of the "Polaris" source. (1) Dissociator, (2) teflon pipe nozzle, (3) heater, (4) skimmer, (5) pole tips, (6) superconducting sextupole, (7) RF transitions, (8) superconducting solenoid, (9) ionizer electrodes, (10) Faraday cup, (11) electrostatic mirror, (12) processor solenoid, (13) ion optics, (14) accelerating column. T_1 = 120 K, T = 20-60 K.

Source of Atoms

To form an atomic beam, a dissociator with a nozzle, a skimmer, and a superconducting sextupole magnet are placed in the cold bore of the first liquid helium cryostat. Deuterium is passed pulsewise to the dissociator through an electromagnetic valve. The dissociator consists of a Pyrex glass bulb and an RF coil. The discharge of the gas is performed by means of an RF coil which is excited by a 80 MHz oscillator. The dissociator is cooled due to the thermal conductivity of a teflon body, which is in thermocontact with a nitrogen shield. The nozzle, in contact with the helium cryostat, is cooled down to 20-60 K. Upon leaving the nozzle, a cold atomic flow produces a high-density free expanding jet. The central part of the jet passes through a conical skimmer 4 mm in diameter and a collimator 9 mm in diameter, forming a beam. The atomic beam is injected into an inhomogeneous field of the sextupole magnet, where the electron spin separation of atoms takes place.

Atoms with electron spin projection 1/2 are focused and produce a polarized beam; atoms with projection -1/2 are defocused and pumped out. The walls of the cryostat surrounding the nozzle, the skimmer, and the sextupole tips are at the temperature of 4.2 K and serve as cryopanels to pump out the gas. Molecules are captured better than atoms at the same wall temperature. In this case the speed of pumping also depends on the process of atom recombination, i.e., an additional number of collisions with the walls is necessary.

The sextupole magnet is made so that its yoke, poles, and coils are placed in the liquid helium and the pole tips in vacuum. The poles and tips are separated by a pipe of the cryostat. A part of the pipe surface serves as a cryopump for the defocused component of the atomic beam. The pole tips are made of sets, 100 and 140 mm long. The first set is a taper with aperture diameter 10 mm at the input and 18 mm at the output. The second set is a cylinder 24 mm in diameter. The yoke and poles are made of steel and the pole tips of FeCo alloys. The coils are made of NbTi superconducting cable 0.5 mm in diameter. The magnetic field on the pole tips is 1.0 T.

Behind the sextupole magnet the beam of atoms enters the RF cells for vector or tensor nuclear polarizations.

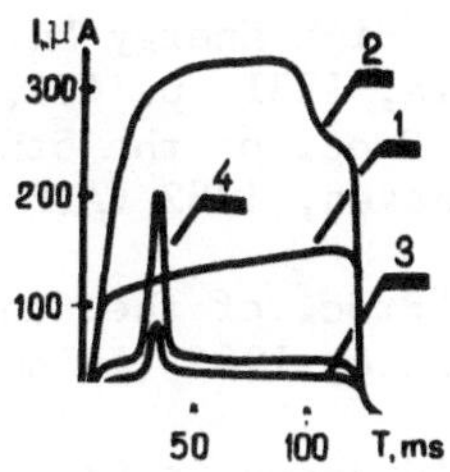

Fig. 2. Typical oscillogram of the ion current of the source. (1,2) Background and atomic signals measured by the Faraday cup, (3,4) the same signals measured downstream of the ion optics (Fig. 1, 13).

Cryogenic Ionizer

The atomic beam is ionized in a Penning ionizer. The ionizer consists of a liquid helium cryostat, a superconducting solenoid, and electronic optics. The ionization of the polarized atomic beam occurs due to the electrons oscillating in the magnetic field along the axis of the electronic optics.

The superconducting solenoid is composed of three series windings, made of a NbTi cable, 1.1, 0.93, and 0.78 mm in diameter, and two correcting coils. The internal section of the winding is wound on a cryostat pipe 61 mm in diameter. The length of the solenoid is 300 mm. The maximum field at the center is 7.0 T. The solenoid of the ionizer as well as the sextupole magnet operate in a persistent current mode.

To match the source and the linac, ion optics and a spin-processor consisting of a 90° electrostatic mirror and a solenoid are used.

Experimental Results

The "Polaris" source is located in a high-voltage terminal and connected to a preaccelerator of the linac with an energy of 5 MeV/nucleon. During injection, the preaccelerator and the source are under a 700 kV potential pulse (~100 μs). For this reason the experimental set-up is insulated from the ground. The helium from the cryostats is recovered through plastic pipes. Power is supplied by an autonomous generator. The source is controlled by a microprocessor. The data are read out from the high-voltage terminal on-line with a microcomputer via fiber-optical lines.[7]

The source operates in a pulsed mode with a pulse of 5-7 s. The pressure of deuterium in the dissociator varies within 5-6 Torr. The flow rate of deuterium is 50 Torr cm^3/pulse. The pressure in the ionizer is usually $3\cdot10^{-8}$ Torr. A typical oscillogram of the ion current of the polarized deuteron beam $\vec{D}^+$ is shown in Fig. 2. In order to measure the vector and tensor polarization of the beam, a polarimeter has been developed using the reactions $^4He(\vec{d},d)^4He$ and $^3He(\vec{d},p)^4He$. The vector polarization of the beam is $p_z = 0.5$, the tensor polarization is $p_{zz} = 0.7\text{-}0.8$. The experimental set-up can operate for 2-3 weeks without deuterium sublimation. Its operation requires 200 liters of liquid helium for a week.

REFERENCES

1. A.A. Belushkina et al. in: "High Energy Physics with Polarized Beams and Polarized Targets," Argonne 1978 (AIP Conf. Proc. No. 51), New York, 1979, p. 351.

2. A.A. Belushkina et al., in: "High Energy Physics with Polarized Beams and Polarized Targets," Basle, 1981, p. 429.
3. N.G. Anishenko et al., in: "Proc. of the 5th Intern. Symp. on High Energy Spin Physics," Brookhaven, 1982 (AIP Conf. Proc. No. 95), New York, 1983, p. 445.
4. N.G. Anishenko et al., in: "Proc. of the 6th Intern. Symp. on High Energy Spin Physics," Marseille, 1984 (J. de Phys., Colloque C2, Tome 46, 1985, p. C2-703).
5. V.A. Bartenev et al., in: "Advances in Cryogenic Engineering," Vol. 18, Plenum, New York, 1973, p. 460.
6. D.D. Yovanovitch, in: "High Energy Physics with Polarized Beams and Polarized Targets," ANL, 1984, XVI-I.
7. Yu.I. Romanov, "Multichannel Fiber Optical Lines for the Control System of the Polaris Source," JINR, 13-82-279, Dubna, 1982.

AUTHOR INDEX

Abdelsalam, M.K. 69, 203
Adachi, Y. 1031
Ageyev, A.I. 19
Anderson, J.E. 909
Andrews, D.E. 1
Andriishchin, A.M. 19
Araoka, O. 551
Arrendale, H.G. 219
Asano, K. 33

Bailey, R.E. 219
Barclay, J.A. 719, 751, 767, 791
Barron, R.F. 355
Barton, Jr., H.R. 569
Beaudry, B.J. 785
Beduz, C. 363
Belushkina, A.A. 1167
Berndt, H. 1147
Bernhard, D.P. 983
Beurer, P. 559
Bilton, J.R. 203
Bodegom, E. 349, 999
Bonneton, M. 591
Boom, R.W. 69, 203
Boroski, W.N. 243, 251, 323
Boyle, R.F. 487
Brandreth, B.H. 497
Brindza, P. 623
Brodie, L.C. 349, 999
Brooks, W.F. 901
Brown, D.P. 663
Buchanan, D.S. 97
Burmeister, H. 313

Cain, C.L. 383
Campenni, C.K. 791
Canavan, E.R. 195
Carson, J.A. 235
Castles, S.H. 631, 799, 819
Chaussy, J. 1023
Chen, W.E.W. 479
Chesebrough, R. 791
Chronis, W.C. 585, 623
Chwalek, J.M. 1097

Clausen, M. 559, 569, 1113
Colwell, J. 53
Cross, C.R. 767
Crowley, C.J. 371
Crunkleton, J.A. 809
Cui, G.-D. 1161

Dacus, M.W. 757
Daney, D.E. 515, 1071
DeGregoria, A.J. 767
Dergham, A.R. 355
Dinaburg, L.B. 183
DiPirro, M.J. 487, 885, 893, 909
Doi, Y. 33, 533, 543, 551, 1105
Doll, R. 1147
Donnelly, C. 699
Dresner, L. 167
Duband, L. 1023
Dustmann, C. 25
Dykaar, D.R. 1097

Eckels, P.W. 1129
Egolf, H. 559
Eisele, P. 991
England, J.E. 341
Ershov, V.P. 1167
Eschricht, W. 313
Eyssa, Y.M. 69, 187, 203

Fast, R.W. 291, 299
Fehling, D.T. 259
Ferri, D. 1039
Fester, D.A. 909
Fietz, W.A. 259
Fimushkin, V.V. 1167
Fouaidy, M. 407
Francois, M.X. 407, 425, 441
Frederking, T.H.K. 431, 479
Fuerst, J.D. 655
Fuge, C. 991
Fukutsuka, T. 125

Gaj, G.I. 1167
Gallagher, T.A. 203
Garrity, T.F. 203

Gerke, C. 559
Gerke, Chr. 1113
Gibbs, R.J. 663
Giraudi, D. 1039
Gistau, G.M. 591, 675
Gonczy, J.D. 227, 243, 251, 323
Gray, K.E. 707
Green, G. 777
Green, M.A. 149, 631
Gridasov, V.I. 19
Guinan, M.W. 639

Hackworth, D. T. 1129
Hahn, P.A. 639
Hamada, M. 125
Han, J. 1161
Hara, K. 615
Harding, D. 791
Hart, H.L. 291, 299
Haruyama, T. 33, 533, 543, 551, 1105
Hashimoto, T. 733, 743
Hassenzahl, W.R. 143
Heiden, C. 845
Helvensteijn, B.P.M. 399
Henning, C.D. 41
Hepler, W.A. 479
Hermanson, L.A. 497
Herzog, H. 559
Hilal, M.A. 143, 203
Hirabayashi, H. 323
Hiyama, T. 599
Hofmann, A. 471
Hong, G.-T. 1161
Hopkins, R.A. 925
Horlitz, G. 313, 559
Hoshino, M. 599
Hosoyama, K. 615
Hsiang, T.Y. 1097
Huang, X. 187
Hubbell, R.H. 383
Hubmann, M. 333

Ibaraki, Y. 551
Ibrahim, E.A. 143
Igra, R.M. 1005
Inomata, K. 733
Inoue, H. 533
Israelsson, U.E. 713
Ito, T. 743
Iwasa, Y. 809

Jackson, H.W. 713
Jaeger, S.R. 751, 767
Jahn, U. 1147
Jebali, F. 425
Joham, E. 699
Johanson, W. 791
Johnson, J.W. 767
Johnson, K. 699
Jones, J.A. 869

Kalinin, V.V. 183
Kamisada, Y. 61
Kaneko, O. 607
Kang, S.S. 371
Karpenko, V.N. 143
Kashani, A. 417
Kashtanov, E.M. 19
Kataoka, S. 607
Kato, M. 1045
Kato, T. 259, 599
Kawada, M. 837
Kawai, K. 283
Kawai, M. 551
Kawamura, S. 615
Kawano, K. 599
Kaz'min, B.V. 19
Keim, T.A. 1137
Kelley, J.P. 623
Kerney, P.J. 683
Kerns, J.A. 175
Kessler, G. 569
Khalil, A. 471
Kichimi, H. 533
Kimura, H. 533
Kimura, N. 1105
Kimura, Y. 615
Kirol, L.D. 757
Kirschman, R.K. 683
Kitagawa, S. 607
Kittel, P. 465, 893, 955
Kiuchi, T. 1045
Knoll, R.H. 341
Kobayashi, M. 61
Kojima, Yuuji 615
Kojima, Yuzo 615
Kokubun, Y. 283
Kondo, Y. 551, 1105
Konstantinov, A.B. 183
Kotova, L.S. 1167
Krämer, H.P. 471
Krempetz, K.J. 1121
Krevet, B. 25
Krooshoop, H.J.G. 117
Kropatsch, G. 333
Krushev, V.N. 183
Kuchnir, M. 1063
Kun, L.C. 963
Kuroda, T. 607

Kuzuhara, T. 733

Langenecker, F. 559
Larson, E.T. 235
Laumer, H. 267
Lee, J.H. 525, 901
Lee, J.M. 431
Lesyna, L. 955
Leung, E.M.W. 219
Li, R. 733
Lierl, H. 559
Logachev, S.N. 19
Lombard, D.S. 885
Longsworth, R.C. 683, 689
Lu, B.C.-Y. 1031
Lubell, M.S. 259
Ludtke, P.R. 515
Lue, J.W. 259
Luther, R.D. 1121
Luton, J.N. 259

Maa, S. 901
MacNeil, P.N. 341
Maddocks, J.R. 449
Maehata, K. 61
Makida, Y. 159
Maki, A. 33
Maki, M. 551
Martindale, D.L. 885
Mart, J.W. 591
Matsumoto, K. 125, 615, 733, 743
Maza, J. 425
McGuire, K. 1063
McIntosh, G.E. 69, 203, 885
McInturff, A. 1063
Mess, K.H. 1113
Meuris, C. 441
Michels, P.H. 219
Miller, J.R. 175
Mills, G.L. 497
Mito, T. 33, 533, 543, 551, 1105
Mitsunobu, S. 615
Mord, A.J. 497
Morimoto, M. 615
Moser, H.O. 25
Mourou, G.A. 1097
Mulder, G.B.J. 117
Mulholland, G.T. 1121
Murakami, Y. 159
Murase, S. 61
Myznikov, K.P. 19

Nakai, H. 615
Nakamura, S. 61
Nakazato, T. 615
Nakogome, H. 733
Nast, T.C. 457
Ng, Y.S. 525, 901
Nicol, T.H. 227, 235, 243, 251
Niemann, R.C. 227, 235, 243, 251, 323
Nisenoff, M. 77
Nishijima, S. 125, 135
Nissen, J.A. 349, 999
Noguchi, S. 615
Nolen, J.A. 267
Nowakowski, K.-D. 559
Nurusheva, M.B. 19

Ogitsu, T. 615
Ohba, T. 533
Ohi, T. 607
Ohmori, T. 323, 533, 551
Ohtani, T. 305
Ohtu, Y. 607
Ohuchi, N. 159
Oka, Y. 607
Okada, T. 125, 135
Okuda, M. 305
Olsen, C. 791
Olson, J.R. 683
Omori, T. 33
Orlov, A.P. 19
Otavka, J.G. 243, 251
Overton, W.C. 751, 791

Pahade, R.F. 975
Parmley, R.T. 935, 943
Patton, W. 777
Paulson, D. 97
Pavese, F. 1039
Payne, D.A. 925
Peck, S.D. 143
Pecoud, Y. 675
Peterson, T.J. 655
Petrac, D. 713
Phillips, R. 699
Pilipenko, Yu.K. 1167
Pirklbauer, H. 333
Poirier, R.N. 203
Poivilliers, J. 441
Powers, R.J. 251
Pryima, M.V. 19

Quack, H. 647

Radebaugh, R. 851, 1013
Rao, M.G. 1005
Ravex, A.E. 675, 1023
Reich, G. 845
Rettig, S. 559, 569, 1113

Rhodenizer, R.L. 9
Rios, P.A. 9
Roach, P.R. 707
Rode, C.H. 391, 623
Roellig, T. 955
Rogers, J.D. 1129
Rothe, P.H. 371
Rowles, H.C. 983
Ruschman, M.K. 243, 251, 323
Russell, M.P. 1013
Ryba, E. 699

Sahashi, M. 733, 879
Saito, S. 615
Sakamoto, Y. 615
Sakurai, K. 607
Sarwinski, R.E. 87
Sato, A. 61, 879
Sauvage-Boutar, E. 441
Schauer, F. 333
Schein, M.E. 819
Schlafke, A.P. 663
Schurter, R.V. 683
Schwenterly, S.W. 259, 271
Scurlock, R.G. 363, 1005
Self, B. 699
Sellmann, D. 313
Selvaggi, J.A. 53
Semura, J.S. 349, 999
Shah, J.M. 203
Shamichev, A.N. 19
Shea, M. 623
Shen, S.S. 259
Shestak, E. J. 1129
Shiba, E. 533
Shimamoto, S. 599
Shu, Q.S. 291, 299
Shutov, V.B. 1167
Siegwarth, J.D. 1089, 1153
Singh, S.K. 1129
Sixsmith, H. 827
Slack, D.S. 175, 585
Smelyansky, V.V. 1167
Smith, Jr., J.L. 809
Smith, D.R. 1013
Snyder, H.A. 497
Sobolewski, R. 1097
Sondericker, J.H. 663
Sousa, A.J. 363
Spradley, I.E. 935
Stahlschmidt, W. 559
Stamps, R.E. 259
Stanton, R. 699
Stevens, J. 777
Steward, W.G. 515
Stewart, W.F. 791
Storch, P.J. 851
Strait, J. 1063
Sugie, H. 1031
Sun, Y. 861
Suzuki, H. 551
Suzuki, S. 551
Swift, W.L. 827
Sytnik, V.V. 19

Tada, E. 599
Tadano, M. 551, 1105
Taira, T. 323
Takahashi, K. 323
Takahashi, M. 733
Takahata, K. 125, 135
Takao, T. 283
Taneda, M. 305
ten Haken, B. 211
ten Kate, H.H.J. 117, 211
Terada, S. 33, 551
Theilacker, J.C. 391, 577
Thompson, D.H. 259
Tomokiyo, A. 733
Tsuchiya, K. 33, 551
Tsukamoto, O. 283
Tsukuda, J. 305

Urbach, A.R. 497
Uyttewaal, W. 211

Valenzuela, J. 827
Valevich, A.I. 1167
van de Klundert, L.J.M. 117, 211
Van Sciver, S.W. 195, 399, 417, 507
Vander Arend, P. 53
Vasiliev, L.M. 19
Vasiliev, V.A. 19
Vázquez, I. 1013
Veselov, O.M. 19
Vidal, F. 425
Volkov, A.F. 183
Voth, R.O. 1089, 1153

Wade, L. 699
Wake, M. 61
Walker, R.J. 1081
Walstrom, P.L. 449
Wands, R.H. 1121
Wanner, M. 917
Waynert, J. 187
Weber, K.J. 1121
Weisend II, J.G. 507
Werner, M. 955

Whitehead, B.B. 991
Wiedemann, W. 1147
Wikswo, Jr., J.P. 107
Williamson, S.J. 97
Wilson, C.T. 259
Wu, K.C. 663

Yamada, Y. 61
Yamamoto, A. 323, 533, 543, 1105
Yamamoto, J. 159, 879
Yamaoka, H. 533
Yamashita, N. 551
Yamashita, T. 125
Yarygin, N.N. 19
Yayama, H. 733
Yazawa, T. 733
Ye, J.D. 313
Yerokhin, A.N. 19
Yoshimura, H. 837
Yuan, S.W.K. 431, 457

Zeller, A.F. 267
Zhigulin, I.V. 1167
Zhou, Y. 861
Zhu, W.X. 861
Zichy, J.A. 1053
Ziegler, B. 559
Ziemer, J.H. 975
Zimm, C.B. 791
Zinchenko, S.I. 19
Zink, R.A. 235

SUBJECT INDEX

ac or cyclic losses: 119, 159, 214, 618, 624
Acoustic emission, uses of: 135,
Active phase separator: 431
Adiabatic demagnetization: 879, 955
Adsorption: 1013, 1023
Adsorption compressor—see Compressor, adsorption
Air separation: 963
Argon calorimeter: 1121

Baker diagram: 391
Bayonets: 885, 898
Biological effects: 107

Carbon, adsorption on: 1013, 1023
Carbon resistors—see Temperature sensor
Compressors,
- adsorption: 869
- cold: 593, 602, 626, 647, 655, 663, 675
- warm: 559, 629, 834

Conductors, electrical—see Superconductors
Control system for refrigerators—see Refrigeration, control systems
Cooldown,
- of superconducting magnet system: 36, 537, 556
- of superconducting RF cavity: 626

Couplings: see Bayonets
Cryocoolers—see also Refrigerators: 91, 799, 809, 819, 827, 837, 845, 851, 861
Cryoelectronics: 77
Cryogenic containers—see Dewars, cryostats
Cryopumps, cryopumping: 607
Cryostability—see Stability
Cryostats—see also Dewars: 227, 235
Current leads: 259
Current sharing—see Stability

Data acquisition system: 223, 580, 1067
Density, measurement of: 1081
Dephlegmators: 983
Deuterons: 1167
Dewars,
- space: 944
- superfluid: 906, 925

Dilution refrigerators: 707, 713

Economics, of SMES: 74
Electrical power generation—see Generators
Electronics, superconducting: 77
Energy storage systems—see Superconducting magnetic energy storage
Epoxy: 126, 135
Equation of state—see PVT data and LVE data
Expansion cycles—see Refrigerator cycles
Expansion engines, turbine: 832, 963

Fatigue: 44
Film boiling: 355, 425
Flow characteristics in superconducting coils: 168, 175
Flow meters: 898, 1056, 1071
Flow, two-phase: 391, 441, 543
Fountain effect: 457, 465, 471, 479, 487, 497
Fusion applications—see Magnets, fusion

GGG—see Refrigerators, magnetic
Gas compression—see Compressors
Generators, cryogenic: 1137
Gorter—Millink parameter—see Heat transfer, He II

Heat exchangers: 383, 693, 833
Heat load measurements/calculations,
- for current leads: 259
- for dewar supports: 227, 235, 243, 249
- for magnets: 40

Heat pump—see Refrigerators, magnetic
Heat transfer,
- boiling: 355, 363, 371, 425
- to He I—see also Stability: 299, 349

Heat transfer (cont.)
to He II—see also Stability: 187, 195, 399, 407, 417, 425, 449
to nitrogen: 291, 305, 313, 323, 355, 363
to oxygen: 341
transient: 373
Helium,
properties of: 713, 999, 1005
Hydrocarbon recovery: 983

ICCS—see Superconductor, internally cooled
Instrumentation—see also Control systems, power supplies, quench protection system, strain gauges, temperature sensor, thermometry: 87, 97, 267, 1053, 1063, 1071, 1081
Insulation, thermal: 291, 299, 305, 313, 323, 333, 341
Insulators, electrical: 271
Ion beams: 1167

J-T valve: 699

Kapitza conductance—see Heat transfer to He II

Leak detectors: 1081
Liquefied natural gas (LNG) terminals: 991
Liquid level, measurement of: 267, 917, 1055

Magnetic field and force calculations: 26, 55, 149
Magnetic field measurement: 99, 107
Magnetic refrigeration—see Refrigerators, magnetic
Magnetic resonance imaging: 1, 9, 89
Magnetic shielding: 5, 100
Magnetocaloric materials: 777, 785, 791
Magnets, superconducting,
accelerator: 19, 25
energy storage: 69, 203
fusion: 41, 175, 183, 187
high energy physics: 33, 533, 543, 551, 631
high field (> 10T): 61
MRI: 3, 9
protection of: 73, 203, 1059
separator: 53
solenoid: 33, 61, 533
SSC: 143, 149
Magnet supports: 227, 235, 243, 251
Medical applications: 87, 97

Methane, in mixtures: 1033
MRI—see Magnetic resonance imaging
Multilayer insulation: 291, 299, 305, 313, 323, 333

Neon, in mixtures: 1031
Neuromagnetism: 97
Nitrogen, production of: 975

Oxygen deficiency hazard: 1127
Oxygen, purity of: 1039

Paschen curve: 273
Permanent magnets,
for MRI: 12
for refrigeration: 760
Phase separators, He II—see also Porous plug and Active phase separator: 431, 897
Porous plugs—see also Refrigerators, space and helium II in space: 487, 497
Power leads—see Current leads
Power supplies for superconducting systems: 35
Pressure transducers: 1006, 1055, 1064
PVT data: 1046
Pulse tubes: 802, 851, 861
Pumps, helium: 507, 515, 605, 1148
Pumps, fountain effect: 457, 465, 471, 479, 487, 497, 896, 909

Quality, fluid: 391, 441, 1055
Quench behavior,
magnet: 66, 539
normal zone propagation velocity: 148, 217
potted magnets: 125, 135
pressure rise: 39, 143, 209
Quench detection: 167, 283, 1058
Quench protection systems: 73, 203, 1059

Radiation damage: 45, 639
Recovery current—see Stability, superconducting coil
Rectifiers, superconducting: 117, 213
Refrigeration system,
BNL: 663
CEBAF: 623
Fermilab: 577, 655
HERA: 313, 559, 569, 1113
JAERI: 599
KEK: 533, 543, 551, 615, 1105
MFTF: 585

Refrigeration system (cont.)
MRI: 90
rf cavity: 615, 623
Separator magnet: 57
Testing facilities: 577, 639
Tore Supra: 591, 675
UNK: 19
Refrigerator,
cooldown: 695
control systems: 1105, 1113
cycles—see also Liquefiers and refrigeration systems: 647, 828, 976
Refrigerators,
dilution: 707, 713
for electronics: 683
J-T: 689, 699
magnetic: 719, 733, 743, 751, 757, 767, 777, 785, 791
MRI: 696
pulsed tube: 802, 851, 861
sorption: 869
space: 799, 819, 837, 879, 1161

Safety: 636, 899, 1127, 1132
Sampler, electro-optic: 1097
SHOOT: 487, 893, 901, 909
Silicon diodes—see Temperature sensor
SMES—see Superconducting magnetic energy storage
Space simulator: 1161
SQUIDS: 85, 93, 98, 109
SSC—see Superconducting Supercollider
Stability, superconducting coil,
LHe I: 44, 125, 135, 143, 149, 159, 175, 183
LHe II: 187, 195
Stirling cycle coolers—see also Refrigerators, space: 845
Superconducting electronics: 77, 117
Superconducting magnetic energy storage: 69, 203
Superconducting magnets—see Magnets, superconducting
Superconducting Supercollider,
cryostats for: 227, 235, 243, 251, 323
magnets for: 143, 149
refrigeration for: 577
Superconductors,
high T_C: 69
internally cooled: 159, 167, 175, 195
Nb-Ti: 27, 61
Superfluid—see also Helium II and Heat transfer to He II,
cryostats for: 63
heat transfer to: 189, 407, 425

Superfluid (cont.)
metrology: 1071
pumps: 457, 465, 471, 479, 487, 497, 507, 515, 525
thermohydraulics in: 200, 399, 417, 431, 449
transfer: 885, 893, 901, 909, 917
Support system,
dewar: 227, 235, 243, 251, 935, 1124
magnet: 35
Synchrotron light source: 25

Tanks—see Dewars and storage tanks
Temperature sensors: 1054, 1065, 1118
Test facilities,
cryogenic: 577, 1097
radiation damage: 639
short sample: 211, 219
Thermal intercepts: 227, 235, 244
Thermodynamic properties—see PVT data
Training—see Quench behavior:
Transfer lines: 574, 621, 628
Transfer of cryogenic fluids: 1147
Transformers,
cryogenic: 1129
superconducting: 219
Turboexpanders—see Expansion engines, turbine
Two-phase flow—see also Flow characteristics: 391, 441, 543

Vacuum technology—see Cryopumps
Valves, cryogenic: 1153
Vapor cooled leads: 257
Vapor-liquid equilibrium—see VLE data
VLE data: 1031, 1045
Voltage breakdown: 271
Vuillenmier cooler: 837

Weight, measurement of: 1089